X-RAY ASTRONOMY

Related Titles from the AIP Conference Proceedings Subseries on Astronomy and Astrophysics

587 Gamma 2001: Gamma-Ray Astrophysics 2001
Edited by Steven Ritz, Neil Gehrels, and Chris R. Shrader, October 2001,
0-7354-0027-X; CD-ROM: 0-7354-0030-X

586 Relativistic Astrophysics: 20th Texas Symposium
Edited by J. Craig Wheeler and Hugo Martel, October 2001, 0-7354-0026-1

566 Observing Ultrahigh Energy Cosmic Rays from Space and Earth: Int'l. Workshop
Edited by Humberto Salazar, Luis Villaseñor, and Arnulfo Zepeda, May 2001,
0-7354-0002-4

565 Young Supernova Remnants: Eleventh Astrophysics Conference
Edited by Stephen S. Holt and Una Hwang, May 2001, 0-7354-0001-6

558 High Energy Gamma-Ray Astronomy: International Symposium
Edited by Felix A. Aharonian and Heinz J. Völk, April 2001, 1-56396-990-4

556 Explosive Phenomena in Astrophysical Compact Objects: First KIAS Astrophysics Workshop
Edited by Heon-Young Chang, Chang-Hwan Lee, Mannque Rho, and Insu Yi,
March 2001, 1-56396-987-4

526 Gamma-Ray Bursts: 5th Huntsville Symposium
Edited by R. Marc Kippen, Robert S. Mallozzi, and Gerald J. Fishman, June 2000,
CD-ROM included, 1-56396-947-5

522 Cosmic Explosions: Tenth Astrophysics Conference
Edited by Stephen S. Holt and William W. Zhang, June 2000, 1-56396-943-2

516 26th International Cosmic Ray Conference: ICRC XXVI, Invited, Rapporteur, and Highlight Papers
Edited by Brenda L. Dingus, David B. Kieda, and Michael H. Salamon, May 2000,
1-56396-939-4

510 The Fifth Compton Symposium
Edited by Mark L. McConnell and James M. Ryan, March 2000, 1-56396-932-7

499 Small Missions for Energetic Astrophysics: Ultraviolet to Gamma-Ray
Edited by Steven P. Brumby, December 1999, 1-56396-912-2

433 Workshop on Observing Giant Cosmic Ray Air Showers from $>10^{20}$ eV Particles from Space
Edited by John F. Krizmanic, Jonathan F. Ormes, and Robert E. Streitmatter, June 1998,
1-56396-788-X

To learn more about these titles, or the AIP Conference Proceedings Series, please visit the webpage **http://proceedings.aip.org**

X-RAY ASTRONOMY

Stellar Endpoints, AGN, and the Diffuse X-ray Background

Bologna, Italy   6–10 September 1999

EDITORS
Nicholas E. White
NASA/Goddard Space Flight Center
Greenbelt, Maryland

Giuseppe Malaguti
ITESRE/CNR
Bologna, Italy

Giorgio G. C. Palumbo
ASI
Rome, Italy

Melville, New York, 2001
AIP CONFERENCE PROCEEDINGS ■ VOLUME 599

Editors:

Nicholas E. White
NASA/Goddard Space Flight Center
Laboratory for High Energy Astrophysics
Code 660
Greenbelt, MD 20771
USA
E-mail: nwhite@lheapop.gsfc.nasa.gov

Giuseppe Malaguti
ITESRE/CNR
via P. Gobetti, 101
40129 Bologna
ITALY
E-mail: malaguti@tesre.bo.cnr.it

Giorgio G. C. Palumbo
ASI
Viale Liegi, 26
00198 Rome
Italy

The articles on pp. 150-159, 377-386, 387-395, 482-485, and 1027-1030 were authored by U.S. Government employees and are not covered by the below mentioned copyright.

Authorization to photocopy items for internal or personal use, beyond the free copying permitted under the 1978 U.S. Copyright Law (see statement below), is granted by the American Institute of Physics for users registered with the Copyright Clearance Center (CCC) Transactional Reporting Service, provided that the base fee of $18.00 per copy is paid directly to CCC, 222 Rosewood Drive, Danvers, MA 01923. For those organizations that have been granted a photocopy license by CCC, a separate system of payment has been arranged. The fee code for users of the Transactional Reporting Service is: 0-7354-0043-1/01/$18.00.

© 2001 American Institute of Physics

Individual readers of this volume and nonprofit libraries, acting for them, are permitted to make fair use of the material in it, such as copying an article for use in teaching or research. Permission is granted to quote from this volume in scientific work with the customary acknowledgment of the source. To reprint a figure, table, or other excerpt requires the consent of one of the original authors and notification to AIP. Republication or systematic or multiple reproduction of any material in this volume is permitted only under license from AIP. Address inquiries to Office of Rights and Permissions, Suite 1NO1, 2 Huntington Quadrangle, Melville, N.Y. 11747-4502; phone: 516-576-2268; fax: 516-576-2450; e-mail: rights@aip.org.

L.C. Catalog Card No. 2001096945
ISBN 0-7354-0043-1
ISSN 0094-243X
Printed in the United States of America

CONTENTS

Preface xvii

INVITED PAPERS

X-ray Sources as Tracers of the Large-Scale Structure in the Universe 3
    X. Barcons, F. J. Carrera, M. T. Ceballos, and S. Mateos
An X-ray View of Millisecond Pulsars................................... 13
    W. Becker
Narrow-Line Seyfert 1 Galaxies as an Extreme of Seyfert Activity 25
    T. Boller
X-ray Transients Monitored by the All-Sky Monitor on RXTE:
A Tabulation 35
    H. Bradt, A. Levine, R. Remillard, and D. A. Smith
X-ray Absorption in Radio-Quiet QSOs 53
    W. N. Brandt, S. C. Gallagher, A. Laor, and B. J. Wills
X-ray Transients in Quiescence 63
    S. Campana
What's Wrong with AGN Models for the X-ray Background? 73
    A. Comastri
Coronal Heating and Emission Mechanisms in AGN 83
    T. Di Matteo
The Energy Output of the Universe...................................... 93
    A. C. Fabian
Relativistic Jets from X-ray Binaries: Recent Advances..................... 101
    R. P. Fender
The BeppoSAX HELLAS Survey: On the Nature of Faint
Hard X-ray Selected Sources.. 111
    F. Fiore, L. A. Antonelli, P. Ciliegi, A. Comastri, P. Giommi, F. La Franca,
    R. Maiolino, G. Matt, S. Molendi, G. C. Perola, and C. Vignali
Relativistic Flows in Blazars 120
    G. Ghisellini
BeppoSAX Observations of Bright Radio Galaxies........................ 130
    P. Grandi
Probing the Cosmic Dark Age in X-rays................................. 140
    Z. Haiman
High-Energy Emission from Isolated Pulsars............................. 150
    A. K. Harding
The 4 and One-Half (Plus or Minus One Half) Soft Gamma
Repeaters in Review... 160
    K. Hurley
X-ray Properties of Ultraluminous Infrared Galaxies (ULIGS) 169
    K. Iwasawa

Spectral Signatures of Kilohertz Quasiperiodic Oscillations from
Accreting Neutron Stars 179
    P. Kaaret
Deep ROSAT Surveys and the Contribution of AGNs to the
Soft X-ray Background 189
    I. Lehmann, G. Hasinger, M. Schmidt, J. E. Gunn, D. P. Schneider,
    R. Giacconi, M. McCaughrean, J. Trümper, and G. Zamorani
Obscured Active Galactic Nuclei 199
    R. Maiolino
The Broad Band Spectrum and Variability of Seyfert 1 209
    G. Matt
The Anomalous X-ray Pulsars 219
    S. Mereghetti
Beat-Frequency Models of Kilohertz QPOs 229
    M. Coleman Miller
Jets in Quasars, Microquasars, and Gamma-ray Bursts 239
    I. F. Mirabel
Isolated Neutron Stars Discovered by ROSAT............................. 244
    C. Motch
Transient Be Star Binary Systems 254
    F. Nagase
Iron K Line Emission in AGN: Observations.............................. 264
    K. Nandra
Photoionized Gas in Starburst and Active Galaxies 274
    H. Netzer
Hard X-ray Tails and Cyclotron Features in X-ray Pulsars................. 283
    M. Orlandini and D. Dal Fiume
X-ray Afterglows of Gamma-ray Bursts 295
    L. Piro
Time Lags in Compact Objects: Constraints on the Emission Models......... 310
    J. Poutanen
Low-Luminosity AGN and Normal Galaxies 326
    A. Ptak
The USA X-ray Timing Experiment 336
    P. S. Ray, K. S. Wood, G. Fritz, P. Hertz, M. Kowalski, W. N. Johnson,
    M. N. Lovellette, M. T. Wolff, D. Yentis, R. M. Bandyopadhyay,
    E. D. Bloom, B. Giebels, G. Godfrey, K. Reilly, P. Saz Parkinson,
    G. Shabad, P. Michelson, M. Roberts, D. A. Leahy, L. Cominsky,
    J. Scargle, J. Beall, D. Chakrabarty, and Y. Kim
Observing the Effects of Strong Gravity with Future X-ray Missions 346
    C. S. Reynolds
Radio-Loud AGNs: The X-ray Perspective................................ 355
    R. M. Sambruna and M. Eracleous
The Relativistic Precession Model for QPOs in Low Mass
X-ray Binaries... 365
    L. Stella

Oscillations during Thermonuclear X-ray Bursts: A New Probe of
Neutron Stars 377
    T. E. Strohmayer
Early Results from the Chandra Observatory 387
    H. Tananbaum
Results from X-ray Surveys with ASCA 396
    Y. Ueda
Kilohertz Quasi-periodic Oscillations—Observational Overview 406
    M. van der Klis
What Have We Learned about Gamma-ray Bursts from Afterglows? 416
    M. Vietri

ORAL PRESENTATIONS

Keck K-Band Observations of Low Mass X-ray Binaries 433
    P. J. Callanan, A. V. Filippenko, and M. R. Garcia
Intensive HST/RXTE Monitoring of NGC 3516: Evidence against
Thermal Reprocessing 437
    R. Edelson
Synchrotron and Compton Components and Their Variability in
BL Lac Objects 441
    P. Giommi, G. Ghisellini, P. Padovani, and G. Tagliaferri
A New Pulsar/SNR Pair: AX J1845-0258 in G29.6+0.1 445
    E. V. Gotthelf, G. Vasisht, K. Torii, and B. M. Gaensler
The X-ray Source Populations of the Magellanic Clouds 449
    F. Haberl
Discovery of Two High-Magnetic-Field Radio Pulsars 453
    V. M. Kaspi, F. Camilo, A. G. Lyne, R. N. Manchester, J. F. Bell,
    N. D'Amico, N. P. F. McKay, and F. Crawford
Evidence for Photon Bubble Oscillations (PBO) and Turbulence in
Centaurus X-3 457
    R. I. Klein, J. G. Jernigan, and J. Arons
ASCA Observations of the New Jet System XTE J1748-288 462
    T. Kotani, N. Kawai, F. Nagase, M. Namiki, M. Sakano, T. Takeshima,
    Y. Ueda, and M. Matsuoka
A BeppoSAX Observation of Nova Velorum 1999: A Very Bright
Classical Nova 466
    M. Orio, A. N. Parmar, L. Amati, R. Benjamin, M. Della Valle, F. Frontera,
    J. Greiner, T. Mineo, H. Ögelman, S. Starrfield, and E. Trussoni
A Self-consistent Test of Comptonization Models Using a Long
BeppoSAX Observation of NGC 5548 470
    P. O. Petrucci, F. Haardt, L. Maraschi, P. Grandi, G. Matt, F. Nicastro,
    L. Piro, G. C. Perola, and A. De Rosa
A ROSAT HRI Survey of Bright Nearby Galaxies 474
    T. P. Roberts and R. S. Warwick

A New Event Analysis Method for the X-ray Photon Count CCD 478
 T. G. Tsuru, H. Awaki, K. Koyama, K. Hamaguchi, H. Murakami,
 M. Nishiuchi, M. Sakano, and H. Tsunemi

Probing Dense Matter in the Cores of AGN: Observations with
RXTE and ASCA 482
 K. A. Weaver

POSTERS

Warm Comptonization in AGN: Effect on the Iron $K\alpha$ Line and the
Lyman Edge...489
 A. Abrassart and A. M. Dumont

Average Properties of Gamma-ray Bursts Spectra from 2 to 700 keV
with BeppoSAX..493
 L. Amati, F. Frontera, D. Dal Fiume, M. Orlandini, E. Palazzi, E. Costa,
 M. Feroci, P. Soffitta, M. N. Cinti, J. Heise, J. in't Zand, J. M. Muller,
 L. Nicastro, and M. Tavani

High-Energy Spectra of Seyfert 2 Galaxies..........................497
 L. Bassani, G. Malaguti, A. Malizia, J. B. Stephen, E. Caroli, G. Di Cocco,
 F. Frontera, and M. Trifoglio

New Results from the HRX BL Lac Sample.............................502
 V. Beckmann and A. Wolter

RXTE Monitoring of Centaurus A.....................................506
 S. Benlloch, R. E. Rothschild, J. Wilms, C. S. Reynolds, W. A. Heindl,
 K. Pottschmidt, A. Orr, I. Kreykenbohm, and R. Staubert

Electron Injection Break and Pair Content of Quasar Jets 510
 M. Błazejowski, M. Sikora, R. Moderski, and T. Bulik

Observing the Unobservable: Coronal Line Data from Narrow-Line
Seyfert 1s as a Test of EUV Continuum Models 514
 P. J. Bleackley and P. T. O'Brien

Low- and High-Frequency Quasi-periodic Oscillations in 4U1915-05:
Relation with Source State 518
 L. Boirin, D. Barret, and J. F. Olive

ROSAT Discovered Soft X-ray Intermediate Polars: UU Col and
RX J0806.3+1527 522
 V. Burwitz and K. Reinsch

A New BL Lac Sample from the REX Survey 526
 A. Caccianiga, T. Maccacaro, A. Wolter, R. Della Ceca, and I. M. Gioia

On the Fate of Stars near a Supermassive Black Hole 530
 A. Čadež and A. Gomboc

Deriving the Emissivity Law in Accretion Disks from X-ray
Iron Emission Lines 534
 A. Čadež, M. Calvani, C. Di Giacomo, and P. Marziani

Variability of the Relativistic Iron K-Line in NGC 3516 538
 A. Čadež and M. Calvani

An EUVE Observation of the Globular Cluster NGC 1851...............542
 P. J. Callanan, J. J. Drake, A. Fruscione, and D. Christian

BeppoSAX Observations of Cen X-4 in Quiescence 546
    S. Campana, L. Stella, S. Mereghetti, and D. Cremonesi
Multiple BeppoSAX Observations of IC 4329A to Probe the Origin of
the Compton Reflection Component in Seyfert 1 Galaxies 550
    M. Cappi, G. Di Cocco, M. Dadina, G. Malaguti, M. Matsuoka, G. Matt,
    G. C. Perola, and L. Piro
Hard Sources, Their Colours, and the X-ray Background 554
    F. J. Carrera, J. P. D. Mittaz, and M. J. Page
PKS 2155-304—A Source of Very High Energy Gamma Rays 558
    P. M. Chadwick, K. Lyons, T. J. L. McComb, K. J. Orford, J. L. Osborne,
    S. M. Rayner, S. E. Shaw, and K. E. Turver
BeppoSAX Observations of MKN 421: Clues on the
Particle Acceleration? 562
    G. Fossati, A. Celotti, M. Chiaberge, L. Chiappetti, and Y. H. Zhang
Very High Energy Gamma Rays from Cen X-3 566
    P. M. Chadwick, K. Lyons, T. J. L. McComb, K. J. Orford, J. L. Osborne,
    S. M. Rayner, S. E. Shaw, and K. E. Turver
Very High Energy Gamma-ray Observations of
Southern Hemisphere AGNs 570
    P. M. Chadwick, K. Lyons, T. J. L. McComb, K. J. Orford, J. L. Osborne,
    S. M. Rayner, S. E. Shaw, and K. E. Turver
Search for Interactions between Ejections of GRS 1915+105 and Its
Environment 574
    S. Chaty, L. F. Rodríguez, and I. F. Mirabel
EUVE/ASCA/RXTE Observations of NGC 5548 578
    J. Chiang, C. Reynolds, O. Blaes, M. Nowak, N. Murray, G. Madejski,
    H. Marshall, and P. Magdziarz
Super-agile—The X-ray Detector for the Gamma-ray Mission AGILE 582
    E. Costa, L. Barbanera, M. Feroci, M. Frutti, I. Lapshov, B. Martino,
    M. Mastropietro, E. Morelli, M. Rapisarda, A. Rubini, P. Soffitta,
    M. Tavani, S. Mereghetti, S. Vercellone, P. Caraveo, F. Perotti,
    G. Barbiellini, G. Budini, F. Longo, M. Prest, E. Vallazza, A. Morselli,
    P. Picozza, V. Cocco, C. Pittori, G. Di Cocco, and C. Labanti
New Extreme Synchrotron BL Lac Objects 586
    L. Costamante, G. Ghisellini, P. Giommi, G. Tagliaferri, A. Celotti,
    M. Chiaberge, L. Chiappetti, G. Fossati, L. Maraschi, F. Tavecchio,
    A. Treves, and A. Wolter
The Complex and Variable Absorption of NGC 3516 Observed by
BeppoSAX 590
    E. Costantini, C. Salvini, A. Comastri, A. Fruscione, S. Mathur, F. Nicastro,
    G. M. Stirpe, and B. Wilkes
Erratic Variability of LMC X-1 in the 1.5-10 keV Range:
Sporadic Presence of a QPO in a Black Hole Candidate 594
    D. Dal Fiume, F. Haardt, M. R. Galli, A. Treves, and L. Chiappetti
The Parkes Multibeam Pulsar Survey: Preliminary Results 598
    N. D'Amico, A. G. Lyne, R. N. Manchester, F. M. Camilo, V. M. Kaspi,
    J. Bell, I. H. Stairs, F. Crawford, D. Morris, and A. Possenti

The ASCA Hard Serendipitous Survey (HSS): A Progress Update............ 602
    R. Della Ceca, V. Braito, I. Cagnoni, and T. Maccacaro

BeppoSAX Observations of Asynchronous Magnetic Cataclysmic
Variables... 606
    D. de Martino, G. Matt, T. Belloni, K. Beuermann, B. T. Gänsicke,
    F. Haberl, M. Mouchet, K. Mukai, and J. M. Bonnet-Bidaud

The BeppoSAX Long Looks at the Seyfert 1 Galaxies NGC 5548 and
NGC 3783 610
    A. De Rosa, L. Piro, F. Nicastro, P. Grandi, M. Dadina, F. Fiore, F. Haardt,
    J. Kaastra, L. Maraschi, G. Matt, T. Mineo, G. C. Perola, P. O. Petrucci,
    and A. Treves

Discovery of Hard X-ray Emission from Type II Bursts of the
Rapid Burster 614
    F. Frontera, N. Masetti, M. Orlandini, L. Amati, E. Palazzi, D. Dal Fiume,
    S. Del Sordo, G. Cusumano, A. N. Parmar, G. Pareschi, I. Lapidus, and
    L. Stella

X-ray Beaming in the High Magnetic Field Pulsar GX 1+4 618
    D. K. Galloway and K. Wu

Variable Fe-Kα Line Profiles in Sy 1 Galaxies........................... 622
    J. Gelbord and K. A. Weaver

AGN X-ray Absorption vs. Optical Classification: Hints from the
XRB Models 626
    R. Gilli, G. Risaliti, and M. Salvati

IBIS/INTEGRAL Galactic Plane Observations: X-ray Novae 630
    P. Goldoni, A. Goldwurm, P. Laurent, and F. Lebrun

RXTE Studies of the Starburst Galaxies M82 and NGC253 634
    D. E. Gruber and Y. Rephaeli

Comptonization in X-ray Bright Neutron Star Globular Cluster
Systems... 638
    M. Guainazzi, A. N. Parmar, and T. Oosterbroek

Transient Taxonomy 642
    C. A. Haswell and A. R. King

The VSOP Mission and the Possible Future................................... 646
    H. Hirabayashi, Y. Murata, P. G. Edwards, and H. Kobayashi

Digitized Astronomical Plates: Optical Data for X-ray Astronomy............ 650
    R. Hudec

X-ray Astronomy 2000: Wide Field X-ray Monitoring with
Lobster-Eye Telescopes... 654
    A. Inneman, R. Hudec, L. Pina, and P. Gorenstein

Spectral Evolution during Dipping in X 1624-490 from the BeppoSAX
Observation 658
    P. J. Humphrey, M. J. Church, M. Bałucińska-Church, and A. N. Parmar

Optical Observations of Supersoft Source 0925-47 662
    J. B. Hutchings, A. P. Cowley, D. Crampton, and P. Schmidtke

X-ray Emission from Supernovae... 666
    S. Immler and W. Pietsch

Detection of Supernova Remnant and Black Hole Candidates in M83
with ROSAT 670
    S. Immler, M. Ehle, W. Pietsch, and A. Vogler

A Systematic Search for new X-ray Pulsators in Public ROSAT HRI
and BeppoSAX SMC Fields 674
    G. L. Israel, S. Campana, S. Covino, D. Dal Fiume, D. Lazzati,
    T. Oosterbroek, M. Orlandini, M. R. Panzera, A. N. Parmar, D. Ricci,
    G. Tagliaferri, and L. Stella

The Relativistic Astrophysics Explorer: A New Mission for X-ray
Timing 678
    P. Kaaret, J. Grindlay, F. K. Lamb, E. H. Morgan, J. H. Swank, and
    W. Zhang

Luminous Supersoft X-ray Emission from the Recurrent
Nova U Scorpii 682
    P. Kahabka

An Optical and X-ray Study of the Peculiar Narrow-Line Quasar
QSO 0117-2837 686
    S. Komossa, D. Grupe, V. Burwitz, and M. Janek

Simultaneous X-ray/Optical Burst from GS 1826-24 690
    A. Kong, L. Homer, E. Kuulkers, P. Charles, and A. Smale

An X-ray/TeV Gamma-ray Study of Mkn 501 during Its
Extraordinary Outburst of 1997 694
    H. Krawczynski, P. S. Coppi, T. Maccarone, and F. A. Aharonian

Wind Accretion in HMXRB 698
    I. Kreykenbohm, J. Wilms, P. Kretschmar, R. Staubert, R. E. Rothschild,
    W. A. Heindl, and D. E. Gruber

Two Cyclotron Lines in Vela X-1? 702
    I. Kreykenbohm, P. Kretschmar, J. Wilms, R. Staubert, W. A. Heindl,
    D. E. Gruber, and R. E. Rothschild

Her X-1 X-ray Turn-On Monitored by RXTE 706
    M. Kuster, J. Wilms, R. Staubert, D. E. Gruber, R. E. Rothschild, and
    W. A. Heindl

RXTE Observations of Seyfert Galaxies: Evidence for Reflection from
Disk and Torus 710
    G. Lamer, P. Uttley, and I. M. McHardy

Ginga Observations of the Short High State in Her X-1 714
    D. Leahy

Two Photon Bremsstrahlung in Strong Magnetic Fields 718
    D. Leahy and L. Semionova

A Series of Eclipses of Her X-1 Observed with RXTE 722
    D. Leahy and D. M. Scott

The Optically Bright REX Sample 726
    D. Lentini, R. Della Ceca, T. Maccacaro, A. Wolter, A. Caccianiga, and
    I. M. Gioia

The REX Survey: The Catalog 730
    T. Maccacaro, R. Della Ceca, A. Wolter, A. Caccianiga, and I. M. Gioia

Measurements of Fluctuations in the Hard X-ray Background with RXTE 734
    D. R. MacDonald, D. E. Gruber, and E. A. Boldt

The Radio to Hard X-ray Spectral Energy Distribution of BeppoSAX Emission Line AGN 738
    A. Malizia and F. Fiore

Time Dependent Comptonisation 742
    J. Malzac and E. Jourdain

Spectral Features from X-ray Illuminated Accretion Discs as Diagnostics of the Black Hole Angular Momentum 746
    A. Martocchia, G. Matt, and V. Karas

EPIC/XMM Calibrations 750
    P. Marty and J.-P. Bernard

BeppoSAX Spectra of Five Low Mass X-ray Binaries 754
    N. Masetti, E. Pian, R. Frontera, E. Palazzi, L. Amati, M. Orlandini, and D. Dal Fiume

X-ray Evidence of an AGN in M82 758
    H. Matsumoto and T. G. Tsuru

Resolving the 10-40 keV Cosmic X-ray Background with Constellation-X 762
    G. Matt, F. Pompilio, and F. La Franca

Correlated Variability in LMC X-2 766
    K. E. McGowan, P. A. Charles, D. O'Donoghue, and A. P. Smale

The Multicolour Disc Model Parameter for Black Hole Candidates: Can We Trust the Inner Disc Measurements? 770
    A. Merloni, A. C. Fabian, and R. R. Ross

Spatial Distribution of Spectral Characteristics of the Supernova Remnant Cas A 774
    T. Mineo, M. C. Maccarone, A. Preite-Martinez, J. Vink, and J. S. Kaastra

Variability of an Iron Line Profile and a Power-Law Continuum ... 778
    K. Misaki, H. Kunieda, and Y. Terashima

Why Do Broad Line AGN Show Up among Faint Hard X-ray Sources? 782
    J. P. D. Mittaz, M. J. Page, and F. J. Carrera

Hysteretic Behavior and Coherence of the Black Hole Candidate X-ray Binaries 786
    S. Miyamoto and S. Kitamoto

An Extended Multi-zone Model for the MCG-6-30-15 Warm Absorber 790
    R. Morales, A. C. Fabian, and C. S. Reynolds

Dust Scattered X-ray Halo of 4U 1538-52 Observed with ASCA ... 794
    F. Nagase, T. Dotani, T. Endo, H. Ozawa, S. Uno, T. Kotani, and T. Mihara

Short-Term Spectral Variations during X-ray Flares in Black Hole Candidates and AGNs 798
    H. Negoro

High-resolution Observations of X-ray Absorbers/Emitters ... 802
    F. Nicastro, M. Elvis, F. Fiore, G. Matt, and S. Savaglio

Intermediate Polar Spin Pulse Profiles 806
    A. J. Norton

Resonant Truncation of Be Disks and Type I Outbursts in Be/X-ray Binaries 810
    A. T. Okazaki and I. Negueruela

Time Lags in Low Mass X-ray Binaries 814
    J.-F. Olive and D. Barret

The X-ray Spectra of AGN Accretion Disk Coronae: First Results 818
    A. Orr, G. Torricelli-Ciamponi, and P. Pietrini

A Survey for Hard Spectrum ROSAT Sources. 822
    M. J. Page, J. P. D. Mittaz, and F. J. Carrera

Automatic Search for Periodic Sources in the ROSAT Database 826
    S. Paltani and P. Bartholdi

The X-ray Spectrum of [$OIII$]λ5007 Selected Seyfert 2 Galaxies 830
    A. Pappa, I. Georgantopoulos, G. C. Stewart, and A. L. Zezas

The 0.5-10 keV Spectra of Broad-Line Quasars and the X-ray Background 834
    A. Pappa, G. C. Stewart, I. Georgantopoulos, R. E. Griffiths, B. J. Boyle.
    T. Shanks, and O. Almaini

BeppoSAX Observations of the Her X-1 Short-On and Anomalous Low-States 838
    A. N. Parmar, T. Oosterbroek, D. Dal Fiume, M. Orlandini, A. Santangelo,
    S. Del Sordo, and A. Segreto

XEUS—The X-ray Evolving Universe Spectroscopy Mission. 842
    A. N. Parmar, T. Peacock, M. Bavdaz, G. Hasinger, M. Arnaud,
    X. Barcons, D. Barret, A. Blanchard, H. Böhringer, M. Cappi, A. Comastri,
    T. Courvoisier, A. C. Fabian, R. Griffiths, G. Malaguti, K. O. Mason,
    T. Ohashi, F. Paerels, L. Piro, J. Schmitt, M. van der Klis, and M. Ward

Spectral Evolution of GRS 1915+105 during the Regular Flares 846
    B. Paul and A. R. Rao

Detection of a Pulsating Soft Component in the X-ray Pulsar XTE J0111.2-7317 850
    B. Paul, J. Yokogawa, M. Ozaki, F. Nagase, D. Chakrabarty, and
    T. Takeshima

BeppoSAX Detection of Highly Ionized Emission and Absorption Features in M81 854
    S. Pellegrini, M. Cappi, L. Bassani, G. Malaguti, G. G. C. Palumbo, and
    M. Persic

The BeppoSAX GRATIS Survey 858
    M. Perri, P. Giommi, F. Fiore, and M. Capalbi

Pair Creation at Shocks: A Possible Origin of the High-Energy Variability of AGNs 862
    P. O. Petrucci, G. Henri, and G. Pelletier

BeppoSAX Observations of Markarian 501 in June 1999 866
    E. Pian, L. Chiappetti, P. Giommi, F. Tavecchio, L. Maraschi, E. Palazzi,
    F. Aharonian, M. Catanese, A. Celotti, B. Degrange, A. Djannati-Atai,
    G. Fossati, G. Ghisellini, H. Krawczynski, C. M. Raiteri, R. M. Sambruna,
    D. Smith, G. Tagliaferri, G. Tosti, A. Treves, C. M. Urry, and M. Villata

Accretion Disk Boundary Layers in Low-Mass X-ray Binaries.............870
      R. Popham and R. Sunyaev

Weighing the Black Hole in Narrow-Line Seyfert 1 Galaxy,
RE J1034+396 874
      E. M. Puchnarewicz, K. O. Mason, A. Siemiginowska, I. Cagnoni,
      A. Comastri, F. Fiore, and A. Fruscione

Spectral Signatures of Reprocessing on Hercules X-1/Hz Herculis............878
      H. Quaintrell, M. D. Still, S. D. Vrtilek, B. Boroson, and P. Roche

X-rays from the Seyfert Galaxy IC 4329A, Its Neighbours, and Its
Galaxy Group 886
      A. M. Read and W. Pietsch

Nature of "Passive" Elliptical Galaxies....................................890
      T. A. Rector and J. T. Stocke

PDS 456: An Extreme Accretion Rate Quasar?............................894
      J. Reeves, P. O'Brien, S. Vaughan, D. Law-Green, M. Ward, C. Simpson,
      K. Pounds, and R. Edelson

The X-ray Properties of Luminous Infrared Galaxies 898
      G. Risaliti, R. Gilli, R. Maiolino, and M. Salvati

The Orbital Light Curve of Aquila X-1 902
      E. L. Robinson, W. F. Welsh, and P. Young

Magnetic Field Limit on SGR 1900+14 906
      R. E. Rothschild, D. Marsden, and R. E. Lingenfelter

X-ray Iron Line Variability for the Model of an Orbiting Flare above
a Black Hole Accretion Disc 910
      M. Ruszkowski

The Influence of Resonant Absorption on the Fe Emission Line
Profiles from Accreting Black Holes 914
      M. Ruszkowski and A. C. Fabian

Morphological Analysis of a Statistically Complete X-ray Selected
Sample of Seyfert Galaxies 918
      M. Salvato, P. Böhm, J.-U. Fischer, G. Hasinger, I. Lehmann, and
      P. Rafanelli

Irradiation of the Secondary Star in X-ray Nova Scorpii 1994
(=GRO J1655-40)..922
      T. Shahbaz, P. Groot, S. N. Phillips, J. Casares, P. A. Charles, and
      J. van Paradijs

Faint-Source Contributions to the Extragalactic X-ray Background in
an XMM Deep Field..927
      R. Shirey, F. Cordova, J. Kennea, D. Pandel, T. Sasseen, and J. West

BeppoSAX Observations of the Galactic Center Region: Soft X-rays
from the Radio Halo of SGRA East.....................................931
      L. Sidoli, S. Mereghetti, A. Treves, L. Chiappetti, G. L. Israel, and
      M. Orlandini

On Pair Content of Quasar Jets..935
      M. Sikora and G. Madejski

CCD Photometry of Outbursts in GK Per 939
      V. Šimon and Z. Velič

Color Variations during Activity States of the Binary V SGE 943
    V. Šimon and S. Shugarov
Activity of the Dwarf Nova CH Uma 947
    V. Šimon
ROSAT Reveals the Large Scale Distribution of Matter 951
    A. Sołtan, M. Freyberg, G. Hasinger, T. Miyaji, M. Treyer, and J. Trümper
The IBIS Data Handling System 955
    M. Stuhlinger, E. Goehler, C. Dreischer, R. Volkmer, R. Staubert,
    E. Kendziorra, N. v.Krusenstiern, P. Risse, and R. Weiss
The Population of Faint X-ray Sources in the Galaxy and Their
Contribution to the Galactic Ridge X-ray Emission 959
    M. Sugizaki, K. Matsuzaki, H. Kaneda, S. Yamauchi, K. Mitsuda, and the
    ASCA Galactic Plane Survey Team
An H-R Diagram for AGN? 963
    J. W. Sulentic, P. Marziani, and M. Calvani
On the Modelling of Time-Dependent Phenomena around Black Holes 967
    E. Szuszkiewicz and J. C. Miller
Flaring Blazars with BeppoSAX 971
    G. Tagliaferri, G. Ghisellini, P. Giommi, L. Chiappetti, L. Maraschi,
    A. Celotti, M. Chiaberge, A. Comastri, L. Costamante, G. Fossati,
    E. Massaro, R. Nesci, E. Pian, C. M. Raiteri, M. Ravasio, F. Tavecchio,
    G. Tosti, A. Treves, M. Villata, and A. Wolter
Gamma-Loud Quasars: A View with BeppoSAX 975
    F. Tavecchio, L. Maraschi, G. Ghisellini, A. Celotti, L. Chiappetti,
    A. Comastri, G. Fossati, P. Grandi, F. Haardt, E. Pian, G. Tagliaferri,
    A. Treves, C. M. Raiteri, R. Sambruna, and M. Villata
Constraints to the SSC Model for MKN 501 979
    F. Tavecchio and L. Maraschi
Variability of the Iron Line in the Low-Luminosity AGN NGC 4579 983
    Y. Terashima, T. Yaqoob, P. J. Serlemitsos, H. Kunieda, K. Misaki,
    A. F. Ptak, and L. C. Ho
X-ray Properties of FR I Radio Galaxies 987
    E. Trussoni, L. Feretti, A. Capetti, A. Celotti, and M. Chiaberge
Tartarus—An ASCA AGN Database 991
    T. J. Turner, K. Nandra, D. Turcan, and I. M. George
X-ray Spectral Complexity in Narrow-Line Seyfert 1 Galaxies 995
    S. Vaughan, J. Reeves, R. Warwick, R. Edelson, and K. Pounds
ASCA View on High-Redshift Radio-Quiet Quasars 999
    C. Vignali, A. Comastri, M. Cappi, G. G. C. Palumbo, and M. Matsuoka
The Warm Scattering Medium in NGC 4151 1003
    R. S. Warwick and R. G. Griffiths
The X-ray Spectra of Symbiotic Stars 1007
    P. J. Wheatley
Hard Synchrotron BL Lacs: The Case of 1ES 1101-232 1011
    A. Wolter, G. Ghisellini, G. Tagliaferri, F. Tavecchio, and A. Caccianiga
ASCA/ROSAT Observations of PKS 2316-423: Spectral Properties of
a Low Luminosity Intermediate-Type BL Lac Object 1015
    S.-J. Xue and Y.-H. Zhang

**Energy Dependent X-ray Variability of the TeV Blazars
PKS 2155-304 and MKN 421** 1019
    Y. H. Zhang, A. Celotti, A. Treves, G. Fossati, L. Maraschi, E. Pian,
    S. Paltani, F. Tavecchio, L. Chiappetti, G. Ghisellini, G. Tagliaferri, and
    M. Chiaberge

**Structure of the Circumnuclear Region of Seyfert 2 Galaxies—Clues
from RXTE Observations of NGC 4945** 1023
    P. Życki, G. Madejski, and C. Done

The Constellation X-ray Mission 1027
    N. E. White and H. Tananbaum

List of Participants 1031
Author Index 1035

PREFACE

This conference took place at a time when results from the new and powerful X-ray observatories Chandra and XMM-Newton were keenly anticipated. The meeting marked an opportunity to look back on a successful decade of results from the ROSAT, ASCA, RXTE and BeppoSAX missions. This meeting was a successor to a similar meeting held ten years previously, also in Bologna, the 23rd ESLAB Symposium on "Two Topics in X-ray Astronomy" (ESA SP-296). The two topics covered at both meetings were: 1) X-ray Binaries and 2) AGN and the X-ray background. X-ray binaries provide a natural laboratory for probing the extreme environments of black holes and neutron stars. A substantial fraction of the X-ray background is the superposition of active galactic nuclei (AGN), which are powered by super-massive black holes.

The Meeting was organized jointly by ITESRE/CNR and NASA/GSFC with support from ASI. The program was developed and selected by the Scientific Organizing Committee (SOC). The SOC was chaired by N. E. White and split into two parts to deal with the two topics covered by the meeting:

Stellar Endpoints:

D. Barret, L. Bildsten, G. Di Cocco, F. Haberl, N. Kawai, M. van der Klis, F. Nagase, F. Paerels, A. Parmar, L. Stella, R.A. Sunyaev

AGN & the X-ray background:

N. Brandt, X. Barcons, T. Courvoisier, G. Hasinger, A.C. Fabian, T. Maccacaro, G.G.C. Palumbo, L. Piro, H. Tananbaum, J. Turner, M. Urry, M. Ward

The local organizing committee consisted of G. Malaguti (chair), L. Bassani, M. Cappi, A. Comastri, D. Dal Fiume, M. Orlandini, E. Palazzi, E. Pian, and A. Spizzichino who ensured a very smoothly run meeting.

The meeting itself proved to be very timely and gave an excellent summary of the past discoveries, as well as the science challenges awaiting the new missions that would dominate the first decade of the 21st century. The first public presentation of the exciting results from the newly launched Chandra X-ray Observatory were a highlight of the meeting.

INVITED PAPERS

X-RAY SOURCES AS TRACERS OF THE LARGE-SCALE STRUCTURE IN THE UNIVERSE

X. Barcons [1], F.J. Carrera [2,1], M.T. Ceballos [1], S. Mateos [1,3]

1) Instituto de Física de Cantabria (CSIC-UC), 39005 Santander, Spain
2) Mullard Space Science Laboratory, University College London, UK
3) Departamento de Física Moderna, Universidad de Cantabria, 39005 Santander, Spain

ABSTRACT We review the current status of studies of large-scale structure in the X-ray Universe. After motivating the use X-rays for cosmological purposes, we discuss the various approaches used on different angular scales including X-ray background multipoles, cross-correlations of the X-ray background with galaxy catalogues, clustering of X-ray selected sources and small-scale fluctuations and anisotropies in the X-ray background. We discuss the implications of the above studies for the bias parameter of X-ray sources, which is likely to be moderate for X-ray selected AGN and the X-ray background ($b_X \sim 1-2$). We finally outline how all-sky X-ray maps at hard X-rays and medium surveys with large sky coverage could provide important tests for the cosmological models.

KEYWORDS: Large-scale structure of Universe; cosmology; galaxies: active,clustering.

1. INTRODUCTION

In the current cosmological picture, galaxies, clusters and large-scale structures have grown from initial small perturbations in the density of the Universe via gravitational collapse. Cosmological models are required to meet two basic observational constraints: on the one hand the Universe at $z \sim 1500$ was very smooth, as the cosmic microwave background (CMB) is seen to have anisotropies of amplitude $\sim 10^{-5}$; on the other hand local mass inhomogeneities measured through the distribution of galaxies exhibit fluctuations of the order ~ 1 on scales $\sim 10\,\mathrm{Mpc}$. Different cosmologies, however, predict highly discrepant ways in which structures on different scales grow up to the current state from the CMB initial conditions. The largest discrepancies occur at redshifts $z \sim 1-5$ which is when galaxies began to collapse and to form stars. Accessing these intermediate redshifts will provide crucial tests for the cosmological models.

The isotropy of the cosmic X-ray background (XRB) on large angular scales ($\frac{\Delta I}{I}$ less than a few % on scales of degrees and larger) suggests that most of the X-ray photons we receive from the Universe must have been originated in the distant Universe. Surveys at different depths carried out with *ROSAT* have revealed that 50-70% of the (soft) XRB is resolved into point sources, mostly Active Galactic

Nuclei (AGN) of different classes. Although there are still some discrepancies in the determination of the X-ray luminosity function and its redshift evolution, there is no doubt that most of the XRB originates at redshift $z > 1$. Boyle et al (1994) and Page et al (1996) who find their samples of X-ray selected AGN consistent with pure luminosity evolution models, predict a peak in the X-ray volume emissivity around $z \sim 1.5 - 2$. Miyaji et al (1998) instead find better consistency with luminosity dependent density evolution, in which case the X-ray volume emissivity in AGN more luminous than $10^{44.5}$ erg s^{-1} (which for the broken power-law shape of the luminosity function account for most of the X-rays emitted by AGN) rises steeply from $z = 0$ to $z = 1 - 2$ with no evidence for a decline at higher redshifts. In both cases it is clear that soft X-ray emission from the extragalactic sky comes mostly from redshifts $z = 1 - 2$ or larger, in a situation very similar to the star formation in the Universe (Madau et al 1996, Boyle & Terlevich 1998). Studying the X-ray Universe is then likely to provide a major handle to understand the evolution of the Universe at intermediate redshifts and therefore it is an issue of prime cosmological relevance.

There are other reasons to prefer X-rays to carry out cosmological studies. On the one hand the high-latitude X-ray sky is 'clean', at least at photon energies above 2 keV, galactic absorption has negligible effects and the contribution of the Galaxy to the XRB is less than a few % (Iwan et al 1982). A further reason is the small content in stars of high galactic latitude surveys, ranging from 25% at bright fluxes down to probably less than 10% at the faintest fluxes.

In this paper we review the current status of studies of the large-scale structure of the Universe, which up to now has produced relevant but certainly not spectacular results. The two main questions that we address are:

- Do X-ray sources (and the XRB) trace mass in the Universe and what is their bias parameter?

- What are the best observational approaches to obtain information on the large-scale structure of the Universe at intermediate redshifts with X-rays?

Except when otherwise stated we use $H_0 = 100\,h\,\mathrm{km\,s^{-1}\,Mpc^{-1}}$, $q_0 = 0.5$ and $\Lambda = 0$.

2. THE X-RAY SKY ON THE LARGEST SCALES

The distribution of the XRB fluctuations on the largest scales and their link to inhomogeneities in the distribution of matter has been an active field of research for many years. The observational resources have been mostly limited to the HEAO-1 A2 experiment which scanned the sky with a resolution of $3° \times 1.5°$ at photon energies 2-60 keV.

	l (deg)	b (deg)	Err (deg)	XRB ampl (%)	Ref
CMB	264	48		0.15	F96
AGN	318	38	30		MB90
Clusters	260	5	15		PK98
Soft XRB	288	25	19	1.7	PG99
Hard XRB	338	47	25	0.11	S99

TABLE 1. Dipoles of X-ray source populations and the XRB. References are Fixsen et al (1996); MB80: Miyaji & Boldt (1980); PK98: Plionis & Kolokotronis (1998); PG99: Plionis & Georgantopoulos (1999); S99: Scharf et al (1999).

2.1. The dipole of X-ray sources

Since the Galaxy is moving with respect to the frame where the CMB would be isotropic towards $l = 264°, b = 48°$, there must be an overdensity of sources which are pulling us towards that direction. The distribution of X-ray sources in the sky should therefore exhibit an approximate large-scale dipolar distribution pointing towards the same direction.

Using the AGNs in the Piccinotti et al (1982) flux-limited sample of X-ray sources (2-10 keV flux limit $\sim 3 \times 10^{-11}$ erg cm^{-2} s^{-1}), Miyaji & Boldt (1990) and Miyaji (1994) found the dipole of these sources to point towards $l = 318°, b = 38°$ with a large error circle ($\sim 30°$ radius). The dipole appears to saturate at $50-100\,h^{-1}$ Mpc and is roughly aligned with the CMB dipole. Within the framework of linear theory, this allows the bias parameter of the X-ray selected AGN to be estimated, giving a somewhat large value ($b_X \Omega_0^{-0.6} \sim 3-6$). Uncertainties come primarily from the indetermination of the redshift at which the dipole saturates.

Plionis & Kolokotronis (1998) and Kolokotronis et al (1998) have measured the dipole of an X-ray flux-limited sample of galaxy clusters. This is again in rough alignment with the CMB dipole, but it appears to saturate at $\sim 160 h^{-1}$ Mpc. As expected in all popular scenarios where clusters arise in extreme peaks of the underlying dark-matter distribution, they exhibit a large bias parameter ($b_X \sim 4$, see Table 4).

The fact that the dipoles of the two most numerous classes of extragalactic X-ray sources (AGNs and clusters) are roughly aligned with the CMB dipole is encouraging. We note, however, that all-sky deeper samples of these objects (particularly X-ray selected AGN) would enormously help in defining the distance at which the contribution to the dipole saturates and therefore in measuring the bias parameter.

2.2. The dipole of the X-ray background

There are two reasons why the XRB should show a dipole signal: our motion relative to the CMB rest frame (the so-called Compton-Getting effect) and the excess contribution of the sources that cause this motion in the same direction. The XRB dipole is expected to be aligned with the CMB dipole, but the amplitude should be larger than the Compton-Getting effect, allowing for the excess emissivity.

There are two basic problems in measuring the XRB dipole: one is the contribution of the Galaxy and the other one is the integrated nature of the XRB whereby confusion noise dominates on all angular scales. Warwick, Pye & Fabian (1980) realized that even at photon energies > 2 keV and galactic latitudes $\mid b \mid$> 20° a residual galactic contribution $\sim 2-7\%$ is present. Iwan et al (1982) modelled this galactic component in terms of a finite radius disk with thermal spectrum at $T \sim 9$ keV. To emphasize how difficult is to obtain the extragalactic signal, the galactic contribution amounts to a few % at the galactic poles, while the effect it is being looked for is less than 1%.

Attempts to look for singular enhancements of the XRB surface brightness include those by Warwick et al (1980), the Jahoda & Mushotzky (1989) search for emissivity from the great attractor, the Mushotzky & Jahoda (1992) search for XRB negative fluctuations towards the most prominent voids and the unsuccessful detection of X-ray emission from superclusters by Persic et al (1990).

By modelling out the Galaxy, Shafer (1983) and Shafer & Fabian (1983) found a dipole signal significant at $\sim 2\sigma$ level in the HEAO-1 A2 map. Most of the subsequent dipole refinements have used the same data with increasingly finer corrections for detector drifts and other unwanted effects. The latest one is by Scharf et al (1999), who excluded the galactic plane, the Magellanic clouds and also regions around the Piccinotti et al (1982) sources, which leaves less than 50% of the sky for the dipole analysis. Various methods are used to deal with the masked regions (including spherical harmonic reconstruction) and the results are shown in Table 1. The dipole signal is very clearly detected and its intensity appears larger than the Compton-Getting effect. The direction of this extra large-scale structure dipole caused by the fluctuations in the source density is only roughly aligned with the direction of our motion, and its amplitude is similar to the Compton-Getting effect as predicted by theory (Lahav, Piran & Treyer 1997).

In an analysis of the *ROSAT* all-sky data (0.9-2.4 keV), Plionis & Georgantopoulos (1999) also find a dipole component. The Galaxy is modeled according to the Iwan et al (1982) model and they further exclude other regions associated with the Galaxy. The direction of the resulting dipole is in better agreement with the CMB dipole, but the amplitude is almost a factor of 10 larger than the Compton-Getting effect.

There are various reasons for the discrepancy between these measurements. First, an extra residual contribution from the Galaxy is likely to contaminate more strongly the *ROSAT* data than the HEAO-1 A2 data. This would explain why the *ROSAT* dipole points closer to the galactic plane and that its amplitude is larger. A second reason for the discrepancy is the fact that Scharf et al (1999) have excluded regions around the galaxy clusters present in the Piccinotti et al (1982) sample (which are known to have a very large bias parameter and represent 50% of the extragalactic sources in that sample) but Plionis & Georgantopoulos (1999) have not. In fact these last authors note that the contribution from the Virgo cluster alone is of the order of 20% of the detected dipole. A good exercise which could give some insight on the level of the galactic contamination in the $ROSAT$ data would be to exclude the clusters in the $ROSAT$ analysis and not excise the Piccinotti et al (1982) sources from the analysis of the A2 data.

2.3. Higher order multipoles of the X-ray background

Lahav, Piran & Treyer (1997) proposed the use of a multipole expansion of the angular variations of the XRB in order to measure the large-scale structure of the Universe. Under fairly general assumptions, the coefficients a_{lm} of the harmonic expansion would be the sum of a large-scale structure term $a_{lm}^{(LLS)} \propto l^{-0.4}$ and a confusion noise term which is a function of the flux S_{cut} down to which sources have been excised from the maps for the multipole analysis $a_{lm}^{(Noise)} \propto S_{cut}^{\gamma-1}$, where γ is the slope of the *integral* source counts in the energy band used ($N(>S) \propto S^{-\gamma}$).

Treyer et al (1998) performed this analysis on the HEAO-1 A2 all sky data by removing regions around the Piccinotti et al (1982) sample and the galactic plane. They find evidence for a growth of the spherical harmonic coefficients growing at low values of l in a manner roughly consistent with the predictions. The significance of the signal is difficult to assess as the harmonic coefficients are not independent due to cross-talk between different orders introduced by the masking. Assuming a redshift dependent bias parameter for the X-ray sources parametrized as $b_X(z) = b_X(0) + z[b_X(0) - 1]$ (which assumes that all galaxies form at some past epoch, Fry 1996), they estimate a rather modest bias parameter ($1.0 < b_X(0) < 1.6$). In their diagrams it is also seen that the dipole ($l = 1$) has an unusually large amplitude compared to higher harmonics.

The way to go is indeed to have precise measurements of the XRB intensity on large angular scales, but with the possibility of excluding sources down to the faintest possible levels. Treyer et al (1998) suggest that an all-sky map with XRB intensities measured with a 1% precision and with sources excised down to 3×10^{-13} erg cm^{-2} s^{-1} (i.e., 100 times fainter than the Piccinotti et al catalogue) would be ideal for a spherical harmonic analysis.

3. CROSS-CORRELATIONS OF GALAXY CATALOGUES WITH XRB INTENSITIES

An alternative way that has been devised to look for structure in the X-ray sky is to cross-correlate the unresolved XRB intensity with catalogues of galaxies. The amplitude of the cross-correlation function (CCF) between the X-ray intensity I_{XRB} and the galaxy surface density N_g, $W_{Xg}(\theta) = \langle I_{XRB} N_g \rangle_\theta / \langle I_{XRB} \rangle \langle N_g \rangle - 1$ at zero-lag ($\theta = 0$) provides an approximate measurement of the fraction of the XRB arising either in the catalogued galaxies or in sources clustered with them within a scale of the beam with which X-ray observations have been obtained (Lahav et al 1993).

Positive signals have been found for W_{Xg}, typically of the order of 1% when the galaxies are optically or infrared selected, and up to $> 10\%$ when active galaxies

XRB data	Galaxy catalogue	$W_{XG}(0)$	Ref
HEAO-1 A2	IRAS 2Jy	7×10^{-3}	M94
GINGA NGP	IRAS 0.7Jy	1.4×10^{-2}	C95
GINGA NGP	UGC	1.1×10^{-2}	C95
HEAO-1 A2	IRAS 12μ all	$< 9.6 \times 10^{-3}$	B95
HEAO-1 A2	IRAS 12μ Sy1	1.1×10^{-1}	B95
HEAO-1 A2	IRAS 12μ Sy2	3.1×10^{-2}	B95

TABLE 2. Cross-correlation signal at zero-lag (W_{Xg}) of galaxy catalogues with XRB data either from the HEAO-1 A2 experiment or the GINGA North Galactic Pole scans. References are M94: Miyaji et al (1994); C95: Carrera et al 1995; (B95): Barcons et al (1995).

are selected (see table 2). The interpretation of this signal requires to model the clustering of X-ray sources around the catalogued galaxies, which is indeed modulated by the bias parameter b_X. Using $b_X = 1$, it is found that the local volume emissivity of optically selected galaxies amounts to $\sim 10^{39}\, h\,\mathrm{erg\,s^{-1}\,Mpc^{-3}}$ (Lahav et al 1993; Miyaji et al 1994; Carrera et al 1995), most of which is contributed by Seyfert galaxies and QSOs (Barcons et al 1995).

When this volume emissivity is extrapolated to higher redshifts, the fraction of the XRB intensity due to the precursors of the catalogued galaxies can be predicted (Lahav et al 1993). Carrera et al (1995) find that 10-30% of the hard X-ray background might be produced by optically selected galaxies without exceeding the upper limits on the autocorrelation function of the XRB. This value is similar to the result of cross-correlation analyses of deep *ROSAT* X-ray images with deep optical images in the same fields (Almaini et al 1997).

A constraint on the bias parameter of X-ray sources from the CCF results can be derived by taking into account that a fraction $f \approx 2/3$ of the CCF signal arises from sources clustered with the catalogued galaxies. As that contribution scales linearly with b_X, the local volume emissivity scales $\propto \frac{3}{1+2b_X}$. Since the AGN-only local emissivity is also $\sim 10^{39}\,\mathrm{erg\,cm^{-2}\,s^{-1}}$, we can safely derive that $b_X < 2$ as otherwise the total volume emissivity will be significantly less than the AGN emissivity.

4. CLUSTERING OF X-RAY SELECTED SOURCES

In the recent years large, complete samples of X-ray selected AGN have been built. This has allowed the direct measurement of the 3D spatial correlation function $\xi(r)$ for these objects and its comparison with the spatial correlation function of galaxies selected at other wavebands.

Carrera et al (1998) used two complete samples of X-ray selected AGN in pencil beam survey regions, spanning a wide redshift range ($0 < z < 2$) to search for clustering signals and deriving its amplitude and redshift evolution. Clustering is found to be 99% significant at $z < 1$. When the spatial correlation function is fitted to a standard power-law form $\xi(r) = (1 + z)^p (\frac{r}{r_0})^{-1.8}$ (for comoving r), it is seen that comoving or slower clustering evolution is excluded, and that even for stable or linear growth the values of r_0 permitted by the data are of the same order as the ones derived from clustering of $IRAS$ galaxies. Carrera et al (1998) conclude that X-ray selected AGN are not significantly biased $0.7 < b_X < 2$.

Akylas, Plionis & Georgantopoulos (1999) have used the $ROSAT$ All Sky Survey sources to derive a local angular correlation function from which they estimate a somewhat higher correlation length ($r_0 \sim 7 - 9\,\mathrm{Mpc}$), consistent with optically selected QSO clustering (La Franca et al 1998) and comoving clustering evolution. The obvious weakness of this method is that it is not based on 3D but 2D data.

5. FLUCTUATIONS AND ANISOTROPIES IN THE XRB

The method of auto-correlating the XRB intensity at various separations has been extensively used in an effort to detect small scale structure in the XRB attributable to source clustering (Barcons & Fabian 1989, De Zotti et al 1990, Jahoda & Mushotzky 1991, Carrera et al 1991, Carrera & Barcons 1992, Carrera et al 1993, Chen et al 1994). These works produced a set of upper limits for the auto-correlation function of the XRB $W_{XX}(\theta) = \langle I_{XRB} I_{XRB} \rangle_\theta / \langle I_{XRB} \rangle^2 - 1$ on different angular scales (except for the Jahoda & Mushotzky 1991 work, which claimed a detection at separations $\sim 10°$) of the order of $10^{-3} - 10^{-4}$ which constrained the clustering properties of the underlying source population (see, e.g., Fabian & Barcons 1992).

Under the assumption of comoving clustering evolution, the sources of the XRB cannot be more strongly clustered than optically selected galaxies (see Carrera & Barcons 1992), in which case $b_X \sim 1$. However, as explained above, Carrera et al (1998) found marginal evidence for faster clustering evolution in samples of X-ray selected AGN. This means that b_X could be higher without violating the upper limits on the autocorrelation function, as the sources that produce the bulk of the XRB at high redshift could be very weakly clustered.

A further method employing the XRB angular variations has been to search for fluctuations in the XRB intensity distribution in excess of the ones expected from confusion noise produced by unresolved sources. These excess fluctuations should then be attributed to source clustering if all remaining noises (counting noise, systematics, etc.) could be removed. Studies of this kind have invariably lead to upper limits summarized in Table 3. What actually limits the sensitivity of this method is the statistics: it scales as $N_{obs}^{-1/2}$ where N_{obs} is the number of independent measurements of the XRB intensity that have been used to derive the excess fluctuations.

Excess fluctuations are related to the power spectrum of the density field of the Universe, weighted with the X-ray volume emissivity as a function of redshift (Barcons, Fabian & Carrera 1998). The method is potentially very powerful as it reflects the clustering properties of the sources that produce the bulk of the XRB at redshifts $z > 1$.

XRB data	Beam	$\left(\frac{\Delta I}{I}\right)_{excess}$	Ref
HEAO-1 A2	$5° \times 5°$	< 0.02	S83
Ginga	$2° \times 1°$	< 0.04	B97
ROSAT	$\pi \times (2.5')^2$	< 0.07	CFB97
ROSAT	$\pi(10'^2 - 5'^2)$	< 0.12	CFB97

TABLE 3. Upper limits to excess fluctuations. References are S83: Shafer (1983); B97: Butcher et al (1997); CFB97: Carrera, Fabian & Barcons (1997)

Measurement	Reference	Scale (Mpc)	b ($\Omega_0 = 1$)
X-ray cluster dipole	PK98	10-100	4
X-ray AGN dipole	MB90	1000	3-6
XRB-galaxy CCF		10-100	< 2
XRB dipole vs bulk motions	S99	1000	2-7
XRB multipoles vs bulk motions	T98	100-1000	1-2
Clustering of distant AGN	C98	10-100	1-2
Clustering of nearby AGN	APG99	100	2-3

TABLE 4. Bias parameters as inferred from various measurements. references are: PK98 Plionis & Kolokotronis (1998); MB90: Miyaji & Boldt (1990); S99: Scharf et al (1999); T98: Treyer et al (1998); C98: Carrera et al (1998); APG99: Akylas, Plionis & Georgantopoulos (1999).

6. THE BIAS PARAMETER OF X-RAY SOURCES

Over the past sections we have discussed various approaches to detect and measure the clustering properties of X-ray sources. Table 4 summarizes the inferred bias parameter b_X from these studies. Measurements are carried out with a variety of methods, correspond to different objects, are sensitive to different redshifts and also to different scales. Besides that, all dynamical estimates actually measure the combination $b_X \Omega_0^{-0.6}$.

Measurements of the correlation function are also affected by the cosmological parameters in the computation of the distances at significant redshifts, beyond the obvious linear dependence on H_0. If we live in an accelerating Universe, the Carrera et al (1998) correlation length would have to be scaled up by 30-50%, resulting in a subsequent increase of almost a factor of 2 in the bias parameter. Given the uncertainties in the values of q_0 and Λ (even for a flat Universe), the Carrera et al (1998) and Akylas et al (1999) results cannot be considered inconsistent.

As expected, clusters are a largely biased population ($b_X \sim 4$) compared to AGN ($b_X \sim 1-2$). The multipoles of the XRB are expected to be dominated by AGN, as these objects are the main sources of the XRB. The bias parameter derived from the XRB multipoles is consistently in agreement with the bias parameter derived from AGN clustering ($b_X \sim 1-2$). The exception to this is the XRB dipole which implies a larger value of b_X. This could be partly due to a larger cluster contribution, as the lowest order multipoles are most sensitive to nearest (and brightest) sources, where the cluster contribution to the source counts ($\sim 10\%$ on average in the deep extragalactic surveys) is $\sim 50\%$ for the Piccinotti et al (1982) sample.

7. FUTURE PROSPECTS

X-ray astronomy is now in a position to address cosmological studies. X-ray selected AGN which produce most of the X-rays in the Universe, appear to trace mass with a moderate bias parameter $b_X \sim 1-2$, but that has to be better defined as a function of scale and redshift. *Chandra* and XMM will carry out several deep 'pencil beam' surveys which, after subsequent identification of the serendipitous sources discovered, will define the redshift evolution of the AGN X-ray luminosity function at photon energies > 2keV and therefore the X-ray volume emissivity as a function of redshift. However, these surveys will not map sufficiently large areas of the sky which are necessary to trace the large-scale structure of the Universe at the redshifts where the XRB was produced.

The obvious way to go would be to survey very large areas of the sky (the whole sky even better) for X-ray sources, in order to have a most complete picture. Unless hard X-rays are produced at significantly lower redshifts than soft X-rays (which is doubtful in view of the *ASCA* and *BeppoSAX* surveys), to reach $z \sim 1$ where a significant fraction of the X-ray emissivity in the Universe resides, these surveys will have to go at least down to $\sim 10^{-14}$ erg cm^{-2} s^{-1}.

There is an alternative which is to perform high sensitivity observations of the XRB with a beam corresponding to the linear scale to be probed (Barcons, Fabian & Carrera 1998). As the peak of the power spectrum of the density field of the Universe occurs at comoving wavenumbers $\sim 0.01 - 0.1\,h\,\text{Mpc}^{-1}$, for a standard geometry a $1°$ resolution is well matched to this at $z \sim 1-3$. All-sky measurements of the XRB intensity on that angular scale with a precision of a few % could then be used to detect the excess fluctuations due to source clustering which are expected to be just below 1% in amplitude. Controlling all other possible sources of excess fluctuations well below that level requires a stable large-area detector (to reduce photon counting noise) and probably an X-ray monitor which images simultaneously the brightest sources in the field.

ACKNOWLEDGEMENTS

Partial financial support for this work was provided by the DGESIC under project PB95-0122.

REFERENCES

Akylas, T., Plionis, M., Georgantopoulos, 1999, in: MPA/ESO conference on Evolution of Large Scale Structure, in the press
Almaini, O. et al 1997, MNRAS, 291, 372
Barcons, X., Fabian, A.C., 1989, MNRAS, 237, 119
Barcons, X., Franceschini, A., De Zotti, G., Danese, L., Miyaji, T., 1995, ApJ, 455, 480
Barcons, X., Fabian, A.C., Carrera, F.J., 1998, MNRAS, 293, 60
Boyle, B.J. et al, 1994, MNRAS, 271, 639
Boyle, B.J., Terlevich, R.J., 1996, MNRAS, 293, L49
Butcher, J.A. et al, 1997, MNRAS, 291, 437
Carrera, F.J. et al 1991, MNRAS, 249, 698
Carrera, F.J., Barcons, X., 1992, MNRAS, 257, 507
Carrera, F.J. et al 1993, MNRAS, 260, 376
Carrera, F.J. et al 1995, MNRAS, 275, 22
Carrera, F.J., Fabian, A.C., Barcons, X., 1997, MNRAS, 285, 820
Carrera, F.J. et al 1998, MNRAS, 299, 229
Chen, L.-W. et al, 1994, MNRAS, 266, 846
De Zotti, G., et al 1990, ApJ, 351, 22
Fixsen, D.J. et al, 1996, ApJ, 473, 576
Fabian, A.C., Barcons,X., 1992, ARAA, 30, 429
Fry, J.N., 1996, ApJ, 461, L65
Iwan, D. et al, 1982, ApJ, 260, 111
Jahoda, K., Mushotzky, R.F., 1989, ApJ, 346, 638
Kolokotronis, V., Plionis, M., Coles, P., Borgani, S., 1998, MNRAS, 295, 19
La Franca, F., Andreani, P., Cristiani, S., 1998, ApJ, 497, 529
Lahav, O. et al 1993, Nat, 364, 693
Lahav, O., Piran, T., Treyer, M.A., 1997, MNRAS, 284, 499
Madau, P. et al 1996, MNRAS, 283, 1388
Miyaji, T., Boldt, E., 1990, ApJ, 353, L3
Miyaji, T., 1994, PhD Thesis, Univ of Maryland
Miyaji, T., Hasinger, G., Schmidt, M., 1999, A&A, in press
Mushotzky, R., Jahoda, K., 1992, in: The X-ray background, Barcons, X., Fabian, A.C. eds, CUP
Page, M.J. et al 1996, MNRAS, 281, 597
Persic, M. et al, 1990, ApJ, 364, 1
Piccinotti, G. et al 1982, ApJ, 253, 485
Plionis, M., Kolokotronis, V., 1998, ApJ, 500, 1
Plionis, M., Georgantopoulos, I., 1999, MNRAS, in the press
Scharf, C., Jahoda, K., Treyer, M., Lahav, O., Boldt, E., Piran, T., 1999, ApJ, submitted
Shafer, R.A., 1983, PhD thesis, Univ of Maryland
Shafer, R.A., Fabian, A.C., 1983, in: IAU Symposium 104, Early evolution of the Universe and its present structure, Reidel, p. 333
Treyer, M., Scharf, C., Lahav, O., Jahoda, K., Boldt, E., Piran, T., 1998, ApJ, 509, 531
Warwick, R.S., Pye, J.P., Fabian, A.C., 1980 , MNRAS, 190, 243

AN X-RAY VIEW OF MILLISECOND PULSARS

Werner Becker

Max-Planck-Institut für extraterrestrische Physik, Giessenbachstrasse 1, 85740 Garching, Germany

ABSTRACT The satellite observatories ROSAT, ASCA, RXTE and BeppoSAX have brought important progress in neutron star and pulsar astronomy. With significantly higher sensitivities compared with previous X-ray satellites they allowed for the first time to detect X-ray emission from objects as faint as millisecond (ms) pulsars. In this paper we summarize the current observational status and discuss the empirical constraints on the pulsars' X-ray emission mechanisms.

1. INTRODUCTION

Millisecond pulsars are distinguished from the group of ordinary rotation-powered pulsars by their small spin periods ($P \leq 20$ ms) and high spin stability ($dP/dt \approx 10^{-18} - 10^{-21}$ s/s). Consequently, they are very old objects with spin-down ages of typically $P/2\dot{P} \sim 10^9 - 10^{10}$ years and magnetic dipole components $B_\perp \propto (P\dot{P})^{(1/2)}$ of the order of $10^8 - 10^{10}$ G. More than $\sim 75\%$ of the known disk millisecond pulsars are in binaries with a compact companion star, compared with the $\cong 1\%$ of binary pulsars found in the general population. This gives support to the idea that their fast rotation has been acquired by angular momentum transfer during a past mass accretion phase.

Before the launch of ROSAT, nothing was known on the X-ray emission properties of ms-pulsars. According to the standard models of cooling neutron stars, ms-pulsars are too old to expect detectable thermal emission related with the star's heat content at birth (see Tsuruta 1998 and references therein). Thermal X-ray emission, however, may be emitted from polar caps heated up to temperatures of few million degrees by high energy secondary electrons/positrons, streaming back from the outer magnetosphere to the neutron star's polar cap regions. A non-thermal process which may account for X-rays is the emission from relativistic charged particles accelerated in the pulsar magnetosphere. The interaction of a relativistic pulsar wind with a nearby companion star or with the interstellar medium is another likely scenario. The latter process causes a DC component in the observed emission, and, if the physical conditions are appropriate, will manifest in an extended synchrotron nebula.

Eleven of the 35[1] detected rotation-powered pulsars belong to the small group

---
[1] An updated version of Table 1 from Becker & Trümper 1997 is available online. See http://www.xray.mpe.mpg.de/~web/bt97/bt97_update.html.

of millisecond pulsars (see Fig.1). Although this is almost 1/3 of all X-ray detected rotation-powered pulsars, the origin of their X-ray emission is only known for few of them.

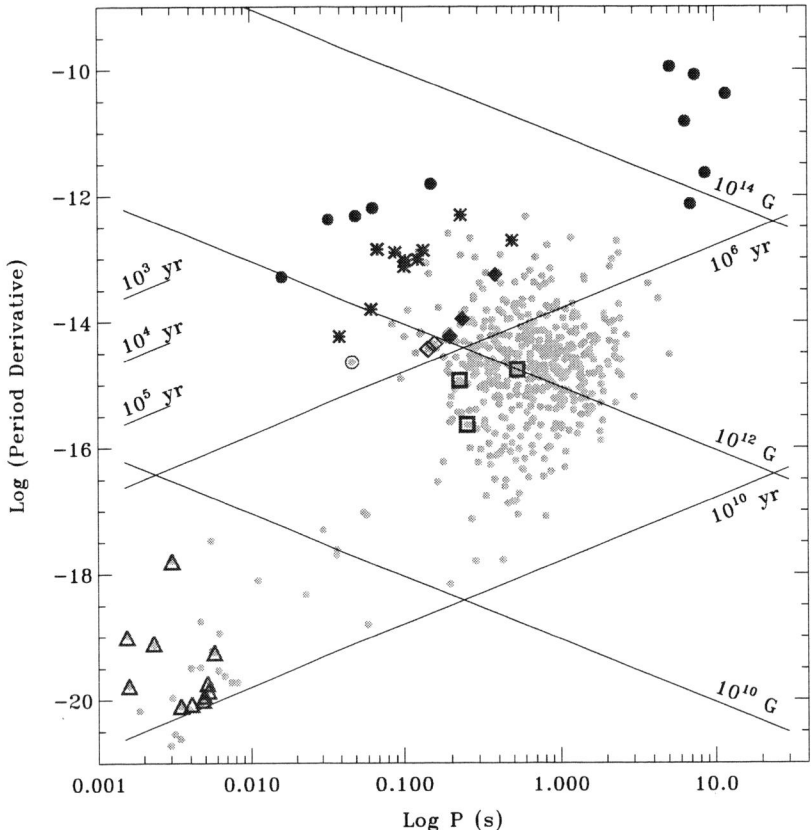

FIGURE 1. The sample of rotation-powered pulsars plotted with respect to their spin parameters $P, \dot{P}$ (small gray dots). Separate from the majority of ordinary field pulsars are the ms-pulsars in the lower left corner. X-ray detected pulsars are indicated by symbols. Lines representing constant ages $\tau = P/2\dot{P}$ and magnetic field strength $B_\perp = 3.3 \times 10^{19}(P\dot{P})^{1/2}$ are indicated.

Five of the eleven detected ms-pulsars (PSR B1957+20, J1012+5307, B0751+18, J1744-1134 and J1024-0719) are identified only by their positional coincidence with the radio pulsar (see also Table 1), and in view of the low number of detected counts do not provide much more than a rough flux estimate. More detailed results are found for the other six ms-pulsars which all provide important empirical information on the X-ray emission mechanisms operating in these objects. The following sections present these six pulsars in greater detail.

2. THE GLOBULAR CLUSTER PULSAR 1821-24 IN M28

The 3 ms-pulsar PSR 1821-24 in the globular cluster M28 shows Crab-like X-ray pulses up to ~ 17 keV (Saito et al. 1997; Kawai & Saito 1999). The two sharp peaks in the pulse profile (see Fig.3) and the power-law nature of the spectrum argue without doubt for a magnetospheric (non-thermal) origin of the detected emission. The alignment between a radio and X-ray pulse component adds further support to this interpretation and implies a common emission site for the main X-ray and radio pulse component observed at 800 MHz. The photon-index for the *pulsed* emission is found to be $\alpha \sim 1.2$ (Kawai & Saito 1999). Imaging M28 with the ROSAT HRI (Fig. 2) has shown two separate sources: a point source (RX J1824.5-2452P) consistent with the ms-pulsar position and a brighter but extended source RX J1824.5-2452E, whose nature is not yet clear. The spectral results obtained from ASCA for RX J1824.5-2452E favors the model in which RX J1824.5-2452E is a pulsar-powered synchrotron nebula, similar to what is seen for the Crab. An alternative and maybe more likely interpretation, however, is that RX J1824.5-2452E is made of a number of point sources, e.g. accreting binaries containing white dwarfs (cataclysmic variables, CV) or neutron stars (low mass X-ray binaries, LMXB), which have not been spatially resolved by the HRI. An interpretation in terms of low accretion LMXBs is also supported from a long term X-ray luminosity study of M28. Becker & Trümper (1999) found recently that the ROSAT PSPC and HRI data taken in 1991 and 1995 suggest an X-ray flux variability of M28 on time scales of years. The total HRI energy flux taken in March 1995 from M28 is about a factor of 3 higher than the total energy flux deduced from the March 1991 PSPC observation. This behavior is in agreement with the results by Gotthelf & Kulkarni (1997) who discovered an X-ray burst from M28 in the 1995 ASCA data.

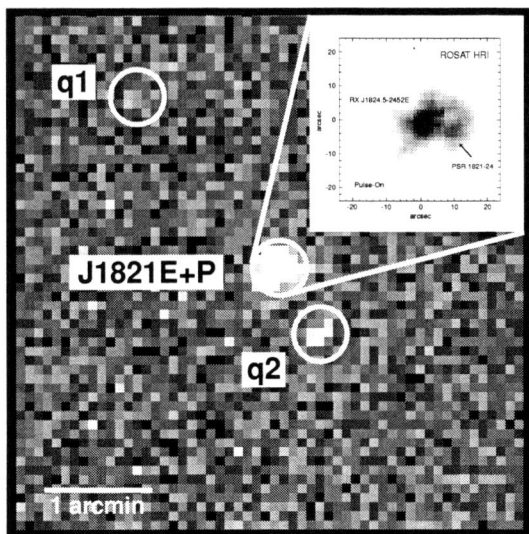

FIGURE 2. $5' \times 5'$ ROSAT HRI image of M28, indicating positions of four x-ray sources (J1824E, J1824P, q1, and q2). The circle q1 indicates the PSPC position for this source, which is undetected by the HRI. The upper-right inset magnifies the core encompassing J1824E+P. Here, the HRI data are oversampled at $1''$ bins and temporally phased to emphasize "pulse-on" events from the millisecond pulsar (arrow).

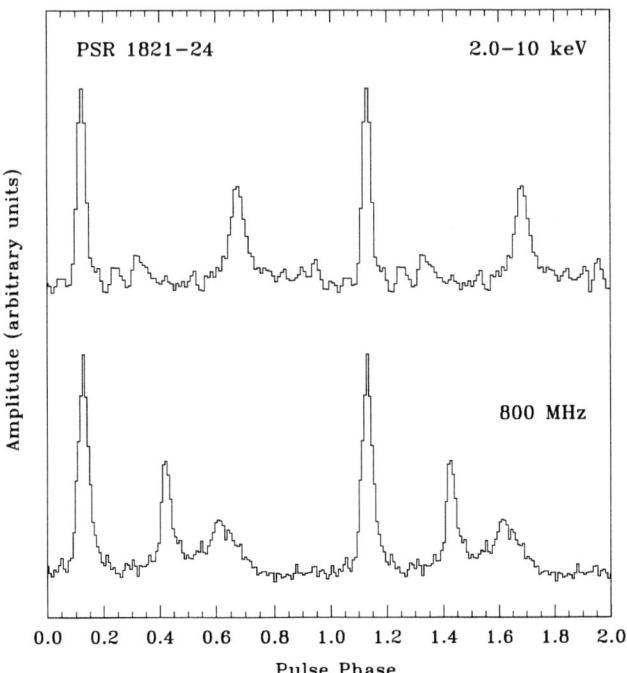

FIGURE 3. Integrated pulse profiles of the globular cluster pulsar PSR 1821−24 as observed with RXTE (Kawai & Saito 1999) and with the NRAO at 800 MHz (*bottom*) by Backer & Sallmen (1997). Two phase cycles are shown for clarity. The X-ray pulse profile is characterized by a double peak structure with a phase separation of ∼ 0.6 between the two peaks. The radio profile at 800 MHz depicts three pulse components. At this frequency, the dominating radio pulse is nearly phase aligned with the primary X-ray pulse.

3. PSR 1937+21

A recent result based on ASCA observations is the detection of X-ray pulses from PSR 1937+21 (Takahashi et al. 1998), an isolated 1.56 ms pulsar which is the fastest spinning pulsar known so far. The pulsar has a dispersion measure based distance of 3.6 kpc and a rather high column absorption of $N_h \sim 6 \times 10^{21}$ cm^2. The latter prevented the detection of soft X-rays in a ∼ 30 ksec ROSAT PSPC observation (Verbunt et al. 1996; Becker & Trümper 1997). Most of the flux detected by ASCA (probably up to 80%) is in the 1.7−6.5 keV band. The X-ray detection is remarkable because the dominating X-ray pulse has a peak width of only ≲ 100 μs (i.e. ∼ 0.08 in phase) and a pulsed fraction of ∼ 75% (see Fig. 4). The narrow peak infers a very small beaming factor which is difficult to explain in terms of thermal emission from hot polar caps. A non-thermal origin is further suggested by the high pulsed fraction and the spectral analysis. The latter, although strongly limited by photon statistics (less than 100 cts were recorded with the ASCA GIS), yields a photon-index of 0.9 ± 1.

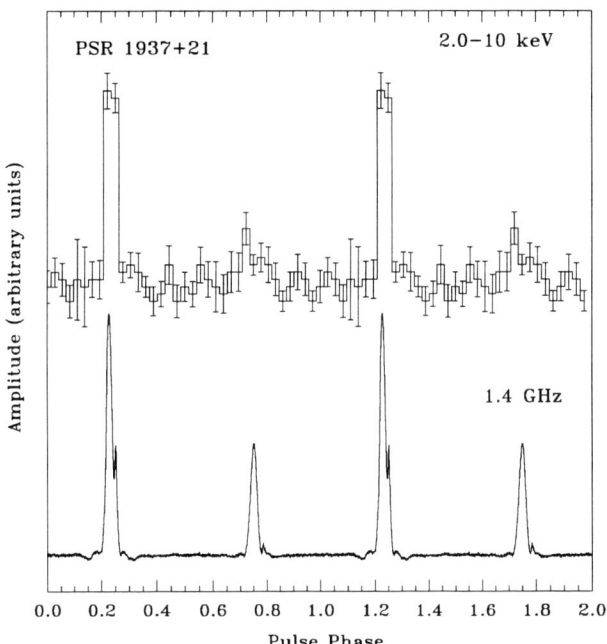

FIGURE 4. Integrated pulse profiles of the 1.56 ms pulsar PSR 1937+21. The X-ray profile (top) was observed by Takahashi et al. (1998) using ASCA. The radio profile (bottom) was taken with the Arecibo radio telescope at 1.4 GHz (D. Backer, priv. com.). The radio profile is characterized by two narrow peaks, which both have X-ray counterparts (although the counterpart for the second radio peak is only marginal detected). Because the relative phase between the X-ray and radio profile is unknown both profiles have been aligned arbitrarily.

4. PSR J0218+3242

The 2.3 ms pulsar PSR J0218+4232 is in a two day binary orbit with a low-mass white dwarf companion and shows significant *unpulsed* radio emission throughout the pulse period. The latter has been taken as an indication that the magnetic dipole is almost aligned with the rotation axis.

X-ray pulses from the 2.32 ms pulsar PSR J0218+3242 have been detected by Kuiper et al. (1998) in ROSAT HRI data and subsequently with BeppoSAX by Mineo et al. (2000). The BeppoSAX data are best fitted by a power-law spectrum with photon-index $\sim 1 \pm 0.2$ (phase averaged). The pulse profile shows two peaks which change their relative intensity with increasing photon energy. The flux in the energy band $2-10$ keV is $f_x = 4 \times 10^{-13}$ erg^{-1}s^{-1}cm^2 (Mineo et al. 2000). ROSAT data have revealed the existence of a compact synchrotron nebula surrounding the pulsar (see Fig. 5). The HRI data imply a source extend of about 14 arcsec. Although this is close to the HRI's attitude solution, the BeppoSAX data seem to confirm the existence of a soft nebula component, showing an energy dependent pulsed fraction of $37 \pm 13\%$ ($0.1-2.4$ keV), $67 \pm 12\%$ ($1.6-4$ keV) and $89 \pm 10\%$ ($4-10$ keV). A summary of spectral fits based on the BeppoSAX data and pulse profiles for different energy bands are shown in Figure 6.

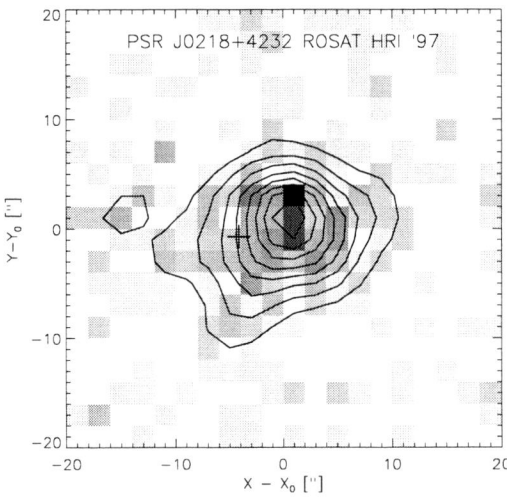

FIGURE 5. PSR J0218+3242 as observed with the ROSAT HRI. The data imply a source extend of about 14 arcsec, associated with a compact synchrotron nebula surrounding the pulsar. (Image from Kuiper et al. 1998).

Noting the spatial coincidence of PSR J0218+4232 with the EGRET source 2EG J0220+4228, Verbunt et al. (1996) tentatively identified the pulsar with the high-energy γ-ray source. Using some additional EGRET observations, applying a combination of spatial and timing analysis, Kuiper et al. (1999) conclude that 2EG J0220+4228 is probably multiple: between 0.1 and 1 GeV PSR J0218+4232 is the most likely counterpart, and above 1 GeV the bright AGN 3C 66A is the best candidate counterpart. The third EGRET catalog (Hartman et al. 1999), which is based on more viewing periods than the 2EG catalog, identifies 3EG J0222+4253 (2EG J0220+4228) with 3C 66A, rather than with the ms-pulsar. However, in a note on this source, Hartman et al.(1999) indicate that the identification with 3C 66A stems from the catalog position based on the > 1 GeV map. Furthermore they confirm that for lower energies (100-300 MeV) the EGRET map is consistent with all the source flux coming from the pulsar.

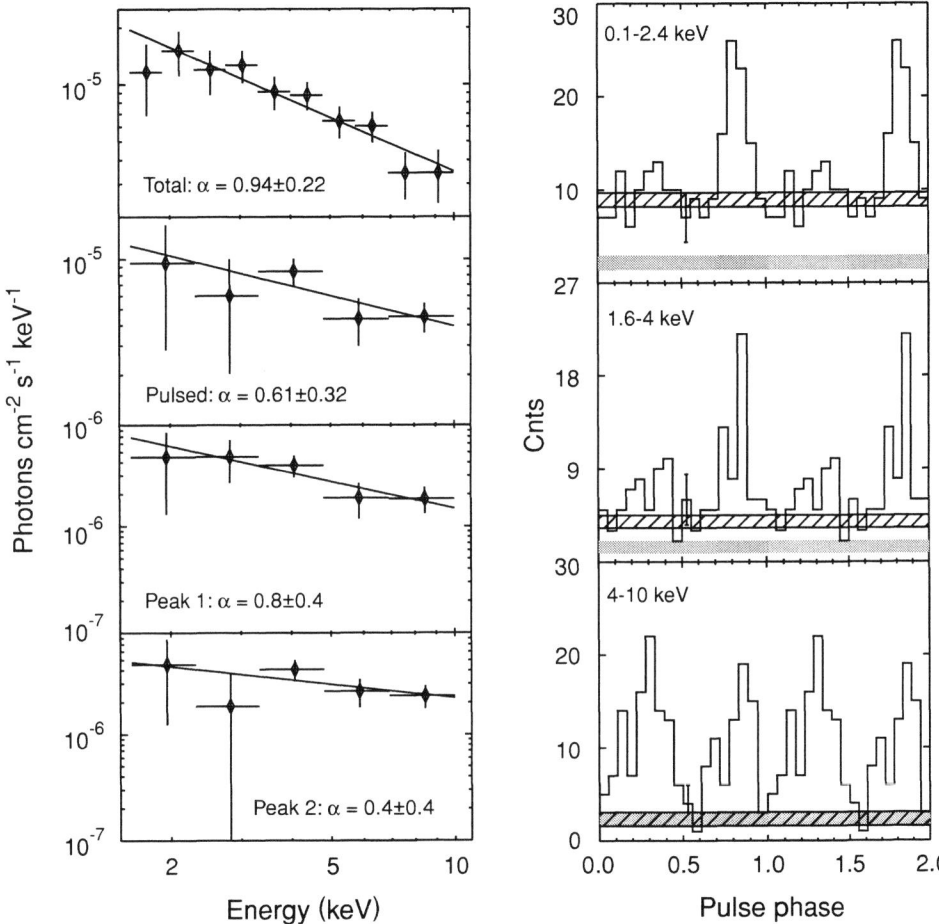

FIGURE 6. **Left:** Spectral fits based on the BeppoSAX data of PSR J0218+3242 (from Mineo et al. 2000). The upper top panel shows the phase averaged data fitted with a power-law spectrum. Fits to the total pulsed emission and to the separate peaks are shown in the other panels. Although the small photon-statistics results in large uncertainties, the data suggest a spectral hardening of the pulsed emission with increasing energy. **Right:** pulse profiles as observed by ROSAT (top panel) and BeppoSAX. While in the soft band the profile is in agreement with a single pulse, a second peak shows up in the hard band. The gray-shaded area indicates the background level whereas the hatched area indicates the DC-level. The difference between both levels and its decrease with increasing photon energy supports the idea that there is a compact nebula contribution in the soft X-ray band which fades away with increasing photon energy.

5. PSR 0437-4715 AND J2124−3358

The case for the 5.75 ms binary pulsar PSR J0437-4715 and the isolated 4.93 ms pulsar J2124−3358 is not as certain as for 1821−24, 1937+21 and J0218+3242 for which a non-thermal origin of the radiation is constrained.

There is no common agreement on the origin of the X-rays detected from PSR J0437-4715. Although this ms-pulsar is the closest and brightest of all ms-pulsars detected so far, the small bandwidth of ROSAT and the limited photon-statistic of the ASCA and SAX data did not allow to discriminate between various possible emission scenarios (Fig. 7). Multi-component thermal spectra (thermal polar-cap emission) and non-thermal spectra (power-law and broken power-law) fit the data equally well (see Becker & Trümper 1999 for a more detailed discussion and further references). The X-ray pulse profile is broad with a single peak stretching almost the entire phase cycle. The pulsed fraction is $\sim 30\%$ in the $0.1 - 2.4$ keV band.

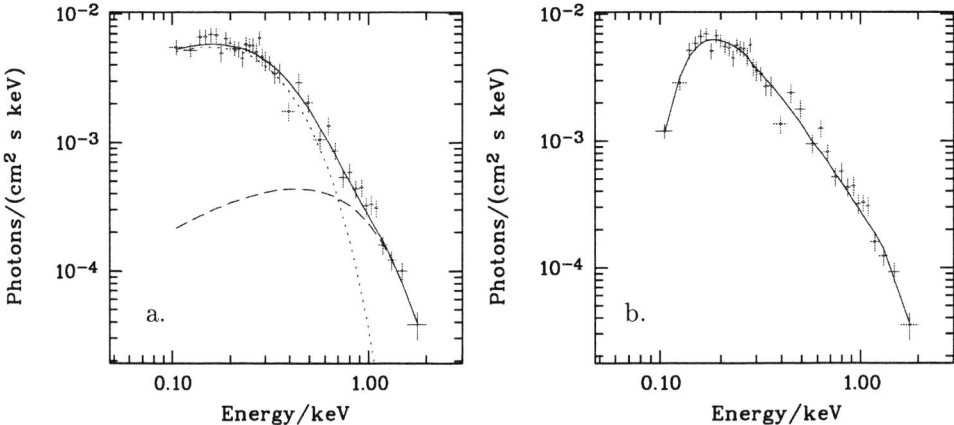

FIGURE 7. **a.** The ROSAT PSPC data of PSR J0437−4715 fitted with a two component black body model. The harder component (dashed line) represents the blackbody spectrum as deduced from spectral fits to ASCA data. The dotted line represents the soft component as obtained from a spectral fit to the ROSAT data. **b.** The PSPC data fitted with a broken power-law which also fits the ASCA data. The break energy in this model is 1.4 keV. The slopes of the first and second power-law components are -2.35 and -4.5, respectively.

PSR J2124−3358 was the first isolated galactic ms-pulsar detected at X-ray energies (Becker and Trümper 1999). Its lightcurve shows indication for a double pulse structure, which in terms of pulsed components and pulse phase separation implies a similarity between the X-ray and radio profile observed at 436 MHz. This similarity points towards a non-thermal origin of the X-ray emission (spectral information from this pulsar is not yet available). However, the shape of a pulse profile itself is not a strong indicator for the origin of the detected emission if the profile (as observed for J2124−3358) is broad: a radiation cone which yields sharp peaks at one aspect angle may well be seen as hardly modulated away from this angle, so that sharp peaks may indicate non-thermal emission processes whereas the reversal – soft modulated emission originates from thermal processes – must not be true. The fraction of pulsed X-ray photons observed from this pulsar is ∼ 33% in the ROSAT band .

6. PSR J0030+0451

The solitary ms-pulsar PSR J0030+0451 was discovered with the Arecibo radio observatory by Lommen et al. (2000). The pulsar has a rotation period of $P = 4.86$ ms and a period derivative of $\dot{P} \leq 1.0(2) \times 10^{-20}$ s s^{-1}, implying a pulsar age of $P/2\dot{P} \geq 8 \times 10^9$ yrs. The rotational energy loss is $\dot{E} \leq 3.4 \times 10^{33}$ erg s^{-1}. The radio dispersion measure of $DM = 4.33$ pc cm^{-3} implies a distance of ∼ 230 pc and a column density of only $N_H \sim 10^{20}$ cm^{-2}. In terms of the spin-parameters, distance and column absorption PSR J0030+0451 is thus a "twin" of the solitary millisecond pulsar J2124−3358 (Bailes et al. 1997; Becker & Trümper 1999) and is very similar to PSR J0437−4715 (see Table 1). The pulsar was one of the last targets observed with ROSAT in 1998 December, just prior to turning off the observatory (Becker et al. 2000). A 7500 sec PSPC exposure detected roughly 120 source counts at the position of the pulsar. A timing analysis found pulsations at the 4.86 ms period (Becker et al. 2000). According to the conventional classification scheme discussed above, PSR J0030+0451 should clearly belong to the class of objects for which thermal emission and a sinusoidal pulse shape are expected. However, the pulsar shows a ∼ 70% modulated, Crab-like X-ray pulse profile with two narrow peaks, similar to the radio pulse profile seen at 1.4 GHz (see Fig. 8)

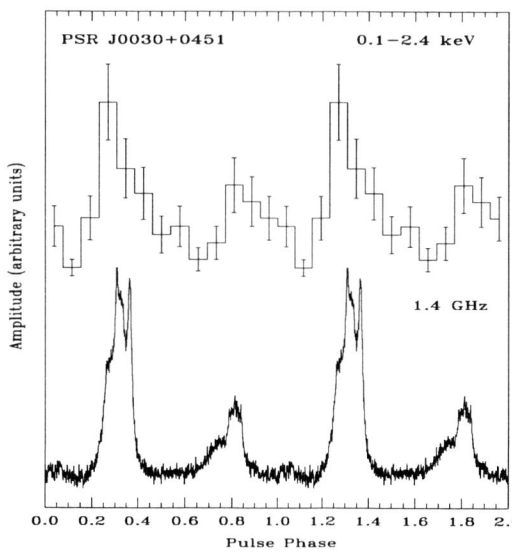

FIGURE 8. X-ray and radio pulse profile of PSR J0030+0451 as observed with the ROSAT PSPC in the 0.1−2.4 keV band (top) and the Arecibo radio telescope at 1.4 MHz (bottom). Two phase cycles are shown for clarity. The phase alignment is arbitrary (Becker et al. 2000).

Unfortunately, because of the small number of detected events and the difficulty of determining an accurate response function for the "dying" PSPC, no X-ray spectral parameters could be determined from the ROSAT data. The high pulsed fraction and the gross similarity in the pulse morphology between the radio and X-ray pulses, however, argue for a non-thermal origin of the detected X-rays.

7. SUMMARY AND DISCUSSION

The X-ray emission observed from PSR 1821-24, 1937+21 and J0218+4232 is dominated by non-thermal processes. This is constrained by power-law spectra and pulse profiles with narrow peaks and high pulsed fractions. For the other ms-pulsars like PSR 0437−4715 and J2124−3358 the existing data do not allow to unambiguously identify the emission process. It is not excluded that in these pulsars thermal polar-cap emission dominates whereas the X-rays from PSR J0030+0451 are likely non-thermal. Hot polar caps are predicted by many magnetospheric emission models as a consequence of pair-creation by the vacuum gap discharge, after which a significant amount of highly energetic charged particles is expected to stream back to the neutron star, heating the surface to a few million degrees.

Two of the eleven X-ray detected ms-pulsars are found to show diffuse extended emission: PSR 1821-24 in M28 and PSR J0218+4232. While for 1821-24 the extended emission possibly comes from unresolved globular cluster sources, the faint diffuse emission found around J0218+4232 might be plerionic. Chandra and XMM observations will constrain the existence and origin of such nebulae.

Comparing the stellar parameters of PSR 1821−24, 1937+21 and J0218+4232 with those of the young Crab-like pulsars (for which similar X-ray emission properties are observed), it can be seen that the intensity of the magnetic field at the pulsar's light-cylinder radius[2] $R_L = cP/(2\pi)$ may play a key role in determining the emission properties and the efficiency of $\dot{E} \rightarrow L_X$ conversion (see Fig. 9). All three pulsars have a period in the range $\sim 1.5 - 3$ ms. As a consequence, in spite of its surface magnetic field $B_S \sim 10^8$ G, its magnetic field at the light-cylinder is large ($B_L = B_S(R_S/R_L)^3 \simeq 10^6$). By comparison for the Crab it is $\simeq 2 \times 10^6$, even though B_S is 5 orders of magnitude larger! If this correlation reflects intrinsic dependences of the emission process on P and $\dot{P}$, it implies that the X-ray emission in these pulsars is emitted from a location close to the light-cylinder rather than from a place close to the star's surface.

Another argument in favor of a non-thermal origin for the bulk of the observed X-rays from ms-pulsars is the close correlation between the pulsars spin-down energy $\dot{E}$ and the X-ray luminosity within 0.1-2.4 keV. Becker & Trümper (1997) found that from the young Crab-like pulsars down to the ms-pulsars all X-ray luminosities from ROSAT detected rotation-powered pulsars correlate very closely with the pulsar's spin-down energy; implying $L_X = 10^{-3}\dot{E}$. So in particular the ms-pulsars obey the same linear relationship with the same X-ray efficiency as the Crab type pulsars,

---

[2] the radius at which the magnetosphere, co-rotating with the neutron star, will have a speed equal to the speed of light

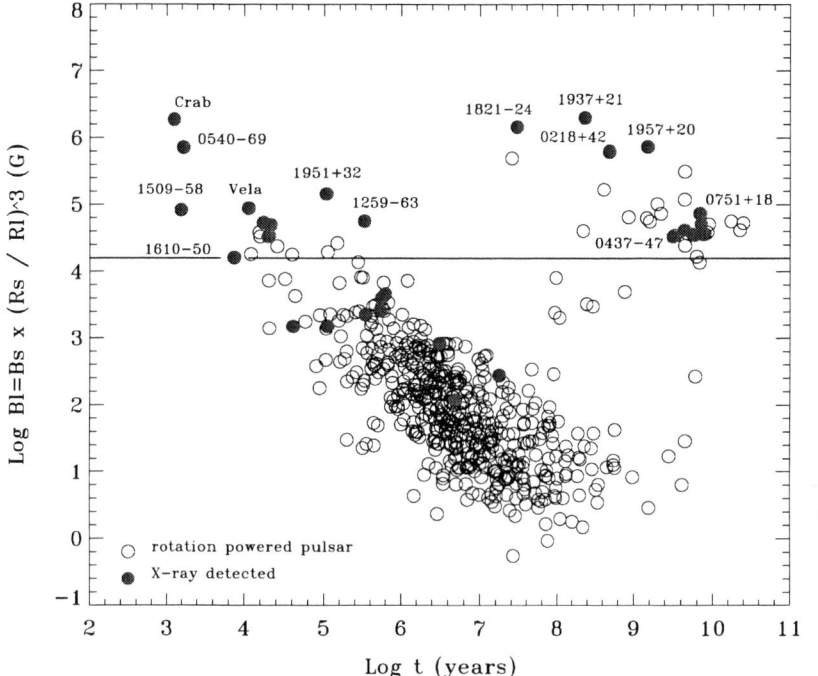

FIGURE 9. Magnetic field at the pulsar's light-cylinder vs. spin-down age for the radio pulsars of the Princeton pulsar catalog. The 35 X-ray detected pulsars are plotted as filled circles (some are labeled). The line at Bl=4.2 indicates the level of Bl for the radio pulsar PSR 1610-50 for which (by its young age, i.e. P and $\dot{P}$) Crab- or Vela-like X-ray emission is expected. Because of their small light-cylinder radius, all ms-pulsars have a Bl comparable with those of the young pulsars, for which non-thermal emission is found to be the dominating process.

indicating that their emission is mainly due to magnetospheric processes. It will be a key question for XMM and Chandra to test this correlation and to find out on what level thermal/non-thermal processes contribute. The main progress in respect to this is expected from those ms-pulsars for which no detailed spectral and/or temporal information could be obtained by sensitivity limitations. Table 1 lists the eleven X-ray detected ms-pulsars. All pulsars in the list are promising targets, worth to be studied further with XMM and Chandra in order to better understand the X-ray emission mechanisms working in this kind of pulsars.

Note	Name PSR	P ms	$\dot{P}$	log age yr	$\dot{E}$ erg/s	$\dot{E}/4\pi d^2$ erg/s/cm^2	Nh $10^{21} cm^{-2}$	dist kpc
p	J0437−4715	5.75	5.7e-20	9.50	33.62	1.0e-09	∼ 0.08	0.18
p	B1937+21	1.5	1.1e-19	8.37	36.04	7.1e-10	∼ 2	3.6
p	B1821−24	3.05	1.6e-18	7.48	36.35	6.2e-10	∼ 4	5.5
p	J2124−3358	4.93	1.1e-20	9.86	33.55	4.7e-10	∼ 0.5	0.25
p	J0030+0451	4.86	1.0e-20	9.88	33.53	5.4e-10	∼ 0.13	0.23
d	J1744−1134	4.07	8.6e-19	9.86	33.28	5.5e-10	∼ 0.09	0.17
d	B1957+20	1.60	1.2e-20	9.32	35.06	4.1e-10	∼ 4.5	1.53
d	J1024−0719	5.16	1.8e-20	9.76	33.72	3.6e-10	∼ 0.2	0.35
d	J1012+5307	5.25	1.4e-20	9.76	33.60	1.2e-10	∼ 0.07	0.5
p	J0218+4232	2.32	8.0e-20	8.66	35.29	6.5e-11	∼ 1.8	5.7
d	J0751+1807	3.47	8.0e-21	9.84	33.88	1.5E-11	∼ 0.4	2.0

TABLE 1. Properties of the X-ray detected millisecond pulsars. The list depicts the pulsars' spin parameters $P, \dot{P}$, their spin-down energy $\dot{E}$ and their characteristic age $P/2\dot{P}$. The column densities N_H and the pulsar distances are estimated from the radio dispersion measure. Pulsars which are identified by positional coincidence with a radio pulsar (d) and by pulsed emission (p) are indicated.

ACKNOWLEDGEMENTS

We thank D. Backer, Y. Saito, M. Takahashi and S. Shibata for their support with the radio and X-ray pulse profiles used to produce the Figures 3 and 4.

REFERENCES

Backer, D.C., Sallmen, S., 1997, ApJ, 114, 1539
Becker, W., Trümper, J., 1997, A&A, 326, 682
Becker, W., Trümper, J. 1999, A&A, 341, 803
Becker, W., Trümper, J., Lommen, A.N., Backer, D.C., 2000, Apj, 545, 1015
Danner, R., Kulkarni, S.R., Saito, Y., Kawai, N., 1997, Nat, 388, 751
Gotthelf, E.V., Kulkarni, S.R. 1997, ApJ, 490, L161
Hartman R.C., Bertsch D.L., Bloom S.D., et al. 1999, ApJS, 123,79
Kawai, N., Saito, Y., 1999, 1999, Astro. Lett. and Communications, 38, 1
Kuiper et al. 1998, A&A, 336, 545
Kuiper L., Hermsen W., Verbunt F., et al. 1999, 1999, Astro. Lett. and Communications, 38
Mineo, T., Cusumano, G., Kuiper, L., Hermsen, L., Massaro E., Becker, W., Nicastro, L., Sacco, B., Verbunt, F., Lyne, A.G., Stairs, I.H., Shibata, S. 2000, A&A, 355, 1053
Saito, Y., Kawai, N., Kamae, T., et al., 1997, ApJ, 477, L37
Takahashi M., Shibata S., Torii K., Saito Y., Kawai N. 1998, IAU circ. 7030
Tsuruta S., 1998, Physics Reports, 292, 1
Verbunt F., Kuiper L., Belloni T., et al. 1996, A&A, 311, L9

NARROW-LINE SEYFERT 1 GALAXIES AS AN EXTREME OF SEYFERT ACTIVITY

Th. Boller

Max-Planck-Institut für extraterrestrische Physik, Postfach 1603, D-85748 Garching bei München, Germany

ABSTRACT

The recent progress in the study of narrow-line Seyfert 1 galaxies is summarized. A clear correlation between the strength of the soft X-ray excess and the optical line width of the Hβ line is detected. A similar relation is detected between the hard 2−10 keV continuum slope and the Hβ line width. The most extreme X-ray variability in radio-quiet active galaxies is seen to date in narrow-line Seyfert 1 galaxies. These newly discovered observational properties give rise to requirements on their physical interpretation. The extreme X-ray variability in some narrow-line Seyfert 1 galaxies is probably due to relativistic effects. The extreme soft X-ray excess emission detected in narrow-line Seyfert 1 galaxies may be due to a more highly ionized accretion disk, with respect to broad line Seyfert 1 galaxies. Narrow-line Seyfert 1 galaxies are therefore thought to form an extreme of Seyfert activity. Because of their extreme properties, the study of narrow-line Seyfert 1 galaxies should allow us to progress in solving the diverse problems more generally posed by the Seyfert phenomenon.

KEYWORDS: galaxies: active–Galaxies: Seyferts–quasars: emission lines–X–rays: galaxies

1. INTRODUCTION

Narrow-line Seyfert 1 galaxies (see Osterbrock & Pogge 1985) form a peculiar group of Seyfert 1 galaxies with **(i)** line widths of hydrogen comparable to the line widths of the forbidden lines, like [O III], [N II] and [S II] (typical values of the line widths at FWHM range between about 500 − 2000 km s^{-1}); **(ii)** emission lines from Fe II, i.e. the optical multiplet emission lines between 4500−4600 Å and 5250−5350 Å, or higher ionized lines like [Fe VII] 6087 Å and [Fe X] 6375 Å; **(iii)** the ratio of the [O III] to the Hβ line being smaller than 3, a value used to distinguish between Seyfert 1 and Seyfert 2 galaxies (Shuder & Osterbrock 1981). As an example, the optical spectrum of the narrow-line Seyfert 1 galaxy IRAS 13224−3809 is shown in Figure 1. ROSAT, ASCA and BeppoSAX observations have shown many narrow-line Seyfert 1 galaxies to have characteristic, unique and extreme X-ray properties. These include strong soft X-ray excess emission, extremely rapid and large-amplitude X-ray variability, steep 2−10 keV power-law continua, and unusual discrete spectral

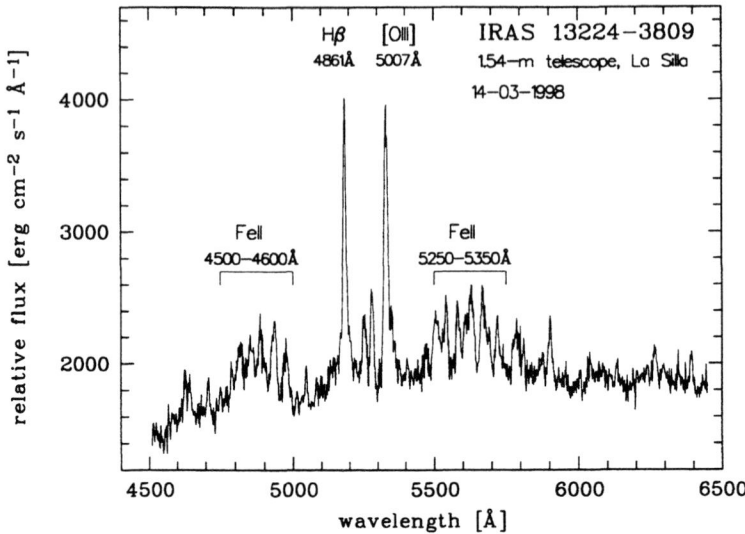

FIGURE 1. Optical spectrum of IRAS 13224−3809 obtained on March 14, 1998, with the Danish 1.54-m telescope at La Silla, Chile. The characteristic features of narrow-line Seyfert 1 galaxies, comparable line widths of Hβ and [O III] as well as strong optical Fe II multiplet emission, are clearly visible.

features.

2. X-RAY PROPERTIES OF NLS1

The observational properties of narrow-line Seyfert 1 galaxies can be summarized as follows:

Narrow-line Seyfert 1 galaxies are those Seyfert 1 galaxies with the steepest X-ray continua in the 0.1−2.4 keV energy range and the strongest soft X-ray excess components. A significant correlation between the steepness of the X-ray continua and the FWHM Hβ line widths is detected through measurements with the X-ray satellite ROSAT (see Fig. 2 (a)) by Boller, Brandt & Fink (1996). The distribution of the measured values, the slope of the spectral energy distribution in the soft X-ray range and the FWHM Hβ line widths, indicate that there exists a continuous distribution between both quantities. A region in the Γ - Hβ plane (Fig. 2 (a)) with $\Gamma > 3$ and FWHM H$\beta > 3000$ km s^{-1} is not occupied by Seyfert 1 galaxies. This region is sometimes be called the 'zone of avoidance'. In the 0.1−2.4 keV energy range, the soft X-ray excess, as well as the underlying power law component, contribute to the measured spectral energy distribution. The slope of the spectral energy distribution is therefore a measure of the relative contributions of the emission from the accretion disk and the emission from the accretion disk corona. The

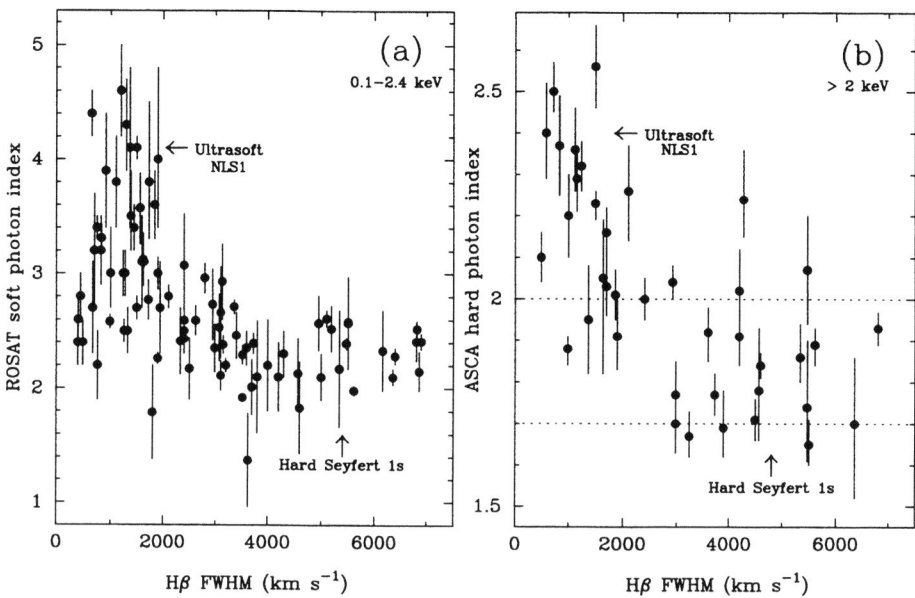

FIGURE 2. **Left Figure (a):** Photon index in the 0.1−2.4 keV energy range as a function of the FWHM line width of the Hβ line. The photon index serves as a measurement of the steepness of the X-ray continuum. All object in the diagram are Seyfert 1 galaxies. A significant correlation between the slope of the X-ray continuum and the FWHM Hβ line width is found. Seyfert 1 galaxies with large line widths (FWHM Hβ > 2000 km s^{-1}) show a relatively small dispersion in their values of the photon index with a mean of about 2.3. Narrow-line Seyfert 1 galaxies show a large dispersion in their values of the photon index. Some of these objects exhibit values of the photon index up to about 5. A region of the diagram, with values of the photon index larger than 3 and optical line widths larger than about 3000 km s^{-1}, is not occupied by Seyfert galaxies. This region is sometimes called as 'zone of avoidance'. **Right Figure (b):** Photon index in the 2−10 keV energy range as a function of the FWHM Hβ line width discovered by (Brandt, Mathur & Elvis 1997). The measurements were obtained with the Japanese X-ray satellite ASCA. Also in the hard X-ray energy range, narrow-line Seyfert 1 galaxies show a stronger dispersion in their photon indices than previously thought. Typical values of the photon index in the hard X-ray energy band range between about 1.7 and 2.6. In this energy range, the spectrum is dominated by the power law component. The correlation between the slope of the X-ray spectral energy distribution and the optical line width is expected to allow statements on the physical conditions in the accretion disk corona.

distribution of the data points in Fig. 2 (a) suggests that the strength of the soft X-ray excess emission, as a measure of the strength of the ionizing radiation from the nuclei of active galaxies, determines the size of the broad-line region.

Similar to the soft X-ray spectral slopes, the 2–10 keV continua of narrow-line Seyfert 1 galaxies show a larger dispersion than previously thought. Characteristic values for the photon index Γ in the 2–10 keV energy range, resulting from power law fits, lie in the range between about 1.9 and 2.6, respectively. A significant correlation between the slopes of the 2–10 keV continua and the FWHM of the Hβ line was found (see Fig. 2 b) by Brandt, Mathur & Elvis (1997). In the 2–10 keV energy range, the spectral energy distribution is dominated only by the power law component, which is thought to originate in the accretion disk corona. Therefore, the distribution of data points in Fig. 2 (b) may provide constraints on the physical conditions in the accretion disk corona.

As a class, narrow-line Seyfert 1 galaxies show shorter doubling times of their X-ray flux variations than broad-line Seyfert 1 galaxies (see Fig. 3, left panel). The rapid variations indicate that the X-ray emission originates in a compact region with a size of only a few thousand light seconds, probably in the inner parts of the accretion disk. In addition, extreme (amplitude of variability above 10) and rapid (time scales of less than about one day) variability has been detected in the narrow-line Seyfert 1 galaxy IRAS 13224−3809 (Boller et al. 1997), the narrow-line quasar PHL 1092 (Brandt et al. 1999) and the narrow-line Seyfert 1 galaxy 1ES 1927+654 (c.f. Section 4).

As a class, narrow-line Seyfert 1 galaxies show no significant amount of absorption by neutral hydrogen above the galactic value (see Fig. 3, right panel). The X-ray emitting inner parts of the accretion disk are therefore directly visible.

The newly discovered relations between the slopes of the soft and hard X-ray continua and the optical line widths in narrow-line Seyfert 1 galaxies may allow us to make statements on the physical conditions in the accretion disk and the accretion disk corona. X-ray observations of narrow-line Seyfert 1 galaxies are therefore a new and effective tool in the study of the innermost parts of active galactic nuclei.

3. RELATION OF THE X-RAY PROPERTIES TO OTHER OPTICAL EMISSION LINE PARAMETERS

A whole set of optical emission line parameters has been found recently (e.g. Brandt & Boller 1998) that correlate with the X-ray 0.1–2.4 and 2–10 keV continuum slopes. The optical emission line parameters include the Fe II strength, [O III] strength, the Hβ FWHM and the Hβ asymmetry. Many NLS1 tend to have strong Fe II emission, weak [O III] emission, small Hβ FWHM and large Hβ asymmetry. These correlations suggest that NLS1 may have an extreme value of an underlying physical parameter. The relations discussed above suggest that the primary physical driver is located in the energetically important circumnuclear region, within about 50 Schwarzschild

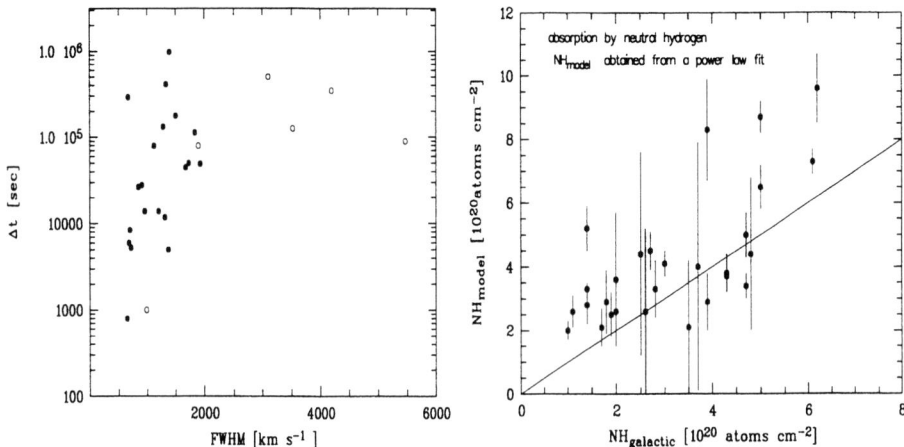

FIGURE 3. **Left panel:** X-ray doubling time as a function of the FWHM Hβ line width. The filled circles represent narrow-line Seyfert 1 galaxies investigated by Boller, Brandt & Fink (1996). The open circles are taken from Wandel & Mushotzky (1986). The doubling time correlates with the optical line width. Narrow-line Seyfert 1 galaxies exhibit more rapid X-ray variability than broad-line Seyfert 1 galaxies. **Right panel:** Column density of neutral hydrogen along the line of sight, obtained from power law fits to ROSAT PSPC measurements of narrow-line Seyfert 1 galaxies, as a function of the Galactic neutral hydrogen column density. The majority of narrow-line Seyfert 1 galaxies do not show significant absorption above the Galactic value.

radii of the supermassive black hole. Models with a high ratio of the accretion rate to the black hole mass in NLS1 seem to be able to explain the observational properties of broad and narrow-line Seyfert 1 galaxies (Boller, in preparation). Other possible interpretations include the black hole spin and the orientation effects.

4. X-RAY TIMING PROPERTIES OF NLS1

In this Section, the discovery of extreme and rapid X-ray variability in the narrow-line Seyfert 1 galaxy IRAS 13224−3809 (Boller et al. 1997) the narrow-line quasar PHL 1092 (Brandt et al. 1999) and the narrow-line Seyfert 1 galaxy 1ES 1927+654 (Boller, in preparation) are described and physically discussed.

4.1. The long-term light curve of IRAS 13224-3809 from 1992 to 1998

In this Section, the X-ray variability of IRAS 13224−3809 between 1992 and 1998 is discussed. In Figure 4, the count rate as a function of time is shown for the ROSAT observations in the years 1992, 1994, 1995, 1996, and 1998. In all observations, expect that of 1995, which covers a time scale of only one day, strong and rapid

X-ray variability is detected. The strongest variations are detected during the 1996 observations. The X-ray variability of IRAS 13224−3809 is characterized by periods of relative quiescence, interrupted by strong flaring events on time scales of only a few hours. Whenever IRAS 13224−3809 was observed, strong and rapid X-ray variations were detected. This is of relevance with respect to the analogy discussed between narrow-line Seyfert 1 galaxies and galactic black hole candidates.

The resulting maximum amplitude variability for the 1996 observations is a factor of 61. The most extreme amplitude variability on a short time scale is detected between day 16.0160 and 17.9861. The count rate increases from $(5.0 \pm 1.9) \cdot 10^{-3}$ photons s^{-1} to 0.287 ± 0.019 photons s^{-1}. This corresponds to an amplitude variability by a factor of about 57 within only two days.

In this context it is of importance to determine the changes in the luminosity, assuming isotropic emission. To convert ROSAT HRI count rates into fluxes and luminosities, the best fitting spectral model (power law plus two black body components) to the ROSAT PSPC data of IRAS 13224−3809 are used. From this model, we see that 1 HRI count per second corresponds to a flux value, corrected for absorption by neutral hydrogen along the line of sight, of $1.15 \cdot 10^{-10}$ erg cm^{-2} s^{-1}. The resulting luminosity [1] for one HRI count per second is $2.9 \cdot 10^{45}$ erg s^{-1}. Assuming isotropic emission from IRAS 13224−3809, the resulting increase in luminosity from $1.5 \cdot 10^{43}$ erg s^{-1} to $8.3 \cdot 10^{44}$ erg s^{-1} within about 2 days, and the corresponding change in luminosity of $\Delta L \approx 8.2 \cdot 10^{44}$ erg s^{-1} would be remarkable. The most probable explanation is that the extreme amplitude variability is caused by relativistic flux amplification in the inner accretion disk, and that there are no significant changes in the energy production within IRAS 13224−3809. In this model, the extreme flux variations are only detected in the observers frame (Sunyaev 1973, Guilbert, Fabian & Rees 1993, Boller et al. 1997).

4.2. Discovery of extreme and rapid X-ray variability in the narrow-line quasar PHL 1092

PHL 1092 (z = 0.396) is one of the most X-ray luminous narrow-line Seyfert 1 objects. In comparison with the mean X-ray luminosity of IRAS 13224−3809 of $1.45 \cdot 10^{44}$ erg s^{-1}, the X-ray luminosity of PHL 1092 is between 4 and 48 times higher (Brandt et al. 1999). A power law fit to the ROSAT PSPC observation of PHL 1092 from 1992 results in a photon index of $\Gamma = 4.2 \pm 0.5$ (Forster & Halpern 1996). Within this observation, flux variations by a factor of about 4 were observed. For an object of that luminosity, such a rapid variation is remarkable. Because of its steep X-ray spectrum and observed rapid variability, PHL 1092 was proposed for a 18-day ROSAT HRI monitoring observation. The results of this monitoring observation are reported below.

The most extreme combination of amplitude variability with time is observed at

---

[1] The conversion between flux and luminosity was done using the relations given by Schmidt & Green (1986) for a value of Hubble constant of 50 km s^{-1} Mpc^{-1}, a value of q_0 of 0.5 and for the redshift of IRAS 13224−3809 of z = 0.0067.

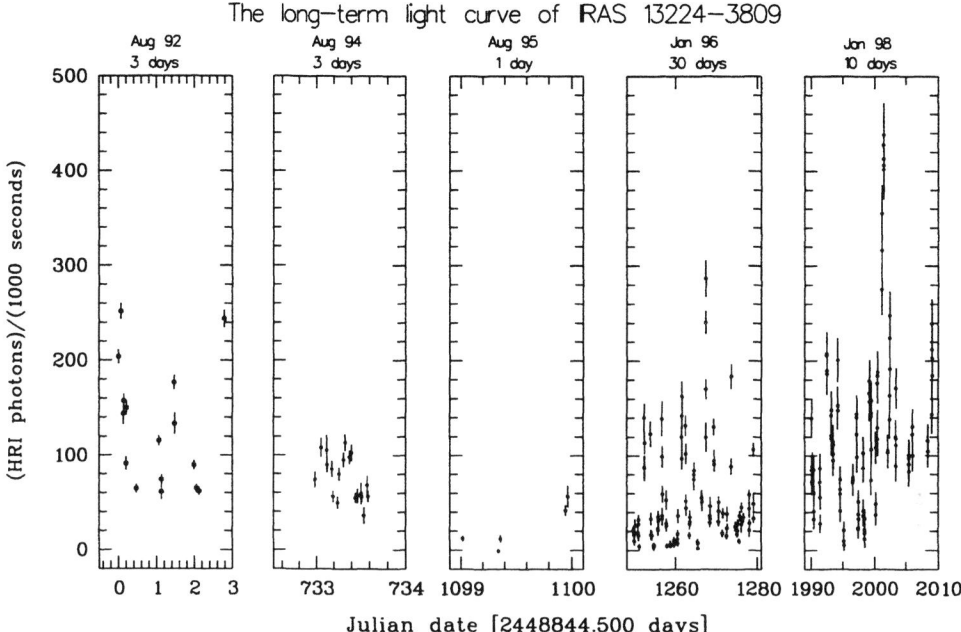

FIGURE 4. Compilation of all ROSAT observations of IRAS 13224−3809. The x-axis gives the Julian date minus 244844.500 days, the y-axis shows the count rate per 1000 seconds in the ROSAT HRI detector. IRAS 13224−3809 shows, during the full observing time interval, from 1992 to 1998, strong and rapid X-ray variability.

day 8.6 (cf. Fig. 5). The count rate increases by a factor of 3.8 from 0.0346 ± 0.0098 to 0.1310 ± 0.0190 photons per second within 3580 seconds. In order to determine the conversion factor between HRI count rate, flux and luminosity, a power law model combined with two black body models to the PSPC data is used. From this model, it follows that one HRI count per second corresponds to a luminosity of $5.2 \cdot 10^{46}$ erg s^{-1}. During the 18-day observation of PHL 1092, the luminosity therefore varies between $(5.9 \pm 1.9) \cdot 10^{44}$ erg s^{-1} and $(8.2 \pm 0.8) \cdot 10^{45}$ erg s^{-1}. For the most extreme combination of amplitude variability with time, at day 8.6, we obtain a change in luminosity of $\Delta L = (5.0 \pm 1.1) \cdot 10^{45}$ erg s^{-1} within a time interval of $\Delta t = 3580$ seconds. The observed variability per time of $\frac{\Delta L}{\Delta t} = 1.40 \cdot 10^{42}$ erg s^{-2} is presently the largest value measured in a radio-quiet galaxy. This value exceeds the efficiency limit for accretion onto a Kerr black hole (Fabian 1979), and further supports the model of relativistic flux boosting for the rapid and extreme variability detected in PHL 1092 and IRAS 13224−3809.

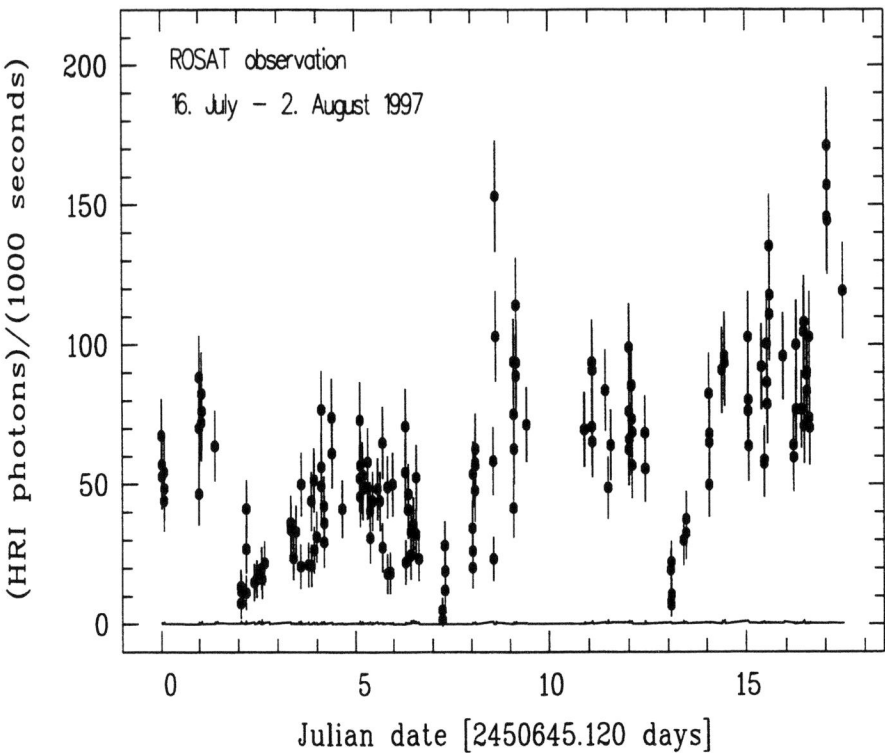

FIGURE 5. ROSAT HRI light curve of PHL 1092, obtained within a 18-day observation between July 16, 1997 (02:47:15 UT) and August 2, 1997 (13:53:58 UT). The x-axis gives the Julian date minus 2450645.120 days). The dashed line gives the background count rate as function of time. The maximum factor of the amplitude variability is 13.9. PHL 1092 shows the most extreme change of luminosity per time interval so far found in a radio-quiet galaxy.

4.3. Discovery of persistent and rapid X-ray variability in 1ES 1927+654

With a mean count rate of 1.2 counts s^{-1} in the ROSAT 0.1–2.4 keV energy range, 1ES 1927+654 is one of the brightest AGN detected in the ROSAT All-Sky Survey. More interestingly, the source shows unusual rapid and extreme X-ray variability (cf. Fig. 6). The maximum factor of amplitude variability is at about 20. The source is found to be persistently variable on time scales of fractions of a day. Due to the scanning law during the All-Sky Survey observations, the data points shown in Fig. 6 are separated by 5600 seconds and we are not able to search for faster variations. NED gives a redshift for 1ES 1927+654 of z = 0.017. By inspecting the APM finding charts we estimate an optical B-magnitude of about 15 (no optical magnitude is given in NED). The 1.4 GHz radio flux density is 0.0039 Jansky, resulting to a ratio

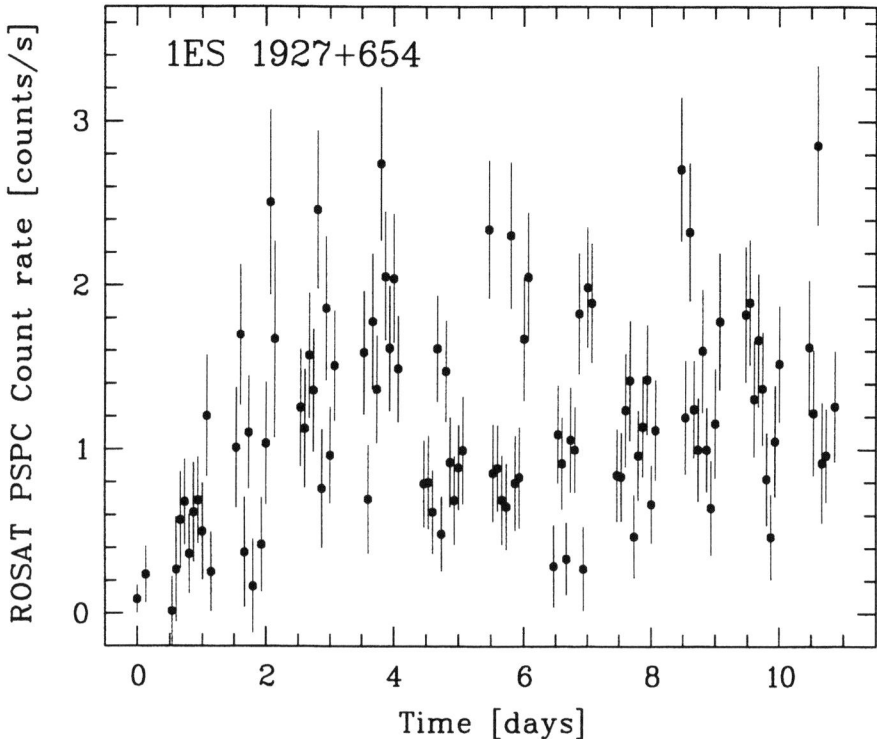

FIGURE 6. ROSAT All-Sky Survey light curve of 1ES 1927+654. The maximum factor of amplitude variability is about 20.

between the radio to optical flux density of about 0.1. Following the Kellermann formalism (Kellermann et al. 1989), 1ES 1927+654 can be considered as a radio-quiet object. From the arguments given above it seems plausible, that relativistic flux boosting, rather than a strong relativistic jet is causing the X-ray variability detected during the ROSAT All-Sky Survey observations. The type of variability seems to be similar to the persistent giant and rapid variability found in the narrow-line Seyfert 1 galaxy IRAS 13224−3809 (Boller et al. 1997) and the narrow-line quasar PHL 1092 (Brandt et al. 1999). Relativistic flux amplification from X-ray hot spots, orbiting the central black hole at close distances (less than about 10 Schwarzschild radii) seems to be a plausible explanation for the extreme variability. Assuming a simple power law for the continuum emission with a photon index Γ, the boosting factor for the emission from a X-ray hot spot is: $f/f' = \delta^{3+\Gamma}$, where δ is the relativistic Doppler factor, f is the observed flux and f' is the emitted flux by a X-ray hot spot. In the relativistic flux boosting scenario, the X-ray hot spot emits a significant amount of flux, comparable to that of the accretion disk, in order to achieve the observed factors of amplitude variability. If accretion disk emission lines are emitted by X-ray hot spots, moderate line shifts with Doppler factors up to about 2 (for a X-ray hot spot orbiting the black hole at 3 R_S) are expected to appear.

REFERENCES

Boller, Th., Brandt W.N., Fink H., 1996, A&A 305, 53
Boller Th., Brandt W.N., Fabian A.C., Fink H. 1997, MNRAS 289, 393
Brandt W.N., Mathur S., Elvis M., 1997, MNRAS 285, 25P
Fabian A.C., 1979, Proc. Roy. Soc. London A, 366, 449
Forster K., Halpern J.P., 1996, ApJ 468, 565
Guilbert P.W., Fabian A.C., Rees M.J., 1983, MNRAS 205, 593
Kellermann K.I., Sramek R., Schmidt M., Shaffer D.B., Green R., 1989, AJ 98, 1195
Osterbrock D.E., Pogge R.W., 1985, ApJ 297, 166
Schmidt M., Green R.F., 1983, ApJ 269, 352
Shuder J.M., Osterbrock D.E., 1981, ApJ 250, 55
Sunyaev R.A., 1973, Soviet Astronomy AJ 16, 941
Wandel A., Mushotzky R.F., 1986, ApJ 306, L61

X-RAY TRANSIENTS MONITORED BY THE ALL-SKY MONITOR ON RXTE: A TABULATION

Hale Bradt, Alan Levine, Ronald Remillard, and Donald A. Smith

CSR and Physics Dept., MIT

ABSTRACT We present a tabulation of 46 transient x-ray sources monitored with the All-Sky Monitor (ASM) on the Rossi X-ray Timing Explorer (RXTE). They fall into four broad categories: short (~ 1 d), intermediate, and long (> 500 d) duration of outbursts, and long period binary systems that flare up at periastron (e.g., Be systems). The mixture of outburst/quiescent cycles and low-level persistent emission in a few systems could indicate conditions are near the limit for stable mass flow in the accretion disk. The two short-time-scale systems, CI Cam and V4641 Sgr, are within 1 kpc of the sun, and hence many more such systems may await discovery.

KEYWORDS: X-rays: stars

1. ASM SKY SURVEY

The All-Sky Monitor (ASM; Levine et al. 1996) on RXTE has been monitoring the entire sky for new (uncataloged) transient x-ray sources while also recording the intensities of the known sources. The current catalog contains about 325 source positions of which about 180 have yielded positive detections on some occasion. The monitoring has been reasonably continuous except for times when the sun is relatively close to a source and except for a period of ~ 7 weeks shortly after launch when the detectors were turned off due to a temporary breakdown problem. The detected sources include many well known persistent sources as well as a substantial number of transient sources. Some of these are recurrent and others are in their first known outburst. Most of the latter were discovered in the RXTE era, either with other satellites, *e.g.* CGRO and BSAX, or with RXTE. Some were discovered prior to the launch of RXTE.

Of the 180 positive detections, approximately 150 reached 15 mCrab on at least some occasion and 30 are detected at levels 2 to 15 mCrab in averages over long periods, up to 6 months. For sources with known positions, the detection threshold (3 sigma) away from the galactic center is about 30 mCrab in a single sweep of the ASM cameras across the source. A sweep usually consists of four 90-s integrations or "snapshots" as the cameras step across the source. The one-day threshold (typically 5 – 8 sweeps) can reach down to ~ 10 mCrab.

The data are routinely searched for new (*i.e.*, not in the ASM catalog) sources with a cross-correlation search of the entire FOV. Confidence in the detection of a new persistent source arises through multiple detections that yield crossed lines of position. In one day, a 50-mCrab source is solidly established. Fainter sources to about 7 mCrab can be retrieved from cross-correlation maps that integrate one week of data. These thresholds apply to positions reasonably removed from bright sources.

The list of detections include about 50 sources we call "transients". Another 23 objects are extragalactic (14 Sy1 and QSOs, 4 BL Lacs, and 5 clusters). About 40 objects exhibit periodicities in the ASM data from the spin period of X Per (837 s) to the 164-d precession period of SS 433.

2. TRANSIENT DETECTIONS

We have collected a list of the 46 brighter transients monitored with the ASM (Table 1). We further tabulate comments about the sources in Table 2. The criterion for inclusion on this list is that the source be known to have been below Uhuru/HEAO-1 thresholds (few mCrab) for sustained periods and that the source was found in a bright state of at least 25 mCrab, as measured by the ASM. The fainter objects omitted include, for example, some of those detected in the galactic plane scans with the sensitive PCA instrument on RXTE (Valinia, Kinzer, & Marshall 2000) or from observations with the Wide-Field Camera on BeppoSAX (Jager et al. 1997).

The tabulated sources are divided into several groups that depend on the temporal character of their variability:

1. two sources with very short outbursts (hours to a few days),
2. transients of intermediate durations which have are further divided into
    (a) the thirteen monitored with the ASM in the process of their first known outburst (which may have occurred before the launch of RXTE) and
    (b) nineteen that are known to be recurrent,
3. six sources with very long outbursts (> 500 d), and finally
4. six periodic systems that typically flare up when the compact object in an elliptical orbit approaches periastron.

The definition of a transient can be rather elusive. For example, the existence of long-duration transients (Table 1C) suggests that there may be no clear boundary between transients and persistent sources. Conversely, the close binary sources X 2129+47 and X 1755−338, long considered to be persistent sources, have disappeared both optically and in X rays (see, *e.g.*, refs in van Paradijs 1995). Neither of these sources have been detected with the ASM to levels of a few mCrab since the Dec. 1995 launch.

3. THE TABULATION

The tabulation describes each transient in terms of the outburst profile shape, the peak flux, the hardness ratio, the first date of outburst, the rise and decay times and finally the duration. The light curves exhibit much more richness than these few parameters indicate. Sample X-ray light curves for six neutron-star systems are shown in Fig. 1 and for six black hole systems in Fig. 2. Plots on expanded time-scales reveal even more detailed structure than is evident in these figures.

3.1. Description of data in Table 1

Column 1: Source name. Sources are listed in RA order within each category. Satellite prefixes are given for objects discovered in the past two decades, but longer-known objects are designated with the prefix "X".
Column 2: Type indicates black hole candidate or neutron star system.
Column 3: Outburst profiles are categorized as fast-rise-slow-decay (frsd), symmetric, or irregular.
Column 4: The peak count rate is given in mCrab. Note that 1 Crab is 75 ASM cts/s at 1.5–12 keV.
Column 5: The hardness ratio HR2 is the ratio of counting rate in the 5 – 12 keV band to that in the 3–5 keV band.
Column 6: The start date (MJD) is the date of the first positive detection at the onset of an outburst, or the onset of the first outburst in Table 1D. "pre-XTE" indicates the source was first detected above threshold when RXTE observations began after launch. MJD conversions are:
   1996 Jan. 0.0 = MJD 50082.0
   1997 Jan. 0.0 = MJD 50448.0
   1998 Jan. 0.0 = MJD 50813.0
   1999 Jan. 0.0 = MJD 51178.0
   2000 Jan. 0.0 = MJD 51543.0
Columns 7, 8, 9: The rise and decay times and the durations are approximately the total time for the full rise, the exponential time constant for the decay, and the total duration above threshold, respectively. Some outbursts are still in progress (IP) at this date (2000 Feb. 28).

3.2. Description of Table 2

The notes give descriptive features of the light curves and hardness ratios that complement the tabulated values and also reference recent cogent results. They are not meant to be complete; refereed publications are favored as are later works as they ease entry into the literature. Results from before the RXTE era may be found in the reviews by van Paradijs (1995) and Bradt & McClintock (1983). References to earlier catalogs may also be found in these works. The references to the table are coded based on the author and source names.

4. HIGHLIGHTS OF THE TABULATION

The nature of a given source is well correlated with the ASM hardness ratio, HR2 as follows: neutron-star low-mass binaries have HR2 = 1.0–1.5, pulsars (neutron-star high-mass binaries) have HR2 = 2–4, and black-hole candidates exhibit large temporal variations of HR2 from extremely soft to higher values (0.3–1.5).

The outburst profiles exhibit several types of wave forms as indicated in the table. Similarities exist from source to source and from outburst to outburst in one source. However, there are substantial differences also. In general, the profiles should shed light on the disk accretion instabilities that give rise to the episodes of high accretion luminosity.

One notable effect is the presence of long ($\sim$ 1 year) marginally-on states after a major outburst, e.g. in 1630–47 and 1608–522, and "failed" outbursts in Aql X–1. These states may indicate that the conditions for outburst are marginal. In fact, Aql X–1 lies on the on the thermal-viscous disk-instability boundary (van Paradijs 1996).

The range of detected outburst durations is extremely wide as noted above. The listed intermediate outbursts range from about $\sim$ 10 to $\sim$ 200 days. The two fast x-ray novae (CI Cam and V4641 Sgr) were only recently discovered. These two objects are both quite close to the sun, at distances inferred from 21 cm absorption profiles of 1.0 and 0.5 kpc respectively. It is thus possible that infrequent such outbursts from other sources could have been missed because of the intermittency of coverage or limited solid angles of past and present x-ray monitoring missions. The long-duration transients are by definition "quasi-persistent". These too may help reveal the factors that lead to instability.

5. FUTURE WORK

The ASM instrument continues to operate with most of its initial capability, so another 1–3 years or more of useful data are expected. The archival ASM data have recently been reprocessed with improved *a posteriori* calibrations, increased temporal coverage, and improved analysis algorithms. With these we may retrieve additional transients. The final data base should be useful for the determination of rates of transients, the nature of accretion processes, and possibly may reveal new distinctions between neutron stars and black holes.

ACKNOWLEDGEMENTS

We are grateful to the RXTE teams at MIT, GSFC, and UCSD.

REFERENCES

Bradt, H. & McClintock, J. 1983, ARA&A 21, 63
Jager, R., et al. 1997, A&AS, 125, 557
Levine, A., Bradt, H., Cui, W., Jernigan, J. G., Morgan, E., Remillard, R., Shirey, R., & Smith, D. A. 1996, ApJL, 469, L33
Valinia, A., Kinzer, R., & Marshall, F. E. 2000, ApJ 534, in press.; also ApJ, 505, 13
van Paradijs, J. 1995, in *X-ray Binaries*, eds. W. H. G. Lewin, J. van Paradijs & E. P. J van den Heuvel (Cambridge: Cambridge Univ. Press), 536
van Paradijs, J. 1996, ApJL, 464, L1

Table 1. RXTE ASM: Transients above 25 mCrab

Source Name	type	profile	peak mCrab	ASM HR2	start date	Rise days	Decay days	Duration days
A. Fast X-ray Novae: Decay $\tau < 1$ day									
XTE J0421+560	bhc?	frsd	1885	0.8–2.2	50903	0.3	0.5	7.7
SAX J1819.3–2525	bhc	irr	12200	0.8–2.1	51436	0.2	0.01	0.6
B. Intermediate-Duration X-ray Transients (Nonperiodic)									
Recent Initial Outbursts									
XTE J0111–733	ns	irr?	50	3.2	51119	–	–	53
XTE J1550–564	bhc	frsd	6800	0.3–1.6	51062	4.2	11	246
XTE J1723–376	ns	irr?	100	1.5	51108	–	–	182
GRS 1737–310	bhc?	frsd?	26	1.8	50497	–	–	46
GRS 1739–278	bhc	frsd	805	0.6	pre XTE	12	9	> 400
XTE J1739–285	??	frsd?	193	1.4	51471	6	22	~50
XTE J1748–288	bhc	frsd	485	1.4	50966	1.4	15	63
SAX J1750.8–2900	ns	frsd	117	1.3	50515	<1	9	28
XTE J1755–324	bhc	frsd	188	0.3	50653	3	30	104
XTE J1858+034	ns	sym	26	3.2	50842	–	–	28
XTE J1859+226	bhc	frsd	1045	0.4–1.4	51460	10	29	IP
XTE J2012+381	bhc	frsd	209	0.3	50956	3.5	32	182
XTE J2123–058	ns	frsd	84	1.2	50987	<1	31	52
Known Recurrent Transients									
EXO 0748–676	ns	frsd	>50	0.9	pre XTE	–	40	–
X 1246–588	ns?	sym?	35	1.0	51271	–	–	IP
X 1354–644	bhc	sym	52	1.3	50744	70	40	>85
X 1608–522	ns	frsd	??	1.1	pre XTE	–	15	–
"		frsd	911	1.1	50842	> 8	13	IP
X 1630–472	bhc	irr	336	1.1	50153	17	16	150
"		frsd	416	1.6	50846	60	15	122
"		irr	215	1.2	51295	60	17	98
GRO J1655–40	bh	irr	3138	0.6	50198	15	140	484
X 1658–298	ns	frsd	45	0.9	51265	70	50	82

Continued Next Page.

Table 1. (continued) RXTE ASM: Transients above 25 mCrab

Source Name	type	profile	peak mCrab	ASM HR2	start date	Rise days	Decay days	Duration days
\multicolumn{9}{c}{*Known Recurrent Transients (cont.)*}							
X 1704+241	ns?	sym	33	1.3	50707	110	35	160
RX J1709.5−266	ns	frsd	210	0.9	50448	< 50	50	86
X 1711−339	ns?	sym	50	1.2	51016	10	20	280
X 1730−333	ns	frsd	377	1.5	mult.	−	−	25
GRO J1744−28	ns	frsd	1291	2.5	pre XTE	−	65	>120
”	ns	frsd	1291	2.5	50433	40−60	40	>120
X 1803−245	ns?	frsd	740	1.2	50904	20	25	75
SAX J1808.4−3658	ns	frsd	108	0.9	50333	8	8	19
”		frsd	79	1.1	50911	4	12	21
GS 1843+009	ns	frsd	30	2.5	50480	30	50	104
XTE J1856+053	bhc	sym	75	0.4	50189	20	10	27
”		frsd	79	0.4	50328	20	40	70
X 1908+005	ns	sym	515	1.0	mult.	−	−	78
” (failed)		sym	40	0.9	50232	−	−	∼60
GS 2138+568	ns	sym	45	2.0	50630	12	10	27
SAX J2103.5+4545	ns	sym	25	−	50487	75	30	135
”		sym	25	−	51471	12	12?	IP
\multicolumn{9}{c}{C. Long Duration Transients (Duration > 500 d)}							
X 1210−64	ns?	qp	30	1.2	pre XTE	−	−	> 675
KS J1716−389	??	qp	50	1.6	pre XTE	−	−	> 1500
KS 1731−260	ns	qp	356	1.1	pre XTE	−	−	> 1500
GRS 1758−258	bhc	qp	54	1.6	pre XTE	−	−	> 1500
GS 1826−238	ns	qp	35	1.2	pre XTE	−	−	> 1500
GRS 1915+105	bhc	wild	2497	1.3	pre XTE	−	−	> 1500
\multicolumn{9}{c}{D. Periodic, Hard Transients}							
X 0115+634	ns	sym	400	3.2	pre XTE	−	−	21
RX J0812.4−3114	ns	sym	25	2.0	50926	−	−	20
X 1145−619	ns	sym	93	1.9	50166	−	−	30
X 1845−024	ns	sym	25	2.5	50345	−	−	30
X J1946+274	ns	mult	80	2.6	51055	−	−	85
EXO 2030+375	ns	mult	25	1.6	pre XTE	−	−	20

Table 2. Notes and References for ASM Transients

A. Fast X-ray Novae: Decay $\tau < 1$ day

XTE J0421+560 (CI Cam, radio jets)
- X-ray outburst (Smi99)
- Rapid rise (few hours) and decay time scale 0.5 d to 2.3 d (Bel99)
- X-ray properties, unusual spectrum (Ued98, Fro98, Orr98, Rev99)
- Distance 1.0 kpc (Orl00)
- Radio jets (Hje99)
- Optical outburst: (Wag99, Bar98)
- Opt spectrum (Dow84)
- IR spectrum, dense circumstellar wind (Cla99)

SAX J1819.3–2525 (V4641 Sgr, radio jets)
- Discovery: Feb 99, (Int99), Sept. 99 (Smi99)
- Five brief X-ray outbursts in 6 days in Sept. 99 (Smi99, Wij99, McC99)
- Rapid 1-s variability (Wij99)
- Radio jets (Hje99)
- Distance 0.5 kpc (Hje99)
- Opt counterpart (Gor90, Gre99)
- Optical outburst (Stu99, Gar99)

B. Intermediate-Duration Transients (Nonperiodic)
Recent Initial Outbursts

XTE J0111–733 (31-s pulsar in SMC)
- Pulsations (Cha98)
- Hard X-ray profile and spinup (Wil98)
- Optical counterpart (Isr99)

XTE J1550–564 (bright bhc transient)
- Acquired early in its rise (Smi98)
- Reached 6.8 Crab brightness in brief flare (Rut98)
- Detected to 200 keV with BATSE (Wil98)
- Evolution of spectra (Sob99)
- QPO 0.05 — 285 Hz (Cui99, Rem99, Hom99)
- Optical counterpart K star at 2.5 kpc (Jai99, San99)
- Hard lags in X-ray QPO and broad band var. (Cui00)
- Likely radio counterpart (Cam98)

XTE J1723–376
- X-ray outburst w. 816 Hz osc. (Mar99a)
- X-ray position and Type I bursts (Mar99b)

GRS 1737–310
- Weak X-ray outburst: (Tru99, Mar97)
- Similarity to Cyg X–1 spectrum (Cui97)
- BSAX intensity and position (Hei97)
- Spectrum and distance of 8500 pc (Ued97)

GRS 1739–278 (radio emitter)
- Multiple X-ray sub-peaks (Asm00)
- X-ray outburst; black-hole candidate (Var97)
- Radio emission (Hje96)
- Candidate optical/IR object at radio position (Mar97)
- X-ray spectra variations (Bor98)
- 5-Hz QPO (Bor00)

XTE J1739–285
- X-ray outburst (Mar99)

XTE J1748–288 (radio jets, shock in ISM)
- Single outburst w. 2-d rise (Smi98)
- Detected to 100 keV (Har98)
- QPO at 0.5 and 32 Hz (Fox98)
- Spectral and QPO evolution (Rev99)
- Transient radio with jet that shocked in ISM (Hje98, Fen98)

SAX J1750.8–29
- Bursting transient (Nat99)

Table 2. (continued) Notes and References for ASM Transients

XTE J1755–324 (extremely soft spectrum)
Steep soft spectrum with hard component, (Rem97)
Temporal/spectral evol. similar to Nova Muscae 1991 (Rev98)

Hard X-ray flux (Gol99)

XTE J1858+034 (221-s pulsar)
X-ray outburst, hard spectrum (Rem98)
221-s pulsar (Tak98)

QPO at 0.11 Hz (Pau98)
Celestial position (Mar98)

XTE J1859+226 (radio source)
X-ray outburst (Woo99)
Oscillations from 0.5 Hz to 5.5 Hz (Mar99, Dal99)
Detected to 200 keV w. variable cutoff (McC99, Dal99, Foc99)
Radio outburst (Poo99)

Optical counterpart R = 15.1 (Gar99)
Opt IR consistent w. short period soft transients (Hyn99)
Possible optical orbital modulation 0.28 d (Uem99)

XTE J2012+381 (radio source)
X-ray outburst (Rem98)
Hard initial spike and later becomes very soft, bhc (Asm00)
Ultra soft comp. w. hard tail in ASCA, bhc (Whi98)

Radio counterpart (Hje98, Poo98)
Optical counterpart (tentative) V = 21.3 (Hyn99)

XTE J2123–058 (high-lat. LMXB)
High galactic latitude –36.2 (Lev98)
Atoll LMXB, bursts, twin kHz (Hom99, Tom99)
Optical counterpart w. 6-h orbit (Tom99, Sor99)

Optical outbursts (Gne99)
Precursor activity may be solar contamination (Asm00)

Known Recurrent Transients

EXO 0748–676 (eclipsing LMXB)
Soft x-ray excess (Bri97)
Eclipse Timings (Her97)
Progressive covering of disk corona (Chu98)

Quiescent properties (Gar99)
QPO 0.6 – 2.4 Hz (Hom99)

X 1246–588
Probable X-ray Burster (Pir97)

Probable ROSAT source 1RXS J124938.0–590525 (Bol97)

X 1354–644 (LMXB, BW Cir)
Modest outburst (Rem97)
Detected to 200 keV (Har97)

Ginga detection (Kit90)
Low/hard state; rapid var. (Rev00)

X 1608–522 (bright recurrent LMXB)
Sustained one-year low states after each outburst (Asm00)
KHz QPO (Men98)
KHz QPO freq. dependence on position in color diagram (Men99a)
Quiescent luminosity (Asa96) possibly thermal (Rut99)

KHz QPO peak separation not constant (Men99b)
Island state kHz QPO (Yu97)
Outburst with hard spectrum (Zha96)

Table 2. (continued) Notes and References for ASM Transients

X 1630−472 (bright recurrent bhc, 184 Hz, radio source)
Three outbursts w. intervals of 700 d and 450 d (Asm00)
Double-peaked and flat-topped profiles (Asm00)
Sustained (1 year) low state after 2nd outburst (Asm00)
Historical outbursts behavior (Kuu97)
Absorption dips (Kuu98)
Evolution of spectral components (Oos98)
QPOs 0.06 – 14 Hz (Die00)
QPO 184 Hz (Rem99)
Radio and Hard X-rays (Hje99)

GRO J1655−40 (rel. radio jets; 300 Hz QPO)
Black hole, radio jets, "microquasar" (Tin95)
Mass 6 – 7 $M_\odot$ (Oro97a, Sha99)
Optical turn-on precedes X-ray by 5 d (Oro97b)
Low freq. QPOs; 300 Hz when source hard (Rem99)
Spectral evolution (Men98, Tom99, Sob99)
Echo mapping (X-ray to optical) (Hyn98)

X 1658−298 (X-ray burster)
Recovery by BSAX and X-ray burst (Hei99)

X 1704+241 (HD 154791)
Peculiarities in M Giant spectrum (Gau99)

RX J17095−26
Hard X-ray outburst (Mar97)
Possible radio counterpart (Hje97)
X-ray bursts (Coc98)

X 1711−339
Recovered Ariel-5 and SAS-3 source (Rem98)

X 1730−333 (Rapid Burster)
Seven outbursts w. intervals 210 d (Asm00)
Outbursts last 5 weeks w. two phases: Type I thermonuclear bursts
followed by Type II accretion bursts (Gue99)

GRO J1744−28 (bursting pulsar)
Hard X-ray pulsations 0.47 s (Fin96)
Pulsar phase changes associated with bursts (Kos98 and refs therein)
QPOs (Zha96, Kom97)
Super-Eddington fluxes imply beaming (Gil96)
Possible near IR counterpart (Aug97)
Propeller effect (Cui97)
HEXE/Mir-Kvant observations (Bor97)
Hard X-ray bursts with Konus and Mir- Kvant (Apt98, Ale98)
X-ray properties from BATSE and ASCA (Woo99, Nis99)

X 1803−245 (XTE J1806−246)
X-ray outburst (Mar98)
QPOs (Rev99, Wij99)
Possible burst source (Mul98)

SAX J1808.4−36 (401-Hz accreting pulsar)
X-ray outburst (Int98)
401-Hz pulsations and 2-h orbit (Wij98a, Cha98)
Soft phase lags (Cui98, Vau98)
X-ray spectrum (Gil98, Hei98)
Broad-band power spectrum (Wij98b)
Optical counterpart (Roc98)
Transient radio emission (Gae99)
Renewed activity Feb. 00 (vdK00, Wac00)

Table 2. (continued) Notes and References for ASM Transients

GS 1843+009 (30-s pulsar)
100-d flare followed by weak activity (Asm00)
X-ray recovery, 30-s pulse period, and spectrum (Wil97, Tak97)

XTE J1856+053 (bhc)
X-ray outburst (Mar96)
Soft spectrum (Asm00)
Hard X-ray flux (Bar96)

X 1908+005 (Aql X–1; bright recurrent transient)
Five strong and two failed outbursts (Asm00)
Optical counterpart clarified: V = 21.6, late K (Che99)
KHz oscillations change freq after burst (Yu99)
Low-energy phase lags (For99)
Propeller effect (Cam98)
Outside-in outburst (opt-IR-X-ray) (Sha98)
At thermal-viscous instability boundary (vaP96)

GS 2138+568 (Cep X–4(?), 66-s pulsar)
Be star optical counterpart: (Bon98)
Spindown rate (Wil99)
X-ray pulse profile changes (Muk00)

SAX J2103.5+4545 (359-s pulsar)
Faint transient, 359-s pulsar (Hul98)
Second outburst (Bay00)

C. Long Duration Transients (Duration > 500 d)

X 1210–64 (quasi-persistent)
On until 50763 (Asm00)
Uhuru and OSO–7 source

KS J1716–389 (100-d dipper)
Galactic center source (Ale95)
On until MJD 50763 (Asm00)
Quasi persistent source with periodic dips (Rem99)
Periodicity $\sim$ 100 d (Wen99)

KS 1731–260 (524 Hz during bursts)
KHz QPO at 524 Hz during burst (Smi97)
Two KHz QPO at 898 and 1159 Hz (Wij97)
ROSAT observations, celestial position, persistent source? (Bar98)

GRS 1758–258 (bright hard galactic center source)
Radio jets, x-ray spectral var. and similarity to 1E 1740.7–2942 (Smi97)
ASCA spectrum, soft excess (Mer97)
Optical candidates (Mar98)
Long-term monitoring (Mai99, Kuz99)

GS 1826–238 (burster)
Bursts at reg. intervals (Ube97)
Spectrum and distance from bursts (Int99)
Possibly steady accretor since 1988 (Ube97, Int99)

GRS 1915+105 (microquasar)
Superluminal jets (Mir94, Fen99)
Ten distinct accretion states, some oscillatory (Gre97, Mun99)
Variable low freq. QPOs and persistent 67 Hz when spectrum hard (Mor97)
Coincident X-ray, IR and radio outbursts (Poo97, Eik98, Mir98)
Disk emptying episodes (Bel97, Poo97, Fer99)
Hard phase lags for 67 Hz QPO (Cui99)
Interplay between QPOs and spectral components (Mun99, Mar99, Fer99)

Table 2. (continued) Notes and References for ASM Transients

D. Periodic, Hard Transients

X 0115+634 (P = 24-d, 3.61 s)
~ 8 maxima detected through 2/00 (Asm00)
Major outburst Feb. 99 (Asm00)
Mini outbursts May – July 96 at multiples of

24-d orbital period (Asm00)
Four cyclotron lines, (Hei99, San99)
Optical counterpart reclassified (Ung99)

RX J08124–3114 (P = 81.4 d, 32 s)
~ 7 maxima detected through 2/00 (Asm00)
Optical counterpart is Be star LS992 (Mot97)

Orbital period 80 d (Cor00)
X-ray pulsar 31.9 s (Rei99)

X 1145–619 (P = 189 d, 292 s)
~ 4 maxima detected through 2/00 (Asm00)
Outburst (Cor96)

Multiwavelength observations, 13-yr review (Ste97)

X 1845–024 (P = 242 d, 95 s)
(= GRO J1849–03 = GS1843–02)
~5 maxima detected through 2/00 (Asm00)

BATSE outbursts w. 242-d period (Fin99)
GRO source identified as Ariel-5/SAS-3/Ginga source (Sof98, Fin99)

X J1946+274 (= 3A 1942+274; P = 80 d, 16-s)
~ 7 maxima detected through 2/00 (Asm00)
First detections since 1976 (Asm00)

Pulsar, P = 16 s, (Smi98)
Orbital period 80 d, (Cam99)

EXO 2030+375 (P = 46 d, 42 s)
~ 30 maxima detected through 2/00 (Asm00)
Pulse period dependence on luminosity (Rey96)
Timing properties (Rei98a)
Long-term variability and IR spectroscopy (Rei98b)

Thirteen outbursts at 46-d intervals, orbit from pulse phases (Sto99)
Spectra at low luminosities (Rei99)

REFERENCES FOR TABLE 2

Ale95_1716: Aleksandrovich, N. L., Aref'ev, V. A., Borozdin, K. N., Sunyaev, R. A., & Skinner, G. K. 1998, Astron. Letters, 21, 431
Ale98_1744: Aleksandrovich, N. L., Borozdin, K. N., Aref'ev, V. A., Sunyaev, R. A., & Skinner, G. K. 1998, Astron. Letters, 24, 7
Apt98_1744: Aptekar', R. L., et al. 1998, ApJ, 493, 404
Asa96_1608 Asai, K., Dotani, T., Mitsuda, K., Hoshi, R., Vaughan, B., Tanaka, Y., & Inoue, H. 1996, PASJ, 48, 257
Asm00_xxxx: The ASM Teams at MIT and GSFC, public data.
Aug97_1744: Augusteijn, T., et al. 1997, ApJ, 486, 1013
Bar96_1856: Barret, D., et al. 1996, IAUC 6519
Bar98_1731: Barret, D., Motch, C., & Predehl, P. 1998, A&A, 329, 965
Bar98_0421: Barsukova, E.A., et al. 1998, Bull. Spec. Astrophys. Obs., 45, 14 (astro-ph/9905338)
Bay00_2103: Baykal, A., Stark, M., Swank, J. H 2000, IAUC 7355
Bel97_1915: Belloni, T., Mendez, M., King, A. R., van der Klis, M., & van Paradijs, J. 1997, ApJL, 488, L109
Bel99_0421: Belloni, T., et al. 1999, ApJ, 527, 345
Bol97_1246: Boller, T., Haberl, F., Voges, W., Piro, L., & Heise, J. 1997, IAUC 6546
Bon98_2138: Bonnet-Bidaud, J. M., & Mouchet, M. 1998, A&A, 332, L9

Table 2. (continued) Notes and References for ASM Transients

Bor97_1744: Borkus, V. V., et al. 1997, Astron. Letters, 23, 421

Bor98_1739−278: Borozdin, K. N., Revnivtsev, M. G., Trudolyubov, S. P., Aleksandrovich, N. L., Sunyaev, R. A., & Skinner, G. K. 1998, Astron. Letters, 24, 435

Bor00_1739−278: Borozdin, K. N. & Trudolyubov, S. P. 2000, ApJL (submitted), astroph/9911290

Bri97_0748: Brian, T, Corbet, R., Smale, A., Asai, K., & Dotani, T. 1997, ApJL, 480, L21

Cam98_1550: Campbell-Wilson, D., et al. 1998 IAUC 7010

Cam98_1908: Campana, S., et al. 1998, ApJL, 499, L65

Cam99_1946: Campana, S. Israel, G., & Stella, L. 1999, A&A, 352, L91

Cha98_0111: Chakrabarty, D., Levine, A. M., Clark, G. W., & Takeshima, T. 1998, IAUC 7048

Cha98_1808: Chakrabarty, D., & Morgan, E. H. 1998, Nature 394, 346

Cla99_0421: Clark, J. S., Steele, I. A., Fender, R. P., Coe, M. J. 1999, A&A, 348, 888

Che99_1908: Chevalier, C., Ilovaisky, S. Leisy, P., & Patat, F. 1999, A&A, 347, L51

Chu98_0748: Church, M. J., Balucinska-Church, M., Dotani, T., & Asai, K. 1998, ApJ, 504, 516

Coc98_1709: Cocchi, M., et al., ApJ, 508, L163

Cor96_1145−619: Corbet, R., & Remillard, R. 1996, IAUC 6486

Cor00_0812: Corbet, R., & Peele, A. G. 2000, ApJL, 530, L33

Cui97_1737: Cui, Wei., Heindl, W. A., Swank, J. H., Smith, D. M., Morgan, E. H., Remillard, R., & Marshall, F. E. 1997, ApJL, 487, L73

Cui97_1744: Cui. W. 1997, ApJL, 482, L163

Cui98_1808: Cui, W., Morgan, E. H., & Titarchuk, L. G. 1998, ApJL, 504, L27

Cui99_1550: Cui, E., Zhang, S. N., Chen, W., & Morgan, E. H. 1999, ApJL, 512, L43

Cui99_1915: Cui, W. 1999, ApJL, 524, L59

Cui00_1550: Cui, W., Zhang, & S. N., Chen, W. 2000, ApJL, 531, L45

Dal99_1859+226: Dal Fiume, D., et al. 1999, IAUC 7291

Die00_1630: Dieters et al. 2000, ApJ, in press (astro-ph 991202)

Dow84_0421: Downes, R. 1984, PASP, 96, 80

Eik98_1915: Eikenberry, S.S., Matthews, K., Morgan, E. H., Remillard, R. A., & Nelson, R. W. 1998, ApJL, 494, L61

Fen98_1748: Fender, R. P., & Stappers, B. W. 1998, IAUC 6937

Fen99_1915: Fender, R. P., et al. 1999, MNRAS, 304, 865

Fer99_1915: Feroci, M., Matt, G., Pooley, G., Costa, E., Tavani, M., & Belloni, T. 1999, A&A, 351, 985

Fin96_1744: Finger, M. H., Koh, d. T., Nelson, R. W., Prince, T. A., Vaughan, B. A., & Wilson, R. B. 1996, Nature, 381, 291

Fin99_1845: Finger, M. H., et al. 1999, ApJ, 517, 449

Foc99_1859+226: Focke, W. B., Markwardt, C. B., Swank, J. H., & Taam, R. E. 1999, BAAS, 31, 1555

For99_1908: Ford, E. C. 1999, ApJL, 519, L73

Fox98_1748: Fox, D., & Lewin, W. 1998, IAUC 6964

Fro98_0421: Frontera, F., et al.1998 A&A, 339, L69

Gae99_1808: Gaensler, B. M., Stappers, B. W., & Getts, T. J. 1999, ApJL, 522, L117

Gar99_0748: Garcia, M. R., & Callanan, P. J. 1999, AJ, 118, 1390

Gar99_1819: Garcia, M.R., & McClintock, J.E. 1999, IAUC 7271

Gar99_1859+226: Garnavich, P, M., Stanek, K. Z., & Berlind, P. 1999, IAUC 7276

Gau99_1704: Gaudenzi, S., & Polcari, V. F. 1999, A&A, 347, 4

Gol99_1755: Goldoni, P., et al. 1999, ApJ, 511, 847

Gor90_1819: Goranskij 1990, IBVS 346

Gil96_1744: Giles, A. B., Swank, J. H., Jahoda, K., Zhang, W., Strohmayer, T., Stark, M. J., & Morgan, E. H 1996, ApJL, 469, L25

Gil98_1808: Gilfanov, M., Revnivtsev, M., Sunyaev, R., & Churazov, E. 1998, A&A, 338, L83

Gne99_2123: Gneiding, C. D., Steiner, J. E., & Cieslinski, D. 1999, A&A, 352, 543

Table 2. (continued) Notes and References for ASM Transients

Gre97_1915: Greiner, J., Morgan, E. H., & Remillard, R. A. 1997, ApJL, 473, L79

Gre99_1819: Green, D.W.E. 1999, IAUC 7277

Gue99_1730: Guerriero, R., et al. 1999, MNRAS, 307, 179

Har97_1354: Harmon, B. A., & Robinson, C. R. 1997, IAUC 6774

Har98_1748: Harmon, B. A., McCollough, M. L., Wilson, C. AS., Zhang, S. N., & Paciesas, W. S. 1998, IAUC 6933

Hei97_1737: Heise, J. IAUC 6606

Hei98_1808: Heindl, W. A., & Smith, D. M. 1998, ApJL, 506, L35

Hei99_0115: Heindl, W. A., et al. 1999, ApJL, 521, L49

Hei99_1658: Heise, J., et al. 1999, IAUC 7263

Her97_0748: Hertz, P., Wood, K., & Cominsky, L. 1997, ApJ, 486, 1000

Hje96_1739−278: Hjellming, R. M., Rupen, M. P., Marti, J., Mirabel, F., & Rodriguez, L. F. IAUC 6383

Hje97_1709: Hjellming, R. M., & Rupen, M. P. 1997, IAUC 6547

Hje98_1748: Hjellming, R. M., et al. 1998, Paper in preparation; see Abstract #103.08, AAS Mtg #193

Hje98_2012: Hjellming, R. M., & Rupen, M. P. 1998, IAUC 6924; see also IAUC. 6932

Hje99_0421: Hjellming, R.M., & Mioduszewski, A.J. 1999, IAUC 6862

Hje99_1608: Hjellming, R. M., et al. 1999, ApJ, 514, 383

Hje99_1819: Hjellming., R.M., et al. 1999, IAUC 7265; see also 2000, BAAS, Meeting 195, late papers

Hom99_0748: Homan, J., Jonker, P. G., Wijnands, R., van der Klis, M., & van Paradijs, J. 1999, ApJL, 516, L91

Hom99_1550: Homan, J., Wijnands, R., & van der Klis 1999, IAUC 7121

Hom99_2123: Homan, J., Mendez, M., Wijnands, R., van der Klis, M., & van Paradijs, J. 1999, ApJL, 513, L119

Hul98_2103: Hulleman, F., In't Zand, J. J. M., & Heise, J. 1998, A&A, 337, L25

Hyn98_1655: Hynes, R. I., O'Brien, K., Horne, K., Chen, W., & Haswell, C. A. 1998, MNRAS, 299, L37

Hyn99_1859+226: Hynes, R. I., Haswell, A. J., & Chaty, S. 1999, IAUC 7294

Hyn99_2012: Hynes, R. I., Roche, P., Charles, P. A., & Coe, M. J. 1999, MNRAS, 305, L49

Int98_1808: In 't Zand, J. J. M., Heise, J., Muller, J. M., Bazzano, A., Cocchi, M., Natalucci, L., Ubertini, P. 1998, A&A, 331, L25

Int99_1819 In't Zand, J., et al. 1999, IAUC 7119

Int99_1826: In 't Zand, J. J. M., Heise, J., Kuulkers, E., Bazzano, A., Cocchi, M., Ubertini, P. 1999, A&A, 347, 891

Isr99_0111: Israel, G. L., Stella, L., & Mereghetti, S. 1999, IAUC 7101

Jai99_1550: Jain, R. K., Bailyn, C. D., Orosz, J. A., Remillard, R. A., & McClintock, J. E. 1999, ApJL, 517, L131

Kit90_1354: Kitamoto, S., et al. 1990, ApJ, 361, 590

Kom97_1744: Kommers, J. M., Fox, D. W., Lewin, W. H. G., Rutledge, R. E., van Paradijs, J., & Kouveliotou, C. 1997, ApJL, 482, L53

Kos98_1744: Koshut et al. 1998, ApJL, 496, L101

Kuu97_1630: Kuulkers, E., Parmar, A. N., Kitamoto, S., Cominsky, L. R., & Sood, R. K. 1997, MNRAS, 291, 81

Kuu98_1630: Kuulkers, E., Wijnands, R., Belloni, T., Mendez, M., van der Klis, M., & van Paradijs, J. 1998, ApJ, 494, 753

Kuz99_1758: Kuznetsov, S. I. et al. 1999, Astron. Letters, 25, 351

Lev98_2123 Levine, A., Swank, J., & Smith, D.A. 1998, IAUC 6955

Mai99_1758: Main, D. S., Smith, D. M., Heindl, W. A., Swank, J., Leventhal, M., Mirabel, I. F., & Rodriguez, L. F. 1999, ApJ, 525, 901

Mar96_1856: Marshall, F. E., Ebisawa, K., Remillard, R., & Valinia, A. 1996, IAUC 6504

Mar97_1709 Marshall, F. E., Swank, J. H., Thomas, B., Angelini, L., Valinia, A., & Ebisawa, K. 1997, IAUC 6543

Mar97_1737: Marshall, F. E., & Smith, D. M. 1997, IAUC 6603

Mar97_1739−278: Marti, J., Mirabel, I. F., Duc, P.-A, Rodriguez, L. F. 1997 A&A, 323, 158

Table 2. (continued) Notes and References for ASM Transients

Mar98_1803: Marshall, F. E., Strohmayer, T., & Remillard, R. 1998, IAUC 6891

Mar98_1858: Marshall, F. E. & Chakrabarty, D., A&A, 337, 815

Mar98_1758: Marti, J., Mereghetti, S., Chaty, S., Mirabel, I. F., Goldoni, P., & Rodriguez, L. F. 1998, A&A, 338, L95

Mar99a_1723: Marshall, F. E., & Markwardt, C. B. 1999, IAU Circ., 7103

Mar99b_1723: Marshall, F.E., Ueda, Y. & Markwardt, C. B. 1999, IAUC 7133

Mar99_1739−285: Markwardt, C. B., Marshall, F. E., Swank, J. H., & Cui, W. 1999, IAUC 7300

Mar99_1859+226: Markwardt, C. B., Focke, W. B., Swank, J. H., & Taam, R. E. 1999, BAAS, 31, 1555

Mar99_1915: Markwardt, C. B., Swank, J. H., & Taam, R. E. 1999, ApJL, 513, L37

McC99_1819: McCollough, M. L., Finger, M. H., &Woods, P. M. 1999, IAUC 7257

McC99_1859+226: McCollough, M. L., & Wilson, C. A. 1999 IAUC 7282

Men98_1655: Mendez, M., Belloni, T., & van der Klis, M. 1998, ApJL, 499, L187

Men98_1608: Mendez, M., et al. 1998, ApJ, 494, L65

Men99a_1608 Mendez, M., van der Klis, M., Ford, E. C., Wijnands, R., & van Paradijs, J. 1999 ApJL, 511, L49

Men99b_1608: Mendez, M., van der klis, M., Wijnands, R., Ford, F.C., van Paradijs, J., & Vaughan, B.A. 1998, ApJ, 505, L23

Mer97_1758: Mereghetti, S., Cremonesi, D. I., Haardt, F., Murakami, T., Belloni, T., & Goldwurm, A. 1997, ApJ, 476, 829

Mir94_1915: Mirabel, I. F., & Rodriguez, L. F. 1994, Nature, 371, 46

Mir98_1915: Mirabel, I. F., et al. 1998, A&A, 330, L9

Mor97_1915: Morgan, E. H., Remillard, R., & Greiner, J. 1997, ApJL, 482, L155

Mot97_0812: Motch, C., Haberl, F., Dennerl, K., Pakull, M.,& Janot-Pacheco, E. 1997, A&A, 323, 853

Muk00_2138: Mukerjee, K., Agrawal, P. C., Paul, B., Rao, A. R., Seetha, S., Kasturirangan, K. 2000, A&A, 353, 239

Mul98_1803: Muller, J. M., et al. 1998, IAUC 6867

Mun99_1915: Muno, M., Morgan, E.,& Remillard, R. 1999, ApJ, 527, 321

Nat99_1750: Natalucci, L., Cornelisse, R., Bazzano, A., Cocchi, M., Ubertini, P., Heise, J., In't Zand, J. J. M., & Kuulkers, E. 1999, ApJL, 523, L45

Nis99_1744: Nishiuchi, M., et al. 1999, ApJ, 517, 436

Oos98_1630: Oosterbroek et al. 1998, A&A, 340, 431

Orl00_0421: Orlandini, M., et al. 2000, A&A (in press, astro-ph 0001530)

Oro97a_1655: Orosz, J. A., & Bailyn, C. D. 1997, ApJ, 477, 876

Oro 97b_1655: Orosz, J. A., Remillard, R. A., Bailyn, C. D., & McClintock, J. E. 1997, ApJL, 478, L83

Orr98_0421: Orr, A., et al. 1998, A&A, 340, L190

Pau98_1858: Paul, B., & Rao, A. R. 1998, A&A, 337, 815

Pir97_1246: Piro, L., et al. 1997, IAUC 6538

Poo97_1915: Pooley, G. G., & Fender, R. P. 1997, MNRAS, 292, 925

Poo98_2012: Pooley, G. G. 1998, IAUC 6926

Poo99_1859+226: Pooley, G. G., & Hjellming, R. M. 1999, IAUC 7278

Rei98a_2030: Reig, P., & Coe, M. J. 1998, MNRAS, 294, 118

Rei98b_2030: Reig, P., Stevens, J. B., & Fabregat, J. 1998, MNRAS, 301, 42

Rei99_0812: Reig, P., & Roche, P. 1999, MNRAS, 306, 95

Rei99_2030: Reig, P., & Coe, M. 1999, MNRAS, 302, 700

Rem97_1354: Remillard, R., Marshall, F., & Takeshima, T. 1997, IAUC 6772

Rem97_1755: Remillard, R., Levine, A., Swank, J., & Strohmayer, T. 1997, IAUC 6710

Rem98_1711: Remillard, R. 1998, IAUC 6893

Rem98_1858: Remillard, R., & Levine, A. 1998 IAUC 6826

Rem98_2012: Remillard, R., Levine, A., & Wood, A. 1998, IAUC 6920

Table 2. (continued) Notes and References for ASM Transients

Rem99_1550: Remillard, R. A., McClintock, J. E., Sobczak, G. J., Bailyn, C. D., Orosz, J. A., Morgan, E. H., & Levine, A. M. 1999, ApJL, 517, L127

Rem99_1630: Remillard, R., & Morgan, E. 1999, BAAS, 31, 1421

Rem99_1655: Remillard, R. A., Morgan, E. M., McClintock, J. E., Bailyn, C. D., & Orosz, J. A. 1999, ApJ, 522, 397

Rem99_1716: Remillard, R. A. 1999, Mem. Soc. Astron. Ital., 70, 881 (astro-ph 9805224)

Rev98_1755: Revnivtsev, M., Gilfanov M., & Churazov, E. 1998, A&A, 339, 483. See also Astron. Letters 25, 493 and ApJ, 511, 847

Rev99_0421: Revnivtsev, M. G., Emel'yanov, A. N., & Borozdin, K. N. 1999, Astron. Letters, 25, 294

Rev99_1748: Revnivtsev, M. G., Trudolyubov, S. P., & Borozdin, K. N., 2000, MNRAS, 312, 151

Rev99_1803: Revnivtsev, M., Borozdin, K., Emelyanov, A. 1999, A&A, 344 L25

Rev00_1354: Revnivtsev, M., Borozdin, K., Priedhorsky, W. & Vikhlinin, A. 2000, ApJ 530, 955

Rey96_2030: Reynolds, A. P., Parmar, A. N., Stollberg, M. T., Verbunt, F., Roche, P., Wilson, R. B., Finger, M. H. 1996, A&A, 312, 872

Roc98_1808: Roche, P., Chakrabarty, D., Morales-Rueda, L., Hynes, R., Slivan, S. M., Simpson, C., & Hewett, P. 1998, IAUC 6885

Rut98_1550: Rutledge, R., Fox, D., & Smith, D. A. 1998, Astr. Tel. 36

Rut99_1608: Rutledge, R. E., Bildsten, L., Brown, E. F., Pavlov, G. G., & Zavlin, V. E. 1999, ApJ, 514, 945

San99_0115: Santangelo, A., et al. 1999, ApJL, 523, L85

San99_1550: Sanchez-Fernandez, C., et al. 1999, A&A, 348, L9

Sha98_1908: Shahbaz, T., Bandyopadhyay, R. M., Charles, P. A., Wagner, R. M., Muhli, P., Hakala, P., Casares, J., & Greenhill, J. 1998, MNRAS, 300, 1035

Sha99_1655: Shabaz, T., van der Hooft, F., Casares, J., Charles, P. A., & van Paradijs, J. 1999 MNRAS, 306, 89

Smi97_1731: Smith, D. A., Morgan, E. H., & Bradt, H. 1997, ApJL, 479, L137

Smi97_1758: Smith, D. M., Heindl, W. A., Swank, J., Leventhal, M., Mirabel, I. F., & Rodriguez, L. F. 1997, ApJL, 489, L51

Smi98_1550: Smith, D. A. 1998, IAUC 7008

Smi98_1748: Smith, D.A., Levine, A., & Wood, A. 1998 IAUC 6932

Smi98_1946: Smith, D. A., & Takeshima, T. 1998 IAUC 7014

Smi99_1819: Smith, D.A., Levine, A. M., & Morgan, E. H. 1999, IAUC 7253

Smi99_0421: Smith, D., Remillard, R., Swank, J., Takeshima, T., & Smith, E. 1998, IAUC 6855

Sob99_1550: Sobczak, G. J., McClintock, J. E., Remillard, R. A., Levine, A. M., Morgan, E. H., Bailyn, C. D., & Orocz, J. E. 1999, ApJL, 517, L121

Sob99_1655: Sobczak, G. J., McClintock, J. E., Remillard, R. A., Bailyn, C. D., & Orosz, J. A. 1999, ApJ, 520, 776

Sof98_1845: Soffitta, P., et al. 1998, ApJL, 494, L203

Sor99_2123: Soria, R., Wu, K., & Galloway, D. 1999, MNRAS, 309, 528

Ste97_1145-619: Stevens, J. B., Reig, P., Coe, M. J., Buckley, D. A. H., Fabregat, J., & Steele, I. A. 1997, MNRAS, 288, 988

Sto99_2030: Stollberg, Mark T., Finger, Mark H., Wilson, R. B., Scott, D. M., Crary, D. J., & Paciesas, W. S. 1999, ApJ, 512, 313

Stu99_1819: Stubbings, R., & Pearce, A. 1999, IAUC 7253

Tak97_1843: Takeshima, T. 1997, IAUC 6595

Tak98_1858: Takeshima, T. Corbet, R., Marshall, F., Swank, J., & Chakrabarty, D. 1998, IAUC 68

Tin95_1655: Tingay, S. J., et al. 1995, Nature, 374, 141

Tom99_1655: Tomsick, J. A., Kaaret, P., Kroeger R. A., & Remillard, R. A. 1999, ApJ, 512, 892

Tom99_2123: Tomsick, J. A., Halpern, J., Kemp, J., & Kaaret, P. 1999, ApJ, 521, 341

Tru99_1737: Trudolyubov, S., et al. 1999, A&A, 342, 496; see also IAUC 6599

Table 2. (continued) Notes and References for ASM Transients

Ube97_1826: Ubertini, P., Bazzano, A., Cocchi, M., Natalucci, L., Heise, J., Muller, J. M., & In 't Zand, J. J. M. 1999, ApJL, 514, L27

Ued97_1737: Ueda et al. 1997, IAUC 6627

Ued98_0421: Ueda, Y., Ishida, M., Inoue, H., Dotani, T., Greiner, J., & Lewin, W. H. G.. 1998, ApJ, 508, L167

Uem99_1859+226: Uemura, Kato, T., Pavlenko, E., Shugarov, S., & Mitskevich, M., 1999 IAUC 7303

Ung99_0115: Unger, S. J. Roche, P., Negueruela, I., Ringwald, F. A, Lloyd, C., & Coe, M. J., et al. 1998, A&A, 336, 960

Var97_1739-278: Vargas, M., et al. 1997, ApJL, 476, L23

vdK00_1808: van der Klis, M., Chakrabarty, D., Lee, J. C., Morgan, E. H.,Wijnands, R., Markwardt, C. B., & Swank, J. H. 2000, IAUC 7358

VaP96_1908: vanParadijs, J. 1996, ApJL, 464, L1

Vau98_1808: Vaughan, B. A., et al. 1998, ApJL, 509, L145; also ApJL, 483, L115

Wac00_1808: Wachter, S., & Hoard, D. W. 1998, IAUC 7363

Wag99_0421: Wagner, R.M., & Starrfield, S.G. 1999, IAUC 6857

Wen99_1716: Wen, L., Levine, A., & Bradt, H. 1999, BAAS 31, 1427

Whi98_2012: White, N. E., Ueda, Y., Dotani, T., & Nagase, F. 1998, IAUC 6927

Wij97_1731: Wijnands, R. A. D., & van der Klis, M. 1997, ApJL, 482, L65

Wij98a_1808: Wijnand, R., & van der Klis, M. 1998, Nature 294, 344

Wij98b_1808: Wijnands, R., & van der Klis, M. 1998, ApJL, 507, L63

Wij99_1803: Wijnands, R., van der Klis, M. 1999, ApJ, 522, 965

Wij99_1819: Wijnands, R., & van der Klis, M. 2000, ApJL, 528, L93

Wil97_1843: Wilson, R., Harmon, B., Scott, D., Finger, M., Robinsin, C., Chakrabarty, D., & Prince, T. A. 1997, IAUC 6586

Wil98_0111: Wilson, C. A., & Finger, M. H. 1998, IAUC 7048

Wil98_1550: Wilson, C. A., Harmon, B. A., Paciesas, W. S., & McCollough, M. L. 1998, IAUC 7010

Wil99_2138: Wilson, C. A., Finger, M. H., & Scott, D. M. 1999, ApJ, 511, 367

Woo99_1859+226: Wood, A., Smith, D. A., Marshall, F. E., & Swank, J. 1999, IAUC 7274

Woo99_1744: Woods, P. M., et al. 1999, ApJ, 517, 431

Yu97_1608: Yu, W., et al. 1997, ApJL, 490, L153

Yu99_1908: Yu, W., Li, P., Zhang, W., & Zhang. S. N. 1999, ApJL, 512, L35

Zha96_1608: Zhang, S. N., et al. 1996 A&A Suppl. Ser., 120, 279

Zha96_1744: Zhang, W., Morgan, E. H., Jahoda, K., Swank, J. H., Strohmayer, T. E., Jernigan, J. G., & Klein, R. I. 1996 ApJL, 469, L2

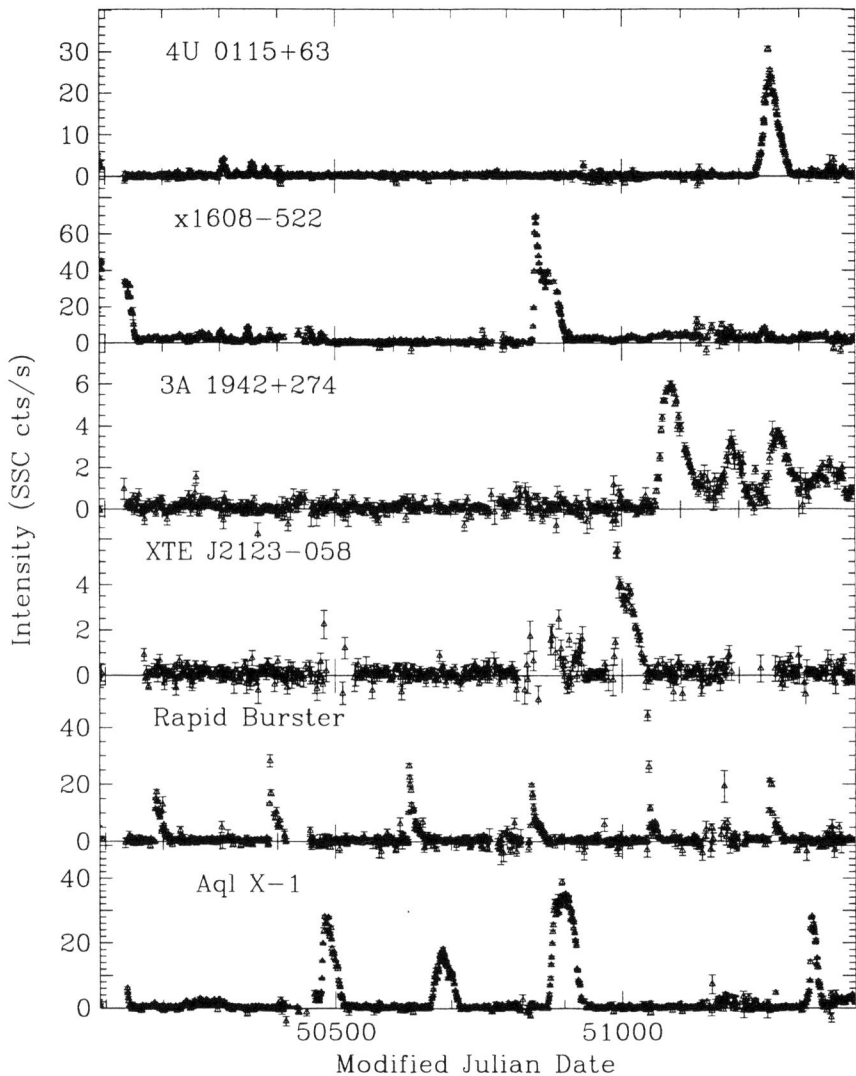

FIGURE 1. RXTE/ASM light curves for six neutron-star binary-system transients for the period early January 1996 through mid August 1999. The data points represent 1-day averages of the 10 – 20 (typical) daily measurements in the 1.5–12 keV band. 75 ct/s corresponds to 1.0 Crab. MJD 50082 = 1996 Jan. 0.0.

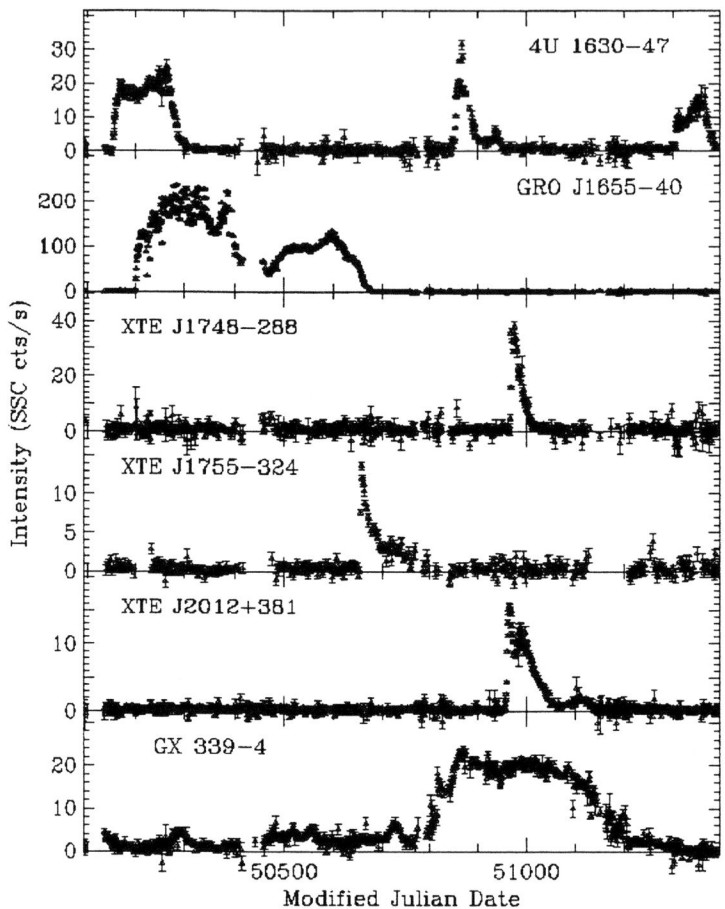

FIGURE 2. RXTE/ASM light curves for six black-hole binary-system transients. GX 339–4 shows a transient bright soft state; it is not listed as a transient in Table 1. See also caption to Fig. 1.

X-RAY ABSORPTION IN RADIO-QUIET QSOS

W.N. Brandt,[1] S.C. Gallagher,[1] A. Laor[2] and Beverley J. Wills[3]

1) Department of Astronomy & Astrophysics, 525 Davey Lab, The Pennsylvania State University, University Park, Pennsylvania 16802 USA
2) Physics Department, Technion, Haifa 32000 Israel
3) Department of Astronomy, University of Texas at Austin, Austin, Texas 78712 USA

ABSTRACT Major flows of ionized gas are thought to be present in the nuclei of most luminous QSOs, and absorption by these flows often has a significant effect upon the observed X-ray continuum from the black hole region. We briefly discuss X-ray studies of this gas and attempts to determine its geometry, dynamics, and ionization physics. Our focus is on X-ray warm absorber QSOs, Broad Absorption Line QSOs, and soft X-ray weak QSOs. We also discuss some prospects for further study with the next generation of X-ray observatories.

KEYWORDS: QSOs: absorption — QSOs: general — galaxies: active — X-rays: galaxies.

1. INTRODUCTION

X-ray absorption studies of active galaxies are proving to be one of the most powerful ways of probing material in the immediate vicinity of supermassive black holes. Rapid X-ray variability suggests that the nuclear X-ray source is the most compact emitter of continuum radiation, and it thus provides a point-like and luminous 'flashlight' right at the heart of the active galaxy. Because X-rays are highly penetrating, X-ray spectra can be used to probe column densities over the wide range 10^{19}–10^{25} cm^{-2}. X-ray absorption is produced by the innermost electrons of metals, and it provides a probe of matter in nearly all forms (i.e., neutral gas, ionized gas, molecular gas and dust).

Here we shall briefly review several types of X-ray absorption seen in luminous, radio-quiet QSOs. We will discuss X-ray warm absorber QSOs, Broad Absorption Line QSOs, and soft X-ray weak QSOs. We will also discuss some future prospects for radio-quiet QSO X-ray absorption studies. Due to lack of space, we will not be able to cover the important red QSO and type 2 QSO debates or the exciting recent results on X-ray absorption in radio-loud QSOs.

2. X-RAY WARM ABSORBERS IN RADIO-QUIET QSOS

Warm absorption by ionized nuclear gas is familiar from the lower-luminosity Seyfert 1 galaxies where it has been intensively studied (e.g., Reynolds 1997; George et al.

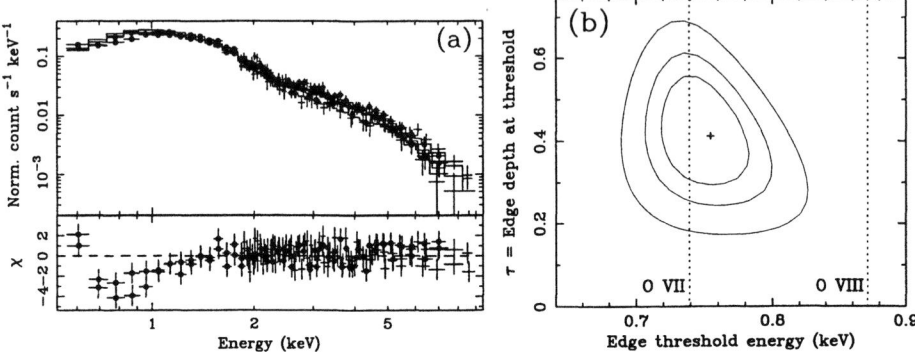

FIGURE 1. (a) *ASCA* SIS (solid dots) and GIS (plain crosses) observed-frame spectra of the warm absorber QSO IRAS 13349+2438 ($z = 0.107$). A power-law model has been fit to the 2–9 keV data and extrapolated back to show the deviations at low energies. The ordinate for the lower panel (labeled χ) shows the fit residuals in terms of σ with error bars of size one. Note the systematic absorption residuals at low energies due to ionized oxygen. (b) Confidence contours for the edge parameters. Contour levels are for 68.3, 90.0 and 99.0% confidence. The fitted edge energy has been corrected for cosmological redshift. From Brandt et al. (1997).

1998). Warm absorbers imprint moderately strong edges (e.g., O VII and O VIII) on the continuum, but this absorption is not so strong that it completely extinguishes the soft X-ray flux. Assuming photoionization equilibrium, the column density and ionization parameter of the ionized gas can be obtained via X-ray spectral fitting.

To our knowledge, only five luminous radio-quiet QSOs have been shown to have X-ray warm absorbers: MR 2251-178 (e.g., Halpern 1984; Pan, Stewart & Pounds 1990; Reeves et al. 1997), IRAS 13349+2438 (Brandt, Fabian & Pounds 1996; Brandt et al. 1997; Siebert, Komossa & Brinkmann 1999; see Figure 1), PG 1114+445 (Laor et al. 1994; George et al. 1997), IRAS 17020+4544 (e.g., Leighly et al. 1997), and IRAS 12397+3333 (e.g., Grupe et al. 1998). It has been difficult to perform detailed studies of the warm absorbers in these QSOs due to limited photon statistics, but the basic physical properties of their warm absorbers appear similar to those seen in Seyfert 1s. Edges from O VII and O VIII seem to be the strongest spectral features, and column densities of $\approx 10^{21}$–10^{23} cm^{-2} and ionization parameters of $\xi \approx 20$–160 erg cm s^{-1} are inferred. In three cases (IRAS 12397+3333, IRAS 13349+2438 and IRAS 17020+4544), the warm absorber probably contains dust which causes significant reddening of the optical continuum. Dust will not be rapidly sputtered at warm absorber temperatures (the gas temperature is $\sim 10^5$ K for a photoionized warm absorber), and it will not be sublimated if the warm absorber is located outside the Broad Line Region (BLR). Two of the QSOs with X-ray warm absorbers (as well as most of the Seyfert 1 galaxies; Crenshaw et al. 1999) show UV absorption lines (PG 1114+445: Mathur, Wilkes & Elvis 1998; MR 2251-

178: Mathur et al., in preparation), and it has been argued that the X-ray and UV absorption arise in the same gas. While there is still debate over the extent to which the X-ray and UV absorbers can be unified, they are likely to have qualitatively similar dynamics. The UV absorbing gas is measured to be outflowing from the nucleus at speeds of several hundred km s^{-1}.

The incidence of warm absorbers in luminous, radio-quiet QSOs is difficult to address at present (e.g., George et al. 1999). Warm absorbers are detected in $\gtrsim 50\%$ of Seyfert 1s, while a much smaller percentage of radio-quiet QSOs have *detected* warm absorbers. However, the X-ray spectra of most radio-quiet QSOs have significantly lower signal-to-noise than those of Seyfert 1s, and cosmological redshifting also moves the main warm absorber edges down to regions of low effective area and often poor calibration. Seyfert-like warm absorbers could be lurking undetected in the noisy X-ray spectra of many radio-quiet QSOs, and the only clear conclusion that can be drawn at present is that better data are needed (although Laor et al. 1997 suggest that warm absorbers are relatively rare in radio-quiet QSOs based upon *ROSAT* PSPC data).

3. BROAD ABSORPTION LINE (BAL) QSOS

Luminous radio-quiet QSOs show another type of absorption that is not familiar from Seyfert 1s: UV Broad Absorption Lines (BALs) that are created in an outflowing 'wind' with velocities up to $\sim 0.1c$. BALs have been intensively studied in the UV for many years, and it is likely that most QSOs create BAL outflows (e.g., Weymann 1997). The BAL region is thought to be major part of the nuclear environment with a covering factor of 10-50% (e.g., Goodrich 1997; Krolik & Voit 1998), and the BAL phenomenon may be fundamentally connected to the QSO 'radio volume control' (e.g., Weymann 1997). In addition, BAL outflows may clear gas from QSO host galaxies and thereby affect star formation and QSO fueling over long timescales (e.g., Fabian 1999).

Ideally, one would like to use X-rays to study the absorption properties, nuclear geometries, and continuum shapes of BAL QSOs. X-ray absorption studies would constrain the column density, ionization state, abundances, and covering factor of the BAL gas, and the nuclear geometry could be constrained using the iron Kα line and X-ray variability. Regarding the continuum shape, it is important to establish that, underneath their absorption, BAL QSOs emit like normal radio-quiet QSOs.

ROSAT observations found BAL QSOs to be very weak in the soft X-ray band with few X-ray detections (e.g., Kopko, Turnshek & Espey 1994; Green & Mathur 1996, hereafter GM96). This was an important and surprising result since, if BAL QSOs indeed have normal underlying X-ray continua, large neutral column densities of $\gtrsim 4 \times 10^{22}$ cm^{-2} are required to extinguish the X-ray emission. Ionization of the absorbing gas led to even larger required column densities. The *ROSAT* column densities were at least an order of magnitude larger than those inferred from UV data. Subsequently, it was realized that the UV absorption lines are severely saturated (e.g., Hamann 1998), leading to much larger inferred UV column densities.

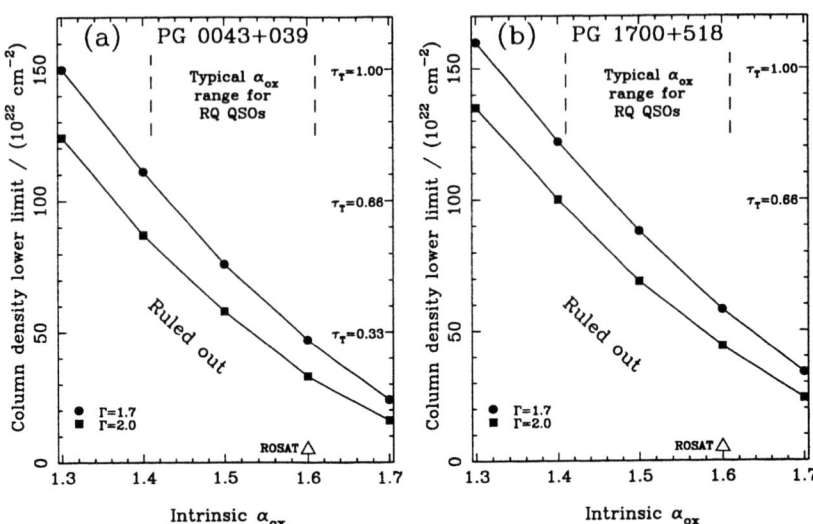

FIGURE 2. Column density lower limits for (a) PG 0043+039 and (b) PG 1700+518 derived using our *ASCA* SIS0 data. We show the inferred column density lower limit as a function of the intrinsic (i.e., absorption-corrected) value of α_{ox} (the slope of a nominal power law connecting the rest-frame flux density at 2500 Å to that at 2 keV). The square data points are for an X-ray photon index of $\Gamma = 2.0$, and the circular dots are for an X-ray photon index of $\Gamma = 1.7$. The open triangle at $\alpha_{\mathrm{ox}} = 1.6$ illustrates the typical BAL QSO column density lower limit found by GM96 based on *ROSAT* data. The numbers along the right-hand sides of the panels show the Thomson optical depth of the corresponding column density; note that our inferred column densities are within a factor of ≈ 3 of being optically thick to Thomson scattering. The column density lower limits shown in this plot are for absorption by neutral gas with solar abundances. In reality, the absorbing gas is probably ionized, and this can significantly raise the required column density. We have made similar plots to those above using α_{ix} (between 1.69 μm and 2 keV), and we find similarly large column densities are required. From Gallagher et al. (1999).

These column densities are then consistent with the X-ray data.

Hard X-rays are much more penetrating than soft X-rays, and if BAL QSOs are absorbed by column densities of a few times 10^{22} cm^{-2} they should be much brighter above 2 keV than below this energy. Using *ASCA*, Mathur, Elvis & Singh (1995) detected the famous BAL QSO PHL 5200 ($z = 1.98$) and claimed to measure a large column density for this object via spectral fitting. Our independent analysis of these data confirms the detection, but the claim for a large column density in this object is not reliable at present due to extremely limited photon statistics (Gallagher et al. 1999). The nearby BAL QSO Mrk 231 ($z = 0.042$) has also been studied by *ASCA* (Iwasawa 1999; Turner 1999) and appears to show absorption with a column density of $\gtrsim 2 \times 10^{22}$ cm^{-2}, although precise constraints are difficult due to the complex X-ray spectrum of this object (e.g., there appears to be a significant starburst contribution in X-rays).

Recently, we have been performing an exploratory BAL QSO survey using *ASCA* and *BeppoSAX* (Gallagher et al. 1999; Gallagher et al., in preparation). We chose these satellites because they provide access to penetrating 2–10 keV X-rays. We performed moderate-length ($\approx$ 20–30 ks) exploratory observations to learn about the basic X-ray properties of as many BAL QSOs as possible without being too heavily invested in the uncertain results from any one object. Our goals were to define the 2–10 keV properties (e.g., fluxes) of the class, to discover good objects for follow-up studies with *Chandra* and *XMM*, and to set absorption, geometry and continuum constraints (to the greatest extent possible with exploratory observations). We proposed many of the optically brightest BAL QSOs known since the optical and X-ray fluxes are generally correlated for QSOs. Most of our objects should have been easily detected if they have normal QSO X-ray continua absorbed by column densities of several times 10^{22} cm^{-2}. We focused on bona-fide BAL QSOs (no mini-BALs; see §3.1 of Weymann et al. 1991), and we also tried to sample a few objects with extreme properties (e.g., optical continuum polarization) to look for correlations.

We have performed new *ASCA* and *BeppoSAX* observations for 8 BAL QSOs in total, and we have also analyzed the archival data for 4 BAL QSOs. Our objects have $z = 0.042$–3.505 and $B = 14.5$–18.5; PHL 5200 ($B = 18.5$) is the optically faintest member of our sample. We detect 5 of our 12 BAL QSOs, with our most distant and most luminous detected BAL QSO being CSO 755 ($z = 2.88$, $M_V = -27.4$; Brandt et al. 1999). Our detection fraction is higher than in soft X-rays, consistent with the idea that heavy absorption is present in these objects. However, we find that BAL QSOs are still generally faint 2–10 keV sources, and several of them are strikingly faint. For example, we did not detect the optically bright BAL QSOs PG 0043+039 ($B = 15.9$, $z = 0.384$, 24 ks *ASCA* exposure) and PG 1700+518 ($B = 15.4$, $z = 0.292$, 21 ks *ASCA* exposure). If these objects have normal underlying X-ray continua, then large neutral column densities of $\gtrsim 5 \times 10^{23}$ cm^{-2} are needed to explain their X-ray non-detections (see Figure 2). Because of our access to more penetrating X-rays, our column density lower limits for some objects are about an order of magnitude larger than those set by *ROSAT*. Ionization of the absorbing gas raises our required column densities to the point where they are almost 'Compton-thick' ($N_\mathrm{H} \gtrsim 1.5 \times 10^{24}$ cm^{-2}; compare with Murray et al. 1995). These large column densities increase the inferred mass outflow rate and kinetic luminosity. If the X-ray absorption arises in gas at $\gtrsim 3 \times 10^{16}$ cm that is outflowing with a significant fraction of the terminal velocity measured from the UV BALs, one derives extremely large mass outflow rates ($\dot{M}_\mathrm{outflow} \gtrsim 5$ M$_\odot$ yr^{-1}) and kinetic luminosities ($L_\mathrm{kinetic} \gtrsim L_\mathrm{ionizing}$). While such powerful winds are perhaps not impossible, the mass outflow rate and kinetic luminosity can be reduced if much of the X-ray absorption occurs at velocities significantly smaller than the BAL terminal velocity. Note that the X-ray and UV absorbers in BAL QSOs have not yet been shown to be identical, and the X-ray and UV light paths may differ.

Our exploratory observations demonstrate that it is risky to attempt long X-ray spectroscopic observations of BAL QSOs that do not have established X-ray fluxes, and we find that optical flux is *not* a good predictor of X-ray flux for BAL QSOs. We fail to detect some of our optically brightest objects, while some of our optically faintest are clearly detected. We have empirically searched for other predictors of X-ray brightness, and while the data are limited there is a tentative connection between high optical continuum polarization and X-ray brightness (see Brandt et al. 1999 for details). Such a connection could be physically understood if the direct lines of sight into the X-ray nuclei of BAL QSOs are usually blocked by Compton-thick matter, and we can only see X-rays when there is substantial electron scattering in the nuclear environment by a 'mirror' of moderate Thomson depth. Further studies of uniform, well-defined BAL QSO samples are needed to avoid biases and check this potential connection better. It can also be checked with detailed X-ray studies of highly polarized BAL QSOs. Iron Kα lines with large equivalent widths could be formed if most of the X-ray flux is scattered, and one would also not expect rapid ($\lesssim 1$ day) X-ray variability.

4. SOFT X-RAY WEAK (SXW) QSOS

BAL QSOs are generally weak in the soft X-ray band, presumably due to heavy X-ray absorption. One can also address the converse questions: Do all Soft X-ray Weak QSOs (SXW QSOs) suffer from absorption? Do all SXW QSOs have BALs? Alternative possible causes of soft X-ray weakness include unusual intrinsic spectral energy distributions (SEDs) and extreme X-ray or optical variability (e.g., changes in α_{ox} over time). The presence of QSOs with relatively weak soft X-ray emission was recognized at least as early as the mid-1980s, with some observed to be $\gtrsim 20$ times weaker than expected given their optical fluxes (e.g., Elvis & Fabbiano 1984; Avni & Tananbaum 1986; Elvis 1992). For example, Avni & Tananbaum (1986) discussed a 'skew tail' towards soft X-ray weak objects for the α_{ox} distribution of the PG QSOs. Many new SXW QSOs were found in *ROSAT* samples (e.g., Laor et al. 1997; Yuan et al. 1998), and *ROSAT* was also able to place significantly tighter constraints upon α_{ox}. This sparked further detailed studies of these objects (e.g., Wang et al. 1999; Wills, Brandt & Laor 1999), and we have recently completed the first systematic study of a well-defined SXW QSO sample (Brandt, Laor & Wills 1999). Our goals for this study were (1) to determine the origin of soft X-ray weakness in general, (2) to discover relations between SXW QSOs, BAL QSOs, and X-ray warm absorber QSOs, and (3) to search for correlations between soft X-ray weakness and other interesting observables.

We selected all SXW QSOs from the Boroson & Green (1992, hereafter BG92) sample of 87 $z < 0.5$ PG QSOs. The BG92 sample is well defined and representative of the optically selected QSO population, and there is already a large amount of high-quality and uniform data available for it. We computed our own α_{ox} values for the BG92 objects using data mainly from *ROSAT* but also from *ASCA* and *Einstein* as needed, and our resulting α_{ox} values were substantially more complete

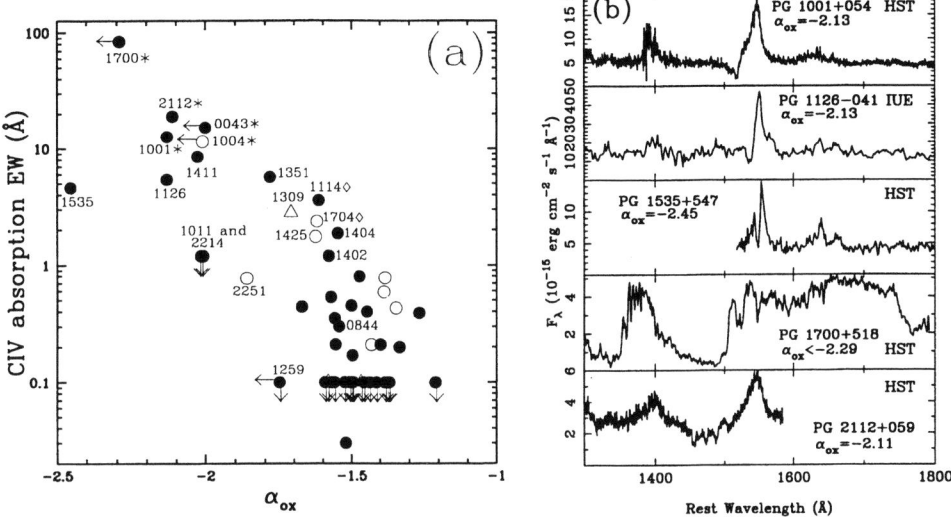

FIGURE 3. (a) Plot of α_{ox} versus C IV $\lambda 1549$ absorption-line EW (in the rest frame) for the BG92 QSOs with UV coverage. Following BG92, solid dots are radio-quiet QSOs, open triangles are core-dominated radio-loud QSOs, and open circles are lobe-dominated radio-loud QSOs. SXW QSOs (those QSOs with $\alpha_{ox} \leq -2$) and other interesting QSOs are labeled by the right ascension part of their name. An asterisk to the right of an object's name indicates that it is a BAL QSO or probable BAL QSO, and a diamond indicates that it is known to have an X-ray warm absorber. (b) UV spectra of the regions of the broad C IV emission line for some of the SXW QSOs. Note the strong, blueshifted C IV absorption. From Brandt, Laor & Wills (1999).

and constraining than those previously available (especially for the SXW QSOs). We used $\alpha_{ox} \leq -2$ as our criterion for soft X-ray weakness (note in this section we take α_{ox} to be a *negative* quantity). Thus, given their optical fluxes, our SXW QSOs were ≥ 25 times weaker than 'usual' in soft X-rays. We found 10 SXW QSOs with $\alpha_{ox} \leq -2$, and thus SXW QSOs appear to comprise $\approx 11\%$ of the optically selected QSO population. Nine of our SXW QSOs are radio-quiet, and one is radio-loud.

We compared the continuum and line properties of our 10 SXW QSOs to those of the other 77 BG92 non-SXW QSOs using nonparametric tests. The properties compared included those listed in Tables 1 and 2 of BG92 as well as the optical continuum polarization, the optical continuum slope, and the radio structure. We also compared the C IV $\lambda 1549$ absorption-line properties for the 55 QSOs from BG92 that have high-quality UV coverage in this spectral region. All C IV measurements were made by B. J. Wills with particular effort toward ensuring consistency and uniformity. We found that the SXW QSOs and non-SXW QSOs have consistent distributions of M_V, z, radio loudness (R), optical continuum slope, optical continuum polarization, and Hβ FWHM. In addition, they have consistent EW distributions of Hβ, He II and Fe II. SXW QSOs were found to have significantly lower [O III] luminosities than those of non-SXW QSOs; low [O III] luminosities have similarly been noted for low-ionization BAL QSOs (e.g., Turnshek et al. 1997). Since [O III] emission is likely to be a reasonably isotropic property for radio-quiet PG QSOs (e.g., Kuraszkiewicz et al. 1999), this result is significant as it suggests there may be an *intrinsic* difference between SXW QSOs and non-SXW QSOs (see Brandt, Laor & Wills 1999). In addition, SXW QSOs appear to have 'peaky' Hβ line profiles and large Hβ line shifts (either to the blue or the red relative to the rest frame defined by [O III]).

The most striking difference between the SXW QSOs and non-SXW QSOs is their UV absorption. SXW QSOs show greatly enhanced C IV absorption (see Figure 3). We find blueshifted C IV absorption with EW > 4.5 Å in 8 of our 10 SXW QSOs, while only 1 of 45 non-SXW QSOs had EW > 4.5 Å. The two SXW QSOs without clear UV absorption, 1011−040 and 2214+139, have UV spectra of only limited quality. Given that UV and X-ray absorption have a high probability of joint occurrence in Seyfert galaxies and QSOs, we consider Figure 3 to be evidence that absorption is the primary cause of soft X-ray weakness in QSOs. Only one of our SXW QSOs, 1411+442, has a broad-band X-ray spectrum at present, and it indeed shows evidence for strong X-ray absorption with $N_H \gtrsim 10^{23}$ cm^{-2} (Brinkmann et al. 1999). We can argue against unusual SEDs as the primary cause of soft X-ray weakness by noting that our SXW QSOs have normal Hβ, He II and Fe II EWs (see Korista, Ferland & Baldwin 1997). We also do not find general evidence for strong α_{ox} variability when we compare our α_{ox} values with the limited historical data available. The fact that we find no evidence for QSOs with intrinsically weak soft X-ray emission underscores the universality of QSO X-ray production.

The general correlation between α_{ox} and C IV absorption EW shown in Figure 3 provides a useful overall view of QSO absorption. Unabsorbed QSOs and BAL QSOs lie at opposite extremes of the correlation, while X-ray warm absorber QSOs and moderate SXW QSOs lie at intermediate positions. We find 4–5 bona-fide BAL QSOs in the BG92 sample; 3 were already known (0043+039, 1700+518 and 2112+059) and 2 are new (1001+054 and probably 1004+130; see Wills, Brandt & Laor 1999 for detailed discussion of 1004+130). If all BAL QSOs are SXW QSOs, then we should have found all the BAL QSOs in the BG92 sample. The incidence of BAL QSOs in the BG92 sample appears to be statistically consistent with the $\approx 11\%$ observed for the LBQS (e.g., Weymann 1997), although this issue could be examined more reliably with complete UV coverage of the BG92 QSOs.

The UV results for our SXW QSOs imply that selection by soft X-ray weakness is an effective ($\sim 80\%$ successful) way to find low-redshift QSOs with strong UV absorption. This is important from a practical point of view because, for bright QSOs, the optical and X-ray flux densities needed to establish soft X-ray weakness can often be obtained from publicly available data. This method has already been exploited in several cases for individual objects (e.g., Fiore et al. 1993; Mathur et al. 1994), and it could be profitably applied to larger QSO samples.

5. SOME FUTURE PROSPECTS

With the next generation of X-ray observatories, it should be possible to find many more X-ray warm absorbers in radio-quiet QSOs or demonstrate that few are present. This will allow study of their basic physical properties as well as a reliable determination of their incidence. Detailed X-ray spectroscopy and modeling should be possible for a few of the X-ray brightest sources, although this will require a significant investment of observation time.

For BAL QSOs, further exploratory observations are needed to look for correlations with optical continuum polarization and other properties. These would be most effective if performed on uniform and well-defined samples. Moderate-quality X-ray spectroscopy should be possible for a few of the X-ray brightest BAL QSOs to study their absorption properties, nuclear geometries, and continuum shapes. For BAL QSOs with enough X-ray flux, the widths and amplitudes of X-ray bound-free edges can constrain the dynamics and metallicity of the absorber. It is also important to study the radio-loud BAL QSOs (e.g., Becker et al. 1999) in X-rays to determine if they follow the same patterns as radio-quiet objects, and deep X-ray surveys over moderate areas may be able to constrain the BAL covering factor (see Krolik & Voit 1998). Studies of SXW QSOs more generally would benefit from a focused X-ray and UV study of a complete sample. The PG SXW QSOs are probably a good starting point, but a larger complete sample would be even better. Intense studies of particularly interesting objects (e.g., 1004+130) are important as well.

Finally, it is crucial to test models that propose to unify the different types of X-ray absorption into a coherent physical picture. Such testing should provide an exciting challenge for even the next generation of X-ray observatories.

ACKNOWLEDGEMENTS

We acknowledge the support of NASA LTSA grant NAG5-8107 and the Alfred P. Sloan Foundation (WNB), NASA grant NAG5-4826 and the Pennsylvania Space Grant Consortium (SCG), the fund for the promotion of research at the Technion (AL), and NASA LTSA grant NAG5-3431 (BJW). We thank C.S. Reynolds for a careful reading and J. Chiang for a helpful discussion.

REFERENCES

Avni, Y., Tananbaum, H. 1986, ApJ, 305, 83
Becker, R.H., et al. 1999, ApJ, in preparation
Boroson, T.A., Green, R.F. 1992, ApJS, 80, 109 (BG92)
Brandt, W.N., Fabian, A.C., Pounds, K.A. 1996, MNRAS, 278, 326
Brandt, W.N., Mathur, S., Reynolds, C.S., Elvis, M. 1997, MNRAS, 292, 407
Brandt, W.N., Laor, A., Wills, B.J. 1999, ApJ, in press (astro-ph/9908016)

Brandt, W.N., Comastri, A., Gallagher, S.C., Sambruna, R.M., Boller, Th., Laor, A. 1999, ApJ, in press (astro-ph/9909284)
Brinkmann, W., Wang, T., Matsuoka, M., Yuan, W. 1999, A&A, 345, 43
Crenshaw, D.M., Kraemer, S.B., Boggess, A., Maran, S.P., Mushotzky, R.F., Wu, C. 1999, ApJ, 516, 750
Elvis, M., Fabbiano, G. 1984, ApJ, 280, 91
Elvis, M. 1992, in Frontiers of X-ray Astronomy, ed. Tanaka, Y., Koyama, K. (Universal Acad. Press, Tokyo), p. 567
Fabian, A.C. 1999, MNRAS, in press (astro-ph/9908064)
Fiore, F., Elvis, M., Mathur, S., Wilkes, B.J., McDowell, J.C. 1993, ApJ, 415, 129
Gallagher, S.C., Brandt, W.N., Sambruna, R.M., Mathur, S., Yamasaki, N. 1999, ApJ, 519, 549
George, I.M., et al. 1997, ApJ, 491, 508
George I.M., Turner T.J., Netzer H., Nandra K., Mushotzky R.F., Yaqoob T. 1998, ApJS, 114, 73
George, I.M., et al. 1999, ApJ, in press (astro-ph/9910218)
Goodrich, R.W. 1997, ApJ, 474, 606
Green, P.J., Mathur, S. 1996, ApJ, 462, 637 (GM96)
Grupe, D., Wills, B.J., Wills, D., Beuermann, K. 1998, A&A, 333, 827
Halpern, J.P. 1984, ApJ, 281, 90
Hamann, F. 1998, ApJ, 500, 798
Iwasawa, K. 1999, MNRAS, 302, 96
Kopko, M., Turnshek, D.A., Espey, B.R. 1994, in Multi-Wavelength Continuum Emission of AGN, ed. Courvoisier, T., Blecha, A. (Kluwer, Dordrecht), p. 450
Korista, K., Ferland, G., Baldwin, J. 1997, ApJ, 487, 555
Krolik, J.H., Voit, G.M. 1998, ApJ, 497, L5
Kuraszkiewicz, J., Wilkes, B.J., Brandt, W.N., Vestergaard, M. 1999, ApJ, submitted
Laor, A., Fiore, F., Elvis, M., Wilkes, B.J., McDowell, J.C. 1994, ApJ, 435, 611
Laor, A., Fiore, F., Elvis, M., Wilkes, B.J., McDowell, J.C. 1997, ApJ, 477, 93
Leighly, K.M., Kay, L.E., Wills, B.J., Wills, D., Grupe, D. 1997, ApJ, 489, L25
Mathur, S., Wilkes, B.J., Elvis, M., Fiore, F. 1994, ApJ, 434, 493
Mathur, S., Elvis, M., Singh, K.P. 1995, ApJ, 455, L9
Mathur, S., Wilkes, B.J., Elvis, M. 1998, ApJ, 503, L23
Murray, N., Chiang, J., Grossman, S.A., Voit, G.M. 1995, ApJ, 451, 498
Pan, H.C., Stewart, G.C., Pounds, K.A. 1990, MNRAS, 242, 177
Reeves, J.N., Turner, M.J.L., Ohashi, T., Kii, T. 1997, MNRAS, 292, 468
Reynolds, C.S. 1997, MNRAS, 286, 513
Siebert, J., Komossa, S., Brinkmann, W. 1999, A&A, in press (astro-ph/9909323)
Turner, T.J. 1999, ApJ, 511, 142
Turnshek, D.A., Monier, E.M., Sirola, C.J., Espey, B.R. 1997, ApJ, 476, 40
Wang, T.G., Brinkmann, W., Wamsteker, W., Yuan, W., Wang, J.X. 1999, MNRAS, 307, 821
Weymann, R.J., Morris, S.L., Foltz, C.B., Hewett, P.C. 1991, ApJ, 373, 23
Weymann, R.J. 1997, in Mass Ejection from AGN, ed. Arav, N., Shlosman, I., Weymann, R.J. (ASP Press: San Francisco), p. 3
Wills, B.J., Brandt, W.N., Laor, A. 1999, ApJ, 520, L91
Yuan, W., Brinkmann, W., Siebert, J., Voges, W. 1998, A&A, 330, 108

X-RAY TRANSIENTS IN QUIESCENCE

Sergio Campana

Osservatorio astronomico di Brera, Via E. Bianchi 46, Merate (LC), I-23807 Italy; e-mail: campana@merate.mi.astro.it

ABSTRACT Transient X-ray binaries remain in their quiescent state for a long time (months to hundred years) and then bright up as the most powerful sources of the X-ray sky. While it is clear that, when in outbursts, transient binaries are powered by accretion, the origin of the low luminosity X-ray emission that has been detected in the quiescent state has different interpretations and provides the unique opportunity for testing different accretion regimes. In this paper we concentrate on the various aspects of the accretion physics at low rates onto compact objects. We describe the observational panorama of quiescent emission for the three classes of X-ray transients and try to interpret these data in light of the different regimes accessible at such low mass inflow rates.

KEYWORDS: X-ray: stars – Accretion, accretion disks – Black hole physics – Stars: neutron

1. INTRODUCTION

X-ray transients are classified based on their outburst spectral properties (e.g. White et al. 1984): hard X-ray transients (HXRTs), soft X-ray transients (SXRTs) and ultra soft X-ray transients (USXRTs). This classification is successful since it reflects the true nature of these systems: HXRTs host a high magnetic field neutron star (NS) accreting from a high mass, usually Be, companion star [1] (for a review see Bildsten et al. 1997); SXRTs host a low field NS accreting from a late type, usually K-M, star (e.g. Campana et al. 1998a) and USXRTs consisting of a black hole candidates (BHCs) accreting from a low mass companion too (e.g. Tanaka & Shibazaki 1996). Transient binaries are characterised by an X-ray luminosity that varies over many orders of magnitude, allowing to probe different conditions and accretion regimes that are unaccessible to persistent (bright) sources.

In this paper I will first review the observational properties of transient X-ray sources and then challenge these with simple theoretical models.

---

[1] Here I consider only sources containing fastly spinning NS ($P \lesssim 10$ s), which share several properties with SXRTs.

2. BLACK HOLE TRANSIENTS

BHCs are usually faint in their X-ray quiescent state and strong upper limits exist for a number of them (Menou et al. 1999; Campana & Stella 2000). The first, and only short orbital period (7.8 hr), BHC detected in quiescence to date is A 0620–00. This is the prototype BHC: after a very bright outburst peaking at ~ 7 Crab, the source slowly returned to quiescence and only ROSAT, years later, was able to reveal it. The 0.1–2.4 keV luminosity is $\sim 6 \times 10^{30}$ erg s^{-1} (for a distance of 1.2 kpc and by fixing the column density to $N_H = 1.2 \times 10^{21}$ cm^{-2}; McClintock et al. 1995). Due to the small number of photons (~ 40) however, the spectrum is very poor and can be well fit by a variety of single component models. In particular, it cannot be excluded that such a low luminosity arises from the K dwarf companion (see also Bildsten & Rutledge 2000).

The other two BHCs detected in quiescence are characterised by longer orbital periods (> 2.5 d) and therefore higher average mass inflow rates, based on evolutionary models. GS 2023+338 (V 404 Cyg) was detected at $L \sim 2 \times 10^{33}$ erg s^{-1} with ASCA and BeppoSAX (Narayan et al. 1997; Campana et al. 2000c). The spectrum is well fit by either a power law (photon index $\Gamma \sim 1.5 - 2$) or a bremsstrahlung ($kT_{\rm br} \sim 5-10$ keV). GRO J1655–40 was detected at $L \sim 2 \times 10^{32}$ erg s^{-1} (Hameury et al. 1997), with a spectrum that can be described by a power law model with $\Gamma \sim 1.5$.

2.1. ADAF

The very low luminosity of BHCs in quiescence has stimulated a number of works. The paradigm is now represented by advection-dominated accretion flow (ADAF) models, where the radiative efficiency is very low ($\sim 10^{-4} - 10^{-3}$) and most of the gravitational energy of the inflowing matter is stored as thermal and/or bulk kinetic energy and advected towards the collapsed star (e.g. Narayan et al. 1996; Narayan et al. 1997; Menou et al. 1999). Solutions of this type exist for sub-Eddington mass accretion rates ($\dot M < 0.1 - 0.01~\dot M_{\rm Edd}$). In this regime the bolometric luminosity scales approximately as $\dot M^2$ (as opposed to the $\dot M$ scaling of standard accretion; cf. Fig. 1). These models (which where modified under way to include the optical/UV luminosity as produced by the ADAF itself; cf. Narayan et al. 1997) provide good fit to the multi-wavelength spectra of quiescent BHCs.

3. HARD X-RAY TRANSIENTS

Despite a considerable increase in the number of new HXRTs discovered in the last few years thanks to RXTE, observational data on the transition to quiescence of HXRTs are still sparse. For only a few systems (V 0332+53 Stella et al. 1986; 4U 0115+63 Tamura et al. 1992) there were indications of a sudden turn off of the X-ray luminosity when the sources achieve a level of $\sim 10^{36}$ erg s^{-1}.

Even more rare are the observations of HXRTs in quiescence. In the last few years, only the HXRT A 0538–66 in the Large Magellanic Cloud (containing the

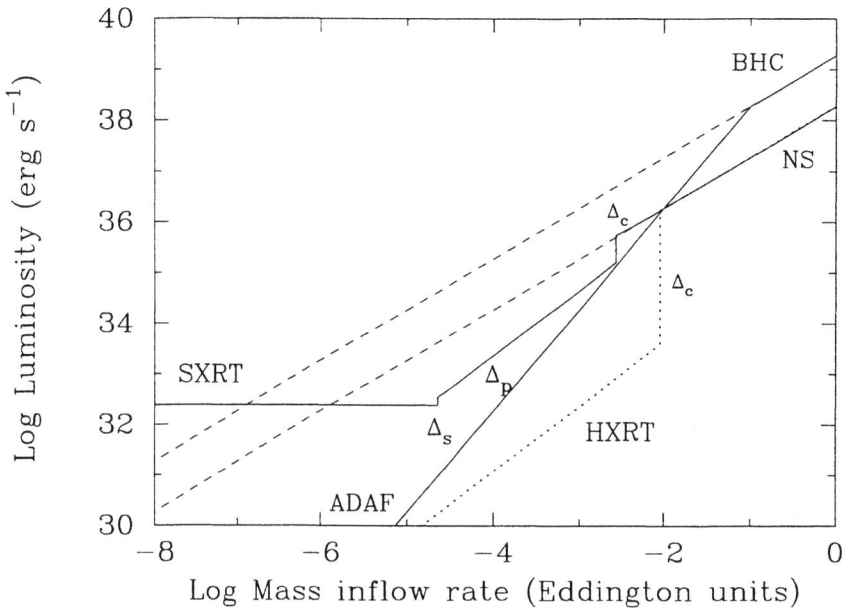

FIGURE 1. Luminosity versus the mass inflow rate (in Eddington units $\dot{M} \sim 1.4 \times 10^{18} M/M_\odot$ g s^{-1}) for different accretion regimes onto BHs and NSs. The upper line refers to a 14 $M_\odot$ BH. The dashed line marks the standard accretion regime (efficiency $\epsilon = 0.1$) and the continuous line the ADAF model. The lower lines refer to accretion onto a 1.4 $M_\odot$ NS. The continuous line gives the luminosity produced by a 2.5 ms spinning, $B = 10^8$ G NS in different regimes, the dotted line by a 4 s, $B = 10^{12}$ G NS. The lower dashed line refers to accretion onto the NS surface.

fastest accreting NS at 69 ms) has been positively detected in quiescence. During the ROSAT all-sky survey two weak outbursts were detected, with peak luminosities of ~ 4 and $\sim 2 \times 10^{37}$ erg s^{-1} in the 0.1–2.4 keV range (Mavromatakis & Haberl 1993) and a similar weak outburst was detected by ASCA at $\sim 6 \times 10^{36}$ erg s^{-1} (1–10 keV; Corbet et al. 1997; Corbet 1996). A 0538–66 was detected several times at a level of about $10^{34} - 10^{35}$ erg s^{-1} during ROSAT PSPC serendipitous pointings (Campana 1997). ASCA and ROSAT observations gave a first indication of the spectrum at such low rates: the ASCA spectrum is well fit by a power law (photon index $\Gamma \sim 2$) plus a soft component, e.g. a black body with temperature ~ 3 keV and equivalent radius of ~ 2 km. The ROSAT PSPC spectrum at a factor of 10–100 lower luminosity can be fit by a black body model with much smaller temperature (~ 0.2 keV) and larger radii (~ 70 km). The presence of a hard power law however cannot be excluded. Recently, we obtained BeppoSAX observations of three fast spinning HXRTs (A 0538–66, V 0332+53 and 4U 0115+63) during their quiescent states (Gastaldello et al. 2000). The most striking results comes from the observation

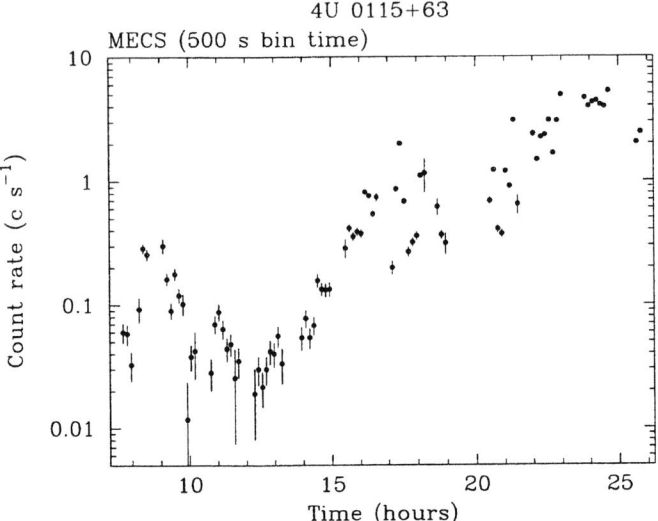

FIGURE 2. BeppoSAX MECS light curve of 4U 0115+63. The orbital phase is just before periastron, based on the ephemerides of Cominsky et al. (1992).

of 4U 0115+63. A 15 hr BeppoSAX observation shows a variation in the count rate by a factor of ~ 250 (cf. Fig. 2; Campana et al. 2000a). A time-resolved spectral analysis reveals that this variation is intrinsic to the source, which does not change its spectrum (hard power law with photon index $\Gamma \sim 1$) nor its column density (a few 10^{22} cm^{-2}, washing out any soft component). The mean 0.1–200 keV luminosity in each interval varies from $\sim 2 \times 10^{34}$ erg s^{-1} to $\sim 2 \times 10^{36}$ erg s^{-1} (at 4 kpc). Pulsations were detected all the way down to the smaller fluxes.

3.1. Propeller regime

In the relatively slow ($P \geq 1$ s) and high magnetic field ($B \sim 10^{12}$ G) NSs of HXRTs accretion onto the surface can take place as long as the magnetospheric radius (r_m, at which the NS magnetic field starts dominating the motion of the infall matter) is smaller than the corotation radius (r_cor, at which matter in Keplerian rotation orbits at the same angular frequency of the NS). In this regime, the accretion luminosity is given by $L(R) = GM\dot{M}/R$. As the mass inflow rate decreases, r_m expands till it reaches the corotation radius. For smaller rates, matter getting attached to the NS field lines at r_m experiments a centrifugal force stronger than gravity and gets expelled. The source starts to be centrifugally inhibited (propeller regime) at a luminosity of [2]

$$L_{cb} \simeq 2 \times 10^{36} B_{12}^2 M_{1.4}^{-2/3} R_6^5 P_{4s}^{-7/3} \text{ erg s}^{-1}$$
$$\simeq 5 \times 10^{35} B_8^2 M_{1.4}^{-2/3} R_6^5 P_{2.5\,\text{ms}}^{-7/3} \text{ erg s}^{-1} \quad (1)$$

($B = 10^{12} B_{12}$ G - $B = 10^8 B_8$ G, $P = 4 P_{4s}$ s - $P = 2.5 \times 10^{-3} P_{2.5\,\text{ms}}$ s, $R = 10^6 R_6$ cm and $M = 1.4 M_{1.4} M_\odot$ are the magnetic field, spin period, radius and mass of the NS, respectively). Matter, being stopped at r_m rather than R, releases a lower accretion luminosity. The accretion luminosity gap, Δ_c, across the centrifugal barrier is (Corbet 1996; Campana & Stella 2000)

$$\Delta_c = \frac{r_{cor}}{R} = \left(\frac{GMP^2}{4\pi^2 R^3}\right)^{1/3} \simeq 620 P_{4s}^{2/3} M_{1.4}^{1/3} R_6^{-1}$$
$$\simeq 3 P_{2.5\,\text{ms}}^{2/3} M_{1.4}^{1/3} R_6^{-1}. \quad (2)$$

Δ_c depends almost exclusively on the spin period P and is basically a measure of how deep r_{cor} is in the potential well of the NS.

This simple picture is challenged by the recent observations of 4U 0115+63. The centrifugal gap that has been modeled as a step-like transition from the accretion to the propeller (or viceversa) regimes has likely been observed. Assuming a typical mass inflow rate variation as derived from BeppoSAX observations of 4U 0115+63 in outburst, one can derive a relation between the observed luminosity and the mass accretion rate $L \propto \dot{M}^\alpha$ which in the case of standard accretion implies $alpha = 1$ and in the propeller $\alpha = 9/7$. We derive a value of $\alpha \sim 50$, indicating that a very small variation in $\dot{M}$ induces a huge variation in luminosity. As a confirmation of this, disk and wind accretion model for a system like 4U 0115+63 allows for a mass inflow rate variation by a factor of a few in 15 hr, at most (e.g. Raguzova & Lipunov 1998).

At variance with simple model predictions, X–ray pulsations are detected all the way down the lower part of the gap (even if with decreasing pulsed fractions). The most straightforward interpretation is that some matter still leaks the centrifugal barrier likely from the highest magnetic latitudes. We conclude that we are observing for the first time the transition from the propeller regime to the standard accretion onto the NS surface. This transition is very fast and opens the possibility to study in detail how the centrifugal mechanism works (see also Campana et al. 2000a, Pizzolato et al. 2000).

4. SOFT X-RAY TRANSIENTS

SXRTs in quiescence were the target of early X–ray astronomy missions such as Einstein (Petro et al. 1981) and EXOSAT (van Paradijs et al. 1987). These data

---

[2] We use here simple spherical accretion theory. This is a reasonably accurate approximation when the accretion disk at the magnetospheric boundary is dominated by gas pressure. For a more general approach see e.g. Campana et al. (1998a).

however provide very poor spectral information. ROSAT gave for the first time a clear assessment in the field, revealing the SXRT prototype source Aql X-1 on three occasions at a level of $\sim 10^{33}$ erg s^{-1} (0.4–2.4 keV) with a very soft spectrum (e.g. a black body with a temperature of $kT_{bb} \sim 0.3$ keV and an equivalent radius of $\sim 10^5$ cm; Verbunt et al. 1994).

The number of SXRTs detected in quiescence is now increasing thanks to BeppoSAX and ASCA: Aql X-1 (Campana et al. 1998b; Asai et al. 1998), Cen X-4 (Asai et al. 1996; Campana et al. 2000b), 4U 1608–522 (Asai et al. 1998), 4U 2129+47 and EXO 0748–676 (Garcia & Callanan 1999) and most recently SAX J1808.4–3658 (Stella et al. 2000) and X 1732–304 (Guainazzi et al. 1999). All these sources have X-ray luminosities in the $10^{32} - 10^{33}$ erg s^{-1} range (e.g. Campana et al. 1998a). Together with the soft spectral component which is present in all SXRTs in quiescence observed to date (usually modeled as a black body of $kT_{bb} \sim 0.1 - 0.3$ keV), a hard power law can be revealed in the best studied sources (Aql X-1, Cen X-4 and X 1732–304). This power law (with photon index $\sim 1.5 - 2$) makes up $\sim 50\%$ of the 0.5–10 keV luminosity. This two components spectrum is becoming the 'canonical' spectrum for SXRTs in quiescence and further confirmations will come from Chandra and XMM-Newton pointings.

4.1. Shock emission

It had long been suspected that the NSs of persistent and transient LMXRBs have been spun up to very short rotation periods by accretion torques however, conclusive evidence has been been gathered only recently. The best example is SAX J1808.4–3658, a bursting transient source discovered with BeppoSAX in 1996 (in't Zand et al. 1998). In April 1998, RossiXTE observations revealed a coherent ~ 401 Hz modulation, testifying to the presence of magnetic polar cap accretion onto a fast rotating magnetic NS (Wijnands & van der Klis 1998; Chakrabarty & Morgan 1998). Millisecond rotation periods have also been inferred for 7 other LMXRBs of the Atoll (or suspected members of the) group through the oscillations that are present for a few seconds during type I X-ray bursts (for a review see van der Klis 2000). Regarding the NS magnetic field, Psaltis & Chakrabarty (1999) estimate for SAX J1808.4–3658 a value of $B \sim 10^8 - 10^9$ G, by adopting different models for the disk-magnetosphere interaction. Indirect evidence for fields of $B \sim 10^8$ G derives also from the steepening in the X-ray light curve decay and marked change of the X-ray spectrum when the luminosity reaches a level of $\sim 10^{36}$ erg s^{-1} in Aql X-1 (Campana et al. 1998b; Zhang et al. 1998) SAX J1808.4–3658 (Gilfanov et al. 1998), XTE 2123–058 (Tomsick et al. 1999) and Rapid Burster (Masetti et al. 2000), once these changes are interpreted in terms of the onset of the centrifugal barrier. The spin period and the inferred magnetic field strength of SAX J1808.4-3658 provide the first direct evidence for the long suspected Low Mass X-Ray Binaries millisecond pulsars connection.

As the mass inflow rate decreases further the magnetosphere expands until the light cylinder radius, $r_{lc} = cP/2\pi$, is reached; beyond this point the radio pulsar dipole radiation will turn on and begin pushing outward the inflowing matter, due to a flatter radial dependence of its pressure compared to that of disk or radial accretion inflows (Illarionov & Sunyaev 1975; Stella et al. 1994; Campana et al. 1995). The equality $r_\mathrm{m} = r_\mathrm{lc}$ defines the lowest mass inflow rate (and therefore accretion luminosity) in the propeller regime. An accretion luminosity ratio of

$$\Delta_\mathrm{p} = \left(\frac{r_\mathrm{lc}}{r_\mathrm{cor}}\right)^{9/2} \simeq 440\, P_{2.5\,\mathrm{ms}}^{3/2}\, M_{1.4}^{-3/2} \qquad (3)$$

characterises the range over which the propeller regime applies. Note that also this ratio depends mainly on the spin period P.

Once in the rotation powered regime, a fraction η of the spin down luminosity, L_sd, converts to shock emission in the interaction between the relativistic wind of the NS and the companion's matter flowing through the Lagrangian point. Theoretical models indicate that the material lost by the companion star may take somewhat different shapes, ranging from a bow shock to an irregular annular region in the Roche lobe of the NS, depending on radio pulsar wind properties and the rate and angular momentum of the mass loss from the companion star (Tavani & Brookshaw 1993). η may be as large as 0.1 (Tavani 1991) and the shock luminosity can be expressed as $L_\mathrm{shock} = \eta\, L_\mathrm{sd} \sim 2 \times 10^{32}\, \eta_{-1}\, B_8^2\, P_{2.5\,\mathrm{ms}}^{-4}$ erg s^{-1} ($\eta \sim \eta_{-1}\, 0.1$). The luminosity ratio across the transition from the propeller to the rotation powered regime can be approximated as (Stella et al. 1994; Campana et al. 1998a)

$$\Delta_\mathrm{s} = \frac{3}{2\sqrt{2}\,\eta}\left(\frac{r_\mathrm{g}}{r_\mathrm{lc}}\right)^{1/2} \simeq 2\,\eta_{-1}^{-1}\, P_{2.5\,\mathrm{ms}}^{-1/2}\, M_{1.4}^{1/2}\,, \qquad (4)$$

where $r_\mathrm{g} = GM/c^2$ is the gravitational radius. The energy spectrum due to shock emission is expected to be a power law with photon index of $\sim 1.5 - 2$ that extends from a ~ 10 eV to ~ 100 keV, with both energy boundaries shifting as $B_8\, P_{2.5\mathrm{ms}}^{-3}$ (Tavani & Arons 1997; Campana et al. 1998a).

Observationally, the nature of the two spectral components observed during the quiescent state of SXRTs is a matter of debate. One possibility is that some matter leaks through the centrifugal barrier accreting onto the NS surface and produces in turn the observed soft component, whereas the hard component arises in an ADAF (Zhang et al. 1998; Menou et al 1999). However a clear assessment of this spectral model has not been carried out yet, nor self-consistent ADAF models for NSs exist. The other possibility complains a cooling NS[3] (which contributes to the soft component) working as a radio pulsar, the relativistic wind of which generates a shock power law spectrum (Campana et al. 1998a). This is, at least

---

[3]Concerning the soft X-ray thermal-like component, we note that the effective black body radii inferred from spectral fitting are substantially smaller than the NS radius. At the same time, radiative transfer calculations for the NS atmosphere indicate that the emergent thermal-like X-ray spectrum is complex and simple black body fits are likely to underestimate the effective emission radius and overestimate the temperature by a factor of 3-10 and 2-3, respectively (Rutledge et al. 1999; Brown et al. 1998). Consequently, it cannot be ruled out yet that thermal emission from the whole NS surface powers the soft X-ray component of quiescent NS SXRTs.

qualitatively, in agreement with the hard power law like X–ray component observed in the quiescent X–ray spectrum of Cen X-4, Aql X-1 and X 1732–304. The extended power law spectrum expected from shock emission is also in agreement with the recently determined residual UV spectrum of Cen X-4, which shows no evidence for a turnover down to lowest measured UV energies ($\sim$ 7.5 eV; McClintock & Remillard 2000) and matches quite well the extrapolation of the (power law) X–ray spectrum.

5. CONCLUSIONS

In the last few years a large wealth of new data on quiescent transient sources has been obtained thanks to BeppoSAX and ASCA. These observations give us the opportunity to study in some details the different regimes that can be expected at such low X–ray fluxes, such as the ADAF models for BHCs, the propeller regime for HXRTs and the shock emission for SXRTs. New and better data will be collected in the next few years thanks to Chandra and XMM-Newton, answering to some basic questions that are still open.

Here I list some topics that can be addressed in the next years:

- **Quiescent emission of BHCs (1).** ADAF models are now very popular in explaining the BHC quiescent emission. However these models face problems since their optical/UV luminosity (after subtraction of the contribution from the companion star) dominates by a factor of $\gtrsim 10$ over the X–ray luminosity, at variance with SXRTs in which the two luminosities are comparable, at the most. Based on the latest ADAF models, this optical/UV luminosity should be ascribed to the ADAF itself, arising from synchrotron processes. These results indicate that the luminosity swing from the outburst peak to quiescence, which has been claimed to be different in BHCs and SXRTs (Narayan, Garcia & McClintock 1997), does not provide evidence for a significant difference between the two classes. Therefore, one of the key motivations for considering ADAF models, i.e. that most of the energy is hidden beyond the BH event horizon, is considerably weakened (see Campana & Stella 2000).

- **Quiescent emission of BHCs (2).** The only short orbital period BHC detected in quiescence is A 0620–00 at a level of $\sim 10^{31}$ erg s^{-1} (0.5–10 keV, e.g. Menou et al. 1999). In analogy with the X–ray emission from K stars in RS CVn type binaries (which emit up to $\sim 10^{32}$ erg s^{-1} in the ROSAT band; Dempsey et al. 1993) one may argue that, in short orbital period BHC systems, such as A 0620–00, the low level X–ray quiescent luminosity ($\sim 10^{31} - 10^{32}$ erg s^{-1}) might also arise from due to coronal activity of the companion star (see also Bildsten & Rutledge 2000).

- **Centrifugal gap in HXRTs.** BeppoSAX observations of 4U 0115+63 provide for the first time the opportunity to study the transition from the accretion regime to the propeller state. Good data have been obtained and their potential is being exploited, with the aim of understanding how and when the centrifugal barrier closes (and/or opens; Campana et al. 2000a; Gastaldello et al. 2000). These data also demonstrate that the monitoring of HXRTs in quiescence, especially near periastron, is worthy.

- **Quiescent emission of SXRTs.** Two facing models have been proposed to explain the quiescent emission of SXRTs: one is based on ADAFs and included the leaking of matter through the centrifugal barrier to explain the soft X–ray component; the other one envisages the presence of an active radio pulsar. One of the most straightforward prediction of the ADAF scenario is that pulsations should be detected in the soft component. In the case of an active radio pulsar, a radio signal might be detected if the circumstellar material does not absorb completely the signal. Doppler maps can also be useful to pinpoint unusual geometries not related to an accretion disk but to a shock front.

ACKNOWLEDGEMENTS

I thank useful discussions with a number of people, together with part of this work has been carried out. In particular, I would like to thank L. Stella, S. Mereghetti, M. Colpi, G.L. Israel, M. Tavani, F. Gastaldello, D. Ricci, T. Belloni, D. Lazzati, G. Tagliaferri, S. Covino, G. Ghisellini. This work was partially supported through ASI grant.

REFERENCES

Asai K. et al., 1998, PASJ 50 611
Bildsten L. & Rutledge R.E., 2000, ApJ in press (astro-ph/9912304)
Bildsten L. et al., 1997, ApJS 113 367
Brown E.F., Bildsten L. & Rutledge R.E., 1998, ApJ 504 L95
Campana S., 1997, A&A 320 840
Campana S., Colpi M., Mereghetti S., Stella L. & Tavani M., 1998a, A&A Rev. 8 279
Campana S. & Stella L., 2000, ApJ in press (astro-ph/0005118)
Campana S., Stella L., Mereghetti S. & Colpi M., 1995, A&A 297 385
Campana S. et al., 1998b, ApJ 499 L65
Campana S. et al., 2000a, in preparation
Campana S. et al., 2000b, A&A 358 583
Campana S. et al., 2000c, in preparation
Chakrabarty D. & Morgan E.H., 1998, Nat 394 346
Cominsky L., Roberts M. & Finger M.H., 1994, 1st Compton Symposium, AIP, p. 294
Corbet R.H.D., 1996, ApJ 457 L31
Corbet R.H.D. et al., 1997, ApJ 476 833
Dempsey R.C., Linsky J.L., Flaming T.A. & Schmitt J.H.M.M., 1993, ApJS 86 599
Garcia M.R. & Callanan P.J., 1999, AJ 118 1390
Gastaldello F. et al., 2000, in preparation
Gilfanov M., Revnivtsev M., Sunyaev R. & Churazov E., 1998, A&A 338 L83

Guainazzi M., Parmar A.N. & Oostebroek T., 1999, A&A 349 819
Hameury J.-M., Lasota J.-P., McClintock J.E. & Narayan R., 1997, ApJ 489 234
Illarionov A.F. & Sunyaev R.A., 1975, A&A 39 185
in't Zand J.J.M. et al., 1998, A&A 331 L25
Masetti N. et al., 2000, A&A submitted
Mavromatakis F. & Haberl F., 1993, A&A 274 304
McClintock J.E., Horne K. & Remillard R.A., 1995, ApJ 442 358
McClintock J.E. & Remillard R.A., 2000, ApJ 531 956
Menou K. et al., 1999, ApJ 520 276
Narayan R., Barret D. & McClintock J.E., 1997, ApJ 482 448
Narayan R., Garcia M.R. & McClintock J.E., 1997, ApJ 478 L79
Narayan R., McClintock J.E. & Yi I., 1996, ApJ 457 821
Petro L.D., Bradt H.V., Kelley R.L., Horne K. & Gomer R., 1981, ApJ 251 L7
Pizzolato F. et al., 2000, in preparation
Psaltis D. & Chakrabarty D., 1999, ApJ 521 332
Raguzova N.V. & Lipunov V.M., 1998, A&A 340 85
Rutledge R.E., Bildsten L., Brown, E.F., Pavlov, G.G. & Zavlin V.E., 1999, ApJ 514 945
Stella L., Campana S., Colpi M., Mereghetti S. & Tavani M., 1994, ApJ 423 L47
Stella L., White N.E. & Rosner R., 1986, ApJ 308 669
Stella L. et al., 2000, ApJ 537 L115
Tamura K., Tsunemi H., Kitamoto S., Hayashida K. & Nagase F., 1992, ApJ 389 67
Tanaka Y. & Shibazaki N., 1996, ARA&A 34 607
Tavani M., 1991, ApJ 379 L69
Tavani M. & Arons J., 1997, ApJ 477 439
Tavani M. & Brookshaw L., 1993a, A&A 267 L1
Tomsick J.A. et al., 1999, ApJ 521 341
van der Klis M., 2000, ARA&A in press (astro-ph/0001167)
van Paradijs J., Verbunt F., Shafer R.A. & Arnoud K.A., 1987, A&A 182 47
Verbunt F., Belloni T., Johnston H., van der Klis M. & Lewin W.H.D., 1994, A&A 285 903
White N.E., Kaluzienski J.L. & Swank J.H., 1984, in "High Energy Transients in Astrophysics", ed. S.E. Woosley, AIP Conf. Proc. 115, p 31
Wijnands R.A.D. & van der Klis M., 1998, Nat 394 344
Zhang S.N., Yu W. & Zhang W.W., 1998, ApJ 494 L71

WHAT'S WRONG WITH AGN MODELS FOR THE X-RAY BACKGROUND ?

Andrea Comastri

Osservatorio Astronomico di Bologna, via Ranzani 1, I–40127 Bologna, Italy

ABSTRACT

The origin of the hard X-ray background (XRB) as a superposition of unabsorbed and absorbed Active Galactic Nuclei is now widely accepted as the standard model. The identification of faint X-ray sources in ROSAT, ASCA, and BeppoSAX medium–deep surveys and their average spectral properties are in broad agreement with the model predictions. However, AGN models, at least in their simplified version, seem to be at odds with some of the most recent findings calling for substantial revisions. I will review the recent XRB "best fit" models and discuss how the foreseen XMM and Chandra surveys will be able to constrain the allowed parameter space.

KEYWORDS: Cosmology, Diffuse emission, X-rays, Active Galactic Nuclei

1. INTRODUCTION

It has been recognized, already a few years ago, that a self-consistent AGN model for the XRB requires the combined fit of several observational constraints in addition to the XRB spectral intensity such as the number counts, the redshift and absorption distribution in different energy ranges, the average spectra and so on (Setti & Woltjer 1989). First attempts towards a "best fit" solution relied on simplified assumptions for the AGN spectral properties and for the evolution of their luminosity function (Madau, Ghisellini & Fabian 1994 (MGF94), Comastri et al. 1995 (CSZH95), Celotti et al. 1995 (CFGM95)). A three step approach has been followed to build the so-called baseline model: the first step is to assume a single average spectrum for the type 1 objects which is usually parameterized as a power law plus a reflection component from a face-on disk and a high-energy cut-off at a few hundreds of keV. A distribution of absorbing column densities for type 2 objects is then added in the second step. Finally the template spectra are folded with an evolving XLF which, in the framework of unified models, does not depend on the amount of intrinsic obscuration. The number density and absorption distribution of obscured sources are then varied until a good fit is obtained. The baseline model led to a successful description of most of the observational data available before 1995 and to testable predictions for the average properties of the sources responsible for the bulk of the XRB. The increasing amount of data from soft and hard X-ray surveys combined with the study of nearby bright sources has been used to obtain a more detailed description of the AGN X-ray spectra and absorption distribution. In addition, the optical identification of sizeable samples of faint AGNs discovered in the ROSAT, ASCA and BeppoSAX surveys has shed new light on the evolution of the AGN luminosity function opening the possibility to test in more detail the AGN synthesis model predictions. As a consequence, the modelling of the XRB has attracted renewed attention and several variations/improvements with respect to the baseline model have been proposed. However, despite the increasing efforts, a coherent self-consistent picture of "the" XRB model has yet to be reached, as most of its ingredients have to be extrapolated well beyond the present limits. Besides the interest in a best-fit model it is by now clear that the problem of the origin of the XRB is closely related to the evolution of accretion and obscuration in AGN. As a consequence, the XRB spectrum should be considered as a useful tool towards a better understanding of the history of black hole formation and evolution in the Universe (Fabian & Iwasawa 1999) and the interplay between AGN activity and star-formation (Franceschini et al. 1999; Fabian this volume).

2. RECENT OBSERVATIONAL CONSTRAINTS

2.1. The XRB spectrum

The low energy (below 10 keV) XRB spectrum has been measured with the imaging detectors onboard ROSAT, ASCA, and BeppoSAX and a summary of the results is given in Figure 1 together with a compilation of recent re-analysis of the HEAO1 A2 and A4 experiments data. The comparison between the different datasets in the overlapping $\sim$ 1–8 keV energy range points to a systematic difference in the normalization of the XRB flux while the average spectrum is similar among all the observations. The largest deviation is of the order of $\sim$ 40 % between the HEAO1 A2 and BeppoSAX data (see Vecchi et al. 1999 for a more detailed discussion). Such a discrepancy could be due to residual, not fully understood, cross-calibration errors among the different detectors and/or to field-to-field fluctuations. These findings cast shadows on the intensity and the location of the XRB peak as measured by HEAO1 A2 ($\sim$ 43 keV cm^{-2} s^{-1} sr^{-1} at 30 keV; Gruber et al. 1999). Indeed a simple extrapolation of the BeppoSAX and HEAO1 A4 MED best fit spectra imply that the XRB spectrum peaks at $\sim$ 23 keV with a much higher intensity introducing an extra-degree of freedom in AGN models parameter space. A new measurement of the 10–100 keV spectrum would be extremely important. Unfortunately such observations are not foreseen in the near future.

2.2. The AGN spectrum

As far as the model of the XRB is concerned, the most important parameters to deal with are a good estimate of the average continuum slope and of the absorption column density. The broad-band energy range exploited by BeppoSAX turned out to be extremely useful to probe column densities as high as 10^{24-25} cm^{-2}, to assess the strength of the reflection component which peaks around 20–30 keV, and the

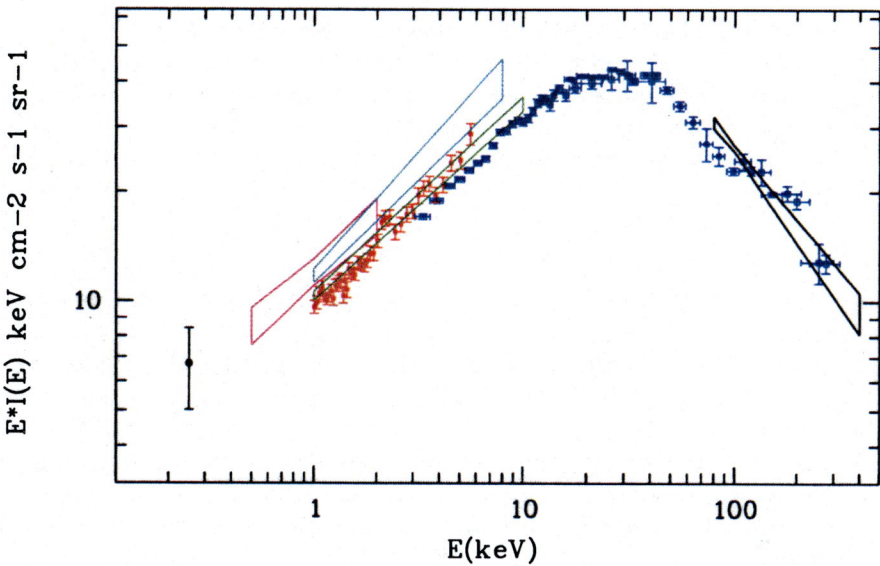

FIGURE 1. The XRB spectral energy density from 0.25 to 400 keV. The red points in the 1-6 keV range are from ASCA (Gendreau et al. 1995), the blue points in the 3-300 keV range are from HEAO1 A2 and A4 LED experiments (Gruber et al. 1999), while the best-fit spectrum from the HEAO1 A4 MED experiment (Kinzer et al. 1997) is reported as a black bow-tie contour between 80 and 400 keV. The best-fit estimates from recent observations at low energies (< 10 keV) are also displayed as bow-ties contours. Green (1-10 keV): joint ROSAT/ASCA analysis (Miyaji et al. 1998). Purple (0.5-2.0 keV): ROSAT results from Georgantopoulos et al. (1996). Cyan (1-8 keV): BeppoSAX data from Vecchi et al. (1999). Finally an estimate of the extragalactic background intensity at 0.25 keV (Roberts & Warwick 1998) is also plotted.

shape of the low-energy soft-excess emission below ~ 1 keV. In addition ASCA observations of sizeable samples of relatively faint AGNs have allowed to probe the spectral properties of high-luminosity high-redshift objects. The most important new results emerging from these observations can be summarized as follows:
• The fraction of heavily obscured ($24 < \log N_H < 25$) and Compton thick ($\log N_H > 25$) sources in the local Universe is much higher than previously thought (Risaliti, Maiolino & Salvati 1999) and a fraction as high as 50% of the Seyfert 2 in the local Universe could be obscured by these large column densities.
• Soft excess emission is uncommon among bright Seyfert 1 galaxies (Matt this volume) and nearby quasars (George et al. 2000) and estimated to be present in less than ~ 30 % of AGN.

- First observations of high redshift quasars suggest a flattening of the power law slope which cannot be ascribed to the reflection component (Vignali et al. 1999).
- Despite intensive searches for high luminosity highly absorbed objects (the so-called type 2 quasars) these sources appear to be elusive and only a few bona-fide examples have been reported in the literature (i.e. Barcons et al. 1998; Georgantopoulos et al. 1999).

2.3. The evolution of the AGN X-ray luminosity function

The evolution of the AGN XLF has been extensively studied mainly in the soft X-rays and usually parametrized with a pure luminosity evolution (PLE) model (i.e. Boyle et al 1994). A major step forward in the determination of the soft XLF has been recently achieved by Miyaji et al. (2000). Combining the results of several ROSAT surveys it has been possible to explore the low-luminosity high-redshift tail of the XLF in much greater detail than before. The results favour a luminosity dependent density evolution (LDDE) as the best description of the available data. In agreement with previous studies, X-ray selected AGN undergo strong evolution up to a redshift $z_c = 1.5$–2.0 and a levelling-off or a weak negative evolution up to $z_{max} \simeq 4$–5. Two parametric descriptions (LDDE1 and LDDE2) encompassing the statistically acceptable fits to the soft XLF have been worked out by Miyaji and collaborators. The integration of the LDDE1 and LDDE2 XLF up to $z \simeq 5$ accounts for about 60 % and 90 % of the soft XRB respectively.

3. THE AGN MODELS PARAMETER SPACE

3.1. Warnings

Before discussing and comparing the various models, it is important to stress the strong coupling between the input spectral parameters and those describing the XLF evolution, which instead are often uncorrectly considered to be independent in the models. Indeed the X-ray luminosities are usually computed converting count rates into fluxes assuming a single valued (relatively steep) slope. This procedure might easily lead to a wrong estimate of the intrinsic luminosity for a very hard absorbed spectrum or if the soft X-ray flux is due to a component not directly related to the obscured nucleus (as in the case of a thermal spectrum from a starburst or scattered emission). According with the XRB baseline model, absorbed AGN become progressively more important towards faint fluxes and thus an additional spurious density evolution term can be introduced in the derivation of the XLF. It turns out that not only the evolution and the space density of obscured AGN are highly uncertain, but also the common practice to consider the soft XLF as representative of the properties of type 1 objects is likely to contain major uncertainties especially when extrapolated to higher energies. Unfortunately our present knowledge of the AGN spectral and evolutive properties does not allow to disentangle the spectral and evolutionary parameters, leaving this ambiguity in all the XRB synthesis models.

3.2. An incomplete tour of the parameter space

The baseline model (cfr § 1) has been recently extended, taking into account some of the new observational findings described in §2, by several authors: Gilli, Risaliti & Salvati 1999 (GRS99); Miyaji, Hasinger & Schmidt 1999 (MHS99); Wilman & Fabian 1999 (WF99); Pompilio, La Franca & Matt 2000 (PLM00). A good agreement among the various models has been reached on the high energy cut–off in the input spectrum (300–500 keV), which is basically fixed by the XRB shape above 40 keV (Comastri 1999), and on the z_c and z_{max} values. GRS99 and MHS99 adopted the LDDE model for the evolution of the XLF and also introduced a cut–off in the luminosity distribution of absorbed AGN for $L > 10^{44}$ erg s^{-1} to cope with the lack of type 2 QSO. The absorption distribution has been fixed according to the recent BeppoSAX results only in the GRS99 model. PLM00 and WF99 both stressed that a proper treatement of the high energy spectrum of heavily obscured ($24 < \log N_H < 25$) objects has important consequences for the modelling. In particular the evolution of the obscured to unobscured ratio as a function of redshift (PLM00) or the need of super–solar abundances to better fit the XRB peak at 30 keV (WF99; but see § 2.1) have been invoked.

Table. 1 - Comparison of model parameters

Model	XLF^a	Evolution	QSO2^b	N_H^c	α_E^d	SE^e	CT^f
MGF 94	2-10	PLE	Yes	Fitted	0.9	No	Yes
CSZH 95	0.3-3.5	PLE	Yes	Fitted	0.9	Yes	Yes(*)
CFGM 95	2-10	PLE	Yes	Fitted	0.9	No	Yes
MHS 99	0.5-2.0	LDDE1	No	Fitted	0.7	Yes	No
GRS 99	0.5-2.0	LDDE1	No	Fixed	0.9	Yes	Yes(*)
WF 99	2-10	PLE	Yes	Fitted	0.9	No	Yes
PLM 00	0.3-3.5	PLE	Yes	Fitted	0.9	Yes/No	Yes

^{a} Energy range of the adopted XLF; ^{b} Presence of type 2 quasars; ^{c} Absorption distribution; ^{d} Spectral energy slope; ^{e} Presence of soft excess emission in the model spectrum; ^{f} Presence of Compton thick sources (The * indicates that a simplified treatment has been employed)

A comparison between the various models (all of them providing a fairly good description of the present data) is made difficult by the large dispersion in the starting assumptions among the different authors (see Table 1) and also by the relatively large uncertainty in the XRB spectrum normalization (see § 2.1).

The most up-to-date treatment of the XLF evolution has been adopted only by GRS99 and MHS99 who also made an attempt to correct for the biases described in § 3.1. In both cases the model predictions fall short the hard X–ray (2–10 keV and especially 5–10 keV) counts at relatively bright 10^{-13}–10^{-12} cgs fluxes. This effect, which is less severe for MHS99 given the very hard input spectrum ($\alpha = 0.7$ plus reflection), can be explained by the relatively low average luminosity of

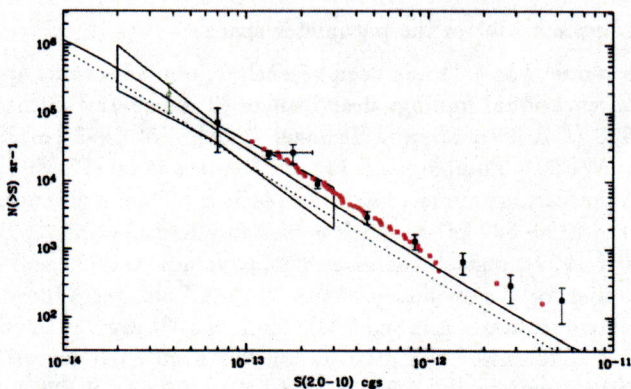

FIGURE 2. The CSZH95 (solid line) and the GRS99 (dotted line) AGN model predictions are compared with the total 2–10 keV counts. Data points at bright fluxes are from Cagnoni et al. 1998 (single sources plotted); Ueda et al. 1999 (points with error bars); Ogasaka et al. 1998 (at the faintest flux). The bow–tie contour is from a fluctuation analysis of ASCA data by Gendreau, Barcons & Fabian 1998.

absorbed sources which show up only at fainter fluxes. The hard X–ray counts are better accounted for in PLE models (Fig. 2), which however are based on a less appropriate description of the XLF and include high luminosity highly absorbed sources. It is worth noting that source counts at fluxes $> 10^{-13}$ cgs, both in the soft and hard bands, should not be entirely accounted for by AGN models as a non negligble fraction of these relatively bright sources are not AGN. The 2–10 keV and 5–10 keV counts are best fitted by those models without soft excess emission in type 1 objects. However in this case the predicted average spectrum of faint sources in the ROSAT band ($\alpha_E \simeq 0.5$–0.6) is much harder than the observed value ($\alpha_E \simeq 1.0$, Hasinger et al. 1993).

Another inconsistency of most of these models concerns the relatively small expected percentage of type 1 unobscured AGN at the 2–10 keV fluxes currently sampled. Indeed optical identifications of medium–deep ASCA surveys (Boyle et al. 1998; Akiyama et al. 2000) suggest that the fraction of unabsorbed broad line AGN is of the order of 60–70 % while only one third of the sources should be type 1 AGN on the basis of the models predictions (but see § 4). The fraction of type 1 AGN can be increased assuming an LDDE2 model for the evolution of the XLF and a flat $\alpha_E = 0.7$ spectrum for high luminosity objects (Vignali et al. 1999). With these parameters a good fit to the hard XRB spectrum can be obtained even without including heavily obscured ($N_H > 10^{24}$) sources. As a result the relative ratio between absorbed and unabsorbed objects at relatively bright fluxes (Fig. 3) decreases significantly. However also within this model the hard X–ray counts are seriously underestimated (being consistent with the dotted line in Fig. 2) owing to

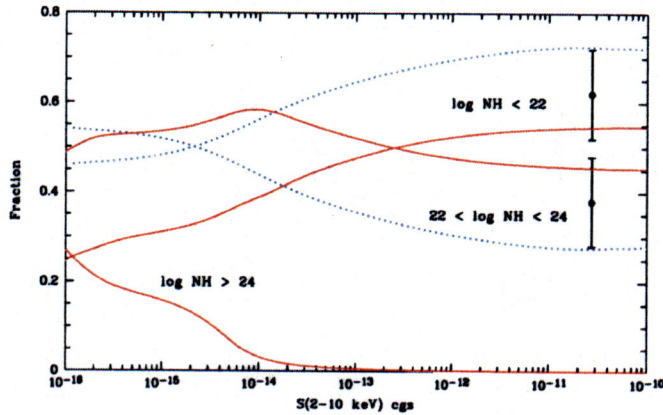

FIGURE 3. The relative fraction of unobscured and obscured sources for a LDDE2 model (dotted line) and a PLE model (solid line). The points with error bars represent the fraction of unabsorbed (about 60%) and absorbed (about 40%) sources in the Piccinotti et al. (1982) sample.

the decreased emissivity of hard absorbed sources.

As an example of the link between the various parameters I have computed three different models which provide a good fit to the overall XRB spectrum but differ in the choice of the input spectra and XLF evolution (Fig. 4). Assuming a high fraction of type 1 objects as in the LDDE2 scenario a soft excess component cannot be accomodated as the 1/4 keV background would be overpredicted. On the other hand the class of PLE models without soft excess (which better reproduce the hard X-ray counts, but with the caveats discussed above) suggest a possible contribution from other, steep spectrum, sources to the 0.25 keV background.

4. IS THERE A WAY OUT ?

The main message emerging from what discussed above is that a self-consistent description of all the observational constraints is still lacking. The major problem is the discrepancy between the predictions of those models computed assuming the most up-to-date results, and the high energy (> 2 keV) source counts. One obvious possibility is a substantial contribution from non-AGN, flat spectrum sources. Extremely hard ($\alpha_E \simeq 0.2$) power-law tails above a few keV, possibly originating in advection dominated accretion flows, have been recently discovered in a small sample of nearby elliptical galaxies (Allen, Di Matteo & Fabian 2000). It has been proposed (Di Matteo & Allen 1999) that these objects constitute the missing population needed to fill the gap between the hard counts and the AGN model predictions. However in this case elliptical galaxies should be a non-negligible fraction of the already identified X-ray sources in ASCA and BeppoSAX surveys at variance with the present breakdown of optical identifications.

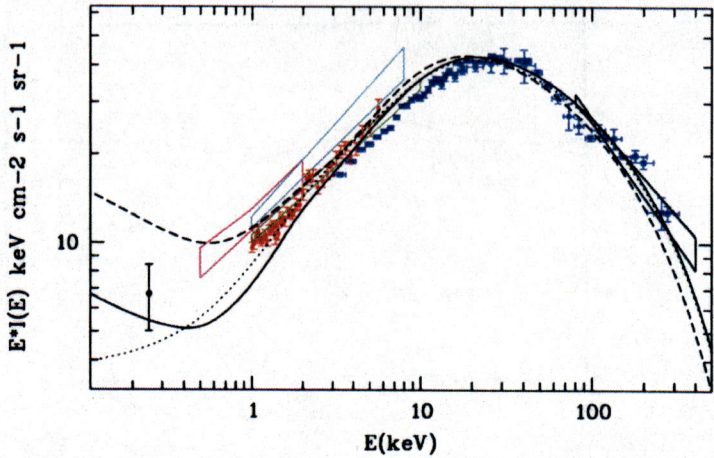

FIGURE 4. XRB fits normalized at 30 keV for 3 different assumptions on the input spectral shape and XLF evolution. Solid line : PLE with Soft Excess. Dotted line: PLE without soft excess. Dashed line: LDDE2 with soft excess. The observational data are the same of Fig. 1.

Another interesting possibility, which would allow to include high luminosity highly absorbed AGN in the models and at the same time reproduce most of the observational constraints, is that the optical properties of X-ray obscured AGN are different from what expected (i.e. narrow lined AGN). In this respect the identification of the first High Energy LLarge Area Survey (HELLAS) carried out with BeppoSAX in the 5–10 keV band (Fiore et al. 1999, and this volume) is providing new and unexpected results. In particular, X-ray absorbed AGN are identified with objects which show a large variety of optical classification, such as intermediate type 1.5–1.9 objects, red quasars (Vignali et al. 2000) and even broad line "blue" quasars. A similar behaviour has been also reported for a sample of ROSAT AGN (Mittaz, Carrera, Page this volume). It is also interesting to note that large columns of cold gas have been detected in Broad Absorption Line quasars (Brandt et al. this volume) and in several Broad Line Radio galaxies and radio quasars observed by ASCA (Sambruna, Eracleous & Mushotzky 1999). Although the statistics is not yet good enough to reach firm conclusions, it is quite possible that the correlation between X-ray absorption and optical appearance of AGN change with redshift and/or luminosity (Fig. 5). A decreasing value of the dust–to–gas ratio as a function of the X-ray luminosity would provide a possible explanation of this effect.

5. CONCLUSIONS

In order to achieve a major improvement in the exploration of XRB models parameter space, the resolved fraction of its energy density should be of the order of 50–60

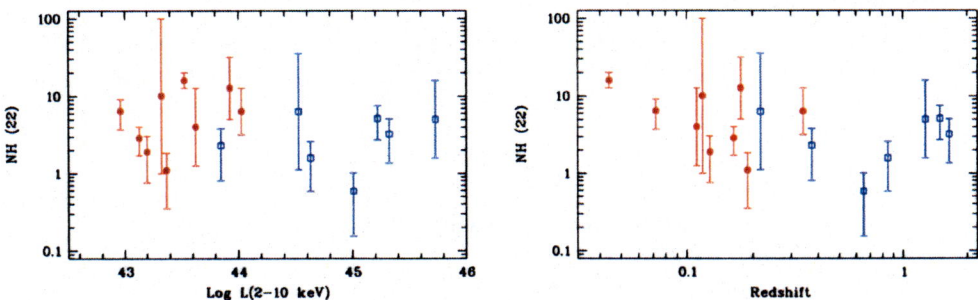

FIGURE 5. The absorption column density (in units of 10^{22} cm^{-2} versus the 2-10 keV luminosity (left panel) and redshift (right panel) of obscured AGN in hard X-ray selected samples. Broad lined objects are plotted with open (blue) symbols, narrow lined ones with filled (red) symbols.

% or higher. The expected contribution of AGN to the 2–10 keV XRB is reported

Table. 2 - Resolved fraction of the XRB

2-10 keV Flux interval	Relative %	Integral %
$> 10^{-11}$	0.5	0.5
10^{-12}-10^{-11}	2	2.5
10^{-13}-10^{-12}	8	10.5
10^{-14}-10^{-13}	32.5	43
10^{-15}-10^{-14}	39	82
10^{-16}-10^{-15}	16	98
$< 10^{-16}$	2	100

in Table 2 as a function of flux. The model parameters are such to account for an intensity of $\sim 7 \times 10^{-8}$ erg cm^{-2} s^{-1} sr^{-1} (in between the ASCA and BeppoSAX measurements) at $\sim 10^{-17}$ cgs. The predictions are model dependent and should be considered as indicative. Nevertheless it is clear that at the fluxes sampled by the foreseen Chandra and XMM medium–deep surveys most of the XRB will be resolved allowing to unveil the nature of the sources making the bulk of its energy density. The most important challenge for XRB models will be the study of X-ray absorption and luminosity distribution for 2–10 keV fluxes $< 10^{-13}$ cgs, the search for heavily obscured AGN which according to the predictions are expected to show up in a substantial fraction below $< 10^{-14}$ cgs (cfr. Fig. 3), and the optical–infrared follow–up of X-ray obscured sources.

ACKNOWLEDGEMENTS

Partial support from ASI contract ARS-98-119 and MURST grant Cofin98-02-32 is acknowledeged. I thank G. Zamorani and R. Gilli for useful discussions.

REFERENCES

Allen S.W., Di Matteo T., Fabian A.C., 2000, MNRAS 311, 493
Akiyama M., et al., 2000, ApJ in press (astro-ph/0001289)
Barcons X., et al., 1998, MNRAS 301, L25
Boyle B.J., et al., 1994, MNRAS 271, 639
Boyle B.J., et al., 1998, MNRAS 296, 1
Cagnoni I., Della Ceca R., Maccacaro T., 1998, ApJ 493, 54
Celotti A., Fabian A.C., Ghisellini G., Madau P., 1995, MNRAS 277, 1169
Comastri A., Setti G., Zamorani G., Hasinger G., 1995, A&A 296, 1
Comastri A., 1999, Astr. Lett. & Comm. 39, 181
Di Matteo T., Allen S.W., 1999, ApJ 527, L21
Fabian A.C., Iwasawa K., 1999, MNRAS 303, L34
Fiore F., et al., 1999, MNRAS 306, L55
Franceschini A., et al., 1999, MNRAS 310, L5
Gendreau K.C., et al., 1995, PASJ 47, 5
Gendreau K.C., Barcons X., Fabian A.C., 1998, MNRAS 297, 41
Georgantopoulos I., et al., 1996, MNRAS 280, 276
Georgantopoulos I., et al., 1999, MNRAS 305, 125
George I.M., et al., 2000, ApJ 531, 52
Gilli R., Risaliti G., Salvati M., 1999, A&A 347, 424
Gruber D.E., et al., 1999, ApJ 520, 124
Hasinger G., et al., 1993, A&A 275, 1
Kinzer R.L., et al., 1997, ApJ 475, 361
Madau P., Ghisellini G., Fabian A.C., 1994, MNRAS 270, L17
Miyaji T., et al. 1998, A&A 334, L13
Miyaji T., Hasinger G., Schmidt M., 1999, Adv. Space Res. in press
Miyaji T., Hasinger G., Schmidt M., 2000, A&A 353, 25
Ogasaka Y., et al., 1998, AN 319, 43
Piccinotti G., et al., 1982, ApJ 253, 485
Pompilio F., La Franca F., Matt G., 2000, A&A 353, 440
Risaliti G., Maiolino R., Salvati M., 1999, ApJ 522, 157
Roberts T.P., Warwick R.S., 1998, AN 319, 34
Sambruna R.M., Eracleous M., Mushotzky R.F., 1999, ApJ 526, 60
Setti G., Woltjer L., 1989, A&A 224, L21
Ueda Y., et al., 1999, ApJ 524, L11
Vecchi A., Molendi S., Guainazzi M., et al., 1999, A&A 349, L73
Vignali C., Comastri A., Cappi M., et al., 1999, ApJ 516, 590
Vignali C., Mignoli M., Comastri A., et al., 2000, MNRAS in press (astro-ph/0002279)
Wilman R.J., Fabian A.C., 1999, MNRAS 309, 862

CORONAL HEATING AND EMISSION MECHANISMS IN AGN

Tiziana Di Matteo*

Harvard-Smithsonian Center for Astrophysics, 60 Garden St., Cambridge, MA 02138;
tdimatteo@cfa.harvard.edu

ABSTRACT

Popular models for the formation of X-ray spectra in AGN assume that a large fraction of the disk's angular momentum dissipation takes place in a hot corona that carries a small amount of the accreting mass. Here I discuss the formation of a magnetically-structured accretion disk corona, generated by buoyancy instability in the disk and the heating of localized flare regions up to the canonical X-ray emitting temperatures. I also examine the analogy between accreting disk-coronae and ADAFs and discuss the relevant emission mechanism in these two accretion models and how observational constraints can allow us to discriminate between them.

KEYWORDS: accretion, accretion disks; magnetic fields; radiation mechanisms: thermal; X-rays: general

1. INTRODUCTION

Observations of the central regions of AGN show that a significant fraction of their bolometric luminosity comes out in hard X-rays (from ~ 0.1 keV all the way up to a few 100 keV) and sometimes up to 1 GeV.

According to the standard paradigm, AGN are powered by accretion onto their central black hole. An accretion disk around a supermassive black hole (in an AGN) leads to the production of a strong optical/ultraviolet continuum, the so-called 'blue bump'. Such a component is attributed to quasi-blackbody emission (e.g. see Koratkar & Blaes 1999 for relevant modifications to the blackbody spectrum for an accretion disk). The effective absorptive optical depth in a disk is typically $\tau \gg 1$ which implies that photons are close enough to being in thermal equilibrium with the electrons to produce a blackbody-like spectrum. The luminosity of this component scales as $L \sim \pi r_g \sigma T^4$ where r_g is the Schwarzschild radius. This implies gas temperatures in the disk of the order of

$$T \sim 5 \times 10^5 L_{44}^{-1/4} \left(\frac{L}{L_{Edd}}\right)^{1/2} K, \tag{1}$$

Chandra Fellow

where $L = 10^{44} L_{44}$. L_{Edd} is the Eddington luminosity and the temperature decreases with increasing luminosity (or increasing black hole mass).

It is evident from Eqn. 1 that if AGN generated their energy solely by accretion of matter in thermodynamic equilibrium, the highest temperatures achieved would be of the order of 10^5 K and negligible X-ray emission would be expected. Phenomenologically, therefore, we know that there must be an efficient mechanism for transferring the energy released in an accretion disk into a plasma component that is far from thermodynamic equilibrium with the ambient radiation and that radiates the high energy portion of AGN spectrum.

Although there are many uncertainties concerning how such energy transfer occurs, we know there must be mechanisms that can sustain the presence of a very hot plasma near an accretion disk: e.g. the Sun which has a surface temperature of only 5500 K, is surrounded by a magnetically-dominated corona with a temperature of $2 - 3 \times 10^6$ K.

Here, I address the issue of why we expect hot electrons to be present in AGN. I will discuss how AGN coronae formation can be understood as a direct consequence of the internal dynamics of an accretion disk where shock-like events (magnetic reconnection and MHD processes) are responsible for heating the coronal plasma. I will examine the relevant radiative processes in AGN that are responsible for the production of the X-ray emission that we observe. In particular I will discuss the relevance of these processes for both AGN coronae and hot advection-dominated accretion flows (ADAFs) and their relative importance for different regimes of source luminosities.

1.1. The X-ray emission

Before discussing in more detail the proposed picture of coronae formation I will briefly review the observed characteristics of the X-ray emission in AGN and the information that these give when trying to construct a model for coronae.

Assuming for now the existence of a hot plasma (see Section 2), it is well established that the X-ray continuum in AGN can be explained by thermal Comptonization of the soft UV radiation (e.g Haardt & Maraschi 1993). There is evidence also that this X-ray continuum is reprocessed in a cold medium (e.g. the accretion disk) and gives rise to a reflection bump at around 30 keV and a broad, Fe $K\alpha$ emission line at 6.4 keV. The presence of these features in the spectrum place constraints on the geometry of the X-ray emitting region and tell us that the hot plasma has to be situated above the colder accretion disk. Also, the different ratios of soft luminosity (attributed to the accretion disk) to hard X-ray luminosity imply that the hot coronal plasma is not a slab but consists of localized active regions (e.g. Haardt, Ghisellini & Maraschi 1994). This is also consistent with the characteristically short X-ray variability timescales observed in Seyfert galaxies (as short as a few hours) which imply that enormous amounts of energy are released in a very short time in flare-like events.

Finally, the average X-ray spectra of Seyfert galaxies show a high-energy cutoff

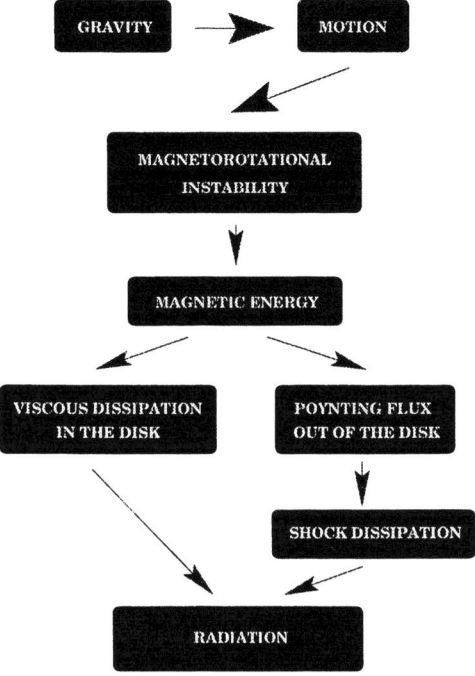

FIGURE 1. Energy flow and angular momentum transport in accretion disks

usually above 100 keV which can be reproduced quite well by models of thermal Comptonization. The absence of conspicuous electron pair annihilation line indicates that most of the hot electrons in a corona are thermal. Whatever the processes that operate in coronae to heat the plasmas are, they do not accelerate a large number of electrons. Alternatively, mechanisms exist for efficiently thermalizing the electron population (e.g. Svensson & Ghisellini 1998).

2. ACCRETION DISK CORONAE

In recent years significant progress has been made in understanding accretion disks and how angular momentum transport operates with the identification of the Balbus–Hawley instability (e.g. Balbus & Hawley 1997) for weakly magnetized disks. Thanks to this fundamental progress, we can now think more confidently of coronae formation as a direct consequence of the internal dynamics of an accretion disk; much the same way the solar corona is thought to be heated by dynamical processes lower in the Sun's atmosphere. More specifically, Balbus & Hawley have identified an instability in weakly magnetized accretion flows that is responsible for the transport of angular momentum. The way such a magneto-rotational instability works is by producing strong amplification of the seed magnetic fields and in this way channeling the energy present in the system into magnetic energy (see Figure 1). The formation of a corona can be understood as an efficient way for a disk to saturate the Balbus–

Hawley instability and to dissipate the accretion energy/angular momentum into particles, which can then radiate it away. The built-up magnetic energy is dissipated into particles locally in the disk and partly builds-up strong magnetic flux tubes leading to a net vertical flux of magnetic energy which inevitably escapes from the disk to form a magnetically-dominated corona.

The idea that buoyancy of strong flux tubes in the disk and their expulsion from the disk form magnetic coronae has been proposed in the past (Stella & Rosner 1984, Coroniti 1981, Galeev, Rosner & Vaiana 1979), but can only now be integrated in a deeper understanding of accretion phenomena.

2.1. Coronal heating: magnetic reconnection

Within the context of such a model, the question of how the coronal plasma heats up to X-ray emitting temperatures can be assessed. Such coronae (e.g. ensembles of flux tubes) contain a very small amount of mass and are magnetically dominated. By definition, the magnetic flux tubes become buoyant when $\beta \sim 8\pi P/B^2 \gtrsim 1$ where B is the magnetic field strength and P the gas pressure in the disk. The typical speed of the rising flux tubes is then given by their Alfvén speed e.g. $V_A = B/\sqrt{4\pi\rho} \gtrsim c_s$ and is by definition always larger then the relevant sound speed (c_s) implying (in a simple view) that the buoyant magnetic energy has to be dissipated in shocks. So, whereas the core of the disk is usually dominated by subsonic turbulence the coronal gas above the disk is, inevitably, supersonic.

More realistically we would expect this energy to be dissipated in shocks in reconnection sites where strong impulsive heating occurs when magnetic field lines are brought together. Reconnection can occur either 'spontaneously' in a given magnetic loop or can be 'driven' when more than one magnetic tubes are brought together. A reconnection site is thought to be a collection of particle acceleration and heating (e.g. direct Joule heating near X-point, slow shock acceleration, Fermi magnetic mirroring in turbulent outflows, conduction, downstream fast shocks etc..) but the detailed physics of how it occurs is still an unsolved MHD problem (for the case Petschek reconnection).

Although the general physical picture of accretion disk coronae described above provide us with an understanding of why we expect to find hot plasmas above accretion disks, there remain many uncertainties. These include the question of which pressure is relevant for magnetic field amplification and buoyancy. It is not clear whether magnetic fields build up to equipartition such as $B^2/8\pi \sim P_{tot}$ or $B^2/8\pi \sim P_{gas}$ Also, it is uncertain what fraction of the magnetic energy is dissipated into e^- and p. It is clear that when energy dissipation occurs one needs to treat the plasma as a 2-temperature medium: different wave-particle interactions will heat electrons and protons differently. One can construct 2-T AGN coronae if the protons contain most of the energy (Di Matteo, Blackman & Fabian 1997), but no clear-cut arguments can be made to support their plausibility over one temperature models. The same problems exist in the case of ADAF plasma where the 2-T condition is a crucial assumption but, at this stage, yet to be proven. Coronal plasmas are often collisionless. It is not clear, therefore, whether electrons are thermalized or dissipative processes cannot accelerate particles efficiently. In other words the importance of direct heating versus acceleration in either coronal or ADAFs plasma, cannot be determined. In AGN coronae or ADAFs, V_A can approach c, and one should really consider the effects of relativistic MHD. Such effects are usually not taken into account.

Important recent results from numerical simulations (Miller & Stone 1999) do indeed show the formation and heating of magnetized coronae above accretion disks. In particular, Miller & Stone have shown that when weak B fields are amplified in the disk via MHD turbulence driven by the Balbus–Hawley instability some of the magnetic energy is dissipated locally but a good fraction escapes due to buoyancy and forms a strongly magnetized corona above the disk. Most of the energy in their simulations is dissipated at a few scale heights above the disk, and strong shocks are continuously produced making the corona hot up to X-ray emitting temperatures. Their results on the impulsive heating of coronal plasmas, are in accordance with simple analytical estimates (Di Matteo 1998) on the occurrence of an ion–acoustic instability, associated with slow shocks in Petschek magnetic reconnection in flare-like events in a magnetically-dominated corona. The occurrence of an ion–acoustic instability, associated with slow shocks in Petschek magnetic reconnection, can be shown to result in a violent release of energy and heat the coronal plasma to canonical X-ray emitting temperatures (of a few $\times 10^9$K).

3. EMISSION MECHANISMS

In the previous sections I have discussed the vertical structure of an accretion disk and how its internal dynamics can lead to the formation of a highly-dynamic, magnetically dominated and heated corona.

Both in AGN coronae and in ADAFs (also magnetized and with hot $\sim 10^9$ K electrons; see Narayan, Quataert & Mahadevan 1999 for a recent review) the relevant interactions and relative emission mechanisms are: particle-photon $\to$ Compton processes; particle-magnetic field $\to$ cyclo/synchrotron emission and particle-particle $\to$ bremsstrahlung emission.

Inverse-Compton scattering of disk photons off the hot electrons is usually the dominant process in most AGN. The importance of Inverse-Compton processes scales $\propto U_{rad} \exp(y)$ where $y \sim 4(kT/m_e c^2)\tau$, $\tau = n_e \sigma_T r$ and the energy density $U_{rad} \sim L/(R^2 c)\tau$ is usually attributed to the external soft photon flux coming from the disk. Bremsstrahlung instead scales as $n_e^2 T^{-1/2}$, where n_e is the electron number density, and dominates IC only in very low luminosity objects e.g.

$$IC > BREM \to \frac{L}{L_{Edd}} > \frac{10^{-5}}{\sqrt{\theta}} r_s \qquad (2)$$

(see also Section 3.2), where θ is the dimensionless electron temperature.

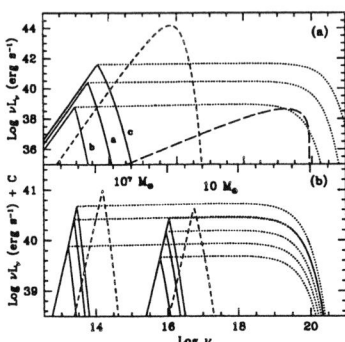

FIGURE 2. (a) The solid line is the synchrotron component, the dotted line its Comptonization, small dashed line the blackbody component from the accretion flow. (b) The shift in frequency of the synchrotron peak between supermassive and galactic black holes.

3.1. Synchrotron emission and Comptonization in coronae and ADAFs

Both in the case of an AGN corona or an ADAF, magnetic fields are close to their equipartition values and synchrotron emission should be taken into account.

In both cases electrons are considered to be thermal. Thermal synchrotron is heavily self-absorbed up to a frequency $\nu_s \propto T^2 B$. Equipartition arguments (in the case of a supermassive black hole with $M \sim 10^7 \, M_\odot$) imply values of $B \sim P_{gas} + P_{rad} \sim 10^{3-5}$ Gauss and for canonical corona temperatures of 10^9 K, synchrotron emission peaks in the Infrared/Optical bands (Di Matteo, Celotti & Fabian 1997; see Figure 2a).

The synchrotron soft photon flux is Inverse Compton scattered up to X-ray energies by the hot electrons (dotted line in Figure 2a). In most cases though, synchrotron self-Compton does not dominate the X-ray emission because the energy density due to the soft disk photon field dominates the scattering. Due to the high self-absorption, the synchrotron energy density $U_{syn} < B^2/8\pi$ which, given the equipartition arguments, implies $U_{syn} < U_{disk} \sim P_{rad}$ (Fig. 4a) and Comptonization of the soft disk photons dominates the X-ray emission. Given the strong dependences of thermal synchrotron emission on both temperature and B, and the very dynamical structure of the corona, estimates of an 'average' T and B, which are usually employed in these calculations are likely to be unrealistic. As shown by the above relations, the importance of synchrotron and its Inverse Compton might be highly enhanced if flares are at different temperatures and some are hotter and/or with higher magnetic fields than the values usually assumed from global arguments. It is plausible that a non-thermal population of particles could be present which would also significantly enhance the synchrotron and its IC component but this has not been taken into account in current models).

In contrast, in an ADAF the synchrotron photons are, in most cases, the only source of soft photons for Comptonization (even if the ADAF is matched to a thin

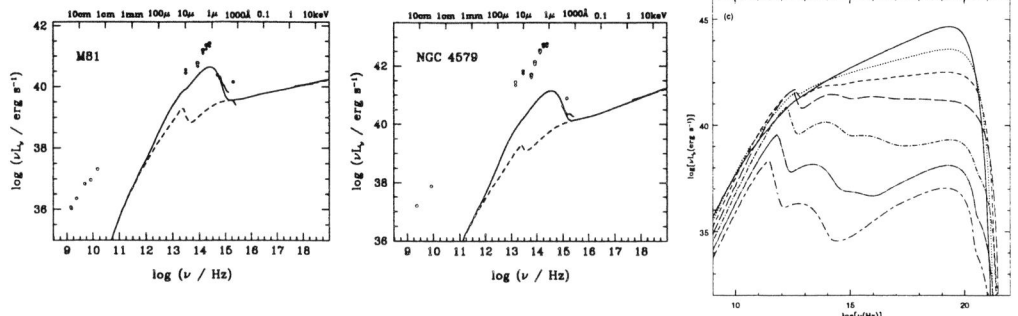

FIGURE 3. Model for M81 NGC 4579 in which a thin disk is truncated at $r \sim 100 r_S$, inside of which there is an ADAF. The solid line shows the total "disk + ADAF" emission, while the dashed line shows the ADAF contribution (Quataert et al. 1999). On the rightmost panel the spectra of ADAFs for a $10^9 \, M_\odot$ black hole with $\dot{m}$ decreasing by about 3 orders of magnitude from the top curve to the bottom one. The high energy spectrum changes from Comptonized synchrotron to bremsstrahlung.

disk at large distances, as in models by Esin et al. 1997; Quataert et al 1999 its contribution is negligible; e.g. see Figure 3).

Comptonization of the synchrotron component in an ADAF can explain the observed X-ray emission in some low-luminosity AGN. Figure 3 shows the case for M81 and NGC 4579 both of which have an estimated mass for the central black hole, detected hard X-ray emission, and optical/UV emission too low to allow for the presence of a geometrically thin, optically thick accretion disk close to the black holes (Quataert et al. 1999). In general, in a standard ADAF, Comptonization becomes important for $\dot{m} \lesssim \dot{m}_{crit}$ above which the hot flow cannot exist. In the high $\dot{m}$ regime considered here, the characteristic electron scattering optical depth τ of the ADAF becomes of order unity since $\tau \propto \dot{m}$. As τ decreases with decreasing $\dot{m}$, bremsstrahlung becomes the dominant process (see Figure 3).

3.2. Bremsstrahlung emission in elliptical galaxy nuclei

Equation 2 shows that bremsstrahlung emission can only be important in sources with very low luminosities (or low radiative efficiencies). The nuclear regions of elliptical galaxies provide excellent environments in which to study the physics of low-luminosity accretion. There is now strong evidence, from high-resolution optical spectroscopy and photometry, that black holes with masses of $10^8 - 10^{10} \, M_\odot$ reside at the centers of bulge dominated galaxies, with the black hole mass being roughly proportional to the mass of the stellar component (e.g. Magorrian et al. 1998). X-ray studies of elliptical galaxies also show that they possess extensive hot gaseous halos, which pervade their gravitational potentials. Given the large black hole masses inferred, some of this gas must inevitably accrete at rates which can be estimated from Bondi's spherical accretion theory. Such accretion should, however, give rise to far more nuclear activity (e.g. quasar-like luminosities) than is observed, if the radiative efficiency is as high as 10 per cent (e.g. Fabian & Canizares 1988), as is generally postulated in standard accretion theory.

Accretion with such high radiative efficiency need not be universal, however. As suggested by several authors (Rees et al. 1982; Fabian & Rees 1995), the final stages of accretion in elliptical galaxies may occur via an advection-dominated accretion flow (ADAF; Narayan & Yi 1995, Abramowicz et al. 1995) at roughly the Bondi rates. Within the context of such an accretion mode, the quiescence of the nuclei in these systems is not surprising; when the accretion rate is low, the radiative efficiency of the accreting (low density) material will also be low. Other factors may also contribute to the low luminosities observed. As discussed by Blandford & Begelman (1999; and emphasized observationally by Di Matteo et al. 1999a), and shown numerically by Stone, Pringle & Begelman (1999), winds may transport energy, angular momentum and mass out of the accretion flows, resulting in only a small fraction of the material supplied at large radii actually accreting onto the central black holes.

If the accretion from the hot interstellar medium in elliptical galaxies (which should have relatively low angular momentum) proceeds directly into the hot, advection-dominated regime, and low-efficiency accretion is coupled with outflows (Di Matteo et al. 1999a), the question arises of whether *any* of the material entering into the accretion flows at large radii actually reaches the central black holes. The present observational data generally provide little or no evidence for detectable optical, UV or X-ray emission associated with the nuclear regions of these galaxies.

The discovery of hard X-ray emission from a sample of six nearby elliptical galaxies (Allen, Di Matteo & Fabian 1999), including the dominant galaxies of the Virgo, Fornax and Centaurus clusters (M87, NGC 1399 and NGC 4696, respectively), and NGC 4472, 4636 and 4649 in the Virgo cluster, has important implications for the study of quiescent supermassive black holes. The ASCA data for all six sources provide clear evidence for hard, power-law emission components, with photon indices in the range $\Gamma = 0.6 - 1.5$ and intrinsic $1 - 10$ keV luminosities of $2 \times 10^{40} - 2 \times 10^{42}$ erg s^{-1} (Allen et al. 1999). This potentially new class of accreting X-ray source has X-ray spectra significantly harder than Seyfert nuclei and bolometric luminosities relatively dominated by their X-ray emission.

We argue that the X-ray power law emission is most likely to be due to accretion onto the central supermassive black holes, via low-radiative efficiency accretion (Allen et al. 1999, Di Matteo et al. 1999b).

The broad band spectral energy distributions for these galaxies, which accrete from their hot gaseous halos at rates comparable to their Bondi rates, can be explained by low-radiative efficiency accretion flows in which a significant fraction of the mass, angular momentum and energy are removed from the flows by winds. The observed suppression of the synchrotron components in the radio band (Di Matteo

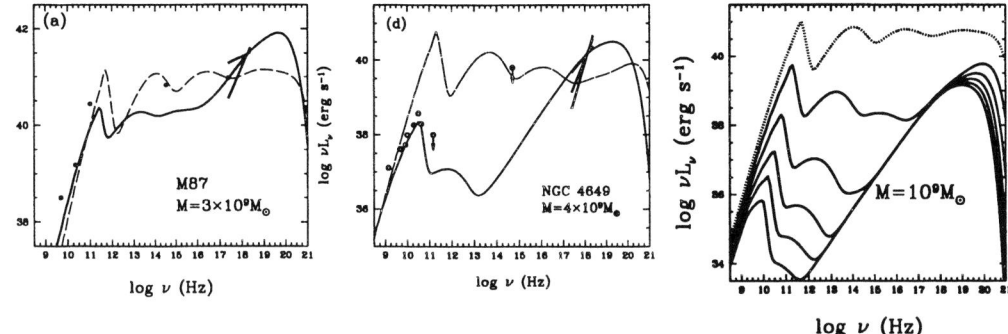

FIGURE 4. Spectral models calculated for ADAF with outflows and without outflows (dashed lines) are shown for two representative cases (for the other objects see Di Matteo et al. 1999b). The solid dots are the best constraints on the core emission. The thick solid lines the slopes and fluxes obtained from the ASCA analysis. The leftmost panel shows explicitly the effects of winds on the spectra of ADAFs, The mass outflow scales as $\dot{M} = \dot{M}_{out} \left(\frac{R}{R_{out}} \right)^p$. Models are calculated for the same $\dot{M}_{out}$ and R_{out} with p increasing from 0 to 1 (in steps of 0.2) from the top dashed curve to the lower one

et al. 1999a; excluding the case of M87) and the systematically hard X-ray spectra, which are interpreted as thermal bremsstrahlung emission, support the conjecture that significant mass outflow is a natural consequence of systems accreting at low-radiative efficiencies (see the representative cases of NGC 4649 and M87 in Figure 4 and for all of the objects Di Matteo et al. 1999b).

The presence of outflows in the hot flows suppresses completely the importance of Comptonization in ADAF flows and bremsstrahlung becomes (irrespectively of the accretion rate, c.f. Figure 3) the dominant X-ray emission mechanism. A representation of the effects on the ADAF spectra of outflows is shown in Figure 4.

ACKNOWLEDGEMENTS

TDM acknowledges support for this work provided by NASA through Chandra Fellowship grant number PF8-10005 awarded by the Chandra Science Center, which is operated by the Smithsonian Astrophysical Observatory for NASA under contract NAS8-39073.

REFERENCES

Abramowicz M.A., Chen X., Kato S., Lasota J.P., Regev O., 1995, ApJ, 438, L37
Allen, S.W., Di Matteo, T., Fabian, A.C., 1999, MNRAS, in press

Balbus, S. A., Hawley, J. F., in Accretion Processes in Astrophysical Systems: Some Like it Hot! Eighth Astrophysics Conference, College Park, MD, October 1997. Edited by Stephen S. Holt and Timothy R. Kallman, AIP Conference Proceedings 431., p.79

Blandford R.D., Begelman M.C., 1999, MNRAS, 303, L1

Coroniti F.V., 1981, ApJ, 244, 587

Di Matteo T., 1998, MNRAS, 299, L15

Di Matteo T., Blackman E.G., Fabian A.C., 1997, MNRAS, 291, L23

Di Matteo T., Celotti A., Fabian A.C., 1997, MNRAS, 291, 805

Di Matteo, T., Fabian, A. C., Rees, M. J., Carilli, C. L., Ivison, R. J. 1999a, MNRAS, 305, 492

Di Matteo, T., Quataert, E., Allen, S. W., Narayan, R., Fabian, A. C., 1999b, MNRAS, in press

Fabian A.C., Canizares C.R., 1988, Nature, 333, 829

Fabian A.C., Rees M.J., 1995, MNRAS, 277, L55

Ghisellini, G., Haardt, F.,Svensson, R. 1998, MNRAS, 297, 348

Galeev A.A., Rosner R., Vaiana G.S., 1979, ApJ, 229, 318

Haardt F., Maraschi L., ApJ, 1993, 413, 507

Haardt F., Maraschi L., Ghisellini G., 1994, ApJL, 432, L92

Koratkar A., Blaes O., 1999, Publ. Astron. Soc. of Pacific, 111, 1

Magorrian J. et al., 1998, AJ, 115, 2285

Lee J.C,, Fabian A.C., Reynolds C.S., Brandt W.N., Iwasawa K., 1999, MNRAS, submitted

Miller K. A., Stone J.M., 1999, ApJ, submitted

Narayan R., Yi I., 1995, ApJ, 444, 231

Narayan R., Mahadevan R., Quataert E., 1998, Theory of Black Hole Accretion Disks, edited by Marek A. Abramowicz, Gunnlaugur Bjornsson, and James E. Pringle. Cambridge University Press, 1998., p.148

Quataert E., Di Matteo T., Narayan R., Ho L., 1999, ApJL, 525, L89

Rees M.J., Phinney E.S., Begelman M.C., Blandford R.D., 1982, Nature , 295, 17

Reynolds C.S., Fabian A.C., Inoue H., MNRAS, 276, 1311

Stone J.M., Pringle J.E., Begelman M.C., 1999, MNRAS, in press

Stella L., Rosner R., 1984, ApJ, 277, 312

THE ENERGY OUTPUT OF THE UNIVERSE

A.C. Fabian

Institute of Astronomy, Madingley Road, Cambridge CB3 0HA, U.K.

ABSTRACT   The total energy emitted by the growth of massive black holes is large and can be 10-50 per cent of that emitted by stars in the universe. I show how the X-ray Background provides a good measure of this energy and also why most accretion power is absorbed and re-emitted in the far infrared band. A model for the obscured growth of massive black holes is presented which accounts for both the high absorption and the observed black hole to galaxy spheroid mass correlation. Future missions should detect the obscured X-ray sources associated with the growth of massive black holes.

KEYWORDS: galaxies:active – quasars:general – X-rays:general

1. INTRODUCTION

The X-ray Background (XRB) is the integrated emission from all X-ray sources. Its hard spectrum has proved difficult to explain since, in the 2–10 keV band, it is flat with a power-law of energy index 0.4. This is flatter than the spectrum of any known common population of objects. For the last decade the most popular explanation has been that the XRB intensity is dominated by many absorbed sources (Setti & Woltjer 1989), with ranges of absorbing column density and redshift causing the observed spectrum to be a power-law. The absorption model has been extensively studied by Madau et al (1994), Matt & Fabian (1994), Comastri et al (1995), Celotti et al (1995), and Wilman & Fabian (1999). The most complete studies include Compton down-scattering in the estimation of the observed spectrum of the Compton-thick sources.

The absorption model is adopted here and is used in a simple way to show that black holes grow by radiatively efficient accretion and to determine a) the local mean density of black holes, b) the fraction of accretion power which has been absorbed, and c) constraints on the fraction of power in the Universe due to accretion (see also Fabian & Iwasawa 1999). After some discussion of how so much obscuring material can surround most sources, and how the nuclei might be fuelled, I then outline a model of obscuration in a forming, isothermal galaxy spheroid (Fabian 1999). The XRB is shown to be a key diagnostic of the accretion power of the Universe.

2. ACCRETION AND THE XRB

I assume that the underlying active galactic nuclei (AGN) which power the XRB have a quasar-like spectrum with an energy photon index of one. The spectrum is then constant in a νF_ν sense (Fig. 1). The action of photoelectric absorption by increasing amounts of material, characterised by a column density N_H, is (Fig. 2) to cut out the lower energy emission from the observed spectrum up to an (approximate) energy $E \sim 10 N_H^{8/3}$ keV, where N_H is in units of 10^{24} cm^{-2}. As the column density exceeds about 1.5×10^{24} cm^{-2} so the absorber becomes Compton thick and Compton (electron) scattering causes the residual spectrum above this cutoff to decrease in intensity. This means that the intensity observed above about 30 keV is close to the intrinsic unattenuated intensity from Compton-thin sources, and is a lower limit for Compton-thick ones. Therefore the intensity of the XRB at 30 keV equals the normalization of the XRB after correction for absorption by Compton-thin sources. This normalization can be increased by a factor of f^{-1} if only a fraction f of all the power emerges from sources which are Compton thin. f is at most 3/4 (Maiolino et al 1998) and could be less than one half.

The absorption-corrected XRB spectrum can then be extended into the ultra-violet band assuming the mean quasar spectral energy distribution of Elvis et al (1994). This shows that about 3 per cent of the power from a typical quasar is emitted in the 2–10 keV band. The total absorption-corrected AGN background can now be converted into an energy density ϵ_{AGN} and hence, through the use of $E = Mc^2$ or rather $\epsilon(1 + \bar{z}) = \eta \rho c^2$ with an accretion efficiency factor η and a mean redshift $\bar{z}$ (since photons lose energy in the expansion of the Universe but mass does not), we have the mean density in black holes ρ_{bh}.

The resulting value of $\rho_{bh} = 6 \times 10^5$ M$_\odot$ Mpc^{-3} is about half the value found by Magorrian et al (1998) from a study of ground-based optical data of the cores of nearby galaxies, and in rough agreement with an HST photometric study made by van der Marel (1999). Similar agreement has been obtained by Salucci et al (1999) from a detailed considerations of source counts etc. This close agreement emphasises that most of the mass in black holes has been accreted by a radiatively efficient (but obscured) process, and not by some inefficient process such as an advective flow. The correction required for absorption is extensive and requires that most, about 85%, of the accretion power has been absorbed.

3. AGN, THE FIR BACKGROUND AND THE ENERGY FROM STARS

The absorbed power is assumed to be emitted in the Far Infrared (FIR) bands, and when redshifted it should contribute to the sub-mm background. The total predicted is about 3 nW m^{-2} sr^{-1} which is several tens percent of total the sub-mm background (Fixsen et al 1998; see also Almaini et al 1999 for estimates of the AGN contribution to the sub-mm background). This suggest that to within a factor of two the total integrated power (ie the total energy released) from accretion onto black holes is about one quarter of that from stars (mostly starlight but including

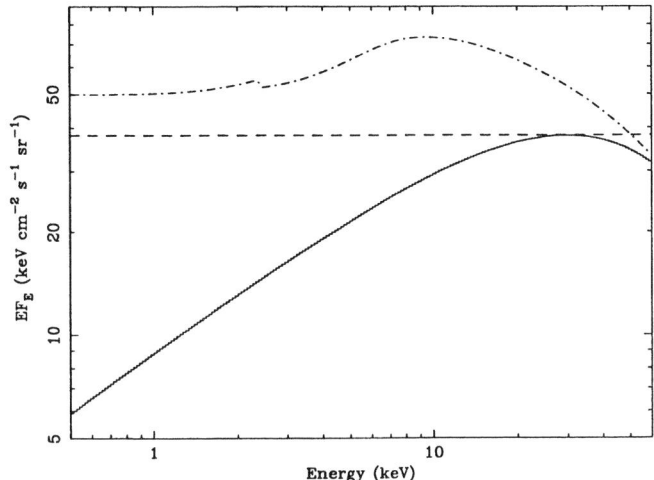

FIGURE 1. XRB spectrum (solid line) with the assumed unabsorbed spectrum of photon index 2 (dotted line). A typical AGN spectrum with reflection, in which the direct emission is a power law of photon index 2 with an exponential cutoff of 300 keV is shown by the dot-dash line, matching around the XRB peak. If unabsorbed quasars contribute 50 per cent of the XRB at 1 keV, then their contribution lies along the bottom of the figure.

supernovae), i.e.
$$E_{AGN}/E_\star \sim 0.25.$$
The details of any comparison depend upon the history of the starlight and of the accretion. No estimate of the contribution to the NIR and optical backgrounds, which could lower the above value, has been made here.

A simple check on this is obtained from an argument due to G. Hasinger (see Fabian & Iwasawa 1999). Magorrian et al (1998) find the following relation between the black hole mass M_{bh} and spheroid mass M_{sph} of a galaxy:
$$M_{bh} \approx 0.005 M_{sph},$$
so if the total energy radiated
$$E_{AGN} \approx 0.1 M_{bh} c^2$$
then
$$E_{AGN} \approx 0.1 \times 0.005 M_{sph} c^2.$$
But the total energy radiated by stars
$$E_\star \approx 0.1 \times 0.005 a^{-1} M_{sph} c^2,$$

where the first term is the fraction of a star which undergoes nuclear fusion and the second is the efficiency (in a $E = mc^2$ sense) of that fusion. a is the ratio of the present mass of the spheroid to its original mass (many of the stars have evolved) and for a Salpeter mass function is about 20 per cent. Therefore

$$E_{AGN}/E_\star \approx a \approx 0.2.$$

4. UNCERTAINTIES

The above estimate reduces to 0.1 if the scaling relation of van der Marel (1999), which agrees better with the XRB intensity, is used, but can increase towards unity if stellar mass loss is recycled into new stars, so that $a \sim 1$. A mass-to-energy efficiency of 0.1 has been used but it can be 0.06 if the black hole is not spinning. or 0.42 if it becomes a maximally spinning, Kerr, black hole.

An even more extreme possibility which defines an upper limit on the efficiency relative to the final (dead) black hole mass is to assume that the black hole was maximally spinning during the accretion phase and then spun down by, say, the Blandford-Znajek (1977) mechanism. The total energy released relative to the final black hole mass can then approach unity and allows for an order of magnitude uncertainty in η and thus E_{AGN}. Of course a high value here, which maximises $E_{AGN}/E_\star$, overpredicts the XRB intensity unless most of the growing phase of black holes is Compton thick. It is also possible that a significant fraction of the power from an AGN is in the form of a wind and not directly in radiation. As discussed later, growing black holes may be both Compton thick and powering winds. If this is correct, then $E_{AGN}/E_\star$ may be significantly higher than the estimate in the last section.

5. OBSCURATION, METALLICITY AND FUELLING

As outlined above, at least 85 per cent of accretion power is absorbed. Since about ten per cent is in quasars which show very little absorption, this means that most lines of sight out of the remaining objects are highly absorbed. This is difficult for the standard obscuring torus model, which could absorb perhaps one half to two thirds of all sight lines. Even then it is unclear what inflates the torus, which is supposed to be cold and molecular. Dissipation in a system of orbiting clouds should cause it to flatten into a disc, with low covering factor.

Energy must be continuously injected into any cold absorbing cloud system to keep it inflated and so sky covering. One plausible solution is that a gas-rich star cluster surrounds the black hole and it is the massive stars (winds and supernovae) which supply the energy (Fabian et al 1998). The surrounding starburst can thereby obscure the active nucleus.

The starburst should enhance the metallicity of the absorbing gas. This makes a given mass of gas more efficient at absorbing X-rays and indeed increases the effect of absorption before Compton down-scattering comes into play. This is important

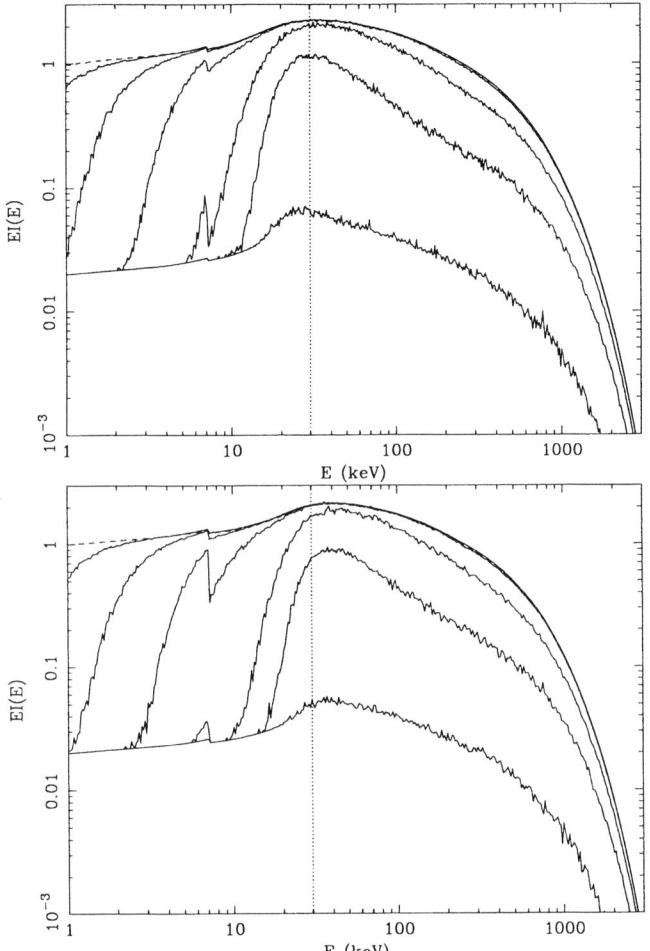

FIGURE 2. Monte-Carlo simulations of of an accretion disc spectrum (a power-law of unit energy index with cold reflection and an exponential cutoff at 360 keV) propagated through a solar abundance spherical cloud of column densities ranging from $10^{21.25}$ to $10^{25.25}$ cm^{-2} in steps of a factor 10. Note that the spectra peak around 30 keV, indicated by the vertical line. The lower panel shows the effect of increasing the iron abundance by 5, which causes the peak to shift to 40–50 keV (from Wilman & Fabian 1999).

in opening up the parameter space for model-fitting of the XRB spectrum (Wilman & Fabian 1999).

Fuelling of the nucleus is an old problem (see e.g. Shlosman et al 1990). Although there may be lots of gas around the nucleus, angular momentum may prevent it from rapidly accreting to the centre. In this respect, a hot phase in the surrounding medium may be important, with Bondi accretion from this phase being the dominant mechanism (see e.g. Nulsen & Fabian 1999). Angular momentum may be transported outward by turbulence within such a hot phase, so allowing rapid accretion to proceed.

6. THE MEAN LUMINOSITY OF THE DISTANT AGN DOMINATING THE XRB INTENSITY

Since the mass of the black hole in nearby galaxies appears to be proportional to the spheroid mass, the mass function of black holes must be similar in shape to the spheroid mass function. The mean black hole mass is therefore that appropriate to an L^* galaxy, or about 3×10^8 $M_\odot$. The Eddington limit of such a black hole is about 3×10^{46} erg s^{-1} and its mass doubling (Salpeter) time is about 3×10^7 yr. If the typical mass black hole has therefore grown from say a million solar mass one in 3×10^9 yr (i.e. by $z \sim 2$), then we probably need $L > 0.05 L_{Edd} \sim 10^{45}$ erg s^{-1}. This means that the typical growing black hole was powerful and of quasar-like luminosity (indeed housing a quasar at the centre).

Such an obscured powerful object would locally be classified as a ULIRG (see Sanders & Mirabel 1996), although the distant ones need not be the same as the local ones, which are perhaps mainly fuelled by mergers.

Of course it is possible that massive black holes grew inside galaxies which themselves were merging back at $z \sim 2$. Nevertheless, unless they were all assembled from smaller holes just before accretion switched off, it is probable that they emit for a reasonable fraction of the last doubling time as a single object.

7. OBSCURATION IN A GROWING, ISOTHERMAL GALAXY SPHEROID

Consider an isothermal galaxy in which a significant fraction f_c of cooled gas remains as cold dusty clouds instead of rapidly forming stars. At the centre a black hole grows by accretion from the surrounding cold (and hot) gas. Assume that the nucleus also blows a wind of velocity v_w which has a power $L_w = \alpha L_{Edd}$. Eventually the wind becomes powerful enough to blow away the surrounding gas and so shut off the accretion and further growth to the black hole and spheroid. The Magorrian et al (1998) black-hole – spheroid mass relation can then be obtained (Silk & Rees 1998; Fabian 1999; Blandford 1999).

The kinetic power of a wind at which it ejects cold gas of column density N_H from a spheroid is given by

$$L_W \approx 2\pi G M_{sph} m_p N_H v_w$$

or
$$L_w \approx f_c \sigma^4 v_w G^{-1},$$
where σ is the velocity dispersion within the spheroid. (I have used a force argument here, see Fabian 1999; Silk & Rees 1998 use an energy argument to obtain a limit of σ^5/G, which is a factor σ/v_w smaller than the above L_w.) Ejection occurs when
$$M_{bh} \approx \frac{\sigma^4 \sigma_T}{4\pi G^2 m_p} \frac{v_w}{c} \frac{f_c}{\alpha}.$$
Using the Faber-Jackson relation for spheroids ($M_{sph} \propto \sigma^4$) then yields, if $\frac{v_w}{c} \frac{f_c}{\alpha} \sim 1$
$$M_b \sim 0.005 M_{sph},$$
close to the Magorrian et al (1998) relation.

At that point the column density in to the accretion radius $N_H \sim N_T = \sigma_T^{-1}$, so the growth is (just) Compton thick. The growth of massive black holes is radiatively efficient, highly obscured and gives rise to much of the XRB. It is also intimately linked with the growth of galaxy spheroids, the main evolution of which is terminated by a quasar wind. X-ray observations probe best the underlying obscured nucleus at (rest frame) energies of about 30 keV. Indeed X-rays are the best diagnostic of the black hole accretion history of the Universe.

The optically bright quasar phase (from an outside observer's point of view) follows over the next few million years as the accretion disc around the black hole empties. The early phase as the wind clears the gas away can be identified with BAL quasars. The central engine is only revived after the quasar phase if a merger or other event brings in sufficient low angular momentum gas to fuel it.

The prospects of testing the above scenario and absorption models of the XRB are close at hand, with Chandra and XMM. They should detect large numbers of faint, but powerful absorbed sources in the 3–10 keV band, due to the negative K correction involved (see Fig. 3) and identify them with luminous FIR/sub-mm–emitting young galaxy spheroids.

ACKNOWLEDGEMENTS

I am grateful to Kazushi Iwasawa, Paul Nulsen and Richard Wilman for continued collaboration and the organisers for an interesting conference. The Royal Society is thanked for support.

REFERENCES

Almaini O Lawrence A Boyle B 1999 MNRAS 305 59
Blandford RD 1999 astro-ph 9906025
Blandford RD Znajek RL 1977 MNRAS 179 433
Celotti A Fabian AC Ghisellini G Madau P 1995 MNRAS 277 1169
Comastri A Setti G Zamorani G Hasinger G 1995 A&A 296 1

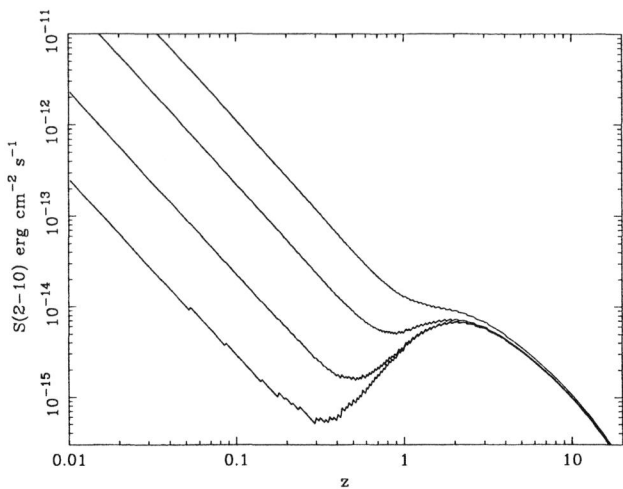

FIGURE 3. The observed 2–10 keV flux as a function of redshift from a source of intrinsic (unabsorbed) 2–10 keV luminosity of 10^{45} erg s^{-1} with a column density of $10^{24.5}$ cm^{-2}. Scattering fractions (by thin ionized gas) of 5, 1, 0.1 and 0.01 per cent are included (top to bottom). Note that the negatice K-correction means that sources at $z \sim 0.1, 0.8$ and 7 can have the same observed 2–10 keV flux. From Wilman & Fabian (1999).

Elvis M et al 1994 ApJS 95 1
Fabian AC 1999 MNRAS 308 L39
Fabian AC Barcons X Almaini O Iwasawa K 1998 MNRAS 297 L11
Fabian AC Iwasawa K 1999 MNRAS 303 L34
Fixsen D Dwek E Mather JC Bennet CL Shafer RA 1998 ApJ 508 123
Madau P Ghisellini G Fabian AC 1994 MNRAS 270 L17
Magorrian J et al 1998 AJ 115 2285
Maiolino R et al 1998 A&A 338 781
Matt G Fabian AC 1994 MNRAS 267 187
Nulsen PEJ Fabian AC MNRAS in press
Salucci P Szuskiewicz E Monaco P Danese L 1999 MNRAS
Sanders DB Mirabel IF 1996 ARAA 34 749
Silk J Rees MJ 1998 A&A 331 L1
Setti G Woltjer L 1989 A&A 224 L21
Shlosman I Begelman MC Frank 1990 Nature 345 679
van der Marel RP 1999 AJ 117 744
Wilman RJ Fabian AC 1999 MNRAS 309 862

RELATIVISTIC JETS FROM X-RAY BINARIES : RECENT ADVANCES

R. P. Fender

Astronomical Institute 'Anton Pannekoek' and Center for High-Energy Astrophysics, University of Amsterdam, Kruislaan 403, 1098 SJ Amsterdam, The Netherlands

ABSTRACT

I summarise important recent advances in the observation of relativistic jets from X-ray binary systems. Specifically I concentrate on a comparison of jet production in neutron star and black hole candidate X-ray binaries, and on the relation between black hole state and jet formation. It is concluded that jet formation is probably broadly similar in low magnetic field neutron stars and black holes, and that there is a clear empirical relation between jet formation and black hole state. Furthermore the strong correlations observed between hard X-rays and radio emission point to a physical connection between the Comptonising corona and the jet.

KEYWORDS: X-rays: stars – Radio continuum: stars – ISM:jets and outflows – stars:neutron – black hole physics

1. INTRODUCTION

Radio emission, sometimes spectacularly resolved into jet-like outflows, is becoming recognised (by some) as an increasingly important diagnostic of accretion flows in the vicinity of black holes and neutron stars in X-ray binary systems. Recent reviews can be found in Hjellming & Han (1995), Mirabel & Rodríguez (1999), Fender (2000) and references therein.

Several aspects of the radio emission are by now well-established :

- The emission mechanism is the synchrotron process, produced by relativistic electrons (+ positrons ?) spiralling around magnetic field lines.

- Major ejections, such as those observed from Cyg X-3, GRS 1915+105, GRO J1655-40 etc. correspond to the transient formation of radio-emitting outflows.

- These outflows propagate at bulk relativistic ($\gtrsim 0.9c$; exceptionally the jets of SS 433 are observed to propagate with a lower, nearly constant $v = 0.26c$) velocities along well collimated (opening angle in all cases $\lesssim 20°$), sometimes curved (SS 433, Cyg X-3 ?, GRS 1915+105 ?) paths.

- The energetics of these jet-formation episodes are extreme, rivalling both the accretion luminosity and mass-flow (the latter point is valid only if each synchrotron-emitting electron is accompanied by a proton).

In this paper I will highlight two new areas of research in which significant progress has been made in recent years, namely:

1. The relation between jet formation in neutron-star and black-hole X-ray binaries;

2. The relation between jet formation and X-ray 'state' in black-hole X-ray binaries.

2. BLACK HOLES AND NEUTRON STARS

2.1. Transients

Bright X-ray transients are generally accompanied by transient radio emission which follows the pattern outlined in the introduction. Bright transients can contain both neutron stars (e.g. Aql X-1) and black holes (e.g. GS 1124-684); do the radio properties of these two populations differ?

The answer is yes and no. Firstly, it is clear that, excluding bright, exotic objects for which classification of the compact object type has proved impossible to date (e.g. Cyg X-3, SS 433, LSI +61° 303), the black hole transients are, at the peak of outburst, the brightest radio sources associated with X-ray binaries. However, this is also the case for their X-ray emission – i.e. the brightest transients in the X-ray band are also the black holes. It is unclear at present whether this simply reflects the larger average masses of the black holes, or differences in the accretion flows onto the two types of compact accretor. On the other hand, the ratio of radio to X-ray peak fluxes is comparable for both neutron star and black hole X-ray binaries. As an example, the neutron star transient Cen X-4 reached peak fluxes of $\sim$ 4 Crab and $\sim$ 10 mJy at soft X-ray and radio wavelengths respectively during its 1979 outburst. For comparison, the black hole transient A 0620-00 reached peak fluxes of $\sim$ 45 Crab and $\sim$ 200 mJy during its 1975 outburst. While the ratios are not exactly the same (but bear in mind there is likely to be a scatter in *observed* radio fluxes due to beaming – see Kuulkers et al. 1999), their order-of-magnitude correspondence indicates that in both neutron star and black hole systems the ratio of X-ray to radio luminosities is comparable. Furthermore, the ratio is similar for most other systems (Fender & Kuulkers, in prep). This in turn implies that the accretion and jet formation mechanisms are broadly the same, during outburst, for both types of system.

2.2. Persistent sources

Only four persistently bright X-ray sources in our Galaxy are believed to contain black holes : Cyg X-1, GX 339-4, 1E 1740.7-2942 and GRS 1758-258. The latter two

Source type	$S_\nu/(\text{kpc}^2)$ (mJy)	Inferred physical characteristics		
		$(\dot{m}/\dot{m}_{\text{Edd}})$	B (Gauss)	inner disc radius (km)
BHC (low/hard state)	55 ± 13	≤ 0.1	–	few $\times 100$
Z (horizontal branch)	"	$0.1 - 1.0$	$10^9 - 10^{10}$	few $\times 10$
Atoll	$\leq 10 \pm 2$	$0.01 - 0.1$	$10^9 - 10^{10}$	few $\times 10$
X-ray pulsar	$\leq 6 \pm 2$	≤ 1.0	$\geq 10^{12}$	≥ 1000

TABLE 1. Comparison of derived mean intrinsic radio luminosities for the BHC, Z, Atoll and X-ray pulsar classes of persistent X-ray binary, plus simple interpretations of their physical differences. From Fender & Hendry (2000).

are too faint for regular radio monitoring, however for Cyg X-1 and, especially, GX 339-4, we have a good idea of how radio emission is related to X-ray state. This is explored in detail in Sect. 3. Discrete, bright, outbursts correspond to major state transitions and/or the Very High State; the formation of a steady outflow seems to occur in the Low/Hard and, more weakly, Off states.

Neutron star systems do not display analogs of the black hole states, but can nonetheless be classified into different groups. These are the 'Z' and 'atoll' X-ray binaries (believed to contain low magnetic field neutron stars), and the high-field X-ray pulsars. Thus the population of neutron star X-ray binaries allows us to explore the effects of mass accretion rate and accretor magnetic field on the production of radio jets. The Z sources (and the unusual atoll source GX 13+1) are all regularly detected as variable radio sources.

Penninx et al. (1988), in observations of the Z-source GX 17+2, established a link between radio emission and location on the Z track (corresponding to the soft X-ray colours, and hence presumably the state of the accretion disc, at the time of observation – see e.g. van der Klis 1995), such that radio emission is strongest on the 'horizontal branch', weak on the 'normal branch' and absent on the 'flaring branch'. This relation is in agreement with (most) subsequent studies, and at face value appears to demonstrate an anti-correlation between radio emission and accretion rate in these sources, but this is almost certainly an oversimplification.

The atoll sources are in general not detected except during outbursts (e.g. Aql X-1 is an atoll source and a transient, which had a radio outburst) and/or at very high mass accretion rates (which may be the cause in the case of GX 13+1). No X-ray pulsar has ever been detected as a radio source.

These results, and their inferences, are summarised in Table 2 (adapted from Fender & Hendry 2000). It seems clear that an inner (≤ 1000 km) accretion disc, as a result of a low ($\leq 10^{10}$ G) accretor magnetic field and, most importantly and intuitively, a high mass accretion rate, are required for jet formation. In addition, it is demonstrated in Table 1 that the Z sources (plus GX 13+1) have approximately the same radio luminosity as the persistent black hole candidates when in the Low/Hard X-ray state (although the Z sources may be a litte more variable). Once again we are forced to conclude that the accretion and jet formation processes in neutron star and black hole X-ray binaries are similar (although it is noted that the Z sources are apparently a little more luminous in X-rays on average than the black holes in the Low/Hard state).

2.3. Discussion

This comparison of the radio properties of the neutron-star and black-hole X-ray binaries has revealed that whether the systems are *transient*, as a result (in most cases) of low average accretion rates and disc instability mechanisms, or *persistent*, as a result of higher average accretion rates, the coupling between radio jet formation and accretion luminosity is similar for both classes of accretor. It appears that the disc : jet coupling does not really care too much about the nature of the accretor (as long as the magnetic field is not strong enough to disrupt the accretion disc, as in the case of the X-ray pulsars).

However, this is an oversimplification of the situation, and many important questions remain. For example, the atoll sources, amongst the neutron star X-ray binaries, appear to show the strongest X-ray spectral evidence for Comptonising coronae, yet they are weak radio sources, while in the black holes the presence of the corona is (nearly) always associated with observable radio emission. Is it just a question of accretion rate, or is another factor allowing only one of the classes to form jets ? This and other questions will only be resolved by future coordinated radio and X-ray observations of neutron-star, as well as black-hole, X-ray binaries.

3. RADIO JET FORMATION AND BLACK HOLE STATE

Black hole candidate (BHC) X-ray binaries (Tanaka & Lewin 1995) are believed to consist of an accreting black hole with a mass in the range $3M_\odot \leq M_{BH} \leq 20M_\odot$ (see e.g. Charles 1999 for a review of the dynamical evidence for black holes in these systems). Four systems in our Galaxy (Cyg X-1, GX 339-4, 1E1740.7-2942 and GRS 1758-258) and two in the LMC (LMC X-1 and LMC X-3) are persistent X-ray sources, in the sense that their X-ray emission is more or less constant over time (which means, in effect, over several decades since their discoveries). Several more systems, the soft X-ray transients (SXTs), are known only to have undergone a handful of X-ray outbursts during the history of X-ray astronomy, but these systems offer the most convincing dynamical evidence for black holes. Charles (1999) lists several such systems, which include A 0620-00, GRO J0422+32, GS 2000+25 and GRS 1124-68. Generally speaking, these X-ray outbursts of the SXTs are accompanied by radio outbursts (Hjellming & Han 1995; Kuulkers et al. 1999).

It was initially realised that black holes displayed at least two spectral states, 'high' (in this paper High/Soft) and 'low' (Low/Hard), based upon the strength of their soft ($\leq$ few keV) X-ray emission (Tanaka & Lewin 1995). Additionally, the 'Very High' (Miyamoto et al. 1991) and 'Intermediate' (Méndez & van der Klis 1997) states have been identified, and an 'Off' state, corresponding to sources at very low X-ray flux levels ('quiescence' in the case of transients), introduced. In addition

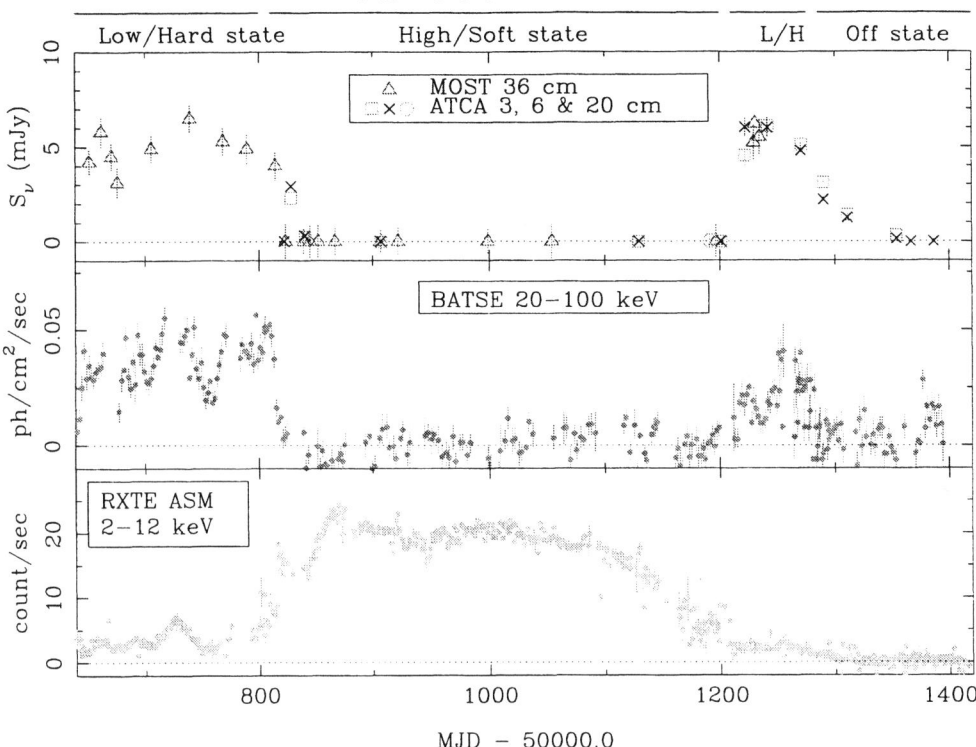

FIGURE 1. Radio, soft- and hard-X-ray observations of the Low/Hard, High/Soft and Off states in GX 339-4. From Fender et al. (1999) and Corbel et al. (2000).

Méndez, Belloni & van der Klis (1998) have shown that even unusual systems like the 'superluminal' transient GRO J1655-50 display these 'canonical' states.

The physical interpretation of the different states is based upon an origin for the soft X-ray component in an accretion disc and the hard X-ray component via Comptonisation of softer photons in a corona of high-energy electrons. The most popular current models invoke a truncated accretion disc in the Low/Hard X-ray state, inside of which there may be an advection-dominated flow (e.g. Esin et al. 1998), and the presence of a strong Comptonising corona. In the High/Soft state the disc component dominates and the corona is smaller and cooler. In the Very High state both the disc and corona are strong; the Intermediate state may be similar but at a lower overall luminosity.

3.1. State transitions

How does radio emission relate to black hole state and/or state transitions? For the transient sources the radio emission is generally (although not always) associated with one or more discrete ejections, the first (and generally foremost) of which is associated with the rapid rise (corresponding to something like a Off→High/Soft state transition in the space of a day or so) of X-ray emission at the time of the outburst. The evidence for discrete ejections comes from both direct imaging and the radio light curves (discrete peaks) and radio spectra (rapidly evolving to optically thin and subsequently decaying) observed (e.g. Kuulkers et al. 1999). As well as the transients, the persistent sources Cyg X-1 (Zhang et al. 1997) and maybe also GX 339-4 (Corbel et al. 2000) show flaring at times of state transitions.

3.2. Off and Low/Hard states

We know from the X-ray transients that radio emission is pretty weak (often undetectable) during X-ray quiescence ≡ Off state. This is supported by weak radio detections of GX 339-4 when at very low X-ray flux levels (Corbel et al. 2000; Fig. 1).

Cyg X-1 and GX 339-4 are generally observed in the Low/Hard X-ray state, and reveal very similar characteristics, namely weak (few mJy at cm wavelengths) but steady radio emission with a flat/inverted radio spectrum (i.e. spectral index $\alpha = \Delta \log S_\nu / \Delta \log \nu \gtrsim 0$) which is correlated with both soft and hard X-ray emission over approximately a factor of two in intensity (Brocksopp et al. 1999 and references therein; Hannikainen et al. 1998; Corbel et al. 2000). The radio emission from Cyg X-1 is additionally modulated at the 5.6-day orbital period (Pooley, Fender & Brocksopp 1999; Brocksopp et al. 1999), probably due to free-free absorption in the dense stellar wind of the massive companion star. Models for flat/inverted radio emission generally invoke a partially self-absorbed outflow of the kind originally envisaged for AGN cores by Blandford & Königl (1979). This is supported by the large size scale required (a sphere with radius greater than the binary separation of Cyg X-1 is required to generate the observed cm-wavelength emission). Final confirmation of the jet hypothesis has come from VLBA observations of an extension on mas-scales from Cyg X-1 (Stirling et al. 1998; Stirling et al. in prep).

As well as the persistent sources Cyg X-1 and GX 339-4, some X-ray transients also spend extended periods in the Low/Hard state (e.g. GRO J0422+32, GS 2023+338/V404 Cyg). These systems also display a flat radio spectrum with comparable luminosity to Cyg X-1 and GX 339-4, when in the Low/Hard state (Fender, in prep).

3.3. The High/Soft state

It was already suspected that the radio flux density in Cyg X-1 dropped significantly when the system was in the High/Soft X-ray state (Brocksopp et al. 1999 and references therein). Dramatic confirmation that this was a phenomenon shared by other BHCs came from long-term monitoring of GX 339-4 at radio, soft- and hard-X-ray energies, in which it was found that the radio emission from the system was reduced by a factor ≥ 25 during a year-long period in the High/Soft state (Fender et al. 1999). Figure 1 demonstrates this result, encompassing Low/Hard,

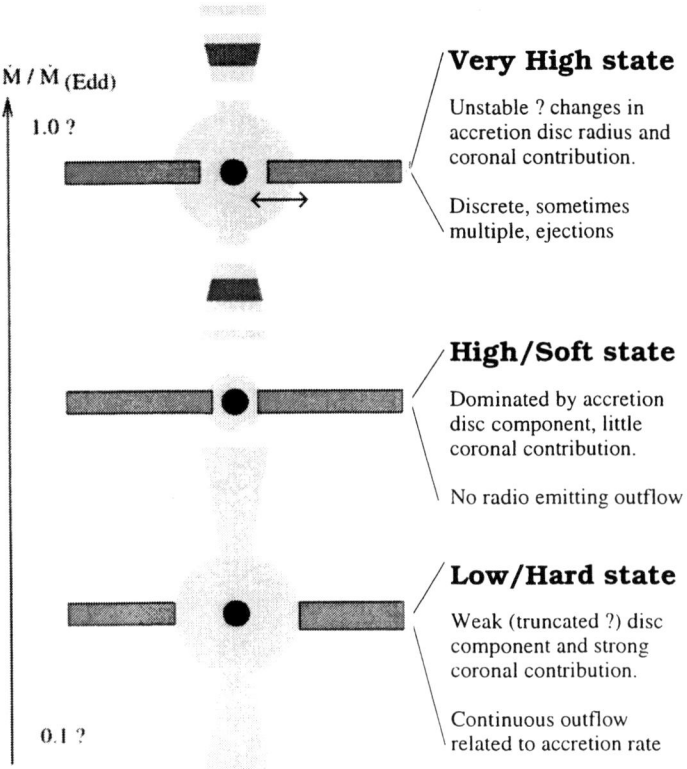

FIGURE 2. A summary of the empirical relation between models for BHC X-ray states and observed radio emission

High/Soft, Low/Hard and finally Off states during the period of our monitoring program (Corbel et al. 2000), and dramatically reveals the extremely strong correlation between hard X-ray and radio emission. Is the outflow physically suppressed during this state or cooled (via inverse Compton losses) so rapidly by the more luminous disc that the electrons no longer produce synchrotron emission ? The answer to this is unclear, but at least one (naive) argument, the evidence for radio emission during the (even more luminous) Very High State (see below), suggests physical suppression of the jet.

Note that while many X-ray transients appear, at face value, to be radio-bright whilst in the High/Soft state, more careful scrutiny reveals that the radio emission is usually just the decaying tail of emission from the flare event at the start of the outburst (i.e. at the state transition), which is by then physically decoupled from the system.

State	Radio properties
Very High	Bright ejections, spectral evolution from absorbed $\to$ opt. thin
High/Soft	Radio suppressed by factor ≥ 25
Intermediate	?
Low/Hard	Low level, steady, flat spectrum extending to at least sub-mm
Off	Weak; similar to Low/Hard but reduced $\propto \dot{m}$

TABLE 2. The relation of radio emission to black hole X-ray state.

3.4. The Very High (and Intermediate) state(s)

The Very High State is much rarer than the Low/Hard or High/Soft states, and few clear examples exist of radio observations of this state. GX 339-4, while having previously entered this state (Miyamoto et al. 1991), has not done so since radio monitoring began. GS 1124-683 also spent an extended period in the Very High State (Miyamoto et al. 1993), during which time there *was* radio coverage (for at least 30 days – Kuulkers et al. 1999 and references therein), revealing fairly bright and variable emission. However, the radio coverage does not appear to have been good enough to distinguish between rapid radio variability (as in GRS 1915+105, see below), or simply two large, discrete ejections.

Belloni (1998) has suggested that GRS 1915+105, when exhibiting its highly variable 'dipping' behaviour, may be in the Very High State. This pattern of behaviour is associated with bright and repeated flare events with a flat spectrum from the radio – mm – infrared, which are most likely associated with discrete ejection events $\equiv$ transient jet formation (Fender & Pooley 1998 and references therein). The intermediate state may be similar to the Very High State, but at lower luminosity levels, but again there has been no good radio coverage of this state.

Belloni et al. (1996) suggested that the soft state transition of Cyg X-1 in 1996 was only to the 'Intermediate' state, and that Cyg X-1 may never have achieved the 'genuine' High/Soft state observed in transients. This is of interest as there is evidence that in soft states Cyg X-1 undergoes suppression of the radio emission similar to that observed in GX 339-4 (Brocksopp et al. 1999 and references therein).

3.5. Discussion

We are now in a position to be able to characterise the relation of radio emission from BHC X-ray binaries to their 'state' as defined observationally by their X-ray spectral and timing properties, except for the Intermediate state, for which there is no clear case of simultaneous radio observations. This relation is summarised in Fig. 2 and Table 2. Several implications are quite clear:

- The Low/Hard state supports a quasi-continuous outflow whose contribution to the overall energetics of the system is likely to be significant. In this state there is a three-way correlation between radio, soft- and hard-X-rays, which may reflect small changes in the accretion rate. The Off state may simply be the Low/Hard state 'turned down' to lower accretion rates.

- In the High/Soft state the radio emission drops below detectable levels, probably corresponding to the physical disappearance of the jet. It is this state, in both radio and X-ray emission, which is most different to the other states.

- Rapid changes in the accretion disc radius (or whatever it is that is physically changing which is currently modelled as an inner disc radius!) correspond to discrete ejections, which appear as radio (and sometimes mm and infrared) flares. This mechanism may be operating in analogous ways in both general state transitions and in the Very High State.

- There is an extremely strong correlation between radio and hard X-ray emission. The radio emission has been *directly observed* to arise in outflows and to originate in synchrotron emission from a population of high-energy electrons. The hard X-ray emission is *inferred* to arise via Comptonisation by a similar population of electrons (albeit the low-energy tail of the distribution). Furthermore both components, the jet and corona, are believed to originate at the centre of the accretion disc, in the vicinity of the black hole. By far the simplest interpretation therefore is that the Comptonising corona, at least in the case of BHC systems, is simply the base of the jet.

A simple observational relation between black hole state and radio emission, and hence accretion and jet formation, has now been established for the BHC systems. Furthermore all the evidence points to the Comptonising corona in these systems being physically related to the presence of a jet. The next stage will be to investigate these relations quantitatively.

ACKNOWLEDGEMENTS

RPF would like to thank Mariano Méndez, Eric Ford, Jeroen Homan, Erik Kuulkers and Michiel van der Klis for stimulating discussions.

REFERENCES

Belloni T., 1998, New Astronomy Reviews, 42, 585
Belloni T., Mendez M., van der Klis M., Hasinger G., Lewin W.H.G., van Paradijs J., 1996, ApJ, 472, L107
Blandford R., Königl A., 1979, ApJ, 232, 34
Brocksopp C., Fender R.P., Larionov V., Lyuty V.M., Tarasov A.E., Pooley G.G., Paciesas W.S., Roche P., 1999, MNRAS, 309, 1063
Charles P.A., 1999, 'Black Holes in our Galaxy : observations', In : Abramowicz, M. A., Björnsson, G., Pringle, J. E. (Eds), Theory of Black Hole Accretion Discs, Cambridge Contemporary Astrophysics, CUP, 1998, p.1

Corbel S., Fender R.P., Tzioumis A.K., McIntyre V., Nowak M., Sood R., Durouchou P., Campbell-Wilson D., 2000, A&A, submitted

Esin A.E., Narayan R., Cui W., Grove J.E., Zhang S.N., 1998, ApJ, 505, 854

Fender R.P., 2000, 'Relativistic Jets from X-ray binaries' (**astro-ph/9907050**)

Fender R.P., Pooley G.G., 1998, MNRAS, 300, 573

Fender R.P., Hendry M.A., 2000, MNRAS, 317, 1

Fender, R., Corbel, S., Tzioumis, T., McIntyre, V., Campbell-Wilson, D., Nowak, M., Sood, R., Hunstead, R., Harmon, A., Durouchoux, P., Heindl, W., ApJ, 1999, 519, L165

Hannikainen D.C., Hunstead R.W., Campbell-Wilson D., Sood R.K., 1998, A&A, 337, 460

Hjellming, R. M., Han, X., 1995, Radio properties of X-ray binaries, In : Lewin, W. H. G., van Paradijs, J., van der Heuvel, E. P. J. (Eds.), X-ray binaries, Cambridge University Press, Cambridge, 308–330

Kuulkers, E., Fender, R. P., Spencer, R. E., Davis, R. J., Morison, I., 1999, MNRAS, 306, 919

Méndez M., van der Klis M., 1997, ApJ, 479, 926

Méndez M., Belloni T., van der Klis M., 1998, ApJ, 499, L187

Mirabel, I.F., Rodríguez, L.F., 1999, ARA&A, **37**, 409

Miyamoto S., Kimura K., Kitamoto S., Dotani T., Ebisawa K. 1991, ApJ, 383, 784

Miyamoto S., Iga S., Kitamoto S., Kamado Y., 1993, ApJ, 403, L39

Penninx W., Lewin W.H.G., Zijlstra A.A., Mitsuda K., van Paradijs J., 1988, Nature, 336, 146

Pooley G.G., Fender R.P., Brocksopp C., 1999, MNRAS, 302, L1

Stirling, A., Spencer, R., Garrett, M., 1998, New Astronomy Reviews, 42, 657

Tanaka Y., Lewin W.H.G., 1995, Black-hole binaries, In : Lewin, W. H. G., van Paradijs, J., van der Heuvel, E. P. J. (Eds.), X-ray binaries, Cambridge University Press, Cambridge, 308–330

Van der Klis M., 1995, Rapid aperiodic variability in X-ray binaries, In : Lewin, W. H. G., van Paradijs, J., van der Heuvel, E. P. J. (Eds.), X-ray binaries, Cambridge University Press, Cambridge, 252–307

Zhang S.N., Mirabel I.F., Harmon B.A., Kroeger R.A., Rodríguez L.F., Hjellming R.M., Rupen M.P., 1997, Galactic Black Hole Binaries : Multifrequency Connections, In : Dermer C.D., Strickman M.S., Kurfess J.D. (Eds.), Proc. 4th Compton symposium, AIP, p.141

THE BEPPOSAX HELLAS SURVEY: ON THE NATURE OF FAINT HARD X-RAY SELECTED SOURCES

F. Fiore[1,2,3], L.A. Antonelli[2], P. Ciliegi[4], A. Comastri[4], P. Giommi[1],
F. La Franca[5], R. Maiolino[6], G. Matt[5], S. Molendi[7], G.C. Perola[5], C. Vignali[4]

1) BeppoSAX Science Data Center, Via Corcolle 19, I–00131 Roma, Italy;
2) Osservatorio Astronomico di Roma, Via Frascati 33, I-00044 Monteporzio, Italy;
3) Harvard-Smithsonian Center of Astrophysics, 60 Garden Street, Cambridge MA 02138 USA;
4) Osservatorio Astronomico di Bologna, via Ranzani 1, I40127 Bologna, Italy;
5) Dipartimento di Fisica, Università degli Studi "Roma Tre", Via della Vasca Navale 84, I–00146 Roma, Italy;
6) Osservatorio Astrofisico di Arcetri, p. E. Fermi 5, Firenze, Italy;
7) IFCTR/CNR, via Bassini 15, Milano, I20133, Italy

ABSTRACT The BeppoSAX 4.5-10 keV High Energy Large Area Survey has covered about 80 deg^2 of sky down to a flux of $F_{5-10keV} \sim 5 \times 10^{-14}$ erg cm^{-2} s^{-1}. Optical spectroscopic identification of $\lesssim$ half of the sources in the sample (62) shows that many ($\approx 50\%$) are highly obscured AGN, in line with the predictions of AGN synthesis models for the hard X-ray background (XRB, see e.g. Comastri et al. 1995). The X-ray data, complemented by optical, near-IR and radio follow-up, indicate that the majority of these AGN are "intermediate" objects, i.e. type 1.8-1.9 AGN, 'red' quasars, and even a few broad line, blue continuum quasars, obscured in X-rays by columns of the order of $10^{22.5-23.5}$ cm^{-2}, but showing a wide dispersion in optical extinction. The optical and near-IR photometry of the obscured objects are dominated by galaxy starlight, indicating that a sizeable fraction of the accretion power in the Universe may actually have been missed in optical color surveys. This also implies that multicolor photometry techniques may be efficiently used to assess the redshift of the hard X-ray selected sources.

KEYWORDS: X–ray: selection – background – galaxies – AGN

1. INTRODUCTION

Hard X-ray observations are the most efficient way of tracing emission due to accretion mechanisms, such in Active Galactic Nuclei (AGN). Hard X-ray selection is less affected by strong biases present at other wavelengths. For example, a column of a few times 10^{22} cm^{-2} has negligible effect in the 5-10 keV band, while it reduces by ~ 100 times nuclear emission below 2 keV. Soft X-ray surveys (e.g. Hasinger et al. 1998, Schmidt et al. 1998) are also often contaminated by non nuclear components, like emission from binaries and/or from optically thin plasmas in star-formation regions surrounding the nucleus. Optical and UV color selection is biased against objects with even modest extinction or an intrinsically 'red' emission spectrum (see e.g. Vignali et al. 2000). Sensitive hard X-ray surveys are therefore powerful tools to select large samples of AGN less biased against absorption and extinction. Our approach consists in taking advantage of the large field of view and good sensitivity of the BeppoSAX MECS instrument (Boella et al. 1997a,b) to survey tens to hundreds of square degrees at fluxes $\gtrsim 5 - 10 \times 10^{-14}$ erg cm^{-2} s^{-1} (Fiore et al. 2000a), and using higher sensitivity XMM-Newton and Chandra observations to extend the survey down to $\sim 10^{-14}$ erg cm^{-2} s^{-1} on several deg^2 (at this flux the majority of the hard XRB is resolved in sources). The results on the optical identification of a sample of faint Chandra sources discovered over the first 0.14 deg^2 have been published by Fiore et al. (2000b). This approach is complementary to deep pencil beam surveys ($\sim$ 0.1 deg^2, see e.g. Mushotzky et al. 2000, Hornschemeier et al. 2000), and we cover a different portion of the redshift–luminosity plane. Our purpose is to study cosmic source populations at fluxes where a reasonably large fraction of the hard XRB is resolved (20-30% at the BeppoSAX flux limit), but where the X-ray flux is high enough to provide X-ray spectral information in higher sensitivity follow-up observations. This would allow the determination of the distribution of absorbing columns in the sources making the hard XRB, providing strong constraints on AGN synthesis models for the XRB. Furthermore, large area surveys allow the search for previously 'rare' AGN, like 'red' quasars or other minority AGN populations (Kim & Elvis 1999) and quantify their fractional contribution to the AGN family. Finally, at our X-ray flux limits the optical counterparts are bright enough to allow relatively high quality optical spectroscopy, useful to investigate the physics of the sources.

2. THE HELLAS SURVEY

The High Energy Large Area Survey (HELLAS) has been performed in the 4.5-10 keV band because: a) this is the band closest to the maximum of the XRB energy density which is reachable with the current imaging X-ray telescopes, and b) the BeppoSAX MECS Point Spread Function (PSF) greatly improves with energy, providing a 95 % error radius of 1' in the hard band (Fiore et al. 2000a), allowing optical identification of the X-ray sources. About 80 deg^2 of sky have been surveyed so far using 142 BeppoSAX MECS fielda at $|b| > 20$ deg. (Fields centered on bright extended sources and bright Galactic sources were excluded from the survey, as well as fields close to LMC, SMC and M33.)

A robust detection algorithm has been used in coadded MECS1+MECS2+MECS3 (or MECS2+MECS3 for observations performed after the failure of the MECS1 unit) 4.5-10 keV images. The method consists in first convolving the X-ray image with a wavelet function, to smooth the image and increase contrast, and then in running a standard slide-cell detection method on the smoothed image, to locate count excesses above the local background. The statistics of each candidate detection is then accurately studied and the final net counts are estimated from the original (un-smoothed) image to preserve Poisson statistics. The background is calculated using source-free boxes near the source region. The detection has been run several time for each field, changing the size of the wavelet function, to a) take into account the variation of the MECS PSF with the offaxis angle, and b) to detect efficiently sources with variable extension. The quality of the detection has always been checked interactively. We used a probability threshold of 99.94 % (about 3.5 σ). Sources detected in regions of radius 4, 6 or 8 arcmin around targets (depending on the target brightness) have been excluded from the sample, which includes a total of 147 sources down to a 5-10 keV flux of $\sim 5 \times 10^{-14}$ erg cm^{-2} s^{-1}. Count rates were converted to fluxes using a fixed conversion factor equal to 7.8×10^{-11} erg cm^{-2} s^{-1} (5-10 keV flux) per one "3 MECS count" (4.5-10 keV). This factor is appropriate for a power law spectrum with $\alpha_E = 0.6$, but due to the narrow band it is not strongly sensitive to the spectral shape: for $\alpha_E = 0.4$ and 0.8 it is 8.1 and 7.6 $\times 10^{-11}$ erg cm^{-2} s^{-1} respectively. The skycoverage varies from ~ 1 deg^2 at 5×10^{-14} erg cm^{-2} s^{-1} to 6.6, 50 and 84 deg^2 at fluxes of $10^{-13}, 3 \times 10^{-13}$ and $> 10^{-12}$ erg cm^{-2} s^{-1} respectively. After correcting for this skycoverage we find 16.9±3.0(6.4) sources deg^{-2} at $F_{5-10keV} = 4.8 \times 10^{-14}$ erg cm^{-2} s^{-1} (Fiore et al. 2000a, Comastri et al. 2000). First quoted errors are the 1σ statistical confidence interval. Errors in brackets include systematic uncertainties, due to the lack of knowledge of the real spectrum of the faint sources. This logN-logS corresponds to a resolved fraction of the 5-10 XRB equal to 20-30 %, depending on the XRB normalization (Comastri this meeting, Vecchi et al. 1999). The observed number counts are consistent, within the errors, with the extrapolations from the BeppoSAX (Giommi et al 2000) and ASCA (Cagnoni et al. 1998, Ueda et al. 1999, Della Ceca et al. 1999) 2-10 keV number counts, assuming an average power law spectrum with $\alpha_E \sim 0.6$.

3. HARDNESS RATIO ANALYSIS

To study the spectral variety of the HELLAS sources we have calculated for each source the softness ratio (S-H)/(S+H), S=1.3-4.5 keV, H=4.5-10 keV. Figure 1 plots (S-H)/(S+H) as a function of source total 5-10 keV flux. Sources under or close to the berillium strongback supporting the MECS window have been excluded from this analysis, because their observed softness may be systematically lower than real. The number of the remaining sources in the sample is 126. Many of the HELLAS sources have a low (S-H)/(S+H), indicating a hard spectrum. Assuming $\alpha_E = 0.6$, we find that 36(5) of the 126 sources have 1.3-4.5 keV count rates lower than that expected at confidence level $\gtrsim 95\% (\gtrsim 99.7\%)$. Large absorbing columns densities are likely responsible for the hard spectrum of these sources. A deficit of very hard sources at fluxes $\lesssim 1 - 2 \times 10^{-13}$ erg cm^{-2} s^{-1} is also evident in figure 1. This can be due to both an astrophysical and a technical reason. The first is a redshift effect: the observed softness ratio of sources with similar intrinsic absorbing column density increases with the redshift, as the observed cut-off energy moves toward lower energies. The second is the MECS reduced sensitivity to hard sources, due to the rapid increase of the vignetting of the telescopes with the energy and with

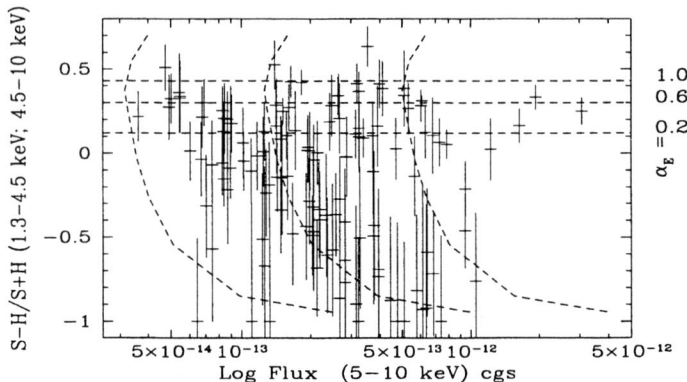

FIGURE 1. The softness ratio (S-H)/(S+H) versus the 5-10 keV flux for the 126 sources in the HELLAS sample not covered by the strongback support of the MECS window. Dashed lines mark loci of equal 4.5-10 keV count rate

the off-axis angle. To quantify the latter effect we have computed loci of equal 4.5-10 keV count rate for a given flux, shown by dashed lines in figure 1. The strong curvature of these lines toward low values of (S-H)/(S+H) indicate that a large part of the deficit of faint hard sources is probably due to this effect. The curves are bent toward high flux values at high (S-H)/(S+H) too, because the MECS sensitivity is reduced for very soft sources by the berillium window, which absorbs most photons below ~ 2 keV. The 4.5-10 keV sensitivity is maximum for an unabsorbed power law spectrum of $\alpha_E = 0.6 - 0.8$.

4. OPTICAL IDENTIFICATIONS

Correlations of the HELLAS source catalog with catalogs of cosmic sources provide 26 coincidences (7 radio-loud AGN, 13 radio-quiet AGN, 6 clusters of galaxies), suggesting that most of the HELLAS sources are AGN. Optical spectroscopic follow-ups have been performed on about 45 HELLAS error-boxes, providing 36 new identifications (Fiore et al. 1999, La Franca et al. 2000 in preparation). The radio-quiet AGN sample includes:
  i) 28 broad line blue continuum quasars
  ii) 2 broad line 'red' continuum quasars
  iii) 14 type 1.8-1.9-2 AGN
  iv) 5 sources have optical spectra typical of LINERS of starburst galaxies (all have strong [OII] emission).
• The number of chance coincidences in 45 error boxes is $< 0.8 - 1.5$ for broad line quasar and < 4 for narrow line AGN (including in this category strong [OII]

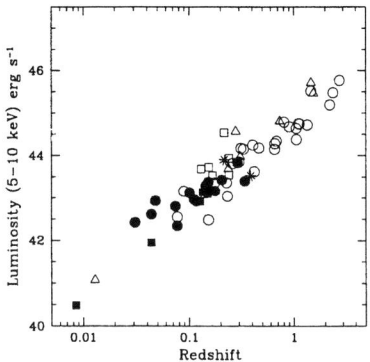 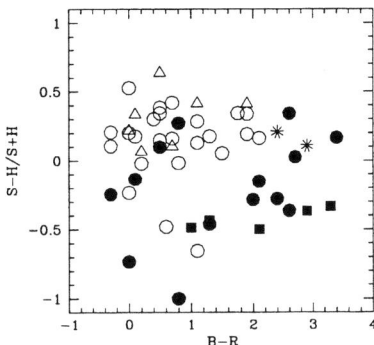

FIGURE 2. a) The luminosity of the HELLAS sources as a function of their redshift. Different symbols mark different sources: open circles = broad line, 'blue' continuum quasars and Sy1; stars= broad line 'red' continuum quasars; filled circles= type 1.8-1.9-2.0 AGN; filled squares= starburst galaxies and LINERS; open triangles= radio-loud AGN; open squares= clusters of galaxies. A cosmology with $H_0 = 70$ and $q_0 = 0.5$ has been used. b) The (S-H)/(S+H) softness ratio versus the B–R color

TABLE 1. Mean (S-H)/(S+H) and B–R of HELLAS radio-quiet AGN

	$< B - R >$	σ_{B-R}	$< (S-H)/(S+H) >$	$\sigma_{(S-H)/(S+H)}$
total AGN (49)	1.2±0.2	1.0	-0.03±0.05	0.34
broad line AGN (28)	0.8±0.1	0.7	0.11±0.05	0.26
narrow line AGN (19)	1.7±0.3	1.1	-0.26±0.08	0.34

emission line galaxies like LINERS and starburst galaxies).
• At least 5 of the narrow emission line AGN lie in small groups and/or in interacting couples.
• About 1/3 of the error-boxes studied in detail contains no 'reasonable' (in terms of X-ray to optical ratio of known classes of sources) counterpart to the X-ray source down to R=20.5
• Optical spectroscopy indicates a wide variety of spectra, with a large fraction of "intermediate" objects (type 1.8-1.9 AGN, 'red' quasars).

Figure 2a) plots the luminosity of the identified HELLAS sources as a function of their redshift. We have identifications of broad line quasars up to z=2.76 and luminosity of $\sim 10^{46}$ erg s^{-1}, and of narrow line AGN up to z=0.4 and luminosity of $\sim 10^{44}$ erg s^{-1}. Figure 2b) plots the (S-H)/(S+H) of the identified sources as a function of their B – R color. Note as the broad line AGN are relatively well separated from the narrow line AGN in the diagram, although the scatter in both

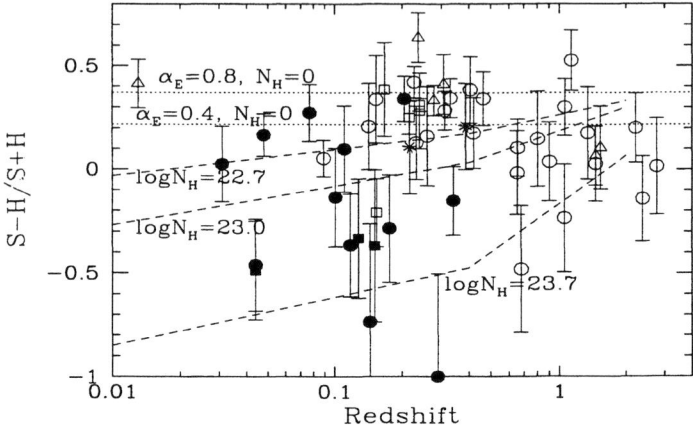

FIGURE 3. (S-H)/(S+H) versus the redshift for the identified sources. Symbols as in figure 2. Dotted lines show the expected softness ratio for a power law model with $\alpha_E = 0.4$ (lower line) and $\alpha_E = 0.8$ (upper line). Dashed lines show the expectations of absorbed power law models (with $\alpha_E = 0.8$ and $\log N_H = 23.7, 23.0, 22.7$, from bottom to top) with the absorber at the source redshift.

B − R and (S-H)/(S+H) is large for both class of AGN (also see Table 1). Both (S-H)/(S+H) and B − R of broad line AGN are different (at the ∼ 3σ level) from that of narrow line AGN, which, on average, have a lower (S-H)/(S+H), and are therefore likely to be more X-ray absorbed, and an higher B − R, and are therefore subjected to a greater extinction.

(S-H)/(S+H) is plotted as a function of the redshift in figure 3 for the 53 identified sources detected far from the berillium strongback supporting the MECS window. The dotted lines represent the expectation of unabsorbed power law with $\alpha_E = 0.4$ and 0.8. The dashed lines represent the expectations of a power law absorbed by columns of 5×10^{22}, 10^{23} and 5×10^{23} cm^{-2} respectively, in the source frame. Note that the softness ratios of constant column density models strongly increases with the redshift. Most of the narrow line AGN have (S-H)/(S+H) inconsistent with that expected from a power law model with $\alpha_E = 0.4$. Absorbing columns, of the order of $10^{22.5-23.5}$ cm^{-2}, are most likely implied. Note also that some of the broad line AGN have (S-H)/(S+H) inconsistent with that expected for a $\alpha_E = 0.8$ power law, in particular at high redshift. The (S-H)/(S+H) of the 24 broad line AGNs is marginally anticorrelated with z (Spearman rank correlation coefficent of -0.364 for 22 degrees of freedom, corresponding to a probability of 92%). The number of sources is not large enough to reach a definite conclusion,

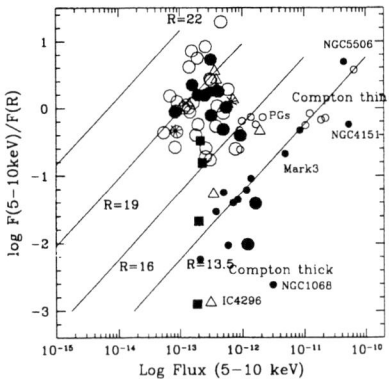 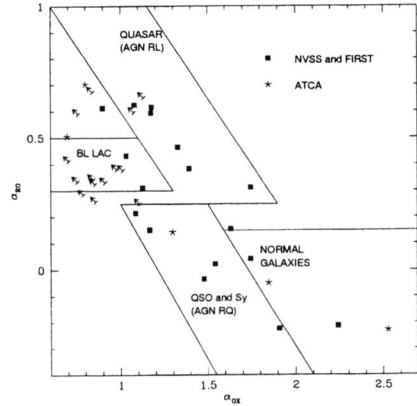

FIGURE 4. a) X-ray to optical ratio versus the 5-10 keV flux. HELLAS AGN: big symbols; known AGN: small symbols (symbols as in figure 2). In Compton thick AGN ($\log N_H > 24.3$), the nuclear emission is completely blocked in the 5-10 keV band. Thick lines identify regions of constant apparent R magnitude. b) The α_{ro} versus the α_{ox} for the HELLAS sources with a radio counterpart.

but it is interesting to note that this correlation goes in the opposite direction than expected. In fact, the ratio of the optical depth in the optical band, due to dust extinction, to that in the X-ray band, due to photoelectric absorption, should scale as $(1+z)^4$. Highly X-ray obscured broad line blue continuum quasar can exist only if their dust to gas ratio or their dust composition strongly differs from the Galactic one (also see Maiolino, this meeting). Similar results have been recently found in ASCA samples by Akiyama et al. (1999) and Della Ceca et al. (this meeting). XMM-Newton and Chandra follow-up observation may easily confirm or disregard a significant absorbing column in these high z broad line quasars.

5. SPECTRAL ENERGY DISTRIBUTION

5.1. X-ray to optical color

Figure 4 shows the hard X-ray (5-10 keV) to optical (R band) flux ratio as a function of the X-ray flux for the identified HELLAS sources and a sample of relatively bright, nearby AGN observed by BeppoSAX (Seyfert 1 galaxies, Seyfert 2 galaxies, PG quasars with z<0.4). The X-ray to optical ratio of the HELLAS sources is similar to that of the X-ray brightest objects in the local universe (with the exception of X-ray selected blazars, like the HBL, which have higher X-ray to optical ratio, but also a relatively strong radio emission). While supporting the robustness of our identifications, this suggests that roughly one third of the hard X-ray background is due to sources similar to local Seyferts and quasars.

5.2. Near infrared to optical colors

Photometric infrared and optical observations of 10 HELLAS sources have been carried out using the Telescopio Nazionale Galileo (TNG; Maiolino et al. 2000). The sample includes 4 broad line 'blue' quasars, 2 broad line 'red' quasars, 3 type 1.9 AGN and 1 LINER. The B, R and J photometry of the 2 'red' quasars and the 4 narrow line galaxies is dominated by the emission from the host galaxy. AGN contribution is observed in the K band, especially in the 2 'red' quasar. This means that many, if not most of the objects making the hard X-ray background cannot be distinguished from normal galaxies using optical and near-IR photometry. In fact only the 4 broad line blue quasar would have passed the color criterion of the PG catalog, thus indicating that a large fraction of the accretion power in the Universe may actually have been missed in optical color survey such as the BQS (U − B < −0.44). Multicolor photometry techniques based on galaxy templates (e.g. Giallongo et al. 1998) may be efficiently used to assess the redshift of the hard X-ray selected sources. This will be more and more important when large samples of faint X-ray sources will be available from Chandra and XMM observations, and optical spectroscopic identification of all of them will not be feasible.

5.3. Radio to optical to X-ray broad band spectral indices

Nearly all the 147 HELLAS sources have been observed in either the NVSS and FIRST surveys or by our collaboration using ATCA. In particular, we observed with ATCA 20 sources obtaining 8 5σ detections (40 % of the sample) at a 5 GHz flux limit of 0.5 mJy. This fraction is higher than in Radio follow-ups of Einstein and ROSAT X-ray surveys (Ciliegi et al. 2000 in preparation). This is mainly due to the lower radio-to-X-ray flux limit ratio in our survey. A similar fraction of radio detections was found by Akiyama et al. (2000) in their correlation of the ASCA Large Sky Survey sources with the FIRST catalog, which have a radio-to-X-ray flux limit ratio similar to ours. Figure 4b) shows the radio-to-optical broad band spectral index as a function of the X-ray-to-optical index (arrows identify sources without a R<20 counterpart in the small Radio error box). Note that many of the HELLAS detections have α_{RO} and α_{XO} consistent with that of radio quiet AGN.

6. CONCLUSIONS

A large area, hard X-ray survey performed with BeppoSAX has found a population of faint obscured AGN. The X-ray data, complemented by optical, near-IR and radio follow-ups, indicate that the majority of these sources are "intermediate" AGN i.e. type 1.8-1.9 AGN, 'red' quasars, obscured in X-ray by columns of the order of $10^{22.5-23.5}$ cm^{-2}, but showing a wide dispersion in optical extinction. The sample of identified HELLAS AGN contains higher redshift analogs of nearby Seyfert galaxies and quasars. Furthermore, we find marginal evidence for a population of X-ray obscured quasars at $z \gtrsim 0.5$ showing broad lines in their optical spectra.

ACKNOWLEDGEMENTS

We thank the BeppoSAX SDC, SOC and OCC teams for the successful operation of the satellite and preliminary data reduction and screening. This research has been partly supported by ASI ARS/99/75 contract and MURST Cofin-98-032 contract.

REFERENCES

Akiyama et. al. 2000, ApJ, 532, 700
Boella, G. et al. 1997, A&AS, 122, 299
Boella, G. et al. 1997, A&AS, 122, 327
Comastri, A., Setti, G., Zamorani, G. & Hasinger, G. 1995, A&A, 296, 1
Comastri, A., Fiore, F., Vignali, C., La Franca, F. & Matt, G. 2000, in "Large Scale Structure in the X-ray Universe", M. Plionis & I. Georgantopoulos eds., Atantisciences 2000, astro-ph/0001037
Cagnoni, I., Della Ceca, R. & Maccacaro, T. 1998, ApJ, 493, 54
Della Ceca, R., Castelli, G., Braito, V., Cagnoni, I., & Maccacaro, T. 1999 ApJ, 524, 674
Fiore, F. et al. 1999, MNRAS, 306, L55
Fiore, F. et al. 2000a, MNRAS, submitted
Fiore, F. et al. 2000b, New Astronomy in press, astro-ph/0003273
Giallongo, E. et al. 1998, AJ, 115, 2169.
Giommi, P., Perri, M. & Fiore, F. 2000, A&A in press, astro-ph/0006333
Hasinger, G. et al. 1998, A&A, 329, 482
Hornschemeier, A.E. et al. 2000, ApJ in press, astro-ph/0004260
Kim, D.-W. & Elvis, M. 1999, ApJ, 516, 9
Maiolino, R. et al. 2000, A&A, 335, L47
Mushotzky, R.F., Cowie, L.L., Barger, A.J., Arnaud, K.A. 2000, Nature, 404, 459
Schmidt, M. et al. A&A 329 495
Ueda, Y. et al., 1999, ApJL, 524, L11
Vecchi, A., Molendi, S. Guainazzi, M., Parmar, A. & Fiore F. 1999 A&A, 349, L73
Vignali, C. et al. 2000, MNRAS, 313, L11

RELATIVISTIC FLOWS IN BLAZARS

Gabriele Ghisellini

Osservatorio Astronomico di Brera, Via Bianchi 46, I–23807 Merate, Italy

ABSTRACT The radiation we observe from blazars is most likely the product of the transformation of bulk kinetic energy into random energy. This process must have a relatively small efficiency (e.g. 10%) if jets are to power the extended radio–structures. Recent results suggest that the average power reaching the extended radio regions and lobes is of the same order of that produced by accretion and illuminating the emission line clouds. Most of the radiative power is produced in a well localized region of the jet, and, at least during flares, is mainly emitted in the γ–ray band. A possible scenario qualitatively accounting for these facts is the internal shock model, in which the central engine produces a relativistic plasma flow in an intermittent way.

KEYWORDS: Jets, AGNs, blazars, radiation processes: synchrotron, inverse Compton, electron–positron pairs

1. INTRODUCTION

We believe that the continuum radiation we see from blazars comes from the transformation of bulk kinetic energy, and possibly Poynting flux, into random energy of particles, which quickly produce beamed emission through the synchrotron and the inverse Compton process. This is analogous to what we believe is happening in gamma–ray bursts, although the bulk Lorentz factor of their flow is initially larger.

Evidences for bulk motion in blazars with Lorentz factors between 5 and 20 have been accumulated along the years, especially through the monitoring of superluminally moving blobs on the VLBI scale (Vermeulen & Cohen 1994), and, more recently, through the detection of very large variable powers emitted above 100 MeV (see the third EGRET catalogue, Hartman et al., 1999), which require beaming for the source to be transparent to photon–photon absorption (e.g. Dondi & Ghisellini, 1995).

The explanation of intraday variations of the radio flux, leading to brightness temperatures in excess of $T_B = 10^{18}$ K (much exceeding the Compton limit) are instead still controversial (Wagner & Witzel 1995). Interstellar scintillation is surely involved, but it can work only if the angular diameter of the variable sources is so small to nevertheless lead to $T_B = 10^{15}$ K, which requires either a coherent process to be at work (e.g. Benford & Lesch 1998) or a Doppler factor of the order of a thousand.

Another controversial issue is the matter content of jets. We still do not know if they are dominated by electron–positron pairs or by normal electron–proton plasma (see the reviews by Celotti, 1997, 1998).

Part of our ignorance comes from the difficulty of estimating intrinsic quantities, such as the magnetic field and the particle densities, using the observed flux, which is strongly modified by the effects of relativistic aberration, time contraction and blueshift, all dependent on the unknown plasma bulk velocity and viewing angle. Furthermore it is now clear (especially thanks to multiwavelength campaigns) that the blazar phenomenon is complex.

On the optimistic side, we have for the first time a complete information of the blazar energy output, after the discovery of their γ–ray emission, and some hints on the acceleration process, through the behaviour of flux variability detected simultaneously in different bands (see the review by Ulrich, Maraschi & Urry 1997). Also, blazar research can now take advantage of the explosion of studies regarding gamma–ray bursts, which face the same problem of how to transform ordered to random energy to produce beamed radiation (for reviews: Piran 1999; Meszaros 1999).

2. ACCRETION = ROTATION?

Despite the prediction that jets carry plasma in relativistic motion dates back to 1966 (Rees, 1966), and intense studies over the last 20 years (Begelman, Blandford & Rees, 1984), quantitative estimates of the amount of power transported in jets have been done only relatively recently, following new observational results.

One important point is that the extended (or lobe) radio emission of radiogalaxies and quasars traces the energy content of the emitting region. Through minimum energy arguments and estimates of the lobe lifetime by spectral aging of the observed synchrotron emission and/or by dynamical arguments, Rawlings & Saunders (1991) found a nice correlation between the average power that must be supplied to the lobes and the power emitted by the narrow line region. Although one always expects some correlation between powers (they both scales with the square of the luminosity distance) it is the ratio of the two quantities to be interesting, being of order of 100. Since we also know that, on average, the total luminosity in narrow lines is of the order of one per cent of the ionizing luminosity, we have the remarkable indication that the power carried by the jet (supplying the extended regions of the radio–source) and the power produced by the accretion disk (illuminating the narrow line clouds) are of the same order.

Celotti, Padovani and Ghisellini (1997) later confirmed this by calculating the kinetic power of the jet at the VLBI scale (see Celotti & Fabian 1993) and the broad line luminosity (assumed to reprocess $\sim 10\%$ of the ionizing luminosity).

A possible explanation involves the magnetic field being responsible for both the extraction of spin energy of a rotating black hole and the extraction of gravitational energy of the accreting matter. Assume in fact that the main mechanism to power the jet is the Blandford–Znajek (1977) process:

$$L_{\rm jet} \simeq \left(\frac{a}{m}\right)^2 U_{\rm B}(3R_{\rm s})^2 c \qquad (1)$$

where (a/m) is the specific black hole angular momentum (~ 1 for a maximally rotating Kerr hole), $U_{\rm B}$ is the magnetic energy density and $R_{\rm s}$ is the Schwarzschild radius. Note that Eq. 1 has the form of a Poynting flux. Assume now that most of the luminosity of the accretion disk is produced at $3R_{\rm s}$. The corresponding radiation energy density is then $U_{\rm r} = L_{\rm disk}/(36\pi R_{\rm s}^2 c)$, leading to

$$L_{\rm disk} = U_{\rm r}(3R_{\rm s})^2 c \qquad (2)$$

Therefore a magnetic field in equipartion with the radiation energy density of the disk would lead to $L_{\rm jet} \sim L_{\rm disk}$.

3. MASS OUTFLOWING RATE

We can estimate the ratio of the outflowing (in the jet) to the inflowing mass rate, since

$$L_{\rm disk} = \eta \dot{M}_{\rm in} c^2; \quad L_{\rm jet} = \Gamma \dot{M}_{\rm out} c^2; \quad \rightarrow \quad \dot{M}_{\rm out} = \frac{\eta}{\Gamma} \frac{L_{\rm disk}}{L_{\rm jet}} \dot{M}_{\rm in} \qquad (3)$$

If jets carry as much energy as the one produced by the accretion disk, we then obtain that the mass outflow rate is $\sim 1\%$ of the accreting mass rate (if $\eta = 10\%$ and $\Gamma = 10$).

4. THE BLAZAR DIVERSITY

BL Lac objects and Flat Spectrum Radio Quasars (FSRQ) are characterized by very rapid and large amplitude variability, power law spectra in restricted energy bands and strong γ–ray emission. These common properties justify their belonging to the same blazar class. However they differ in many other respects, such as the presence (in FSRQ) or absence (in BL Lacs) of broad emission lines, the radio to optical flux ratio, the relative importance of the γ–ray emission, the polarization degree, and the variability behavior. Within the BL Lac class, Giommi & Padovani (1994) have subdivided the objects according to where (i.e. at what frequency) the first broad (synchrotron) peak is located. Low energy peaked BL Lacs (LBL) show a peak in the IR–optical bands, while in High energy peaked BL Lacs (HBL) this is in the X–ray band (see, in this volume, the contributions of Costamante et al., Giommi et al., Pian et al., Tagliaferri et al., Tavecchio & Maraschi, Wolter et al.).

As the emission of all blazars is beamed towards us, so there must be a parent population of objects pointing in other directions. The parent populations of BL Lacs and FSRQs are believed to be FR I and more powerful FR II radio galaxies, respectively (see the review by Urry & Padovani 1995). The absence of broad emission lines in BL Lacs is shared by FR I radio galaxies, whose nuclei are well visible by Hubble Space Telescope observations (Chiaberge, Capetti & Celotti 1999). This suggests that in FR I and BL Lac objects broad emission lines are intrinsically weaker than in more powerful objects.

5. THE RE-UNITED BLAZARS

Fossati et al. (1998) found that the SED of all blazars is related to their observed luminosity. There is a rather well defined trend: low luminosity objects are HBL–like, and furthermore their high energy peak is in the GeV–TeV band. As the bolometric luminosity increases, both peaks shift to lower frequencies, and the high energy emission is increasingly more dominating the total output. [1] Ghisellini et al. (1998), fitted the SED of all blazars detected in the γ–ray band for which the distance and some spectral information of the high energy radiation were available. They found a correlation between the energy $\gamma_{\rm peak} m_e c^2$ of the electrons emitting at the peaks of the spectrum and the amount of energy density U (both in radiation and in magnetic field), as measured in the comoving frame: $\gamma_{\rm peak} \propto U^{-0.6}$. This indicates that, at $\gamma_{\rm peak}$, the radiative cooling rate $\dot\gamma(\gamma_{\rm peak}) \propto \gamma_{\rm peak}^2 U \sim$const. It also suggests that this may be due to a "universal" acceleration mechanism, which must be nearly independent of γ and U: in less powerful sources with weak magnetic field and weak lines the radiative cooling is less severe and electrons can be accelerated up to very high energies, producing a SED typical of a HBL. The paucity of photons produced externally to the jet leaves synchrotron self–Compton as the only channel to produce high energy radiation. At the other extreme, in the most powerful sources with strong emission lines, electrons cannot be accelerated to high energies because of severe cooling. Their spectrum is therefore peaked in the far IR and in the MeV band. In these sources the inverse Compton scattering off externally produced photons is the dominant cooling mechanism, producing a dominant γ–ray luminosity.

5.1. Powers

For the same sample of blazars fitted in Ghisellini et al. (1998) we can estimate the powers radiated and transported by jets in the form of cold protons, magnetic field and hot electrons and/or electron–positron pairs. Since the model allows us to determine the bulk Lorentz factor, the dimension of the emitting region, the value of the magnetic field and the particle density, we can then determine

$$L_{\rm p} = \pi R^2 \Gamma^2 \beta c\, n'_{\rm p} m_{\rm p} c^2; \quad L_{\rm e} = \pi R^2 \Gamma^2 \beta c\, n'_{\rm e} \langle \gamma \rangle m_e c^2; \quad L_{\rm B} = \pi R^2 \Gamma^2 \beta c\, \frac{B^2}{8\pi} \quad (4)$$

where $n'_{\rm p}$ and $n'_{\rm e}$ are the comoving proton and lepton densities, respectively, R is the cross section radius of the jet, and $\langle \gamma \rangle m_e c^2$ is the average lepton energy. These

---

[1] A note of caution: the limited sensitivity of EGRET (onboard CGRO) and ground based Cherenkov telescopes allows us to detect sources which are in high states. Therefore the trend of more high energy dominated spectra as the total power increases strictly refers to high states.

powers can be compared with the radiated one estimated in the same frame (in which the emitting blob is seen moving). The power radiated *in the entire solid angle* is thus $L_r = L'_r \Gamma^2$ (the same holds for the power $L_{\rm syn}$ emitted by the synchrotron process). All these quantities are plotted in Fig. 2 (Celotti & Ghisellini 2000, in prep.). In this figure hatched areas correspond to BL Lac objects. Several facts are to be noted:

- If the jet is made by a pure electron–positron plasma, then the associated kinetic power is L_e. However, we note that $L_e \ll L_r$ posing a serious energy budget problem.

- If there is a proton for each electron, the bulk kinetic power $L_p \sim 10 L_r$. This corresponds to an efficiency of $\sim 10\%$ in converting bulk into random energy. The remaining 90% is therefore available to power the radio lobes, as required.

- The power in the Poynting flux, L_B, is of the same order of L_e, indicating that the magnetic field is close to equipartition with the electron energy density. This suggests that, on these scales, the magnetic field is not a prime energy carrier, but is a sub-product of the process transforming bulk into random energy.

6. INTERNAL SHOCKS

The central engine may well inject energy into the jet in a discontinuous way, with individual shells or blobs having different masses, bulk Lorentz factors and energies. If this occurs there will be collisions between shells, with a faster shell catching up a slower one. This idea has become the leading model to explain the emission of gamma–ray bursts, but it was born in the AGN field, due to Rees (1978) (see also Sikora 1994).

- **Location** — The γ–ray emission of blazars and its rapid variability imply that there must be a preferred location where dissipation of the bulk motion energy occurs. If it were at the base of the jet, and hence close to the accretion disk, the produced γ–rays would be inevitably absorbed by photon–photon collisions, with associated copious pair production, reprocessing the original power from the γ–ray to the X–ray part of the spectrum (contrary to observations). If it were far away, in a large region of the jet, it becomes difficult to explain the observed fast variability, even accounting for the time–shortening due to the Doppler effect. The region where the radiation is produced is then most likely located at a few hundreds of Schwarzschild radii ($\sim 10^{17}$ cm) from the base of the jet, within the broad line region (see Ghisellini & Madau 1996 for more details). The extra seed photons provided by emission lines enhance the efficiency of the Compton process responsible for the γ–ray emission. This is indeed the typical distance at which two shells, initially separated by $R_0 \sim 10^{15}$ cm (comparable to a few Schwarzschild radii) and moving with $\Gamma \sim 10$ and $\Gamma \sim 20$ would collide.

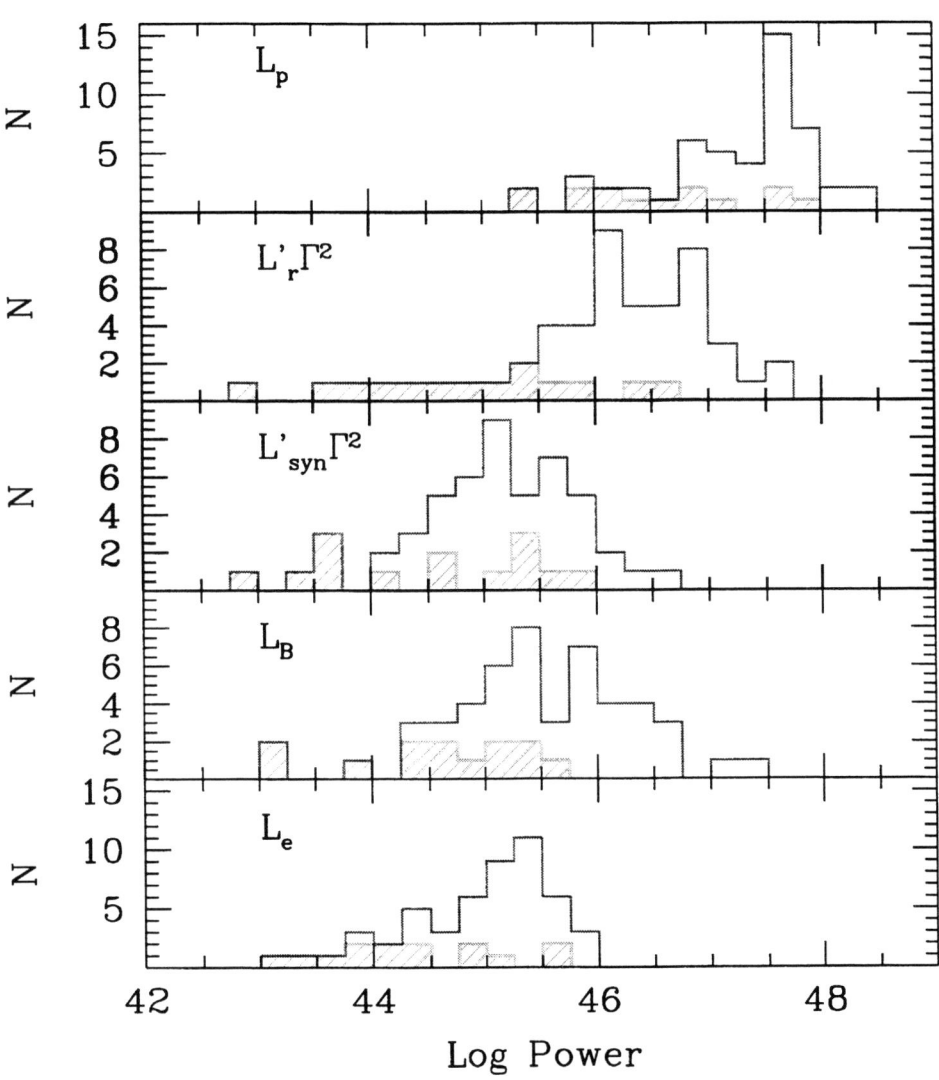

FIGURE 1. Histograms of the powers carried by the jet in protons, total radiation, synchrotron radiation, magnetic field and relativistic electrons, from top to bottom. Hatched areas correspond to BL Lac objects. The electron distribution was assumed to extend down to $\gamma_{min} \sim 1$. From Celotti & Ghisellini (2000, in prep.)

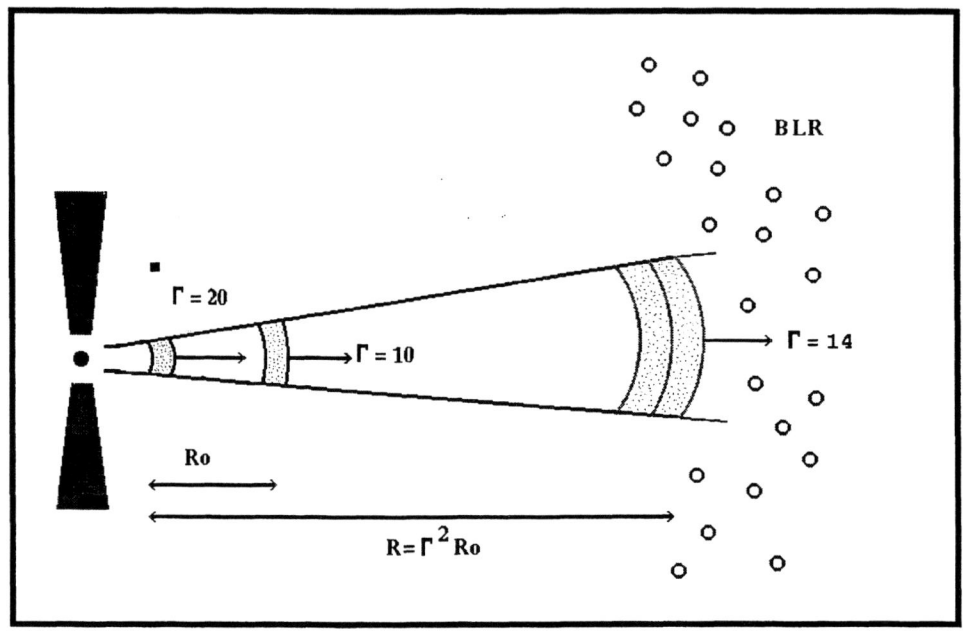

FIGURE 2. Cartoon illustrating the internal shock scenario. The intermittent activity of the central engine produces two shells, initially separated by R_0. The faster one will catch up the slower one at $R \sim \Gamma^2 R_0$.

- **Variability timescales** — In fact if the initial separation of the two shells is R_0 and if they have Lorentz factors Γ_1, Γ_2, they will collide at

$$R = \frac{2\Gamma_1^2}{1 - (\Gamma_1/\Gamma_2)^2} R_0 \qquad (5)$$

If the shell widths are of the same order of their initial separation the time needed to cross each other is of the order of R/c. The observer at a viewing angle $\theta \sim 1/\Gamma$ will see this time Doppler contracted by the factor $(1-\beta\cos\theta) \sim \Gamma^{-2}$. The typical variability timescale is therefore of the same order of the initial shell separation. If the mechanism powering GRB and blazar emission is the same, we should expect a similar light curve from both systems, but with times appropriately scaled by the different R_0, i.e. the different masses of the involved black holes.

- **Efficiencies** — As most of the power transported by the jet must reach the radio lobes, only a small fraction can be radiatively dissipated. The efficiency η of two blobs/shells for converting ordered into random energy depends on their masses m_1, m_2 and bulk Lorentz factors Γ_1, Γ_2, as

$$\eta = 1 - \Gamma_f \frac{m_1 + m_2}{\Gamma_1 m_1 + \Gamma_2 m_2} \qquad (6)$$

where $\Gamma_f = (1-\beta_f^2)^{-1/2}$ is the bulk Lorentz factor after the interaction and is given by (see e.g. Lazzati, Ghisellini & Celotti 1999)

$$\beta_f = \frac{\beta_1\Gamma_1 m_1 + \beta_2\Gamma_2 m_2}{\Gamma_1 m_1 + \Gamma_2 m_2} \qquad (7)$$

The above relations imply, for shells of equal masses and $\Gamma_2 = 2\Gamma_1 = 20$, $\Gamma_f = 14.15$ and $\eta = 5.7\%$.

Efficiencies η around 5–10% are just as needed for blazar jets.

- **Peak energies?** — In the rest frame of the fast shell, the bulk kinetic energy of each proton of the slower shell is $\sim (\Gamma' - 1)m_p c^2$, where $\Gamma' \sim 2$. This is what can be transformed into random energy. Assume now that the electrons share this available energy (through an unspecified acceleration mechanism). In the comoving frame, the acceleration rate can be written as $\dot{E}_{heat} \sim (\Gamma' - 1)m_p c^2/t'_{heat}$. The typical heating timescale may correspond to the time needed for the two shells to cross, i.e. $t'_{heat} \sim \Delta R'/c \sim R/(c\Gamma)$, where $\Delta R'$ is the shell width (measured in the same frame). The heating and the radiative cooling rates will balance for some value of the random electron Lorentz factor γ_{peak}:

$$\dot{E}_{heat} = \dot{E}_{cool} \rightarrow \frac{\Gamma m_p c^3}{R} = \frac{4}{3}\sigma_T c U \gamma_{peak}^2 \rightarrow \gamma_{peak} = \left(\frac{3\Gamma m_p c^2}{4\sigma_T R U}\right)^{1/2} \qquad (8)$$

The agreement of the above simple relation with what can be derived from model fitting the SED of blazars is surprisingly good (see Ghisellini 2000).

- **Radio flares** — Collisions between shells may (and should) happen in a hierarchical way. As an illustrative example, assume that one pair of shells after the collision moves with a final Lorentz factor $\Gamma_1 = 14$ (this number corresponds to $\Gamma = 10$ and 20 for the two shells before the interaction). The collision produces a flare –say– in the optical and γ-ray bands. After some observed time Δt two other shells collide and another flare is produced. Assume that the final Lorentz factor is now $\Gamma_2 = 17$ (corresponding to an initial $\Gamma = 10$ and 30 before collision). Since the second pair is faster, it will catch up the first one after a distance (from eq. 5) $R \sim 1200 c\Delta t$. A time separation of $\Delta t \sim$ a day between the two flares then corresponds to $R \sim 1$ pc, i.e. the region of the radio emission of the core. Due again to Doppler contraction, this radio flare will be observed ony a few days after the second optical flare. Since the ratio Γ_2/Γ_1 is small, the efficiency is also small (at least a factor 10 smaller than the firsts shocks). There is then the intriguing possibility of explaining the birth of radio blobs after intense activity (i.e. more than one flare) of the higher energy flux. Radio light–curves should have *some* memory of what has happened days–weeks earlier at higher frequencies.

7. CONCLUSIONS

Here I will dare to assemble different pieces of information gathered in recent years in a coherent, albeit still preliminary, picture.

There is a link between the extraction of gravitational energy in an accretion disk and the formation and acceleration of jets, since both have the same power. Objects of low luminosity accretion disks also lack strong emission lines, suggesting that it is the paucity of ionizing photons, not of gas, the reason for the lack of strong lines in BL Lacs. Correspondingly, this implies that, if FR I are the parents of BL Lacs, they also have intrinsically weak line emission (i.e. no need for an obscuring torus). Despite the fact that the jet power in blazars spans at least four orders of magnitude, the average bulk Lorentz factor is almost the same, suggesting a link between the power and the mass outflowing rate: their ratio is constant. In the region where most of the radiation is produced, the jet is heavy, in the sense that protons carry most of the bulk kinetic energy. There the jet dissipates $\sim 10\%$ of its power and produces beamed radiation. The power dissipated at larger distances is much less, and therefore the jet can transport $\sim 90\%$ of its original power to the radio extended regions. One way to achieve this is through internal shocks, which can explain why the major dissipation occurs at a few hundreds Schwarzschild radii, why the efficiency is of the order of 10%, and give clues on the observed variability timescales and even on why electrons are accelerated at a preferred energy. The spectral energy distribution of blazars depends on where shell–shell collisions take place, and on the amount of seed photons present there. Even in a single source it is possible that the separation of two consecutive shells is sometimes large, resulting in a collision occurring outside the broad line region. In this case the corresponding spectrum should be produced by the synchrotron self–Compton process only, without the contribution of external photons: we then expect a simultaneous optical–γ–ray flare of roughly equal powers (but with the self–Compton flux varying quadratically, see Ghisellini & Maraschi 1966). This is what should always happen in lineless BL Lac objects. On the other hand, if the initial separation of the two shells is small (or the Γ–factor of the slower one is small), the collision takes place close to the disk. X–rays produced by the disk would then absorb all the produced γ–rays and a pair cascade would develop, reprocessing the power originally in the γ–ray band mainly into the X–ray band. We should therefore see an X–ray flare without accompanying emission above $\Gamma m_e c^2$.

Pairs of shells which have already collided can interact again between themselves, at distances appropriate for the radio emission. This offers the interesting possibility to explain why the radio luminosity is related with the γ–ray one, and why radio flares are associated with flares at higher frequencies. Work is in progress in order to quantitatively test this idea against observations.

ACKNOWLEDGEMENTS

I thank Annalisa Celotti for very insightful discussions.

REFERENCES

Begelman M.C., Blandford R.D. & Rees M.J., 1984, Rev. Mod. Phys., 56, 255
Benford G. & Lesch H., 1998, MNRAS 301 414
Blandford R.D. & Znajek R.L., 1977, MNRAS, 176, 465
Celotti A., 1997, in Relativistic jets in AGNs, eds. M. Ostrowski, M. Sikora, G. Madejski & M. Begelman, p. 270
Celotti A., 1998, in Astrophysical jets: open problems, (Gordon & Breach Science publ.), eds. S. Massaglia & G. Bodo (Amsterdam), p. 79
Celotti A. & Fabian A.C. 1993, MNRAS, 264, 228
Celotti A., Padovani P. & Ghisellini G., 1997, MNRAS, 286, 415
Chiaberge, Capetti & Celotti, 1999, A&A, 349, 77
Dondi L. & Ghisellini G., 1995, MNRAS, 273, 583
Fossati G., Celotti A., Comastri A., Maraschi L. & Ghisellini G., 1998, MNRAS, 299, 433
Ghisellini G. & Madau P., 1996, MNRAS, 280, 67
Ghisellini G. & Maraschi L., 1996, in Blazar Continuum Variability, ASP Conference series, Vol. 110, 1996, eds. H.R. Miller & J.R. Webb, p. 436
Ghisellini G., Celotti A., Fossati G., Maraschi L. &Comastri A., 1998, MNRAS, 301, 451
Ghisellini G., 2000, in The Asca Symposium, Tokio, March 1999, in press
Giommi P. & Padovani P., 1994, ApJ, 444, 567
Hartman, R.C. et al., 1999, ApJS, 123, 79
Lazzati D., Ghisellini G. & Celotti A., 1999, MNRAS, 309, L13
Mészáros, P., 1999, Nuclear Phys. B, in press (astro-ph/9904038)
Piran, P., 1999, Phys. Rep., 314, 575
Rawlings S.G. & Saunders R.D.E., 1991, Nature, 349, 138
Rees M.J., 1966, Nature, 211, 468
Rees, M.J., 1978, MNRAS, 184, P61
Sikora M., 1994, ApJS, 90, 923
Ulrich M.-H., Maraschi L. & Urry C.M., 1997, ARA&A, 35, 445
Urry M.C. & Padovani P., 1995, PASP, 107, 803
Vermeulen R.C. & Cohen M.H., 1994, ApJ, 430, 467
Wagner S.J. & Witzel A., 1995, ARA&A, 33, 163

BEPPOSAX OBSERVATIONS OF BRIGHT RADIO GALAXIES

Paola Grandi

Istituto di Astrofisica Spaziale - CNR, Via Fosso del Cavaliere 100, I-00133, Roma, Italy

ABSTRACT BeppoSAX observations of Broad Line Radio Galaxies (BLRGs) have shown that they have a considerable variety of spectral properties and important differences with respect to their radio-quiet counter-part Seyfert 1s. In radio galaxies the soft photons are often absorbed by cold material. In contrast, in Seyfert 1s the soft photons are generally absorbed by warm gas. The iron lines, always detected in Seyfert 1s, are not always present in BLRGs and generally are weak. In addition, small iron line equivalent widths seem to correspond to weak reflection components in radio galaxies.

The emerging picture of BLRGs is complex. Probably several X-ray components, jet, accretion flow and molecular torus, mixed in different way in different objects, contribute to the production of their X-ray spectrum and determine the observed variety. The weakness of the reprocessed features can be explained either by a dilution of the Seyfert-like continuum from non-thermal (jet) radiation or by an accretion gas that is hot and geometrically thick close to the black hole and cold geometrically thin (i.e. able to reprocess the primary X-ray radiation) at larger radii.

1. INTRODUCTION

The most basic classification of Active Galactic Nuclei (AGN) consists in dividing them in two classes: radio-quiet and radio-loud. This represents not only an observational distinction but a basic physical difference whose basis is still not understood. Radio morphologies (lobes, jets etc.) are obviously the product of some physical mechanisms at work in the nuclei of radio-loud objects. Rees et al. (1982) suggested, for example, that the primary engine in radio galaxies was a spinning black hole fed by thick, and hot, accretion flow. Later, Blandford (1990) and Meier (1999) have explicitly identified the black hole spin as a possible physical parameter responsible for the radio-loud and radio-quiet dichotomy. If the hole rotates faster, it is more efficient in producing the jets observed in radio-loud objects.

X-ray photons seem to be the best probe to investigate the radio loudness issue as they are produced (and reprocessed) in the inner regions of AGNs where accretion occurs. X-ray observations of radio-loud objects therefore provide unique data to test the model hypotheses. In particular, two questions can be addressed: 1) Does the jet contribute to the X-ray emission in non-blazar radio-loud AGNs? 2) Are the same accretion processes at work in radio galaxies (and in general in radio-loud AGNs) and Seyfert galaxies?

TABLE 1. BeppoSAX Sample of Broad Line Radio Galaxies

Source	z	N_{Gal} 10^{21} cm^{-2}	Radio^a Morphology	$L^b_{2-10 keV}$ $\times 10^{44}$
PKS2152-69	0.027	0.25	FRI/FRII	0.2
Pictor A	0.035	0.42	FRII	1.0
3C120	0.033	1.1	FRI/S	2.4
3C111	0.048	3.2	FRII/S	2.9
3C390.3	0.057	0.42	FRII/S	3.3
3C382	0.059	0.88	FRII	9.4

^{a} – FR = Fanaroff-Riley, S = superluminal source
^{b} – Luminosity corrected for absorption in ergs sec^{-1}

BeppoSAX is a broad band (0.1-200 keV) X-ray satellite (Scarsi 1993) which has ideal characteristics for this purpose. Here we present a BeppoSAX analysis of 6 Broad Line Radio Galaxies and compare our sources with a sample of 13 Seyfert 1 galaxies also observed by BeppoSAX (Matt these proceedings, hereafter M00). Based on the properties of optical-UV spectra, BLRGs are considered the radio-loud counterpart of Seyfert 1s (Urry and Padovani 1995). The comparison of the X-ray (nuclear) properties of these two classes of objects is then particularly appropriate to investigate the radio-loud and radio-quiet dichotomy.

2. OBSERVATIONS AND RESULTS

The radio galaxies presented here are part of a larger sample of Radio-Loud Emission-Line AGNs observed by BeppoSAX, whose analysis is still on-going. In this paper, we restrict the discussion to a subsample of 6 BLRGs, which can be directly compared with the sample of Seyfert 1s discussed by M00. The sample is shown in Table 1. All the 3C sources have been observed for about 100 ksec. PKS2152-69 and Pictor A were observed for $\sim$ 17 and 30 ksec, respectively.

2.1. BLRG Spectral Analysis

The spectral analysis results, obtained by simultaneously fitting the LECS (0.1-4 keV) MECS (1.5-10 keV) and PDS (15-100 keV) instruments are reported in Table 2.

A simple power law plus Galactic absorption gave good fits to the data of PKS2152-69 and Pictor A. The iron line is absent in Pictor A in agreement with the ASCA measurement (Eracleous et al. 1998). For PKS2152-69, the exposure time was too short to allow the detection of even a strong feature. For both, we could obtain only an upper limit for the iron line equivalent width. In 3C111 the continuum was also a simple power law and the iron line only marginally detected (in spite of the

TABLE 2. BLRG SPECTRAL FIT RESULTS

Source	Γ	N_H^b	Refl.	Cutoff keV	E_{Fe} keV	σ keV	EW eV
PKS2152-69^a	1.79±0.1	0.25 (f)	...	...	6.4 (f)	0.0 (f)	< 429
Pictor A	1.63±0.06	0.42 (f)	...	...	6.4 (f)	0 (f)	< 102
3C120*	$1.80^{+0.12}_{-0.30}$	$2.4^{+1.8}_{-1.3}$	0.7±0.4	109^{+125}_{-77}	6.2±0.4	$0.27^{+0.40}_{-0.27}$	59^{+82}_{-42}
3C111†	1.65±0.04	$7.1^{+0.9}_{-0.8}$	< 0.3	> 90	$6.6^{+0.4}_{-0.2}$	0 (fixed)	58^{+31}_{-55}
3C390.3	$1.80^{+0.05}_{-0.04}$	1.3±0.2	$1.2^{+0.4}_{-0.3}$	> 123	6.4±0.1	$0.07^{+0.21}_{-0.07}$	136^{+40}_{-36}
3C382*	1.79±0.04	0.88 (f)	$0.4^{+0.3}_{-0.2}$	155^{+148}_{-59}	$6.5^{+0.9}_{-0.2}$	0.00 (f)	31^{+44}_{-15}

^{a} – MECS data only; ^{b} – in unit of 10^{21} cm^{-2}
* – Sources with detected soft excess; † – Iron line marginal detection

long exposure time). The soft photons appeared strongly absorbed. It is however possible that the cold absorber is not intrinsic to the source but associated to our Galaxy. 3C111 is in fact behind the Galactic dark cloud, Taurus B (see Reynolds et al. 1998 for more details).

More complex models were necessary to fit the other radio galaxies. Absorption in excess of the Galactic column density, iron lines, reflection (Ref.) humps, soft excesses and bendings of the high energy spectrum (Cutoff) are common spectral features, although not always simultaneously present in each source.

The soft excess was detected in two sources, namely 3C120 and 3C382. It was parameterized with a steep power law ($\Gamma^{soft} = 3 - 4$) and contributed to the total emission in the 0.1-2 keV band more than 60% in both the sources. The origin of this excess is not clear. It could be related to thermal emission from a cold thin disk, to radiation coming by extended thin plasma as well as to a very soft jet emission.

All the BLRGs with detected iron line show reflection components. Although the strength of the reflection (Ref.) is not very well constrained, data seem to suggest that radio-galaxies with weak iron lines have also weak reflections (Figure 1). If confirmed, this would imply that the line is generated in the same material which reflects the X-ray primary photons.

2.2. Comparison between BLRGs and Seyfert 1s

We compared the spectral properties of our BLRGs with a sample of 13 Seyfert 1s observed by BeppoSAX (M00). The 2-10 keV luminosities of Seyferts (ranging from $\sim 10^{42}$ to 10^{44} erg sec^{-1}) and BLRGs (see Table 1) partially overlap.

The X-ray primary power law of BLRGs is generally flat ($< \Gamma^{BLRG} = 1.73 >$, rms dispersion $\sigma^{BLRG}_{rms} = 0.08$). The Seyfert 1 sample is characterized by a steeper average spectral slope. However, the larger spread of the spectral indices in radio-quiet AGNs ($< \Gamma^{Sey\ 1} >=1.87$, $\sigma^{Sey\ 1}_{rms} = 0.24$), does not allow to statistically confirm any difference between Seyfert 1s and BLRGs.

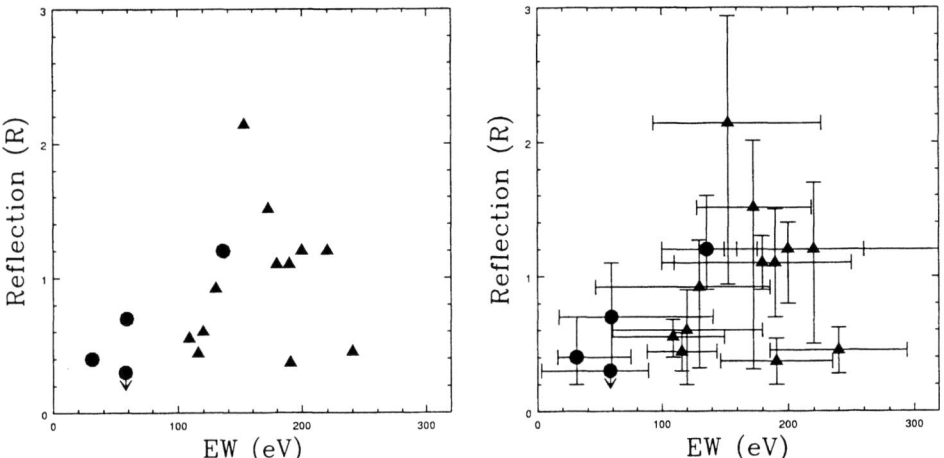

FIGURE 1. (*left panel*) – The amount of reflected radiation (Reflection) is plotted as a function of the strength of the iron line (EW) for the BLRG sample (circles). For comparison the Seyfert 1 (triangles) sample of M00 is also shown. Note that the radio-loud AGN are characterized by generally weaker iron lines and reflection components. (*right panel*) – Same plot with error bars.

Half the 3C sources in our sample shows a bending of the X-ray spectrum at high energies. When it is modeled with an exponential cutoff, the cutoff energies are similar to those observed in Seyferts.

While in radio-quiets the iron lines and the reflection components are always present, in BLRGs the reprocessed features are detected in only 3 sources. It should be noted the absence of the lines cannot be only attributed to poor statistics. In addition in radio-louds, the iron line equivalent widths are significantly smaller than in radio-quiets ($\langle EW^{BLRG} \rangle = 71$ eV, $\sigma_{rms}^{BLRG} = 45$ eV; $\langle EW^{Sey1} \rangle = 174$ eV, $\sigma_{rms}^{Sey1} = 57$ eV).

A general weakness of the reprocessed features in BLRGs is also confirmed by the XTE results presented by Sambruna and Eracleous (these proceedings). Note that Wozniak et al. (1998) have already discussed this possibility analyzing ASCA, GINGA and OSSE data.

A cold absorbing column in excess of Galactic is observed in 3 (including 3C111) out of 5 BLRGs with LECS data. No source shows absorption edges typical of warm absorber, which, on the contrary, is rather common in Seyfert 1s. It is possible that the absorbing material is different in BLRGs and Seyfert 1s, being warm in radio-quiets and cold in radio-louds (see also Sambruna Eracleous and Mushotzky 1999). This is also supported by a historical study of the column density changes in 3C390.3. The long-time (years) variability of the intrinsic N_H does not appear to be correlated to the flux intensity at 1 keV (see Fig. 2 in Grandi et al. 1999). It is possible that variations in the geometry of the absorber rather than changes of its ionization state (as expected in the case of a warm absorber) are responsible for the N_H long term variability.

3. DISCUSSION

Two important points arise from the BeppoSAX study of BLRGs: 1) there is not a unique type of BLRG X-ray spectrum, but a variety of cases; 2) there are several differences between BLRGs and Seyfert 1 galaxies. The most impressive difference concerns the reprocessed features that are weaker in radio-louds than in radio-quiets (Fig. 1). How can these results be explained?

BLRGs are complex objects, in which at least three X-ray components, a jet, an accretion flow and a molecular torus, can contribute to the formation of the spectrum.

Jet – In some BLRGs the radio jet shows superluminal motion. It is then reasonable to suppose that Doppler-enhanced radiation also contaminates the X-ray spectra and dilutes the reprocessed spectral features when the Doppler factor of the jet ($\delta=[\gamma(1-\beta\cos)]^{-1}$) is sufficiently large. If, in agreement with the AGN Unified Schemes, BLRGs and intrinsically powerful blazars are the same objects seen at different angles of view, also the jet output of radio galaxies should be Inverse Compton in the PDS band (Fossati et al. 1998, Ghisellini 1998). If this is the case, it is rather improbable that the observed high energy steepening in BLRGs is related to the jet. In powerful blazars, the Compton break (i.e. the Compton peak in the Spectral Energy Distribution, SED) is usually revealed at MeV-GeV energies. In order to occur at 100-200 keV as observed in BLRGs, the Doppler factor ($\delta=[\gamma(1-\beta\cos)]^{-1}$) should be smaller by about a factor 100 or more. This would imply a strong de-amplification of the non-thermal radiation ($I^{obs}(\nu)=\delta^3 I^{intr.}(\nu')$) and, in turn, a difficult detection of the jet.

Accretion Flow – It is not really clear how gas accretion occurs in radio-loud objects. The simplest approach is to extend the physical models developed for Seyferts to Broad Line Radio Galaxies. As in the case of radio-quiet AGNs, the accretion gas flow could be a cold geometrically thin disk with a hot corona above it (Haardt and Maraschi 1991, 1993; Petrucci et al. 2000). The role of the corona is to transform into X-ray photons, via inverse Compton scattering, the UV photons generated by the disk. Down-scattered X-rays, in turn, hit the disk and are reprocessed, generating an iron line and a reflection hump above 10 keV. A Seyfert-like accretion disk should then produce the following signatures in X-ray BLRG spectra: a soft excess (related to the thermal disk emission), a high energy thermal cutoff (related to the corona temperature), a broad red-shifted iron line (which suffers the vicinity of the strong black hole gravitational field) and a strong reflection component (produced by a disk subtending a $\sim 2\pi$ solid angle to the X-ray primary source, i.e. Refl$\sim \Omega/2\pi \sim 1$).

Alternatively, one can assume that different physical/geometrical accretion configurations occur in AGNs which discharge large amount of energy in jets. Rees

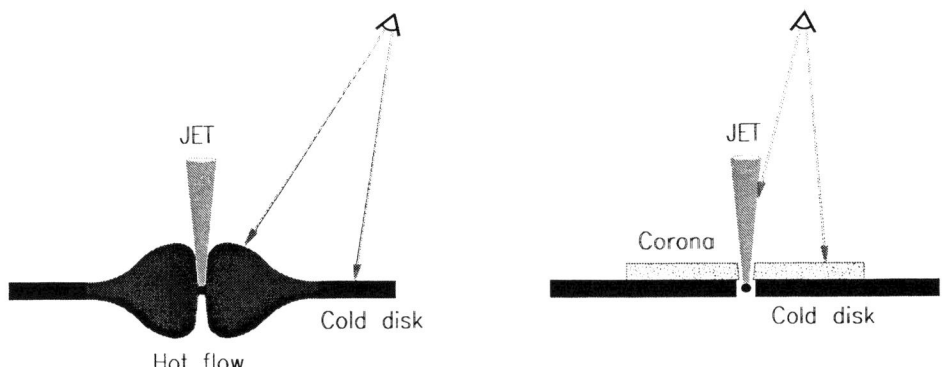

FIGURE 2. A schematic diagram of possible radio-loud accretion flows. The accretion can be hot geometrically thick near the black hole and cold geometrically thin in the outer regions (left panel). The weakness of the reprocessed features is assured by the small solid angle subtended by cold material. In this model, the jet radiation can be negligible. If accreting processes are the same in radio-louds and radio-quiets AGNs, the accretion flow is a unique cold and optically thin disk (right panel). In this case, a jet contribute is necessary to explain the weak iron lines and reflections components observed in BLRGs.

et al. (1982) speculated on the possibility that in radio-galaxies the hot accreting gas is in the shape of an ion-supported torus characterized by low radiative efficiency. The Advection Dominated Accretion Flow (ADAF) models (Narayan e al. 1998), recently proposed, follow similar lines of though. Shapiro Lightman and Eardley (SLE, 1976) found another solution for the hot flow, that resembles the ion-supported torus. However, in the SLE model the energy produced in the accreting gas by viscosity is locally radiated and not advected radially.

Then the accretion flow in BLRGs could be hot and geometrically thick in the inner region and become cold and geometrically thin only at larger radii (Chen and Halpern 1989). Given the smaller covering factor of the cold (and reprocessing) matter to the X-ray primary source, this accretion configuration predicts less prominent soft excesses, narrower weaker iron lines and smaller reflection components than the Seyfert case. In addition, a lack of correlation between the variations of the primary X-ray source and the reprocessed features is expected; the entity of the temporal delay depending on the inner radius of the cold disk.

Molecular Torus – It is probable that a thick wall of absorbing material (perhaps in a toroidal shape) surrounds the accretion flow. The torus can produce the iron line and reflect the X-ray primary photons (Ghisellini Haardt and Matt 1994) further

TABLE 3. BLRG NUCLEAR COMPONENTS

Source	BeppoSAX Spectral characteristics	Dominant Spectral Components
3C120	soft excess weak (broad?) Fe line reflection high energy cutoff	Jet + Seyfert-like disk
3C111	iron line (?) no reflection no cutoff	Jet
3C390.3	narrow strong Fe line strong reflection no soft excess	Hot flow + cold disk + Torus
3C382	narrow weak Fe line weak reflection energy cutoff soft excess	Hot flow + Torus (+ external cold disk?)

complicating the X-ray spectrum. In addition, if the opening angle of the torus is small, its (less thick) upper layers could be intercepted by the observer and cause the soft X-ray depletion sometimes observed in BLRGs.

Since its distance from the X-ray source is large (≥ 1 pc), iron line and reflection components should respond with a considerable delay (light-years) to the continuum variations.

It is possible and also probable that all these components are present in BLRGs. If they are mixed in different ways in different objects, the variety of X-ray spectra observed with BeppoSAX would be easily explained. As shown in Table 3 the main BeppoSAX spectral features of the BLRGS with high signal-to noise spectra (i.e. pointed for about 100 ksec) can be opportunely reproduced choosing a plausible combination of the nuclear components.

This tentative table shows that a Seyfert-like disk alone can not reproduce the observations. The weakness of the reprocessed features requires the presence either of a non-thermal radiation from a jet or of a cold disk sub-tending a small solid angle to the X-ray primary source (see figure 2).

4. SEYFERT-LIKE OR ADAF-LIKE ACCRETION?

The sources in Table 3 which seem better to fit the Seyfert-like+jet model are 3C111 and 3C120. If the model is correct, the weak (but detected) iron line in 3C120 assures that Seyfert and Blazar components have comparable intensity around 6 keV, i.e. the non-thermal power law is important and can effect the continuum shape. We then tested whether the 3C120 spectrum can be fitted by a mix of beamed and un-beamed radiation on the entire BeppoSAX band (i.e. on about 3 decades in energies).

A simple power law was assumed to represent the jet (blazar) spectrum and a power law with cutoff plus iron line and reflection component were utilized to mimic the Seyfert emission. The Seyfert equivalent width was assumed $EW^{sey} = 174$ eV, the average value from the Matt sample. Different combinations of spectral slopes for the Seyfert and blazar models were tested. In all the cases, the relative normalization at 6.4 keV of the two power laws was fixed in order to reproduce the observed Fe equivalent width reported in Table 2. As shown in Figure 3 (*left panel*), 3C120 data can be fitted by a blazar and a Seyfert-like power law (in Figure $\Gamma^{Blazar} = 2$, $\Gamma^{Sey} = 1.7$). However the model requires very low energy cutoff ($E_{Cut} \sim 70$ keV) and a strong reflection (Ref~ 2). These best fit values are rather extreme if compared to those of Seyferts. This slightly disfavors the jet model, although it cannot be completely rejected given the large uncertainties associated to the parameters.

A geometry that assumes a hot inner flow and cold disk surrounding it seems to be particularly appropriate for 3C390.3. In this source the UV bump is weak, there is not soft excess and the iron line is narrow. In this case a molecular torus (or a warped disk) also contributes to increase the strength of the iron line and the reflection component (Grandi et al. 1999).

The idea that the iron line emitting regions cannot be located near the X-ray is also supported by recent BeppoSAX observations of the very bright radio-galaxy Centaurus A (Cen A) ($F_{2-10\ keV} \sim 10^{-10}$ erg cm^{-2} sec^{-1}). It is a FR I optically classified as Narrow Line Radio Galaxy and shows a radio-optically-X-ray jet pointed far away from us ($i \sim 60-70^0$). It is characterized by a X-ray spectrum rather complex. However above 3 keV it is dominated by the nuclear point-like source (Turner et al. 1998).

Cen A was observed on 1997 February 20-21 (30 ksec), on 1998 January 6-7 (50 ksec) and twice for 40 ksec during the 1999 summer on July 10-11 and August 2-3. The repeated observations of this radio-galaxy, thanks to their high statistics, have allowed a detailed study of the nuclear flux variations versus the iron line flux changes. As clearly shown in Figure 3 (*right panel*), line and continuum do not vary together, in particular the iron line was more intense when the source was weaker (compare the 1997 and 1998 observations). Since the inclination of the jet is large in Cen A, the contribution of non-thermal radiation to the total X-ray nuclear continuum is expected to be negligible. Then Figure 3 (*right panel*) simply indicates that the line emitting region responds with a significant delay to the continuum variations, as it is expected if the reprocessing gas is located far from the primary X-ray source.

We conclude that, although, in principle, the two proposed accretion scenarios are both viable, the idea that a jet reduces the reprocessed features is lightly disfavored.

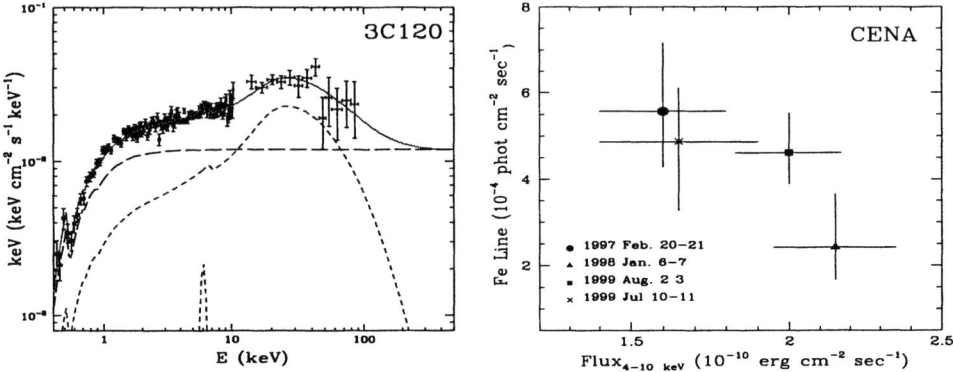

FIGURE 3. (*left panel*) – 3C120: the 3C120 data can be fitted with a mix of Seyfert-like (red line) and blazar-like (green line) spectra. (*right panel*) – CenA: the variations of the Fe line intensity are reported as a function of the 4-10 keV nuclear flux changes. The iron line and the continuum do not change together.

5. CONCLUSIONS

BeppoSAX analysis of 6 BLRGs has shown that a variety of X-ray spectra exits. A mix of different components, a jet, an accretion flow and a molecular torus, could explain the observations. The comparison between our 6 BLRG and a sample of 13 Seyfert 1s has pointed out important differences between radio-loud and radio-quiet AGNs. In particular, it has been shown that BLRG reprocessed features are often absent or weaker than in Seyfert 1s. This result has important implications. If the accretion flow in BLRGs and Seyferts is identical (i.e a cold thin optically thin disk with a hot above it) a strong contribution by Doppler-enhanced radiation is necessary to explain the weakness of the reprocessed features. Alternatively, if the jet emission is not important, the cold re-processing gas has to subtend a smaller solid angle to the X-ray primary source. A possibility is that the X-ray continuum is produced by a hot geometrically thick ion-supported torus that illuminates a cold thin disk at larger radii.

Choosing between the two scenarios is still premature. The new satellites (XMM, Chandra, INTEGRAL) will play an important role in solving the problem. Detailed studies of the iron line profiles in a large (and well selected) sample will allow to confirm the presence of cold matter very near to the black hole (if lines are broad and redshifted) or at large distances (if lines are narrow).

Variability studies will be also crucial. An anti-correlation is expected between the EW and the X-ray continuum flux, if the jet is dominant: the observed continuum (disk + jet) should be more variable than the Fe line (from disk). Alternatively, if no correlation is found, the Fe variations do not follow the continuum flux variations and a temporal delay effect (as observed in CenA case) is present. Finally, detailed studies of the hard X-ray continuum will be able to detect any beamed radiation. At very high energies, where the Seyfert-like power law drops, the jet, if intense, should emerge and become directly detectable (see Figure 3 *left panel*).

ACKNOWLEDGEMENTS

I would like to thank all the people who, working with me on the BeppoSAX radio-galaxy project, have allowed the writing of this paper: L. Maraschi, C. M. Urry, E. Massaro, G. Matt, M. Guainazzi, F. Haardt, P. Giommi, G.G. Palumbo, G. Malaguti. An acknowledgment also go the L. Piro and the BeppoSAX CDC team for their support to this project. I also thank G. Ghisellini, L. Ferretti and C. G. Perola for the useful comments and stimulating discussions and A. Bazzano and A.J. Bird for critical reading of the manuscript. I am very grateful to M. Frutti for invaluable help in realizing Figure 2.

REFERENCES

Blandford, R. D., 1990 in Active Galactic Nuclei, Saas-Fee Advanced Course, pag. 264
Chen, K. and Halpern, J. P., 1989, ApJ, 344, 115
Eracleous et al. 1998 ApJ, 550,557.
Fossati G., Maraschi L., Celotti A., Comstri A., Ghisellini, G., 1998, MNRAS, 299, 433
Ghisellini G., Haardt F., and Matt G., 1994, MNRAS, 267, 743
Ghisellini G., Celotti A., Fossati G., Maraschi L., Comastri A., 1998, MNRAS, 301, 451
Grandi, P., et al. 1999, A&A, 343, 40
Haardt, F., Maraschi, L., 1991, ApJ, 380, L51
Haardt F., Maraschi L., 1993, ApJ, 413, 507
Meier D. L., 1999, New Astronomy Reviews in press (astro-ph/9908283)
Morganti R., Killeen N., Tadhunter C.N. 1993, MNRAS, 263, 1023
Narayan, R., Mahadevan, R. and Quataert, E. 1998, *The Theory of Black Hole Accretion Disk*, eds M.A. Abramowicz, G. Bjornsson, J.E. Pringle, pag.148
Nandra P., et al. 1997, ApJ, 477, 602
Petrucci P.O. et al. these proceedings
Reynolds C. S., 1997, MNRAS, 286, 513
Reynolds C. S., Iwasawa K., Crawford C.S., Fabian A.C., 1998, MNRAS, 299, 410
Sambruna M. R., Eracleous M., Mushotzky R. F., 1999, ApJ, 256, 60
Sambruna M. R., Eracleous M., these proceedings
Rees M.J., Begelman M.C., Blandford R.D., Phinney E.S., 1982, Nature, 295, 17
Shakura, N. I., Sunyaev, R. A. , 1973, A&A, 24, 337
Shapiro, S. L., Lightman A. P., Eardley D. M., 1976, ApJ, 204, 187
Urry C.M. and Padovani P., 1995, PASP, 107,803
Wozniak P. R., Zdziarski A. A., Smith D., et al. 1998, MNRAS, 299, 449

PROBING THE COSMIC DARK AGE IN X–RAYS

Z. Haiman[1,2]

1) Department of Astrophysics, Princeton University, Princeton, NJ 08544, USA
2) Fermi National Accelerator Laboratory, P.O. Box 500, Batavia, IL 60510, USA

ABSTRACT Empirical studies of the first generation of stars and quasars will likely become feasible within the next decade in several different wavelength bands. Microwave anisotropy experiments, such as MAP or Planck, will set constraints on the ionization history of the intergalactic medium due to these sources. In the infrared, the Next Generation Space Telescope (NGST) will directly detect sub-galactic objects at $z \gtrsim 10$. In the optical, data from the Hubble Deep Field (HDF) already places a constraint on the abundance of high-redshift quasars. However, the epoch of the first quasars might be first probed in X-ray bands, by instruments such as the Chandra X-ray Observatory (CXO) and the X-ray Multi-mirror Mission (XMM). In a 500 Ksec integration, CXO reaches a sensitivity of $\sim 2 \times 10^{-16}$ erg s^{-1} cm^{-2}. Based on simple hierarchical CDM models, we find that at this flux threshold $\sim 10^2$ quasars might be detectable from redshifts $z \gtrsim 5$, and ~ 1 quasar at $z \sim 10$, in each $17' \times 17'$ field. Measurement of the power spectrum of the unresolved soft X-ray background will further constrain models of faint, high-redshift quasars.

KEYWORDS: cosmology: theory; quasars: general; black hole physics

1. INTRODUCTION

The cosmic dark age ended when the first gas clouds condensed out of the primordial fluctuations at redshifts $z = 10 - 20$ (Peacock 1992; Rees 1996). These condensations are likely the sites where the first clusters of stars and the first quasar black holes appeared, giving birth to the first "mini–galaxies" or "mini–quasars" in the Universe. Despite the lack of observational data, this epoch has become a subject of intense theoretical study in the past few years. The recent interest can be attributed to forthcoming instruments: *NGST* could directly image sub–galactic objects at $z \gtrsim 10$ in the infrared, while microwave satellites such as MAP or Planck could measure signatures from the reionization of the intergalactic medium (IGM). Currently, bright quasars are detected out to $z \sim 5$ (Fan et al. 1999). Although the abundance of optically and radio bright quasars declines at $z \gtrsim 2.5$ (Schmidt et al. 1995; Shaver et al. 1996), a recent determination of the X-ray luminosity function (LF) of quasars from ROSAT data (Miyaji et al. 2000) has not confirmed this decline. In this contribution, we point out that future X-ray observations might provide yet another probe of the first quasars and the end of the dark age at $z \sim 10$, and that X-ray data might be uniquely useful in distinguishing quasars from stellar systems.

2. THE APPEARANCE OF THE FIRST QUASARS AND STARS

In popular Cold dark matter (CDM) cosmologies, the first baryonic objects appear near the Jeans mass ($\sim 10^6$ $M_\odot$) at redshifts as high as $z \sim 30$ (Haiman & Loeb 1999b, and references therein). At any redshift, the mass function of collapsed dark halos is given to within a factor of two by the Press–Schechter formalism. Following collapse, the gas in the first baryonic condensations is virialized by a strong shock (Bertschinger 1985). Provided it is able to cool on a timescale shorter than the Hubble time, the shock–heated gas continues to contract. Depending on the details of the cooling and angular momentum transport, the gas then either fragments into stars, or forms a central black hole exhibiting quasar activity. Although the actual fragmentation process is likely to be rather complex, the average fraction $f_{\rm star}$ of the collapsed gas converted into stars can be calibrated empirically so as to reproduce the average metallicity observed in the Universe at $z \approx 3$. The observed ratio, inferred from CIV absorption lines in Lyα forest clouds, is between 10^{-3} and 10^{-2} of the solar value (Songaila 1997 and references therein). If the carbon produced in the early mini–galaxies is uniformly mixed with the rest of the baryons in the Universe, this implies $f_{\rm star} \approx$ 2–20% for a Scalo stellar mass function.

An even smaller fraction of the cooling gas might condense at the center of the potential well of each cloud and form a massive black hole, exhibiting quasar activity. In the simplest scenario, the peak luminosity of each black hole is proportional to its mass, and the light–curve, expressed in Eddington units, is a universal function of time. Indeed, for a sufficiently high fueling rate, quasars are likely to shine at their maximum possible luminosity, which is some constant fraction of the Eddington limit, for a time which is dictated by their final mass and radiative efficiency. Here we assume that the black hole mass fraction $r \equiv M_{\rm bh}/M_{\rm halo}$ obeys a log-Gaussian probability distribution, $p(r) = \exp[-(\log r - \log r_0)^2/2\sigma^2]$, with $\log r_0 = -3.5$ and $\sigma = 0.5$ (Haiman & Loeb 1999b). These values roughly reflect the distribution of black hole to bulge mass ratios found in a sample of 36 local galaxies (Magorrian et al. 1998) for a baryonic mass fraction of $\sim (\Omega_{\rm b}/\Omega_0) \approx 0.1$. We further postulate that each black hole emits a time–dependent bolometric luminosity in proportion to its mass, $L_{\rm q} \equiv M_{\rm bh} f_{\rm q} = M_{\rm bh} L_{\rm Edd} \exp(-t/t_0)$, where $L_{\rm Edd} = 1.5 \times 10^{38}$ $M_{\rm bh}/M_\odot$ erg s^{-1} is the Eddington luminosity, t is the time elapsed since the formation of the black hole, and $t_0 = 10^6$ yr is the characteristic quasar lifetime. As shown by Haiman & Loeb (1998), this simple quasar-model accurately reproduces the evolution of the luminosity function of optical quasars in the B–band (Pei 1995) at redshifts $z \gtrsim 2.2$.

Finally, we assume that the shape of the emitted spectrum follows the mean spectrum of known quasars (Elvis et al. 1994) up to a photon energy of 10 keV. We extrapolate the spectrum up to ~ 50 keV, assuming a spectral slope of $\alpha=0$ (or a photon index of -1). For reference, Figure 1 shows the adopted spectrum of quasars,

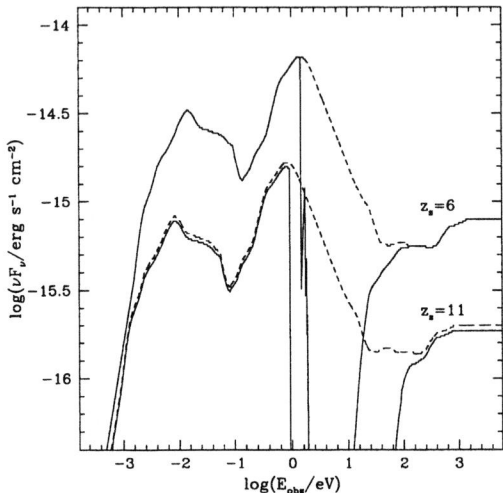

FIGURE 1. The observed spectra of quasars with a central black hole mass of $M_{bh} = 10^8$ M$_\odot$. The upper curves correspond to a source redshift of $z_s = 6$ and the lower curves to a source redshift of $z_s = 11$. In both cases we assume sudden reionization at $z_r = 10$. The dashed curves show the assumed intrinsic spectral shape, and the solid curves show the spectra after absorption by neutral intergalactic H and He.

assuming a black hole mass $M_{bh} = 10^8 M_\odot$, placed at two different redshifts, $z_s = 11$ and $z_s = 6$, and processed through the IGM, and assumed that reionization occurred at $z_r = 10$ and that at higher redshifts the IGM was homogeneous and fully neutral. At lower redshifts, $0 < z < z_r$, we included the hydrogen opacity of the Lyα forest given by Madau (1995), extrapolating his fitting formulae for the evolution of the number density of absorbers beyond $z = 5$ when necessary. As Figure 1 shows, the minimum black hole mass detectable by the $\sim 2 \times 10^{-16}$ erg s^{-1} cm^{-2} flux limit of *CXO* (see below) is $M_{bh} \sim 10^8$ M$_\odot$ at $z = 10$ and $M_{bh} \sim 2 \times 10^7$ M$_\odot$ at $z = 5$. In our model, the corresponding halo masses are $M_{halo} \sim 3 \times 10^{11}$ M$_\odot$, and $M_{halo} \sim 6 \times 10^{10}$ M$_\odot$, respectively. Although such massive halos are rare, their abundance is detectable in wide-field surveys.

3. INFRARED: EXPECTED COUNTS WITH NGST

The Next Generation Space Telescope[1] (*NGST*) will be able to detect the early population of mini–galaxies and mini–quasars. *NGST* is scheduled for launch in 2008, and is expected to reach an imaging sensitivity of ~ 1 nJy (S/N=10 at spectral resolution $\lambda/\Delta\lambda = 3$) for extended sources after several hours of integration in the wavelength range of 1–3.5μm. Figure 2 shows the predicted number counts in

---

[1] see http://www.ngst.nasa.gov

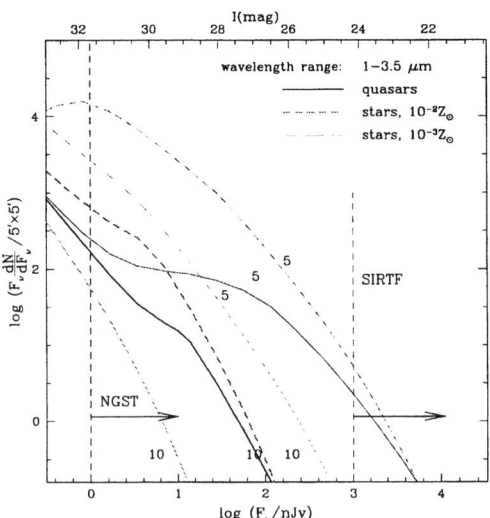

FIGURE 2. Infrared Number Counts. The solid curves refer to quasars, while the long/short dashed curves correspond to galaxies with low/high normalization for the star formation efficiency. The curves labeled "5" or "10" show the cumulative number of objects with redshifts above $z = 5$ or 10.

the mini–galaxy and mini–quasar models described above, in a ΛCDM cosmology with $(\Omega_0, \Omega_\Lambda, \Omega_b, h, \sigma_{8h^{-1}}, n)$=(0.35, 0.65, 0.04, 0.65, 0.87, 0.96), normalized to a $5' \times 5'$ field of view. This figure shows separately the number per logarithmic flux interval of all objects with redshifts $z > 5$ (thin lines), and $z > 10$ (thick lines). As the figure shows, NGST will be able to probe about ~ 100 quasars at $z > 10$, and ~ 200 quasars at $z > 5$ per field of view. The bright–end tail of the number counts approximately follows a power law, with $dN/dF_\nu \propto F_\nu^{-2.5}$. The dashed lines show the corresponding number counts of mini–galaxies, assuming that each halo undergoes a starburst that converts a fraction of 2% (long–dashed) or 20% (short–dashed) of the gas into stars. These lines indicate that NGST would detect $\sim 40 - 300$ mini–galaxies at $z > 10$ per field of view, and $\sim 600 - 10^4$ mini–galaxies at $z > 5$. Unlike quasars, galaxies could in principle be resolved if they extend over a scale comparable to the virial radius of their dark matter halos (Haiman & Loeb 1997; Barkana and Loeb 2000). The supernovae and γ-ray bursts in these galaxies might outshine their hosts and may also be directly observable (Miralda-Escudé & Rees 1997). Finally, we note that recent data in the J and H infrared bands from deep NICMOS observations of the HDF (Thompson et al. 1999) could already be useful to constrain mini–quasar and mini–galaxy models.

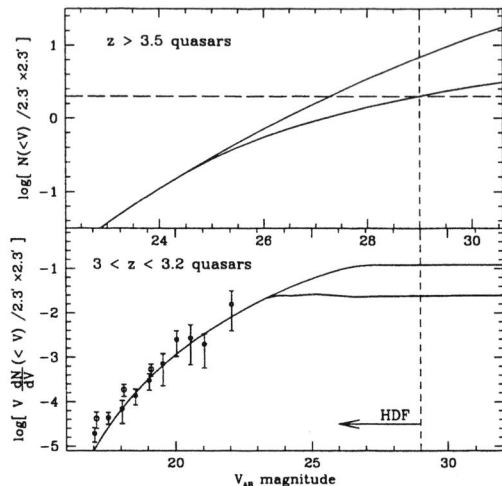

FIGURE 3. V-band number counts for high–redshift quasars. The lower panel shows differential counts for $3 < z < 3.2$, as well as data adapted from Pei (1995). The upper panel shows cumulative number counts for $z > 3.5$. The top curve shows predictions of our standard ΛCDM model with $M_{bh} \propto M_{halo}$. The bottom curve shows the counts in a model with halo circular velocities restricted to $v_{circ} \geq 75$ km s^{-1}. This model is consistent with the lack of detections in the HDF.

4. OPTICAL: CONSTRAINTS FROM THE HUBBLE DEEP FIELD

Although the infrared wavelengths are the best suited to detect the redshifted UV–emission from objects at $z \sim 10$, present data in the optical already yields a constraint on quasar models of the type described above. The properties of faint *extended* sources found in the HDF (Madau et al. 1996) agree with detailed semi-analytic models of galaxy formation (Baugh et al. 1998). On the other hand, the HDF has revealed only a handful of faint *unresolved* sources, but none with the colors expected for high redshift quasars (Conti et al. 1999). The simplest mini–quasar model described above predicts the existence of ~ 10 B–band "dropouts" in the HDF, inconsistently with the lack of detection of such dropouts up to the $\sim 50\%$ completeness limit at $V \approx 29$ in the HDF. To reconcile the models with the data, a mechanism is needed for suppressing the formation of quasars in halos with circular velocities $v_{circ} \lesssim 50 - 75$ km s^{-1} (see Figure 3 for the counts). This suppression naturally arises due to the photo-ionization heating of the intergalactic gas by the UV background after reionization (Thoul & Weinberg 1996; Navarro & Steinmetz 1997). Alternative effects could help reduce the quasar number counts, such as a change in the background cosmology, a shift in the "big blue bump" component of the quasar spectrum to higher energies due to the lower black hole masses in mini–quasars, or a nonlinear black hole to halo mass relation; however, these effects

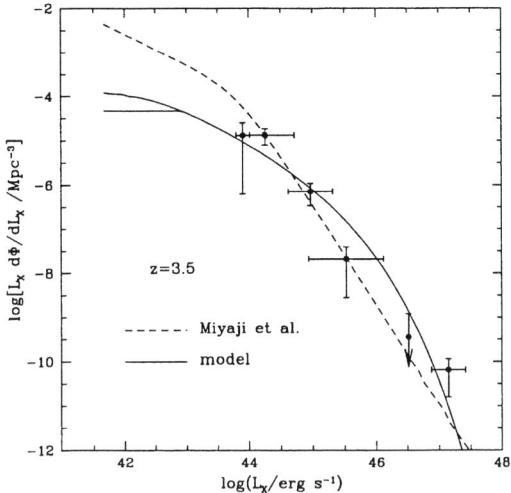

FIGURE 4. The predicted X–ray luminosity function at $z = 3.5$ in our model (solid curves). The lower curve shows the effect of a cutoff in circular velocity for the host halos of $v_{\rm circ} \geq 50$ km s^{-1}. The ROSAT data points are adopted from Miyaji et al. (2000) and the dashed curve shows their fitting formula (for our background cosmology).

are too small to account for the lack of detections in the HDF (Haiman, Madau & Loeb 1999).

5. X–RAYS: PREDICTIONS FOR CXO

Quasars can be best distinguished from star forming galaxies at high redshifts by their X-ray emission. Detections of high-z quasars would therefore be highly valuable: detections, or upper limits would help in answering the important question of whether the IGM at $z \lesssim 6$ was reionized by stars or quasars, by yielding constraints on the ionizing photon rate from high–z quasars. The simple quasar–model described above was constructed to accurately reproduce the evolution of the optical luminosity function in the B–band (Pei 1995) at redshifts $z \gtrsim 2.2$ (Haiman & Loeb 1998). However, it yields good agreement with the data on the X–ray LF, as demonstrated in Haiman & Loeb (1999c), and shown here in Figure 4. We regard this model as a minimal toy model which successfully reproduces the existing data, and use a straightforward extrapolation of this model to predict the X-ray number counts. In Figure 5, we show the predicted counts in the 0.4–6keV energy band of the CCD Imaging Spectrometer (ACIS) of *CXO*. Note that these curves are insensitive to our extrapolation of the template spectrum beyond 10 keV. The figure is normalized to the $17' \times 17'$ field of view of the imaging chips. The solid curves show that of order a hundred quasars with $z > 5$ are expected per field at the *CXO* sensitivity of $\sim 2 \times 10^{-16}$ erg s^{-1} cm^{-2} for a 5σ detection of a point source. Note

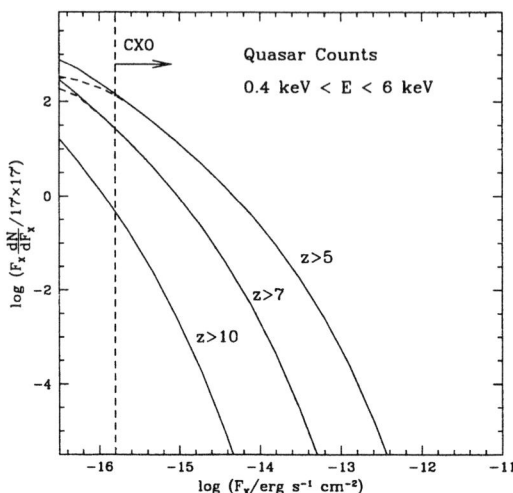

FIGURE 5. The total number of quasars with redshift exceeding $z = 5$, $z = 7$, and $z = 10$ are shown as a function of observed X-ray flux in the CXO detection band. The solid curves correspond to a cutoff in circular velocity for the host halos of $v_{\rm circ} \geq 50$ km s^{-1}, the dashed curves to a cutoff of $v_{\rm circ} \geq 100$ km s^{-1}. The vertical dashed line show the CXO sensitivity for a 5σ detection of a point source in an integration time of 5×10^5 seconds.

that CXO's arcsecond resolution will ease the separation of these point sources from background noise. The abundance of quasars at higher redshifts declines rapidly; however, a few objects per field are still detectable at $z \sim 8$. The dashed lines show the results for a minimum circular velocity of the host halos of $v_{\rm circ} \geq 100$ km s^{-1}, and imply that the model predictions for the CXO satellite are not sensitive to such a change in the host velocity cutoff. This is because the halos shining at the CXO detection threshold are relatively massive, $M_{\rm halo} \sim 10^{11}$ M$_\odot$, and possess a circular velocity above the cutoff. In principle, the number of predicted sources would be lower if we had assumed a steeper spectral slope. For example, as figure 6 shows below, our model falls short of predicting the hard X-ray background, by about an order of magnitude at 10 keV. The difference could be explained by a change in our template spectrum to include a population of quasars with hard, but highly absorbed spectra (caused by the denser, and more gas rich hosts at high redshift; see also the discussion by A. Comastri in these proceedings for models consistent with the ROSAT surveys that also produce the correct XRB spectrum). We note, however, that the agreement between the LF predicted by our model at $z \approx 3.5$ and that inferred from ROSAT observations would be upset by such a change, and require a modification of the model that would in turn tend to counter-balance the decrease in the predicted counts.

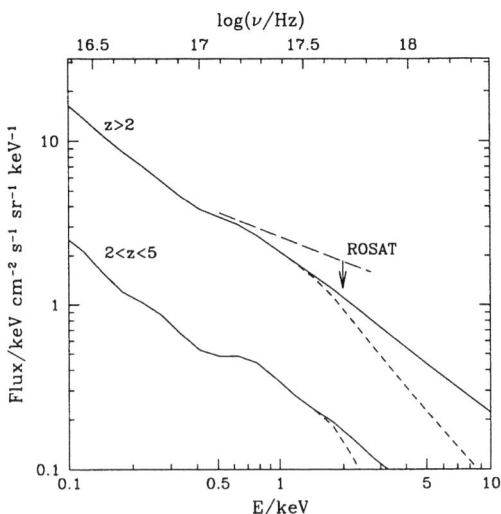

FIGURE 6. Spectrum of the unresolved soft X-ray background in our model. We assume that the median X-ray spectrum of each source follows the mean spectrum of Elvis et al. (1994) up to 10 keV, and has a spectral slope of 0.5 (solid lines) or -0.5 (short-dashed lines) at higher photon energies. The lower curves show the spectra resulting from quasars with redshifts between $2 < z < 5$, and the upper curves include contributions from all redshifts $z > 2$. The long-dashed line shows the unresolved fraction (assumed to be 25%) of the soft X-ray background spectrum from Miyaji et al. (1998).

6. THE X-RAY BACKGROUND

Existing estimates of the X-ray background (XRB) provide another useful check on our quasar model. Figure 6 shows the predicted spectrum of the XRB in our model at $z = 0$ (solid lines). The unresolved background flux is shown, obtained by summing the emission of all quasars whose individual observed flux at $z = 0$ is below the ROSAT PSPC detection limit for discrete sources of 2×10^{-15} erg cm^{-2} s^{-1} (Hasinger & Zamorani 1997). The short dashed lines show the predicted fluxes assuming a steeper spectral slope beyond 10 keV ($\alpha = -0.5$, or a photon index of -1.5). The long dashed line shows the 25% unresolved fraction of the soft XRB observed with ROSAT (Miyaji et al. 1998; Fabian & Barcons 1992). This fraction represents the observational upper limit on the component of the soft XRB that could in principle arise from high-redshift quasars. As the figure shows, our quasar model predicts an unresolved flux just below this limit in the 0.5-3 keV range. The model also predicts that most ($\gtrsim 90\%$) of this yet unresolved fraction arises from quasars beyond $z = 5$. The power spectrum of the unresolved background therefore might carry information on quasars at $z > 5$, and be useful in constraining the models (Haiman & Hui 1999, in preparation). The correlations in the background have recently been measured by Soltan et al. (1999, see also this Proceedings).

7. DISCUSSION

We have demonstrated that state–of–the–art X-ray observations could yield more stringent constraints on quasar models than currently available from the Hubble Deep Field (Haiman, Madau, & Loeb 1999). The X–ray data might provide the first probe of the earliest quasars, complementing subsequent infrared and microwave observations. More specifically, we have found that forthcoming X–ray observations with the *CXO* satellite might detect of order a hundred quasars per field of view in the redshift interval $5 \lesssim z \lesssim 10$. Our numerical estimates are based on the simplest toy model for quasar formation in a hierarchical CDM cosmology, that satisfies all the current observational constraints on the optical and X-ray luminosity functions of quasars. Although a more detailed analysis is needed in order to assess the modeling uncertainties in our predictions, the importance of related observational programs with *CXO* is evident already from the present analysis. Other future instruments, such as the HRC or the ACIS-S cameras on *CXO*, or the EPIC camera on *XMM*, which has a collective area 5–20 times larger than that of *CXO*, will also be useful in searching for high–redshift quasars.

The relation between the black hole and halo masses may be more complicated than linear. With the introduction of additional free parameters, a non–linear (mass and redshift dependent) relation between the black–hole and halo masses can also lead to acceptable fits (Haehnelt et al. 1998) of the observed quasar LF near $z \sim 3$. Such fits, when extrapolated to higher redshift, can result in different predictions for the abundance of high–redshift quasars. From the point of view of selecting between these alternative models, even a non–detection by CXO would be invaluable. It is hoped further that either observations of the clustering properties of $z \sim 3$ quasars in the Sloan Digital Sky Survey, or a measurement of the power spectrum of the soft X-ray background, would break model degeneracies (Haiman & Hui 1999, in preparation).

Quasars emit a broad spectrum which extends into the UV and includes strong emission lines, such as Lyα. For quasars near the *CXO* detection threshold, the fluxes at $\sim 1\mu$m are expected to be relatively high, $\sim$ 0.5–0.8 μJy. Therefore, infrared spectroscopy of X-ray selected quasars with the Space Infrared Telescope Facility *(SIRTF)* or *NGST* can identify the redshifts of the faint X-ray point-like sources detected by the *CXO* satellite. Such an approach could prove to be a particularly useful approach for unraveling the reionization history of the intergalactic medium at $z \gtrsim 5$. At present, the best constraints on hierarchical models of the formation and evolution of quasars originate from the Hubble Deep Field. However, *HST* observations are only sensitive to a limiting magnitude of $V \sim 29$, and cannot probe the earliest quasars, beyond $z \sim 6$. The combination of X-ray data from the *CXO* satellite and infrared spectroscopy from *SIRTF* and *NGST* could potentially resolve one of the most important open questions about the thermal history of the Universe, namely whether the intergalactic medium was reionized by stars or by accreting black holes.

ACKNOWLEDGEMENTS

I thank A. Loeb for his advice and guidance throughout many projects, M. Rees and P. Madau for many useful discussions, and N. White for the invitation to this stimulating conference. Support for this work was provided by the DOE and NASA through grant NAG 5-7092 at Fermilab, and through a Hubble Fellowship.

REFERENCES

Barkana, R., & Loeb, A. 2000, ApJ, 531, 613
Baugh, C. M., Cole, S., Frenk, C. S. & Lacey, C. G. 1998, ApJ, 498, 504
Bertschinger, E. 1985, ApJS, 58, 39
Conti, A., Kennefick, J. D., Martini, P., & Osmer, P. S. 1999, AJ, 117, 645
Elvis, M., et al. 1994, ApJS, 95, 1
Fabian, A. C. & Barcons, X. 1992, ARA&A, 30, 429
Fan, X. et al. (SDSS collaboration) 1999, AJ, 118, 1
Haehnelt, M. G., Natarajan, P. & Rees, M. J. 1998, MNRAS, 300, 817
Haiman, Z., & Loeb, A. 1997, in "Science with the NGST", eds. E. Smith & A. Koratkar, p. 251
———————— 1998, ApJ, 503, 505
———————— 1999a, ApJ, 519, 479
———————— 1999b, in Proc. of "After the Dark Ages: When Galaxies Were Young (the Universe at $2 < z < 5$)", eds. S. Holt & E. Smith, p. 34
———————— 1999c, ApJ, 521, 9
Haiman, Z., Madau, P., & Loeb, A. 1999, ApJ, 514
Hasinger, G., & Zamorani, G. 1997, in "Festschrift for R. Giacconi's 65th birthday", World Scientific Publishing Co., H. Gursky, R. Ruffini, L. Stella eds., in press, astro-ph/9712341
Madau, P. 1995, ApJ, 441, 18
Madau, P., et al. 1996, MNRAS, 283, 1388
Magorrian, J., et al. 1998, AJ, 115, 2285
Miralda-Escudé, J. 1998, ApJ, 501, 15
Miralda-Escudé, J., & Rees, M. J. 1997, ApJ, 478, L57
Miyaji, T., Hasinger, G., & Schmidt, M. 2000, A&A, 353, 25
Miyaji, T., Ishisaki, Y., Ogasaka, Y., Ueda Y., Freyberg, M. J., Hasinger, G., & Tanaka, Y. 1998, A&A 334, L13
Navarro, J. F., & Steinmetz, M. 1997, ApJ, 478, 13
Peacock, J. 1992, Nature, 355, 203
Pei, Y. C. 1995, ApJ, 438, 623
Press, W. H., & Schechter, P. L. 1974, ApJ, 181, 425
Rees, M. J. 1996, preprint astro-ph/9608196
Schmidt, M., Schneider, D. P., & Gunn, J. E. 1995, AJ, 110, 68
Shaver, P. A., et al. 1996, Nature, 384, 439
Songaila, A. 1997, ApJL, 490, 1
Soltan, A., et al. 1999, A&A, 349, 354
Thompson, R., et al. 1999, AJ, 117, 17
Thoul, A. A., & Weinberg, D. H. 1996, ApJ, 465, 608

HIGH-ENERGY EMISSION FROM ISOLATED PULSARS

Alice K. Harding

Laboratory for High Energy Astrophysics, NASA Goddard Space Flight Center, Greenbelt, MD USA

ABSTRACT Rotation-powered pulsars can produce observable high-energy emission by a number of mechanisms. Non-thermal X-rays and gamma-rays are produced by relativistic particles, accelerated in the pulsar magnetosphere at a still unknown location, via curvature, synchrotron and inverse Compton radiation. Thermal X-rays may be produced at the stellar surface by cooling of the neutron star or by a variety of possible heating mechanisms that include the bombardment of the surface by backflowing particles and internal friction. In pulsars with ultra-strong fields, known as magnetars, the magnetic field is the dominant energy source responsible for both the radiation and the spin-down behavior. I will review models for pulsar emission in light of recent X-ray data from ROSAT, ASCA and RXTE.

KEYWORDS: pulsars:general; acceleration of particles; magnetic fields

1. INTRODUCTION

Pulsars are powerful sources of high energy emission, with some pulsed spectra extending above 1 GeV. There are presently about eight rotation powered γ-ray pulsars detected by CGRO (Thompson et al. 1997) and eleven X-ray pulsars detected by ROSAT (Becker & Trumper 1997) and ASCA (Saito 1998). The high energy emission in these pulsars is seen to be a significant fraction of their spin-down power and requires very efficient acceleration and radiation mechanisms. I will discuss the mechanisms that are currently being studied in the two major classes of models: polar cap models where acceleration and radiation occurs close to the neutron star surface, and outer gap models where these processes take place in the outer magnetosphere.

There is recent evidence for a new class of pulsars that derive their primary radiation power from decaying, ultra-strong magnetic fields. Such "magnetars" include four or five soft gamma-ray repeaters and about six anomalous X-ray pulsars. Their measured periods and period derivatives imply surface magnetic fields in the range $10^{14} - 10^{15}$ G. Unlike the pulsars having "normal" field strengths, the magnetars appear to be radio quiet.

2. SPIN-POWERED PULSARS

Rotating, magnetized neutron stars are unipolar inductors, generating huge $\mathbf{v} \times \mathbf{B}$ electric fields capable of pulling charges out of the star against the force of gravity (Goldreich & Julian 1969). It is believed that the resulting charge density that builds up in a neutron star magnetosphere is able to short out the electric field parallel to the magnetic field ($E_\parallel$) (thus allowing the field to corotate with the star) everywhere except at a few locations. These spots where $\mathbf{E} \cdot \mathbf{B} \neq 0$ are thought to occur above the surface at the polar caps and along the null charge surface, $\mathbf{\Omega} \cdot \mathbf{B} = 0$, where the corotation charge changes sign. These are the purported sites of particle acceleration and have given rise to the two classes of high energy emission models.

2.1. Polar cap model

2.1.1. Acceleration and non-thermal emission

Polar cap models of pulsar high energy emission date from the early work of Sturrock (1971) and Ruderman & Sutherland (1975), who proposed particle acceleration and radiation near the neutron star surface at the magnetic poles. There is a large variation among polar cap models, with the primary division being whether or not there is free emission of particles from the neutron star surface. This question is still somewhat subject to debate, due to our incomplete understanding of the neutron star surface composition and physics. The subclass of polar cap models based on free emission of particles of either sign, called space-charge limited flow models, assumes that the surface temperature of the neutron star (many of which have now been measured in the range $T \sim 10^5 - 10^6$ K) exceeds the ion and electron thermal emission temperatures. Although $E_\parallel = 0$ at the neutron star surface in these models, the space charge along open field lines above the surface falls short of the corotation charge, due to the curvature of the field (Arons 1983) or to general relativistic inertial frame dragging (Muslimov & Tsygan 1992). The $E_\parallel$ generated by the charge deficit accelerates particles, which radiate inverse Compton (IC) photons by resonant scattering (when they reach energies (in units of mc^2) of $\gamma \sim 10^2 - 10^6$) and curvature (CR) photons (at energies $\gamma \lesssim 10^6$). Both IC and CR photons can produce e^+e^- pairs in the strong magnetic field. The e^+e^- pairs from IC photons are capable of screening the $E_\parallel$ before the particles can produce many CR photons, which would limit the particle energies to only $\gamma \lesssim 10^6$ (Harding & Muslimov 1998a [HM98]). But the field is screened by the positrons which turn around and return to the surface (see Figure 1). The pairs from these returning positrons may screen the $E_\parallel$ above the surface, because the IC scattering occurs above the resonance producing more energetic photons, disrupting the stability of particle acceleration near the surface. HM98 find that stable acceleration zones can form at 0.5 - 1 stellar radii above the surface, where the density of soft X-rays from the neutron star surface decreases and CR photons from both primary electrons and returning positrons produce stable pair formation fronts. The primary particle energies can

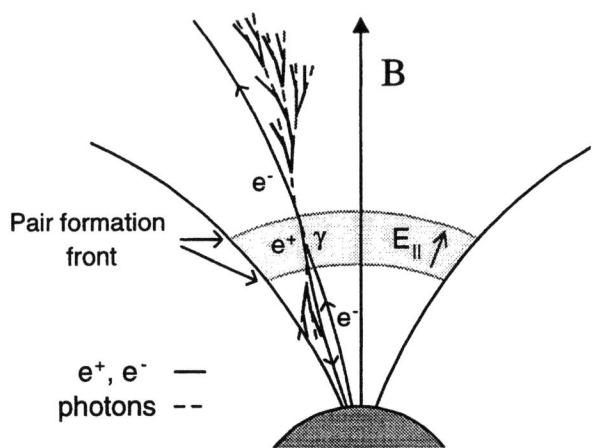

FIGURE 1. Acceleration zone above a pulsar polar cap, showing the cascades from upward moving electrons which produce the upper pair formation front, and cascades from downward moving positrons which produce the lower pair formation front.

then reach $\sim 10^7$ before pair production screens the field.

In the original version of the polar cap model (Sturrock 1971, Daugherty & Harding 1982, 1996) curvature radiation from the primary particles initiates a pair/synchrotron cascade. The resulting cascade spectrum is dominated by synchrotron radiation from the pairs and has a very sharp high energy cutoff at several GeV due to pair production attenuation. While this model spectrum agrees quite well with pulsar spectra observed by CGRO at energies above 100 keV, the pair synchrotron spectrum terminates at around 50 - 100 keV, the local cyclotron energy (~ 1 keV) blueshifted by the pair's parallel energy ($\gamma_\pm \sim 50$) (Harding & Daugherty 1999). Such turnovers may have been observed in the X-ray spectra of several middle-aged pulsars such as Vela (Strickman et al. 1999) and PSR1951+32 (Kuiper et al. 1998). However, this model falls short of explaining the strong X-ray emission observed in the spectra of young pulsars like the Crab, PSR0540-69 and PSR1509-58.

Recently, Zhang & Harding (1999 [ZH99]) noted that the pairs produced in polar cap cascades may resonant scatter the soft thermal photons from the neutron star surface, losing most of the remaining parallel energy they could not lose via synchrotron emission. Previously, polar cap models had considered only IC emission from the primary particles. The spectrum of this resonant scattering from pairs will extend down to at least the soft X-ray (ROSAT) band and possible even to

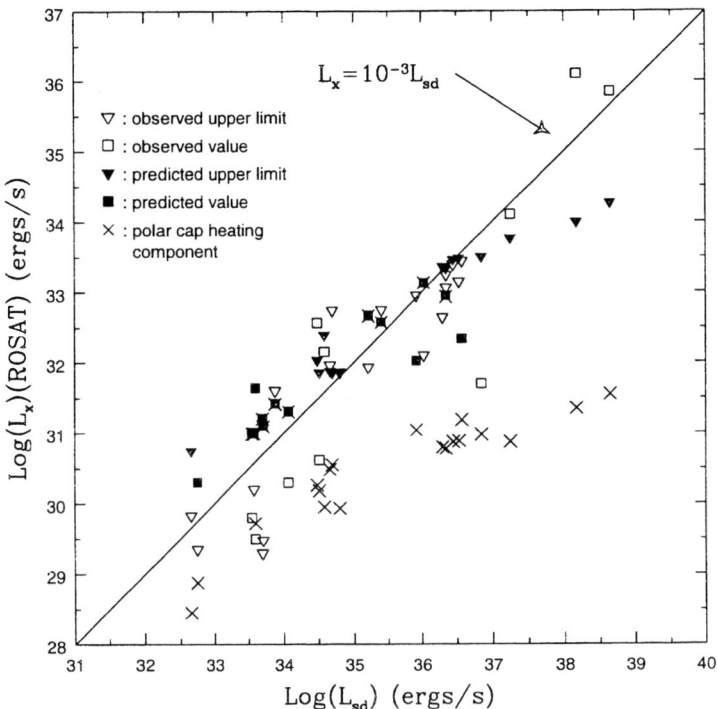

FIGURE 2. Comparison of X-ray luminosities predicted for the full polar cap cascade (resonant scattering of pairs and observed in the ROSAT band (0.1 - 2 keV) vs. spin-down luminosity for 29 pulsars.

the optical band. The IC luminosity will depend on the temperatures of two soft thermal photon sources, one due to cooling of the whole surface and one due to heating of the polar cap by returning positrons (see section 2.1.2). Figure 2 shows luminosities of pulsars detected by ROSAT in the soft X-ray band (0.2 - 1 keV) along with the predicted luminosities which include three components (non-thermal IC emission, full-surface thermal emission and polar cap thermal emission). The relation $L_x = 10^{-3} L_{sd}$ proposed by Becker & Trumper (1997) to fit the ROSAT luminosities, where L_{sd} is the spin-down luminosity, is roughly reproduced by the ZH99 model. However, the predicted luminosities of several young pulsars, the Crab and PSR0540-69 (the two highest open squares in the upper-right corner), are more than an order of magnitude smaller than the measured values.

2.1.2. Thermal Emission

Neutron star thermal emission can arise from several sources. Surface cooling will be the dominant source in young pulsars with ages $\tau \lesssim 10^5$ yr, and the X-ray emission of many young pulsars are consistent with standard cooling models (Becker & Trumper 1997). For older pulsars, any residual cooling will be dominated by surface heating due to backflowing positrons or internal friction (Shibazaki & Lamb 1989).

The positrons that are turned around at the upper pair formation front will accelerate downward to the neutron star surface through the same potential and heat the polar cap. In a space-charge limited flow model, the trapped positrons needed to screen the $E_\parallel$ are only a small fraction of the number of primary electrons because this field was created by a small charge imbalance in the first place. One can estimate the maximum returning positron fraction by evaluating $(\nabla \cdot E)_\parallel$ at the upper pair formation front. The maximum thermal luminosity due to polar cap heating can then be estimated in two regimes (ZH99)

$$L_{PC}^{max}(I) = 9.1 \times 10^{29} \text{erg s}^{-1} B_{12}^{2/7} P^{-9/7} r_{e,6}^{-8/7} (\cos \alpha)^{5/7}$$
$$L_{PC}^{max}(II) = 9.1 \times 10^{26} \text{erg s}^{-1} B_{12} P^{-13/4} r_{e,6}^{3} (\cos \alpha)^{5/4}, \quad (1)$$

where $B_{12} \equiv B/10^{12}$ G is the surface field strength, P is the period, $r_{e,6} \sim 1.5$ is the radius of the accelerator in units of the stellar radius and α is the pulsar inclination angle. A pulsar is in regime I when $B_{12}^{1/7} P^{-11/28} r_{e,6}^{10/7} (\cos \alpha)^{3/28} > 6$ and in regime II otherwise. Note that L_{PC}^{max} is not proportional to L_{sd}, but has a much weaker dependence on pulsar parameters. This polar cap heating luminosity L_{PC}^{max} is plotted in Figure 2. The contribution of L_{PC}^{max} to the total predicted X-ray luminosity is negligible for young pulsars having high L_{sd}, but is significant for several old pulsars, PSR B0950+08 and PSR B1929+10, and dominant for some millisecond pulsars, PSR J0437-4715 and PSR J2124-3358, having low L_{sd}.

Polar cap models predict that the high-energy non-thermal emission is radiated in a (nearly) hollow cone centered on the magnetic pole. One thus expects the pulse profile of the non-thermal emission to in general be double peaked, with the phase of closest approach to the magnetic pole situated midway between the two peaks. Since the thermal emission from either a hot polar cap or the full surface is centered on the magnetic pole, one important prediction of polar cap models is that the peak of the thermal X-ray emission should lie between the peaks of the non-thermal emission (Harding & Muslimov 1998b).

2.2. Outer gap model

The outer gap models for γ-ray pulsars are based on the existence of a vacuum gap in the outer magnetosphere which may develop between the last open field line and the null charge surface ($\Omega \cdot \mathbf{B} = 0$) in charge separated magnetospheres. The gaps arise because charges escaping through the light cylinder along open field lines above the null charge surface cannot be replenished from below. The first outer gap γ-ray pulsar models (Cheng, Ho & Ruderman 1986 [CHR]) assumed that emission is seen from gaps associated with both magnetic poles, but this picture, although successful in fitting the spectrum of the Crab and Vela pulsars, did not reproduce the observed pulsar light curves. More recent outer gap models assuming emission from one pole can more successfully reproduce the observed light curves (Romani & Yadigaroglu 1995).

2.2.1. Acceleration and non-thermal emission

The electron-positron pairs needed to provide the current, and therefore allow particle acceleration, in the outer gaps are produced by photon-photon pair production. In young Crab-like pulsars, the pairs are produced by curvature photons from the primary particles interacting with non-thermal synchrotron X-rays from the same pairs. In older Vela-like pulsars, where non-thermal X-rays emission is much lower, the pairs were assumed to come from interaction of primary particle inverse Compton photons with infra-red photons. However, this original Vela-type model (CHR) predicted large fluxes of TeV emission, from inverse Compton scattering of the infrared photons by primary electrons, which violates the observed upper limits (Nel et al. 1993) by several orders of magnitude. Cheng (1994) revised the outer gap model for Vela-type pulsars by proposing another self-sustaining gap mechanism where thermal X-rays from the neutron star surface interact with primary particle radiation to produce pairs, replacing the infra-red radiation (which has also never been observed). Some of the accelerated pairs flow downward to heat the surface and maintain the required thermal X-ray emission. The modern outer gap Vela-type models (Romani 1996, Zhang & Cheng 1997) all adopt this picture.

The observed non-thermal radiation in Crab-like pulsars is a combination of synchrotron emission and synchrotron self-Compton emission from pairs. In Vela-type pulsars, the non-thermal radiation is a combination of curvature and curvature self-Compton emission from the primaries at γ-ray energies, and synchrotron emission from the pairs at optical through X-ray energies. The high-energy spectra in both types of model have cutoffs around 10 GeV, due to the radiation-reaction cutoff in the primary particle spectrum, which are much less sharp than the attenuation cutoffs in polar cap model spectra.

2.2.2. Thermal Emission

The pairs produced in the outer gaps that are accelerated back toward the neutron star surface radiate curvature photons that can produce pairs in the strong magnetic field near the surface, initiating pair cascades (Zhang & Cheng 1997, Wang et al. 1998). But the pairs still have enough residual energy to heat the surface at the footpoints of open field lines that thread the outer gaps, producing thermal emission at a temperature around 1 keV, which is much higher than the temperatures observed in thermal emission from pulsars. According to the model, this emission is not observed directly (except right along the poles) but only through the blanket of pairs produced by the downward-going particle cascades. The 1 keV photons are reflected back to the surface by the pair blanket through cyclotron resonance scattering (Halpern & Ruderman 1993), and are re-radiated from the entire surface at a temperature around 0.1 keV. Thus, these outer gap models predict three X-ray emission components from the return particles from the outer gaps: hard thermal emission from direct heating of the polar caps (seen only along the poles), soft thermal emission reflected from the pair blanket, and non-thermal emission from the downward pair cascades (Cheng & Zhang 1999). The components actually observed from a particular pulsar depend on inclination and viewing angle. The predicted X-ray luminosities for pulsars in the ROSAT band can account for the observed $L_x = 10^{-3} L_{sd}$ relation.

3. MAGNETICALLY-POWERED PULSARS (MAGNETARS)

Until a little over a year ago, neutron stars having supercritical fields in the range $10^{14} - 10^{15}$ G lay in the realm of theoretical fantasy. They were first proposed to explain a number of properties of soft gamma-ray repeaters (SGRs), including the highly super-Eddington luminosities (Paczynski 1992), the young age and long period of SGR 0526-66 (the Mar 5, 1979 source) and the tremendous energy of the SGR giant bursts (Thompson & Duncan 1995). Recently (see review by Hurley 2000), long ($P > 5$ s) periods and high period derivatives ($\dot{P} \sim 10^{-10}$ s s^{-1}), the major prediction of the magnetar model, have been detected in both SGR sources and in anomalous X-ray pulsars (AXPs). The derived surface magnetic fields of these sources, assuming dipole spin-down such that $B_0 = 6.4 \times 10^{19} (P\dot{P})^{1/2}$, are indeed in the range $10^{14} - 1.6 \times 10^{15}$ G. The rotational energy loss rate of these sources is too low to account for their observed radiation, but the magnetic energy released through field decay is more than adequate.

Isolated neutron stars with magnetar-strength fields spin down much faster than do neutron stars with fields in the normal pulsar range of $10^{11} - 10^{13}$ G. The spin-down timescale, or characteristic age,

$$\tau_{SD} = \frac{P}{2\dot{P}} = 1600 \left(\frac{P}{B_{14}}\right)^2 \text{ yr}, \qquad (2)$$

from integrating the rotational energy loss rate for magnetic dipole radiation, $\dot{E}_D = I\Omega\dot{\Omega} \propto \Omega^4$, where B_{14} is the surface field in units of 10^{14} G, I is the neutron star moment of inertia and P and Ω are the present period and rotation frequency, predicts that magnetars will reach relatively long periods approaching 10 s in an age of several thousand years. Particle wind flows may also contribute to the rotational energy loss of some magnetars, affecting both the characteristic age and the derived surface magnetic field (Harding et al. 1999, Thompson & Blaes 1998).

The magnetic energy of a magnetar dominates over rotational energy at an early age, $\tau_B \sim 1.5 \times 10^5$ yr $(B_0/10^{14} \text{ G})^{-4}$, and it becomes a plentiful supply of free energy to power both the bursting and quiescent emission observed in the SGRs.

3.1. Burst radiation

The large magnetic flux diffusing out of the neutron star core, and eventually through the crust, creates imbalances and stresses. TD95 postulate that a large-scale rearrangement of the field, either through an interchange instability or both the internal and external field or shearing of the external field, can cause a massive reconnection event. These events could release a significant portion of the magnetic field energy,

$$\frac{B_{core}^2}{8\pi} R^3 \simeq 4 \times 10^{46} \,\text{erg} \left(\frac{B_{core}}{10^{15}\,\text{G}}\right)^2 \quad (3)$$

and may be responsible for the giant SGR flares observed from SGR 0526-66 and SGR 1900+14 having energies $\sim 10^{45}$ erg.

The small SGR bursts having average energies of 10^{41} ergs may be due to cracking of the neutron star crust. Ambipolar diffusion of the magnetic flux through the crust will built up stresses which can cause it to crack, producing small horizontal displacements of the magnetic footpoints. The sudden motion of the field can drive transient Alfven waves into the magnetosphere (TD95) or induce parallel electric fields (Miller et al. 1992), either of which can accelerate particles. Interestingly enough, the size spectrum of the SGR bursts is a power spectrum with the same slope as Earth quakes (Cheng et al. 1995).

The radiation from particles trapped on closed field lines is expected to fall in the soft γ-ray range. TD95 suggested that the radiation would form an optically thick pair plasma at temperatures around 20-30 keV, the peak energies of the observed SGR bursts. However, Baring (1995) pointed out that photon splitting, a QED process in which a single photon splits into two lower energy photons, dominates over one-photon pair production in magnetar fields. If all three modes of photon splitting allowed by CP invariance in QED operate at higher field strengths (which is not certain), so that both photon polarization states are allowed to split, then pair production would be completely suppressed. This would have several important consequences for SGRs. First, the complete suppression of pair production would cause these sources to be radio quiet (Baring & Harding 1998), according to standard pulsar models which require electron-positron pairs for the production of coherent radio emission. Second, photon splitting cascades would efficiently degrade energies of radiated photons, producing a quasi-thermal spectrum peaking around 30 keV (Harding & Baring 1996). In addition, the model spectra from photon splitting cascades when $B \geq 4B_{cr}$ do not vary with emission location in the magnetosphere, if the emission occurs in equatorial regions. This characteristic seems to be born out in the data and would favor burst emission in the closed-field region of the neutron star magnetosphere. The particles accelerated on open field lines may produce the harder emission seen in the giant bursts, and would drive a powerful wind.

4. FUTURE ISSUES FOR PULSAR X-RAY ASTRONOMY

In the next few years, X-ray measurements with Chandra and XMM will play a key role in understanding pulsar acceleration and radiation. The energy range from 0.1 - 100 keV is very important because it is the transition region for the different radiation mechanisms predicted by the models. Polar cap models for rotation-powered pulsars predict low energy turnovers of the synchrotron spectrum of the pair cascades around 10 - 100 keV, and a transition to inverse Compton radiation below this energy. Outer cap models also predict a transition from synchrotron to inverse Compton (for Crab-like pulsars) or curvature radiation (for Vela-like pulsars) in the X-ray range. In many old and middle-aged pulsars, the thermal radiation component from the hot neutron star surface will dominate at energies 0.1 - 1 keV, standing out above the non-thermal component. Measurements with energy resolution, spectral range and sensitivity a step beyond what is available with ROSAT, ASCA and RXTE will be necessary to seperate the different spectral components. Absolute phases of the pulse profiles of the thermal and non-thermal components will be a powerful discriminator between the different models.

The study of magnetars and radio quiet pulsars has just begun. More sensitive searches for (and more detections of) long period pulsars with ultra-strong fields may answer the important questions: how many radio-quiet pulsars exist? and what distinguishes magnetars from radio pulsars?

ACKNOWLEDGEMENTS

I thank Bing Zhang for valuable help in preparation of this paper.

REFERENCES

Arons, J. 1983, ApJ, 266, 215.
Baring, M. G. 1995, ApJ, 440, L69.
Baring, M. G. & Harding, A. K., 1998, ApJ, 507, L55.
Becker, W. & Trumper, J. 1997, 326, 682.
Cheng, B. et al. 1995, Nature, 382, 518.
Cheng, K. S., 1994, Proc. Toward a Major Atmospheric Cherenkov Detector, ed. T. Kifune (Tokyo: Universal Academy, 25.
Cheng, K. S., Ho, C., & Ruderman, M. A. 1986, ApJ, 300, 500.
Cheng, K. S. & Zhang, L. 1999, ApJ, 515, 337.
Daugherty, J. K. & Harding, A. K., 1982, ApJ, 252, 337.
Daugherty, J. K. & Harding, A. K., 1996, ApJ, 458, 278.
Goldreich, P. & Julian, W. H. 1969, ApJ, 157, 869.
Halpern, J. P. & Ruderman, M. A. 1993, ApJ, 415, 286.
Harding, A. K. & Baring, M. G. 1996, Proc. 3rd Huntsville Workshop on Gamma-Ray Bursts, ed. Kouveliotou, C., M. Briggs & Fishman, G. (AIP, New York), p. 941.
Harding, A. K., Contopoulos, I. & Kazanas, D. 1999, ApJ Letters, in press.

Harding, A. K. & Daugherty, J. K. 1999, Proc. of 3rd Integral Workshop, Ap. Lett. & Comm., 38, 25.
Harding, A. K. & Muslimov, A. G. 1998a, ApJ, 508, 328 [HM98].
Harding, A. K. & Muslimov, A. G. 1998b, ApJ, 500, 862.
Hurley, K. 2000, these proceedings.
Kuiper, L. et al. 1998, A & A, 337, 421.
Muslimov, A. G. & Tsygan, A. I. 1992, MNRAS, 255, 61.
Nel, H. I. et al. 1993, ApJ, 418, 836.
Paczynski, B. 1992, Acta Astron., 42, 145.
Romani, R. W. 1996, ApJ, 470, 469.
Romani, R. W. & Yadigaroglu, I.-A. 1995, ApJ, 438, 314.
Ruderman, M.A. & Sutherland, P. G. 1975, ApJ, 196, 51.
Saito, Y. 1998, Ph.D. Thesis, Univ. of Tokyo.
Shibazaki, N. & Lamb, F. K. 1989, ApJ, 346, 808.
Strickman, M., Harding, A. K. & De Jager, O. C., 1999, ApJ, 524, 373.
Sturrock, P.A. 1971, ApJ, 164, 529.
Thompson, C. & Blaes, O. 1998, Phys. Rev. D, 57, 3219.
Thompson, C. & Duncan, R. C. 1995, MNRAS, 275, 255 [TD95].
Thompson, D. J., Harding, A.K., Hermsen, W. & Ulmer, M.P. 1997, in Proc. of the 4th Compton Symposium, ed. C.D. Dermer, M.S. Strickman & J.D. Kurfess (AIP 410: New York), 39.
Wang, F. Y.-H. et al. 1998, ApJ, 498, 373.
Zhang, L. & Cheng, K. S. 1997, ApJ, 487, 370.
Zhang, B. & Harding, A. K. 1999, ApJ, in press [ZH99].

THE 4 AND ONE-HALF (PLUS OR MINUS ONE HALF) SOFT GAMMA REPEATERS IN REVIEW

K. Hurley

UC Berkeley
Space Sciences Laboratory
Berkeley, CA 94720-7450

ABSTRACT Four Soft Gamma Repeaters (SGRs) have now been identified with certainty, and a fifth has possibly been detected. I will review their X-ray and gamma-ray properties in both outburst and quiescence. The magnetar model accounts fairly well for the observations of SGR1806-20 and SGR1900+14, but data are still lacking for SGR1627-41 and SGR0525-66. The locations of the SGRs with respect to their supernova remnants suggest that they are high velocity objects.

KEYWORDS: stars: neutron; stars: magnetic fields; gamma rays: bursts

1. INTRODUCTION

The Soft Gamma Repeaters are sources of short, soft-spectrum ($\leq$ 100 keV) bursts with super-Eddington luminosities. They undergo sporadic, unpredictable periods of activity, sometimes quite intense, which last for days to months, often followed by long periods (up to years or decades) during which no bursts are emitted. Very rarely, perhaps every 20 years, they emit long duration *giant flares* which are thousands of times more energetic than the bursts, with hard spectra ($\sim$ MeV). The SGRs are quiescent, and in some cases periodic, 1-10 keV soft X-ray sources as well. They all appear to be associated with supernova remnants, and a good working hypothesis is that they are all *magnetars*, i.e. highly magnetized neutron stars for which the magnetic field energy dominates all other sources, including rotation (Duncan and Thompson 1992; Thompson and Duncan 1995). Figure 1 shows the time histories of bursts from SGR1900+14, and figure 2 shows a typical energy spectrum.

In this paper, I will mainly review the radio, X-ray, and gamma-ray properties of the SGRs in outburst and in quiescence, and indicate how the magnetar model accounts for these properties.

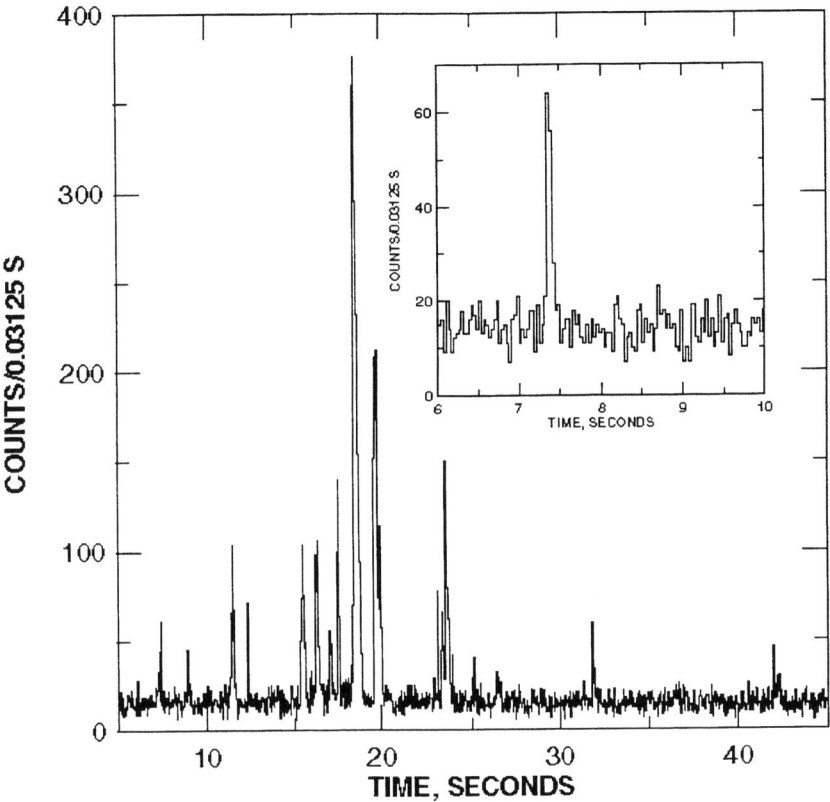

FIGURE 1. From Hurley et al. (1999a). Inset: a typical burst from SGR1900+14 as observed in the 25-150 keV range by *Ulysses* . Main figure: bursts during a period of intense activity.

2. SGR1806-20

Kulkarni and Frail (1993) suggested that this SGR was associated with the Galactic supernova remnant (SNR) G10.0-0.3, based on its localization to a $\sim 400\,\mathrm{arcmin}^2$ error box by the old interplanetary network (IPN) (Atteia et al. 1987). This was confirmed when ASCA observed and imaged the source *in outburst* , localizing it to a 1' error circle (Murakami et al. 1994). A quiescent soft X-ray source was also detected by Cooke (1993) using the ROSAT HRI. Based on more recent observations, Kouveliotou et al. (1998) have found that the quiescent source is periodic (P=7.48 s) and is spinning down rapidly ($\dot{P}=2.8 \times 10^{-11}$s/s). If this spindown is interpreted as being due entirely to magnetic dipole radiation, the implied field strength is B=8 $\times 10^{14}$G. The 2-10 keV X-ray luminosity of the source is 2×10^{35}erg/s, and the low energy X-ray spectrum may be fit by a power law with index 2.2.

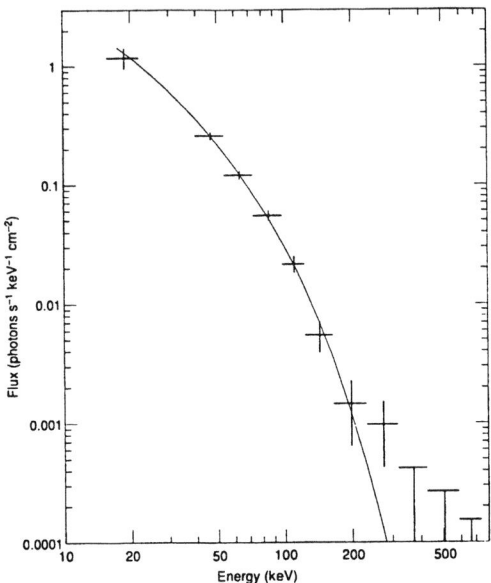

FIGURE 2. Reprinted by permission from Nature (Kouveliotou et al. 1993) copyright 1993 Macmillan Magazines Ltd.. Typical spectrum of a burst from SGR1900+14 as observed by BATSE. The spectrum is fit here with an optically thin thermal bremsstrahlung function, with kT=39 keV.

The SNR G10.0-0.3 has a non-thermal core, and Frail et al. (1997) have detected changes in the radio contours of the core on $\sim$ year timescales. Van Kerkwijk et al. (1995) have found an unusual star at the center of this core, which they identify as a luminous blue variable (LBV). The presence of this object has been a mystery up to now, because it was thought that the SGRs were single neutron stars. Recent work from the 3rd IPN has shed some light on this issue (Hurley et al 1999b). Figure 3 shows the location of the SGR superimposed on the radio contours of the SNR. It can be seen that the SGR is in fact offset from the LBV. The LBV may be powering the non-thermal core of the SNR, and causing the changes in the radio contours. It is also possible that the SGR progenitor was once bound to the LBV, but that it became unbound when it exploded as a supernova. A transverse velocity of $\sim$100 km/s would then be required to explain the displacement between the two. Alternatively, it is possible that the apparent SGR-SNR association is due to a chance alignment of these two objects along the line of sight.

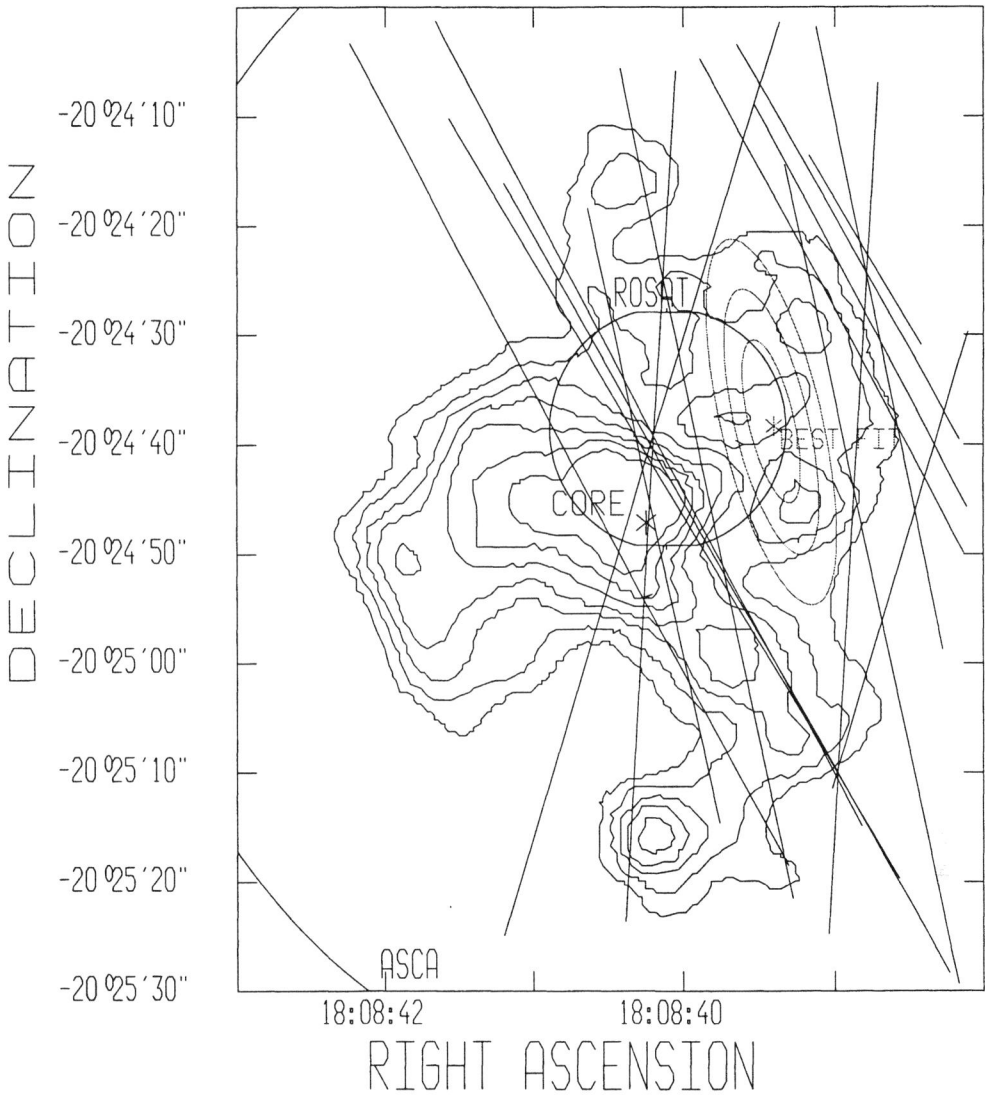

FIGURE 3. From Hurley et al. (1999b). Eight IPN annuli (lines), and the 1, 2, and 3 σ equivalent confidence contours (ellipses) for SGR1806-20. The best fit position and the position of the non-thermal core are indicated. The ASCA error circle is just visible in the lower left and upper left hand corners; its radius is 1', quoted as a systematic error, with no confidence limit given (Murakami et al. 1994). The ROSAT PSPC error circle is at the center; its radius is 11", with no confidence limit quoted (Cooke et al. 1993). We have reanalyzed the ROSAT data and confirm the presence of a weak source at this position, but are unable to establish confidence limits for its position. The 3.6 cm radio contours of G10.0-0.3 are also shown, from Vasisht, Frail, & Kulkarni (1995).

3. SGR1900+14

SGR1900+14 was discovered by Mazets et al. (1979) when it burst 3 times in two days. A precise localization by the IPN (Hurley et al. 1999a) showed that this source lay just outside the Galactic SNR G42.8+0.6, with an implied proper motion >1000 km/s. The SGR is associated with a quiescent soft X-ray source (Vasisht et al. 1994; Hurley et al. 1999c; Kouveliotou et al. 1999). The quiescent source has a period 5.16 s, and a period derivative 6.1 $\times 10^{-11}$s/s; again, assuming purely dipole radiation, $B \sim 8 \times 10^{14}$G. The 2-10 keV luminosity is 3 $\times 10^{34}$erg/s, and the spectrum may be fit with a power law of index 2.2.

On 1998 August 27, the SGR emitted a giant flare which was probably the most intense burst ever detected at Earth (Hurley et al. 1999d). Its luminosity was 2 $\times 10^{43}$ erg/s in >25 keV X-rays, or $10^5 L_E$ (the Eddington luminosity). The time history of this burst clearly displayed the 5.16 s periodicity of the quiescent source (figure 4). The magnetic field strength required to contain the electrons responsible for the X-ray emission is $> 10^{14}$G; this constitutes an independent argument for the presence of strong fields in SGRs. From measurements of the ionospheric disturbance which this burst caused, Inan et al. (1999) have estimated that there must have been one order of magnitude more energy in 3-10 keV X-rays than in >25 keV X-rays, bringing the total energy to $\sim 4 \times 10^{44}$erg. Frail et al. (1999) detected a transient radio source with the VLA at the SGR position following the giant flare. This is the only case where a radio point source is present at an SGR position.

4. SGR0525-66

This SGR was discovered when it emitted the giant flare of 1979 March 5 (Cline et al. 1980, Golenetskii et al. 1979). It was localized by the IPN to a 0.1 arcmin^2 error box within the N49 supernova remnant (Evans et al. 1980). For an LMC distance of 55 kpc, this burst had a luminosity of 5×10^{44} erg/s in X-rays >50 keV, or $2 \times 10^6 L_E$; the total energy emitted was $\sim 7 \times 10^{44}$erg in >50 keV X-rays. The time history displayed a clear 8 s periodicity (Barat et al. 1979). Paczynski (1992) was the first to suggest a strongly magnetized neutron star as the origin of this burst. Although the source remained active through 1983 (Golenetskii et al. 1987), it has not been observed to burst since then.

Rothschild et al. (1994) found a quiescent soft X-ray point source in the SGR error box with a ROSAT HRI observation. As no energy spectra are obtained from the HRI, the soft X-ray luminosity can only be estimated by assuming various spectral shapes. The 0.1-2.4 keV luminosity is in the range $10^{36} - 10^{37}$ erg/s, depending on the assumed spectrum. No periodicity was detected in this observation, but the upper limit to the pulsed fraction is only 66%. If the age of the N49 SNR is taken to be 5 kyr (Vancura et al. 1992), the implied transverse velocity of the SGR is several thousand km/s. *Chandra* observations of the SNR are scheduled, and are bound to reveal more about this interesting object.

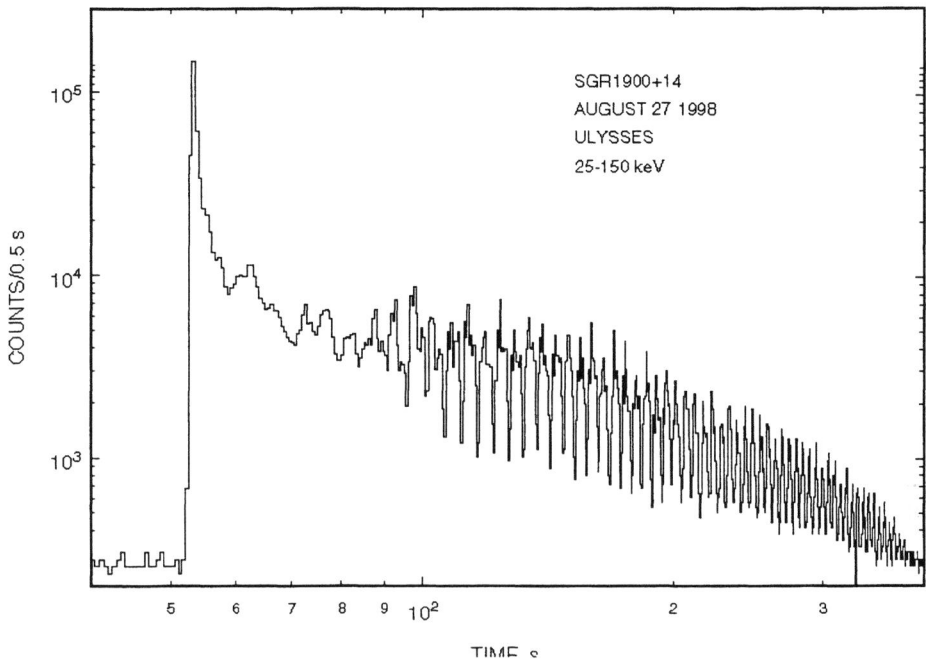

FIGURE 4. Reprinted by permission from Nature (Hurley et al. 1999d) copyright 1999 Macmillan Magazines Ltd.. The *Ulysses* 25-150 keV time history of the 1998 August 27 giant flare from SGR1900+14. Note the 5.16 s periodicity.

5. SGR1627-41

SGR1627-41 burst about 100 times in June-July 1998, and has not been observed to burst since then. During that period, observations by BATSE (Woods et al. 1999a), *Ulysses* (Hurley et al. 1999e), KONUS-*Wind* (Mazets et al. 1999), and RXTE (Smith et al. 1999) led to a precise source localization. The SGR lies near the SNR G337.0-0.1, at a distance of ~ 11 kpc. The implied transverse velocity of the SGR is in the range 200 - 2000 km/s. Although no giant flare has been observed from this source, there is a KONUS-*Wind* observation of an extremely energetic event (Mazets et al. 1999). The luminosity and total energy of the burst in the >15 keV range were $\sim 8 \times 10^{43}$ erg/s and $\sim 3 \times 10^{42}$ erg/s, respectively.

Like the other SGRs, this one also appears to be a quiescent soft X-ray source. *BeppoSAX* observations revealed a variable source with spectral index 2.1 and luminosity $\sim 10^{35}$ erg/s (Woods et al. 1999b). Although the *BeppoSAX* observations gave weak evidence for a possible 6.4 s periodicity, this was not confirmed in later ASCA observations of the source with better statistics (Hurley et al. 1999f).

6. SGR1801-23

The latest SGR to be discovered is 1801-23 (Cline et al. 1999). It was observed to burst just twice, on June 29, 1997, by *Ulysses*, BATSE, and KONUS-*Wind*. The burst spectra were soft, and could be fit by an optically thin thermal bremsstrahlung function with a kT of ~ 25 keV. The time histories were short. In both respects, then, the source properties resemble those of the other SGRs. However, because only two bursts were observed, and they occurred on the same day, the IPN localization is not very precise. The error box is 3.8° long, and has an area of $\sim 80\,\text{arcmin}^2$. The source lies in the general direction of the Galactic center, and the error box crosses numerous possible counterparts (figure 5). The source would have a super-Eddington luminosity for any distance > 250 pc; at the approximate distance of the Galactic center, its luminosity would be $1200 L_E$. At present, the best hypothesis is that this source is indeed an SGR; recall that SGR1900+14 was similarly detected when it burst 3 times in two days, and it remained quiescent for many years. Like SGR1900+14, the identification of SGR1801-23 may have to await a new period of bursting activity.

Table 1 summarizes the essential properties of all the SGRs.

TABLE 1. Essential properties of the SGRs

SGR	Super-Eddington Bursts?	Giant Flare?	Periodicity Observed in Burst?	Quiescent Soft X-ray Source? (erg/s)	Periodicity in Quiescent Source?	$\dot{P}$ 10^{-11} s/s
1806-20	1000×	No	No	2×10^{35}	7.47 s	2.8
1900+14	1000×	270898	5.16 s	3×10^{34}	5.16 s	6.1
0525-66	20000×	050379	8 s	10^{36-37}	No	—
1627-41	400000×	No	No	10^{35}	6.4 s?	—
1801-23	?	No	No	?	—	—

7. THE MAGNETAR MODEL

Briefly, the magnetar model (Duncan and Thompson 1992; Thompson and Duncan 1995) explains the short, soft bursts by localized cracking on the neutron star surface, with excitation of Alfven waves which accelerate electrons. Every 20–100 y, a massive, global crustquake takes place. Regions of the neutron star with magnetic fields of opposite polarity suddenly encounter one another, resulting in magnetic field annihilation and energization of the magnetosphere, giving rise to a giant flare. Magnetars are thought to be born in ~ 1 out of 10 supernova explosions, and remain active for perhaps 10,000 y. Thus there should be about 10 active magnetars in the Galaxy at any given time. So far, we have found 4.5 ± 0.5. Stay tuned for more!

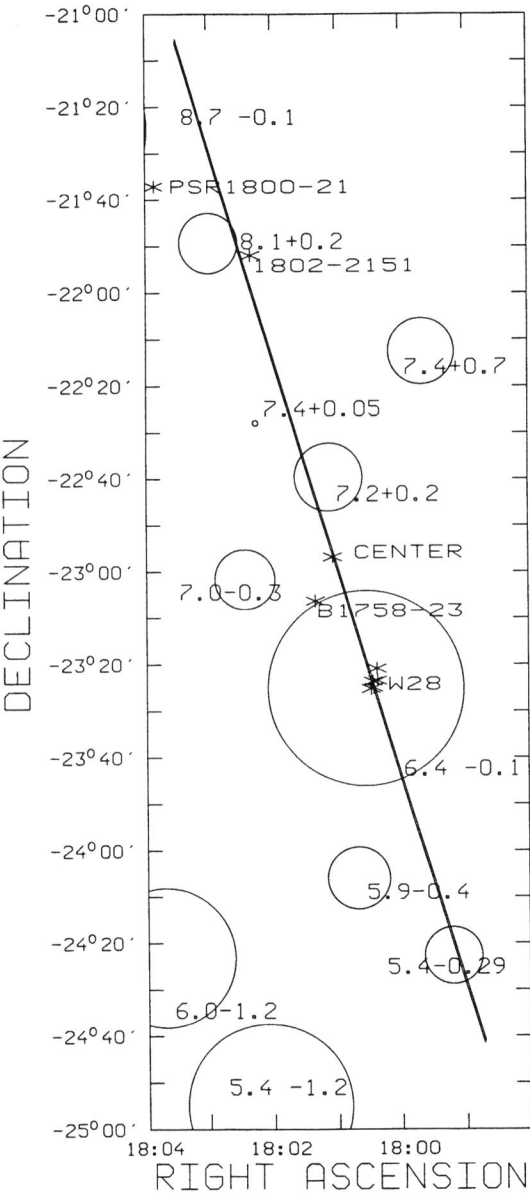

FIGURE 5. From Cline et al. (1999). IPN error box for SGR1801-23 (the lines are too closely spaced to distinguish). The center is indicated with an asterisk. Circles give the approximate locations of confirmed and suspected SNRs; the radii have been taken as half the size given in the catalogs. Asterisks give the positions of ROSAT X-ray sources, and two pulsars, PSR1800-21 and B1758-23, probably associated with SNRs 8.7-0.1 and 6.4-0.1. Coordinates are J2000.

ACKNOWLEDGEMENTS

We are grateful to JPL support of *Ulysses* operations under Contract 958056 and to NASA for grant NAG5-7810.

REFERENCES

Atteia, J.-L. et al. 1987, Ap. J. 320, L105
Barat, C. et al. 1979, Astron.Astrophys. 79, L24
Cline, T. et al. 1980, Ap. J. 237, L1
Cline, T. et al. 1999, Ap. J., accepted, astro-ph/9909054
Cooke, B. 1993, Nature 366, 413
Duncan, R., and Thompson, C. 1992, Ap. J., 392, L9
Evans, W. D. et al. 1980, Ap. J. 237, L7
Frail, D. et al. 1997, Ap. J. 480, L129
Frail, D. et al. 1999, Nature 398, 127
Golenetskii, S. et al. 1979, Sov. Astron. Lett. 5, 340
Golenetskii, S. et al. 1987, Sov. Astron. Lett. 13(3), 166
Hurley, K. et al. 1999a, Ap. J., 510, L107
Hurley, K. et al. 1999b, Ap. J., 523, L37
Hurley, K. et al. 1999c, Ap. J. 510, L111
Hurley, K. et al. 1999d, Nature 397, 41
Hurley, K. et al. 1999e, Ap. J. 519, L143
Hurley, K. et al. 1999f, Ap. J., submitted, astro-ph/9909355
Inan, U. et al. 1999, GRL, 26(22), 3357
Kouveliotou, C. et al. 1993, Nature, 362, 728
Kouveliotou, C. et al. 1998, Nature, 393, 235
Kouveliotou, C. et al. 1999, Ap. J. 510, L115
Kulkarni, S., and Frail, D. 1993, Nature 365, 33
Mazets, E. et al. 1979, Sov. Astron. Lett. 5(6), 343
Mazets, E. et al. 1999, Ap. J. 519, L151
Murakami, T. et al. 1994, Nature 368, 127
Paczynski, B. 1992, Acta Astronomica 42, 145
Rothschild, R. et al. 1994, Nature 368, 432
Smith, D. et al. 1999, Ap. J. 519, L147
Thompson, C., and Duncan, R. 1995, Mon. Not. R. Astron. Soc., 275, 255
van Kerkwijk, M. et al. 1995, Ap. J. 444, L33
Vancura, O. et al. 1992, Ap. J. 394, 158
Vasisht, G. et al. 1994, Ap. J. 431, L35
Vasisht, G. et al. 1995, Ap. J. 440, L65
Woods, P. et al. 1999a, Ap. J. 519, L139
Woods, P. et al. 1999b, Ap. J. 519, 139

X-RAY PROPERTIES OF ULTRA-LUMINOUS INFRARED GALAXIES (ULIGS)

K. Iwasawa

Institute of Astronomy, Madingley Road, Cambridge CB3 0HA, UK

ABSTRACT Recent ASCA results on ULIGs are reviewed. A search for obscured active nuclei in 24 ASCA ULIGs is reported. X-ray emission from ULIGs are, in general, very faint and a further X-ray sensitivity is required to test whether they contain heavily obscured AGN.

KEYWORDS: galaxies: infrared — galaxies: X-Ray

1. INTRODUCTION

The population of galaxies characterized by a large infrared excess in their spectral energy distribution (SED) has been recognized as an important subject to investigate in understanding galaxy evolution, the genesis of active galactic nuclei (AGN), background radiation in various wave bands (see the review by Sanders & Mirabel 1996 and references are therein). Particularly, objects emitting infrared luminosity larger than $10^{12} L_\odot$ are termed as ultra-luminous infrared galaxies (ULIGs). The origin of the large energy output in the infrared band has been discussed extensively since the discovery by IRAS. Whether a starburst or AGN is a major power source is not entirely clear.

One important aspect of ULIGs for X-ray astronomy is their potential importance in producing the diffuse X-ray background (XRB). Generally accepted AGN synthesis models for the XRB require a significant number of heavily absorbed sources (Setti & Woltjer 1989). Fabian & Iwasawa (1999) has pointed out, from a study of the XRB, that a large fraction (as large as 85 per cent) of accretion-powered X-ray radiation in the universe might be absorbed. ULIGs are found to contain a large amount of gas and dust in their nuclear regions and considered to be an important phase for forming AGN (e.g., Sanders et al 1988). Given the nature of ULIGs, one could expect heavily obscured AGN in them. Earlier X-ray studies of ULIGs showed that they are faint in X-ray (Rieke 1988). However, this is consistent with being strongly absorbed X-ray sources. With the sensitivity available from ASCA and BeppoSAX, we are limited to a handful of nearby objects for studying their X-ray spectral properties in detail.

Bulk of IRAS-detected ULIGs are located at moderate redshifts as a result of the limited sensitivity of IRAS (see Kim & Sanders 1998). Since the fraction of objects classified as AGN increases as the bolometric luminosity goes up (Sanders & Mirabel

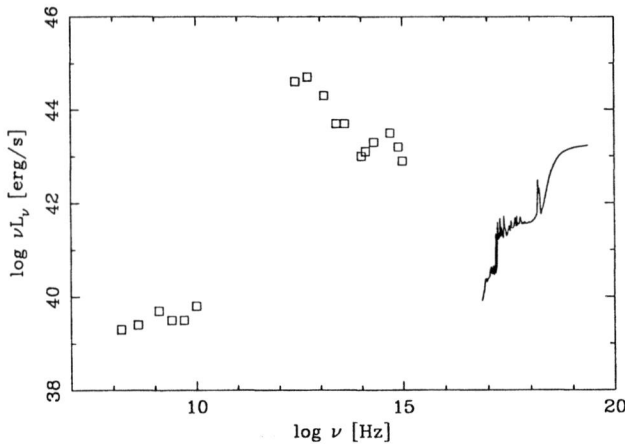

FIGURE 1. The spectral energy distribution of NGC6240. The data are taken from NED and Vignati et al (1999).

1996), a survey of ULIGs at higher redshifts, where a majority of the XRB sources are usually located ($z = 1$–2), is required. While powerul missions such as Chandra and XMM will provide more appropriate X-ray sensitivity to study the ULIGs lying in the important redshift range, it is worth summarising the current status of X-ray results on ULIGs obtained from ASCA and BeppoSAX (for BeppoSAX results, see Risaliti et al in this proceeding).

2. A BURIED ACTIVE NUCLEUS IN THE PROTOTYPE FIRG, NGC6240

Optical and infrared spectroscopic surveys of ULIGs suggest the importance of starburst in the energetics (Genzel et al 1998; Lutz, Veilleux & Genzel 1999). One advantage of X-ray technique in searching for heavily obscured sources is the penetrating power of hard X-rays. An intensive starburst surrounding the nuclear region of a ULIG could easily mask the AGN activity in the central nucleus, as suggested by Fabian et al (1998). This is certainly the case for a nearby far-infrared galaxy NGC4945 which shows clear evidence for AGN in the X-ray band above 10 keV, first revealed by a Ginga observation (Iwasawa et al 1993), but no sign of AGN in optical and infrared. A similar case is found in the higher luminosity object NGC6240, which is one of the prototype far-infrared galaxies. This galaxy possesses typical characteristics of ULIGs (although its L_{ir}, just below $10^{12} L_\odot$, is not enough for NGC6240 to qualify as a ULIG): two nuclei in this merging system, large infrared excess in the SED, "superwind" signatures, large amount of molecular gas in the central region of the system. Optical spectroscopy classifies this object as a LINER and the ISO SWS observation classifies it as a starburst (Genzel et al 1998). However, the ASCA and BeppoSAX observations of this object have revealed an active nucleus deeply buried in an obscuration that are only accessible with hard X-rays.

The spectral energy distribution of NGC6240 is shown in Fig. 1. Luminous thermal emission which is likely due to the starburst Ais detected in the soft X-ray band. An extended X-ray emission has been found with the ROSAT HRI (Komossa et al 1998). The ASCA spectrum above 3 keV is flat and shows a strong iron K line feature (Mitsuda 1995; Iwasawa & Comastri 1998; Netzer, Turner & George 1998), indicating that this energy range is dominated by reflection. A subsequent BeppoSAX observation shows a further strong excess above 10 keV, which is attributed to a primary X-ray source absorbed by Thomson-thick matter with $N_H \sim 2 \times 10^{24} \text{cm}^{-2}$ (Vignati et al 1999). Although this component is significantly suppressed by the large column density (for which Compton down-scattering becomes important), it is possible to estimate the insrinsic luminosity of the primary continuum. The absorption-corrected 2–10 keV luminosity is $2 \times 10^{44} \text{erg s}^{-1}$ which is near the higher luminousity end of Seyfert nuclei or low luminosity end of QSO. Assuming a bolometric correction for a typical QSO (see e.g., Elvis et al 1994), the bolometric luminosity of the obscured AGN in NGC6240 is comparable with that observed in the infrared band. Therefore the AGN contribution to the total energetics in this object appears to be significant.

3. ASCA ULIG SAMPLE

3.1. Sample

X-ray data of ULIGs were collected from the ASCA public archive, thus the sample is by no means complete. The selection criteria applied here are 1) the 8–1000μm luminosity exceeding $10^{12} L_\odot$ (H_0=75km s^{-1} Mpc^{-1} and q_0=0); 2) detection at the IRAS 60μm band in the Faint Source Catalogue (FSC); 3) being observed by either the ASCA SIS or GIS within 20 arcmin from the centre of the field of view. Details of the sample galaxies are listed in Table. 1. Applying the strict luminosity limit, several objects that have sometimes been regarded as ULIGs are dropped (e.g., Mrk463E).

The number of ULIGs selected are 24, including four hyper-luminous infrared galaxies of which bolometric luminosity in excess of $10^{13} L_\odot$. The infrared luminosities are calculated using the formula given in Kim & Sanders (1998) and the IRAS measurements. The optical classifications were taken mainly fom Kim, Veilleux & Sanders (1998): S1 (quasar or Seyfert 1); S2 (Seyfert 2); L (LINER); SB (starburst). Number distribution of the sample against redshift and L_{ir} are shown in Fig. 2. Four objects at redshifts higher than 0.2 are shown together on the right end of the figure.

X-ray images are created by adding the same type of detectors together, e.g., SIS0+SIS1 in the 0.5–10 keV band, and GIS2+GIS3 in the 0.7–10 keV band, for an inspection of X-ray detection. Pointing errors of the satellite have been corrected using the Look-up table provided by the ASCA Guest Observer Facility (GOF) so that positional uncertainty of the SIS image should be smaller than 20 arcsec (Gothelf & Ishibashi 1997). Occasionally, this pointing error is as large as 1 arcmin. In a few cases, this correction made a difference in detection status, i.e., some targets which have been reported to be no detection turned out to be detected positively

TABLE 1. The ASCA ULIG sample. The optical classes are mainly taken from Kim, Veilleux & Sanders (1998): S1:(Seyfert1 and QSO), S2 (Seyfert 2), L (LINER), H (HII type, or starburst). The infrared and X-ray luminosities are calculated using ($H_0 = 75$km s^{-1} Mpc^{-1}, q_0=0) for the wave bands of 8–1000μm and *observed* 2–10 keV, respectively. When a galaxy is not detected or too weak to perform a spectral analysis, the X-ray spectral shape of F05189–2524 is assumed to derive the X-ray luminosity from the count rate obtained from the imaging analysis. The 2σ of the local background is given as the X-ray upper limit.

FSC name	Other name	z	log L_{ir} $L_\odot$	Optical class	log L_X [erg/s]
00235+1024		0.575	12.83	H	< 43.73
01355−1814		0.192	12.39	H	< 42.96
05189−2524		0.042	12.07	S2	43.26
07599+6508		0.149	12.46	S1	42.88
08572+3915		0.058	12.11	L	< 42.32
09105+4108		0.442	13.14	S2	45.04
09320+6134	UGC5101	0.039	12.01	L	41.98
10214+4724		2.286	14.00	S2	< 45.00
12265+0219	3C273	0.159	12.73	S1	45.79
12540+5708	Mrk231	0.042	12.50	S1	42.28
13349+2438		0.108	12.25	S1	44.15
13428+5608	Mrk273	0.038	12.10	S2	42.30
13451+1232	PKS1345+12	0.122	12.28	S2	43.43
13454−2956		0.129	12.21	S2	< 42.86
14394+5332		0.105	12.04	S2	< 42.54
15307+3252		0.926	13.64	S2	< 44.36
15327+2340	Arp220	0.018	12.11	L	41.18
16348+7037	PG1634+706	1.337	14.00	S1	46.00
17207−0014		0.043	12.35	H	41.81
19254−7245	Super Antennae	0.062	12.04	S2	42.51
20460+1925		0.181	12.43	S2	44.15
22560−3501		0.172	12.07	H	< 43.20
22204−0214		0.140	12.10	H	< 42.94
23060+0505		0.174	12.44	S2	43.90

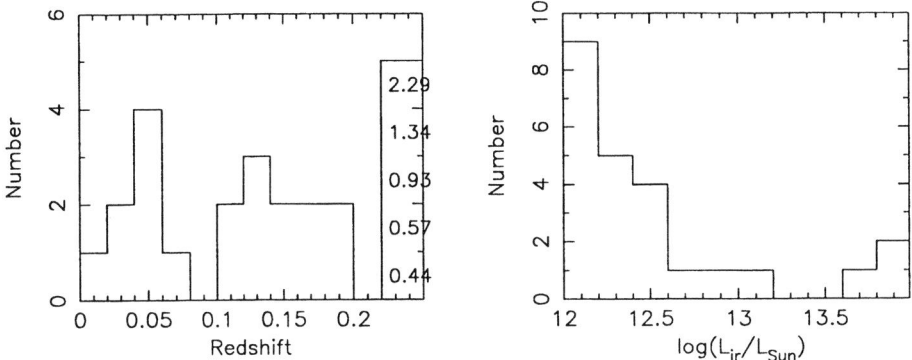

FIGURE 2. Redshift and infrared (8–1000μm) luminosity distribution of the ASCA ULIG sample. Objects with redshifts higehr than 0.2 are indicated together on the the right-end of the plot in the left panel.

TABLE 2. X-ray detection for the ASCA ULIG sample. Objects detected with 3σ or larger significance are counted for each optical class: (S1: Seyfert1 or QSO; S2: Seyfert 2; L: LINER; H: HII or starburst).

Opt. class	N_{Total}	N_{Det}
S1	5	5
S2	11	7
L	3	2
H	5	1

(UGC5101, for example). Objects deteted above 3σ of the local background are decided as positive detection. For objects of no detection, upper limits of the 2–10 keV flux are given, converted from the 2σ upper limit in count rate obtained from the SIS image when it is available (otherwise the GIS result is used) assuming the spectral shape of the bright, one of the nearest ULIG, F05189 (see the next section).

The X-ray detectability for objects in each optical class is shown in Table 2. A spectral analysis was performed only for bright enough objects to search for X-ray absorption or strong iron K line as signatures of reflection-dominated spectrum in a heavily obscured AGN. Unabsorbed spectra in the S1 objects, and a few absorbed spectra are observed in S2 objects whilst no reflection-dominated spectra are found in the sample.

4. THE NEAREST FIVE ULIGS

The nearest five ULIGs, F05189-2524, Mrk231, Arp220, Mrk273, and UGC5101, are all detected with ASCA. X-ray emission probably due to AGN is found in F05189, Mrk273 and Mrk231. These objects are classified as Seyfert 1 or Seyfert 2 based on the optical emission line properties. The ASCA spectra show that the observed X-ray emission is due to an absorbed primary source definitely in IRAS05189, probably in Mrk231 and Mrk273. However, the absorption-corrected X-ray luminosities (in the 2–10 keV band) is about 3 orders of magnitude below the infrared luminosities. The ASCA SIS spectrum of the brightest object F05189 is shown in Fig. 3. The ASCA spectra of Mrk231, Mrk273 and Arp220 (see Fig. 4) and their implications are disscussed in Iwasawa (1999). For the two LINERs, there is no significant hard X-ray emission indicating presence of AGN in Arp220 while UGC5101 is too faint to perform a spectral analysis. It should be noted that, in Arp220, not only the AGN-related X-ray emission is faint but the thermal soft X-ray emission due to the starburst is also notably weak ($L_{0.5-2\text{keV}} \sim 4 \times 10^{40}\text{erg s}^{-1}$) relative to the luminous infrared emission.

5. OBSCURED AGN IN THE ASCA ULIG SAMPLE

In the rest of the ASCA ULIG sample, a few more objects (PKS1345+12, Super Antennae, F23060, F20460) show evidence of X-ray absorption with column densities of order of 10^{23}cm^{-2}. The ASCA results for these sources have been published by Brandt et al (1997), Ogasaka et al (1997) and Imanishi & Ueno (1999). Two objects F20460 and F23060, in which evidnce for hidden BLRs have been found, have the absorption-corrected 2–10 keV luminosities exceeding 10^{44}erg s^{-1}, well in the range of quasars, hence they can be called as type 2 quasars. The inferred column desnities of X-ray absorption in these objects are order of 10^{23}cm^{-2}, much smaller than that in Compton-thick objects like NGC6240. Results from ASCA and ROSAT HRI observations of F15307 have been reported by Fabian et al (1996) and Ogasaka et al (1997).

F09105+4108 is considered as a type 2 quasar at $z = 0.442$ based on the results

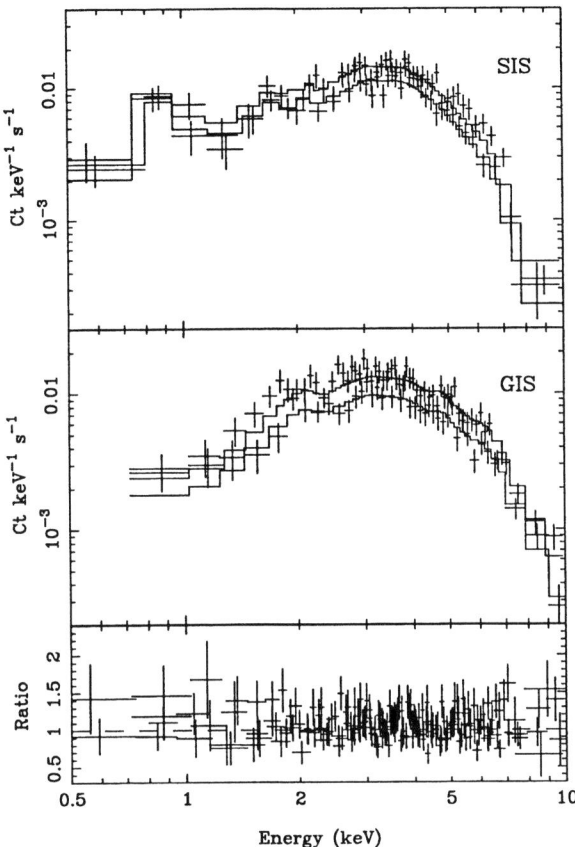

FIGURE 3. The ASCA spectrum of F05189−2524. The power-law continuum is absorbed by $N_H \sim 5 \times 10^{22}\,\mathrm{cm}^{-2}$.

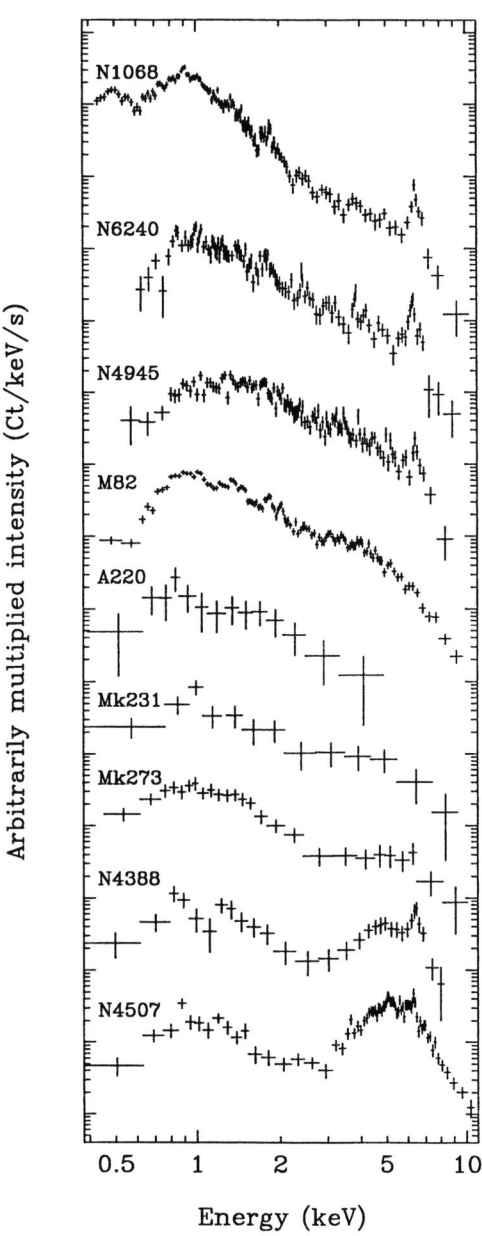

FIGURE 4. The ASCA SIS spectra of some ULIGs, NGC6240, nearby Seyfert 2 galaxies and starburst galaxy M82. Reflection-dominated objects (NGC1068, NGC6240, NGC4945) are shown in the upper part while absorbed sources (NGC4388, NGC4507) are in the lower part of the figure.

from optical spectroscopy and polarimetry (Kleinmann et al 1988; Hines et al 1999). Since this galaxy lies in a cooling flow of a rich cluster which is a luminous X-ray source (Fabian & Crawford 1995), the detection of a strong iron K line with ASCA (Fabian et al 1994) does not necessarily imply the presence of an hidden AGN. However, a recent BeppoSAX observation of this object has detected a hard X-ray excess above 10 keV and it is a strong case for a heavily absorbed quasar nucleus in this hyper-luminous infrared galaxy (Franceschini et al in this proceeding).

6. SUMMARY

This ASCA survey confirms that ULIGs are weak in X-ray, as previously suggested. Some fraction of the sample galaxies which have evidence for AGN in optical show absorbed X-ray sources. No significant detection of hard X-ray emission in Arp220 apparently supports the idea that a starburst is a major energy source, as concluded from the ISO results, although the weak thermal X-ray emission due to the starburst is also puzzling. No reflection-dominated spectrum source (or a Compton-thick source), apart from F09105, is found. This probably means that present ASCA observations are not deep enough to detect sources absorbed by a large column density at the distance farther than the nearby objects like NGC4945 and NGC6240. So far the detection of reflection-dominated spectrum sources are limited to the objects in the nearby universe, although the systematic BeppoSAX survey increased the number of Compton-thick sources (e.g, Risaliti et al 1999). Indeed, the strong iron-K AGN observed with ASCA are nearby Seyfert 2s and the 2–10 keV flux is usually order of 10^{-13}erg cm^{-2} s^{-1}. Given the dusty/gaseous environment of the nuclear regions of ULIGs, even the reflected light could also be absorbed significantly. The small fraction of reflected light relative to the primary emission naturally explains the difficulty with detecting Compton-thick sources at the redshift range where most ULIGs are populated.

As mentioned in the Introduction, there is a demands of a population of strongly absorbed X-ray sources to account for the XRB. The faint X-ray sources such that have been found in the deep ROSAT HRI survey (Hasinger et al 1998), show similar SED to that of NGC6240 (Hasinger, in this proceeding). Therefore, ULIGs still remain a strong candidate to harbour heavily obscured AGN at moderate to high redshifts. Recently discovered faint SCUBA sources (e.g., Barger et al 1998) could be the higher redshift counterpart of ULIGs. Further X-ray surveys of ULIGs with much improved X-ray sensitivity provided by Chandra and XMM will be a significant step towards understanding the XRB and the nature of ULIGs themselves.

ACKNOWLEDGEMENTS

The author thanks the organisers of the conference. Data taken from the NASA/IPAC Extragalactic Database (NED) are used. PPARC is acknowledged for support.

REFERENCES

Barger A.J., et al, 1998, Nat, 394, 248
Brandt W.N., et al, 1997, MNRAS, 290, 617
Elvis M., et al 1994, ApJS, 95, 1
Fabian A.C., et al, 1994 ApJ, 436, L51
Fabian A.C., Crawford C.S., 1995, MNRAS, 274, L75
Fabian A.C., Cutri R.M., Smith H.E., Crawford C.S., Brandt W.N., 1996, MNRAS, 283, L95
Fabian A.C., Barcons X., Almaini O., Iwasawa K., 1998, MNRAS, 297, L11
Fabian A.C., Iwasawa K., 1999, MNRAS, 303, L34
Genzel R., et al, 1998, ApJ, 498, 579
Gothelf E.V., & Ishibashi K., 1997, ASCA GOF ASCA News Letter
Hasinger G., Burg R., Giacconi R., Schmidt M., Trumper J., Zamorani G., 1998, A&A, 329, 482
Hines et al 1999, ApJ, 512, 145
Imanishi M., Ueno S., 1999, MNRAS, 305, 829
Iwasawa K., et al, 1993, ApJ, 409, 155
Iwasawa K., 1999, MNRAS, 302, 96
Iwasawa K., Comastri A., 1998, MNRAS, 297, 1219
Kim D.-C., Sandsers D.B., 1998, ApJ, 119, 41
Kim D.-C., Veilleux S., Sanders D.B., 1998, ApJ, 508, 627
Kleinmann S.G., et al, 1988, ApJ, 328, 161
Komossa S., Shultz H., Greiner J., 1998, A&A, 334, 110
Lutz D., Veilleux S., Genzel R., 1999, ApJ, 517, L13
Mitsuda K., 1995, 17thTexas Symposium, eds Böhringer, Morfill, Trümper, NY Acad. Sci., New York, p213
Netzer H., Turner T.J., George I.M., 1998, ApJ, 504, 680
Ogasaka Y., et al, 1997, PASJ, 49, 179
Rieke G., 1988, ApJ, 331, L5
Risaliti G., Maiolino R., Salvati M., 1999, A&A, 522, 157
Sanders D.B., et al. 1988, ApJ, 325, 74
Sanders D.B., Mirabel I.F., 1996, ARAA, 34, 749
Setti G., Woltjer I., 1989, A&A, 224, L21
Vignati P., et al, 1999, A&A, 349, L57

SPECTRAL SIGNATURES OF KILOHERTZ QUASI-PERIODIC OSCILLATIONS FROM ACCRETING NEUTRON STARS

Philip Kaaret

Harvard-Smithsonian Center for Astrophysics

ABSTRACT

Correlations discovered between millisecond timing properties and spectral properties in neutron star x-ray binaries are described and then interpreted in relation to accretion flows in the systems. Use of joint timing and spectral observations to test for the existence of the marginally stable orbit, a key prediction of strong field general relativity, is described and observations of the neutron star x-ray binary 4U 1820-303 which suggest that the signature of the marginally stable orbit has been detected are presented.

KEYWORDS: accretion, accretion disks — gravitation — relativity — stars: individual (4U 1820-303) — stars: neutron — X-rays: stars

1. INTRODUCTION

The orbital periods associated with the innermost orbits around solar mass compact objects are in the millisecond range. The millisecond quasi-periodic oscillations (QPOs) discovered (Strohmayer et al. 1996; van der Klis et al. 1996) with the *Rossi X-Ray Timing Explorer* (RXTE; Bradt, Rothschild, & Swank 1993) in the x-ray emission from accreting neutron stars are most likely associated with orbital motion near the neutron star, and, thus, potentially offer a new probe of accretion flows and strong gravitational fields near neutron stars in x-ray binaries (for a brief review, see Kaaret & Ford 1997). The key to exploiting the kHz QPOs as probes of strong gravity is to understand the QPOs at a sufficient level so that those aspects of the QPOs which depend on the properties of the gravitational field can be separated from the aspects which depend on details of the accretion physics. It may not be necessary to construct a full theory of the QPO generation mechanism, just as a binary pulsars are excellent test beds for the study of relativity even in the absence of an adequate theory for the generation of the radio pulses.

Here, I present some initial steps of attempts to understand the key quantitative features of the QPOs, such as the centroid frequency, and their relation to the properties of the accretion flow. I consider several key questions:

1. Is the x-ray spectral state of a source correlated with QPO frequency?

2. Is the QPO frequency related to the geometry of an accretion disk?

3. Is the QPO frequency determined by the mass accretion rate?

In the following, I present and then interpret observational data to attempt to answer these three questions.

If an adequate understanding of the QPOs can be obtained, we can then extend our study to address the properties of the gravitational field surrounding the neutron star. One important question is whether there exists an innermost radius, the marginally stable orbit also referred to as the innermost stable circular orbit, inside of which there exist no stable circular orbits. The lack of stability of circular orbits sufficiently close to a compact object is a key prediction of strong field general relativity (e.g. Misner, Thorne, & Wheeler 1970). The relevance of the existence of the marginally stable orbit to to the physics of accretion around black holes has long been understood (e.g. Shakura & Sunyaev 1973). More recently, it was pointed out by Kluzniak and Wagoner (1985) that, for most of the preferred equations of state of nuclear matter, the marginally stable orbit lies outside the neutron star radius and thus should be dynamically important for the accretion flow on to neutron stars. The kHz QPOs appear to provide an observational means to test these predictions.

2. CORRELATION OF QPO FREQUENCY WITH SPECTRAL STATE

We wish to use the kHz QPOs as kinematic probes and, thus, must understand the relation of the QPO parameters, particularly the centroid frequency, to the geometry of the accretion flow. We begin by searching for correlations in the time variations of QPO parameters and x-ray spectral properties.

The most obvious spectral parameter with which to correlate the QPO frequency is the total X-ray flux. In a neutron star system, accreted matter must eventually come to rest on the neutron star surface, and, therefore, the total luminosity, L, of a neutron star system must be directly proportional to the total mass accretion rate onto the stellar surface $\dot{M}$ via $L = GM\dot{M}/R$, where G is the gravitational constant, and M and R are the mass and radius of the neutron star. If the emission is isotropic, or, more generally, if the beaming pattern of the emission does not change with time, then the bolometric flux from a source must also be proportional to $\dot{M}$. In Fig. 1, we show the relation between QPO frequency and X-ray flux in the 2.5–25 keV band for the neutron star binary 4U 1728-34 (GX 354-0). The figure includes data from observations spanning three years. It is apparent from the figure that, while the QPO frequency and X-ray flux may be correlated for subsets of the data, the overall behavior shows no correlation. The lack of correlation between QPO frequency and X-ray flux may be due to one or more of several different factors: time variable beaming of the X-rays, the contribution of significant flux outside the measurement band, an outflow from the system, or a fundamental lack of physical correlation between QPO frequency and total mass accretion rate.

Spectral shape has long been employed as an indicator of mass accretion rate in neutron star systems. For sources with x-ray spectra which can be adequately

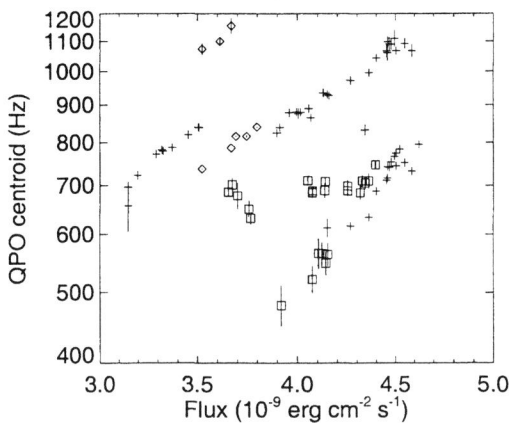

FIGURE 1. QPO centroid frequency versus x-ray flux in the 2.5–25 keV band for 4U 1728-34 (GX 354-0). Included are data from observations spanning three years. The different symbols indicate data from different years.

described with a simple power law model (often used for black hole candidates and low luminosity neutron star binaries), a natural indicator of spectral state is the power law photon index. Kaaret et al. (1998) showed that the QPO frequency is well correlated with photon index, but not with x-ray flux, in the neutron star x-ray binaries 4U 0614+091 and 4U 1608-52. This suggests that the spectral state and QPO frequency are related to a common physical parameter. Kaaret et al. (1998) suggested that common parameter is the mass accretion rate through the accretion disk.

A more general indicator of spectral state is an x-ray color, defined as the ratio of count rates in two different energy bands. X-ray colors can be applied to any source and do not rely on an assumed spectral model. However, they are strongly instrument dependent, making it difficult to compare colors obtained from different instruments, and they have no direct interpretation in terms of physical parameters of the source. X-ray colors have been used as spectral state indicators both individually and in pairs forming a "color-color" diagram. When their time varying spectral state is plotted on a color-color diagram, each individual neutron star x-ray binary tends to follow a well-defined track, see Fig. 2 where we present a color-color diagram for 4U 1728-34. This collapse of a potentially two dimensional pattern to a one-dimensional track suggests that a single parameter determines the spectral state. This parameter has usually been interpreted as the total mass accretion rate in the system (Hasinger & van der Klis 1989). However, the lack of correlation between total flux and spectral state may indicate that the parameter is not the total mass accretion rate.

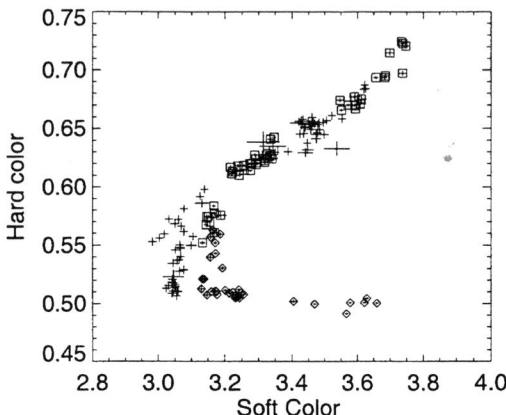

FIGURE 2. Color-color diagram for 4U 1728-34 (GX 354-0). The soft color is defined as the ratio of counts in the 3.5–6.4 keV band to counts in the 2.0–3.5 keV band and the hard color as 9.7–16.0 keV to 6.4–9.7 keV. The plot symbols are the same as in Fig. 1.

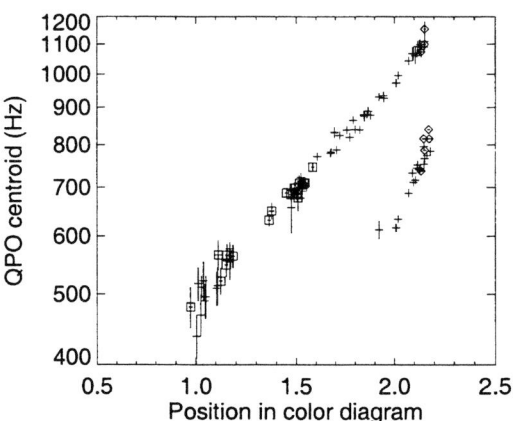

FIGURE 3. QPO centroid frequency versus position in color-color diagram for 4U 1728-34. Larger values of the position in the color-color diagram indicate higher inferred mass accretion rates. The upper and lower branches correspond to the upper and lower frequency kHz QPOs which often appear simultaneously. The plot symbols are the same as in Fig. 1.

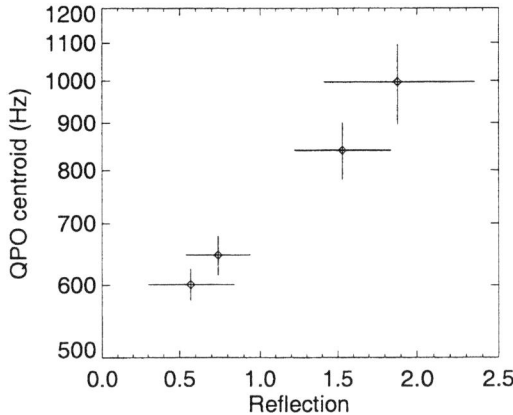

FIGURE 4. QPO frequency versus magnitude of reflection for 4U 0614+091. The inclination angle for the reflection component was fixed to 0°.

Position in the color-color diagram can be parameterized by position along a fiducial track drawn through the diagram (Hasinger et al. 1990; Hertz et al. 1992). In Fig. 3, we show the QPO frequency versus position in the color-color diagram (Fig. 2) for the same observations presented in Fig. 1. The correlation between QPO frequency and position in the color-color diagram appears robust across several years of observations. A robust correlation is found between QPO frequency and x-ray colors, either position in a color-color diagram or simply a hard x-ray color, for most neutron star x-ray binaries (e.g. Mendez 1999).

3. QPO FREQUENCY AND REFLECTION

The robust correlation between QPO centroid frequency and various indicators of spectral state, photon index or x-ray colors, suggest that a single physical parameter determines both spectral state and QPO frequency. To determine the physical nature of this parameter, we performed a detailed analysis of high quality x-ray spectra of 4U 0614+091 obtained with BeppoSAX (Piraino et al. 1999). When fit with a simple power-law model, the spectra showed strong and systematic residuals near 10–30 keV with a shape characteristic of that expected for a reflection component in the spectrum. Addition of reflection to the spectral model greatly improved the quality of the fits and gave residuals with no systematic variations. Thus, there is strong evidence for a reflection component in the spectra of 4U 0614+091. We found that the magnitude of reflection is well correlated with photon index, confirming a relation suggested by Zdziarski et al. (1999).

The reflection component found in the spectra requires the presence of cool matter located close to the primary x-ray source, presumably the neutron star, and subtending a large solid angle as viewed by the primary source. The most natural interpretation is that the reflection occurs in an accretion disk surrounding the neutron star. In this case, the magnitude of reflection should be related to the properties of the disk. In particular, if the disk has a variable inner radius, then higher magnitudes of reflection will result for smaller disk inner radii.

Fig. 4 shows the QPO centroid frequency plotted versus magnitude of reflection for 4U 0614+091. The QPO parameters are from RXTE observations while the magnitude of reflection is from BeppoSAX observations. For the two points with lower frequency, we had simultaneous RXTE and BeppoSAX observations. For the upper two points, we inferred the QPO frequency based on the QPO frequency versus photon index relation from Kaaret et al. (1998) and the photon index measured with BeppoSAX and allowing for a systematic offset in photon indices between BeppoSAX and RXTE measured using simultaneous observations. The large error bars for these two points are due to the uncertainty in this extrapolation. The QPO frequency appears correlated with the magnitude of reflection. The correlation is consistent with that expected if the QPO frequency is determined by the orbital frequency at the inner edge of the disk and the variation in reflection is due to changes in the inner disk radius.

4. KHZ QPOS AND ACCRETION GEOMETRY

I suggest the following physical picture to explain the correlations of QPO frequency with spectral state and magnitude of reflection presented in the previous sections. Mass accretion can occur in neutron star x-ray binaries both through the accretion disk and radially (e.g. Ghosh & Lamb 1978). The total mass accretion rate, disk plus radial, determines the total luminosity of the system. The mass accretion rate through the disk determines the radius of the inner edge of the disk and, via the dependence of the overall spectrum on the soft photon flux emitted from the disk, the spectral state of the system. The QPO frequency is determined by the radius of the inner edge of the disk. The data presented above are fully consistent with this picture.

To answer the questions posed earlier:

1. The spectral state of a source is correlated with QPO frequency as demonstrated by the robust correlations found between QPO frequency and photon index or x-ray color.

2. The QPO frequency is related to the geometry of the disk as demonstrated by the correlation between QPO frequency and magnitude of reflection in the x-ray spectrum.

3. The QPO frequency is determined by the mass accretion rate, but the important parameter is the mass accretion rate through the disk and not the total mass accretion rate.

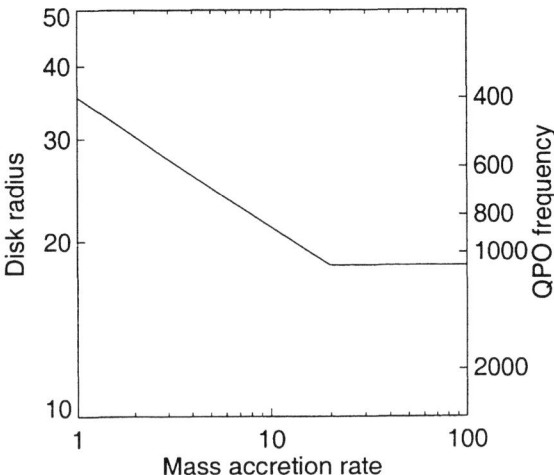

FIGURE 5. Disk radius versus mass accretion rate through the disk showing the effect of the marginally stable orbit. The disk radius, QPO frequency, and mass accretion rate are all in arbitrary units.

5. KHZ QPOS AND THE MARGINALLY STABLE ORBIT

The existence of the marginally stable orbit will have a strong effect on the configuration of the accretion disk near the neutron star as the lack of stable orbits inwards of the marginally stable orbit implies that a stable disk can not exist in that region. The inner radii of disks around neutron stars appear to be variable. The truncation of the disk may be caused by the neutron star magnetic field, by radiation forces acting on the disk, or by a disk instability. In general, the inner disk radius decreases with increasing mass accretion rate through the disk. The marginally stable orbit will modify this behavior by limiting the minimum possible inner disk radius, see figure 5. Thus, saturation at a minimum disk radius for large mass accretion rates is a signature of the marginally stable orbit (Miller, Lamb, & Psaltis 1998). As described in Kaaret, Ford, & Chen (1997), when the disk reaches the marginally stable orbit, the inner radius will not approach a single value, but instead wander over some range due to the properties of the transonic flow near the marginally stable orbit.

If one is willing to accept the assertions made in the previous section that the QPO frequency is determined by the radius of the inner edge of the accretion disk and that the x-ray spectral state is an indicator of mass accretion rate through the disk, then the relation of QPO frequency versus spectral state can be used to probe the relation of inner disk radius versus mass accretion rate through the disk. Thus, the shape of the QPO versus spectral state diagram, i.e. whether or not the QPO frequency saturates at high mass accretion rates, provides a test for the presence of

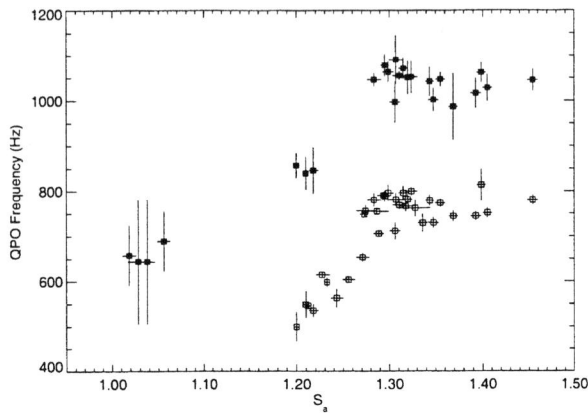

FIGURE 6. QPO centroid frequency versus position in color-color diagram for 4U 1820-30 from Bloser et al. (1999). Larger values of the position in the color-color diagram indicate higher inferred mass accretion rates. Filled squares indicate the upper kHz QPO while open squares indicate the lower kHz QPO.

the marginally stable orbit.

The QPO frequency versus spectral state diagram for 4U 1728-34, Fig. 3, shows no evidence for a saturation of QPO frequency at high mass accretion rates. This is also true for almost all neutron star binary for which kHz QPOs have been detected. There is only one source which does show a clear saturation of QPO frequency at high inferred mass accretion rates: 4U 1820-30. Zhang et al. (1998) demonstrated a saturation of QPO frequency versus x-ray count rate for 4U 1820-30. However, as QPO frequency is poorly correlated with count rate in most sources (the relation tends to look similar to the QPO frequency versus flux relation in Fig. 1), this approach came under significant criticism. Kaaret et al. (1999) showed that the QPO frequency saturates when plotted versus a spectral state indicator derived from a hard x-ray color. As QPO frequency is generally well correlated with hard x-ray color (e.g. Mendez 1999), this result is compelling evidence that the QPO frequency saturates at high disk mass accretion rates. Recently, Bloser et al. (1999) performed a similar analysis using position in a color-color diagram as a spectral state indicator, see Fig. 6, and again find a saturation of QPO frequency.

The fact that only 4U 1820-30 exhibits a saturation of the QPO frequency versus inferred disk mass accretion rate, and thus appears to be the only source for which the disk reaches the marginally stable orbit, merits some consideration. Interestingly, 4U 1820-30 also holds a unique position as the stellar binary system with the shortest known orbital period, see Fig. 7, and is thought to be a double degenerate binary which evolved through a common envelope phase (Rappaport et al. 1987). These unique features of 4U 1820-30 suggest that the neutron star may

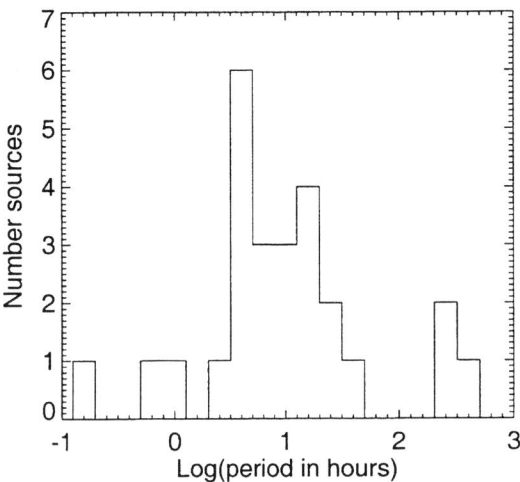

FIGURE 7. Period distribution of low-mass x-ray binaries from the catalog of van Paradijs (1995). 4U 1820-30 has the shortest known orbital period at 11 minutes.

have accreted more matter and thus is more massive than other neutron stars in x-ray binaries.

For the marginally stable orbit to be dynamically important, it must lie outside the neutron star surface. If there is a boundary layer on the neutron star surface formed by the accretion flow, then to be observed via kHz QPOs, the marginally stable orbit must lie outside the top surface of the boundary layer. Inogamov & Sunyaev (1999) have recently calculated the properties of such boundary layers and found typical equatorial thicknesses near 1 km. This is comparable to the separation between the marginally stable orbit and neutron star surface for a 1.4 $M_\odot$ neutron star, rotating at 250–400 Hz, for many of the currently favored equations of state for nuclear matter. The separation increases as the neutron star mass increases since the surface generally moves inward and the marginally stable orbit moves outward. Thus, the signature of the marginally stable orbit is most likely to be observed from the most massive neutron stars.

6. CONCLUSIONS

Millisecond x-ray timing is a promising probe of accretion flows in x-ray binaries. Robust correlations exist between millisecond timing properties and spectral properties. Study of these correlation may help improve our understanding of the physics and geometry of accretion flows in neutron star x-ray binaries and, perhaps, also of strong field gravity.

ACKNOWLEDGEMENTS

I thank Peter Bloser for use of Fig. 6 before publication.

REFERENCES

Bloser, P.F., Grindlay, J.E., Kaaret, P., Zhang, W., Smale, A.P., & Barret, D. 2000, ApJ to appear, astro-ph/0005496
Bradt, H.V., Rothschild, R.E., & Swank, J.H. 1993, AAS, 97, 355
Ghosh, P., & Lamb, F. K. 1979, ApJ, 234, 296
Hasinger, G. & van der Klis, M. 1989, A&A, 225, 79
Hasinger, G., van der Klis, M., Ebisawa, K., Dotani, T., & Mitsuda, K. 1990, A&A, 235, 131
Hertz, P., Vaughan, B., Wood, K.S., Norris, J.P., Mitsuda, K., Michelson, P.F., & Dotani, T. 1992, ApJ, 396, 201
Inogamov, N.A. & Sunyaev, R.A. 1999, Astron. Lett. 25, 269
Kaaret, P., Ford, E. C. & Chen, K. 1997, ApJ, 480, L27
Kaaret, P. & Ford, E.C. 1997, Science, 276, 1386
Kaaret, P., Yu, W., Ford, E.C., & Zhang, S.N. 1998, ApJ, 497, L93
Kaaret, P. et al. 1999, ApJ, 520, L37
Kluzniak, W. & Wagoner, R.V. 1985, ApJ, 297, 548
Mendez, M. 1999, Proceedings of the 19th Texas Symposium in Paris, astro-ph/9903469
Miller, M. C., Lamb, F. K., & Psaltis, D. 1998, ApJ, 508, 791
Misner, C.W., Thorne, K.S., & Wheeler, J.A. 1970, Gravitation (San Francisco: Freeman)
Piraino, S., Santangelo, A., Ford, E.C., & Kaaret, P. 1999, A&A, 349, L77
Rappaport, S., Ma, C.P., Joss, P.C., Nelson, L.A. 1987, ApJ, 322, 842
Shakura, N.I. & Sunyaev, R.A. 1973, A&A, 24, 337
Strohmayer, T.E. et al. 1996, ApJ, 469, L9
van der Klis, M. et al. 1996, ApJ, 469, L1
van Paradijs, J. 1995, in X-Ray Binaries, ed. W.H.G. Lewin, J. van Paradijs, & P.J. van den Heuvel, (Cambridge: Cambridge University Press), 536
Zdziarski, A.A., Lubiski, P., Smith, D.A. 1999, MNRAS, 303, L11
Zhang, W., Smale, A.P., Strohmayer, T.E., & Swank, J.H. 1998, ApJ, 500, L171

DEEP ROSAT SURVEYS & THE CONTRIBUTION OF AGNS TO THE SOFT X-RAY BACKGROUND

I. Lehmann [1], G. Hasinger [1], M. Schmidt [2], J.E. Gunn [3], D.P. Schneider [4], R. Giacconi [5], M. McCaughrean [1], J. Trümper [6] and G. Zamorani [7,8]

1) Astrophysikalisches Institut Potsdam (AIP), An der Sternwarte 16, D-14482 Potsdam, Germany
2) California Institute of Technology, Pasadena, CA 91125, USA
3) Princeton University Observatory, Peyton Hall, Princeton, NJ 08540, USA
4) Department of Astronomy and Astrophysics, The Pennsylvania State University, University Park, PA 16802, USA
5) Associated Universities, Inc. 1400 16th Street, NW, Suite 730 Washington, DC 20036, USA
6) Max-Planck-Institut für extraterrestrische Physik, Karl-Schwarzschild-Str. 2, D-85748 Garching bei München, Germany
7) Osservatorio Astronomico, Via Ranzani 1, I-40127, Bologna, Italy
8) Istituto di Radioastronomia del CNR, via Gobetti 101, I-40129, Bologna, Italy

ABSTRACT   The ROSAT Deep Surveys in the Lockman Hole have revealed that AGNs are the main contributors ($\sim$75 %) to the soft X-ray background in the 1–2 keV band. Using new optical/infrared and radio observations we have obtained a nearly complete identification (93 %) of the 91 X-ray sources down to a limiting flux of $1.2 \cdot 10^{-15}$ erg cm^{-2} s^{-1} in the 0.5–2.0 keV band. We present the optical colours and the emission line properties of our AGNs in comparison with other X-ray selected AGN samples. Furthermore we discuss the fraction of red AGNs found in the ROSAT Deep Surveys. From the ROSAT Deep Surveys we see no evidence for a new class of X-ray bright galaxies, which significantly contributes to the soft X-ray background.

KEYWORDS: Surveys–galaxies: active–Galaxies: Seyferts–quasars: emission lines–X–rays: galaxies

1. INTRODUCTION

Deep X-ray surveys and the optical/infrared identification of the detected faint X-ray sources are the key to understand the origin of the X-ray background, discovered more than thirty years ago (Giacconi et al. 1962). We have performed the most sensitive soft X-ray survey (to date) in the Lockman Hole (Hasinger et al. 1993), a region of extremely low galactic hydrogen column density (Lockman et al. 1986).

The Deep ROSAT Surveys in the Lockman Hole consist of a 207 ksec ROSAT

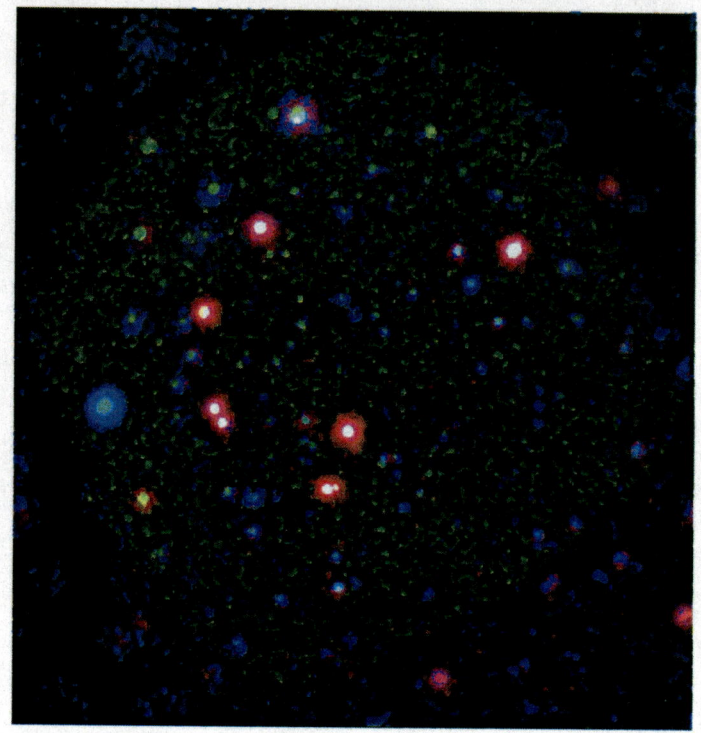

FIGURE 1. Combined ROSAT PSPC/HRI image of the Lockman Hole region. The HRI sources are shown in green colour. The red and blue colours indicate PSPC sources in the 0.1–0.5 keV and 0.5–2.4 keV energy bands. The HRI sources appear less extended due to the better spatial resolution of the HRI ($\sim$ 5 arc–sec) compared with that of the PSPC ($\sim$ 20 arc–sec). The field size is about 30 arc-min.

PSPC exposure, a 205 ksec ROSAT HRI raster scan and a series of HRI images of a fixed area totalling 1112 ksec (see Fig. 1). The ROSAT Deep Survey (RDS) based on the PSPC image includes a statistically complete sample of 50 X–ray sources with fluxes in the 0.5–2.0 keV band larger than $5.5 \cdot 10^{-15}$ erg cm^{-2} s^{-1} (Hasinger et al. 1998). The optical identification of these faint X–ray sources has shown that over 75 % of the RDS sources are AGNs (Schmidt et al. 1998). This study identified a much larger fraction of AGNs than did any previous X–ray survey (Boyle et al. 1995, Georgantopoulos et al. 1996, Bower et al. 1996 and McHardy et al. 1998).

The HRI images are the basis for the Ultradeep HRI Survey (Hasinger et al. 1999), which reaches now a surface density of $\sim$1000 sources deg^{-2} at 10^{-15} erg cm^{-2} s^{-1}. At this flux limit 70–80 % of the soft X-ray background has been resolved into discrete sources. Both surveys contain in total 91 X-ray sources with fluxes in the 0.5–2.0 keV band larger than $1.2 \cdot 10^{-15}$ erg cm^{-2} s^{-1}.

Here we present the optical identification status of the ROSAT Deep Surveys in the Lockman Hole. We will discuss the population of red sources, which we have analyzed in the RDS. Finally, we compare the emission line properties of the RDS AGNs with those of several optical/UV and X-ray selected AGN samples.

2. IDENTIFICATION OF THE ROSAT DEEP/ULTRADEEP SURVEYS

The optical identification of the counterparts of most X–ray sources was obtained during the last five years using the Low Resolution Imaging Spectrometer (LRIS: Oke et al. 1995) on the Keck I and II telescopes on Mauna Kea, Hawaii. R-band images of the Lockman Hole field were taken to identify the faint optical counterparts ($R > 24$ mag) of the weakest X–ray sources. Fig.2 shows the finding charts of some RDS sources. The HRI error circle is always smaller than the PSPC circle due to the higher spatial resolution of the ROSAT HRI detector as mentioned before. The HRI detection was crucial for the optical identification effort, because we found in nearly all cases a unique optical object inside the HRI error circle. Confusion in some RDS sources is resolved by the HRI, where the PSPC image shows only a single source (Schmidt et al. 1998, Lehmann et al. 2000a; hereafter Paper III).

The majority of optical spectra were taken with the LRIS instrument on the Keck I and II telescopes in the long-slit and multi-slit mode. The typical integration times were 1800 sec for long-slit spectra and 3600 sec for multi-slit masks. The wavelength coverage of the long–slit spectra is between 3800 and 8200 Å using a 300 lines mm^{-1} grating. The wavelength range of the multi-slit spectra depends on the position of the objects on the masks (typical 4500–8900 Å). The resolution of the spectra is about ∼10-15 Å (FWHM). The high signal to noise Keck spectra are very important to identify the faint optical counterparts of the X–ray sources.

For several relatively bright objects we have obtained spectra with the 4-Shooter spectrograph (Gunn et al. 1987) at the 5-m Hale telescope. Using a 1.5 × 100″ slit and a 200 lines mm^{-1} transmission grating the spectra cover a wavelength range of 4500-9500 Å at a resolution of 25 Å. The optical spectra of some bright stars were taken with the Boller & Chivens Cass Twin spectrograph at the 3.5-m telescope on Calar Alto. A 1.5″ wide slit and the 600 lines mm^{-1} transmission grating (T13) was used resulting in a wavelength coverage of 3500-5500 Å and a resolution of about 4.4 Å.

In Fig. 2 we present the Keck spectra of the counterparts of faint X–ray sources, which are contributors to the soft X–ray background (type I AGNs, type II AGNs and groups/clusters of galaxies). The flux scale is normalized so that one count corresponds to an AB magnitude 28 ($f_\nu = 2.29 \cdot 10^{-31}$ erg s^{-1} cm^{-2} Hz^{-1}). The location of typical AGN emission lines (eg. Lyα, C IV, C III], Mg II) and galaxy absorption lines (Ca H+K λ3934/3968, CH G λ4304, Mg I λ5175, Na I λ5890) are overplotted. For the whole set of optical spectra of RDS sources see Paper III.

The existence of broad emission lines reveals the AGN nature of many faint optical counterparts (eg. 37G in Fig.2). But the optical spectra of several objects

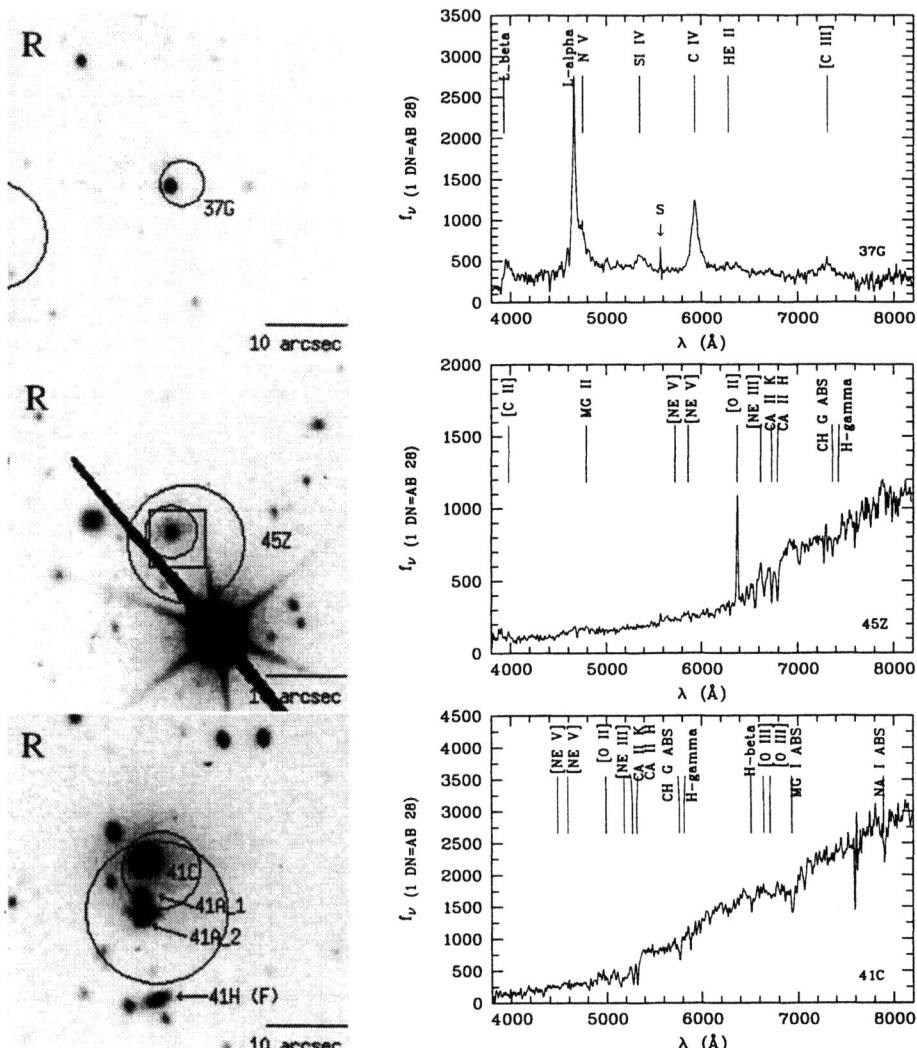

FIGURE 2. Examples of Keck R-band images and low-resolution spectra of optical counterparts (37G – type I AGN, 45Z – type II AGN, 41C – a member of a galaxy group). North is up, and east to the left. The small and large circles show the ROSAT HRI and PSPC error circles (80 % error radius). The 2σ error box indicates a VLA 20 cm detection (DeRuiter et al. 1997). Residuals of night sky emission lines are indicated with "s".

contain no broad emission lines. Some of them show high ionization [Ne V] $\lambda3426$ emission lines, typical for AGNs. Without these high quality Keck spectra and the accurate HRI positions they may have been identified with narrow-line X–ray galaxies (NLXG) as found in several other less deep soft X–ray surveys (see Boyle et al. 1995, Geogantopoulous et al. 1996, McHardy et al. 1998). Furthermore two of the RDS sources have X–ray luminosities in the 0.5–2.0 keV energy band above 10^{43} erg s^{-1}, show relatively strong [Ne III] $\lambda3869$ emission lines, which are typically quite strong in Seyfert galaxies and quasars (Schmidt et al. 1998). The problem of missidentification of faint X–ray sources is discussed in more detail in Paper III.

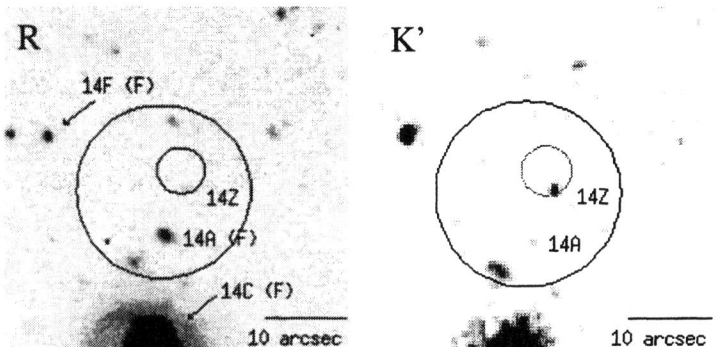

FIGURE 3. Keck R and Calar Alto Omega-Prime K' images of source 14. The counterpart 14Z is clearly visible on the K' image (inside the HRI error circle), whereas it is only marginal detected on the R image (~25.5 mag) resulting in a large $R - K'$ colour of 5.4 mag. Field galaxies with known redshift are marked with "(F)".

Some X-ray sources exhibit only a very faint ($R \sim 25$ mag) or no optical counterpart at the limiting R-magnitude of 25.5, which is clearly below the spectroscopic limit of our survey ($R \sim 23.5$ mag). Therefore we started a deep broad-band K' (1.9244–2.292 μm) survey of the Lockman Hole region with the Omega-Prime camera (Bizenberger et al. 1998) on the Calar Alto 3.5-m telescope. About half of the Ultradeep HRI Survey area is already covered. The camera uses a 1024×1024 pixel HgCdTe HAWAII array with an image scale of 0.396 arcsec/pixel, and covers a field-of-view of 6.7×6.7 arcmin. Point sources are well detected at the 5σ level at K'=19.7 mag in a 45 min (net) exposure.

K' and R images are presented in Fig. 3 for the RDS source 14. A faint K' source is recognizeable within the HRI error circle ($K' = 19.6$ mag), whereas in the R-band the source is only marginally detected. The resulting $R - K'$ colour is about ~ 5.4 mag. The RDS contains a further source with such a large $R - K'$ colour (84Z: $R - K' \geq 6.2$). The most probable identifications of such red objects are either with high redshift ($z \geq 1$) ellipticals or dust-reddened starbursts or AGNs. Since the source 14 has a relatively hard spectrum (Schmidt et al. 1998) we prefer the identification with a quasar reddened by absorption, since no additional red object is seen close to it in the K' image (the object 14A, inside the PSPC error box, but outside the HRI circle, is a narrow emission line galaxy (NELG) at $z = 0.546$, which is typical for the population of field galaxies (see Paper III).

There are several examples for faint X-ray sources identified with red objects (eg. Newsam et al. 1997). In our total sample of 91 X-ray sources we found that all X-ray counterparts with red colours ($R - K' > 4.5$) are either members of high redshift clusters of galaxies (z>1) or obscured AGNs at various redshifts. Recently, we have detected a number of red galaxies (Lehmann et al. 2000b) indicating one of the most distant X-ray selected clusters of galaxies with a gravitationally lensed arc (Hasinger et al. 1999b).

objects class	number of object	total content in [%]
AGNs type I	54	59.3
AGNs type II	15	16.5
galaxies	1	1.1
groups/clusters	10	11.0
stars	5	5.5
unidentified sources	6	6.6
total	91	

TABLE 1. Spectroscopic identification status of the ROSAT Deep Surveys.

In Table 1. we present the identification status of the 91 X-ray sources above the limiting flux of $1.2 \cdot 10^{-15}$ erg cm^{-2} s^{-1} in the 0.5–2.0 keV energy band. The large majority of the X-ray sources in this sample (~ 75 %) turned out to be AGNs. Most of these objects are classified as AGNs type I (quasars and Seyfert galaxies), because they exhibit at least one broad emission line in their optical spectra. Among these objects is the most distant X-ray selected quasar (z=4.45) found to date (Schneider et al. 1998). A non-negligible fraction of AGNs is classified with AGNs type II showing only narrow emission lines or broad Hα/Hβ lines with a large Balmer decrement. For some of them only the existence of high excitation [Ne V] emission lines and/or the high X-ray luminosity ($L_X > 10^{43}$ erg s^{-1}) in the 0.5–2.0 keV energy band reveal the AGN nature of these objects. From the nearly complete (95 %) identification of our ROSAT Deep Surveys we find no indication for new classes of NLXG or of galactic CVs (Maoz & Grindlay 1995), which have been claimed to dominate the soft X-ray counts at fainter fluxes (cf. Paper III).

3. POPULATION OF RED AGNS IN THE RDS

The optical spectra of most RDS AGNs with redshift below 1.1 show absorption lines typical for galaxies, eg. Ca H+K λ3934/3968, CH Gλ4304, Mg I λ5175, Na I λ5890. This suggests a significant contribution by the host galaxy in the optical wavelength range. Therefore we have measured the break index D(4000) in order

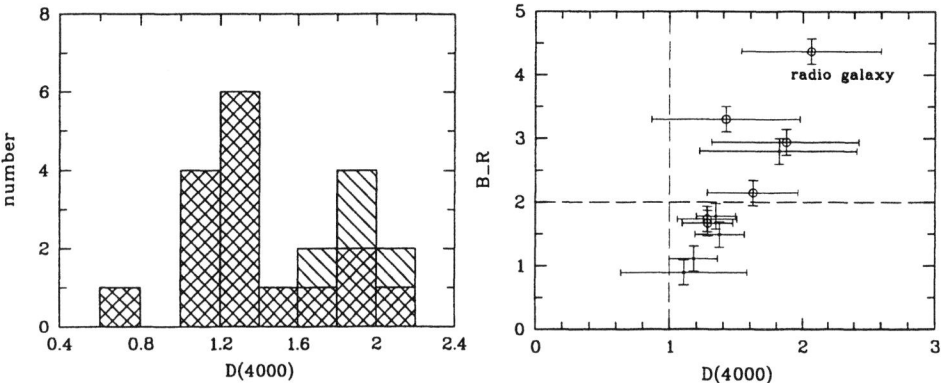

FIGURE 4. a) Distribution of the D(4000) index of 16 RDS AGNs. In addition, the values of the three group members 41C, 58B, 67B(s) and of the RDS galaxy 53A are overplotted with a different hatch. b) $B - R$ colour derived from the optical spectra vs. the D(4000) index for 11 RDS AGNs. The dots with error bars are the broad-line AGNs, whereas the circles indicate the narrow-line AGNs. In the other cases the optical spectra could not be used to obtain the $B - R$ values. The horizontal line marks the region of red AGNs ($B - R > 2.0$). The AGNs right of the vertical line with D(4000)>1 show a continuum component originated from the host galaxy.

to have an estimate of the main stellar population in the host galaxies. Early-type galaxies show a pronounced break around 4000 Å in their optical spectra. The 4000Å break index D(4000) is defined by Bruzual (1983). Due to the wavelength range of our spectra we were able to derive the D(4000) values for all objects with redshift between 0.01 and 0.93.

About 63% of the RDS AGNs for which the D(4000) index could be derived from the optical spectra, have D(4000) > 1 at more than 1 sigma level (cf. Paper III). One object, whose spectrum shows a very blue continuum, is the only AGN with D(4000) < 1, consistent with a power law continuum ($f_\nu = \nu^\alpha$ with $\alpha < 0$). The histogram of the D(4000) index of the RDS AGNs is shown in Fig. 4a. This distribution peaks at D(4000)~1.2, with most of the objects in the range 1.0≤D(4000)≤1.4, plus some objects at larger values (1.6≤D(4000)≤2.0). These two ranges of D(4000) correspond to galaxies with a sustained amount of star formation and galaxies with a dominant old stellar population, respectively. The D(4000) values of the three RDS group members (41C, 58B and 67B) are all consistent with those of elliptical galaxies.

As a comparison, the mean D(4000) values measured from the Canada-France Redshift Survey (CFRS; Lilly et al. 1995) in a redshift interval similar to that of our objects (z<1) are 1.43 for 256 emission line galaxies and 1.92 for 85 "quiescent" galaxies, defined as galaxies with no detected [OII] line (from Table 1 in Hammer et al. 1997). Similar average values (1.40 and 1.83 for emission line galaxies and no-emission line galaxies) have recently been measured from the large, more local (z<0.15) Stromlo-APM Survey (Tresse et al. 1999).

The D(4000) values do not appear to correlate either with the absolute magnitudes of the AGNs, which cover the range $-23.9 < M_V < -20.7$, or with the presence of broad emission lines. In order to check if the D(4000) index is correlated with the colours of the objects we have determined the $B - R$ colours of the RDS AGNs. Unfortunately we have no good photometric catalogue yet in the B-band. Therefore we have derived the $B - R$ colour from the optical spectra for those objects, where the spectra were taken at the paralactic angle to avoid atmospheric refraction. Fig. 4.b shows the $B - R$ colours of 11 RDS AGNs versus the D(4000) index. Despite of the relatively large errors and the small number of objects there is a trend that the $B - R$ colour of the AGNs increases with increasing D(4000) value. Five of the 11 RDS AGNs have red colours ($B - R \geq 2.0$). The AGNs with the largest $B - R$ value is a radio galaxy, were the optical spectrum reveals an elliptical galaxy. All of these red AGNs have relatively large D(4000) values indicating a significant contribution of the host galaxy to the continuum emission. The red colour of some RDS AGNs could therefore originate from their host galaxies.

Recently, Fiore et al. (1999) have found two red AGNs from the first optical identification of the BeppoSAX HELLAS Survey in the 5–10 keV energy band. Their optical spectra are as well dominated by starlight.

4. EMISSION LINE PROPOERTIES OF RDS AGNS

In order to derive a reliable optical identification of the RDS X–ray sources we have determined the emission line properties of the most prominent AGN and galaxy emission lines. Due to the large redshift range of the RDS AGNs from 0.080 to 2.832 with its mean redshift of $\langle z \rangle = 1.2$ (cf. Fig 5. in Paper III), any given emission line is present only in a small number of objects.

The emission lines have been measured by fitting with single or double Gaussian profiles applying the Levenberg-Marquardt algorithm (Press et al. 1992). The Levenberg-Marquardt algorithm provides one sigma errors of the parameters. In some cases a single Gaussian gave a poor fit to the data, so an additional Gaussian was used. From the best set of parameters we have calculated the redshift, the rest frame EW_{rest} in Å and the FWHM in km s^{-1}. The redshift of each object was derived as the mean value from the strongest features (eg. Mg II λ 2798, [O II] λ 3727). Its mean error is around ~ 0.001. The FWHM has been corrected for intrumental resolution. The FWHM and EW_{rest} of all emission lines of the RDS AGNs are given in Paper III.

The emission lines are divided into narrow emission lines (<1500 km s^{-1}) and broad emission lines (>1500 km s^{-1}). In addition, a few spectra required very broad line components (>8000 km s^{-1}) of the semi–permitted line C III] λ1908 and of the permitted lines C IV λ1548 and Mg II λ2798.

We compared the mean FWHM of the RDS AGNs with the properties of the UV/optically selected, low–redshift quasar sample ($\langle z \rangle = 0.54$) from Green 1996 and with the FWHM of a QSO sample at intermediate redshift ($0.9 \leq z \leq 2.2$) from

Brotherton et al. 1994. In addition we used for comparison the emission line properties derived from two X –ray surveys: the RIXOS AGN X–ray sample with $\langle z \rangle = 0.82$ (Puchnariwicz et al. 1997) and the AGN data of the Cambridge–Cambridge ROSAT Serendipity Survey (CRSS) of Boyle et al. (1997), which has a mean redshift of $\langle z \rangle = 0.88$. In general the mean FWHM of the RDS AGN broad emission lines (Si IV $\lambda1397$, C IV $\lambda1548$, C III] $\lambda1908$, Mg II $\lambda2798$, Hβ $\lambda4861$ and Hα $\lambda6563$) are consistent with those values found for other X–ray selected AGN samples as well as with the data obtained from the mentioned optical/UV selected AGN samples. Only the FWHM of the broad Lyα $\lambda1216$ emission lines (detected only in two AGNs) are smaller than the mean FWHM of the Green quasar sample.

For a comparison of the mean EW_{rest} of RDS AGNs we have used in addition the range of the mean EW derived for several emission line surveys (Osmer 1980, Vaucher & Weedman 1980) and optical/radio quasar surveys (Neugebauer 1979, Richstone & Schmidt 1980, Oke & Korycanski 1982, Wampler et al. 1984, Tytler et al. 1987) given in Schmidt et al. (1986) and the sample of high-redshift, bright QSO $\langle z \rangle = 1.6$ from Steidel & Sargent (1991). Further, we have calculated the mean EW (neglecting uncertain values) from the X–ray selected AGN sample ($\langle z \rangle = 0.19$) of Stephens 1989.

The mean RDS AGN EW of the broad emission lines are consistent with the data from several X–ray selected AGN samples at lower mean redshift (eg. RIXOS: $\langle z \rangle = 0.8$ and CCRS: $\langle z \rangle = 0.9$) compared to its mean redshift of $\langle z \rangle = 1.2$. Further they are in general in good agreement with the mean EW from UV/optical selected AGN samples at low and high redshift. Although our results are derived from a small number of emission lines they seem to indicate in comparison with the other samples nearly the same properties of AGN broad emission lines.

ACKNOWLEDGEMENTS

We acknowledge the support in part by NASA grants NAG5–1531 (M.S.), NAG8–794, NAG5–1649, and NAGW–2508 (R.G.), and NSF grant AST–9509919 (D.S.). G.H. acknowledges the grant FKZ 50 OR 9908 0 by the Deutsches Zebtrum für Luft- und Raumfahrt e.V. (DLR). G.Z. acknowledges partial support by the Italian Space Agency (ASI) under contract ARS–96–70 and the Italian Ministry for University and Research (MURST) under grant Cofin 98–02–32.

REFERENCES

Bizenberger P., McCaughrean M. J., Birk C., et al., 1998, Omega Prime: the wide-field near-infrared camera for the 3.5m telescope of the Calar Alto Observatory. In A. M. Fowler, ed., Infrared astronomical instrumentation, SPIE vol. 3354, p. 825

Bower R.G., Hasinger G., Castander F.J., et al., 1996, MNRAS 281, 59

Boyle B.J., McMahon R.G., Wilkes B.J., Elvis M., 1995a, MNRAS 272, 462

Boyle B.J., Wilkes B.J., Elvis M., 1997, MNRAS 285, 511

Brotherton M.S., Wills B.J., Steidel C.C., Sargent W.L.W., 1994, APJ 423, 131

Bruzual A.G., 1983, ApJ 273, 105
De Ruiter H.R., Zamorani G., Parma P., et al., 1997, A&A 319, 7
Georgantopoulos I., Stewart G.C., Shanks T., et al., 1996, MNRAS 280, 276
Giacconi R., Gursky H., Paolini F.R., et al., 1962, Phys. Rev. Lett. 9, 439
Green P.J., 1996, ApJ 467, 61
Gunn J.E., Carr M.L., Danielson G.E., et al., 1987, Opt. Eng. 26, 779
Fiore F., La Franca F., Giommi P., Elvis M., Matt G., et al., 1999, MNRAS 306, 55
Hammer F., Flores H., Lilly S.J., et al., 1997, ApJ 481, 49
Hasinger G., Burg R., Giacconi R., et al., 1993, A&A 275, 1 (erratum A&A, 291, 348)
Hasinger G., Burg R., Giacconi R., et al., 1998, A&A 329, 482 (Paper I)
Hasinger G., Lehmann I., Giacconi R., et al., 1999a, In Proceedings of the Symposium "Highlights in X-ray Astronomy in honour of Joachim Trümper's 65th birthday", eds. B.Aschenbach & M.J.Freyberg, MPE Report 272, p. 199
Hasinger G., Giacconi R., Gunn J. E., et al., 1999b, A&A 340L, 27 (Paper IV)
Lockman F.J., Jahoda K., McCammon D., 1986, ApJ 302,432
Lehmann I., Hasinger G., Schmidt M., et al., 2000a, A&A 354, 35 (Paper III)
Lehmann I., Hasinger G., Giacconi R., et al., 2000b, In Proceedings of the VLT Opening Symposium, held at Antofagasta, Chile, 1-4 March 1999, eds. J. Bergeron & A. Renzini, Springer, p. 121
Lilly S.J., Le Fèvre O., Crampton D., Hammer F., Tresse L., 1995, ApJ 455, 50
Maoz E., Grindlay J.E., 1995, ApJ 444, 183
McHardy I., Jones L.R., Merrifield M.R., et al., 1998, MNRAS 295, 641
Neugebauer G., Oke J.B., Becklin E.E., Matthews K., 1979, ApJ 230, 79
Newsam A.M., McHardy I.M., Jones L.R., Maison K.O., 1997, MNRAS 292, 378
Oke J.B., Cohen J.G., Carr M., et al., 1995, PASP 107, 375
Osmer P.S., 1980, ApJ Suppl. 42, 523
Puchnarewicz E.M., Mason K.O., Carrera F.J., et al., 1997, MNRAS 291, 177
Press H.W., Teukolski S.A., Vetterling W.T., Flannery B.P., 1992, Receipes (in FORTRAN), sec. edition, Cambridge Univ. Press,
Richstone D.O. & Schmidt M., 1980, ApJ 235, 377
Schmidt M., Schneider D.P., Gunn J.E., 1986, ApJ 306, 411
Schmidt M., Hasinger G., Gunn J.E., et al., 1998, A&A 329, 495 (Paper II)
Schneider D.P., Schmidt M., Hasinger G., et al., 1998, AJ, 115, 1230
Steidel C.C., Sargent W.L.W., 1991, ApJ 382, 433
Stephens S.A., 1989, AJ 97, 10
Tresse L., Maddox S., Loveday J., Singleton C., 1999, MNRAS 310, 262
Vaucher B.G. & Weedman D.W., 1980 ApJ 240, 10
Wampler E.J., Gaskell C.M., Burke W.L., Baldwin J.A., 1984, ApJ 276, 403

OBSCURED ACTIVE GALACTIC NUCLEI

R. Maiolino

Osservatorio Astrofisico di Arcetri, Firenze, Italy

ABSTRACT   The properties of the absorption in type 2, narrow line AGNs are reviewed by focusing on the X-ray indicators. I discuss the properties of the cold absorbing medium (the putative torus) and of the reprocessed components, as well as their implications for the unified model. The relation between optical classification and X-ray absorption is examined. The case of "fossil" AGNs, whose type 2 classification is not due to absorption effects, is also discussed. Although this review is mainly focused on nearby Seyfert 2 galaxies, I also shortly discuss the effects of absorption at higher luminosities and higher redshift and the implications for the X-ray background.

KEYWORDS: galaxies: active – galaxies: Seyfert – galaxies: ISM – X-rays: galaxies

1. INTRODUCTION.

Several observational data indicate that type 2, narrow line Seyfert nuclei suffer significant obscuration along our line of sight. The unified model (Antonucci 1993) ascribes this obscuration to a gaseous parsec-scale circumnuclear torus. According to this model, broad line Seyfert 1 (Sy1) and narrow line Seyfert 2 (Sy2) galaxies would be identical physical objects, while the orientation of the line of sight with respect to the torus is responsible for the obscuration of the Broad Line Region (BLR) and of the nuclear engine (X–UV source) in Sy2s.

In this review I show how X-ray observations, especially at high energies (>2 keV), provide a wealth of information which constrain the properties of the absorbing medium in Seyfert 2s, i.e. the putative circumnuclear gaseous torus. I also discuss the implications of these X-ray studies on related issues, such as the fuelling of active nuclei and the X-ray background. Due to the limited space, I do not discuss warm absorbers, although evidence for this component is found in some obscured Seyferts (eg. Komossa & Fink 1997), neither I discuss the soft excess that characterizes most Sy2s (Maiolino et al. 1998, Wilson & Elvis 1997).

2. X-RAY ABSORPTION AS A DIAGNOSTIC OF THE TORUS.

X-ray observations have probably provided the most direct test of the unified model. Indeed, the X-ray spectrum of many Sy2s is characterized by a powerlaw similar to that observed in Sy1s and a photoelectric cutoff due to absorbing, cold gas along our line of sight. Since the absorbing column in Sy2s is generally larger than $10^{22}\mathrm{cm}^{-2}$, these studies were mostly obtained with satellites sensitive in the high energy band above $\sim 1 keV$ such as Ginga, ASCA and BeppoSAX.

If the absorbing column is larger than $\sim 10^{24} cm^{-2}$, i.e. the medium is thick to Compton scattering, then the transmitted component is completely suppressed below 10 keV and the spectrum observed in the 2–10 keV band is dominated by reprocessed components. More specifically, in this case the hard X-ray spectrum is characterized by a flat Compton reflection component, ascribed to the inner surface of the torus, and/or a steeper component due to a ionized, warm scattering medium (eg. Matt et al. 1997). If the absorbing medium has a column in the range $10^{24} - 10^{25} cm^{-2}$, then the transmitted component can be still observed in the 10–300 keV band as an emission excess (eg. Done et al. 1996); for larger absorbing columns even the 10–300 keV band is dominated by the reflection component. The cold reflector also produces a prominent fluorescence iron line at 6.4 keV (EW$\sim$1–2 keV with respect to the reflected continuum). Yet, Netzer et al. (1998) pointed out that also the Narrow Line Clouds can contribute significantly to the observed 6.4 keV line, provided that their column density is large enough. Instead, the warm scattering medium emits lines of He– and H–like iron at 6.7 and 7 keV, which can reach large equivalent widths as well (Matt et al. 1996). Until a few years ago only a couple of Compton thick sources were known. This is because the reflection efficiency of both the cold and warm mirrors is low ($\sim 10^{-2} - 10^{-3}$) and, therefore, this class of objects is more difficult to detect. This highlights one of the major problems in past hard X-ray surveys, as discussed in the following section.

2.1. The distribution of N_H.

Past hard X-ray surveys of Sy2s were strongly biased in favor of X-ray bright sources (according to all-sky surveys), which tend to be the least absorbed ones. More recently, X-ray observations have probed X-ray weaker Sy2s, partly removing the selection against heavily obscured objects (eg. Turner et al. 1997). We (Salvati et al. 1997, Maiolino et al. 1998) used BeppoSAX to observe an [OIII]-selected sample of previously unobserved (X-ray weak) Sy2s. These observations discovered a large fraction of Compton thick objects, a result which confirms the bias against heavily obscured systems affecting previous surveys. However, none of these surveys is actually complete or free from biases and, therefore, none of them is suitable (if taken separately) to derive the distribution of absorbing column densities among Sy2s. To accomplish this goal, we merged all the available hard X-ray observations of Sy2s and extracted a complete subsample limited in *intrinsic* (i.e. unabsorbed) luminosity as inferred from the [OIII] narrow emission line (Risaliti et al. 1999). This subsample is composed of 45 objects and the corresponding N_H distribution, shown in Fig. 1, can be considered the best estimate of the true distribution that can be obtained with the available data. The most interesting result is that this distribution is significantly shifted toward large columns with respect to past estimates: most ($\sim$75%) of the Sy2s are heavily obscured ($N_H > 10^{23} \mathrm{cm}^{-2}$) and about half are

Compton thick. The N_H does not appear to correlate with the (intrinsic) luminosity, at variance with early results. This N_H distribution has various implications some of which will be discussed in the following.

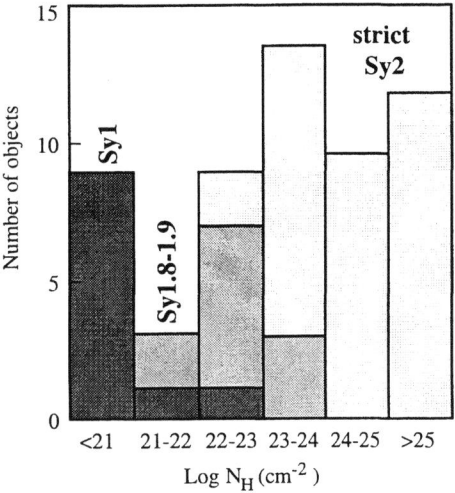

FIGURE 1. The distribution of absorbing column densities among Seyfert galaxies. (from Risaliti et al. 1999, with modifications).

2.2. X-ray absorption as a probe of the parsec scale torus.

The large fraction of Compton thick sources constrains the size of the obscuring medium. Indeed, assuming an axisymmetric geometry, the mass of the gas responsible for the obscuration is about $M_{gas} \approx f\, N_H R^2$, where R is the external radius of the torus and f is the covering factor which, for the Compton thick medium, must be of the order of 0.5. As discussed in Risaliti et al. (1999), the requirement that the gas mass does not exceed the dynamical mass constrains the external radius of the torus to R < 10 pc. Enhanced metallicity could help to relax this constraint. Metallicities a few times solar have actually been inferred for the BLR and in the warm absorber of several Sy1s, but an order of magnitude higher metallicities would be required to relax the mass constraints enough to match the size of the torus of a few 100 pc proposed by some models (sect. 2.3).

A parsec-scale size of the obscuring torus is also supported by the variability of N_H. Indeed, in several Seyfert galaxies the photoelectric cutoff is observed to change significantly on time scales of a few years (Malizia et al. 1997). Fig. 2 shows the specific case of NGC7582 (from a compilation of results in Turner et al. 2000, Xue et al. 1999 and references therein, where the same model was adopted to fit the data). Although in some cases differences in N_H measured at different epochs might be ascribed to the different satellite used, in some cases variation of N_H are observed even with the same satellite. Finally, I recall that a parsec-scale component of the obscuring torus has been directly observed in radio VLBI images, both in continuum (free-free) and H_2O maser emission, in a few active nuclei (eg. Gallimore et al. 1997, Greenhill & Gwinn 1997).

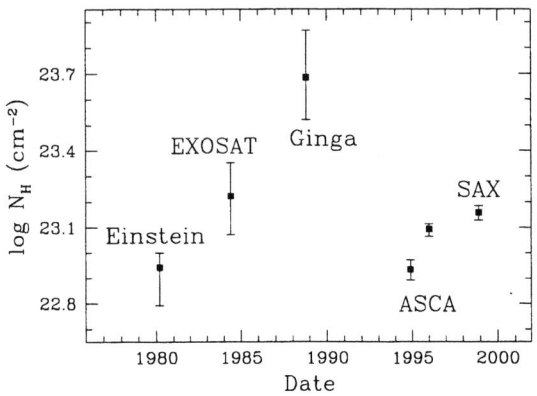

FIGURE 2. Variation of the absorbing N_H in the Seyfert 2 galaxy NGC7582.

2.3. X-ray absorption as a probe of the 100 pc scale torus.

X-ray studies have also provided evidence for obscuring medium on larger scales (~100 pc). Indeed, samples of AGNs selected in the soft X-rays ($\leq$ 2 keV) are characterized by a shortage of edge-on galaxies (Lawrence & Elvis 1982, McLeod & Rieke 1995, Simcoe et al. 1997). This effect is ascribed to circumnuclear material, associated to the galactic gaseous disk, which obscures the AGN when the host galaxy is oriented edge-on. Only a moderate absorbing column ($N_H \geq 10^{22} \text{cm}^{-2}$) is required to suppress significantly the soft X-ray emission.

A 100 pc scale component of the obscuring torus is also required to fit the broad IR spectral energy distribution (Granato et al. 1997) and it is also directly observed in the HST images of some objects (eg. Ford et al. 1997, Malkan et al. 1998).

2.4. Intermediate type Seyferts.

Very likely, the obscuring torus has both a Compton thick parsec-scale component and a larger 100 pc-scale component which is coplanar with the galactic disk and characterized by lower N_H (which does not violates the mass constraints). The internal torus obscures completely the BLR along the intercepting lines of sight (Sy2 case), while the obscuration due to the external torus (in edge-on systems) is moderate and the Seyfert nucleus might show weak broad lines and, therefore, might be classified as intermediate type 1.8–1.9. Indeed, intermediate type Sys are more commonly found in edge-on galaxies (Maiolino & Rieke 1995). This model is also supported by the distribution of absorbing column of intermediate Seyferts, which is significantly shifted towards lower values of N_H than strict Sy2s (Risaliti et al. 1999, Fig. 1). Although, this model might apply to many intermediate type Seyferts, there are Seyfert galaxies whose intermediate properties are not to ascribe to moderate obscuration but to photoionization and variability effects or to more complex forms of absorption (Goodrich et al. 1995, Komossa & Fink 1997).

2.5. Dual absorbers.

A simple, absorbed power law does not provide an adequate description of the observed X-ray spectrum for some of the Compton thin Sy2s. In these cases there is evidence for a second photoelectric cutoff at higher energies due to a medium that only partially covers the X-ray source (Malaguti et al. 1999, Hayashi et al. 1996, Weaver et al. 1994, Vignali et al. 1998). While the photoelectric cutoff at low energies is very likely due to the torus, the partial covering observed at higher energies can only be obtained if the second absorbing medium is close to the X-ray source and has a similar size. A most likely candidate is the BLR. Generally, the column density of the partial covering medium is of the order of $\sim 10^{23} \mathrm{cm}^{-2}$, in agreement with estimates for the N_H of broad line clouds. The covering factor of the partial absorber is generally found to be larger than $\sim 30\%$, that is significantly higher than the covering factor of the broad line clouds which, based on the broad lines equivalent width and on the absence of the Ly-edge cutoff in the UV spectrum of QSOs, is expected to be about 10%. However, I do not consider this a major caveat since the dual absorbers discovered so far are not representative of the true distribution of the partial covering absorption systems, but only sample the tail with high covering factor. Partial covering systems with a covering factor of 10% probably remained undetected in the past observations. I expect Chandra and XMM to discover a large number of dual absorbers with low partial covering.

Finally, thanks to the extended spectral coverage of BeppoSAX some cases of dual absorber with partial covering characterized by an absorbing column as high as $N_H \sim 10^{24} \mathrm{cm}^{-2}$ are being found (eg. Turner et al. 2000). Such a high N_H is still consistent with that expected for the broad line clouds. Indeed, the estimated column of $10^{22} - 10^{23} \mathrm{cm}^{-2}$ for the broad line clouds is only a lower limit that is required to produce the low ionization broad emission lines (MgII, FeII, etc...).

3. X-RAY ABSORPTION AND GALAXY MORPHOLOGY.

Although the column density of the cold obscuring material does not depend on the properties of the AGN (eg. its luminosity) it does depend on the properties of the host galaxy. Indeed, the absorbing N_H strongly correlates with the presence of a stellar bar in the host galaxy (Maiolino et al. 1999). In particular, while non-barred Sy2s are characterized by an average log $N_H \sim 22$ (cm^{-2}), most of the strongly barred Sy2 galaxies are Compton thick. This finding indicates that stellar bars are effective in driving gas into the nuclear region to obscure the AGN. This result is in line with other studies, as reviewed by Sakamoto (1999), which indicate that bars are effective in driving gas into the nuclear region, though these other studies are not specifically focused on AGNs. We might speculate that, more generally, non-axisymmetric potentials (eg. distorted morphologies and galaxy interactions) drive gas into the nuclear region. Although there are some observational indications in this direction, a systemic study, similar to that on barred systems, has not been performed yet. This will be possible with the new Chandra and XMM data.

4. THE RELATION BETWEEN OPTICAL AND X-RAY ABSORPTION.

If the obscuring torus has the same gas-to-dust ratio as in the Galactic ISM, and the dust is characterized a Galactic extinction curve, then the nuclear region of Sy2s should suffer a visual extinction that is related to the gaseous column density by the formula $A_V = 5 \times 10^{-22} N_H (cm^{-2})$. In general this is not the case: A_V is lower than expected from the N_H measured in the X-rays. This was first pointed out by Maccacaro et al. (1982). A visual extinction lower than that expected from the N_H measured in the X-rays is also required to fit the IR spectrum of AGNs (Granato et al. 1997). We have collected a sample of Seyferts which both show X-ray (cold) absorption and whose optical or IR broad lines are not completely suppressed. The ratios between the broad lines provide information on the dust reddening towards the nucleus; however, the broad emission lines must be used with much care, since the extreme conditions of the broad line clouds can affect the intrinsic line ratios through radiative transport effects. By assuming the standard extinction curve we can estimate the visual extinction. The resulting distribution for the A_V/N_H ratio, relative to the Galactic standard value, is shown in Fig. 3. Most of the AGNs in our sample are characterized by a deficit of dust absorption with respect to what expected from the N_H measured in the X-rays, in agreement with early claims. At higher, quasar-like luminosities there are even more extreme examples of this effect: objects that, although absorbed in the X-rays, do not show significant dust absorption in the optical and appear as type 1, broad line AGNs have been recently discovered in hard X-ray and radio surveys (Sambruna et al. 1999, Akiyama et al. 2000, Reeves et al. 1997). Puzzling enough, the early Chandra surveys presented to date have found only a few type 1 QSOs absorbed in the hard X-rays; this issue will be shortly discussed in Sect. 7.

The origin of the reduced A_V/N_H ratio is not clear. An obvious explanation is that the dust-to-gas ratio is much lower than Galactic or that in the inner part of the obscuring torus the dust is sublimated by the strong UV radiation field. However, if the dust content in the absorbing medium is significantly reduced, especially at the inner face, then most of the UV ionizing photons are absorbed by the atomic gas. This should create a huge HII region, which would emit strong ($\sim$ narrow) hydrogen lines corresponding to a large covering factor, i.e. much brighter than the emission lines from the NLR (see also Netzer & Laor, 1993). Also, a simple shortage of dust grains with respect to the gas mass would not explain other peculiar properties of the dust in AGNs, such as the absence of the silicate absorption feature in the mid-IR spectra of most Sy2s (Clavel et al. 2000) and the absence of the carbon dip in the UV spectra of some reddened Sy1s.

Another interesting possibility is that the dust extinction curve is much flatter than the standard Galactic. The high density of the gas in the circumnuclear region of AGNs is likely to favor the growth of large grains (probably through coagulation) which, in turn, should flatten the extinction curve and make it featureless. This effect is directly observed in the dense clouds of our Galaxy (Draine 1995). Within the context of the optical versus X-ray absorption, the effect of a flat extinction curve (due to grain coagulation) is twofold: 1) given the same dust mass, the effective visual extinction is lower, and 2) the broad lines ratio gives a deceiving (low) measure of the extinction. A more thorough discussion of the whole issue is given in Maiolino et al. (2000b).

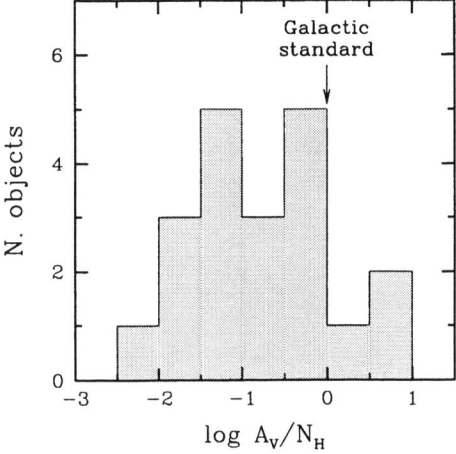

FIGURE 3. Distribution of the A_V/N_H ratio, relative to the Galactic standard value, for a sample of absorbed AGNs.

5. PIERCING COMPLETELY HIDDEN AGNS.

So far I have discussed X-ray absorption in AGNs which were discovered and classified in the optical. However, an increasing number of obscured powerful AGNs has been discovered by means of hard X-ray observations in galaxies which are optically classified as starburst or LINER. Probably, in these objects the obscuring medium hides also the NLR. Alternatively, the nuclear ionizing source might be completely embedded and obscured in all directions. The most spectacular case is the nearby (4 Mpc) edge-on galaxy NGC4945, which hosts one of the brightest AGNs at 100 keV (Done et al. 1996) The powerful X-ray nucleus is obscured by a column of $\sim 5 \times 10^{24} cm^{-2}$ along our line of sight. However, it seems that the nucleus is heavily obscured in all directions. Indeed, optical to mid-IR observations were unable to detect any indication of the AGN activity: at these wavelengths the central region is characterized by a starburst with a spectacular superwind cavity where LINER-like lines are produced (Maiolino et al. 2000a, Marconi et al. 2000). A case quite similar to NGC4945, but at much higher luminosities, is NGC6240. The optical spectrum of this strongly interacting system shows only weak LINER-like emission lines and the mid-IR properties are similar to starburst galaxies (Lutz et al. 1999). However, hard X-ray observations have detected the presence of a heavily obscured AGN whose intrinsic luminosity is in the QSO range (Vignati et al. 1999). Other luminous IR galaxies, optically classified as starburst or LINER, might also host completely hidden AGNs (Risaliti et al. 2000).

Some of the luminous hard X-ray sources recently discovered by deep Chandra surveys have optical and near-IR counterparts that show little or no evidence for AGN activity (eg. Mushotzky et al. 2000, Fiore et al. 2000). Possibly, many of these objects are the analogous at higher redshift of the local, completely hidden AGNs discussed above.

6. FOSSIL ACTIVE NUCLEI.

Not all of the narrow line, type 2 Seyfert galaxies are associated to absorption along our line of sight. A fraction of the Seyfert galaxies have shown a strong decrease of their X-ray flux, on time scales of several years, that is not associated to an increased absorption but to an intrinsic drop of their activity (eg. Bassani et al. 1999b). One of the best studied cases is NGC2992 (Weaver et al. 1996). The fraction of these "fossil" AGNs is about 10% (Maiolino et al. in prep.). However, this is actually a lower limit to their real fraction since the identification of this class of objects requires X-ray observations at different epochs and, in particular, the source had to be in a high, bright state during the observations performed by the early, low sensitivity missions. These conditions are met only for a limited number of Seyferts. Obviously, the conservation of the number of active nuclei requires fossil AGNs to revive after a certain period, as it is actually observed (eg. Gilli et al. 2000a).

Shortly after that an AGN turns off, the BLR fades and the nucleus appears as a type 2 Seyfert whose X-ray emission is dominated by the cold reflection component due to the circumnuclear torus (Guainazzi et al. 1998). Although the AGN appears as a Compton thick Sy2, these features are *not* due to obscuration. About 10 years later also the echo from the torus should fade. If the nucleus remains in a quiescent state for an even longer period then also the gas in NLR clouds can recombine. Yet, the high ionization regions of the clouds should recombine much faster and, therefore, the observed narrow line spectrum should be characterized by low ionization lines similar to LINERs (Eracleous et al. 1995). In this phase the X-ray emission should be dominated by the *non-variable* warm reflection component with, possibly a highly ionized Fe line at 6.7–7 keV, as indeed observed in several LINERs. Some of the LINER nuclei might actually be fossil AGNs.

7. IMPLICATIONS FOR THE X-RAY BACKGROUND.

Obscured AGNs are thought to be a key ingredient of the hard X-ray background (XRB, Setti & Woltjer 1991, Comastri et al. 1995). This is discussed in detail in this volume by other authors (eg. Comastri). However, I wish to emphasize that some of the issues discussed in this review might have implications for our understanding of the origin of the hard XRB.

The distribution of N_H represents the main set of free parameters in the XRB synthesis models. The N_H distribution presented in sect. 2.1 can be used to freeze this set of parameters, under the assumption that the distribution does not evolve with redshift. Detailed models that take into account this constraint are presented in Gilli et al. (1999, 2000b): although the shape and power of the XRB is well reproduced, the observed number counts in the hard X-rays seem to require an evolution of the obscured AGNs faster than for the unobscured population. This result is also related to the finding that non–axisymmetric morphologies increase the obscuration of the active nuclei (sect. 3). Indeed, the latter effect should have an impact on the XRB, since galaxies at high redshift are characterized by a higher rate of distorted/irregular morphologies.

The deficiency of dust absorption, with respect to the X-ray absorption, especially at high luminosities, implies a possible mismatch between optical and X-ray classification of the sources contributing to the hard X-ray background. In particular, some of the type 2 QSOs, which are expected to make most of the hard XRB, could be optically "masked" as type 1 QSOs and already present in optical surveys. As mentioned in Sect. 4, the early Chandra surveys have identified only a few objects of this class, at variance with what found in radio and past hard X-surveys (Sect. 4). Possibly the discrepancy is due to the sensitivity of Chandra which peaks in the soft X-ray band and, therefore, probably biases the surveys in favor of little absorbed sources. To properly tackle this issue we should wait for the identification of a larger number of Chandra sources (especially at fainter fluxes, where the fraction of absorbed sources is higher) and the results of the surveys that are being performed with XMM, whose sensitivity is much more uniform up to ~ 7 keV.

ACKNOWLEDGEMENTS

Many of the new results presented in this paper were obtained in collaboration with R. Gilli, A. Marconi, G. Risaliti and M. Salvati. This work was partially supported by the Italian Space Agency (ASI) through the grant ARS-99-15 and by the Italian Ministry for University and Research (MURST) through the grant Cofin-98-02-32.

REFERENCES

Antonucci, R.R.J., 1993, ARA&A 31, 473
Akiyama, M., et al., 2000, ApJ, 532, 700
Bassani, L., et al. 1999b, ApJS, 121, 473
Clavel, J., et al., 2000, A&A, 357, 839
Comastri, A., Setti, G., Zamorani, G., Hasinger, G., 1995, A&A, 296, 1
Done, C., Madejski, G.M., Smith, D.A., 1996, ApJ, 463, L63
Draine, B.T., 1995, in *The physics of the interstellar medium and intergalactic medium*, ASP Conf. Ser. 80, p.133
Eracleous, M., Livio, M., Binette, L., 1995, ApJ 445, L1
Fiore, F., et al., 2000, New Astr. in press (astro-ph/0003273)
Ford, H., et al., 1997, in *IAU Coll. 163*, ASP Conf. Ser. 121, p.620
Gallimore, J.F., Baum, S.A., O'Dea, C.P., 1997, Nat., 388, 852
Gilli, R., Risaliti, G., Salvati, M., 1999, A&A, 347, 424
Gilli, R., et al. 2000a, A&A, 355, 485
Gilli, R., 2000b, in "Fourth Italian Conference on AGNs", MemSAIt, in press, (astro-ph/0007207)

Goodrich, R.W., 1995, ApJ, 440, 141
Granato, G.L., Danese, L., Franceschini, A., 1997, ApJ, 486, 147
Greenhill, L.J., Gwinn, C.R., 1997, Ap&SS, 248, 261
Guainazzi, M, et al., 1998, MNRAS, 301, L1
Hayashi, I., Koyama, K., Awaki, H., 1996, PASJ, 48, 219
Iwasawa, K., 1999, MNRAS, 302, 96
Komossa, S., Fink, H., 1997, A&A 327, 555
Lawrence, A., Elvis, M., 1982, ApJ, 256, 410
Lutz, D., Veilleux, S., Genzel, R., 1999, ApJ, 517, L13
Maccacaro, T., Perola, G.C., Elvis, M., 1982, ApJ, 257, 47
Malaguti, G., et al., 1999, A&A, 342, L41
Malizia, A., et al. 1997, ApJ 113, 311
Maiolino, R., Rieke, G.H., 1995, ApJ, 454, 95
Maiolino, R., et al., 1998, A&A, 338, 781
Maiolino, R., Risaliti, G., Salvati, M., 1999a, A&A, 341, L35
Maiolino, R., et al. 2000a, in *Imaging the Universe in Three Dimensions*, Eds. W. van Breugel & J. Bland-Hawthorn, ASP Conf. Ser. 195, p. 307 (astro-ph/9906038)
Maiolino, R., et al. 2000b, A&A submitt
Malkan, M.A., Gorjian, V., Tam., R., 1998, ApJS, 117, 25
Marconi, A., et al., 2000, A&A, 357, 24
Matt, G., Brandt, W.N., Fabian, A.C., 1996, MNRAS, 280, 823
Matt, G., et al. 1997, A&A, 325, L13
McLeod, K.K., Rieke, G.H., 1995, ApJ, 441, 96
Mushotzky, R.F., Cowie, L.L., Barger, A.J., Arnaud, K.A., 2000, Nat, 404, 459
Netzer, H., Laor, A., 1993, ApJ, 404, 51
Netzer, H., Turner, T.J., George, I.M., 1998, ApJ, 504, 680
Reeves, J.N., Turner, M.J.L., Ohashi, T., Kii, T., 1997, MNRAS, 292, 468
Risaliti, G., Maiolino, R., Salvati, M., 1999, ApJ, 522, 157
Risaliti, G., Gilli, R. Maiolino, R., Salvati, M., 2000, A&A, 357, 13
Sakamoto, K., 1999, in Galaxy Dynamics: from Early Universe to the Present, ASP Conf. Ser., in press (astro-ph/9910226)
Salvati, M., et al., 1997, A&A, 323, L1
Sambruna , R.M., Eracleous, M., Mushotzky, R.F., 1999, ApJ, 526, 60
Setti, G., Woltjer, L., 1989, A&A, 224, 21
Simcoe, R., McLeod, K.K., Schachter, J., Elvis, M., 1997, ApJ, 489, 615
Turner, T.J., et al., 2000, ApJ, 531, 245
Vignali, C., et al., 1998, A&A, 333, 411
Vignati, P., et al., 1999, A&A, 349, L57
Weaver, K.A., et al. 1996, ApJ, 458, 160
Wilson, A.S., Elvis, M., 1997, Ap&SS, 248, 141
Xue, S.-J., et al., 1998, PASJ, 50, 519

THE BROAD BAND SPECTRUM AND VARIABILITY OF SEYFERT 1

Giorgio Matt

Dipartimento di Fisica, Università degli Studi Roma Tre, via della Vasca Navale 84, I–00146 Roma, Italy

ABSTRACT Recent results on the X–ray spectrum and variability of Seyfert 1 galaxies are reviewed. New spectral results from BeppoSAX observations are also presented and briefly discussed.

KEYWORDS: X–rays:galaxies; galaxies:nuclei

1. INTRODUCTION

With *Chandra* just launched, and XMM and ASTRO-E due to follow suit soon, it is the right time to review what we have learned on Seyfert 1 galaxies from previous generation satellites like ASCA and BeppoSAX, and point out the still open questions, hopefully to be answered or at least addressed by the new generation missions. In particular, I will concentrate here on BeppoSAX results, not only because I worked myself on them but also because it is only now that a general picture from these observations is emerging. Before discussing the BeppoSAX observations, however, let me at least mention briefly the recent results from ASCA and XTE which I consider most significant.

2. RECENT RESULTS ON SEYFERT 1

2.1. Variability

The most exciting result in recent years on variability in Seyfert Galaxies is probably the discovery of periodicity in the flux of the Compton–thin Seyfert 2 galaxy IRAS 18325-5925 (Iwasawa et al. 1998). The discovery has been obtained with an imaging satellite, ASCA; when the field is observed with a better spatial resolution instrument (ROSAT), no evidence for a confusing source is found. The chance of the periodicity being due to a cataclysmic variable (like in the infamous case of NGC 6814, Madejski et al. 1993) or some other confusing source is therefore very low. As periodicity has been observed so far only in one (or two, see below) sources, it is possible that it is a transient event, e.g. a flaring region orbiting around the black hole with a lifetime longer than usual (or being so strong to outshine all other flares simultaneously present). IRAS 18325-5925 has many characteristics typical of the subclass of Narrow Line Seyfert 1 Galaxies (see Boller, 1999, for a review):

large amplitude variation, steep spectrum, possibly ionized disc. One can then argue that periodicities are better to be searched for in NLS1s rather than in 'classical' Seyferts. However, a recent claim of a 33 h periodicity in MCG–6-30-15 by Lee et al. (2000) could indicate that periodicity may be a not so rare phenomenon, after all, and that it has been generally missed so far due to lack of long enough observations.

Thanks mostly to RXTE, long monitoring campaigns in X–rays simultaneously with optical and UV have been performed on a few selected objects, with quite surprising and important results. In NGC 7469 (Nandra et al. 1998), NGC 5548 (Chiang et al. 1999) and NGC 3516 (Maoz et al. 2000), optical/UV and X–ray emissions seem to be uncorrelated or, if any correlation exists, is the optical/UV that leads X–rays, opposite to what expected in the reprocessing scenario. This scenario (Collin-Souffrin 1991), in which optical and UV emissions arises as reprocessing in the accretion disc of the hard X–rays, was suggested to explain the too tight correlation (and lack of delays) between optical and UV fluxes (Courvoisier & Clavel 1991) in some Seyfert 1s (NGC 5548 being one of the best examples). So, the situation is rather confused and puzzling, as in the same source both the viscous dissipation and the reprocessing scenario appear to be unviable.

Regarding even longer term variability, the most spectacular result recently has been the switching off of the "swan's song" galaxy, NGC 4051 (Guainazzi et al. 1998). When observed by BeppoSAX in May 1998, only the Compton reflection component and the iron line were visible, while the primary radiation had disappeared. When put in the context (Uttley et al. 1999), this observation demonstrated that the observed reflected emission occurred in matter distant at least a few light–months from the nucleus. So far, this is the most convincing evidence for the presence of large amount of cold circumnuclear matter in Seyfert 1 galaxies.

2.2. Warm absorbers

Warm absorbers are detected in at least half of Seyfert 1s (Reynolds 1997, George et al. 1998), indicating that the covering factor of the ionized material is fairly large. The reason why warm absorbers are not observed in all Seyferts is not clear. It may be due either to a spread in column density and/or ionization structure, or to a covering factor of the ionized matter less than unity. In the first case, an improved sensitivity, as for instance that provided by XMM, is necessary to search for small edges due either to small columns of oxygen or to less abundant elements (e.g. Neon). In the second case, the ionized material may be observed via emission lines (e.g. Netzer 1993), and high resolution detectors (gratings, *Chandra* and XMM, or calorimeters, ASTRO-E) are needed.

The most interesting recent discovery possibly related to warm absorbers is the 1 keV absorption feature observed in some NLS1s (Vaughan et al., 1999a; Leighly 1999 and references therein). When interpreted as a blend of several resonant lines, mainly from iron L (Nicastro, Fiore & Matt 1999), this implies that the ionization structure of narrow and broad lines Seyferts is rather different, not surprisingly given the different ionizing continuum, especially in soft X–rays. Other possible explanations include blueshifted absorption edges (Leighly et al. 1997) and oxygen smeared edge from ionized discs (Fabian, private communication; see also Vaughan et al. 1999b). XMM should easily distinguish between competing models.

2.3. The iron line

The iron line in Seyfert 1 galaxies is usually broad (Nandra et al. 1997), likely originating in a relativistic disc (e.g. Fabian et al. 1995). In the two best studied sources, MCG–6-30-15 (Tanaka et al. 1995; Guainazzi et al. 1999) and NGC 3516 (Nandra et al. 1999), the observation of a double–horned and asymmetric line profile, expected in presence of fast orbital motion and strong gravity effects (e.g. Fabian et al. 1989; Matt et al. 1992), makes this interpretation quite strong.

In principle, from the line profile and short term variability in response to variations of the continuum (i.e. the so–called reverberation mapping), it is possible to determine the spin of the black hole and the geometry of the accretion disc and the illuminating source (e.g. Martocchia, Karas & Matt 1999 and references therein). Large sensitivity, coupled with good energy resolution, it is required for these goals. It is possible that even the next missions will not sufficient for this purpose, and we have probably to wait for Constellation–X (e.g. Young & Reynolds 2000) and XEUS. It is worth noting, however, that already with ASCA it has been possible to start studying variations of the iron line in MCG–6-30-15 (Iwasawa et al. 1996; 1999). Even if the results are not yet conclusive, these works demonstrated the potentiality of the method.

While most of the iron lines observed in Seyfert 1s can be explained in terms of relativistic discs, it is not clear whether emission from more distant matter (let us call it the "torus") is also present. Evidence for this component is scanty, and the only unambiguous cases so far are those of NGC 4151 (e.g. Piro et al. 1999) and of NGC 4051 discussed in Sec.2.1. In the latter case, the discovery was due more to a stroke of luck rather than to systematic searches, and the probability to find other sources switched off seems not very high. More promising, in view of missions like XMM and expecially ASTRO–E with large sensitivity coupled with good energy resolution, is to search for narrow iron lines (as the torus should be at a distance such that relativistic effects are negligible).

2.4. The continuum

While there is wide consensus on the fact that the power law index (assuming that a power law is actually a good enough description of the primary continuum: see Petrucci et al., 1999) is typically around 1.8–2 (Nandra & Pounds 1994) and that the Compton reflection is a common ingredient, there are still several open questions: is there a correlation between the photon index and the amount of Compton reflection, as suggested by Zdziarski et al (1999)? How common is the soft excess in Broad Lines Seyfert 1s (there is actually no doubt that it is common in Narrow Lines Seyfert 1s)? What is the typical (if any) cut–off energy in the hard X–ray spectra?

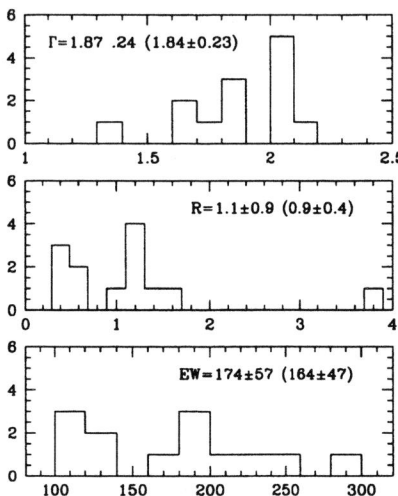

FIGURE 1. The distributions of power law indices, Γ, of the amount of Compton reflection, R and of the iron line EW. The average values and standard deviations are labeled. In parentheses, the value excluding Mkn 841 are also given.

To address these questions, let me discuss the BeppoSAX observations of a dozen of moderately bright Seyfert 1s.

3. A BEPPOSAX SPECTRAL SURVEY OF SEYFERT 1

In Table 1, the observed sources and the main spectral parameters are listed. When the sources have been observed more than once, the results refer to the combined spectrum if there is not spectral variability, or otherwise to the observation when the source was brightest.

For all sources, the spectral model includes: a power law with exponential cut-off; a gaussian line (or a relativistic line, when it provides a significantly better fit); the Compton reflection continuum. A warm absorber and/or a soft excess have been added if required by the data. The spectral parameters reported in the Table refer to the best fit model, whatever its complexity.

In Fig. 1 the distributions of Γ, R (the solid angle, in units of 2π, subtended by the reflecting matter, assumed to be observed face–on) and of the equivalent width of the iron line, are given. The results are in agreement with those found by Nandra & Pounds (1994) and Nandra et al. (1997) and based on *Ginga* and ASCA observations, respectively.

TABLE 1. The best fit parameters for the Seyfert 1s observed by BeppoSAX. For the case of Mkn 841, see the caveats in the text. All errors refer to 90% confidence level for two interesting parameters, i.e. $\Delta\chi^2 = 4.6$.

Source	S.E.	Γ	R	EW (eV)	E_c (keV)
NGC 5548[1,2]	NO	$1.63^{+0.04}_{-0.03}$	$0.44^{+0.18}_{-0.18}$	120 ± 40	160^{+50}_{-70}
NGC 3783[1]	NO	$1.71^{+0.08}_{-0.08}$	$0.37^{+0.22}_{-0.22}$	190 ± 65	160^{+90}_{-60}
NGC 7469[1]	NO?	$2.04^{+0.05}_{-0.05}$	$1.1^{+0.4}_{-0.3}$	180 ± 90	> 230
NGC 4151[1]	?	$1.35^{+0.20}_{-0.20}$	$0.45^{+0.20}_{-0.20}$	240 ± 70	70 ± 20
Fairall 9	?	$2.05^{+0.11}_{-0.09}$	$1.2^{+0.7}_{-0.5}$	220^{+60}_{-100}	> 220
NGC 5506	?	$2.03^{+0.09}_{-0.08}$	$1.5^{+1.2}_{-0.5}$	175 ± 45	> 380
NGC 4593[3]	NO	$1.87^{+0.05}_{-0.05}$	$1.1^{+0.4}_{-0.4}$	190^{+90}_{-6}	> 150
IC 4329A[4]	NO	$1.86^{+0.03}_{-0.03}$	$0.55^{+0.15}_{-0.13}$	110^{+50}_{-40}	270^{+170}_{-80}
Mrk 509[5]	Yes	$1.58^{+0.08}_{-0.09}$	$0.6^{+0.3}_{-0.3}$	110^{+130}_{-60}	70^{+50}_{-20}
MCG-8-11-11[5]	NO	$1.84^{+0.05}_{-0.05}$	$1.0^{+0.6}_{-0.4}+$	130^{+70}_{-50}	170^{+300}_{-80}
MCG-6-30-15[6]	NO	$2.06^{+0.03}_{-0.03}$	$1.2^{+0.4}_{-0.2}$	200^{+50}_{-60}	160^{+130}_{-60}
Mrk 841	Yes?	$2.16^{+0.07}_{-0.05}$	$3.9^{+1.2}_{-1.2}$	290^{+210}_{-220}	> 150
NGC 3516[7]	NO	$2.04^{+0.03}_{-0.03}$	$1.4^{+0.4}_{-0.4}$	100^{+70}_{-60}	> 350

[1] Piro et al. 1999; DeRosa et al. 1999; [2] Nicastro et al. 2000; [3] Guainazzi et al. 1999a; [4] Perola et al. 1999; [5] Perola et al. 2000; [6] Guainazzi et al. 1999b; [7] Costantini et al. 2000.

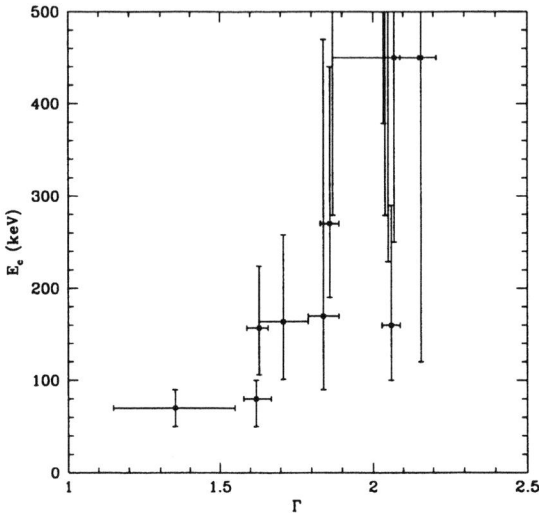

FIGURE 2. The high energy exponential cut–off, E_C, versus the photon index Γ.

3.1. The high energy cut–off

Before the launch of BeppoSAX, a high energy cut–off was unambiguously detected only in one source, NGC 4151 (Jourdain et al., 1992). From the analysis of the spectrum obtained summing several sources observed with OSSE, Gondek et al. suggested an average cut–off energy of a few hundreds keV.

From Table 1, it can be seen that a high energy cut–off is positively detected in about half of the sample. The detected values range from 70 keV (NGC 4151) to 270 keV (IC4329A). When only a lower limit can be obtained, the values are generally consistent with the measured ones. The plot of the e-folding energies vs. the power law index is shown in Fig. 2. While there is no obvious correlation between the two parameters, there is a tendency for flat sources to avoid high values of the cut–off energy. Whether this reflects something intrinsic to the spectra, or simply the fact that it is easier to determine high energy cut–offs in flat sources, cannot be said with the present data.

3.2. The soft excess

Only two out of 13 sources (namely Mkn 509 and Mkn 841) in the BeppoSAX sample discussed here present evidence for a soft excess. Even if for three sources (NGC 5506 and NGC 4151, which have a large intrinsic absorption, and Fairall 9, for which the LECS instrument was switched off, for technical problems, for almost all the observation) we cannot say anything, the soft excess appears to be the exception rather than the rule.

For Mkn 509, the inclusion of a soft excess (in the form of a steeper power law dominating below about 1 keV) improves significantly the goodness of the fit and provides a more astrophysically sound global solution (Perola et al. 2000). For Mkn 841, a warm absorber fits the spectrum almost as well as a black body soft excess, but from ASCA it seems that the warm absorber is not very relevant in this object (see e.g. Reynolds 1997). In both sources, the shape and luminosity of the soft component remains rather undetermined, and what can be said is that a steepening of the spectrum occurs below 1 keV or so. Whether there is a curvature in the primary X–ray emission, or there are two different components, it is impossible to establish.

3.3. The iron line and Compton Reflection

In Fig. 3 the iron line EW is plotted against R. In the reprocessing scenario, the two quantities are expected to be strongly correlated (George & Fabian 1991; Matt, Perola & Piro 1991). The error bars, however, are too large to really check this hypothesis. The relatively small probability (3.4%) that they are not correlated is mainly due one point, i.e. Mkn 841 (but see next paragraph).

More interesting and instructive is the comparison between the average values of the two quantities (Fig. 1), as it can be used to estimate the iron abundance, which turns out to be consistent (say, within a factor of 2 or so) with the cosmic value.

3.4. Are Γ and R correlated?

In Fig. 4 R is shown against the photon index. A correlation between the two quantities is evident, the probability that they are not correlated being only 1.3% if Mkn 841 is excluded, and less than 0.1% when it is included. However, a word of caution is necessary here. The two parameters are strongly correlated in the fit procedure (see also the discussion in Nandra et al. 2000). If, for some reason, the photon index is miscalculated, and e.g. a value steeper than real is found, a larger R is immediately obtained. Let me use the case of Mkn 841 as an example. If the soft excess is modeled with a black body, the results for Γ, R and the iron EW listed in Table 1 and presented in the figures are obtained. An even steeper spectrum ($\Gamma=2.27$) and, correspondingly, larger R (4.6) is obtained with the warm absorber model. Conversely, if the soft excess is modeled by a power law, the photon index is significantly flatter (($\Gamma=1.75$) and R reduces to 1.5 (but the line EW remains fairly large, e.e. 350 eV).

Therefore, I believe that the issue of the reality of the Γ–R correlation is still open, and can be further addressed only by much more sensitive missions like XMM (but the energy band may be not broad enough) or Constellation–X.

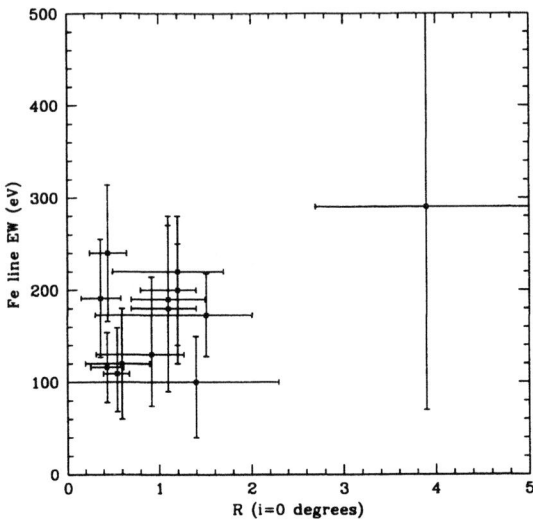

FIGURE 3. The iron line EW vs. R.

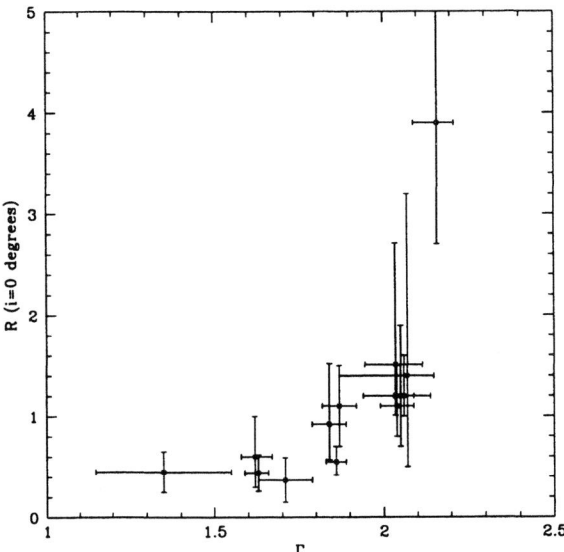

FIGURE 4. R vs. Γ.

4. CONCLUSIONS

The main results from the BeppoSAX observations of a sample of Seyfert 1 Galaxies can be summarized as follows:

- In most sources there is no evidence for a soft excess. In only two sources, Mkn 509 and Mkn 841, a steepening of the spectrum below about 1 keV seems required by the data. Whether it is due to a curvature of the primary emission, or to an altogether distinct component, is impossible to say.

- A high energy cut–off with e–folding energy typically in the range $\sim$100–200 keV is observed, when the quality of the data is good enough.

- The power law index, Γ, and the amount of reflection component are clearly correlated, confirming the findings of Zdziarski et al. (1999). However, it is possible that such a correlation is not real, but partly or entirely due to the fact that the two parameters are strongly correlated in the fit procedure. Let us suppose that some values of Γ are miscalculated, either because there are other components not taken into account, or because a power law is a too simple model (e.g. Petrucci et al. 1999). In this case, a correlation between Γ and R would be immediately found, as the quality of present (and of course past) data is not sufficient to measure the amount of Compton reflection independently of the shape of the primary continuum.

ACKNOWLEDGEMENTS

The BeppoSAX results have been presented on behalf of large teams led by G.C. Perola and L. Piro. I thank E. Costantini, A. De Rosa, F. Nicastro, L. Piro and especially G.C. Perola for many useful discussions, and for kindly providing the results of their analysis before publication. I acknowledge financial support from the Italian Space Agency, and from MURST (grant COFIN98–02–32).

REFERENCES

Boller, T., 1999, this conference
Chiang J., et al., 1999, ApJ, in press (astro–ph/9907114)
Collin–Souffrin S., 1991, A&A, 249, 344
Costantini E., et al., 2000, ApJ, submitted
Courvoisier T.J.L., Clavel J., 1991, A&A, 248, 389
De Rosa A., et al., 1999, this conference
Fabian A.C., Rees M.J., Stella L., White N.E., 1989, MNRAS, 238, 729
Fabian A.C., et al., 1995, MNRAS, 277, L11
George I.M., Fabian A.C., 1991, MNRAS, 249, 352
George I.M., Turner T.J., Netzer H., et al., 1998, ApJSS 114, 73
Gondek D., et al., 1996, MNRAS 282, 646
Guainazzi M., et al., 1998, MNRAS, 301, L1

Guainazzi M., et al., 1999, A&A, 341, L27
Iwasawa K. et al., 1996, MNRAS 282, 1038
Iwasawa K., et al., 1998, MNRAS, 295, L20
Iwasawa K., et al., 1999, MNRAS, 306, L19
Jourdain E., et al. 1992, 256, L38
Lee J., et al., 2000, MNRAS, submitted (astro–ph/9909239)
Leighly K., Mushotzky R.F., Nandra K., Forster K., 1997, ApJ, 489, L25
Leighly K., 1999, ApJS, 125, 317
Madejski G.M., et al., 1993, Nature, 365, 626
Maoz D., Edelson R., Nandra K., 2000, AJ, 119, 199
Martocchia A., Karas V., Matt G., 1999, MNRAS, 312, 817
Matt G., Perola G.C., Piro L., 1991, A&A, 245, 25
Matt G., Perola G.C., Piro L., Stella L., 1992, A&A, 257, 63
Nandra K., Pounds K. A., 1994, MNRAS 268, 405
Nandra K., George I. M., Mushotzky R. F., Turner T. J., Yaqoob T., 1997, ApJ 477, 602
Nandra K., et al., 1998, ApJ, 505, 594
Nandra K., et al., 1999, ApJ, 523, 17
Nandra K., et al., 2000, ApJ, in press (astro-ph/0006339)
Netzer H., 1993, ApJ, 411, 594
Nicastro F., Fiore F., Matt G., 1999, ApJ, 517, 108
Nicastro F., et al., 2000, ApJ, 536, 718
Perola G.C., et al., 1999, A&A, 351, 937
Perola G.C., et al., 2000, A&A, 358, 117
Petrucci P.O., et al., 1999, this conference
Piro L., 1999, Astron. Nach., in press
Reynolds C. S., 1997, MNRAS 286, 513
Tanaka Y., et al., 1995, Nature, 375, 659
Uttley M., et al., 1999, MNRAS, 307, L6
Vaughan S., Reeves J., Warwick R., Edelson R., 1999a, MNRAS, 309, 113
Vaughan S., Pounds K.A., Reeves J., Warwick R., Edelson R., 1999b, MNRAS, 308, L34
Young A.J., Reynolds C.S., 2000, ApJ, 529, 101
Zdziarski A.A., Lubinski P., Smith D.A., 1999, MNRAS, 303, L11

THE ANOMALOUS X-RAY PULSARS

S. Mereghetti

Istituto di Fisica Cosmica G.Occhialini - CNR via Bassini 15, I-20133 Milano, Italy

ABSTRACT The Anomalous X-ray Pulsars (AXP) are a small group of pulsars characterized by periods of several seconds and by the absence of massive companion stars. Three of them are associated with supernova remnants.

It is unclear whether they are accreting from very low mass companions or consist of isolated neutron stars. In the latter case, their rotational energy loss is too small to provide the observed luminosity, and other powering mechanisms must be invoked.

The discovery of similar periodicities and spin-down values in the Soft Gamma-Ray repeaters suggests that also the AXP could be strongly magnetized neutron stars in which the X-ray emission is powered by the dissipation of the magnetic energy. However, also models based on accretion onto isolated neutron stars have been proposed.

KEYWORDS: Neutron Stars; X-ray binaries; Pulsars

1. INTRODUCTION

The Anomalous X-ray Pulsars (AXP, Mereghetti & Stella 1995) are a small group of pulsars with peculiar properties that distinguish them from the more common "normal" pulsars. The latter are rotating, magnetized neutron stars powered either by the loss of their rotational energy (as in the more than 1000 known radio pulsars) or by the accretion of matter from a companion star (as in the $\sim$70 pulsars in High Mass X-ray Binaries).

The AXP show periodic pulsations in the 6-12 s range, which are secularly slowing-down on timescales of $\sim 10^4$ - 4×10^5 years. This fact has two important consequences: First, it allows to establish the very likely presence of a neutron star. Second, it is clear that the rotational energy loss ($\sim 10^{45}$ Ω $\dot{\Omega}$ erg s^{-1} for a neutron star) is not sufficient to power the luminosities of these objects, that are typically in the range 10^{34} - 10^{36} erg s^{-1}. Therefore, it would seem natural to explain the AXP as X-ray binaries, in which accretion of matter from a companion star is responsible for the observed luminosity. However, the presence of a massive companion star can be excluded in most AXP and there are no signatures of orbital motion in the AXP light curves. Furthermore, AXP have other peculiar properties that sets them apart from the majority of accreting pulsars found in High Mass X-Ray Binaries (HMXRB).

There are now six members of the AXP class (see Table 1). Before reviewing the main models for AXP, I briefly summarize their observational properties in Section 2 (for an extensive review of the observations see Mereghetti 2000). Most of the models proposed for these objects belong to two main classes: models based on accretion (Section 3) and those invoking highly magnetized neutron stars powered by the decay of the magnetic field and/or internal heat dissipation (Section 4).

2. PROPERTIES OF THE AXP

The absence of massive companion stars in AXP is supported by two independent observational results: the lack of orbital motion signatures and the limits on the brightness of their possible optical and infrared counterparts.

In fact, no periodic intensity variations (like eclipses or dips), nor orbital Doppler shifts in the pulse frequency that might indicate the presence of a binary system, have been detected in AXP. The most sensitive searches for orbital Doppler shifts have been carried out with the RXTE satellite. Searches for orbital periods between a few minutes and one day gave negative results, yielding upper limits on the projected semi-major axis $a_x \sin i$ of $\sim$30 and $\sim$60 light-ms for 1E 2259+586 and 1E 1048.1−5937 respectively (Mereghetti, Israel & Stella 1998). Similar results were obtained by Wilson et al. (1998) for 4U 0142+61 .

Though in general the limits on the possible optical/IR counterparts of AXP allow to rule out the presence of non-degenerate massive stars, low mass companion stars cannot be ruled out. More sensitive searches for optical/IR counterparts are needed, but due to the crowding of these low galactic latitude fields, this also requires more precise localizations for the AXP.

Three AXP are found at the center of shell-like SNR: 1E 2259+586 is close to the geometrical center of G109.1–0.1 (CTB 109), a partial radio/X-ray shell with an angular diameter of $\sim$30' (see, e.g., Rho & Petre 1997). The other AXP clearly associated to a SNR is 1E 1841−045 , located at the center of Kes 73, a young ($\sim$2000 yr) SNR at a distance of $\sim$7 kpc (Helfand et al. 1994). Gaensler et al. (1999) have recently reported the discovery of a radio SNR around AX J1845.0−0300 . These three AXP are found close to the geometrical center of the respective SNR, implying transverse velocities of a few hundreds km s^{-1} at most. No reliable associations with SNRs have been reported for the three remaining AXP (4U 0142+61 , 1E 1048.1−5937 and 1RXS J170849−400910).

The AXP are characterized by soft X-ray spectra, clearly different from those of the pulsars in HMXRB. The latter have relatively hard spectra in the 2-10 keV range (power law photon index α_{ph} $\sim$1) that steepen with an exponential cut-off above $\sim$20 keV. Recent observations with the ASCA and BeppoSAX satellites, have shown that in most cases a single power law is not sufficient to describe the spectra of AXP. All the AXP for which good quality observations are available (White et al. 1996, Parmar et al. 1998, Oosterbroek et al.1998, Israel et al. 1999a) require the combination of a blackbody-like component with kT$\sim$0.5 keV, accounting up to $\sim$40-50% of the observed luminosity, and a steep (α_{ph} $\sim$3–4) power law (see

Table 1 - Properties of the Anomalous X-ray Pulsars[a]

SOURCE (SNR)	P (s) $\dot{P}$ (s s^{-1})	L_x (erg s^{-1}) d^b (kpc)	SPECTRUM kT_{BB}/α_{ph}	Ref.[c]
1E 1048.1−5937	6.45 $[1.5-4] \times 10^{-11}$	2×10^{34} 5	BB+PL ∼0.64 keV / ∼2.5	1,2,3
1E 2259+586 (G109.1−0.1)	6.98 ∼5×10^{-13}	5×10^{34} 5	BB+PL ∼0.44 keV / ∼3.9	4,5,6
4U 0142+61	8.69 ∼2×10^{-12}	8×10^{34} 1	BB+PL ∼0.4 keV / ∼4	7,8,9
RXSJ170849−4009	11.00 2×10^{-11}	9×10^{35} 8	BB+PL ∼0.41 keV/ 2.92	10,11
1E 1841−045 (Kes 73)	11.77 4.1×10^{-11}	3×10^{35} 7	PL − / ∼3.4	12,13
AX J1845.0−0300 (G29.6+0.1)	6.97 −	5×10^{34} 8	BB ∼0.7 keV / −	14,15

(a) We exclude from the AXP group the following sources:
4U 1626−67: was originally included among AXP, however it has a hard spectrum, an optical identification, is clearly in a binary system, and showed an extended period of spin-up.
RXJ 1838.4−0301: the pulsations at 5.45 s (Schwentker 1994) have not been confirmed; it is probably a normal late type star (Mereghetti et al. 1997).
RX J0720.4−3125: this 8.4 s pulsar (Haberl et al. 1997) has some similarity with the AXP, but its spectrum is much softer and its luminosity (∼3×10^{31} erg s^{-1}) much smaller.

(b) Distance assumed to compute L_X.

(c) References: [1] Seward et al. 1986; [2] Mereghetti 1995; [3] Oosterbroek et al. 1998; [4] Fahlman & Gregory 1981; [5] Kaspi et al. 1999; [6] Parmar et al. 1998; [7] Israel et al. 1994; [8] White et al. 1996; [9] Israel et al. 1999a; [10] Sugizaki et al. 1997; [11] Israel et al. 1999b; [12] Gotthelf & Vasisht 1997; [13] Gotthelf et al. 1999; [14] Torii et al. 1998; [15] Gaensler et al. 1999.

Table 1). The emitting area inferred from the blackbody components ($R_{BB} \sim$ 1-4 km) corresponds to a large fraction of the neutron star surface.

It is possible that this two component model be an oversimplified description of the true spectra of AXP, resulting from the limitations of the current instruments. Future observations with XMM should resolve this issue, possibly leading to the discovery of narrow spectral features that so far escaped detection. In particular, the energy of cyclotron lines from ions lies in the 0.1 - 10 keV range for the high values of the magnetic field (B$\sim 10^{14}$ G) expected for the magnetar model (see Section 4).

Despite the lack of optical identifications, some indications on the distances of AXP can be derived from their location in the Galaxy, and more reliable estimates are possible for the three AXP associated with SNR. The narrow distribution in the galactic plane ($< |b| > = 0.35°$), indicates that, as a population, AXP are unlikely to be nearby ($\lesssim$ 1 kpc) objects. This is also consistent with the relatively high column density derived from the X-ray spectral fits ($N_H \gtrsim 10^{22}$ cm^{-2}).

Even taking into account conservative errors in the AXP distances, the observed fluxes translate to luminosities in the $\sim 10^{34}$-10^{36} erg s^{-1} range. It is therefore well established that AXP have X-ray luminosities smaller than those typically observed in persistent HMXRB pulsars.

In general, AXP have relatively steady X-ray fluxes, compared with the kind of variability displayed by other classes of accreting compact objects. Most AXP have always been detected at similar flux levels. There are, however, some interesting exceptions. AX J1845.0−0300 showed a flux decrease greater than a factor 14 in observations spaced 3.5 years apart (Torii et al. 1998, Gaensler et al. 1999). Variability was also observed in 1E 2259+586 (a factor $\sim$2 with GINGA observations; Iwasawa et al. 1992).

Another distinctive peculiarity of AXP is their long-term period evolution. In general, accreting neutron stars are expected to spin up, due to the angular momentum transferred from the accreting material and in the interaction between the accretion disk and the neutron star magnetosphere (see, e.g. Henrichs 1983). Indeed this is observed in many HMXRB pulsars in which there is evidence for an accretion disk. Other pulsars show alternating episodes of spin-up and spin-down, the origin of which is not completely understood. On the contrary, the spin periods of AXP are increasing at a nearly constant rate (on timescales ranging from $\sim$5,000 to $\sim 4 \times 10^5$ yrs). This behavior has now been observed in a few AXP for a very long period, spanning more than two decades. Accurate timing measurements have shown that the spin-down of AXP is not constant, but is subject to small fluctuations (see, e.g., Iwasawa et al. 1992, Mereghetti 1995). Recent observations of 1E 2259+586 and 1RXS J170849−400910 with XTE have shown that, at least on timescales of a few years, some AXP can be very stable rotators, with a timing noise as small as that of radio pulsars (Kaspi et al. 1999).

3. MODELS BASED ON ACCRETION

3.1. Accretion from a companion star

Mereghetti & Stella (1995) originally proposed that the AXP are weakly magnetized neutron stars (B$\sim 10^{11}$ G) rotating close to their equilibrium period. This requires accretion rates of the order of a few 10^{15} g s^{-1}, consistent with the AXP luminosities.

The possible nature of the companion stars is constrained by the optical/IR limits on the AXP counterparts and by the absence of orbital Doppler modulations of the pulses. The former allow to exclude bright massive companions, while the latter are now beginning to exclude also main sequence stars for large regions of the orbital parameter space. Except for the unlikely possibility that these systems are seen face-on, main sequence companions can be ruled out in the three best studied AXP (1E 2259+586 , 1E 1048.1−5937 , 4U 0142+61 ; Mereghetti, Israel & Stella 1998, Wilson et al. 1998). Helium burning stars with M $\lesssim 0.8$ $M_\odot$ cannot be excluded, but the accretion rate resulting from Roche lobe overflow would produce a much greater luminosity than the observed one. A possibility is that the He companion underfill its Roche lobe and a smaller accretion rate be provided by a stellar wind. White dwarf companion stars are compatible with the $a_x \sin i$ limits and yield consistent values of accretion. For example, a white dwarf with mass of ~ 0.02 $M_\odot$ and $P_{orb} \sim 30$ min would give the $\dot{M}$ of a few $\times 10^{-11}$ $M_\odot$ yr^{-1} required by the observed luminosity of 1E 2259+586.

3.2. Accretion from the interstellar medium

Accretion from the interstellar medium (ISM) cannot provide the luminosities observed in AXP for typical ISM parameters and neutron star velocities. In fact, the accretion luminosity is given by $L_{acc} \sim 10^{32}$ v_{50}^{-3} n_{100} erg s^{-1}, where v_{50} is the relative velocity between the neutron star and the ISM in units of 50 km s^{-1} and n_{100} is the gas density in units of 100 atoms cm^{-3}. Unless all the AXP lie within nearby (~ 100 pc) molecular clouds, which seems very unlikely considered their distribution in the galactic plane, the accretion rate is clearly insufficient to produce the observed luminosities.

3.3. Isolated neutron stars with residual accretion disks

The scenario of isolated neutron stars fed from a residual accretion disk was first advanced by Corbet et al. (1995) for 1E 2259+586 . van Paradijs et al. (1995) proposed that AXP could be one possible outcome of the common envelope evolutionary phase of close HMXRB systems. The connection with massive binaries is supported by the fact that the AXP seem to be relatively young objects, being located at small distances from the galactic plane and, in at least 50% of the cases, found within SNRs. A residual accretion disk could be formed after the complete spiral-in of a neutron star in the envelope of a giant companion (a Thorne-Zytkow object, TZO, Thorne & Zytkow 1977). According to van Paradijs et al. (1995), the estimated birthrate of AXP is consistent with that of TZO.

These ideas were further developed by Ghosh et al. (1997), who put this model in the broader context of the evolution of close massive binaries. A HMXRB undergoing common envelope evolution can produce two kinds of objects, depending on the (poorly known) efficiency with which the envelope of the massive star is lost. Relatively wide systems have enough orbital energy to led to the complete expulsion of the envelope before the settling of the neutron star at the center of the massive companion. This results in the formation of binaries composed of a neutron star and a Helium star, like 4U 1626–67 and Cyg X–3. Closer HMXRB, on the other hand, produce TZO, due to the complete spiral in of the neutron star in the common envelope phase, and then evolve into AXP.

According to Ghosh et al. (1997), this model can also explain the two component spectra of AXP, as well as their secular spin-down. The accretion flow is supposed to consist of two distinct components: one in form of a disk and one spherically symmetric, resulting from the part of the envelope with less angular momentum. The hot (kT∼1 keV) and ionized spherically symmetric flow forms a shock at the magnetospheric boundary, cools, and enters into the magnetosphere through a Rayleigh-Taylor instability. This results in accretion over a large fraction of the neutron star surface, producing the observed blackbody emission. The power law spectral component is instead produced by the conventional, field-aligned accretion onto the polar caps resulting from the disk component. The AXP are supposed to rotate close to their equilibrium periods, which increase due to the decreasing mass accretion rate.

Another possibility for the formation of a disk around an isolated neutron star is through fallback of some material from the progenitor star after the supernova explosion. According to Chatterjee et al. (2000), for appropriate values of the neutron star magnetic field, initial spin period, and mass of the residual disk, these systems can evolve into AXP with luminosities, periods and lifetimes consistent with the observed values. Due to the steadily declining mass accretion rate, the rotating neutron star evolves through different states. During an initial "propeller" phase, lasting a few thousand years, the spin period increases up to values close to those observed in AXP. In this phase, the AXP progenitors are very faint, undetectable X–ray sources, since accretion down to the neutron star surface is inhibited (or greatly reduced) by the magnetospheric centrifugal barrier. In the following phase, the spin frequency approaches the Keplerian frequency at the inner edge of the disk $\Omega(r_m)$, most of the mass flow is accreted, and the star becomes visible as an AXP. During this quasi-equilibrium phase, the neutron star spins down trying to match $\Omega(r_m)$, which decreases with the declining mass accretion rate in the disk. To explain the narrow range of spin periods observed in AXP, Chatterjee et al. (2000) propose that an advection-dominated accretion flow (ADAF) ensues when the accretion rate further decreases. This would result in a much lower X–ray luminosity, explaining the lack of old AXP ($\gtrsim 5 \ 10^4$ yrs) with long spin periods.

4. MAGNETAR MODELS

Models based on strongly magnetized (B$\sim 10^{14}$–10^{15} G) neutron stars, or "magnetars" (Duncan & Thompson 1992; Thompson & Duncan 1995,1996), were originally developed to explain the peculiar properties of the Soft Gamma-ray Repeaters (SGR). SGRs are remarkable transient events characterized by brief ($<$ 1 s) and relatively soft (peak photon energy $\sim$20-30 keV) bursts of super-Eddington luminosity. Only four (or possibly five) SGRs are currently known (see, e.g., Hurley 2000 for a review). Several authors (e.g. Thompson & Duncan 1995) pointed out some analogies between the prototype AXP 1E 2259+586 and the soft repeater SGR 0526–66, located in the Large Magellanic Cloud SNR N49 and for which pulsations at 8 s were reported during the famous super-burst of March 5, 1979.

In the last two years the possible connection between AXP and SGR, received renewed attention, following the discovery of periodicities also in SGR 1806-20 (Kouveliotou et al. 1998) and SGR 1900+14 (Hurley et al. 1999, Kouveliotou et al. 1999). The values of P and $\dot{P}$ ($\sim$ (5-15) 10^{-11} s s^{-1}) observed in SGRs are very similar to those of AXP. Other similarities are the luminosities of the quiescent counterparts of SGR, ($L_X \sim 10^{34}$-10^{35} erg s^{-1}) and the fact that all of them appear to be associated with SNRs.

If the spin-down in AXP and SGR is interpreted as due to magnetic dipole radiation losses, the neutron star magnetic field can be estimated as B $\sim 3.2 \times 10^{19}$ $(P\dot{P})^{1/2}$ G. The observed values of P and $\dot{P}$ lead to values of B $\gtrsim 10^{14}$-10^{15} G. In the "magnetar" model the magnetic field is the main energy source, powering both the persistent X–ray (and particle) emission and the soft gamma-ray bursting activity. This involves internal heating, due to the magnetic field dissipation, and the generation of seismic activity. The latter is responsible for the soft γ-ray bursts, when the magnetic stresses in the neutron star crust shake the magnetosphere and accelerate particles.

Heyl & Hernquist (1997) showed that, if the magnetic fields in AXP are $\gtrsim 10^{15}$ G, their residual thermal energy can be sufficient to power for a few thousand years the observed X-ray luminosity. This requires the presence of an envelope of hydrogen and helium (an iron envelope is much more efficient in insulating the core, resulting in a lower luminosity and effective temperature at the neutron star surface). The envelope of light elements, with a required mass of $\sim 10^{-11}$-10^{-8} $M_\odot$, could be due to fallback material after the supernova explosion and/or to accretion from the interstellar medium if the neutron star is born in a sufficiently dense environment ($\gtrsim 10^4$ cm^{-3}).

Actually, the AXP magnetic fields derived through the dipole radiation formula are very likely overestimated. In fact, the particle wind outflow, either continuous or in the form of strong episodic outbursts, also contributes significantly to the spin-down (Thompson & Blaes 1998). Harding et al. (2000) derived the relations to estimate magnetic field and spin-down age as a function of the particle wind duty cycle and luminosity. In the case of continuous particle outflows, the AXP and SGR magnetic fields do not need to be much greater than those of conventional radio pulsars.

Colpi et al. (2000) noted that, in the context of the magnetar scenario, the period clustering of AXP can be explained only if the magnetic field decays on a timescale of $\sim 10^4$ years. Models without a significant field decay would lead to the presence of AXP with longer periods, which have not been observed.

Different authors discussed the kind of spin-down irregularities expected in the magnetar model. Melatos (1999) described the oscillation in $\dot{P}$ caused by radiative precession, an effect due to the star asphericity induced by the very strong magnetic field. He fitted the observed evolution of rotation frequency of the AXP 1E 2259+586 and 1E 1048.1−5937 in terms of the periodic ($\sim$5-10 yrs) behavior of $\dot{P}$ resulting from this effect. Unfortunately, the sparse period measurements available for AXP do not allow, for the moment, to discriminate against alternative possibilities.

For instance, Heyl & Hernquist (1999) fitted the same data with a constant spin-down interrupted by a few glitches. The magnitude of these glitches is similar to that observed in radio pulsars. Earlier analysis of the same 1E 2259+586 data showed that the level of $\dot{P}$ fluctuations was similar to that observed in accreting X-ray binaries (Baykal & Swank 1996), and was therefore taken as support to accretion based models. More recently, Kaspi et al. (1999) could obtain a phase-coherent timing solution thanks to RXTE observations of 1E 2259+586 spanning 2.6 years. These data show a very low level of timing noise, contrary to the previous results that were based on sparse (not phase-coherent) observations spanning $\sim$20 years. Also 1RXS J170849−400910 , monitored with RXTE for 1.4 yrs, was found to have a similar level of timing noise (Kaspi et al. 1999), while an even more stable rotator is 1E 1841−045 (Gotthelf et al. 1999).

5. CONCLUSIONS

The absence of a massive companion and the presence of a neutron star are observationally well established, but the AXP remain one of the more enigmatic classes of galactic X-ray sources. The current data are unable to discriminate between the different models proposed for these sources, that do not seem to fit with other "standard" manifestations of neutron stars.

Most of the AXP models involve very peculiar objects, the existence of which is inferred from theoretical considerations or evolutionary scenarios, but not yet supported by compelling observational data. This is what makes the study of AXP so exciting.

Another reason for the great interest in AXP is the growing evidence that a large fraction of neutron stars are born with properties very different from that of the "canonical" Crab-like and Vela-like pulsars. This might explain why only very few SNRs are associated with energetic, rapidly spinning radio pulsars.

Due to their relatively low luminosity and soft spectrum (critically affected by the interstellar absorption) AXP are not easy to find. Several of the known X-ray sources, too faint for sensitive pulsations searches, could be AXP and we can expect that, thanks to the *Chandra* and XMM X-ray satellites, many more will be discovered in the near future.

ACKNOWLEDGEMENTS

I thank my collaborators and friends Luigi Stella, Gianluca Israel and Sergio Campana for their outstanding contribution and continuous help in the study of AXP.

REFERENCES

Baykal A. & Swank J.H. 1996, ApJ 460, 470.
Chatterjee P., Hernquist L., & Narayan R. 2000, ApJ 534, 373.
Colpi M., Geppert U., & Page D., 2000, ApJ 529, L29.
Corbet R.H.D, Smale A.P., Ozaki M. et al. 1995, ApJ 443, 786.
Duncan R.C. & Thompson C. 1992, ApJ 392, L9.
Fahlman G.G. & Gregory, P.C. 1981, Nature 293, 202.
Gaensler B.M., Gotthelf E.V. & Vasisht G. 1999, ApJ 526, L37.
Ghosh P., Angelini L. & White N.E. 1997, ApJ 478, 713.
Gotthelf E.V. & Vasisht G. 1997, ApJ 486, L133.
Gotthelf E.V., Vasisht G. & Dotani 1999, ApJ 522, L49.
Haberl F., et al. 1997, A&A 326, 662.
Harding A.K., Contopoulos I., & Kazanas D. 2000, ApJ 525, L125.
Helfand D. et al. 1994, ApJ 434, 627.
Henrichs H.F. 1983, in "Accretion Driven Stellar X-ray Sources", ed. W.H.G. Lewin & E.P.J. van den Heuvel (Cambridge University Press).
Heyl J.S. & Hernquist L. 1997, ApJ 489 L67.
Heyl J.S. & Hernquist L. 1999, MNRAS 304, L37.
Hurley K. et al. 1999, ApJ 510, L111.
Hurley K. 2000, Proceedings 5th Huntsville GRB Symposium, in press, astro-ph/9912061
Israel G.L., Mereghetti, S., Stella L. 1994, ApJ 433, L25.
Israel G.L., et al. 1999a, A&A 346, 929.
Israel G.L., et al. 1999b, ApJ 518, L107.
Iwasawa K., Koyama, K., & Halpern, J.P. 1992, PASJ, 44, 9.
Kaspi V.M., Chakrabarty D. & Steinberger J. 1999, ApJ 525, L33
Kouveliotou C. et al. 1998, Nature 393, 235.
Kouveliotou C. et al. 1999, ApJ 510, L115.
Melatos A. 1999, ApJ 519, L77.
Mereghetti S. 1995, ApJ 455, 598.
Mereghetti S. 2000, Proceedings NATO ASI "The Neutron Star - Black Hole Connection", in press, astro-ph/9911252
Mereghetti S. & Stella L. 1995, ApJ 442, L17.
Mereghetti S., Belloni T. & Nasuti F. 1997, A&A 321, 835
Mereghetti S., Israel G.L. & Stella L. 1998, MNRAS 296, 689.
Oosterbroek T., Parmar A.N., Mereghetti S. & Israel G.L. 1998, A&A 334, 925.

Parmar A. et al., 1998, A&A 330, 175.
Rho J. & Petre R. 1997, ApJ 484 828.
Schwentker O. 1994, A&A 286, L47.
Seward F., Charles, P.A., Smale, A.P. 1986, ApJ 305, 814.
Sugizaki M. et al. 1997, PASJ, 49, L25.
Thompson C. & Duncan R.C. 1995, MNRAS 275, 255.
Thompson C. & Duncan R.C. 1996, ApJ 473, 322.
Thompson, C. & Blaes O. 1998, Phys. Rev. D 57, 3219.
Thorne K.S. & Zytkow A.N. 1977, ApJ 212, 832.
Torii K., et al. 1998, ApJ 503, 843.
van Paradijs J., Taam R.E. & van den Heuvel E.P.J. 1995, A&A 299, L41.
White N.E. et al. 1996, ApJ 463, L83.
Wilson C.A., Dieters S., Finger M.H., Scott D.M. & van Paradijs J., 1998, ApJ 513, 464.

BEAT-FREQUENCY MODELS OF KILOHERTZ QPOS

M. Coleman Miller

University of Maryland

ABSTRACT Kilohertz QPO sources are reasonably well-characterized observationally, but many questions remain about the theoretical framework for these sources and the consequent implications of the observations for disk physics, strong gravity, and dense matter. We contrast the predictions and implications of the most extensively studied class of kilohertz QPO models, the beat-frequency models, with those of alternative classes of models. We also discuss the expected impact of new observations of these sources with satellites such as Chandra, XMM, Astro-E, and Constellation-X.

KEYWORDS: neutron stars; general relativity; X-ray

1. INTRODUCTION

Soon after the launch of the *Rossi* X-ray Timing Explorer (*RXTE*) in late 1995, observations with it of neutron star low-mass X-ray binaries revealed kilohertz quasi-periodic brightness oscillations (QPOs) in the accretion-powered emission from many of these sources (see, e.g., van der Klis 1998 for a review). These oscillations had high frequencies (up to ~1200 Hz, the highest-frequency astrophysical oscillations ever observed), high amplitudes (up to 15% rms in the 2-60 keV band of the Proportional Counter Array on *RXTE*), and high coherences (with quality factors $Q \equiv \nu/\text{FWHM} > 100$ in many cases), and often appeared as two (but no more) high-frequency oscillations in a single power density spectrum. Beat-frequency models (BFMs) were quickly proposed for this phenomenon (Strohmayer et al. 1996; Miller, Lamb, & Psaltis 1998). In these models, the higher-frequency of the two oscillations is attributed to the orbital frequency of gas at some special radius near the star, and the lower-frequency oscillation is a beat between this orbital frequency and the stellar spin frequency. These models were consistent with many of the trends evident in the early data, including the approximate constancy of the frequency difference between the two simultaneous kilohertz oscillations and the close match in four sources of this frequency difference with the stellar spin frequency inferred from brightness oscillations during thermonuclear X-ray bursts.

It has, however, been established recently that in several sources the frequency difference is *not* constant, and indeed can vary by more than 50 Hz. Moreover, this variation is systematic: the higher the lower peak frequency, the lower the frequency difference. The explicability of this behavior in the beat-frequency picture has direct bearing on some of the most important inferences drawn from the kilohertz QPOs.

For example, only in BFMs is the leveling-off of the frequency of both kilohertz QPOs possibly observed from 4U 1820–30 (Zhang et al. 1998) a signature of the presence of the innermost stable circular orbit (ISCO), a crucial prediction of strong-gravity general relativity. Hence only in BFMs can one infer a high gravitational mass $M > 2.1 M_\odot$ for this source, which constrains strongly the equation of state of the high-density matter in the core of neutron stars.

The changing separation frequency observed in several sources provided part of the motivation for the development of other models of the kilohertz QPOs, in particular the relativistic precession models (e.g., Stella & Vietri 1998). In these models the close match between the separation frequency and the spin frequency inferred from burst brightness oscillations is a coincidence, but they do predict the qualitative effect of a separation frequency that drops with increasing kilohertz QPO frequency.

Here we discuss the beat-frequency model in light of these new developments and contrast it with alternate pictures. In § 2 we describe the observational trends that motivated the development of beat-frequency models. We then elaborate on these models, in particular the sonic-point beat-frequency model. In § 3 we discuss the evidence for a changing difference frequency in the four sources Sco X-1, 4U 1608–52, 4U 1728–34, and 4U 1735–44. We show that an aspect of the sonic-point beat-frequency model, included in the dynamics but originally omitted from the frequency estimates, naturally accommodates the changing difference frequency and can quantitatively fit the data. Finally, in § 4 we contrast some of the predictions of the beat-frequency model with the predictions of the relativistic precession model, and discuss analysis that might be done with current data to help discriminate between the two interpretations. We also explore the impact of future observations, both with the upcoming generation of high spectral resolution satellites (such as Chandra, XMM, and Astro-E) and with longer-term projects such as Constellation-X and a hypothetical high-area follow-on to *RXTE*.

2. MOTIVATION FOR BEAT-FREQUENCY MODELS

As discussed in the introduction, soon after the discovery of kilohertz QPOs it was established that these oscillations have (1) high frequency, (2) high amplitude, and (3) high coherence, and that there are always two or fewer kilohertz QPOs in a given power density spectrum. In addition, the separation frequency appeared consistent with constant in many sources and close to the spin frequency inferred from burst brightness oscillations in the four sources where this could be tested. It is now known (see § 3) that in several, and perhaps all, sources, the separation frequency is *not* constant (although it is still close to the inferred spin frequency), and in fact decreases systematically with increasing lower peak frequency. In § 3 we discuss how this new result may be interpreted within the beat-frequency model.

The high frequency indicates that the source of the brightness oscillations is close to the neutron star. A natural candidate for these oscillations is the orbital frequency at some special radius. Given that the burst oscillation frequency is most convincingly interpreted as the stellar spin frequency or its first overtone (see, e.g., Strohmayer & Markwardt 1999), the close match of the separation frequency with the inferred stellar spin frequency suggests a sideband relation between the two simultaneous kilohertz QPOs. The lack of a third kilohertz QPO suggests a beat-frequency relation, because most other mechanisms will produce both an upper sideband and a lower sideband. For example, amplitude modulation of one frequency by another will produce two sidebands of equal strength. For all of these reasons, beat-frequency mechanisms are natural to propose for the kilohertz QPO phenomenon.

The qualitative match of beat-frequency expectations with the data is not enough to accept this model: it is also important to establish a reason why a particular radius in the accretion disk would be selected. The high coherence of the oscillations requires, in this model, that the radial range from which the oscillations are generated has a fractional width less than $\sim 1/Q$, or less than $\sim 1\%$ in the most stringent cases. Moreover, the high amplitudes observed in some sources indicates that the luminosity cannot be generated in just the small range of radii where the frequency is determined.

For these reasons, the sonic-point beat-frequency model was proposed (Miller, Lamb, & Psaltis 1998). In this model, the special radius at which the frequencies are generated is where the inward radial velocity of the gas in the disk increases rapidly with decreasing radius. As described in Miller, Lamb, & Psaltis (1998), this rapid increase in inward radial velocity is usually caused by radiation drag, which removes angular momentum from the accreting gas and hence causes it to spiral inwards. If for some reason radiation drag is not effective in a particular instance, as might happen if the accretion rate is so high that the optical depth from the stellar surface is large, then a similar rapid increase in velocity will nonetheless occur near the ISCO, where the inspiral of gas will open up simply because of gravitational effects. The radial velocity typically goes from subsonic to supersonic in this transition, and it is therefore convenient to label it $r_{\rm sonic}$, the sonic point. However, in this model it is only the rapid change in velocity, and not the specific fact that the gas crosses a sonic point, that is important.

In this model we assume that there are dense clumps of gas orbiting at many radii near the star, and as we show in Miller, Lamb, & Psaltis (1998) only the clumps near $r_{\rm sonic}$ will produce sharp brightness oscillations. As gas streams from the clump onto the star, it produces a bright pattern of impact that rotates with the clump (see Miller, Lamb, & Psaltis 1998), and hence a distant observer sees a modulation of the flux from the system as the bright impact spot passes into and out of view. This occurs at approximately the orbital frequency of the clumps at $r_{\rm sonic}$. If the intensity of radiation from the stellar surface is also modulated at the stellar spin frequency, as it will be if a weak stellar magnetic field funnels extra matter towards one or two magnetic poles, then this variation in intensity modifies the mass accretion rate from the clumps, and hence changes the total luminosity from the impact spots. As shown in Miller, Lamb, & Psaltis (1998), this creates an observed luminosity modulation from the system at approximately the beat frequency between the orbital frequency of the clumps and the stellar spin frequency. In this model the luminosity in the oscillations is generated at the stellar surface, where most of the gravitational energy is released, and hence the oscillations can have high amplitudes.

We now demonstrate that in this model the slow inward drift of the clumps, included in the dynamical calculations but not the frequency estimates of Miller, Lamb, & Psaltis (1998), generally produces a separation frequency between the pair of kilohertz QPO peaks that is less than the stellar spin frequency.

3. CHANGING DIFFERENCE FREQUENCY

3.1. Observational Evidence

By late 1996, observations of Sco X-1 by van der Klis and colleagues (van der Klis et al. 1997) demonstrated that the separation frequency between the kilohertz QPO peaks in this source is definitely not constant, and in fact that the separation clearly decreases with increasing lower peak frequency. This is contrary to the simplest expectations of the beat frequency model, in which the clumps generating the frequency are orbiting at a constant radius and hence the separation is extremely close to the stellar spin frequency.

Many explanations for this behavior were discussed in the community. Most of them centered on the apparently unique nature of Sco X-1 in this respect and hinged on other of its properties such as its very high (near-Eddington) luminosity at the point when the change in the difference frequency was most pronounced. These explanations included jet models (van der Klis et al. 1997) and other unpublished ideas that involved effects such as the finite thickness of the accretion disk. These models all had different physical bases and even different mathematical models for the changing difference frequency, but were all able to fit the relatively smooth $\Delta\nu$ vs. ν_{lower} curve for Sco X-1.

In the last year or so, however, careful analyses by Méndez and colleagues have shown that Sco X-1 is *not* unique in having a changing difference frequency. Specifically, the low-luminosity sources 4U 1608−52 (Méndez et al. 1998), 4U 1728−34 (Méndez & van der Klis 1999), and 4U 1735−44 (Ford et al. 1998) also have difference frequencies that drop with increasing lower peak frequency, and the separation frequency in 4U 1636−536 is slightly but significantly less than the spin frequency inferred from burst brightness oscillations (Méndez, van der Klis, & van Paradijs 1998). More generally, Psaltis et al. (1998) showed that, within the observational uncertainties, all of the kilohertz QPO sources have separation frequencies that are consistent with this trend. This is therefore clearly not an effect that requires a high luminosity, and more general explanations must be sought.

3.2. Explanation Within the Beat-Frequency Model

As described in detail in Lamb & Miller (1999), and as we now discuss, in the sonic-point model the inward drift of the clumps tends to increase the frequency of the lower kilohertz peak and decrease the frequency of the upper kilohertz peak. This produces a difference frequency that is generally less than the stellar spin frequency, in agreement with observations.

If a clump were to orbit in a perfect circle, then the inspiral time and phase traversed during inspiral of gas from the clump to the stellar surface would be the same regardless of when the gas separated from the clump. In this case, consider two successive "beats", i.e., two successive maxima of the mass flow rate from the clump. These occur a time $\Delta T = 1/(\nu_{\rm orb} - \nu_{\rm spin})$ apart. Because the inspiral time to the surface is the same for each of the two beats, the arrival time at the surface is separated by exactly that same ΔT, and the observed frequency of the beat is therefore just $1/\Delta T = \nu_{\rm orb} - \nu_{\rm spin}$.

Now consider clumps that are drifting in slowly. Then the inspiral time decreases as time progresses. If we again think of two successive beats, suppose that the inspiral time for the second beat is a small time δt less than the inspiral time for the first beat. Then the interval between releases of gas from the clump in the first and second beats is still $\Delta T = 1/(\nu_{\rm orb} - \nu_{\rm spin})$, but the arrival times at the surface are now separated by $\Delta T - \delta t$. Therefore the observed frequency, which depends on surface arrival times because the surface is where the luminosity is generated, is $1/(\Delta T - \delta t)$, which is higher than it would have been without the inward drift. This is somewhat analogous to Doppler blueshifting. Further analysis and comparison with numerical simulations shows (Lamb & Miller 1999) that the upper peak frequency is also affected. In fact, the inward drift decreases the upper peak frequency, typically by a fractional amount that is $\sim 50\%$ of the fractional amount by which the lower peak frequency is increased compared to the circular orbit approximation. Both the increase in the lower peak frequency and the decrease in the upper peak frequency act to decrease the frequency separation.

As described in Lamb & Miller (1999), the magnitude of the change in the frequencies depends on the details of the inspiral, and in particular on how the total phase and time of the inspiral depend on the radial location of the clump. However, much of the physics is captured in a simple model in which the frequency change depends only on the ratio of two radial velocities: the velocity $v_{\rm clump}$ of the clump near the sonic point and a characteristic velocity $v_{\rm gas}$ of the gas when it leaves the clump (see Lamb & Miller 1999 for more details). The observed frequencies are then approximately $\nu_{\rm low} = \nu_{\rm beat}/(1 - v_{\rm clump}/v_{\rm gas})$ and $\nu_{\rm high} = \nu_{\rm orb}(1 - \frac{1}{2}v_{\rm clump}/v_{\rm gas})$, where $\nu_{\rm beat} \equiv \nu_{\rm orb} - \nu_{\rm spin}$ and $\nu_{\rm orb}$ is the orbital frequency at the sonic point. Numerical simulations show that the characteristic velocity $v_{\rm gas}$ of the gas is fairly insensitive to radius, and hence for simplicity we assume that $v_{\rm gas}$ is independent of radius.

What remains is to produce a physical model for $v_{\rm clump}$ as a function of radius. The radial velocity of a clump with angular momentum J under the influence of a torque N is $v_r = N/(\partial J/\partial r)$. In a Schwarzschild spacetime, for clumps whose mass does not depend on the distance from the star,

$$\partial J/\partial r \propto \frac{1}{2}\frac{r-6M}{(r-3M)^{3/2}}, \qquad (1)$$

where we use geometrized units in which $G \equiv c = 1$. The singularity at $r = 6M$ in the equation for the radial velocity is only apparent. In reality, the velocity deviates from this simple formula near the ISCO. Numerical simulations show (Lamb & Miller 1999) that for plausible torques the radial velocity of the clump can be small enough that even near the ISCO the inspiral time of the clumps is large, and hence the coherence of the brightness oscillations can be as high as observed. The simulations also show that for small enough torques the separation frequency can remain almost constant even near the ISCO. Therefore, in this model the observations of 4U 1820–30, which indicate an approximately constant separation frequency even when the kilohertz QPO frequencies may be leveling off, are still consistent with the asymptotic frequency being approximately the orbital frequency at the ISCO (Lamb & Miller 1999).

As a particular model of the torque, assume that the clumps interact only weakly with each other and with any background gas in the accretion disk, so that the primary torque is exerted by the stellar magnetic field. The general form of the torque at radius r is then $N \sim B^2 Ar$, where A is the cross-sectional area of the clump and B is the strength of the stellar magnetic field at the location of the clump. Assume also that the clump mass is approximately independent of the distance from the star, and that the clumps are in pressure equilibrium with the external magnetic field. Suppose that the clumps are kept at the local Compton temperature, which is roughly independent of the radius. Then $nkT \sim B^2 \sim r^{-6}$ for a dipolar field, meaning that the diameter of the clumps scales as r^2 and the cross-sectional area scales as r^4. Then the torque scales as $N \sim r^{-1}$.

This model can then be fit to the data, using the relations between radial velocity and upper and lower QPO frequency given above. For a source such as Sco X-1 which has an unknown spin frequency, there are three parameters: the stellar gravitational mass M, the spin frequency $\nu_{\rm spin}$, and the coefficient of the torque. For a source such as 4U 1728–34 with a known $\nu_{\rm spin}=364$ Hz, there are two free parameters. Figure 1 shows the results of the fitting. It is encouraging that one physical model can fit both of these sources (this model also fits 4U 1608–52 and 4U 1735–44).

4. TESTS OF MODELS WITH CURRENT AND FUTURE DATA

With the immense archive of data accumulated with *RXTE*, there are a number of further analyses that could in principle be performed to discriminate between models or test predictions of models. In this section we discuss such tests, particularly as they might discriminate between beat-frequency models and relativistic precession models, and also look to the future to consider the qualitatively new types of observations that may be available with the upcoming generation of X-ray satellites.

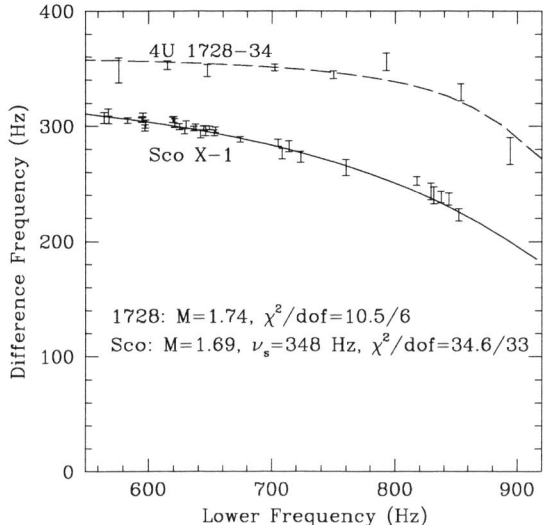

FIGURE 1. Difference frequency data and fit using the drifting clump model described in this section. The data for 4U 1728-34 and Sco X-1 were kindly provided by Mariano Méndez. The spin frequency for 4U 1728-34 was fixed at 364 Hz to correspond to the frequency of burst brightness oscillations in that source.

4.1. Tests With Current Data

An important difference between the predictions of the beat-frequency model and the relativistic precession model concerns expectations for the value of the stellar spin frequency. In the beat-frequency model it is expected that the stellar spin frequency is close to (although slightly larger than) the separation frequency, and is also very close to the burst oscillation frequency or half the burst frequency, depending on the symmetry. In contrast, as recently emphasized by Psaltis & Norman (1999), in the simplest version of the relativistic precession model the data on the lower frequency horizontal branch type brightness oscillations combined with the kilohertz oscillations requires that the stellar spin frequency exceed ~ 500 Hz in all sources, and be especially large (up to ~ 900 Hz) in the Z sources. Simple estimates of corrections due to fluid viscosity (Psaltis & Norman 1999) suggest that these corrections would increase further the required spin frequency. The predictions for $\nu_{\rm spin}$ therefore differ greatly in the two pictures, and if a clear spin frequency is detected in the persistent, accretion-powered emission then this will have strong bearing on the modeling. In addition, if it happens that the spin frequency is not equal to the burst oscillation frequency or half of it, this will completely overturn the apparently compelling arguments that the coherence and stability of the burst oscillation frequency in a given source can only be explained via a rotational modulation mechanism.

The two models also differ in their predictions of the frequencies of other, weak brightness oscillations. In the relativistic precession model, there is an upper sideband of the orbital frequency expected (Psaltis & Norman 1999). The amplitude of this oscillation need not be as great as the amplitude at the lower sideband; for example, if the underlying perturbations of the accretion disk cut off sharply at frequencies exceeding the observed orbital frequency then the upper sideband could be weaker. Nonetheless, the oscillation is expected to be there at some level. In the beat frequency model there is an oscillation expected at twice the observed lower peak frequency. This is expected for several reasons. For example, the waveform generated by the beat is not expected to be perfectly sinusoidal, and hence overtones will be generated. In addition, if the radiation pattern rotating with the star, which helps produce the beat, is itself not perfectly sinusoidal, then it will generate weak overtones of the lower peak frequency.

The predicted frequencies discussed above depend only on the observed frequencies. If the observed frequencies are ν_{low} and ν_{high}, then the predicted frequency in the relativistic precession model is $2\nu_{\text{high}} - \nu_{\text{low}}$ and the predicted frequency in the beat frequency model is $2\nu_{\text{low}}$. The frequencies will change with time for a given source, but these predictions are ideally suited for testing by use of the shift-and-add method first used by Méndez et al. (1998) to discover a weak brightness oscillation in 4U 1608−52.

4.2. Tests With Future Data and Satellites

The launch of a new generation of high spectral resolution satellites such as Chandra, XMM, and Astro-E opens up new ways to probe the strong gravity and dense matter of neutron stars. For example, suppose that simultaneous observations of a kilohertz QPO source are performed with *RXTE* and Astro-E. If an Fe Kα profile from the inner edge of the nearly-circular flow is detected and characterized with Astro-E, this profile gives the radius of that inner edge in units of the gravitational mass of the neutron star. If at the same time a pair of kilohertz QPOs is detected with *RXTE*, then the frequency of the upper peak, combined with the radius in units of the gravitational mass, yields the mass of the star. This is true independent of where the inner edge is; that is, it need not be at the innermost stable circular orbit. Therefore, such simultaneous measurements could provide a clean way to measure the gravitational masses of neutron stars in low-mass X-ray binaries.

Future, high-area missions such as Constellation-X or a hypothetical high-area follow-up timing mission to *RXTE* could provide even more information, by allowing characterization of the waveforms of the brightness oscillations seen in accretion-powered and burst-powered emission. An example is shown in the left panel of Figure 2, which shows theoretical waveforms for burst brightness oscillations. The curve for the more compact star is broader, as is expected due to the extra gravitational light deflection. In addition, the curve for the larger (less compact) star is more asymmetric, due to the greater Doppler shifts from the surface rotation velocity. The waveform therefore encodes information about both the mass and radius of the star, meaning that repeated observations of bursts from a single source can constrain the mass and radius tightly, with consequent constraints on the equation

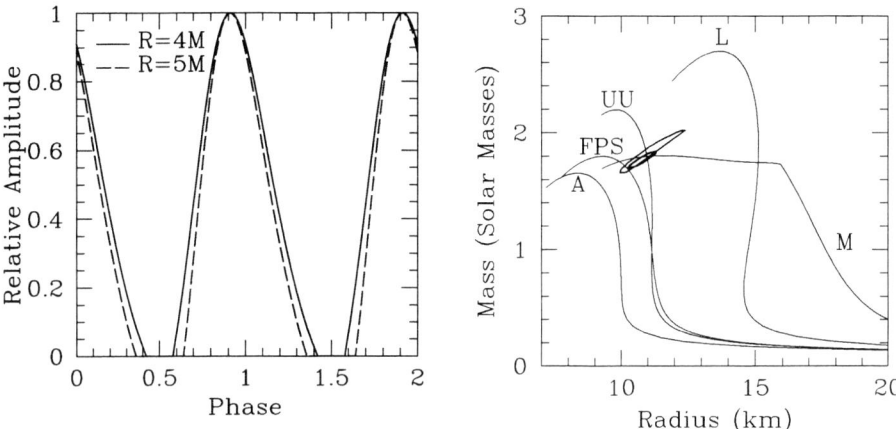

FIGURE 2. (left panel) Theoretical waveforms for burst brightness oscillations. Here we assume a small bright spot on the rotational equator, as seen by a distant observer in the rotational plane. We show simulated light curves over two cycles for a star with a gravitational mass $M = 1.8 M_\odot$ and stellar radii $R = 4GM/c^2$ (solid curve) and $R = 5GM/c^2$ (dashed curve), with a stellar spin frequency as seen at infinity of 364 Hz (the spin frequency of 4U 1728–34, the source most frequently observed with *RXTE* to have burst brightness oscillations). (right panel) Constraints on mass and radius possible with waveform fitting of burst brightness oscillations. We simulated the bolometric waveform from a 1.8 $M_\odot$ star rotating at 364 Hz, with high-density equation of state UU, a small emitting spot on the rotational equator, and no scattering after emission from the surface. We then performed a likelihood analysis using this simulated waveform. We combined the constraints from five bursts, assuming 5 seconds each of 5% rms amplitude, and a flux typical of the bursts observed with *RXTE*. The light outer contour shows the 1σ confidence region expected from observations with Constellation-X, and the dark inner contour shows the 1σ confidence region expected for a hypothetical future 10 m^2 timing satellite. The light solid curves show the mass-radius relations given by different high-density equations of state, labeled as in Miller, Lamb, & Psaltis (1998).

of state of the high-density matter in the core of neutron stars.

The right panel of Figure 2 shows an example of the constraints possible with Constellation-X (outer contour, at 1σ) and a hypothetical 10 m^2 future timing instrument (inner contour, at 1σ). Clearly, waveform fitting can in principle yield very precise information about the mass and radius of individual neutron stars, and therefore about the equation of state of matter at supranuclear densities.

High-area timing missions also will potentially allow us to do qualitatively new types of tests of the strong-gravity predictions of general relativity. For example, we have just discussed two independent ways of estimating the gravitational mass of a neutron star: by combining Fe Kα profiles with kilohertz QPOs, and by waveform fitting of burst brightness oscillations. If either of these is successful for an individual source, then we can predict precisely the orbital frequency at the innermost stable circular orbit. With high-area timing instruments we expect to see more cases like 4U 1820–30, in which there is leveling-off of the QPO frequency and thus evidence for the ISCO. The match of this asymptotic frequency with the frequency predicted from the gravitational mass and general relativity will provide us with unprecedented quantitative tests of general relativity in strong gravity. In conclusion, therefore, the continued qualitative and quantitative agreement of the beat-frequency model with observations of kilohertz QPOs has not only yielded important new constraints on the equation of state of the dense matter in the core of neutron stars and, possibly, the first direct evidence for unstable orbits around neutron stars, a key prediction of general relativity. It also indicates strongly that future observations of these sources, especially with high-area timing missions, will allow us to continue to make qualitative leaps in our observational understanding of strong gravity and dense matter.

ACKNOWLEDGEMENTS

We thank Michiel van der Klis and Tod Strohmayer for discussions about possible information from high-area timing missions. This research was supported in part by NASA ATP grant number NRA-98-03-ATP-028.

REFERENCES

Ford, E. C., van der Klis, M., van Paradijs, J., Méndez, M., Wijnands, R. A. D., & Kaaret, P. 1998, ApJ, 508, L155
Lamb, F. K., & Miller, M. C. 1999, in preparation
Méndez, M., & van der Klis, M. 1999, ApJ, 517, L71
Méndez, M., van der Klis, M., & van Paradijs, J. 1998, ApJ, 506, L117
Méndez, M., van der Klis, M., Wijnands, R. A. D., Ford, E. C., van Paradijs, J., & Vaughan, B. A. 1998, ApJ, 505, L23
Miller, M. C., Lamb, F. K., & Psaltis, D. 1998, ApJ, 508, 791
Psaltis, D. et al. 1998, ApJ, 501, L95
Psaltis, D., & Norman, C. 1999, ApJ, submitted
Stella, L. & Vietri, M. 1998, ApJ, 492, L59
Strohmayer, T. E., & Markwardt, C. B. 1999, ApJ, 516, L81
Strohmayer, T. E., Zhang, W., Swank, J. H., Smale, A., Titarchuk, L., & Day, C. 1996, ApJ, 469, L9
van der Klis, M. 1998, in 'The many faces of Neutron Stars', R. Buccheri, J. van Paradijs & M.A. Alpar (Eds), NATO ASI Series C, Vol. 515, pp. 337-368 (Kluwer Academic Publishers)
van der Klis, M., Wijnands, R. A. D., Horne, K., & Chen, W. 1997, ApJ, 481, L97
Zhang, W., Smale, A. P., Strohmayer, T. E., & Swank, J. H. 1998, ApJ, 500, L171

JETS IN QUASARS, MICROQUASARS AND GAMMA-RAY BURSTS

I.F. Mirabel

Centre d'Etudes de Saclay/ CEA/DSM/DAPNIA/SAP
91911 Gif/Yvette, France &
Intituto de Astronomía y Física del Espacio. Bs As, Argentina

ABSTRACT

Relativistic outflows are common in accreting and forming black holes. Despite the enormous differences in scale, stellar-mass black holes in X-ray binaries and supermassive black holes in Galactic Nuclei produce jets with analogous properties. In both are observed two types of relativistic outflows: 1) quasi-steady compact jets with flat-spectrum, and 2) episodic large-scale ejections with steep-spectrum and apparent superluminal motions. Because of the short time scale of the phenomena in black hole binaries, the formation of synchrotron jets is associated to changes in the X-ray thermal emission from the accretion disk. Besides, the most common class of gamma-ray bursts can be conceived as extreme microquasars, since they are afterglows from ultra-relativistic jets associated to the formation of black holes at cosmological distances.

KEYWORDS: Black holes, jets, quasars, microquasars, gamma-ray bursts

1. THE MICROQUASAR ANALOGY

The discovery of *microquasars* (Margon, 1994; Mirabel et al. 1992) with apparent superluminal motions (Mirabel & Rodríguez, 1994) has opened new perspectives for the astrophysics of black holes (Mirabel & Rodríguez, 1999 for a review). These scaled-down versions of quasars are believed to be powered by spinning black holes with masses of up to a few tens that of the Sun. The word *microquasar* was chosen to suggest that the analogy with quasars is more than morphological, and that there is an underlying unity in the physics of accreting black holes over an enormous range of scales, from stellar-mass black holes in binary stellar systems, to supermassive black holes at the centre of distant galaxies (Rees, 1998).

However, in microquasars the black hole is only a few solar masses instead of several million solar masses; the accretion disk has mean thermal temperatures of several million degrees instead of several thousand degrees; and the particles ejected at relativistic speeds can travel up to distances of a few light-years only, instead of the several millon light-years as in some giant radio galaxies. In quasars matter can be drawn into the accretion disk from disrupted stars or from the interstellar medium of the host galaxy, whereas in microquasars the material is being drawn from the companion star in the binary system. In quasars the accretion disk has sizes of $\sim 10^9$ km and radiates mostly in the ultraviolet and optical wavelenghts, whereas in microquasars the accretion disk has sizes of $\sim 10^3$ km and the bulk of the radiation comes out in the X-rays. It is believed that part of the spin energy of the black hole can be tapped to power the collimated ejection of magnetized plasma at relativistic speeds. This analogy between quasars and microquasars resides in the fact that in black holes the physics is essentially the same irrespective of the mass, except that the linear and time scales of phenomena are proportional to the black hole mass. Because of the relative proximity and shorter time scales, in microquasars it is possible to firmly establish the relativistic motion of the sources of radiation, and to better study the physics of accretion flows and jet formation near the horizon of black holes.

At first glance it may seem paradoxical that relativistic jets were first discovered in the nuclei of galaxies and distant quasars and that for more than a decade SS433 was the only known object of its class in our Galaxy (Margon 1984). The reason for this is that disks around supermassive black holes emit strongly at optical and UV wavelengths. Indeed, the more massive the black hole, the cooler the surrounding accretion disk is. For a black hole accreting at the Eddington limit, the characteristic black body temperature at the last stable orbit in the surrounding accretion disk will be given approximately by $T \sim 2 \times 10^7 \, M^{-1/4}$ (Rees 1984), with T in K and the mass of the black hole, M, in solar masses. Then, while accretion disks in AGNs have strong emission in the optical and ultraviolet with distinct broad emission lines, black hole and neutron star binaries usually are identified for the first time by their X-ray emission. Among these sources, SS 433 is unusual given its broad optical emission lines and its brightness in the visible. Therefore, it is understandable that there was an impasse in the discovery of new stellar sources of relativistic jets until the recent developments in X-ray astronomy. Strictly speaking and if it had not been for the historical circumstances described above, the acronym *quasar* would have suited better the stellar mass versions rather than their super-massive analogs at the centers of galaxies.

2. COUPLING BETWEEN ACCRETION DISK AND JET

Since the characteristic times in the flow of matter onto a black hole are proportional to its mass, variations with intervals of minutes in a microquasar correspond to analogous phenomena with durations of thousands of years in a quasar of 10^9 M$_\odot$, which is much longer than a human life-time Sams et al. (1996). Therefore, variations with minutes of duration in microquasars could be sampling phenomena that we have not been able to study in quasars. The repeated observation of two-sided moving jets in a microquasar (Rodríguez & Mirabel, 1999) has led to a much greater acceptance of the idea that the emission from quasar jets is associated with moving material at speeds close to that of light.

On the other hand, simultaneous multiwavelength observations of GRS 1915+105 are revealing the connection between the sudden disappearance of matter through the horizon of the black hole, with the ejection of expanding clouds of relativistic plasma. Radio, infrared, and X-ray light curves of GRS 1915+105 at the time of quasi-periodic oscillations on 1997 September 9 (Mirabel et al. 1998) have shown that the infrared flares occur during the recovery from X-ray dips. These simultaneous multiwavelength observations have shown the connection between the rapid disappearance and follow-up replenishment of the inner accretion disk seen in the X-rays (Belloni et al. 1997), and the ejection of relativistic plasma clouds observed as synchrotron emission at infrared wavelengths first and later at radio wavelengths.

3. COMPACT JETS IN X-RAY BINARIES AND GALACTIC NUCLEI

The class of stellar-mass black holes that are persistent X-ray sources (e.g. Cygnus X-1, 1E 1740-2942, GRS 1758-258, etc.) and some supermassive black holes at the centre of galaxies (e.g. Sgr A* and many AGNs) do not exhibit luminous outbursts with large-scale sporadic ejections. However, despite the enormous differences in mass, steadily accreting black holes have analogous radio cores with steady, flat ($S_\nu \propto \nu^\alpha$; $\alpha \sim 0$) emission at radio wavelengths. The fluxes of the core component in AGNs are typically of a few Janskys (e.g. Sgr A* ~1Jy) allowing VLBI high resolution studies, but in stellar mass black holes the cores are much fainter, typically of less than a few mJy, which makes difficult high resolution observations of the core.

Although there have been multiwavelength studies and speculation about the nature of the faint and steady compact radio emission in X-ray black hole binaries (e.g. Rodríguez et al. 1995; Fender et al. 1999, 2000), GRS 1915+105 is the black hole binary where the core has been succesfully imaged at AU scale resolution (Dhawan, Mirabel & Rodríguez, 2000). GRS 1915+105 is the only X-ray binary where both, a compact core with steady fluxes ≥ 20 mJy, as well as large-scale superluminal ejections are *unambigously* observed. VLBA images during different states of the source (quiescent and QPO states) always show compact jets with sizes $\sim 10\lambda_{cm}$ AU along the same position angle as the superluminal large-scale jets. The length of the compact jet and the period of the oscillations are consistent with bulk motions $\geq 0.9c$, comparable with the velocities of the large-scale superluminal ejecta (Mirabel & Rodríguez, 1994). As in the radio cores of AGNs, the brightness temperature of the compact jet in GRS 1915+105 is $T_B \geq 10^9$ K. The VLBA images of GRS 1915+105 are consistent with the conventional model of a conical expanding jet with syncrotron emission (Hjellming & Johnston, 1988; Falke & Biermann, 1999) in an optically thick region of solar system size.

4. MICROBLAZARS AND GAMMA-RAY BURSTS

In all three galactic microquasars where θ (the angle between the line of sight and the axis of ejection) has been determined, a large value is found (that is, the axis of ejection is close to the plane of the sky). This result is not inconsistent with the statistical expectation since the probability of finding a source with a given θ is proportional to $sin\,\theta$. We then expect to find as many objects in the $60° \leq \theta \leq 90°$

range as in the $0° \leq \theta \leq 60°$ range. However, this argument suggests that we should eventually detect objects with a small θ. For objects with $\theta \leq 10°$ we expect the timescales to be shortened by $2\gamma^2$ and the flux densities to be boosted by $8\gamma^3$ with respect to the values in the rest frame of the condensation. For instance, for motions with $v = 0.98c$ ($\gamma = 5$), the timescale will shorten by a factor of ~50 and the flux densities will be boosted by a factor of ~10^3. Then, for a galactic source with relativistic jets and small θ we expect fast and intense variations in the observed flux. These microblazars may be quite hard to detect in practice, both because of the low probability of small θ values and because of the fast decline in the flux.

There is increasing evidence that the central engine of the most common form of gamma-ray burst (GRBs), those that last longer than a few seconds, are afterglows from ultra-relativistic jets produced during the formation of black holes (McFaden & Woosley, 1999). Mirabel & Rodríguez (1999) propose that ultra-relativistic bulk motion and beaming are needed to explain: 1) the enormous energy requirements of $\geq 10^{54}$ erg if the emission were isotropic (e.g. Kulkarni et al. 1999; Castro-Tirado et al. 1999); 2) the statistical correlation between time variability and brightness (Ramirez-Ruiz & Fenimore, in Vth Compton workshop on GRBs 1999), and 3) the statistical anticorrelation between brightness and time-lag between hard and soft components (Norris et al. 1999). Beaming reduces the energy release by the beaming factor $f = \Delta\Omega/4\pi$, where $\Delta\Omega$ is the solid angle of the beamed emission. Additionally, the photon energies can be boosted to higher values. Extreme flows from collapsars with bulk Lorentz factors > 100 have been proposed as sources of γ-ray bursts (Mészáros & Rees 1997). High collimation (Dar 1998; Pugliese et al. 1999) can be tested observationaly (Rhoads, 1997), since the statistical properties of the bursts will depend on the viewing angle relative to the jet axis.

Recent multiwavelength studies of gamma-ray afterglows suggest that they are highly collimated jets. The brightness of the optical transient associated to GRB 990123 showed a break (Kulkarni et al. 1999), and a steepening from a power law in time t proportional to $t^{-1.2}$, ultimately approaching a slope $t^{-2.5}$ (Castro-Tirado et al. 1999). The achromatic steepening of the optical light curve and early radio flux decay of GRB 990510 are inconsistent with simple spherical expansion, and well fit by jet evolution. It is interesting that the power laws that describe the light curves of the ejecta in microquasars show similar breaks and steepening of the radio flux density (Rodríguez & Mirabel, 1999). In microquasars, these breaks and steepenings have been interpreted (Hjellming & Johnston 1988) as a transition from slow intrinsic expansion followed by free expansion in two dimensions. Besides, linear polarizations of about 2% were recently measured in the optical afterglow of GRB 990510 (Covino et al. 1999), providing strong evidence that the afterglow radiation from gamma-ray bursters is, at least in part, produced by synchrotron processes. Linear polarizations in the range of 2-10% have been measured in microquasars at radio (e.g. Rodríguez et al. 1995), and optical (Scaltriti et al. 1997) wavelengths.

In this context, the jets in microquasars of our own Galaxy seem to be less extreme local analogs of the super-relativistic jets associated to the more distant gamma-ray bursters. However, there are caveats to this analogy and gamma-ray bursters are different to the microquasars found so far in our own Galaxy. The former do not repeat, seem to be related to catastrophic events, and have much larger super-Eddington luminosities. Therefore, the scaling laws in terms of the black hole mass that are valid in the analogy between microquasars and quasars do not seem to apply in the case of gamma-ray bursters.

REFERENCES

Belloni, T, Méndez, M, King, AR, van der Klis, M, van Paradijs, J. 1997, *Ap. J.* 479, L145-48

Castro-Tirado, AJ. et al. 1999, *Science* 283, 2069-73

Covino, S. et al. 1999, *Astron. Astrophys.* 348, L1-4

Dar, A. 1998, *Ap. J.* 500, L93-96

Dhawan, V, Mirabel, IF, Rodríguez, LF. 2000, To be submitted to *Ap. J.*

Falke, H. & Biermann, P.L. 1999, *Astron. Astrophys.* 342, 49

Falke, H. et al. 1999, *astro-ph/9912436*

Fender, R.P. et al. 1999, *Ap. J.* 519, 165

Fender, R.P., Pooley, G.G., Durouchoux, P., Tilanus, R.P.J. & Brocksop, C. 2000, *MNRAS* in press

Hjellming, RM, Johnston, KJ. 1988, *Ap. J.* 328, 600-09

Kulkarni, S.R. et al. 1999, *Nature* 398, 389-94

Margon, BA. 1984, *Annu. Rev. Astr. Astrophys.* 22, 507-36

MacFayden, A.I. & Woosley, S.E. 1999, *Ap. J.* 524, 262

Mészáros, P, Rees, MJ. 1997, *Ap. J.* 482, L29-32

Mirabel, IF, Dhawan, V, Chaty, S, Rodríguez, LF, Robinson, C, Swank, J, Geballe, T. 1998, *Astron. Astrophys.* 330, L9-12

Mirabel, IF, Rodríguez, LF., Cordier, B., Paul, J., Lebrun, F. 1992, *Nature* 358, 215-17

Mirabel, IF, Rodríguez, LF. 1994, *Nature* 371, 46-48

Mirabel, IF, Rodríguez, LF. 1999, *Annu. Rev. Astr. Astrophys.* 37, 409

Norris, J.P. Marani, G.F. & Bonell, J.T. 1999, *submitted to Ap. J.* astro-ph/9903233

Pugliese, G, Falcke, H, Biermann, PL. 1999, *Astron. Astrophys.* 344, L37-40

Rees, MJ. 1966, *Nature* 211, 468-70

Rees, MJ. 1984, *Annu. Rev. Astr. Astrophys.* 22, 471-506

Rees, MJ. 1998, in *Black Holes and Relativistic Stars*, ed. Wald, RM, University of Chicago, 79-101

Rhoads, JE. 1997, *Ap. J.* 487, L1-4

Rodríguez, LF, Gerard, E., Mirabel, IF, Gómez, Y., & velázquez, A. 1995, *Ap. J. Supp.* 101, 173-79

Rodríguez, LF, Mirabel, IF. 1999, *Ap. J.* 511, 398-404

Sams, BJ, Eckart, A, Sunyaev, R. 1996 *Nature* 382, 47-49

Scaltriti, F, Bodo, G, Ghisellini, G, Gliozzi, M, Trussoni, E. 1997, *Astron. Astrophys.* 327, L29-31

ISOLATED NEUTRON STARS DISCOVERED BY ROSAT

C. Motch [1]

1) CNRS, Observatoire de Strasbourg, 11 rue de l'Université, 67000 Strasbourg, France

ABSTRACT

ROSAT has discovered a new group of isolated neutron stars characterized by soft blackbody like spectra (kT $\sim$ 50-120 eV), apparent absence of radio emission and no association with supernovae remnants. So far only six such sources are known. A small fraction of these stars exhibit X-ray pulsations with relatively long periods of the order of 10 sec. Two very different mechanisms may be envisaged to explain their properties. The neutron stars may be old and re-heated by accretion from the ISM in which case their population properties could provide information on past stellar formation and secular magnetic field decay. Alternatively, this group may at least partly be made of relatively young cooling neutron stars possibly descendant from magnetars. We review the last observational results and show how they can shed light on the evolutionary path of these new objects within the whole class of isolated neutron stars.

KEYWORDS: neutron stars; pulsars; magnetars

1. INTRODUCTION

The vast majority of the 10^8-10^9 isolated neutron stars (INS) present in the Galaxy should be virtually undetectable using current observational means. Young cooling neutron stars might emit thermal X-rays during the first $\sim 10^6$ yr and their pulsed radio emission will reveal them up to ages of $\sim 10^8$ yr. Recycled millisecond pulsars are old objects but their previous accreting binary phase may have altered their physical properties in particular the magnetic field strength. In this context, the possibility that a sizeable fraction of the entire 'fossil' population is re-heated by accretion from interstellar medium and becomes detectable in the EUV / X-ray domain is exciting. This could allow an observational study of old neutron stars and give access to information on past stellar formation, heating and cooling mechanisms and magnetic field decay. This idea was first proposed by Ostriker, Rees & Silk (1970). Population synthesis models were later computed by Treves & Colpi (1991), Blaes & Madau (1993) and Madau & Blaes (1994). A recent and extensive review on current issues can be found in Treves et al. (1999). Early models predicted that a rather large number of ROSAT XRT all-sky survey (RASS) sources could be accreting INS, initiating several optical identification campaigns. In this paper we shall describe the general observational properties of the INS discovered by ROSAT, discuss the possible X-ray powering mechanisms, consider what iden-

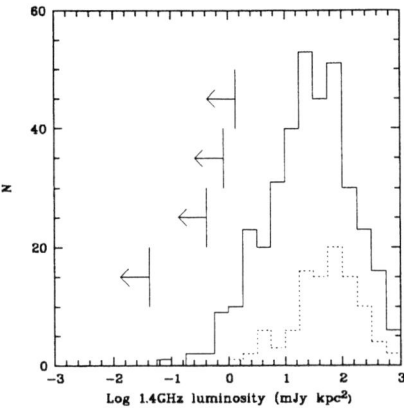

FIGURE 1. Upper limits on 1.4GHz radio luminosity of ROSAT discovered INS. Vertical displacement is there only for readability. Histograms represent the distribution of observed luminosities of radio pulsars as in Taylor et al. (1993) (Total population is the continuous line, pulsars with ages below 10^6 yr is the dashed line)

tification campaigns can tell us so far on the total population and discuss in more details the pulsating sources.

2. GENERAL PROPERTIES

The group of INS discovered by ROSAT shares a rather well defined set of properties. Although each of these features may be present individually in sub groups of INS (e.g. some radio pulsars have very soft X-ray spectra and some X-ray emitting radio quiet INS are found in SNR environments), the properties characterizing this new group are never encountered together in previously known classes of INS.

Soft X-ray spectra. All ROSAT discovered INS exhibit soft X-ray spectra which at the energy resolution of the PSPC can be very well fitted by blackbody models with temperatures in the range of $\sim$ 50 to 120 eV. No bright X-ray hard component seems to be present, at least at the level encountered in most X-ray detected radio pulsars. Another common spectral feature is the low N_H towards these sources ($\sim 10^{20}$ cm^{-2}) which implies distances of the order of 100 to 1000 pc at most.

No strong radio emission. The absence of detected bright radio emission seems to be also a common feature although not all members have been surveyed yet. At present, constraints on radio emission are mostly based on the FIRST and NVSS surveys. We list in Table 1 various upper limits on luminosity gathered for a subset of INS candidates. In Fig. 1 we show the position of these upper limits with respect to observed 1.4 GHz luminosity distribution of radio pulsars.

ROSAT discovered INS are undoubtedly less radio luminous than most known radio pulsars. The difference is stricking if one assumes that these are young cooling

TABLE 1. Upper limits on radio emission from ROSAT discovered INS

ROSAT source	Frequency GHz	Assumed distance (kpc)	Max L mJy kpc^2	Survey
RX J1856−3754	0.43	0.13	0.058	Parkes survey
	1.4		0.042	NVSS
RX J0720−3125	1.4	0.41	0.420	NVSS
RX J1605+3249	1.4	0.70	0.400	FIRST
RX J1308+2127	1.4	1.20	1.350	FIRST

objects with ages of less than about 10^6 yrs. However, targeted radio observations should easily improve on current upper limits.

No association with SNR. ROSAT and Einstein observatories have discovered a number of X-ray bright but apparently radio quiet neutron stars associated with SNRs (Gotthelf et al. 1997, Brazier & Johnston 1999). Their X-ray spectra, although often thermal like are in general significantly hotter than those of the group of INS discussed here. In contrast, the fact that none of the INS considered here lies close to an SNR suggests ages older than $\sim 10^5$ yr.

No long term variability. ROSAT was able to monitor the mean X-ray luminosity of some candidates over several years, basically from the all-sky survey time till the last operational phases. EXOSAT and Einstein have also serendipitously observed some of these targets further extending the time base. In the cases investigated in details no evidence for X-ray variability over months or years time scale exist with very stringent upper limits of the order of few percents (Walter et al. 1996, Haberl et al. 1997, Motch et al. 1999).

Long rotation periods. At least one of the INS candidates, RX J0720-3125 exhibits X-ray pulsations with a period of 8.39 s (Haberl et al. 1997). Another case is RX J0420−5022 which is possibly pulsating at 22.7 s (Haberl et al. 1999). The spin period of other candidates is not known but this issue may be settled by forthcoming XMM observations. These long spin periods are most unusual for radio pulsars. We shall see below that the long periods have profound impact on the possible evolutionary status of these objects.

2.1. Optical identifications

Only the two X-ray brightest candidates have a secure optical identification. The ROSAT source RX J1856.5−3758 was identified by Walter et al. (1996) with a V = 25.6, U-V = −1.2 object. A faint B = 26.1-26.5 blue star is also the likely counterpart of RX J0720-3125 (Motch & Haberl 1998, Kulkarni & van Kerkwijk 1998). In these two cases, the optical continuum lies only 1.4 and 1.7 magnitude respectively above the Rayleigh-Jeans tail of the blackbody seen in soft X-rays. For the remaining candidates, the identification with an INS eventually rests on the absence of optical

TABLE 2. The catalogue of isolated neutron stars discovered by their X-ray emission

ROSAT source	PSPC cts/s	kT eV	N_H 10^{20} cm^{-1}	P s	B mag	References
RX J1856−3754	3.64	57 ± 1	1.4 ± 0.1	-	25.8	1
RX J0720−3125	1.64	79 ± 4	1.3 ± 0.3	8.39	26.5	2
RX J1605+3249	0.90	92 ± 6	1.1 ± 0.4	-	>25	3
RX J0806−4123	0.38	78 ± 7	2.5 ± 0.9	-	>24	4
RX J1308+2127	0.29	113 ± 14	2.0 ± 0.4	-	>26	5
RX J0420−5022	0.14	<85	2.0	22.7:	>25.2	6
MS 0317.7−6647	~0.03	180 ± 30	40 ± 20	-	>21.4	7

References: 1) Walter et al. (1996), 2) Haberl et al. (1997), 3) Motch et al. (1999), 4) Haberl et al. (1998), 5) Schwope et al. (1999), 6) Haberl et al. (1999), 7) Stocke et al. (1995).

signatures of other possible soft X-ray emitters such as magnetic CVs or soft AGNs.

2.2. The catalogue

We list in Table 2 the main properties of the INS discovered so far on the basis of their X-ray emission. In almost all cases, their nature was established by ROSAT observations although some of them already had Einstein or EXOSAT detections. MS 0317.7-6647 (Stocke et al. 1995) is a particular case as its X-ray energy distribution undergoes very high absorption. Also, its INS nature is not entirely clear since an identification with a very massive accreting black hole in the field spiral galaxy NGC 1313 is not excluded.

3. X-RAY POWERING MECHANISMS

A number of possible mechanisms leading to production of X-rays from INS may be envisaged. Basically, the X-ray luminosity may be extracted from spin down as 'normal' radio pulsars do, from cooling, or the neutron star may be re-heated by accretion of interstellar material. We consider below in some details each of these mechanisms.

3.1. Rotation

Becker & Trümper (1997) showed that for 'normal' radio pulsars, the X-ray luminosity is tightly linked to rotational energy loss with $L_X \sim 10^{-3}$ Ė. Wang et al. (1999) pointed out that at least in the case of RX J0720−3125 this mechanism could not work since the observed X-ray luminosity was more than twice the upper limit on spin down power ($\dot{P} \leq 0.8 \ 10^{-12}$, Haberl et al. 1997). However in the absence of spin period measurements and derivatives this mechanism cannot be ruled

TABLE 3. Brightest X-ray pulsars in the ROSAT PSPC instrumental system after Becker & Trümper (1997)

Pulsar name	PSPC cts/s	P ms	Log(P/2Ṗ) years	Note
Crab	17.8	33.4	3.10	
Vela	3.40	89.29	4.05	
B0656+14	1.92	384.87	5.05	cooling pulsar
B1055-52	0.35	197.10	5.73	cooling pulsar
J0437-47	0.20	5.75	9.50	

out for other INS candidates. All radio pulsars bright enough to have a constraining ROSAT PSPC spectrum exhibit a rather hard power law component which is thought to be the signature of intense magnetospheric activity (Becker & Trümper 1997). The absence of a luminous hard X-ray tail above the thermal component in ROSAT discovered INS is a general argument against a rotationally driven origin of X-rays.

3.2. Cooling

Blackbody temperatures in excess of 50 eV imply ages younger than $\sim 10^6$ yr for standard cooling curves while the absence of nearby SNR sets a lower limit of $\sim 10^5$ yr. Among the X-ray emitting radio pulsars listed by Becker & Trümper (1997) two middle aged (10^5-10^6 yr) pulsars (PSR 0656+14 and PSR 1055-52) exhibit in addition to the power law a clear thermal component believed to be due to cooling from the neutron star surface. A luminous thermal component is probably also present in the younger (10^4-10^5 yr) Vela type pulsars. In PSR 0656+14 the soft blackbody dominates the 0.1-2.4 keV energy distribution and has characteristics ($T_{bb} = 87$ eV, $N_H = 1.9 \pm 0.4\ 10^{20}$ cm^{-2}) quite similar to those of the INS discussed here. In the case of RX J1605+3249 for instance, Motch et al. (1999) show that owing to the lower statistics, a hard component of similar intensity as in PSR 0656+14 would not have been detected nor would the 0.384 s period with a 9% pulsed fraction have been seen. In this context, a natural and simple explanation for the absence of radio emission from an otherwise X-ray bright and nearby source is that the radio beam does not sweep the earth. The proportion of the sky swept by the radio beam decreases with increasing periods and is of the order of 0.3 for the period range of cooling pulsars (Biggs 1990). In the range of PSPC count rates covered so far there are two cooling radio pulsars (see Table 3). Although this is small number statistics, the picture is in fact consistent with the entire ROSAT discovered population being cooling radio pulsars whose beam remains undetected because it does not cross the earth.

3.3. Accretion from interstellar medium

Accretion of interstellar matter onto old neutron stars can produce X-ray luminosities large enough to detect a sizeable fraction of the total galactic population. Assuming Bondi-Hoyle accretion, the mass accretion rate is $\dot{M} \sim 10^{10}$ n $(V/40\,\mathrm{km s^{-1}})^{-3}$ g s^{-1} with V the neutron star velocity with respect to interstellar medium and n the mean density. The temperature of the polar cap heated by accretion is $T_{bb} \sim 20\,(\dot{M}/10^{10}\mathrm{g s^{-1}})^{1/4}\,f^{-1/4}$ eV with f the relative surface area of the polar cap. However, before matter can reach the neutron star surface a number of conditions must be fulfilled. The first one is that ram pressure at accretion radius exceeds pulsar momentum flux otherwise the star is in the ejector state. This requires the spin period to be less than $\sim 9\,(B/10^{12}\mathrm{G})^{1/2}\,n^{-1/4}\,(V/40\,\mathrm{km s^{-1}})^{1/2}$ s. Magnetic dipole braking allows to reach this period in $\sim 5\,(B/10^{12}\mathrm{G})^{-1}\,n^{-1/2}\,(V/40\,\mathrm{km s^{-1}})$ Gyr (Blaes & Madau 1993). The most constraining condition is that the corotation radius must be larger than the Alfven radius. For the low accretion rates prevailing here, this condition requires very slow rotation $P_{rot} \geq 3000\,(B/10^{12})^{6/7}\,n^{-3/7}$ $(V/40\,\mathrm{km s^{-1}})^{9/7}$ s. Dipole magnetic radiation is not efficient enough to slow down the neutron star on a reasonable time scale. Braking is probably achieved by the propeller mechanism which acts on a short time scale of $\sim 0.8\,(B/10^{12}\mathrm{G})^{-11/14}$ $n^{-17/28}\,(V/40\,\mathrm{km s^{-1}})^{29/14}$ Gyr (Blaes & Madau 1993).

4. CONSTRAINTS ON POPULATIONS OF X-RAY EMITTING ISOLATED NEUTRON STARS

The first constraints came from the general identification campaigns which aimed at a complete census of the RASS source population at different flux levels and galactic latitudes. In addition a number of projects specifically searching for INS were initiated. Zickgraf et al. (1997), Motch et al. (1997), Belloni et al. (1997) and Danner et al. (1998) reported on deep surveys in small areas including molecular clouds. Shallow surveys of large areas were performed by Manning et al. (1996), Bade et al. (1998) and Thomas et al. (1998), among others. These campaigns led to the discovery of the handful of candidates discussed here. They also yielded the unexpected result that the number of INS present in the RASS was far below that predicted by early population synthesis models which assumed that a few thousands neutron stars should be detected in the RASS. The results of these observational constraints have been summarized by Neuhäuser & Trümper (1998). We show an updated version of their figure 1 in Fig. 2. At the faint flux end, upper limits on ROSAT INS densities are incompatible with model predictions from Treves & Colpi (1991) and Blaes & Madau (1993) (not shown) but still consistent with the revised models of Madau & Blaes (1994). The most recent large sky area searches (labeled A,B and C in Fig. 2) yield even stronger constraints. The results (A) of Danner et al. (1998) is restricted to dark clouds area and is therefore not directly comparable with model predictions. Constraints labeled B and C cover large unbiased sky areas. (C) results from the identification of very bright soft ROSAT sources with $|b| \geq 20°$ by Thomas et al. (1998). The upper limit shown in (B) is derived from an identification campaign of soft sources over $\sim 1/4$ of the whole sky conducted at ESO (Motch et al. 2000). Optical identification of a total of 116 sources down to a PSPC count rate of 0.17 cts/s yielded only 3 possible cases among which 2 are already known INS. Spectral selection is discussed in Motch et al. (1999).

Contrary to other studies, surveys (B) and (C) assume that accreting INS are intrinsically soft sources and preselect candidates on the basis of ROSAT PSPC hardness ratios. The brightest INS found in the RASS without spectral selection are indeed soft and a soft spectrum is also theoretically expected (e.g. Zampieri et al. 1995). However, the possibility remains that because of the spectral preselection a fraction of the population, for instance INS accreting in very dense medium, escapes the search.

The painstaking optical identification campaigns have yielded the very significant and unexpected result that the number of accreting old INS shining in X-rays is a factor 10 to 1000 below that predicted by population synthesis models. This discrepancy points to a fundamental error in one or more of the model assumptions used. Constraints on an accreting population may be even stronger since; i) the observed LogN-LogS is almost consistent with a young cooling neutron star population only and ii) for none of the INS candidates discovered so far can the accreting model be unambiguously established by the discovery of X-ray variability or the measurement of a negative $\dot{P}$ for instance. In two cases INS are found in regions of particularly low mean densities contrary to naive expectations ($n \leq 0.4$ and 0.3 cm^{-3} for RX J0720$-$3125 and RX J1605+3249 respectively). Very low relative velocities of less than $10 \, \mathrm{km \, s^{-1}}$ must be assumed to account for the observed luminosities. However, selection effects and ISM patchiness on very small scales may solve this paradox. Finally, accretion should enrich neutron star atmosphere in H and He and could cause a brighter optical continuum than observed (Pavlov et al. 1996).

5. ARE OLD NEUTRON STARS ACCRETING ?

Newly born neutron stars must be spun down efficiently before they can start accreting from the interstellar medium. The evolution of magnetic field with time seems a key parameter as it both determines braking strength during ejector and propeller phases and the minimum period allowing accretion. Treves et al. (1999) argue that for certain values of initial magnetic field and decay time scales, the life time of the ejector phase may be longer than the age of the Galaxy. Colpi et al. (1998) and Livio et al. (1998) also showed that field decay significantly increases the duration of the propeller phase with the consequence that most old neutron stars could still not be able to accrete. Because of the steep dependence of Bondi-Hoyle accretion rate with relative velocity, the velocity distribution of old neutron stars is another key parameter. Early models were based on the work of Narayan & Ostriker (1990). Since then new observations by Lyne & Lorimer (1994) suggest a typical mean velocity at birth of the order of 450 km s^{-1} about twice that of Narayan & Ostriker (1990). These two effects may cooperate to explain the small number of

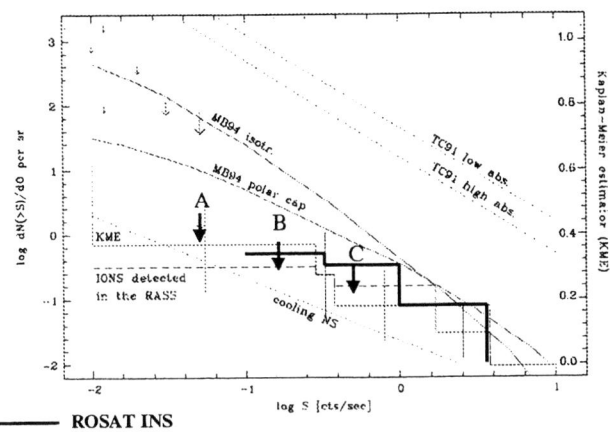

FIGURE 2. LogN-LogS curves for X-ray detected isolated neutron stars after Neuhäuser & Trümper (1998). Lines labelled TC91 amd MB94 represent model predictions from Treves & Colpi (1991) and Madau & Blaes (1994). Thin arrows are upper limits derived from various identification campaigns. The thick curve is the revised INS LogN-LogS curve. Thick arrows marked A,B and C represent upper limits resulting from A) dark clouds (Danner et al. 1998), B) Soft sources on 1/4 of the sky (see text), C) RASS soft sources (Thomas et al. 1998).

X-ray detections.

6. PULSATING SOURCES

Two INS, RX J0720−3125 and possibly RX J0420−5022 have long rotation periods which modulate their X-ray light curve by 20 to 80 %. These spin period are significantly longer than those generally measured in radio pulsars. In the accretion scenario, the spin period yields an estimate of the surface magnetic field which has to be weaker than $\sim 10^{10}$ G to allow for accretion (Haberl et al. 1997). If confirmed, this would clearly point to secular magnetic field decay. Unfortunately, alternative explanations exist. Based on the similarity of spin periods Haberl et al. (1997) proposed that RX J0720−3125 is related to class of anomalous (braking) X-ray pulsars (AXPs). The nature of AXPs itself remains a mystery. Two models often mentioned are i) isolated neutron stars evolving from Thorne-Żytkow objects (van Paradijs et al. 1995) and accreting from a remnant disk and ii) magnetars (Thompson & Duncan 1996). An interpretation in terms of a cooling neutron star faces the difficulty to reconcile thermal and rotational ages. For RX J0720−3125, $T_{bb} = 8 \; 10^5$ K implies ages in the range of 1 to 4 10^5 yr (Heyl & Hernquist 1998) whereas the time required to spin down from a short birth period to 8.39 s with dipole radiation is $\sim 1.2 \; 10^9$ $(B/10^{12} G)^{-2}$ yr. This discrepancy may be solved assuming the star was born with a long period. Alternatively, Heyl & Hernquist (1998) proposed the exciting possibility that RX J0720−3125 is a magnetar with B∼ 10^{14} G, similar to those proposed to explain soft-γ repeaters and AXPs. Because radio emission is quenched by the strong magnetic field magnetars remain undetected by classical radio means. Their birth rate may be ∼ 10% of that of ordinary pulsars (Kouveliotou et al. 1994). Gotthelf & Vasisht (1999) also argue that a large fraction of SNR contains previously unrecognized slowly rotating, radio-quiet and X-ray bright pulsars which could be the natural progenitors of the long period INS discovered by ROSAT. The high magnetic field allows efficient magnetic dipole spin down and its decay provides an additional source of heat which allows magnetars to remain detectable in X-rays over longer times than ordinary pulsars. Consequently, although less numerous than normal pulsars, cooling magnetars may show up in comparable numbers in X-ray surveys.

7. CONCLUSIONS

All optical identification campaigns carried out so far confirm the scarcity of X-ray emitting isolated neutron stars in the ROSAT all sky survey. Clearly, the large population of old galactic neutron stars does not accrete from interstellar medium at the rate predicted by early models, or the accretion energy is re-radiated outside the ROSAT energy band. This unexpected result tells us something fundamental on the kinematics and/or magnetic field evolution of old neutron stars and points to a secular magnetic field decay or a large mean velocity. ROSAT has however discovered a handful of INS sharing as common properties a soft blackbody like X-ray spectrum, absence of luminous radio emission and absence of long term variability. It is not yet proven that any of the 6-7 confirmed INS accretes from ISM. Their small number could be compatible with cooling normal pulsars whose radio beam does not cross the earth or with magnetars evolved from Soft-γ repeaters, AXPs and more generally from the slowly rotating X-ray bright and radio quiet pulsars found in several SNRs. Future observations with VLTs and Chandra/XMM should help to solve the puzzle of the evolutionary status of this population by measuring proper motions, rotation periods and period derivatives which are key parameters to distinguish between young cooling and old accreting INS. Finally these INS should allow the detailed observation of a rather unperturbed neutron star atmosphere and the accurate measurement of fundamental neutron star properties.

ACKNOWLEDGEMENTS

I would like to thank F. Haberl for enlightening discussions.

REFERENCES

Bade, N., Engels, D., Voges, W., et al. , 1998, A&A 127, 145
Becker, W., Trümper, J., 1997, A&A 326, 682
Belloni, T., Zampieri, L., Campana, S., 1997, A&A 319, 525
Biggs, J.D., 1990, MNRAS 245, 514

Blaes, O., Madau, P., 1993, ApJ 403, 690
Brazier, K.T.S., Johnston, S., 1999, MNRAS 305, 671
Colpi, M., Turolla, R., Zane, S., Treves, A., 1998, ApJ 501, 252
Danner, R., 1998, A&AS 128, 349
Gotthelf, E.V., Petre, R., Hwang, U., 1997, ApJ 487, L175
Gotthelf, E.V., Vasisht, G., 1999, in "Pulsar Astronomy 2000 and Beyond, ASP Conf. Series, M. Kramer, N. Wex and R. Wielebinski, eds
Haberl, F., Motch, C., Buckley, D.A.H., Zickgraf, F.J., Pietsch, W., 1997, A&A 326, 662
Haberl, F., Motch, C., Pietsch, W., 1998, Astronomische Nachrichten, 319, 97
Haberl, F., Pietsch, W., Motch, C., 1999, A&A 351, L53
Heyl, J.S., Hernquist, L., 1998, MNRAS 297, L69
Kouveliotou, C., et al., 1994, Nat 368, 125
Kulkarni, S.R., van Kerkwijk, M.H., 1998, ApJ 507, L49
Livio, M., Xu, C., Frank, J., 1998, ApJ 492, 298
Lyne, A.G., Lorimer, D.R., 1994, Nat 369, 127
Madau, P., Blaes, O., 1994, ApJ 423, 748
Manning, R.A., Jeffries, R.D., Willmore, A.P., 1996, MNRAS 278, 577
Motch, C., Guillout, P., Haberl, et al., 1997, A&A 318, 111
Motch, C., Haberl, F., 1998, A&A 333, L59
Motch, C., Haberl, F., Zickgraf, F.J., Hasinger, G., Schwope, A.,D., 1999, A&A in press,
Motch, C., et al., 2000, in preparation
Narayan, R., Ostriker, J.P., 1990, ApJ 352, 222
Neuhäuser, R., Trümper, J., 1999, A&A 343, 151
Ostricker, J., Rees, M.J., Silk, J., 1970, Astrophys. Lett., 6, 179
Pavlov, G.G., Zavlin, V.E., Trümper, J., Neuhäuser, R., 1996, ApJ 472, L33
Schwope, A.D., Hasinger, G., Schwarz, R., Haberl, F., Schmidt, M., 1999, A&A 341, L51
Stocke, J.T., Wang, Q.D., Perlman, E.S., Donahue, M.E., Schachter, J.F., 1995, AJ 109, 1199
Taylor, J.H., Manchester, R.N., Lyne, A.G., 1993, ApJS 88, 529
Thomas, H.-C., Beuermann, K., Reinsch, K., Schwope, A.D., Trümper, J., Voges, W., 1998, A&A 335, 467
Thompson, C., Duncan, R.C., 1996, ApJ 473, 322
Treves, A., Colpi, M., 1991, A&A 241, 107
Treves, A., Turolla, R., Zane, S., Colpi, M., 1999, PASP, in press
van Paradijs, J., Taam, R.E., van den Heuvel, E.P.J., 1995, A&A 299, L41
Walter, F.M., Wolk, S.J., Neuhäuser, R., 1996, Nat 379, 233
Wang, J.C.L., Link, B., Van Riper, K., Arnaud, K.A., Miralles, J.A., 1999, A&A 345, 869
Zampieri, L., Turolla, R., Zane, S., Treves, A., 1995, ApJ 439, 849
Zickgraf, F.-J., Thiering, I., Krautter, J. et al. 1997, A&AS 123, 103

TRANSIENT BE STAR BINARY SYSTEMS

F. Nagase [1]

1) *The Institute of Space and Astronautical Science, 3-1-1 Yoshinodai, Sagamihara, Kanagawa 229-8510, Japan*

ABSTRACT   In this paper we review current status of X-ray observations of transient Be star binary systems. An overview of remarkable discoveries of transient X-ray pulsars in the last few years is given and general properties common to transient Be star binary pulsars are summarized. A few interesting topics revealed from recent observations are also presented.

KEYWORDS: eclipsing binaries; interstellar dust; X-ray pulsars; X-ray sources.

1. INTRODUCTION

X-ray binaries containing neutron stars or black holes are classified into two types, the High Mass X-ray Binaries (HMXB) or Low Mass X-ray Binaries (LMXB) depending on the mass of the companion star. Most of the HMXBs show pulsations due to strong dipole magnetic field and mass accretion onto the magnetic poles at the neutron star surface. Bildsten et al. (1997) tabulated 44 X-ray pulsars in their review article. However, within a few years after their review, a dramatic rush of pulsar discoveries has been continuing, and the total number of accreting X-ray pulsars is about 80 at present (see e.g., Nagase 1999). Among them about a dozen wind-fed high mass X-ray binaries, four low mass X-ray binary pulsars and six anomalous X-ray pulsars are included. The rest three quadrants show a transient nature and more than half of them are identified to be massive Be star binaries. Thus the Be star X-ray binaries represent the largest sub-class of massive X-ray binaries. It seems a good assumption that all the transient X-ray pulsars are Be star binary systems, even though identification for some of them has not been done yet and occasionally it is difficult due to large extinction for the sources located in the Galactic plane near the center (see e.g., Coe 2000 for review of optical observations of Be star systems). Hereafter in this paper I designate this class of transient (Be star) X-ray binary pulsars as a TrXBP, regardless of the identification of the Be star companion.

The pulse periods of TrXBPs (and every classes of X-ray pulsars) distribute uniformly in the range of a few seconds to a thousand seconds with a few exceptions as shown in Figure 1. This is in good contrast to the radio pulsars, the pulse periods of which are more narrowly distributed below a few seconds. It will be worthy to note that more than 20 TrXBP were discovered from Magellanic Clouds in the last

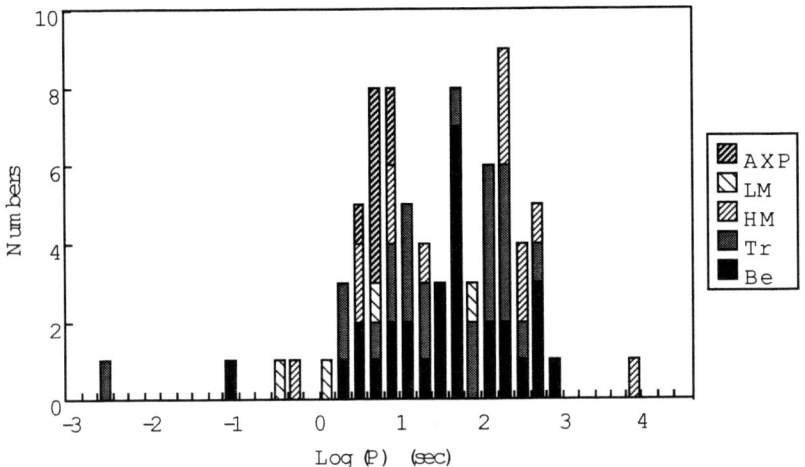

FIGURE 1. Distribution of pulse periods in accreting X-ray pulsars.

few years, in contrast to the fact that only three accreting pulsars were known for the past two decades. This rush of pulsar discoveries in LMC/SMC owes to the ideal line-up of X-ray observatories, ROSAT, ASCA, BeppoSAX, and XTE.

In the next section I summarize the general X-ray properties of TrXBPs. In section 3, I present some recent topics, such as (1) pulsar survey in Magellanic Clouds and investigation of X-ray source population in LMC/SMC, (2) Discoveries of TrXBPs lead by ROSAT Galactic plane survey, and (3) detection of soft-excess feature in the spectra of TrXBPs and consideration of the origin of the soft-excess feature.

2. X-RAY PROPERTIES OF TRANSIENT X-RAY BINARY PULSARS

Main X-ray properties common to TrXBPs, which are mostly Be star binaries, are

- Complex X-ray pulse profiles, depending on the energy bands and X-ray luminosity,
- Random pulse period change with the rate of change often correlating to the luminosity change,
- Simple, featureless power law spectrum with a weak iron fluorescent line, and
- Correlation between the binary orbital period and the pulsar spin period.

Some of the TrXBPs with a Be-star companion show regularly recurrent outbursts which take place at the periastron passage of the neutron star. Occasionally some TrXBPs exhibit giant outbursts that occur regardless of the orbital phase

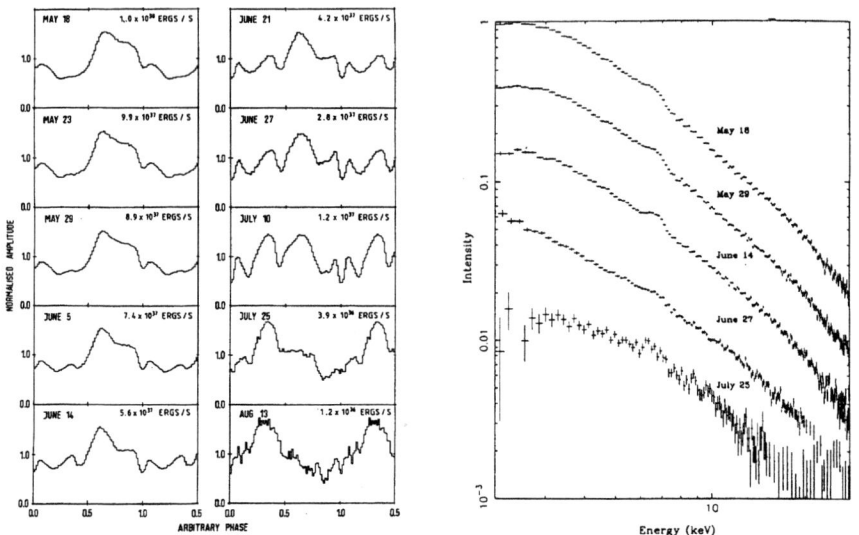

FIGURE 2. Changes of pulse profile and energy spectrum in a Be star X-ray binary pulsar EXO 2030+375 along with an outburst decay (taken from Parmar et al. 1989b and Renoulds et al. 1993).

which may cause an extremely large expansion of the circumstellar disk. As mentioned in the previous section, pulse periods of TrXBPs distribute widely from ∼ 3 s to 1000 s. The pulse profiles are fundamentally broad with a sinusoidal shape. However, they often exhibit complex change of features, including a multi-peak feature, a sharp dip at some pulse phase, according to the change of observed energy bands and X-ray luminosity during the observation. An example is shown in the left panel of Figure 2 where the complex change of pulse profile is clearly seen along the decay of an outburst in EXO 2030+375 (Parmar et al. 1989b). This figure shows almost every faces of pulse profiles so far observed in TrXBPs.

The trends of pulse period change in TrXBPs are complex and differ from source to source (see Bildsten et al. 1997). In general, there is a tendency that the neutron star spin rate is accelerated during outbursts, when the mass accretion rate increases and an accretion disk might be formed surrounding the neutron star. On the other hand the spin rate is decelerated during quiescent phases, when mass accretion from the Be star companion ceases and the disk formed shrinks (see e.g., Nagase 1989). Occasionally a correlation between the rate of change in the pulse periods and the X-ray luminosity is observed in an outbursts along with the evolution of the outburst. During the declining phase of the 1985 outburst of EXO 2030+375, the correlation of $-\dot{P}_{spin} \propto L^{1.08-1.35}$ was measured (Parmar et al. 1989a), which

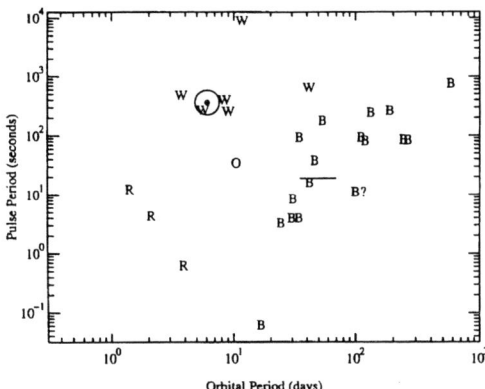

FIGURE 3. Pulse period versus orbital period of X-ray binary pulsars taken from Corbet et al. (1999).

was approximately consistent with the theory of accretion torque in disk-fed pulsars developed at the time (e.g., Ghosh and Lamb 1979).

Energy spectra of TrXBPs exhibit a hard power law of a photon index $\Gamma \sim 1$ with an exponential high energy cutoff at 15-30 keV. This feature in spectrum is common to accreting X-ray pulsars where mass accretion is halted by a strong magnetic field and the mass flow is channelled to the magnetic poles thus leading X-ray emission from accretion columns on the magnetic poles. There is a tendency that the spectral photon index changes with luminosity, usually hard at low luminosity. It has been known that wind-fed pulsars, such as Vela X-1, Cen X-3, and GX 301−2, exhibit a strong fluorescent line at 6.4 keV due to reprocessing by circumstellar matter, either stellar winds or an accretion disk. On the contrary, the flouorescent iron line in TrXBPs is relatively weak and often show broad line features (see e.g., White et al. 1983, Nagase 1989). An example of such spectral features is shown in the right panel of Figure 2 which was derived from the EXOSAT observation of an outburst in 1985 (Reynolds et al. 1993)

A correlation between the spin period and the orbital period of TrXBPs was discovered by Corbet (1986). The correlation diagram, so called "Corbet diagram" was useful at that time to discriminate the TrXBPs ("B" in Fig. 3) from Roche lobe overflow sources ("R" in Fig. 3) and wind accretion sources ("W" in Fig. 3) when both the spin and orbital periods are determined. With new orbital period measurements available mostly with RXTE/ASM monitoring, the boundary between TrXBPs and high mass wind-fed pulsars tend to become unclear (see Fig. 3 taken from Corbet et al. 1999). For a TrXBP, XTE J1855−026 (marked by a circled dot in Fig 3), Corbet et al. (1999) prefer the interpretation of a wind accretion pulsar rather than a Be star binary pulsar from the position on the Corbet diagram, although the nature of the companion star is not identified yet.

3. RECENT TOPICS

Several interesting findings on TrXBPs have been reported recently from observations with ROSAT, ASCA, BeppoSAX, and RXTE. The topics presented in this section are:

- Survey of Magellanic Clouds with ROSAT and discoveries of many TrXBPs with ASCA, BeppoSAX, and RXTE,

- Discoveries of TrXBPs lead by the systematic Galactic plane survey with ROSAT and ASCA and temporarily from RXTA/ASM all sky monitoring, and

- Detection of a soft-excess feature in the energy spectra of TrXBPs, especially from the sources in SMC.

3.1. Survey of Magellanic Clouds

Extensive surveys of the LMC and SMC regions were performed with ROSAT and more than 500 X-ray sources in LMC (Pietsch and Kahabka 1992) and 248 sources in SMC (Kahabka et al. 1999) were catalogued. Variabilities of 27 LMC X-ray sources were studied by Haberl and Pietsch (1999) using the ROSAT observations between 1990 and 1994. Details of the results of ROSAT LMC survey were presented by Haberl (1999) in this conference. Complete compilation and optical identification for the X-ray sources in LMC/SMC regions were performed in a series of papers by Schmidtke et al. (1994), Cowley et al (1997), and Schmidtke et al. (1999). Using spectral parameters estimated by two colors hardness ratios, Kahabka et al. classified 60 % of the detected 248 SMC sources into super-soft sources (SSS), X-ray binaries, supernova remnants (SNR), foreground stars,and background AGNs. About half of them were identifies to be X-ray binaries.

Recently more than 20 TrXBPs were discovered in the LMC/SMC regions (see e.g., Nagase 1999) from ROSAT, ASCA, BeppoSAX and RXTE observations. Yokogawa et al. (2000a) conducted a systematic analysis of the ASCA survey data of SMC region and found coherent pulsations from 12 sources among 39 sources detected with ASCA. Most of them were found to exhibit long-term flux variability, suggesting that they are TrXBPs. They also found that these TrXBPs can be clearly seperated from SNRs and other class of sources using the hardness ratio analysis (see Fig. 4).

It is remarkable that very few LMXBs were discovered so far from the SMC in contrast to the recent tremendous discoveries of TrXBPs. Thus the number ratio of HMXBs to LMXBs in SMC is strikingly different to our Galaxy where the population of LMXBs is larger than that of HMXBs as suggested by Schmidtke et al. (1999) and Yokogawa (2000a). This implies that the SMC has been more active than our Galaxy in massive star formation. Since HMXBs are relatively young it is suspected that strong star formation activity in the SMC took place in the recent past, several million years ago.

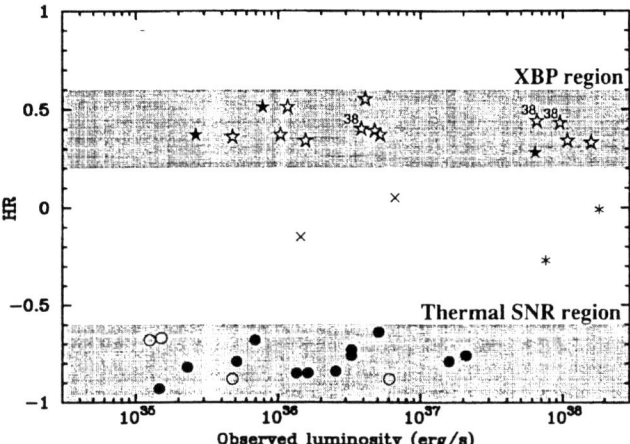

FIGURE 4. Hardness ratio versus luminosity of SMC/LMC sources taken from Yokogawa et al. (2000a). Solid symbols represent LMC sources and open symbols represent SMC sources. The stars represent identified X-ray pulsars and the circles represent identified supernova remnants. Two asterisks in between XBP region and SNR region are black hole candidates and two crosses are Clab-like pulsars.

3.2. New Transients in Galactic Plane

In addition to the TrXBPs listed in the table of Bildsten et al. (1997), more than a dozen of TrXBPs were discovered from the Galactic plane observations with ASCA, BeppoSAX and RXTE. These sources are listed in Table 1 together with references.

It will be worthy to note that the discoveries of pulsations from three ROSAT sources among the TrXBPs in Table 1 were based on the ROSAT Galactic plane survey and optical identifications of these sources by Motch et al. (1997). The three ROSAT sources, RX J0812.4−3114, RX J0440.9+4431, and RX J1037.5−564 were readily identified to the Be binary systems, LS 992, BSD 24-491, and LS1698, respectively in the table compiled by Motch et al. (1997). From the RXTE observations of these sources, Reig and Roche (1999a, 1999b) discovered coherent pulsations. These sources show pulse profiles and energy spectra typical to TrXBPs as were described in section 2. An outburst recurrence period of about 80 d period was recently observed from RX J0812.4−3114/LS 992 system (Corbet & Peele 2000). Assuming that the recurrent outbursts are due to the periastron passage of the neutron star in the eccentric orbital around the Be star companion, this source is a typical TrXBP with the correlation between the spin period and the orbital period on the Corbet diagram (Corbet 1986).

TABLE 1. X-ray pulsars recently discovered in the Galaxy and not cited in the table of Bildsten et al. (1997).

Source	l	b	Period(s)	Ref.[a]
SAX J1808.4−3658	355.395	−8.161	0.00249	1
SAX J0635+0533	206.151	−1.043	0.0338	2
XTE J1946+274	63.204	1.392	15.83	3
RX J0812.4−3114	249.578	1.543	31.8851	4
XTE J1906+09	42.561	1.047	89.17	5
AX J1820.5−1434	16.478	0.060	152.26	6
1SAX J1324.4−6200	306.793	0.609	170.84	7
RX J0440.9+4431	159.849	−1.2791	202.5	8
AX J1749.2−2725	1.707	0.106	220.38	9
XTE J1858+034	36.789	−0.068	221.0	10
SAX J2103.5+4545	87.130	−0.691	358.6	11
XTE J1855−026	31.102	−2.088	361.1	12
1SAX J1452.8−5949	317.645	−0.463	437.4	13
AX J170006−4157	344.0	0.2	714.5	14
RX J1037.5−5647	285.349	1.433	860	8

[a] References for pulsation discoveries: 1, Wijnands and van der Klis 1998, 2, Cusumano et al. 2000, 3, Smith et al. 1998, 4. Reig & Roche 1999a, 5, Marsden et al. 1998, 6, Kinugasa et al. 1998, 7, Angelini et al. 1998, 8, Reig & Roche 1999b, 9, Torii et al. 1998, 10, Remillard et al. 1998, 11, Hulleman et al. 1998, 12, Corbet et al. 1999, 13, Oosterbroek et al. 1999, 14, Torii et al. 1999.

3.3. Soft-Excess in Spectra of TrXBPs

A feature of excess intensities at energy below 1 keV in the spectrum over the extrapolation of a power low spectrum fitted to the higher energy has been reported from various sub-class of X-ray pulsars except for TrXBPs. The soft excess feature is considered to be a common propertiy in the spectra of anomalous X-ray pulsars (e.g., Mereghetti 1999) and the excess emission is usually fitted by a blackbody of temperature $kT = 0.4 - 0.7$ keV. Some LMXBs, such as Her X-1 and 4U 1626−67, show a soft excess feature in their spectra (e.g., McCray et al. 1982, Dal Fiume et al. 1998, Endo et al. 2000 for Her X-1 and Angelini et al. 1995, Orlandini et al. 1998 for 4U 1626−67). Evidence of such a soft excess feature has also been reported from HMXBs, LMC X-4 (Woo et al. 1996) and SMC X-1 (Wojdowski et al. 1998) from the combined analyses of ROSAT and Ginga data.

Such a clear soft excess feature, however, has not been reported so far from TrXBPs, because most of Galactic TrXBPs are subjected to heavy soft X-ray absorption and furthermore X-ray missions prior to ASCA did not have agood efficiency for a wide energy range from below 1 keV to 10 keV. Recently, clear exam-

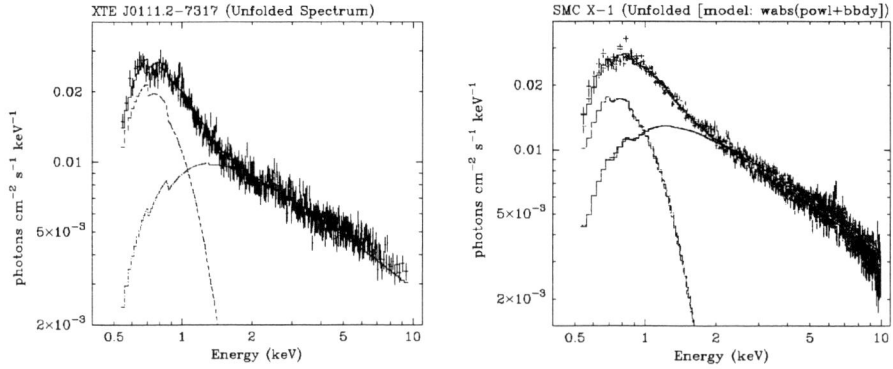

FIGURE 5. Energy spectra of a Be transient XTE J0111.2−7313 and a massive X-ray binary pulsar SMC X-1 observed with ASCA.

ples of such a soft excess feature were obtained from ASCA observations of TrXBPs in SMC, such as RX J0059.2−7138 (Hughes 1994, Kohno et al. 2000) and XTE J0111.2−7317 (Paul et al. 1999, Yokogawa et al. 2000b).

The energy spectrum with soft excess feature observed with ASCA from a TrXBP, XTE J0111.2−7317 is compared in Figure 6 with that observed with ASCA from a HMXB pulsar SMC X-1. Both can be fitted by a model of power law plus blackbody emission. The best fit parameters are $\Gamma = 0.8$, $kT = 0.15$ keV, and $N_H = 2.7 \times 10^{21}$ cm^{-2}, and $\Gamma = 0.8$, $kT = 0.18$ keV, and $N_H = 2.2 \times 10^{21}$ cm^{-2}, respectively for XTE J0111.2−7317 and SMC X-1. Thus, the spectral model that involves a power law and a soft blackbody emission, which is widely adopted to fit the soft-excess spectra observed from X-ray pulsars, can be adopted also to interpret the soft-excess feature seen in the spectra of some TrXBPs.

However, the total X-ray luminosities of these sources are 1.8×10^{38} erg s^{-1} and 2.4×10^{38} erg s^{-1}, respectively at the distance of SMC. If the soft excess emission is really blackbody, the luminosity fraction of the blackboly emission is about one thenth of the total luminosity. This requires blackbody radii of several hundred km if the surces are really located in the SMC.

The energy dependent pulse profiles of XTE J0111.2−7317 and SMC X-1 are shown in Figure 7. As can be seen, the pulse shape and pulse fraction do not change drastically with energies in these pulsars, although the pulse amplitude in XTE J0111.2−7317 changes slightly. This means that the soft excess component is pulsating in the same manner as the power law component. Thus the blackbody interpretation seems implausible, because the corresponding size of blackbody emission is several 10 times the neutron star radius at the distance of SMC and moreover this soft excess component should be pulsating.

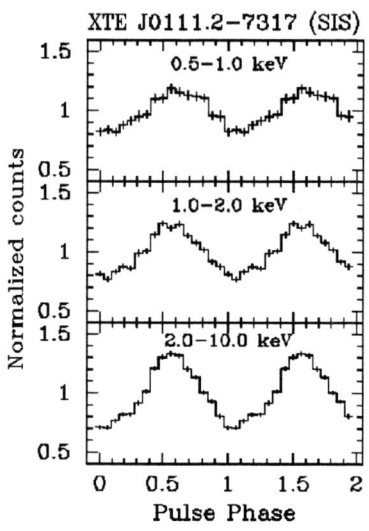 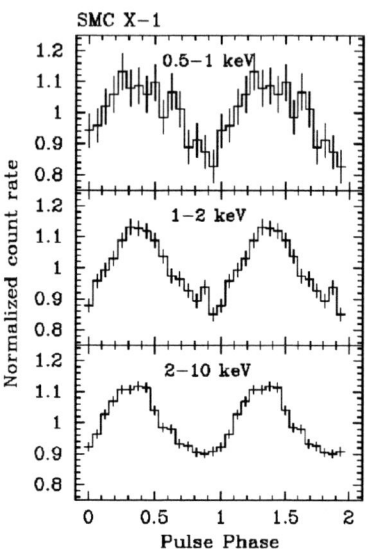

FIGURE 6. Pulse profiles of a Be transient XTE J0111.2−7313 and a massive X-ray binary pulsar SMC X-1 observed with ASCA.

Hence, it is likely that all the emission in 0.5-10 keV, including the soft excess emission, originate from the same site, for instance from accretion columns on the neutron star magnetic poles. The soft excess feature in spectra would be a feature intrisically common to X-ray pulsar emission from neutron star accretion columns, because many pulsars in all sub-class of X-ray pulsars exhibit such a feature.

ACKNOWLEDGEMENTS

The author thanks B. Paul for the useful comments and discussion.

REFERENCES

Angelini, L. et al. 1995, ApJ, 449, L41
Angelini, L. Church, M.J., Parmar, A.N., Balucinska-Church, M., Mineo, T. 1998, A&A, 339, L41
Bildsten, L., Chakrabarty, D., Chiu, J., et al. 1997, ApJS, 113, 367
Coe, M.J., 2000, in ASP Conf. ser. 214, *The Be Phenomenon in Early-Type Stars*, eds. M.A. Smith, H.F. Henrichs & J. Fabregat (San Francisco: ASP), p.656.
Corbet R.H.D. 1986, MNRAS, 220,1047
Corbet R.H.D., Marshall, F.E., Peele, A.G, Takeshima, T. 1999, ApJ, 517, 956
Corbet R.H.D., Peele, A.G. 2000, ApJ, 530, L33
Cowley, A.P., et al. 1997, PASP, 109, 21

Cusumano G., Maccarone, M.C., Nicastro, L., Sacco, B., Kaaret, P. 2000, ApJ, 528, L25
Dal Fiume, D., et al. 1998, A&A, 329, L41
Endo, T., Nagase, F., Mihara, T. 2000, PASJ, 52, 223
Ghosh, P., Lamb, F.K. 1979, ApJ, 234, 296
Haberl, F. 1999, in this Proceedings
Haberl, F., Pietsch, W. 1999, A&A, 344, 521
Hughes, J.P. 1994, ApJ, 427, L25
Hulleman F., in't Zand, J.J.M., Heise, J. 1998, A&A, 337, L25
Kahabka, P., Pietsch, W., Fillipovic, M.D., Haberl, F. 1999, A&AS, 136, 81
Kohno, M. Yokogawa, J., Koyama, K., 2000, PASJ, 52, 299
Kinugasa K., et al. 1998, ApJ, 495, 435
Marsden D., Gruber, D.E., Heindl, W.A., Pelling, M.R., Rothschild, R.E. 1998, ApJ, 502, L129
McCray, R.A., et al. 1982, ApJ, 262, 301
Mereghetti S., 1999, in *The Neutron Star - Black Hole Connection*, NATO Advanced Study Institute, in press.
Motch C., Haberl, F., Dennel, K., Pakull, M., Janot-Pacheco, E. 1997, A&A 323, 853
Nagase, F. 1989, PASJ, 41, 1
Nagase, F. 1999, in the Proceedings of *Japanese-German Workshop on High Energy Astrophysics*, eds. W. Becker and M. Itoh, (MPE Report 270), p. 160
Oosterbroek, T., et al. 1999, A&A, 351,
Orlandini, M., et al. 1998, ApJ, 500, L163
Parmar, A.N., et al. 1989a, ApJ, 338, 359
Parmar, A.N., White, N.E., Stella, L. 1989b, ApJ 338, 373
Paul, B., et al. 1999, in this Proceeding
Pietsch, W., KahabkaP. 1992, in *Lecture Notes in Physics 416: New Aspects of Magellanic Cloud Research* eds. B. Baschek, G. Klare, K. Beuermann, p. 59
Reig, P., Roche, P. 1999a, MNRAS, 306, 95
Reig, P., Roche, P. 1999b, MNRAS, 306, 100
Remillard, R., et al. 1998, IAUC 6826
Reynolds, A.P. Parmar, A.N., White, N.E. 1993, ApJ, 414, 302
Schmidtke, P.C., et al. 1994, PASP, 106, 843
Schmidtke, P.C., et al. 1999, AJ, 117, 927
Smith D.A., et al. 1998, IAUC 7014
Torii K., et al. 1998, ApJ, 508, 854
Torii K., Sugizaki, M., Kohmura, T., Endo, T., Nagase, F. 1999, ApJ, 523, L65
White N.E., Swank J.H., Holt S.S. 1983, ApJ, 270, 711
Wijnands R., van der Klis, M. 1998, Nature, 394, 344
Wojdowski, P., Clark, G.W., Levine, A.M., Woo, J.W., Zhang, S.N. 1998, ApJ, 502, 253
Woo, J.W., Clark, G.W., Levine, A.M., Corbet, R.H.D., Nagase, F. 1996, ApJ, 467, 811
Yokogawa, J., et al. 2000a, ApJS, 128, 491
Yokogawa, J., et al. 2000b, ApJ, 539, 191

IRON K LINE EMISSION IN AGN: OBSERVATIONS

K. Nandra [1,2]

1) NASA/Goddard Space Flight Center, Mail Code 662, Greenbelt, MD 20771, USA
2) Universities Space Research Association

ABSTRACT   Iron Kα lines are key diagnostics of the central regions of AGN. Their profiles indicate that they are formed deep in the potential well of the central black hole, where extreme broadening and red shift occur. The profiles are most easily reproducible in an accretion disk: the lack of significant emission blue-ward of the rest energy is difficult produce in other geometries. In one source an apparent (and perhaps variable) absorption feature in the red wing of the line may represent rare evidence for inflow onto the black hole. Sample analysis has defined the mean properties, showing a strong concentration of the emission in the central regions and face-on accretion disks, at least in Seyfert 1 galaxies. Surprising results have been obtained from examination of the line variability. Strong profile changes may be accounted for by changes in the illumination pattern of the central, relativistic part of the disk. In at least the case of MCG-6-30-15, there is evidence for emission from within $6R_g$, possibly indicating a spinning black hole. Developing an understanding of these complex changes has the potential to reveal the geometry and kinematics of the inner few gravitational radii around extragalactic black holes.

KEYWORDS: accretion, accretion disks; line:profiles; galaxies: active; galaxies: Seyfert

1. INTRODUCTION

The first iron Kα lines were discovered in NGC 4151, and a few source with large absorbing columns, in which the line was thought to originate ((Mushotzky, Holt & Serlemitsos 1978; Mushotzky 1982). The first unobscured AGN to show line emission was MCG-6-30-15 (Nandra et al. 1989; Matsuoka et al. 1990) and *Ginga* subsequently found iron Kα emission to be extremely common in Seyfert galaxies (Pounds et al. 1990; Nandra & Pounds 1994). Line emission had been predicted from optically-thick material close to the nucleus (Guilbert & Rees 1988), including the accretion disk (Fabian et al. 1989). Detailed predictions of the line strength from the disk (George & Fabian 1991; Matt, Perola & Piro 1991) were found to be in excellent agreement with the observations (Nandra & Pounds 1994), but the *Ginga* data were unable to determine the width or profile of these lines. This is of clear importance, as the profiles allow the location and geometry of the material to be constrained. Specifically, in the case of an accretion disk, large widths and distinctive profiles are expected due to the rotation and gravitational effects of the black hole (Fabian et al. 1989; Stella 1990; Laor 1991; Matt et al. 1992). The launch

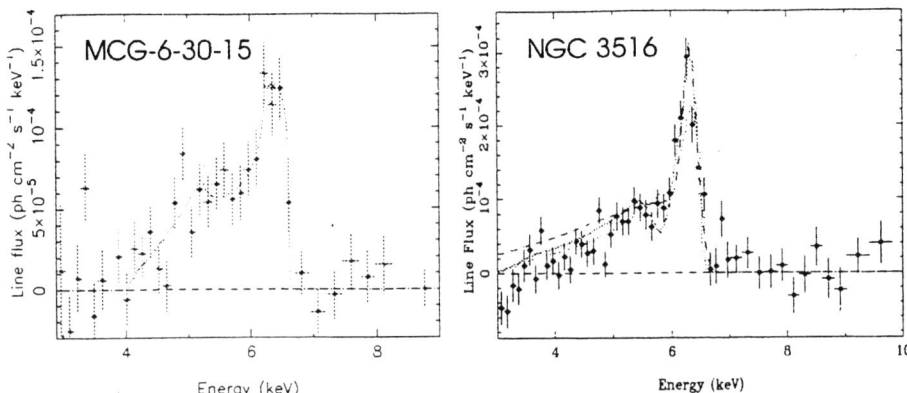

FIGURE 1. Iron Kα profiles for MCG-6-30-15 (left panel; Tanaka et al. 1995) and NGC 3516 (right panel; Nandra et al. 1999). The ASCA SIS data, derived from interpolating a local continuum, are shown as the crosses. The profiles for both sources are very similar, and are extremely broad. They exhibit a relatively-narrow core peaked at the rest energy of near-neutral iron (∼ 6.4 keV) and a very broad wing to lower energies. These profiles are characteristic of Doppler and gravitational effects in an accretion disk orbiting a black hole and the lines show various models of such emission, which fit the data extremely well. In NGC 3516, the best model includes an absorption line around 5.8 keV. This feature may be due to resonance scattering in material inflowing into the central regions.

of *ASCA* offered an opportunity to test these models, with the SIS detectors having good sensitivity and a factor ∼ 4 improvement in energy resolution compared to *Ginga*.

2. PROFILES OF INDIVIDUAL SOURCES

The early ASCA data did indeed show evidence that the iron Kα lines in AGN were broad with velocity widths of ∼ 50,000 km s^{-1}, characteristic of material extremely close to the central black hole (Fabian et al. 1994; Mushotzky et al. 1995). Uncertainties in calibration, limited sensitivity above ∼ 7 keV and the short exposures of these early observations made it difficult to determine the profiles, however. Long observations have provided the best constraints:

MCG-6-30-15: The first high signal-to-noise ratio profile was obtained in a ∼ 150 ks observation of the Seyfert 1 galaxy MCG-6-30-15 (Tanaka et al. 1995). These data provided strong confirmation of the hypothesis that the iron Kα line arises from the inner accretion disk (Fig. 1). The profile is extremely broad, with FWZI∼ 0.3c, and is skewed to the red. This is characteristic of accretion disk models in which the disk is observed close to face-on, where the dominant broadening process is the strong gravitational field of the black hole, rather than Doppler motions. Indeed,

Parameter	Symbol	Typical range
Rest Energy	E	6.4-6.9 keV
Inclination	i	0-90 degree
Inner radius	R_i	1-6 R_g
Outer radius	R_o	20-1000 R_g
Emissivity Index	q	0-3
Equivalent Width	EW	100-500 eV

TABLE 1. Disk line parameters

the particular characteristics of the line in MCG-6-30-15 are extremely difficult to explain without the invocation of a black hole and accretion disk (see, e.g. Fabian et al. 1995). The line profile of MCG-6-30-15 has therefore rightly received much attention and scrutiny, as it presents arguably the most direct evidence we have for the existence of black holes in active galaxies. The broad profile in MCG-6-30-15 has been confirmed by BeppoSax (Guainazzi et al. 1999).

NGC 4151: Yaqoob et al. (1995) presented a profile very similar to MCG-6-30-15 for NGC 4151, based on another long ASCA exposure. In this case the origin of the line is less clear-cut, as the complexity of the continuum in NGC 4151 makes the line difficult to model (Zdziarski et al. 1996). Nonetheless, the similarity of the two profiles is highly suggestive of a common origin.

NGC 3516: A third example (Nandra et al. 1999; Fig. 1), which again shows a profile remarkably similar to MCG-6-30-15. Once more, an origin in a face-on accretion disk orbiting a black hole is indicated and in this case there is also evidence for an absorption feature in the red wing of the line. This feature may be due to resonance scattering by iron, redshifted from the rest energy. If the redshift is due to kinematic effects, this feature presents rare evidence for material inflowing into the black hole and could be an important tracer of accretion. The interpretation is not unique, however, as it is possible that redshift is gravitational, in which case it indicates that there may be an ionized "skin" above or around the accretion disk. Ruzkowski & Fabian (these proceedings) show that this also fits the data.

3. ACCRETION DISK MODELS

Fig. 1 shows various models of line emission from an accretion disk, which are used to fit the ASCA data. The disk line models such as those of Fabian et al. (1989) and Laor (1991) are characterized by a number of parameters, which can in principle be constrained by the data (Table 1). It has already been mentioned that the very red profiles of MCG-6-30-15 and NGC 3516 favor low inclinations for the disk. Because the emission tends to be centrally-concentrated, the inner radius of the disk is usually better constrained than its outer radius. The former is of particular interest because it can help constrain the black hole spin. The innermost stable orbit around the black hole metric: for a Schwarzschild (non-rotating) hole this occurs at 6 R_g;

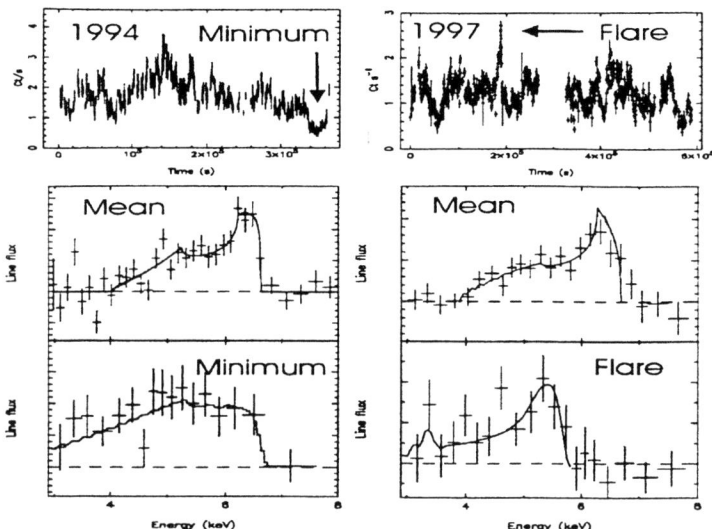

FIGURE 2. The top panels shows the light curves of MCG-6-30-15 from the long ASCA observations in 1994 and 1997 (Iwasawa et al. 1996, 1999). The integrated line profiles of both observations are very similar (middle panels). In both observations, however, the line profile was found to vary. In the earlier data, a very broad and redshifted profile was observed during a deep minimum in the flux. In 1997, a similarly extreme profile was observed, but this time during a flare. The profile variations may be attributed to changes in the illumination pattern of the disk due to localized flares.

for a rapidly rotating hole, however, the stable orbits exist close to the gravitational radius. Another crucial parameter is the line emissivity law, which parameterizes the X-ray illumination of the disk. In the models of Fabian et al. (1989) and Laor (1991), a power-law emissivity assumed, which is a useful parameterization, if somewhat unphysical. The true emissivity depends on the geometry of the X-ray source and accretion disk and their relationship, modified by relativistic effects and ionization. As the geometries are very poorly known, what is really required is to formulate specific physical models and compare them with the data. Alternatively, one can attempt to "invert" the problem and derive the emissivity from the line profile (Dabrowski et al. 1997; Cadez & Calvani, these proceedings).

4. AGN SAMPLES

The profiles of individual sources provide strong constraints, but studying samples has also been extremely informative. The widespread applicability of the disk line models has been demonstrated by the fact that they fit the data better than symmetric profiles, such as a gaussian (Nandra et al. 1997a hereafter N97; Reynolds 1997). The parameters can also be constrained.

N97 presented ASCA iron Kα data for 18 Seyfert 1 galaxies and found good constraints on the inclination of a number of these. Low inclinations are very strongly preferred, with a mean of $\sim$ 30 deg. As these are Seyfert 1 galaxies, this may not be considered surprising, as in standard unification schemes, highly inclined sources would be expected to be obscured, and seen as Seyfert 2 galaxies (Lawrence & Elvis 1982; Antonucci & Miller 1995). It is puzzling then, that Turner et al. (1998) also found low inclinations to be preferred for a sample of Seyfert 2s and NELGs. A possible solution to this is suggested by Weaver & Reynolds (1998), who fitted the spectra with an additional, narrow line at 6.4 keV, presumed to arise from the obscuring torus (Ghisselini, Haardt & Matt 1994; Krolik, Madau & Zycki 1994). If such a line is allowed, then so is a higher inclination of $\sim$ 50 deg.

The mean emissivity was found by N97 to be proportional to $R^{-2.5}$ for Seyfert 1 galaxies. This rather steep function implies that the illumination - and therefore the line emission - is strongly concentrated in the inner regions of the AGN, with $\sim$ 50 % of the line coming from within $20R_g$ and 80 % from within $100R_g$. Such an emissivity is roughly consistent with that of a centrally-illuminated disk. No universal emissivity law was found, however, suggesting that there is no single geometry.

The emission line usually peaks at 6.4 keV, and for low-inclination disks this implies a low state of ionization, typically <Fe XX. If the peak is due to a contribution from a narrow line from another source, however, no constraint can be placed on the line rest energy (N97), and therefore ionization. Standard accretion disks are expected to be very dense, and can therefore remain cool despite the intense illumination. Nonetheless, significant ionization might be expected especially in the central regions, where the X-rays are most strongly concentrated (e.g. Matt, Fabian & Ross 1993). A range of ionization sates throughout the disk is certainly consistent with the current data. The inner disk could be highly ionized, with the gravitational and Doppler shifts dominating and making it difficult to tie down the rest energy. The observed profiles are well-fit with a disk line alone, and in Seyfert 1s there is no strong requirement from line emission from other regions (N97). In Seyfert 2s, however, there is evidence for an additional component (Weaver & Reynolds 1998) and it is quite plausible that the optical BLR or obscuring torus could contribute to the line emission in all AGN. The difficulty is in distinguishing such emission from that of the low-velocity outer disk, which requires high spectral resolution.

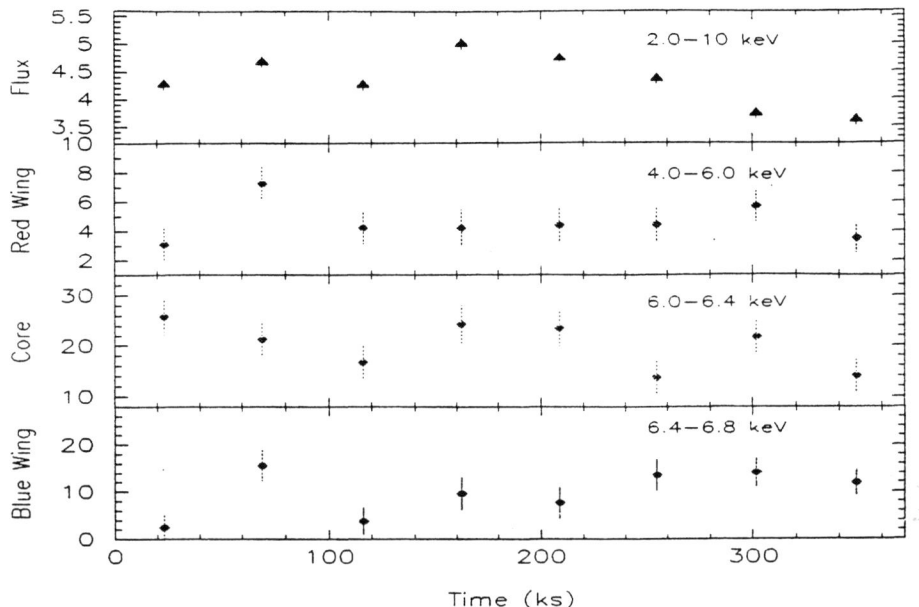

FIGURE 3. Light curves of NGC 3516. In descending order they are the 2-10 keV continuum, and the excess flux above the continuum in three line bands. Neither the core nor the blue wing flux is consistent with a constant and though the red wing is formally consistent with no variability, it appears strongly correlated (at 95% confidence) with the blue wing.

4.1. Differences and similarities

All of the high quality profiles obtained so far show profiles which are remarkably similar, and similar to the composite profile of Seyfert 1s. There are a few cases where it has been claimed that the line is narrow (e.g. NGC 4051 Mihara et al. 1994; NGC 7469 Guainazzi et al. 1994; Mrk 766 Leighly et al. 1996) but it is very difficult to exclude a broad component with the ASCA data, and none of these case is clear cut. Nonetheless there do appear to be differences in the profiles comparing different sources. The lack of a universal emissivity law is one suggestion of this (N97). Good quality data for more sources is required to confirm this, and investigate the origin. They may, for example, represent differences in geometry. Alternatively, they could be as simple as variations in the relative contributions of narrow and disk-line components.

5. VARIABILITY STUDIES

Variability studies often contribute fundamentally to our understanding of AGN. The line emission is no exception, and here some of the observations are reviewed.

- **MCG-6-30-15:** During the observation reported by Tanaka et al. (1995) the profile of the line was found to be variable (Fig. 3). Iwasawa et al. (1996) reported an unusually broad and redshifted profile during a "deep minimum" in continuum flux. Another long observation in 1997 showed a similar mean profile but this time exhibited the extreme broadening during a flare (Iwasawa et al. 1999). The profile variability was interpreted as being due to changes in the illumination pattern of the disk. If, instead of a single coherent source, localized flares produce the X-rays, then at certain times a few or even a single flare could dominate the emission. If that flare occurred in the very innermost regions, the line profile would temporarily appear more redshifted than the average. In MCG-6-30-15 the line is so broad during the deep minimum that it implies that the line emission arises within ~ 6 R_g thus implying a Kerr black hole (Iwasawa et al. 1996; see also Reynolds & Begelman 1997; Dabrowski et al. 1997; Young et al. 1999).

- **NGC 3516:** Nandra et al. (1999) have noted profile variability in NGC 3516 also (Fig. 3). The absorption feature at 5.8 keV (Fig. 1) was found to be prominent only in the middle part of the observation, where the flux was high. Furthermore, while the core of the line appeared to follow the variations in the continuum, the red and blue wings wings did not. They were well correlated with each other, however, and showed show strong variability (factor ~ 2). This was in excess of the variations in the driving continuum and therefore very difficult to explain in standard models. An interpretation similar to MCG-6-30-15 remains valid, where a localized flare beamed towards the inner disk causes the variation. Alternatively, or additionally, flares could cause an increase in the ionization of the inner disk. This could cause on over-response because the effective fluorescence yield increases sharply in the helium-like and hydrogen-like ionization states. Nandra et al. (1997b) also noted significant variability of the line flux in NGC 3516 over a 1 year baseline, with no obvious change in profile.

- **Other sources:** Yaqoob et al. (1996) found evidence for rapid variability of the broad wing in *NGC 7314* but no obvious changes were seen in the core. This is consistent with the simple disk model, where the red wing is expected to come from close in. In *NGC 4051* Wang et al. claimed a significant flux change comparing two ASCA observations, though some portion of the line flux is constant and comes from a more distant region (Uttley et al. 1999). Chiang et al. (1999) found no evidence for changes in the line flux of *NGC 5548* despite large variability of the continuum.

6. PROBLEMS AND OPEN ISSUES

Some of the iron $K\alpha$ line observations - particularly in the realm of variability - have been surprising in the context of the standard disk-line model. Depending on ones viewpoint, this may be interpreted as a further demonstration of their diagnostic power (see above) or as being problematic for the disk line model (e.g., Sulentic, Marziani & Calvani 1998a). Alternatives to the relativistic disk have been suggested and we review those briefly here. Many of these issues have been discussed by Fabian et al. (1995). They include:

Alternative geometries: in the original paper predicting the iron Kα lines, Guilbert & Rees (1988) suggested that they may come from optically thick clouds, sheets or filaments, rather than the disk. Such material can produce the observed line strengths, as long as the covering fraction is high, but less than unity to avoid obscuring the continuum (e.g. Bond & Matsuoka 1993; Nandra & George 1994). Detailed calculations of the profiles in a pseudo-spherical geometry have not been carried out and it is therefore difficult to make a definite statement as to whether the observations can be reproduced. It does, however, seem unlikely. In MCG-6-30-15 and NGC 3516 (Fig. 1), there is very little evidence for emission blue-ward of the rest energy, and it is hard to envisage a non-disk geometry in which the blue emission is suppressed. The details depend on the geometry, kinematics and self-covering of the cloud system but, for example rotating blobs are almost certainly excluded, given that the high velocities would cause Doppler boosting of the blue wing. Inflowing or outflowing cloud distributions are another possibility, but once again it seems hard to suppress the emission from material moving towards us without a special geometry.

Complex continuum: As mentioned above, Zdziarski et al. (1996) have suggested that an apparently-broad line in NGC 4151 can be accounted for by a complex continuum - in particular complex absorption. Many AGN have material in their lines-of-sight, such as "warm absorbers" (e.g. Reynolds 1997; George et al. 1998) but typically these are not thought to affect the spectrum above about 3 keV. As in all other wavebands, a great challenge in determining the line strengths and profiles is determining the underlying continuum. Although different continuum-deconvolution methods tend to produce similar results, unknown complexities add additional uncertainty to the profile determination.

Comptonization: Kinematics and gravitation are not the only mechanisms by which lines can be broadened. A particular alternative suggestion has been that Compton scattering can broaden the line (e.g., Misra & Khembavi 1998). Fabian et al. (1995) argued against such a model, as the Comptonizing medium must be finely-tuned to generate the observed profile. In particular a very compact, high optical depth ($\tau \sim 5$), cool (~ 0.2 keV) medium is required. Such a medium would down-scatter the X-ray continuum also, producing a spectral break at ~ 20 keV which is not observed.

Line blending: Combinations of numerous lines can appear broad when observed at low resolution. Differently-ionized species of iron-K, for example, result in emission from 6.4-6.9 keV. This cannot explain the observed profiles, as the broad emission occurs below 6.4 keV. No abundant element produces line emission in the 4-6 keV region where the red wing is observed, but Skibo (1997) has suggested that spallation of iron nuclei by > 10 MeV photons could result in significant amounts of V, Ti, Cr and Mn. There are numerous difficulties with such a model. For example, the observed profile changes (see above) argue forcefully against such a suggestion, as all the fluorescence lines should vary together.

6.1. Open issues

None of these alternatives is as compelling as the standard disk, but some open questions remain. The broad lines have been confirmed by BeppoSax (Guainazzi et al. 1999), but we await further confirmation and definition of the profiles. In particular, it is important to quantify and deconvolve the disk line contribution from narrower components from the BLR and obscuring torus. ASCA observations have shown some differences comparing objects, but the high signal-to-noise profiles all show common features, and in particular the derived inclinations are often very similar. This, and possible disagreements with other inclination indicators (Sulentic et al. 1998b) are not yet significant problems, but it will be important in the future to relate the properties derived from iron $K\alpha$ with other AGN observables. With large collecting area, it may be possible to discover the weak and very broad lines expected from edge-on accretion disks, which have so far eluded us. Deconvolution from a (potentially-complex) continuum is the key here.

The $K\alpha$ observations so far have been of great importance, but have shown unexpected complications. The interpretation of future observations is therefore likely to be challenging. Two particular issues are the emissivity and ionization of the disk as a function of radius. These are both arbitrary from an observational standpoint, and difficult to predict theoretically. Detailed interpretation of, e.g., variability data - including reverberation mapping (e.g. Reynolds et al. 1999) - will require an understanding of these effects. Iron $K\alpha$ observers can therefore look forward to developing the kind of complex physical models and advanced data analysis techniques that many other astronomers have been enjoying for many years. We hope and expect the rewards to be substantial.

ACKNOWLEDGEMENTS

I am grateful to Andy Fabian, Ian George, Kazushi Iwasawa, Richard Mushotzky, Chris Reynolds, Jane Turner and Tahir Yaqoob for much discussions and data. Financial support is provided by NASA grant NAG 5-7067, through USRA.

REFERENCES

Antonucci, R.R., Miller, J.S, 1985, ApJ, 297, 621
Bond, I.A., Matsuoka, M., 1993, MNRAS, 265, 619
Chiang, J., et al., ApJ, submitted
Dabrowski, Y., et al. 1997, MNRAS, 288, L11
Fabian, A.C., Rees, M.J., Stella, L., White, N.E., 1989, MNRAS, 238, 729
Fabian, A.C., et al., 1994, PASJ, 46, 59
Fabian, A.C., 1995, MNRAS, 277, L11
George, I.M., Fabian, A.C., 1991, MNRAS, 249, 352

George, I.M., et al. 1998, ApJS, 114, 73
Ghisellini, G., Haardt, F., Matt, G., 1993, MNRAS, 267, 743
Guilbert, P.W., Rees, M.J., 1988, MNRAS, 233, 475
Guainazzi, M., Matsuoka, M., Piro, L., Mihara, T., Yamauchi, M., 1994, ApJ, 436, L35
Guainazzi, M., et al., 1999, A&A, 341, L27
Iwasawa, K., et al., 1996, MNRAS, 282, 1038
Iwasawa, K., Fabian, A.C., Young, A.J., Inoue, H., Matsumoto, C., 1999, MNRAS, 306, L19
Krolik, J.H., Madau, P., Zycki, P., 1993, ApJ, 420, L57
Laor, A., 1991, ApJ, 376, 90
Lawrence, A., Elvis, M., 1982, ApJ, 256, 410
Leighly, K., Mushotzky, R.F., Yaqoob, T., Kunieda, H., Edelson, R., 1996, ApJ, 469, 147
Matt, G., Perola, G.C., Piro, L., 1991, A&A, 245, 75
Matt, G., Perola, G.C., Piro, L., Stella, L., 1992, A&A, 257, 63
Matt, G., Fabian, A.C., Ross, R.R., 1993, MNRAS, 262, 179
Mihara, T., et al. 1994, PASJ, 46, L137
Misra, R., Kembhavi, A., 1998, ApJ, 499, 205
Mushotzky, R.F., 1982, ApJ, 256, 92
Mushotzky, R.F., et al., 1995, MNRAS, 272, 9P
Mushotzky, R.F., Holt, S.S., Serlemitsos, P.J., 1978, ApJ, 225, L115
Nandra, K., George, I.M., 1994, MNRAS, 267, 974
Nandra, K., George, I.M., Mushotzky, R.F., Turner, T.J., Yaqoob, T., 1997a, ApJ, 477, 602 (N97)
Nandra, K., George, I.M., Mushotzky, R.F., Turner, T.J., Yaqoob, T., 1999, ApJ, 523, L17
Nandra, K., Mushotzky, R.F., Yaqoob, T., George, I.M., Turner, T.J., 1997b, MNRAS, 284, L7
Nandra, K., Pounds, K.A., 1994, MNRAS, 268. 405
Nandra, K., Pounds, K.A., Stewart, G.C., Fabian, A.C., Rees, M.J., 1989, MNRAS, 236, L39
Pounds, K.A., Nandra, K., Stewart, G.C., George, I.M., Fabian, A.C., 1990, Nat, 344, 132
Reynolds, C.S., 1997, MNRAS, 286, 513
Reynolds, C.S., Begelman, M.C., 1997, ApJ, 487, 135
Reynolds, C.S, Young, A.J., Begelman, M.C., Fabian, A.C., 1999, ApJ, 514, 164
Skibo, J., ApJ, 478, 522
Stella, L., 1990, Nat, 344, 747
Sulentic, J.W., Marziani, P., Calvani, M., 1998a, ApJ, 497, L65
Sulentic, J.W., Marziani, P., Zwitter, T., Calvani, M., Dultzin-Hacyan, D., 1998b, ApJ, 501, 54
Tanaka, Y., et al., 1995, Nat, 375, 659
Turner, T.J., George, I.M., Nandra, K., Mushotzky, R.F., 1998, ApJ, 493, 91
Uttley, P., McHardy, I.M., Papadakis, I.E., Guainazzi, M., Fruscione, A., 1999, MNRAS, 307, L6
Wang, J.X., Zhou, Y.Y., Xu, H.G., Wang, T.G., 1999, ApJ, 516, L65
Weaver, K., Reynolds, C.S., 1998, ApJ, 503, L39
Yaqoob, T., et al., 1995, ApJ, 453, 81
Yaqoob, T., Serlemitsos, P.J., Turner, T.J., George, I.M., Nandra, K., 1996, ApJ, 470, L27
Young, A.J., Ross, R.R., Fabian, A.C., 1998, MNRAS, 300, L11
Zdziarski, A.A., Johnson, W.N., Magdziarz, P., 1996, MNRAS, 283, 193

PHOTOIONIZED GAS IN STARBURST AND ACTIVE GALAXIES

Hagai Netzer

School of Physics and Astronomy and the Wise Observatory, Tel Aviv University, Tel Aviv 69978, Israel

ABSTRACT The coming observations by *Chandra* and *XMM* will probably revolutionize the area of X-ray spectroscopy. This review discusses the expected X-ray spectrum of starburst and active galaxies and suggest ways by which the new observations will help understand the nature of such sources and the AGN-starburst connection.

KEYWORDS: galaxies:active - galaxies:abundances - galaxies:nuclei - galaxies:starburst - X-rays

1. INTRODUCTION

ASCA, RXTE and BeppoSax observations of starburst galaxies and active galactic nuclei (AGN) shed new light on the role of the central X-ray source as a major excitation mechanism for the gas in the inner kpc of these sources. The study of this gas will, no doubt, become a major area of investigation in the near future, in particular the study of X-ray emission lines with the *Chandra* and *XMM* grating instruments. It is therefore time to consider in detail some of the more important physical processes contributing to the X-ray line emission.

Photoionized and collisionally ionized plasmas are expected to produce rich emission line spectrum that reflect the level of ionization of the gas, its temperature and composition. The aims of this review are to assess the importance of excitation by the hard, central radiation field in sources hosting active AGN as well as star forming regions, to estimate the level of excitation of the various gas components, and to make simple predictions about the 0.3–10 keV spectrum and its use as a major diagnostics tool for the understanding the physics of extragalactic sources. The main observational results are summarized in various recent publications, in particular Reynolds (1997), George et al. (1998) and Nicastro, Fiore and Matt (1999) for the properties of warm absorbers; Turner et al. (1997), Netzer, Turner and George (1998) and Guanazzi et al. (1999) for emission lines in type-II AGN; Netzer et al. (1998), Iwasawa (1999), Nakagawa et al. (2000), Vignati et al. (1999) and Matt (2000) for the spectrum of obscured AGN and ultra luminous IR galaxies (ULIRG), and Ptak et at. (1998) for X-ray observations of starburst galaxies.

2. WHERE IS THE HOT GAS?

High energy photons from AGN accretion disks, from nonthermal sources in the vicinity of massive black holes, and from other continuum emitting processes, are capable of interacting with low density gas and ionize it sufficiently to emit strong emission lines. The relative intensity of the lines depend both on the central radiation field as well as on the gas properties. It is therefore important to review the "nuclear gas inventory" in starburst and active galaxies and to estimate the location, density and perhaps mass of the various components. The major known nuclear components in AGN are:

1. The accretion disk, with dimensions of several hundred gravitational radii. The gas density in this component is unknown but theoretical calculations suggest it to be in the range of $10^{12} - 10^{17}$ cm^{-3}.

2. The BLR with typical dimensions of

$$R_{BLR} \simeq 0.03 \times \left(\frac{\lambda L_\lambda(5000\text{Å})}{10^{44} \text{ erg/sec}}\right)^{0.7} \text{ pc,} \qquad (1)$$

where L_λ is the monochromatic luminosity in erg/sec/Å (Kaspi et al. 1999). The typical density is $10^9 - 10^{11}$ cm^{-3}.

3. The torus with typical dimensions of 1–10 pc and densities similar to those observed in galactic molecular clouds.

4. The warm absorber with yet unknown location and density. A very large density range is likely, perhaps $10^5 - 10^{11}$ cm^{-3}.

5. The NLR with typical dimensions of 0.1–1 kpc and density of the order of 10^4 cm^{-3}.

6. The starburst region, with dimension exceeding 1 kpc and mean volume density of the order of 1 cm^{-3}. However, optical and IR studies suggest that much of the gas, that originate in supernovae driven shocks and stellar winds, is in condensations of much higher densities.

The following table summarizes the properties of the various components.

Component	location	density
Accretion disk	10^{-3} pc	$\sim 10^{15}$ cm^{-3}
BLR	~ 0.1 pc	$\sim 10^{10}$ cm^{-3}
Torus	1–10 pc	10^{3-6} cm^{-3}
Warm absorber	??	10^{5-11} cm^{-3}
NLR	0.1–1 kpc	$\sim 10^4$ cm^{-3}
Starburst	~ 1 kpc	10^{0-4} cm^{-3}

3. THE EMITTED SPECTRUM

The presence of a powerful central X-ray source, and a large amount of low density gas in its vicinity, suggest that some of the gas will be sufficiently ionized to produce X-ray emission lines. The most important indicator of the level of excitation is the so called "ionization parameter" (denoted here by U)

$$\text{Ionization parameter} = \frac{\text{photon density}}{\text{gas density}}. \quad (2)$$

The above ratio is energy dependent and a more meaningful relation is obtained when calculating the photon density over the range of energies where the gas is expected to emit the strongest emission lines. Thus, the level of ionization of an optical-UV emitting gas is best given by the UV ionization parameter (e.g. using the density of the Lyman continuum photons) and that of an X-ray excited gas by considering the density of the X-ray photons, i.e. only those photons energetic enough to ionized and excite gas such that the strongest emission is over the X-ray range. With this definition, $U = 1$ generally produces high level of ionization and strong emission lines over the energy range of interest.

A useful definition for X-ray gas is obtained by considering the density of the 0.1–10 keV photons. This "X-ray ionization parameter" is given by

$$U_x = \int_{0.1\text{ keV}}^{10\text{ keV}} \frac{L_\nu d_\nu/h\nu}{4\pi r^2 c N_H} \quad (3)$$

For a "typical AGN" with a power-law continuum in the optical-UV (energy slope of ~ -0.7) and the X-ray (energy slope of ~ -1), and with $\alpha_{ox} \sim 1.3$, we obtain the following estimate

$$U_x \simeq L_{45} R_{pc}^{-2} N_5^{-1}, \quad (4)$$

where L_{45} is the bolometric luminosity in units of 10^{45} erg/sec (some 30 times the 2–10 keV luminosity given the above α_{ox}), R_{pc} the distance to the center in pc and N_5 the density in units of 10^5 cm^{-3}.

Given the above definition, and the location and density of the gas as specified in Table 1, we can obtain crude estimates for the value of U_x in the various components. This enable us to predict and to calculate the X-ray spectrum which will eventually be compared with the coming *Chandra* and *XMM* observations. The following description of the expected spectrum is based on pre-*Chandra* observations.

- The accretion disk: The value of the ionization parameter is highly uncertain, $U_x = 10^{-3} - 10$, i.e. both neutral disk lines and highly ionized disk lines are possible (see Ross, Fabian and Young 1999).

- The BLR: Standard size and dimension estimates (see eqn. 1) give $U_x = 10^{-2} - 10^{-4}$, i.e. there is no significant X-ray emission, except for a sharp, "neutral" Fe-Kα line with EW which depends on the gas column density and covering fraction.

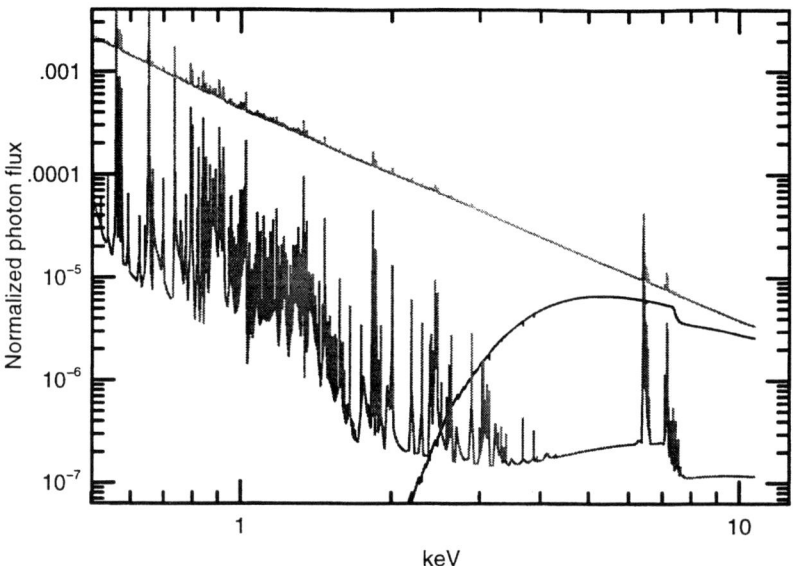

FIGURE 1. The predicted X-ray spectrum of a central torus showing the various possible views of the system. The gas is assumed to have a column density of $10^{23.5}$ cm^{-2}, the photon index of the central X-ray source is $\Gamma = 2$ and the ionization parameter at the illuminated face is U_x=0.75. The pure absorption curve, on the bottom right, is the spectrum seen by an observer looking through the absorber. This observer cannot see the inner torus walls. The bottom spectrum, with the numerous emission lines, is produced by the inner torus walls. Such a spectrum is seen by an observer who has a direct view of the torus walls but not the torus center. The top curve is the spectrum seen by an observer with a direct view of the central source (i.e. the one seeing a Seyfert 1 galaxy) assuming there is no more ionized gas. Note that the broad disk line is not included in the calculations.

- The torus: We roughly know the expected dimension and column density of this component form observations of Seyfert 2 galaxies. These can be combined to guess the density. The results is $U_x=10^{-2}-1$. The torus can thus be relatively neutral but can also be highly ionized. This raises the interesting possibility that some of the strong X-ray lines observed in Seyfert 2 galaxies originate in the torus. Fig. 1 shows the calculation of a hypothetical torus spectrum as would be observed from various viewing angles. The model assumes $U_x=0.75$ which corresponds to an inner torus dimension of 1 pc and a source luminosity of 10^{45} erg/sec (i.e. $L_{2-10 \text{ keV}} = 3 \times 10^{43}$ erg/sec).

- The warm absorber: The ionization parameter of this component is constrained to the range $U_x=0.1-1$ (George et al. 1998). However the location and density are not known. An important issue is whether the same component is responsible for the 0.7–1 keV absorption features in type-I sources and the strong emission lines in type-II sources. The Netzer et al. (1998) analysis of Seyfert 2 spectra suggest that type-I and type-II AGN differ in terms of their X-ray level of ionization, a result that needs to be confirmed by high resolution spectroscopy. An interesting possibility is that the warm absorbing gas is located inside the BLR in which case it is unlikely to be observed in Seyfert 2 galaxies.

  Given the known warm-absorber properties, we can predict its emission line spectrum. This is shown in Fig. 2 for a solar metallicity gas cloud assumed to be located just outside the BLR, with a particle density of 10^8 cm^{-3} and a column density of $10^{22.5}$ cm^{-2}.

- The NLR: The conditions in this gas are similar to those in the BLR except that the column density and perhaps also the covering fraction are lower. No significant X-ray emission is expected except for a narrow Fe-Kα line.

- The starburst region: These regions are located outside the NLR and the resulting X-ray flux is considerably lower. However, the gas density is likely to be very low and the covering factor large. We therefore expect $U_x=10^{-3}-10^0$, i.e. some nuclear star-forming regions are likely to produce strong X-ray lines due to excitation by the central X-ray sources.

  The situation regarding nuclear starburst regions is thus very interesting. Observations of "pure" starburst galaxies (if such exist, see Ptak and Griffith 1999) suggest that they are dominated by thermal X-ray emission. The expected spectroscopic signature of such gas is typical of a narrow range of excitations, even for a multi-temperature gas. In particular, emission from highly excited ions, such as Fe xxv and Fe xxvi, must be very weak. However, starburst regions near powerful AGN (e.g. NGC 6240 and other ULIRGs, NGC 1068 etc.) can be excited by both collisional and radiation processes. The resulting spectrum can differ significantly from the spectrum of pure collisionally ionized or pure photoionized plasma. Figs. 3–6 illustrate this case

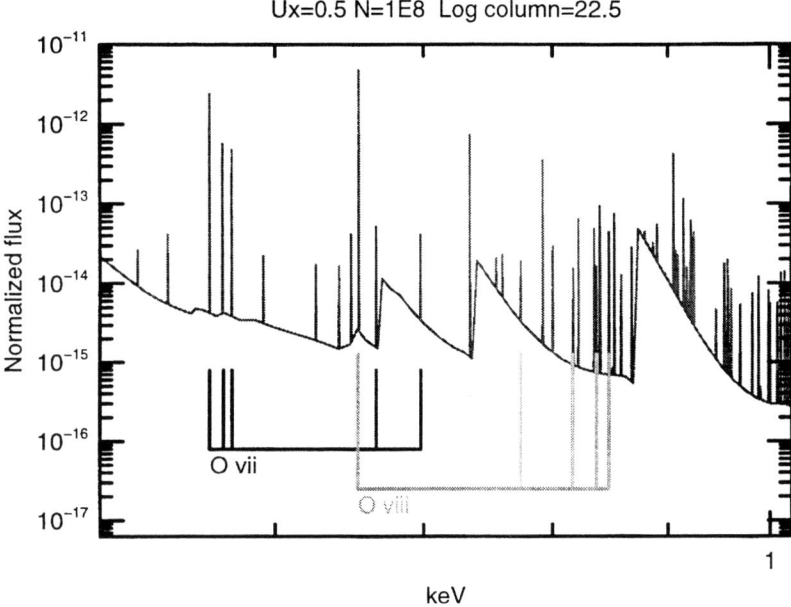

FIGURE 2. The predicted X-ray spectrum of a "typical" warm absorbing gas assumed to be located just outside the BLR. Note the rich spectrum of H-like and He-like lines of oxygen, the large number of Fe L-shell lines, and the relatively strong recombination emission features. Such gas is likely to produce a rich absorption spectrum with the strongest lines coinciding with the strongest emission by showing the calculated spectrum of a 10^7 K, low density gas subjected to illumination by a central power-law continuum source with a range of luminosities. The diagram shows the remarkable spectral changes caused by the increasing intensity of the central radiation source (i.e. the increasing U_x). The spectrum of such gas reflects both the local temperature as well as the local radiation density. The analysis of such spectra has never been attempted and will require new computational tools.

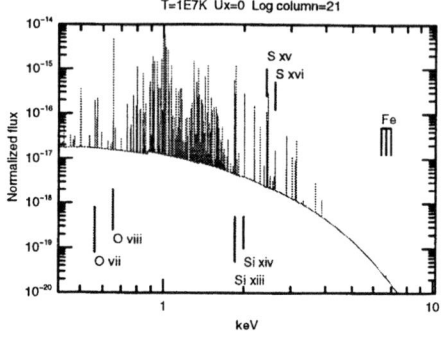

FIGURE 3. The spectrum of a 10^7 K solar composition plasma, similar to the one expected to be found in "pure starburst" regions

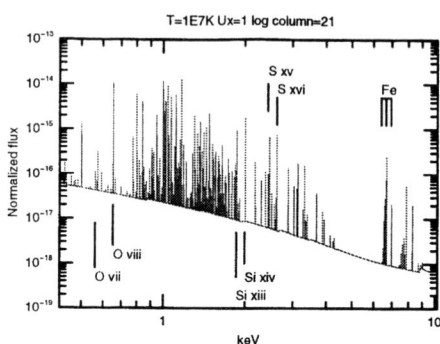

FIGURE 5. As in the previous case but for $U_x=1$.

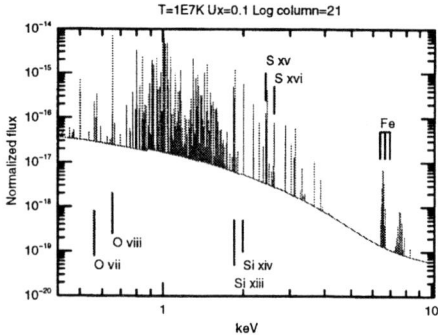

FIGURE 4. The spectrum of the same gas when illuminated by a medium luminosity AGN source. The combination of source luminosity, gas location and density gives, in this case, $U_x=0.1$

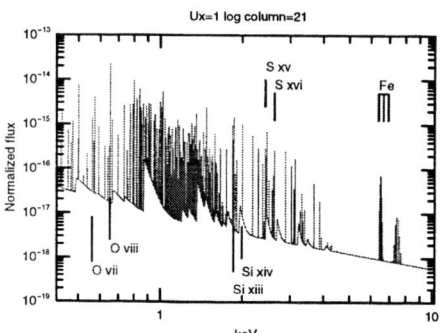

FIGURE 6. The spectrum of a "pure photoionized" gas with $U_x=1$.

4. SPECTRAL DIAGNOSTICS

Chandra and *XMM* will probably revolutionize X-ray spectroscopy. Many new spectroscopic diagnostics will be available to address various questions of interest regarding the physics of starburst and active galaxies. The following is a short list of some of the more important spectral features that are likely to be used, in various ways, to help understand the starburst-AGN connection.

He-like and H-like lines of the more abundant elements: A major spectroscopic issue is to be able to separate the various components of the He-like triplet. This is important since the calculation of two of the three components are model dependent (George and Netzer 2000). The forbidden line is the simplest component to model and its comparison with the H-like line of the same element will enable an abundant independent estimate of U_x. The comparison of He-like lines of the various elements, once U_x is known, will enable a clean abundance determination.

Low ionization fluorescence lines of O, Ne, Mg, Si, S, Ar & F: Some of these are likely to be strong in low ionization plasma (Netzer et al. 1998) and will provide diagnostics for the low ionization components such as the central disk and the torus.

Fe L-shell lines: Currently there is a large scale effort to model these features (e.g. Kallman and Liedahl 1996) and to understand their nature. The key spectroscopic issue is to be able to separate them from the H-like and He-like lines and to enable a clean abundance and ionization determination.

Recombination continua: These are potentially important in determining the temperature of a photoionized gas - see Fig. 2 for an example.

Absorption lines: Large column density plasma, with large (few×100 km/sec) turbulent or expansion velocity, seen against the central continuum, can show a rich X-ray absorption spectrum. Preliminary studies (Nicastro et al. 1999; Netzer 2000) suggests this to be a complementary tool for addressing issues such as the motion and dynamics of the warm absorbing gas.

A useful method of comparing the measured line intensities, and using them to probe the physical conditions in the various components, is to construct "diagnostic diagrams" (Netzer et al. 1998; Netzer 1999). Such diagrams are useful in determining U_x, metallicity and other parameters in photoionized and collisionally ionized gas. Three lines (two line ratios) are required to construct one such diagram and the expectation is that several of those will be available, for a sample of starburst and active galaxies, in a few years.

X-ray spectroscopy is entering a new area and the coming observations are likely to provide important clues about the nature of starburst galaxies, AGN and the link between the two classes of sources.

ACKNOWLEDGEMENTS

I am grateful to RIKEN (Japan), Goddard (USA) and the Institute of Astronomy, Cambridge (UK) for hospitality and support during my 1999 sabbatical, when this review was written. Astronomical research at Tel Aviv University is supported by the Israel Science Foundation and the Reymond and Beverly Sackler Institute of Astronomy.

REFERENCES

George, I.M., et al.: 1998, Astrophys. J. Suppl. 114, 73
George, I.M., Netzer, H.: 2000 (in preparation)
Guanazzi, M, et al., 1999, Mon. Not. R. Astron. Soc. 310, 10
Iwasawa, K.: 1999, Mon. Not. R. Astron. Soc. 302, 961
Kallman, T., & Liedahl , D., 1996, Osterheld, A., Goldstein, W., & Kahn, S, Astrophys. J. 465, 994
Kaspi, S., Smith, P.S., Netzer, H., Maoz, D., Jannuzi, B.T., & Giveon, U., 1999, Astrophys. J. (in press)
Matt, G., Fabian, A.C., Guanazzi, M, Iwasawa, K, Bassani, L & Malaguti, G, Mon. Not. R. Astron. Soc. (in press)
Nakagawa et al.: 2000, Astrophys. Sp. Sci. (in press)
Netzer, H..: 1996, Astrophys. J. 473, 781
Netzer, H..: 1999, (in the proceedings of the ASCA99 symposium, in press)
Netzer, H..: 2000, (in preparation)
Netzer,H., Turner,T.J.,George,I.M. 1998.: Astrophys. J. 504, 680
Nicastro, F., 1999, Fabrizio, F, & Matt, G, Astrophys. J. 517, 108
Ptak, A., Serlemitsos, P., Yaqoob, T., Mushotzky, R..: 1998, Astrophys. J. Suppl. 120, 179
Ptak, A., & Griffiths, R., 1999, Astrophys. J. Lett. 517, L85
Reynolds, C., 1997, Mon. Not. R. Astron. Soc. 286, 513
Ross, R., Fabian, A., & Young, A., 1999, Mon. Not. R. Astron. Soc. 306, 461
Turner, T.J., et al.: 1997, Astrophys. J. 488, 164
Vignati et al., 1999, Astron. Astrophys. 349, L57

HARD X-RAY TAILS AND CYCLOTRON FEATURES IN X-RAY PULSARS

Mauro Orlandini and Daniele Dal Fiume

TeSRE Institute, CNR, Bologna, Italy

ABSTRACT

We review the physical processes occurring in the magnetosphere of accreting X-ray pulsars, with emphasis on those processes that give rise to observable effects in their high (E>10 keV) energy spectra. In the second part we compare the empirical spectral laws used to fit the observed spectra with theoretical models, at the light of the BeppoSAX results on the broad-band characterization of the X-ray pulsar continuum, and the discovery of new (multiple) cyclotron resonance features.

KEYWORDS: Magnetic fields — Stars: magnetic fields — Stars: neutron — pulsars: general — X-rays: stars

1. INTRODUCTION

An X-ray pulsar is, by definition, a celestial source showing pulsed emission when observed in X-rays. The very first pulsating X-ray source, Centaurus X-3, was discovered in 1971 by the first scientific X-ray satellite *Uhuru* [6]. Its 4.8 s pulse period implied a small emitting region, and because the object responsible for the pulsation is not destroyed by the centrifugal force it is necessary that at its surface the gravitational force is greater than the centrifugal one. This implies $\Omega_p \lesssim \sqrt{G\rho}$, where Ω_p is the pulse frequency, G the gravitational constant, and ρ the object mean density. The observed value of Ω_p implies $\rho \gtrsim 10^7$ g/cm^3 and therefore the compact nature of the object responsible of the pulsed emission was established. The binary nature of Cen X-3 was soon after recognised by the observation of Doppler modulation in the observed pulse period [26]. The 2.1 day modulation was coincident with the periodic disappearing of the source X-ray flux, interpreted as eclipse of the compact object by the companion. Finally the optical counterpart was discovered as an early-type O star [10]. With all these elements it was possible to determine the mass of the compact object, which resulted to be 1.4 M$_\odot$: a neutron star (NS).

2. PHYSICAL PROCESSES IN X-RAY PULSARS

The physical scenario able to explain the production of pulsed X-ray emission was elaborated by Shklovskii [27] *before* the discovery of Cen X-3. X-rays are produced in the conversion of the kinetic energy of the accreted matter (coming from the intense stellar wind of an early-type star — wind-fed binaries, or coming from an accretion disc due to Roche-lobe overflow — disk-fed binaries) into radiation, because of the interaction with the strong magnetic field of the NS, of the order of 10^{11}–10^{13} gauss[1]. The dipolar magnetic field of the NS drives the accreted matter onto the magnetic polar caps, and if the magnetic field axis is not aligned with the spin axis, the NS acts as a "lighthouse", giving rise to pulsed emission when the beam (or the beams, according to the geometry) crosses our line of sight.

For a detailed description of the spectral properties of accreting X-ray pulsars (AXPs) it is therefore necessary to describe the interactions of the X-rays produced at the NS surface with the highly magnetized plasma forming the magnetosphere. This is a formidable task because we cannot use a linearized theory for the radiative transfer equations but we have to deal with the fully magnetohydrodynamical system. This is due to the fact that the coupling constants among the interactions are so large that a series expansion is impossible.

This is the reason why there is not a *parametrized* description of AXP spectra in terms of physical quantities, but only empirical laws to fit the observed spectra. An alternative method is the numerical solution of the radiative transport equations assigning particular values to the physical parameters, comparing the obtained spectra with the observed ones, and varying the parameters until a match is reached.

2.1. Cyclotron resonant features

At some distance from the NS, that we will call magnetospheric radius r_m, the motion of the accreted matter will be dominated by its intense magnetic field. We define magnetosphere the region around the NS delimited by r_m. The electrons present in the magnetosphere will have an helicoidal motion along the magnetic field lines, with gyromagnetic (Larmor) frequency given by

$$\omega_c = \frac{eB}{\gamma mc} \qquad (1)$$

where γ is the Lorentz factor. For the magnetic field strength B expected in the NS magnetosphere, the motion of the electron in the direction perpendicular to B is quantized in the so-called Landau levels (see e.g. [13]). In the nonrelativistic case, the energy associated to each level is given by

$$\hbar\omega_n = n\,\hbar\omega_c \qquad (2)$$

---
[1] Obtained from conservation of magnetic flux during the process of collapse from a "normal" star ($B \sim 10$–100 gauss, $R \sim 10^6$ Km) to a NS ($R \sim 10$ Km)

where ω_c is the Larmor gyrofrequency given in Eq. 1. As an aside, from Eq. 2 we have that $E_n = 11.6 \cdot B_{12}$ keV, where B_{12} is the magnetic field strength in units of 10^{12} gauss. Therefore we expect to observe cyclotron features in the hard ($E > 10$ keV) energy range. As we have seen, in the non-relativistic case the energy levels are harmonically spaced. When relativistic corrections are taken into account a slight anharmonicity is introduced in the Landau levels. Indeed, we have

$$\hbar\omega_n = mc^2 \frac{\sqrt{mc^2 + 2n\hbar\omega_c \sin^2\theta} - 1}{\sin^2\theta} \tag{3}$$

where θ is the angle between the line of sight and B.

Another consequence of the existence of the Landau levels is that an electromagnetic wave propagating in such a plasma will have well defined polarization normal modes, i.e. the medium will be birifringent [7]. It is not our intention to enter into the details of the propagation of waves in the magnetospheric plasma; we will develop a semi-quantitative approach by highlighting the plasma properties that have observable consequences in the AXP spectra.

If we introduce the complex refraction index N, with its real part the geometric refraction index and with its imaginary part the absorption coefficient, then the dispersion relation in the non-relativistic case can be written as a bi-quadratic equation in N. The solution for N will have the form [13]

$$N_1^2 \propto \frac{1}{\omega - \omega_c} \qquad N_2^2 \propto \frac{1}{\omega + \omega_c} \ . \tag{4}$$

The wave with N_1 presents resonance and is right-handed circularly polarized (that is in the same sense as the electron gyration). This wave is called *extraordinary*, in opposition to the *ordinary* wave — described by N_2, which is left-handed circularly polarized. By introducing the complex refractive index is straightforward to obtain the cyclotron absorption cross section. By means of the optical theorem we obtain

$$\sigma_c = 4\pi^2 \alpha_f \frac{\hbar}{m} |e_1|^2 \delta(\omega - \omega_c) \tag{5}$$

where $\alpha_f = e^2/\hbar c$ is the fine structure constant, and $\vec{e_1}$ is the polarization versor of the extraordinary wave.

Up to now, we worked neglecting both relativistic corrections and thermal motions (cold plasma approximation). The release of the latter condition allows an electron to absorb waves not only of frequency $\omega = \omega_c$, but in the interval $\omega_c \pm \Delta\omega_D$, where the Doppler width is given by

$$\Delta\omega_D = \omega_c \sqrt{\frac{2kT}{mc^2}} |\cos\theta| \tag{6}$$

where T is the electron temperature (we assumed a Maxwell-Boltzmann distribution for the electrons).

Once the electron absorbs a photon it (almost) immediately de-excitates on a time scale $t_r \sim 2.6 \times 10^{-16} B_{12}^{-1}$ sec [13]. This has important consequences for the scattering cross sections. Indeed, while a scattering process involves two photons (one going in, one going out), absorption (or emission) processes involve only one photon. Therefore one expect that the two cross section are different. This is not true just because an absorbed photon is immediately re-emitted, and therefore the absorption-emission process is equivalent to a scattering. It is possible to show [13] that the cyclotron scattering cross section σ_{res} has the same form as the cyclotron absorption cross section σ_c (Eq. 5) with the prescription

$$\delta(\omega - \omega_c) \to \frac{\Gamma_r/2\pi}{(\omega - \omega_c)^2 + \Gamma_r^2/4} \qquad (7)$$

where $\Gamma_r = \gamma_r \omega_c$, and $\gamma_r = (2/3)(e^2/mc^3)\omega$ is the radiative damping.

Therefore photons with frequency close to ω_c will be scattered out of the line of sight, creating a drop in their number. Cyclotron "lines" observed in the spectra of AXPs are therefore *not* due to absorption processes, but are due to scattering of photons resonant with the magnetospheric electrons (as it occurs for the Fraunhofer lines in the Solar spectrum). This is why we will not use the term cyclotron lines but the more appropriate "cyclotron resonant features" (CRFs).

When relativistic effects are taken into account, it is possible to show that also ordinary waves show resonance, and are scattered out of the line of sight. Another important effect on the radiative properties of the plasma is due to a pure quantum effect: the so-called vacuum polarization. The magnetic field at which a quantum mechanical treatment of the plasma is necessary can be defined when the classical cyclotron energy $\hbar \omega_c$ becomes equal to the electron rest mass mc^2. That is

$$B_{cr} = \frac{m^2 c^3}{e\hbar} = 4.414 \times 10^{13} \text{ gauss} \quad . \qquad (8)$$

We call B_{cr} the "critical" magnetic field strength. For B not far from B_{cr} virtual electron-positron pairs can be created. These virtual photons dominate the polarization properties of the plasma for frequencies in the range $\omega_{v1} \lesssim \omega \lesssim \omega_{v2}$, where [14, 15]

$$\omega_{v1} \simeq 3 \text{ keV} \frac{\sqrt{n_{22}}}{B/0.1 B_{cr}} \quad ; \qquad \omega_{v2} \simeq \omega_c \qquad (9)$$

and therefore affect the scattering cross sections (n_{22} is the electron density in units of 10^{22} cm^{-3}). In Fig. 1 we show the effects of the inclusion of vacuum polarization on the opacity (cross section times density) as obtained by [32].

2.2. Continuum emission

The main physical process responsible for the continuum emission in AXPs is Compton scattering. We will not enter into the details of the problem of repeated scatterings in a finite, thermal medium (see e.g. [23]). Let us only summarize that an

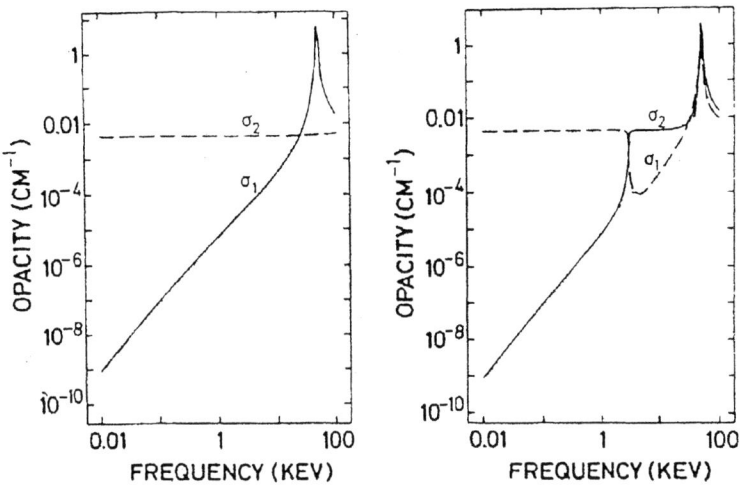

FIGURE 1. Angle-averaged scattering opacities (cross section times density) for the extraordinary (σ_1) and ordinary (σ_2) modes, computed for $n_{22} = 1$ and $\hbar\omega_c = 50$ keV. *Left:* without vacuum polarization; *Right:* with vacuum polarization. From Ventura et al. (1979).

input photon of energy E_i will emerge from a cloud of non-relativistic electrons (at a temperature T) with an average energy $E_f \sim E_i \, e^y$ (this is valid in the regime $E_f \ll 4kT$). The comptonization parameter y therefore gives a measure of the photon energy variation in traversing the plasma, and is given by

$$y = \begin{cases} \dfrac{4kT}{mc^2} \max(\tau, \tau^2) & \text{Nonrelativistic} \\ \left(\dfrac{4kT}{mc^2}\right)^2 \max(\tau, \tau^2) & \text{Relativistic} \end{cases} \tag{10}$$

where $\max(\tau, \tau^2)$ is nothing else but the average number of scattering suffered by the photons (τ is the optical depth of the medium). Note that if $E_i < 4kT$ then photons can increase their energy at the expense of the electrons: this is *inverse Compton scattering*.

The detailed description of the spectrum of the emergent photons requires the solution of the Kompaneets equation, but it is possible to obtain qualitative information for special cases:

- $y \ll 1$ In this case only coherent scattering is important, and the emergent spectrum will be a blackbody spectrum or a "modified" blackbody spectrum according whether the photon frequency is lower or greater than the frequency at which scattering and absorption coefficients are equal [23].

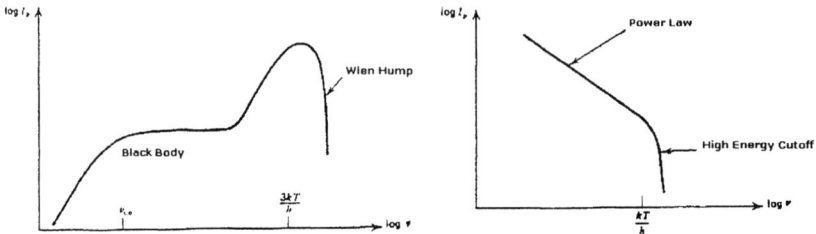

FIGURE 2. Spectrum due Compton scattering in a thermal, nonrelativistic medium. *Left:* At low frequency the spectrum is blackbody, while at higher frequencies develops a Wien hump due to saturated inverse Compton. *Right:* Spectrum produced by unsaturated inverse Compton scattering (adapted from [23]).

- $y \gg 1$ Inverse Compton scattering can be important. If we define a frequency ω_{co} such that $y(\omega_{co}) = 1$, then for $\omega \gg \omega_{co}$ the inverse Compton scattering is saturated and the emergent spectrum will show a Wien hump, due to low-energy photons up-scattered up to $\hbar\omega \sim 3kT$ [23]. In the case in which there is not saturation a detailed analysis of the Kompaneets equation shows that the spectrum will have the form of a power law modified by a high energy cutoff [23, 29]. These two regimes are qualitatively depicted in Fig. 2.

3. SPECTRAL X-RAY OBSERVATIONS OF AXPS

3.1. Before *BeppoSAX*

The very first observation of a CRF in a spectrum of an X-ray pulsar was performed in 1978 when Trümper et al. [31] observed a ~35 keV CRF in the spectrum of Hercules X-1. A while later it was observed not only the fundamental but also the first harmonics in the spectrum of the transient X-ray pulsar 4U0115+63 [33]. Observations of CRFs in other AXPs showed that they are a quite common phenomenon in this class of objects. But it was with the advent of the Japanese satellite Ginga that a systematic analysis of the spectra of AXPs was performed in search of CRFs. Mihara [16] analysed the spectra of 23 AXPs and found that 11 among them showed CRFs.

3.1.1. Continuum characterization

Because CRFs are broad features, the exact determination of the continuum is of paramount importance. From the analysis of the HEAO-1/A2 spectra of AXPs White et al. [34] found an empirical law that was able to fit their energy spectra

$$\text{POHI}(E) = \begin{cases} E^{-\alpha} & E < E_{cut} \\ E^{-\alpha} \exp\left(-\dfrac{E - E_{cut}}{E_f}\right) & E > E_{cut} \end{cases} \quad (11)$$

It is evident that this model tries to simulate the unsaturated inverse Compton process shown in Fig. 2. But this model suffers the problem of a too abrupt break around the cutoff energy E_{cut}, problem enhanced by following more sensitive instruments. Therefore Tanaka [30] introduced a "smoother" cutoff of the form

$$\text{FDCO}(E) = \frac{1}{1 + \exp\left(\frac{E - E_{cut}}{E_f}\right)} \quad (12)$$

that he called Fermi-Dirac cutoff because of it resemblance with the Fermi-Dirac distribution function. It is important to stress that the FDCO model does not have any physical meaning: it only gives a better description of the break in the AXP spectra.

Makishima and Mihara [11] were the first to note that in the AXPs showing CRFs there was a correlation between the cutoff energy E_{cut} and the CRF energy E_c, namely $E_c \simeq (1.2 - 2.5) \cdot E_{cut}$. Therefore it seemed that the cutoff was in some way due to the presence of the CRF. The next step was performed by Mihara, who introduced the so-called NPEX (Negative Positive EXponential) model

$$\text{NPEX}(E) = (AE^{-\alpha} + BE^{+\beta}) \exp\left(-\frac{E}{kT}\right) \quad . \quad (13)$$

This model is quite successful in describing the AXP spectra observed by Ginga in the 3–30 keV. Its components have also a physical meaning, because it mimics the saturated inverse Compton spectrum shown in Fig. 2 if $\beta = 2$. Furthermore, because the (non relativistic) energy variation of a photon during Compton scattering is [23]

$$\frac{\Delta E}{E} = \frac{4kT - E}{mc^2} \quad (14)$$

then when $E = E_c$ the medium is optically thick and therefore $E_c \sim 4kT$.

3.1.2. CRF characterization

Mihara, besides the introduction of the NPEX model to describe the AXP continuum, introduced a new form for the CRF

$$\text{CYAB}(E) = \exp\left(-\frac{\tau(WE/E_c)^2}{(E - E_c)^2 + W^2}\right) \quad (15)$$

which has the form of a Lorenzian of width W, and depth τ. From a physical point of view, this is the form assumed by the cyclotron scattering cross section described by Eq. 7.

Source	Obs Date	E_{cyc} (keV)	FWHM (keV)	References
4U0115+63 (M)	20 Mar 1999	12.78 ± 0.08	3.58 ± 0.33	[25]
4U1538−52 (M)	29 Jul 1998	21.5 ± 0.4	6.7 ± 1.2	[22]
Cen X−3 (M?)	27 Feb 1997	28.5 ± 0.5	7.3 ± 1.9	[24]
XTE J1946+27	09 Oct 1998	33 ± 4	16 ± 2	[26]
OAO1657−415	04 Sep 1998	36 ± 2	10	[17]
4U1626−67	06 Aug 1996	38.0 ± 0.9	11.8 ± 1.7	[19]
4U1907+09 (M)	29 Sep 1997	38.3 ± 0.7	9.7 ± 2.3	[1]
Her X−1	27 Jul 1996	42.1 ± 0.3	14.7 ± 1.1	[2]
GX301−2 (M)	24 Jan 1998	49.5 ± 1.0	17.9 ± 2.5	[20]
Vela X−1 (M)	14 Jul 1996	54.8 ± 0.9	25.0 ± 2.1	[18]
GX1+4	25 Mar 1997	...	...	[9]
GS1843+00	04 Apr 1997	...	...	[21]
X Persei	09 Sep 1996	...	...	[4]

M stands for multiple lines detected/suspected

TABLE 1. BeppoSAX observations of X−ray binary pulsars

3.2. *BeppoSAX observations of AXPs*

With the advent of *BeppoSAX* the study of energy spectra of AXPs received new impulse because it was now possible to characterize with unprecedent detail the continuum on a broader energy range (0.1−200 keV), and the two high energy instruments aboard *BeppoSAX*, namely HPGSPC (sensitive in 5−60 keV; [12]) and PDS (15-200 keV; [5]), are the best suited for the detailed spectroscopy of CRFs. *BeppoSAX* observed all the persistent AXPs, plus a couple of transient ones (see Table 1). As a first result, we found that the NPEX model, successfully used to fit the Ginga data, is not adequate to describe the broad AXP continuum [3]. In particular we find that their continuum can be described in terms of (i) a blackbody component with temperature of few hundreds eV; (ii) a power law of photon index ~ 1 up to ~ 10 keV; and a (iii) a high energy ($\gtrsim 10$ keV) cutoff that makes the spectrum rapidly drop above ~ 40−50 keV.

Furthermore, the CRFs observed with *BeppoSAX* are better described in terms a Gaussian in absorption, defined as [28]

$$\text{GAUABS}(E) = \left[1 - I \exp\left(-\frac{(E-E_c)^2}{2W^2}\right)\right] \quad . \tag{16}$$

In order to better characterize the CRF we introduced a new tool, the so-called normalized Crab ratio. The Crab ratio is simply the ratio between the source count rate spectrum and the count rate spectrum of the Crab Nebula. As this second spectrum is known, with great accuracy, to be free of features and to be modeled

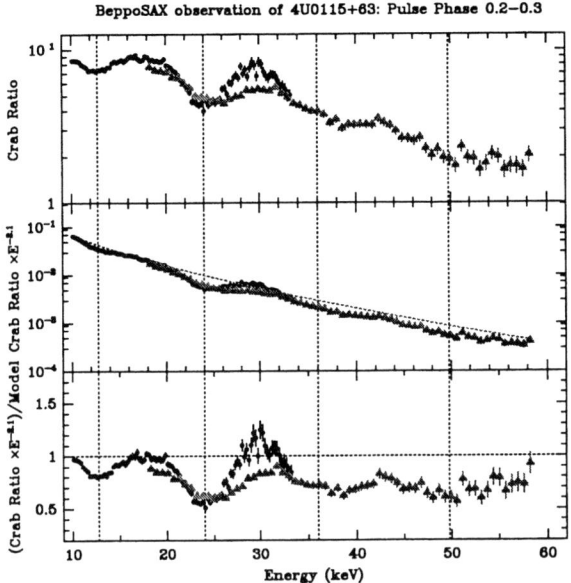

FIGURE 3. Normalized Crab ratio for the *BeppoSAX* observation of the transient AXP 4U0115+63. Both the two high energy instruments, HPGSPC (black marks) and PDS (grey marks), are shown. Note the presence of at least four cyclotron harmonics [25].

at first order with a power law in a very broad energy range, this ratio is quite well suited to enhance the presence of features in the spectrum. Furthermore the ratio is in first approximation independent from the calibration of the instrument.

In order to enhance the deviations from the continuum we multiply the ratio by a $E^{-2.1}$ power law, that is the functional form of the Crab Nebula spectrum, and we divide by the functional describing the continuum shape of the source (from this the name normalized Crab ratio). The procedure is described in Fig. 3 where we plot the result of each different step used to obtain the final result in the case of 4U0115+63. Note the presence of up to four cyclotron harmonics in the spectrum [25].

By performing the normalized Crab ratio on all the AXPs listed in Table 1 it is immediate to observe that higher the CRF energy, broader the feature [3]. The correlation between E_c and the CRF FWHM is quite evident in Fig. 4, and is easily understood in terms of Doppler broadening of the electrons responsible of the resonance, and holds for all the sources displaying single CRFs (see Eq. 6)[2].

---

[2] We assume that CRFs are produced quite close to the NS surface, therefore neglecting the gravitational redshift, which shifts the centroid energy by a factor $(1 + z)^{-1} = (1 - 2GM/Rc^2)$. See discussion in [3].

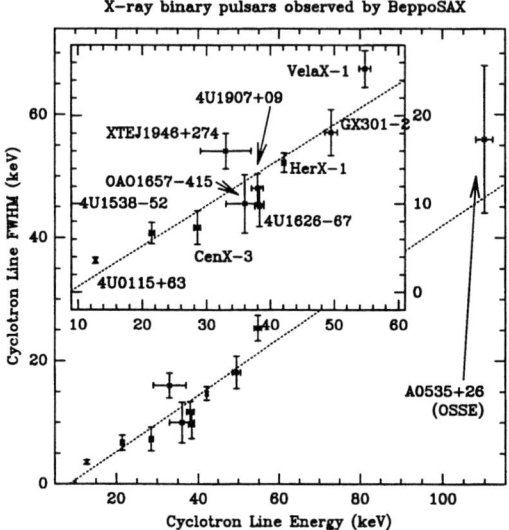

FIGURE 4. FWHM vs centroid energy for the cyclotron features observed by *BeppoSAX* and listed in Table 1. We include the OSSE measurement on the 1994 outburst of A0535+26 [8]. Note that the linear correlation has been computed *without* taking into account the A0535+26 point.

It is important to stress that this relation does not hold in presence of multiple harmonics. In other words, it seems that the temperature of the electrons responsible of higher CRF harmonics is different from that of the electrons responsible of the fundamental CRF. From Fig. 4 and by means of Eq. 6 we derived that the electron temperature responsible for the resonance is in the range $\sim$15–30 keV. This energy range is somehow "critical", because in some AXPs (and the effect is particularly evident in OAO1657–40 [17]) we observe actually *two* changes of slope in the high energy part of their spectra: a first change of slope occurs in the $\sim$10–20 keV range, while a second steepening occurs for higher energies. This leds to our last issue: "anomalous" multiple CRFs.

There are three sources, namely 4U1907+09, Vela X–1, and GX301-2 that require two CRFs in their energy spectra. The anomaly is that (i) the two CRFs are not harmonically spaced; (ii) the depth of the "fundamental" is much smaller than that of its "harmonics", (iii) their width does not correlate with their centroid energy and, more importantly, (iv) there is no trace of the "fundamental" in the normalized Crab ratio. Because of this last point we have some doubt about the interpretation of them as CRF, expecially because they are all in the critical energy range 10–30 keV where we observe the change of slope in the continuum. We are therefore inclined to interpret them as due to a not correct modelization of the continuum. Another possible explanation could be that they are due to vacuum polarization effects (see Fig. 1), but this interpretation requires a more quantitative analysis. For the three sources discussed above, we did not plot in Fig. 4 the "anomalous" CRF but the one obtained from the normalized Crab ratio.

4. CONCLUSIONS

The broad-band capabilities of *BeppoSAX* have shown that the simple phenomenological spectral laws used to describe AXP spectra in narrow energy ranges are inadequate to fit broad-band spectra. The study of AXPs as a class has shown that there is a critical region between ~ 10 and ~ 30 keV in which we observe a change of slope in the continuum. If not well modeled this could give rise to extraneous features that could be interpreted as CRF. Probably a detailed treatment of Compton scattering taking into account the effects of the magnetic field could help to solve this issue. Also vacuum polarization effects could alter the emergent energy spectra and explain the observed "anomalous" CRF harmonics.

Doppler broadening of the electrons responsible for the CRF is able to explain the observed correlation between CRF FWHM and centroid energy, showing that the electron temperature is in the range 15–30 keV for all the observed AXPs. This correlation does not hold for sources showing multiple CRFs, implying that the temperature of the electrons giving rise to higher harmonics could be different.

Acknowledgements We wish to thank the "X-ray pulsar fans" working group, formed by the friends at the TeSRE, IFCAI and ESTEC/SSD institutes in Bologna, Palermo and Noordwijk, who produced a good wealth of results on *BeppoSAX* observations and without whom this work would not have been possible.

REFERENCES

[1] Cusumano, G., et al. 1998, A&A, 338, L79
[2] Dal Fiume, D., et al. 1998, A&A, 329, L41
[3] Dal Fiume, D., et al. 2000, ASR, 25, 399
[4] Di Salvo, T., et al. 1998, ApJ, 509, 897
[5] Frontera, F., et al. 1997, A&AS, 122, 357
[6] Giacconi, R., et al. 1971, ApJ, 167, L67
[7] Ginzburg, V.L. 1970, *The Propagation of Electromagnetic Waves in Plasmas*, Pergamon Press, Oxford
[8] Grove, J.E., et al. 1995, ApJ, 438, L25
[9] Israel, G.L., et al. 1998, Nucl. Phys. B (Proc. Suppl.), 69, 141,
[10] Krzeminski, W. 1974, ApJ, 192, L135
[11] Makishima, K., & Mihara, T. 1992, in Frontiers of X-ray Astronomy, eds. Tanaka, Y., & Koyama, K. Universal Academy Press, Tokyo, p. 23
[12] Manzo, G., et al. 1997, A&AS, 122, 341
[13] Mészáros, P. 1992, *High-Energy Radiation from Magnetized Neutron Stars*, Chicago University Press
[14] Mészáros, P., & Ventura, J. 1978, Phys.Rev.Lett., 41, 1544
[15] Mészáros, P., & Ventura, J. 1979, Phys.Rev., D19, 3565

[16] Mihara, T. 1995, PhD thesis, RIKEN
[17] Orlandini, M., et al. 1999, A&A, 349, L9
[18] Orlandini, M., et al. 1998, A&A, 332, 121
[19] Orlandini, M., et al. 1998, ApJ, 500, L163
[20] Orlandini, M., et al. 2000, ASR, 25, 417
[21] Piraino, S., et al. 2000, A&A, 357, 501
[22] Robba, N.R. et al. 1999, in preparation
[23] Rybicki, G.B., & Lightman, A.P. 1975, *Radiative Processes in Astrophysics*, John Wiley & Sons
[24] Santangelo, A., et al. 1998, A&A, 340, L55
[25] Santangelo, A., et al. 1999, ApJ, 523, L85
[26] Schreirer, E., et al. 1972, ApJ, 172, L79
[26] Segreto, A. et al. 1999, in preparation
[27] Shklovskii, I.S. 1967, ApJ, 148, L1
[28] Soong, Y., et al. 1990, ApJ, 348, 641
[29] Sunyaev, R.A., & Titarchuk, L.G. 1980, A&A, 86, 121
[30] Tanaka, Y. 1986, in Radiation Hydrodynamics in Stars and Compact Objects, eds. Mihalas, D., & Winkler, K.H. Springer, Berlin, p. 198
[31] Trümper, J., et al. 1978, ApJ, 219, L105
[32] Ventura, J., et al. 1979, ApJ, 233, L125
[33] Wheaton, W.A., et al. 1979, Nat, 282, 240
[34] White, N.E., et al. 1983, ApJ, 270, 711

X-RAY AFTERGLOWS OF GAMMA-RAY BURSTS

Luigi Piro

Istituto Astrofisica Spaziale, C.N.R., Via Fosso del Cavaliere, 00133 Roma, Italy

ABSTRACT

The afterglow emission has become the main stream of Gamma-Ray burst research since its discovery three years ago. With the distance-scale enigma solved, the study of the late-time GRB emission is now the most promising approach to disclose the origin of these explosions and their relationship with the environment of the host galaxy in the early phase of the Universe. In this contribution I will review X-ray observations and their implication on our undertstanding on the GRB phenomenon. These measurements are providing a direct probe into the nature of the progenitor and a measurement of the GRB beaming properties, crucial to establish the total energy output. Some evidence of iron lines connects the GRB explosion with massive progenitors, thence with star-forming regions. Furthermore a comparison of the spectral properties with the temporal evolution indicates that the fireball expansion should not be - on average - highly collimated, with a jet angle $> 10°$.

KEYWORDS: Gamma rays: bursts; X-rays: general; Cosmology: early Universe

1. INTRODUCTION

Gamma-Ray Bursts (GRB) were discovered in 1969 (Klebesadel *et al.* 1973) by the Vela satellites, deployed by USA to verify the compliance of USSR to the nuclear test ban treaty. In the following 28 years thousands of events have been observed by several satellites, leading to a good characterization of the global properties of this phenomenon. A big step in this area was achieved with BATSE (Fishman *et al.* 1994) The isotropical distribution of the events in the sky (Fishman & Meegan 1995) was suggestive of an extragalactic origin, but a direct measurement of the distance in a single object was not available. What was lacking was a *fast AND precise* position, where the *Holy Grail* of GRB scientists, i.e. the *counterpart*, could have been searched for at all wavelenghts with more chances to catch it. This was achieved in 1996, with observations of GRB by BeppoSAX.

2. GAMMA-RAY BURSTS IN THE AFTERGLOW ERA

2.1. The first afterglow: GB970228

Before BeppoSAX (Piro *et al.* 1995, Boella *et al.* 1999) GRB astronomy has proceeded on a statistical approach and the only information gathered was limited to the tens of seconds of the GRB: the subsequent evolution was completely unknown. The hunt for GRB by BeppoSAX is based on the full complement of instruments. The Gamma-Ray Burt Monitor (GRBM) provides the temporal signature of the event (i.e. the trigger). If the event falls in the field of view (20° FWHM) of the the Wide Field Cameras (WFC: 2-26 keV) a burst in coincidence with the GRB trigger should be present. In such a case the position (with an error of $\approx 3'$) is derived from the WFC image. This position is distributed to the world-wide community by BeppoSAX mails & GCN. At the same time a follow-up with BeppoSAX Narrow Field Instruments is programmed to search for the X-ray afterglow. The operations for a prompt follow up of GRBs became operative on December 1996, after an off-line analysis of a GRB (GRB960720: Piro et al. 1998a) had demonstrated the designed capability of the mission. The first opportunity was on January 11, 1997: GRB970111. The field was pointed with the NFI 16 hours after the GRB. The possible association of one of the faint sources found in the error box with the GRB was under scrutiny (Feroci et al. 1998), when on February 28, 1997, another event, GRB970228, was detected by BeppoSAX GRBM and WFC. The NFI were pointed to the GRB location 8 hours after burst. A previously unknown X-ray source was detected in the field of view of the LECS and MECS instruments with a flux in the 2-10 keV energy range of 3×10^{-12} erg cm^{-2} s^{-1}. The new source appeared to be fading away during the observation. On March 3 we performed another observation that confirmed that the source was quickly decaying : at that time its flux was a factor of about 20 lower than the first observation (Figure 1). This was the first detection of an "afterglow" of a GRB (Costa et al. 1997).

The flux of the source appeared to decrease following a power law dependence on time ($\sim t^{-\delta}$) with index $\delta = (1.3 \pm 0.1)$. Further X-ray observation with the X-ray satellites ASCA and ROSAT detected the source about one week later with a flux consistent with the same law (Yoshida et al. 1997, Frontera et al. 1998a). This kind of temporal behaviour agrees with the general predictions of the fireball models for GRBs (e.g.Mészáros P. & Rees 1997, Vietri 2000). A backward extrapolation of this power law decay (Fig. 2) is consistent with the X-ray flux measured during the burst, suggesting that the afterglow started soon after the GRB. Another important result came from the spectral analysis of the X-ray afterglow. It excluded a black body emission, therefore arguing against a model in which the radiation comes from the cooling of the surface of a neutron star (Frontera et al. 1998b).

While the X-ray monitoring of GRB970228 was going on, an observational campaign of the same object was simultaneously started with the most important optical telescopes. This campaign led to the discovery (van Paradijs et al. 1997) of an optical transient associated with the X-ray afterglow. As in the X-ray domain, the optical flux of the source showed a decrease well described by a power law with index -1.12 (e.g. Garcia et al. 1998), again in agreement with the general predictions of the fireball model. The images taken with the Hubble Space Telescope (HST) (Fruchter et al. 1997, Sahu et al. 1997) showed the presence of a nebulosity around the optical transient. However the nebulosity was very weak, and it was not possible to disentangle whether it was associated with the host galaxy (extragalactic

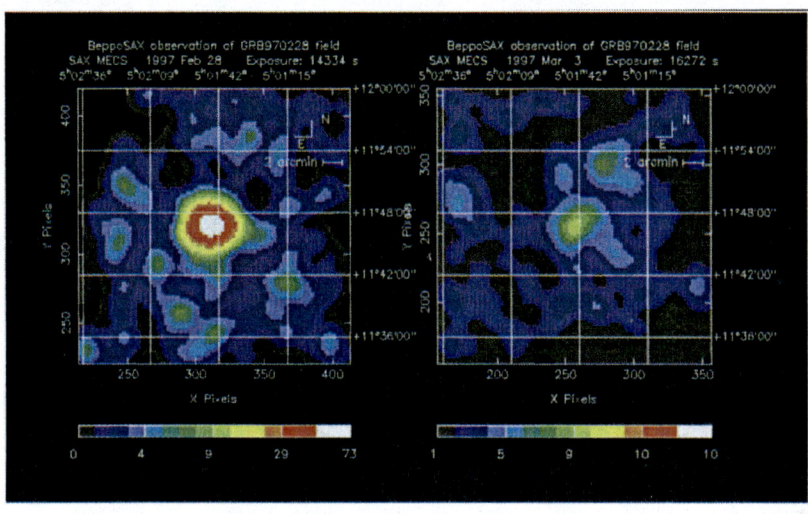

FIGURE 1. BeppoSAX MECS images of the GRB970228 afterglow, 8 hours after the GRB (left) and 3 days after the GRB (right) (from Costa *et al.* 1997)

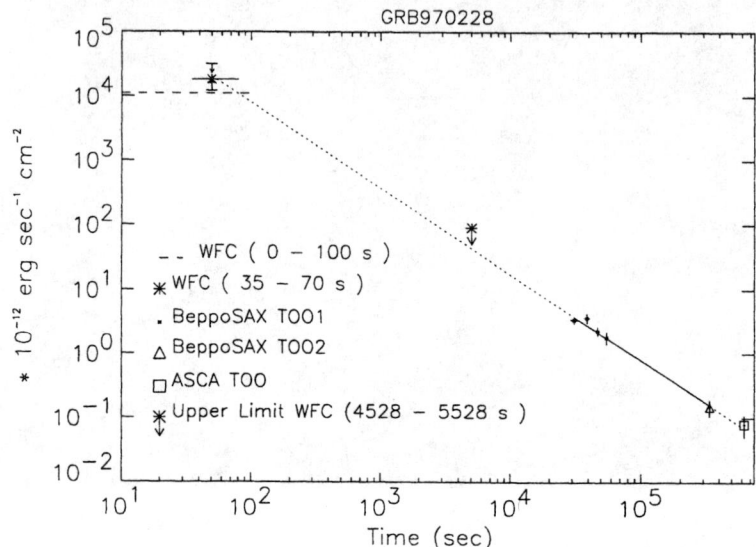

FIGURE 2. BeppoSAX decay curve of the GRB970228 afterglow in the 2-10 keV energy range, obtained with the WFC and the NFI. The result of the ASCA observation is also shown at the bottom right(from Costa *et al.* 1997).

origin) or with a transient diffuse emission representing the residual of the explosion (galactic origin).

2.2. The first measurement of redshift: GRB970508

On 8 May 1997 the second breakthrough arrived with GRB970508 (Piro *et al.* 1998b), detected just few minutes before the satellite was passing over the ground station in Malindi. This opportunity and the experience gained from previous events allowed to point the BeppoSAX NFI on source 5.7 hours after the burst, while optical observations started 4 hours after the burst.

The early detection of the optical transient (Bond 1997), and its relatively bright magnitude allowed a spectroscopical measurement of its optical spectrum with the Keck telescope (Metzger *et al.* 1997). The spectrum revealed the presence of FeII and MgII absorption lines at a redshift of $z = 0.835$, attributed to the presence of a galaxy between us and the GRB, and therefore demonstrated that GRB970508 was at a cosmological distance.

2.3. GRB are long-lasting phenomena ... after all!

The BeppoSAX observation of GB970508 has also changed our view of the GRB *phenomenon*. The old concept of a brief - sudden release of luminosity concentrated in few seconds does not stand the new information provided by BeppoSAX. Indeed

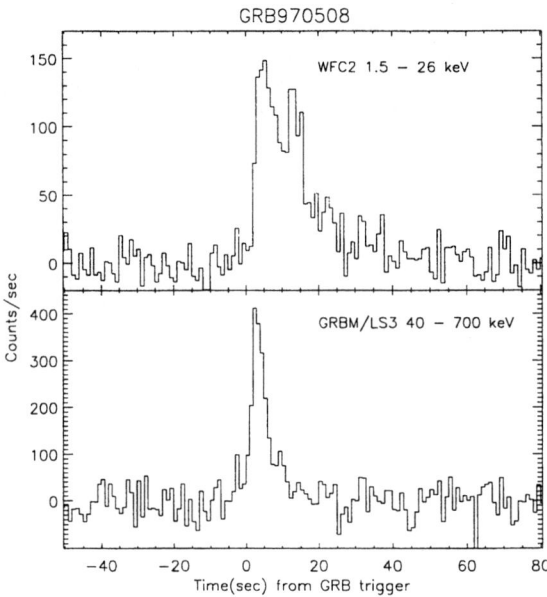

FIGURE 3. WFC and GRBM light curves of GRB970508 (from Piro *et al.* 1998b)

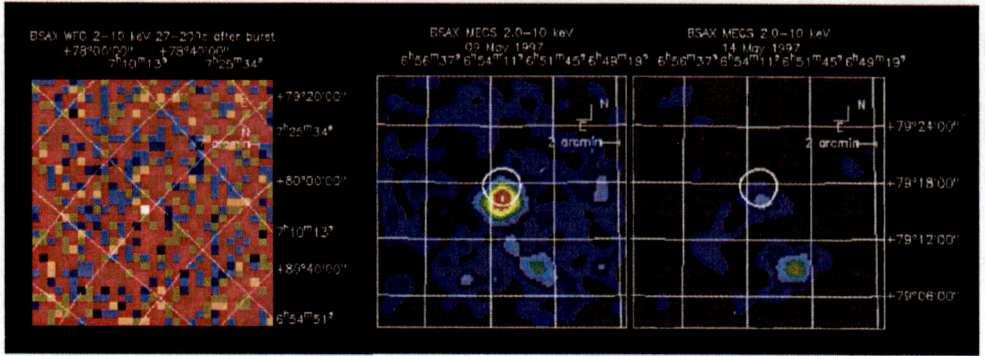

FIGURE 4. The X-ray afterglow of GB970508 (from Piro et al. 1998b): Time sequence of images of the field of GB970508 observed by the WFC2 (left image, 27–200 s after the burst), MECS(2+3) on May 9 (center image, 6 hours after the GRB), and MECS(2+3) on May 14 (after 6 days). The WFC2 show the presence of the afterglow that was then detected by the LECS and MECS (1SAXJ0653.8+7916 visible in the 99% error circle of the WFC). Note the decrease in intensity between the two MECS observations, as compared to 1RXSJ0653.8+7916, the source in the lower right corner

the name *afterglow* attributed to the X-ray emission observed after the event is somewhat misleading. This is clear when one considers the energy produced in the afterglow phase, which turns out to be comparable to that of the GRB. In order to compute the energy emitted in the afterglow phase it is necessary to integrate in time its luminosity and it is then crucial to know *when* the afterglow starts. A detailed analysis of the data of the WFC of GB970508 (Piro et al. 1998b) shows that the X-ray emission is present (Fig. 4) even when the signal of the light curve disappears in to the noise at $\sim$ 30 s (Fig.3) and remains visible for at least 2000 seconds, when the flux goes below the sensitivity of the WFC (Fig. 5). The conclusion is then that the afterglow phase starts immediately after (or even before) than the prompt emission settles down. The energy emitted in the afterglow in X-rays turns out to be a substantial fraction ($\sim 40 - 50\%$) of the energy produced by the GRB. Furthermore the light curve in Fig. 5 shows a rebursting event starting 1 day after the initial burst, an evidence that indicates that the source of the energy can re-ignite on long time scales (Piro et al. 1998b).

3. THE LARGEST AND MOST DISTANT EXPLOSIONS SINCE THE BIG BANG

The GRB observed by BeppoSAX on December 14, 1997 if, on one hand, has consolidated the extragalactic origin of GRB, on the other has underlined the problem of the energy budget and, ultimately, of the nature of the "central engine".

The chain of steps leading to the identification of the counterpart of GRB971214 (Dal Fiume et al. 1999) and its distance was the same of previous BeppoSAX observations. With a redshift z=3.42 (Kulkarni et al. 1998a) this GRB and its host galaxy are at a distance that corresponds to a look-back time of about 85% of the present age of the Universe. At this distance, the luminosity would be about $3 \times 10^{53} erg\ s^{-1}$, were the emission isotropic. This was the highest luminosity ever observed from any celestial source. Initially the huge luminosity appeared to be not compatible with the energy available in the coalescence of neutron-star mergers (Kulkarni et al. 1998a), unless beaming were invoked. Other alternative energetic models are based on the death of extremely massive stars, leading to an explosion orders of magnitude more energetic than a supernova, hence named *hypernova* (Paczynski 1998, Woosley 1993). However it was shown (Mészáros & Rees 1998a) that all these progenitors, whether Neutron Star - Neutron Star mergers or hypernovae, eventually go through the formation of a same Black Hole/torus system, from which the energy is extracted to form the GRB. The radiation physics and energy of all mergers and hypernovae are then, to order of magnitude the same, and still compatible with the luminosity observed in GRB971214.

4. TO BEAM OR NOT TO BEAM....

The energy problem became much more severe with GRB990123, again one of the BeppoSAX GRB (Heise et al. 1999, Piro et al. 1999a). It was one of the brightest

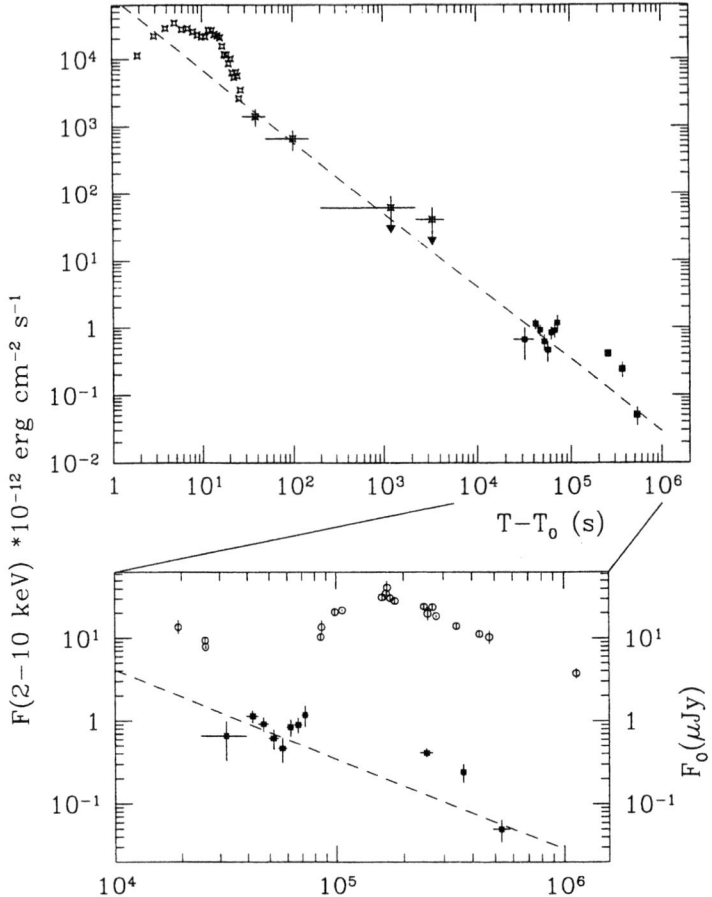

FIGURE 5. Top panel: Decay law of the GRB970508 afterglow as detected by the BeppoSAX WFC and NFI. The WFC provided the data up to 5.000 seconds after the burst. Bottom panel: enlargement of the X-ray decay law and comparison with the simultaneous time history of the optical transient (open circles) (from Piro *et al.* 1998b)

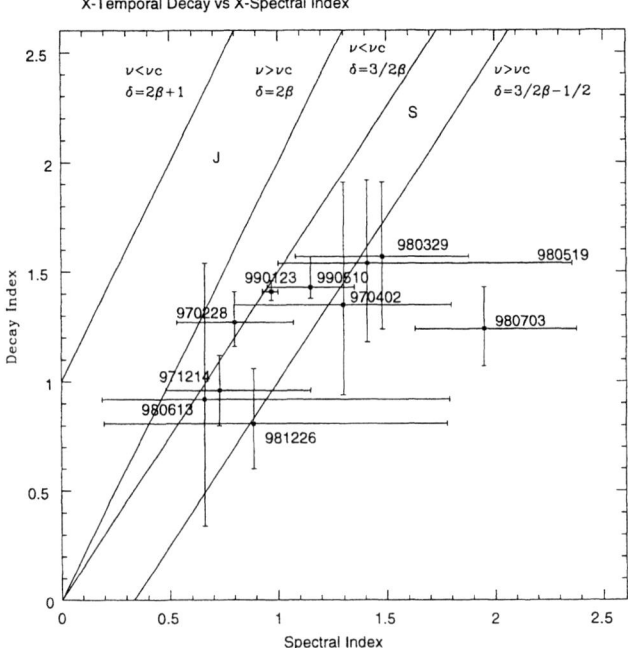

FIGURE 6. Decay vs spectral (photon index) power law slopes for 11 X-ray afterglows observed by BeppoSAX. The continuous lines trace the expected relationship in the cases of spherical (S) and jet (J) expansion in a uniform density medium

GRB ever observed, ranking in the 0.3% top of the BATSE flux distribution. Its distance (z=1.6, Kulkarni et al. 1999) would imply a total energy of $1.6 \times 10^{54} erg$, assuming isotropical emission. This corresponds to $\sim 2 M_\odot c^2$, at the limit of all models of mergers (Mészáros & Rees 1998a).

This piece of evidence is lending support to the idea that, at least in some case, the emission is collimated. This would reduce the energy budget by $\sim \theta^2/4\pi$, where θ is the angle of the jet. A typical feature of a jet expansion (vs spherical) would be the presence of an achromatic (i.e. energy-independent) break in the light curve, that appears when the relativistic beaming angle $1/\Gamma$ becomes $\approx \theta$ (e.g. Rhoads 1997, Sari et al. 1999). The presence of such a break has been claimed in GRB990123 (Kulkarni et al. 1999) and in another more recent GRB, GRB990510 (Harrison et al. 1999). With an angle $\theta \approx 10°$, the total energy would be reduced by $\approx 10^3$, within the limits of current models. So far evidence of an achromatic break is limited to the optical range and an independent measurement confirming its presence in different regions of the spectrum is lacking or not conclusive (e.g. in X-rays Kuulkers et al. 1999).

Very important indications on the geometry of the expansion can be derived by comparing the prediction of the standard scenario (i.e. fireball expansion with synchrotron emission) on the spectral and temporal behaviour of the afterglow with observations. In particular, the spectral and temporal slopes of the afterglow emission $F \sim t^{-\delta} \nu^{-\beta}$ are linked together by a relationship that depends on the geometry of the fireball expansion (Sari et al. 1997, 1999). In the assumption of an adiabatic spherical expansion in a constant density medium we have $\delta = 3\beta/2$ and $\delta = 3\beta/2 - 1/2$ below and above the cooling frequency ν_c respectively. In the case of a jet expansion the relations are $\delta = 2\beta + 1$ ($\nu < \nu_c$) and $\delta = 2\beta$ (($\nu > \nu_c$). These relationship are plotted in fig.6 along with the measured slopes we have derived for a first sample of X-ray afterglows. The data refer to a sample of 11 afterglows, among the brightest pointed by BeppoSAX upto May 1999, observed from few hours to about 2 days after the GRB (Stratta et al. 1999, Piro et al. 2000). The average property of the sample are fully consistent with a spherical expansion and deviates substantially from a jet expansion *in the first two days*. We stress that our sample is not biased against steep spectral slopes ($\beta > 1.5$), because $\gtrsim 90\%$ of the GRB detected by BeppoSAX and followed on with a fast observation, do show an X-ray afterglow.

The disagreement with the jet prediction does not imply that the geometry is spherical, because deviations from the emission pattern of a spherical expansion are expected only when the beaming angle of the relativistic emission Γ^{-1} becomes comparable to the opening angle of the jet θ_0. This happens at a time $t_{jet} \approx 6.2(E_{52}/n_1)^{1/3}(\theta_0/0.1)^{8/3} hr$, which should then be $\gtrsim 48\ hr$. This is compatible with the breaks detected in the optical range mentioned above, considering that the transition from one regime to the other can be rather smooth. We then derive $\theta_0 \gtrsim 12°(n_1/E_{52})^{1/8}$. Hence collimation, if present, cannot be very high.

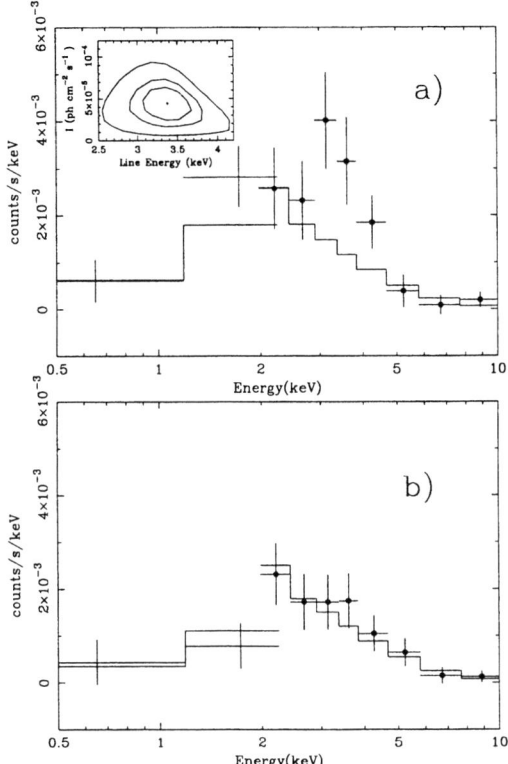

FIGURE 7. Spectra (in detector counts) of the X-ray afterglow of GRB970508 taken in the periods 0.2-0.6 day, (panel a) and 0.6-1 day (panel b) after the burst The continuous line represents the best fit power law. Note the feature around 3.5 keV in the first part (contour plots at 68%, 90%, 99% in the inset) that, at a redshift of 0.83, corresponds to the energy of the iron line complex (6.4-6.8 keV). The line disappears in the second part of the observation (panel b) (from Piro *et al.* 1999)

5. THE NATURE OF THE PROGENITOR

Information on the nature of the progenitor can be drawn from the GRB environment. In the case of hypernova the massive star should die young ($\approx 10^6$ years) and therefore GRB should be preferentially hosted in regions near the center of star-forming galaxies. On the contrary, NS-NS coalescence happens on much longer time scales and the kick velocity given to the system by two consecutive supernova explosions should bring a substantial fraction of these systems away from the the parent galaxy. So far the angular displacement of 5 optical counterparts indicates that GRB are located within their host galaxies (Bloom et al. 1999a), favoring the association with star-forming regions. We note also that those events are not located in the very center of their galaxies, that excludes an association of GRB with AGN activity.

The other diagnostics of the progenitor is based on spectral measurements of broad and narrow features imprinted by a dusty - gas rich environment expected in the hypernova scenario (e.g. Perna & Loeb 1998, Mészáros & Rees 1998b, Bottcher et al. 1998). The absence of an optical transient in about 50% of well localized BeppoSAX GRB (25 as of Dec. 99), in which instead an X-ray afterglow has been found in almost all the cases, may be explained by heavy absorption by dust in the optical range, which leaves almost unaffected the X-rays (Owens et al. 1998).

An exciting possibility is opened by the possible detection of X-ray iron line features in two different GRB, one by BeppoSAX (GRB970508 Piro et al. 1999b; Fig. 7) and the other by ASCA (GRB970828, Yoshida et al. 1999), associated with rebursting on time scales of the order of a day. It should be remarked that the presence of rebursting appears to be an uncommon feature of X-ray afterglows, whose temporal behaviour is very well described by power laws (Fig.8) at least until 2-3 days, when the X-ray flux goes below the sensitivity of current X-ray instruments. Both the temporal and spectral features betray the presence of dense ($n \sim 10^{10} cm^{-3}$) medium of $\approx 1 M_\odot$ near the site of the explosion ($\approx 10^{16} cm$) (Piro et al. 1999b). Such a medium should have been pre-ejected before the GRB explosion, but the large value of the density excludes stellar winds. A possible, intriguing explanation is that the shell is the result of a SN explosion preceding the GRB (Piro et al. 1999b, Vietri et al. 1999, Vietri & Stella 1998).

Other evidences argue in favour of the association GRB-SN. In the BeppoSAX error box of GRB980425 (Pian et al. 1999) two groups (Kulkarni et al. 1998b, Galama et al. 1998) found a supernova (SN1998bw) that had exploded at about the same time of the GRB. The probability of a chance coincidence of the two events is $\approx 10^{-4}$. Since the majority of GRB are not associated with SN (e.g. Graziani et al. 1999), this event (if the association is true) should apparently represent an uncommon kind of GRB. However it is also possible that the two families are indeed associated: this scenario would require that the GRB are emitted by collimated jets. The majority of GRB and afterglow we see are beamed towards us, so that the contribution of the supernova to the total emission is negligible. The case of SN1998bw was then particular in that the jet producing the GRB was collimated

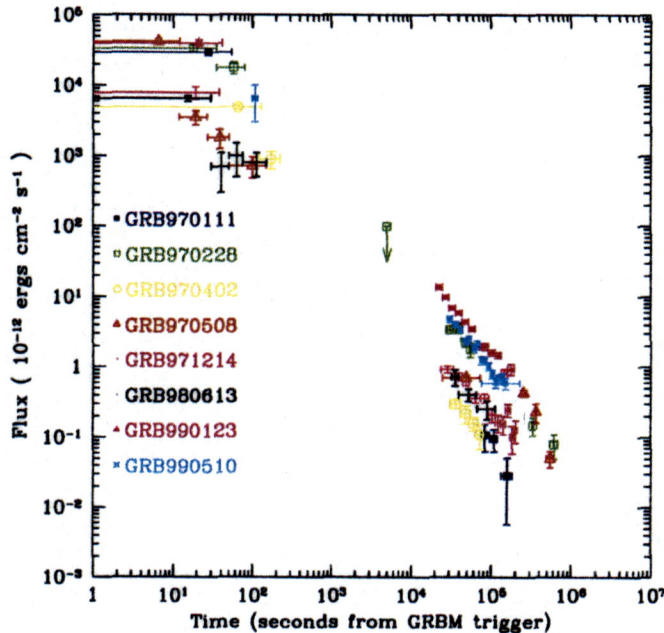

FIGURE 8. X-ray (2-10 keV) light curves of a sample of afterglows by BeppoSAX

away from our line of sight, allowing the detection of the (isotropic) SN emission at an early phase. This scenario also explains why GRB980425 was not particularly bright, notwithstanding its redshift (z=0.0085), much lower than the typical value of the other GRB ($z \approx 1$). Since the afterglow decays as a power law, it is possible that at late times the emission of the SN becomes detectable. Evidence of such emission has been claimed in at least two cases (GRB990326: Bloom et al. 1999b; GRB970228: Reichart et al. 1999)

6. CONCLUSIONS

Several bits of evidence supporting the association of GRB to star-forming regions have been gathered so far. The potential perspectives of this link are extremely exciting. Being GRB the most powerful and distant sources of ionizing photons, we can think of using them as probes of heavy elements and star/galaxy formation in the early Universe. A direct proof of this association is still missing but the near future appears very promising in this respect. BeppoSAX is discovering and localizing GRB and X-ray afterglows at a pace of 1 per month. Other satellites have also set up with success procedures for rapid GRB localization (BATSE, XTE, ASCA and IPN). The launch of HETE2, foreseen in early 2000 will increase substantially the number of well localized GRB. Furthermore present and near-future big X-ray satellites, like Chandra, XMM, ASTRO-E will allow detailed spectral studies of X-ray afterglows and provide (Chandra) arcsec position of X-ray counterparts and, possibly, a direct redshift determination.

ACKNOWLEDGEMENTS

The BeppoSAX results presented here were obtained through the joint effort of all the components of the BeppoSAX Team. BeppoSAX is a major program of the Italian space agency (ASI) with participation of the Netherlands agency for aerospace programs (NIVR)

REFERENCES

Bloom J.S. et al. . 1999a, ApJ, in press
Bloom J. et al. 1999b, Nature, in press
Boella G. et al. 1997 A&AS, 122 299.
Bond H., IAU Circular n. 6654, May 1997.
Bottcher M., Dermer C.D., Crider A. W. & Liang E. D. 1998 A&A, 343, 111
Costa E. et al., 1997 Nature, 387, 783.
Dal Fiume D. et al. 1999, A&A, in press
Feroci M. et al. 1998, A&A, 332, L29
Fishman J. et al. 1994 Astrophys. Journal Suppl. Ser., 92, 229
Fishman J. and C.A. Meegan 1995 , Annual Review Astron. Astrophys., 33, 415
Frontera F. et al., 1998a A&A, 334, L69
Frontera F. et al., 1998b ApJ 493, L67

Fruchter A. et al., IAU Circular n. 6747, September 1997.

Galama T. *et al.* 1998, Nature, 395, 670

Graziani C., D. Lamb & G.H,. Marion 1999, A&AS, 138, 469, Proc. of Gamma-Ray Bursts in the Afterglow Era, F. Frontera & L. Piro ed.s.

Garcia M. et al., 1998 ApJ, 500, L105

Harrison F.A. *et al.* 1999, Apj 523, L21

Heise J.*et al.* 1999, submitted to Nature

Klebesadel R.W. et al., 1973 Astrophys. Journal Letters 182, L85.

Kulkarni S. R. *et al.* 1998a, Nature, 393, 35

Kulkarni S. R. *et al.* 1998b, Nature, 395, 663

Kulkarni S. *et al.* 1999, Nature 398, 389

Kuulkers E. *et al.* 2000, ApJ, 538, 638

Mészáros P. & Rees M. J. 1997 ApJ 476, 319

Mészáros P. & Rees M. J. 1998a, New Astronomy, (astro-ph/9808106)

Mészáros P. & Rees M. J. 1998b MNRAS, 299, L10.

M.R. Metzger et al. 1997, Nature, 387, 878.

Owens A. *et al.* 1998, A&A, 339, L37

Paczynski, B. 1998 ApJ, 494, L45

Perna R. & Loeb A., 1998, ApJ, 501, 467

E. Pian *et al.* 1999, A&A, in press

Piro L., Scarsi L. & Butler R.C. 1995, in X-Ray and EUV/FUV Spectroscopy and Polarimetry, (ed. S. Fineschi) SPIE 2517, 169-181

Piro L. et al., 1998a A&A, 329, 906

Piro L. *et al.* 1998b, A&A, 331, L41

Piro L. *et al.* 1999a , GCN 199,203;

Piro L. *et al.* 1999b, ApJ, 514, L73

Piro L. *et al.* 2000, in preparation

D. Reichart *et al.* 1999, Ap.J., in press

Rhoads J.E. 1997, ApJ, 478, L1

Sahu K.C. et al. 1997, Nature, 387, 476

Sari,R.,Piran,T. & Narayan,R., 1997 ApJ, 497 ,L17

Sari R., Piran T, Halpern J.P. 1999, ApJ, 497, L17

Stratta G., Piro L., Gandolfi G. *et al.* , 1999, Proc.s of the 5th Huntsville Symposium on GRB

J. van Paradijs et al., 1997 Nature, 386, 686.

Vietri M., Perola G.C., Piro L. & Stella L. 1999, MNRAS, 308, P29

M. Vietri & L. Stella 1998, ApJ, 507, L45

Vietri M. 2000, this conference

A. Yoshida et al., IAU Circular n. 6593, 19 March 1997.

A. Yoshida *et al.* 1999 A&AS, 138, 433, Proc. of the Gamma-Ray Bursts in the Afterglow Era, F. Frontera &L. Piro ed.s.

Woosley S. 1993, ApJ, 405, 273

TIME LAGS IN COMPACT OBJECTS: CONSTRAINTS ON THE EMISSION MODELS

Juri Poutanen

Stockholm Observatory, SE-133 36, Saltsjöbaden, Sweden

ABSTRACT Accreting black holes and neutron stars in their hard (low) state show not only very similar X/γ-ray spectra but also that the behaviour of their light curves is quite similar which can be quantified as having similar power-density spectra and Fourier-frequency-dependent time/phase lags. Taken together this argues for a common mechanism of the X/γ-ray production in these objects. This mechanism is probably a property of the accretion flow only since it does not depend on the nature of the compact object. In this paper, I review the observational data paying most attention to the properties of the temporal variability such as the time/phase lags that hopefully can help us to discriminate between different theoretical models. I also discuss the models developed to account for the basic observational facts. Particularly, I show that the commonly used Compton cloud models with constant temperature cannot explain variable sources without violating the energy conservation law. Alternative models where time lags are related to the spectral evolution during X-ray flares are discussed and compared with observations. Compton reflection from the outer edge of the accretion disc is shown to affect markedly the time lag Fourier spectrum.

KEYWORDS: accretion, accretion discs; black hole physics; stars: neutron; stars: flare; stars: individual (Cygnus X-1); X-ray: stars.

1. MODELS FOR THE FORMATION OF X/γ-RAY SPECTRA

X-ray and gamma-ray spectra of accreting black holes and neutron stars are deconvolved into (at least) two components: a soft component interpreted as emission from an optically thick accretion disc, and a hard tail associated with a hot (10–100 keV) "corona". Reviews of the spectral properties of Galactic black hole candidates (GBHs) can be found in Gilfanov et al. (1995), Tanaka & Lewin (1995), Grebenev et al. (1993, 1997), Grove et al. (1997), and Poutanen (1998). X/γ-ray properties of radio-quiet active galactic nuclei (AGN) are reviewed by Zdziarski et al. (1997), Johnson et al. (1997), and Zdziarski (1999). Recent results on the broad-band spectra of accreting neutron stars are presented by Barret et al. (2000).

An amusing fact is that super-massive black holes in AGN, GBHs and accreting neutron stars in their hard (low) states (see Tanaka & Lewin 1995; Gilfanov et al. 1995 for the definition of the spectral states) show very similar X/γ-ray spectra (see Zdziarski 1999; Barret et al. 2000). Furthermore, properties of their rapid temporal variability are also similar (van der Klis 1995b; Wijnands & van der Klis 1999; Psaltis, Belloni & van der Klis 1999; Ford et al. 1999; Edelson & Nandra 1999; Chiang et al. 2000). All this argues for a common mechanism of the X-ray production in all these sources.

There are good reasons to believe that the main radiative mechanism for the production of the hard X-rays is Comptonization of soft photons (e.g., Shapiro, Lightman, & Eardley 1976; Sunyaev & Trümper 1979; Sunyaev & Titarchuk 1980). However, it is not completely clear what determines the observed spectral slopes. The geometry of the X-ray emitting region and the source of soft photons is still a matter of debate (see Svensson 1996; Poutanen 1998; Beloborodov 1999b; Wardzinski & Zdziarski 2000).

An important clue to our understanding of the X-ray production came from the discovery of Fe lines (at ~ 6.4 keV) and the hardening of the spectra above 10 keV in AGN (Pounds et al. 1990; Mushotzky, Done, & Pounds 1993; Nandra & Pounds 1994), Cygnus X-1 (e.g., Done et al. 1992; Gierliński et al. 1997), and neutron stars (e.g., Yoshida et al. 1993). These features are associated with the reflection of hard X-rays from cold material (Basko, Sunyaev, & Titarchuk 1974; George & Fabian 1991; Magdziarz & Zdziarski 1995; Poutanen, Nagendra, & Svensson 1996). These observations gave support to the so called two-phase accretion disc-corona models. In such models, X-rays are emitted by a hot rarified corona above the cold accretion disc (Haardt & Maraschi 1993; Haardt, Maraschi, & Ghisellini 1994; Stern et al. 1995; Poutanen & Svensson 1996). Hard X-rays from the corona, being reprocessed in the cold disc, produce the reflection hump as well as most of the seed soft photons that are subsequently Comptonized to produce the hard X-rays. This is the *feedback* mechanism. The geometry of the corona determines the feedback factor which in its turn determines the spectral slope of the escaping radiation. The temperature of the emitting plasma (or to be more exact, the Kompaneets y-parameter) is determined by the energy balance between heating (by magnetic reconnection?) and cooling (by Comptonization of soft photons).

Further support for the feedback models was recently given by Zdziarski, Lubinski, & Smith (1999) (see also Zdziarski 1999; Gilfanov, Churazov, & Revnivtsev 1999) who found a correlation between the amount of reflection ($R \equiv \Omega/(2\pi)$, where Ω is a solid angle subtended by cold material as viewed from the X-ray source) and the intrinsic photon spectral index, Γ, of the hard X-ray component. Such a correlation can easily be explained if there is overlap between the hot corona and the cold disc (Poutanen, Krolik, & Ryde 1997). The further the cold disc penetrates into the corona, the larger is the cooling, the smaller is the temperature of the corona, the softer is the spectrum, and, finally, the larger is the amplitude of the reflection. The model, however, appears to have trouble giving reflection amplitudes above $R_{\max} \sim 0.5$ (if the coronal optical depth $\tau_T \sim 1$, see Zdziarski et al. 1997) due to partial smearing of the reflection component by the hot corona.

Alternatively, the observed $R - \Gamma$ correlation can be reproduced by variations of the bulk velocity of the X/γ-ray emitting plasma (Beloborodov 1999a,b). If the emitting regions are sufficiently compact to produce electron-positron pairs, the pressure of the radiation reflected and reprocessed in the disc accelerates pairs to mildly relativistic velocities away from the disc. On the other hand, for proton dominated plasmas, a small anisotropy in the energy dissipation mechanism can result in the ejection of particles away or towards the disc. Ejection away from the disc reduces R below 1 and leads to hard spectra, while ejection towards to the disc can result in apparent $R > 1$ as is observed in some objects.

The physical possibility of the corona formation was studied by Galeev, Rosner, & Vaiana (1979). They showed that the magnetic fields, being amplified in the cold disc due to turbulent motions and differential rotation, do not have time to annihilate inside the disc on the inflow time scale. Instead, the field loops are expelled from the disc by buoyancy (the Parker instability) and they annihilate in the tenuous corona. Beloborodov (1999a) showed that the mechanism studied by Galeev et al. is able to produce a corona of limited luminosity which is h/r (the ratio of the disc height to its radius) times smaller than the disc luminosity. By contrast, in some sources most of the energy escapes in the form of hard X-rays. Beloborodov also argued that the magneto-rotational instability (Velikhov 1959; Chandrasekhar 1960; Balbus & Hawley 1991) increases the rate of the magnetic field generation (as compared with the Galeev et al. model) thus producing an active magnetic corona where a large fraction of the gravitational energy can finally be dissipated in magnetic flares. These qualitative arguments were recently supported by numerical three-dimensional magnetohydrodynamical simulations of Miller & Stone (2000) who showed that about 25 % of the total energy dissipation can occur in the rarified corona.

An alternative to the magnetic corona is the hot disc model (Shapiro et al. 1976; Ichimaru 1977; Narayan, Mahadevan, & Quataert 1998; Zdziarski 1998; Esin et al. 1998) which is also able to explain the observed X/γ-ray spectra. In order to distinguish between the models, it would be helpful to compare the predictions of different models with the temporal variability data (see van der Klis 1995a,b and Cui 1999a for recent reviews). Unfortunately, most of the papers on the spectral models do not consider the temporal variability. On the other hand, most of the models designed to explain the temporal variability data do not pay enough attention to the emission processes and the physics of the spectral formation.

In this review, we will discuss the variability data keeping in mind recent advances in modelling broad-band X/γ-ray spectra of accreting black holes and neutron stars. Most attention will be paid to the time lags that can shed light on the mechanism of the X-ray production. Then we discuss simple phenomenological models that are able to explain some of the observational facts. After that, we switch to the physical models. In particular, the properties of the Comptonizing regions will be discussed. We will point out the flaws in models that do not consider the energy balance in the "Compton cloud", and then discuss models that satisfy the energy conservation law and confront them with the available data.

2. OBSERVING TIME LAGS IN ACCRETING BLACK HOLES AND NEUTRON STARS

The standard temporal characteristics that are usually computed are the power-density spectra (PDS) in different energy channels, auto/cross-correlation functions (ACF/CCF), the time/phase lags between the variability in different energy channels, the coherence function, etc. Time lags have been studied by two methods, by constructing the CCF and by cross-spectral analysis (for details, see Lewin et al. 1988; van der Klis 1989; Nowak et al. 1999a).

2.1. Lags in Black Hole Sources

The observations of Cygnus X-1 from sounding rockets and *HEAO 1* (Priedhorsky et al. 1979; Nolan et al. 1981) showed that the CCF between different energy channels peaks very close to zero lag (delay ≲ 40 ms), but it is slightly asymmetric. Similar asymmetry was found in the *EXOSAT* data by Page (1985) who claimed a 6 ± 1 ms shift of the peak of the CCF between the 5-14 keV and the 2-5 keV bands. Recent *RXTE* observations clearly show asymmetries of the CCFs, which, however, peak within ∼ 1 ms from zero lag (see Fig. 1 and Maccarone, Coppi, & Poutanen 2000). This suggests that the relation between the variation in the two bands are not simply a time delay. Asymmetry is also observed in the *RXTE* data of GX 339-4, where the CCFs are offset by ≲ 5 ms from zero (using the 2-5 and 10-40 keV bands, Smith & Liang 1999). The CCFs of AGN also display similar properties (e.g., Papadakis & Lawrence 1995; Lee et al. 2000).

The CCFs cannot be fitted with simple exponentials at any time scale. A reasonably good description of the CCF is in terms of a stretched exponential $CCF(t) = \beta \exp[-(|t-t_0|/\tau)^\nu]$, where the normalisation $\beta \approx 1$, the time where the CCFs peak $t_0 < 10^{-3}$ s, $\nu \sim 2/3$ in the range $|t| < 0.3$ s, and the time constant τ is different for rising and decaying part of the CCF. Such a behaviour is probably the result of self-similarity of the light curve. It is interesting to note that the ACF of gamma-ray bursts also show a similar stretched exponential behaviour (see Stern & Svensson 1996; Beloborodov 1999c).

Since the CCF does not show which frequencies contribute most to the observed lags, van der Klis et al. (1987) suggested to use instead the cross-spectrum for such an analysis.[1] In the hard state of Cyg X-1 observed by the *Ginga* satellite, the time lags between the variability in the 1.2-4.7 and 4.7-9.3 keV energy bands reached 0.1 s and had a strong Fourier-frequency dependence $\delta t(f) \sim f^{-1}$, i.e., the phase lag $\delta\phi(f) \approx$ const (Miyamoto et al. 1988; Miyamoto & Kitamoto 1989). Similar lags were observed in other GBHs, GX 339-4 and GS 2023+338 (≡V404 Cyg), in their hard state (Miyamoto et al. 1992). The analysis of the *RXTE* data

---
[1] The cross-spectrum $C(f) \equiv S^*(f)H(f)$, where $S(f)$ and $H(f)$ are the Fourier transforms of the light curves in the soft and hard energy channels, respectively. The phase lag, $\delta\phi(f) \equiv \arg[C(f)]$, and the time lag, $\delta t(f) \equiv \delta\phi(f)/(2\pi f)$. The lags are positive when hard photons are lagging the soft ones.

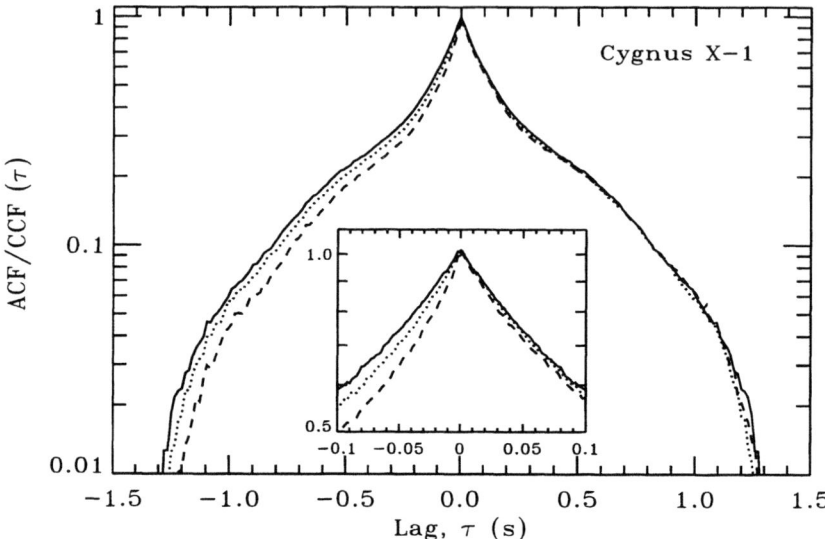

FIGURE 1. The cross-correlation functions of Cyg X-1 in the hard state (*RXTE* observations from October 22, 1996). The solid curves are the autocorrelation function for the 2-3.9 keV energy channel, the dotted and dashed curves are the CCFs for the 6-8.2 and 14-70 keV vs. the 2-3.9 keV energy channel, respectively. Note, that CCFs in the hard state sometimes have much broader wings extending to ~ 8 s (e.g., Nolan et al. 1981; Maccarone et al. 2000).

for Cyg X-1 (Nowak et al. 1999a), GX 339-4 (Nowak et al. 1999b), 1E 1740.7-2942 and GRS 1758-258 (Smith et al. 1997), and GS 1354-644 (Revnivtsev et al. 2000) confirming the general features seen in the *Ginga* data, showed a complicated behaviour of the phase lags which have a number of shelves and breaks (see Fig. 2).

Grove et al. (1998) extended this analysis to higher energies with the data from *CGRO*/OSSE. The time lags at low frequencies reached 0.3 s between the 50-70 and the 70-100 keV photons in the light curve of GRO J0422+32 ($\equiv$Nova Persei 1992). The breaks detected in the time lag spectrum at 0.1 Hz may be related to the quasi-periodic oscillation (QPO) observed at 0.23 Hz. The *CGRO*/BATSE data of Cyg X-1 (Crary et al. 1998), GRO J0422+32 and GRO J1719-24 ($\equiv$Nova Oph 1993) (van der Hooft et al. 1999a,b) show very similar time lag spectra.

The time lags of GBHs in their soft state (when $\Gamma \approx 2.5$) turn out to be somewhat different. *Soft* lags were observed between the 1.2-2.3 keV and the 2.3-4.6 keV bands in GX 339-4 (Miyamoto et al. 1991) and GS 1124-68 ($\equiv$Nova Muscae 1991; see Miyamoto et al. 1993; Takizawa et al. 1997), while the higher energy photons were lagging the variability in the 2.3-4.6 keV band. In one observation of GS 1124-68, the variability in the 4.6-9.2 keV was the most advanced. The phase lag reached ~ 1 rad which is much larger than the lags in the broad band noise observed in the GBHs in their hard state. Rapid time variations were mostly due to the harder power-law

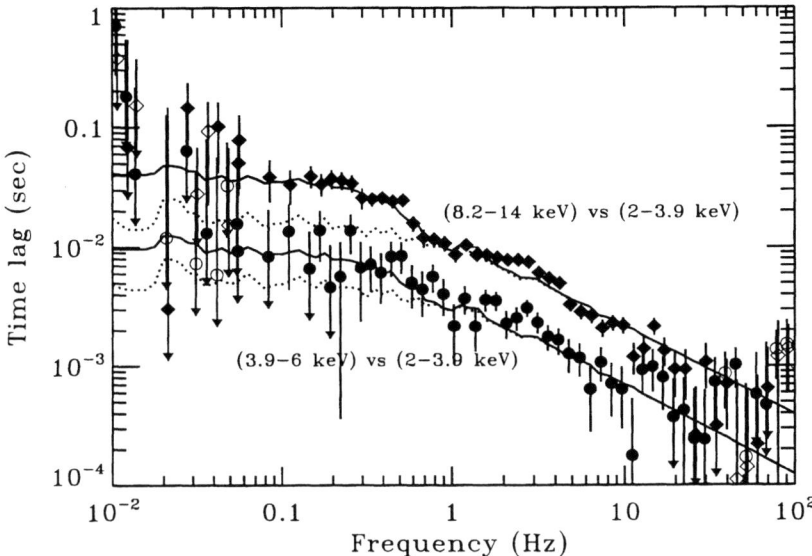

FIGURE 2. Time lags between signals in the 8.2-14 keV and the 3.9-6 keV bands vs the 2-3.9 keV band in Cyg X-1 (*RXTE* observations from October 22, 1996). The dotted curves are the model of Poutanen & Fabian (1999b) with the maximum flare time scale of $\tau_{\max} = 0.2$ s. The solid curves show the same model with 70% of the observed Compton reflection produced at a distance of $t_{\mathrm{refl}} \sim 1$ light seconds from the central X-ray source. The reflector acts as a low pass filter so that additional delays appear only at frequencies $f \lesssim 1/t_{\mathrm{refl}}$, and only at the energies where reflection is significant. See § 4.2 for details.

component which is clearly seen from the rms amplitude. It is interesting that the lags seem to saturate above 10 keV. It is worth pointing out that the largest lags here are observed at the QPO frequency. On the other hand, the time lag spectrum of Cyg X-1 in the soft state looks quite similar to that in the hard state (Cui et al. 1997).

Time lags have been observed in GRS 1915+105 in the 67 mHz QPO by Cui (1999b) and in the broad-band noise and QPOs by Reig et al. (2000). The lags show a very complicated structure, sometimes changing signs from one harmonic to another, and the sign also depends on the frequency of the QPO. Wijnands, Homan, & van der Klis (1999) and Cui, Zhang, & Chen (2000) observed lags in the broad-band noise and the QPOs of XTE J1550-564.

2.2. Lags in Neutron Star Sources

Hard time lags in neutron star sources were discovered by Hasinger (1987) in the CCF of Cyg X-2 (comparing the 1-5 and the 5-17 keV bands) in its horizontal branch (HB, see Lewin et al. 1988 for definitions of branches). The CCF also had sinusoidal oscillations due to the QPO. The time lags showed anti-correlation with the QPO frequency, dropping from 4 ms to 1.5 ms when the QPO changed from 20 Hz to 50 Hz. Hasinger interpreted the lags as delays due to scattering (Comptonization) in the hot cloud and the anti-correlation as an indication of a change in the system size. Associating the QPO frequency with the Keplerian frequency at some radius gives the relation, $\delta t(f) \propto f_{QPO}^{-2/3}$, while the actual data are much better described by $\delta t \propto f_{QPO}^{-1}$, i.e. $\delta\phi = 2\pi f \cdot \delta t = $ const.

Using the cross-spectrum techniques van der Klis et al. (1987) confirmed the existence of $\sim$ 3 ms hard lags in the 20-40 Hz QPO of Cyg X-2 (and GX 5-1) and discovered 8 ms *soft* lags in the low-frequency noise, which were interpreted as a softening of the spectrum during the shots that cause QPOs. These results were confirmed by Vaughan et al. (1994) who also showed (from the analysis of the *Ginga* data of GX 5-1 on the HB) that the time lags increase with photon energy.

In both Cyg X-2 and GX 5-1 in their normal branch, the lags at the $\sim$ 5 Hz QPO showed energy dependence (Mitsuda & Dotani 1989; Vaughan et al. 1999) reaching $\delta\phi \sim \pi$ rad (i.e., $\delta t \sim 0.2$ s) for 10 keV photons vs 2 keV. At the same time, the rms amplitude of the QPO in Cyg X-2 had a minimum at 5 keV and in GX 5-1 it increased above 2.5 keV. This behaviour can be interpreted as a pivoting of the spectrum around 3-5 keV.

With the larger effective area of *RXTE*, Ford et al. (1999) and Olive & Barret (2000) discovered phase lags in the broad-band noise of three atoll sources, 4U0614+09, 4U1705-44, and 4U1728-34. These lags are very similar to those in GBHs like Cyg X-1 and GX 339-4, which tells us that the mechanism responsible for the lags does not depend on the presence or absence of the hard surface of the neutron star, magnetosphere, boundary layer, etc., but instead is a property of the accretion flow.

A number of neutron stars show kHz QPOs in their light curves as revealed by *RXTE*. Kaaret et al. (1999) find 25 μs soft lags between the 4-6 keV and the > 9 keV photons in the 800 Hz QPO in the atoll source 4U1636-536. Analysing the 550 Hz oscillations of Aquila X-1, Ford et al. (1999) found soft lags, $\delta\phi \sim 1$ rad, between the 3-6 keV and the > 6 keV photons. Similar lags were found in the accreting millisecond pulsar SAX J1808.4-3658 (Cui, Morgan, & Titarchuk 1998; Ford 2000). The lags in the 830 Hz QPO in 4U 1608-52 (Vaughan et al. 1997, 1998) reach 60 μs between the 5 and the 25 keV photons.

3. PHENOMENOLOGICAL ENERGY DEPENDENT SHOT NOISE MODELS

It is clear that the observed zoo of time lags cannot be explained by any single model. Different mechanisms should be involved in producing the lags at different Fourier frequencies, the lags in the QPOs and the coherent pulsations, and the lags in the broad-band noise.

Let us first consider the simplest possible model that produces time lags: a shot noise model (Terrell 1972), where shots (=flares) are uncorrelated with each other. We assume that the shot time profiles at different energies have the same shape, but slightly different time constants. As an example we take a shot profile at soft energies $s(t) = [t/\tau]^p \exp[-t/\tau]$ and at hard energies $h(t) = [t/(\eta\tau)]^p \exp[-t/(\eta\tau)]$, where $t > 0$ is measured from the beginning of the shot and p is positive. The Fourier transforms, $S(f)$ and $H(f)$, are $S(f) \propto \tau\Gamma(p+1)/(1-i2\pi f\tau)^{p+1}$ and $H(f) \propto \eta\tau\Gamma(p+1)/(1-i2\pi f\eta\tau)^{p+1}$. The PDSs are $\propto |S(f)|^2$ and $|H(f)|^2$. For small frequencies, $f \ll 1/(2\pi\tau)$, the PDSs have a flat dependence on frequency, $\propto f^0$, while for large frequencies, $f \gg 1/(2\pi\tau)$, the PDSs decay as $f^{-2(p+1)}$. The power per logarithm of frequency (i.e., $f \times PDS(f)$) peaks for soft photons at $f_{s,\max} = 1/[2\pi\tau\sqrt{2p+1}]$ and at $f_{h,\max} = 1/[\eta 2\pi\tau\sqrt{2p+1}]$ for hard photons. The phase lags $\delta\phi(f) = (p+1)[\arctan(\eta 2\pi f\tau) - \arctan(2\pi f\tau)]$. The lag is positive when hard photons are lagging soft ones (i.e., for $\eta > 1$). For small frequencies, $\delta\phi(f)$ rises as $\approx (p+1)(\eta-1)2\pi f\tau$, while for large frequencies, it decays as $(p+1)(\eta-1)/(\eta 2\pi f\tau)$. The lag reaches a maximum of $\delta\phi_{\max} = 2(p+1)(\arctan\sqrt{\eta} - \pi/4)$ at $f = 1/(2\pi\tau\sqrt{\eta})$ close to $f_{s,\max}$ and $f_{h,\max}$, the frequencies where $f \times PDS_{s,h}(f)$ peak.

One can also consider a modified shot noise model, where the shot time scales are distributed according to a power law, $\rho(\tau) \propto \tau^{-p}$ between $\tau_{\min}$ and $\tau_{\max}$ (see, e.g., Miyamoto & Kitamoto 1989; Lochner, Swank, & Szymkowiak 1991), with the same ratio η. Physically this could correspond to, for example, the situation when flares of different durations appear at different radii from the central object (Poutanen & Fabian 1999a). A power-law distribution of τ assures that the PDS is also a power-law $\propto f^{-(3-p)}$ (Lochner et al. 1991). If the flares are self-similar, then the phase lag is constant $\approx \delta\phi_{\max}$ for $f \gg f_{\min} \equiv 1/(2\pi\tau_{\max})$ and $f \ll f_{\max} \equiv 1/(2\pi\tau_{\min})$, decays as $1/f$ at $f > f_{\max}$, and rises linearly at $f < f_{\min}$. The corresponding time lags are constant, $\delta t_{\max} = 2\pi\tau_{\max}\delta\phi_{\max}$, for $f < f_{\min}$, and decay approximately as $1/f$ between $f_{\min}$ and $f_{\max}$. Note that in this model the coherence function (Vaughan & Nowak 1997; Nowak et al. 1999a) is close to unity, since the light curves at different energies are almost perfectly synchronised.

If we assume that $\eta > 1$, there are hard lags and the predicted behaviour of the time lags and coherence function is in a very good agreement with the observations of GBHs (Poutanen & Fabian 1999a). However, this model contradicts the CCF of Cyg X-1 (see Fig. 1 and Maccarone et al. 2000). The CCF becomes narrower at larger energies which requires the shots to be narrower at larger energies. If one, however, reverses the time profiles of the shots, so that they rise slower and decay faster (e.g., $s(t) = (-t/\tau)^p \exp(t/\tau), t < 0$), and one assumes $\eta < 1$ (i.e., hard shots are narrower), the CCFs and the time lags can be reproduced, simultaneously.

Much more complicated models which also account for lags in the QPOs sources were developed by Shibazaki et al. (1988). We just note here that if a signal consists of shots appearing almost periodically and if shots at different energies are shifted in time one against another (or, e.g., the minima are reached at the same time and the peaks are not), the phase lag has a very complicated dependency on frequency (e.g., changes sign from one harmonic to another) depending on the shot profiles.

4. PHYSICAL MECHANISMS FOR PRODUCING LAGS

4.1. Static Compton Cloud Models

Since Comptonization is the most probable mechanism for X-ray production in compact objects, it is natural to attribute the time delays between hard and soft photons to this process. Hard photons are the result of more scattering and so emerge after, or lag behind, softer ones. Consider a static "Compton cloud" with fixed Thomson optical depth, τ_T, and electron temperature $\Theta = kT_e/m_e c^2$. A soft seed photon of energy E_0 injected into the cloud increases its energy by a factor of $A_1 = 1 + 4\Theta + 16\Theta^2$ on average after each scattering, so that after N-scatterings its energy $E_N = A_1^N E_0$. The photon mean free path is $\lambda \approx R/\max(1, \tau_T)$ (where R is the size of the X-ray producing region, and where we accounted for the fact that we are interested only in those photons that actually have undergone scatterings in the cloud). The time between successive scatterings is then $t_c = R/(c \max[1, \tau_T])$, so the time needed to reach the energy E_N is (Sunyaev & Titarchuk 1980; Payne 1980)

$$t_N = N t_c = \frac{R/c}{\max(1, \tau_T)} \frac{\ln(E_N/E_0)}{\ln A_1},$$

which translates to $t_N \sim 10^{-4}$ s for $kT_e \sim 50$ keV, $\tau_T \sim 1$, $R = 10$ km, and $E_N/E_0 \sim 10$.

This model was criticised by Miyamoto et al. (1988), Miyamoto et al. (1991) and Vaughan et al. (1994). First, the large size of the cloud ($10^3 - 10^5 R_g$, where R_g is the Schwarzschild radius, $2GM/c^2$) is needed to produce large delays observed in GBHs and neutron stars. Such cloud is physically unrealistic, since most of the gravitational energy is dissipated within $10 R_g$. Second, the lags predicted by the model are independent of the Fourier frequency (Miyamoto et al. 1988) contrary to the observed $\sim 1/f$ dependence. Third, it is assumed that the soft photons produce the variability, while the hot cloud is not variable. Observationally, it is well established that when a soft black body spectrum is observed in GBHs it is much *less* variable than the hard X-rays (e.g. Miyamoto et al. 1991), so that the hard X-ray variability is most probably intrinsic to the hot cloud itself.

Finally, we would like to point out that due to the requirement of energy conservation the whole concept of a static Compton cloud with a constant temperature is physically unrealistic. The total emitted X-ray luminosity (produced by Comptonization of soft radiation) is $L_{tot} = L_h + L_s$, where L_h is the heating rate in the hot cloud and L_s is the luminosity of seed soft photons. For hard spectra (i.e. $L_h \gg L_s$), $L_{tot} \approx L_h$. The total X-ray luminosity is thus a function of the heating rate *only* and it does not depend on the amount of seed soft photons. By changing L_s, one effectively changes the spectral slope of the emergent X-ray radiation which is a function of the Compton amplification factor $A \equiv L_h/L_s$. This results in the pivoting of the spectrum (see Poutanen 1998; Beloborodov 1999b) without a noticeable increase in L_{tot}. The larger the L_s, the smaller the equilibrium temperature of the emitting electrons, and the softer the spectrum. By contrast, in static

Compton cloud models no changes in the electron temperature are considered in reaction to the changes in the number of soft photons, violating thus the energy conservation law. The increase of the emitted luminosity requires changes in the energy dissipation rate, which will be then driving the variability.

Recently, Kazanas, Hua, & Titarchuk (1997) (see also Böttcher & Liang 1998; Hua, Kazanas, & Cui 1999) modified the Comptonization model. Instead of a homogeneous Compton cloud, the density profile, $n(r) \sim 1/r$, was assumed (in this case, one has equal optical depth per logarithm of radius). The variability is still driven by changing the rate of soft photon injection in the center of the cloud. Then larger radii produce lower frequency variability (filtering out high frequency signal) and larger lags, while the smaller radii produce higher frequency variability and smaller lags (see also Nowak et al. 1999a,c). This model solves only one of the problem mentioned above ($1/f$ time lag dependence), while the other problems remain unsolved. Furthermore, in this model the width of the ACF should increase with photon energy contrary to what is observed (Maccarone et al. 2000). Another modification of the model was considered by Böttcher & Liang (1999) based on an earlier suggestion by Miyamoto et al. (1988). Here, small cold clouds are assumed to free-fall into the hot central cloud thus changing the input of soft photons. Again, this model has problems with energy conservation unless one assumes that the drift of cold blobs in correlated with the energy dissipation in the hot cloud.

4.2. The Dynamic Compton Cloud

Miyamoto et al. (1988) pointed out that some modulation mechanism must be invoked to produce the strongly frequency-dependent time lags. Poutanen & Fabian (1999a,b) proposed a model where the time lags are produced by the evolution of the flare spectrum. They assumed that the energy dissipation varies in time. For a small (of the order of R_g) emitting region (ER) one can consider the spectral evolution as a sequence of steady-states as long as the characteristic time scale of variability is $\tau \gg R/c$. Any changes in the amplification factor A would cause spectral variability and, specifically, a continuous increase of A with time during the course of the flare would cause a soft-to-hard spectral evolution producing hard time lags of the order of the flare time scale (see § 3). Poutanen & Fabian (1999b) considered three mechanisms that can increase A.

(1) A flare starts in the background of soft photons. The X-ray spectrum is soft as long as $L_h < L_{s,\text{bkg}}$. With increasing L_h, the soft photon input gets dominated by reprocessed photons. The spectral slope is then determined by the feedback parameter $D \approx 1/A$ (the fraction of L_h returned to the ER after the reprocessing in the disc into soft seed photons, see Stern et al. 1995; Svensson 1996; Beloborodov 1999b) and the spectrum becomes hard.

(2) The dissipation is accompanied by pumping net momentum into the hot plasma of the emission region (Beloborodov 1999a,b). The resulting bulk velocity increases with increasing luminosity. It leads to a lower feedback and higher A (if the velocity is directed away from the disc).

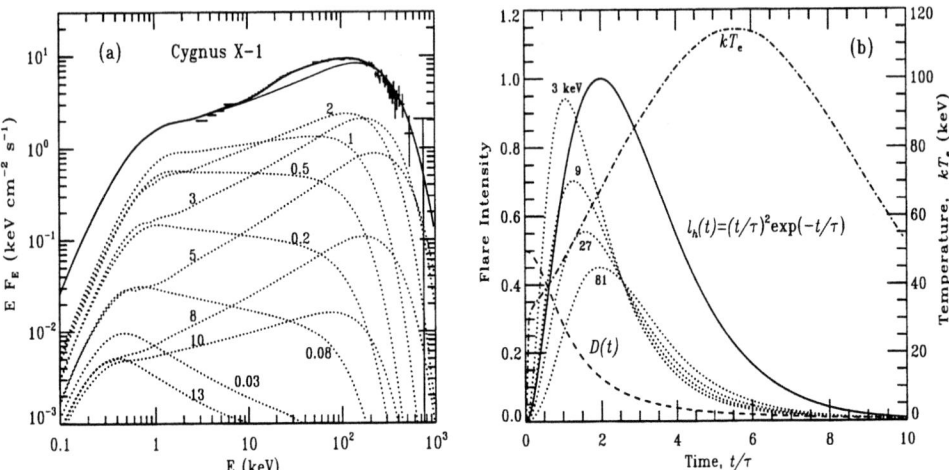

FIGURE 3. (a) Spectral evolution of a magnetic flare. Time resolved spectra (without Compton reflection) are presented by *dotted* curves; marks are times, t/τ, from the beginning of the flare. The time-averaged Comptonized spectrum is shown by a *thin solid* curve. The time-averaged spectrum of Cygnus X-1 (simultaneous *Ginga* and OSSE data from June 1991, the data set # 1 in Gierliński et al. 1997) is plotted with crosses and the best fit with the flare model with a *solid* curve (χ^2/dof= 50.0/75). Interstellar absorption is removed when plotting the model spectrum. (b) The flare light curves at 3, 9, 27, and 81 keV are presented by *dotted* curves. *Solid* curve - the heating rate, $l_h(t) \propto (t/\tau)^2 \exp(-t/\tau)$; *dashed* curve - the feedback factor $D(t)$; *dot-dashed* curve - the temperature of the emission region. See Poutanen & Fabian (1999a,b) for details.

(3) The differential rotation of the footpoints of a magnetic loop at the disc surface causes a twisting and elevation of the loop (Romanova et al. 1998). The time scale of the evolution is of the order of the Keplerian time-scale. When the emission region moves away from the disc, the feedback decreases and A increases (see Fig. 3).

In all these cases the spectral evolution proceeds from soft to hard. The *hard* time lags between energies E and E_0 are $\propto \tau \ln(E/E_0)$. If there is a distribution of time scales τ between, say, 1 ms and 0.3 s (e.g., Keplerian time scales at the radii between the innermost radius of the accretion disc and $\sim 50 R_g$), the time lags $\propto 1/f$ at the characteristic frequencies of the variability (see § 3 and Fig. 2).

If $\tau \sim$ a few light crossing time of the ER, the spectrum in the beginning of the flare is hard because of photon starvation (one needs a few R/c to get reprocessed soft photons into the ER) and softens towards the end of the flare (Poutanen & Fabian 1999a; Malzac & Jourdain 2000) producing *soft* time lags. Hard lags constrain the size of the ER that produces variability at a given frequency, f, $R \lesssim c/(10\pi f)$. For Cyg X-1 and $f = 20$ Hz, we get $R \lesssim 15 R_g$.

4.3. Small Scale Spectral Transitions

The models considered above can explain the time lags in the broad-band noise. In some QPO sources the time lags show a very complicated behaviour (see § 2) and may require different explanation. In radiation-hydrodynamic model (see, e.g., Lamb 1989; Miller & Lamb 1992), QPOs appear as a result of oscillations in the optical depth of the radial flow due to the radiation feedback from the neutron star surface. The resulting spectral pivoting produces phase lags and the increase of the rms amplitude variability above the pivoting point. This model, however, is not applicable to the QPOs and the lags in black holes sources (see, e.g., Takizawa et al. 1997) because of the absence of a hard surface.

The galactic microquasar GRS 1915+105 shows large amplitude oscillations with periods varying from <1 up to 100 s. These time scales are a few orders of magnitudes larger than Keplerian time scales and up to 10^6 time larger than the light crossing time of the ER. The best candidate for producing the spectral variability that causes the lags in GRS 1915+105 is the oscillation of the inner radius of the cold disc on viscous time scales. Such oscillations are similar to the spectral transitions, but have smaller amplitude and occur at shorter time scales. Changes of the relative geometry of the hot corona and the cold disc (with or without changes of the total luminosity) cause spectral pivoting at a few keV (see, e.g., Poutanen, Krolik, & Ryde 1997; Esin et al. 1998). The fluxes below and above the pivot point oscillate then with a phase shift of $\sim \pi$. The rms amplitude of the QPO increases with the energy. The phase lags between the energies above the pivot point can then be produced if the oscillations are time asymmetric (see Fig. 11 in Morgan, Remillard, & Greiner 1997; Vilhu & Nevalainen 1998).

Similar (but aperiodic) changes in the inner disc radius can be responsible for the broad-band variability observed at $f < 1$ Hz in, e.g., Cyg X-1. Associated spectral changes can manifest themselves in time lags observed at these frequencies.

4.4. Delays due to Compton Reflection and Reprocessing

The spectra of accreting GBHs and neutron stars show signatures of Compton reflection (see § 1). Some fraction of the X-ray photons can be reflected from the outer edge of a flared accretion disc, a wind from the companion, etc. Such a reflector acts as a low pass filter smearing out the high frequency variations and produces lags corresponding to the light travel time to the reflector only at lower frequencies. Such processes can explain the break in the time lag spectra observed in Cyg X-1 (see Fig. 2) and in other GBHs at $f \lesssim 1$ Hz.

Reprocessed soft radiation which accompanies Compton reflection is emitted in the optical and UV spectral bands if the reprocessing occurs far away from the central X-ray source. The time delays can then be measured between the optical/UV and the X-ray radiation (e.g., Hynes et al. 1998). On the other hand, reprocessing in the vicinity of the X-ray emitting region, produces time delays of the order of the time scale of the spectral evolution in the hard X-ray band. This may be one reason for the observed soft lags in the soft state of GX 339-4 and GS 1124-68 (§ 2.1).

4.5. Hot Spots on the Neutron Star Surface

Some neutron star sources show lags in their periodic oscillations at kHz frequencies. Ford (1999, 2000) interpreted the soft lags in Aquila X-1 and in the accreting millisecond pulsar SAX 1808.4-3658 using a model of a rotating hot spot with a black body spectrum at the surface of a neutron star where the lags appear due to Doppler effects. The weak energy dependence of the rms amplitude of the oscillations reported by Cui et al. (1998) rules out the black body model for the spectrum. Detailed analysis of the pulsations in the time domain by folding techniques by Revnivtsev (1999) revealed that the pulse profile is distorted at different energies, while the minima are reached at the same time (i.e., there are *no* lags in the normal meaning of this word).

5. SUMMARY

Time lags and other temporal variability data provide strong constraints on the models of the X-ray production. It was demonstrated that static Compton cloud models are based on physically unrealistic assumptions. The models invoking spectral evolution of the flare spectrum can fit both the CCF and the time lag Fourier spectra only if (1) the energy dissipation rate increases slowly and decreases rapidly and (2) the flare spectrum evolves from soft to hard. If soft seed photons are produced by reprocessing the hard ones, the change of sign in the time lag spectrum is expected at high frequencies corresponding to the light crossing time of the emission region. The absence of such a change constrains on the size of the emitting region.

We also argued that the reflection of hard X-rays from the outer part of the accretion disc produces time delays that we already might have observed in GBHs. If so, the disc should be flared and the break in the time lag Fourier spectra then corresponds to the size of the accretion disc. Of course, such an interpretation is not unique. Alternatively, small scale spectral transitions (e.g., oscillations of the inner radius of the accretion disc at viscous time scales) might produce time lags observed at lower frequencies.

In the case of (quasi-) periodic oscillations from the neutron star sources, we argued that in order to reproduce both the time lags and the energy dependent rms amplitude, the spectrum of the hot spots should not be close to a black-body.

ACKNOWLEDGEMENTS

This work was supported by the Swedish Natural Science Research Council and the Anna-Greta and Holger Crafoord Fund. I thank Katja Pottschmidt for providing the time lag Fourier spectra and the light curves of Cyg X-1 used in the calculations of the cross-correlation functions. I am grateful to Andrei Beloborodov and Roland Svensson for valuable comments.

REFERENCES

Balbus, S.A., Hawley, J.F. 1991, ApJ, 376, 214
Basko, M.M., Sunyaev, R.A., Titarchuk, L.G. 1974, A&A, 31, 249
Barret, D., Olive, J.F., Boirin, L., Done, C., Skinner, G.K., Grindlay, J.E. 2000, ApJ, 533, 329
Beloborodov, A.M. 1999a, ApJ, 510, L123
Beloborodov, A.M. 1999b, in High Energy Processes in Accreting Black Holes, ASP Conf. Series Vol. 161, ASP, San Francisco, p.295 (astro-ph/9901108)
Beloborodov, A.M. 1999c, in Gamma-Ray Bursts: The First Three Minutes, ASP Conf. Series Vol. 190, ASP, San Francisco, p.47 (astro-ph/9911122)
Böttcher, M., Liang, E.P 1998, ApJ, 506, 281
Böttcher, M., Liang, E.P. 1999, ApJ, 511, L37
Chandrasekhar, S. 1960, Proc. Natl. Acad. Sci. USA, 46, 253
Chiang, J. et al. 2000, ApJ, 528, 292
Crary, D.J. et al. 1998, ApJ, 493, L71
Cui, W., Zhang, S.N., Focke, W., Swank, J.H. 1997, ApJ, 484, 383
Cui, W., Morgan, E.H., Titarchuk, L.G. 1998, ApJ, 504, L27
Cui, W. 1999a, in High Energy Processes in Accreting Black Holes, ASP Conf. Series Vol. 161, ASP, San Francisco, 97 (astro-ph/9809408)
Cui, W. 1999b, ApJ, 524, L59
Cui, W., Zhang, S.N., Chen, W. 2000, ApJ, 531, L45
Done, C., Mulchaey, J.S., Mushotzky, R.F., Arnaud, K.A. 1992, ApJ, 395, 275
Edelson, R., Nandra, K. 1999, ApJ, 514, 682
Esin, A.A., Narayan, R., Cui, W., Grove, E.C., Zhang, S.-N. 1998, ApJ, 505, 854
Ford, E.C. 1999, ApJ, 519, L73
Ford, E.C. 2000, ApJ, 535, L119
Ford, E.C. et al. 1999, ApJ, 512, L31
Galeev, A.A., Rosner, R., Vaiana, G.S. 1979, ApJ, 229, 318
George, I.M., Fabian, A.C. 1991, MNRAS, 249, 352
Gierliński, M. et al. 1997, MNRAS, 288, 958
Gilfanov, M. et al. 1995, in The Lives of the Neutron Stars, NATO C 450. Kluwer Academic Publishers, Dordrecht, p.331
Gilfanov, M., Churazov, E., Revnivtsev, M. 1999, A&A, 352, 182
Grebenev, S.A. et al. 1993, ApJS, 97, 281
Grebenev, S.A., Sunyaev, R.A., Pavlinsky, M.N. 1997, Adv. Space Res., 19, (1)15
Grove, J.E. et al. 1997, in Proceedings of 4th Compton Symposium, AIP Conf. Proc. Vol. 410, AIP, New York, p.122
Grove, J.E. et al. 1998, ApJ, 502, L45
Haardt, F., Maraschi, L. 1993, ApJ, 413, 507
Haardt, F., Maraschi, L., Ghisellini, G. 1994, ApJ, 432, L95
Hasinger, G. 1987, in The Origin and Evolution of Neutron Stars, IAU Symp. 125, D. Reidel Publ. Co., Dordrecht, p.333
Hua, X.-M., Kazanas, D., Cui, W. 1999, ApJ, 512, 793
Hynes R.I., O'Brien, K., Horne, K., Chen, W., Haswell, C.A. 1998, MNRAS, 299, L37

Ichimaru, S. 1977, ApJ, 214, 840
Johnson, W.N. et al. 1997, in Proceedings of 4th Compton Symposium, AIP Conf.Proc. Vol. 410, AIP, New York, p.283
Kaaret, P., Piraino, S., Ford, E.C., Santangelo, A. 1999, ApJ, 514, L31
Kazanas, D., Hua, X.-M., Titarchuk, L. 1997, ApJ, 480, 735
Lamb, F.K. 1989, in Proc. 23rd ESLAB Symp. on Two-Topics in X-ray Astronomy, ESA SP-296, p.215
Lee, J.C. et al. 2000, MNRAS, in press (astro-ph/9909239)
Lewin, W.H.G., van Paradijs, J., van der Klis, M. 1988, Space Sci. Rev., 46, 273
Lochner, J.C., Swank, J.H., Szymkowiak, A.E. 1991, ApJ, 376, 295
Maccarone, T., Coppi, P.S., Poutanen, J. 2000, ApJ, 537, L107
Magdziarz, P., Zdziarski, A.A. 1995, MNRAS, 273, 837
Malzac, J., Jourdain, E. 2000, A&A, 359, 843
Miller, G.S., Lamb, F.K. 1992, ApJ, 388, 541
Miller, K.A., Stone, J.M. 2000, ApJ, 534, 398
Mitsuda, K., Dotani, T. 1989, PASJ, 41, 557
Miyamoto, S., Kitamoto, S. 1989, Nature, 342, 773
Miyamoto, S., Iga, S., Kitamoto, S., Kamado, Y. 1993, ApJ, 403, L39
Miyamoto, S., Kitamoto, S., Mitsuda, K., Dotani, T. 1988, Nature, 336, 450
Miyamoto, S. et al. 1991, ApJ, 383, 784
Miyamoto, S. et al. 1992, ApJ, 391, L21
Morgan, E.H., Remillard, R.A., Greiner, J. 1997, ApJ, 482, 993
Mushotzky, R.F., Done, C., Pounds, K.A. 1993, Ann. Rev. Astron. Astrophys., 31, 717
Nandra, K., Pounds, K.A. 1994, MNRAS, 268, 405
Narayan, R., Mahadevan, R., Quataert, E. 1998, in Theory of Black Hole Accretion Discs, Cambridge Univ. Press, Cambridge, p.148
Nolan, P.L. et al. 1981, ApJ, 246, 494
Nowak, M.A. et al. 1999a, ApJ, 510, 874
Nowak, M.A., Wilms, J., Dove, J.B. 1999b, ApJ, 517, 355
Nowak, M.A. et al. 1999c, ApJ, 515, 726
Olive, J.-F., Barret, D. 2000, these proceedings (astro-ph/0007325)
Page, C.G. 1985, Space Sci. Rev., 40, 387
Papadakis, I.E., Lawrence, A. 1995, MNRAS, 272, 161
Payne, D.G. 1980, ApJ, 237, 951
Pounds, K.A. et al. 1990, Nature, 344, 132
Poutanen, J. 1998, in Theory of Black Hole Accretion Discs, Cambridge Univ. Press, p.100
Poutanen, J., Fabian, A.C. 1999a, MNRAS, 306, L31
Poutanen, J., Fabian, A.C. 1999b, in High Energy Processes in Accreting Black Holes, ASP Conf. Series Vol. 161, ASP, San Francisco, p.135
Poutanen, J., Svensson, R. 1996, ApJ, 470, 249
Poutanen, J., Krolik, J.H., Ryde, F. 1997, MNRAS, 292, L21
Poutanen, J., Nagendra, K.N., Svensson, R. 1996, MNRAS, 283, 892
Priedhorsky, W. et al. 1979, ApJ, 233, 350
Psaltis, D., Belloni, T., van der Klis, M. 1999, ApJ, 520, 262

Reig, P. et al. 2000, ApJ, in press (astro-ph/0001134)
Revnivtsev, M. 1999, PhD thesis, Space Research Institute, Moscow (astro-ph/9912556)
Revnivtsev, M., Borozdin, K., Priedhorsky, W.C., Vikhlinin, A. 2000, ApJ, 530, 955
Romanova, M.M. et al. 1998, ApJ, 500, 703
Shapiro, S.L., Lightman, A.P., Eardley, D.N. 1976, ApJ, 204, 187
Shibazaki, N. et al. 1988, ApJ, 331, 247
Smith, D.M. et al. 1997, ApJ, 489, L51
Smith, I.A., Liang, E.P. 1999, ApJ, 519, 771
Stern, B.E., Poutanen, J., Svensson, R., Sikora, M., Begelman, M.C. 1995, ApJ, 449, L13
Stern, B.E., Svensson, R. 1996, ApJ, 469, L109
Sunyaev, R.A., Titarchuk, L.G. 1980, A&A, 86, 121
Sunyaev, R.A., Trümper, J. 1979, Nature, 279, 506
Svensson, R. 1996, A&AS, 120C, 475
Takizawa, M. et al. 1997, ApJ, 489, 272
Tanaka, Y., Lewin, W.H.G. 1995, in X-ray binaries, Cambridge Astrophysics Series, vol. 26, Cambridge University Press, Cambridge, p.126
Terrell, N. J. Jr. 1972, ApJ, 174, L35
van der Hooft, F. et al. 1999a, ApJ, 513, 477
van der Hooft, F. et al. 1999b, ApJ, 519, 332
van der Klis, M. et al. 1987, ApJ, 319, L13
van der Klis, M. 1989, in Timing Neutron Stars, NATO ASI C 262. Kluwer, 27
van der Klis, M. 1995a, in The Lives of the Neutron Stars, NATO ASI C 450. Kluwer, 301
van der Klis, M. 1995b, in X-ray binaries, Cambridge Astrophysics Series, vol. 26, Cambridge University Press, Cambridge, p.252
Vaughan, B.A., Nowak, M.A. 1997, ApJ, 474, L43
Vaughan, B. et al. 1994, ApJ, 421, 738
Vaughan, B. et al. 1997, ApJ, 483, L115 (erratum 1998, ApJ, 509, L145)
Vaughan, B.A. et al. 1999, A&A, 343, 197
Velikhov, E.P. 1959, Sov. Phys. JETP, 36, 995
Vilhu, O., Nevalainen, J. 1998, ApJ, 508, L85
Wardzinski, G., Zdziarski, A.A. 2000, MNRAS, 314, 183
Wijnands, R., van der Klis, M. 1999, ApJ, 514, 939
Wijnands, R., Homan, E., van der Klis, M. 1999, ApJ, 526, L33
Yoshida, K. et al. 1993, PASJ, 45, 605
Zdziarski, A.A. 1998, MNRAS, 296, L51
Zdziarski, A.A., Johnson, W.N., Poutanen, J., Magdziarz, P., Gierliński, M. 1997, in The Transparent Universe, ESA SP-382, p.373
Zdziarski, A.A. 1999, in High Energy Processes in Accreting Black Holes, ASP Conf. Series Vol. 161, p.16
Zdziarski, A.A., Lubiński, P., Smith, D.A. 1999, MNRAS, 303, L11

LOW-LUMINOSITY AGN AND NORMAL GALAXIES

A. Ptak

Carnegie Mellon University, Dept. of Physics, 5000 Forbes Ave., Pittsburgh, PA 15213

ABSTRACT Low-luminosity AGN (with $L_X < 1 \times 10^{42}$ ergs s^{-1}) far outnumber ordinary AGN, and are therefore perhaps more relevant to our understanding of AGN phenomena and the relationship between AGN and host galaxies. Many normal galaxies harbor LINER and starburst nuclei, which, together with LLAGN, are a class of "low-activity" galaxies that have a number of surprisingly similar X-ray characteristics, despite their heterogenous optical classification. This strongly supports the hypothesis of an AGN-starburst connection. Further, X-ray observations of normal galaxies without starburst or AGN-like activity in their nuclei offer opportunities to study populations of X-ray binaries, HII regions, and warm or hot ISM under different conditions than is often the case in the Milky Way. The results of recent X-ray observations of these types of galaxies are reviewed, and what we hope to learn about both nearby and high redshift galaxies of each type from observations with forthcoming and planned satellites is discussed.

KEYWORDS: galaxies; abundances; galaxies:active; galaxies:starburst

1. INTRODUCTION

Since active galactic nuclei (AGN) comprise only a small fraction of all galaxies, it is more relevant for our general understanding of galactic processes to observe normal galaxies. With the availability of *imaged* X-ray observations, galaxies not necessarily dominated by a point source in their nucleus have now been studied in detail. This review will give an overview of these results, with an emphasis on galaxies exhibiting moderate or low amounts of activity in their nuclei.

In this paper, a low-luminosity AGN (LLAGN) is considered to be a galaxy with Seyfert-like optical spectra and an X-ray luminosity less than 10^{42} ergs s^{-1}. A starburst galaxy is a galaxy in which evidence of enhanced star-formation rates are observed, particularly in kpc-sized nuclear regions. Most often this is observed by the presence of HII-region like optical emission lines, however other signs include high IR luminosities (i.e., $L_{IR} > 10^{10} L_\odot$; c.f., Telesco 1988). In LINER galaxies the optical emission lines are observed with line diagnostic ratios that differ from both starburst and LLAGN ratios. The most likely scenarios for the ionization flux in LINERs are a LLAGN (albeit with different physical properties than LLAGN with Seyfert-like spectra), shocks (Dopita et al. 1996) and hot stars (c.f., Shields 1992). Interestingly, 15% of LINERs exhibit *broad* $H\alpha$ emission (Ho, Filippenko, Sargent, & Chen 1997). These "LINER 1" galaxies are almost certainly LLAGN, and the

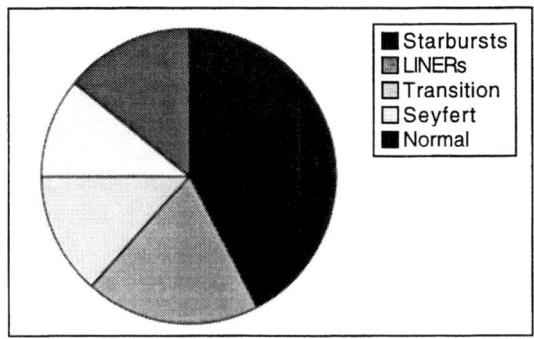

FIGURE 1. Demographics of activity in nearby galaxies.

fact that the ratio of LINER 1 to LINER 2 galaxies is similar to the fraction of Seyfert 1/Seyfert 2 galaxies is suggestive that most LINERs are indeed LLAGN.

1.1. Demographics

In order to properly access the prevalence of LLAGN, LINER and starburst activity, it is necessary to perform a careful survey of nearby galaxies in which the galactic light is subtracted from the nuclear spectrum in order for the often-weak optical emissions line to be detected. Such as survey was recently completed by Ho, Filippenko & Sargent (1997). The demographics of these types of activity is shown in Figure 1. Clearly, "low-activity" galaxies (starburst, LINER, and LLAGN, and transition nuclei containing both LLAGN and starburst emission), 86% of the total, dominate over "normal" galaxies. Furthermore, "normal" AGN only comprise $\sim$ 10% of all galaxies (Ho, Filippenko, & Sargent 1997). This emphasizes the importance of studying these types of galaxies: the usual state of affairs is for a galaxy to exhibit starburst or LLAGN activity (or both) in its nucleus, and most AGN in the local universe are in a low-luminosity state.

2. NORMAL GALAXIES

In elliptical/early-type galaxies the X-ray emission is dominated by hot gas (in some cases similar to group or cluster of galaxies X-ray emission, with a King-like radial surface brightness profile, occasionally including a cooling flow), with a temperature of $0.8 - 1.0$ keV (Trinchieri, Fabbiano, & Kim 1998). Some early-type galaxies also contain a contribution from a LLAGN (see Di Matteo, et al.; these proceedings) and low-mass X-ray binaries (LMXB, Irwin & Bregman, 1998).

The X-ray emission of spiral galaxies is generally dominated by X-ray binaries (Fabbiano 1989). Spiral arms often contain HII regions, particularly regions of local density enhancements (i.e., knots). These HII regions are often X-ray bright as a result of the products of enhanced star formation: hot stars, supernovae (SN) and

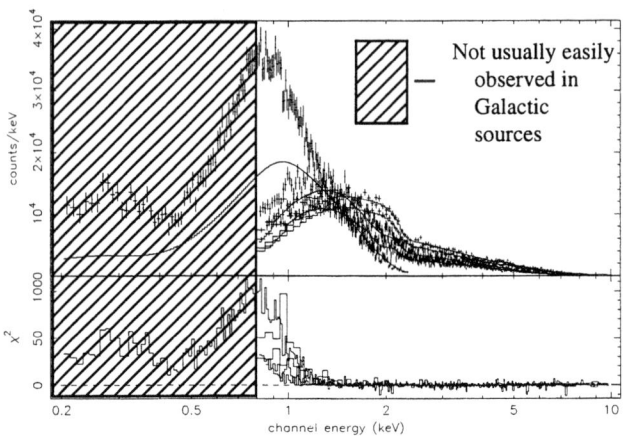

FIGURE 2. ASCA + PSPC spectra from the bulge of M31. Figure taken from Irwin & Bregman (1999), with a shaded area added to delineate the region typically absorbed in Galactic X-ray sources.

supernova remnants (SNR), high-mass X-ray binaries and black hole candidates (BHC) (c.f. NGC 1313 in Colbert et al. 1997). Not surprisingly, "bluer" galaxies, which tend to be later-type galaxies with higher star-formation rate (with the blue colors being a result of a higher proportion of massive stars than is observed in galaxies dominated by older stellar populations) are X-ray bright (Fabbiano, Feigelson, & Zamorani 1982). There is also a strong correlation between X-ray luminosity and IR luminosity, with the IR emission being produced by dust that has been heated by massive stars (David, Jones & Foreman 1992; Green, Anderson & Ward 1992). In cases of very high star formation rates occurring near the edge of galaxies, blow-outs can occur when the local pressure resulting from outflows from hot stars and SN exceeds the ambient interstellar medium (ISM) pressure, as observed by ROSAT in galaxies such as NGC 55 (Schlegel et al. 1997). There is also X-ray evidence that normal galaxies (Cui et al. 1996), including the Milky Way, possess hot ($T \sim 10^6$K) "halos" or "coronae" (Spitzer 1956).

Spiral galaxies also typically contain a bulge component, which may be counterparts to elliptical galaxies. A substantial fraction of the X-ray sources found in the nearest spiral galaxy, M31, are concentrated in the bulge. These sources are presumably mostly LMXB associated with the older stellar populations in globular clusters. The X-ray spectrum of the bulge region of M31 as observed by the ROSAT PSPC and ASCA (Irwin & Bregman 1999) and BeppoSAX (Trinchieri et al. 1999). This spectrum has a power-law component consistent with X-ray binaries in our galaxies, plus a soft component (see Figure 2) that is similar to that considered to be due to X-ray binaries in early-type galaxies. Note that it would be difficult to observe this soft component in Galactic binaries since most lie in the disk of the Milky Way and hence are highly absorbed.

Colbert and Mushotzky (1999) discuss a survey of X-ray sources in normal

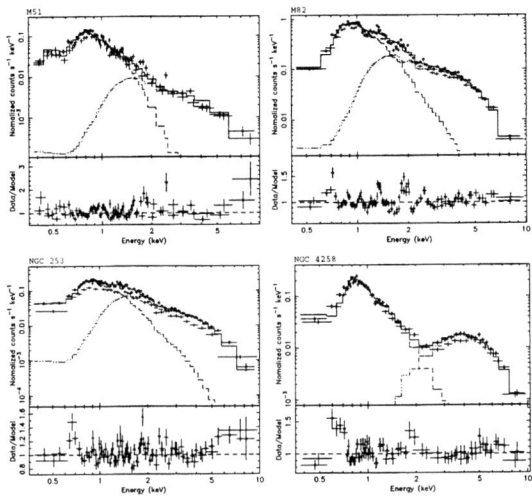

FIGURE 3. ASCA SIS spectra of M51 (starburst + LLAGN), M82 (starburst), NGC 253 (starburst) and NGC 4258 (LLAGN).

galaxies with luminosities in excess of 1.3×10^{38} ergs s^{-1}, the Eddington luminosity for accretion onto a solar-mass object. These intermediate-luminosity X-ray objects (IXO) have X-ray spectra consistent with accretion onto black holes with masses of $\sim 10^{1-4} M_\odot$. IXOs are typically significantly displaced from the galactic centers, and hence they are not AGN. This implies that IXOs may be cases of "intermediate" mass black holes, and may be precursors to some modern-day AGN (see also M82 below and in Ptak & Griffiths 1999).

3. LOW-ACTIVITY GALAXIES

3.1. Spectral Properties

Despite their heterogeneous optical classification, low-activity galaxies usually have a similar spectral shape (Serlemitsos, Ptak & Yaqoob 1996). Specifically, in general these galaxies exhibit at least two spectral components: a soft, thermal component with kT ~ 0.7 keV and a hard component well-modeled by either a thermal bremsstrahlung with kT $\sim 5 - 10$ keV or a power-law with $\Gamma \sim 1.8$ (Ptak et al. 1999). The hard (soft) component is typically absorbed by a column density of $\sim 10^{22}$ cm^{-2} ($\sim 10^{20-21}$ cm^{-2}). The fact that the hard component tends to be more spatially compact (see below) and absorbed than the soft component implies that the hard component is emanating from further within the galaxies (i.e., the nuclei). The fact that starburst galaxies exhibit a hard component, likely due accreting sources, and LINER and LLAGN galaxies exhibit soft emission, likely due to starburst activity, strongly supports the idea of a starburst/AGN connection.

The luminosity of the hard component tends to be on the order of 10^{40-41} ergs s^{-1} and the luminosity of the soft component 10^{39-40} ergs s^{-1}, with the relative intensity varying from galaxy to galaxy (see Figure 3). It is evidently rare for starburst activity, the origin of the soft component, to achieve luminosities in excess of 10^{40-41} ergs s^{-1} (Halpern, Helfand, & Moran 1995; however see Moran, Lehnert & Helfand 1999 for a counter-example in NGC 3256). Accordingly, sources with X-ray luminosities $> 10^{41}$ ergs s^{-1} (e.g., NGC 3998 and NGC 3147) only require a power-law component since the starburst component is overwhelmed. Conversely, in Seyfert 2s where the AGN is highly absorbed, starburst emission is often observed below 2 keV (see Turner et al. 1998).

The abundances inferred from the soft component tend to be sub-solar (on the order of 10^{-1} solar). This is surprising since starburst emission is presumably the result of massive star evolution and accordingly should be highly enriched, however many effects might be contributing to this. For example, it is probably too simple to be fitting the multi-temperature starburst emission with a single component (c.f., Breitschwerdt & Komossa 1999; Dahlem, Weaver & Heckman 1998), or other sources of continuum may be present such as soft emission from X-ray binaries (c.f., Figure 2). In brighter sources (see the residuals in Figure 3 and Ptak et al. 1997), it appears that there is a deficiency of Fe relative to α-process elements (e.g., Ne, Mg, Si, S, etc. produced in massive stars), although the effect is diminished somewhat when more complex models are invoked (see Dahlem, Weaver, & Heckman 1999).

3.2. Fe-K Emission

Fe-K emission is an important diagnostic in AGN studies since its energy, physical width, and equivalent width (EW) are functions of the physical conditions in the accretion region. In the case of an obscured nuclear region, the EW of an Fe line is expected to be greatly enhanced since the direct continuum is diminished. Although most low-activity galaxies are too faint for Fe-K to detected, in several cases Fe-K *is* detected and is often complex. For example, the Fe-K EW in NGC 3147 (Ptak et al. 1996), NGC 1365 (Iyomoto et al. 1997), M51 (Terashima et al. 1998a), NGC 4736 (Roberts, Warwick, & Ohashi 1999) and NGC 1052 (Weaver et al. 1999) are high (> 100 eV) relative to Seyfert 1 EW values ($\sim 100 - 200$ eV; Nandra et al. 1997), consistent with an obscured nucleus. In M81, the Fe-K line appears to contain several components or may be broad, possibly due to an accretion disk (Ishisaki et al 1996; Serlemitsos, Ptak & Yaqoob, 1996). NGC 4579 (Terashima et al. 1998b; Terashima et al. 2000ab) is a particularly interesting case where the line was observed to be due to *ionized* material in a 1995 ASCA observation, but was observed to be due to neutral material in 1998.

(Ionized) Fe-K emission was detected marginally by ASCA in M82 (Ptak et al. 1997) and in M82 and NGC 253 significantly by BeppoSAX (Cappi et al. 1999). This emission stongly suggests that very hot gas (T $\sim 10^8$ K) is present in these starburst galaxies. However, the EW of the Fe-K lines is only on the order of $100 - 200$ eV, considerably less than that expected from solar-abundance hot gas (EW ~ 600 eV).

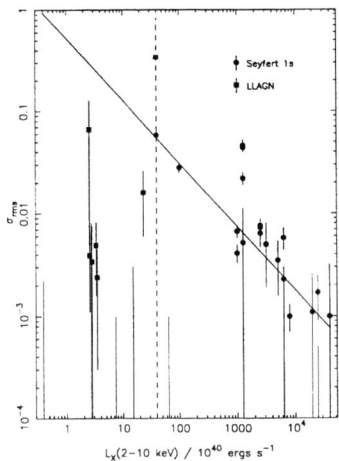

FIGURE 4. The trend of "excess variance", a measure of short-term variability, with X-ray luminosity in Seyfert 1s and low-activity galaxies, from Ptak et al. (1998).

Again, the Fe abundance may be depressed in these galaxies, however it is also likely that other sources of continuum are present, diluting the thermal Fe-K emission.

3.3. Temporal Properties

Low-activity galaxies tend to vary on long (months to year) time scales (c.f., M81 in Ishisaki et al. 1996; NGC 4579 in Serlemitsos, Ptak & Yaqoob 1996), but not on short time scales as observed in Seyferts (Ptak et al. 1998; however note that M81 has been observed to vary at the 30% level on day time scales; Pellegrini et al. 2000). Suprisingly, some of the most variable low-activity galaxies have been starbursts. For example, the nuclear source in the starburst NGC 3628 "shut off", varying by a factor of ~ 40 (Dahlem, Heckman, & Fabbiano 1995), and M82 has varied considerably in the 2-10 keV bandpass (Ptak & Griffiths 1999; Matsumoto & Tsusru 1999; Gruber et al., these proceedings). This implies that at least some of the contribution to the hard component in starburst galaxies is due to accreting sources. The lack of variability on short times scales is demonstrated in Figure 4. As argued in Ptak et al. (1998), this marked break from the temporal behavior of Seyfert 1s implies a large source extent for the X-ray producing regions in low-activity galaxies. While in some cases this might be due to a multiple sources of X-ray emission, it may also be due to the prevelance of advection-dominated accretion disks, in which the entire disk contributes to the X-ray emission, as opposed to the "α"-disks in Seyferts in which the X-rays are most likely produced by flares. On the other hand, the short-term variability observed in M82 may be our first look at the hard X-ray light curve of an IXO, assuming that the source of the 2-10 keV variability is the off-nuclear point source observed by the ROSAT HRI (Collura et al. 1994).

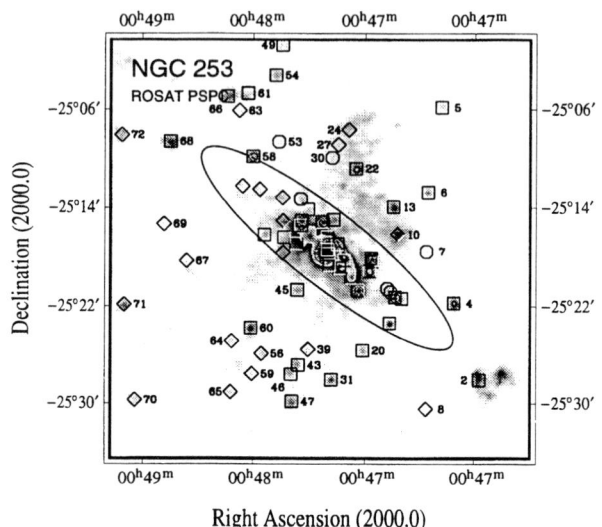

FIGURE 5. ROSAT PSPC image of NGC 253, from Vogler & Pietsch (1999).

3.4. Spatial Properties

Spatially most low-activity galaxies have 2-10 keV emission which is concentrated in the nucleus (see Ptak 1997, although the statistics are limited in many cases) and are extended over kpc scales in addition to being concentrated in multiple point sources. This tends to be particularly true of starburst galaxies (c.f., Read, Ponman, & Strickland 1997; Dahlem, Weaver, & Heckman 1998) but these phenomena are also seen in LINERs and LLAGN (c.f., the LINER NGC 4594 in Fabbiano & Juda 1997; the LINER NGC 3079 in Pietsch, Trinchieri, & Vogler 1998; the LLAGN M51 in Marston et al. 1995, and the LLAGN NGC 4258 in Cecil, Wilson, & De Pree 1998). In one of the nearest starburst galaxies NGC 253 (at $\sim$ 2.5 Mpc), $\sim$ 73 point sources (some of which are background QSOs) have been detectedby ROSAT (Vogler & Pietsch 1999; see Figure 5), in addition to the extended emission associated with the nuclear region, disk, and halo of NGC 253. Interestingly, the point source distribution in NGC 253 is consistent with that of M31 (with the older-population buldge sources removed) and M33 (Figure 6). A similar result is observed more generally by Roberts et al. (these proceedings), which provides motivation that a *universal* luminosity distribution of X-ray binaries can be found, and the high-luminosity end of which would be the IXOs.

4. CONCLUSIONS AND FUTURE PROSPECTS

Both "normal" galaxies and low-activities provide very rich data sets for studying phenomena that is difficult to study in the Milky Way (i.e., due to extinction) and/or is not present in the Milky Way (i.e., starburst regions with very high star

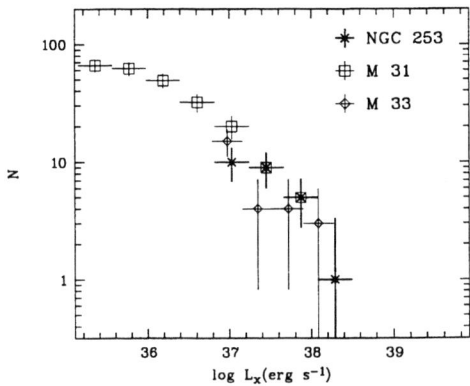

FIGURE 6. The luminosity distribution of point sources in NGC 253, M33 and M31 (excluding buldge sources), from Vogler & Pietsch (1999).

formation rates resulting in superwinds). The X-ray emission occurs in both point sources, which tend to be the most luminous known X-ray binaries and supernovae, and in complex diffuse emission that is the result of heating of the ISM by hot star, supernovae and, in some cases, superwind outflows. The brightest X-ray binaries (IXOs) are black hole candidates that potentially have masses on the order of 10^{1-4} $M_\odot$, intermediate to that of Galactic BHC and AGN. Both starburst and accretion emission tend to be observed whenever either type of activity is present, strongly supporting the notion of a starburst/AGN connection. There appears to be a natural upper-limit to the luminosity of starburst processes on the order of 10^{41} ergs s^{-1}, and so starburst emission is usually overwhelmed by AGN emission that is not absorbed and exceeds $L_X \sim 10^{42}$ ergs s^{-1}. Observed abundances tend to be absurdly low, but that is almost certainly due to "contamination" of the continuum from multiple temperature gas emission and unresolved point sources. Chandra and XMM will be able to resolve the starburst emission spatial and extract CCD resolution spectra, which will resolve this issue and allow abundance and temperature enhancements to be observed in individual regions within the galaxies.

The X-ray properties of nearby normal and low-activity galaxies suggest that they will contribute less $\sim 10\%$ of the X-ray background (c.f., Griffiths & Padovani 1991; Yaqoob et al. 1995). This fraction can increase if these galaxies become harder and more luminous at earlier epochs, i.e., due to the enhanced starburst activity associated with the peak in the star formation history of the universe at redshifts of $\sim 1-2$ (Hughes et al. 1998). It may be possible to directly observe evolution in low-activity galaxies with ultra-deep XMM surveys or with telescopes such as XEUS that promise effective areas on the order of m^2 (see Figure 7).

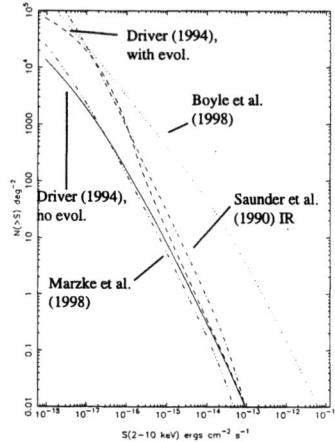

FIGURE 7. The expected logN-logS distributions in the 2-10 keV bandpass obtained by converting the optical luminosity functions of Driver (1994) and the IR luminosity function of Saunders et al. (1990) to the X-ray bandpass using the optical and IR to X-ray correlations of David, Jones & Forman (1992) and Green, Anderson, & Ward (1992), respectively. Note that for the mean power-law spectral slope of low-activity galaxies described in Ptak et al. (1999) of $\Gamma \sim 1.7 - 1.8$, $F_{2-10 \text{ keV}} = F_{0.5-4.0 \text{ keV}}$.

REFERENCES

Breitschwerdt, D. & Komossa, S. 1999, Ap&SS, in press
Cappi, M. et al. 1999, A&A, 350, 777
Cecil, G., Wilson, A., & De Pree, C. 1995, ApJ, 440, 181
Colbert, E. et al. 1995, ApJ, 446, 177
Colbert, E., & Mushotzky, R. 1999, ApJ, 519, 89
Collura, A., Schulman, E., Reale, F., & Bregman, J. 1994, ApJ, 420, L63
Cui, W., Sanders, W., McCammon, D., Snowden, S., & Womble, D. 1996, ApJ, 468, 102
David, L., Jones, C., Forman, W. 1992, ApJ, 388, 82
Dahlem, M., Heckman, T., & Fabbiano, G. 1995, ApJ, 442, 49L
Dahlem, M., Weaver, K., & Heckman, T. 1998, ApJS, 118, 401
Dopita, M., Koratkar, A., Evans, I., Allen, M., Bicknell, G., Sutherland, R., Hawley, J., & Sadler, E. 1996, in "The Physics of LINERS", ASP Conference Series, Vol. 103, ed. M. Eracleous, A. Koratkar, C. Leitherer, and L. Ho, p.44
Driver, S. 1994, Ph.D. thesis, University of Wales
Fabbiano, G., Feigelson, E., Zamorani, G. 1982, ApJ, 256, 397
Fabbiano, G. 1989, ARA&A, 27, 87
Fabbiano, G. & Juda, J. 1997, ApJ, 476, 666
Green, P., Anderson, S. & Ward, M. 1992, MNRAS, 254, 30
Griffiths, R. & Padovani, P. 1990, ApJ, 360, 483

Halpern, J., Helfand, D., & Moran, E. 1995, ApJ, 453, 61
Ho, L., Filippenko, A., Sargent, W. 1997, ApJ, 487, 568
Ho, L., Filippenko, A., Sargent, W., Peng, C. 1997, ApJS, 112, 391
Hughes, D. et al. 1998, Nature, 395, 47
Irwin, J. & Bregman, J. 1999, ApJ, 527, 125
Ishisaki, Y. et al. 1996, PASJ, 48, 237
Iyomoto, N., Makishima, K., Fukazawa, Y., Tashiro, M., & Ishisaki, Y. 1997, PASJ, 49, 425
Marston, A. et al. 1995, ApJ, 438, 663
Marzke, R., Da Costa, L., Pellegrini, P., Willmer, C., & Geller, M. 1998, ApJ, 503, 617
Matsumoto, H. & Tsuru, T. 1999, PASJ, 51, 321
Moran, Lehnert, & Helfand 1999, ApJ, 526, 649
Nandra, K., George, I., Mushotzky, R., Turner, T. & Yaqoob, T. 1997, ApJ, 476, 70
Pellegrini, S. et al. 2000, A&A, 353, 447
Pietsch, W., Trinchieri, G., & Vogler, A. 1998, A&A, 340, 351
Ptak, A., Yaqoob, T., Serlemitsos, P., Kunieda, H. & Terashima, Y. 1996, ApJ, 459, 542
Ptak, A., Serlemitsos, P., Yaqoob, T., & Mushotzky, R. 1997, AJ, 113, 1286
Ptak, A. 1997, Ph.D. Thesis, The University of Maryland
Ptak, A., Yaqoob, T., Mushotzky, R., Serlemitsos, P. & Griffiths, R. 1998, ApJ, 501, L37
Ptak, A., Serlemitsos, P., Yaqoob, T., & Mushotzky, R. 1999, ApJS, 120, 179
Ptak, A. & Griffiths, R. 1999, ApJ, 517, 85L
Read, A., Ponman, T. & Strickland, D. 1997, MNRAS, 286, 626
Saunders, W. et al. 1990, MNRAS, 242, 318
Schlegel, E., Barrett, P., & Singh, K. 1997, AJ, 113, 1296
Serlemitsos, P., Ptak, A., & Yaqoob 1996, in "The Physics of LINERS", ASP Conference Series, Vol. 103, ed. M. Eracleous, A. Koratkar, C. Leitherer, and L. Ho, p.70
Shields, J. 1992, ApJ, 399, 27L
Spitzer, L. 1956, ApJ, 124, 20
Telesco, C. 1988, ARA&A, 26, 343
Terashima, Y. et al. 1998a, ApJ, 496, 210
Terashima, Y. et al. 1998b, ApJ, 503, 212
Terashima, Y. et al. 2000, ApJ, 535, 79L
Trinchieri, G. et al. 1999, A&A, 348, 43
Trinchieri, G., Fabbiano, G, & Kim, D. 1997, A&A, 318, 361
Vogler, & Pietsch, W. 1999, A&A, 342, 101
Weaver, K., Wilson, A., Henkel, C. & Braatz, J. 1999, ApJ, 520, 130
Weaver, K., Heckman, H. & Dahlem, M. 2000, ApJ, 534, 684
Yaqoob, T. et al. 1995, ApJ, 455, 508

THE USA X-RAY TIMING EXPERIMENT

P. S. Ray[1], K.S. Wood[1], G. Fritz[1], P. Hertz[1], M. Kowalski[1], W.N. Johnson[1], M.N. Lovellette[1], M.T. Wolff[1], D. Yentis[1], R. M. Bandyopadhyay[2], E.D. Bloom[3], B. Giebels[3], G.Godfrey[3], K. Reilly[3,4], P. Saz Parkinson[3,4], G. Shabad[3,4], P. Michelson[4], M. Roberts[4], D.A. Leahy[5], L. Cominsky[6], J. Scargle[7], J. Beall[8], D. Chakrabarty[9], Y. Kim[10]

1) E. O. Hulburt Center for Space Research, Naval Research Laboratory, 2) NRC Research Associate, 3) Stanford Linear Accelerator Center, 4) Stanford University, 5) University of Calgary, 6) Sonoma State University, 7) NASA Ames, 8) Saint John's College, 9) MIT, 10) Saddleback College

ABSTRACT   The USA Experiment is a new X-ray timing experiment with large collecting area and microsecond time resolution capable of conducting a broad program of studies of galactic X-ray binaries. USA is one of nine experiments aboard the Advanced Research and Global Observation Satellite which was launched February 23, 1999. USA is a collimated proportional counter X-ray telescope with about 1000 cm^2 of effective area per detector with two detectors sensitive to photons in the energy range 1-15 keV. A unique feature of USA is that photon events are time tagged by reference to an onboard GPS receiver allowing precise absolute time and location determination. We will present an overview of the USA instrument, capabilities, and scientific observing plan as well as the current status of the instrument.

1. INTRODUCTION

The Unconventional Stellar Aspect (USA) Experiment is a low-cost X-ray timing experiment with the dual purpose of timing X-ray binary systems and exploration of applications of X-ray sensor technology. USA was launched on February 23, 1999 on the Advanced Research and Global Observation Satellite (ARGOS). It is a reflight of two proportional counter X-ray detectors that performed excellently on the NASA Spartan-1 mission (Kowalski *et al.* 1993). The primary targets are bright Galactic X-ray binaries that are used simultaneously for both scientific and applied objectives. X-ray photon event times are measured to high precision using the GPS receiver on ARGOS. USA has the effective area, precise timing ability, and data throughput capability to probe these sources at the timescales of processes near neutron star surfaces or the innermost stable orbits around black holes. A second objective of the experiment is to conduct experiments involving applied uses of X-ray detectors in space and with reliable computing in space. These will not be discussed here but descriptions are available elsewhere (Wood 1993).

Key characteristics of the experiment and mission that facilitate this overall program include (i) a mission concept that allows long observing times on bright X-ray objects, (ii) large-area detectors with high time resolution capability (effective area: 2000 cm^2; telemetry: 40 kbps, with 128 kbps available for short periods; 2 μs time resolution), (iii) good low energy response (down to 1 keV), and (iv) a high flexibility in data handling. Other special features include absolute time-tagging (to 2 μs) using a GPS receiver.

2. SCIENTIFIC PROGRAM

The principal targets for USA are X-ray binaries whose X-ray emitting members are neutron stars, black holes, or white dwarfs. Study of physical processes in these systems have been among the main thrusts of X-ray astronomy since the founding of the field. Today it remains true that many of the most important results on these systems are found by studying their X-ray variability, and the push to shorter (millisecond) timescales is proving highly fruitful. If the source is bright ($>$ milliCrabs) such short timescales are more readily reached with non-imaging instruments having large collecting apertures than with imaging instruments. Physics issues studied in these sources are generally related to the fact that parameters such as magnetic field strength, mass and energy densities, and gravitational fields reach extreme values, hence providing the preferred testing grounds for physical theories. X-ray timing is a cornerstone of relativistic astrophysics.

USA, in turn, is one of the two main resources at the present epoch for X-ray timing experiments, the other being the Proportional Counter Array (PCA) on RXTE. USA has its own special areas of emphasis, one of which is its observing plan. Present plans call for the observation of about 30 primary targets, with each being observed for about 1 month over a nominal mission life of 3 years; selected targets will be observed for shorter periods of time. Sources observed to date (through 31 August 1999) include Cyg X-1 (700 ks on target), Aql X-1 (100 ks), Cen X-3 (65 ks), X1630-472 (60 ks), Cyg X-2 (50 ks), X1636-536 (45 ks), GX 1+4 (40 ks), 1E2259+586 (40 ks), X1820-30 (40 ks), X1630-472 (35 ks), 1E1048.1-5937 (30 ks), and GRS 1915+105 (25 ks). The total time on each source is typically scheduled as a number of $\sim$1 ks observations distributed over weeks or months. Simultaneous observation with other observatories, such as the Compton Gamma Ray Observatory and Rossi X-ray Timing Explorer, and with ground based telescopes are also being undertaken.

Figure 1 shows two sample light curves taken with USA. The first is an X-ray burst from the burster X1735-444, and the second is an observation of a flaring state of the Galactic microquasar GRS1915+105. In 1735-44 the instrument is on the source throughout the interval displayed while in GRS 1915+105 the steep rise at the beginning of the plot is the instrument slewing onto the source; the earliest seconds represent the background for this observation.

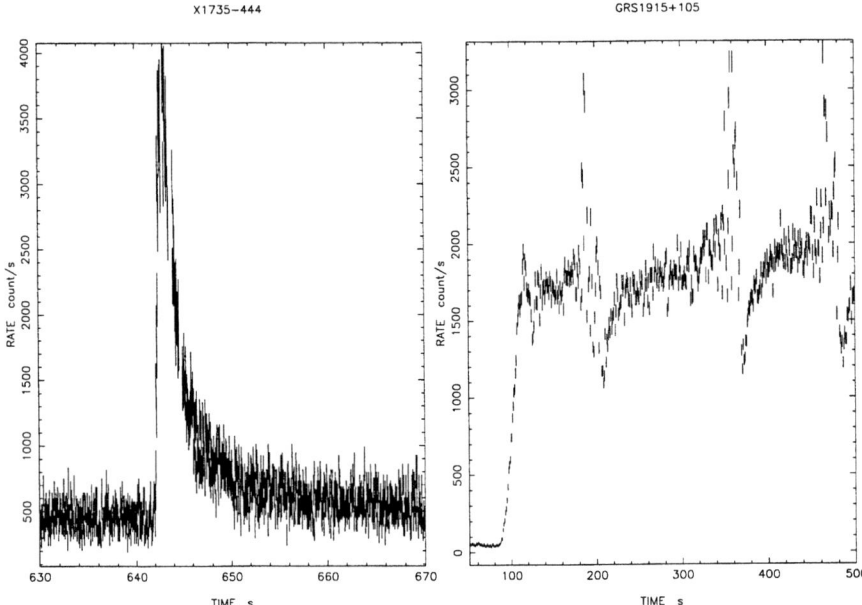

FIGURE 1. Left: A burst from X1735-444. Right: An observation of GRS1915+105

2.1. Low-Mass X-ray Binaries

The special importance of the low mass X-ray binaries (LMXBs) arises from their comparatively weak magnetic fields which allows the disk to penetrate very close to the star. This gives rise to fast timing effects that can be used to probe the extreme conditions in the neutron star vicinity. Major gains in the understanding of these phenomena have been made since the launch of the Rossi X-ray Timing Explorer (RXTE) in late 1995. High frequency quasiperiodic oscillations (QPOs) and short strings of coherent pulsations during bursts have been used to argue convincingly that effects associated with inner disk edges and the innermost stable orbits predicted by General Relativity are being seen (van der Klis 1998). Another milestone is the establishment of the spinup evolution of neutron stars through the discovery of the first accretion powered millisecond pulsar (SAX J1808.4-3658).

USA will make further contributions to the study of LMXBs with the application of its unique strengths. In some cases this will mean exploiting the ability to dedicate large blocks of time to a key source, e.g., to refine understanding of SAX J1808.4-3658 or to observe transitions between modes or states. Significant time is being devoted to searches for coherent periods, both on and off bursts. Off burst work is carried out using coherence recovery searches for periods. Observations can also be carried out in various ways to detect or refine orbital periods in LMXBs.

Overall, LMXBs are sources that stand to bring major rewards including advances in understanding the role of General Relativity in the dynamics of inner disk regions, but past experience has also shown that these rewards are achieved only through major investments of observing time and analysis, chiefly because of the elusiveness and short timescale of the spin periods.

2.2. High-Mass X-ray Binaries

USA will also accumulate significant time on a number of high-mass X-ray binaries. Many of these systems have accretion rates that far exceed the Eddington limit locally in the accretion column. This means that radiation pressure has a significant influence on the flow. Recently Jernigan *et al.* (1999) reported the discovery of Photon Bubble Oscillations (PBOs) in Cen X-3. USA will observe this source and other bright HMXBs to help characterize their high frequency power spectra independently from RXTE. Outstanding puzzles in these systems are the details of angular momentum transfer from the disk to the star (including possible reversals of the sense of disk rotation) and understanding in detail the photohydrodymics of the accretion column in which the super Eddington accretion funneled flow is converted to the observed X-ray emission. Bright binary pulsars such as Her X-1, Cen X-3, and Vel X-1 will be observed to gain insight into these issues. Monitoring over both short and long time periods allows the correlation between period, period derivative, and luminosity to be probed which addresses the angular momentum transfer issue. USA observes at significantly lower energies than the BATSE instrument on CGRO, which has gathered much of the data on this topic in recent years.

2.3. Black Hole Candidates

USA will pursue several investigations into the nature of black holes. Chief among these are (1) The characterization of high frequency ($\nu > 1$ kHz) variability in Cyg X-1. USA has begun a long term study of Cyg X-1 in which more than 1 Ms of exposure will be accumulated on Cyg X-1 in each of the spectral states it exibits during the USA three year mission. Using techniques developed for calibrating high frequency systematic effects in HEAO A-1 and RXTE (Chaput *et al.* 1999), we will constrain the sub-millisecond variability of Cyg X-1. (2) Simultaneous X-ray and infrared observations of the galactic microquasar GRS 1915+105. USA will, for the first time, determine the 1–3 keV behavior of this interesting source. Figure 1 does not begin to convey the range of modulation patterns seen in this object. A new window at lower energies from USA, gathered in simultaneity with other space-based and ground facilities, may contribute to modeling the fluid dynamical processes near the black hole. The model that can account for the wealth of variability effects seen in GRS 1915+105 may bring us closer to understanding how plasma behaves near black holes, including the relativistic effects on orbits.

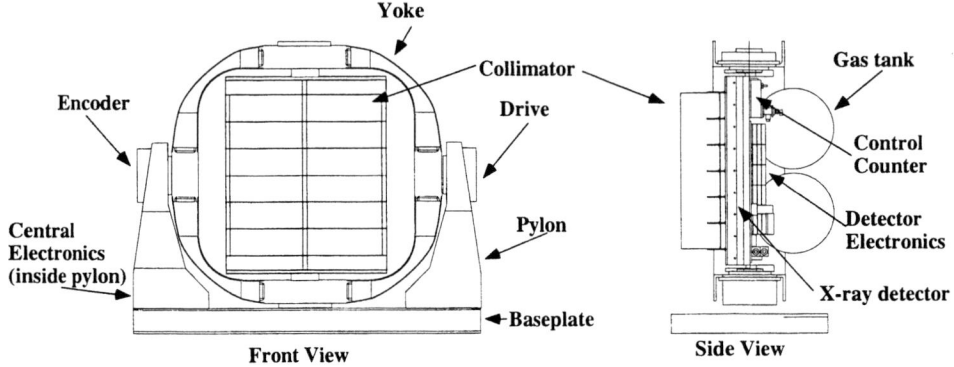

FIGURE 2. Two Views of the USA Experiment

2.4. Other Sources

The Anomalous X-ray Pulsars (AXPs) appear to be a distinct subpopulation whose observed emission is powered by spin-down. Timing these pulsars over a period of years will place strong constraints on their nature (whether they are magnetars or accreting neutron stars) and emission mechanism. Subtle effects of (currently unsuspected) binary companions may show up or else derivatives that provide insight into source dynamics may be measured. There are several good candidates and the result will not necessarily be the same in all instances. Long-term monitoring of AXPs is made feasible by the soft response of USA and the good absolute timing. Rotation-powered (radio) pulsars are also observed by USA to validate the time transfer between USA and RXTE, and to measure the radio to X-ray offset of the pulses.

Cataclysmic variables (CVs) exhibit a wide range of timing phenomena,including QPOs, X-ray transients, and complex light curves. While CVs are typically ~ 100 times fainter than LMXBs, their dynamical time scales are ~ 1000 times longer. Moreover the magnetic CVs, which will be USA's prime targets among CVs are distinguished by having the largest magnetic moments among known stellar populations, including even magnetars. Curiously, accretion-induced QPOs were predicted historically to result from this flow before they were observed. The QPOs have been seen repeatedly in optical wavelengths but never in X-rays, despite searches. Highly correlated optical and X-ray luminosity variations are predicted in current hydrodynamic models (Wolff, Wood & Imamura 1991)

Finally, USA can exploit its great flexibility to observe targets of opportunity that are deemed important by the science working group. Already, USA has observed Aql X-1, the Rapid Burster, and X1630-472 during outbursts. Of course, the accreting millisecond pulsar SAXJ1808.4-3658 would be of particular interest if it becomes active during the USA mission.

TABLE 1. Operational Features of the USA Detector System

Gas:	P-10 at 1.1 atmosphere
Flow rate:	0.1 cc/min allows for 3–6 year life
Window:	5.0 μm Mylar + ~ 30–40Å Nichrome
Energy range:	1–15 keV
Field of view:	collimation of 1.2° × 1.2° (FWHM)
Energy resolution:	0.17 (1 keV @ 5.9 keV), 128 raw PHA channels
Aperture (effective):	2000 cm² @ 3 keV
On-orbit calibration:	solenoid operated ^{55}Fe source

3. INSTRUMENT DESCRIPTION

3.1. Proportional Counter X-ray Detectors

The detector (Table 1) consists of two multiwire constant flow proportional counters equipped with a 5.0 μm Mylar window and an additional 1.9 μm thick aluminized Mylar heat shield. The detector is filled with a mixture of 90% argon and 10% methane (P-10) at a pressure of 16.1 psia (at room temperature). The detector interior contains an array of wires which provides two layers of nine 2.8 cm square cells, each containing one anode wire, running the length of the counter. An additional wire runs around the periphery of the array as part of the cosmic ray veto system. The electronics are designed to accept primarily X-ray events, which arise in one cell only. Events registered in two or more wires by a cosmic ray track are vetoed with an efficiency of about 99%.

The high voltage on the anode wires is adjusted continuously to stabilize the gain, using a feed-back loop which monitors the pulse-height distribution of X-ray events in a small separate proportional counter. Two discriminators provide a normalized value independent of the absolute source intensity of the ^{55}Fe source used in the feed-back counter.

The collimators serve to support the window as well as to define the field of view. To place reasonable requirements on the pointing system the collimator was constructed with a field of view of approximately 1.2°×1.2° and a flat top of approximately 0.05°. Each collimator consists of 8 modules 7.5 cm × 28 cm × 11 cm high filled with copper hexcell formed from 25 μm sheet stock with a 2.5 mm altitude for each hexagon. The sides and ends of each module are formed from 1 mm Cu sheet which provides stiffness across the width of the collimator to support a 98% transmission nickel mesh that in turn supports the Mylar window. The response function of each collimator module was measured with X-rays before the modules were assembled into the collimator frames.

TABLE 2. Pointing System Characteristics

Pitch/Yaw drive capability:	$\sim 3.6°$/min (track), ~ 20 °/min (slew)
Field of Regard:	2π Sr
Common pitch/yaw pivot design:	180° travel in each axis
Drive system:	1.8 deg stepper motor
	800:1 gear reduction, ~ 20 ″/step
Position sensing system:	16-bit optical shaft encoder

3.2. Support Hardware

The ARGOS spacecraft is three-axis-stabilized and nadir-pointed. The X-ray detectors are mounted on a 2-axis gimballed platform to permit inertial pointing at celestial objects. The pointer is configured as an equatorial mount looking aft (away from the velocity vector of the spacecraft). Pointing is accomplished by a yaw rotation to acquire the target followed by a continuous slew in the pitch to track the target as it rotates about the orbital pole.

The primary tasks of the central command and control electronics (CE) are command and data interface to the ARGOS spacecraft MIL-1553 bus, data acquisition from the detector modules, control of the pointer system, and interface with USA's RH3000 and IDT3081 processors. The command and control processor is a Harris radiation hardened 80C86 microprocessor.

The structural elements of the pointer (Table 2), the support pylons, and the yoke which serves as the inner gimbal form the primary structure of the experiment. Each axis has a drive unit which forms the pivot on one end of the axis and a position encoder unit which supports the opposite end. Actual alignment is measured by rastering through sources in flight.

The detector interface board (DIB) in the CE performs time tagging and data formatting for the X-ray science data as well as formatting detector housekeeping data. The microprocessor used is an Analog Devices ADSP2100. The DIB receives a fast photon arrival signal from each detector which enables the timing to ~ 1 microsecond accuracy. A 1 Hz clock (with corresponding GPS time tag) is received directly from the spacecraft to synchronize the event time tagging clock. Pulse height data for each photon are transmitted from the detector to the DIB upon completion of the analog to digital conversion. There are two standard telemetry modes: event and spectral. Event mode is the "workhorse" telemetry mode for USA; for moderate count rates, it allows the maximal amount of information to be preserved on each photon. In event mode, the arrival time and some energy information is stored for each photon detected. There are two submodes of event mode providing 32 μs time and 16 pulse height channels in a 12 bit word and 2 μs time with 8 pulse height channels in a 15 bit word respectively. Data may be output in event mode at either 40 or 128 kbps providing maximum count rates of 3060 or 9940 events per second for 32 μs time or 2448 or 7952 events per second for μs time. In spectral mode, a full resolution energy spectrum (48 channels) is generated every 10 milliseconds for each detector.

The USA experiment also provides space for two "ride-along" processor boards, the RH3000 and the IDT3081. The RH3000 board is built around a pair radiation hardened Harris Semiconductor version of the MIPS R3000 configured as a shadow pair with 2 MB of memory. The IDT3081 board incorporates the commercial-off-the-shelf IDT3081 processor and 2 MB of DRAM without any special error correcting hardware. Both computer boards have access to the downlink science telemetry stream. These processors will be used to conduct experiments in fault-tolerant computing, autonomous spacecraft navigation, and to perform special data analysis functions which are beyond the scope of the normal science telemetry modes, or which require bandwidths greater than 128 kbps.

3.3. Instrument Status

The USA instrument has been performing well since activation began on 30 April 1999, but the USA mission has not been without its difficulties. Approximately two weeks after launch the detector heat shields suffered from degradation which has imposed additional constraints on USA pointing with respect to the Sun. On 8 June 1999 Detector 2 suffered an event which increased the gas leak rate to a very high level and exhausted the P-10 supply leaving only Detector 1 to complete the mission, halving the effective area. Two spacecraft performance issues which are described in more detail below have also impacted USA operations.

4. ARGOS MISSION DESCRIPTION

USA exploits the flight opportunity provided by the ARGOS mission under the DoD Space Test Program (STP). STP was established in 1965 as an activity under the executive management of the Air Force Systems Command with the objective of providing spaceflight for DoD research and development experiments which are not authorized to fund their own flight. Both engineering/technology development and scientific payloads have been flown with great frequency under this program. ARGOS is the only Delta-class STP free-flyer mission to be launched in the 1990s.

The 5000 lb ARGOS satellite was launched from Vandenberg AFB at 10:30 UT on 23 February 1999 aboard a Boeing Delta-II rocket. The prime satellite contractor was Boeing who built and tested the satellite at their Seal Beach, CA facility. ARGOS carries a complement of 9 experiments which address such topics as ionospheric remote sensing, space dust, advanced electric propulsion, and high temperature superconductivity.

Spacecraft downlink telemetry bandwidth is done at 1, 4, or 5 Mbps. Data are stored in a 2.4 Gbit solid state recorder and downlinked at station passes to AFSCN ground stations. The spacecraft is operated in a 3-axis stabilized mode, with Z-axis of the spacecraft always pointed to nadir. Attitude control is based on a system of gyros and horizon sensors feeding into reaction wheels and CO_2 thrusters. The orbit is nearly circular with a 830 km altitude and a 98.7° inclination. It is Sun-synchronous with a beta-angle of 25–45 degrees, i.e., it crosses the equator at approximate local times of 14:00 on the day side and 02:00 on the night side. This nearly polar orbit means that USA encounters a high radiation environment multiple times per orbit as it passes through the Earth's radiation belts at latitudes above 50°. This forces USA to take data at lower duty cycle, turning off the detectors in the radiation belts and the South Atlantic Anomaly to prevent detector breakdowns.

4.1. Mission Operations and Data Processing

The satellite mission operations are handled by the Air Force SMC/TEO at Kirtland AFB, NM. They are responsible for uplinking commands to and for receiving data from the satellite. Individual experiment command uploads are delivered to TEO via FTP and uploaded during ground contacts. Data downlinked during the pass is recorded at the ground station and mailed to TEO because the AFSCN does not support real time links of > 1 Mbps. This results in a delay of 7–21 days in getting science data back to the experimenters.

USA operation is largely automatic. Twice daily command uploads contain timed execution commands to slew the insrument, command it to track the source, switch on the high voltage, select telemetry rate and perform calibrations. These command sequences are generated by a highly automated observation scheduling system which optimizes source selection, manages solid state recorder space, and builds the command uploads.

The USA data processing system is also highly automated. As data appear at Kirtland, files are automatically retreived via FTP and the first several processing steps are performed. Quicklook data are checked for anomalous conditions and the USA team is alerted by e-mail if problems occur. Subsequently, observations are extracted from the Level 0 archive, converted to FITS, and distributed to the scientific analysis centers, including NRL and SLAC.

A Science Working Group (SWG) has been established to help optimize the scientific potential of USA. The SWG determines scientific priorities for observing targets, subject to certain constraints. Scheduling of targets during the USA mission will be consistent with experiment science objectives, priorities, and mission operations capabilities. Telemetry formats will be selected to support overall objectives. USA has the flexibility to respond quickly to some targets of opportunity with approximately a 1–3 day turnaround after the decision to revise the observing plan. The SWG decides whether to respond to potential targets of opportunity and also identifies instances when coordinated observations with ground-based observers, CGRO, RXTE, or other ARGOS instruments are scientifically advantageous. The USA team is receptive to collaborations to make better use of the data, but the small size of the group does not allow us to operate a conventional guest observer facility.

4.2. ARGOS Mission Anomalies and Events

The ARGOS launch and deployment went flawlessly, but since then several problems have surfaced with various subsystems. Generally, the spacecraft has been very robust and has autonomously safed itself when presented with dramatic disturbances, such as a battery exploding on the electric propulsion experiment and during the USA Detector 2 gas leak. Here we will just summarize the issues and describe how they affect the operation of USA.

Shortly after launch, it was discovered that the GPS receiver is unable to stay locked on to the GPS solution and provide good navigation information. This problem was traced to an unexpectedly large input level to the receiver which causes cross correlation errors which disrupt the solution. Generally the receiver will lock on, then oscillate between navigation and acquisition mode for a period of a few minutes to a few hours before losing the solution completely. To recover the time resolution required for many of the USA objectives, new software was uploaded to the satellite to make it safer to initialize the receiver repeatedly. Currently the receiver is initialized 4 times per day. Software is being developed to be able to interpolate times using the onboard clock to recover precise absolute times between periods when the receiver is locked on to a good solution.

A problem was discovered with the Scanning Horizon Sensors which are used to control the pitch and roll of the satellite. They are more radiation sensitive than expected and experience data dropouts or return incorrect data during most passages through the South Atlantic Anomaly (SAA). This causes the spacecraft to respond and produces attitude disturbences in the satellite when it is in the SAA. This does not affect USA because USA never operates in the SAA.

At the time of the software upload to work around the GPS receiver problem, a problem with the offset pointing of USA from the satellite appeared. It appears that the navigation message sent to USA every second no longer represents the true attitude of the spacecraft. It appears, after numerous scanning observations using USA, that the satellite is out of alignment in the roll direction by about 1°. The cause of this is currently unknown, and work is ongoing to troubleshoot this problem and design a workaround.

Research in X-ray Astronomy at NRL is supported by the Office of Naval Research.

REFERENCES

Chaput, C., Bloom, E., Cominsky, L., et al. 1999, ApJ, submitted (astro-ph/9901131)

Jernigan, J. G., Klein, R. I., Arons, J. 1999, ApJ, in press (astro-ph/9909133)

Kowalski et al. 1993, ApJ, 412, 489

van der Klis, M. 1998, in Proceedings of the Third William Fairbank Meeting

Wolff, M. T., Wood, K. S., & Imamura, J. N. 1991, ApJ, 375, L31

Wood, K. S., et al. 1991, ApJ, 379, 295

Wood, K.S. 1993, SPIE 1940, 105

OBSERVING THE EFFECTS OF STRONG GRAVITY WITH FUTURE X-RAY MISSIONS

Christopher S. Reynolds*

JILA, University of Colorado, Campus Box 440, Boulder, CO 80309, USA

ABSTRACT Spectroscopy of the broad iron iron with *ASCA* and *BeppoSAX* has up opened the innermost regions of accreting black hole systems to detailed study. In this contribution, I discuss how observations with future X-ray missions will extend these studies and all us to observationally address issues which are currently only in the realm of the theorists. In particular, high-throughput spectroscopy with *XMM* and, eventually, *Constellation-X* will allow the full diagnostic power of iron line variability to be realized. Instabilities of the inner accretion flow, the geometry of the variable X-ray source, and the black hole mass and spin will all be open to study. Eventually, X-ray interferometry will allow direct imaging of the black hole region in nearby active galaxies, thereby providing the ultimate probe of black hole astrophysics.

KEYWORDS: accretion, accretion discs – black hole physics – galaxies: active – X-rays: general – line: profiles

1. INTRODUCTION

As we have heard in this meeting, X-ray spectroscopy with *ASCA* and *BeppoSAX* are providing probes of the region very close to the supermassive black holes in active galactic nuclei (AGN). In particular, detailed observations and modeling of the broad $K\alpha$ fluorescence emission line of iron, which is thought to originate from the surface layers of the inner accretion disk, allow us to probe the inner disk structure and strong-field gravity in completely new ways. The current observational status of this field has been summarized in Dr. Nandra's contribution in this volume. In this paper, I will discuss what there still is to learn, and how observations with future X-ray missions will help us understand the environment near an accreting supermassive black hole (SMBH).

As one might expect, the region close to an accreting SMBH is complex, with many basic issues still unknown to us. At a fundamental level, the mass of most *active* SMBHs is very uncertain. Furthermore, there are essentially no robust indicators of black hole spin. Many models for the radio-loud/radio-quiet dichotomy of AGN postulate that the black hole mass and, especially, the spin are the control parameters that determine the radio-loudness of the object. However, without

---
* Hubble Fellow

observational signatures of black hole masses and spins, it will be difficult or impossible to test such models. In additional, the physics governing the interaction of the accreting matter with the SMBH is far from clear. Some of the outstanding questions are:

1. Does the inner accretion disk in some objects become hot and geometrically-thick (see Dr. Sambruna's contribution in this volume for a suggestion that this might be the case in broad line radio galaxies)?

2. Is the violently variable X-ray emission due to magnetic flares on the accretion disk surface, or changes within a central corona sitting within the cold accretion disk?

3. What happens within the radius of marginal stability? Does this region have observational relevance? For example, Krolik (1999) recently suggested that the magnetic field becomes very strong in this region, and as a result Alfén waves might plausibly transport significant amounts of energy from this region into an inner corona or the rest of the disk.

4. How are jets launched from the black hole region and collimated, and what contribution do they make to the emissions observed from non-blazar AGN.

This article describes how future X-ray observations may attempt to disentangle these phenomena.

2. CURRENT UNCERTAINTIES AND PURE SPECTRAL STUDIES

The accretion disk model is highly successful at explaining the X-ray reprocessing spectrum observed in many AGN. A small number of AGN (MCG–6-30-15, Tanaka et al. 1995; NGC 3516, Nandra et al. 1999; NGC 4151, Wang et al. 1999) have been the subject of very long integrations with *ASCA* yielding high quality iron line profiles which match the predictions of the accretion disk model well (Fabian et al. 1995). However, there are still ambiguities present.

Firstly, a time-averaged iron line profile contains no information about the mass of the central black hole. All parameters relevant to determining the line profile scale with the gravitational radius. Secondly, and more interesting from an astrophysics point of view, the line profile is sensitive to the X-ray source geometry, accretion disk structure (including the region inside the innermost stable orbit), and the spin of the SMBH. Degeneracies exists in the sense that different astrophysical assumptions and space-time geometries can produce very similar iron line profiles. The best studied example of this degeneracy is the case of the very-broad state of the iron line in MCG–6-30-15 found by Iwasawa et al. (1996). Making the standard assumptions that the line emission is axisymmetric, and there is only emission from outside of the radius of marginal stability, Iwasawa et al. (1996) suggested that the SMBH in this object must be rapidly rotating to produce a line as broad and redshifted as that seen. Dabowski et al. (1996) computed grids of iron line profiles for various values

FIGURE 1. Density structure in a slice through an MHD simulation of disk accretion in a pseudo-Newtonian potential. The inner edge of the simulated wedge is at $r = 4GM/c^2$ and the outer edge is at $r = 12GM/c^2$. Strong clumping can be seen at all radii, and especially within the radius of marginal stability at $r = 6CM/c^2$. From Armitage & Reynolds (2000).

of the SMBH spin with the same assumptions and set a formal limit of $a > 0.94$ on the spin of this SMBH. However, Reynolds & Begelman (1997) showed that the same iron line profile can result from a non-rotating SMBH if a high-latitude X-ray source illuminates disk material within the radius of marginal stability. This is an explicit demonstration of how uncertainties in the assumed astrophysics (e.g. the X-ray source geometry) leads to the degeneracy between models with very different space-time geometries (i.e. Schwarzschild vs. extremal Kerr). In a rather different vain, Weaver & Yaqoob (1998) showed that non-axisymmetric obscuration of the line emitting region could also reproduce these data.

The first question to address is whether better spectroscopy with much higher signal-to-noise and/or larger bandpass than *ASCA* and *BeppoSAX* will remove these degeneracies. Returning to the example of MCG–6-30-15, Young, Fabian & Ross (1998) showed that iron fluorescence from material within the radius of marginal stability would be accompanied by a large iron edge. While it is questionable whether the current *ASCA* data are of sufficient quality to rule out the presence of such an edge, one might think that this would be a tell-tale signature that could be used to distinguish the Schwarzschild and extremal-Kerr models for this object. However, it is important to realize that such conclusions are at the mercy of extra epicycles of astrophysical theory. Both the Reynolds & Begelman (1997) and Young et al. (1998) models assume a smooth accretion flow within the radius of marginal stability. But strong magnetic fields in that region will inevitably produce clumping of the material which will in turn lower the ionization parameter of the material which produces the X-ray reflection signatures (Armitage & Reynolds 2000; also see Fig. 1). This, in turn, may diminish the depth of the iron edge that one would expect in the spectrum.

3. IRON LINE VARIABILITY

Spectral variability, and in particular variability of the broad iron line, is a powerful probe of AGN central engines. Many of the degeneracies described above can be broken by considering line variability. In this section, I shall distinguish three types of line variability and discuss how the study of each may help unravel the complexities of these systems.

3.1. Structural changes in the source

As has already been mentioned above, *ASCA* has already seen broad iron line variability in several objects, e.g. MCG–6-30-15 (Iwasawa et al. 1996) and NGC 4051 (Wang et al. 1999). Figure 2 shows the line variability in MCG–6-30-15 in which the line changed from its 'normal' state (shown with open squares) to a very broad and strong state (shown by filled circles). This change in line profile accompanied a sharp drop in the continuum flux level during an event that lasted at least 60 ksec (which is greater than the dynamical timescale $t_{\rm dyn}$ for the inner accretion disk by a factor of ~ 100 or more for any plausible SMBH mass; Reynolds 1999). Unless the occultation scenario of Weaver & Yaqoob (1998) is correct, some dramatic change in the structure of the accretion disk and/or the geometry of the illuminating X-ray source is required to produce such dramatic and long-lived line changes. Changes in the thermal structure of the disk, which occur on a timescale of $t_{\rm th} \sim t_{\rm dyn}/\alpha$ (where $\alpha \sim 0.1 - 0.01$ is the standard viscosity parameter), may produce this type of variability.

Even given the long-lived nature of these events, *ASCA* cannot produce high signal-to-noise line profiles in the different states. This hampers our ability to probe details of the disk/corona variability using these line changes. *XMM* will completely change this situation. With an effective area at iron line energies more than a factor of 10 greater than *ASCA*, very high quality iron line profiles will be obtained at different times as a source such as MCG–6-30-15 undergoes one of these events. While I dare not predict what these observations will find, these studies will undoubtedly revolutionize our understanding of the kind of instabilities suffered by the inner accretion disk and X-ray emitting corona.

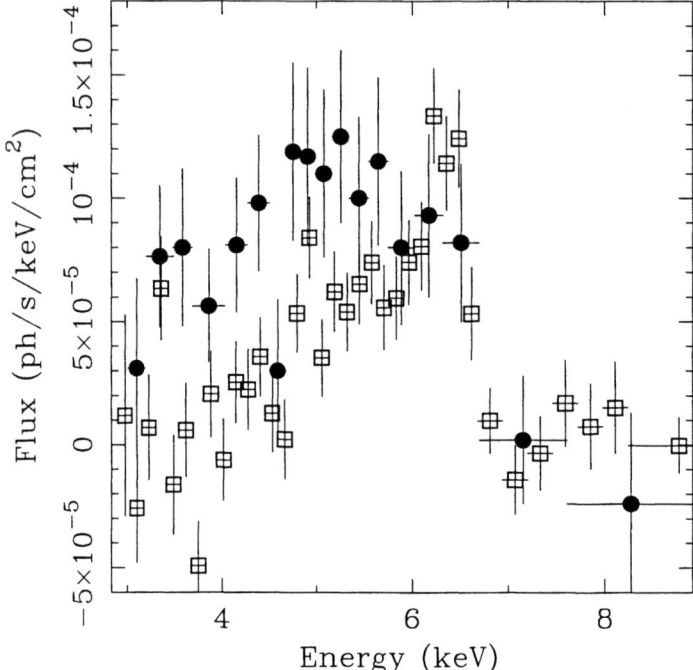

FIGURE 2. Iron line variability in MCG–6-30-15 detected by *ASCA* by Iwasawa et al. (1996). The open squares show the 'normal' state of the line whereas the filled circles show the 'very-broad' state of the line, during which time the continuum level was seen to drop dramatically.

3.2. Orbiting flares

The X-ray emission from most AGN is observed to be highly variable on timescales down to (our best estimate for) the dynamical timescale. Whether the X-ray emission is due to magnetic flares exploding out of the accretion disk or some other instability in a hot disk corona, the instantaneous X-ray emission is likely to be non-axisymmetric. If these non-axisymmetric structures are long lived (i.e. survive at least a couple of dynamical timescales), the iron line will be observed to undergo distinct profile changes as the system orbits the central SMBH.

The computation of observables from an orbiting hot-spot on an accretion disk around a black hole is a classical problem and has been worked on my many authors (e.g. Ginzburg & Ozernoi 1977, Bao et al. 1994, Bromley et al. 1997). Most recently, Ruszkowski (1999; also see contribution in this volume) has computed the observed iron line variability when it is powered by an X-ray flare that is co-rotating with the disk. *XMM* should be able to track these profile changes and measure several key parameters. Firstly, the period and amplitude of energy variations in the peak energy of the iron line are an easy and robust way of determining the black hole mass. Note that the inclination can be measured from the time-averaged iron line profile and so is a known quantity in this calculation. Secondly, departures from sinusoidal time-dependence of the iron line peak can be attributed to relativistic effects and used to probe, for example, the spin parameter of the black hole. Such observations may yield signatures of a spinning black hole: if iron line variations are found that imply a flare orbiting on a circular orbit at a radius less the Schwarzschild radius of marginal stability ($r = 6GM/c^2$), a rapidly rotating black hole is will be implied.

3.3. Reverberation

If some X-ray flares are very short lived, or activate rapidly (as compared to the light-crossing time of the inner accretion disk), line profile changes due to the finite speed of light will occur. This then raises the possibility of performing 'reverberation mapping' of the central regions of the SMBH accretion disk (Stella 1990; Reynolds et al. 1999).

In principal, reverberation provides powerful diagnostics of the space-time geometry and the geometry of the X-ray source. When attempting to understand reverberation, the basic unit to consider is the point-source transfer function, which gives the response of the observed iron line to an X-ray flash at a given location. As a starting point, one could imagine studying the brightest flares in real AGN and comparing the line variability to these point-source transfer functions in an attempt to measure the SMBH mass, spin and the location of the X-ray flare. By studying such transfer functions, it is found that a characteristic signature of rapidly rotating black holes is a 'red-tail' on the transfer function. This feature corresponds to highly redshifted and delayed line emission that originates from an inwardly moving ring of illumination/fluorescence that asymptotically freezes at the horizon (see Reynolds et al. 1999 for a discussion of this feature).

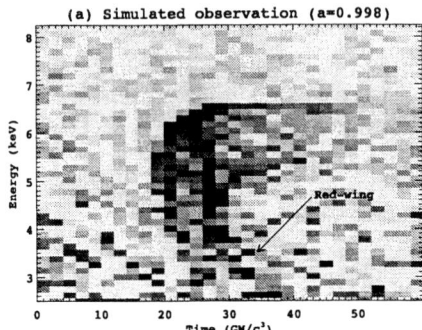
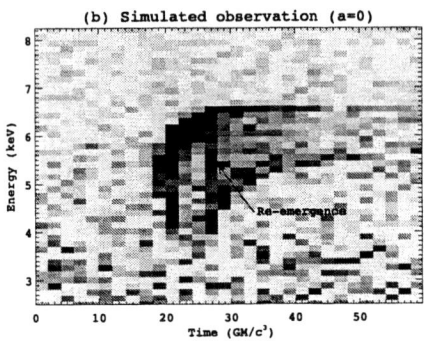

FIGURE 3. *Constellation-X* simulations of iron line reverberation. Panel (a) shows the case of a rapidly rotating SMBH whereas panel (b) shows a non-rotating SMBH. In both cases, an X-ray flash on axis at a height of $10GM/c^2$ has been assumed and the iron line response calculated for an accretion disk inclination (away from normal) of 30°. Sequential 1000s *Constellation-X* observations of the time varying iron line are then simulated, continuum subtracted, and stacked in order to make an observed transfer function. Figure from Young & Reynolds (1999).

The primary observational difficulty in characterizing iron line reverberation will be obtaining the required signal-to-noise. One must be able to measure an iron line profile on a timescale of $t_{\rm reverb} \sim GM/c^3 \approx 500 M_8$ s, where we have normalized to a mass of $10^8 \, M_\odot$. This requires an instrument such as *Constellation-X*. Figure 3 shows that *Constellation-X* can indeed detect reverberation from a bright AGN with a mass of $10^8 \, M_\odot$. Furthermore, the signatures of black hole spin may well be within reach of *Constellation-X* (Young & Reynolds 1999). Although these simulations make the somewhat artificial assumption that the X-ray flare is instantaneous and located on the axis of the system, it provides encouragement that reverberation signatures may be observable in the foreseeable future.

Of course, the occurrence of multiple, overlapping flares will also hamper the interpretation of iron line reverberation. The best way to disentangle these flares is still the subject of current work. However, *Constellation-X* may have the required signal to noise to allow the direct fitting of multiple transfer functions to real data (see Young & Reynolds 1999).

4. DIRECT IMAGING OF BLACK HOLE ACCRETION DISKS

I will end by briefly discussing an exciting idea which will allow us to image the central regions of nearby AGN with sufficient angular resolution to probe structure on scales smaller than the size of the event horizon. By combining diffraction limited X-ray optics with the interferometric technologies that are currently being developed for the Space Interferometer Mission (*SIM*), it is within our technological

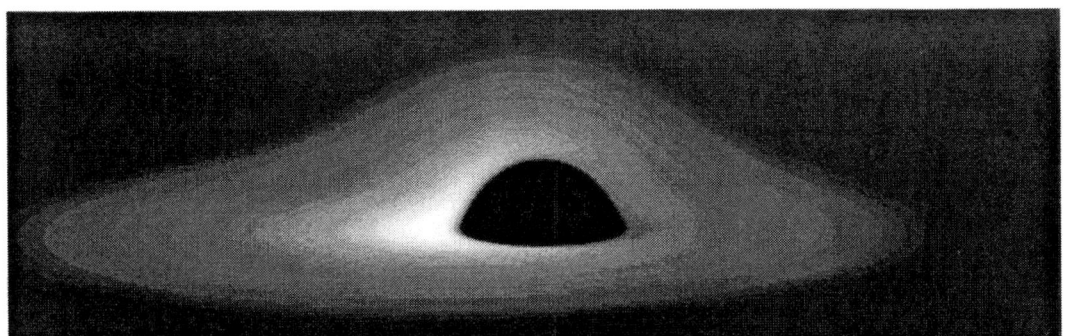

FIGURE 4. Theoretical image of a nearly edge-on accretion disk around a Schwarzschild black hole. The hole in the center of the image corresponds to the radius of marginal stability at $r = 6GM/c^2$. The distortions in the image of the far side of the accretion disk are due to strong light bending effects. In the future, X-ray Interferometry will allow us to obtain such images for real systems.

reach to construct an X-ray interferometer capable of achieving sub-microarcsecond resolution (this concept has become known as *MAXIM*, the Micro-arcsec X-ray Interferometer Mission; see *http://maxim.gsfc.nasa.gov/*).

As well as the obvious appeal of directly imaging an accreting black hole, an observatory capable of achieving 0.1μ arcsec would yield major scientific return. The geometry of the X-ray source (and the spatial nature of the X-ray flares) would be open to direct imaging studies. X-ray activity or fluorescence from within the radius of marginal stability could be easily seen (this region would be well resolved). We might also expect there to be X-ray emission from the base of the jet in the region where the magnetic field couples to the black hole spin via the Blandford-Znajek process. Such emission could be imaged, thereby providing the first look at these exotic physical mechanisms at work. If an interferometer can be constructed with sufficient effective area, we will be able to use the fluorescent iron line to make detailed velocity maps across the image. These velocity fields would provide direct constraints of the black hole mass and spin, and implicitly provide a stringent test of strong field General Relativity.

5. CONCLUSIONS

The immediate environment of an accreting supermassive black hole is extremely exotic. Broad iron lines provide us with the best tool to date for studying these regions. *ASCA* and *BeppoSAX* observations have already shown us that the accretion disk in at least some AGN extends very close to the black hole (and maybe so close as to suggest that the black hole must be rotating). Furthermore, the detection of broad iron line variability by *ASCA* is most likely tracking structural changes in the accretion disk and/or X-ray emitting corona. However, large effective area detectors are required to make further progress. *XMM* will allow these structural changes to be characterized in detail, thereby probing the instabilities that affect the inner accretion disk/corona. Furthermore, *XMM* will allow us to study iron line variability caused by the accretion disk rotation, allowing us to measure the mass of the black hole and constrain the location/lifetime of the X-ray flares. Eventually, *Constellation-X* will allow us to search for iron line reverberation. The detection of reverberation will give robust signatures of black hole spin and provide the tools to study the inner disk structure in unprecedented detail.

Further in the future, direct imaging of the inner disk and black hole region in nearby AGN will be possible using X-ray interferometry. This will provide the ultimate observational probe of black hole astrophysics.

ACKNOWLEDGEMENTS

CSR appreciates support from Hubble Fellowship grant HF-01113.01-98A. This grant was awarded by the Space Telescope Institute, which is operated by the Association of Universities for Research in Astronomy, Inc., for NASA under contract NAS 5-26555.

REFERENCES

Armitage P. J., Reynolds C. S., submitted
Bao G., Hadrava P., Ostgaard E., 1994, ApJ, 423, 63
Bromley B. C., Chen K., Miller W. A., 1997, ApJ, 475, 57
Dabrowski Y., Fabian A. C., Iwasawa K., Lasenby A. N., Reynolds C. S., 1997, MNRAS, 288, L11
Fabian A. C. et al. 1995, MNRAS, 277, L11
Ginzburg V. L., Ozernoi L. M., 1977, Ap&SS, 48, 401
Krolik J. H., 1999, ApJL, 515, L73
Iwasawa K. et al., 1996, MNRAS, 282, 1038
Nandra K., George I. M., Mushotzky R. F., Turner T. J., Yaqoob T., 1999, ApJL, 525, 17
Reynolds C. S., 1999, ApJ, submitted
Reynolds C. S., Begelman M. C., 1997, ApJ, 488, 109
Reynolds C. S., Young A. J., Begelman M. C., Fabian A. C., 1999, ApJ, 514, 164
Ruszkowski M., 1999, MNRAS, submitted (astro-ph/9906397)
Stella L., 1990, Nat, 344, 747
Tanaka Y. et al., 1995, Nat, 375, 659
Wang J. X, Zhou Y. Y., Wang T. G., 1999, ApJL, 523, 129
Wang J. X., Zhou Y. Y., Xu H. G., Wang T. G., 1999, ApJL, 516, 65
Weaver K. A., Yaqoob T., 1998, ApJ, 502, L139
Young A. J., Reynolds C. S., 1999, ApJ, in press
Young A., Fabian A. C., Ross R. R., 1998, MNRAS, 300, L11

RADIO-LOUD AGNS: THE X-RAY PERSPECTIVE

Rita M. Sambruna and Michael Eracleous

Pennsylvania State University, Dept. of Astron. & Astrophys., 525 Davey Lab, University Park 16802 (emails: rms@astro.psu.edu, mce@astro.psu.edu)

ABSTRACT

The X-ray emission of radio-loud (RL) AGNs is a powerful tool for probing the structure of the accretion flow in these objects. We review recent spectral and variability studies of RL AGNs, which show that these systems have systematically different X-ray properties than their radio-quiet (RQ) counterparts. Specifically, RL AGNs have weaker and narrower Fe Kα lines and weaker Compton reflection components above 10 keV. The nuclear continuum of RL AGNs in the 2-10 keV band is well described by a power law with photon indices $\sim$ 1.8, similar to RQ AGNs of comparable X-ray luminosity. RL AGNs have little or no flux variability on short time scales ($\lesssim$ 0.5 days); however, flux and spectral variations are observed on time scales of weeks in two well-monitored objects, 3C 390.3 and 3C 120. These properties strongly suggest that the central engines of the two AGNs classes are different. We discuss the implications of these observational results, in particular the possibility that the central engines of RL AGNs are harbor an ion torus (also known as an Advection-Dominated Accretion Flow or ADAF). We show that a beamed component from the jet is unlikely in the majority of sources. Moreover, the X-ray data provide evidence that the circumnuclear environs of RL and RQ AGNs also differ: large amounts of cold gas are detected in BLRGs and QSRs, contrary to Seyfert galaxies of similar X-ray luminosity where an ionized absorber seems to be the norm. The role of future X-ray missions in advancing our understanding of the central engines of RL AGNs is briefly highlighted.

KEYWORDS: Radiogalaxies; X-rays; black hole; AGNs

1. THE X-RAY ADVANTAGE

The spectra and variability properties of AGNs are a diagnostic of the conditions of the matter in the inner parts of the accretion flow. Recent studies of X-ray-bright, radio-quiet (RQ) Seyfert galaxies have provided evidence that their X-ray emission is complex. At high energies, the most prominent features are the Fe Kα emission line between 6 and 7 keV and the Compton reflection hump at $\lesssim$ 10 keV, originating from a cold reprocessor near the central black hole. At soft X-rays, many Seyferts exhibit absorption from partially ionized gas with column densities $N_W \sim 10^{21}$–10^{24} cm^{-2} along the line of sight. The rapid flux variations of Seyferts galaxies on timescales of hours and even minutes (e.g., Edelson 2000) indicate that the X-ray source is compact and located very close to the central black hole.

In contrast, the X-ray spectra and variability of radio-loud (RL) AGNs[1] are not as well known. Past X-ray studies with *Einstein, EXOSAT* and *GINGA* showed that RL AGNs had systematically harder (i.e, flatter) X-ray continua than their radio-quiet counterparts (Wilkes & Elvis 1987; Shastri et al. 1993; Lawson & Turner 1997). These studies, however, were plagued by the low sensitivity and resolution of the instruments, in addition to the heterogeneity of the samples.

Fundamental progress was recently achieved with the advent of *ASCA* and *RXTE* (for a review of the *BeppoSAX* results, see Grandi 2000). The wide-band coverage of these instruments, combined to their higher sensitivity and/or resolution compared to older instruments, enable us to disentangle the various spectral components and study flux and spectral variability. We started a program of systematic study of the X-ray properties of RL AGNs, with the ultimate goal of elucidating the structure of the accretion flow in these systems and comparing it to RQ AGNs. Here we review the results of our *ASCA* and *RXTE* surveys (Sambruna, Eracleous, & Mushotzky 1999; Eracleous, Sambruna, & Mushotzky 2000; also reporting references to previous works), and present new observations of selected objects.

2. THE *ASCA* AND *RXTE* DATABASES FOR RL AGNS

In the following we concentrate on AGNs with lobe dominated radio morphologies. We define radio-loud AGN as those objects with either a 5 GHz radio power of $P_{5\,\mathrm{GHz}} > 10^{25}$ W Hz^{-1} or with rest-frame 5 GHz-to-4400 Å flux-density ratios of $\mathcal{R}_{\mathrm{ro}} > 10$, following Kellermann et al. (1994). There are 39 objects in the *ASCA* archive up to September 1998 which satisfy these criteria, of which 9 are Broad Line Radio Galaxies (BLRGs), 6 are Quasars (QSRs), 12 are Narrow Line Radio Galaxies (NLRGs), and 11 are Radio Galaxies (RGs). This subdivision depends on their optical spectral properties, namely, the presence of broad, permitted lines in their optical spectra and the luminosity of the [O III] $\lambda 5007$ line. Clearly the sample is not statistical or complete, and most likely is biased toward the brightest sources of each type.

Because of its high background rate, only the brightest sources, typically BLRGs, were observed by *RXTE*. Our sample (Eracleous et al. 2000) includes four BLRGs, namely 3C 120, 3C 111, 3C 382, and Pictor A, observed for typically ~ 40 ks during AO1 and AO2. These have fluxes $F_{2-10\,keV} \sim 1 - 4 \times 10^{-11}$ erg cm^{-2} s^{-1}, ensuring that their spectra could be measured adequately with both the PCA and HEXTE. The NLRG Centaurus A was also observed with *RXTE*; the results are presented by Rothschild et al. (1999).

The BLRGs 3C 390.3 and 3C 120 were intensively monitored with *RXTE* in 1997 May and January for 134 and 150 ks, respectively (P.I.: R. Remillard), as part of multiwavelength campaigns. Here we present the preliminary results of our analysis of the archival data. We also discuss simultaneous *ASCA* 100 ks and *RXTE* 60 ks observations of 3C 382 obtained by us in 1999 March.

---

[1] Here we will exclude blazars, as their continuum is entirely dominated by the the beamed jet emission.

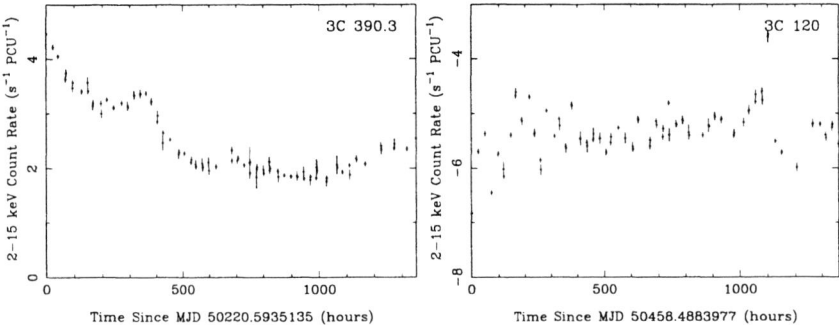

FIGURE 1. Light curves from *RXTE* monitoring observations of the BLRGs 3C 390.3 and 3C 120 in 1997 May and 1997 January, respectively. The light curves are binned at 4600 s. Flux variability is apparent, with different behaviors for the two sources. While 3C 390.3 shows a long-term trend, with large-amplitude variations on timescales of $\lesssim 2$ weeks, "flickering" variations are observed for 3C 120.

3. THE NUCLEAR X-RAY EMISSION OF RL AGNS

3.1. Continuum Shape

In the *ASCA* sample, a power law component is detected in 100% of BLRGs and QSRs and in 90% of NLRGs and RGs above 2 keV. The average photon index is $\langle \Gamma_{2-10\ keV} \rangle \sim 1.7 - 1.9$ for all the four subclasses, in agreement with unification models, which postulate that both type-1 and type-2 AGNs sport the same type of central engine, although viewd from a different direction. Indeed, in NLRGs and RGs the nuclear power law spectrum is heavily absorbed. The column densities detected by *ASCA* are $N_H \sim 10^{21-24}$ cm^{-2} and are most likely due to the putative obscuring torus on parsec-scales (e.g., Urry & Padovani 1995). More surprising is the detection of similar columns of *cold* gas in a fraction of BLRGs and QSRs, which we discuss further below.

We compared the photon index distributions of the RL objects of our sample to those of RQ AGNs of matching X-ray luminosity observed with *ASCA* (Nandra et al. 1997; Turner et al. 1997), and found that the distributions of X-ray continuum slopes for the two classes are not demonstrably different. This finding is consistent with previous results that a flat X-ray component in RL AGNs is due to the beamed contribution of the jet (see above). This conclusion is bolstered by the fact that no correlation is found between the nuclear X-ray and core radio luminosities. This was expected since the AGNs in our sample are are associated with lobe-dominated radio sources.

3.2. Continuum Flux and Spectral Variability

While X-ray variability on short-timescales (from hours to minutes) is common in Seyfert 1s (e.g., Edelson 2000), no pronounced flux variations are observed in the

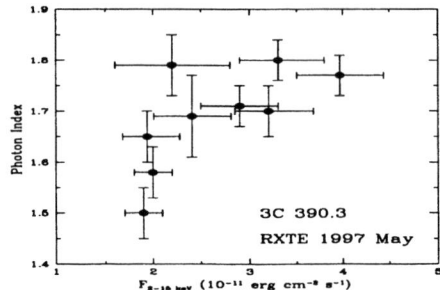

FIGURE 2. Spectral variability of 3C 390.3 during the $RXTE$ intensive monitoring in 1997 May (see Figure 1). The photon index from fits to time-resolved $RXTE$ spectra is plotted versus the observed flux. There is a trend of steeper slopes for increasing flux, an a large and probably intrinsic dispersion in spectral indices at lower fluxes.

BLRGs and QSRs of our $ASCA$ and $RXTE$ samples on timescales < 10 hours. The only exception is 3C 120, where flux changes with amplitude $\sim 20\%$ are present in the archival $ASCA$ 40 ks observation. The lack of pronounced variability may be a consequence of the high luminosity of these objects since Seyferts of comparable luminosities have similar variability properties.

In contrast, significant X-ray flux and spectral variations on longer timescales are observed in two well-monitored BLRGs, 3C 390.3 and 3C 120. Figure 1 shows the $RXTE$ 2–15 keV light curves obtained during the intensive monitorings in 1996 May and 1997 January, with a sampling characterized by short pointings roughly once a day for 60 days. Flux variability is readily apparent in both cases, with qualitatively different behaviors. 3C 390.3 exhibits a trend of decreasing flux by a factor 2 in ~ 20 days, with smaller changes ($\sim$ a few percent) on timescales of $\lesssim 4$ days superposed. Non-linear variability was found in the soft X-ray light curve of this object from previous $ROSAT\ HRI$ observations (Leighly & O'Brien 1998). In contrast, 3C 120 exhibits more erratic flux variations similar to "flickering", with small-amplitude intra-day flares superposed on a constant baseline.

We investigated accompanying spectral variability by integrating PCA spectra during selected intervals in the light curves in Figure 1 corresponding to high, intermediate, and low states. We fitted the spectra in 4–20 keV with a model consisting of a power law, absorbed by the Galactic absorbing column, plus the Fe Kα line at 6.4 keV. This model was chosen because it describes well the average PCA+HEXTE spectrum. Figure 2 shows the plot of the best-fitting photon index versus the 2–10 keV flux in the case of 3C 390.3. Spectral variability is readily apparent, with the spectrum becoming steeper as the flux increases. Note, however, the large dispersion of slopes at low flux levels ($\lesssim 3 \times 10^{-11}$ erg cm^{-2} s^{-1}), which is larger than the typical error bar, indicating a true intrinsic dispersion of values. A similar correlation between the X-ray slope and flux was previously observed in 3C 120 (Halpern

1985) and is consistent with the general trend of steeper-when-brighter observed in Seyfert 1s (Grandi et al. 1992).

3.3. The Fe Kα Line

At high energies, the Fe Kα line is detected with *ASCA* in 67% of BLRGs, 20% of QSRs, 33% of NLRGs, and 30% of RGs. It is generally broad in type-1 sources, with FWHM $\lesssim$ 50,000 km $^{-1}$s, and unresolved in NLRGs and RGs. However, in most cases it is difficult to study the line profile due to the limited sensitivity of *ASCA* and the fact the line is weak in RL AGNs. More stringent constraints are provided by *RXTE* thanks to its larger collecting area. The Fe Kα line is detected in the BLRGs of the *RXTE* sample with an Equivalent Width of EW $\lesssim$ 100 eV, lower than what was measured by *ASCA* in most Seyferts (Figure 3a). Unfortunately, because of the poor *RXTE* resolution, the lines are unresolved, with $FWHM \lesssim 50,000$ km s^{-1} at 90% confidence.

3C 390.3. Interesting constraints on the Fe Kα line profile are obtained for the nearby, bright BLRG 3C 390.3. The Fe Kα line profile is better fitted by a disk-line model than by a Gaussian model, with EW=132^{+40}_{-48} eV; however, the inner disk radius is not constrained. A deep (134 ks) *RXTE* observation shows that the line is unresolved, with width FWHM $\lesssim$ 24,000 km s^{-1} (Figure 3c). These findings lend support to the idea that the Fe Kα line in RL AGNs comes from the cold, thin accretion disk exterior to an ADAF. No reflection component is required to fit the *RXTE* data above 10 keV, with an upper limit to the covering factor $\Omega/2\pi \lesssim 0.5$ at 90% confidence.

3C 382. We observed 3C 382 simultaneously with *ASCA* and *RXTE* for 100 ks and 60 ks, respectively, in 1999 March, in order to study in detail the profile of the Fe Kα line. This source was selected because it showed an extremely broad (FWHM $\sim$ 170,000 km s^{-1}) line in a 40 ks *ASCA* exposure. However, our previous *RXTE* exposure showed an unresolved line, with FWHM $\leq$ 56,000 km s^{-1} at 90% confidence (Eracleous et al. 2000). The combination of spectra from *ASCA* and *RXTE* lends itself to studying the line profile because it provides wide-band coverage and good resolution and collecting area in the energy range 6–7 keV. Thus in principle it allows a reliable determination of the continuum underlying the emission line. Unfortunately, it was later discovered that there are calibration discrepancies between the two instruments, which hamper the determination of the continuum. In practice, during the joint *ASCA+RXTE* fits the slope and normalization of the *ASCA* and *RXTE* continua were left untied.

The 0.6–50 keV continuum is well described by a power law with photon index $\Gamma = 1.8 - 1.9$ and observed flux $F_{2-10\ keV} \sim 5 - 6 \times 10^{-11}$ erg cm^{-2} s^{-1}, plus a bremsstrahlung plasma component with $kT \sim 0.4$ keV below 1 keV. The power law flux is a factor of $\sim$ 2 higher than during the previous *ASCA* and *RXTE* observations. The simultaneous *ASCA* and *RXTE* observations confirm that the Fe Kα line is quite broad compared to other BLRGs: we measure a a width of FWHM $78,000^{+31,000}_{-25,000}$ km s^{-1} at 90% confidence, and with EW=228^{+131}_{-81} eV, resolved at

> 99% confidence. No improvement to the fit is obtained when a disk-line profile is used, nor when the reflection component is added. The upper limit to the solid angle subtended by any medium producing Compton reflection $\Omega/2\pi \lesssim 0.5$ at 90% confidence.

The large width of the Fe Kα line in our observations of 3C 382 is puzzling and requires explanation. One possibility is that the line is a blend of different components originating in various parts of the accretion flow and/or the jet. We tested this idea adding a narrow unresolved Gaussian to the fit to the *ASCA+RXTE* data; while the fit is not improved, the degraded SIS resolution at the time of these observations does not allow us to rule out the presence of multiple components to the Fe Kα line in 3C 382. However, calibration problems with *ASCA* could be responsible for the observed width of the line, which is unresolved in deep *BeppoSAX* measurements (Grandi 2000). Future observations with *XMM* will provide the final solution to this puzzle.

3.4. The Reflection Component

The wide energy range (2–250 keV) covered by the PCA and HEXTE instruments on *RXTE* is well suited to the study of the Compton reflection in RL AGNs at energies above 10 keV. The strength of this component is parameterized in terms of $R = \Omega/2\pi$, i.e., the fraction of solid angle subtended by the reprocessor to the illuminating source. We find that the reflection component is generally weak or undetected in the BLRGs observed with *RXTE*, with strengths $R \lesssim 0.4$–0.5. This is much weaker than what is typically observed in Seyfert 1s (Figure 3b).

Interestingly, the spectrum of the NLRG Centaurus A, includes an unresolved Fe Kα line and no reflection component (Rothschild et al. 1999), consistent with unification models.

4. POSSIBLE INTERPRETATIONS

In the currently accepted accretion scenario for RQ AGNs, X-ray emission is produced in a hot "corona" overlaying and illuminating a standard, Shakura-Sunyaev accretion disk (e.g., Haardt, Maraschi, & Ghisellini 1994). In this picture the reprocessor (disk) subtends a solid angle of $\Omega = 2\pi$ to the illuminating source, as observed by *Ginga* and *RXTE* in Seyfert galaxies. The Fe Kα line is emitted within a few gravitational radii from the black hole (Fabian et al. 1989), and has a skewed profile with a broad red wing. Its equivalent width depends on the solid angle subtended by the disk to the X-ray source.

In BLRGs we observe weaker reflection components, indicating $\Omega \approx \pi$. From this perspective, one possible scenario is that the inner parts of the disk (below some transition radius) inflate under the pressure of the ions which are much hotter than the electrons, thus forming an ion torus or ADAF. The ion torus responsible for the emitted continuum radiation from radio to IR via synchrotron, bremsstrahlung, and/or inverse Compton emission. The cold, thin disk exterior to the ADAF produces the reflection component and the Fe Kα line; both of these features would be

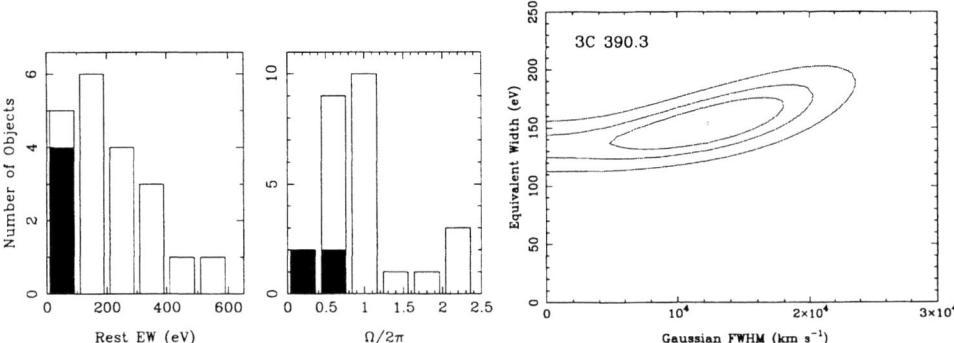

FIGURE 3. *(a) Left and (b) Center:* Histograms of the rest-frame EW of the Fe Kα line and covering fraction of the reflection component, respectively, for the BLRGs of our *RXTE* survey (shaded) and for Seyfert 1s studied by *ASCA* and GINGA (open), from the literature (Nandra et al. 1997; Nandra & Pounds 1994). Radio-loud AGNs have weaker lines and Compton humps than radio-quiet AGNs. *(c) Right:* Contours at 68%, 90% and 99% confidence for the EW of the Fe Kα line versus its velocity width from the fit to an archival 134 ks *RXTE* observation of 3C 390.3. The line is marginally resolved, with an interesting upper limit to its width of FWHM $\lesssim$ 24,000 km s^{-1}.

weaker than in Seyferts since the disk subtends a solid angle $\Omega \approx \pi$ to the X-ray source.

For an assumed black hole mass of $M = 10^8 M_\odot$, the lack of X-ray flux variability on timescales $\lesssim 0.4$ days constrains the light-crossing radius of the emitting region to be $\gtrsim 100~R_g$ (where $R_g \equiv GM/c^2$ is the gravitational radius, with M the mass of the black hole). This is consistent with the expected light-crossing size of an ADAF, which is somewhere between a few light-hours and a few light-days. The ADAF scenario can also accommodate X-ray slopes similar to Seyferts, depending on the transition radius between the inner ADAF and the disk.

An alternative scenario was proposed by Woźniak et al. (1998) on the basis of similar results to ours, obtained from non-simultaneous, archival *ASCA*, *CGRO*/OSSE, and *Ginga* data. In that scenario the central engines of BLRGs harbor a standard, geometrically thin disk but the bulk of the X-ray continuum is produced by the inner parts of a *mildly* relativistic jet. The X-rays from the jet are emitted in a cone that is wide enough to illuminate the obscuring torus but not wide enough to illuminate the accretion disk. Thus, these authors proposed that the Compton reflection component and the Fe Kα line originate in the obscuring torus (the former via Thomson scattering and the latter via fluorescence). However, our result that RL AGN have similar X-ray slopes to RQ AGNs argues against a beamed component in the X-rays. Moreover, the scenario suffers from the additional drawback that X-rays from the jet are unlikely to be beamed in a wide-angle cone: since jet Lorentz factors are thought to have values around 10, the opening angle of the beam should be less than 10°. Therefore, the obscuring torus will not be illuminated by this beam.

These scenarios can in principle be tested further by studying the profile of the Fe Kα line in detail with instruments such as *XMM* and *Astro-E2*. If the ADAF scenario is correct, the line will be double peaked with FWHM $\lesssim$ 20,000 km s^{-1} (as indicated in the case of 3C 390.3), corresponding to an inner radius of a few hundred R_g or more. If the Fe Kα line is produced in the obscuring torus, i.e., at a very large distance from the central black hole, then its profiles will be rather narrow (FWHM $\sim$ 300 km s^{-1}), and thus unresolved by the *XMM* detectors and possibly even unresolved by the *Astro-E2* calorimeter. Finally, if the Fe Kα lines of BLRGs have the same profiles as those of Seyfert 1 galaxies, then these will be easily measurable in high signal-to-noise ratio spectra obtained with *XMM*.

It has also been suggested (Blandford & Znajek 1977; Meier 1999) that RL AGNs harbor more rapidly spinning black holes than RQ sources. While this mechanism would provide a ready explanation for the formation and collimation of the jets, its implications for the broad-band and X-ray properties of the two AGN classes are less clear. Nevertheless, it should be possible at least in principle to test this model through the shape of the Fe Kα line (see Reynolds 2000 and references therein). This awaits the large collecting areas and resolutions of future X-ray missions, such as the proposed *Constellation-X* (NASA) and *XEUS* (ESA).

5. X-RAY ABSORPTION IN RL AGNS

X-ray observations of RL AGNs also provide constraints on the gas surrounding their central engines. While ionized absorption is common in Seyfert 1s (Reynolds 1997; George et al. 1998), no analogous evidence is found in the BLRGs of our sample, except in 3C 390.3. Instead, large columns of *cold* gas are detected in 50% of BLRGs and QSRs. In some case the absorbing columns are as high as those found in NLRGs and RGs. This is in contrast to unification models where the line of sight to type-1 sources should be devoid of cold material.

The columns detected in BLRGs and QSRs of our sample are similar to the columns observed in more distant RL QSRs with *ASCA* and *ROSAT* (Elvis et al. 1998; Cappi et al. 1997). This is apparent from Figure 4, where the excess X-ray column $N_{\rm H^{exc}}$ (in the source's rest-frame) is plotted versus the nuclear 2–10 keV luminosity. While there is a large dispersion in the values of the column density at lower luminosity, no trend is present over more than five decades: the more distant sources have similar intrinsic absorbing columns to the nearby sources. Interestingly, while the BLRGs and QSRs of our sample are lobe-dominated, the more distant objects are core-dominated. This suggests that the absorber, common to both low- and high-redshift sources, must be isotropic in the sources' rest-frame, or subtend a relatively large solid angle to the central illuminating source. As discussed by Halpern (1997), the nature of the absorbing medium and its location relative to

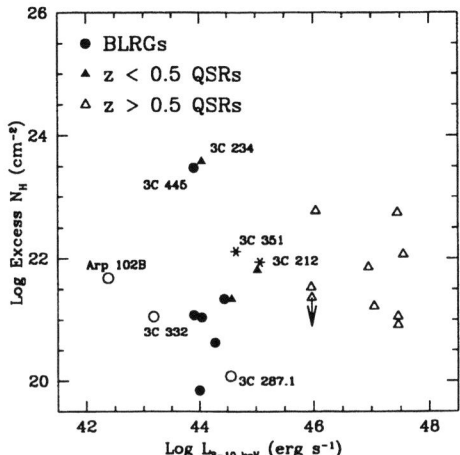

FIGURE 4. Plot of the intrinsic (rest-frame) excess N_H versus luminosity for the BLRGs and QSRs of our study, together with more distant sources studied with *ROSAT* and *ASCA* in the literature. No trend of the column density with luminosity is apparent over more than five decades.

the emission line regions surrounding the central engine is unclear. It is plausible to place the X-ray absorber between the broad-line region and the central engine, as long as it does not contain a significant amount of dust. In this context the absorber could be a wind outflowing from the central engine (e.g., Blandford & Begelman 1999).

6. CONCLUSIONS

Recent observations at high sensitivity and resolution with *ASCA* and *RXTE* have shown that systematic differences in the X-ray properties of RL and RQ AGNs. The X-ray results suggest that the origin of the RL/RQ AGN dichotomy must be sought in the intrinsic properties of the central engines of these systems.

Much remains to be done to further our understanding of RL AGNs. The three major X-ray observatories of the next century, *Chandra*, *XMM*, and *Astro-E2*, will have a central role as they will allow us to study in detail the Fe Kα line profile, the nature of the mysterious X-ray absorber, and X-ray variability on the shortest timescales. Future X-ray observations at high sensitivity and spectral resolution will provide a giant leap forward in our understanding of RL AGNs (comparable to the one *ASCA* already provided for RQ AGNs), opening a new perspective on the origin of the RL/RQ AGN dichotomy.

ACKNOWLEDGEMENTS

We acknowledge support from NASA contract NAS-38252 and NASA grants NAG5-9133, NAG5-7733, and NAG5-8369.

REFERENCES

Blandford, R. D. & Znajek, R. L. 1977, MNRAS, 179, 433
Blandford, R.D . & Begelman, M. C. 1999, MNRAS, 303, L1
Cappi, M. et al. 1997, ApJ, 478, 492
Edelson, R. 2000, these proceedings
Elvis, M. et al. 1998, ApJ, 492, 91
Eracleous, M., Sambruna, R. M., & Mushotzky, R. F. 2000, ApJ, in press (astro-ph/0002265)
Fabian, A. C. et al. 1989, MNRAS, 238, 729
George, I. M. et al. 1998, ApJS, 114, 73
Grandi, P. 2000, these proceedings
Grandi, P. et al. 1992, ApJS, 82, 93
Haardt, F., Maraschi, L., & Ghisellini, G. 1994, ApJ, 432, L95
Halpern J. P. 1985, ApJ, 290, 130
Halpern, J. P. 1997, in "Mass Ejection from AGNs", ASP Conference Series, No. 128 (San Francisco: ASP), 41
Kellerman, K. et al. 1994, AJ, 108, 1163
Lawson, A. J. & Turner, M. J. L. 1997, MNRAS, 288, 920
Leighly, K.M. & O'Brien, P. T. 1997, ApJ, 481, L15
Meier, D. L. 1999, ApJ, 522, 753
Nandra, K. et al. 1997, ApJ, 477, 602
Nandra, K. & Pounds, K. A. 1994, MNRAS, 268, 405
Reynolds, C. S. 1997, MNRAS, 286, 513
Reynolds, C. S. 2000, these proceedings
Rothschild, R. et al. 1999, ApJ, 510, 651
Sambruna, R. M., Eracleous, M., & Mushotzky, R. F. 1999, ApJ, 526, 60
Shastri, P., et al. 1993, ApJ, 410, 29
Turner, T.J. et al. 1997, ApJS, 113, 23
Urry, C. M. & Padovani, P. 1995, PASP, 107, 803
Wilkes, B. J. & Elvis, M. 1987, ApJ, 323, 293
Woźniak, P. et al. 1998, MNRAS, 299, 449

THE RELATIVISTIC PRECESSION MODEL FOR QPOS IN LOW MASS X-RAY BINARIES

L. Stella

Osservatorio Astronomico di Roma, Via Frascati 33, I-00040 Monteporzio Catone (Roma), Italy, e-mail stella@coma.mporzio.astro.it; affiliated to I.C.R.A.

ABSTRACT

The relativistic precession model for quasi periodic oscillations, QPOs, in low mass X-ray binaries is reviewed. The behaviour of three simultaneous types of QPOs is well matched in terms of the fundamental frequencies for geodesic motion in the strong field of the accreting compact object for reasonable star masses and spin frequencies. The model ascribes the higher frequency kHz QPOs, the lower frequency kHz QPOs and the horizontal branch oscillations to the Keplerian, periastron precession and nodal precession frequencies of matter inhomogeneities orbiting close to the inner edge of the accretion disk. The remarkable correlation between the centroid frequency of QPOs in both neutron star and black hole candidate low mass X-ray binaries is very well fit by the model. QPOs from low mass X-ray binaries might provide an unprecedented laboratory to test general relativity in the strong field regime.

KEYWORDS: accretion – black hole physics – relativity – stars: neutron – X-rays: stars

1. INTRODUCTION

Old accreting neutron stars, NSs, in low mass X-ray binaries, LMXRBs, display a complex variety of quasi-periodic oscillation, QPO, modes in their X-ray flux. The *low frequency* QPOs ($\sim 1 - 100$ Hz) that were discovered and studied from high luminosity Z-sources in the eighties are further classified into horizontal, normal and flaring branch oscillations (HBOs, NBOs and FBOs, respectively), depending on the simultaneous position occupied by a source in the X-ray colour-colour diagram (for a review see van der Klis 1995). The kHz QPOs (~ 0.2 to ~ 1.3 kHz) that were revealed and investigated with RXTE in a number of NS LMXRBs (see van der Klis 1998, 1999, 2000 and references therein) involve timescales similar to the dynamical timescales close to the NS. A common phenomenon is the presence of a pair of kHz QPOs (centroid frequencies of ν_1 and ν_2) which drift in frequency while mantaining their frequency difference $\Delta\nu \equiv \nu_2 - \nu_1 \approx 250 - 360$ Hz roughly constant. Detailed studies showed that in four sources $\Delta\nu$ decreases significantly (by up to ~ 100 Hz) as ν_2 increases; these are Sco X-1 (van der Klis et al. 1997), 4U1608-52 (Mendez et al. 1998a,b), 4U1735-44 (Ford et al. 1998) and 4U1728-34 (Mendez & van der Klis 1999). Owing to poor statistics, a similar variation of $\Delta\nu$ in other sources would have remained undetected (Psaltis et al. 1998).

kHz QPOs show remarkably similar properties across NS LMXRBs of the Z and Atoll groups, the luminosity of which differs by a factor of ~ 10 on average. During type I bursts from six Atoll sources, a nearly coherent signal at a frequency of $\nu_{burst} \sim 290 - 580$ Hz has also been detected (for a review see Strohmayer 2000). In a few cases ν_{burst} is consistent, to within the errors, with the frequency separation of the kHz QPO pair $\Delta\nu$ or twice its value $2\Delta\nu$. Yet there are currently two sources (4U1636-53, Mendez et al. 1999, and 4U1728-34, Mendez & van der Klis 1999) for which ν_{burst} is significantly different from $\Delta\nu$ and its harmonics.

The presence of HBOs has been firmly established in both Atoll and Z-sources. Their frequency, ν_{HBO} (~ 15 to ~ 60 Hz) shows an approximately quadratic dependence ($\sim \nu_2^2$) on the higher kHz QPO frequency that is observed simultaneously in a number of sources. The frequency changes of the kHz QPOs and HBOs are positively correlated with the instantaneous accretion rate. Some evidence has also been found for an equivalent of the NBOs and FBOs of Z-sources (Wijnands et al. 1999; Psaltis, Belloni & van der Klis 1999).

A remarkable correlation between the centroid frequencies of QPOs (or peaked noise components) from LMXRBs has been recently discovered (Psaltis, Belloni & van der Klis 1999). This correlation extends over nearly 3 decades in frequency and encompasses both NS and black hole candidate, BHC, systems.

The frequencies of these QPOs, despite their quasi-periodic nature, provide the most accurately measured observables of LMXRBs. A primary goal of any QPO model is therefore to explain the frequency range and dependence of the different QPO types of these sources. The basic features of the relativistic precession model, RPM, are reviewed here (Stella & Vietri 1998a, 1999; Stella, Vietri & Morsink 1999). In the RPM the QPO signals arise from the fundamental frequencies of the motion of matter in the vicinity of the NS. The corresponding orbits are supposed to be slightly eccentric and tilted. As in other models, the higher frequency kHz QPOs at ν_2 are produced by the ϕ-motion (*i.e.* the Keplerian motion) of inhomogeneities orbiting the inner disk boundary, while the lower frequency QPO signal at ν_1 originates from their periastron precession, which is primarily determined by strong-field effects. The HBOs are due to the nodal precession in the orbits of the same inhomogeneities, an effect which is dominated by frame-dragging around fast-rotating collapsed stars. The RPM can be applied to BHCs as well.

2. PERIASTRON PRECESSION AND KHZ QPOS

We consider here only infinitesimally eccentric and tilted orbits, under the assumption that the motion of matter in the innermost disk regions is dictated by the star's gravity alone. In the case of a circular geodesic in the equatorial plane ($\theta = \pi/2$) of a Kerr black hole of mass M and specific angular momentum a, the coordinate frequency measured by a static observer at infinity is

$$\nu_\phi = \pm M^{1/2} r^{-3/2} [2\pi(1 \pm aM^{1/2} r^{-3/2})]^{-1} \qquad (1)$$

(we use units such that $G = c = 1$). The upper sign refers to prograde orbits. If we slightly perturb a circular orbit in the r and θ directions, the coordinate frequencies of the small amplitude oscillations within the plane (the epicyclic frequency ν_r) and in the perpendicular direction (the vertical frequency ν_θ) are given by (see Stella & Vietri 1999 and references therein)

$$\nu_r^2 = \nu_\phi^2(1 - 6Mr^{-1} \pm 8aM^{1/2}r^{-3/2} - 3a^2r^{-2}) \ , \qquad (2)$$

$$\nu_\theta^2 = \nu_\phi^2(1 \mp 4aM^{1/2}r^{-3/2} + 3a^2r^{-2}) \ . \qquad (3)$$

In the Schwarzschild limit ($a = 0$) ν_θ coincides with ν_ϕ, such that the nodal precession frequency $\nu_{nod} \equiv \nu_\phi - \nu_\theta$ is identically zero. ν_r, instead, is always lower than the other two frequencies, reaching a maximum for $r = 8M$ and going to zero at $r_{ms} = 6M$. This qualitative behaviour of ν_r is preserved in the Kerr field ($a \neq 0$). Therefore the periastron precession frequency $\nu_{per} \equiv \nu_\phi - \nu_r$ is dominated by a "Schwarzschild" term over a wide range of parameters (Stella & Vietri 1999).

In the RPM the higher and lower frequency kHz QPOs are identified with $\nu_2 = \nu_\phi$ and $\nu_1 = \nu_{per}$, respectively. Therefore $\Delta\nu \equiv \nu_2 - \nu_1 = \nu_\phi - (\nu_\phi - \nu_r) = \nu_r$. For $a = 0$, Eqs. 1-2 give

$$\nu_r = \nu_\phi(1 - 6M/r)^{1/2} = \nu_\phi[1 - 6(2\pi\nu_\phi M)^{2/3}]^{1/2} \ . \qquad (4)$$

The curves in Fig. 1A show ν_r vs. ν_ϕ for $a = 0$ and selected values of M, the only free parameter in Eq. 4. The measured $\Delta\nu$ vs. ν_2 for eleven NS LMXRBs is also plotted. It is apparent that for NS masses in the 2 $M_\odot$ range, the simple model outlined above is in qualitative agreement with the measured values, including the decrease of $\Delta\nu$ for increasing ν_2 seen in Sco X-1, 4U1608-52, 4U1735-44 and 4U1728-34. The model above is only an approximation: first, the spacetime around a fast rotating NS is different from a Kerr spacetime (due to the star's oblateness induced by rotation); second, the orbits might possess a finite (though small !) eccentricity. Analytical formulae to partly correct for these effects were derived by Stella & Vietri (1999); Fig. 1B shows the fit to the observed $\Delta\nu$ vs. ν_2 relationship in Sco X-1 that was obtained through them. The orbital eccentricity, in particular, was varied in order to obtain different frequencies, while keeping the periastron distance $r_p = a(1-e)$ fixed. The best model for a non-rotating NS is shown in Fig. 1B. The model reproduces fairly accurately the data with a minimum number of free parameters, the NS mass ($M \sim 1.9$ $M_\odot$) and periastron distance ($r_p \simeq 6.2$ M). The latter value is close to the marginally stable orbit radius. When the NS spin is allowed a finite value (say $\nu_s \sim 300 - 600$ Hz), fits of very similar quality are obtained, the parameters of which differ only slightly from those given above. In essence, the effects induced by the NS rotation on ν_r are small, though non-negligible.

The behaviour of the curves in Fig. 1A,B , and therefore the ability of the model to match the observations, reflects the properties of the strong field Schwarzschild metric, since lower order expansions fail to reproduce the observed frequencies (see Stella & Vietri 1999).

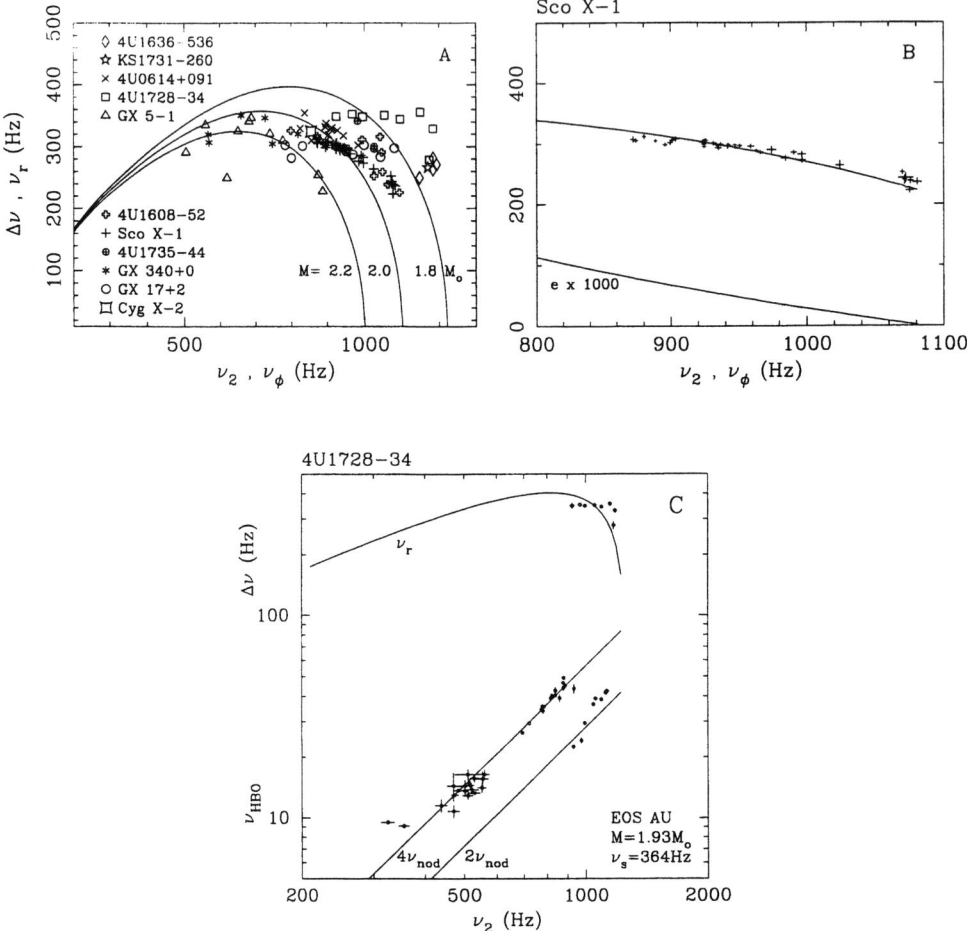

FIGURE 1. (A) kHz QPO frequency difference $\Delta\nu$ versus higher QPO frequency ν_2 for eleven LMXRBs. Error bars are not plotted for the sake of clarity. The curves give the r- and ϕ-frequencies of matter in nearly circular orbit around a non-rotating neutron star, of mass 2.2, 2.0 and 1.8 $M_\odot$. (B) $\Delta\nu$ versus ν_2 in Sco X-1. The best fit model corresponds to the r- and ϕ-frequencies of matter orbiting a non-rotating 1.90 $M_\odot$ neutron star at a periastron distance of 6.25 M (17.5 km). The line marked with e gives the orbital eccentricity ($\times 1000$) as a function of the ϕ-frequency. (C) kHz QPO frequency difference $\Delta\nu$ and (double-branched) HBO frequency versus higher QPO frequency ν_2 in 4U1728-34 (Strohmayer et al. 1996; Ford & van der Klis 1998; Mendez & van der Klis 1999). The solid lines give the r-frequency and the 2nd and 4th harmonics of the nodal precession frequency ν_{nod} as a function of the ϕ-frequency for infinitesimally eccentric and tilted orbits in the spacetime of a 1.93 $M_\odot$ neutron star spinning at 364 Hz (EOS AU; Wiringa et al. 1988).

Within the RPM, the maximum value of $\nu_r = \Delta\nu$ depends mainly on the mass of the compact object. The NS masses deduced from the simple modelling in Fig. 1A,B are in the $\sim 1.8 - 2.0$ M$_\odot$ range, in agreement with the only relatively accurate mass measurement from optical spectro-photometry in any of these systems (Cyg X-2; $M = 1.78 \pm 0.23$ M$_\odot$; Orosz & Kuulkers 1999). In general within the RPM, $\Delta\nu$ should not be obviously related to the NS spin frequency ν_s. Therefore, it seems natural to identify ν_s with ν_{burst}, i.e. the stable frequency seen during type I X-ray bursts (for a review see Strohmayer 2000). The distribution of NS spins inferred in this way is fairly wide ($\sim 290 - 580$ Hz) and compares well with that of millisecond radio pulsars, MSPs, in agreement with evolutionary scenarios in which LMXRBs are the progenitors of MSPs. The 401 Hz spin of SAX J1808.4-3658, the only bursting LMXRB displaying coherent pulsations in its persistent emission (Wijnands & van der Klis 1998; Chakrabarty & Morgan 1998), is also in the range of spin frequencies deduced from ν_{burst}. None of the LMXRBs of the (high luminosity) Z-class has yet displayed burst oscillations; therefore their spin period is still to be measured. Cyg X-2 and GX 17+2, the only type I X-ray bursters in the group, might provide this important piece of information.

3. NODAL PRECESSION AND HBOS

If the orbits giving rise to the signals at ν_ϕ and ν_{per} are slightly tilted relative to the equatorial plane, nodal precession will take place around the spin axis. In the RPM the HBO frequency is related to the nodal precession frequency. From Eqs. 1 and 3 this can be written in the slow rotation limit ($a/M \ll 1$)

$$\nu_{nod} \simeq 4\pi a \nu_\phi^2 \simeq 6.2 \times 10^{-5} (a/M) m \nu_\phi^2 \text{ Hz} \simeq 4.4 \times 10^{-8} I_{45} m^{-1} \nu_\phi^2 \nu_s \text{ Hz}, \quad (5)$$

where $M = m$ M$_\odot$. This is the well known Lense-Thirring nodal precession formula. The latter equality refers to a rotating NS, where $aM = 2\pi\nu_s I$ with $I = 10^{45} I_{45}$ g cm^2 its moment of inertia.

If ν_ϕ and ν_s are measured, the only parameter in Eq. 5 that is not identified from observations is $I_{45} m^{-1}$; this can vary over a limited range, $0.5 < I_{45} m^{-1} < 2$, for virtually any mass and EOS (see the rotating NS models of Friedman, Ipser & Parker 1986 and Cook, Shapiro & Teukolsky 1992). The stellar oblateness induced by the star's rotation gives rise to correction terms in the nodal precession frequency also (see Morsink & Stella 1999 for a post-Newtonian formula). Their relative importance increases for high ν_s and ν_ϕ. Yet the Lense-Thirring term dominates over a wide range of parameters, such that a $\sim \nu_\phi^2$ dependence is expected for ν_{nod}.

An approximately quadratic dependence of ν_{HBO} on the higher frequency kHz QPOs has been measured in a number of LMXRBs. This dependence was originally suggested on the basis of a few power spectra of the Atoll source 4U1728-34 (Stella & Vietri 1998a). Ford & van der Klis (1998) analysed a large set of power spectra from the same source and determined that the frequency ν_{low} of the $\sim 10 - 50$ Hz QPOs scales as $\nu_2^{2.11 \pm 0.06}$. Note that the low frequency QPO vs. ν_2 relation of this source appears to be double-branched, with the centroid frequency shifting by a factor of ~ 2 across different observations. This suggests that on occasions the 2nd harmonic of ν_{HBO} is excited instead of the fundamental. Stella & Vietri (1998b) first noticed that the HBO frequency of the Z-source GX 17+2 displays a nearly quadratic dependence on ν_2. Psaltis et al. (1999) carried out a systematic study of Z-sources and determined that the HBO frequency is consistent with a ν_2^2 scaling (Cyg X-2 and Sco X-1 show evidence for a somewhat flatter dependence). In essence these results confirmed one of the basic features of the RPM, namely the nearly quadratic dependence of the nodal precession frequency on the ϕ-frequency.

If the NS spin frequency is measured, then for any value of ν_ϕ the model yields a predicted nodal precession frequency which is uncertain only by a factor of a few, mainly due to the allowed range of $I_{45}m^{-1}$ (Stella & Vietri 1998a). Only in the Atoll source 4U1728-34 burst oscillations and simultaneous kHz QPOs and HBOs have so far been detected unambiguously (Strohmayer et al. 1996; Ford & van der Klis 1998; Mendez & van der Klis 1999). Therefore its QPO frequencies can be used to test both the ν_{HBO} and $\Delta\nu$ versus ν_2 relationships predicted by the RPM, when the NS spin derived from burst oscillations is used ($\nu_{burst} \simeq 364$ Hz). In order to take fully into account of all the effects that contribute determining geodetic motion in the vicinity of the NS, we adopted a numerical approach and computed the spacetime metric of the star using Stergioulas' (1995) code, an equivalent of that of Cook et al.(1992); see also Stergioulas and Friedman (1995). From this, ν_r and ν_{nod} were derived as a function of ν_ϕ for infinitesimally small tilt angles and eccentricities (Morsink & Stella 1999; Stella, Vietri & Morsink 1999).

Fig. 1C shows the measured values of $\Delta\nu$ and ν_{HBO} versus ν_2 in 4U1728-34. Relatively high NS masses (see also Sect. 2) and stiff EOSs such as AU and UU (Wiringa et al. 1988) are required in this application of the RPM. The solid lines in Fig. 1C are for a 1.93 $M_\odot$ NS with EOS AU and $\nu_s = 364$ Hz. A good agreement is obtained if the HBO frequency, the lower of the two branches seen in 4U1728-34, is identified with the 2nd harmonics of ν_{nod} (*i.e.* $2\nu_{nod}$; see also Morsink & Stella 1999; Stella & Vietri 1999a). Correspondingly the upper HBO branch is well fit by $4\nu_{nod}$. The geometry of tilted orbits in the innermost disk regions might be such that a stronger signal is produced at the even harmonics of the nodal precession frequency (e.g. Psaltis et al. 1999). The frequency range and trend of the epicyclic frequency ν_r in this model are also in reasonable agreement with the $\Delta\nu$ measurements; a more complex model is clearly required in order to fit these data more accurately.

In summary, the model presented here is capable of reproducing the salient features of both the $\Delta\nu$ versus ν_2 and ν_{HBO} versus ν_2 relationships, with just two free parameters (M and the EOS), the allowed range of which is fairly limited (moreover the EOS cannot even be varied continuously !). Concerning Z-sources and all other Atoll sources for which burst oscillations have not been detected yet, the NS spin can still be regarded as a free parameter. The application of the RPM to the HBOs of these sources can therefore be used to constrain the spin of their NSs. Conversely, in those Atoll source in which ν_{burst} is measured, but HBOs have not been detected yet, the RPM can be used to predict ν_{HBO}. These issues are briefly addressed in the next Section.

4. THE RELATIVISTIC PRECESSION MODEL AND THE PBV CORRELATION

Psaltis, Belloni & van der Klis (1999) recently identified two QPOs and peaked noise components the frequency of which follows a tight correlation over nearly three decades. This correlation (hereafter PBV correlation) involves both NS and BHC LMXRBs spanning different classes and a wide range of luminosities (see the points in Fig. 2). In kHz QPO NS systems, these components are the lower frequency kHz QPOs, ν_1, and the low frequency, HBO or HBO-like QPOs, ν_{HBO}. For BHC systems and lower luminosity NS LMXRBs the correlation involves either two QPOs, or a QPO and a peaked noise component. In all cases the frequency separation is about a decade and an approximate linear relationship ($\nu_{HBO} \sim \nu_1^{0.95}$) holds. The QPO frequencies from the peculiar NS system Cir X-1 varies over nearly a decade while closely following the PBV correlation and bridging its low and high frequency ends. Psaltis, Belloni & van der Klis (1999) noted also that the ν_2 vs. ν_1 relations of different Atoll and Z-sources line-up with good accuracy.

The RPM matches precisely the PBV correlation, without resorting to any additional assumption (see Stella, Vietri & Morsink 1999). We assume that in all QPO sources, including BHCs, $\nu_{HBO} \simeq 2\nu_{nod}$ as in 4U1728-34 (see Sect.3). Fig. 2A shows $2\nu_{nod}$ and ν_ϕ obtained from Eqs. 1-3 as a function of ν_{per} for corotating orbits and selected values of M and a/M. The high frequency end of each line is dictated by the orbital radius reaching the marginally stable orbit.

The separation of the lines in Fig. 2A testifies that while ν_{nod} depends weakly on the mass and more strongly on a/M, the opposite is true for ν_ϕ. By taking the weak field ($M/r \ll 1$) and slow rotation ($a/M \ll 1$) limit of Eqs. 1-3 the relevant first order dependence is made explicit,

$$\nu_\phi \simeq (2\pi)^{-2/5} 3^{-3/5} M^{-2/5} \nu_{per}^{3/5} \simeq 33\, m^{-2/5} \nu_{per}^{3/5}\ \text{Hz}\ , \qquad (6)$$

$$\nu_{nod} \simeq (2/3)^{6/5} \pi^{1/5} (a/M) M^{1/5} \nu_{per}^{6/5} \simeq 6.7 \times 10^{-2}\, (a/M) m^{1/5} \nu_{per}^{6/5}\ \text{Hz}\ . \qquad (7)$$

For the case of rotating NSs we adopt the numerical approach outlined in Sect. 3. Results are shown in Fig. 2B for a NS mass of 1.95 M$_\odot$, EOS AU and $\nu_s = 300, 600, 900$ and 1200 Hz (corresponding to $a/M = 0.11, 0.22, 0.34$ and 0.47, respectively). Note that the approximate scalings in Eqs. 6-7 remain valid over a wide range of frequencies. Only for the largest values of ν_{per} and ν_s, ν_{nod} departs substantially from the $\sim \nu_{per}^{6/5}$ dependence.

The measured QPO and peaked noise frequencies giving rise to the PBV correlation are also plotted in Fig. 2B. Higher kHz QPO frequencies from NS systems (ν_2) are included (for the sake of clarity NBOs and FBOs were excluded). The agreement over the range of frequencies spanned by each kHz QPO NS system should not be surprising: together with the accurate matching of the corresponding $\nu_1 - \nu_2$ relationship in Z-sources, this is indeed part of the evidence on which the RPM model was proposed. However the fact that the dependence of ν_{nod} on ν_{per} matches the observed $\nu_{HBO} - \nu_1$ correlation to a good accuracy over ~ 3 decades in frequency (down to ν_1 of a few Hz), encompassing both NS and BHC systems, provides additional independent evidence in favor of the RPM. The observed variation of ν_{HBO} and ν_1 in individual sources (Cir X-1 is the most striking example, see Fig. 2B) further supports the scaling predicted by the RPM. The matching of the observed ν_2 vs. ν_1 relation in terms of ν_ϕ vs. ν_{per} is also quite accurate.

For EOS AU and $m = 1.95$, the ν_{HBO} vs. ν_1 values of most NS LMXRBs are best matched for ν_s in the ~ 600 to 900 Hz range. It is apparent from Fig. 2B that ν_s as low as ~ 300 Hz are required for the Atoll sources with ν_{HBO} somewhat below the main PBV correlation: these are 4U1728-34 (see Sect. 3) and 4U1608-52 (from which burst oscillations have not been detected yet). The values above are close to the range of ν_s inferred from ν_{burst} in a number of other Atoll sources (van der Klis 1999). Z-type LMXRBs appear to require ν_s in the ~ 600 to 900 Hz range, a possibility that is still open since for none of these sources there exists yet a ν_s measurement. Note that the upper HBO branch of 4U1728-34 matches well the main PBV correlation. In the interpretation of Sect. 3 the lower branch corresponds to $2\nu_{nod}$ and the upper branch to $4\nu_{nod}$. One could further speculate that sources following the main PBV correlation, Z-sources in particular, are also in the upper HBO branch at $4\nu_{nod}$; in this case their ν_s might be expected in the $\sim 300-400$ Hz range.

For BHC LMXRBs the scatter around the PBV correlation implies values of a/M of $\sim 0.1-0.3$ (see Fig. 2B). The points from XTE J1550-564, while inconsistent with any single value of a/M, might lie along two distinct branches separated by a factor of ~ 2 in ν_{HBO}, similar to the case of 4U1728-34. In the RPM the high frequency BHC QPOs (e.g. the ~ 300 Hz QPOs of GRO1655-40) are interpreted terms of $\nu_1 = \nu_{per}$. This is at variance with the $\nu_1 = \nu_{nod}$ interpretation of Cui et al. (1998), which requires high values of a/M (~ 0.95 in GRO1655-40), in contrast with BHC accretion-driven spinup scenarios (King & Kolb 1999). From Fig. 2A it is apparent that the ~ 300 Hz QPOs from GRO 1655-40 lie close to the high frequency end of the $a/M = 0.1$, $m = 7$ line. Since the mass of the BHC in GRO 1655-40 determined through optical observations is ~ 7 M$_\odot$ (Shahbaz et al. 1999), we conclude that, according to the RPM, $\nu_{per} \simeq 300$ Hz close to the marginally stable orbit, where by definition $\nu_\phi = \nu_{per}$. This suggests that any additional QPO signal at $\nu_2 = \nu_\phi$ would be very close to or even blended with the QPO peak at ν_1. The detection of two closeby or even partially overlapping QPO peaks close to ~ 300 Hz in GRO 1655-40 would therefore provide further evidence in favor of the RPM interpretation.

5. DISCUSSION

5.1. Beat Frequency Models

In the alternative scenario provided by beat frequency models, BFMs, disk inhomogeneities at the magnetospheric boundary (r_m) and the sonic point radius (r_s) are accreted at the beat frequency between the local Keplerian frequency, ν_ϕ, and the NS spin frequency, ν_s, giving rise to the HBOs ($\nu_{HBO} = \nu_\phi(r_m) - \nu_s$) and the lower frequency kHz QPOs ($\nu_1 = \nu_\phi(r_s) - \nu_s$), respectively (Alpar & Shaham 1985; Lamb

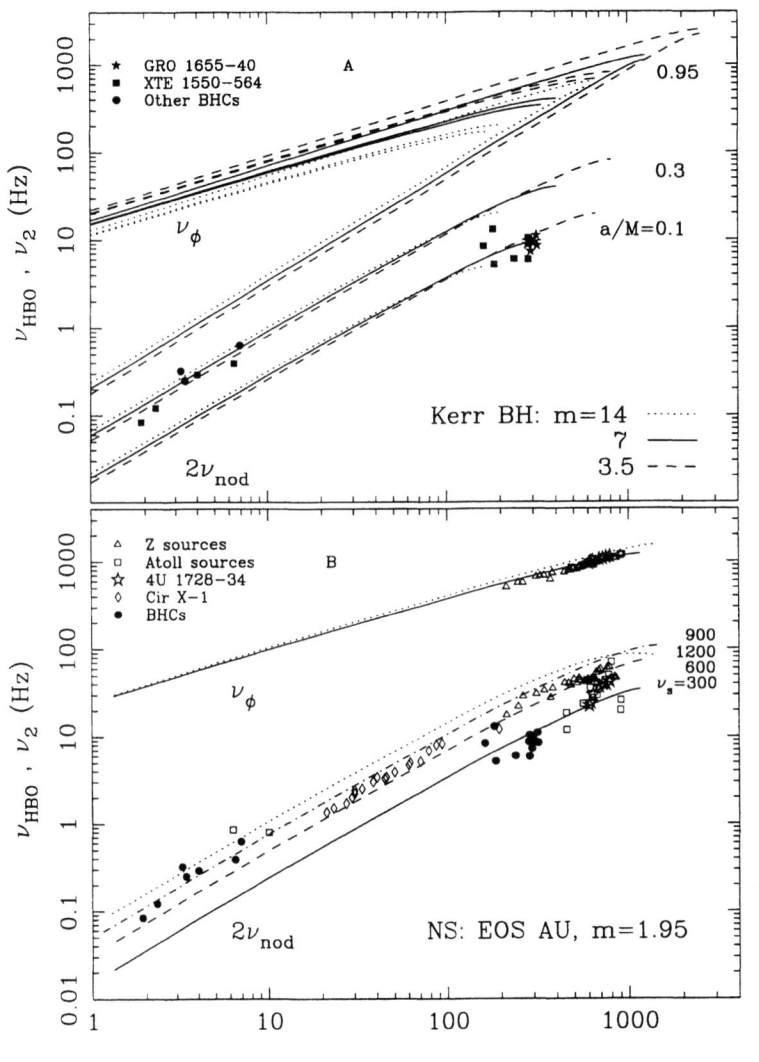

FIGURE 2. Twice the nodal precession frequency, $2\nu_{nod}$, and ϕ-frequency, ν_ϕ, vs. periastron precession frequency, ν_{per}, for black hole candidates of various masses and angular momenta (panel A) and rotating neutron star models (EOS AU, $m = 1.95$) with selected spin frequencies (panel B). The measured QPO (or peaked noise) frequencies ν_1, ν_2 and ν_{HBO} giving rise to the PBV correlation are also shown in panel B for both BHC and NS LMXRBs and in panel A for BHC LMXRBs only; errors bars are not plotted (see Psaltis, Belloni & van der Klis 1999 for a complete list of references). We included only those cases in which QPOs at ν_1 were unambiguously detected. NBO and FBO frequencies are not plotted.

et al. 1985; Miller et al. 1998). The higher frequency kHz QPOs are attributed to the Keplerian motion at the sonic point radius ($\nu_2 = \nu_\phi(r_s)$). The frequency separation $\Delta\nu$ therefore, yields the NS spin frequency, ν_s. The narrow distribution of ν_s inferred in this way ($\sim 250-360$ Hz) is far from the mass shedding limit of any NS model and considerably less extended than that of fast MSPs (~ 600 Hz). Accordingly the NSs of LMXRBs should be equilibrium rotators with a narrow frequency range despite their different average mass transfer rates, magnetic field strenghts and evolutionary histories. Moreover, if LMXRBs are the progenitors of MSPs, then BFMs would require a different evolutionary path leading to the formation of radio pulsars with $\nu_s > 400$ Hz.

The fact that in some Atoll LMXRBs $\nu_{burst} \simeq \Delta\nu$ (or $\simeq 2\Delta\nu$) is readily interpreted, because in BFMs $\nu_s \equiv \Delta\nu$. Yet, $\Delta\nu$ does vary and is significantly different from ν_{burst} in several sources (see Sect. 1). This in contrast with the expectations of simple BFMs.

Attempts at fitting the PBV correlation within BFMs by using the range of spin frequencies inferred from $\Delta\nu$ fail to produce a power-law like behaviour over a sufficiently large range of frequencies (see van der Klis 1999). This is because both ν_1 and ν_{HBO} result from the difference of a variable frequency (ν_ϕ at r_s and r_m, respectively) and a fixed frequency (ν_s). Moreover, BFMs are not applicable to BHCs, since the *no hair theorem* excludes the possibility that an offset magnetic field or radiation beam can be stably anchored to the black hole, as required to produce the beating with the disk Keplerian frequency.

5.2. The Relativistic Precession Model

The RPM naturally explains the frequency range and dependence of the kHz QPOs and HBOs in NS LMXRBs, as well as the PBV correlation, which involves both NS and BHC systems. The model has a minimum number of free parameters. Predictions that can be tested through future analyses and/or observations include:

(a) The frequency difference $\Delta\nu = \nu_r$ is expected to decrease also for low values of $\nu_2 = \nu_\phi$ (see Eq. 4 and Fig. 1). Moreover if the highest $\nu_2 = \nu_\phi$ frequencies do originate from nearly circular orbits (see Fig. 1A), then $\Delta\nu$ should quickly decrease as ν_2 increases further and the orbital radius approaches the marginally stable orbit.

(b) $\nu_2 = \nu_\phi$ is expected to scale as $\nu_1^{3/5} = \nu_{per}^{3/5}$. Extending the ν_2 vs. ν_1 correlation in NS systems toward lower frequencies and detecting the signal at ν_2 in BHC systems would provide important new tests.

In the RPM the QPO signals are produced at $r/M \sim 100\, m^{-2/5} \nu_{per}^{-2/5}$, which, for individual sources, must decrease for increasing mass accretion rates (as $\dot{M}$ is positively correlated with e.g. ν_2). The inferred radii range from close to the marginally stable ($r/M \sim 6$) to $r/M \sim 30$ over the frequency span covered by the PBV correlation. Many NS and BHC LMXRBs display two component X-ray spectra consisting of a soft thermal component, usually interpreted in terms of emission from an optically thick accretion disk, and a harder, often power-law like component, likely due to a hot inner disk region. The QPOs might originate at the transition radius between the optically thick disk and the hot inner region as a result of occultation or modulated emission by inhomogeneities (Stella, Vietri & Morsink 1999; see also Di Matteo & Psaltis 1999).

Mechanisms that can induce a finite (though small) eccentricity and tilt in the motion of matter in the innermost disk regions are currently under investigation. In the case of NSs, some kind of resonance between the star spin and the motion of matter inhomogeneities (see *e.g.* Vietri & Stella 1998) might be responsible for the close commensurability of ν_{burst} and $\Delta\nu$ (or $2\Delta\nu$) observed in a few sources. These issues will be addressed elsewhere.

A simple test-particle approximation has been adopted so far within the RPM. Much needed hydrodynamical calculations are still in their infancy; among other things these are hampered by uncertainties concerning the physics of the innermost disk regions. Yet we note that in the hydrodynamical approach explored by Psaltis & Norman (2000), the test particle frequencies (the same as in the RPM plus an additional frequency at $\nu_\phi + \nu_r$) are selected by the response of an annulus in the disk, when this is subject to a wide-band input noise.

If confirmed, the RPM will provide an unprecedented opportunity to measure GR effects in the strong field regime, such as the periastron precession in the vicinity of the marginally stable orbit and the radial dependence of the Lense-Thirring nodal precession frequency. In principle, accurately measured kHz QPO and HBO frequencies would yield crucial information on the compact object such as its mass and angular momentum (*e.g.* by solving Eqs. 1-3 for m, a/M and r). Should suitable, additional observables be found, it might become possible to obtain a self-consistency check of the RPM, together with tests of GR in the strong field regime.

ACKNOWLEDGEMENTS

I am grateful to my close collaborators in this project, S. Morsink and M. Vietri.

REFERENCES

Alpar, A. & Shaham, J. 1985, Nature, 316, 239
Chakrabarty D. & Morgan E.H. 1998, Nature, 394, 346
Cook, G.B., Shapiro, S.L., & Teukolsky, S.A. 1992, ApJ, 398, 203
Cui, W., Zhang, S.N., Chen, W. 1998, ApJ, 492, L53
Di Matteo, T. & Psaltis, D. 1999, ApJ, 526, L101
Ford, E.C. & van der Klis, M. 1998, ApJ, 506, L39
Ford, E.C. et al. 1998, ApJ, 508, L155
Friedman, J.L, Ipser, J.R. & Parker, L. 1986, ApJ, 304, 115
King, A.R. & Kolb, U. 1999, MNRAS, 305, 654
Lamb, F.K., Shibazaki, N., Alpar, A. & Shaham, J. 1985, Nature 317, 681
Mendez, M. et al. 1998a ApJ, 494, L65
——— 1998b ApJ, 505, L23
Mendez, M. & van der Klis, M. 1999, ApJ, 517, L51

Mendez, M., van der Klis, M., Ford, E.C., Wijnands, R. & van Paradijs, J. 1999, ApJ, 511, L49
Miller, M.C., Lamb, F.K. & Psaltis, D. 1998, ApJ, 508, 791
Morsink, S. & Stella, L. 1999, ApJ, 513, 827
Orosz, J.A. & Kuulkers, E. 1999, MNRAS, 305, 132
Psaltis, D. & Norman, C. 2000, ApJ, in press (astro-ph/0001391).
Psaltis, D., et al. 1998, ApJ, 501, L95
——— 1999, ApJ, 520, 763
Psaltis, D., Belloni, T., & van der Klis, M. 1999, ApJ, 520, 262
Shahbaz, T. et al. 1999, MNRAS, 306, 89
Stella, L. & Vietri, M. 1998a, ApJ, 492, L59
——— 1998b, in "The active X-ray sky", Nucl Phys B (Proc. Suppl.), 69, 135
——— 1999, Phys. Rev. Lett, 82, 17
Stella, L., Vietri, M. & Morsink, S.M. 1999, ApJ, 524, L63
Stergioulas, N. 1995, code available via ftp from `pauli.phys.uwm.edu` in directory `/pub/rns`
Stergioulas, N., & Friedman, J.L. 1995, ApJ 444, 306
Strohmayer, T.E. 2000, this volume
Strohmayer, T.E. et al. 1996, ApJ, 469, L9
van der Klis, M. 1995, in "X-ray Binaries", Eds. W. H. G. Lewin, J. van Paradijs & E. P. J. van den Heuvel (Cambridge University Press), p. 252
——— 1998, in: Proc. NATO ASI "The many faces of neutron stars", Series C, Vol. 515, p. 337
——— 1999, in: "The Lense-Thirring effect", Proc. of the 3rd William Fairbank meeting, Ed. R. Ruffini, C. Sigismondi (World Scientific: Singapore) in press; astro-ph/9812395
——— 2000, this volume.
van der Klis, M. et al. 1997, ApJ, 481, L97
Vietri, M. & Stella, L. 1998, ApJ, 503, 350
Wijnands R.A.D. & van der Klis M. 1998, Nature, 394, 344
Wijnands, R., van der Klis, M. & Rijkhorst, E.J. 1999, ApJ, 512, L39
Wiringa, R.B., Fiks, V. & Fabrocini, A. 1988, Phys. Rev. C, 38, 1010

OSCILLATIONS DURING THERMONUCLEAR X-RAY BURSTS: A NEW PROBE OF NEUTRON STARS

Tod E. Strohmayer

NASA's Goddard Space Flight Center, Greenbelt, MD 20771

ABSTRACT

Observations of thermonuclear (also called Type I) X-ray bursts from neutron stars in low mass X-ray binaries (LMXB) with the Rossi X-ray Timing Explorer (RXTE) have revealed large amplitude, high coherence X-ray brightness oscillations with frequencies in the 300 - 600 Hz range. Substantial spectral and timing evidence point to rotational modulation of the X-ray burst flux as the cause of these oscillations, and it is likely that they reveal the spin frequencies of neutron stars in LMXB from which they are detected. Here I review the status of our knowledge of these oscillations and describe how they can be used to constrain the masses and radii of neutron stars as well as the physics of thermonuclear burning on accreting neutron stars.

KEYWORDS: stars: neutron – stars: rotation – X-rays: bursts – equation of state

1. INTRODUCTION

During the past 25 years X-ray astronomers have expended a good deal of effort in an attempt to identify the spin frequencies of neutron stars in LMXB (see for example Wood et al. 1991; Vaughan et al. 1994). These efforts were largely prompted by the discovery in the radio band of rapidly rotating neutron stars, the millisecond radio pulsars (see Backer et al. 1984), and subsequent theoretical work suggesting their origin lie in an accretion-induced spin-up phase of LMXB (see the review by Bhattacharya 1995 and references therein). However, up to the mid-90's there was little or no direct evidence to support the existence of rapidly spinning neutron stars in LMXB. This situation changed dramatically with the launch of the *Rossi X-ray Timing Explorer* (RXTE) in December, 1995. Within a few months of its launch RXTE observations had provided strong evidence suggesting that neutron stars in LMXB are spinning with frequencies $\geq$ 300 Hz. These first indications came with the discovery of high frequency (millisecond) X-ray brightness oscillations, "burst oscillations," during thermonuclear (Type I) X-ray bursts from several neutron star LMXB systems (see Strohmayer et al. 1996; Smith, Morgan & Bradt 1997; Zhang et al. 1996).

At present these oscillations have been observed from six different LMXB systems (see Strohmayer, Swank & Zhang 1998). The observed frequencies are in the range from $\approx$ 300 – 600 Hz, similar to the observed frequency distribution of binary millisecond radio pulsars (Taylor, Manchester & Lyne 1993), and consistent with some theoretical determinations of spin periods which can be reached via accretion-induced spin-up (Webbink, Rappaport & Savonije 1983). In this contribution I will review our observational understanding of these oscillations, with emphasis on how they can be understood in the context of spin modulation of the X-ray burst flux. I will discuss how detailed modelling of the oscillation amplitudes and harmonic structure can be used to place interesting constraints on the masses and radii of neutron stars and therefore the equation of state of supranuclear density matter. Inferences which can be drawn regarding the physics of thermonuclear burning will also be discussed. I will conclude with some outstanding theoretical questions and uncertainties and where future observations and theoretical work may lead.

2. OBSERVATIONAL PROPERTIES OF BURST OSCILLATIONS

Burst oscillations with a frequency of 363 Hz were first discovered from the LMXB 4U 1728-34 by Strohmayer et al. (1996). Since then an additional five sources with burst oscillations have been discoverd. The burst oscillation sources and their observed frequencies are given in table 1. In the remainder of this section I review the important observational properties of these oscillations and attempt to lay out the evidence supporting the spin modulation hypothesis.

2.1. Oscillations at burst onset

Many bursts show detectable oscillations during the $\approx 1 - 2$ s risetimes typical of thermonuclear bursts. For example, Strohmayer, Zhang & Swank (1997) showed that some bursts from 4U 1728-34 have oscillation amplitudes as large as 43 % within 0.1 s of the observed onset of the burst. They also showed that the oscillation amplitude decreased monotonically as the burst flux increased during the rising portion of the the burst lightcurve. Figure 1 shows this behavior in a burst from 4U 1636-53. This burst had an oscillation amplitude near onset of $\approx$ 80 %, and then showed an episode of radius expansion beginning near the time when the oscillation became undectable (see Strohmayer et al. 1998a). The presence of modulations of the thermal burst flux approaching 100 % right at burst onset fits nicely with the idea that early in the burst there exists a localized hot spot which is then modulated by the spin of the neutron star. In this scenario the largest modulation amplitudes are produced when the spot is smallest, as the spot grows to encompass more of the neutron star surface, the amplitude drops, consistent with the observations.

X-ray spectroscopy during burst rise also suggests that the emission is localized near the onset of bursts. Prior to RXTE few instruments had the collecting area and temporal resolution to study spectral evolution during the short rise times of thermonuclear bursts. Day & Tawara (1990) used GINGA observations of 4U 1728-34 in an attempt to constrain the e-folding spreading time of the burning front to ≈ 0.1 s in two bursts. With RXTE Strohmayer, Zhang & Swank (1997) investigated the spectral evolution of bursts from 4U 1728-34. They fit a black body model to

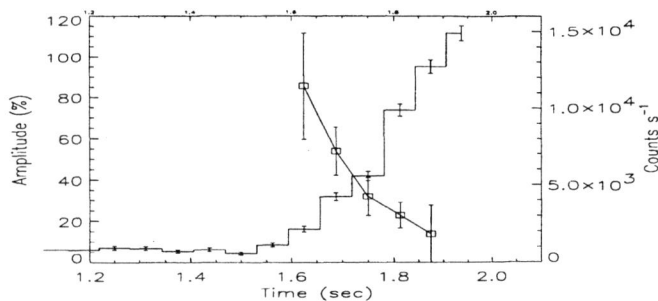

FIGURE 1. The amplitude of oscillations at 580 Hz during the rising phase of a burst from 4U 1636-53. The amplitude is greatest near the onset of the burst and decreases as the burst flux increases.

TABLE 1. Burst oscillation sources and frequencies

Object	Frequency (Hz)
4U 1728-34	363
4U 1636-53	580 (290)
4U 1702-429	330
KS 1731-26	526
Aql X-1	549
Gal. Center	589

intervals during several bursts and plotted the flux F_{bol} versus $F_{bol}^{1/4}/kT_{BB}$. For black body emission from a spherical surface this ratio is a constant proportional to $(R/d)^{1/2}$, where d and R are the source distance and radius, respectively. Figure 2 shows such a plot for a burst from 4U 1728-34. In this plot the solid line connects successive time intervals, with the burst beginning in the lower left and evolving diagonally to the upper right and then across to the left. This evolution indicates that the X-ray emitting area is *not* constant, but *increases* with time during the burst rise. The spectra of type I bursts are not true black bodies (see London, Taam & Howard 1986; Ebisuzaki 1987; Lewin, van Paradijs & Taam 1993), however, the argument here concerns the energetics and not the detailed shape of the spectrum. Since the effect is seen in bursts that do not show photospheric radius expansion the atmosphere is always geometrically thin compared with the stellar radius, so the physics of spectral formation depends only on conditions locally. The most straightforward interpretation is that the deficit in $F_{bol}^{1/4}/kT_{BB}$ evident at burst onset reflects a localized region of X-ray emission which then spreads to encompass the neutron star.

2.2. Expectations from the theory of thermonuclear burning

The thermonuclear instability which triggers an X-ray burst burns in a few seconds the fuel which has been accumulated on the surface over several hours. This $> 10^3$ difference between the accumulation and burning timescales means that it is unlikely that the conditions required to trigger the instability will be achieved simultaneously over the entire stellar surface. This realization, first emphasized by Joss (1978), led to the study of lateral propagation of the burning over the neutron star surface (see Fryxell & Woosley 1982, Nozakura, Ikeuchi & Fujimoto 1984, and Bildsten 1995). The subsecond risetimes of thermonuclear X-ray bursts suggests that convection plays an important role in the physics of the burning front propagation, especially in the low accretion rate regime which leads to large ignition columns (see Bildsten (1998) for a review of thermonuclear burning on neutron stars). Bildsten (1995) has shown that pure helium burning on neutron star surfaces is in general inhomogeneous, displaying a range of behavior which depends on the local accretion rate, with low accretion rates leading to convectively combustible accretion columns and standard type I bursts, while high accretion rates lead to slower, nonconvective propagation which may be manifested in hour long flares. These studies emphasize that the physics of thermonuclear burning is necessarily a multi-dimensional problem and that *localized* burning is to be expected, especially at the onset of bursts. The properties of oscillations near burst onset described above fit well into this picture of thermonuclear burning on neutron stars.

Miller (1999) has recently found evidence for a significant 290 Hz subharmonic of the strong 580 Hz oscillation seen in bursts from 4U 1636-53. Evidence for the subharmonic was found by adding together in phase data from the rising intervals of 5 bursts. This result suggests that in 4U 1636-53 the spin frequency is 290 Hz, and that the strong signal at 580 Hz is caused by nearly antipodal hot spots. If correct this result has interesting implications for the physics of nuclear burning, in particular, how the burning can be spread from one pole to the other within a few tenths of seconds, and how fuel is pooled at the poles, perhaps by a magnetic field.

2.3. The coherence of burst oscillations

One of the most interesting aspects of the burst oscillations is the frequency evolution evident in many bursts. The frequency is observed to increase by $\approx 1 - 3$ Hz in the cooling tail, reaching a plateau or asymptotic limit (see Strohmayer et al. 1998a). An example of this behavior in a burst from 4U 1702-429 is shown in figure 3. However, increases in the oscillation frequency are not universal, Strohmayer (1999) and Miller (1999) have recently reported on an episode of *spin down* in the cooling tail of a burst from 4U 1636-53. Frequency evolution has been seen in five of the six burst oscillation sources and appears to be commonly associated with the physics of the modulations. Strohmayer et. al (1997) have argued this evolution results from angular momentum conservation of the thermonuclear shell. The burst expands the shell, increasing its rotational moment of inertia and slowing its spin rate. Near burst onset the shell is thickest and thus the observed frequency lowest.

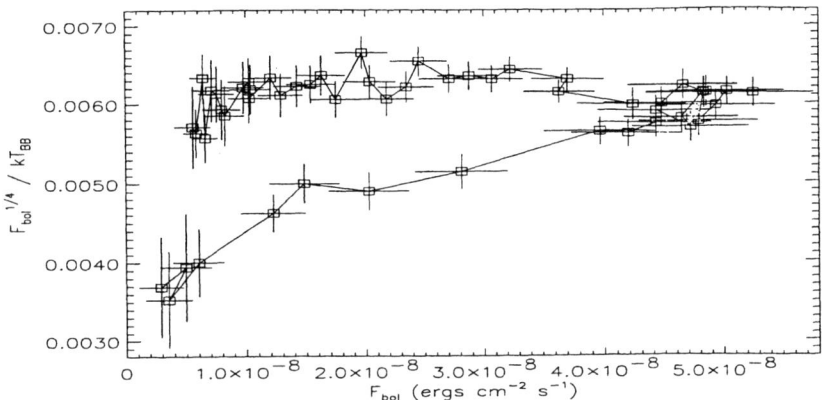

FIGURE 2. Bolometric flux F_{bol} versus $F_{bol}^{1/4}/kT_{BB}$ for a burst from 4U 1728-34. The burst evolves from lower left to upper right and then to the left. This behavior is strong evidence for an increasing X-ray emission area during the burst rise.

The shell then spins back up as it cools and recouples to the bulk of the neutron star. Calculations indicate that the ~ 10 m thick pre-burst shell expands to ~ 30 m during the flash (see Joss 1978; Bildsten 1995), which gives a frequency shift due to angular momentum conservation of $\approx 2\,\nu_{spin}(20\text{ m}/R)$, where ν_{spin} and R are the stellar spin frequency and radius, respectively. For the several hundred Hz spin frequencies inferred from burst oscillations this gives a shift of ~ 2 Hz, similar to that observed.

In bursts where frequency drift is evident the drift broadens the peak in the power spectrum, producing quality factors $Q \equiv \nu_0/\Delta\nu_{FWHM} \approx 300$. In some bursts a relatively short train of pulses is observed during which there is no strong evidence for a varying frequency. Recently, Strohmayer & Markwardt (1999) have shown that with accurate modeling of the frequency drift quality factors as high as $Q \sim 4,000$ are achieved in some bursts. They modelled the frequency drift and showed that a simple exponential "chirp" model of the form $\nu(t) = \nu_0(1 - \delta_\nu \exp(-t/\tau))$, works remarkably well. The resulting quality factors derived from the frequency modelling are very nearly consistent with the factors expected from a perfectly coherent signal of finite duration equal to the length of the data trains in the bursts. These results argue strongly that the mechanism which produces the modulations is a highly coherent process, such as stellar rotation, and that the asymptotic frequencies observed during bursts represent the spin frequency of the neutron star.

2.4. The long-term stability of burst oscillation frequencies

The accretion-induced rate of change of the neutron star spin frequency in a LMXB is approximately 1.8×10^{-6} Hz yr^{-1} for typical neutron star and LMXB parameters.

The Doppler shift due to orbital motion of the binary can produce a frequency shift of magnitude $\Delta\nu/\nu = v\sin i/c \approx 2.05 \times 10^{-3}$, again for canonical LMXB system parameters. This doppler shift easily dominates over any possible accretion-induced spin change on orbital to several year timescales. Therefore the extent to which the observed burst oscillation frequencies are consistent with possible orbital Doppler shifts, but otherwise stable over $\approx$ year timescales, provides strong support for a highly coherent mechanism which sets the observed frequency.

At present, the best source available to study the long term stability of burst oscillations is 4U 1728-34. Strohmayer et al. (1998b) compared the observed asymptotic frequencies in the decaying tails of bursts separated in time by ≈ 1.6 years. They found the burst frequency to be highly stable, with an estimated time scale to change the oscillation period of about 23,000 year. It was also suggested that the stability of the asymptotic periods might be used to infer the X-ray mass function of LMXB by comparing the observed asymptotic period distribution of many bursts and searching for an orbital Doppler shift.

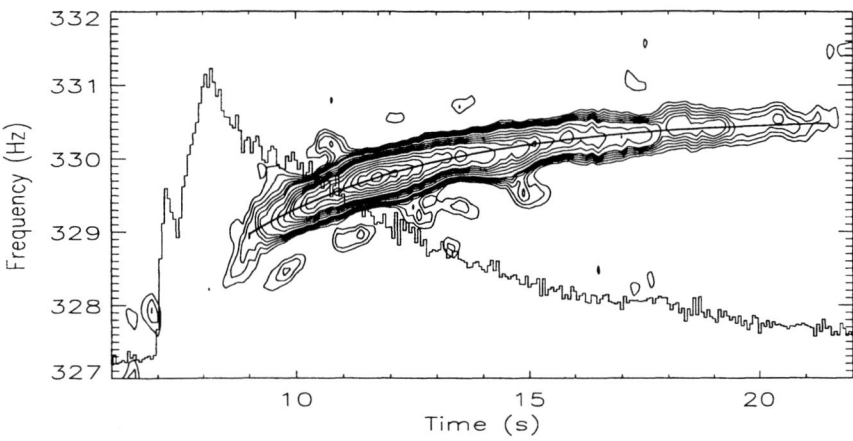

FIGURE 3. A dynamic power density spectrum of a burst from 4U 1702-429 showing frequency drift toward an asymptotic limit. The solid curve is a best fit using an exponential recovery.

3. BURST OSCILLATIONS AS PROBES OF NEUTRON STARS

Detailed studies of the burst oscillation phenomenon hold great promise for providing new insights into a variety of physics issues related to the structure and evolution of neutron stars. In particular, the burst oscillations have given astronomers their first direct method to investigate the two dimensional nature of nuclear flame front propagation. In this section I will outline how the burst oscillations can be used to probe neutron stars.

3.1. Mass - Radius constraints and the EOS of dense matter

Using the rotating hot spot model it is possible to determine constraints on the mass and radius of the neutron star from measurements of the maximum observed modulation amplitudes during X-ray bursts as well as the harmonic content of the pulses. The physics that makes such constraints possible is the bending of photon trajectories in a strong gravitational field. The strength of the deflection is a function of the stellar compactness, GM/c^2R, with more compact stars producing greater deflections and therefore weaker spin modulations. An upper limit on the compactness can be set since a star more compact than this limit would not be able to produce a modulation as large as that observed. Complementary information comes from the pulse shape, which can be inferred from the strength of harmonics. Information on both the amplitude and harmonic content can thus be used to bound the compactness. Detailed modelling, during burst rise for example, can then be used to determine a confidence region in the mass - radius plane for neutron stars. Miller & Lamb (1998) have investigated the amplitude of rotational modulation pulsations as well as harmonic content assuming emission from a point-like hot spot. They also show that knowledge of the angular and spectral dependence of the emissivity from the neutron star surface can have important consequences for the derived constraints. More theoretical as well as data modelling in this area are required.

3.2. Doppler shifts and pulse phase spectroscopy

Stellar rotation will also play a role in the observed properties of spin modulation pulsations. For example, a 10 km radius neutron star spinning at 400 Hz has a surface velocity of $v_{spin}/c \leq 2\pi\nu_{spin}R \approx 0.084$ at the rotational equator. This motion of the hot spot produces a Doppler shift of magnitude $\Delta E/E \approx v_{spin}/c$, thus the observed spectrum is a function of pulse phase (see Chen & Shaham 1989). Measurement of a pulse phase dependent Doppler shift in the X-ray spectrum would provide additional evidence supporting the spin modulation model and also yields a means of constraining the neutron star radius, perhaps one of the few direct methods to infer this quantity for neutron stars.

The rotationally induced velocity also produces a relativistic aberration which results in asymmetric pulses, thus the pulse shapes also contain information on the spin velocity and therefore the stellar radius (Chen & Shaham 1989). The component of the spin velocity along the line of site is proportional to $\cos\theta$, where θ is the latitude of the hotspot measured with respect to the rotational equator. The modulation amplitude also depends on the latitude of the hotspot, as spots near the rotational poles produce smaller amplitudes than those at the equator. Thus a correlation between the observed oscillation amplitude and the size of any pulse phase dependent Doppler shift is to be expected. Dectection of such a correlation in a sample of bursts would provide strong confirmation of the rotational modulation hypothesis.

Searches for a Doppler shift signature are just beginning to be carried out. Studies in single bursts have shown that spectral variations with pulse phase can be detected (see Strohmayer, Swank, & Zhang 1998). The varations with pulse phase show a 4-5 % modulation of the fitted black body temperature, consistent with the idea that a temperature gradient is present on the stellar surface, which when rotated produces the flux modulations. Ford (1999) has analysed data during a burst from Aql X-1 and finds that the softer photons lag higher energy photons in a manner which is qualitatively similar to that expected from a rotating hot spot. Strohmayer & Markwardt (1999) have shown that signals from multiple bursts can be added in phase by modelling the frequency drifts present in individual bursts. This provides a stronger signal with which to test for Doppler shift effects. So far, burst oscillation signals from 4U 1702-429 have been added in phase in an attempt to identify a rotational Doppler shift. A difficulty in analysing the phase resolved spectra from bursts is the systematic change in the black body temperature produced as the surface cools. A simpler measure of the spectral hardness, rather than the black body temperature, is the mean energy channel of the spectrum. I computed the distribution of mean channels in the RXTE proportional counter array (PCA) as a function of pulse phase using spectra from 4 different bursts from 4U 1702-429. Figure 4 shows the results. A strong modulation of the mean PCA channel is clearly seen. There is a hint of an asymmetry in that the leading edge of the pulse appears harder (as expected for a rotational Doppler shift) than the trailing edge, but the difference does not have a high statistical significance. More data will be required to decide the rotational Doppler shift issue.

3.3. Physics of thermonuclear burning

The properties of burst oscillations can tell us a great deal about the processes of nuclear burning on neutron stars. The amplitude evolution during the rising phase of bursts contains information on how rapidly the flame front is propagating. If the anitpodal spot hypothesis to explain the presence of a subharmonic in 4U 1636-53 is correct, then it has important implications for the propagation of the instability from one pole to another in ≈ 0.2 s (see Miller 1999). In addition, a two pole flux anistropy suggests that the nuclear fuel is likely pooled by some mechanism, perhaps associated with the magnetic field of star. Further detections and study of the subharmonic in 4U 1636-53 could shed more light on these issues.

Until recently, much of the work concerning burst oscillations has concentrated on studies of the pulsations themselves and their relation to individual bursts. With the samples of bursts growing it is now possible to concieve of more global studies which correlate the properties of oscillations with other measures of these sources, for example, their spectral state and mass accretion rate. This will allow researchers to investigate the system parameters which determine the likelihood of producing bursts which show oscillations. Such investigations will provide insight into how the

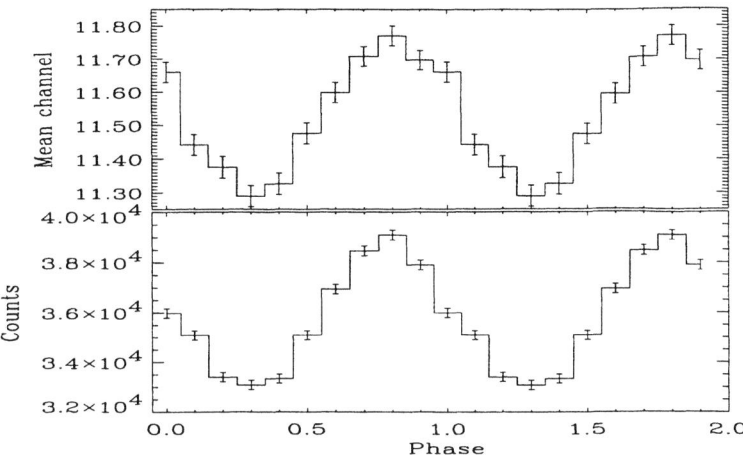

FIGURE 4. Pulse phase spectral variations in bursts from 4U 1702-429. The top panel shows the mean PCA energy channel as a function of pulse phase for 4 bursts which were co-added in phase. The bottom panel shows the pulse profile in the 2 - 24 keV band.

properties of thermonuclear burning (as evidenced in the presence or absence of oscillations) are influenced by other properties of the system. Furthermore, we can test if theoretical predictions of how the burning should behave are consistent with the hypothesis that the oscillations result from rotational modulation of nonuniformities produced by thermonuclear burning. Initial work in this regard suggests that bursts which occur at higher mass accretion rates show stronger burst oscillations more often (see Franco et al. 1999). Although preliminary this result appears roughly consistent with theoretical descriptions of the thermonuclear burning which indicates an evolution from vigorous, rapid (thus uniform) burning at lower mass accretion rates (lower persistent count rates) to weaker, slower burning (thus more non-uniform) at higher mass accretion rates (see Bildsten 1995, for example).

4. REMAINING PUZZLES AND THE FUTURE

Although much of the burst oscillation phenomenology is well described by the spin modulation hypothesis several important hypotheses need to be confronted with more detailed theoretical investigations. Perhaps the most interesting is the mechanism which causes the observed frequency drifts. Expressed as a phase slip the frequency drifts seen during the longest pulse trains correspond to about 5 - 10 revolutions around the star. Whether or not a shear layer can persist that long needs to be further investigated. The recent observation of a, so far unique, spin down in the decaying tail of a burst from 4U 1636-53 (see Strohmayer 1999), which might be the first detection of the spin down caused by thermal expansion of the burning layers, needs to be better understood in the context of thermonuclear energy release at late times in bursts. Another perplexing issue is the mechanism which allows flux asymmetries to both form and then persist at late times in bursts.

Although RXTE provided the technical advancements required to discover the burst oscillations, it may take future, larger area instruments such as Constellation-X or a successor timing mission to RXTE to fully exploit their potential for unlocking the remaining secrets of neutron stars.

REFERENCES

Backer, D. C., Kulkarni, S. R., Heiles, C., Davis. M. M. & Goss, W. M. 1982, Nature, 300, 615
Bhattacharya, D. 1995,in *X-ray Binaries*, ed. W. H. G. Lewin, J. Van Paradijs, & E. P. J. Van den Heuvel, (Cambridge: Cambridge Univ. Press), p 233
Bildsten, L. 1995, ApJ, 438, 852
Bildsten, L. 1998, in "The Many Faces of Neutron Stars", ed. R. Buccheri, A. Alpar & J. van Paradijs (Dordrecht: Kluwer), p. 419
Chen, K. & Shaham, J. 1989, ApJ, 339, 279
Day, C. S. R. & Tawara, Y. 1990, MNRAS, 245, 31P
Ebisuzaki, T. 1987, PASJ, 39, 287
Ford, E. C. 1999, ApJ, 519, L73
Franco, L. et al. 1999, in preparation
Fryxell, B. A., & Woosley, S. E. 1982, ApJ, 261, 332
Joss, P. C. 1978, ApJ, 225, L123
Lewin, W. H. G., van Paradijs, J. & Taam, R. E. 1993, Space Sci. Rev., 62, 233
London, R. A., Taam, R. E. & Howard, W. M. 1986, ApJ, 306, 170
Miller, M. C. 1999a, ApJ, 515, L77
Miller, M. C. 1999b, ApJ, submitted
Miller, M. C., Lamb, F. K. & Psaltis, D. 1998, ApJ, 508, 791
Miller, M. C. & Lamb, F. K. 1998, ApJ, 499, L37
Nozukura, T., Ikeuchi, S., & Fujimoto, M. Y. 1984, ApJ, 286, 221
Smith, D., Morgan, E. H. & Bradt, H. V. 1997, ApJ, 479, L137
Strohmayer, T. E. 1999, ApJ, 523, L51
Strohmayer, T. E. & Markwardt, C. B. 1999, ApJ, 516, L81
Strohmayer, T. E., Swank, J. H., & Zhang, W. 1998, Nuclear Phys B (Proc. Suppl.) 69/1-3, 129-134
Strohmayer, T. E., Zhang, W., Swank, J. H., White, N. E. & Lapidus, I. 1998a, ApJ, 498, L135
Strohmayer, T. E., Zhang, W., Swank, J. H. & Lapidus, I. 1998b, ApJ, 503, L147
Strohmayer, T. E., Zhang, W. & Swank, J. H. 1997, ApJ, 487, L77
Strohmayer, T. E. et al. 1996, ApJ, 469, L9
Taylor, J. H., Manchester, R. N. & Lyne, A. G. 1993, ApJS, 88, 529
Vaughan, B. A. et al. 1994, ApJ, 435 362
Webbink, R. F., Rappaport, S. A. & Savonije, G. J. 1983, ApJ, 270, 678
Wood, K. S. et al. 1991, ApJ, 379, 295
Zhang, W., Lapidus, I., Swank, J. H., White, N. E. & Titarchuk, L. 1996, IAUC 6541

EARLY RESULTS FROM THE CHANDRA OBSERVATORY

Harvey Tananbaum

Harvard-Smithsonian Center for Astrophysics

KEYWORDS: X-rays: general, instrumentation: detectors and telescopes, Chandra

1. INTRODUCTION

The launch of the Chandra Observatory on July 23, 1999 has opened a new era of high spatial resolution and high spectral resolution X-ray astronomy. Here we present a description of the Observatory and an overview of the results from the first three weeks of science calibration observations.

2. DESCRIPTION OF CHANDRA

Figure 1 is a diagram of the Chandra Observatory showing the High Resolution Mirror Assembly (HRMA), the Aspect Camera, the two Transmission Gratings, and the Optical Bench connecting the telescope to the two focal plane detectors 10m away. The latter are: the Advanced Charge-Coupled Imaging Spectrometer (ACIS) and the High Resolution Camera (HRC). Overall the observatory is ~ 14m long, measures ~ 20m across from one tip of the solar array to the other, and weighs nearly 5000kg.

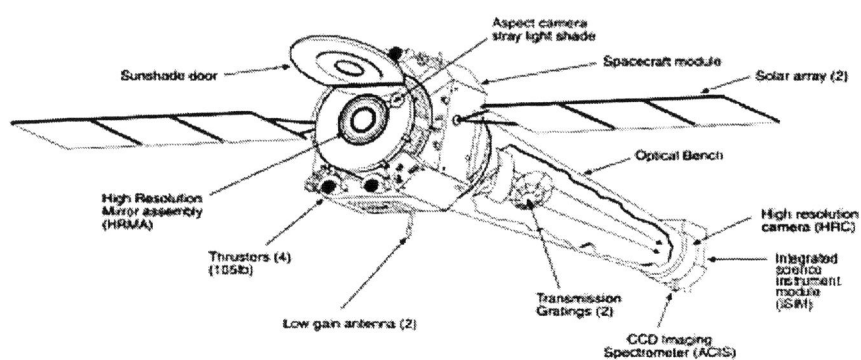

FIGURE 1. Chandra - Schematic

The four nested pairs of paraboloids and hyperboloids which comprise the HRMA are made from zerodur blanks (supplied by Schott-Glaswerke), shaped and polished at Hughes Danbury Optical Systems, coated with Iridium by Optical Coating Laboratories, Inc., and assembled into a single telescope by Eastman Kodak (see Figure 2). Leon Van Speybroeck (Smithsonian Astrophysical Observatory) is the Telescope Scientist. The ACIS consists of two sets of charge-coupled devices, a 2x2 array called

FIGURE 2: Mirror Assembly at EK   FIGURE 3: ACIS Hardware

ACIS-I primarily intended for imaging studies and a 1x6 array called ACIS-S primarily intended for readout of the High Energy Transmission Grating (HETG). Illustrated in Figure 3, the ACIS was built by Pennsylvania State University (PI - Gordon Garmire), MIT, Lincoln Laboratories, and Lockheed-Martin.

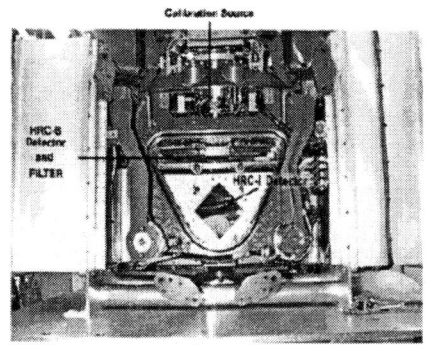

FIGURE 4: HRC Hardware   FIGURE 5: Grating Hardware

Figure 4 shows the High Resolution Camera built by the Smithsonian Astrophysical Observatory (PI - Steve Murray) using microchannel plates supplied by Galileo Electro-Optics and Photonis. There are two HRC units, a diamond-shaped detector for imaging and a rectangular unit consisting of three MCPs primarily intended for readout of the Low Energy Transmission Grating (LETG). The HRC and ACIS are somewhat complementary in that the HRC has a larger field of view, higher time resolution, and slightly better spatial resolution; while ACIS has higher quantum efficiency (above 0.5 keV) and much better intrinsic spectral resolution. The two gratings are pictured in Figure 5 which shows two structures, each comprised of a set of four concentric rings which can be inserted and retracted into the optical path behind the mirror. The LETG (PI - Bert Brinkman) was built by Space Research Organization of the Netherlands (SRON) and Max Planck Institut (MPI), and the HETG (PI - Claude Canizares) was built by Massachusetts Institute of Technology (MIT).

FIGURE 6: Chandra and IUS in the Shuttle Bay

FIGURE 7: Launch

The spacecraft (pointing and altitude control system, on-board computers, command and communication system, electrical power system, solar arrays, propulsion, etc.) was provided by TRW and their contractor team. Ball Aerospace & Technology designed and built the Integrated Science Instrument Module which houses and moves the HRC and ACIS. Figure 6 shows the integrated Chandra atop the Inertial Upper Stage in the Shuttle Orbiter Bay, and Figure 7 shows the actual launch.

3. LAUNCH, TRANSFER ORBIT, AND INITIAL ACTIVATION

About 7 hours after liftoff by STS-93 (Columbia), Chandra was deployed from the orbiter bay. Then the two-stage Inertial Upper Stage built by Boeing boosted Chandra into a highly elliptical orbit, the solar arrays deployed, and Chandra successfully separated from the IUS (within 10 hours of Columbia's liftoff). Over the next several days, five burns of Chandra's Integral Propulsion System (built by TRW) boosted the observatory to the intended 10,000 km x 140,000 km orbit. The HRC vacuum door was released and then the ACIS protective door was also opened. Contamination covers were retracted, starting with those at the back of the HRMA (so that ACIS could check for buildup of molecular contaminants over the preceding two years by viewing radioactive sources built into the back of the forward mirror covers - no contaminants were detected). When the front HRMA cover (also serving as the sunshade door) was opened on August 12, celestial X-rays were detected for the first time and the Aspect Camera (built by Ball Aerospace) was able to detect and lock on to guide stars thereby providing stable 3-axis pointing (and pointing knowledge) for the Observatory.

4. SCIENCE CALIBRATION - FIRST IMAGES AND SPECTRA

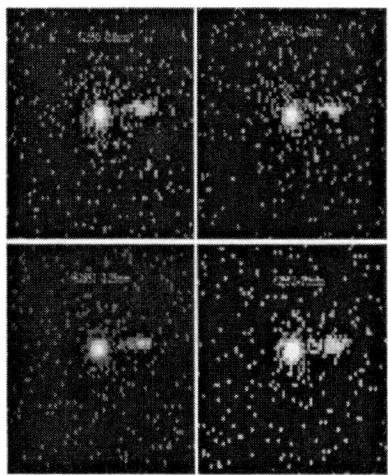
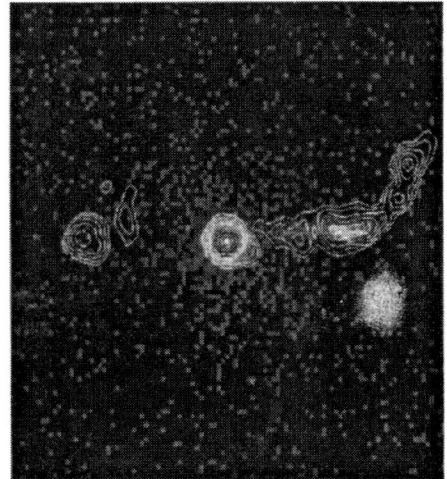

FIGURE 8: Plate focus for PKS 0637-752

FIGURE 9: PKS 0637 X-ray Image and Radio Contours

One of the first celestial targets was the field containing the radio loud quasar PKS0637-752 at a redshift of 0.654. The plan was for Chandra to observe this "point-like" source at a series of different focal plane settings, by moving the ACIS-S detectors forward and aft of the expected system best focus. In fact, the four images displayed in Figure 8 illustrate that the approximately round central image does go through a minimum size. However, the data also provide a surprise with an X-ray jet extending $\sim 10''$ outwards from the central source. At the distance of PKS0637-752, this jet has a linear dimension of ~ 80 kpc, more than large enough to fit the Milky Way Galaxy into the space between the nucleus and the bright spot in the jet some $8''$ away.

Figure 9 shows radio contours from a nearly concurrent observation with the Australia Telescope (J. Lovell, private communication) showing a peak in the radio jet coincident with the peak in the X-ray jet. The sharpness of the Chandra Point Spread Function (PSF) is invaluable, allowing us to unambiguously resolve the peak in the jet from the nuclear source at the $8''$ separation, even though the peak in the jet is only $\sim 7\%$ of the intensity of the central source.

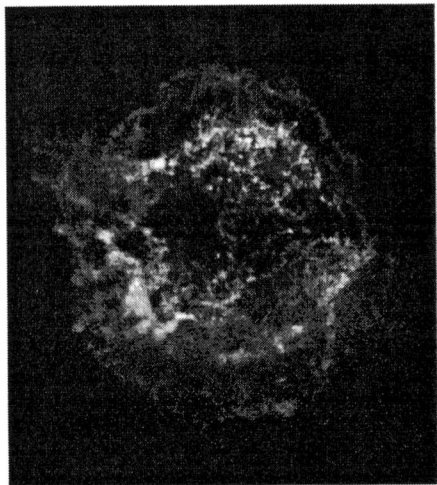

FIGURE 10: Cas A - First Light Image   FIGURE 11: Cas A - Energy Spectrum

Having focussed the telescope system, notwithstanding the presence of the jet in the quasar, we observed the supernova remnant Cas A for an official "First Light Picture". Figure 10 shows the incredibly detailed structure seen with Chandra, and also provides another discovery with the point-like source seen almost in the center of the remnant. It is very likely that this point source is the collapsed star (neutron star or black hole) formed at the time of the explosion. The point source has a counting rate of $\sim 1/3000$ that of the overall X-ray remnant. Assuming a similar energy spectrum for the point source and the rest of the remnant and not correcting for interstellar absorption, we obtain a luminosity of order 10^{32} erg s^{-1} for the point source. Noting that interstellar absorption in this direction is ~ 1.2 x

10^{22} cm^{-2}, we could fit a black body spectrum which corrected for absorption and extended down to 0.01 keV would have a luminosity of order 10^{34} erg s^{-1}. Clearly, a larger exposure (First Light was 5000 sec with telemetry saturation resulting in an effective exposure time of $\sim$ 3000 sec) will provide the photons needed for an accurate spectral fit, with the Chandra spatial resolution allowing the spectrum of the point source to be separated from the extended remnant emission. Figure 11 shows the ACIS S-3 (Back-Illuminated CCD) spectrum of Cas A integrated over the source with prominent lines of Mg, Si, S, Ar, Ca, and Fe. Again, longer exposures should provide comparable spectra spatially resolved across and around the remnant.

Figures 12 and 13 illustrate the results from a subsequent plate focus on another radio loud quasar PKS 0312-77 (redshift=0.223), without a jet this time. The images show a clear cut minimum size. The best focus data (no longer discernible from the other curves in Figure 13 without color) show that 65% of the X-rays fall within a circle of 1" diameter (1 pixel radius) and 90% fall within a 2" diameter. The performance actually exceeds the Chandra requirements including the HRMA, ACIS, and Aspect Camera specifications.

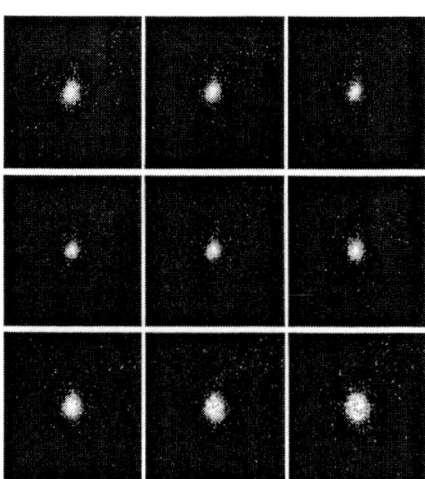
FIGURE 12: Plate focus for PKS 0312-77

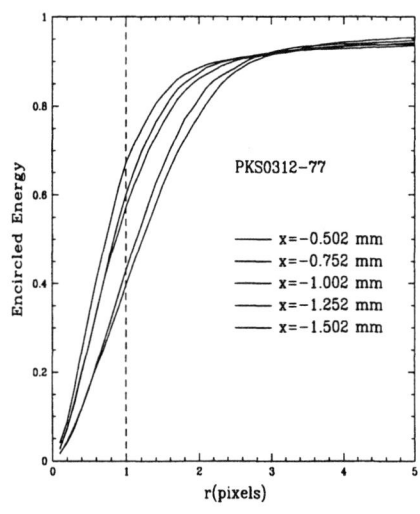
FIGURE 13: Encircled Energy vs radius for PKS 0312-77

Figure 14 shows an ACIS image of the supernova remnant E0102-72 in the Small Magellanic Cloud. The remnant is $\sim 40''$ in diameter and is thought to be 1000-2000 years old. The outer edge of the remnant seems to be quite sharp in most directions suggesting that the shock is expanding into a relatively uniform medium. ACIS and subsequent HETG observations will be used to generate spatially resolved spectra tracing the oxygen, neon, and other abundant elements over the face of the remnant. Figure 15 shows an ACIS image of the supernova remnant G21.5-0.9 in the plane of

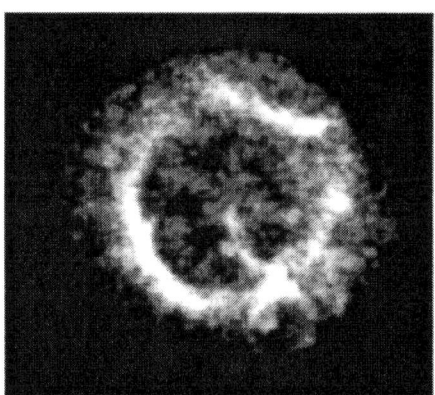
FIGURE 14: E0102-72 Image

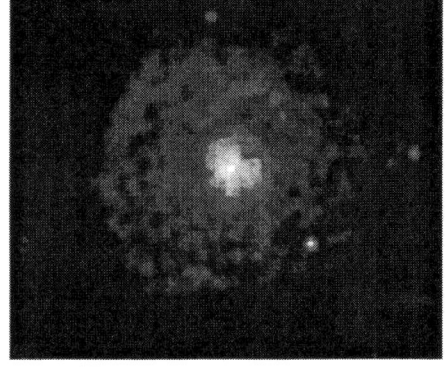
FIGURE 15: G21.5-0.9 Image

our Galaxy. Previously reported as a "Crab-like" plerion $\sim 1.5'$ in diameter, G21.5-0.9 shows for the first time here a point-like central source (possibly the neutron star powering the plerion). We also see that the plerionic emission is embedded in a fainter disk or shell $\sim 4'$ across with an integrated X-ray flux about 2% of that of the plerionic component. Most likely we are seeing a composite shell and plerion source perhaps similar to that previously reported for the remnant 0540-69 in the Large Magellanic Cloud. Spectral comparisons of the plerion and the more extended emission may be quite enlightening as should a search for pulsations from the point-like source embedded in the plerion.

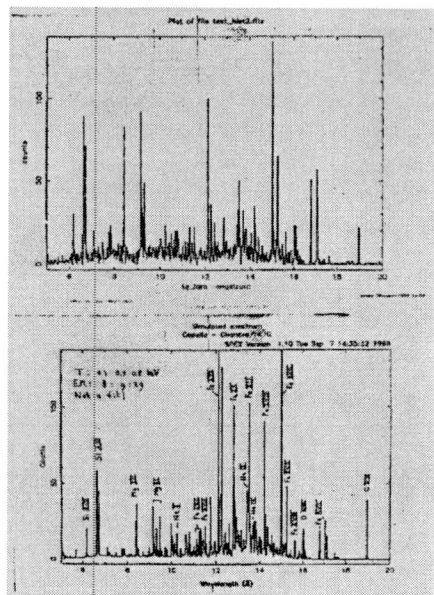
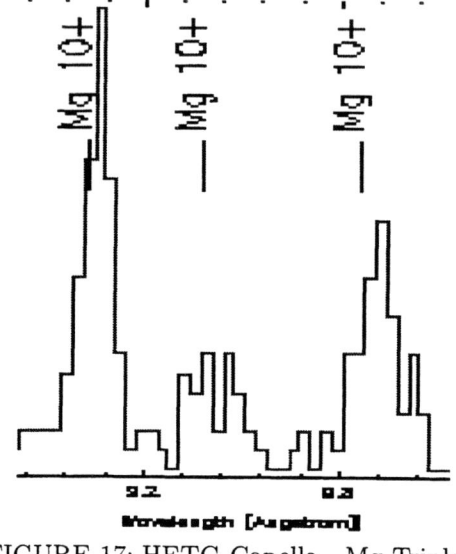

FIGURE 16: (Top) HETG Capella Spectrum; (Bottom) 3-component Model

FIGURE 17: HETG Capella - Mg Triplet

In Figures 16 and 17 we see the "First Light Spectrum" of Capella obtained with the High Energy Transmission Grating. The upper half of Figure 16 shows a line-dominated coronal spectrum with copious emission from Si, Mg, Ne, Fe, and O, and very little underlying continuum. The lower half shows a model with three temperature components capable of predicting most of the observed lines but not fitting the details of the spectrum particularly well. Figure 17 shows an expanded view of the Mg 10^+ triplet with the resonance, intercombination, and forbidden lines all clearly detected and available for use in determining the density of the emitting region(s). The resolution $E/\Delta E$ at 9Å exceeds 500 (reaching 1000 at the longest wavelengths observed with the HETG and ~ 2000 at the longest wavelengths subsequently observed with the LETG), meeting the Chandra specifications and clearly demonstrating the feasibility and value of high resolution X-ray spectroscopy.

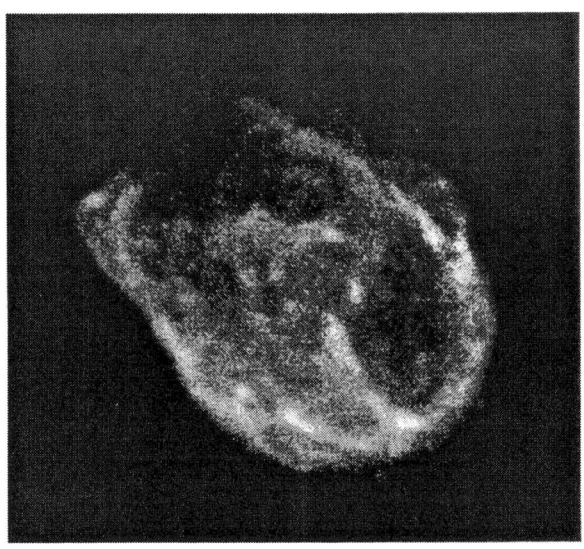

FIGURE 18. HRC Image of N132D

Shortly before this paper was presented the High Resolution Camera obtained its first celestial image, the supernova remnant N132D in the Large Magellanic Cloud, which is shown in Figure 18. The detailed structures seen in this image challenge interpretation - it may be that we are seeing more than one shell suggesting the possibility of two supernova explosions with interacting (or superimposed) ejecta. Or it may be that the surrounding circumstellar and interstellar material has a very complicated structure and is imprinting some of that information onto the X-ray image.

As this paper was presented, the first Low Energy Transmission Grating spectra of Capella were obtained, again demonstrating expected performance and completing the successful activation of the Observatory.

ACKNOWLEDGEMENTS

The Chandra project would not have been possible without the incredible efforts and talents of thousands of people who worked long and hard to make Chandra a success. In addition to the PIs named in the text, I would like to single out the contributions of Martin Weisskopf, the NASA Project Scientist, and Fred Wojtalik, the NASA Program Manager. The Chandra Program at SAO was supported under Contract-NAS8 39073.

REFERENCES

Additional information on the Chandra instrumentation, images, and spectra can be located at: http://chandra.harvard.edu, and NASA and other science websites listed in these pages.

RESULTS FROM X-RAY SURVEYS WITH ASCA

Yoshihiro Ueda

Institute of Space and Astronautical Science, Kanagawa 229-8510, Japan

ABSTRACT

We present main results from X-ray surveys performed with *ASCA*, focusing on the *ASCA* Large Sky Survey (LSS), the Lockman Hole deep survey, and the *ASCA* Medium Sensitivity Survey (AMSS or the GIS catalog project). The Log N - Log S relations, spectral properties of sources, and results of optical identification are summarized. We discuss implications of these results for the origin of the CXB.

KEYWORDS: diffuse radiation — surveys — galaxies: active — X-rays: galaxies

1. INTRODUCTION

Understanding the origin of the Cosmic X-ray Background (CXB or XRB) and cosmological evolution of X-ray extragalactic populations is one of the main goals of X-ray astronomy. In the soft X-ray band, the *ROSAT* satellite resolved 80% of the 0.5–2 keV CXB into individual sources (Hasinger et al. 1998) and optical identification revealed that the major population is type-I AGNs (Schmidt et al. 1998). Because of the technical difficulties, imaging sky surveys in the hard X-ray band (above 2 keV), where the bulk of the CXB energy arises, were not available until the launch of *ASCA*. The sensitivity limits achieved by previous mission such as *HEAO1* (Piccinotti et al. 1982) and *Ginga* (Kondo et al. 1991) are at most $\sim 10^{-11}$ erg s^{-1} cm^{-2} (2–10 keV), and the sources observed by them only account for 3% of the CXB intensity in the 2–10 keV band. In particular, there is a big puzzle on the CXB origin, called the "spectral paradox": bright AGNs observed with *HEAO1*, *EXOSAT* and *Ginga* have spectra with an average photon index of $\Gamma = 1.7$–1.9 (e.g., Williams et al. 1992), which is significantly softer than that of the CXB itself ($\Gamma \simeq 1.4$; e.g., Gendreau et al. 1995). Furthermore, the broad band properties of sources at fluxes from $\sim 10^{-11}$ to $\sim 10^{-13}$ erg s^{-1} cm^{-2} (2–10 keV) are somewhat puzzling according to previous studies. The extragalactic source counts in the soft band (0.3–3.5 keV) obtained by *Einstein* Extended Medium Sensitivity Survey (EMSS; Gioia et al. 1990) is about 2–3 times smaller than that in the hard band (2–10 keV) obtained by the *Ginga* fluctuation analysis (Butcher et al. 1997) when we assume a power-law photon index of 1.7.

The *ASCA* satellite (Tanaka, Inoue, & Holt 1994), launched in 1993 February, was expected to change this situation. It is the first imaging satellite capable of

TABLE 1. Summary of *ASCA* Surveys

Survey Project	Area (deg^2)	Sensitivity (2–10 keV) (erg s^{-1} cm^{-2})
Large Sky Survey (LSS)	7.0	1.5×10^{-13}
Deep Sky Survey (DSS)	0.3	4×10^{-14}
Lockman Hole Deep Survey	0.2	4×10^{-14}
Survey of deep *ROSAT* fields	1.0	5×10^{-14}
ASCA Medium-Sensitivity Survey (AMSS)	110	7×10^{-14}

study of the X-ray band above 2 keV with a sensitivity up to several 10^{-14} erg s^{-1} cm^{-2} (2–10 keV) and covers the wide energy band from 0.5 to 10 keV, which allows us to directly compare results of the energy bands below and above 2 keV with single detectors, hence accompanied with much less uncertainties than previous studies. By taking these advantages, several X-ray surveys have been performed with *ASCA* to reveal the nature of hard X-ray populations: the *ASCA* Large Sky Survey (LSS; Ueda et al. 1998), the *ASCA* Deep Sky Survey (DSS; Ogasaka et al. 1998; Ishisaki et al. 1999 for the Lockman Hole), the *ASCA* Medium-Sensitivity Survey (AMSS or the GIS catalog project: Ueda et al. 1997, Takahashi et al. 1998, Ueda et al. 1999b; see also Cagnoni, Della Ceca, & Maccacaro 1998 and Della Ceca et al. 1999), a survey of *ROSAT* deep fields (Georgantopoulos et al. 1997; Boyle et al. 1998), and so on. The sensitivity limits and survey area are summarized in Table 1. In this paper, we present main results of the *ASCA* surveys, focusing on the LSS (§ 2), the Lockman Hole deep survey (§ 3), and the AMSS (§ 4). In § 5, we summarize these results and discuss their implications for the origin of the CXB.

2. THE LARGE SKY SURVEY

2.1. X-ray Data

The survey field of the *ASCA* Large Sky Survey (LSS; Ueda et al., 1998) is a continuous region near the north Galactic pole, centered at $RA(2000) = 13^h14^m$, $DEC(2000) = 31°30'$. Seventy-six pointings have been made over several periods from Dec. 1993 to Jul. 1995. The total sky area observed with the GIS and SIS amounts to 7.0 deg^2 and 5.4 deg^2 with the mean exposure time of 56 ksec (sum of GIS2 and GIS3) and 23 ksec (sum of SIS0 and SIS1), respectively. From independent surveys in the total (0.7–7 keV), hard (2–10 keV), and soft (0.7–2 keV) bands, 107 sources are detected with sensitivity limits of 6×10^{-14}, 1×10^{-13}, and 2×10^{-14} erg s^{-1} cm^{-2}, respectively. The Log N - Log S relations derived from the LSS are summarized in Ueda et al. (1999a) together with a complete X-ray source list. At these flux limits, 30($\pm$3)% of the CXB in the 0.7–7 keV band and 23($\pm$3)% in the 2–10 keV band have been resolved into discrete sources. The 2–10 keV Log N - Log S relation combined with the AMSS result (§4) is plotted in Figure 3.

The spectral properties of the LSS sources suggest that contribution of sources with hard energy spectra become significant at a flux of $\sim 10^{-13}$ erg s^{-1} cm^{-2} (2–10 keV), which are different from the major population in the soft band. The average 2–10 keV photon index is 1.49±0.10 (1σ statistical error in the mean value) for 36 sources detected in the 2–10 keV band with fluxes below 4×10^{-13} erg s^{-1} cm^{-2}, whereas it is 1.85±0.22 for 64 sources detected in the 0.7–2 keV band with fluxes below 3×10^{-13} erg s^{-1} cm^{-2}. The average spectrum of 74 sources detected in the 0.7–7 keV band with fluxes below 2×10^{-13} shows a photon index of 1.63±0.07 in the 0.7–10 keV range: this index is consistent with the comparison of source counts between the hard and the soft band.

To investigate the X-ray spectra of these hard sources, we made deep follow-up observations with *ASCA* for the five hardest sources in the LSS, selected by the apparent 0.7–10 keV photon index from the source list excluding very faint sources. The results are summarized in Ueda et al. (1999c); see also Sakano et al. (1998) and Akiyama et al. (1998) for AX J131501+3141, the hardest source in the LSS. Three sources in this sample are optically identified as narrow-line AGNs and one is a weak broad-line AGN by Akiyama et al. (2000); one is not identified yet. We found that spectra of these sources are most likely subject to intrinsic absorption at the source redshift with column densities of $N_\mathrm{H} = 10^{22} \sim 10^{23}$ cm^{-2}.

2.2. Optical Identification

Akiyama et al. (2000) summarize the results of optical identification for a subsample of the LSS sources, consisting of 34 sources detected in the 2–7 keV band with the SIS. The major advantage of this sample compared with other *ASCA* surveys is good position accuracy; it is 0.6 arcmin in 90% radius from the *ASCA* data alone, thanks to superior positional resolution of the SIS. To improve the position accuracy further, we made follow-up observations with *ROSAT* HRI over a part of the LSS field in Dec. 1997. Optical spectroscopic observations were made using the University of Hawaii 88″ telescope, the Calar Alto 3.5m telescope, and the Kitt Peak National Observatories Mayall 4m and 2.1m telescopes.

Out of the 34 sources, 30 are identified as AGNs, 2 are clusters of galaxies, 1 is a Galactic star, and only 1 object remains unidentified. The identification as AGNs is based on existence of a broad emission line or the line ratios of narrow emission lines ([NII]6583Å/Hα and/or [OIII]5007Å/Hβ); see Akiyama et al. 2000 and references therein. Figure 1(a) shows the correlation between the redshift and the apparent photon index in the 0.7–10 keV range, which is obtained from a spectral fit assuming no intrinsic absorption, for the identified objects. The 5 sources that have an apparent photon index smaller than 1.0 are identified as 4 narrow-line AGNs and 1 weak broad-line AGN, all are located at redshift smaller than 0.5. On the other hand, X-ray spectra of the other AGNs are consistent with those of nearby type 1 Seyfert galaxies. Four high redshift broad-line AGNs show somewhat apparently hard spectra with an apparent photon index of 1.3 ± 0.3, although it may be still marginal due to the limited statistics.

To avoid complexity in classifying the AGNs by the optical spectra, we divide the identified AGNs into two using the X-ray data: the "absorbed" AGNs which show intrinsic absorption with a column density of $N_H > 10^{22}$ cm^{-2} and the "less-absorbed" AGNs with $N_H < 10^{22}$. Correcting the *flux* sensitivity for different X-ray spectra, we found the contribution of the absorbed AGNs is almost comparable to that of less-absorbed AGNs in the 2–10 keV source counts at a flux limit of 2×10^{-13} erg s^{-1} cm^{-2}. Figure 1(b) shows the correlation between the redshift and the 2–10 keV luminosity of the identified AGNs. The redshift distribution of the 5 absorbed AGNs is concentrated at $z < 0.5$, which contrasts to the presence of 15 less-absorbed AGNs at $z > 0.5$. This suggests a deficiency of AGNs with column densities of $N_H = 10^{22-23}$ at $z = 0.5-2$, or in the X-ray luminosity range larger than 10^{44} erg s^{-1}, or both. Note that if the 4 broad-line AGNs with hard spectra have intrinsic absorption instead of other hardening mechanism such as Compton reflection, it could complement this deficiency.

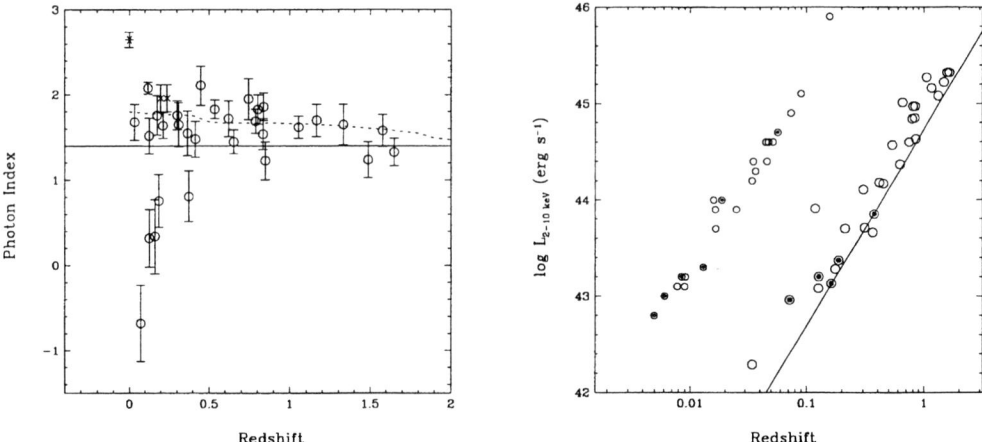

FIGURE 1. (a) left: the correlation between the redshift and the apparent 0.7–10 keV photon index for the identified objects in the LSS (Akiyama et al. 2000). The open circles, crosses, and asterisk represent AGNs, clusters of galaxies, and a Galactic star, respectively. The dotted curve shows the expected apparent photon index in the observed 0.7–10 keV band as a function of redshift, for a typical spectrum of type-1 Seyfert galaxies with a Compton reflection component. (b) right: The 2–10 keV luminosity versus redshift diagram for the LSS AGNs (with large open circles, Akiyama et al. 2000), and for the *HEAO1A2* AGNs (with small marks, Piccinotti et al. 1982). The "absorbed" AGNs are plotted with dots. Lines indicate detection limits of the LSS for a source with an photon index of 1.7 with no intrinsic absorption.

3. THE LOCKMAN HOLE DEEP SURVEY

Deep surveys were performed with *ASCA* over several fields (Ogasaka et al. 1998), although optical identification is more difficult than the LSS because of faint flux levels and source confusion problem. To overcome this difficulty, we have been conducting a deep survey of the Lockman Hole, where the *ROSAT* deep survey was performed (Hasinger et al. 1998). The advantage of selecting this field is that we already have a complete soft X-ray source catalog down to a flux limit of 5.5×10^{-15} erg s^{-1} cm^{-2} (0.5–2 keV), most of which have been optically identified (Schmidt et al. 1998). In addition, we utilized an X-ray source list at even fainter flux limits (G. Hasinger, private communication). Since the flux limits of the *ROSAT* surveys are extremely low, we can expect most of *ASCA* sources could have *ROSAT* counterparts within reasonable range of spectral hardness. Utilizing positions of the *ROSAT* sources, we can determine the hard-band flux for individual sources, which would otherwise have been difficult to separate, to a flux limit of 3×10^{-14} erg s^{-1} cm^{-2} (2–10 keV). Preliminary results are reported in Ishisaki et al. (1999).

Up to present, we have made 3 pointings in the direction of the Lockman Hole with *ASCA* on 1993 May 22–23, 1997 April 29–30, and 1998 November 27 for a net exposure of 63 ksec (average of the 8 SIS chips), 64 ksec, and 62 ksec, respectively. The pointing positions were arranged so that the superposed image of the SIS field of views (FOVs) covers the PSPC and HRI FOVs as much as possible. We here used only the SIS data considering its superior positional resolution. Analysis was made through the 2-dimensional maximum-likelihood fitting to a raw, superposed image in photon counts space in the sky coordinates, with a model consisting of source peaks (point spread functions) and the background. As a first step, we put sources into the model at the positions of the *ROSAT* catalogs. Then, after checking the residual image of the fit, we added remaining peaks that were missing in the *ROSAT* catalogs. Thus, we determined the significance and flux of each source in three energy bands, 0.7–7 keV, 2–7 keV, and 0.7–2 keV, including new sources detected with *ASCA*. We corrected for the degradation of detection efficiency caused by the radiation damage using the CXB intensity. Note that the *ASCA* sensitivity limits strongly depends on position due to the multiple pointings and the vignetting of the XRT.

We detected 27 sources altogether with significances higher than 3.5σ in either of the three survey bands. Two sources were newly detected with *ASCA*. One object is a variable source having a 0.7–7 keV photon index of about 1.7, which was very faint during the *ROSAT* observations. The other shows a very hard spectrum and is detected only in the 2–7 keV band. In the combined SIS FOVs, 43 sources out of 50 sources in the Schmidt et al. (1998) catalog are located. Identification of *ASCA* sources using the *ROSAT* catalog is summarized in Table 2. Since the number of sources detected in the 2–7 keV band is limited due to poor photon statistics, we here use the results for 25 sources detected in the 0.7–7 keV band for comparison with the *ROSAT* survey. Four unidentified sources in the *ASCA* survey have *ROSAT* counterparts in the deeper X-ray source catalog (G. Hasinger, private

TABLE 2. Summary of optical identification of the *ASCA* Lockman Hole deep survey by the *ROSAT* catalog (Schmidt et al. 1998)

Population	ROSAT	ASCA (0.7–7 keV)
Total	43	25
Type-1 AGN (a-c)	26	13
Type-2 AGN (d-e)	7	6
Group/Galaxies	3	0
Star	3	1
Unidentified	4	4+1

communication) and remaining one is the variable source detected only with *ASCA*. For AGNs identified by Schmidt et al. (1998), we divided them into two according to their optical spectra: (1) type-1 AGNs, corresponding to either of class a, b, or c, showing broad emission lines, and (2) type-2 AGN, class d or e, showing only narrow emission lines. As noticed from the table, 6 out of 7 type-2 AGNs were detected, whereas only half of the 26 type-1 AGNs were detected with *ASCA*, which covers much harder band than the *ROSAT*. This suggests that contribution of type-2 AGNs are more dominant in higher energy bands than in the soft band at similar flux levels.

4. THE *ASCA* MEDIUM-SENSITIVITY SURVEY

Because these surveys are limited in sky coverage, the sample size is not sufficient to obtain a self-consistent picture about the evolution of the sources over the wide fluxes, from $\sim 10^{-11}$ erg s^{-1} cm^{-2} (2–10 keV) which is the sensitivity limit of *HEAO1* A2 (Piccinotti et al. 1982), down to $\sim 10^{-13}$ erg s^{-1} cm^{-2} (2–10 keV), that of *ASCA* . To complement these shortcomings, we have been working on the project called the "*ASCA* Medium Sensitivity Survey (AMSS)", or the GIS catalog project. In the project, we utilize the GIS data from the fields that have become publicly available to search for serendipitous sources. The large field of view and the low-background characteristics make the GIS instrument ideal for this purpose.

Main results from the AMSS are reported in Ueda et al. (1999b), which were obtained from selected GIS fields of $|b| > 20°$ observed from 1993 to 1996, covering the total sky area of 106 deg^{-2}. The sample contains 714 serendipitous sources, of which 696, 323, and 438 sources are detected in the 0.7–7 keV (total), 2–10 keV (hard), and 0.7–2 keV (soft) band, respectively. This is currently the largest X-ray sample covering the 0.7–10 keV band. Figure 2(a) shows the correlation between the 0.7–7 keV flux and the hardness ratio between the 2–10 keV and 0.7–2 keV count rates. We also plot the average hardness ratio in several flux ranges, separated by the dashed curves, with crosses. It is clearly seen that the average spectrum becomes harder with a decreasing flux: the corresponding photon index (assuming

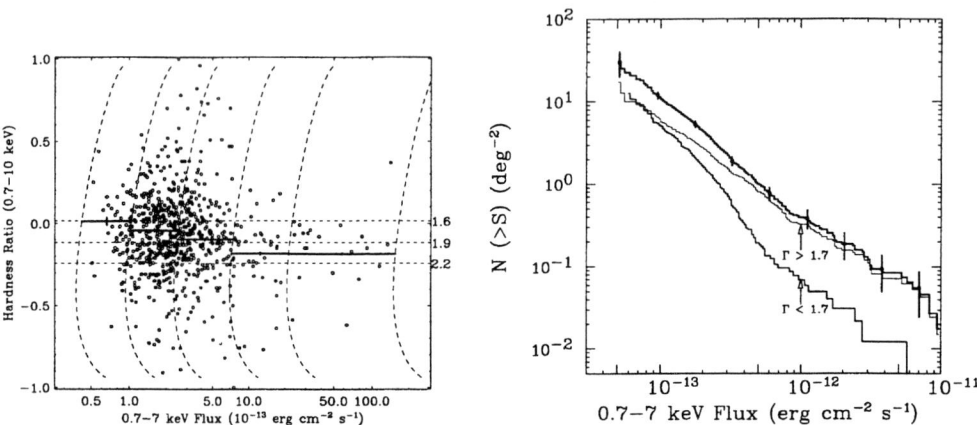

FIGURE 2. (a) left: The correlation between the 0.7–7 keV flux and the hardness ratio between the 0.7–2 keV and 2–10 keV count rates for sources detected in the 0.7–7 keV survey in the AMSS sample (Ueda et al. 1999b). The crosses show the average hardness ratios (with 1σ errors in the mean value) in the flux bin separated by the the dashed curves, at which the count rate hence the sensitivity limit is constant. The dotted lines represent the hardness ratios corresponding to a photon index of 1.6, 1.9, and 2.2 assuming a power law spectrum. (b) right: The integral Log N - Log S relations in the 0.7–7 keV survey band, derived from the AMSS sample. The medium-thickness curve represents the result for the hard source sample, consisting of sources with an apparent 0.7–10 keV photon index Γ smaller than 1.7, the thin curve represents that for the soft source sample (with Γ larger than 1.7), and the thick curve represents the sum. The 90% statistical errors in source counts are indicated by horizontal bars at several data points.

a power law over the 0.7–10 keV band with no absorption) changes from 2.1 at the flux of $\sim 10^{-11}$ erg s^{-1} cm^{-2} to 1.6 at $\sim 10^{-13}$ erg s^{-1} cm^{-2} (0.7–7 keV). Similar hardening are also reported in the 2–10 keV range by Della Ceca et al. (1999) using 60 serendipitous sources. Figure 2(b) shows the integral Log N - Log S relations in the 0.7–7 keV survey band for the soft source sample, consisting of sources with an apparent 0.7–10 keV photon index larger than 1.7, and for the hard source sample, with an index smaller than 1.7. This demonstrates that sources with hard energy spectra in the 0.7–10 keV range are rapidly increasing with decreasing fluxes, compared with softer sources.

5. SUMMARY

The *ASCA* surveys have brought a clear, self-consistent picture about statistical properties of sources that constitute about 30% of the CXB in the broad energy band of 0.7–10 keV. Figure 3 summarizes the 2–10 keV Log N - Log S relation obtained from the *ASCA* surveys together with the results from previous missions. The direct source counts from combined results of the LSS (Ueda *et al.* 1999b) and the AMSS (Ueda *et al.* 1999c; these contain the data used by Cagnoni, Della Ceca, & Maccacaro 1998) give the tightest constraints so far over a wide flux range from $\sim 10^{-11}$ to $\sim 7 \times 10^{-14}$ erg s^{-1} cm^{-2}: $N(>S) = 16.8\pm7.2$ (90% statistical error), 11.43±2.4, 3.76±0.42, 1.08±0.17, and 0.33±0.09 deg^{-2}, at $S = 7.4 \times 10^{-14}$, 1.0×10^{-13}, 2.0×10^{-13}, 4.0×10^{-13}, and 1.0×10^{-12} erg s^{-1} cm^{-2}, respectively. The DSS gives a direct source counts at the faintest flux, 3.8×10^{-14} erg s^{-1} cm^{-2} (Ogasaka *et al.* 1998), whereas the fluctuation analysis of deep SIS fields constrains the Log N - Log S relation at fluxes down to 1.5×10^{-14} (Gendreau, Barcons, & Fabian 1998). As seen from the figure, the *ASCA* direct source counts smoothly connect the two regions constrained by the *Ginga* and *ASCA* fluctuation analysis.

The AMSS/LSS results demonstrate that the average spectrum of X-ray sources becomes harder toward fainter fluxes: the apparent photon index in the 0.7–10 keV range changes from 2.1 at the flux of $\sim 10^{-11}$ to 1.6 at $\sim 10^{-13}$ erg s^{-1} cm^{-2} (2–10 keV). This fact can be explained by the rapid emergence of population with hard energy spectra, as is clearly indicated in Figure 2(b). The evolution of broad-band properties of sources solves the puzzle of discrepancy of the source counts between the soft (EMSS) and the hard band (*Ginga* and *HEAO1*). If we compare the *ASCA* Log N - Log S relations (including Galactic objects) between above and below 2 keV, the hard band source counts at $S \sim 10^{-13}$ erg s^{-1} cm^{-2} (2–10 keV) matches the soft band one when we assume a photon index of 1.6 for flux conversion, whereas at brighter level of $S = 4 \times 10^{-13} \sim 10^{-12}$ erg s^{-1} cm^{-2} (2–10 keV), we have to use a photon index of about 1.9 to make them match. The latter fact is consistent with the average 0.7–10 keV spectrum at the same flux levels, and can be connected the "soft" spectrum of the fluctuation observed with *Ginga*, which shows a photon index of 1.8±0.1 in the 2–10 keV range (Butcher *et al.* 1997).

The optical identification revealed that the major population at fluxes of 10^{-13} erg s^{-1} cm^{-2} are AGNs. The population of hard sources, which are most responsible for making the average spectrum hard, are X-ray absorbed sources. They are mostly identified as narrow line (type-2) AGNs. The contribution of these type-2 AGNs is larger in the hard band than in the soft band at the same flux limit. Recent results of the 5–10 keV band survey by *BeppoSAX* confirms this tendency (Fiore *et al.* 1999). These results support the scenario that the CXB consists of unabsorbed AGNs and absorbed AGNs, whose contribution becomes more significant with decreasing fluxes and in harder energy band.

We found, however, possible evidence that is not consistent with the "unified scheme" of AGNs (e.g., Awaki *et al.* 1991), on which many CXB synthesis models are based. The LSS results may imply deficiency of X-ray luminous, absorbed AGNs, with $N_H = 10^{22-23}$ at $z = 0.5-2$, or in the X-ray luminosity range larger than 10^{44} erg s^{-1}, although we cannot rule out the possibility, for example, that there are many luminous AGNs at $z > 2$ with extreme heavy absorption of $N_H > 10^{24}$. On the other hand, there is another implication that there could be a population of

AGNs at high redshifts ($z > 1$) that are optically identified as type-1 AGNs but have apparently hard X-ray spectra, although the origin of the hardness is not clear yet. Future surveys by *Chandra* and *XMM* together with optical identification of the AMSS sources will reveal the luminosity, number, and spectral evolutions of extra-galactic populations including absorbed AGNs, which will eventually lead us to full understanding of the origin of the CXB.

ACKNOWLEDGEMENTS

I thank all the collaborators of our ASCA survey projects, especially, M. Akiyama, G. Hasinger, H. Inoue, Y. Ishisaki, I. Lehmann, K. Makishima, Y. Ogasaka, T. Ohashi, K. Ohta, M. Sakano, T. Takahashi, T. Tsuru, W. Voges, T. Yamada, and A. Yamashita.

REFERENCES

Akiyama, M., *et al.* 1998, ApJ, 500, 173
Akiyama, M., *et al.* 2000, ApJ, in press
Awaki, H., Koyama, K., Inoue, H., & Halpern, J.P. 1991, PASJ, 43, 195
Boyle, B.J., *et al.* 1998, MNRAS, 296, 1
Butcher, J.A., *et al.* 1997, MNRAS, 291, 437
Cagnoni, I., Della Ceca, R., & Maccacaro, T. 1998, ApJ, 493, 54
Della Ceca, R., *et al.* 1999, ApJ, 524, 674
Fiore, F., *et al.* 1999, MNRAS, 306, L55
Gendreau, K.C., *et al.* 1995, PASJ, 47, L5
Gendreau, K.C., Barcons, X., & Fabian, A.C. 1998, MNRAS, 297, 41
Georgantopoulos, I., *et al.* 1997, MNRAS, 291, 203
Gioia, I.M. *et al.* 1990, ApJS, 72, 567
Hasinger, G. 1998, AN, 319, 37
Hasinger, G., *et al.* 1998, A&A, 329, 482
Ishisaki, Y., *et al.* 1999, Advanced in Space Research, in press
Kondo, H. *et al.* 1982, in "Frontiers of X-ray Astronomy", Universal Academy Press, Tokyo, p.655
Ogasaka, Y., *et al.* 1998, Astro. Nachr., 319, 43
Piccinotti, G., *et al.* 1982, ApJ, 253, 485
Sakano, M. *et al.* 1998, ApJ, 505, 129
Schmidt, M., *et al.* 1998, A&A, 329, 495
Takahashi, T., Ueda, Y., Ishisaki, Y., Ohashi, T., & Makishima, K. 1998, AN, 319, 91
Tanaka, Y., Inoue, H., & Holt, S.S. 1994, PASJ, 46, L37
Ueda, Y., Takahashi, T., Ishisaki, Y., Makishima, K. & 1997, in "All-Sky X-Ray Observations in the Next Decade", RIKEN, Saitama, p55
Ueda, Y., *et al.* 1998, Nature, 391, 866
Ueda, Y., *et al.* 1999a, ApJ, 518, 656
Ueda, Y., Takahashi, T., Ishisaki, Y., Ohashi, T., & Makishima, K. 1999b, ApJ, 524, L11
Ueda, Y., *et al.* 1999c, Advanced in Space Research, in press
Williams, O.R. *et al.* 1992, ApJ, 389, 157

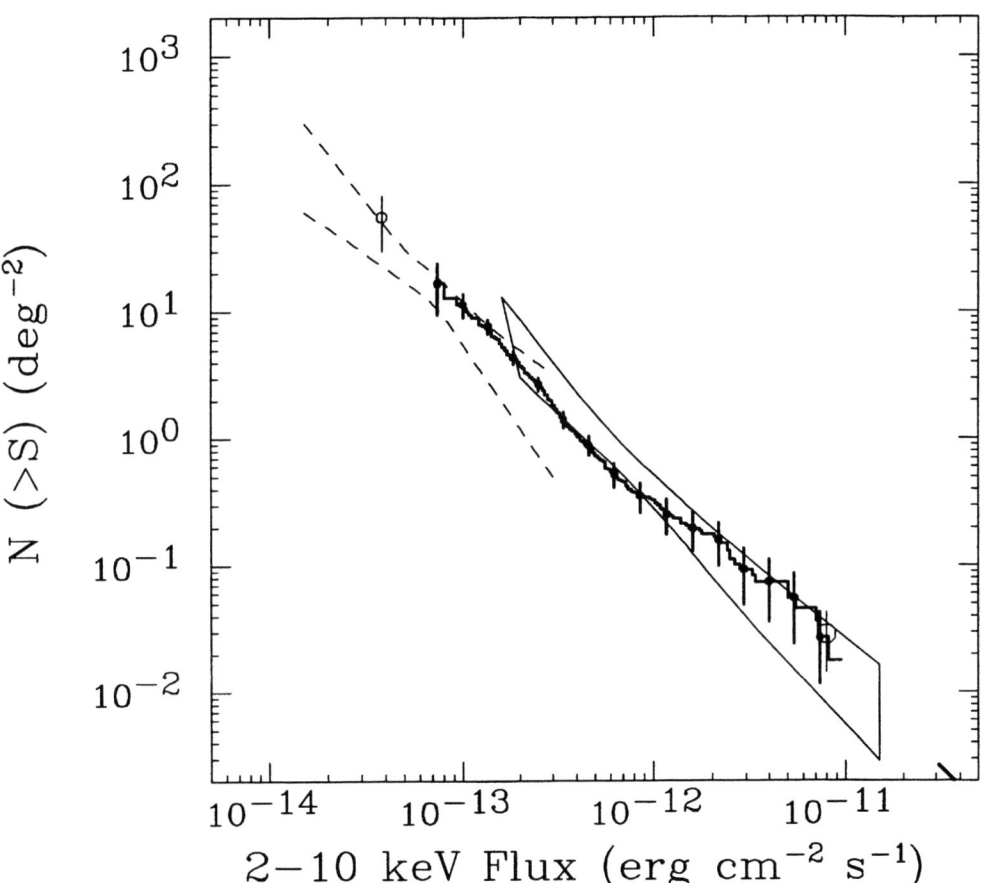

FIGURE 3. Summary of the 2–10 keV Log N - Log S relation obtained by the *ASCA* surveys, compared with previous results. The steps are the combined results from the LSS (Ueda *et al.* 1998) and the AMSS (Ueda *et al.* 1999b). The faintest point at 4×10^{-14} erg s^{-1} cm^{-2} is derived from the DSS utilizing the SIS data (Ogasaka *et al.* 1998). The trumpet shape between two dashed lines indicates 1σ error region from the fluctuation analysis of *ASCA* SIS deep fields (Gendreau, Barcons & Fabian 1998). The contour at $10^{-13} \sim 10^{-11}$ erg s^{-1} cm^{-2} represents the constraints by the *Ginga* fluctuation analysis at 90% confidence level (Butcher *et al.* 1997). The open circle at 8×10^{-12} erg s^{-1} cm^{-2} corresponds to the source count by *Ginga* survey (Kondo *et al.* 1991), and the thick-line above 3×10^{-11} erg s^{-1} cm^{-2} is the extragalactic Log N - Log S relation determined by *HEAO1* A2 (Piccinotti *et al.* 1982). All the horizontal bars represent 90% statistical errors in source counts.

KILOHERTZ QUASI-PERIODIC OSCILLATIONS — OBSERVATIONAL OVERVIEW

M. van der Klis

Astronomical Institute "Anton Pannekoek" and Center for High-Energy Astrophysics, University of Amsterdam

ABSTRACT Twin kilohertz quasi-periodic oscillation peaks have been discovered with the Rossi X-Ray Timing Explorer in some 20 neutron-star low-mass X-ray binaries. The frequencies of these peaks vary between ~500 and ~1200 Hz and they likely diagnose the motion of the accreting matter in the accretion flow just a few kilometers from the neutron star surface. In these regions strong field gravity theory is essential to correctly describe the motion of matter, and the precise equation of state of supranuclear density matter, by determining the radius of and the mass distribution in the neutron star, becomes a crucial ingredient in understanding the observations. Models for kHz QPOs therefore explicitly include these fundamental theories and point the way as to how to test general relativity in the strong field limit and proposed equations of state. In this paper I review the observational situation with respect to this phenomenon.

KEYWORDS: stars:neutron; binaries; X-rays:stars; relativity; dense matter; equation of state; black hole physics

1. INTRODUCTION

Twin kilohertz quasi-periodic oscillation (QPO) peaks began to be discovered with the Rossi X-ray Timing Explorer (RXTE) early 1996, briefly after its launch (van der Klis et al. 1996a,b; Fig. 1, Strohmayer et al. 1996a,c). The peaks had associated rms amplitudes of $\lesssim 1\%$ up to several 10% of the total source flux, peak separations of ~250–360 Hz, coherences between $Q \sim 1$–10^2, and their frequencies (of the higher-frequency peak of the two) varied between 500 and 1200 Hz with mass accretion rate $\dot{M}$. Only a short time later another series of RXTE discoveries began, of burst oscillations (Strohmayer et al. 1996b,c). These came from some of the same sources, but only during type 1 X-ray bursts, and had frequencies between 330 and 590 Hz, depending on the source. They were interpreted in terms of the spin frequency of a layer in the neutron star atmosphere that is presumably close to the neutron star spin frequency itself (see Stromayer, these proceedings). Finally, the first true spin frequency was detected (at 401 Hz) in an accreting low-magnetic field neutron star (Wijnands and van der Klis 1998; see Chakrabarty, these proceedings).

In several sources it was found that the twin kHz peak separation was equal to the frequency of the burst oscillations, or half that, to within a few percent. This strongly suggested a beat-frequency interpretation, where the higher-frequency peak (hereafter the upper peak, at a frequency ν_2) is identified with orbital motion at

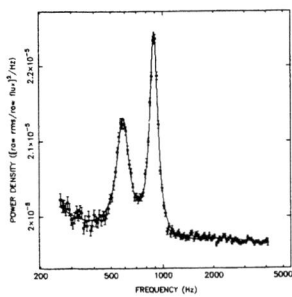

FIGURE 1. Twin kHz peaks in Sco X-1. (van der Klis et al. 1997b)

some preferred radius in the disk and the lower-frequency kHz QPO peak (hereafter the lower peak, at ν_1) is at the beat frequency between this orbital frequency and the spin frequency of the neutron star: $\nu_1 = \nu_2 - \nu_s$ (Strohmayer et al. 1996c, Miller et al. 1998; see Miller, these proceedings). As the peak separations $\Delta\nu = \nu_2 - \nu_1$ are then an indication for the neutron star spin frequency, and as $\Delta\nu$ is similar in all sources (~250–250 Hz), the conclusion would be that all these neutron stars spin at similar frequencies. Possibly, this is evidence for strong gravitational radiation from these objects limiting their spins and preventing further spin-up by the accretion torques (Bildsten 1998, Andersson et al. 1999).

However, in several sources $\Delta\nu = \nu_2 - \nu_1$ does not remain constant when ν_2 varies as its identification with ν_s would predict, but decreases. While some beat-frequency models can accomodate this, another interpretation, where ν_2 is also the orbital frequency but ν_1 is the apsidal motion frequency of slightly eccentric orbits in the accretion disk, also became attractive (Stella and Vietri 1999; see Stella, these proceedings).

It was immediately realized that if we are indeed seeing the signature of orbital motion around neutron stars with orbital frequencies ν_2 of up to 1200 Hz, then we are probing a region of space-time where strong-field general relativity is required to describe orbital motion, and constraining the mass-radius relation of neutron stars and thereby the equation of state (EOS) of supranuclear-density matter. This spurred a large amount of observational work on these phenomena, some of which I will summarize here.

2. PHENOMENOLOGY

The number of sources that have shown kilohertz quasi-periodic oscillations (kHz QPOs) is now twenty. Together with the one source that showed burst oscillations but no kHz QPOs (probably MXB 1743−29), and the millisecond pulsar SAX J1808.4−3658, which so far showed neither, this brings the total number of sources that has shown kilohertz periodic or quasi-periodic phenomena (here defined as phenomena at frequencies exceeding $10^{2.5}$ Hz) to 22. All but one of the kHz

QPO sources have shown twin kHz peaks in their power spectrum (Fig. 1); the exception with only a single peak so far is EXO 0748−676. Tables 1 and 2 summarize the results on these sources.

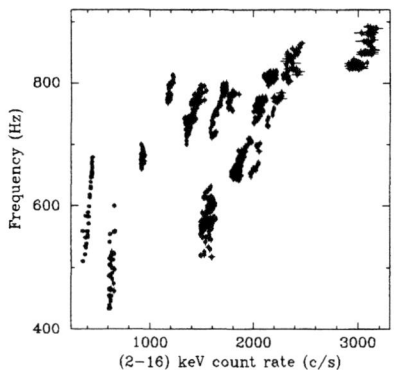

FIGURE 2. In 4U 1608−52 a QPO frequency vs. count rate plot shows parallel lines. (Méndez et al. 1999)

The two kHz QPO peaks increase in frequency with inferred mass accretion rate $\dot{M}$ both in Z and in atoll sources (see Hasinger and van der Klis 1989 or the review by van der Klis 1995 for the introduction of these subtypes of LMXBs). In Z sources, the QPOs are nearly always seen down to the lowest inferred $\dot{M}$ levels these sources reach, in atoll sources the QPOs tend to occur in the middle of the $\dot{M}$ range of each source. Kilohertz QPOs are seen in a similar frequency range (500–1200 Hz for the upper peak) in sources that differ in average X-ray luminosity L_x by 2.5 orders of magnitude (where L_x is here defined simply as $4\pi d^2 f_x$ with f_x the X-ray flux and d the distance), and the kHz QPO frequency seems to be determined more by the difference between some average, and instantaneous L_x than by L_x itself (Ford et al. 2000). Individual sources observed at different epochs show a similar pattern of parallel lines in the X-ray flux vs. QPO frequency plot (Fig. 2). These findings are unexplained and must mean that another parameter than $\dot{M}$, related in some way to the average L_x, affects the QPO frequency (van der Klis 1997).

In 4 sources, both twin kHz peaks and burst oscillations are seen, and the fact that the burst oscillation frequency is near 1 or 2 times the kHz peak separation frequency is the main argument for the beat frequency interpretation of kHz QPOs. The evidence for this is summarized in Table 3. In 4U 1636−53, and 4U 1728−34 Méndez et al. (1998c, Méndez and van der Klis 1999) have shown that the correspondence is not exact. Yet, although we would dearly like to have a few more examples, the preponderance of the evidence still seems to indicate that at least an approximate beat-frequency relation exists between the three frequencies in those sources where they are all observed.

TABLE 1. Observed frequencies of kilohertz QPOs in Z sources

Source	ν_1 (Hz)	ν_2 (Hz)	$\Delta\nu$ (Hz)	ν_{burst} (Hz)	References
Sco X-1	565	870	307±5		Van der Klis et al. 1996a,b,c,1997b
	845	1080	237±5		
		1130			
GX 5−1	215	505	298±11		Van der Klis et al. 1996e
	660	890			Wijnands et al. 1998c
	700				
GX 17+2		645			Van der Klis et al. 1997a
	480	785	294±8		Wijnands et al. 1997b
	780	1080			
Cyg X-2		730			Wijnands et al. 1998a
	530	855	346±29		
	660	1005			
GX 340+0	200	535	339±8		Jonker et al. 1998, 1999
	565	840			
	625				
GX 349+2	710	980	266±13		Zhang et al. 1998a
		1020^a			Kuulkers & van der Klis 1998

Values for ν_1 and ν_2 were rounded to the nearest 5, for ν_{burst} to the nearest 1 Hz. Entries in one column not separated by a horizontal line indicate ranges over which the frequency was observed or inferred to vary; ranges from different observations were combined assuming the ν_1, ν_2 relation in each source is reproducible (no evidence to the contrary exists). Entries in one uninterrupted row refer to simultaneous data (except for ν_{burst} values). Values of $\Delta\nu$ straddling two rows, or adjacent to a vertical line refer to measurements made over the range of frequencies indicated. Note: ^{a} Marginal detection.

In an interpretation where the burst oscillations and the kHz peak separation are both close to the neutron star spin, this peak separation is predicted to be approximately constant. It is clear that this is not always the case: in several sources this separation decreases systematically when the kHz QPO frequencies increase (Fig. 3; see also Table 2).

The possibility to detect evidence in kHz QPOs for the existence of the innermost stable circular orbit (ISCO) predicted by general relativity, which would constitute the first direct detection of a strong-field general-relativistic effect, has fascinated since the beginning (Kaaret, Ford and Chen 1997; Zhang, Strohmayer and Swank 1997b). It has been conjectured that when the inner edge of the accretion disk reaches the ISCO, the QPO frequency might level off and remain constant while $\dot{M}$ continues rising. For this reason, the measurement of an apparent leveling off of the increase of QPO frequency with X-ray count rate in 4U 1820−30 (Zhang et al. 1998b; Fig. 4) attracted considerable attention. The leveling off is also observed as a function of X-ray flux and color (Kaaret et al. 1999) and position along the atoll track (Bloser et al. 1999). However, the frequency vs. flux relations are known in other sources to be variable, and in 4U 1820−30 the leveling off seems not to be reproduced in the same way in all data sets (Méndez et al. 2000). It may also be more gradual in nature than originally suggested. No evidence for a similar saturation in frequency was seen in other sources, and most reach higher frequencies.

A number of intriguing correlations has been found between the kHz QPOs and the phenomena at lower frequencies. Wijnands and van der Klis (1999) point out,

TABLE 2. Observed frequencies of kilohertz QPOs in atoll sources

Source	ν_1 (Hz)	ν_2 (Hz)	$\Delta\nu$ (Hz)	ν_{burst} (Hz)[d]	References
4U 0614+09	418 825	450 765 1160 1215 1330	312±2		Ford et al. 1996, 1997a,b; Van der Klis et al. 1996d; Méndez et al. 1997; Vaughan et al. 1997,1998; Kaaret et al. 1998; van Straaten et al. 1999
EXO 0748−676		695			J. Homan 1999, in prep.
4U 1608−52	415 440 475 865 895	765 800 1090[b]	325±7 326±3 225±12[b]		Van Paradijs et al. 1996; Berger et al. 1996; Yu et al. 1997; Kaaret et al. 1998; Vaughan et al. 1997,1998; Méndez et al. 1998a,b, 1999; Méndez 1999; Markwardt et al. 1999b
4U 1636−53	830 900 950 1070	1150 1190 1230	251±4[b]	291,582	Zhang et al. 1996a,b,1997a; Van der Klis et al. 1996d; Vaughan et al. 1997,1998; Zhang 1997; Wijnands et al. 1997a; Méndez et al. 1998c; Méndez 1999; Markwardt et al. 1999b
4U 1702−43	625 655 700 770 902	1000[b] 1040[b] 1085[b]	344±7[b] 337±7[b] 315±11[b]	330	Markwardt et al. 1999a,b
4U 1705−44	775 870	1075[a]	298±11		Ford et al. 1998a
XTE J1723−376		815			Marshall & Markwardt 1999
4U 1728−34	510 875 920	325 845 1160	349±2[c] 279±12[c]	364	Strohmayer et al. 1996a,b,c; Ford & van der Klis 1999; Méndez & van der Klis 1999; Méndez 1999; Markwardt et al. 1999b; di Salvo et al. 1999
KS 1731−260	900	1160 1205	260±10	524	Wijnands & van der Klis 1997
4U 1735−44	630 730 900[a]	980 1025 1150 1160	341±7 296±12 249±15		Wijnands et al. 1996, 1998b; Ford et al. 1998b
4U 1820−30	500 795	655 860 1075 1100	358±42 278±11		Smale et al. 1996, 1997; Zhang et al. 1998b; Kaaret et al. 1999; Bloser et al. 1999
Aql X-1	670 930	1040[a]	241±9[a]	549	Zhang et al. 1998c; Cui et al. 1998; Yu et al. 1999; Reig et al. 1999; Méndez et al. 2000 in prep.
4U 1915−05	515 560 655 705[a] 880	820 925 1005 1055 1265[a]	348±11		Barret et al. 1997,1999; Boirin et al. 1999
XTE J2123−058	845 855 870[a]	1100 1130 1140	255±14 276±9 270±5[a]		Homan et al. 1998,1999; Tomsick et al. 1999

Caption: see Table 1. Notes: [a] Marginal detection. [b] Shift and add detection method, cf. Méndez et al. (1998a). [c] See Fig. 3.

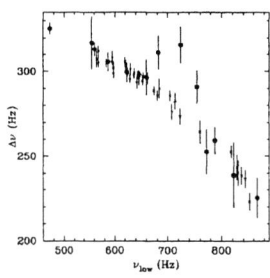 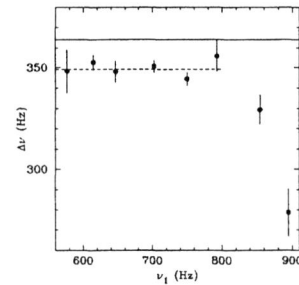

FIGURE 3. The variations in kHz QPO peak separation in Sco X-1 and 4U 1608−52 (left) and 4U 1728−34 (right) as a function of the lower peak frequency. (Méndez et al. 1998b, Méndez and van der Klis 1999)

TABLE 3. Commensurability of kHz QPOs and burst oscillation frequencies

Source	Burst oscillation frequency (Hz)	kHz QPO separation frequency (Hz)	Ratio (burst/separation frequency)
4U 1636−53	581.8	254±5	2.29±0.04
4U 1702−43	330.6	344±7	0.96±0.02
4U 1728−34	364.2	349.3±1.7	1.043±0.005
KS 1731−260	523.9	260±10	2.015±0.08
Aql X-1	548.9	241±9	2.28±0.09

Largest observed burst oscillation frequency and kHz QPO separation frequency is listed for each source. References see Table 2.

that the low $\dot{M}$ power-spectral similarity between BHCs and low-luminosity low-magnetic field neutron stars found with EXOSAT and Ginga (van der Klis 1994) also holds for the millisecond pulsar SAX J1808.4-3658, and may even extend to Z sources in *their* lowest $\dot{M}$ states (the left end of the so-called horizontal branch), where by the way these sources are still quite luminous. The power spectra look very similar (Fig. 5), and with the better RXTE data it is now evident that nearly always, in addition to the break at low frequency, a QPO-like feature above the break is present. The correlation between the break frequency and the QPO frequency is excellent (Fig. 6), and encompasses both neutron stars and black-hole candidates. This suggests that both the band-limited noise component with the 0.03–30 Hz break frequency, and the 0.2–70 Hz QPO (with the possible exception of the Z sources, which are slightly off the main relation) are found in both neutron stars and black holes. This would exclude spin-orbit beat-frequency models and any other models requiring a material surface, an event horizon, a magnetic field, or their absence, for their explanation, essentially implying these phenomena are generated in the accretion flow towards *any* low-magnetic field compact object and are most likely

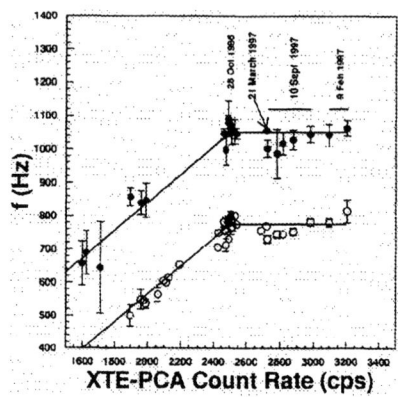

FIGURE 4. Evidence for a leveling off of the kHz QPO frequency with count rate in 4U 1820−30. (Zhang et al. 1998b)

disk variability features. way.

A coincidence of QPO properties pointed out by Psaltis, Belloni and van der Klis (1999) may provide an answer to this question. Power spectra of many Z and atoll sources, of Cir X-1 and of a few low luminosity neutron stars and BHCs show two QPO or broad noise phenomena whose centroid frequencies, when plotted vs. each other, seem to line up (Fig. 7). This would not only identify the low-frequency (few 10 Hz) QPOs in Z and atoll sources with even lower-frequency QPOs in Cir X-1 (3–10 Hz) and BHCs (0.3–1 Hz) as also suggested by the Wijnands and van der Klis (1999) results, but would *also* indicate that the *lower* kHz peak in Z and atoll sources (the same argument does not hold for the upper kHz peak), the broad 20–100 Hz bumps found in Cir X-1, and even lower-frequency (0.2–1 Hz) bumps in some low luminosity neutron stars and BHCs are due to the same physical phenomenon. While this conjecture remains to be confirmed, the currently available data are suggestive. The interpretation would be similar to that mentioned above, particularly, it would imply that the lower kHz QPO is not unique to neutron star systems either, and therefore presumably a general feature of the accretion flow onto a compact object. Of course, *orbital motion* in the disk is such an accretion flow phenomenon and remains an attractive interpretation for some of the observed frequencies. The phenomenology is quite complex; further detailed work will show to what extent these various suggestions will stand the test of additional data. Recent work by di Salvo et al. (1999) suggests how the Wijnands and van der Klis (1999) and Psaltis et al. (1999) work may be related. In 4U 1728−34 they see a feature that at low inferred accretion rate appears as a break in the noise and at higher accretion rate as a QPO peak, whose frequency relates to the other low-frequency QPO peak in the spectra according to the Wijands and van der Klis (1999) relation, and which falls on a second, lower branch in the Psaltis et al. (1999) plot that had

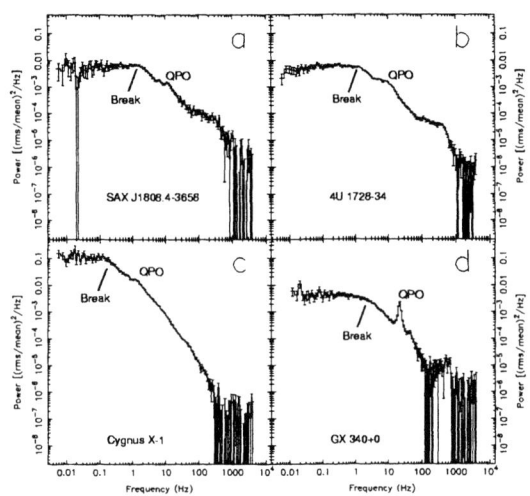

FIGURE 5. Broad-band power spectra of, respectively, the millisecond pulsar, an atoll source, a black-hole candidate and a Z source. (Wijnands and van der Klis 1999)

already been noted by those authors. So, it may be that the two branches in the Psaltis plot, when plotted vs. each other, produce the Wijands plot.

REFERENCES

Andersson, N., Kokkotas, K.D., Stergioulas, N., 1999, ApJ 516 307
Barret, D., et al., 1997, $IAU\ Circ.\ No.$ 6793
Barret, D., et al., 1999, ApJ in press. astro-ph/9911042
Berger, M., et al., 1996, ApJ 469 L13
Bildsten., L., 1998, ApJ 501 L89
Bloser, P.F., 1999, ApJ submitted
Boirin, L., 1999, $A\&A$ submitted
Cui, W., 1998, ApJ 502 L49
Di Salvo, T., 1999, ApJ submitted
Ford, E.C., van der Klis, M., 1998, ApJ 506 L39
Ford, E., et al., 1996, $IAU\ Circ.\ No.$ 6426
Ford, E., et al., 1997a, ApJ 475 L123
Ford, E.C., et al., 1997b, ApJ 486 L47
Ford, E.C, van der Klis, M., Kaaret, P., 1998a, ApJ 498 L41
Ford, E.C., et al., 1998b, ApJ 508 L155
Ford, E.C., et al. 2000, ApJ submitted
Hasinger, G., van der Klis, M., 1989, $A\&A$ 225 79
Homan, J., et al., 1998, $IAU\ Circ.\ No.$ 6971
Homan, J., et al., 1999, ApJ 513 L119
Jonker, P.G., et al., 1998, ApJ 499 L191
Jonker, P.G., et al., 1999, ApJ submitted
Kaaret, P., Ford, E., Chen, K., 1997, ApJ 480 L27

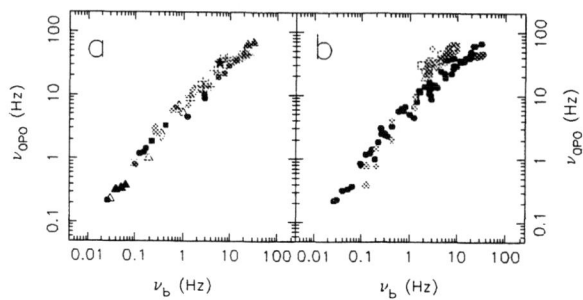

FIGURE 6. Relation between noise break frequency and QPO frequency for sources of the types shown in the previous figure. (Wijnands and van der Klis 1999)

Kaaret, P., et al., 1998, *ApJ* 497 L93
Kaaret, P., et al., 1999, *ApJ* 520 L37
Kuulkers, E., van der Klis, M., 1998, *A&A* 332 845
Markwardt, C.B., Strohmayer, T.E., Swank, J.H., 1999a, *ApJ* 512 L125
Markwardt, C.B., Lee, H.C., Swank, J.H., 1999b, *AAS HEAD* 31 15.01 (Abstr.)
Marshall, F.E., Markwardt, C.B., 1999, *IAU Circ. No.* 7103
Méndez, M., 1999, *Relativistic Astrophysics and Cosmology, Texas Symp., 19th, Paris.* in press. astro-ph/9903469
Méndez, M., van der Klis, M., 1999, *ApJ* 517 L51
Méndez, M., et al., 1997, *ApJ* 485 L37
Méndez, M., et al., 1998a, *ApJ* 494 L65
Méndez, M., et al., 1998b, *ApJ* 505 L23
Méndez, M., van der Klis, M., van Paradijs, J., 1998c, *ApJ* 506 L117
Méndez, M., et al., 1999, *ApJ* 511 L49
Méndez, M., et al., 2000, in prep.
Miller, M.C., Lamb, F.K., Psaltis, D., 1998, *ApJ* 508 791-
Psaltis, D., Belloni, T., van der Klis, M., 1999, *ApJ* 520 262
Reig, P., et al., 1999, *ApJ* submitted. astro-ph/9910139
Smale, A.P., Zhang, W., White, N.E., 1996, *IAU Circ. No.* 6507
Smale, A.P., Zhang, W., White, N.E., 1997, *ApJ* 483 L119
Stella, L., Vietri, M., 1999, *Phys. Rev. Lett.* 82 17
Strohmayer, T., Zhang, W., Swank, J., 1996a, *IAU Circ. No.* 6320
Strohmayer, T., et al., 1996b, *IAU Circ. No.* 6387
Strohmayer, T.E., et al., 1996c, *ApJ* 469 L9
Tomsick, J.A., et al., 1999, *ApJ* 521 341
Van der Klis, M., 1994, *ApJS* 92 511
Van der Klis, M., 1995, in Lewin, WHG, van Paradijs, J, van den Heuvel, EPJ. *X-Ray Binaries.* Cambridge: Cambridge University Press pp. 252
Van der Klis, M., 1997, *Astronomical Time Series, Wise Obs. Anniv. Symp., 25th, Tel Aviv, Kluwer Astroph. and Space Science Library* 218 121. Dordrecht: Kluwer
Van der Klis, M., et al., 1996a, *IAU Circ. No.* 6319
Van der Klis, M., et al., 1996b, *ApJ* 469 L1
Van der Klis, M., et al., 1996c, *IAU Circ. No.* 6424
Van der Klis, M., et al., 1996d, *IAU Circ. No.* 6428
Van der Klis, M., et al., 1996e, *IAU Circ. No.* 6511
Van der Klis, M., et al., 1997a, *IAU Circ. No.* 6565

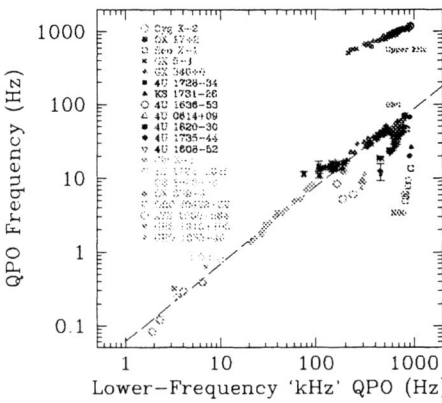

FIGURE 7. Relation between two selected QPO frequencies across a wide variety of sources. (Psaltis, Belloni and van der Klis 1999)

Van der Klis, M., et al., 1997b, *ApJ* 481 L97
Van Paradijs, J., et al., 1996, *IAU Circ. No.* 6336
Van Straaten, S., et al., 1999, *ApJ* submitted
Vaughan, B.A., et al., 1997, *ApJ* 483 L115
Vaughan, B.A., et al., 1998, *ApJ* 509 L145
Wijnands, R.A.D., van der Klis, M., 1997, *ApJ* 482 L65
Wijnands, R., van der Klis, M., 1998, *Nature* 394 344
Wijnands, R., van der Klis, M., 1999, *ApJ* 514 939
Wijnands, R.A.D., et al., 1996, *IAU Circ. No.* 6447
Wijnands, R.A.D., et al., 1997a, *ApJ* 479 L141
Wijnands, R., et al., 1997b, *ApJ* 490 L157
Wijnands, R., et al., 1998a, *ApJ* 493 L87
Wijnands, R.A.D., et al. 1998b, *ApJ* 495 L39
Wijnands, R.A.D., et al., 1998c, *ApJ* 504 L35-
Yu, W., et al., 1997, *ApJ* 490 L153
Yu, W., et al., 1999, *ApJ* 512 L35
Zhang, W., 1997, presented at *AAS Meeting, 190th, Winston-Salem*
Zhang, W., et al., 1996a, *ApJ* 469 L17
Zhang, W., et al., 1996b, *ApJ* 473 L135
Zhang, W., et al., 1997a, *IAU Circ. No.* 6541
Zhang, W., Strohmayer, TE, Swank, JH., 1997b, *ApJ* 482 L167
Zhang, W., Strohmayer, TE, Swank, JH., 1998a, *ApJ* 500 L167
Zhang, W., et al., 1998b, *ApJ* 500 L171
Zhang, W., et al., 1998c, *ApJ* 495 L9

WHAT HAVE WE LEARNED ABOUT GAMMA RAY BURSTS FROM AFTERGLOWS?

Mario Vietri

Università di Roma III

ABSTRACT The discovery of GRBs' afterglows has allowed us to establish several facts: their distance and energy scales, the fact that they are due to explosions, that the explosions are relativistic, and that the afterglow emission mechanism is synchrotron radiation. On the other hand, recent data have shown that the fireball model is wrong when it comes to the emission mechanism of the true burst (which is unlikely to be synchrotron again) and that shocks are not external. Besides these relatively tame points, I will also discuss the less well established physics of the energy deposition mechanism, as well as the possible burst progenitors.

KEYWORDS: gamma rays: bursts – stars: neutron – black holes – relativity – hydrodynamics – emission mechanisms

1. INTRODUCTION

Gamma ray bursts (GRBs) were discovered in 1969 (Klebesadel, Strong and Olson 1973) by American satellites of the *Vela* class aimed at verifying Russian compliance with the nuclear atmospheric test ban treaty. Though the discovery was made in 1969, the paper appeared only four years later because the authors had lingering doubts about the reality of the effects they had discovered. Since then, several thousands of bursts have been observed by a more than a dozen different satellites, but it is remarkable that the basic burst features outlined in the abstract of the 1969 paper (photons in the range $0.2 - 1.5\ MeV$, durations of $0.1 - 30\ s$, fluences in the range $10^{-5} - 2 \times 10^{-4}\ ergs\ cm^{-2}$) have remained substantially unchanged.

Current evidence (Fishman and Meegan 1995) has highlighted a wide ($0.01 - 100\ s$) duration distribution, with hints of a bimodality which is claimed to correlate (at the 2.5σ level) with spectral properties. All bursts' spectra observed so far are strictly non–thermal, and there has never been any confirmation by BATSE of a supposed thermal component (nor of cyclotron lines or precursors, for this matter) claimed in previous reports. A remarkable feature reported by BATSE is the bewildering diversity of light curves, ranging from impulsive ones (a spike followed by a slower decay, nicknamed FREDs for Fast Rise-Exponential Decay), to smooth ones, to long ones with amazingly sharp fluctuations, including even some with a strongly periodic appearance (two such examples are the 'hand' and the 'comb', so nicknamed from the number of high–Q, regularly repeating sharp spikes).

2000 BATSE Gamma-Ray Bursts

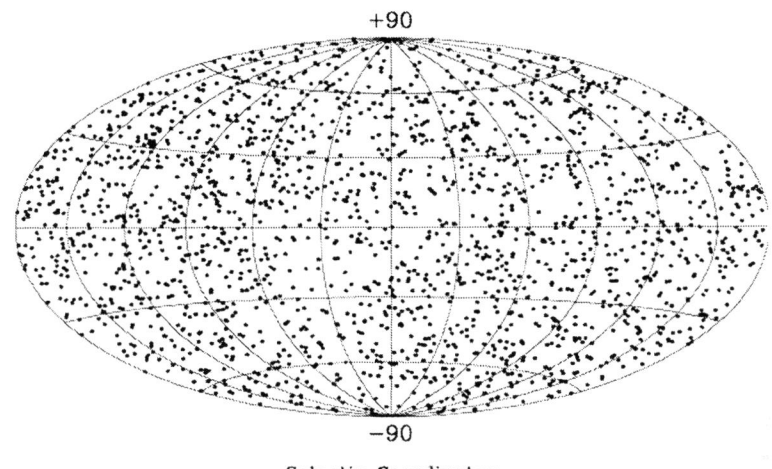

FIGURE 1. Burst distribution on the plane of the sky,

The most exceptional result from BATSE, though, was the sky distribution of the bursts (Fig.1). It was obvious from it that the bursts *had* to be extragalactic, as already discussed by theorists (Usov and Chibisov 1975, Paczyǹski 1986).

2. AFTERGLOWS

The next major step was triggered by BeppoSAX: in the summer of 1996, L. Piro and his coworkers located in archival data of the satellite the soft X-ray counterpart of a GRB (GRB 960720). They immediately conceived the idea of implementing a procedure to follow the next burst in real time, by re-orienting the whole satellite, after the initial detection by the Wide Field Cameras, so that the more sensitive Narrow Field Instruments could pinpoint the burst location to within 45 arcsecs, a feat never achieved in such short times, and by a single satellite. After an initial snafu (GRB 970111), the gigantic effort paid off with the discovery of the X-ray afterglow of GRB 970228 (Costa *et al.*, 1997), immediately followed by the discovery of its fading optical counterpart (van Paradijs *et al.*, 1997), obtained through a search inside the WFC error box, in perfect agreement with theoretical predictions (Vietri 1997a, Mèszàros and Rees 1997)[1].

After the detection of the optical counterpart, the door was open to find the bursts' redshifts: Table I summarizes the status of our current knowledge (september

---

[1] There is an interesting lesson to be drawn from this: in case one should wonder why a soft X-ray telescope was not placed onboard Compton to track GRBs, it was because of rivalries between different NASA subsections, the X-ray and the γ-ray divisions.

GRB	z	E_{iso}
970228	0.695	5×10^{51} erg
970508	0.835	2×10^{51} erg
971214	3.4	3×10^{53} erg
980703	0.93	3×10^{53} erg
990123	1.7	4×10^{54} erg
990510	1.6	2×10^{53} erg
990712	0.43	

1999); bursts' luminosities are for isotropic sources. Two comments are in order. First, the bursts have *prima facie* a redshift distribution not unlike that of AGNs and of the Star Formation Rate (SFR). The initial hope that they might trace an even more distant and elusive Pop III, triggered by the fact that the second redshift detected was also the largest so far (GRB 971214, $z = 3.4$), has now vanished. Second, in order to place the energy release of GRB 990123 in context, one should notice that 4×10^{54} *ergs* is the energy obtained by converting the rest–mass of two solar masses, or, alternatively, the energy emitted by the whole Universe out to $z \approx 1$ within the burst duration. So, a single (perhaps double) star outshines the whole Universe.

Besides the distance and energy scales, the major impact of the discovery of afterglows has been the establishment of some key features of the fireball model (Rees and Mèszàros 1992):

1. bursts are due to explosions, as evidenced by their power–law behaviour;

2. the explosions are relativistic, as proved by the disappearence of radio flares;

3. the burst emission is due to synchrotron emission, as shown by the afterglow spectrum, and its optical polarization.

I will illustrate these points in the following, but, lest we become too proud, we should also remember that the fireball model has met some failures. The original version of the model (Mèszàros and Rees 1993) advocated the dissipation of the explosion energy at external shocks (*i.e.*, those with the interstellar medium). Sari and Piran (1997), following a point originally made by Ruderman (1975) showed that these shocks smooth out millisecond timescale variability, which can only be maintained by the internal shocks proposed by Paczyǹski and Xu (1994). Also, the fireball model originally ascribed even the emission from the burst proper (as opposed to the afterglow) to optically thin synchrotron processes; I will discuss in the section *Embarrassments* why this is exceedingly unlikely. Furthermore, even the last tenet of mid–90s common wisdom, *i.e.*, that bursts are due to neutron binary mergers, does not look too promising at the moment (since some bursts seem to be located inside star forming regions, incompatible with the long spiral–in time), though of course it is by no means ruled out yet.

2.1. The fireball model

Here, one may assume that an unknown agent deposits $10^{51}-10^{54}$ $ergs$ inside a small volume of linear dimension $\approx 10^6 - 10^7$ cm. The resulting typical energy density corresponds to a temperature of a few MeVs, so that electrons and positrons cannot be bound by any known gravitational field. In these conditions, optical depths for all known processes exceed 10^{10}. The fluid expands because of its purely thermal pressure, converting internal into bulk kinetic energy. Parametrizing the baryon component mass as $M_b \equiv E/\eta c^2$, it can be shown that, for $1 \leq \eta \leq 3 \times 10^5$ (Mèszàros, Laguna, Rees 1993) the fluid achieves quickly (the fluid Lorenz factor increases as $\gamma \propto r$) a coasting Lorenz factor of $\gamma \approx \eta$.

The requisite asymptotic Lorenz factor is dictated by observations: photons up to $\epsilon_{ex} \approx 18$ GeV have been observed by EGRET from bursts (Fishman and Meegan 1995). For these photons to evade collisions with other photons, and thus electron/positron pair production, it is necessary that, in the reference frame in which a typical burst photon (with $\epsilon \approx 1$ MeV) and the exceptional photon are emitted, they appear as below pair production threshold: thus we must have $\epsilon' \epsilon'_{ex} \leq 2 m_e c^2$. Since $\epsilon' \approx \epsilon/\gamma$, and similarly for the other photons, we find (Baring 1993)

$$\gamma \approx 300 \left(\frac{\epsilon}{1\ MeV} \frac{\epsilon_{ex}}{10\ GeV}\right)^{1/2}. \qquad (1)$$

From what we said above, we thus require a maximum baryon contamination, in an explosion of energy E, of $M_b \lesssim 10^{-6} M_\odot (E/10^{51}\ erg)(300/\eta)$.

The energy release is now assumed to be in the form of an inhomogeneous wind, with parts having a Lorenz factor larger than parts emitted previously. This leads to shell collisions (the internal shock model) at radii r_{sh} which allow a time–scale variablity $\delta t \approx r_{sh}/2\gamma^2 c$; for $\delta t = 1$ ms, $r_{sh} \approx 10^{13}$ cm, which fixes the internal shock radii. Particle acceleration at these internal shocks and ensuing non–thermal emission is thought to lead to the formation of the burst proper. At larger radii, a shock with the surrounding ISM forms, and shell deceleration begins at a radius $r_{ag} = (3E/4\pi n m_p c^2 \gamma^2)^{1/3} \approx 10^{17}$ cm for a $n = 1$ cm^{-3} particle density typical of galactic disks. It is thought that the afterglow begins when the shell begins the slowdown, as this drives a marginally relativistic shock into the ejecta, thusly extracting a further fraction of their bulk kinetic energy.

2.2. Why explosions

The success of the fireball model lies in this, that it decouples the problem of the energy injection mechanism from the following evolution, which is, furthermore, an essentially hydrodynamical problem. It can be shown, in fact (Waxman 1997) that the evolution of the external shock is adiabatic, that the shock Lorenz factor

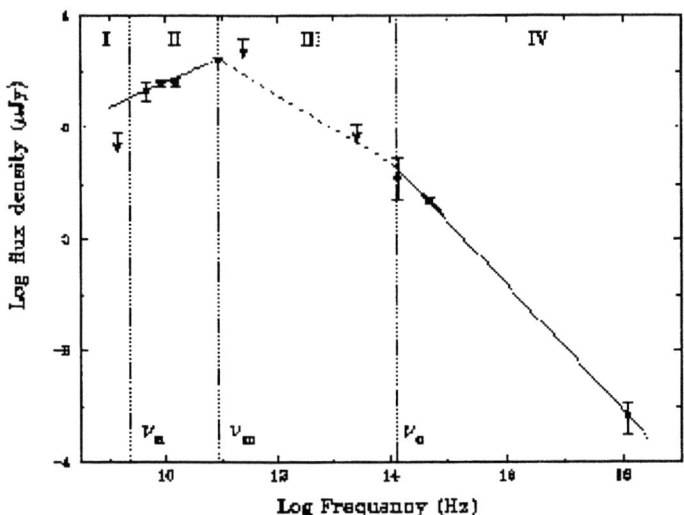

FIGURE 2. Simultaneous spectrum of the afterglow of GRB 970508, from Galama et al., 1998.

decreases as $\gamma \propto r^{-3/2}$ because of the inertia of the swept-up matter, and thus r scales with observer's time as $t = r/\gamma^2 c \to \gamma \propto t^{-3/8}$ (for a radiative solution $\gamma \propto r^{-3/7}$, Vietri 1997b). If afterglow emission is due to optically thin synchrotron in a magnetic field in near-equipartition with post-shock energy density, it can be shown that $B \propto \gamma$, that the typical synchrotron frequency at the spectral peak $\nu_m \propto \gamma B \gamma_e^2 \propto \gamma^4$ (where $\gamma_e \propto \gamma$ is the lowest post-shock electron Lorenz factor), and that $F(\nu_m) \propto t^{-3\beta/2}$, where β is the afterglow spectral slope. As it can be seen, these expectations are based exclusively upon the hydrodynamical evolution (and the synchrotron spectrum), and are thus reasonably robust.

We thus expect power-law time decays, a characteristic of strong explosions (see the Sedov analogy!), with time- and spectral-indices closely related. This is what is observed everywhere, from the X-ray through the optical to the radio, (see Piro and Fruchter, this volume), the few exceptions being discussed later on. In fact, the equality of the time-decay index of the X-ray and optical data in afterglows of individual sources has been taken as the key element to show that emission in the different bands is due to the same source. Time indices in the X-ray are in the range $0.7 - 2.2$ (Frontera et al., in preparation).

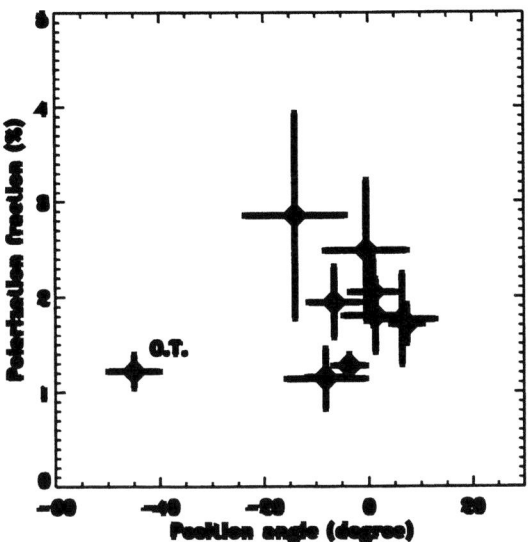

FIGURE 3. Polarization amplitude and position angle for optical afterglow of GRB 990510, from Covino et al., 1999

2.3. Why synchrotron spectrum in the afterglow

After having established that bursts are due to explosions, we happily learn that afterglows emit through synchrotron processes. In fig. 2 (Galama et al., 1998), we show the superposition of theoretical expectations for an optically thin synchrotron spectrum (including the cooling break at $\nu \approx 10^{14}\ Hz$) with observations for GRB 970508. The remarkable agreement is even more exciting as we remark that observations are not truly simultaneous, but are scaled back to the same time by means of the theoretically expected laws for time–decay, thus simultaneoulsy testing the correctness of our hydro. Another piece of evidence comes from the discovery of polarization in the optical afterglow of GRB 990510 (Fig. 3, Covino et al., 1999, Wijers et al., 1999). This polarization may appear small ($\approx 2\%$), but it is surely not due to Galactic effects: stars in the same field show a comparable degree of polarization, but along an axis different by about 50°. Also, polarization in the source galaxy is unlikely, because of a very stringent upper limit on the reddening due to this galaxy (Covino et al., 1999). The only remaining question mark is emission from an anisotropic source, but this would require a disk of $10^{18}\ cm$ to survive the intense γ ray (and X, and UV) flash: though not excluded, it does not look likely.

2.4. Why relativistic expansion

Radio observations of the first burst observed so far (GRB 970508, Frail et al., 1997) showed puzzling fluctuations by about a factor of 2 in the flux, over a time–scale

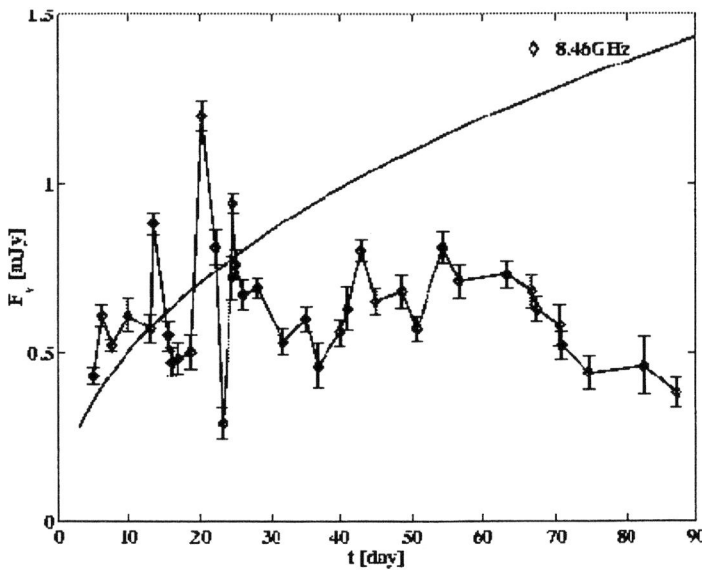

FIGURE 4. VLA observatons at 8.46 GHz of the afterglow of GRB 970508, from Waxman,Frail and Kulkarni 1998.

of days, disappearing after about 30 days from the burst (Fig. 4). This extreme, and unique behaviour, was explained by Goodman (1997), who showed that it is due to interference of rays travelling along different paths through the ISM, and randomly deflected by the spatially varying refractive index of the turbulent ISM. The wonderful upshot of this otherwise marginal phenomenon, is that these effects cease whenever the source expands beyond a radius

$$R = 10^{17} \; cm \frac{\nu_{10}^{6/5}}{d_{sc,kpc} h_{75}} \left(\frac{SM}{10^{-2.5} m^{-20/3} \; kpc} \right)^{-3/5}, \qquad (2)$$

where ν_{10} is the radio observing frequency in units of 10^{10} Hz, $d_{sc,kpc}$ is the distance of the ISM from us (assumed to be a uniform scattering screen), and SM is the Galactic scattering measure, scaled to a typical Galactic value. The existence of interference effects is made more convincing by the amplitude of the average increase (a factor of 2, as observed), the correctness in the prediction of the time–interval between different peaks, and of the decorrelation bandwidth. Since flares disappear after about 30 days, it means that the average speed of the radio source is $R/30 \mathrm{days} = 3 \times 10^{10}$ $cm \; s^{-1}$. So we see directly that GRB 970508 expanded at an average speed of c over a whole month, giving us a direct observational proof that the source is highly relativistic. This proof is completely equivalent to superluminal motions in blazars, and is the strongest evidence in favor of the fireball model.

2.5. GRB 970508: our best case

The afterglow of GRB 970508 is our best case so far: it is in fact a burst for which not only do we know the redshift, but also a radio source that has been monitored for more than 400 days after the explosion (Frail, Waxman and Kulkarni 2000). Through these observations we can see the transition to a sub–relativistic regime at $t \approx 100\ d$, measure the total energetics of the following Sedov phase (unencumbered by relativistic effects!) $E_{New} = 5 \times 10^{50}\ ergs$, determine two elusive parameters, $\epsilon_{eq} = 0.5$ and $\epsilon_B = 0.5$ (the efficiencies with which energy is transfered to post–shock electrons by protons, and with which an equipartition field is built up), and the density of the surrounding medium $n \approx 0.4\ cm^{-3}$. All of these values look reasonable (perhaps ϵ_{eq} and ϵ_B exceed our expectations by a factor of 10, a fact that could be remedied by introducing a slight density gradient which would keep the shock more efficient), so that our confidence in the external–shock–in–the–ISM model is boosted.

Another precious consequence of these late–time observations is that they yield information on beaming and energetics. In fact, GRB 970508 appeared to have a kinetic energy of $E_{rel} = 5 \times 10^{51}\ erg$ when in the relativistic phase, a measurement which can be reconciled with E_{New} (remember that the expansion is adiabatic, so that we must have $E_{New} = E_{rel}$!) only if the unknown beaming angle, assumed $= 4\pi$ in deriving E_{rel}, is smaller than 4π by the factor E_{New}/E_{rel}; we thus have the only measurement of $\delta\Omega/4\pi = 0.1$, so far. This already rules out all classes of models requiring unplausibly large amounts of beaming, 10^{-8} or even beyond. Hopefully, more such measurements will come in the future, since this observationally heavy method is subject to many fewer uncertainties than the competing method of trying to locate breaks in the time–decay of afterglows. Also, the radiative efficiency of the burst can be estimated: correcting the observed burst energy release $E_{GRB} = 2 \times 10^{51}\ erg$ for the same beaming factor, the radiative efficiency is $E_{GRB}\delta\Omega/4\pi/(E_{New}+E_{rel}\delta\Omega/4\pi) = 0.3$, again a unique determination. Notice however that this figure is subject to a systematic uncertainty: we do not know whether the beaming fraction is the same for the burst proper and for the afterglow.

3. EMBARRASSMENTS

Something is rotten in the fireball kingdom as well, namely, departures from pure power–law behaviours, and the spectra of the bursts proper.

3.1. Unpowerlawness

Departures from power–laws are expected when one considers the extremely idealized character of the solutions discussed so far: perfect spherical symmetry, uniform surrounding medium, smooth wind from the explosion, ϵ_{eq} and ϵ_B constant in space and time. The tricky point here is to disentangle these distinct factors. In GRB 970508 and GRB 970828 (Piro et al., 1999, Yoshida et al., 1999) a major departure was observed in the X–ray emission, within a couple of days from the burst; they constitute the single, largest violations observed so far, in terms of number of photons. It is remarkable that spectral variations were simultaneously observed, and that both bursts showed traces (at the 2.7σ significance level) of an iron emission line. The similarity of the bursts' behaviour argues in favor of the reality of these spectral features, which have been interpreted as thermal emission from a surrounding stellar–size leftover, pre–expelled by the burst's progenitor (Lazzati et al., 1999, Vietri et al., 1999). Clearly, these departures hold major pieces of information on the bursts' surroundings, and the nature of bursts' progenitors.

It has been argued (Rhoads 1997) that, whenever the afterglow shell decelerates to below $\gamma \approx 1/\theta$, where θ is the beam semi–opening angle, emission should decrease because of the lack of emitting surface, compared to an isotropic source. But, in view of the existence of clear environmental effects (GRB 970508 and GRB 970828), it appears premature to put much stock in the interpretation of time–power–law breaks as due to beaming effects. And equally, it appears to this reviewer that the same comment applies to the interpretation of a resurgence of flux as due to the appearance of a SN remnant behind the shell. The major uncertainty here is the non–univoqueness of the interpretation: Waxman and Draine (2000) have shown that effects due to dust can mimic the same phenomenon.

3.2. Bursts' spectra

A clear prediction of the emission of optically thin synchrotron is that the low–photon–energy spectra should scale like $dN_\nu/d\nu \propto \nu^\alpha$, with $\alpha = -3/2$, since the emission is in the fast cooling regime. Within thin synchrotron, there is no way to obtain $\alpha > -3/2$. This early–recognized requirement (Katz 1994) is so inescapable that it has been dubbed the 'line of death'. Observations are notoriously discordant with this prediction. Preece et al.(1999) have shown that, for more than 1000 bursts, α is distributed like a bell between -2 and 0, with mean $\bar{\alpha} \approx -1$. The tail of this distribution also contains a few tens of objects with $\alpha \approx +1$. An example of these can be found in Frontera et al., 1999 (GRB 970111), which is instructive since BeppoSAX has better coverage of the critical, low–photon–energy region. In particular, BATSE seems to loose sensitivity below ≈ 30 keV, but this is still not enough to explain away the discrepancy with the theory. Also, Preece et al., 1999, showed that the time–integrated spectral energy distribution has a peak at a photon energy $\epsilon_{pk} \approx 200$ keV, and that ϵ_{pk} has a very small variance from burst to burst. Again, this does not seem dependent upon BATSE's lack of sensistivity above 700 keV, and again this has no explanation within the classic fireball model.

Any theorist who worked on blazars will say that the root of the disagreement is the neglect of Inverse Compton processes, but the trick here is not to identify the culprit, on which everyone agrees, but to devise a fireball model that smoothly incorporates it. One should remember that the details of the fireball evolution are *generic*, *i.e.*, they do not depend upon any detailed property of the source, so that things like the radius at which the fireball becomes optically thin (to pairs or baryonic electrons), the radius at which acceleration ends, the equipartition magnetic field, and so on, are all reliably and inescapably fixed by the outflow's global properties. A step toward the solution has been made by Ghisellini and Celotti (1999) who remarked that at least some bursts have compactness parameters $l = 10(L/10^{53} \, erg \, s^{-1})(300/\gamma)^5 \gg 1$. Under these conditions, a pair plasma will form, nearly thermalized at $kT \approx m_e c^2$, and with Thomson optical depth $\tau_T \approx 10$. The modifications which this plasma will bring to the burst's spectrum are currently unknown, but it may be remarked that this configuration will be optically thick to both high–energy synchrotron photons due to non–thermal electrons accelerated at the internal shocks, and to low–energy cyclotron photons emitted by the thermal plasma, but it will be optically thin in the intermediate region reached by cyclotron photons upscattered via IC processes off non–thermal electrons. A model along this line (*i.e.*, upscattering of cyclotron photons by highly relativistic electrons) is in preparation (Vietri 2000a), but it remains to be seen whether it (like any other model, of course) can simultaneously explain the spectral shape and the narrow range of the spectral distribution peak energy ϵ_{pk}.

4. ON THE CENTRAL ENGINE

As remarked several times already, the fireball evolution is independent of the source nature. The only exisiting constraint is the maximum amount of baryon contamination, which is

$$M_b = \frac{E}{\eta c^2} = 10^{-6} M_\odot \frac{E}{10^{51} \, erg} \frac{300}{\gamma} \, . \tag{3}$$

This is a remarkably small value: since the inferred luminosities exceed the Eddington luminosity by 13 orders of magnitude, they clearly have all it takes to disrupt a whole star, no matter how compact. Yet, the energy deposition must somehow occur outside the main mass, lest the explosion be slowed down to less relativistic, or even possibly Newtonian speeds. In order to satisfy this constraint, it has emerged that the most favorable configuration has a stellar–mass black hole ($M_{BH} \approx 3 - 10 M_\odot$) surrounded by a thick torus of matter ($M_t \approx 0.01 - 1 M_\odot$, with $\rho \approx 10^{10} \, g \, cm^{-3}$). The presence of a black hole is *not* required by observations in any way: models involving neutron stars are still admissible, the advantage of having a black hole being only the deeper potential well: you get more energy out per unit accreted mass. The configuration thusly envisaged has a cone surrounding the symmetry axis devoid of baryons, since all models leading to this configuration have large amounts of specific angular momentum, and thus baryons close to the rotation axis either are not there, or have accreted onto the black hole due to their lack of centrifugal support.

4.1. Energy release mechanism

There are two major mechanisms for energy release discussed in the literature, the first one to be proposed (Berezinsky and Prilutskii 1986) being the reaction $\nu + \bar{\nu} \to e^- + e^+$. Neutrinos have non–negligible mean free paths in the tori envisaged here, so that this annihilation reaction will take place not inside tori themselves, where they are preferentially generated because densities are highest, but in a larger volume surrounding the source. This is both a blessing and a disgrace: by occupying a larger volume, the probability that every neutrino finds its antiparticle to annihilate decreases, but then the energy is released in baryon–cleaner environments. The problem, though complex, is eminently suitable for numerical simulations, showing (Janka et al., 1999, and references therein) that about 10^{50} $ergs$ can be released this way, above the poles of a black hole where less than $10^{-5} M_\odot$ are found.

Highly energetic bursts cannot be reproduced by this mechanism, due to its low efficiency: the second mechanism proposed involves the conversion of Poynting flux into a magnetized wind. The basic physical mechanisms are well–known (Usov 1992) since they have been studied in the context of pulsar emission: electrons are accelerated by a motional electric field $\vec{E} = \vec{v} \wedge \vec{B}/c$ due to the rotation of a sufficiently strong magnetic dipole, attached either to a black hole, or to the torus. Photons are then produced by synchrotron or curvature radiation, and photon/photon collisions produce pairs, to close the circle and allow looping. In order to carry away 10^{51} $erg\ s^{-1}$, a magnetic field of $\approx 10^{15}$ G is required. This is not excessive, since it is about three orders of magnitude below equipartition with torus matter, and because such fields already exist in nature, see SGR 1806-20 and SGR 1900+140: the key point is to understand whether some kind of dynamo effect can lead to these high values within the short allotted time.

Depending upon whether the open magnetic field lines extending to infinity are connected to the black hole or to the torus, the source of the energy of the outflow will be the rotational energy of the black hole (the so–called Blandford–Znajek effect) or of the torus. The first case is traditionally discussed in the context of AGNs (Rees, Blandford, Begelman and Phinney 1984), but it is harshly disputed whether the energy outflow may be actually dominated by the black hole rather than by the disk (Ghosh and Abramowicz, 1997, Livio, Ogilvie and Pringle 1998). On the other hand, the torus looks ideal as the source of a dynamo: its large shear rate, the presence of the Balbus-Hawley instability to convert polidal into toroidal flux, and the possible presence of the anti–floating mechanism inhibiting ballooning of the magnetic field (Kluzniak and Ruderman 1998), all seem to favor the existence of a fast dynamo. It should also be remarked that the configuration of the magnetic field in this problem is known: in fact, the configuration discussed in Thorne et al., 1986 for black holes, only uses the assumptions of steady–state and axial symmetry, and is thus immediately extended to magnetic fields anchored to the torus. What is really required here is a first order study, of the sort published by Tout and Pringle (1992) on angular momentum removal from young, pre–main–sequence stars via magnetic stresses, and on the associated $\alpha - \omega$ dynamo. Until such studies are made, it will be premature to claim that neutrino annihilations are responsible for the powering of GRBs.

4.2. Progenitors

There is no lack of proposed progenitors, but I will discuss only binary neutron mergers (Narayan, Paczyński and Piran 1992), collapsars (Woosley 1993, Paczyński 1998) and SupraNovae (Vietri and Stella 1998, 1999).

Clearly, NS/NS mergers is the best model on paper: it involves objects which have been detected already, orbital decay induced by gravitational wave emission is shown by observations to work as per the theory, and numerical simulations by Janka's group show that a neutrino–powered outflow in baryon–poor matter can be initiated. The major theoretical uncertainties here concern bursts' durations and energetics: all numerical models produce short bursts ($\approx 0.1\ s$) with modest energetics, $E < 10^{51}\ erg$. This is a direct consequence of the mechanism for powering the burst: large, super–Eddington luminosities are carried away by neutrinos, leading to a large mass influx, but only a small fraction, $1 - 3\%$, can be harnessed for the production of the burst. Furthermore, we cannot invoke large beaming factors in this case: the outflow is only marginally collimated, in agreement with expectations that an accretion disk with inner and outer radii $R_{out}/R_{in} \approx$ a few (for the case at hand) can only produce a beam semi–opening angle of R_{in}/R_{out}. So, perhaps, this model may account for the short bursts, but it should be remembered that nothing of what was discussed above pertains to this subclass: BeppoSAX (and thus all BeppoSAX-triggered observations) can only detect long bursts.

On the other hand, future space missions, whether or not able to locate short bursts, can provide a decisive test of this model, provided they can follow with sufficient sensitivity a given burst for several hours. This model, in fact, is the only one proposed so far according to which some explosions should take place outside galaxies: according to Bloom, Sigurdsson and Pols (1999), about 50% of all bursts will be located more than 8 kpc from a galaxy, and 15% in the IGM. This characteristic is testable without recourse to optical observations. In fact, the afterglow begins with a delay (as seen by an outside observer) of $t_d = (r_{ag} - r_{sh})/\gamma^2 c \approx r_{ag}/\gamma^2 c$, which varies greatly depending upon the environment in which the burst takes place:

$$t_d = \begin{cases} 15\ s & \text{ISM, n} = 1\ cm^{-3} \\ 5\ min & \text{galactic halo, n} = 10^{-4}\ cm^{-3} \\ 4\ h & \text{IGM, n} = 10^{-8}\ cm^{-3} \end{cases} \quad (4)$$

Between the burst proper and the beginning of the power–law–like afterglow, thus a silence of recognizable duration is expected (Vietri 2000b).

Collapsars are currently in great vogue as a possible source of GRBs: the large amount of energy available as the core of a supermassive star collapses directly to a black hole is in fact very attractive, even though (again!) the limited efficiency of the reaction $\nu + \bar\nu \rightarrow e^- + e^+$ makes most of this energy unavailable. Here too there is some evidence that these objects must exist (Paczyński 1998), and numerical simulations again showing energy preferentially deposited along the hole rotation axis are also available (McFayden and Woosley 1999). Here however, what is truly puzzling is how the outflow can pierce the star's outer layers without loading itself with baryons: we should remember that at most $10^{-6} M_\odot$ can be added to 10^{51} erg: more baryons imply a proportionately slower outflow. The argument is that the dynamical timescale of the outer layers of a massive stars is of order of a few hours, so that, even if the core collapses and pressure support is removed, nothing will happen during the energy release phase: the outflow must pierce its way through. Two processes seem especially dangerous: Rayleigh–Taylor instability of the fluid heated–up by neutrino annihilations as it is weighed upon by the colder, denser outer layers, and Kelvin–Helmholtz instability after the hot fluid has pierced the outer layers and is passing through the hole. It is well–known that the non–linear development of these instabilities leads to mass entrainment, and that the time–scale for the development of these instabilities is very fast. Furthermore, the baryon–free outflow may be 'poisoned' by baryons to a deadly extent, even if numerical simulations, with their finite resolution, were to detect nothing of the kind.

The third class of models, SupraNovae, concerns supramassive neutron stars which are stabilized against self–gravity by fast rotation, to such an extent that they cannot be spun down to $\omega = 0$ because they implode to a black hole. As the star's residual magnetic dipole sheds angular momentum, this is exactly the fate to be expected for the whole star, except for a small equatorial belt , whose later accretion will power the burst. It is easy to show that this implosion must take place in a very baryon–clean environment. The major uncertainties here concern the channels of formation and the existence of this equatorial belt. Two channels of formation have been proposed: direct collapse to a supramassive configuration (Vietri and Stella 1998) and slow mass accretion in a low–mass X–ray binary (Vietri and Stella 1999). Both are possible, though none yet is supported by observations. The existence of the left–over belt has recently been questioned by Shibata, Baumgarte and Shapiro (1999), who however simulated the collapse of neutron stars with intermediate equations of state, which are entirely (or nearly exactly so) contained inside the marginally stable orbit even before collapse: clearly, these must be swallowed whole by the resulting black hole. Soft equations of state are free of this objection, and are thus much more likely to leave behind an equatorial belt. The soft EoSs are especially favored since the neutron stars must survive the r–mode instability, and thus soft EoSs (Weber 1999) would be in any case required. So one might say that the existence of these stars hinges on one uncertainty only, the EoS of nuclear matter. Besides the baryon–clean environments, SupraNovae have another advantage over rival modlels: only the lowest density regions would be left behind, precisely those with the smallest neutrino losses. The powering of the burst can thus occur through accretion caused by removal of angular momentum by magnetic stresses, without the parallel, unproductive, neutrino generation.

5. CONCLUSIONS

It is difficult to end on an upbeat note: we cannot expect in the near future a rate of progress similar to the one we witnessed in the past three years. In particular, it may be expected that the next flurry of excitement will come with the beginning of the SWIFT mission, which promises to collect relevant data (redshifts, galaxy types, location within or without galaxies, absorption or emission features in the optical and in the X–ray) for a few hundred bursts. This data will nail the major characteristics of the environment (at large) in which bursts take place, and we may be able to rule out a few models. On the other hand, the energy release process, shrouded as it is in optical depths $> 10^{10}$, will remain mysterious, our only hope in this direction being gravitational waves.

Judging by the analogy with radio pulsars, this will correspond to the flattening of the learning curve. Aside from this, we may hope to locate the equivalent of the binary radio pulsar, but, differently from Jo Taylor, we have to be awfully quick in grabbing it.

ACKNOWLEDGEMENTS

Thanks are due to Gabriele Ghisellini, who wisely steered me away from synchrotron emission, and toward the true light of Inverse Compton.

REFERENCES

Baring, M., 1993, Ap.J., 418, 391.
Berezinsky, V.S.,, Prilutskii, O.F., 1986, Astr.Ap., 175, 309.
Bloom, J.S., Sigurdsson, S., Pols, O.R., 1999, MNRAS, 305, 763.
Costa, E., *et al.*, 1997, Nature, 387, 783.
Covino, S., *et al.*, 1999, Astr.Ap., 348, L1.
Fishman, G.J., Meegan, C.A., 1995, ARAA, 33, 415.
Frail, D., *et al.*, 1997, Nature, 389, 261.
Frail, D., Waxman, E., Kulkarni, S., 2000, Ap.J., submitted, astro–ph 9910319.
Frontera, F., *et al.*, 1999, Ap.J.S., in press, astro–ph 9911228.
Galama, T., *et al.*, 1998, Ap.J.L., 500, L97.
Ghisellini, G., Celotti, A., 1998, Ap.J.L., 511, L93.
Ghosh, P., Abramowicz, M.A., 1997, MNRAS, 292, 887.
Goodman, J.J., 1997, New Astr., 2, 449.
Janka, T., Eberl, T., Ruffert, M., Fryer, C.L., 1999, Ap.J.L., 527, L39.
Katz, J., 1994, Ap.J., 422, 248.
Klebesadel, R.W., Strong, I.B., Olson, R.A., 1973, Ap.J.L., 182, L85. ,
Kluzniak, W., Ruderman, M., 1998, Ap.J.L., 505, L113.
Livio, M., Ogilvie, G.I., Pringle, J.E., 1999, Ap.J., 512, 100.
Lazzati, D., Campana, S., Ghisellini, G., MNRAS, 1999, 304, L31.
Mc Fayden, A., Woosley, S.E., 1999, Ap.J., 524, 262.
Mészáros, P. Laguna, P., Rees, M.J., 1993, Ap.J., 415, 181.

Mészáros, Rees, M.J., 1993, Ap.J., 405, 278.
Mészáros, Rees, M.J., 1997, Ap.J., 476, 232.
Narayan, R., Paczynski, B., Piran, T., 1992, Ap.J.L., 395, L83.
Paczyǹski, B., 1986, Ap.J.L., 308, L43.
Paczyǹski, B., 1998, Ap.J.L., 494, L45.
Paczyǹski, B., Xu, G., 1994, Ap.J., 427, 708.
Piro, L., *et al.*, 1999, Ap.J.L., 514, L73.
Preece, R.D., *et al.*, 1999, Ap.J.S., in press, astro-ph. 9908119.
Rees, M.J., Begelman, M.C., Blandford, R.D., Phinney E.S., 1984, Nature, 295, 17.
Rees, M.J., Mèszàros, P., 1992, MNRAS, 258, P41.
Rhoads, J., 1997, Ap.J.L., 487, L1.
Ruderman, R., 1975, Ann. NY. Acad. Sci., 262, 164.
Sari, R., Piran, T., 1997, Ap.J., 485, 270.
Shibata, M., Baumgarte, T.W., Shapiro, S.L., 1999, Phys.Rev.D, submitted, astro-ph. 9911308.
Thorne, K.S., Price, R.H., MacDonald, D.A., 1986, *Black holes: the membrane paradigm*, Yale:New Haven, Yale Univ. Press.
Tout, C.A., Pringle, J.E., 1992, MNRAS, 256, 269.
Usov, V.V., 1992, Nature, 357, 472.
Usov, V.V., Chibisov, G.V. Soviet Astr., 19, 115.
van Paradijs, J., *et al.*, 1997, Nature, 386, 686.
Vietri, M., 1997a, Ap.J.L., 478, L9.
Vietri, M., 1997b, Ap.J.L., 488, L105.
Vietri, M., 2000a, in preparation.
Vietri, M., 2000b, Ap.J.L., submitted.
Vietri, M., Perola, G.C., Piro, L., Stella, L., 1999, MNRAS, 308, P29.
Vietri, M., Stella, L., 1998, Ap.J.L., 507, L45.
Vietri, M., Stella, L., 1999, Ap.J.L., 527, L43.
Waxman, E., 1997, Ap.J.L., 489, L33.
Waxman, E., Draine, B.T., 2000, Ap.J., in press, astro-ph. 9909020.
Waxman, E., Frail, D., Kulkarni, D., 1998, Ap.J., 497, 288.
Weber, F., 1999, in *Pulsars as astrophysical laboratories for nuclear and particle physics*, Bristol, U.K.; Institute of Physics.
Wijers, R.A.M.J., *et al.*, 1999, Ap.J.L., 523, L33.
Woosley, S., 1993, Ap.J., 405, 273.
Yoshida, A., *et al.*, 1999, Astr.Ap.S., 138, 433.

ORAL PRESENTATIONS

KECK K-BAND OBSERVATIONS OF LOW MASS X-RAY BINARIES

Paul J. Callanan[1], Alexei V. Filippenko[2], and Michael R. Garcia[3]

1) Department of Physics, University College, Cork, Ireland
2) Department of Astronomy, University of California, Berkeley, CA 94720-3411
3) Center for Astrophysics, 60 Garden Street, Cambridge, MA 02138 USA

ABSTRACT

We present K-band observations of a sample of XRN/LMXB using NIRC on Keck I. For GRO J0422+32, we show that the K-band flux is significantly contaminated by the accretion disk. We present K-band images of 4U 1630-47, Aql X-1, and GX 17+2, and we discuss the implications of the newly discovered IR/optical counterparts for the latter two systems.

KEYWORDS: binaries: close – X-rays: stars

1. INTRODUCTION

The main areas in which near-IR (1–2.5 μm) observations have made significant contributions to our understanding of X-ray binaries are the following:

(1) IR photometry of black hole X-ray novae (XRN) in quiescence has been used to measure the ellipsoidal variation of the secondary star. These observations are thought to be less susceptible to contamination from the accretion disk than those at optical wavelengths, and hence a more reliable ellipsoidal modulation should be observed. Radial velocity studies combined with these measurements allow the mass of the black hole to be inferred (e.g., Shahbaz et al. 1994; Callanan et al. 1996).

(2) IR imaging of highly reddened LMXBs/XRN is often the only method of identifying counterparts at wavelengths other than X-ray. Furthermore, higher resolution images are easier to achieve in the IR than at optical wavelengths.

In what follows we briefly review some recent results we have obtained using the Near-IR Camera (NIRC) on Keck I. NIRC uses a 256×256 pixel InSb detector (0.15″/pixel) and operates in an imaging and low resolution (R $\approx$60–120) spectroscopic mode.

2. THE ACCRETION DISK CONTAMINATION OF THE K-BAND FLUX OF QUIESCENT XRN

Filippenko, Matheson, & Ho (1995), measured the mass function of GRO J0422+32 to be 1.21 ± 0.06 $M_\odot$. We used NIRC in an attempt to measure the K-band ellipsoidal lightcurve of GRO J0422+32 (see Fig. 1). Although an ellipsoidal modulation may exist, it is severely contaminated by additional flickering (c.f. the comparison star plotted in the lower panel). This is the first evidence for significant K-band contamination by an accretion disk in a quiescent XRN. In Fig. 2, we plot the NIRC K-band spectrum: strong Brγ emission is observed. Furthermore, such contamination is not unique to GRO J0422+32: NIRC K-band spectroscopy of A0620-00 also shows Brγ in emission — although in this case the IR contamination is not as significant as in the case of GRO J0422+32.

Hence an uncontaminated K-band flux in quiescent XRN cannot be taken for granted. Such contamination can have a profound effect on the ellipsoidal lightcurves and the compact object masses derived from them. Higher resolution IR spectroscopy of quiescent XRN is required to properly constrain such contamination.

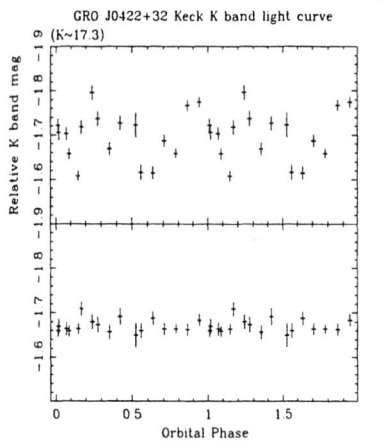 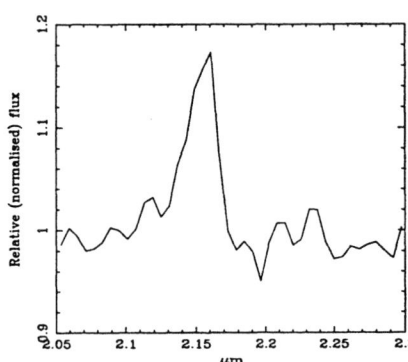

Fig. 1: NIRC lightcurve of GRO J0422+32   Fig. 2: NIRC K-band spectrum

3. THE PECULIAR K-BAND COUNTERPART OF GX17+2

GX 17+2 is one of the brightest of the persistent X-ray sources in the sky. It is a bursting "Z" source, with a time variable radio counterpart. Using *HST* NICMOS observations, Deutsch et al. (1999) showed that the radio position was inconsistent with that of the previously suggested optical counterpart NP Ser, and identified two possible H-band counterparts: the brightest of these had a H-band magnitude of 19.8 during their observations. We plot their image in Fig. 3 (NW is up and NE is to the left): the small and large circles denote the radio and X-ray error circles, respectively. NP Ser is the bright star to the south of the radio error circle.

We observed GX17+2 using NIRC on 1999 June 26 UT: the mean XTE ASM countrate of GX17+2 at this time was very similar to that during the Deutsch et al. observations (~50 counts/s with 10-20% variability superimposed). A bright counterpart was visible in the radio error circle, with a K-band magnitude of ~14.9: NP Ser itself was clearly resolved 0.9" south of this with K~14.5 (Callanan, Filippenko, & Garcia 1999b: Fig. 4). Our observation implies an amplitude of variability of

~3.5-4 mag at K — surprising, considering the stability of the X-ray emission.

Fender (1999, 1997) has suggested this variability may originate in synchrotron emission: the large amplitude *radio* variability observed by Pennix et al. (1988) is entirely consistent with this. Furthermore, the radio flux increases along the normal and horizontal branches of the "Z" curve, regions in which the overall X-ray luminosity changes by only ~10–20% (Pennix et al. 1988). However, the radio data of Pennix et al rarely (if ever) dip to a level consistent with K≈19. Simultaneous radio and IR measurements are required to assess the feasibility of this model.

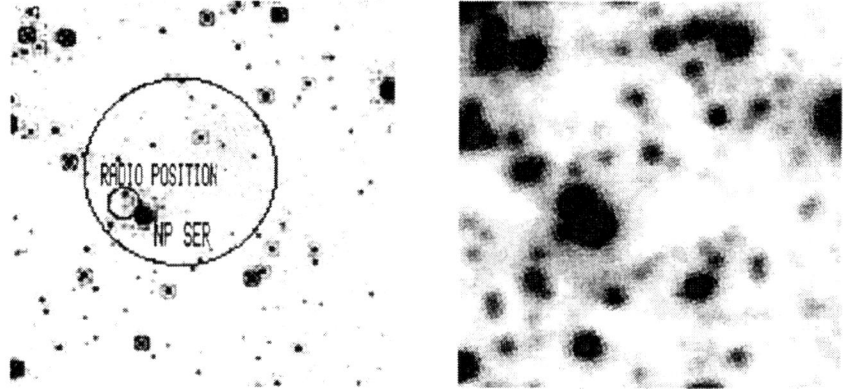

Fig. 3: *HST* NICMOS image of GX17+2   **Fig. 4:** Keck NIRC K-band image

4. THE OPTICAL/IR COUNTERPART OF AQL X-1

Aql X-1 is one of the most active X-ray transients, and one of the few neutron star systems bright enough in quiescence (but see below) for radial velocity studies. Recent optical measurements have yielded puzzling results. During outburst a deep 0.6 mag modulation is observed (with a period of 19 hr), implying a moderately high inclination; however, radial velocity measurements in quiescence yield $i < 30°$ (e.g., Garcia et al. 1999). These conflicting results have confounded attempts to measure the mass of the neutron star primary.

We observed Aql X-1 using NIRC on 1998 September 29 UT; XTE ASM observations show Aql X-1 to be quiescent at this time. Our K-band images clearly resolved Aql X-1 into 2 stars, with a separation of 0.46″ (Callanan, Filippenko & Garcia 1999a, and Fig. 5). Subsequent optical observations by Chevalier et al. (1999) showed that the unresolved quiescent counterpart is dominated by the line of sight interloper. Hence quiescent state ellipsoidal and radial velocity studies have been dominated by this unrelated star, explaining the inconsistent results found previously. Our IR colors for the true Aql X-1 counterpart are $(J - K)_0 \approx 0.8$, implying a K6V secondary (assuming negligible accretion disk contamination). As Chevalier et al. measure a "true" quiescent magnitude of $V \approx 21.6$, *HST* observations will likely be required to accurately measure the mass of the neutron star.

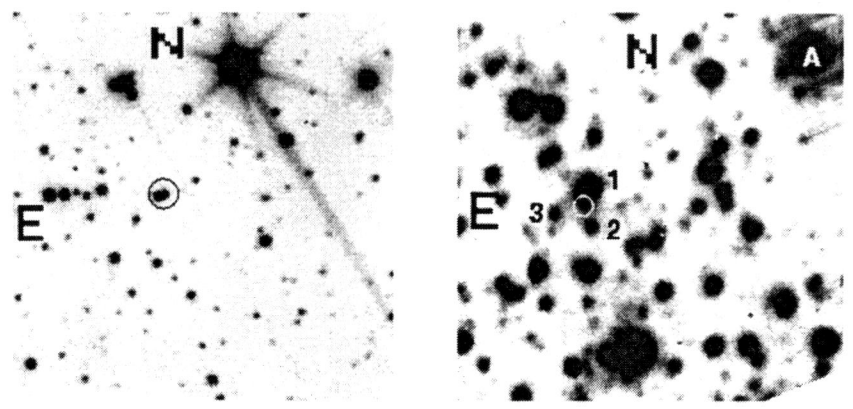

Fig. 5: NIRC image of Aql X-1   **Fig. 6:** NIRC image of 4U 1630-47

5. THE IR COUNTERPART OF 4U 1630-47

4U 1630-47 is a highly obscured ($A_V \geq 13$ mag) X-ray transient. More than a dozen outbursts have been observed so far (e.g., Kuulkers 1998, and references therein). It exhibits rapid X-ray variability akin to that of GRO J1655-40 and GRO J1915+105, the Galactic "superluminal" jet sources. Using the recently discovered radio position (Hjellming et al. 1999), Callanan et al. (2000) searched unsuccessfully for an IR counterpart using CIRIM on the 1.5-m at CTIO. We tried again with NIRC on 1999 June 26 UT, catching 4U 1630-47 in outburst. We show this image in Fig. 6, with the radio position (2σ radius) superimposed. The stars identified are those of Callanan et al. (2000). Of interest here is a star $\sim$ 2 mag fainter than star 1 $\sim$0.8'' to the SE, only 0.5'' away from the radio position: this may be related to 4U 1630-47. Comparison of the Keck photometry with the CTIO data (in which these two stars are completely unresolved) suggest that this star is unlikely to be strongly variable. Deep IR imaging in quiescence is now required.

REFERENCES

Callanan, P. J., Garcia, M. J., Filippenko, A. V., McLean, I., & Teplitz, H. 1996, ApJ, 470, L57
Callanan, P.J., Filippenko, A.V., & Garcia, M.R., 1999a IAUC 7086; and 1999b IAUC 7219
Callanan, P.J., McCarthy, J.F., & Garcia, M.R., 2000, A&A, in press
Chevalier, C. et al. 1999, A&A, 347, L51
Deutsch, E.W. et al. 1999, ApJ, 524, 406
Garcia, M.R. et al. 1999, ApJ, 518, 422
Fender, R. et al. 1997, MNRAS, 290, L65; and Fender 1999, private communication
Filippenko, A. V., Matheson, T., & Ho, L. C. 1995, ApJ, 455, 614
Hjellming, R.M. et al. 1999, ApJ, 514, 383
Kuulkers, E., 1998, New A, 42, 613
Penninx, W. et al. 1988, Nature, 336, 146
Shahbaz, T. et al. 1994, MNRAS, 268, 756

INTENSIVE HST/RXTE MONITORING OF NGC 3516: EVIDENCE AGAINST THERMAL REPROCESSING

Rick Edelson

Astronomy Department; University of California; Los Angeles, CA 90095-1562; USA and X-ray Astronomy Group; Leicester University; Leicester LE1 7RH; United Kingdom

ABSTRACT

We report the results of intensive monitoring of NGC 3516 with HST and $RXTE$, which showed strong (50% peak-peak) variability in the X-rays and much weaker but highly significant (3.5% peak-peak) variability in the optical. The optical variations were highly correlated throughout the observed, band, with no measurable lags to a limit of $\tau \leq 0.15$ d between 3590 and 5510 Å, while they showed no measurable correlation with the X-rays. This presents severe problems for "reprocessing" models in which the X-rays heat a stratified accretion disk that reradiates in the optical.

KEYWORDS: Active Galactic Nuclei — NGC 3516

1. OBSERVATIONS AND DATA REDUCTION

The bright, strongly variable Seyfert 1 galaxy NGC 3516 was monitored almost continuously with HST for 2.8 d at optical wavelengths and simultaneous $RXTE$ monitoring covered the same period. This was the most intensive multiwavelength Seyfert 1 monitoring campaign ever undertaken. The light curves were analyzed by standard techniques. The resultant light curves are plotted in Figure 1.

This shows that the X-rays were strongly variable (~50% peak-to-peak) while the optical continuum showed much smaller but still highly significant variations: a slow ~2.5% rise followed by a faster ~3.5% decline. The fact that the mean emission line light curve (constructed by averaging [OIII], Hβ and Hγ fluxes) was flat with 0.26% RMS dispersion means that the systematic errors were no larger than this.

2. TEMPORAL ANALYSIS

Temoral cross-correlation functions were measured using both the discrete correlation function (DCF; Edelson & Krolik 1989) and the interpolated correlation function (ICF; White & Peterson 1994), and errors on the latter were estimated as in Peterson et al. (1998). The results are plotted in Figure 2. The optical continuum was highly correlated ($r \geq 0.9$), with no measurable interband lag (to a 3σ limit of $\tau \lesssim 0.15$ d between 3590 Å and 5510 Å). However, temporal cross-correlation

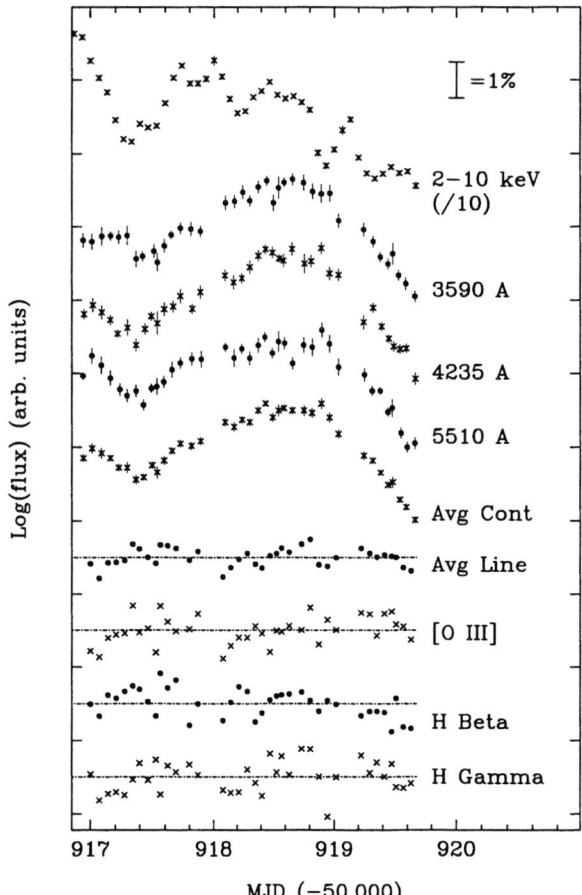

FIGURE 1. Light curves of NGC 3516. At the top is the *RXTE* hard X-ray light curve, with the variability scaled down by a factor of 10, for comparison with the *HST* light curves given below. The optical data are, from the top, 3590 Å, 4235 Å, 5510 Å and average continuum light curves, and the average line, [OIII], Hβ and Hγ light curves. The error bars include only statistical errors. Note specifically that the average optical continuum light curve shows a clear ∼2.5% rise followed by a ∼3.5% decline. During the same period, the average optical emission line light curve is flat with an RMS scatter of 0.26%, indicating that systematic effects are not likely to be a problem above this level.

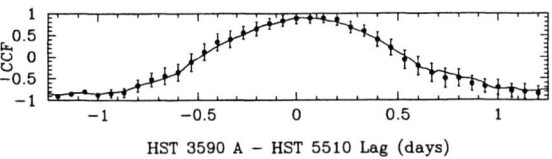

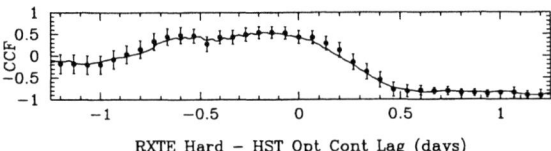

FIGURE 2. Interband temporal cross-correlation functions. The solid line refers to the interpolated cross-correlation function, while the error bars refer to the discrete cross-correlation function. At the top is the *HST* 3590 Å - 5510 Å correlation function, and at the bottom is the *RXTE* X-ray - *HST* mean optical continuum correlation function. A negative lag means that the first listed band leads the second.

functions gave no evidence for a simple relation between the X-ray and optical variations. The most significant value was the anticorrelaton of $r \lesssim -0.8$ for $\tau = -0.8$ to -1.3 d and the maximum positive correlation of $r = +0.53$ at $\tau = -0.21$ d, which was not deemed significant after accounting for interdependence of the data.

3. DISCUSSION: IMPLICATIONS FOR REPROCESSING MODELS

Much attention has been given to reprocessing models for the optical emission from Seyfert 1 galaxies in which an X-ray continuum source irradiates dense, cool material which, in turn, emits thermal radiation at longer (optical/ultraviolet) wavelengths (e.g., Guilbert & Rees 1988, Rokaki et al. 1992, Collier et al. 1998). In many models, the energy absorbed by the disk is greater than that generated internally, so reprocessing is the dominant source of luminosity in the optical/ultraviolet. However, the combination of a lack of measurable correlation between the X-ray and optical light curves and the synchronicity within the optical band reported in this study presents serious general problems for this picture, because each of these results implies a limit on the size of the putative reprocessing region that is incompatible with the other.

The lack of response to the X-rays indicates that the light-crossing time of the optical/ultraviolet reprocessing region is of order or larger than the duration of the simultaneous X-ray/optical monitoring. Otherwise, the light curves should show some correlation if the optical/ultraviolet is in fact driven by the X-rays. Since this experiment ran 2.8 d, this would indicate that the reprocessing region would have to be $\gtrsim 1$ lt-d in size. In fact, much longer term monitoring also failed to show the X-rays leading the optical variations, on delay time scales of weeks to months (Maoz et al. 1999). This in fact indicates that the reprocessing region should be light-weeks across or larger.

On the other hand, the observation of significant optical variability with no lags between the bands down to time scales of $\lesssim 0.15$ d yields an upper limit on the size of the reprocessing region. For the most straightforward geometry, this upper limit in the lag yields an upper limit on the radius of the emitting region of $\lesssim 0.3$ lt-d. Furthermore, if the optical continuum is produced in the same region as the iron Kα line, the relativistic effects observed in the line profile of NGC 3516 (Nandra et al. 1999) would also argue for an origin much closer to the central source.

There are a number of possible model fixes, including anisotropic emission or source geometry, localized flares, or long processing time scales in the disk. However, perhaps the most natural explanation is that the unproven assumption of this reprocessing model, that the X-ray emission powers that at optical/ultraviolet wavelengths, is in error. Clearly, much theoretical work needs to be done to make the reprocessing model fit with this new, emerging picture of Seyfert 1 interband variability.

More observational work is also required: this was an unexpected result in an experiment designed for different purposes. Now that it has been established that *HST* is capable of extremely high precision relative photometry, it is straightforward to design an experiment expressly to search for very small lags between optical and ultraviolet variations. This can eliminate the inherent uncertainties in comparing model predictions to the data and resolve the ambiguity concerning the relationship between the X-ray through optical variations.

This work is discussed in greater detail in Edelson et al. (1999).

REFERENCES

Collier, S. et al. 1998, ApJ, 500, 162
Edelson, R., Krolik, J. 1989, ApJ, 333, 646
Edelson, R., 1999, ApJ, submitted
Gulibert, P., Rees, M. 1988, MNRAS, 233, 47
Maoz, D., Edelson, R., Nandra, K. 1999, AJ, in press
Nandra, K. et al. 1999, ApJL, 523, L17
Peterson, B. M. et al. 1998, PASP, 110, 660
Rokaki, E. et al. 1992, A&A, 253, 57
White, R., Peterson, B. M. 1994, PASP, 106, 879

SYNCHROTRON AND COMPTON COMPONENTS AND THEIR VARIABILITY IN BL LAC OBJECTS

P.Giommi [1], G. Ghisellini [2], P. Padovani [3,4], and G. Tagliaferri [2]

1) BeppoSAX Science Data Center, Via Corcolle 19, 00131 Rome, Italy
2) Osservatorio Astronomico di Brera, Via Bianchi 46, 23807 Merate, Italy
3) Space Telescope Science Institute, San Martin Drive, Baltimore, MD 21218 U.S.A.
4) Astrophysics Division, ESTEC, 2200AG Noordwijk, The Netherlands

1. INTRODUCTION

BL Lacertae objects are extreme extragalactic sources characterized by the emission of strong and rapidly variable nonthermal radiation over the entire electromagnetic spectrum. Synchrotron emission followed by inverse Compton scattering in a relativistic beaming scenario is generally thought to be the mechanism powering these objects (e.g. Kollgaard 1994., Urry & Padovani 1995). BL Lacs can be divided into different subclasses depending on their Spectral Energy Distribution (SED), namely LBL for objects with the synchrotron emission peaking at $\nu_{peak} \approx 10^{13-14} Hz$, intermediate objects ($\nu_{peak} \approx 10^{15-16} Hz$) and HBL or high energy peaked BL Lacs with $\nu_{peak} \approx 10^{17-18} Hz$ (Padovani & Giommi 1995). The wide X-ray band pass of the BeppoSAX satellite (Boella et al. 1997) is well suited for the detailed spectral study of all types of BL Lacs. In fact, direct measurements of the Compton part of the spectrum have been obtained for a number of LBLs (e.g. Padovani et al. 1999), and the very variable tail of the Synchrotron component has been studied in several HBLs (e.g. Pian et al. 1998, Wolter et al. 1998, Giommi, Padovani & Perlman 1999, Chiappetti et al. 1999). In the case of the two intermediate BL Lacs S5 0716+714 and ON 231 BeppoSAX for the first time was able to detect both spectral components within a single instrument (Giommi et al. 1999, Tagliaferri et al. 1999).

The BeppoSAX archive at the Science Data Center (SDC, Giommi & Fiore 1998) presently includes over 100 observations of 56 distinct BL Lacs, about half of which are already publicly available. We have started a project to construct the SED of a large number of all types of BL Lacs by combining a) public BeppoSAX data (0.1-200 keV); b) simultaneous optical and radio data when these are available from monitoring campaigns, or from the University of Michigan Radio Astronomy Observatory (UMRAO) on-line data base (Aller et al. 1999); and c) non-simultaneous photometric data form NED. Here we present the first results of this project.

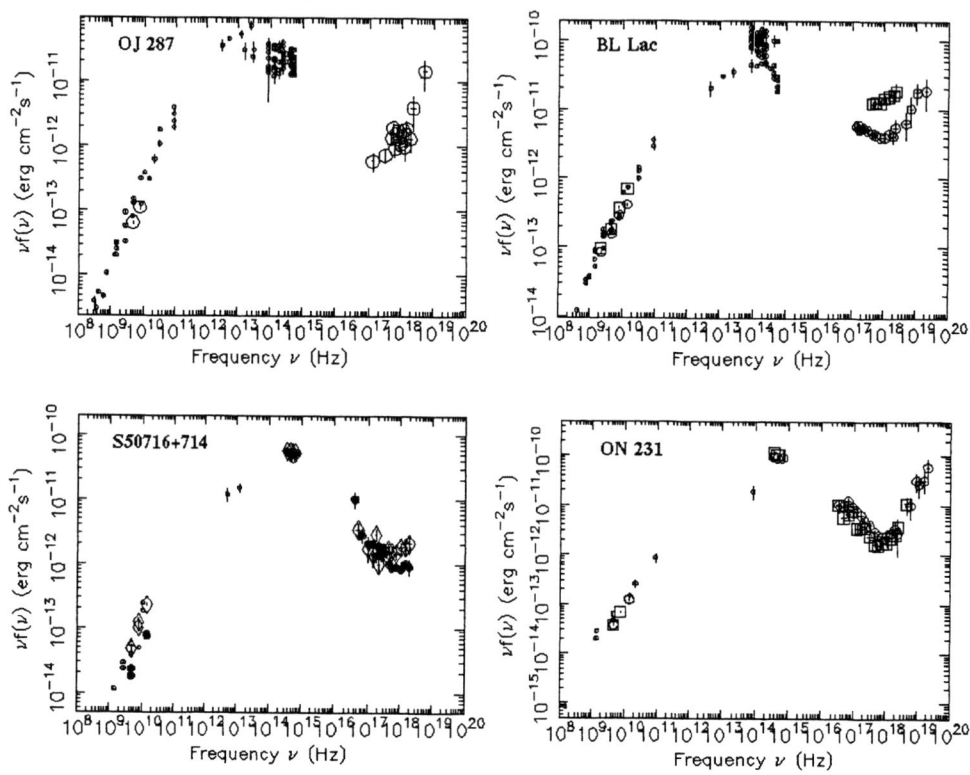

FIGURE 1. Spectral energy distribution of four LBL and intermediate BL Lacs.

2. SPECTRAL ENERGY DISTRIBUTIONS AND VARIABILITY

The SEDs that we have assembled are shown in figure 1 for LBLs and intermediate objects, and in figure 2 for HBL BL Lacs. The X-ray part of the plots have been constructed using data from the LECS, MECS and PDS instruments of the BeppoSAX satellite. The cleaned and calibrated data files have been taken from the SDC on-line archive and have been analyzed using the XSPEC package. Unfolded spectral data have been corrected for low energy absorption assuming N_H equal to the Galactic value. Nearly simultaneous data are plotted with the same symbols used for the X-ray data. Optical monitoring observations are available for S5 0716+714 (Giommi et al. 1999), and ON 231 (Tagliaferri et al. 1999). Nearly simultaneous radio data from the UMRAO database are available for several objects. All other (non-simultaneous) data are plotted as small open circles and are from the photometric data points provided by NED. Strong variability at several frequencies is apparent from Figures 1 and 2. In particular quite spectacular spectral changes are concentrated at or just after the synchrotron peak. BeppoSAX observations of

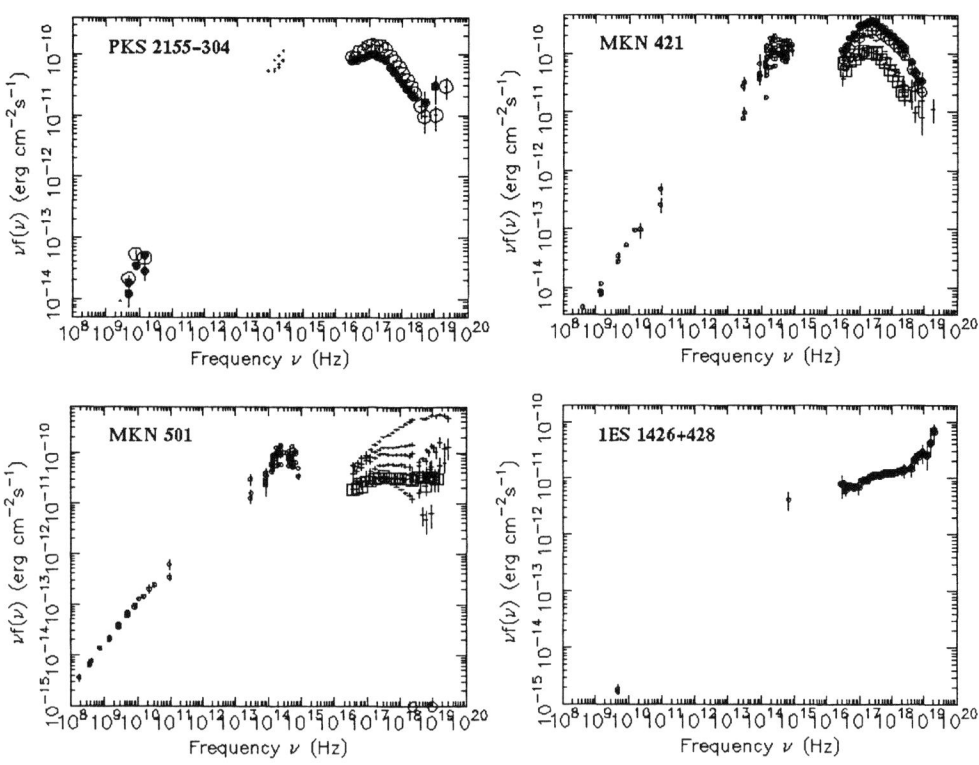

FIGURE 2. Spectral energy distribution of HBL BL Lacs.

intermediate BL Lacs clearly show that the soft X-ray synchrotron radiation vary in a different way compared to the harder Compton components (Giommi et al. 1999, Tagliaferri et al. 1999). The SED shown here indicate that the variability of the Compton component may be correlated with radio flux and not with the optical and soft X-ray synchrotron emission (see figure 1).

3. ULTRA HIGH ENERGY SYNCHROTRON PEAKED BL LACS (UHBLS) ?

Figures 1 and 2 clearly show that the peak frequency of the synchrotron emission ranges from around $10^{13} Hz$ for OJ 287 to well above $10^{19} Hz$ for 1ES1426+428. Ghisellini (1999) argued that this trend could continue to much higher energies. We have thus been searching for BL Lacs with Ultra High synchrotron peak energy (UHBLs). We have selected candidates UHBLs from the sample of extreme BL Lacs of the "Sedentary Multifrequency Survey" (Giommi, Menna & Padovani 1999) by looking for objects within the error circle of unidentified sources in the third EGRET catalog. One such object is 1RXS J23511.1-14033; its finding chart is shown in figure 3 (left). The SED of 1RXS J23511.1-14033, on the right part of figure 3, indicates

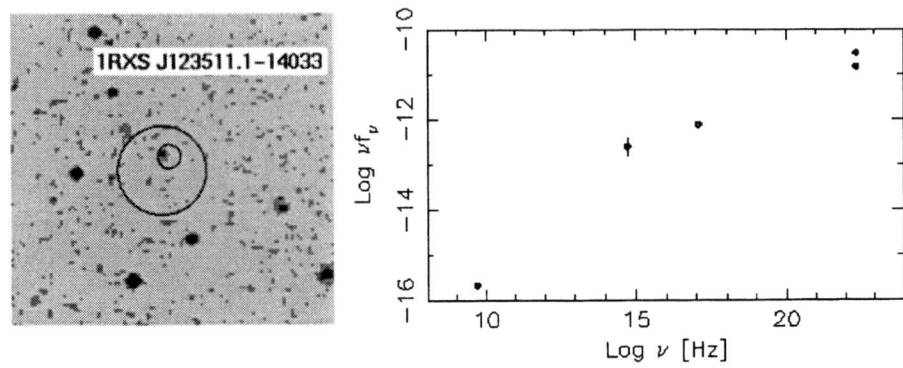

FIGURE 3. Left: the ROSAT and NVSS error circles showing the candidate UHBL 1RXS J123511.1-14033. Right: the SED of 1RXS J123511.1-14033 if this BL Lac is the correct counterpart of the EGRET source 2EGJ1233-1407

that the synchrotron emission could reach the gamma ray band. A first BeppoSAX pointing of this object unfortunately gave inconclusive results since the observation had to be split into three short exposures and the spectrum appears to be variable. Details will be published elsewhere. A second UHBL candidate will be observed by BeppoSAX in a few months. If these observations will confirm the hypothesis that UHBLs exist, this type of sources could be the long sought counterpart of many of the still unidentified high galactic latitude EGRET sources.

REFERENCES

Aller M.F, Aller H.D., Huges P.A., & Latimer G.E., 1999 ApJ 512, 601
Boella G. et al. 1997 A&AS, 122, 299
Chiappetti,L., et al. 1999, ApJ 521, 552
Ghisellini, G., 1999, Proc 3rd Integral Workshop, Taormina, astro-ph/9812419
Giommi P.,& Fiore F. 1998, in Proc. 5th Workshop on Data Analysis in Astronomy, World Scientific, Singapore, p. 93
Giommi, P., Padovani, P. & Perlman, E. 1999, MNRAS in press, astro-ph/9907377
Giommi, P. et al. 1999, A&A, in press, astro-ph/9909241
Giommi, P., Menna, M.T., & Padovani, P. 1999, MNRAS in press, astro-ph/9907014
Kollgaard R.I., 1994 Vistas in Astronomy, 38, 29
Padovani, P. & Giommi, P. 1995, ApJ, 444, 567
Padovani, P. et al. 1999, in preparation
Pian, E. et al. 1998, APJ L,492, L17
Tagliaferri, G., et al. 1999, A&A, submitted
Urry, C.M., & Padovani, P., 1995, PASP, 107, 803
Wolter, A. et al. 1998 A&A 335, 899

A NEW PULSAR/SNR PAIR: AX J1845−0258 IN G29.6+0.1

E.V. Gotthelf[1], G. Vasisht[2], K. Torii[3] & B.M. Gaensler[4]

1) Columbia University, 550 West 120th Street, New York, NY 10027
2) California Institute of Technology, 4800 Oak Grove Drive, Pasadena, CA 91109
3) NASDA TKSC SURP, 2-1-1 Sengen, Tsukuba, Ibaraki 305-8505, Japan
4) Massachusetts Institute of Technology, 70 Vassar Street, Cambridge, MA 02139

ABSTRACT

We present a follow-up X-ray and radio study of the field containing the 7-s X-ray pulsar AX J1845−0258, the serendipitous *ASCA* source whose characteristics are found to be similar to those of the anomalous X-ray pulsars (AXPs). Newly acquired *ASCA* data confirms a dramatic reduction in flux from the pulsar and reveals instead a faint X-ray point source, AX J184453.3−025642, within the pulsar's error circle. This X-ray source is surrounded by a partial shell of emission coincident with a newly discovered young shell-type radio supernova remnant, G29.6+0.1. The central X-ray source is too faint to provide a detection of the expected pulsations which might confirm AX J184453.3−025642 as the pulsar. We argue that this system is similar to that of RCW 103, another AXP-like object whose central source displays low/high flux states (but no pulsations). The alternative interpretation of a binary system, perhaps associated with a supernova remnant, is still possible. In either case, this result may have profound implication on the evolution of young neutron stars.

KEYWORDS: pulsars: individual (AX J1845−0258); supernova remnants: individual(G29.6+0.1); star: individual (AX J184453.3−025642); stars: neutron.

1. INTRODUCTION

The 7-s pulsar, AX J1845−0258, was discovered during an automatic search of the *ASCA* archival data (Gotthelf & Vasisht 1998; Torii et al. 1998). Based on its spectral and timing properties, AX J1845−0258 is likely the latest addition to the class of anomalous x-ray pulsars (AXPs) (Duncan & Thomson 1996; Mereghetti & Stella 1995; van Paradijs et al. 1995). Evidence included a long rotation period, large modulation ($\sim 30\%$), steady short-term X-ray flux during the original *ASCA* observation, steep characteristic spectrum (power-law photon index $\Gamma \sim 5$), location at low Galactic latitude, and the lack of known counterpart. A rough distance estimate derived from the X-ray absorption places the pulsar at distance of $5 - 15$ kpc giving an inferred X-ray luminosity of order $\sim 10^{35}$ erg s^{-1}.

Herein we report on new *ASCA* X-ray and VLA radio observations directed at the pulsar's location. Our goal was to identify the pulsar and confirm or repudiate the AXP hypothesis by measuring the spin-down rate of the pulsar and searching for an associated radio supernova remnant (SNR). We succeeded in finding a young radio SNR within the pulsar's error circle, however the pulsator was not seen again. Instead, we find a faint ASCA point source (Vasisht et al. 2000) at the center of the newly discovered radio SNR G29.6+0.1 (Gaensler et al. 1999). We argue that this faint source is the pulsar AX J1845−0258 in a low state; we consider the pulsar's location at the center of a young SNR and the lack of a radio counterpart as evidence for the AXP interpretation, but with a twist.

2. THE X-RAY OBSERVATIONS

We revisited the field containing the pulsar AX J1845−0258 with the *ASCA* observatory on March 28-29, 1999 UT. Figure 1 reproduced the smoothed and exposure corrected image taken with the gas imaging spectrometers (GISs) aboard *ASCA*. The GIS is sensitive to photon in the $\sim 1 - 10$ keV energy range and has a spatial resolution $\sim 1 - 2'$. All data were edited following the standard *ASCA* reduction procedures resulting in an effective observation time is 49 ks.

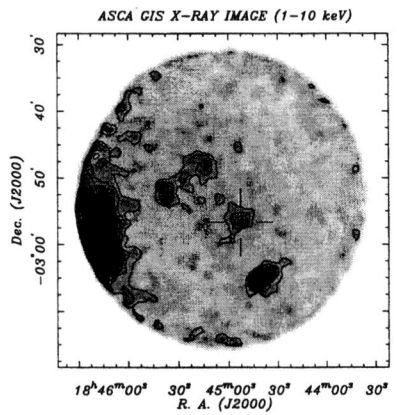

FIGURE 1. The follow-up *ASCA* observation of AX J1845−0258. The full GIS image showing the new X-ray source AX J184453.3−025642 marked by the cross, and several serendipitous sources to the northeast and southwest. The bright emission near the east edge is the SNR Kes 75.

Near the center of the field-of-view we find a faint unresolved point source within the large $\sim 3'$ radius error circle for AX J1845−0258. The pulsar's poor astrometry is due to the extreme off-axis detector location of the discovery observation. The faint source is also detected by *ASCA*'s solid-state imaging spectrometers (SISs) (see Fig 2) with a similar significance of $\sim 5\sigma$. The spatial resolution of the SIS is $\sim 1'$, but the derived coordinates of $18^h 45^m 53.3^s$, $-02°56'42''$ (J2000) have an uncertainty of only $20''$ after correcting for the temperature dependent coordinate offsets (Gotthelf et al. 2000). We refer to this source as AX J184453.3−025642, and consider whether this is indeed the expected pulsar, but at a flux level an order of magnitude less than expected; the dearth of source photon prohibits a proper spectral analysis or search for pulsations, which might allow identification with AX J1845−0258.

3. THE VLA RADIO IMAGES

Radio observations of the field of AX J1845−0258 were made with the D-configuration of the Very Large Array (VLA) on 1999 March 26. The total observing time was 6 hr, of which 4.5 hr was spent observing in the 5 GHz band, and the remainder in the 8 GHz band. At both 5 and 8 GHz a distinct shell of emission is seen, which is designated G29.6+0.1 (see Fig. 2). The shell is clumpy, with a particularly bright clump on its eastern edge. In the east the shell is quite thick (up to 50% of the radius), while the north-western rim is brighter and narrower. Two point sources can be seen within the shell interior.

The shell-like radio emission ($\sim 5\rlap{.}'0$ in diameter) is found to be linearly polarized and non-thermal, which, along with the lack of significant counterpart in 60 μm *IRAS* data, are characteristic properties of supernova remnants (e.g. Whiteoak & Green 1996). G29.6+0.1 is thus classified as a previously unidentified SNR. Its inferred age suggests a young remnant, with an upper limit of 8000 yr. The location of the X-ray source AX J184453.3−025642 at the center of the SNR is highly unlikely to be due to a chance superposition, suggesting that the two are related.

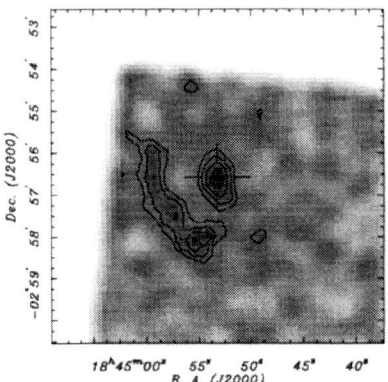 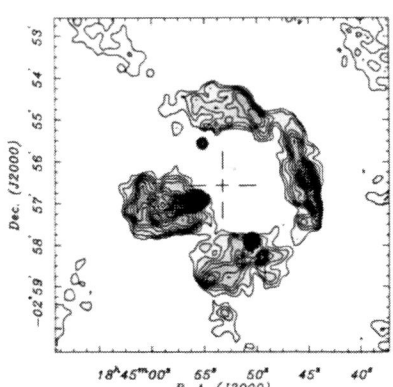

FIGURE 2. Discovery of a new supernova remnant, G29.6+0.1, containing a central X-ray source, AX J184453.3−025642, within the error box of the X-ray pulsar AX J1845−0258. (**LEFT**) The *ASCA* SIS X-ray image centered on the AX J184453.3−025642, marked by the cross. An arc of emission is evident surrounding the point source overlapping the radio shell shown in the next panel. (**RIGHT**) The 5 GHz VLA radio map of the same region, which reveals a clumpy shell, whose spectral index is consistent with a SNR hypothesis.

4. THE NATURE OF AX J1845−0258: A VARIABLE AXP?

The lack of a bright pulsator in the new *ASCA* observation of AX J1845−0258 is quite surprising. The spectral and temporal properties of this pulsar had strongly implied an AXP interpretation. Indeed the discovery of a young radio remnant coincident with the pulsar is consistent with the AXP hypothesis. Conversely, the detection of an X-ray point source in the center of the SNR is in itself indicative of a neutron star candidate associated with the remnant. This new source is exactly where we would expect the AXP to be, to within errors, consistent with this interpretation. We therefore suggest that AX J184453.3−025642 is indeed the pulsar, but at a much reduced (∼ order of magnitude) X-ray flux.

We now consider the interesting possibility that AXP can exhibit extreme, factor of ten, variability. There is some evidence for this already from two well studied AXPs which show large $\gtrsim$ 4 variations in flux on year timescales (e.g. 1E 1048.1-593, Oosterbroek et al. 1998). Most intriguing, the properties of the central, unpulsed, neutron star candidate in SNR RCW 103 are otherwise typical of the AXPs, but its flux has also been found to vary by an order of magnitude at energies > 3 keV (Gotthelf, Petre, & Vasisht 1999), just what is observed for AX J1845−0258. Conversely, this provides further evidence that RCW 103 is an AXP with unseen pulsations, perhaps due to unfavorable beaming geometry.

The identification of another AXP at the center of a young SNR have important implications on the birth properties of pulsars. This result is certainly consistent with AXPs being young, isolated neutron stars, as argued by the magnetar hypothesis. There is the possibility then that AXPs might exhibit periods of enhanced emission. In this case, the population of AXPs might be much greater than previously thought, and we are only detecting a fraction of AXP, those currently in their bright "on" state. Perhaps a duty-cycle (fraction of time the AXP is "on") of only $\lesssim$ 5% would be required to square the known Galactic SNRs population with the detected AXPs, if most young neutron stars manifest themselves as AXPs as some authors suggest (see Gotthelf 1998). Although the magnetar hypothesis is attractive, we cannot reject a a binary system origin, perhaps embedded and associated with a young SNR. Further monitoring of this region is planned.

ACKNOWLEDGEMENTS

This research is support in part by a NASA Hubble Fellowship grant HF-01107.01-98A (B.M.G.) and by a NASA LTSA grant NAG5–7935 (E.V.G. & G.V.).

REFERENCES

Duncan,R. & Thompson,C. 1995, MNRAS, 275, 255
Gaensler,B.M., et al. 1999, ApJ, 526, TBD
Gotthelf,E.V. & Vasisht, G. 1998, NA, 3, 293
Gotthelf,E.V., 1998, Mem Soc Astro It, Vol 69, 4; astro-ph/9809139
...
Gotthelf,E.V. et al. 1999, ApJ, L514, 107
Gotthelf,E.V., 2000 ApJ, submitted.
Mereghetti,S. & Stella,L. 1995, ApJ,442,L17
Oosterbroek,T.et al. 1998 A&A, 334, 925
Torii,K., et al., 1998 ApJ, 503, 843
van Paradijs,J. et al.1995, A&A, 299, L41
Vasisht,G. et al. 2000 ApJ, in prep.
Whiteoak,J. & Green,A.J. 1996, A&ASS, 118, 329

THE X-RAY SOURCE POPULATIONS OF THE MAGELLANIC CLOUDS

F. Haberl

Max-Planck-Institut für extraterrestrische Physik, Gießenbachstraße, 85748 Garching, Germany

ABSTRACT   ROSAT performed about 250 PSPC observations pointed towards the Magellanic Clouds (MCs). A systematic analysis of the data yielded catalogues with more than 750 X-ray sources in the LMC and about 520 in the SMC region. Cross-correlations with catalogues from various other wavelength bands and utilizing the derived X-ray properties allowed to identify a significant fraction of the X-ray sources and to classify many unknown sources as candidates for X-ray binaries, supernova remnants and supersoft sources in the MCs. This makes it possible to study in detail the source populations in these nearby galaxies on a large statistical basis.

KEYWORDS: Catalogues; Galaxies: Magellanic Clouds; Galaxies: stellar content; X-rays: stars

1. INTRODUCTION

The first X-ray surveys of the Magellanic Clouds with imaging instruments were performed with the Einstein satellite (LMC: Long et al. 1981, Wang et al. 1991; SMC: Wang & Wu 1992). These observations revealed large-scale diffuse emission originating from hot gas with temperatures of several 10^6 K in the LMC and first catalogues of discrete X-ray sources were derived.

The ROSAT (Trümper 1982) observations of the MCs performed between 1990 and 1998 build up the most sensitive surveys of these nearby galaxies with imaging instruments in the soft X-ray band (0.1 – 2.4 keV). The coverage of the LMC in the all-sky survey provided a complete view of the diffuse X-ray emission and yielded more than 500 discrete X-ray sources in a $13° \times 13°$ field around the LMC (Pietsch & Kahabka 1993). The large number of pointed observations using the Position Sensitive Proportional Counter (PSPC, Pfeffermann et al. 1986) and the High Resolution Imager (HRI, David et al. 1993) provide an invaluable archive to investigate the X-ray source populations in the MCs. In particular due to the large field of view of the PSPC detector these observations covered 59 square degrees of the LMC and 18 square degrees of the SMC. Here I give a summary of the first results obtained from a systematic study of all PSPC observations of the MCs.

2. ROSAT PSPC CATALOGUES

A first ROSAT PSPC catalogue of the SMC was produced by Kahabka et al. (1999) which was based on a dedicated SMC survey. Using the complete set of PSPC observations from the ROSAT archive catalogues of point and point-like X-ray sources were recently published by Haberl & Pietsch (1999b) and Haberl et al. (1999) for LMC and SMC, respectively. The catalogues comprise of ∼760 and ∼520 sources in the LMC and SMC fields. Together with X-ray position the catalogues contain information on properties as likelihood of existence, count rate, hardness ratios derived from count rates in different energy bands, source extent with likelihood, exposure time and other observational details. From multiple detections of a source the one with the smallest error on the X-ray position entered the catalogue and therefore the X-ray source properties are representative only for a single observation. Flux variability studies of the LMC X-ray sources on time scales of days to years are presented in Haberl & Pietsch (1999a).

3. SOURCE IDENTIFICATION AND CLASSIFICATION

The weakest sources in the MC PSPC catalogues have count rates of a few 10^{-4} counts s^{-1} corresponding to X-ray luminosities down to about 7×10^{33} erg s^{-1} at the distance of LMC and SMC. This allows to detect supernova remnants (SNRs), supersoft sources and X-ray binaries in the MCs, the latter even in low-luminosity state. From the identified Einstein LMC sources about 50% were assigned to background active galactic nuclei (AGN) and foreground stars. Similarly a large fraction of AGN and stars is expected to contribute to the PSPC catalogues.

To investigate the X-ray source populations in the MCs it is not always necessary to identify the sources optically by taking an optical spectrum. Using X-ray properties often allows to classify a source indicating the type of X-ray emitter with high probability. From the information provided in the PSPC catalogues source extent, hardness ratios as indicator for the shape of the X-ray spectrum and time variability can be used to classify the X-ray sources. X-ray flux together with optical magnitudes of the brightest object in the X-ray error circle yield lower limits for the X-ray to optical flux ratio. Regions in parameter space unique to certain source types can be defined from subsamples of already identified sources of known type and used to classify the unknown sources.

3.1. Extended sources

Using source extent and extent likelihood Haberl & Pietsch (1999b) and Haberl et al. (1999) could separate SNRs from unresolved sources. Fig. 1 shows extent likelihood and extent of MC sources. The known SNRs are clearly distinguished from other source classes and some new SNR candidates were classified.

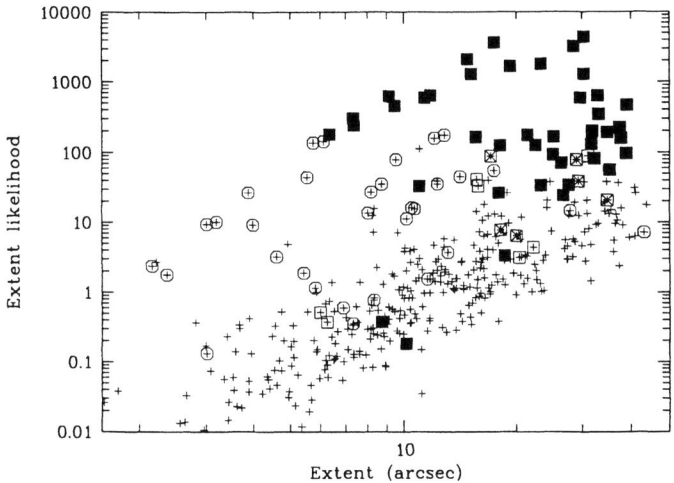

FIGURE 1. Source extent and extent likelihood for PSPC sources in LMC and SMC observed with off-axis angle less than 18' (crosses). SNRs are marked with squares (filled: secure, open: candidates, crossed: classified) and known point sources (X-ray binaries, SSSs, stars and AGN) with hexagons

3.2. Spectral classification

Kahabka et al. (1999) classified PSPC sources in the SMC using the hardness ratios. Similarly Haberl & Pietsch (1999b) and Haberl et al. (1999) used HR1 = (hard + soft)/(hard - soft) and HR2 = (hard2 + hard1)/(hard2 - hard1) with soft, hard, hard1 and hard2 defined as count rates in the 0.1 – 0.4 keV, 0.5 – 2.0 keV, 0.5 – 0.9 keV and 0.9 – 2.0 keV energy bands, to define hardness ratio ranges which allow to uniquely classify PSPC sources in both LMC and SMC. In Fig. 2 HR2 is plotted versus HR1 for PSPC sources with secure identification. Below HR2 of -0.7 one finds only SSSs while above HR2 = -0.7 and below HR1 = 0.25 only foreground stars are located. On the other hand SNRs are detected with HR1 > 0.25 and -0.7 < HR2 < -0.1. The hard sources with HR1 > 0.25 and HR2 > -0.1 can not be uniquely classified by their hardness ratio alone.

4. CONCLUSIONS

From the identification and classification of ROSAT PSPC sources Haberl & Pietsch (1999b) and Haberl et al. (1999) yield likely source types for nearly 400 X-ray sources in the fields around LMC and SMC. This allows for the first time to investigate the source populations in the Magellanic Clouds on a large statistical basis. First results reveal e.g. an inhomogeneous spatial distribution of X-ray binaries and SSSs. The latter are found only around the optical bar of the LMC and around the main body of the SMC. SSSs may exist also inside the dense stellar regions of the MCs but gas

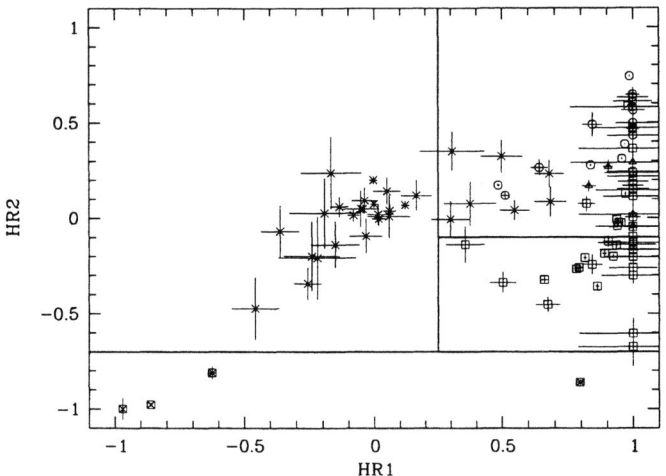

FIGURE 2. Hardness ratios of known PSPC source in the MCs. X-ray binaries are marked with a hexagon, SSSs with crossed square, SNRs with square, stars with x, and AGN with triangle. The thick lines separate areas where only members of a single source class are found

and dust hide these sources with very soft X-ray spectrum.

ACKNOWLEDGEMENTS

I thank my collaborators in this project W. Pietsch, M. Filipović, M. Sasaki and P. Kahabka.

REFERENCES

David L.P., Harnden F.R., Kearns K.E., Zombeck M.V., 1993, The ROSAT High Resolution Imager (HRI), ROSAT Announcement of Opportunity
Haberl F., Pietsch W., 1999a, A&A 344, 521
Haberl F., Pietsch W., 1999b, A&AS 139, 277
Haberl F., Filipović M.D., Pietsch W., Kahabka P., 1999, A&AS submitted
Kahabka P., Pietsch W., Filipović M.D., Haberl F., 1999, A&AS 136, 1
Long K.S., Helfand D.J., Grabelsky D.A., 1981, ApJ 248, 925
Pfeffermann E., Briel U.G., Hippmann H., et al., 1986, Proc. SPIE 733, 519
Pietsch W., Kahabka P., 1993, in Lecture Notes in Physics 416: New Aspects of Magellanic Cloud Research, eds. B. Baschek, G. Klare, J. Lequeux, 59
Trümper J., 1982, Adv. Space Res. Vol. 2, No. 4, 241
Wang Q., Hamilton T., Helfand D.J., Wu X., 1991, ApJ 374, 475
Wang Q., Wu X., 1992, ApJS 78, 391

DISCOVERY OF TWO HIGH-MAGNETIC-FIELD RADIO PULSARS

V. M. Kaspi [1], F. Camilo [2], A. G. Lyne [2], R. N. Manchester [3], J. F. Bell [3], N. D'Amico [4], N. P. F. McKay [2], F. Crawford [1]

1) Department of Physics and Center for Space Research, 70 Vassar Street, MIT, Cambridge, MA 02139
2) University of Manchester, Jodrell Bank Observatory, Macclesfield, Cheshire, SK11 9DL, UK
3) Australia Telescope National Facility, CSIRO, P.O. Box 76, Epping, NSW 1710, Australia
4) Osservatorio Astronomico di Bologna, via Ranzani 1, 40127 Bologna and Istituto di Radioastronomia del CNR, via Gobetti 101, 40129 Bologna, Italy

ABSTRACT

We report the discovery of two isolated radio pulsars having the largest inferred surface dipole magnetic fields yet seen in the population: 4.1×10^{13} G and 5.5×10^{13} G. These pulsars show apparently normal radio emission in a regime of magnetic field strength where some models predict no emission should occur. They have spin parameters and magnetic fields similar to those of some magnetar candidates, but exhibit very different radiative properties. This demonstrates that if the putative magnetars are indeed isolated neutron stars, their unusual attributes cannot be only a consequence of their large inferred magnetic fields.

KEYWORDS: stars: neutron, pulsars: general, pulsars: individual (PSR J1119−6127, J1814−1744)

1. INTRODUCTION

Recently it has been suggested (Thompson & Duncan 1992, Kulkarni & Frail 1993, Vasisht & Gotthelf 1997, Kouveliotou et al. 1998) that, in addition to the radio pulsars, there exists a class of isolated rotating neutron stars with ultra-strong magnetic fields – the so-called "magnetars." The observational properties of radio pulsars and putative magnetars are very different. Known radio pulsars, whose spin periods span the range from 0.0015–8.5 s, rarely have observable X-ray pulsations, and, in all cases, the X-ray power is much smaller than the spin-down luminosity. By contrast, the objects that have been suggested to be magnetars, namely soft gamma repeaters (SGRs) and anomalous X-ray pulsars (AXPs), exhibit pulsations with periods 5–12 s, and high energy emission that is many orders of magnitude stronger than their spin-down luminosity (Mereghetti & Stella 1995). Their pulsations have gone undetected at radio wavelengths. The dichotomy is thought to be a result of the much larger magnetic fields in magnetars, with magnetic field decay heating the neutron star to produce thermal X-ray emission (Thompson & Duncan 1993), or thermal emission from initial cooling enhanced by the large field (Heyl & Hernquist 1997).

Here we report the discovery of two isolated radio pulsars, PSRs J1119−6127 and J1814−1744, which have the largest inferred surface magnetic fields yet seen among radio pulsars. The results reported here will be described in more detail by Camilo et al. (2000).

2. OBSERVATIONS AND RESULTS

PSRs J1119−6127 and J1814−1744 were discovered as part of an ongoing survey of the Galactic Plane using the 64-m Parkes radio telescope (Lyne et al. 2000, see also D'Amico et al., these proceedings). PSR J1119−6127 has $P = 0.41$ s and period derivative $\dot{P} = 4.0 \times 10^{-12}$, the largest known among radio pulsars. We have also measured an apparently stationary period second derivative, $\ddot{P} = -4 \times 10^{-23}$ s^{-1}, making this only the third pulsar for which this has been possible through absolute pulse numbering. PSR J1814−1744 has $P = 4.0$ s and $\dot{P} = 7.4 \times 10^{-13}$. Spin and astrometric parameters for both pulsars are given in Table 1. From the standard equation for the dipolar surface magnetic field $B = 3.2 \times 10^{19} (P\dot{P})^{1/2}$ G, we infer surface magnetic fields of 4.1×10^{13} G and 5.5×10^{13} G for PSRs J1119−6127 and J1814−1744, respectively. These are the highest magnetic field strengths yet observed among radio pulsars.

3. DISCUSSION

Figure 1 is a plot of $\dot{P}$ versus P for the radio pulsar population, with PSRs J1119−6127 and J1814−1744 indicated. Also shown are the sources usually identified as magnetars, namely the five AXPs and two SGRs for which P and $\dot{P}$ have been measured.

Most models of the radio emission physics depend on pair-production cascades above the magnetic poles and hence on the strength of the magnetic field. However, at field strengths near or above the quantum critical field, $B_c \equiv m_e^2 c^3/e\hbar = 4.4 \times 10^{13}$ G, the field at which the cyclotron energy is equal to the electron rest-mass energy, processes such as photon splitting may inhibit pair-producing cascades. It has therefore been argued (Baring & Harding 1998) that a radio-loud/radio-quiet boundary can be drawn on the P–$\dot{P}$ diagram, with radio pulsars on one side, and AXPs and SGRs on the other (see Fig. 1). The existence of PSRs J1119−6127 and J1814−1744, however, demonstrates that radio emission can be produced in neutron stars with surface magnetic fields equal to or greater than B_c.

Especially noteworthy is the proximity of PSR J1814−1744 to the cluster of AXPs and SGRs at the upper right corner of Figure 1. In particular, this pulsar has a nearly identical $\dot{P}$ to the AXP 1E 2259+586 (Fahlman & Gregory 1981, Baykal et al. 1996, Kaspi et al. 1999). The disparity in their emission properties is therefore surprising. The radio emission upper limit (Coe et al. 1994) for 1E 2259+586 implies

Table 1: Parameters for pulsars J1119−6127 and J1814−1744.

Right ascension (J2000)	$11^h19^m14^s.2(1)$	$18^h14^m43^s.0(2)$
Declination (J2000)	$-61°27'48''.3(6)$	$-17°44'47(23)''$
Spin period, P (s)	0.4076034160(1)	3.975823037(1)
Period derivative, $\dot{P}$	$4.022930(4) \times 10^{-12}$	$7.434(4) \times 10^{-13}$
Period second derivative, $\ddot{P}$ (s^{-1})	$-4.1(3) \times 10^{-23}$	...
Period epoch (MJD)	51075.0	51075.0
Dispersion measure (cm^{-3} pc)	713(20)	834(20)
Flux density at 1374 MHz, S (mJy)	0.7(2)	0.5(2)
Surface magnetic field, B (Gauss)	4.1×10^{13}	5.5×10^{13}
Characteristic age, $P/(2\dot{P})$ (kyr)	1.6	85
Spin-down luminosity (erg s^{-1})	2.3×10^{36}	4.7×10^{32}
Distance, d (kpc)	2.4–8	10(2)
Radio luminosity at 1374 MHz, Sd^2 (mJy kpc^2)	∼ 20	∼ 50
Braking index, n	3.0(1)	...

a radio luminosity at 1400 MHz of <0.8 mJy kpc^2. This limit is comparable to the lowest values known for the radio pulsar population. That the radio pulse may be unobservable because of beaming cannot of course be ruled out.

The radio-loud/radio-quiet boundary line displayed in Figure 1 is more illustrative than quantitative. However, the apparently normal radio emission from PSRs J1119−6127 and J1814−1744, and the absence of radio emission from AXP 1E 2259+586, suggests that it will be difficult to delineate any such boundary without fine model-tuning. Furthermore, Pivovaroff et al. (2000) show, from archival data, that PSR J1814−1744 must be significantly less X-ray luminous than 1E 2259+586. The similar spin parameters for these two stars, and in turn the common features between 1E 2259+586 and the other AXPs and SGRs, imply that very high inferred magnetic field strengths cannot be the primary factor governing whether an isolated neutron star is a magnetar.

PSR J1119−6127 is notably young. Only three other pulsars having ages under 2 kyr are known: the Crab pulsar ($\tau = 1.3$ kyr), PSR B1509−58 ($\tau = 1.6$ kyr), and PSR B0540−69 ($\tau = 1.7$ kyr). The age of a pulsar is given by $\tau = [1 - (P_0/P)^{n-1}](P/(n-1)\dot{P}) \simeq P/2\dot{P}$, where P_0 is the spin period at birth (generally assumed to be much smaller than the current spin period) and n is the "braking index," defined via the relation for the spin evolution $\dot{\nu} \propto \nu^n$, where $\nu \equiv 1/P$. In the standard oblique rotating vacuum dipole model, $n \equiv \nu\ddot{\nu}/(\dot{\nu})^2 = 3$. For PSR J1119−6127, the parameters listed in Table 1 imply $n = 3.0 \pm 0.1$. This is the first measured braking index for a pulsar that is consistent with the usually assumed $n = 3$.

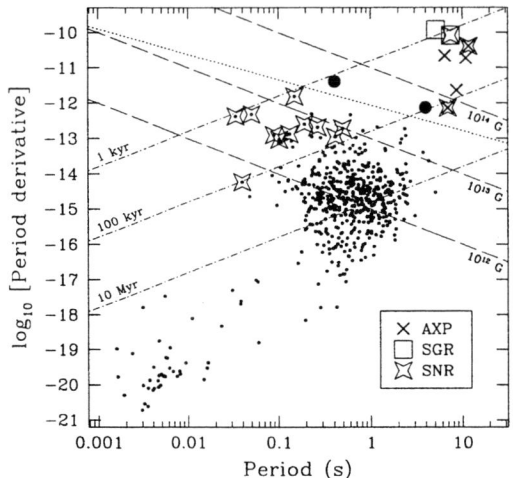

FIGURE 1. $P - \dot{P}$ diagram for radio pulsars, with SGRs and AXPs indicated. PSRs J1119−6127 and J1814−1744 are indicated with large solid circles. Lines of constant magnetic field are shown dashed, and lines of constant characteristic age ($\tau_c \equiv P/2\dot{P}$) are shown dot-dashed. The dotted line is the proposed illustrative radio-loud/radio-quiet boundary (Baring & Harding 1998). Radio pulsars plausibly associated with supernova remnants (SNRs) are indicated with four-point stars.

REFERENCES

Baring, M. G. & Harding, A. K. 1998, ApJ, 507, 55
Baykal, A., Swank, J. H., Strohmayer, T. & Stark, M. J. 1998, A&A, 336, 173
Camilo, F., Kaspi, V. M., Lyne, A. G., Manchester, R. N., Bell, F. J., D'Amico, N., McKay, N. P. F., Crawford, F. 2000, submitted
Coe, M. J., Jones, L. R. & Lehto, H. 1994, MNRAS, 270, 178
Duncan, R. C. & Thompson, C. 1992, ApJ, 392, L9
Fahlman, G. G. & Gregory, P. C. 1981, Nature, 293, 202
Heyl, J. S. & Hernquist, L. 1997, ApJ, 489, L67
Kaspi, V. M., Chakrabarty, D. & Steinberger, J. S. 1999, ApJ, 525, L33
Kouveliotou, C. e. a. 1998, Nature, 393, 235
Kulkarni, S. R. & Frail, D. A. 1993, Nature, 365, 33
Lyne, A. G., Camilo, F., Manchester, R. N., Bell, J. F., Kaspi, V. M., D'Amico, N., McKay, N. P. F., Crawford, F., Morris, D. J., Sheppard, D. C., Stairs, I. H. 2000, MNRAS, in press
Mereghetti, S. & Stella, L. 1995, ApJ, 442, L17
Pivovaroff, M., Kaspi, V. M. & Camilo, F. 2000, ApJ, submitted
Thompson, C. & Duncan, R. C. 1993, ApJ, 408, 194
Vasisht, G. & Gotthelf, E. V. 1997, ApJ, 486, L129

EVIDENCE FOR PHOTON BUBBLE OSCILLATIONS (PBO) AND TURBULENCE IN CENTAURUS X-3

Richard I. Klein [2,3,5], J. Garrett Jernigan [1], Jonathan Arons [2,4,5]

1) Space Sciences Laboratory, Un California, Berkeley, CA 94720
2) Dept Astronomy, 601 Campbell Hall, Un California, Berkeley, CA 94720
3) Lawrence Livermore Nat'l Lab., Un California L-23, P.O. Box 808, Livermore, CA 94550
4) Dept Physics, Un California, Berkeley, CA 94720
5) Theoretical Astrophysics Center, Un California, Berkeley, CA 94720

ABSTRACT We report the discovery of kHz fluctuations, including quasi-periodic oscillations (QPO) at $\sim$ 330 Hz and $\sim$ 760 Hz and a broadband kHz continuum in the power density spectrum of the high mass X-ray binary pulsar Centaurus X-3 (Jernigan, Klein and Arons 2000). These observations of Cen X-3 were carried out with the Rossi X-ray Timing Explorer (RXTE). The fluctuation spectrum is flat from mHz to a few Hz, then steepens to f^{-2} behavior between a few Hz and $\sim$ 100 Hz. Above a hundred Hz, the spectrum shows the QPO features, plus a flat continuum extending to $\sim$ 1200 Hz and then falling out to $\sim$ 1800 Hz.

Multi-dimensional radiation hydrodynamics simulations of optically thick plasma flow onto a magnetized neutron star show that the fluctuations at frequencies above 100 Hz are the likely consequence of the photon bubble turbulence and oscillations (PBO) previously predicted (Klein et al. 1996) to be observable in this source. We show that previous observations of Cen X-3 constrain the models to depend on only one parameter, the size of the polar cap. For a polar cap opening angle of 0.25 radians (polar cap radius $\sim$ 2.5 km and area $\sim$ 20 km^2, for a neutron star radius of 10 km), we show that the spectral form above 100 Hz is reproduced by the simulations, including the frequencies of the QPO and the relative power in the QPO and the kHz continuum. This has resulted in the first measurement of the polar cap size of an X-ray pulsar.

KEYWORDS: X-ray pulsar: Binary X-ray source: QPO: radiation hydrodynamics

1. INTRODUCTION

The binary system composed of the neutron star Cen X-3 and its companion O-type supergiant has been studied extensively by all orbiting X-ray astronomy observatories. The results from the Uhuru satellite identified the system as an eclipsing binary X-ray pulsar with a 2.087 day orbit and a 4.84 s pulse period. The well determined distance to the binary system yields an accurate determination of the luminosity $L_x \simeq 9.4 \times 10^{37}$ erg s^{-1}. Recently a cyclotron feature was detected in the X-ray spectrum of Cen X-3 providing precise knowledge of its surface magnetic

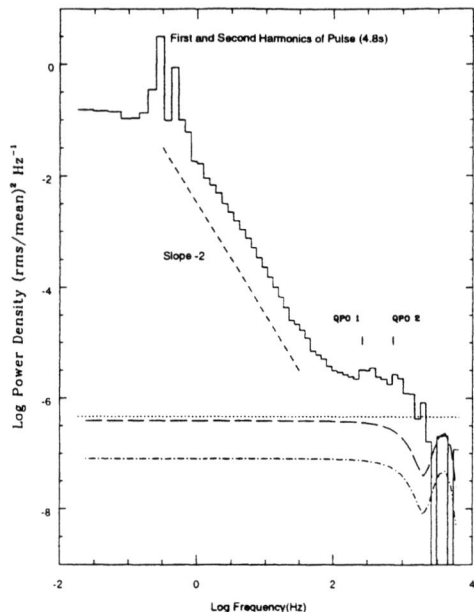

FIGURE 1. The curve is the observed average power density in units $\log_{10}\left[\left(\frac{\text{rms}}{\text{mean}}\right) \text{Hz}^{-1}\right]$ as a function of $\log_{10}$ [frequency (Hz)]. Note the presence of the first and second harmonics of the pulse (0.21 and 0.42 Hz) due to the rotation of the neutron star, the power law continuum below ~ 100 Hz with a slope of ~ -2 (dashed line), the flattening of the curve near 100 Hz, the two kHz QPO peaks, and the continuum power between $100 - 2500$ Hz. This intrinsic power density spectrum has been corrected for the effects of various noise components (dashed curves).

field strength $B \simeq 3.2 \times 10^{12}$ Gauss.

Motivated by earlier predictions of the likelihood of the existence of Photon Bubble Oscillations (PBO) in Cen X-3 (Klein et a. 1996), we have analyzed two consecutive binary cycles of the source to search for PBO. We have also calculated new theoretical models constrained by the observable parameters of Cen X-3. The physics describing the radiation hydrodynamics which governs the accretion of matter onto the highly magnetized polar caps of luminous X-ray pulsars has been described elsewhere (Arons, Klein and Lea 1987; Klein and Arons 1989). Early numerical results (Klein and Arons 1991) and linear stability analysis (Arons 1992) suggested the formation of small scale but large amplitude fluctuations in the matter density and velocity, which form radiation filled pockets almost devoid of plasma ("photon bubbles"). These photon bubbles (PB) result in significant observable fluctuations in the emitted luminosity.

We observed the X-ray emission from Cen X-3 with the RXTE over two binary cycles and clearly detect both kHz QPO and a broadband kHz continuum in the power density spectrum which we identify with PBO and the high frequency power law continuum generated by photon bubble "turbulence". The *only* remaining free parameter in our numerical models is the size of the polar cap. We show that by adopting a reasonable size for the polar cap of Cen X-3, we are able to semi-quantitatively match the frequencies of two observed QPO/PBO peaks and the kHz continuum with the power density spectrum of our calculated radiation-hydrodynamic models; thus giving strong evidence for the existence of PBO and photon bubble turbulence in Cen X-3.

2. OBSERVATIONS

Separate continuous observations of Cen X-3 were carried out with the PCA onboard RXTE for two consecutive binary cycles. Figure 1. is a log–log presentation of the average deadtime corrected power density spectrum for the entire data train of the low energy channel (1–7 keV). Note the smooth plateau at very low frequencies ($10^{-2} - 10^{-1}$ Hz) followed by the first and second harmonics of the 4.8 s pulse ($10^{-1} - 10^0$ Hz). The periodic pulse and its harmonics are superimposed on a broadband power law which falls from an amplitude of $\sim 10^{-1}$ to 10^{-6} with a slope of approximately -2 (see dashed line). At about 100 Hz another power density plateau appears, followed by two QPO peaks (~ 331 and ~ 761 Hz). These two QPO peaks (50% confidence ranges $260 - 407$ Hz and $671 - 849$ Hz) are superimposed on the power density plateau which falls sharply at frequencies above ~ 1200 Hz. The curve falls to a level of about $10^{-7} \left(\frac{\text{rms}}{\text{mean}}\right)^2$ Hz^{-1} at ~ 1800 Hz which has reached the Poisson detection limit of the transform.

3. THEORETICAL MODEL AND RESULTS

We calculate a model appropriate for the surface conditions of the neutron star Cen X-3 as the self-consistent solution of the full two-dimensional, time-dependent equations of the radiation hydrodynamics.

Figure 2. shows the results for the power density spectra of the time series of the emergent transverse luminosity out of the sides of the accretion column for three models (A, B and C) for which the values L_x and B have been set to those measured for Cen X-3. The models A, B and C are distinguished by setting the polar cap size to 0.25, 0.3 and 0.4 radians respectively. Model A ($\theta_c = 0.25$ radians) shows PBO at 350 Hz and a composite PBO at about 700 Hz. The frequencies of these PBO are in good agreement with frequencies of the QPO peaks observed in the power density spectrum. The approximate balance between PBO/QPOs and continuum is also roughly true for all three theoretical models. The additional strong broad composite PBO at ~ 3000 Hz that is seen in models A and B is not seen in the observed power density spectrum of Cen X-3. However it is below the level that could be easily detected with the PCA. The only free parameter in our calculations is the size of the polar cap. If we consider the whole picture these models taken together determine a range for the polar cap radius of ~ 2–3 km.

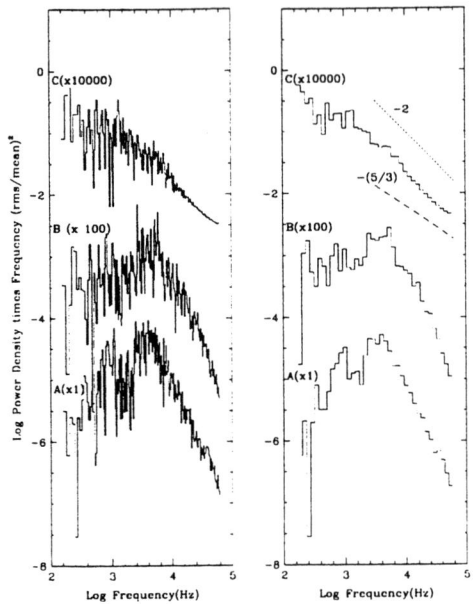

FIGURE 2. The left panel is a log-log plot of power density spectra in units $\log_{10}\left[\text{PowerDensity}\left(\frac{\text{rms}}{\text{mean}}\right)\right]$ of models A,B and C versus $\log_{10}\left[\text{Frequency(Hz)}\right]$. The right panel is a smoothed representation of the data in the left panel and is more appropriate for direct comparison with the observed data. The vertical scale has been shifted up by 2 and 4 decades for models B and C respectively for illustrative purposes. Note that all three models show evidence of PBOs and PB continuum power. Lines that indicate power laws of the form f^{-2} (dotted) and $f^{-(\frac{5}{3})}$ (dashed) are indicated in the right panel.

4. CONCLUSIONS

We have advanced the view that the fluctuations observed in the power spectrum at frequencies exceeding a few hundred Hz have their origin in the intrinsic photon bubble instability of the accretion flow at the stellar surface. Prior to the RXTE launch we predicted the observation of photon bubble phenomena including the specific existence of kHz fluctuations and some power in quasi-periodic oscillations identifiable as PBO in Cen X-3 (Klein et al. 1996). We have now observed these fluctuations in Cen X-3. A careful comparison of the predicted power density from the PB simulations with the observed power density spectrum constrains the size of the polar caps of Cen X-3 to a radius of ∼2–3 km. This has permitted the first model-dependent measurement of the size of the accreting polar cap in an X-ray pulsar.

ACKNOWLEDGEMENTS

This research has been supported by NASA ATP grant NAG5-3809 and a NASA XTE Guest Observer grant NAG5-3385. Part of this work was supported under the auspices of the US Dept. of Energy bu Univ. of Cal LLNL under contract W-7405-Eng-48.

REFERENCES

Arons, J., Klein, R.I. and Lea, S.M. 1987, ApJ, 312, 666

Arons, J. 1992, ApJ, 388, 561

Jernigan, J.G., Klein, R.I. and Arons, J. 2000, ApJ, in Press Feb. 2000

Klein, R.I., and Arons, J. 1989, in Proc. 23rd ESLAB Symposium on Two Topics in X-ray Astronomy, N.E. White and T.D. Guyenne, eds., ESA SP-296 (Paris: European Space Agency), 1, 89

Klein, R.I. and Arons, J. 1991, in Stellar Atmospheres: Beyond Classical Models, L. Crivilari, ed. (Boston: Kluwer), 205

Klein, R.I., Arons, J. Jernigan, J.G. and Hsu, J. 1996, ApJ 457, L85

ASCA OBSERVATIONS OF THE NEW JET SYSTEM XTE J1748-288

T. Kotani [0,1], N. Kawai [1], F. Nagase [2] M. Namiki [1], M. Sakano [3], T. Takeshima [0], Y. Ueda [2], M. Matsuoka [1,4]

0) Laboratory for High-Energy Astrophysics, NASA/GSFC, Greenbelt, MD 20771, USA
1) Cosmic Radiation Laboratory, RIKEN, Saitama 351-0198, Japan
2) High-Energy Astrophysics Division, ISAS, Kanagawa 229-8510, Japan
3) Department of Physics, Kyoto University, Kyoto 606-8502, Japan
4) Space Utilization Research Program, NASDA, Ibaraki 305-8505, Japan

ABSTRACT XTE J1748−288 is the newly discovered transient with a one-sided radio jet. It was observed with ASCA on 1998/09/06 and 1998/09/26, 100 days after the onset of the radio-X-ray outburst. The spectra were expressed with an attenuated power-law model, and the 2–6-keV flux was $4.6^{+1.0}_{-0.8} \times 10^{-11}$ erg s^{-1} cm^{-2} and $2.2^{+0.8}_{-0.6} \times 10^{-12}$ for 09/06 and 09/26, respectively. The light curve shows that the steady exponential decay with an e-folding time of 14 days lasted over 100 days and 4 orders of magnitude from the peak of the outburst. Possible detection of iron line is reported.

KEYWORDS: binaries: close — stars: individual (XTE J1748-288) — X-rays: stars

1. INTRODUCTION

XTE J1748−288 showed an X-ray outburst on 1998/06/03 and detected by the ASM/RXTE and BATSE/CGRO (Smith, Levine, & Wood 1998; Harmon et al. 1998). The 2-10-keV X-ray flux reached to 600 mCrab on 1998/06/05 (Strohmayer, Marshall, et al. 1998), and decayed with an e-folding time of ∼ 20 days (ASM/RXTE Team, 1999). The radio counterpart was located with the VLA at R.A. = 17h48m05s.06, Decl. = −28°28′25".8 (equinox 2000.0; uncertainty 0".6; Strohmayer, et al. 1998). The spectral index of the radio counterpart was 0.2 ∼ 0.6 (Hjellming et al. 1998a; Rupen & Hjellming 1998), and the radio activity reached to the maximum of 350 mJy at 2.25 GHz around 1998/06/16 (GBI Team 1999). Rupen & Hjellming discovered a one-sided jet of 20 mas day^{-1} in the VLA images, which corresponds to a velocity of 0.93 c assuming a distance of 8 kpc. The time of the jet ejection was estimated to be around 1998/06/01, extrapolating the proper motion (Hjellming et al. 1998b). The radio activity lasted 200 days, suggesting a continuous ejection of jet material. Thus this source is considered to be a jet system, probably similar to SS 433, which has a persistent jet. Around 1998/08/09, the expansion of the leading edge slowed to a rate of ∼ 5 mas day^{-1}, or 0.23 c, and the leading edge of the jet

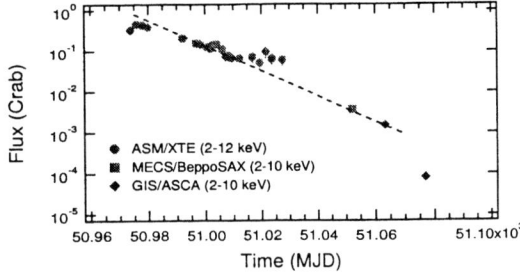

FIGURE 1. Light curve. The data of ASM/RXTE (ASM/RXTE Team 1999), MECS/BeppoSAX (Sidoli et al. 1999), and ASCA are plotted. The convergence of the ASM data to ~ 0.07 Crab is of an artificial effect due to the data selection threshold applied here.

brightened dramatically (Hjellming et al. 1998b). It is interpreted that the ejected jet material has run into external gas and formed a shock which is seen as a "hot spot."

2. LIGHT CURVE

XTE J1748−288 was observed with ASCA from 1998/09/06 09:10 to 17:50 (effective time: 5 ks) and from 1998/09/26 03:30 to 16:30 (20 ks). The source was out of the field of view of the SIS on 09/06. Before the outburst, the source position was observed with ASCA twice during the galactic plane survey, and the result will be reported in forthcoming paper (Kotani et al. 1999). The PH mode with the nominal bit assignment was used for the GIS, and the 1-CCD FAINT mode was used for the SIS (Tanaka, Inoue, & Holt 1994). The standard data-reduction method was applied for both data. There are several contaminating sources near XTE J1748−288, such as 1E 1743.1−2843, Sgr B2, and the galactic ridge emission. We collected photons within 6′ and 2′ of XTE J1748−288 for 09/06 and 09/26, respectively. Assuming that the background count rate is proportional to the sampling area, the background component in the source region of the GIS was estimated from the photons of the annulus region around the source and in the the field of view. The thickness of the annulus was 6′ − 10′ and 6′ − 8′ for 09/06 and 09/26, respectively. Since the field of view of the SIS is not so large as that of the GIS, photons farther than 4.5′ from the source in the CCD chips were used for the background estimation. The 2–6-keV flux was determined to be $4.6^{+1.0}_{-0.8} \times 10^{-11}$ erg s^{-1} cm^{-2} and $2.2^{+0.8}_{-0.6} \times 10^{-12}$ for 09/06 and 09/26, respectively. The light curve is shown in figure 1 together with ASM/RXTE and BeppoSAX data.

In the light curve, the ASCA observation on 09/06 (MJD = 51062) is in the extrapolation line from ASM and BeppoSAX data. It is remarkable that the exponential decay with an e-folding time of 12 $\sim$ 16 days lasted till 09/06 from the peak of the outburst, over 100 days and 4 orders of magnitude. This is a good example of steady exponential decays following outburst which black-hole candidates show. On the other hand, the data point of 09/26 (MJD = 51082) is below the extrapolation line by one order of magnitude. There might be a transition from the exponential-decay phase to another phase, say, off-state phase, in the period between the two ASCA observations.

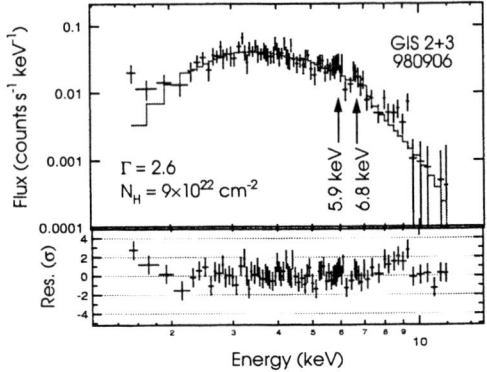

FIGURE 2. Spectrum taken on 1998/09/06 with the GIS. The data of both sensors were combined. The best-fit model of an attenuated power-law with 2 Gaussian lines are plotted.

3. SPECTRUM

Both the spectra of 09/06 and 09/26 were found to be well fitted by an attenuated power-law model. Hydrogen column density was fitted to be $N_H = 9 \times 10^{22}$ cm^{-1} and 6.1×10^{22}, and photon index was 2.6 and 2.7 for 09/06 and 09/26, respectively. In the fit residuals of the data of 09/06, which has a better statistics than the other, there are a hump around the iron-K energy. It can be fitted with two Gaussian lines, and each line passed F test with a significance more than 90 %. The line energies were fitted to be 5.85 ± 0.15 keV and 6.76 ± 0.15 keV. The width of lines were fixed to zero. The best-fit model with two lines and the spectrum are shown in figure 2.

It is interesting to assume that both or either of the lines comes from the jet of XTE J1748−288, and that the energy is Doppler shifted. If they are Fe XXV Kα line, parameters of the jet, e.g., velocity β, inclination θ, and distance to the source D, can be derived. Results are shown in Table 1. If only one of the two lines is of jet origin, parameters of the jet of the other side can be derived, assuming a symmetric bipolar flow. The Doppler parameter of the receding jet would be $z = 0.5 \sim 1$ and consistenet with the non-detection of it in the radio band.

The colliding jet blob observed in radio band may be an X-ray source. It is possible that the observed X ray was due to thermal bremsstrahrung or synchrotron emission. However, neither mecanism can explain the emission lines. Synchrotron spectrum does not have an iron line. And plasma decelerated to 0.23 c from 0.93 c would be heated to $kT \sim$ GeV, and all elements would be fully ionized. So, the colliding jet blob can not be the source of a line, although the blob is still be a candidate X-ray source.

In discussion above, we focus on the possibility that the feature is of XTE J1748−288 origin. Because of the insufficient statistics and the presence of possible contaminating sources, e.g., Sgr B2 and the galactic ridge, it is rather dangerous to conclude that both or either of the lines is intrinsic to XTE J1748−288. The absence of any iron line in the BeppoSAX spectrum of XTE J1748−288 also casts a serious doubt, though the source was at off-center of the MECS, where the energy

TABLE 1. Jet parameters. Four cases are considered. Proper motion of $\mu = 20$ mas day^{-1} was used for estimations (Rupen & Hjellming 1998). For the cases that only one line is of jet origin, distance was assumed to be $D \sim 8$ kpc (Rupen & Hjellming 1998).

	6.8 keV line: Jet origin	6.8 keV line: Not jet origin
5.9 keV line: Jet origin	$\beta = 0.35 \pm 0.05$ $\theta = 78° \pm 4°$ $D = 3.2 \pm 0.7$ kpc	$\beta = 0.747 \pm 0.007$ $\theta = 71.3° \pm 1.1°$ $E_{\rm red} \sim 4.5$ keV
5.9 keV line: Not jet origin	$\beta = 0.714 \pm 0.004$ $\theta = 64.6° \pm 1.1°$ $E_{\rm red} \sim 3.6$ keV	

resolution is not the best (Sidoli et al. 1999). Another question is whether the source formed a jet in such a low state. On the other hand, jet-origin emission lines have been detected from SS 433, even when it was as faint as 10^{35} erg s^{-1} (Kotani 1998). If XTE J1748−288 has a jet with a luminosity $\sim 10^{35}$ erg s^{-1}, it can be observed only when other X-ray emitters in the system becomes fainter than that, i.e., the source is as faint as on 1998/09/06. The estimations above will be checked by future observations of the source in an active state.

ACKNOWLEDGEMENTS

We thank W. B. Focke, J. H. Swank, and L. Sidoli for the useful discussions.

REFERENCES

ASM/RXTE Team 1999, quick-look results provided via http://space.mit.edu/XTE/ASM_lc.html
GBI Team 1999, quick-look results provided via http://www.nrao.edu/~rhjellmi/gbint/plgbi.html
Harmon, B. A., McCollough, M. L., Wilson, C. A., Zhang, S. N., Paciesas, W. S. 1998, IAUC, 6933
Hjellming, R. M., Rupen, M. P., Ghigo, F., Fender, R. P., Stappers, B. W. 1998a, IAUC, 6937
Hjellming, R. M., Rupen, M. P., Mioduszewski, A. J., Smith, D. A., Harmon, B. A., Waltman, E. B., Ghigo, F. D., Pooley, G. G. 1998b, A&A, 193, S10308
Kotani, T. 1998, Doctoral Thesis, University of Tokyo
Kotani, T., Hjellming, R. M., Kawai, N., Nagase, F., Namiki, M., Sakano, M., Takeshima, T., Ueda, Y. 1999, in preparation
Rupen, M. P., Hjellming, R. M. 1998, IAUC, 6938
Sidoli, L., Mereghetti, S., Israel, G. L., Chiappetti, L., Treves, A., Orlandini, M. 1999, ApJ, in press
Smith, D. A., Levine, A., Wood, A. 1998, IAUC, 6932
Strohmayer, T., Marshall, F. E., Hjellming, R. M., Rupen, M. P. 1998, IAUC, 6934
Tanaka, Y., Inoue, H., Holt, S. S. 1994, PASJ, 46, L37

A BEPPOSAX OBSERVATION OF NOVA VELORUM 1999: A VERY BRIGHT CLASSICAL NOVA

M. Orio[1,2], A. N. Parmar[3], L. Amati[4,5], R. Benjamin[2], M. Della Valle[6], F. Frontera[4,5], J. Greiner[7], T. Mineo[8,4], H. Ögelman[2], S. Starrfield[9], E. Trussoni[1]

1) Turin Astronomical Observatory, I-10025 Pino Torinese, To, Italy, 2) Physics Department, University of Wisconsin, Madison, WI, USA, 3) Astrophysics Division, Space Science Dept of ESA, ESTEC, The Netherlands 4) The BeppoSAX Team, 5) TESRE-CNR, Bologna, Italy 6) Arcetri Asstronomical Observatory, Firenze, Italy, 7) Astrophysikalisches Institut, Postdam, FRG, 8) IFCAI-CNR, Palermo, Italy 9) University of Arizona, Tempe, AZ, USA

ABSTRACT

Nova Velorum 1999 (V382 Vel) was detected 12 days after the outburst discovery with the BeppoSAX MECS and LECS with a count rate 0.1537±0.0020 cts s^{-1} and 0.0620±0.0026 cts s^{-1} respectively. It was not detected at higher energies. The LECS-MECS combined spectrum shows huge intrinsic absorption of the nebular ejecta, so the site of X-ray emission must have been deeply inside the nebular material at the time of observation. The spectrum can be fitted with a multiple component model of thermal plasma with at least one component at temperature kT$\geq$10 keV, and the flux is 1.5-2 × 10^{-11} erg cm^{-2} s^{-1}. There is no supersoft emission of the central source due to the heavy absorption of the shell.

KEYWORDS: X-rays: stars; Stars: individual: V382 Vel, novae, cataclysmic variables

1. INTRODUCTION

There are two mechanisms for X-ray emission of classical novae in outburst: they teach us about the *mass ejection* process and the *final outcome* of these systems. "Hard" X-ray emission (thermal bremsstrahlung temperatures of the order 0.5-10 KeV) with luminosities 10^{33-34} erg s^{-1} can be due to *shocks* (interacting winds or interaction ejecta/circumstellar medium).

Luminous *"supersoft"* X-ray emission is also predicted to be due to residual hydrogen burning after the outburst. Classical Novae host white dwarfs with a hydrogen burning shell. If only part of the the burning material m_H accreted in Δm_{env} by a white dwarf with mass m_{WD} is ejected, the rest is exhausted after a time:

$$t_{burn} \propto m_H \propto \Delta m_{env} \propto \frac{1}{m_{WD}}$$

t_{burn} is 2-100 years and critically dependent on m_{WD} and other parameters in the models.

If residual hydrogen is left over at the end of the outburst the secular evolution of the system becomes particularly interesting, because the WD might

- become a type Ia SN

- or undergo an accretion induced collapse.

Out of 105 Galactic and LMC novae, only 4 appeared as *supersoft X-ray sources* as predicted: GQ Mus (Nova Muscae 1983), V1974 Cyg (Nova Cygni 1992), Nova LMC 1995, and the recurrent nova U Sco (see Ögelman et al. 1993, Shanley et al. 1995, Krautter et al. 1997, Orio & Greiner 1999, Orio et al. 1999a, Kahabka et al. 1999). The recurrent nova RS Oph, Novae Her 1991, Pup 1991 and Cyg 1992 and probably LMC 1992 were hard X-ray sources shortly after the outburst (see Orio et al. 1992a and references therein).

2. A VERY BRIGHT NOVA: 382 VELORUM

Nova Velorum 1999 (V382 Vel) was discovered in outburst on May 22 1999 (Williams & Gilmore, 1999). It is the second brightest nova of this half of the century (V=2.6; Seargent & Pearce, 1999). Only 19 classical novae in this century were brighter than 5th magnitude optically. Most are around 8th magnitude at the optical peak.

V382 Vel is a O-Ne-Mg nova (Woodward et al. 1999), and quite a "fast" one ($v \simeq 3500$ km s^{-1}, $t_2 = 6$ d, $t_3 = 10$ d) (Woodward et al. 1999, Della Valle et al. 1999).

3. THE OBSERVATIONS

The nova was observed on June 7-8 1999 for 42.5 ksec in the two MECS and for 13.5 ksec in the LECS; it was detected with a count rate 0.1537±0.0020 cts s^{-1} and 0.0620±0.0026 cts s^{-1} respectively (Orio et al. 1999b). It was *not* detected at higher energies with the PDS : in the band 15-30 keV the 2 σ u.l. is 0.050 cts s^{-1} (flux u.l. 3.9 $\times 10^{-12}$ erg cm^{-2} s^{-1}), at 15-60 KeV the 2 σ u.l. is 0.074 cts s^{-1} (flux u.l. 6.7 $\times 10^{-12}$ erg cm^{-2} s^{-1}).

The LECS-MECS spectrum is characterized by high intrinsic absorption of the ejecta. The peak of energy is poorly constrained, and it might be outside the MECS window. However, the PDS upper limit constraints it to be below $\simeq 15$ keV. A spectral fit with a double component thermal plasma MEKAL model is shown in Fig. 1: fixing solar abundances, the two components have plasma temperatures, kT, 1.7 and $\simeq 20$ keV. The quality of the fit is definitely poor if the second component is assumed to be at lower temperature, primarily because the model predicts excessive Fe-K line emission. A fit to the spectra shown in Figure 1 with the VMEKAL model (in which the abundances of single elements are specified) yields $\chi^2 = 1.8$ per 73 dof with the same parameters used in Figure 1 and reducing the iron abundance from

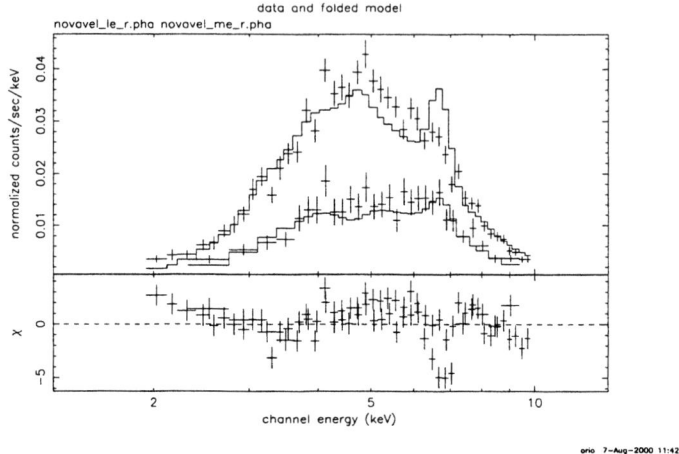

FIGURE 1. The the LECS and MECS spectra (above 1.8 keV: flux below this energy is negligible) and the best fit obtained with a two-component MEKAL model of thermal plasma with solar abundances, kT=1.7 keV and 20 keV for the two components respectively, N(H)=1.7 × 10^{23} cm^{-2}, flux 1.7 × 10^{-11} erg cm^{-2} s^{-1}. The reduced χ^2 is 3 per 73 dof. The emission measure constant is $\frac{\int n_e n_H dV}{4\pi D^2}$=1.89 ×$10^{12}$ × cm^{-5}.

solar to 1/10th the solar value (versus χ^2=3 per 73 dof with solar iron). A very low iron abundance would be exceptional for a nova; we tend to attribute it either to non-equilibrium ionization or a multi-temperature spectrum in which the lines are smoothed out. The excess at ≃2 keV in Figure 1 might be due, rather than to to the emergence of a soft component, to a "leaky absorber", meaning that the ejected material is highly non-homogeneous.

4. CONCLUSIONS

We have learned several facts from this observation:

- The plasma in which the hard X-ray emission begins might have multiple thermal components, at least one of them having kT in the range ≃10-20 keV;

- We find that the peak of hard X-ray emission occurred in the 3rd week post-outburst comparing the flux measured with BeppoSAX and of an ASCA observation done shortly after (Mukai 1999 and Mukai, private communication);

- The mechanism causing emission occurs deeply in the interior of the ejected nebula;

- The *supersoft* X-ray source could NOT be visible in the first month after the outburst due to the *intrinsic* absorption of the ejected shell. Even a *fast* O-Ne-Mg nova like V382 Vel does not turn into a supersoft X-ray source by the end of the first month: comparison with theoretical papers suggests that $m_{WD} < 1.3 M_\odot$.

Since the *intrinsic absorption* of the nebula seems to have dramatically decreased by now (September 1999) judging from the (B-V) color index published in the IAU Circulars, new X-ray observation at t=3-6 months post-outburst will be worthwhile to exploit this rare possibility to study the supersoft central source.

REFERENCES

Della Valle, M. Pasquini, L., Williams, R., 1999, IAUC 7193

Kahabka, P., Hartmann, H.W., Parmar, A.N., Neguerela, I., 1999, A&A 347, L39

Krautter, J., Ögelman, H., Starrfield, S., Wichmann, R., Pfeffermann, E., 1996, ApJ 456, 788

Mukai, K., Ishida, M., 1999, IAUC 7205

Ögelman, H., Orio, M., Krautter, J., Starfield, S., 1993, Nature, 361, 331

Orio, M., Greiner, J., 1999, A&A, 344, L13

Orio, M., Ögelman, H., Covington, J., 1999a, preprint

Orio, M., Torroni, V., Ricci, R., 1999b, IAUC 7196

Seargent, D.A.J., Pearce, A., 1999, IAUC 7177

Shanley, L., Ögelman, H., Gallagher, J., Orio, M., Krautter, J., 1995, ApJ 438, L95

Williams, P., Gilmore, A., 1999, IAUC 7176

Woodward, C.E., Wooden, D.H., Pina, R,K., Fisher, R.S., 1999, IAUC 7220

A SELF-CONSISTENT TEST OF COMPTONIZATION MODELS USING A LONG BEPPOSAX OBSERVATION OF NGC 5548

P.O. Petrucci[1], F. Haardt[2], L. Maraschi[1], P. Grandi[3], G. Matt[6], F. Nicastro[3,4,5], L. Piro[3], G.C. Perola[6], A. De Rosa[3]

1) Osservatorio Astronomico di Brera, Milano, Italy, 2) Universitá dell'Insubria, Como, Italy, 3) IAS/CNR, Roma, Italy, 4) CfA, Cambridge Ma., USA, 5) Osservatorio Astronomico di Roma, Roma, Italy, 6) Universitá degli Studi "Roma 3", Roma, Italy

ABSTRACT We test accurate models of Comptonization spectra using the high quality data of the *Beppo*SAX long look at NGC 5548. The data are well represented by a plane parallel corona with an inclination angle of 30°, a soft photon temperature of 5 eV and a hot plasma temperature and optical depth of $kT_e \simeq 360$ keV and $\tau \simeq 0.1$, respectively. If energy balance applies, such values suggest that a more "photon-starved" geometry (e.g. a hemispheric region) is necessary. The spectral softening detected during a flare appears to be associated with a decrease of the heating-to-cooling ratio, indicating a geometric and/or energetic modification of the disk plus corona system. The hot plasma temperature derived with the models above is significantly higher than that obtained fitting the same data with a power law plus high energy cut off model for the continuum. This is due to the fact that in anisotropic geometries Comptonization spectra show "intrinsic" curvatures which move the fitted high energy cut-off to higher energies.

KEYWORDS: Radiation mechanisms: thermal; Methods: data analysis; *Beppo*SAX ; Galaxies: Seyfert; Galaxies: individual: NGC 5548

1. INTRODUCTION

The X-ray emission of Seyfert galaxies is commonly believed to be produced by Compton scattering of soft photons on a (thermal or non-thermal) population of hot electrons. The non–detection of Seyferts by Comptel and the high energy cut-offs indicated by OSSE have focused attention on thermal models. In this case the X–ray spectral shape is mainly determined by the temperature $\Theta = kT_e/m_e c^2$ and the optical depth τ of the scattering electrons, while the cut–off energy is related essentially to Θ. Moreover, if the Comptonizing region and the source of seed photons are *coupled*, one can write an energy balance equation for the hot coronal plasma which determines a roughly constant value of the Compton parameter $y \simeq 4\Theta\tau(1+4\Theta)(1+\tau)$ (see Petrucci et al. 2000, and references therein) and thus a one to one correpondence between Θ and τ. The required value of y depends on the fraction f of the power dissipated in the corona and on geometry. In the following the limiting case $f = 1$ is considered.

In the present work (Petrucci et al. 2000, hereafter P00), we test Comptonization models using the high quality data of the (8 day) *Beppo*SAX long look at the Seyfert 1 galaxy NGC 5548, deriving constraints on the physical parameters and geometry of the source. These data have already been studied in detail by Nicastro et al. (2000), modelling the continuum with a cut-off power law.

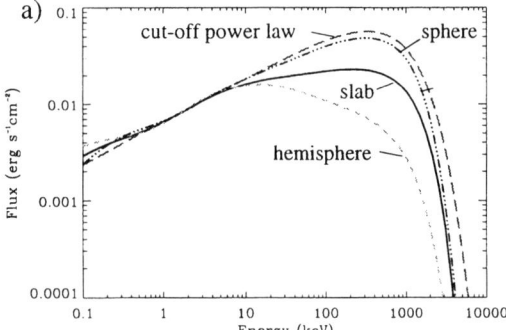 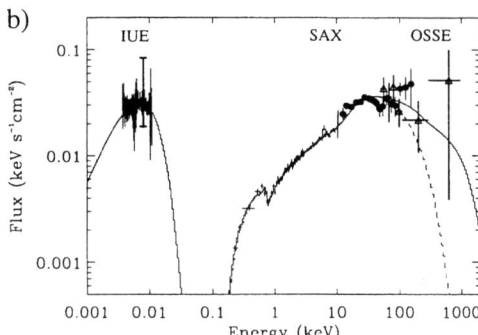

FIGURE 1. **(a)**: Comptonized models for different geometries assuming $\Theta = 0.7$. We have also over-plotted a cut–off power law with $E_c = 2kT_e$ as a dashed line (see the text for details).**(b)**: *Beppo*SAX data set of NGC 5548 with non–simultaneous IUE and OSSE data (from Magdziarzk et al. 1998) with the corresponding best fitting Comptonization model (in slab geometry, solid line) and simple cut–off power law model (dashed line).

2. THE COMPTONIZATION MODEL

2.1. The anisotropy break

In Fig. 1a we show Comptonized spectra computed for different geometries (codes of Haardt, 1994 and Poutanen & Svensson, 1996) for the same value of $\Theta \simeq 0.7$ and for a cut–off power law spectrum (PEXRAV model of XSPEC) with an e–folding energy $E_c = 2kT_e = 720$ keV, as a first order approximation to Comptonization spectral models (for $\tau \lesssim 1$). For the sphere, the soft photons are supposed to be emitted isotropically at the center of the sphere, whereas they come from the bottom for the slab and the hemisphere configurations. In each case, the optical depths have been chosen so as to produce approximately the same spectral index in the 2–10 keV X-ray range (τ=0.09, 0.16 and 0.33 for the slab, hemisphere and sphere geometries, respectively). We see that the spectra are quite different at medium - high energy ($E \gtrsim 10$ keV). In the slab and hemisphere cases, they can be approximately described by broken power laws, the energy of the break E_{break} roughly lying between the second and the third scattering order peaks. The slope at low energies is flat due to the deficiency of photons caused by the anisotropy of the first scattering and is

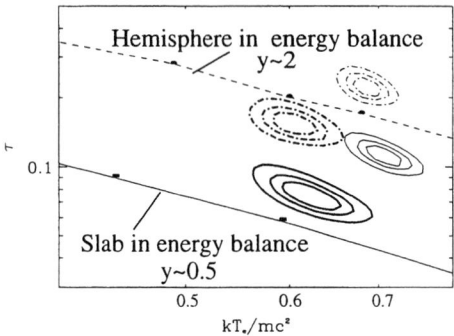

FIGURE 2. Solid and dot-dashed contour plots of τ vs. Θ in the low and high states for the slab and hemisphere configurations respectively. The thick contours correspond to the high states. We have also over-plotted the $\Theta - \tau$ relations predicted when energy balance is achieved with the corresponding (roughly constant) value of the Compton parameter (see Petrucci et al. 2000, and references therein)

State	Geom.	Θ	τ	R	γ	χ^2/dof
Low	Slab	$0.70^{+0.04}_{-0.02}$	$0.12^{+0.01}_{-0.02}$	1.0 ± 0.2	-	90/114
	Hemi.	$0.68^{+0.02}_{-0.02}$	$0.22^{+0.03}_{-0.02}$	1.9 ± 0.3	-	80/114
	PEXRAV	$0.11^{+0.05}_{-0.02}$	-	0.5 ± 0.2	$1.55^{+0.02}_{-0.02}$	93/114
High	Slab	$0.62^{+0.04}_{-0.04}$	$0.07^{+0.02}_{-0.01}$	1.6 ± 0.5	-	137/145
	Hemi.	$0.61^{+0.02}_{-0.03}$	$0.15^{+0.03}_{-0.02}$	2.7 ± 0.6	-	143/145
	PEXRAV	$0.16^{+0.38}_{-0.07}$	-	0.6 ± 0.4	$1.71^{+0.03}_{-0.04}$	142/145

TABLE 1. Best fit parameters for Comptonization models in slab and hemisphere geometries and for PEXRAV. In this last case, we used the approximation $\Theta = E_c/2m_e c^2$ (valid for $\tau \lesssim 1$)

highly angle-dependent. On the other hand, since the higher scattering orders are almost isotropic, the slope above the break is almost angle-independent. Due to the presence of this break, Comptonization spectra (in anisotropic geometry) are softer for $E \gtrsim E_{\text{break}}$ than the cut-off power law one.

2.2. Comptonization model versus cut-off power law

We show in Fig. 1b the best fit models derived using Comptonization (in slab geometry) and a cut–off power law . The two models require different normalizations for the reflection component (larger for the slab) and are roughly in agreement below 200 keV, the upper energy end of our data. However they differ by up to a factor of 10 near 500 keV, since the cut–off energy required by the (harder) power law model is lower than that required by the (intrinsically curved) slab Comptonization spectrum (see Table 1).

2.3. Geometry

Both slab and hemisphere geometries give acceptable fits to the average data (cf. Table 1). The derived parameters are not far from theoretical expectations based on simple energy balance arguments (cf. Fig. 2). For the slab geometry, the data suggest that the hot gas is *photon starved*, i.e., it is undercooled. For the hemispherical geometry the parameters are consistent with the energy balance condition but the required normalization of the reflection component is too large (cf. Table 1). This may suggest that the real geometry is intermediate and/or that the physical situation is more complex possibly involving a non uniform corona and deviations from strict energy balance (Malzac & Jourdain, 2000).

2.4. Variability

Independently of geometry, the low–to–high state transition (the low/high state being the state outside/during the flare) clearly indicates a change of the Compton parameter, i.e., of the Comptonized–to–soft luminosity ratio (cf. Fig. 2). It seems to be most probably due to an increase of the cooling rate, rather than to a decrease of the heating rate, since we observe a pivoting at high energies of the continuum in the two states (cf. P00). If this interpretation is correct, then the spectral softening in the high state is very naturally explained by a drop of the corona temperature, ultimately due to an increase of the UV–EUV soft photon flux.

3. CONCLUSIONS

This *Beppo*SAX observation of NGC 5548 allowed us to show that i) the temperature kT_e of the Comptonizing plasma can be significantly underestimated (up to a factor of 7 here) when derived from simple power law models with high energy cut-offs; ii) the data are well fitted by a plane parallel corona model with an inclination angle of 30°, a soft photon temperature of 5 eV, a hot plasma $kT_e \simeq 360$ keV and an optical depth $\tau \simeq 0.1$. The latter values suggest, however, that the hot Comptonizing gas, if in the shape of slab, is not in energy balance. A better agreement is obtained with a hemispherical geometry; iii) the change of state during the central part of the run clearly indicates a variation of the Compton parameter y, which could be due, as suggested by the data, to an increase of the cooling.

ACKNOWLEDGEMENTS

We gratefully acknowledge J. Poutanen for providing his code. This work was supported in part by the EC under contract number ERBFMRX-CT98-0195 (TMR network "Accretion onto black holes, compact stars and protostars")

REFERENCES

Haardt, F. 1994, PhD, SISSA, Trieste (H94)
Magdziarz, P. et al., 1998, MNRAS, 301, 179
Malzac, J. & Jourdain, E., 2000, A&A, in press (astro-ph/0005523)
Nicastro, F. et al. 2000, ApJ, 536, 718
Petrucci, P. al. 2000, ApJ, in press
Poutanen, J. & Svensson, R. 1996, ApJ, 470, 249

A ROSAT HRI SURVEY OF BRIGHT NEARBY GALAXIES

T.P. Roberts & R.S. Warwick

Department of Physics & Astronomy, University of Leicester, Leicester, LE1 7RH, UK

ABSTRACT The ROSAT public archive contains HRI observations covering more than 100 of the 486 galaxies observed in the 'search for dwarf Seyferts' of Ho et al (1997). We have catalogued a large subset of the best observed galaxies (83 in total), recording discrete X-ray sources coincident with both the galaxy disks and the optical nuclei. We present selected results from this survey. Highlights include a substantial population of very luminous non-nuclear sources. Nuclear X-ray sources are observed in over 50% of galaxies; if these are associated with a super-massive black hole, then they must have a significantly sub-Eddington accretion rate.

KEYWORDS: X-rays:galaxies - galaxies:active - galaxies-general

1. PROJECT OUTLINE

The extensive *ROSAT* archives now contain more than 4400 pointed observations with the High Resolution Imager (HRI), making it an ideal resource for studies of the X-ray properties of particular classes of object. In this paper we outline such a study, concentrating on the discrete X-ray source populations of nearby galaxies[1].

The sample of X-ray observations was drawn from the 486 bright ($B_T \leq 12.5$), predominantly northern galaxies studied by Ho et al. (1997 and references therein, hereafter HFS) in the context of their "search for dwarf Seyferts". The main advantage of using this sample is that Ho et al. have compiled a huge database of both nuclear spectroscopy and host galaxy properties, which is ideal for our purposes. Cross-correlation of the HFS sample and HRI database shows a total of 106 galaxies lying in moderately deep HRI observations (> 10 ks exposure, $< 15'$ off-axis). After optimising our sample for point-source searching by removing fields in which the sensitivity was compromised by the presence of diffuse emission (e.g. giant elliptical galaxies, M 82) we were left with 123 observations of 83 galaxies. We refer to this sample hereafter as the XHFS sample. A comparison with its parent HFS sample shows the XHFS sample is reasonably representative of the galaxy population of the local universe (excluding, of course, giant ellipticals), although possibly subject

---

[1] *ROSAT* HRI data was used in preference to the more sensitive Position Sensitive Proportional Counter (PSPC) data, since superior resolving power is a key requirement for the identification of discrete sources in complicated systems such as nearby galaxies.

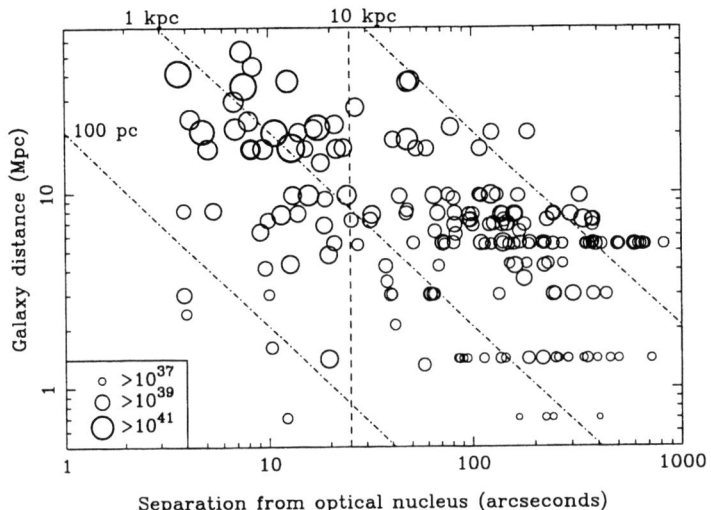

FIGURE 1. The division of the XHFS sources into nuclear and non-nuclear samples, according to whether their positional offset from the optical nucleus of the host galaxy is less than or greater than 25″ (vertical line).

to a slight bias towards high X-ray luminosity systems (e.g. Seyfert nuclei) as may be expected from pointed observation data.

2. THE XHFS SAMPLE

A search for the discrete X-ray sources in the HRI images was conducted using the PSS algorithm (part of the ASTERIX analysis package), and all sources with a detection significance greater than 5σ located within the D_{25} ellipses of the 83 galaxies were catalogued. The initial total of 327 sources was reduced to 187 individual sources after considering multiple observations. The complete source catalogue is available as Appendix A of Roberts & Warwick (2000). The observed count rate of each source was converted to a $0.1 - 2.4$ keV flux, using a standard "AGN-like" spectral form and correcting for Galactic line-of-sight absorption (interpolated from Stark et al. 1992).

Since this study was in part designed to investigate the incidence of low-luminosity active galactic nuclei (LLAGN), this putative population must be separated from the more mundane non-nuclear population. We set a conservative limit on the offset of such sources from the optical nuclear position as 25″, and hence use this limit to define the samples of nuclear X-ray sources (NXS) and non-nuclear sources (Figure 1). A total of 45 NXS are detected in 83 galaxies, a 54% detection rate. The NXS and the non-nuclear source population do not suffer from severe contamination by either foreground or background sources, with upper limits on the contamination set at 12% for the non-nuclear sample, and $< 1\%$ for the NXS.

TABLE 1. X-ray luminosity properties of the nuclear classes.

Nuclear class	Number of objects	Median L_X ($\times 10^{39}$ erg s^{-1})	Detection rate
Seyfert	15	5.4	73%
LINER	15	5.8	73%
Transition	11	5.4	64%
H II	32	< 2.1	41%
NoEL	10	< 0.5	30%

3. RESULTS

3.1. Nuclear X-ray sources

The NXS are observed to be more luminous than their non-nuclear counterparts, with a median luminosity (including 95% upper limits on the non-detections of NXS) a factor 10 higher, implying some sort of enhanced nuclear activity. This is investigated further by considering the HFS spectroscopic classifications of the nuclei. In Table 1 the nuclei potentially holding LLAGN are named in italics; these are clearly more X-ray luminous than the nuclei dominated by star formation activity (H II galaxies), or the inactive (NoEL) nuclei.

The hypothesis that the enhanced X-ray emission may be due to the presence of a LLAGN may be tested by comparison with the properties of "classical" AGN. Specifically, AGN with luminosities in the 10^{40-46} erg s^{-1} regime have an $L_X/L_{H\alpha}$ ratio of the order 1 - 100 (Koratkar et al. 1995). The XHFS nuclei with putative LLAGN all lie in the same parameter range, but extend the relationship down to 10^{38} erg s^{-1}. If this is interpreted as LLAGN activity, then applying the Magorrian et al. (1998) relationships between M_{BH} and L_{bulge} allows a simple estimate of the accretion rate onto the underlying super-massive black hole to be derived for each galaxy, via the fraction of Eddington Luminosity (L_{Edd}) at which the LLAGN is radiating. Using some basic assumptions (e.g. $L_X \sim 0.15$ L_{bol}, Ho (1999); no X-ray absorption), a comparison of the observed luminosity with the implied Eddington value L_{Edd} (Figure 2(a)) demonstrates that the LLAGN are generally accreting at severely sub-Eddington rates (typically $< 10^{-4}$ L_{Edd}), perhaps indicative of ADAF behaviour.

3.2. Non-nuclear X-ray sources

We have investigated the statistical properties of the high L_X discrete source population in nearby spiral galaxies (of Hubble type S0/a – Scd) by constructing a differential luminosity distribution of the non-nuclear X-ray sources in these galaxies in the XHFS sample (Figure 2(b)). Here, the contributions of different galaxies are normalised by their optical blue luminosities (in units of 10^{10} $L_\odot$). XHFS data were used to derive the points above 10^{38} erg s^{-1}, and the M 31 survey of Supper et al. (1997) provides the points at lower L_X values. The dot-dash line shows the

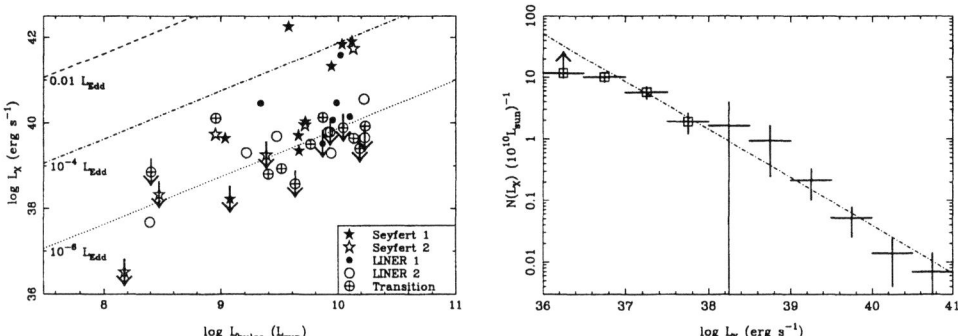

FIGURE 2. (a) Accretion rates of the XHFS LLAGN, shown as fractions of L_{Edd}. (b) The differential luminosity distribution of the non-nuclear X-ray sources in nearby spiral galaxies.

best-fit relationship, which corresponds to $dN/dL_{38} = (1.0 \pm 0.2) L_{38}^{-1.8}$. This implies that there is ~ 1 discrete X-ray source with $L_X > 10^{38}$ erg s^{-1} per 10^{10} $L_\odot$ in the local universe, and since a typical large galaxy may have $L_B \approx 3 \times 10^{10}$ $L_\odot$ then these sources are reasonably common.

A pertinent question is what is the nature of these luminous sources, given that $L_{Edd} \approx 10^{38.3}$ erg s^{-1} for a 1.4 M$_\odot$ neutron star? Possibilities include accreting black holes (e.g. Makishima et al. 2000), X-ray luminous recent supernovae (e.g. Immler & Pietsch 1999), hypernova remnants (Wang 1999), and unresolved source complexes. To address this problem we have initiated a follow-up programme, involving both archival multi-wavelength data and the acquisition of new data. However, since these sources are most readily observed in X-rays, the key to understanding them is likely to emerge from the X-ray variability and spectral data collected by the new generation of X-ray observatories in the forthcoming years.

REFERENCES

Ho L.C., 1999, ApJ, 516, 672
Ho L.C., Filippenko A.V., Sargent W.L.W., 1997, ApJS, 112, 315
Immler S., Pietsch W., these proceedings
Koratkar A., Deustra S.E., Heckman T., et al., 1995, ApJ, 440, 132
Magorrian J., Tremaine S., Richstone D., et al., 1998, AJ, 115, 2285
Makashima K., Kubota A., Mizuno T., et al., 2000, ApJ, 535, 632
Roberts T.P., Warwick R.S., 2000, MNRAS, 315, 98
Stark A., Gammie C.F., Wilson R.W., et al., 1992, ApJS, 79, 77
Supper R., Hasinger G., Pietsch W., et al., 1997, A&A, 317, 328
Wang Q.D., 1999, ApJ, 517L, 27

A NEW EVENT ANALYSIS METHOD FOR THE X-RAY PHOTON COUNT CCD

T.G. Tsuru [1], H. Awaki [1], K. Koyama [1], K. Hamaguchi [1], H. Murakami [1], M. Nishiuchi [1], M. Sakano [1], H. Tsunemi [2]

1) Cosmic Ray Group, Dept. of Physics, Kyoto Univ. Kitashirakawa-Oiwake-Cho, Sakyo, Kyoto, 606-8502, Japan
2) Dept. of Earth and Space Science, Osaka Univ. Machikaneyama-cho, Toyonaka, Osaka, 560-0043, Japan

ABSTRACT   We report here a new event analysis method "two-dimensional Gaussian function fitting" for the X-ray photon count CCD camera. A grade method employed in the ASCA satellite is widely used in the event analysis which excluded the extended event as non X-ray event. Our new method can save extended X-ray events without degradation of the energy resolution and without decrease of the background rejection efficiency. It increases the detection efficiency of X-ray photons by 7.5% at 7 keV and 25% at 12.5 keV, respectively. We will employ this method in the X-ray CCD camera onboard ASTRO-E to be launched in February, 2000.

KEYWORDS: Instrumentation: detectors; Methods: data analysis; Techniques: image processing; Techniques: spectroscopic

1. INTRODUCTION

The solid-state imaging spectrometer (SIS hereafter), on-board the ASCA satellite launched in 1993, was the first X-ray photon count CCD camera in space (Tanaka et al. 1994; Burke et al. 1991). Following ASCA, there are three X-ray astronomical satellites in new generation launched around 2000, Chandra, XMM and ASTRO-E, all of which employ X-ray photon count CCD cameras as their standard detector. These X-ray CCDs have thicker depletion layers (50 μm – 300 μm) than that of SIS (35 μm), allowing us to improve the detection efficiency at high energy X-ray. An X-ray photon absorbed shallow inside the CCD produces a small event size (single pixel event and two pixel split event) while that absorbed deep inside the CCD produces an extended charge cloud in the CCD (multi pixels event) as demonstrated in Figure 1. So far, the ASCA grade method is widely used in the data analysis (Gendreau 1995) in which we regard an event extending over 2×2 pixels as non X-ray. It is useful to apply the data for the CCD with relatively thinner depletion layer. However, the CCD with thicker depletion layer will produce X-ray events which are improper to be analyzed by the grade method. Therefore, we developed a new event analysis method "two-dimensional Gaussian function fitting" to handle

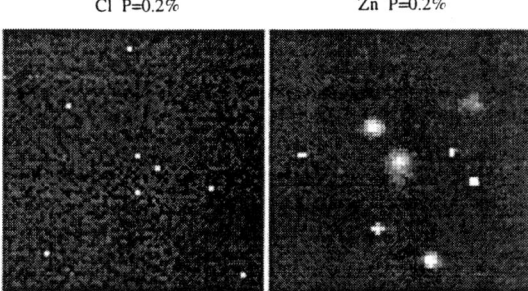

FIGURE 1. The left and right panels show raw frame data of XIS for the chlorine-K line (2.62 keV and 2.82 keV) and the zinc-K line (8.63 keV and 9.57 keV), respectively. Only single pixel events are seen in the figure for the chlorine-K line, while there are some extended events over 2 × 2 pixels for the zinc-K line.

the extended X-ray event properly. This paper describes the introduction and results of this method which will be employed for the X-ray Imaging Spectrometer (XIS hereafter) onboard ASTRO-E (Hayashida et al. 1998; Nishiuchi et al. 1998).

2. THE TWO-DIMENSIONAL GAUSSIAN FUNCTION FITTING METHOD

The new method "two-dimensional Gaussian fitting" is the one in which the 5 × 5 pixel data centered on each event are fitted with a two-dimensional Gaussian function. In this fitting, the center position, normalization and width of the Gaussian function are left as free parameters. The Gaussian function is assumed to be axial symmetric, resulting that the widths in the X- and Y-direction are identical to each other. The pulse height of the event is calculated by integrating the best fit model Gaussian function over the region of 5 × 5 pixels.

2.1. Pulse Height vs. Event Width

The left panel of Figure 2 shows the best fit 1σ width of the Gaussian function (W) as a function of the pulse height (PH) for each illuminated fluorescent X-ray (PH-W diagram hereafter). The distribution for each incident X-ray energy breaks largely at the width of 0.4 pixel (9.6μm).

In the region below the width of 0.4 pixel, the pulse heights of X-ray events are distributed sharply and independent from the width in the horizontal direction, which suggests that the pulse heights of such events are represented well by the fitting method. Figure 2 shows that the relative number of events with large widths for the chlorine-K line is smallest when compared with the other X-ray energies. Since the X-ray energy of the chlorine-K line (2.62 and 2.82 keV) is just above the silicone edge of 1.846 keV, its attenuation length in the CCD is smallest among those of the X-ray energies we illuminated. Therefore, this result suggest that the event with large width is produced by X-ray event absorbed deep inside the CCD.

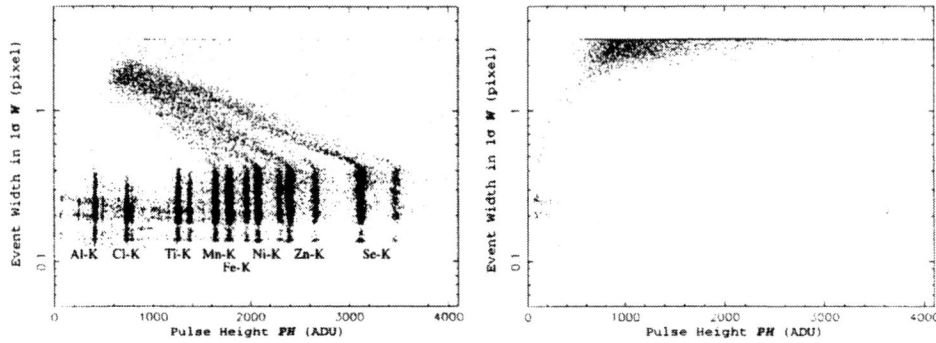

FIGURE 2. The correlation between the event pulse height and the width of the best-fit Gaussian function when illuminated various X-ray energies (left) and ^{60}Co as a non X-ray background simulator (right). The event pulse height is obtained by integrating the best-fit model function over 5 × 5 pixels.

In the region above the width of 0.4 pixel, the pulse height of an X-ray event becomes smaller as the width increases. Expanding the integration region of the Gaussian function from 5 × 5 pixels to 11 × 11 pixels changes this diagram little. Thus, such an largely extended event is thought to be produced by an X-ray absorbed in the field free region and loses some of its electron-hole pairs by recombination (Hopkinson 1983; Hopkinson 1987). The event with the width of 0.4 pixel is produced just at the boundary between the depletion layer and the field free region of the CCD. The properties of these X-ray events will be described in the next paper (Murakami et al. 1999).

The right panel of Figure 2 shows the diagram obtained by illumination of ~ 1 MeV γ-rays from ^{60}Co, which is often used for non X-ray background simulation. The correlation for the γ-rays is very different from that for X-rays. Thus, the PH-W diagram is also useful for non X-ray background rejection.

2.2. Improvement of Quantum Efficiency

We obtained the X-ray energy resolution and quantum efficiency for each X-ray energy from the events with the width smaller than 0.4 pixel. We found that the quantum efficiency increases by 9% at 13 keV without any degradation of energy resolution compared with the ASCA grade method.

Many events are still left unused in the region above the width of 0.4 pixel. We then try to utilize somehow extended events in the region between widths of 0.4 and 0.6 pixel for the improvement of the quantum efficiency. By fitting analytic function to the relationship of PH-W for each incident X-ray energy in this region, we obtained the fraction $f(W, PH)$ of remaining charge after the recombination in the field free region to the initial amount of charge. We next estimated the initial among of charge for each event by dividing PH by $f(W, PH)$. Plotting the

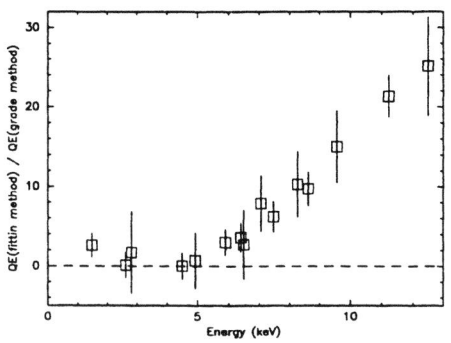
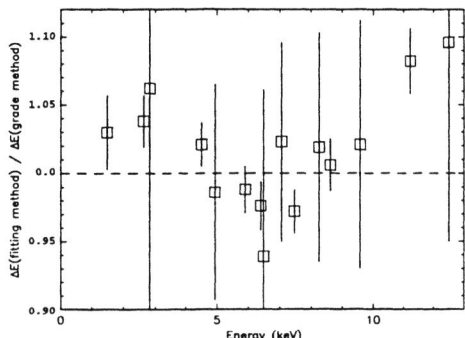

FIGURE 3. The results obtained from events with the width less than 0.6 pixel are displayed. Left: The ratio of the quantum efficiency of the two-dimensional Gaussian fitting method to that of the grade method. Right: The ratio of the energy resolution of the two-dimensional Gaussian fitting method to that of the grade method.

estimated initial amount of charge in the diagram again through this method, we finally obtained X-ray spectra from this "recovered" PH-W diagram.

The ratio of the quantum efficiency obtained with this method to that with the ASCA grade method in the left panel of Figure 3, where the quantum efficiency increases by 7.5% at 7 keV and 25% at 12.5 keV, respectively. On the other hand, there is little degradation of energy resolution (the right panel of Figure 3). Thus, it is concluded that the two-dimensional Gaussian fitting method is useful for the increase of the quantum efficiency of an X-ray CCD camera without degrading its energy resolution.

3. CONCLUSION

The two-dimensional Gaussian fitting method newly introduced in this paper gives us higher quantum efficiency by 7.5% at 7 keV and 25% at 12.5 keV without the degradation of the energy resolution compared with the grade method.

REFERENCES

Burke, B.E., Mountain, R.W., Harrison,D.C., et al. 1991, IEEE Trans. ED-38, 1069
Gendreau, K.C. 1995, Ph D. Theses, MIT
Hayashida, K., Kitamoto, S., Miyata, E., et al. 1998, Proc. SPIE, 3445, 278.
Hopkinson, G.R. 1983, NIM, 216, 423.
Hopkinson, G.R. 1987, Optical Engineering, 26, No.8
Nishiuchi, M., Koyama, K., Tsuru, T. et al. 1998, Proc. SPIE, 3445, 268.
Murakami, H. et al. 1999, in preparetion.
Tanaka, T., Inoue, H., Holt, S.S., 1994, PASJ 46, L37

PROBING DENSE MATTER IN THE CORES OF AGN: OBSERVATIONS WITH RXTE AND ASCA

K. A. Weaver

NASA/GSFC

ABSTRACT

Preliminary results from an X-ray spectral study of Seyfert 1 galaxies with $ASCA$ and $RXTE$ are presented. From an analysis of X-ray reprocessing features of Compton reflection and Fe Kα fluorescence, it is found that iron line strength is not necessarily a good predictor of the amount of reflection. The variability properties of Fe Kα and reflection do not necessarily scale together and substantial decoupling of the behavior of the reprocessed flux with respect to continuum variability is common. Such trends suggest the presence of multiple and/or complex regions of dense matter in AGN cores and that standard, simple accretion disk models drastically oversimplify reality.

KEYWORDS: galaxies: active — galaxies: individual (MCG −2-58-22) galaxies: individual (MCG −5-23-15) — galaxies: nuclei — galaxies: Seyfert — X-rays: galaxies

1. INTRODUCTION

X-ray observations are excellent for studying the accretion process in active galactic nuclei (AGN) because they allow us to see directly into the galaxy core to determine what happens around the black hole, all the way from the gravitational radius out to scales ∼10 billion times larger. X-rays can arise anywhere there is matter to reprocess high-energy continuum photons. Features of X-ray reprocessing such as Fe Kα lines and Compton reflection (Lightman & White 1988, Guilbert & Rees 1988), can arise from an accretion disk (George & Fabian 1991) or the obscuring torus of unified models (Krolik, Madau & Zycki 1994, Ghisellini, Haardt & Matt 1994). Significant Fe Kα emission may also be produced in distributions of clouds such as the UV/optical broad line region.

Spectral variability is critical to understanding where reflection/fluorescence features arise. If the reprocessing arises in an accretion disk, we expect it to respond to the continuum with a lag of only $\sim 3000 M_8$ s, where M_8 is the mass of the central black hole in units of $10^8 M_\odot$. If it comes from as far out as an obscuring torus, the lag should be $\sim 5 \times 10^7 L_{44}^{1/2}$ s, where L_{44} is the luminosity in units of 10^{44} erg s^{-1}. For current favored models of Seyfert 1s where the observed reprocessing occurs in a neutral and Compton-thick accretion disk, the iron line and reflection hump should be physically coupled.

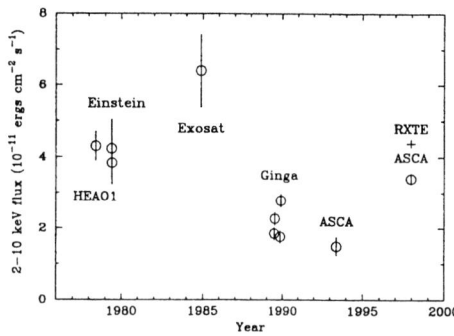

Figure 1: Long-term variability of MCG −2-58-22.

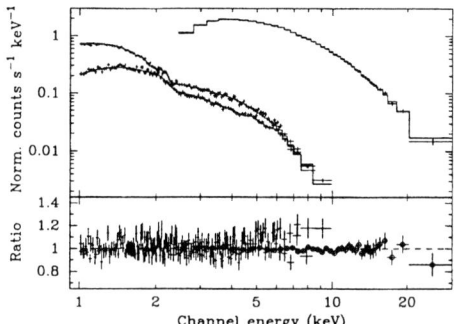

Figure 2: ASCA and RXTE PCA data and best-fitting model folded through the instrumental response. The model consists of a power law and a broad Gaussian ($\sigma = 0.28$ keV). The ratio of the data to the model is shown in the lower panel; circles are RXTE data.

1.1. MCG −2-58-22: A "Reflection-less" Seyfert 1 Galaxy?

MCG −2-58-22 is a bright, unabsorbed Seyfert 1 galaxy with evidence for a Doppler broadened iron line (FWHM of $31,000^{+30,000}_{-18,000}$ km s^{-1}; Weaver et al. 1995). It was observed with ASCA and RXTE for a total of 38 ks on December 15, 1997.

MCG −2-58-22 varies significantly on timescales of years (Figure 1), changing intensity by a factor of three between ASCA observations in 1993 and 1997. The ASCA + RXTE spectrum and best-fitting model are shown in Figure 2. Using the pexrav model in xspec and assuming a face-on disk (where R = 1 for $\Omega/2\pi$ coverage), RXTE detects essentially no reflection, with an upper limit of R = 0.08. Comparing ASCA observations, the iron line normalization stays constant at $\sim 5 \times 10^{-5}$ photons cm^{-2} s^{-1} and the EW drops from ~ 230 eV to ~ 125 eV in four years.

The lack of reflection is puzzling if the broad iron line comes from a disk. Although a large disk inclination or overabundance of iron can suppress reflection, the iron line would be much broader than observed in the former case and much stronger in the latter. Relativistic smearing of the Compton hump is also not likely because this would smear out the line. Plausible explanations include a partly ionized accretion disk, a cutoff in the intrinsic spectrum at energies between 10 and 40 keV, a Compton-thin accretion disk, or a lag in the response of reflection to the continuum (larger than that of the iron line).

The spectrum can be more easily explained if the iron line originates elsewhere. The constant line strength suggests that the line originates at least a few light years away. In fact, the line width is marginally consistent with the inferred velocity of the UV/optical broad line region clouds (16,000 km s^{-1}; Wu, Boggess & Gull 1981). BLR clouds with $N_H \sim 10^{23}$ cm^{-2} would produce fluorescence but no reflection.

1.2. The Contribution of Non-Disk Emission to X-ray Reprocessing

An obvious complication for spectral studies is the contribution of reprocessed emission from non-disk regions. To study this, we have looked at 15 Seyfert 1 galaxies with multiple observations in the ASCA archive. (Gelbord and Weaver, this volume). Modeling the iron line with a narrow Gaussian (to fit the line core), we

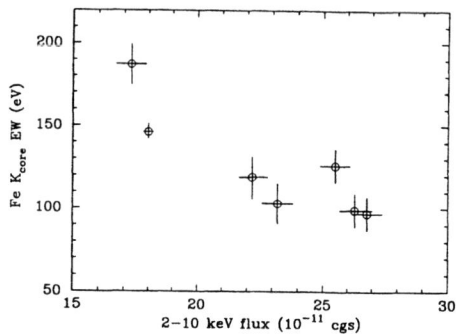

Figure 3: Equivalent width of the Fe Kα line core (measured with a narrow Gaussian) vs. source flux for NGC 4151.

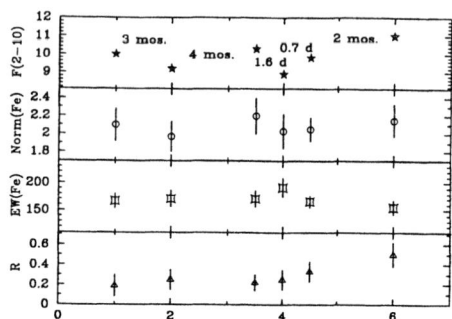

Figure 4: *RXTE* measurements for MCG −5-23-16. Four observations, approx. 2 to 4 months apart are represented, with the third divided into three flux states across ∼ 2 days. Units for panels 1 − 3 from top to bottom are 10^{-11} ergs cm^{-2} s^{-1}, 10^{-4} photons cm^{-2} s^{-1}, and eV.

find that eleven galaxies show some evidence for a change in EW. This can be explained as a significant time delay in flux from the line core, which might be due to emission from the outer disk, the BLR or the torus. One of the best examples of non-variability is NGC 4151, for which the line core flux does not change and the the EW is inversely proportional to the continuum flux (Figure 3).

In a second study, we obtained four *RXTE* observations of MCG−5-23-16 in 1996 and 1997 spaced ∼ 3 months apart. The spectral results are plotted in Figure 4. Between the third (a long-look covering two days) and fourth pointings, the 2 − 10 keV flux decreases while the line normalization remains the same and the EW drops from about 200 eV to 150 eV (Mattson and Weaver, in prep). The lack of response to the continuum suggests that at least some of the line emission occurs from far out. However, the reflection component tracks the continuum during the same time. If real, this would strongly support the idea that MCG −5-23-16 has *two* physically distinct reprocessing regions. Similar behavior of the iron line not tracking reflection is seen by *RXTE* for NGC 5548 (Chiang et al. 1999).

1.3. Compton Reflection and *RXTE*

Preliminary *RXTE* and *ASCA* results suggest that Fe Kα emission and Compton reflection are at least partly decoupled in their behavior. Such a result has strong implications and thus deserves close scrutiny. In particular, it is important to rule out any possible systematic calibration effects. The PCA detector calibration is still being refined; however, up to December 1997, the *ftools* v4.1 matrix for gain epoch 3 (4/15/96 − 3/22/99) provides good fits to the Crab spectrum with less than 4% residuals near 6 − 7 keV and 10 − 20 keV (K. Jahoda, private communication).

Table 1 compares results for a handful of Seyfert 1 galaxies observed during 1996 and 1997. Data were fitted with the *pexrav* plus Gaussian models in *xspec* and response matrices were generated for each observation with *ftools* v4.1. MCG −2-58-22 was observed at the end of 1997, where response deviations from the Crab begin, but there is no clear systematic trend for R to decrease with time. It is possible that R is systematically underestimated (the mean for IC 4329a is 0.43, where R=1

TABLE 1. Compton Reflection Fits to RXTE Seyfert 1s.

Galaxy	Date	PCA rate (count s^{-1})	R	F(2 − 10) (10^{-11} cgs)
NGC 7469	1996 Apr 25 − 26	9	1.10(0.53 − 1.86)	2.9
Mrk 509	1996 Nov 04 − 06	18	0.30(0.07 − 0.45)	5.9
MCG −5-23-16	'96 Apr 24 − '97 Jan 10	33	0.29(0.17 − 0.42)	9.8
IC 4329a	1997 Aug 11	29	0.14(0.00 − 0.80)	10.0
IC 4329a	1997 Aug 22	47	0.46(0.27 − 0.68)	16.0
IC 4329a	1997 Sep 01	36	0.39(0.12 − 0.78)	11.8
IC 4329a	1997 Sep 11	49	0.46(0.22 − 0.76)	17.5
IC 4329a	1997 Sep 18	30	1.04(0.44 − 1.92)	9.9
IC 4329a	1997 Oct 02	43	0.07(0.00 − 0.45)	15.3
MCG −2-58-22	1997 Dec 15 − 17	12	0.02(0.00 − 0.08)*	3.4

Errors are 90% confidence for two parameters. *A 4% systematic error is included.

for an accretion disk), but this cannot account for the non-detection in MCG −2-58-22, even after including a 4% systematic error. We conclude that there is no time-dependent (i.e., calibration-dependent) trend for R to decrease with time.

2. CONCLUSIONS

Results are summarized from an *RXTE* and *ASCA* investigation of spectral variability in Seyfert 1 galaxies. The spectra defy a simple interpretation, with significant time lags and the Fe Kα line and Compton reflection behaving in a decoupled fashion. A possible explanation is that the spectral variability of the accretion-disk is diluted by emission from other, larger regions.

ACKNOWLEDGEMENTS

This work is supported by NASA grants NAG5-7010, NAG5-6917 & NAG5-3504.

REFERENCES

Chiang, J. et al. 2000, ApJ, 528, 292
George, I. M. & Fabian, A. C. 1991, MNRAS, 249, 352
Ghisellini, G., Haardt, F. & Matt, G. 1994, MNRAS, 267, 743
Guilbert, P. W. & Rees, M. J. 1988, MNRAS, 233, 475
Krolik, J., Madau, P. & Zycki, P. 1994, ApJ, 420, L57
Lightman, A. P. & White, T. R. 1988, ApJ, 335, 57
Weaver, K. A., Nousek, J., Yaqoob, T., Hayashida, K. & Murakami, S. 1995, ApJ, 451, 147
Wu, C., Boggess, A. & Gull, T. R. 1981, ApJ, 247, 449

POSTERS

WARM COMPTONIZATION IN AGN: EFFECT ON THE IRON Kα LINE AND THE LYMAN EDGE

A. Abrassart [1], A.M. Dumont[1]

1)Observatoire de Paris, Section de Meudon, Place Janssen, F-92195 Meudon, France

ABSTRACT   It has been suggested that the broad fluorescent X-ray Iron line profile observed in Seyfert 1 galaxies could be explained by Compton scattering in a hot Thomson thick medium, either by reflection (Abrassart 99) or by transmission (Misra and Kembhavi 1998). Here, using our transfer codes TITAN and NOAR (Dumont et al. 1999), we show that for the latter, the iron line and edge can be made consistent with the observed ones, provided the ionization parameter ξ exceeds 10^6 erg.cm.s^{-1}, and the incident spectra has a low enough Compton temperature. This model implies a small black hole mass and an accretion rate close to the Eddington value. A very important effect of such a medium on the line of sight of the central UV/X-ray source is to smooth any intrinsic Lyman edge, thus solving a long standing puzzle.

KEYWORDS: galaxies: Seyfert; X-rays: galaxies; line: profiles; scattering

1. INTRODUCTION.

Active galactic nuclei are thought to be powered by accretion onto a super massive black hole and radiate a high fraction of their luminosity in the UV and X-ray band. Modelling this continuum emission requires at least two phases of plasma with different temperatures. Variability studies have shown that they are both located within hundred gravitational radii from the black hole. Proposed models often involve a hot (10^9 K) corona above a cold (10^5 K), geometrically thin and optically thick accretion disk. The latter could be responsible for the "UV bump", while the former would produce the observed high energy power law via unsaturated, thermal inverse comptonisation of the UV seed photons.

It should be noted that this geometry is not well established, particulary for the inner parts of the disk, because of several instabilities that can arise when the accretion flow is radiation pressure dominated. However, the simplest two phases phenomenogical model, together with the condition that the hot and cold gas must lie within a relatively small volume, makes it likely that each phase reprocess the emission from the other one.

Here we investigate two key features of AGN spectra: the X-ray fluorescent iron Kα line profile and the absence of Lyman break in the UV band. The skewness of the former has been interpreted in terms of relativistic Doppler and gravitationnal shifts affecting line emission from cold disk annuli lying very close to the black hole

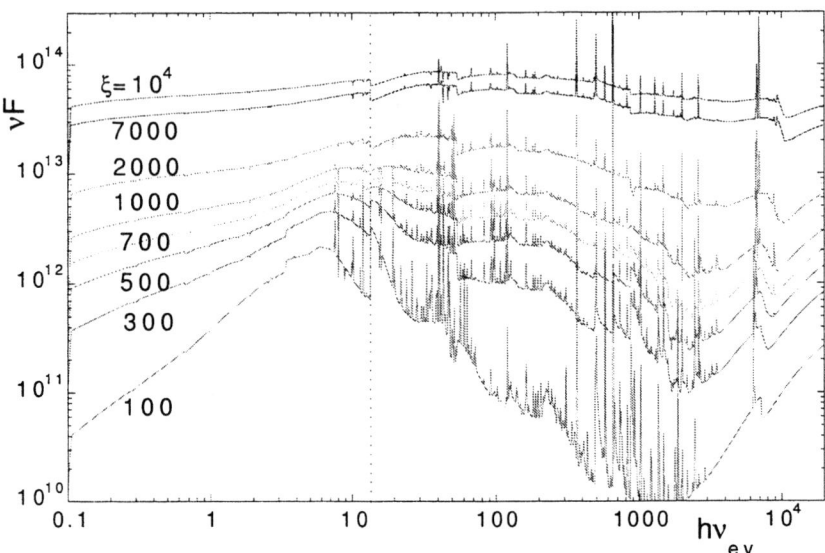

FIGURE 1. Pure reflected spectra from a cosmic abundance slab of gas, computed with the code TITAN for different ξ, using a power law incident spectrum of photon index $\Gamma=2$. Spectral resolution=100, column density=$10^{26} cm^{-2}$, n=$10^{12} cm^{-3}$.

(Fabian et al. 1989; Tanaka et al. 1995). This interpretation of line profile requires the presence of the acccretion disk (at least) down to the radius of marginal stability around a non rotating black hole, i.e. 3 Schwarzshild radii.

We aim at testing an alternative hypothesis, namely that Compton broadening in material on the line of sight is responsible for the asymmetric broadening. We first recall the physical constraints derived under this assumption. We then show our simulation of the transmitted Fe K and Lyman band spectra through such material. Finally, we briefly discuss the ways to discriminate between different models.

2. LYMAN EDGE

There is a strong observational constraint on spectral modeling of AGN: there has been virtually no detection of a significant intrinsic Lyman edge. This is in severe contradiction with photoionization computations, which predict that under general conditions, the reprocessing of hard X-rays by relatively cold matter is associated with a break at 13.6 eV (see Koratkhar and Blaes, 1999, for a review). As shown on fig. 1, this discontinuity can be in emission or in absorption in the reflected spectrum, depending on the ionization parameter ξ ($\xi = 4\pi F/n$, where F is the incident bolometric flux, and n the hydrogen number density of the reprocessing mater).

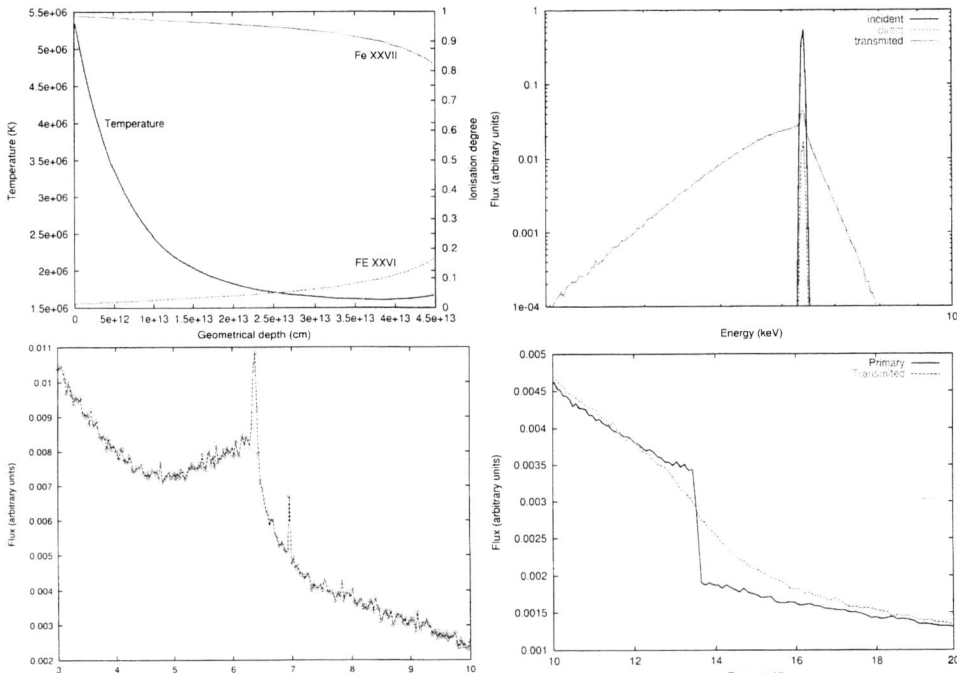

FIGURE 2. Upper left: temperature and ionisation strucure of an extended shell with cosmic abundance, Thomson depth 3, density varying as $\frac{1}{R^2}$, and $\xi = 2\ 10^6$ at the inner, illuminated edge. The incident spectrum is the RQQ composite of Laor et al. (1997), with a high energy cutoff at 100 keV. Upper right: profile of an extrinsic cold narrow (gaussian of width 25 eV) 6.4 keV iron line, transmitted through this same material. Lower left: same, but including a transmitted extrinsic power law continuum ($\Gamma=2$). Note the narrow intrinsic contribution at 6.9 keV. Lower right: effect of the same material on an extrinsic Lyman edge in absorption.

3. COMPTON BROADENING MECHANISM FOR THE FE Kα LINE

There are two basic requirements to reproduce the profile in this scenario. First, there must be a significant Thomson depth of very ionized material on the path of the photons. It must be both Thomson thick and ionized so that enough scattering occurs, as comptonization is in competition with photoelectric absorption. Secondly, because the red wing must be more prominent than any blue wing, the ionized medium must be at a low enough temperature so that direct Compton ($\propto \frac{h\nu}{m_e c^2}$) dominates over inverse Compton ($\propto \frac{4kT_e}{m_e c^2}$) at $h\nu$=6.5 keV. This places an upper limit on the temperature of a few 10^6 K. Such non LTE conditions can be met in a photoionized plasma. The mean RQQ spectrum as derived by Laor et al (1997), for instance, exhibits an ample UV bump, which lowers the Compton temperature in the required range. Fig. 2 shows the temperature and ionization profile of an extended optically thin halo of moderately thick material exposed to such a continuum, as well as its effect on sharp extrinsic features that should be produced via cold reflection on thick matter within it. The smearing of the lyman edge would also operate on a discontinuity in emission, although with reduced efficiency, because inverse Compton dominates at 13 eV.

4. CONCLUSION

A Compton broadening interpretation of the AGN Fe Kα requires that the comptonising matter be exposed to an intense ionizing continuum with a strong UV bump. One asset of this scenario is that it can help to solve the Lyman problem. The very high value of the required ionisation parameter can only be reconciled with variability constraints (namely that line variations should not be washed over timescales greater than a few ks, hence, given the required Thomson depth, a lower limit n~few 10^{10} cm^{-3}) if the material lies close to the central black hole. Indeed, given an observed bolometric luminosity of $\sim 10^{44}$ $erg.s^{-1}$, $\xi = 2.10^6$ $erg.cm.s^{-1}$ then implies that it can not be more than 10^{14} cm away from a central source. We conclude that the Compton mechanism can only be dominant (with respect to gravitational redshift) if the black hole mass does not exceed a few 10^6 $M_\odot$, which in turn requires an Eddington ratio of the order of unity. We note that recent variability studies seems to point to such small masses (Nowak and Chiang 1999).

REFERENCES

Abrassart A. 1999 Proceeding of the 32nd Cospar scientific assembly, AdSR, 25, 465
Dumont, A.M., Abrassart A., Collin S. 1999, A&A, 357, 823
Fabian A.C., Rees M.J., Stella L., White N.E. 1989, MNRAS, 238, 729
Koratkar, A., Blaes, O. 1999 PASP, 111, 1
Laor, A. et al. 1997 ApJ, 477, 93
Misra, R., Kembhavi, A. K. 1998, ApJ, 512, 340
Nowak, MA., Chiang J. 2000, ApJ, 531, L13
Tanaka et al. 1995, Nature, 375, 659

AVERAGE PROPERTIES OF GAMMA-RAY BURSTS SPECTRA FROM 2 TO 700 KEV WITH BEPPOSAX

L. Amati [1], F. Frontera [1], D. Dal Fiume [1], M. Orlandini [1], E. Palazzi [1], E. Costa [2], M. Feroci [2], P. Soffitta [2], M.N. Cinti [2], J. Heise [3], J. in 't Zand [3], J.M. Muller [3], L. Nicastro [4] and M. Tavani [5]

1) Istituto T.E.S.R.E., CNR, Bologna, Italy
2) I.A.S., CNR, Roma, Italy
3) S.R.O.N., Utrecht, Netherlands
4) I.F.C.A.I., CNR, Palermo, Italy
5) I.F.C.T.R., Milano, Italy

ABSTRACT

The co-alignment of two detectors of the BeppoSAX Gamma-Ray Burst Monitor (40–700 keV) with the Wide Field Cameras (2–26 keV) on-board the same satellite has permitted simultaneous detection of several GRBs in X and gamma rays and their localization with high precision. These measurements allow us also to extend the study of GRB spectra down to 2 keV and with good accuracy thanks to the knowledge of source direction. We show and discuss preliminary results of the analysis of X to gamma-rays average spectra based on a sample of 12 BeppoSAX events.

KEYWORDS: gamma-rays:bursts; X-rays:general

1. SCIENTIFIC CASE

Two of the four detection units of the Gamma Ray Burst Monitor (GRBM, 40–700 keV energy band, Frontera et al. 1997) on-board the Italian–Dutch X–ray astronomy mission BeppoSAX (Boella et al. 1997) are co-aligned with the two Wide Field Cameras (WFC, 1.5–26 keV energy band, Jager et al. 1997) on-board the same satellite. This gives to the mission the possibility not only of localizing GRBs entering in the WFC 20 x 20 degrees (FWHM) field of view and detecting associated X–ray afterglow emission (e.g. Costa et al. 1997, Piro et al. 1998a, Nicastro et al. 1998) but also of performing broad band spectral study of these events.

The joint X and gamma ray emission study of GRB is a crucial diagnostic of theoretical models for the main mechanism responsible of the emission, the most popular of which are Synchrotron Shock Models (SSM, e.g. Tavani 1997, Sari et al. 1998), in which the kinetic energy of a relativistic expanding fireball (e.g. Meszaros and Rees 1992) is dissipated through synchrotron emission in the presence of a weak to moderate (in the case of an 'external' shock with the interstellar medium)

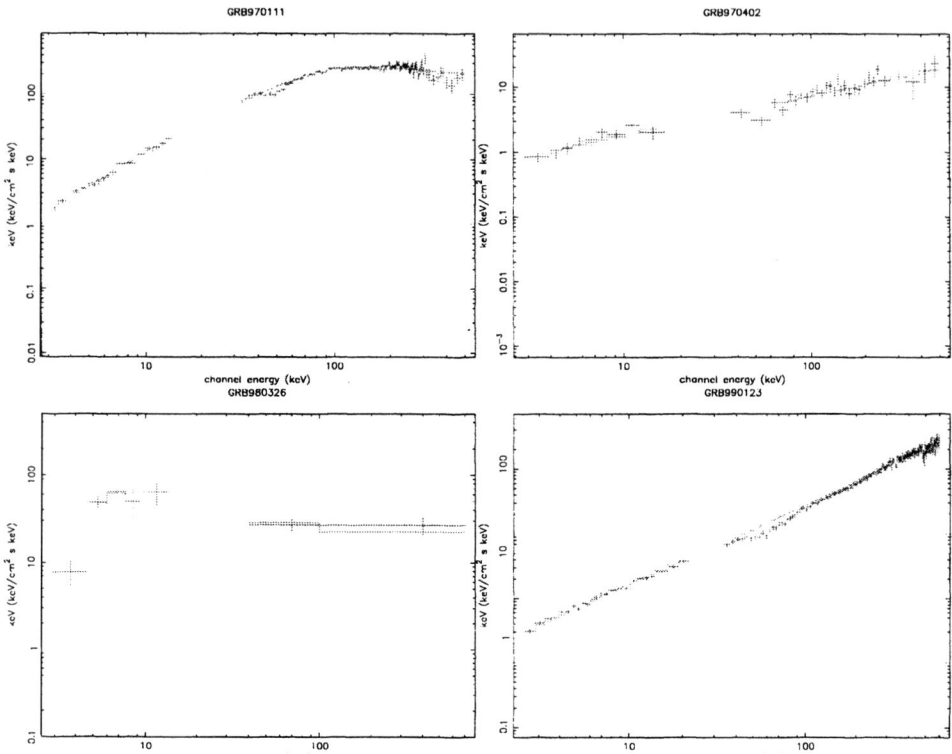

FIGURE 1. Examples of 2–700 keV GRB $\nu F\nu$ spectra with different shapes: GRB970111 (top left) is a typical example of the case in which the peak energy is *inside* the WFC+GRBM energy band, GRB990704 and GRB990123 (top right and bottom right) are outstanding examples of the power–law spectra with the peak energy *above* the WFC+GRBM energy band, and GRB980326 (bottom left) is an example of a power–law spectrum with peak energy *below* the WFC+GRBM energy band and with strong *intrinsic* absorption

or strong (in the case of fireball 'internal shocks') magnetic field (e.g. Waxman 1997).

2. DATA ANALYSIS, RESULTS AND DISCUSSION

The GRBM response function, calibration and spectral data analysis techniques are discussed in Amati et al. (1997) and Amati et al. (1999). For the WFC, details on response function, data reduction and background subtraction can be found in Jager et al. (1997). The two instruments have been cross–calibrated by means of simultaneous observations of the Crab nebula spectrum.
We used XSPEC software package (issue 10) (Arnaud 1996) to deconvolve count

TABLE 1. 2–700 keV average spectra fits results

GRB	nH/nH$_{gal}$	Γ	$\chi^2_{\nu,PL}$	α	β	E$_0$ (keV)	$\chi^2_{\nu,Band}$	$\chi^2_{\nu,BB}$
970111	~1	1.23 ± 0.05	9	−0.21 ± 0.08	−2.18 ± 0.07	78 ± 5	1.1	7.7
970228	~1	1.76 ± 0.13	1.3	−1.34 ± 0.07	−1.91 ± 0.05	43 ± 4	0.98	4.3
970402	~1	1.35 ± 0.08	0.96	~−Γ	~−Γ	>700	0.98	5.5
970508	~1	1.56 ± 0.07	0.89	~−Γ	~−Γ	>700	0.96	8.4
980326	607 ± 151	2.76 ± 0.12	0.29	~−Γ	~−Γ	<2	0.6	15
980329	279 ± 56	1.13 ± 0.03	1.1	−0.54 ± 0.28	−1.33 ± 0.03	78 ± 40	0.98	92
980425	~1	1.82 ± 0.04	1.6	−1.42 ± 0.18	−2.01 ± 0.18	52 ± 20	0.7	17
980613	~1	1.0 ± 0.1	0.73	~−Γ	~−Γ	>700	0.76	1.6
981226	~1	2.2 ± 0.1	2.7	−1.82 ± 0.14	−3.24 ± 0.70	72 ± 35	0.22	9.6
990123	~1	0.78 ± 0.07	1.05	~−Γ	~−Γ	>700	0.87	109
990217	~1	1.63 ± 0.05	1.3	~−Γ	~−Γ	>700	2.6	18
990705	~1	1.01 ± 0.09	1.06	~−Γ	~−Γ	>700	0.96	201

rate spectra, assuming a fixed theoretical model as input function. The average spectra of 12 WFC+GRBM have been fitted with 3 'basic' models:

- a photo–electrically absorbed power–law:

$$N(E) = K \exp(-\sigma n_H) E^{-\Gamma} \qquad (1)$$

where K is in photons cm^{-2}s^{-1}keV^{-1}, $\sigma(E)$ is the photo–electric cross section of a gas with cosmic abundance and n$_H$ is the equivalent hydrogen column density;

- the function proposed by Band et al. (1993), an empirical form which fits well BATSE/CGRO data and also Ginga data (but with lower values of E$_0$):

$$N(E) = \begin{cases} A \left(\frac{E}{100keV}\right)^\alpha \exp\left(-E/E_0\right) & (\alpha - \beta) \cdot E_0 \geq E \\ A \left[\frac{(\alpha-\beta)E_0}{100keV}\right]^{\alpha-\beta} \exp(\beta - \alpha) \cdot \left(\frac{E}{100keV}\right)^\beta & (\alpha - \beta) \cdot E_0 \leq E \end{cases} \qquad (2)$$

;

- a black–body with temperature kT in keV.

As can be seen from Table 1, the data are best fitted by the non–thermal models, the power–law and the Band form, while blackbody emission is ruled out. The reported errors are at 1σ. For the spectra fitted by the Band form, we find that extending GRB spectra study down to X–rays gives systematically lower values of E_0 with respect to those found by fitting BATSE data (> 20 keV) when E_0 is inside the detectors energy band. Indeed, we find that more than half of the events in our sample have 2–700 keV spectra well consistent with a single power–law, indicating that in these cases the bending energy is above (or below, in the case of GRB980326) our energy band. Actually, spectral evolution studies have been performed for some of the GRBs in our sample (Frontera et al. 1999), showing that in many cases E_0 evolves rapidly towards lower values. In some cases, E_0 enters the GRBM band only at the end of the event, when the flux is very low, and thus the average spectrum is dominated by the emission occurred when $E_0 > 700$ keV.

Our broad band average spectral data of GRBs are confirming the general validity of synchrotron shock emission models, which predict average spectra well consistent with the Band form, but remarkably also the need of corrections to them to explain X-ray features like the too low values of the low energy photon index (GRB970111 and GRB980329, see figure on 'the line of death of synchrotron') and the exceptionally high values of n_H found for GRB980326 (for GRB980329 the data do not allow us to discriminate between a strongly absorbed power–law and a Band form). Further data analysis with less simple models, e.g.. internal shocks spectral models (e.g. Pilla and Loeb 1998) and comptonized spectrum models is in progress.

REFERENCES

Amati, L. et al. 1997, SPIE Proceedings, 3114, 176

Amati, L. et al. 1999, A&AS, 138, 403

Arnaud, K.A. 1996, Astronomical Data Analysis Software and Systems V, eds. Jacoby, J. and Barnes, J., ASP Conf. Series 101, 17.

Band, D. et al. 1993, ApJ, 413, 281

Boella, G. et al. 1997a, A&AS, 122, 299

Costa, E. et al. 1997, Nature, 387, 783

Frontera, F. et al. 1997a, A&AS, 122, 357

Frontera, F. et al. 1998, ApJ, 493, L67

Frontera, F. et al. 1999, ApJ, in press

Jager, R., et al. 1997, A&AS, 125, 557

Mészáros, P. and Rees, M.J. 1992, ApJ, 397, 570

Nicastro, L. et al. 1998, A&A, 338, L17

Piro, L. et al. 1998, A&A, 329, 906

Pilla, R.P. and Loeb, A. 1998, ApJ, 494, L167

Sari, R., Piran, T. and Narayan, R. 1998, ApJ, 497, L17

Strohmayer, T.E. et al. 1998, ApJ, 500, 873

Tavani, M. 1997, ApJ, 483, L87

Waxman, E 1997, ApJ, 485, L5

HIGH ENERGY SPECTRA OF SEYFERT 2 GALAXIES

L. Bassani[1], G. Malaguti[1], A. Malizia[2,3], J.B. Stephen[1], E. Caroli[1], G. Di Cocco[1], F. Frontera[1,4], M. Trifoglio[1]

1) Istituto TeSRE/CNR, Via Gobetti 101, 4019 Bologna, ITALY
2) Department of Physics, The University, Southampton, UK
3) BeppoSAX Science Data Centre, Rome, ITALY
4) also at Dipartimento di Fisica, Universit di Ferrara, Ferrara, ITALY

ABSTRACT The PDS instrument onboard BeppoSAX has offered the unprecedented possibility of studying the spectrum of Seyfert 2 galaxies in the 15–100 keV region. Browsing the BeppoSAX public database, we have obtained a catalogue containing 17 Seyfert 2 galaxies detected by the PDS. The 15-100 keV spectra of these sources has been analysed in order to define their high energy continuum (spectral shape, reflection features, and so on), and to look for commonalities or differences among them. The preliminary results of this work are presented.

1. INTRODUCTION

Recent BeppoSAX observations of Seyfert 2 galaxies have demonstrated that these objects can be powerful hard X-ray emitters (above 10 keV) even though their 2–10 keV radiation is severely attenuated by absorption in thick material (Bassani, Cappi and Malaguti 1999). The high sensitivity of the Phoswich Detector System (PDS) instrument (Frontera et al. 1997) provides an opportunity to improve our understanding of the high energy characteristics of this type of objects significantly. Herein we present the results from the systematic data analysis of a large sample of Seyfert 2 galaxies observed with the PDS.

The sample of objects analysed consists of 33 PDS observations of 27 objects; the majority of the sources are of type 2 except for 5 objects of type 1.8/1.9. Two have also Starburst characteristics and 7 are NELG (or narrow emission line galaxies discovered by means of their bright X-ray emission). These data sets are available from the BeppoSAX public archive (www.sdc.asi.it/archive) as of May 1999 except for a few sources for which we have proprietary data. Our intention is to provide a database at high energies and to study the mean properties of Seyfert 2's above 10 keV: thus we do not include a discussion nor do we examine the broad band properties of individual sources. We note however that many of these observations have been published individually and the most relevant reference for each source is given in Table 1.

TABLE 1 : The PDS observation log

Source name	Flux[a]	Obs. Date	Exp. Time ($\times 10^4$ s)	Count rate counts/s	Ref. [b]
NGC 526A	3.8	31 Dec 1998	4.42	0.38±0.02	
NGC 1068	≤1.4	11 Jan 1998	1.77	0.11±0.04	1
NGC 1365	5.3	12 Aug 1997	1.30	0.59±0.05	2
NGC 1386	1.9	10 Dec 1996	1.36	0.21±0.05	2, 3
NGC 2110	6.0	12 Oct 1997	3.87	0.67±0.03	4
MKN 3	10.7	16 Apr 1997	4.48	1.07±0.03	5
NGC 2273	≤3.8	22 Feb 1997	0.99	0.01±0.07	3
ESO 428−G014	≤3.5	21 Nov 1997	0.35	0.11±0.09	3
IRAS 09104+4109	1.3	18 Apr 1998	3.43	0.11±0.03	6
NGC 2992	1.5	01 Dec 1997	3.32	0.16±0.03	7
NGC 3081	≤2.6	20 Dec 1996	0.70	0.10±0.08	3
NGC 3147	≤2.3	15 Nov 1997	1.60	0.06±0.04	
NGC 3393	≤2.9	08 Jan 1997	0.72	0.23±0.08	3
NGC 4388	17.4	09 Jan 1999	5.12	1.75±0.02	
NGC 4507	15.6	26 Dec 1997	2.70	1.64±0.03	
NGC 4507	14.4	02 Jul 1998	1.69	1.47±0.04	
NGC 4507	9.9	13 Jan 1999	2.00	0.99±0.04	
NGC 4941	≤1.5	22 Jan 1997	1.26	0.08±0.05	3
NGC 4939	1.1	27 Jan 1997	1.47	0.13±0.04	3
NGC 5252	0.9	20 Jan 1998	2.92	0.11±0.03	
MKN 463	≤1.7	05 Jul 1998	1.30	0.04±0.04	
MKN 463	≤1.2	28 Jul 1998	2.35	0.05±0.05	
CIRCINUS	18.1	13 Mar 1998	6.32	1.87±0.02	8
NGC 5506	12.3	30 Jan 1997	1.70	1.39±0.04	9
NGC 5506	14.8	14 Jan 1998	1.75	1.45±0.04	9
ESO 273-IG004	≤1.8	02 Mar 1997	1.07	0.08±0.05	
ESO 103-G35	5.7	03 Oct 1996	2.10	0.71±0.04	
ESO 103-G35	5.6	14 Oct 1997	0.59	0.58±0.07	
IRAS 20210+1121	≤1.2	02 Oct 1996	1.86	0.09±0.04	10
NGC 7172	3.9	14 Oct 1996	1.73	0.40±0.04	
NGC 7172	2.2	06 Nov 1997	2.11	0.23±0.04	
NGC 7314	4.4	08 Jun 1999	4.26	0.49±0.02	
IRAS 22017+0319	1.0	28 Nov 1997	2.22	0.12±0.04	

[a] Flux in units of 10^{-11} erg cm^2 s^{-1} (u.l. are at 2σ level); [b] 1) Matt et al. 1997; 2) Risaliti et al. 1999; 3) Maiolino et al. 1998; 4) Malaguti et al. 1999; 5) Cappi et al. 1999; 6) Franceschini et al. 1999; 7) Gilli et al. 1999; 8) Matt et al. 1999; 9) Perola 1998; 10) Ueno et al. 1998.

2. RESULTS

TABLE 2 : Power Law Model

Source name	A[a]	Photon index Γ	χ^2/ν
NGC 526A	0.49	$1.72^{+0.21}_{-0.17}$	86.5/76
NGC 1365	1.47	$1.91^{+0.25}_{-0.23}$	116.9/76
NGC 1386	1.60	$2.21^{+0.65}_{-0.52}$	69.1/76
NGC 2110	1.31	$1.85^{+0.14}_{-0.14}$	89.3/76
MKN 3	0.90	$1.61^{+0.06}_{-0.09}$	68.2/76
IRAS09104+4109	0.03	$1.25^{+0.92}_{-0.81}$	0.59/2
NGC 2992	0.33	$1.85^{+0.75}_{-0.48}$	56.5/76
NGC 4388	1.71	$1.64^{+0.04}_{-0.04}$	88.6/76
NGC 4507	2.68	$1.79^{+0.06}_{-0.06}$	102.7/76
	1.84	$1.71^{+0.09}_{-0.08}$	97.8/76
	0.96	$1.64^{+0.11}_{-0.11}$	88.9/76
NGC 4939	0.22	$1.82^{+2.07}_{-0.96}$	70.8/76
NGC 5252	1.23	$2.35^{+1.02}_{-0.73}$	72.4/76
CIRCINUS	2.79	$1.76^{+0.04}_{-0.04}$	687.8/76
NGC 5506	4.02	$1.95^{+0.10}_{-0.10}$	64.2/76
	3.49	$1.87^{+0.11}_{-0.11}$	101.0/76
ESO 103-G35	4.57	$2.19^{+0.18}_{-0.16}$	87.2/76
	1.08	$1.82^{+0.38}_{-0.33}$	68.2/76
NGC 7172	0.37	$1.64^{+0.32}_{-0.29}$	103.7/76
	0.59	$1.91^{+0.52}_{-0.43}$	63.4/76
NGC 7314	1.29	$1.92^{+0.16}_{-0.15}$	82.0/76
IRAS 22017+0319	0.39	$2.00^{+1.33}_{-0.99}$	68.2/76

[a] Power law flux at 1 keV in units of 10^{-2} ph cm^{-2} s

The PDS observation log is presented in Table 1, where each measurement represents a distinct data set within the archive. The data reduction was performed using the SAXDAS software package with standard screening criteria including a temperature-dependent rise-time threshold which allows a 50% reduction in instrumental background. The resulting exposure time and count rates in the 13 -100 keV range are also listed. Of the 27 objects analysed, 17 were detected at $\geq 3\sigma$ confidence level and further analysed and their 20–100 keV flux is given; for the remainder a 2σ upper limit on the flux is reported. The major drawback of the PDS is the angular resolution which, at 1.3° FWHM is not sufficient to exclude contamination from other sources in the field of view. Indeed, checking the BeppoSAX MECS images as well as searching in public archives for possible 2–10 keV emitting sources (i.e. those most likely to contaminate at high energies) within the instrument field of view, we find that this possibility exists for a number of sources. In particular, the sky region around NGC 1386 and NGC 1365 is particularly confused, as not only may these two sources contaminate each others flux measurement, but also other sources in the field of view (such as the cluster NGC 1399, a QSO and an unidentified object) could contribute to the high energy emission. Another crowded region is that of NGC 4388 in Virgo, where contamination in the low energy channels due to the cluster and to known sources is possible. In the cases of IRAS 22017+0319, NGC 4939 and Circinus galaxy, a single 2–10 keV source is present in the fields of view and identified with the BL Lac PKS 2201+04, the QSO PG 1302-102 and the X-ray binary EXO-TR respectively; whether these objects are a major source of contamination is still to be determined. However, since reporting a high energy source is still interesting, we have included all the above objects in the subsequent analysis but the reader is cautioned to bear in mind that the emission may not be entirely (or even partly) due to the Seyfert galaxy in these cases.

3. SPECTRAL FITTING

In Table 2 we show the results of the spectral analysis of the sources for which we have at least 3σ detection. The best fit parameters are listed in the table for a simple power law fit to the data, while the 20–100 keV fluxes for this model are reported in Table 1. Note that all but one of the sources have a power law index compatible with the canonical value of 1.9 often observed in Seyfert 1's, further confirming the unified model prediction that all Seyferts should look alike when the torus effects are not important. The exception is NGC 4388 which instead shows a flatter spectrum; although this source may be contaminated, the contamination should affect only a few low energy channels (as NGC 4388 is likely to dominate the high energy emission) making the spectrum even flatter. A detailed analysis of the broad-band spectrum of this source is in progress and hopefully will better define its spectral characteristics (Bassani et al. in preparation). The weighted mean value of Γ is 1.74±0.06 considering all objects and becomes 1.79 when all possible contaminated sources are removed. It is evident from table 2 that in cases of highly significant detection a simple power law is not able to fit the data satisfactorily. The observed residual could either be due to high absorption in the source of column density $> 10^{24}$ cm^{-2} and/or to a reflection component. To check this possibility we have fitted the data of those sources detected above 10σ with a power law plus either one of these two extra components. Applying these models and searching for an improvement in χ^2 of at least 6.3 (i.e. 99% fit improvement for 1 extra degree of freedom with respect to a simple power law) we find that these extra components may be at work in NGC 5506 (only reflection is likely), Mkn 3, NGC 4507, NGC 7314 and ESO 103-G35 (both are likely). Note that no further improvement is obtained for NGC 4388 which therefore maintains a flat spectral index. In Mkn 3 and Circinus galaxy both components are probably at work (see for more details Cappi et al. 1999 and Matt et al. 1999) while in NGC 4507, NGC 7314 and ESO 103-G35 we expect reflection to be the dominant component as in these objects the 6.4 iron line EW is in the range 170–180 eV (Bassani et al. 1999) and so unlikely to be due to transmission through column densities much higher than 10^{23} cm^{-2}.

4. CONCLUSIONS

Our preliminary analysis of the PDS data of Seyfert 2 galaxies gives some interesting results: first the high energy output from this class of objects could be intense and similar to that often observed in bright Seyfert 1 objects. The mean photon index observed above 10 keV is around 1.8, i.e. similar to the "canonical value" seen in the type 1 objects. We also found that when the statistical quality of the observation is good, a single power law fit to the data is insufficient requiring extra components in the fitting procedure. We find that either a reflection component or high absorption or both are required to obtain a more acceptable fit. We hope to update and maintain this database in future in order to provide a useful tool for high energy missions such as INTEGRAL and to study on a statistical basis the high energy properties of Seyfert 2 galaxies also in relation to the emission at other wavelengths.

REFERENCES

Bassani L. et al. 1999, Ap. J. Suppl 121, 473

Bassani L., Cappi M., and Malaguti G. 1999, Proc of the 3rd Integral Workshop, in press

Cappi M. et al. 1999, A&A 344,857

Franceschini A. 1999, A&A submitted

Frontera F. et al. 1997 A&A S122, 357

Gilli et al. 1999, A&A submitted

Malaguti G. et al. 1999, A&A 342,L41

Maiolino R. et al. 1998, A&A 338, 781

Matt G. et al. 1997, A&A 325, L13

Matt G. et al. 1999, A&A 341,L39

Perola G.C. 1998, Nucl. Phys. B (Proc. Suppl.) 69/1-3, 477

Risaliti G. et al. 1999, A&A submitted

Ueno S. et al. 1998 Nucl. Phys. B (Proc. Suppl.) 69/1-3

NEW RESULTS FROM THE HRX BL LAC SAMPLE

Volker Beckmann[1,2] and Anna Wolter [2]

1) *Hamburger Sternwarte, Gojenbergsweg 112, 21029 Hamburg, Germany*
2) *Osservatorio Astronomico di Brera, Via Brera 28, 20121 Milano, Italy*

ABSTRACT   We present results for the Hamburg BL Lac sample, based on data provided by the RASS-BSC. By fitting a single power law to the X-ray data we find, in a number of objects, an additional absorbing component to the galactic value of N_H, which might be attributed to intrinsic absorption. A more probable cause seems however to be a curvature in the X-ray spectra in the sense that they are more curved for steeper slopes. The known relation between the X-ray spectral slope and the ratio between optical and X-ray flux (α_{OX}) also applies to this BL Lac sample, even though less significant than in previous works. We also find a dependence of X-ray luminosity on α_{OX}.

KEYWORDS: galaxies: active – BL Lacertae objects: general – X-rays: galaxies

1. INTRODUCTION

BL Lac objects are rare AGN, which are thought to be oriented towards us with their jet, thus showing high polarisation, strong variability at all wavelengths and non-thermal featureless spectra. Still there are several open questions about the nature of the BL Lac objects, e.g. whether there are differences between the BL Lac objects which are selected due to their strong radio-emission (RBL) and those selected because of their X-ray brightness (XBL). The difference between those two classes might be mainly caused by different peak frequencies of the two components in the BL Lac spectrum: XBL have usually a higher peak frequency of the synchrotron branch than RBL and thus we have high-frequency (HBL) and low frequency (LBL) peaked BL Lac objects (see e.g. Fossati et al. 1998). Between those two classes one can find intermediate objects (IBL) whose properties show them to be the link between HBL and LBL. We study the X-ray spectra of IBL and HBL and fit them with a single-power law with absorption by neutral hydrogen to investigate any possible spectral differences between these two samples.

2. THE HAMBURG X-RAY BRIGHT BL LAC SAMPLE

This work is based on the bright sources from ROSAT All Sky Survey (Voges et al. 1996), which have an X-ray flux $f_X > 11 \cdot 10^{-13} \mathrm{erg\ cm^{-2}\ sec^{-1}}$ in the hard PSPC energy band (0.5–2.0 keV). A first complete sample of 39 BL Lac objects (Bade et al.

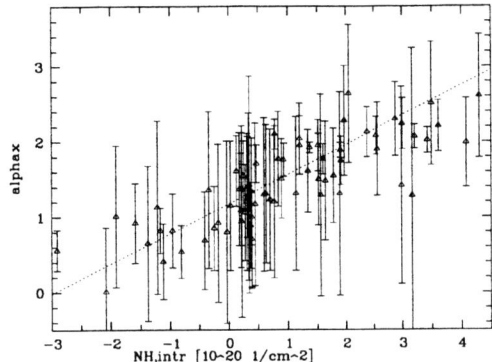

FIGURE 1. The "intrinsic" absorption ($N_{H,intrinsic} = N_{H,free-fitted} - N_{H,galactic}$) versus X-ray spectral slope. The negative values are caused by large errors on the free-fitted N_H so that they are consistent with $N_{H,intrinsic} = 0$. The linear regression takes the errors in N_H and α_X into account.

1998) in an area of 2800 deg^2 is based on the Hamburg/RASS X-ray bright sample (HRX, Cordis et al. in preparation). A larger sample of candidates has been derived by cross-correlating the X-ray sources from RASS in a larger area (4500 deg^2) with radio catalogues (NVSS, FIRST: radio flux limit $\simeq$ 2.5 mJy). These 262 sources, if not already classified, have been included in follow-up spectroscopy with the 3.5m telescope on Calar Alto. Up to now this statistically complete sample is 95 % classified and contains 72 BL Lac objects, both of the HBL and the IBL type. For 61 of them we have already determined the redshift ($z < 0.9$; Beckmann 1999). We will use in the following this sample of 72 BL Lacs.

3. X–RAY SPECTRA

Since the only available spectral measures are the two hardness ratios given in the RASS-BSC, we can only assume a simple spectral shape and apply the method of Schartel et al. (1996) to determine its parameters. We therefore assume that the spectrum is in the form of a single-power law absorbed by two different contributions: the galactic absorption and an "intrinsic" absorption that is given by the difference between the best fit absorption and the Galactic one ($N_{H,intrinsic} = N_{H,free-fitted} - N_{H,galactic}$). The galactic values were taken from Dickey and Lockman (1990). We find a strong correlation of the "intrinsic" absorption with the spectral slope (Fig. 1) on a > 99% confidence level also when taking into account the large errors in $N_{H,intrinsic}$. The negative values of $N_{H,intrinsic}$ seem to be caused by large errors; if we just take into account objects with an error of $N_{H,intrinsic} < 0.3$, we do not have any "negative absorption" at all. Since we do not see any correlation to other observable parameters (such as flux or luminosity), we make the hypothesis that this is an effect of curvature in the X-ray spectra (see e.g. Urry et al. 1996).

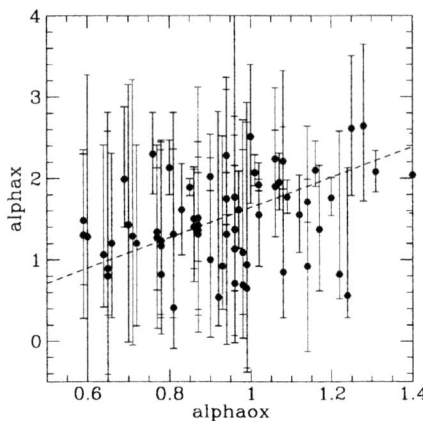

FIGURE 2. The relation between X-ray dominance (α_{OX}) and X-ray spectral slope (α_X) for the whole BL Lac sample. X-ray dominant objects are on the left.

Thus the flat X-ray spectra are well described by a single power law with galactic low energy absorption ($N_{H,intrinsic} \simeq 0$), while for steep spectra the curvature is significant and an additional "absorption" is needed to fit the X-ray data to a single-power law. In this model, a large value of $N_{H,intrinsic}$ would be explained by a convex spectrum. Nevertheless, true intrinsic absorption can not be ruled out in principle.

4. α_{OX} - α_X RELATION FOR HRX-BL LAC

A relation, which is found for BL Lac objects, is the dependence of the spectral slope α_X on α_{OX}[1]. This effect, which shows objects with a higher X-ray dominance having flatter X-ray spectra, was detected in several X-ray bright samples of BL Lac objects (e.g. Wolter et al. 1998, Padovani & Giommi 1995, Comastri et al. 1995). Based on the RASS data we determined this relation for the HRX-BL sample (Fig.2). The effect of spectral flattening in the X-ray region with decreasing α_{OX} is detectable, even though less significant than in previous works. An analysis, which takes into account the errors in α_{OX} and α_X gives a significance for correlation of both values on a > 93% level.

5. DEPENDENCE OF X-RAY LUMINOSITY ON α_{OX}

Another physical parameter which seems to be correlated to the X-ray dominance is the X-ray luminosity of the BL Lac objects. Because we know the redshift and

---

[1] we define α_{OX} as the power law index between 1 keV and 4400 Å with $f_\nu \propto \nu^{-\alpha_{OX}}$, that describes the X-ray dominance over the optical brightness

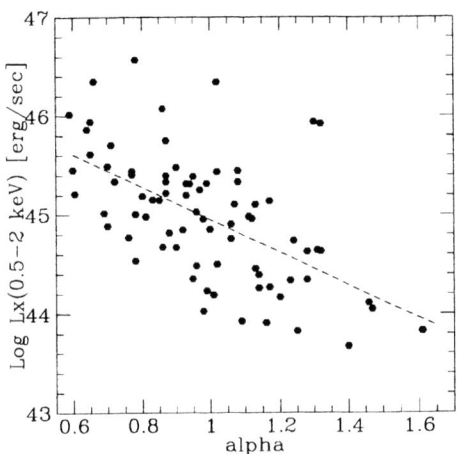

FIGURE 3. The relation between X-ray dominance (α_{OX}) and X-ray luminosity in the hard (0.5 – 2keV) ROSAT-PSPC band. There are no X-ray faint BL Lacs with low α_{OX}.

thus the luminosity for more than 80% of our objects, it is possible to study this relation. Figure 3 shows the dependency of $\log L_X$ on α_{OX}. There seem to be no objects, which are X-ray dominant (low α_{OX}) and have a low X-ray luminosity. This result is solid, because we do not miss any redshift for objects with $\alpha_{OX} < 0.9$ (Beckmann 1999). This effect could be explained in the view of the unified schemes (e.g. Padovani & Giommi 1995). The HBL are more X-ray dominated (low α_{OX}) and the X-ray luminosity is therefore high because the X-ray band is in the vicinity of the peak frequency. On the other hand, LBL (high α_{OX}) have their maximum of the synchrotron emission in the optical region and are X-ray faint.

REFERENCES

Bade, Beckmann, Douglas, et al., 1998, A&A 334, 459
Beckmann, 1999, in: PASPC Vol. 159, eds. Takalo & Silanpää
Comastri, Molendi & Ghisellini, 1995, MNRAS, 277, 297
Dickey and Lockman 1990, ARA&A 28, 215
Fossati, Maraschi, Celotti, et al., 1998, MNRAS 299, 433
Padovani and Giommi, 1995, MNRAS 277, 1477
Schartel, Walter, Fink, and Trümper 1996, A&A 307, 33
Urry, Sambruna, Worral et al. 1996, ApJ 463, 424
Voges, Aschenbach, Boller et al. 1996, IAU Circ. 6420
Wolter, Comastri, Ghisellini et al., 1998, A&A 335, 899

RXTE MONITORING OF CENTAURUS A

S. Benlloch [1], R.E. Rothschild [2], J. Wilms [1], C.S. Reynolds [3], W.A. Heindl [2], K. Pottschmidt [1], A. Orr [1], I. Kreykenbohm [1], R. Staubert [1]

1) Institut für Astronomie und Astrophysik, Astronomie, University of Tübingen, Germany
2) CASS, University of California, San Diego, La Jolla, CA 92093, U.S.A.
3) JILA, University of Colorado, Boulder, CO 80309, U.S.A.

ABSTRACT We analyze a 75 ksec observation of the active galaxy Centaurus A carried out with RXTE in 1998 August. The joint PCA/HEXTE broad band X-ray spectrum can be well described by a heavily absorbed ($N_\mathrm{H} \sim 7 \times 10^{22}\,\mathrm{cm}^{-2}$) power law with a photon index of $\Gamma = 1.8$. We present the interpretation of this spectrum in terms of Monte Carlo computations. No significant temporal variability was found during this observation.

1. INTRODUCTION

At a distance of 3.5 Mpc (Hui et al., 1993), Centaurus A is by far the nearest active radio galaxy. Previous observations have shown Cen A to be a radio-loud object having highly complex X-ray/gamma-ray properties with multi-temperature diffuse flux, a spatially resolved jet in radio and X-rays, and a nuclear component with complex low energy absorption (in part due to the dust-lane that is seen to cross the galaxy). A hard X-ray power-law continuum has been reported for Cen A, with photon index typically ~ 1.7. Iron K-shell absorption and fluorescent emission have been observed between ~ 6–7 keV. Other unusual spectral features of Cen A are the steeping of the spectrum at 140–170 keV reported by Kinzer et al. (1995) and no evidence for a strong Compton reflection component (Woźniak et al. 1998 found $\Omega/2\pi < 0.15$).

In this work we present the results of a 75 ksec observation with the Rossi X-ray Timing Explorer (RXTE) in the range 2.5–240 keV.

2. DATA REDUCTION AND ANALYSIS

RXTE observed Cen A in 1998 August with the Propotional Counter Array (PCA) and the High Energy X-ray Timing Experiment (HEXTE). The standard data from both instruments were used for the accumulation of spectra and light curves with a 16 s time resolution. To account for the imperfect knowledge of the PCA response matrix (Wilms et al., 1999), 1% systematic errors were added to the PCA spectral data. Throughout the RXTE mission, the All-Sky Monitor (ASM) provides estimates of the 1.5–12 keV flux from Cen A (Fig. 2, right).

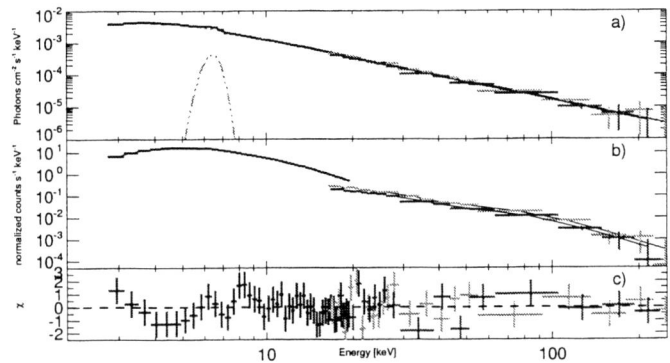

FIGURE 1. Best fit spectrum for an absorbed power law with Gaussian emission line. The HEXTE data have been rebinned for display purposes. a) Unfolded spectrum. b) Folded model spectrum. c) Residuals to the fit in terms of σ with error bars of size one.

2.1. Spectral Analysis

The PCA and HEXTE data were fit simultaneously using XSPEC, v. 10.00ab. We utilize a model that contains a heavily absorbed power law with Gaussian emission line. To account for the difference in the effective areas of the instruments, we included their relative normalization as a fit parameter. We used the photoabsorption cross section of Bałucińska-Church & McCammon (1992), giving a N_H which is $\sim 10\%$ lower than that found from the Morrison & McCammon (1983) cross sections. The best-fit values are given in Table 1. The best fit spectral model is shown in Fig. 1.

TABLE 1. Best Fit Spectral Parameters

Absorption	$N_H = 7.09(30) \times 10^{22}\,\text{cm}^{-2}$
Power law	$\Gamma = 1.88(3)$
	Norm $= 10.5(5) \times 10^{-2}\,\text{ph}\,\text{keV}^{-1}\,\text{cm}^{-2}\,\text{s}^{-1}$ at 1 keV
Iron Line	$E_{Fe} = 6.41(9)\,\text{keV}$
	$\sigma = 0.36(18)\,\text{keV}$
	Norm $= 5(1) \times 10^{-4}\,\text{ph}\,\text{cm}^{-2}\,\text{s}^{-1}$
	EW $= 159(29)\,\text{eV}$
Normalization	cluster A $= 0.95(2)$
(HEXTE vs. PCA)	cluster B $= 0.87(2)$
	Flux(2–10 keV) $= 2.26(3) \times 10^{-2}\,\text{ph}\,\text{cm}^{-2}\,\text{s}^{-1}$
Fit	$\chi^2/\text{d.o.f} = 173.2/188$
	$\chi_\nu^2 = 0.92$

(The uncertainties are 90% for one parameter ($\Delta\chi^2 = 2.7$), and are shown in units of the last digit shown.)

The energy of the iron line indicates that the emission is from cool, i.e., at most weakly ionized, material, and the line is at most weakly broadened. In order to test

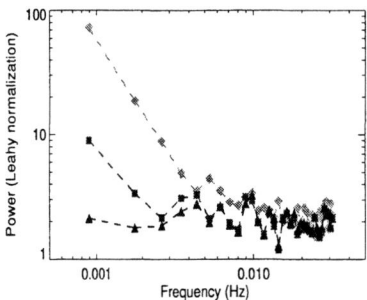
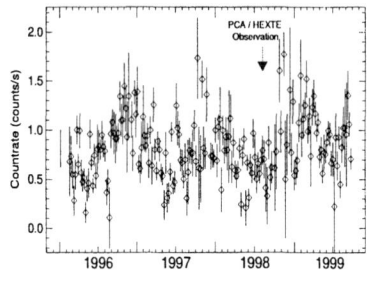

FIGURE 2. Left: Averaged periodogram in (Leahy et al., 1983) normalization of the raw PCA data (squares), of a measured background light curve from the higher energy channels (diamonds), and of the background subtracted source data (triangles). Right: ASM counting rate for Cen A from the beginning of the ASM measurements. Our pointed observation is indicated by the arrow.

for the presence of a Compton reflection component, a power law plus Compton component (XSPEC model PEXRAV) fit was made to the data. The 90% upper limit to a reflection component, expressed as the ratio of the solid angle for primary flux scattered into our line of sight to an infinite slab, was found to be $\Omega/2\pi \leq 0.008$ for an assumed inclination angle of 70°, implying that no significant reflection features are present. This result confirms the earlier OSSE and RXTE results that no reflection component in the spectrum of Cen A was detected (Rothschild et al., 1999; Kinzer et al., 1995). The RXTE data were also tested for the presence of a spectral break at higher energies, using a exponentially cutoff model. The best fit model give a a cutoff-energy of 341 keV with a lower limit of 220 keV and provides an equally adequate fit to the data as the straight power law. This result is consistent with CGRO/OSSE data, where a cutoff energy of 300 keV or higher was found (Kinzer et al., 1995).

2.2. Timing Analysis

Cen A is known to be variable on all time-scales (Jourdain et al., 1993 and references therein). To search for possible short term variability of the spectral parameters we modeled the spectra observed during individual orbits with an absorbed power law with Gaussian emission line model. No significant variation of the spectral parameters is found.

To characterize the variability of Cen A during our observation we computed an average power spectrum (PSD) using time segments of ~ 22 min duration, The analysis was performed for the PCA raw data (i.e., *not* background subtracted) from 4 to 40 keV, where the data are source dominated (Fig. 2 left, squares). Comparison with a PSD measured for the energy band *above* 40 keV, where the data are background dominated, indicates that the power at low frequencies might be due to the PCA background variability. After subtracting a background model from the observed light curve, *no excess variability above the Poisson level is seen* (Fig.2 left, triangles).

3. DISCUSSION AND CONCLUSIONS

A single 10 ks observation of Cen A was made with RXTE in 1996 August. We re-analyzed these data using the new and improved PCA response matrices and background models. Our analysis confirms the previous results of Rothschild et al. (1999) that the spectrum can be characterized by an absorbed power law with $\Gamma = 1.86(2)$, $N_H = 8.03(15) \times 10^{22}$ cm^{-2} (again using the cross sections of Bałucińska-Church & McCammon, 1992), and a Fe line with $\sigma \sim 0.38$ keV. The upper limit for the reflection component in the earlier observation is $\Omega/2\pi < 0.09$. Therefore, with the exception of the absorbing column, the spectral parameters of Cen A during the 1996 and 1998 RXTE campaigns are identical.

From these two RXTE observations we conclude that Cen A *does not exhibit any appreciable evidence for a reflection continuum* in its spectrum. In other words: no cold material is close to the central X-ray source, although a strong Fe line is observable. Therefore, the Fe line does not originate close to the center of Cen A.

A possible source of the Fe line flux would be a molecular torus surrounding the source (similar to Seyfert 2 galaxies) or the dust lane. To test the last hypothesis we have used a Monte Carlo code to compute the iron line flux emerging on the back of a slab of neutral material with $N_H = 7 \times 10^{22}$ cm^{-2} that is irradiated by the unattenuated continuum spectrum of Cen A. The equivalent width of the line resulting from the dust lane is < 30 eV so that most of the contributing Fe line photons do not come from the dust lane region. A natural place would then be the possible torus surrounding the central engine.

ACKNOWLEDGEMENTS

We acknowledge the RXTE Science Operations Center staff to provide the observations and the Guest Observer Facility for providing support in analyzing them. The research has been partially financed by la Caixa/DAAD grant A/98/19182.

REFERENCES

Bałucińska-Church M., McCammon D., 1992, ApJ, 400, 699
Hui X., Ford H.C., Ciardullo R., Jacoby G.H., 1993, ApJ 414, 463
Jourdain E., Bassani L., Roques J.P., et al., 1993, ApJ 412, 586
Kinzer R.L., Johnson W.N., Dermer C.D., et al., 1995, ApJ 449, 105
Leahy D.A., Darbro W., Elsner R.F., et al., 1983, ApJ 266, 160
Morrison R., McCammon D., 1983, ApJ 270, 119
Rothschild R. E., Band D.L., Blanco P.R., et al., 1999, ApJ 510, 651
Wilms J., Nowak M.A., Dove J. B., et al., 1999, ApJ 522, 460
Woźniak P.R., Zdziarski A.A., Smith D., et al., 1998, MNRAS 299, 449

ELECTRON INJECTION BREAK AND PAIR CONTENT OF QUASAR JETS

M. Błażejowski [1], M. Sikora [1], R. Moderski [2,1], T. Bulik [1]

1) N. Copernicus Astronomical Center, Warsaw, Poland
2) JILA, University of Colorado, Boulder, USA

ABSTRACT We study the dependence of nonthermal radiation spectra in OVV quasars on the location of the low energy break in the electron/positron injection function. We show that the high energy spectra produced during outbursts are presumably superposed from two components, one resulting from Comptonization of emission lines, which dominates at MeV-GeV energies, and the other resulting from Comptonization of infrared radiation, which dominates in the X-ray band.

KEYWORDS: quasars, jets, nonthermal radiation

1. ASSUMPTIONS

- Nonthermal radiation in blazars is produced by thin shells, propagating at a constant relativistic ($\Gamma \gg 1$) speed along the conical jet;
- Relativistic electrons are injected in the shells within a distance range $\Delta r = r_0$, starting from r_0. They are injected at a constant rate and with the two power-law energy distribution, $Q = K\gamma^{-p}$ for $\gamma > \gamma_b$ and $Q \propto \gamma^{-1}$ for $\gamma \leq \gamma_b$;
- Radiative energy losses of electrons are dominated by Comptonization of the quasar broad emission lines. This process is responsible for the production of γ-rays. The low energy break at a few MeV results from inefficient radiation cooling of lower energy electrons;
- The intensity of the magnetic field is $B(r) = (r_0/r)B(r_0)$.

2. THE MODEL EQUATIONS

2.1. Electron Evolution

The evolution of the electron energy distribution is given by the continuity equation (Moderski, Sikora & Bulik 2000)

$$\frac{\partial N_\gamma}{\partial r} = -\frac{\partial}{\partial \gamma}\left(N_\gamma \frac{d\gamma}{dr}\right) + \frac{Q}{c\beta\Gamma}, \qquad (1)$$

where

$$\frac{d\gamma}{dr} = \frac{1}{\beta c \Gamma}\left(\frac{d\gamma}{dt'}\right)_{rad} - \frac{2}{3}\frac{\gamma}{r}. \qquad (2)$$

The second term on the rhs of Eq. (2) represents the adiabatic losses. The rate of the radiative energy losses is:

$$\left(\frac{d\gamma}{dt'}\right)_{rad} = -\frac{4\sigma_T}{3m_e c}(u'_B + u'_S + u'_{BEL} + u'_{IR})\gamma^2, \tag{3}$$

where $u'_B = B'^2/8\pi$ is the magnetic energy density, u'_S is the energy density of the synchrotron radiation field, $u'_{BEL} = (4/3)\Gamma^2 L_{BEL}/4\pi r^2 c$ is the energy density of the broad emission lines, $u'_{IR} \simeq (4/3)\Gamma^2 \xi_{IR} 4\sigma_{SB} T^4/c$ is the energy density of the near-IR radiation produced in the molecular torus by hot dust, and ξ_{IR} is the fraction of the accretion disc radiation reprocessed by the dust into the near-IR band.

2.2. Radiation Spectra

The observed spectra as a function of time are computed using the formula

$$\nu L_\nu(t) \equiv 4\pi \frac{\partial(\nu L_\nu(t))}{\partial \Omega_{\vec{n}_{obs}}} = \iint_{\Omega_j} \frac{\nu' L'_{\nu'}[r(\theta,t),\theta]\mathcal{D}^4}{\Omega_j} d\cos\theta d\phi, \tag{4}$$

where $\mathcal{D} = [\Gamma(1-\beta\cos\theta)]^{-1}$ is the Doppler factor, $\nu = \mathcal{D}\nu'$, and $r = c\beta(t-t_0)/(1-\beta\cos\theta) + r_0$. The luminosity $\nu'L'_{\nu'}$ is contributed by: synchrotron radiation

$$\nu' L'_{S,\nu'} \simeq \frac{1}{2}(\gamma N_\gamma)m_e c^2 \left|\frac{d\gamma}{dt'}\right|_S. \tag{5}$$

the synchrotron-self-Compton (SSC) process

$$\nu' L'_{SSC,\nu'} = \frac{\sqrt{3}\sigma_T}{8\Omega_j r^2}\nu'^{3/2} \int N_\gamma \left[\gamma = \sqrt{\frac{3\nu'}{4\nu'_S}}\right] L'_{S,\nu'}\nu'^{-3/2}_S d\nu'_S, \tag{6}$$

and the external-radiation-Compton (ERC) process

$$\nu' L'_{ERC,\nu'}[\theta'] \equiv 4\pi \frac{\partial(\nu' L'_{ERC,\nu'})}{\partial \Omega'_{\vec{n}_{obs}}} \simeq \frac{1}{2}\gamma N_\gamma m_e c^2 \left|\frac{d\gamma}{dt'}\right|_{ERC}[\theta'], \tag{7}$$

where

$$\left|\frac{d\gamma}{dt'}\right|_{ERC}[\theta'] \simeq \frac{4\sigma_T}{3m_e c^2}\gamma^2 \mathcal{D}^2(u_{BEL} + u_{IR}). \tag{8}$$

and

$$\nu' \simeq \mathcal{D}\gamma^2 \nu_{BEL/IR}, \tag{9}$$

where ν_{BEL} and ν_{IR} are the average frequencies of the broad emission lines and of the infrared radiation of hot dust, respectively. Note that in the comoving frame the ERC radiation field is anisotropic, while the synchrotron and SSC radiation fields are isotropic (Dermer 1995).

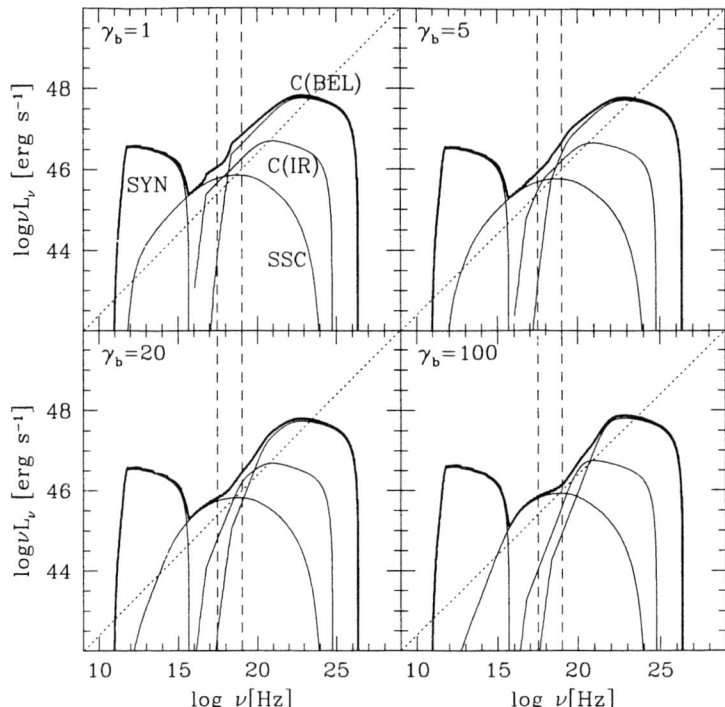

FIGURE 1. The time averaged blazar spectra for four values of γ_b. The dotted line marks the typical slope of the X-ray spectra in blazars ($\alpha = 0.6$); the dashed lines enclose the 1-30 keV X-ray band. In each panel we show four spectral components: synchrotron (SYN), synchrotron-self-compton (SSC), comptonization of broad emission lines [C(BEL)], and comptonization of near-IR dust radiation [C(IR)].

3. RESULTS

In Figure 1 we present the time averaged blazar spectra for four different values of the energy break γ_b. All models are computed for the following set of parameters: $r_0 = 6 \times 10^{17}$ cm; $\Gamma = 15$; $L_{BEL} = 1.4 \times 10^{44}$ erg s^{-1}; $B' = 1.4$ Gauss; $\gamma_{max} = 10^4$; $p = 2.2$; $K = 0.7 \times 10^{50}$ s^{-1}; $\theta_{obs} = \theta_j = 1/15$ rad, $T = 1000$ K, and $\xi_{IR} = 0.08$. For the justification of these choices see Błażejowski et al; in preparation.

4. DISCUSSION

4.1. Low Energy Break in Electron Distribution

We can see from Fig. 1 that for $\gamma_b = 1$ the low energy tail of the C(BEL) component extends down to ~ 2 keV. Thus, any presence of thermal nonrelativistic electrons should be imprinted as a bump, peaking around 2 keV. Since blazar spectra extend down to much lower values without any bump (Comastri et al. 1997; Sambruna 1997; Lawson & McHardy 1998), we exclude the domination of C(BEL) in the soft X-ray band.

In order to get soft X-ray spectra which smoothly join the middle X-ray band, one needs to assume that $\gamma_b > 3$. Then the low energy break of C(BEL) moves above 20 keV and X-ray radiation below this value is dominated by either SSC or C(IR). Noting that the SSC X-ray spectra are much softer than the observed ones ($\alpha_{X,SSC} \sim 1$ vs. $\alpha_{X,obs} \sim 0.6 - 0.7$; Kubo et al. 1998), C(IR) is a better candidate for X-ray production. This, however, can be the case if $\gamma_b \leq 10$. For larger values of γ_b the low energy break of the C(IR) component moves to energies > 20 keV, and, then, at lower energies the C(IR) spectrum becomes too hard in comparison with observations. We conclude that interpreting the blazar X-ray observations within the framework of our model implies that γ_b is enclosed in the range $3 - 10$.

4.2. Pair Content

For a jet dynamically dominated by the energy flux of protons, L_p, and for radiative energy losses of electrons dominated by Comptonization of broad emission lines, the pair content of the jet can be calculated from the formula (Sikora et al., in preparation)

$$\frac{n'_{pairs}}{n'_p} \sim \frac{K}{2(p-1)\gamma_b^{p-1}} \frac{m_p c^2}{L_p} \Gamma^2, \qquad (10)$$

This, for our model parameters and $3 < \gamma_b < 10$ gives

$$6/L_{p,47} < n'_{pairs}/n'_p < 26/L_{p,47}. \qquad (11)$$

Thus, our results suggest that the particle number in quasar jets is dominated by pairs, while the jet inertia is still dominated by protons.

ACKNOWLEDGEMENTS

This project was supported by ITP/NSF grant PHY94-07194, the Polish KBN grant 2P03D00415, and NASA grant NAG-5-6337.

REFERENCES

Comastri, A., Fossati, G., Ghisellini, G., & Molendi, S. 1997, ApJ, 480, 534
Dermer, C.D. 1995, ApJ, 446, L63
Lawson, A.J., & McHardy, I.M. 1998, MNRAS, 300, 1023
Kubo, H., et al. 1998, ApJ, 504, 693
Moderski, R., Sikora, M., Bulik, T., 2000, ApJ, 529, 151
Sambruna, R.M. 1997, ApJ, 487, 536

OBSERVING THE UNOBSERVABLE: CORONAL LINE DATA FROM NARROW-LINE SEYFERT 1S AS A TEST OF EUV CONTINUUM MODELS

P.J. Bleackley and P.T.O'Brien

Department of Physics and Astronomy, University of Leicester

ABSTRACT   We have measured coronal line emission for a sample of 19 Narrow-Line Seyfert 1 galaxies. Kinematic analysis shows that the Coronal Line Region (CLR) consists of gas in a decelerating outflow.

We intend to use these data as a test of the EUV emission of NLS1s. By comparison with photoionization modelling for a number of power-law + black body spectra, we plan to deduce the most likely EUV continua for these objects as well as explore physical conditions in the CLR.

KEYWORDS: galaxies active — galaxies Seyfert — spectroscopy coronal lines — AGN outflows — AGN soft excess

1. INTRODUCTION

Narrow-line Seyfert 1 objects (NLS1s) are a phenomenologically interesting subclass of Seyfert 1s. As their name suggests, they have the narrowest permitted lines of all Seyfert 1s (FWHM(Hβ) $\lesssim$ 2000 km s^{-1}), which appear to be associated with a range of other extremal properties, including strong, steep, soft X-ray excesses, strong permitted Fe II lines, weak forbidden lines, and a high degree of short timescale variability in X-rays. Coronal lines, forbidden transitions from highly ionised states of iron, are often present in their optical spectra. Their properties form a continuum with those of "normal" Seyfert 1s, but as they show extremal properties, they are important for testing AGN unification schemes. It is thought likely that they are Seyfert 1s with low mass black holes and/or high accretion rates.

To further understand the nature of the soft UV/X-ray excess, which is an important component of the energy output of these objects, we wish to study the EUV continuum. As this is not directly observable for extragalactic objects due to Galactic absorption, we use coronal lines (which have ionisation energies $\sim$ a few hundred eV) as diagnostics of this region. This approach also allows us to investigate the physical conditions in the coronal line region, particularly by kinematic methods.

2. OBSERVATIONS AND DATA REDUCTION

Nineteen NLS1s from the soft X-ray observed sample of Boller et al. (1996) were observed between 5 – 11 August 1996, using the Intermediate Dispersion Spectrograph mounted on the Isaac Newton Telescope at La Palma, obtaining dispersions of 1.6 – 3.2 Å pixel^{-1}. The data were reduced with the Starlink project's FIGARO software, and fitted using the ELF routines in DIPSO, which simultaneously fitted gaussian line profiles and a linear continuum. This approach allowed the deblending of neighbouring lines.

The coronal lines [Fe VII] $\lambda 6038.9$, [Fe X] $\lambda 6374.6$, [Fe XI] $\lambda 7891.9$ and [Fe XIV] $\lambda 5302.86$ were identified where possible, and upper limits were placed on them otherwise. Line profiles were then transformed into velocity space relative to the rest-frame of the [O III] $\lambda\lambda 4959, 5007$ doublet.

3. RESULTS

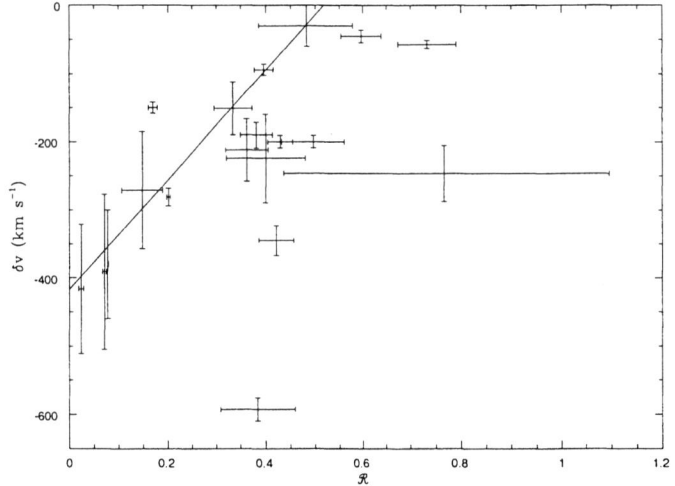

FIGURE 1. The decelerating outflow of the Coronal Line Region

3.1. Kinematic analysis

Erkens et al. (1997) noted a tendency for broader coronal lines to be more blueshifted than narrower ones. To further investigate this correlation in a manner independent of black hole mass, we assume that the FWHM of coronal lines is governed by the virial theorem and define a *radial estimator*

$$\mathcal{R} = \frac{r_{[\text{Fe}]}}{r_{[\text{O III}]}} = \left(\frac{\text{FWHM}_{[\text{O III}]}}{\text{FWHM}_{[\text{Fe}]}}\right)^2$$

where $r_{[Fe]}$ is the distance from the central black hole at which a line, [Fe], is emitted. Plotting the outflow velocity δv against $\mathcal{R}$ shows a decelerating outflow, which was fitted to first order by $\delta v = (806 \pm 93)\mathcal{R} - (417 \pm 25)$ km s^{-1} (see Figure 1). The minimum value of $\mathcal{R}$ observed is 0.024, which constrains the maximum size of the region over which the gas is accelerated.

For those objects for which at least 3 coronal lines had been detected, we plotted $\mathcal{R}$ against the ionisation potential, in order to determine if stratification of the CLR was important. For Arakelian 564, there is a trend towards more highly ionised gas being detected closer to the nucleus, but this trend is not unambiguously detected in other objects (see Figure 2).

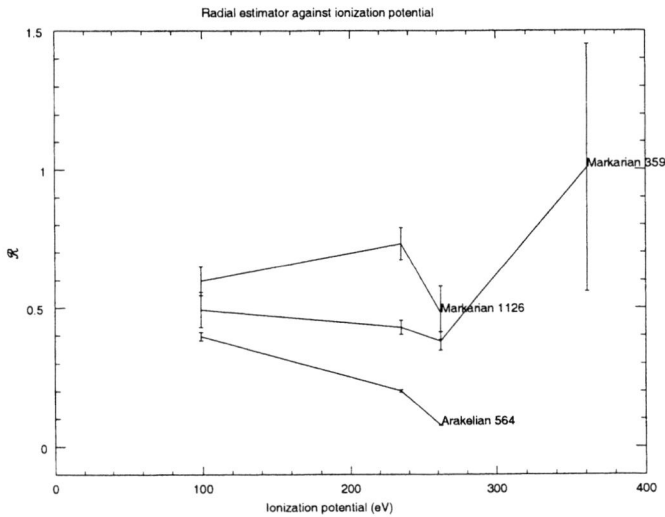

FIGURE 2. Radial estimator against ionisation potential for coronal lines.

3.2. Comparison with X-ray data

Equivalent widths of [Fe VII] and [Fe X] were compared with ROSAT PSPC photon indices obtained from Boller et al. (1996) (see Figure 3). No strong correlation is present for either line.

4. CONCLUSIONS

Gas in the coronal line region is outflowing, and this outflow appears to be decelerating. The coronal line gas has reached an outflow velocity of 420 km s^{-1} within a distance from the nucleus estimated to be $< 1/40$ that of the NLR. Radiation pressure does not appear to drive motion in the CLR, as this would be expected to produce accelerating, rather than decelerating outflows (see Binette (1998) for an example of a radiation pressure driven model).

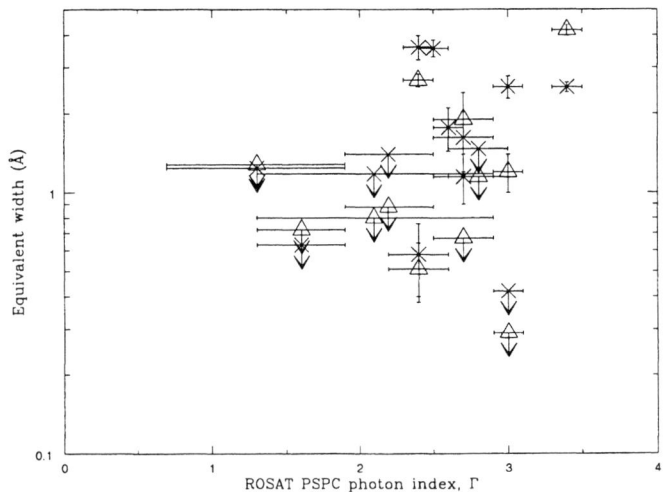

FIGURE 3. Equivalent width of [Fe VII] (crosses) and [Fe X] (triangles) against ROSAT PSPC photon index.

Coronal line strengths do not follow a simple correlation with ROSAT PSPC photon index. To investigate the EUV continuum in more detail, we plan to create photoionisation models based on a fixed power-law, with a black body component of variable luminosity and temperature added, and test the predictions of these models against our observed coronal line strengths. The models will also allow for stratification of the CLR.

ACKNOWLEDGEMENTS

We thank Dr. T. Roberts for his help with the observations. This work uses data obtained using the Isaac Newton Telescope, Observatorio de la Roque de los Muchachos, La Palma. P.J. Bleackley has been supported by a PPARC research studentship.

REFERENCES

Binette, L., 1998, MNRAS, 294, L47–L51
Boller, Th., Brandt, W.N., and Fink, H., 1996, A&A, 305, 53–73
Erkens, U., Appenzeller, I., and Wagner, S., 1997, A&A, 323, 707–716

LOW AND HIGH FREQUENCY QUASI-PERIODIC OSCILLATIONS IN 4U1915-05. RELATION WITH SOURCE STATE.

L. Boirin [1], D. Barret [1], J.F. Olive [1]

1) Centre d'Etude Spatiale des Rayonnements, 9 Av. du Colonel Roche, 31028 Toulouse Cedex 04, France

ABSTRACT

Using Rossi X-ray Timing Explorer (RXTE) observations, we have discovered both low (5-80 Hz, LF) and high (200-1300 Hz, HF) frequency Quasi Periodic Oscillations (QPOs) from the Low Mass X-ray Binary (LMXB) 4U1915-05. We find that the frequencies of the LF and HF QPOs positively correlate, suggesting that both QPOs are produced by related mechanisms. Their properties also depend upon the spectral state of the X-ray source. In this paper, we study the QPOs as a function of the source count rate and its position in the color-color diagram. We conclude that the timing behaviour of the source better correlates with the latter parameter.

KEYWORDS: low mass x-ray binaries; 4U1915-05

1. INTRODUCTION

HFQPOs have been detected in 22 LMXBs since the beginnings of RXTE (Van der Klis 1999 and references therein). It now appears that LFQPOs can be simultaneously present and related to HFQPOs. Thus, this has to be taken into account to interpret the QPO phenomenology and its relation with source states. In this context, we present the properties of the LF and HFQPOs detected from the dipper and burster 4U1915-05 during the RXTE observation.

2. THE RXTE OBSERVATION & RESULTS

4U1915-05 was observed 19 times for a total exposure of 140 ks between 1996 February and October with the Proportional Counter Array (PCA) aboard RXTE. During the observations, the intensity of the persistent emission (dips and bursts filtered out) changed by up to a factor 5 (Fig. 1 *Left*). 4U1915-05 went from a low intensity/hard spectrum ($L_{2-20\,keV} = 3.2 \times 10^{36}$ ergs s^{-1}) to a higher intensity/softer spectrum regime ($L_{2-20\,keV} = 1.4 \times 10^{37}$ ergs s^{-1}) as the accretion rate parameterized with S_a was increasing (Fig. 1 *Right*).

We detect LFQPOs (frequency 5-80 Hz, Full Width at Half Maximum (FWHM) 1-45 Hz, Root Mean Square (RMS) 5-20 %) and HFQPOs (frequency 200-1300 Hz,

FWHM 2-150 Hz, RMS 7-30 %), as shown in Fig. 2, in almost all observations except when 4U1915-05 is in its brighest or in its faintest state (probably a lack of sensitivity in the latter case; the upper limits on the QPOs RMS (7-25 %) are not constraining) (Boirin et al. 2000). In four observations, we detect simultaneously twin HFQPOs separated by 348±12 Hz, interpreted as neutron star spin within Beat Frequency Models (see e.g. Miller et al. 1998). We find that the LF and HF QPO frequencies show a positive correlation that may be compared with the one expected within the relativistic precession model (Stella et al 1999). This correlation, which indeed suggests that LF and HF QPOs are produced by related mechanisms, implies also a common dependence of both QPOs on some quantity that could parameterize the timing behaviour of the source.

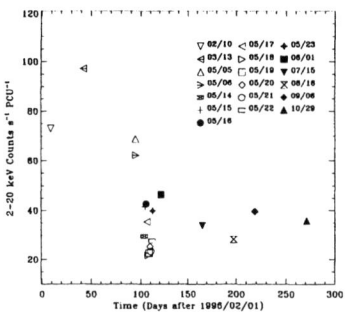 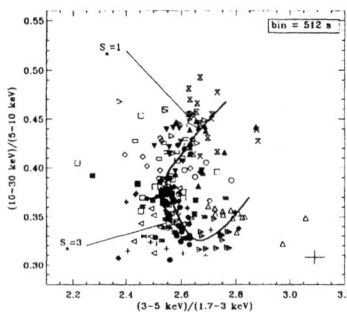

FIGURE 1. *Left* Evolution of the persistent emission of 4U1915-05 during the observations. *Right* Color-color diagram. 4U1915-05 approximately follows the solid spline parameterized with S_a. The accretion rate is inferred to increase with S_a.

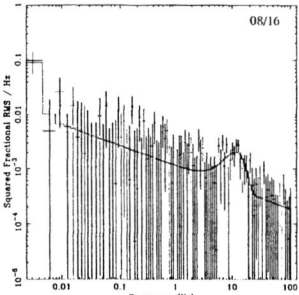 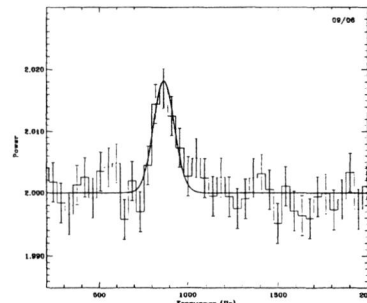

FIGURE 2. *Left* Typical normalized Power Density Spectrum (PDS) computed in the $3 \times 10^{-3} - 100$ Hz frequency range from the 5-30 keV light curve (Aug. 16th observation). The PDS is dominated by red noise (suggesting that 4U1915-05 is in the Banana state) and by a LFQPO at 12 Hz (Gaussian). *Right* 200-2000 Hz PDS showing a HFQPO at 900 Hz (Sept. 6th).

3. LF AND HF QPOS RELATION WITH SOURCE STATE

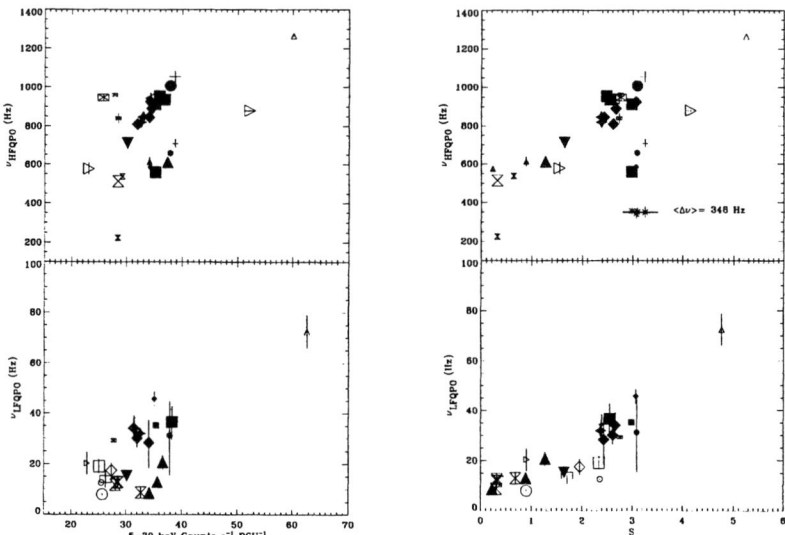

FIGURE 3. LF (*Bot.*) and HF QPO (*Top*) frequency vs intensity (*Left*), vs S_a (*Right*). Small symbols are for 3-4 σ detections, big symbols for detections $\geq 4\ \sigma$. Asterisks represent the frequency separation of the twin HFQPOs.

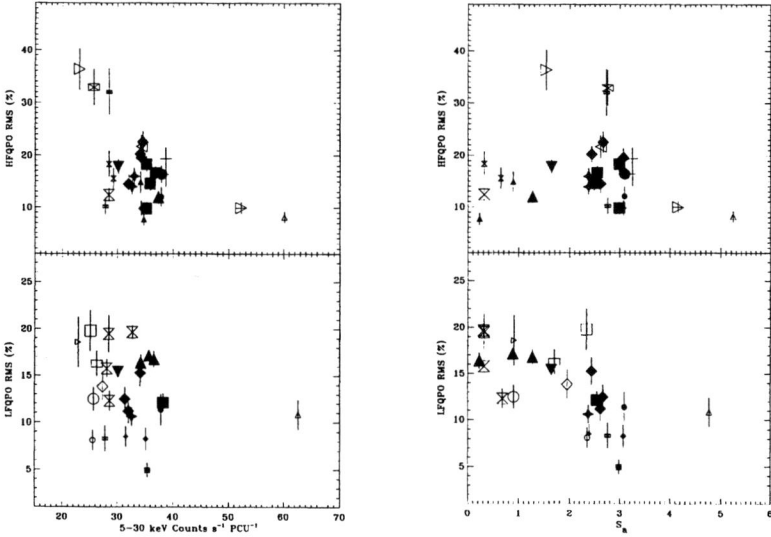

FIGURE 4. LF (*Bot.*) and HF QPO (*Top*) RMS vs intensity (*Left*), vs S_a (*Right*).

Thus, we study the LF and HF QPO properties (frequency, RMS) both as a function of 4U1915-05 intensity and of its position S_a on the color-color diagram (Fig. 1 *Right*). We clearly show a positive correlation between ν_{LFQPO} and S_a (Fig. 3 *Bot. Right*) whereas the correlation with the count rate is not as good, showing larger scatter (*Bot. Left*). The ν_{HFQPO} vs S_a relation (Fig. 3 *Top Right*) also appears much simpler than the ν_{HFQPO} vs count rate relation (Fig. 3 *Top Left*). We can now easily identify the upper and lower HFQPOs. The frequency of the upper HFQPO is well correlated with S_a within its full range. We show that whereas the LFQPO RMS anticorrelates with S_a (Fig. 4 *Bot. Right*), no such anticorrelation is seen with the count rate (Fig. 4 *Bot. Left*). The anticorrelation of the HFQPO RMS with the count rate (Fig. 4 *Top Left*) is stronger than with S_a (Fig. 4 *Top Right*).

4. CONCLUSIONS

The LFQPOs detected for the first time in 4U1915-05 behave (frequency and RMS evolution with S_a) as the Horizontal-Branch QPOs detected in Z sources, suggesting that similar physical processes are at work in Atoll and Z LMXBs.

The HFQPOs behave as the LFQPOs (frequency evolution with S_a), indicating that LF and HFQPOs are produced either by related mechanisms or by the same mechanism seen under different aspects or occuring at different locations in the disk.

The timing behaviour (QPO frequencies) of 4U1915-05 is driven by S_a rather than by the count rate, confirming that the position in color-color diagram is a better indicator of accretion rate than the intensity. However, it is not clear why the HFQPOs RMS amplitude and frequency would be driven by different parameters. Thus, either the accretion rate is better represented by a combination of parameters or by other parameters, or the accretion rate is not the only parameter responsible for timing behaviour evolution.

REFERENCES

Boirin, L., Barret D., Olive J.F. et al., 2000, A&A, in press

Miller, M. C., Lamb, F. K., and Psaltis, D., 1998, ApJ, 508, 791

Stella, L., Vietri M., and Morsink, S.M., 1999, ApJL, 524, 63

Van der Klis, M., 1999, Proceedings of the Third William Fairbank Meeting, Rome June 29 - July 4, 1998

Yoshida, K., 1992, Ph.D. thesis, Tokyo University

ROSAT DISCOVERED SOFT X-RAY INTERMEDIATE POLARS: UU COL AND RX J0806.3+1527

Vadim Burwitz[1,2], Klaus Reinsch[2]

[1] *Max-Planck-Institut für extraterrestrische Physik, Giessenbachstr., D-85748 Garching*
[2] *Universitäts-Sternwarte Göttingen, Geismarlandstr. 11, D-37083 Göttingen*

ABSTRACT We present new results from our ROSAT HRI observations of soft high-galactic latitude ROSAT all-sky survey sources. A 42 ksec pointing on UU Col reveals that the two independent coherent periodicities known from our optical photometry are also present in the soft X-ray flux confirming the soft X-ray intermediate polar nature of this system. In a 13.4 ksec pointing on RX J0806.3+1527 a (321.5393 ± 0.0004) s pulsation with a modulation amplitude of $\sim 100\,\%$ is detected as the only consistent signal in the 0.0002–0.0052 Hz frequency range. Based on its X-ray and optical properties, we suggest that RX J0806.3+1527 is another member of the new class of soft X-ray intermediate polars. Alternatively, it could be a double-degenerate polar.

KEYWORDS: accretion, accretion disks — stars: individual (UU Col, RX J0806.3+1527) — novae, cataclysmic variables — stars: rotation — X-rays: stars

1. INTRODUCTION

Among the ~100 000 X-ray sources detected in the *ROSAT* All-Sky-Survey (RASS) a new class of magnetic cataclysmic variables (mCVs) was discovered: the soft X-ray intermediate polars (IPs). In these systems matter from the secondary star is accreted onto an asynchronously rotating white dwarf. While all IPs discovered before *ROSAT* have hard X-ray spectra, the newly found soft X-ray IPs resemble much the properties of polars and may have field strengths in the lower range of polars (e.g. Mason et al. 1992, Haberl et al. 1994, Haberl & Motch 1995). The similarity of these systems to low-field polars led to the suggestion that soft IPs may be their long sought evolutionary progenitors. We present here new results from *ROSAT* HRI observations of two of these interesting sources.

2. UU COL

UU Col (= RX J0512.2-3241 = 1RXS J051214.5-324140) is contained in the RASS bright-source catalogue (Voges et al. 1999) and has been identified as a soft X-ray IP based on its X-ray and optical spectral characteristics and on the two independent coherent periodicities detected in its optical light curve (Burwitz et al. 1996). The two periods $P_\omega = (863.5 \pm 0.7)$ s and $P_\Omega = (3.45 \pm 0.04)$ h have been interpreted as the spin and orbital periods, respectively, of the magnetic white dwarf in UU Col.

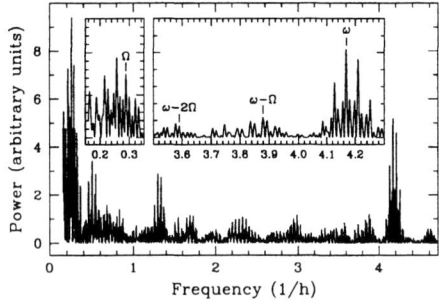
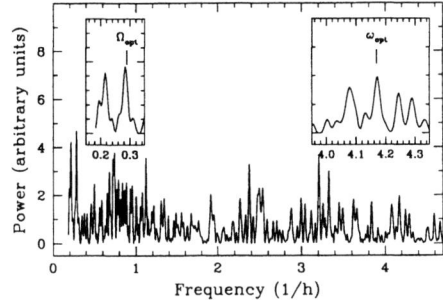

FIGURE 1. UU Col periodogramms (*left*) from optical CCD photometry (from Burwitz et al. 1996) and (*right*) from the September 1997 42 ksec *ROSAT* HRI observation.

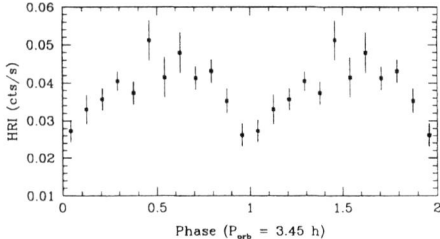
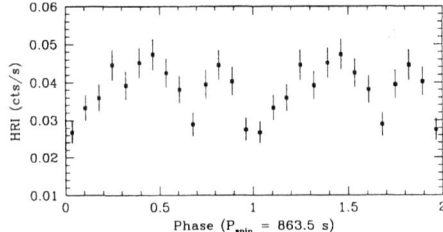

FIGURE 2. Soft X-ray lightcurves of UU Col obtained from the September 1997 42 ksec *ROSAT* HRI observation folded over the (*left*) orbital and (*right*) spin periods. In both cases, the better defined values from our optical observations have been used for the periods.

From the analysis of our 42 ksec pointed *ROSAT* HRI observation of UU Col conducted in September 1997 we have found the X-ray counterparts to the optical spin and orbital periods (Fig. 1). In the Scargle periodogramme of our X-ray data periods are detected at $P_\omega = (863.8 \pm 2.1)\,\mathrm{s}$ and $P_\Omega = (3.53 \pm 0.11)\,\mathrm{h}$. This confirms the presence of the ~3.5 h period tentatively determined from our optical observations and excludes the alias periods seen in our optical power spectra. The optical beat period has not been detected in the X-rays. The average countrate during this HRI observation is ~0.04 cts/s which is similar to that of our previous short HRI observation obtained in September/ October 1995.

The phase-folded X-ray data (Fig. 2) show a pronounced quasi-sinusoidal variation with the orbital period, similar to the lightcurve seen in the optical. The spin modulation of the soft X-ray flux, however, is double-peaked while in the optical it is quasi-sinusoidal. Our HRI data confirm the soft X-ray IP nature of UU Col. The weakness of the beat-period modulation suggests that it is a disk-fed accretor.

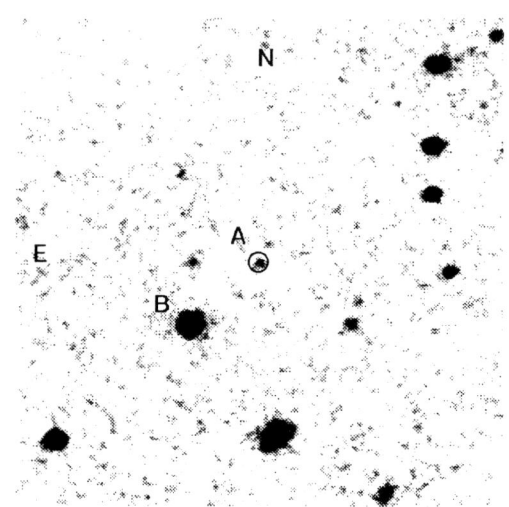

FIGURE 3. V-band CCD image centered on the HRI position of RX J0806.3+1527 (field of view: $120'' \times 120''$). The most likely optical counterpart is the V $\sim$ 21 mag object A located at RA = $08^h06^m22\overset{s}{.}9$, DEC = $+15°27'30.2''$ (2000.0). The V $\sim$ 15.5 mag object B is a G-type star and $23''$ away from the HRI position. The optical image has been taken in April 1998 with the Dutch 0.9 m telescope at ESO, La Silla and an exposure time of 120 s.

3. RX J0806.3+1527

RX J0806.3+1527 is contained in the sample of soft high-galactic latitude RASS sources with PSPC count rates between 0.1 and 0.5 cts/s and has been identified as an IP (Beuermann et al. 1999). The source is not contained in the RASS BSC (Voges et al. 1999) but is present in the photon-event table with an average PSPC count rate of (0.158 ± 0.027) cts/s and a peak count rate of 0.8 PSPC cts/s. The hardness ratio of the source derived from the RASS data is HR1 = -0.67 ± 0.10. The source is also seen in our pointed ROSAT HRI observation with an average countrate of 0.06 cts/s. The error circle of the HRI source position contains only one stellar object ($V \sim 21$ mag) as the most likely optical counterpart (Fig. 3).

From a period analysis of our HRI data we find that the only consistent signal in the frequency range 0.0002–0.0052 Hz is the peak at 0.00311 Hz which is also present in the two subsets of the data (Fig. 4, *left*). The most likely period derived from the combined data set is (321.5393 ± 0.0004) s but an alias period of (321.5465 ± 0.0004) s cannot be excluded. The soft X-ray flux is $\sim 100\%$ modulated with the former period (Fig. 4, *right*). The ephemeris for the heliocentric Julian dates of the pulse maxima is

$$T_{max} = \text{HJD} 2449730.2487(2) + 0.003721520(5) \times E.$$

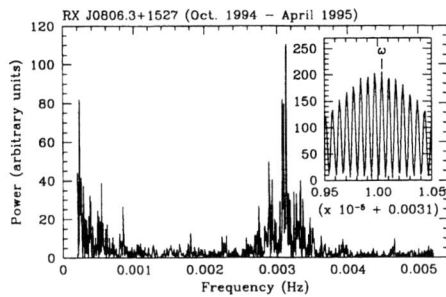
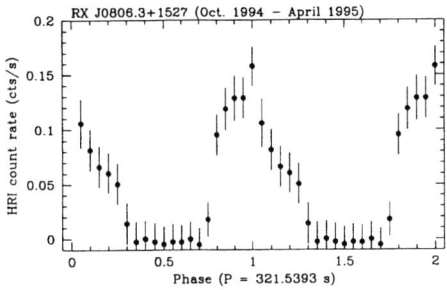

FIGURE 4. *Left:* Scargle periodogramme of the soft X-ray light curve of RX J0806.3+1527 obtained from 13.4 ksec pointed *ROSAT* HRI observations in October 1994 and in April 1995. *Right:* Phase-folded soft X-ray light curve of RX J0806.3+1527 obtained from the *ROSAT* HRI pointing.

where E is the cycle count. The pulse period is in the typical range (5–30 min) of spin periods observed in IPs. Combined with the low hardness ratio this suggests that RX J0806+15 is a further member of the soft X-ray IPs. The X-ray to optical flux ratio $\log(F_X/F_{opt}) = 2.0$ is extremely high but still within the range observed for polars and soft X-ray IPs (Motch et al. 1998). The shape of the lightcurve with practically no flux observed during half of the cycle is, however, difficult to explain with an IP geometry. It resembles much the lightcurve of RX J1914.4+2456 which has been suggested to be a double-degenerate polar (Cropper et al. 1998). It could, therefore, be possible that the 321-s period is the orbital or half the orbital period and that RX J0806+15 is the second example of such an extreme system. Based on the soft X-ray spectrum of the source, we can, however, rule out another possibility discussed by Israel et al. (1999) namely that RX J0806+15 is a nearby old neutron star accreting from the interstellar medium.

ACKNOWLEDGEMENTS

This work has been supported in part by the DLR under grant 50 OR 9210 1.

REFERENCES

Beuermann K., Thomas H.-C., Reinsch K., et al., 1999, A&A 347, 47
Burwitz V., Reinsch K., Beuermann K., Thomas H.-C., 1996, A&A 310, L25
Cropper M., Harrop-Allin M.K., Mason K.O., 1998, MNRAS 293, L57
Haberl F., Motch C., 1995, A&A 297, L37
Haberl F., Thorstensen J.R., Motch C., et al., 1994, A&A 291, 171
Israel G.L., Panzera M.R., Campana S., et al., 1999, A&A 349, L1
Mason K.O., Watson M.G., Ponman T.J., et al., 1992, MNRAS 258, 749
Motch C., Guillout P., Haberl F., et al. 1998, A&AS 132, 341
Voges W., Aschenbach B., Boller Th., et al. 1999, A&A 349, 389

A NEW BL LAC SAMPLE FROM THE REX SURVEY

A. Caccianiga[1], T. Maccacaro[2], A. Wolter[2], R. Della Ceca[2], I.M. Gioia[3,4]

1) Observatório Astronómico de Lisboa, Lisboa, Portugal 2) Osservatorio Astronomico di Brera, Milano, Italy 3) CNR, Bologna, Italy 4) Ifa, Hawaii, USA

ABSTRACT We present and discuss the first well defined BL Lac sample derived from the REX survey. This sample has a size comparable to the existing complete BL Lac samples (i.e. EMSS, 1Jy). A comparison with these samples is also presented and discussed.

KEYWORDS: BL Lacertae objects: general

1. THE SAMPLE

The sample presented here has been selected from the REX survey (Caccianiga et al. 1999; Maccacaro et al. 1999, these proceedings). This survey is the result of a positional cross-correlation between the NRAO VLA Sky Survey (NVSS, Condon et al. 1998) and an X-ray catalog of serendipitous sources detected in about 1200 pointed ROSAT PSPC images. The flux limit in the radio band (at 1.4 GHz) is 5 mJy while the faintest X-ray flux limit is 3.5×10^{-14} erg s^{-1} cm^{-2}. In this paper we consider the subsample defined by the following further constraints:
 1) X-ray flux (0.5-2.0 keV) $\geq 5\times10^{-13}$ erg s^{-1} cm^{-2}
 2) O mag (= blue APM magnitude) ≤ 20
The resulting sample (the X-ray Bright REX sample, XB-REX) contains a total of 147 sources, 133 (90%) of which have been optically classified. Among these sources we have found 33 (25%) BL Lacs, 43 (32%) radiogalaxies (of which 21 in cluster and 22 possibly isolated), 57 (43%) Emission Line AGNs (EL AGNs). We have classified a source as BL Lac if its optical spectrum is featureless (emission lines with EW$\leq$5 Å) and, in case of a "galaxy type" object, if the presence of an extra-source of continuum is evident. To this end, we use the "classical" measurement of the discontinuity at 4000 Å (Ca break) in the optical spectrum. In particular, we consider an object as BL Lac if the Ca break is less than 40%. This limit has been found to be more appropriate than the one (25%) previously used (see the discussions in Browne & Marchã 1993, Laurent-Muehleisen et al. 1997, Caccianiga et al. 1999). The radio-optical (α_{RO}) and optical-X-ray (α_{OX}) spectral indices of the identified sources are shown in figure 1. Note that the majority of the BL Lacs found in this sample falls in the typical region of the High-energy cutoff BL Lacs (HBL), i.e. $0.3\leq \alpha_{RO} \leq 0.6$, $0.5\leq \alpha_{OX} \leq 1.2$.

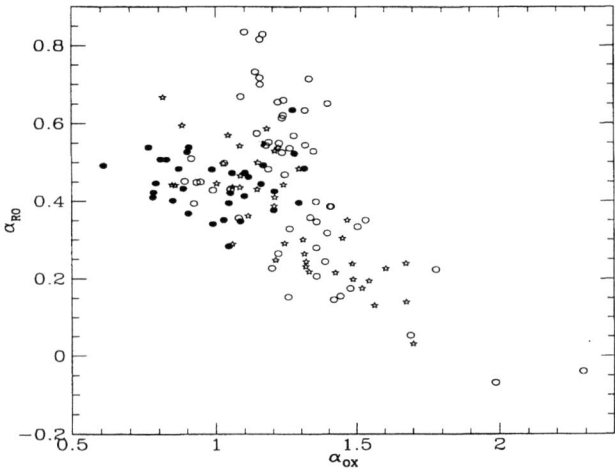

FIGURE 1. The radio-optical (α_{RO}) vs. the optical-X-ray (α_{OX}) spectral indices of the identified sources in the XB-REX sample: filled circles = BL Lacs, open circles = E.L. AGNs, stars = galaxies. Note that some objects, classified in literature as "galaxies" could be low-luminosity AGNs

2. COMPARISON WITH OTHER BL LAC SAMPLES

We want to compare the XB-REX with other BL Lac samples in order to study the origin of the differences between radio and X-ray selected samples. In order to account for the three different flux limits in the X-ray, radio and optical bands, we have performed a number of simulations, using the two statistically complete samples available so far (i.e. the EMSS, Morris et al. 1991, Wolter et al. 1994 and the 1Jy sample, Stickel et al. 1991) to compute the predicted number of BL Lacs in the XB-REX. We integrate the X-ray luminosity function (XLF) between $L=10^{42}$ and 10^{47} erg s^{-1} and from z=0 to z=4, assuming the best-fit evolution reported in Wolter et al. (1994) up to z=2 (negative for the EMSS sample and positive for the 1Jy sample) and no evolution afterwards. We then fold the result with the X-ray sky coverage and, by using the relationships between the different luminosities (radio, optical, X-ray) as presented in Wolter et al. (1994), we simulate a synthetic sample. We can impose to this synthetic sample the same cuts that define the XB-REX i.e., $f_X \geq 5\times 10^{-13}$ erg s^{-1} cm^{-2}, $f_R \geq 5$ mJy and B ≤ 20. We also subtract a 10% from the resulting sample, due to the choice of the tolerance used in the REX cross-correlation (see Caccianiga et al. 1999). If the starting point is the XLF derived from the EMSS sample, the expected number of BL Lacs is 44, while starting from the 1Jy XLF the number is 18. These numbers have to be compared with the actual number (33) of BL Lacs already found in the XB-REX, that can increase by 3-

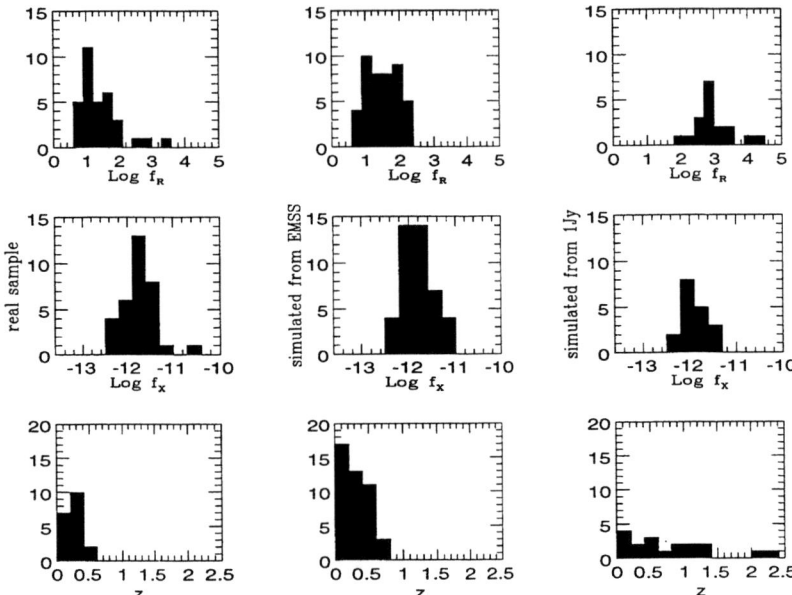

FIGURE 2. Comparison of the observed flux and redshift distributions (first column) of the BL Lacs in the XB-REX with those predicted from the EMSS (second column) and 1Jy (last column) samples. The X-ray fluxes (0.5-2.0 keV) are in erg s^{-1} cm^{-2}. The radio fluxes are in mJy at 1.4 GHz.

4 units, taking into account that 14 sources are still not classified. What is more interesting is to compare the XB-REX distributions of fluxes (in the radio and X-ray band) and of redshift with that predicted by the simulations. Figure 2 shows that the histograms of the XB-REX are consistent with the prediction from the EMSS (KS test probability $\geq 10\%$) while this is not true for the prediction based on the 1Jy. The largest difference is between the radio flux distribution of the XB-REX and the 1Jy predictions. With the KS test we can reject (probability less than 1%) the hypothesis that the two distributions derive from the same population. In particular, almost all of the 18 simulated BL Lacs should have $f_R \geq 100$ mJy, while only 4 in the XB-REX are so bright. Among the unidentified sources, only 2 have radio flux greater than 100 mJy. The 1Jy simulations fail also in reproducing the observed redshift distribution (KS probability<1%). In this case, however, the histogram of the observed redshift reports only the 19 objects for which a determination of z was possible and, thus, it could be not representative of the entire sample.

3. CONCLUSIONS

The conclusions can be summarized as follows:

1) We have presented a new complete sample of 33 BL Lacs selected from the REX survey by imposing an X-ray flux limit of 5×10^{-13} erg s^{-1} cm^{-2}. Ninety percent of this sub-sample is already identified down to a magnitude of O ($\sim$B)=20.

2) The properties of this sample resemble those of a typical X-ray selected sample, with a clear dominance of HBLs (see Fig. 1).

3) The radio and X-ray flux distributions of the BL Lacs found in this sample are in good agreement with those predicted from the XLF and the best-fit (negative) cosmological evolution derived from the EMSS sample. On the contrary, simulations based on the 1Jy sample do not reproduce the basic properties of the XB-REX sample. In particular, the prediction of a large number ($\sim$15-18) of BL Lacs with a radio flux greater than 100 mJy is not confirmed. This result, combined with the fact that almost all the BL Lacs in the XB-REX are HBLs, seems to suggest that Low-energy cutoff BL Lacs (LBLs), found mostly in the 1Jy sample, become rare, in comparison with HBLs, at low radio fluxes.

4) The good agreement between the predicted, based on the EMSS, and the observed properties of the XB-REX sample also confirms that it is possible to take into account the different flux limits in a multi-wavelength selected sample in order to derive the general properties of the BL Lac population. This is particularly important since such multi-wavelength selections are proven to be the most efficient way to select new samples of BL Lacs.

The next step of this research will be the estimate of the LF and cosmological evolution by using the XB-REX sample. Finally, we plan to extend the XB-REX at deeper fluxes.

ACKNOWLEDGEMENTS

This research was partially supported by the Italian Space Agency (ASI), by the European Commission, TMR Programme, Research Network Contract ERBFMR XCT96-0034 "CERES" and by the Italian Ministry for University and Research (MURST) under grant Cofin98-02-32.

REFERENCES

Browne, I.W.A. & Marchã, MJ.M. 1993, MNRAS, 261, 795

Caccianiga, A., Maccacaro, T. Wolter, A., Della Ceca, Gioia, I.M. 1999, ApJ, 513, 51

Condon, J.J. et al. 1998, AJ, 115, 1693

Laurent-Muehleisen, S.A., et al. 1998, ApJS, 118,127

Morris, S.M., et al. 1991, ApJ, 380, 49

Stickel, M., Padovani, P., Urry, C.M., Fried, J.W., Kühr, H. 1991, ApJ, 374, 431

Wolter, A. Caccianiga, A., Della Ceca, R., Maccacaro, T., 1994, ApJ, 433, 29

ON THE FATE OF STARS NEAR A SUPERMASSIVE BLACK-HOLE

Andrej Čadež and Andreja Gomboc

Department of Physics, University of Ljubljana, Slovenia

ABSTRACT Tidal disruption of a star during a close passage by a black hole has been considered as a possible cause of unusual phenomena in centers of some galactic nuclei. Here we discuss a simplified model describing the partial capture of a star by a large (galactic) black hole.

KEYWORDS: black holes, tidal disruption

1. A MASSIVE BLACK HOLE AND A NORMAL STAR

The important feature of this scenario is the similarity between sizes of the black hole and the star; for example a $\approx 3 \times 10^5 M_\odot$ Schwarzschild black hole has the same radius as the Sun. We further note that (Rees 1988; Lacey et al.1982) the Lagrangian point L_1 between the star and the black hole is at the distance $r_D \approx \left(\frac{M_s}{M_{BH}}\right)^{\frac{1}{3}} a$ from the stars center, where M_s and M_{BH} are the masses of the star and of the black hole respectively and a is the distance between them. The Lagrangian point thus plunges below the stellar surface when the star is still quite a few Schwarzschild radii ($R_S = \frac{2GM_{BH}}{c^2}$) away from the black hole. As the approach continues, the Lagrangian point moves below the stellar surface so that the tidal force overcomes stellar gravity in an increasing fraction of stellar envelope. Those parts of the star then begin to effectively free fall in the gravitational field of the black hole. However, the tidal acceleration of the envelope with respect to the center of mass of the star is at first considerably smaller then the velocity of the star divided by the periastron crossing time, so that the stellar surface changes relatively slowly. The circumstances change in an abrupt way as an important fraction of stellar mass with angular momentum $l = 2cR_S M_s$ reaches the radius $2R_S$. Parts of the star that are further away have slightly larger angular momentum and are on an escape orbit, while those closer to the black hole possesses less angular momentum and are bound to cross the horizon of the black hole in a proper time interval only a few times the Schwarzschild radius light crossing time. The relative speed between those parts of the star, therefore, rapidly increases from there on and very rapidly surpasses the speed of sound inside the star. The stellar surface is thus supersonically turning from spherical to thread-like, while the hot dense nucleus still preserves its integrity. While the gas temperature inside the star is unlikely to change, its hot interior is more and more exposed to the outside world as the envelope is sheared away.

Due to circumstances of the scenario just described, we model the star as initially consisting of a large number N of equally massive randomly distributed noninteracting pieces but in such a way that on the average their distribution corresponds to the density and temperature profile of a stellar model. All pieces are assumed to have the same cross section σ so that an external observer sees only the unobstructed ones closest to him. We let such an originally spherical and nonrotating star drop toward a black hole with the parabolic velocity from the point where the Lagrangian point touches the surface of the star. From there on we consider all the pieces as freely falling and preserving their temperature. The simplifying assumption of constant temperature is marginally valid since the star is sheared supersonically and the stellar thermal relaxation time is, at least at the beginning of the process, much longer then the free fall time. The observed luminosity of such a star in the frame comoving with respect to the center of mass of the star is considered to be the sum of luminosities of all the constituent pieces that are seen by the observer.

In our calculations we describe the star with the $n = 5$ polytropic model, so that

$$\rho(r) = \frac{\rho_0}{[1+(\frac{r}{r_0})]^{\frac{5}{2}}} \quad \text{and} \quad T(r) = \frac{T_0}{[1+(\frac{r}{r_0})]^{\frac{1}{2}}} \tag{1}$$

The parameter radius r_0 depends on the visible radius of the star r_s, on the cross section σ and on the number of pieces N that make up the stellar model as follows:

$$r_0 = \frac{r_s^2}{\sqrt{\frac{\sigma}{\pi}}}\left(\frac{32}{27N}\right)^{\frac{1}{2}} \tag{2}$$

Clearly the $n = 5$ polytrope is an oversimplification for the stellar model. Furthermore, combining Eqs. 1 and 2 we can see that with a reasonable σ, say $\sigma = \pi r_s^2/q$, where q is of the order 100, one needs an impractical number of particles ($N >\approx 10^8$) to model a star with the surface to central temperature ratio 1 : 3000 or less characteristic of real stars. However, the results of the $N = 5$ polytrope model are indicative and simple and allow reasonable extrapolation from small to high values of N and σ and this is our reason for considering these models.

2. RESULTS

Numerical calculations show that in the frame comoving with the star the stellar luminosity increases during the passage of the star by the black hole due to the described shearing effect. Results of a simulation are shown in fig.1.

The maximum luminosity change during a flyby or a capture as a function of angular momentum of the star is shown in fig. 2. Note that the ratios should not be taken literally as they were calculated for $N = 2 \times 10^5$ and $\sigma = \frac{\pi r_s^2}{415}$, which does not correspond to a real star, however, the dependence on l is characteristic. We note that only a capture close to critical produces the violent event as described in the previous section. The calculated dependence of the maximum luminosity

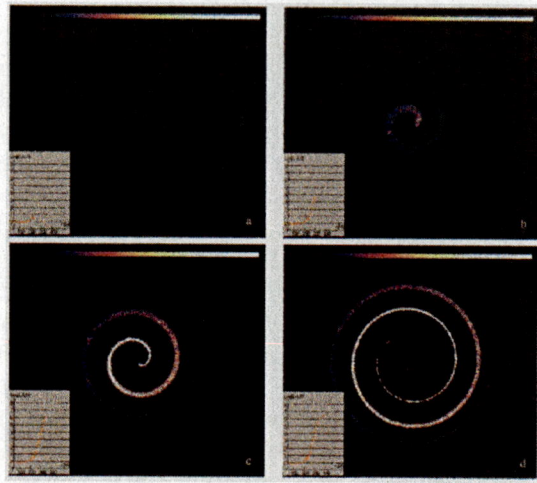

FIGURE 1. A series of snapshots calculated at time steps of $30R_S/c$. The simulation was done with $N = 3 \times 10^5$ and $\sigma = \frac{\pi r_s^2}{415}$, so that for this model the central to surface temperature ratio is only ≈ 14. Graphs in the left bottom corner show the luminosity as a function of time.

increase during a critical capture as a function of the model parameter N is shown in fig. 3. It is apparent that the ratio L_{max}/L_0 increases as N^2. This result can be interpreted as an indication that at the highest luminosity all the parts outside radius $r_s/10$ contribute to radiation irrespective of the number of parts N used in the calculation. Therefore, we further speculate that the star is almost completely disrupted and releases a major fraction of all its energy stored in the form of light just before it is critically captured by the black hole.

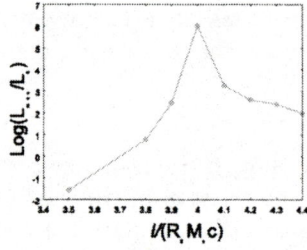

FIGURE 2. The luminosity change as a function of stellar orbital angular momentum, calculated for $N = 2 \times 10^5$

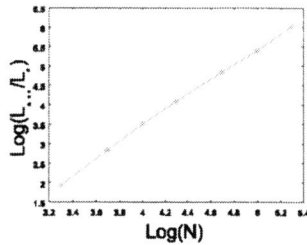

FIGURE 3. The maximum luminosity change as a function of the model parameter N.

3. CONCLUDING REMARKS

Our model, although quite crude, predicts that a medium size galactic black hole can completely disrupt a normal star by critical capture and release a large fraction of stellar thermal energy in a short burst just before half of the star crosses the horizon of the black hole. The appearance of such an event must be calculated using ray tracing in the gravitational field of the black hole. Some related ray-tracing results are in Čadež & Gomboc 1996, Gomboc & Čadež 1996, Gomboc 1999.

ACKNOWLEDGEMENTS

This work has been supported by Slovenian Ministry of Science and Technology under contract number 0790.

REFERENCES

M.J. Rees, *Nature* 333, 523 (1988).

J.H. Lacy, et al., *ApJ* 262, 120 (1982).

M. Cappellari, et al., submitted to *ApJ* (1998).

A. Gomboc, & A. Čadež, *Variable Stars and the Astrophysical Returns of Microlensing Surveys*, R. Ferlet, et al. (eds) 413 (1996).

A. Čadež, & A. Gomboc, *A&A Suppl. Ser.* 119, 293 (1996).

Gomboc et al. 1999, available at http://www.fiz.uni-lj.si/astro/pictures.html

DERIVING THE EMISSIVITY LAW IN ACCRETION DISKS FROM X-RAY IRON EMISSION LINES

A. Čadež [1], M. Calvani [2], C. Di Giacomo [3], P. Marziani [2]

1) Deparment of Physics, University of Ljubljana, Slovenia; 2) Astronomical Observatory, Padova, Italy; 3) Department of Physics, University of Padova, Italy

ABSTRACT The iron Kα fluorescent lines detected in several Seyfert 1 galaxies are usually explained as arising in the innermost regions of an accretion disk around a supermassive black hole. We present a method that allows to derive a free-form fit to the radial emissivity law in the disk. We find that most of the emission comes from quite narrow regions very close to the black hole.

KEYWORDS: Accretion, accretion disks; Galaxies: active; Galaxies: nuclei; X-rays:galaxies;

1. INTRODUCTION

One of the usual assumptions in modeling FeKα line data is that the emissivity in the accretion disk follows a power law with radius. A different approach consists in deriving a free-form fit to the radial emissivity $\epsilon(R)$. The first study of this kind was done by Mannucci et al. (1992) for optical Balmer lines. It was then used in the context of Fe Kα lines by Dabrowski et al. (1997). We here improve upon these results.

2. A FREE-FORM FIT TO $\epsilon(R)$

The relevant integral equation for the radial emissivity profile $\epsilon(r)$:

$$F_{\text{obs}}(E_{\text{obs}}) = \int_{\text{image}} \epsilon(r) g^4 \delta(E_{\text{obs}} - gE_0) d\Xi \quad (1)$$

is written in discrete form as:

$$F_i = \sum_{j=1,N} H_{ij}(a,\theta)\epsilon_j \quad (i = 1\ldots M) \quad (2)$$

where the matrix $\mathbf{H}(a,\theta)$ is evaluated numerically (see Fanton et al. 1997 for details. a is the angular momentum of the black hole, $\mathcal{M}$ its mass, θ the disk inclination to the line of sight); the index $i(=1\ldots M)$ refers to the energy bins into which the spectra are binned, and the index $j(=1\ldots N)$ to the radial bins into which we divide the disk. These equations have no unique solution if $N > M$ and they may not have a solution at all if the rank of $\mathbf{H}(a,\theta)$ is less than M. Following Lucy (1974), we assume that the "best" solution is the one for which the curve $\epsilon(r)$ has the shortest length. We impose the further constraint $\epsilon_i \geq 0$. We therefore treat the above equations as Euler's equations to a minimization problem with the action $S(a,\theta) = \chi^2(a,\theta) + \lambda \Sigma^2$ where

$$\chi^2(a,\theta) = \frac{1}{M-1} \sum_{\substack{i=1,M \\ j=1,N \\ k=1,N}} \frac{[F_i^{obs} - H_{ij}(a,\theta)\epsilon_j][F_i^{obs} - H_{ik}(a,\theta)\epsilon_k]}{\sigma_i^2} \qquad (3)$$

$\Sigma^2 = \sum_{j=1,N} \epsilon_j^2$ where F_i^{obs} is the observed energy flux in the i-th bin. We search for the minimum action S by moving in the direction of negative gradient and in such a way that the Σ part of the gradient is projected on the subspace of the matrix $\mathbf{H}$ with the projector $\mathbf{P}$ such that $\sum_{j=1,N} H_{ij} P_{jk} = 0 \qquad (k=1\ldots N,\ j=1\ldots M)$. The final action is minimum with respect to both χ^2 and Σ^2. This solution is the best one for the disk model corresponding to Eq. 2 for given a and θ. The "best" disk model corresponds to the absolute minimum of χ^2.

3. RESULTS

We applied our inversion procedure to the Fe Kα line residuals (Nandra et al. 1997). The accretion disk is assumed to be geometrically thin and axially symmetric around the hole's axis of rotation. Fig. 1 shows the minimum χ^2 (as a function of a and θ) for some sources; the corresponding best fit radial emissivity profiles are shown in Fig. 2. We find that in most cases the inclination angle is well constrained. The angular momentum of the black hole, however, is not, except in the 4.5d long exposure of MCG-6-30-15 (deep minimum). In practically all cases the source of emissivity is constrained to radii $< 20\mathcal{M}$ and is usually packed in few narrow rings. Note that in the only two available long exposure data (MCG-6-30-15 deep minimum and NGC 3516) the deduced emissivity profiles show a structure that looks like a power law.

ACKNOWLEDGEMENTS

MC and PM acknowledge the financial support of the Italian Ministry for University and Research (MURST) under the grant Cofin98-02-32.

REFERENCES

Dabrowski, Y. et al. 1997, MNRAS, 288, L11
Fanton, C., Calvani, M., de Felice, F. & Čadež, A. 1997, PASJ, 49, 159
Iwasawa, K. et al. 1996, MNRAS, 282, 1038
Lucy, L.B. 1974, AJ, 79
Mannucci, F., Salvati, M. & Stanga, R.M. 1992, ApJ, 394, 98
Nandra, K., George, I.M., Mushotzky, R.F., Turner, T.J., & Yaqoob, T.Y. 1997, ApJ, 477, 602
Nandra, K., George, I.M., Mushotzky, R.F., Turner, T.J., & Yaqoob, T.Y. 1999, Astro-ph 9907193

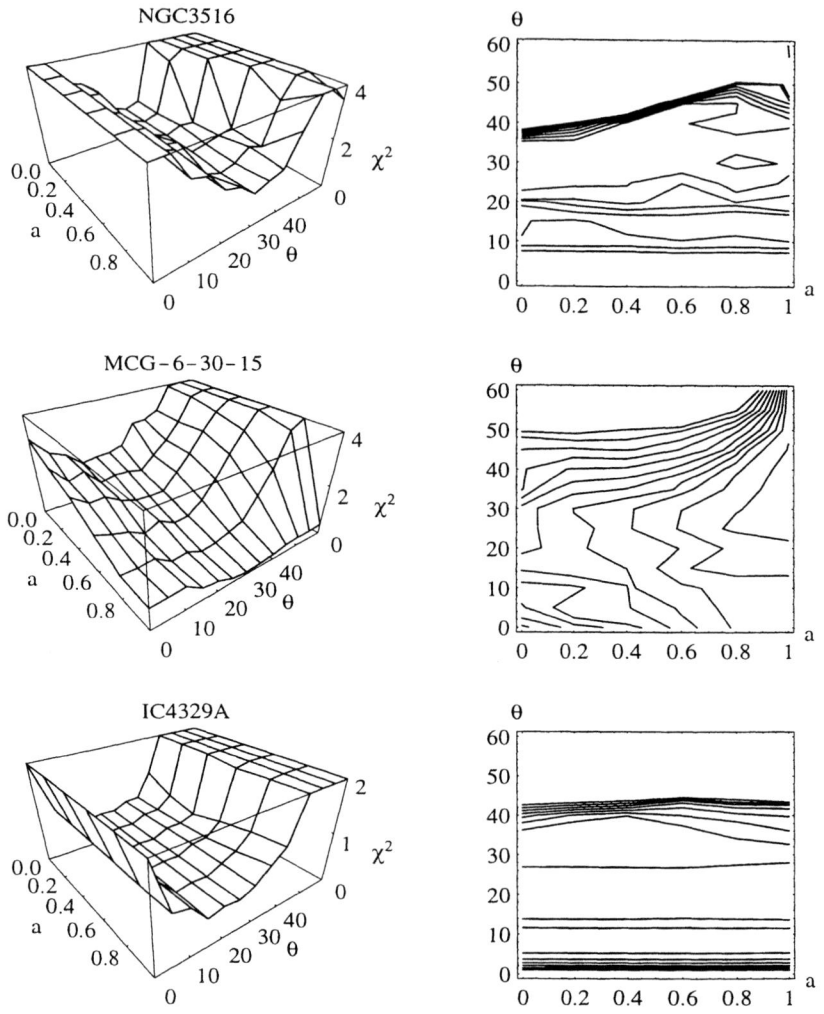

FIGURE 1. *Left columns*: 3D-plots of the minimum χ^2 as function of a and θ. *Right columns*: corresponding χ^2 iso-contours in the (a, θ) plane. Top row: our results for the source NGC 3516 (Nandra et al. 1999). The best fit is for $\theta \approx 30°$ with $\chi^2_{\min} \approx 1.35$; a is not constrained. Isocontours are drawn for $0 \leq \chi^2 \leq 4$ in steps of 0.4. *Middle row*: results for MCG-6-30-15 (Iwasawa et al. 1996). The best fit is for $\theta \approx 30°$ with $\chi^2_{\min} \approx 0.1$; in this case $a/\mathcal{M} \approx 1$ is favoured. Isocontours are drawn for $0 \leq \chi^2 \leq 4$ in steps of 0.4 *Bottom row*: results for IC 4329A (Nandra et al. 1997). The best fit is for $\theta \approx 15°$ with $\chi^2_{\min} \approx 0.59$; a is not constrained. Isocontours are drawn for $0 \leq \chi^2 \leq 2$ in steps of 0.2.

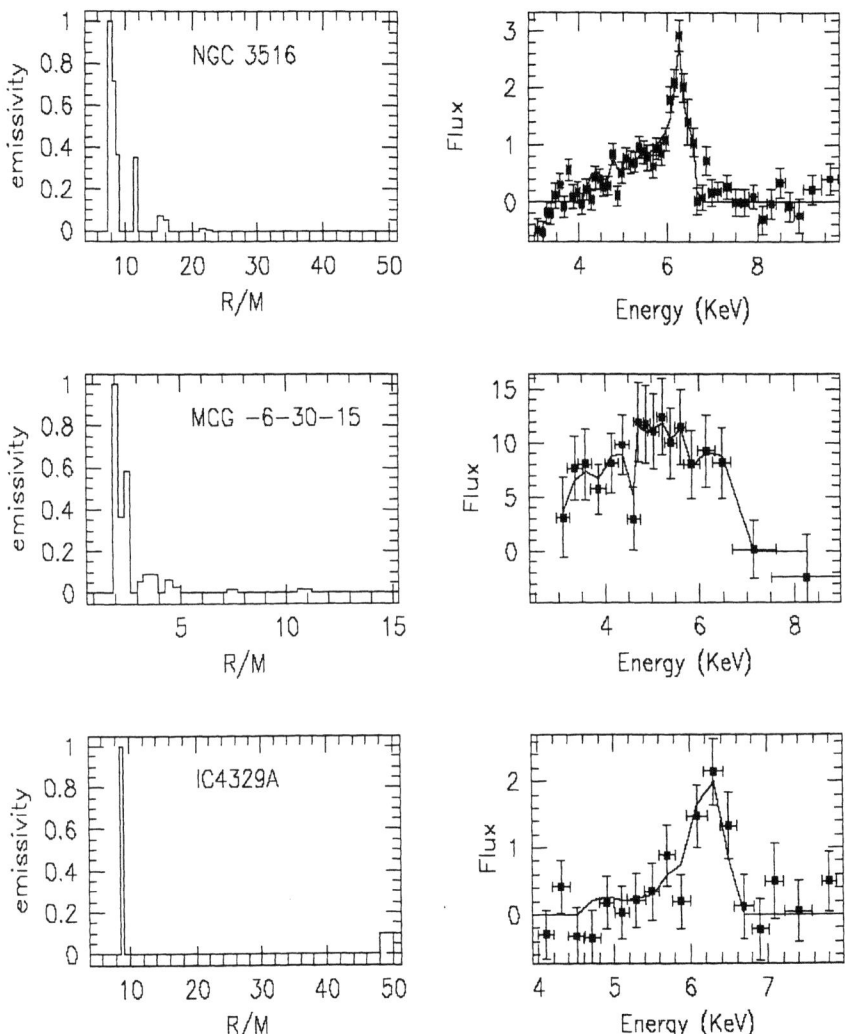

FIGURE 2. *Right columns*: profiles of the FeKα emission lines for the same sources as in Fig. 1; the continuous line is our best fit to the data. Flux is in units of 10^{-4} ph cm^{-2} s^{-1} keV^{-1}. *Left columns*: the corresponding free form emissivity profile that we obtain. Scale in arbitrary units. *Top row*: NGC 3516: $\theta = 30°$, $a = 0$, $R_{\text{out}} = 50\mathcal{M}$, $N = 41$. *Middle row*: MCG-6-30-15: $\theta = 30°$, $a/\mathcal{M} = 0.998$, $R_{\text{out}} = 15\mathcal{M}$, $N = 30$. *Bottom row*: IC 4329A: $\theta = 15°$, $a = 0$, $R_{\text{out}} = 50\mathcal{M}$, $N = 41$.

VARIABILITY OF THE RELATIVISTIC IRON K-LINE IN NGC 3516

A. Čadež [1], M. Calvani [2]

1) Deparment of Physics, University of Ljubljana, Slovenia
2) Astronomical Observatory, Padova, Italy

ABSTRACT A continuos five day ASCA observation of the Seyfert 1 galaxy NGC 3516 (Nandra et al, 1999) has shown that the iron K-line profile is variable. Nandra et al. fitted disk line models to the individual spectra and were not able to identify a single parameter which accounts for the changes in the profile. We show here that the variability can be interpreted in terms of a variable free-form fit to the radial emissivity profile of the accretion disk.

KEYWORDS: Galaxies: active; Galaxies: nuclei; X-rays:galaxies; Accretion, accretion disks

1. INTRODUCTION

An analysis of the relativistic Fe Kalpha line in the Seyfert 1 galaxy NGC 3516, based on a continuus $\approx$4.5 days observation with ASCA has been recently presented by Nandra et al. 1999. Disk line models were fitted to the whole dataset, giving an excellent fit, indicating that the emission is concentrated in the innermost regions of the accretion disk. It was not possible however to distinguish between rotating (Schwarzschild) and non-rotating (Kerr) black hole models.

The data set was then split into 8 segments, with equal duration of 46.4 ks. Nandra et al. fitted disk line models to the individual spectra but were not able to identify a single parameter that accounts for the observed changes in the Fe Kalpha line profiles. The fits were done assuming that the emissivity profile for the accretion disk is a power law $\epsilon(R) \propto R^{-q}$.

2. A FREE-FORM FIT TO $\epsilon(R)$

Rather than assuming a given emissivity law, we developed a procedure that, inverting the line profile, allows to infer the (free-format) emissivity profile, the accretion disk inclination θ and the black hole angular momentum a (details in Čadež et al. 2000). We the applied this procedure to the NGC 3516 data sets of Nandra et al. 1999. The accretion disk is assumed to be geometrically thin and axially symmetric around the hole's axis of rotation.

The results for the whole data set are presented in Čadež et al. 2000. We find that the emissivity profile shows a structure that resembles a power law. Our best fit for the inclination angle to the line of sight is $\theta \approx 30°$. We also are not able to constrain the angular momentum a of the black hole.

3. LINE VARIABILITY

We fixed the inclination angle to be $\theta = 30°$ as derived from fitting the mean spectrum; the rest energy of the line is fixed at 6.4 keV. The free-form emissivity profile was then derived for each of the 8 segments for the Schwarzschild case. The results are shown in Fig. 1.

Each of the 8 segments is well fitted with our procedure. The derived free-form emissivity profile is however different for each of them. Note that the single fits are *not* power laws.

In Fig. 2 the 8 emissivity fits are stack together in order to show the time evolution, while Fig. 2 shows their contour plot in the time vs. R/M plane.

Although 8 segments are not the "optimum" to derive firm conclusions about the line variability, we here argue that there is very likely a pattern developing in time in the emissivity profile. Initially, the emissivity is concentrated in a narrow ring of $\Delta R \approx 0.5M$ at $R \approx 8M$. With time, the ring widens and the emissivity decreases, as if a kind of flash propagates to smaller and larger radii.

4. CONCLUSIONS

We argue that the data can be interpreted in terms of a flare occuring in a narrow ring of the accretion disk (≈ 1 light minute for a 10^7 M$_\odot$ black hole) at $R \approx 8M$ that spreads both towards infinity and down to the black hole.

The flare could be produced locally by the (rotating) magnetic field in the accretion disk produced by tidal winding of the field lines.

ACKNOWLEDGEMENTS

MC acknowledges the financial support of the Italian Ministry for University and Research (MURST) under the grant Cofin98-02-32.

REFERENCES

Čadež, A., Calvani, M., Di Giacomo, C., & Marziani, P. 1999, these proceedings; full details in New Astronomy, 5, 69, 2000.

Nandra, K., George, I.M., Mushotzky, R.F., Turner, T.J., & Yaqoob, T.Y. 1999, ApJ, 523, L17.

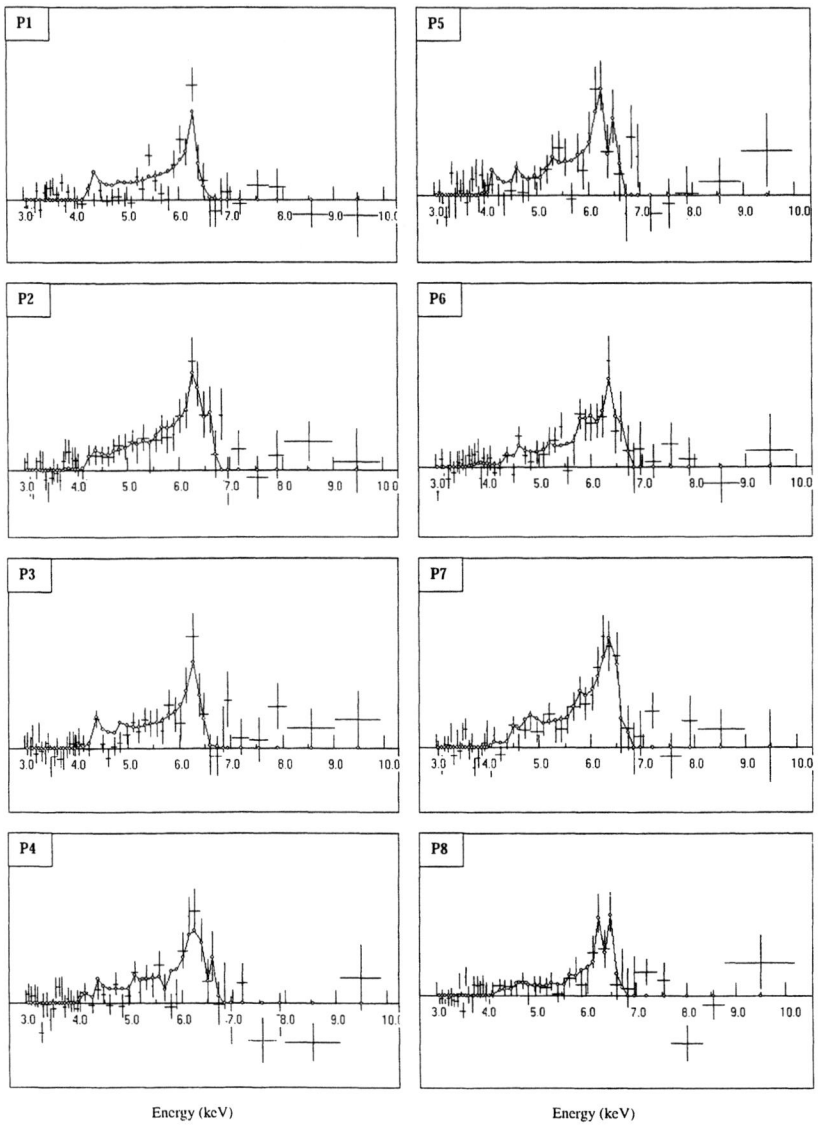

FIGURE 1. The free-format emissivity fits to the 8 segments ($\theta = 30°$, $a = 0$).

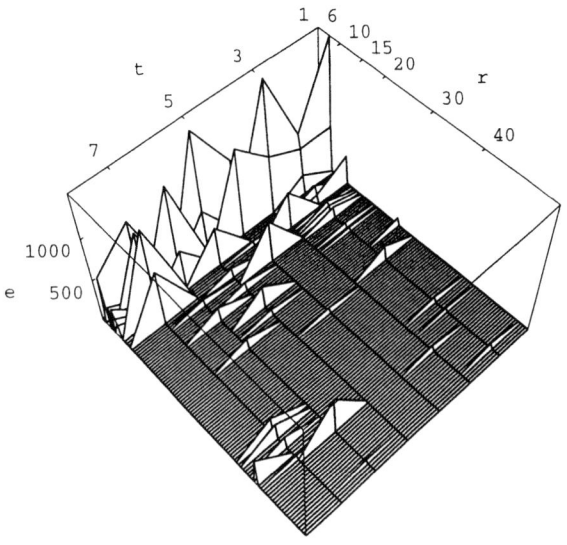

FIGURE 2. 3D graph (time - emissivity - R/M) for the 8 emissivity fits. The emissivity is here multiplied by the appropriate square root of the determinant of the spacial metric to show the power contribution of the various rings. The coordinate t refers to the eight segments into which the data set was split (Nandra et al. 1999), with equal duration (46.4 ks).

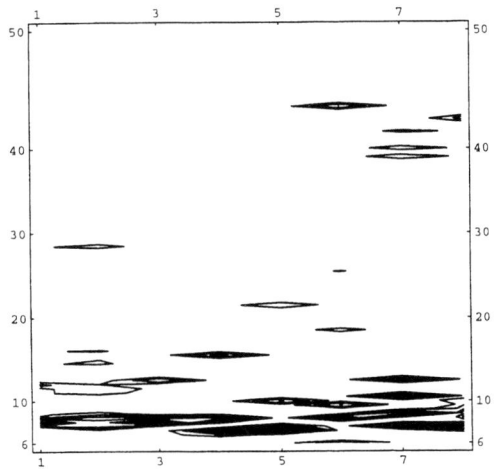

FIGURE 3. Contour plot of the 8 emissivity fits (time vs. R/M).

AN EUVE OBSERVATION OF THE GLOBULAR CLUSTER NGC 1851

Paul J. Callanan[1], Jeremy J. Drake[2], Antonella Fruscione[2] and Damian Christian[3]

1) Department of Physics, University College, Cork, Ireland
2) Center for Astrophysics, 60 Garden Street, Cambridge, MA 02138
3) Space Telescope Science Institute, 3700 San Martin Drive, Baltimore, MD 21218

ABSTRACT   We have detected EUV emission from the globular cluster NGC 1851, using the Deep Survey Photometer aboard the *Extreme Ultraviolet Explorer (EUVE)*. The minimum EUV luminosity implied by our detection is $\sim 10^{35} - 10^{36}$ ergs s^{-1}, for a distance and reddening appropriate to NGC 1851. This is only the second detection of a globular cluster at EUV energies. If this flux is due to the cluster LMXB X0512-401, then the observed EUV luminosity is likely to be comparable to the 2-10 keV X-ray luminosity of the LMXB. With the detection of EUV emission from M15, these observations suggest that the EUV flux of LMXBs in general may represent a sizable fraction of their total bolometric luminosity. However, additional observations are needed to conclusively show that these cluster LMXBs (X0512-401 and AC211) indeed dominate the cluster EUV emission.

KEYWORDS: binaries: close – globular clusters: individual (NGC 1851) – X-rays: stars

1. INTRODUCTION

As discussed by Callanan, Drake & Fruscione (1999), little is known of the EUV properties of Low Mass X-ray Binaries (LMXBs), primarily because most LMXBs are relatively distant (>1 kpc) and lie behind significant optical extinction (>0.5 magnitudes) that dramatically reduces any EUV flux.

The two LMXBs with the lowest reddening are those that reside in the globular clusters M15 (AC211) and NGC 1851 (X0512-401): these systems provide a unique opportunity to study the EUV characteristics of LMXBs in general. Indeed, Callanan, Drake & Fruscione (1999) have found EUV emission from M15, which they argue is most likely due to AC211. This prompted us to search for EUV emission from NGC 1851.

X0512-401 is a moderately bright system (X-ray luminosity $\sim 10^{36}$ ergs s^{-1}), first discovered by Clark et al (1975). Callanan, Penny & Charles (1995) found evidence for only low amplitude variability in the 2-10 keV lightcurve. Recently, Deutsch et al (1996) have found a UV-bright object (B$\sim$21, (U-B) = -0.9), only 2" away from the X-ray position. No orbital period is known, although the faintness of the optical counterpart indicates a short period system: indeed, the only LMXBs in the compilation of van Paradijs & McClintock (1994) of comparable absolute magnitude have periods $\leq$ 1 hr.

2. OBSERVATIONS AND DATA REDUCTION

NGC 1851 was acquired by the *EUVE* Deep Survey and Spectrometer (DSS) telescope on UT 1995 Dec 15.42, and was observed until 1995 Dec 16.73. This telescope, filter and detector combination has significant transmission between about 65 and 170 Å, peaking at approximately 90 Å (Bowyer et al 1994). The total integration time was 33 ksec (see Callanan, Drake & Fruscione 1999 for further details concerning the data reduction).

The EUV source was detected at a position of $RA = 05^h\ 14^m\ 08.4^s$, $Dec = -40°\ 02'\ 58''$ (J2000) with an uncertainty of aproximately ± 25 arc sec. This is in good agreement with the coordinates of the cluster itself (to within 1.2σ: e.g. Djorgovski & Meylan 1993). The average count rate in the DSS was 0.0044 ± 0.0006 count s^{-1}. The lightcurve is shown in Fig 1: there is no evidence for any statistically significant variability.

3. DISCUSSION

At a distance of 12.2 kpc (e.g. Webbink 1985), the EUV source in NGC 1851 lies 1.8 ± 1.5 pc from the centre of the cluster, which corresponds approximately to the half light radius. X0512-401 itself is some $12''$ away from the cluster core: unfortunately, *EUVE* has insufficient positional accuracy to allow us to distinguish between X0512-401 or objects withinin the cluster core as the origin of the EUV emission, on the basis of source location alone.

The reddening towards NGC 1851 has been measured by Walker (1992) to be $E_{B-V} = 0.02 \pm 0.02$: in a subsequent paper, (Walker 1998), a slightly higher value of 0.05 ± 0.02 was derived, although systematic effects may have led to an artifically high value. Using the Savage & Mathis (1979) reddening curve, these correspond to $N_H = 10^{20}$ and 2.5×10^{20} cm^{-2} respectively. However, radio studies (Dickey & Lockman 1990) yield $N_H \sim 3.6 \times 10^{20}$ cm^{-2}, with relatively little scatter ($\sim 10\ \%$ over an area of a few square degrees). For the discussion that follows, we will use a column of $1-4 \times 10^{20}$ cm^{-2}.

Following Callanan, Drake & Fruscione (1999), we model the spectral shape of the source as a blackbody. In Fig 2, we plot the intrinsic bolometric luminosity as a function of effective temperature and line of sight column, corresponding to our DSS countrate: the data imply a bolometric luminosity of $\sim 10^{35} - 10^{36}$ ergs s^{-1}, for the range of columns discussed above.

4. THE LMXB IN NGC 1851 AS THE EUV SOURCE

By analogy with the EUV emission of M15 (Callanan, Drake & Fruscione 1999) we believe that the origin of the observed EUV emission is due to X0512-401 itself.

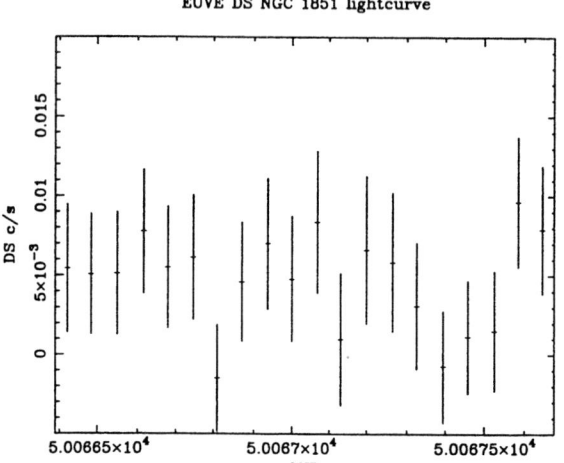

FIGURE 1. The EUVE Deep Survey Photometer light curve of NGC 1851.

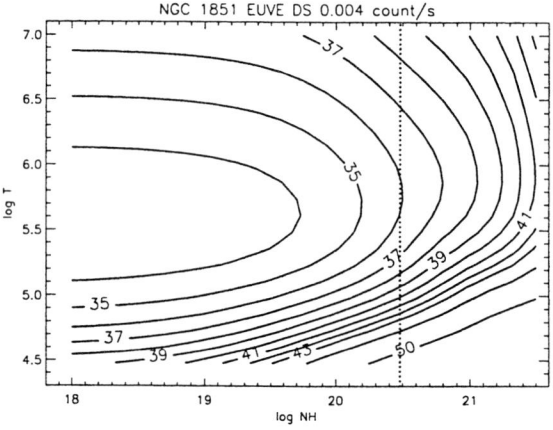

FIGURE 2. A contour plot of the bolometric luminosity of the EUV component, as a function of the line-of-sight neutral hydrogen column density and effective temperature. The contours are logarithmically spaced, for a blackbody spectral shape The luminosity is scaled to our DS countrate of 0.004 counts s^{-1}. The vertical line denotes a column of 3×10^{20} cm^{-2} towards the cluster.

As in the case of M15, our confidence would be bolstered by the detection of any variability in the EUV lightcurve: however, the low S/N of our data precludes any such measurement.

Our simple modelling indicates that the EUV luminosity is likely to be a sizable fraction of the X-ray luminosity of X0512-401, as found in the case of AC211 in M15: indeed because of absorbing column intrinsic to the binary, it could be substantially higher. However, the short period of X0512-401 (as inferred from its low absolute magnitude), coupled with the lack of any obvious X-ray modulation (Callanan, Charles & Penny 1995) suggests that X0512-401 is a lower inclination system. We might therefore expect a lower line-of-sight absorbing column within the binary itself, compared to that of AC211. Hence the correction to the EUV luminosity due to the intrinsic column in the case of X0512-401 is likely to be less than that for AC211. An accurate determination of the NGC 1851 EUV luminosity will require a simultaneous measurement of both the EUV flux and X-ray column.

5. CONCLUSIONS

We have detected EUV emission from the globular cluster NGC 1851. The emission is likely due to the cluster LMXB, and if so represents the second detection at EUV energies of an LMXB, implying in this case a minimum EUV luminosity of $\sim 10^{35}$–10^{36} ergs s^{-1}, and suggesting that the EUV flux of LMXBs in general may represent a sizable fraction of their total bolometric luminosity. However, additional observations are first needed to conclusively show that these cluster LMXBs indeed dominate the cluster EUV emission.

REFERENCES

Bowyer, S., Lieu, R., Lampton, M., Lewis, J., Wu, X., Drake, J.J. & Malina, R.F. 1994, ApJS, 93, 569

Callanan, P.J., Charles, P.A. & Penny, A.J., 1995, MNRAS, 273, 201

Callanan, P.J, Drake, J.J. & Fruscione, A. 1999, ApJL, 521, 125

Clark, G.W., Markert, T.H. & Li, F. 1975, ApJ, 199, L93

Dickey, J.M. & Lockman F.J., 1990, ARAA, 28, 215

Deutsch, E.W., Anderson, S.F., Margon, B. & Downes, R.A. 1996, ApJ, 472, 97

Djorgovski, S. & Meylan, G. 1993, in Structure and Dynamics of Globular Clusters, ed. S. Djorgovski & G. Meylan, ASPCS vol 50, p325

Savage, B.D. & Mathis, J.S., 1979, ARA&A, 17, 73

van Paradijs, J. & McClintock, J.E., 1994, A&A, 290, 133

Walker, A. 1992, PASP, 104, 1063

Walker, A. 1998, AJ, 116, 220

Webbink, R. 1985 in Dynamics of Globular Clusters, IAU Symp 113, ed. J. Goodman & P. Hut (Dordrecht: Reidel)

BEPPOSAX OBSERVATIONS OF CEN X-4 IN QUIESCENCE

Sergio Campana[1], Luigi Stella[2], Sandro Mereghetti[3], Davide Cremonesi[3]

1) Osservatorio Astronomico di Brera, Via Bianchi 46, I–23807 Merate, Italy
2) Osservatorio Astronomico di Roma, Via Frascati 33, I-00040 Monteporzio Catone (Roma), Italy
3) Istituto di Fisica Cosmica "G. Occhialini", CNR, Via Bassini 15, I–20133 Milano, Italy

ABSTRACT   We report on a 61 ks BeppoSAX observation of the Soft X–ray Transient Cen X-4 during quiescence which allowed to study the source X-ray spectrum from ~ 0.3 keV to ~ 8 keV. A two-component spectral model was required, consisting of black body with temperature of ~ 0.1 keV and a power law with photon index ~ 2. These values are compatible with earlier ASCA results indicating that Cen X-4 may be stable, within a factor of a few, over a 5 year baseline.

KEYWORDS: stars: neutron — stars: individual (Cen X-4) — pulsars: general — X–ray: stars

1. INTRODUCTION

Cen X-4 is one of the best studied Soft X–ray Transients (SXRTs; for a review see Campana et al. 1998a). X-ray outbursts have been detected in 1969 and 1979, reaching a peak flux of $L_X \sim 4 \times 10^{37}$ erg s^{-1} ($d \sim 1.2$ kpc; Kaluzienski et al. 1980). Type I bursts were observed, testifying to the presence of an accreting neutron star.

Cen X-4 was observed in quiescence by Einstein IPC (in 1980, ~ 440 d after the 1979 outburst; Petro et al. 1981), EXOSAT CMA (in 1986, van Paradijs et al. 1987) and ROSAT HRI (Campana et al. 1997). ASCA observed twice Cen X-4: in the first observation the luminosity was $L_X \sim 2^{+2}_{-1} \times 10^{32}$ erg s^{-1} (0.5–10 keV, Asai et al. 1996). The X–ray spectrum was well fit by a black body component ($kT_{bb} = 0.16^{+0.03}_{-0.02}$ keV) plus an additional power-law component with photon index $\Gamma = 1.9 \pm 0.3$. The 0.5–10 keV flux from the two spectral components was comparable. The column density was constrained to be $N_H \lesssim 2 \times 10^{21}$ cm^{-2}. The equivalent radius of the black body emission was determined to be ~ 1.8 km, substantially smaller than the radius of a neutron star. The second ASCA observation provided similar results: $L_X \sim 3^{+3}_{-2} \times 10^{32}$ erg s^{-1}, $kT_{bb} = 0.13 \pm 0.02$ keV, $\Gamma = 2.5 \pm 0.5$ and $N_H = (3 \pm 1) \times 10^{21}$ cm^{-2}. Campana et al. (1997) showed all these measurements are consistent with the same luminosity level of the ASCA observations. Yet, during the ROSAT HRI pointing a factor of ~ 3 flux variation was observed on a timescale of a few days (Campana et al. 1997).

A full account of this work is reported in Campana et al. (2000).

2. BEPPOSAX OBSERVATION

A BeppoSAX (Boella et al. 1997a) observation of Cen X-4 took place on 9–11 Feb. 1999. Only the Low Energy Concentrator Spectrometer (LECS; 0.1–10 keV, Parmar et al. 1997) and the Medium Energy Concentrator Spectrometer (MECS; 1.3–10 keV, Boella et al. 1997b) provided 21 ks and 61 ks of useful data, respectively.

The LECS and MECS events were extracted within a radius of $4'$ centered on the source position. We collected 233 photons from the LECS and 632 from the MECS in the full energy range. Background subtraction was applied using the standard BeppoSAX files. The LECS and MECS spectral data were rebinned in order to have at least 40 photons per channel. The spectral analysis was carried out in the 0.2–3 keV energy range for the LECS and 1.7–9 keV for the MECS. The source background subtracted count rates were $(4.7 \pm 0.6) \times 10^{-3}$ (0.1–3.1 keV) and $(2.6 \pm 0.4) \times 10^{-3}$ ct s^{-1} (1.7–9.0 keV) in the LECS and MECS instruments, respectively. We fit the spectral data with the XSPEC (version 10.0) package. All-single component models provided a poor fit to the data ($\chi^2_{red} \gtrsim 2$). Therefore we applied the conventional model for quiescent SXRTs, i.e. a soft black body component plus a hard power-law tail (Campana et al. 1998a,b; Asai et al. 1996, 1998). The fit was good ($\chi^2_{red} = 1.1$). The corresponding black body temperature was $kT_{bb} = 103^{+53}_{-32}$ eV and the power law photon index $\Gamma = 2.01^{+0.65}_{-0.68}$ (uncertainties are 90% confidence for a single parameter). The equivalent black body radius was $R_{bb} = 10.0^{+420}_{-2.2}$ km. The column density was $N_H = 2.6^{+5.2}_{-2.3} \times 10^{21}$ cm^{-2}. The unabsorbed luminosities are 1.5×10^{33} erg s^{-1} (0.1–10 keV) and 4.5×10^{32} erg s^{-1} (0.5–10 keV).

The equivalent black body radius derived from BeppoSAX data ($R_{bb} \gtrsim 6.2$ km) is larger than the ASCA one. Since these authors do not report the uncertainties on the equivalent black body radius, we reanalysed the GIS data (in SIS data Cen X-4 fall across two CCDs), finding, in general, good agreement with previous results. In the first observation the best fit value of the column density is consistent with zero and the other spectral parameters are consistently different from the ones of the second observation. For the second observation count rates in all the instruments are systematically lower than those in the first, even if we derive a factor of ~ 3 higher (unabsorbed) luminosity than Asai et al. (1998). Given the large uncertainties we repeat the analysis keeping fixed the column density to the BeppoSAX value. These fits provide marginally worse χ^2 values and result in a much more homogeneous spectral parameter values

3. DISCUSSION

There is growing evidence that the quiescent spectrum of SXRTs is made of two components, a soft component, usually modeled with a $kT_{bb} \sim 0.1 - 0.3$ keV black

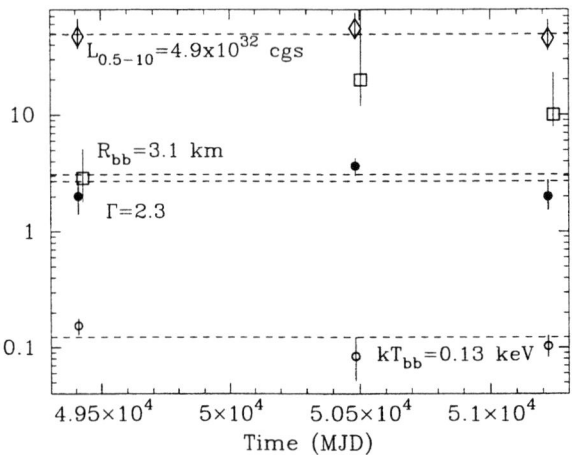

FIGURE 1. Long term behaviour of the 0.5–10 keV unabsorbed luminosity in units of 10^{31} erg s^{-1} (diamonds), of the power law photon index (filled circles), of the equivalent black body radius in km (squares, slightly offset in time) and of the equivalent black body temperature in keV (open circles), taken with ASCA and BeppoSAX imaging instruments. Errors are at the 68% confidence level for three parameters of interest.

body, and a hard power law component with photon index $\sim 1.5 - 2$ (Campana et al. 1998a,b; Asai et al. 1996, 1998; Guainazzi et al. 1999).

The soft component is usually attributed to the radiative cooling of the neutron star warm interior heated up during the accretion episodes of the outbursts (Campana et al. 1998a; Brown, Bildsten & Rutledge 1998). Black body models usually provide an emitting area smaller than the neutron star surface, with $R_{bb} \sim 1 - 2$ km (Verbunt et al. 1994; Campana et al. 1998a,b; Asai et al. 1996, 1998). Neutron star atmosphere models have been used to fit the available data. These models provide a good fit to the soft component and provide substantially larger effective temperatures and radii, consistent with the entire surface, with respect to a pure black body model (Rutledge et al. 1999). The hard component has been interpreted as due to the interaction of the relativistic wind with matter outflowing from the companion (shock emission), powered by a radio pulsar which reactivates in the quiescent state of SXRTs (Campana et al. 1998a; Stella et al. 1994). An alternative explanation identifies the soft component in matter that accretes onto the neutron star surface from polar regions whereas the hard component is produced in an advection-dominated accretion flow (Zhang et al. 1998; Menou et al. 1999).

Recently, an UV spectrum of Cen X-4 has been obtained with HST/STIS (McClintock & Remillard 2000). The main result is that in the ν versus νF_ν plot the unabsorbed flux decreases by only a factor of 2 from X–rays to optical (subtracted from the contribution of the companion star). Such a nearly νF_ν flat spectrum is clearly reminiscent of a shock emission.

The BeppoSAX observation presented here allow to further study the quiescent state of Cen X-4. A comparison in the 0.5–10 keV energy range (i.e. including the hard spectral component) can be carried out only with the two ASCA observations. The long term evolution of these quantities is consistent with a constant (cf. Figure 1). In particular, by averaging the spectral parameters obtained by ASCA and BeppoSAX we derive $\Gamma = 2.36 \pm 0.50$, $kT_{\rm bb} = 0.15 \pm 0.04$ keV and $L_X = (2.7 \pm 1.9) \times 10^{32}$ erg s^{-1}. The difference in the values derived for the equivalent black body radii is not significant due to the large uncertainties. In fact a fit with a constant value of $R_{\rm bb} \sim 1.6 \pm 1.0$ km is statistically acceptable. These observations indicate that the quiescent state of Cen X-4 is stable over a 5 years baseline, at least.

ACKNOWLEDGEMENTS

This work was partially supported through ASI grants.

REFERENCES

Asai K. et al., 1996, PASJ 48 257
Asai K. et al., 1998, PASJ 50 611
Boella G. et al., 1997a, A&AS 122 299
Boella G. et al., 1997b, A&AS 122 327
Brown E.F., Bildsten L., Rutledge R.E., 1998, ApJ 504 L95
Campana S. et al., 1997, A&A 324 941
Campana S. et al., 1998a, A&A Rev. 8 279
Campana S. et al., 1998b, ApJ 499 L65
Campana S. et al., 2000, A&A 358 583
Canizares C.R., McClintock J.E., Grindlay J.E., 1979, ApJ 234 556
Chevalier C., Ilovaisky S.A., van Paradijs J., Pedersen H., van der Klis M., 1989, A&A 210 114
Guainazzi M. et al., 1999, A&A 349 819
Kaluzienski L.J., Holt S.S., Swank J.H., 1980, ApJ 241 779
McClintock J.E., Remillard R.A., 1990, ApJ 350 386
McClintock J.E., Remillard R.A., 2000, ApJ 531 956
Menou K. et al., 1999, ApJ 520 276
Parmar A.N. et al., 1997, A&AS 122 309
Petro L.D., Bradt H.V., Kelley R.L., Horne K., Gomer R., 1981, ApJ 251 L7
Rutledge R.E., Bildsten L., Brown E.F., Pavlov G.G., Zavlin V.E., 1999, ApJ 514 945
Shahbaz T., Naylor T., Charles P.A., 1993, MNRAS 265 655
Stella L., Campana S., Colpi M., Mereghetti S., Tavani M., 1994, ApJ 423 L47
van Paradijs J., Verbunt F., Shafer R.A., Arnoud K.A., 1987, A&A 182 47
Verbunt F., Johnston H., Hasinger G., Belloni T., Bunk W., 1994, A&A 285 903
Zhang S.N., Yu W., Zhang W.W., 1998, ApJ 494 L71

MULTIPLE BEPPOSAX OBSERVATIONS OF IC 4329A TO PROBE THE ORIGIN OF THE COMPTON REFLECTION COMPONENT IN SEYFERT 1 GALAXIES

M. Cappi[1], G. Di Cocco[1], M. Dadina[1], G. Malaguti[1], M. Matsuoka[2], G. Matt[3], G.C. Perola[3], L. Piro[4]

1) ITESRE-CNR, I-40129, Bologna, Italy
2) SURP-NASDA, Tsukuba, Ibaraki 305-3805, Japan
3) Dipartimento di Fisica, Università degli Studi "Roma Tre", I-00146, Roma, Italy
4) Istituto di Astrofisica Spaziale, IAS-CNR, I-00133, Roma, Italy

ABSTRACT

IC 4329A is the brightest known Seyfert galaxy in hard ($\sim$ 2-30 keV) X-rays and is likely to be representative of Seyfert 1 galaxies as a class. A recent 100 ks *BeppoSAX* observation (Perola et al. 1999) clearly confirmed the presence of a warm absorber, a reflection component (R $\sim$ 0.6), and a high-energy cut-off in the power law at $E_c \sim$ 270 keV. Its richness in spectral features, combined with its large flux ($\sim 1.6 \times 10^{-10}$ erg cm^{-2} s^{-1} between 2-10 keV), make this target ideal for multiple observations (in particular with *BeppoSAX*) to search for spectral variations. Results obtained from 3 follow-up observations (40 ks each) are presented here. The first and most important goal of this study was to probe the origin of the Compton reflection component observed in Seyfert galaxies by monitoring the variability of the reflection continuum and Fe Kα line in response to primary continuum variations. The second goal was to search for variability in the high energy cutoff. We obtain however no conclusive results on any of these issues. In fact, all four observations unfortunately caught the source at almost the same flux, showing only little, and marginal, spectral changes between different observations.

KEYWORDS: galaxies: individual (IC 4329A) - galaxies: Seyfert - X-rays: galaxies

1. INTRODUCTION

Ginga, *ASCA* and recent *BeppoSAX* X-ray observations of Seyfert 1 galaxies strongly support a general *observational* framework that includes: ionized absorption by a warm absorber (WA), a steep intrinsic power-law continuum, a Compton reflection hump and associated neutral FeKα line (narrow and/or broad), and a high-energy cutoff (E_{cutoff}). These observational results have raised the following questions:

- What is the location of the X-ray reprocessor? Is it the accretion disk, the broad-line region (BLR) and/or the molecular torus? Is the answer the same for all Seyfert 1s?

- Where is the location of the WA? Is it within the BLR and/or the molecular torus or is it external to them? Is there only one WA?

- Does E_{cutoff} differ from source-to-source (as currently shown by *BeppoSAX*, Piro 1999, Matt 2000) or is it variable with time and/or flux in any single object?

In order to address these questions, we have performed multiple (4) observations of the prototypical Seyfert 1 galaxy IC4329A to obtain time-resolved spectral constrains on the WA, reflection continuum, Fe K_α line and, possibly, E_{cutoff}. This is potentially the best way (see also Weaver 2000) to establish whether these observational features are produced from material near the source (e.g. the accretion disk, in which case short lags, <1000 s, are expected between the reflection and continuum variations) or farther away (e.g. in the BLR or torus, in which case longer lags of at least $\sim 10^{6-7}$ s are expected).

2. THE MULTIPLE *BEPPOSAX* OBSERVATIONS

IC 4329A ($z = 0.014$) is a Seyfert 1 galaxy well studied in X-rays (Miyoshi et al. 1988; Piro, Yamauchi & Matsuoka 1990; Fiore et al. 1992; Cappi et al. 1996) and hard X-rays (Fabian et al. 1993; Madejski et al. 1995; Zdziarski et al. 1995a). It offers the best opportunity to address the above mentioned variability issues because it is among the brightest known Seyferts in X-rays (F(2-10keV)$\sim 1.6 \times 10^{-10}$erg cm^{-2}s^{-1}) and shows only *moderate* variability on short (< day) timescales, but large variability on longer timescales (a factor 2-3 in a few days), both below and above 10 keV (Done, Madejski & Zycki 2000; Fabian et al. 1993).

BeppoSAX observed IC 4329A in 1996 for 100 ks (Perola et al. 1999), and then subsequently 3 times in 1998, with the 3 observations spaced approximately 5 days apart. As expected, the source did *not* vary much *within* each observation. Unfortunately, it did not vary much *between* the 4 observations either (see Fig. 1), despite the large and systematic flux variations typically observed in IC4329A by RXTE (Done, Madejski & Zycki 2000).

Spectral ratios were produced between all observations, in search of any model-independent spectral variations (Fig. 2). We find marginal evidence for variability of the spectral features at $E < 2$ keV and at $E = 8$-10 keV, but we find no clear variations in the FeK_α and reflection component intensities, as well as in the high-E_{cutoff}. The variations at low energies are most likely due to variability of the ionization state of the WA responding to continuum variations, but complications due to the presence of a possible scattered component and/or intrinsic soft (excess) component cannot be ruled out.

Spectral fitting of each observation with a complex model including 2 absorption edges, a steep power-law continuum, a reflection component with associated Fe$K\alpha$, and a high-E_{cutoff} gave best-fit parameters similar to those reported by Perola et al. (1999) with somewhat larger errors (e.g. $\Gamma \sim 1.85 \pm 0.1$, R $\sim 0.6 \pm 0.2$, $E_{cutoff} \sim 250 \pm 100$ keV). Fig. 3 shows the marginal variations of the FeK line intensity and equivalent width. Fig. 4 shows the (similar) best-fit broad-band spectra of two different observations.

The most interesting result of the present analysis is that despite a 30% flux increase between obs. 1 and obs. 2, the line intensity became surprisingly weak (compare Fig. 1 and Fig. 3). During the higher flux obs. 3 the line re-established its typical value of $\sim$ 100-150 eV, though. In other words, there is some evidence (at a $\sim 2\ \sigma$ significance level) that the Fe Kα line *does not* follow "instantaneously" (at least within 40 ks) the continuum variations, but it *does* on a timescale between 4-10 days.

3. CONCLUSIONS

- Moderate variations of the absorption structure, most likely related to the warm absorber, are apparently detected but require further investigation for more quantitative conclusions.

- Despite the unprecedented statistics of these observations, we find *no* significant spectral variations of the Compton reflection continuum, the power law spectral index and the high-enegy cutoff. This is probably because *BeppoSAX* unfortunately caught IC 4329A at a not too different flux level (30% variation) in all four observations hampering, thereby, any spectral variability.

- There is marginal evidence that the FeKα line is being produced by the outer regions of the accretion disk in IC 4329A, thereby suggesting a possible explanation for the lack of a clear broad FeKα line in this object.

ACKNOWLEDGEMENTS

M.C. wishes to thank Martin Elvis and Fabrizio Nicastro for usefull discussions. This analysis has made use of data prepared by the *BeppoSAX* Scientific Data Center.

REFERENCES

Cappi M., Mihara T., Matsuoka M., et al., 1996, ApJ, 458, 149
Done, C., Madejski, G. M. and Życki, P. T. 2000, ApJ, 536, 213
Fabian, A.C. et al., 1993, ApJ, 416, L57
Fiore, F., Perola, G.C., Matsuoka, et al., 1992, A&A, 262, 37
Miyoshi, S., et al., 1988, PASJ, 40, 127
Madejski, G.M., et al. 1995, ApJ, 438, 672
Matt, G. 2000, these proceedings, astro-ph/0007105
Perola, G.C., Matt, G., Cappi, M., et al., A&A, in press
Piro, L., Yamauchi, M., & Matsuoka, M., 1990, ApJ, 360, L35
Piro, L. 1999, in "Heating and Acceleration in the Universe", Astron. Nachr., in press
Weaver, K. 2000, these proceedings, astro-ph/0007327

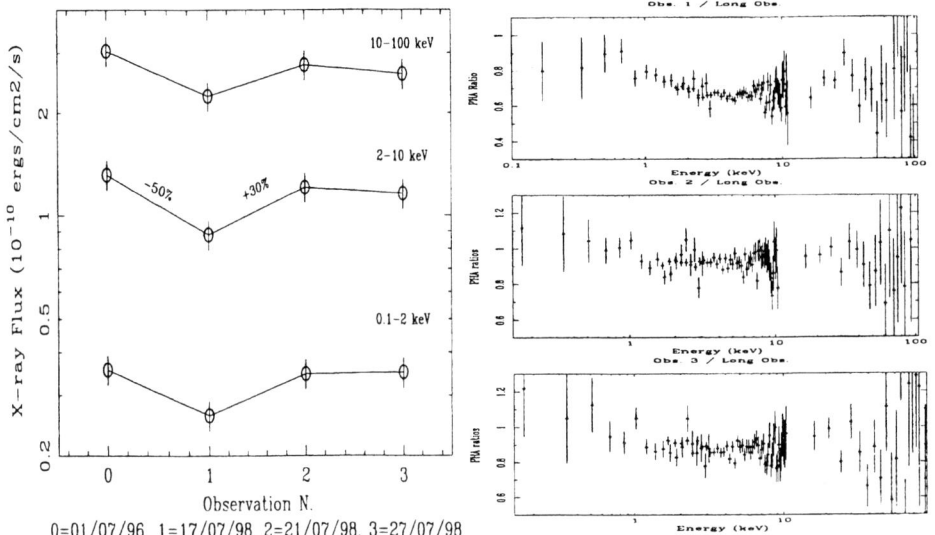

FIGURE 1. 0.1-2 keV, 2-10 keV and 10-100 keV flux of IC 4329A during Obs. 0 (the long 100 ks Obs.), Obs1, 2 and 3.

FIGURE 2. Spectral ratios of $\frac{Obs.1}{LongObs}$ (top pannel), $\frac{Obs.2}{LongObs}$ (mid pannel) and $\frac{Obs.3}{LongObs}$ (lower pannel)

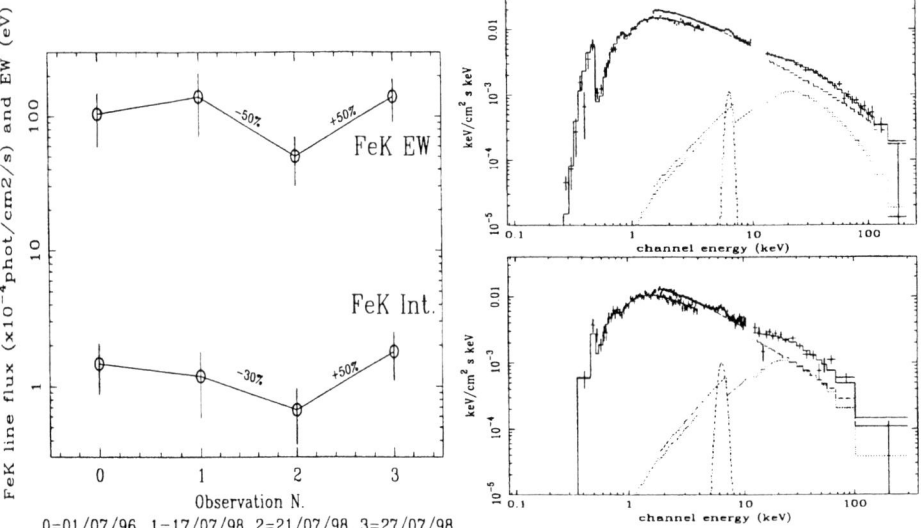

FIGURE 3. FeK line EW (upper points) and Intensity (lower points) during Obs. 0,1,2 and 3

FIGURE 4. Best-fit unfolded spectra during the long 100 Ks Obs. (top) and during Obs.1 (bottom) which illustrate the similarity of the spectra.

HARD SOURCES, THEIR COLOURS, AND THE X-RAY BACKGROUND

F.J. Carrera[1,2], J.P.D. Mittaz[1], M.J. Page[1]

[1] *Mullard Space Science Laboratory-University College London, Surrey RH5 6NT, UK*
[2] *Instituto de Física de Cantabria, CSIC-UC, E-39005 Santander, Spain*

ABSTRACT

The unresolved soft and hard X-ray backgrounds (XRB) are produced by hard extragalactic sources of unknown nature. We have assembled a sample of hard sources in *Rosat* pointings that offer a preview of the properties of these sources. We present here optical and infrared photometry of both unidentified and identified objects in this sample. Our data provide evidence for absorption as the origin of the hard spectra of these sources. Furthermore, the unidentified sources have the same optical and X-ray colours as the identified ones, so they are probably the same population. Since most of the extragalactic identified sources in our sample are broad line AGN, our survey supports the idea of obscured AGN as the main sources of the (soft) XRB.

KEYWORDS: X-rays: galaxies; galaxies: colors; galaxies: active; cosmology: diffuse ratiation

1. INTRODUCTION

The faint extragalactic sources that produce the remaining unresolved (20-40 per cent) soft (0.5-2 keV) X-ray Background (XRB) have to be harder than those already resolved, which have photon indices $\alpha \sim 1$ (Mittaz et al. 1999), and harder than the XRB itself ($\alpha_{1-7\,keV} \sim 0.4$, Miyaji et al. 1998), so that the total integrated emission of the resolved and unresolved sources reproduces the XRB spectral slope.

Most of the sources detected in 2-10 keV are also detected in 0.5-2 keV (Ueda et al. 1999). Therefore, the hard sources that produce the soft XRB also offer a preview of the sources of the hard XRB (2-10 keV), most of which (~ 70 per cent) is still unresolved (e.g. Ueda et al. 1999).

Popular models (based on the Unified Model for active galactic nuclei -AGN-) propose that these hard sources are AGN with intrinsic cold Hydrogen absorption, that modifies their intrinsically soft spectrum ($\alpha \sim 1$) so that their integrated contribution mimics the XRB spectral shape (Comastri et al. 1995, Gilli et al. 1999).

We have assembled a sample of sources significantly harder than the XRB ($\alpha < 0.5$) in about 200 *Rosat* pointings. The breakdown of our identifications shows that they are broad line AGN, narrow emission line galaxies (NELG), "normal" galaxies, clusters of galaxies and Galactic stars (Page et al., 2000a, and these proceedings).

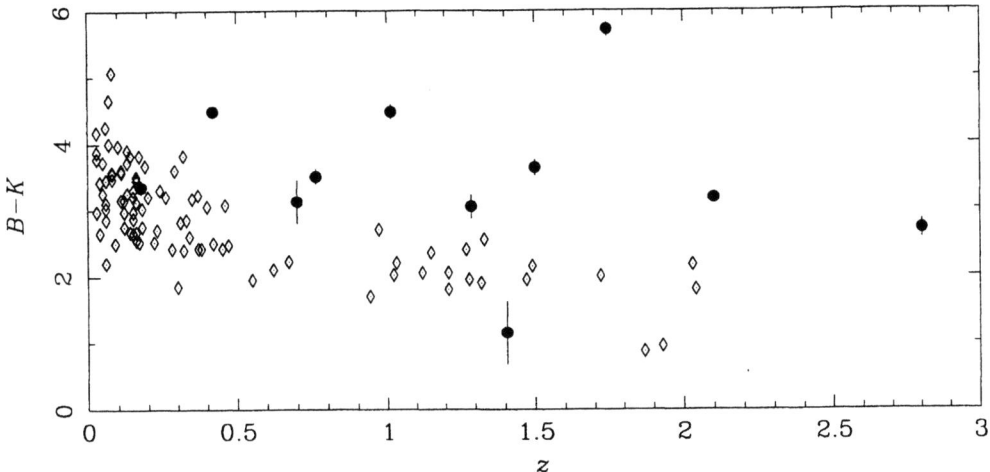

FIGURE 1. $B - K$ vs. z for the broad line AGN in our sample (dots) and for the PG sample of QSO (diamonds).

Most of the extragalactic hard sources are AGN. From the study of those with the highest quality optical spectra we have concluded that they are hard because they are absorbed, but there is a variety of absorption scenarios (Mittaz et al., these proceedings).

We present here the results of optical and infrared photometry of part of our sample. We show further evidence for absorption in some of the AGN from optical and infrared observations in section 2. Comparing the colours of the identified and unidentified objects we have also looked in section 2 for possible differences between them, that might reveal a change of population properties. This comparison is extended to optical-to-X-ray ratios in section 3.

2. OPTICAL AND INFRARED COLOURS

This work has made use of the following facilities: the CAHA 2.2 m ($BVRIK$) and 3.5 m (K') telescopes at Calar Alto (Spain), the OAN 1.52 m ($UBVRI$) telescope at Calar Alto, the ESO 1.5 m ($BVRI$) telescope at La Silla (Chile), and APM data.

We have obtained K filter photometry for almost half of our AGN. Fig. 1 shows that our AGN have "red" spectra, as indicated by $B - K$. Srianand & Kembhavi (1997) showed that, for standard gas-to-dust ratios, absorbing Hydrogen column densities of at least $N_H = 10^{21} - 10^{22}$ cm^{-2} are necessary to obtain $B - K \sim 3 - 5$ from a standard AGN spectrum (Francis et al. 1993). This amount of absorption is similar to that required to harden the X-ray spectra of our sources from a canonical X-ray spectrum with $\alpha \sim 1$ to the observed values.

Fig. 2 shows a ($V-R, V-I$) colour-colour plot, including all the data available for our sources. The unidentified objects are found in the same part of that plane as the

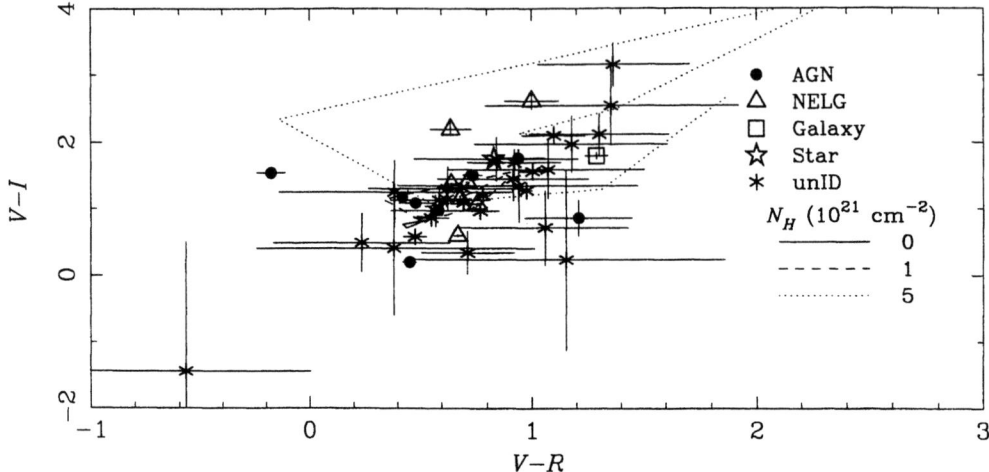

FIGURE 2. $V-I$ vs. $V-R$ for our sample. The lines show the colours of a standard AGN spectrum absorbed by gas with different column densities and standard gas-to-dust ratios. The "steps" in the lines correspond to redshifts between 0 and 3.5 in steps of 0.5. The line for unabsorbed spectra is the little feature at $\sim (0.47, 0.77)$.

identified sources in general, albeit with a few redder exceptions. Also shown in Fig. 2 are the colours expected from an AGN with a standard spectrum, unabsorbed and with different amounts of absorption, and at different redshifts. Again, significant absorbing column densities are required to reproduce the observed distribution of colours of our hard sources.

3. OPTICAL AND X-RAY COLOUR-COLOUR PLOTS

Further information can be gained using both X-ray and optical fluxes simultaneously. We show a $\log(f_X/f_B)$ vs. $O-E$ "colour-colour" plot in Fig. 3, both for the RIXOS sample (Mason et al. 2000) and for our hard sources. We have defined $\log(f_X/f_B) = \log(f_{X,0.5-2\ keV}) + B/2.5 + 5.37$, using the APM E and O magnitudes to obtain $B = O - 0.119(E - O)$ (Evans 1988).

Since AGN are bluer than stars, and brighter in X-rays than them for the same optical flux (Maccacaro et al. 1988), there is a clear segregation between AGN and stars in this colour-colour plot for the RIXOS sample. Most of the sources in our sample (with fainter X-ray fluxes and harder X-ray spectra) share the colours of the RIXOS AGN, but they are on average redder. Interestingly, the brightest unidentified sources (bright enough to have APM magnitudes) also exhibit the same colours. This, together with the results of section 2, indicates that the identified sources are a fair sample of the whole hard source population. In addition, it is reasonable to infer that most of these unidentified sources will turn out to be AGN, because AGN dominate the identified extragalactic component of our survey, and

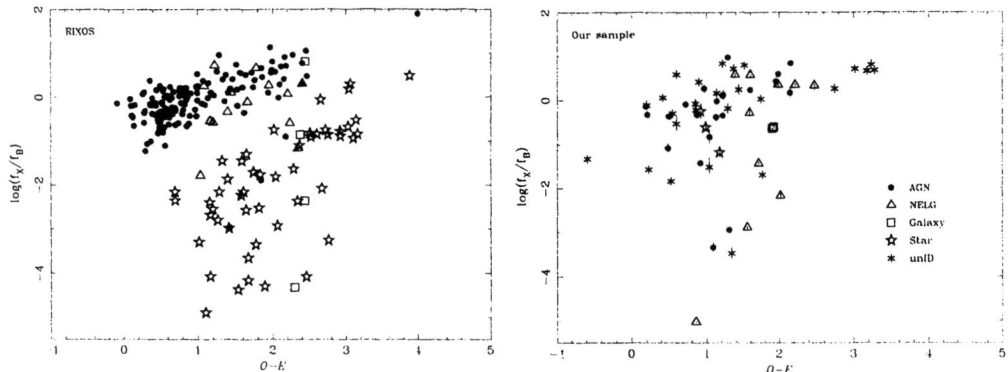

FIGURE 3. log(f_X/f_B) vs. $O-E$ for the RIXOS sources (left) and our hard sources (right).

the unidentified sources have their same colours.

4. DISCUSSION AND CONCLUSIONS

The optical and infrared colours ($B-K$, $V-R$, $V-I$) presented here strongly suggest that the hard spectral appearance of our sources in X-rays is due to absorption by neutral Hydrogen columns $10^{21} - 10^{22}$ cm^{-2}, if standard dust-to-gas ratios are assumed. A possible alternative for the origin of such red optical-infrared colours and hard X-ray spectra would be beaming (Srianand & Kembhavi 1997). However, there is direct evidence for absorption in the optical spectra of the brightest AGN (Mittaz et al., these proceedings). Therefore, absorption is the most likely origin of the hard X-ray spectra of our sample.

The unidentified sources present X-ray-optical colour distributions similar to those of the identified sources, implying that they are the same population. Since AGN are the dominant identified extragalactic population in our survey, AGN are likely to be main fraction of the faint hard source population that produce the unresolved soft XRB.

REFERENCES

Comastri A., et al., 1995, A&A, 296, 1
Evans T., 1988, PhD Thesis, Univ. Cambridge
Francis P.J., et al., 1993, AJ, 106, 417
Gilli R., et al., 1999, A&A, 347, 424
Maccacaro T., et al., 1988, ApJ, 326, 680
Mason K.O., Carrera F.J., et al., 2000, MNRAS, 311, 456
Mittaz J.P.D., et al., 1999, MNRAS, 308, 233
Miyaji T., et al., 1998, A&A, 334, 13
Page M.J., Mittaz J.P.D., Carrera F.J., 2000a, MNRAS, in press (astro-ph/0006347)
Srianand R., Kembhavi A., 1997, ApJ, 478, 70
Ueda Y., et al., 1999, ApJ, 518, 656

PKS 2155-304 — A SOURCE OF VERY HIGH ENERGY GAMMA RAYS

P.M. Chadwick[1], K. Lyons[1], T.J.L. McComb[1], K.J. Orford[1], J.L. Osborne[1], S.M. Rayner[1], S.E. Shaw[1], and K.E. Turver[1]

1) Department of Physics, Rochester Building, Science Laboratories, University of Durham, Durham, DH1 3LE, U.K.

ABSTRACT   The University of Durham Mark 6 very high energy (VHE) γ-ray telescope has been used to observe the X-ray selected BL lac PKS 2155-304 in 1996, 1997 and 1998. The data show evidence for the emission of VHE γ-rays with $E > 300$ GeV in 1996 September and 1997 October/November. There is evidence to support a correlation between the X-ray emission and VHE γ-ray emission. The strongest VHE emission was observed in 1997 November, when the object was producing the strongest high-energy X-ray emission ever recorded, and emission was also detected at $E > 100$ MeV.

KEYWORDS: Gamma rays: observations — BL Lacertae objects: individual (PKS 2155-304)

1. INTRODUCTION

The discovery of PKS 2155-304 as a VHE γ-ray source was made with the University of Durham Mark 6 telescope during observations in 1996-7 which suggested a time variable emitter with the strongest emission in 1997 November at the time of a multiwavelength campaign (Chadwick et al. 1999). Measurements in 1997 November are available from both *RXTE*, *BeppoSAX* and *CGRO*/EGRET (Vestrand and Sreekumar 1999, Chiappetti et al. 1999) which include a short interval (2 hrs) of simultaneous X-ray and TeV observations.

We report here the results of our 1998 measurements and reconsider our 1997 November data in the light of the recently available X-ray results from the multiwavelength campaign. All measurements reported here have been made with the University of Durham Mark 6 γ-ray telescope operating at Narrabri, NSW, Australia. The telescope has been described in detail by Armstrong et al. (1999).

2. NEW MEASUREMENTS IN 1998

Observations in 1998 have involved 9.5 hrs of exposure ON source and an equal amount OFF source. The observing log is in Table 1. All analysis reported here used the method employed previously (Chadwick et al. 1999). We find no evidence

TABLE 1. Observing log for our observations of PKS 2155–304 during 1998.

Date	ON source scans	Date	ON source scans
1998 July 22	1	1998 September 17	4
1998 August 18	5	1998 September 19	3
1998 August 19	7	1998 October 11	2
1998 August 20	4	1998 October 13	2
1998 September 15	5	1998 October 16	1
1998 September 16	2		

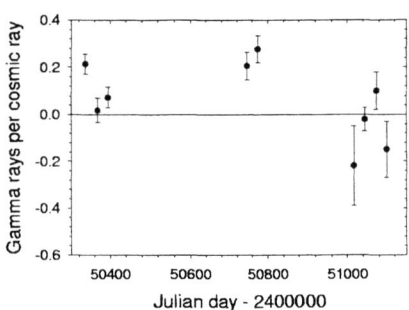

FIGURE 1. The measured VHE γ-ray flux above 300 GeV from PKS 2155–304 averaged over observing periods of approximately 10 days.

for emission of TeV γ-rays throughout the 1998 observations. We calculate a time-averaged 3 σ flux limit of 4.0×10^{-11} photons cm^{-2} s^{-1} at an energy threshold of 300 GeV for 1998 July – October. We show in Figure 1 the time averaged VHE γ-ray fluxes (normalised to the cosmic ray counting rate (Chadwick et al. 1999)) for all dark periods during which we have observed PKS 2155–304. PKS 2155–304 was in an X-ray low state during our measurements in 1998 July – October. Our failure to detect VHE emission during 1998 is thus consistent with the hypothesis that the X-ray and VHE gamma-ray emission from PKS 2155–304 are correlated (Chadwick et al. 1999).

3. MULTIWAVELENGTH OBSERVATIONS IN 1997 NOVEMBER

TeV data are available between 1997 November 17–25 (Chadwick et al. 1999). These observations were made as part of a multiwavelength campaign involving GeV γ-ray and X-ray measurements. The GeV measurements were made using *CGRO*/EGRET from 1997 November 10 – 23 and showed strong emission during the first half of the period (Vestrand and Sreekumar 1999). X-ray observations using PCA and HEXTE on board *RXTE* were made during 1997 November 20 – 22 (Vestrand

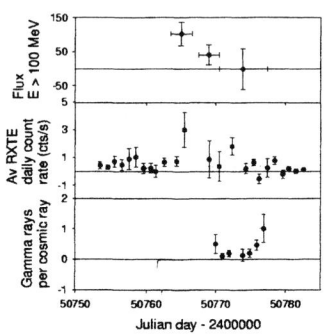

FIGURE 2. The GeV γ-rays recorded with EGRET (Vestrand and Sreekumar 1999 – upper panel), X-rays recorded with ASM on *RXTE* (centre panel) and VHE γ-ray (present work – bottom panel) during the 1997 November observations of PKS 2155-304.

and Sreekumar 1999) and ASM observations are available throughout. *BeppoSAX* observed this object for about 36 hrs during 1997 November 22 – 24 (Chiappetti et al. 1999). The X-ray and γ-ray observations clearly show that PKS 2155-304 was in an active flaring state in mid-November 1997, with X-ray and γ-ray fluxes being as high as ever previously detected — see Figure 2.

Our TeV observations, averaged over the total dataset for 1997 November, indicated the strongest emission during any of the dark periods to date (Chadwick et al. 1999), with a time-averaged flux of $(6.0 \pm 2.0_{stat} \pm 3.0_{sys}) \times 10^{-11}$ photons cm^{-2} s^{-1} at $E > 300$ GeV. The EGRET and X-ray observations suggest that a large outburst occurred in early November prior to the TeV observations (Vestrand and Sreekumar 1999). The TeV observations do not contradict this idea. The only truly contemporaneous multiwavelength data were recorded by *BeppoSAX* and the Mark 6 telescope on 1997 November 23 between 1100 and 1300 hrs UTC. We reproduce the X-ray data from the paper of Chiappetti et al. (1999) — see Figure 3(a).

Our VHE observations occur at the time which Chiappetti et al. (1999) define as a region of low X-ray intensity defined on the basis of the MECS (medium energy) count rate, beginning about 2 hours after the peak of the second X-ray flare detected by *BeppoSAX*. We show in Figure 3(b) the results of our VHE observations for individual 15 min scans on 1997 November 23, along with data taken on 1997 November 22 which were obtained about three hours before the *BeppoSAX* observation started. We have no evidence for strong flaring activity within the VHE data taken during the *BeppoSAX* observations, consistent with the low X-ray state. The VHE data taken on 1997 November 22 are at the same activity level as on November 23. The X-ray data show that an X-ray flare peaked about 2 hrs after our VHE observation finished and that the typical time scale for X-ray flaring is such that the

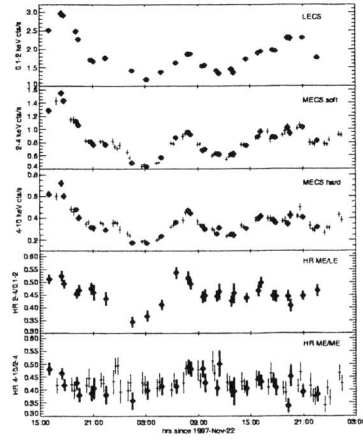
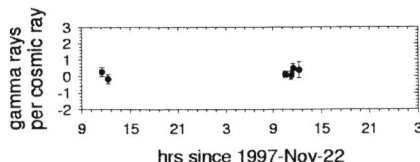

FIGURE 3. (a) The X-ray light curves (upper three panels) and hardness ratios (lower two panels) recorded with *BeppoSAX* during the 1997 November observations of PKS 2155−304 (taken from Chiappetti et al. 1999). Also shown (b) are the VHE γ-ray results from the present work.

flare is likely to have commenced after our observation terminated. The VHE data yield a flux of $(2.0 \pm 5.0_{stat} \pm 1.0_{sys}) \times 10^{-11}$ photons cm^{-2} s^{-1} for the observation on 1997 November 22 and $(7.0 \pm 4.5_{stat} \pm 3.5_{sys}) \times 10^{-11}$ photons cm^{-2} s^{-1} for the observation on 1997 November 23, both at an energy threshold of 300 GeV.

ACKNOWLEDGEMENTS

We are grateful to the UK Particle Physics and Astronomy Research Council for support of the project. This paper uses quick look results provided by the ASM/$RXTE$ team.

REFERENCES

Armstrong, P., et al., Exp. Astron., 9, 51 (1999).
Chadwick, P. M., et al., Astrophys. J., 513, 161 (1999).
Chiappetti, L., et al., Astrophys. J., 521, 552 (1999).
Vestrand, W. T. & Sreekumar, P., Astropart. Phys., 11, 197 (1999).

BEPPO-SAX OBSERVATIONS OF MKN 421: CLUES ON THE PARTICLE ACCELERATION ?

G. Fossati[1], A. Celotti[2], M. Chiaberge[2], L. Chiappetti[3], and Y.H. Zhang[2]

[1] UCSD/CASS, 9500 Gilman Drive, La Jolla, CA 92093-0424, USA – gfossati@ucsd.edu
[2] SISSA, via Beirut 2–4, 34014 Trieste, Italy – celotti, chiab, yhzhang@sissa.it
[3] Istituto di Fisica Cosmica G. Occhialini, via Bassini 15, 20133 Milano, Italy

ABSTRACT   Mkn 421 was repeatedly observed with *Beppo*SAX in 1997–1998. We present highlights of the results of the thorough temporal and spectral analysis discussed by Fossati et al. (1999) and Maraschi et al. (1999), focusing on the flare of April 1998, which was simultaneously observed also at TeV energies. A theoretical picture accounting for all the observational constraints is discussed, where electrons are injected at low energies and then progressively accelerated during the development of the flare.

KEYWORDS: galaxies: active — BL Lacertae objects: individual (Mkn 421)

THE 1998 X–RAY/TEV FLARE

Mkn 421 ($z = 0.031$) is the brightest blazar at X–ray and UV wavelengths and the first extragalactic source discovered at TeV energies (Punch et al. 1992), where dramatic variability has been observed (Gaidos et al. 1996).

In 1998 *Beppo*SAX observed Mkn 421 as part of a long lasting monitoring campaign. *Beppo*SAX observations are dominated by an isolated flare (see Fig. 1), and one of the striking and important results is that in correspondence with the X–ray flare of April 21st a sharp TeV flare was detected by the Whipple Cherenkov Telescope (Figure 1). The peaks in the 0.1–0.5 keV, 4.0–6.0 keV and 2 TeV light curves are *simultaneous within one hour* (see Maraschi et al. 1999).

Here we will focus on the X–ray characteristics of the April 21st flare. We accumulated light curves for different energy bands. The post–flare light curves have been modeled with an exponential decay, superimposed to a steady emission. Four the main results:

<u>Decay Timescales</u>: the timescales range between 30 and 45×10^3 seconds, and *do not* show a clear (if any) relationship with the energy, rather suggesting that the post–flare spectral evolution can be *achromatic*. This result leads to reject the simplest possibility that the decay evolution is driven by the radiative cooling of emitting electrons (this simplest picture would produce a dependence of the timescale with energy, $\tau \sim E^{-1/2}$).

<u>Flaring/Steady components</u>: exponential decay fits require the presence of an underlying less variable component.

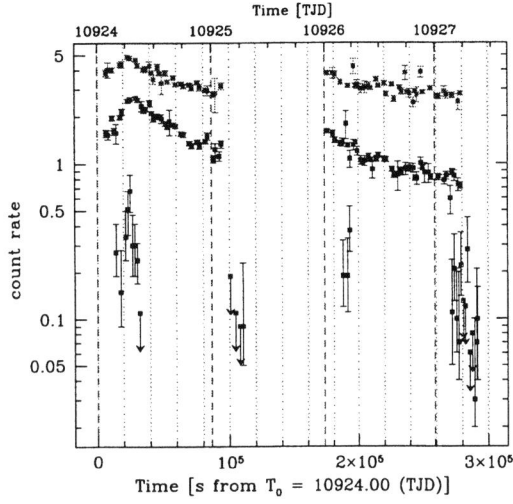

FIGURE 1. Light curves of Mkn 421 at TeV and X-ray energies, during the 1998 campaign. They are shown in order of increasing energy from bottom to top: Whipple E$\geq$ 2 TeV, MECS 4.0–6.0 keV and LECS 0.1–0.5 keV (both with 1500 s bins, multiplied by a factor 4 and 8, respectively). The count rate units are cts/s for *Beppo*SAX data, and cts/min for Whipple data.

<u>Time Lag</u>: the harder X–ray photons lag the soft X–ray ones. We performed a cross correlation analysis using the DCF (Edelson & Krolik 1988) and the MMD (Hufnagel & Bregman 1992) techniques and statistically determined the significance of the time lags using Monte Carlo simulations (Peterson et al. 1998). We refer to Zhang et al. (1999) for the relevant details of such analysis. The result is an average lag of $-2.7^{+1.9}_{-1.2}$ ks for DCF, and $-2.3^{+1.2}_{-0.7}$ ks for MMD (1 σ). This finding is opposite to what is normally found in the best monitored HBL X–ray spectra (e.g. Urry et al. 1993; Kohmura et al. 1994; Takahashi et al. 1996; Zhang et al. 1999) whose hard-to-soft behavior is usually interpreted in terms of cooling of the synchrotron emitting particles.

<u>Rise vs. Fall</u>: possible "asymmetry" of the rise/decay of the flare especially for the higher energy X–rays. The flare seems to be symmetric at the energies corresponding (roughly) to the synchrotron peak, while it might have a faster rise at higher energies. This could be connected to the observed hard–lag.

SPECTRAL VARIABILITY (1997 & 1998)

We accumulated spectra in sub-intervals, and developed an *intrinsically curved spectral model* to be able to estimate the position of the peak of the synchrotron component. In 1997 the source was in a lower brightness state, with a softer ($\Delta\alpha_{97,98} \simeq 0.4$) X-ray spectrum at all energies, and the peak energy 0.5 keV lower. There is a clear relation between the flux variability and the changes in the spectral parameters, both in 1997 and in 1998.

<u>Synchrotron peak energy</u>: the main new result is that we were *able to determine the energy of the peak of the synchrotron component* (with its error). We find a correlation between luminosity and shifts of the peak position (Fig. 2). The source reveals a highly coherent spectral behavior between 1997 and 1998, and through a large flux variability. The peak energies lie along a tight relation $E_{peak} \propto F^{0.55\pm 0.05}$.

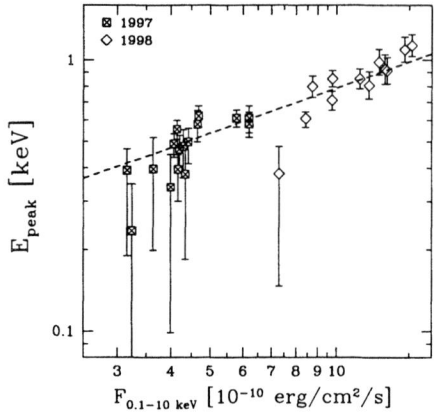
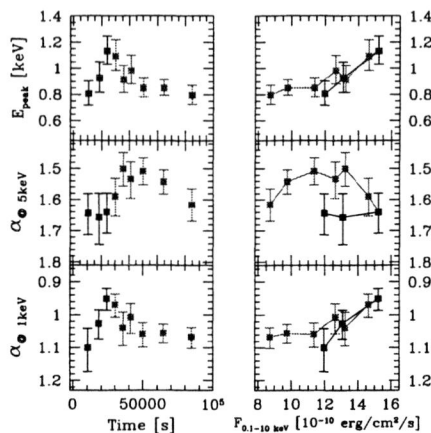

FIGURE 2. Synchrotron peak energy plotted versus "de-absorbed" 0.1–10.0 keV flux. The dashed line represents the best fitting power law, having a slope $\epsilon = 0.55$.

FIGURE 3. The photon spectral indices at 1 keV and at 5 keV, and energy of the peak of the synchrotron component, are plotted versus time and "de-absorbed" 0.1–10.0 keV flux.

Hard Lag in 1998 spectra: the spectral analysis confirms the signature of the hard lag. A blow up of the 1998 flare interval is shown in Figure 3. The main remarkable features are: (a) the synchrotron peak shifts toward higher energy during the rise, and then decreases as soon as the flare is over. (b) The spectral index at 1 keV reflects exactly the same behavior, as expected being computed at the energy around which the peak is moving. (c) On the contrary, the spectral shape at 5 keV does not vary until a few ks after the peak of the flare, and only then –while the flux is decaying and the peak is already receding– there is a response with a significant hardening of the spectrum.

The fact that the spectral evolution at higher energies develops during the decay phase of the flare, produces a nice counter-clockwise loop in the α vs. Flux diagram, i.e. *opposite* way with respect to all the other known cases for HBLs (e.g. Sembay et al. 1993; Kohmura et al. 1994; Takahashi et al. 1996).

PHYSICAL INTERPRETATION

Let us now focus on the possible interpretation of the two main results of this work: the *hard lag* and the *evolution of the synchrotron peak*.

The occurrence of the flare peak at different times for different energies is most likely related to the particle acceleration/heating process.

We therefore *introduced an acceleration term in the time dependent particle kinetic equation* within the model proposed by Chiaberge & Ghisellini (1999), which takes into account the cooling and escape terms and the role of delays in the received photons due to the travel time from different parts of the emitting volume.

The main constraints on the (parametric) form of the acceleration are: [A] particles have to be progressively accelerated from lower to higher energies within the flare rise timescale to produce the hard lag; [B] the emission in the LECS band from the highest energy particles (those radiating initially in the MECS band) should not exceed that from the lower energy ones, as after the peak no further increase of the (LECS) flux is observed; [C] the total decay timescale might be dominated by the achromatic crossing time effects, although the initial phase might be partly determined by the different cooling timescales.

It should be also noted that –within this scenario– the symmetry between the raise and decay of the softer energy light curve seems to suggest that at the same very energies where most of the power is released –possibly determined by the balance between the acceleration and cooling rates– the acceleration timescales are comparable to the region light crossing time.

If the timescales associated with this process are intrinsically linked to the typical size of the emitting region, we indeed expect the observed light curve to be symmetric where the bulk of power is concentrated, and an almost achromatic decay. Indeed, within a single emission region scenario, we have been able to reproduce the sign and amount of lags, postulating that particle acceleration follows a simple law, and stops at the highest particle energies. The same model can account for the spectral evolution (shift of the synchrotron peak) during the flare.

CONCLUSIONS

These results provide us with several *temporal* and *spectral constraints* on any model. In particular, they could possibly be the *first direct signature of the ongoing acceleration process*, progressively "pumping" electrons from lower to higher energies. The measure of the delay provides a tight constraint on the timescale of the acceleration mechanisms.

A last crucial point is that our results support the possibility of the presence and role of quasi–stationary emission. The short-timescale, large-amplitude variability events could be attributed to the development of new individual flaring components (possibly maintaining a quasi-rigid shape), giving rise to a spectrum outshining a more slowly varying emission. The decomposition in these two components might allow to determine the nature and modality of the dissipation in relativistic jets.

REFERENCES

Chiaberge, M., Ghisellini, G. 1999, MNRAS, 306, 551
Edelson, R. A., Krolik, J. H. 1988, ApJ, 333, 646
Fossati, G., et al. 1998, MNRAS, 299, 433
Fossati, G., et al. 1999, submitted to ApJ
Gaidos, J. A., et al. 1996, Nature, 383, 319
Hufnagel, B. R., Bregman, J. N. 1992, ApJ, 386, 473
Kohmura, Y., et al. 1994, PASJ, 46, 131
Maraschi, L., et al. 1999, ApJLetters, in press (vol. 526)
Padovani, P., and Giommi, P. 1995, ApJ, 444, 567
Peterson, B. M., et al. 1998, PASP, 110, 660
Punch, M., et al. 1992, Nature, 358, 477
Sembay, S., et al. 1993, ApJ, 404, 112
Takahashi, T., et al. 1996, ApJ, 470, L89
Urry, C. M., et al. 1993, ApJ, 411, 614
Zhang, Y. H., et al. 1999, ApJ, in press

VERY HIGH ENERGY GAMMA RAYS FROM CEN X-3

P.M. Chadwick[1], K. Lyons[1], T.J.L. McComb[1], K.J. Orford[1], J.L. Osborne[1], S.M. Rayner[1], S.E. Shaw[1], and K.E. Turver[1]

1) *Department of Physics, Rochester Building, Science Laboratories, University of Durham, Durham, DH1 3LE, U.K.*

ABSTRACT Cen X-3 is a well-studied high-mass accreting X-ray binary and a variable source of high energy gamma rays from 100 MeV to 1 TeV. The object has been extensively monitored with the University of Durham Mark 6 telescope. Results of observations, including those taken in 1998 and 1999, are reported. There is no evidence for time variability in the VHE data. There is also no evidence for correlation of the VHE flux with the X-ray flux detected by BATSE and *RXTE*/ASM. A search for periodic emission, at or close to the X-ray spin period, in the VHE data yielded a 3σ upper limit to the pulsed flux of 2.0×10^{-12} cm^{-2} s^{-1}.

KEYWORDS: Stars: individual (Cen X-3) — gamma rays: observations

1. INTRODUCTION

Results of observations of the accreting X-ray binary Cen X-3 using ground based gamma ray telescopes have been reported which have included evidence for sporadic outbursts of strong pulsed emission in the > 1 TeV band (Carraminana at al. 1989, Raubenheimer et al. 1989) and constant but weaker unpulsed emission at > 400 GeV (Chadwick et al. 1998). These results, together with the *CGRO* EGRET measurement of an outburst of pulsed GeV emission (Vestrand, Sreekumar & Mori 1997), indicate that Cen X-3, an accurately measured system containing a 4.8 s pulsar in a 2.1 d orbit around an O-type supergiant, is a sporadic source of high energy gamma rays.

We present the results of a search for a possible correlation between > 400 GeV gamma rays recorded by the University of Durham Mark 6 telescope and X-ray emission according to measurements made with the *RXTE* and *CGRO*/BATSE experiments. We report the results of analysis of data taken during 1998 March and April and 1999 February. We also present the results of searches for variation of the emission at both the orbital and spin periods.

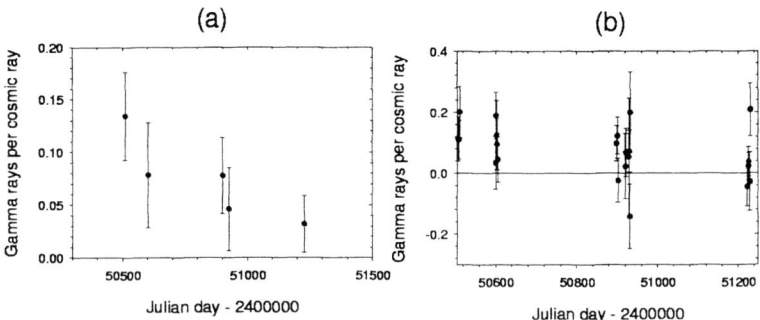

FIGURE 1. (a) The VHE gamma ray flux from Cen X-3 averaged over observing periods. (b) The VHE gamma ray flux from Cen X-3 plotted on a day-by-day basis.

2. RECENT OBSERVATIONS OF VHE GAMMA RAYS

Observations were made with the University of Durham Mark 6 imaging gamma ray telescope operating at Narrabri NSW, Australia. The telescope has been described in Armstrong et al. (1999) and the results of initial observations of Cen X-3 have been reported by Chadwick et al. (1998). Our Cen X-3 dataset now comprises data from 31 hrs of observation during 23 exposures in 1997 March and June (JD 2450508 – JD 2450606), 1998 March and April (JD 2450899 – JD 2450932) and 1999 February (JD 2451220 – JD 2451230). All data were taken during dark moonless conditions.

Our earlier detection (Chadwick et al. 1998) was based on data recorded in 1997 March and June (JD 2450508 – JD 2450606) only. Assuming a collection area of 10^9 cm^2 and that our selection procedure retained $\sim$ 50% of the original gamma ray events, the time averaged flux was estimated to be $(2.0 \pm 0.3) \times 10^{-11}$ cm^{-2} s^{-1} for $>$ 400 GeV. We concluded that the measurements in these two months were consistent with a constant flux. However, ongoing simulations suggest that our current selection procedure retains 20% of the gamma rays. On this basis, the flux for the 1997 March and June (JD2450508 – JD2450606) data would be $(5.0 \pm 0.9) \times 10^{-11}$ cm^{-2} s^{-1}. The additional data taken in 1998 and 1999 provide fewer gamma ray candidates suggesting weaker TeV emission. An analysis of the total data yields a time averaged flux of $(2.8 \pm 1.4_{sys} \pm 0.6_{stat}) \times 10^{-11}$ cm^{-2} s^{-1}.

In the present study, the average gamma ray signal strength from Cen X-3, expressed as a percentage of the cosmic ray background remaining after shape and orientation selection is $(7.0 \pm 1.5)\%$. The most straightforward, but not most powerful, test for constancy of emission is to repeat this process for the data recorded in each of the 5 dark periods as shown in Figure 1(a). On the basis of this test we find no internal evidence for monthly variability of the VHE signal; the data treated this way are consistent with a constant signal ($\chi^2 = 4.5$, 4 df).

3. CORRELATIONS BETWEEN X-RAYS AND VHE GAMMA RAYS

Cen X-3 is a strong but variable X-ray emitter. For example, the average daily rates for X-rays detected with the *RXTE*/ASM during 1997 and 1998 range from 0 to 32 counts s^{-1}; the data are variable on a time scale of days. The daily average for the *RXTE*/ASM count rates are available for 22 of the 23 days when VHE gamma ray observations were made.

The strength of pulsed X-ray emission is also available as a daily average from the BATSE archive for 1997; during the 1998 and some of the 1999 VHE observations, the X-ray flux was less than the threshold for BATSE detection. The BATSE data provide a series of independent X-ray measurements, including a measurement on the single day of the VHE gamma ray observations for which there is no corresponding *RXTE*/ASM measurement.

The VHE gamma ray signal plotted on a day by day basis is shown in Figure 1(b). There is no evidence for outbursts of VHE gamma ray emission on a timescale of days and the data are consistent with a constant VHE gamma ray flux ($\chi^2 = 22.1$, 22 df).

In Figure 2(a) we show the relation between the count rate of the *RXTE*/ASM data and our gamma ray signals. In Figure 2(b) we show a similar plot showing the relation between the individual BATSE pulsed X-ray fluxes and our gamma ray signals. There seems to be no significant evidence for a correlation, although it is interesting to note that the day of highest detected gamma ray flux coincides with the day of most X-ray activity in the dataset (1997 Mar 4).

4. SEARCH FOR PERIODICITY

We have looked for modulation of the gamma ray signal at the orbital period of the binary system. The orbital phase of each of our observations has been calculated using the ephemeris of Kelley et al. (1983). We conclude that there is no modulation of the VHE gamma ray emission at the orbital period.

The data have been subjected to a Rayleigh test for periodicity at a small range of periods around the BATSE pulse period. Phase coherence between observations was not assumed. No significant periodicity was detected, leading to a 3σ upper limit to the pulsed flux of 2.0×10^{-12} cm^{-2} s^{-1} in the total dataset.

5. DISCUSSION

We have detected VHE gamma ray emission from Cen X-3 during each dark moon period in which we have observed this object. The data are consistent with a weak but persistent emission, both when the VHE data is averaged over dark moon periods or when considered observation by observation. Although the observation that yields the strongest gamma ray flux occurs on the day when the daily averaged

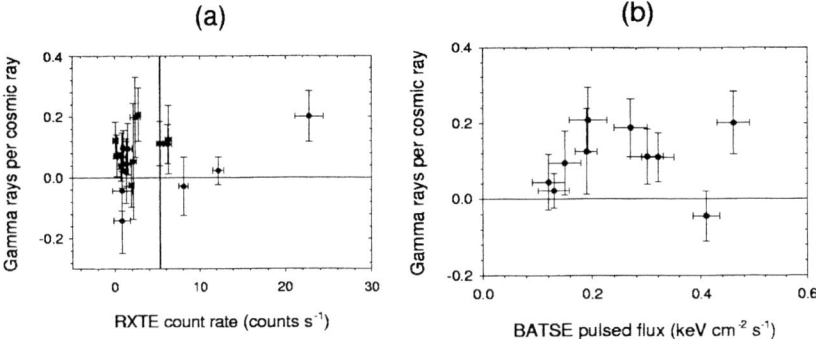

FIGURE 2. The relation between the daily VHE gamma ray flux from Cen X-3 and (a) the X-ray flux detected by ASM/$RXTE$ and (b) the X-ray pulsed flux detected by BATSE.

RXTE X-ray flux was the highest of any day on which we observed Cen X-3, there is no evidence for a formal correlation between the VHE gamma ray and X-ray fluxes.

We have also tested for modulation of the VHE gamma ray flux at the orbital period of the binary system and at the pulsar period. We have no evidence for modulation of the VHE gamma ray emission at either period.

ACKNOWLEDGEMENTS

We are grateful to the UK Particle Physics and Astronomy Research Council for support of the project. The Mark 6 telescope was designed and constructed with the assistance of the staff of the Physics Department, University of Durham. This paper uses quick look results provided by the ASM/$RXTE$ team.

REFERENCES

Armstrong, P., et al. 1999, Exp. Astro., 9, 51

Carraminana, A., et al. 1989, Timing Neutron Stars ed. H. Ögelman & E. P. J. van den Heuvel (Dordrecht: Kluwer Academic Press), p 369

Chadwick, P. M., et al. 1998, Ap. J., 503, 391

Kelley, R. L., et al. 1983, Ap. J., 268, 790

Raubenheimer, B. C., et al. 1989, Ap. J., 336, 349

Vestrand, W. T., Sreekumar, P., Mori, M. 1997, Ap. J., 483, L49

VERY HIGH ENERGY GAMMA RAY OBSERVATIONS OF SOUTHERN HEMISPHERE AGNS

P.M. Chadwick[1], K. Lyons[1], T.J.L. McComb[1], K.J. Orford[1], J.L. Osborne[1], S.M. Rayner[1], S.E. Shaw[1], and K.E. Turver[1]

1) Department of Physics, Rochester Building, Science Laboratories, University of Durham, Durham, DH1 3LE, U.K.

ABSTRACT   A range of AGNs visible from the Southern hemisphere has been observed with the University of Durham Mark 6 very high energy gamma ray telescope. Results of observations of 1ES 0323+022, PKS 0829+046, 1ES 1101–232, Cen A, PKS 1514–24, RKJ 10578–275, 1ES 2316–423, PKS 2005– 489 and PKS 0548–322 are presented.

KEYWORDS: Gamma rays: observations — BL Lacertae objects: individual (PKS 1514–24, PKS 0829+046, 1ES 1011–232, 1ES 2316–423, 1ES 0323+022, RXJ 10578–275, PKS 2005–489, PKS 0548–322) — galaxies: individual (Cen A)

1. INTRODUCTION

One of the most unexpected results in high energy astrophysics in the last decade has been the discovery of high energy and very high energy (VHE) emission from active galactic nuclei (AGNs). The EGRET detector on board the *Compton Gamma Ray Observatory* established that BL Lacs (predominantly radio selected) and flat-spectrum radio sources are strong high energy gamma ray emitters, while X-ray selected BL Lacs (XBLs) have been identified as a source of VHE gamma rays.

The Durham AGN dataset consists of observations of 10 AGNs made with the Mark 6 telescope from 1996 to 1998. The discovery of VHE gamma rays from PKS 2155–304 has already been reported; this is the most distant BL Lac yet detected at these energies (Chadwick et al. 1999). Here we describe observations of PKS 0548–322, PKS 2005–489, 1ES 0323+022, PKS 0829+046, RXJ 10578–275, 1ES 1101–232, Cen A, PKS 1514–24, and 1ES 2316–423, covering a range of classes of AGN. The typical energy threshold for these observations is $\sim$ 300 to 400 GeV. This is $\sim$ 5 times lower than the typical threshold of the CANGAROO telescope, which has also been used to observe Southern hemisphere AGNs (Roberts et al. 1999).

2. OBSERVATIONS

Current VHE γ-ray observations of AGNs support the idea that it is the XBLs which are the most promising sources of VHE emission, as suggested by Stecker et al. (1996). The nine AGNs which are discussed in this paper comprise five XBLs, two RBLs, one intermediate class object and one close radio galaxy (Cen A) which has been detected previously as a VHE γ-ray source. While RBLs are thought to be less promising as VHE γ-ray sources than XBLs, observations in the VHE range will help to confirm the fundamental differences between the XBLs and RBLs. VHE γ-ray observations of BL Lacs have, in general, concentrated on the closest objects ($z \sim 0.07$), but we have sought to extend the current redshift limit of $z = 0.117$ by observing more distant AGNs. With an energy threshold of ~ 300 GeV, the Mark 6 Telescope is well-suited to this task. In the case of one XBL, 1ES 1101–232, the VHE γ-ray observations were made nearly contemporaneously with *BeppoSAX* observations.

The selection criteria applied to these data used a standard set of criteria developed from our successful observations of PKS 2155–304, and include allowance for the variation of image parameters with event size. They are routinely applied to data from all objects recorded at zenith angles less than $45°$, which is the case for all the observations reported here.

3. RESULTS

The dataset for each source has been tested for the presence of gamma ray signals. The flux limits from the nine AGNs are summarised in Table 1. They are all 3 σ flux limits, based on the maximum likelihood ratio test (Gibson et al. 1982, Li and Ma 1983). The threshold energy for the observations has been estimated on the basis of preliminary simulations, and is in the range 300 to 400 GeV for these objects, depending on the object's elevation. The collecting areas which have been assumed, again from simulations, are 5.5×10^8 cm^2 at an energy threshold of 300 GeV and 1.0×10^9 cm^2 at an energy threshold of 400 GeV. These are subject to systematic errors estimated to be $\sim 50\%$. We have assumed that our current selection procedures retain $\sim 20\%$ of the γ-ray signal, which is subject to a systematic error of $\sim 60\%$.

We have also searched our dataset for γ-ray emission on timescales of ~ 1 day. The search for enhanced emission has been conducted by calculating the on-source excess after the application of our selection criteria for the pairs of on/off observations recorded during an individual night. A typical observation comprising 6 on/off pairs of observations (1.5 hours of on-source observations) yields a flux limit of $\sim 1 \times 10^{-10}$ photons cm^{-2} s^{-1} at 300 GeV. Conversely, had any of the objects on which we report here produced a 15-minute flare similar to that seen from Mrk 421 with the Whipple telescope on 1996 May 7 (Gaidos et al. 1996), it would have been detected with the Mark 6 telescope at a significance of around 7 σ. There is no evidence for any flaring activity.

TABLE 1. Flux limits (3 σ) for observations of active galactic nuclei made with the University of Durham Mark 6 Telescope. Also shown are the predictions of the model of Stecker et al. (1996).

Object	Estimated Threshold (GeV)	Flux Limit ($\times 10^{-11}$ photons cm^{-2} s^{-1})	Predicted Flux ($\times 10^{-11}$ photons cm^{-2} s^{-1})
Cen A	300	5.2	
PKS 0829+046	400	4.7	
PKS 1514−24	300	3.7	
1ES 2316−423	300	4.5	0.15
1ES 1101−232	300	3.7	2.0
RXJ 10578−275	300	8.2	0.33
1ES 0323+022	400	3.7	0.40
PKS 2005−489	400	0.79	0.51
PKS 0548−322	300	2.4	1.3

4. DISCUSSION

Whilst the interpretation of VHE upper limits from BL Lacs is complicated by the lack of a complete theory of VHE γ-ray emission from AGNs, Stecker et al. (1996) have predicted the TeV fluxes from a range of objects, three of which (1ES 0323+022, PKS 0548−322, and PKS 2005−489), are included in the present work. The expected fluxes from the other XBLs included in this paper may be estimated on the basis of the work of Stecker et al. (1996) and Stecker (1998) using the simple relation $\nu_x F_x \sim \nu_\gamma F_\gamma$ and the published X-ray fluxes. We estimate that the 300 GeV fluxes of 1ES 1101−232, 1ES 2316−423 and RXJ 10578−275 would be 2.0×10^{-11} photons cm^{-2} s^{-1}, 1.5×10^{-12} photons cm^{-2} s^{-1}, and 3.3×10^{-12} photons cm^{-2} s^{-1} respectively, taking into account photon-photon absorption using the recent determination of γ-ray opacity by Stecker (1998). All these suggested fluxes are lower than the flux limits reported here. However, the lack of contemporaneous X-ray measurements in the case of most of our observations limits the usefulness of these predictions and emphasises the importance of multiwavelength campaigns. In the case of the RBLs, an extended observation of PKS 1514−24, a close RBL, lends support to the suggestion that RBLs are not strong VHE γ-ray emitters.

Our observations of Cen A were made when it was in an X-ray low state, in contrast to the earlier VHE detection of Cen A reported by Grindlay et al. (1975), which was made when Cen A was in X-ray outburst. Further VHE γ-ray observations during an X-ray high state would be desirable.

5. CONCLUSIONS

The Durham University Mark 6 Telescope has been used to make observations of 9 close AGNs: 1ES 0323+022 (XBL, $z = 0.147$), PKS 0548-322 (XBL, $z = 0.069$), PKS 0829+046 (RBL, $z = 0.18$), RXJ 10578-275 (XBL, $z = 0.092$) 1ES 1101-232 (XBL, $z = 0.186$), Cen A (low luminosity radio galaxy, $z = 0.0089$), PKS 1514-24 (RBL, $z = 0.049$), PKS 2005-489 (XBL, $z = 0.071$) and 1ES 2316-423 (transitional BL Lac, $z = 0.055$). We find no evidence for either steady or flaring emission of γ-rays above 300 – 400 GeV in any of these sources. The flux limits are in excess of the fluxes predicted on the basis of the simple model of Stecker et al. (1996). The flux limits derived for 1ES 0323+022 (3.7×10^{-11} photons cm^{-2} s^{-1}), PKS 0548-322 (2.4×10^{-11} photons cm^{-2} s^{-1}), and PKS 2005-489 (0.79×10^{-11} photons cm^{-2} s^{-1}) are not in conflict with the specific predictions of Stecker et al. (1996) (4.0×10^{-12} photons cm^{-2} s^{-1}, 1.3×10^{-11} photons cm^{-2} s^{-1}, and 0.51×10^{-11} photons cm^{-2} s^{-1} respectively).

ACKNOWLEDGEMENTS

We are grateful to the UK Particle Physics and Astronomy Research Council for support of the project and the University of Sydney for the lease of the Narrabri site. The Mark 6 telescope was designed and constructed with the assistance of the staff of the Physics Department, University of Durham. The efforts of Mrs. S. E. Hilton and Mr. K. Tindale are acknowledged with gratitude. We would like to thank Anna Wolter for providing us with information about *BeppoSAX* observations of 1ES 1011-232 in advance of publication. This paper uses quick look results provided by the ASM/*RXTE* team and uses the NASA/IPAC Extragalactic database (NED), which is operated by the Jet Propulsion Laboratory, Caltech, under contract with the National Aeronautics and Space Administration.

REFERENCES

Chadwick, P. M., et al., Ap. J., 513, 161 (1999).
Gaidos, J. A., et al., Nature, 383, 319 (1996).
Gibson, A. I., et al., Proc. Intl. Workshop on Very High Energy Gamma Ray Astro., Bombay: Tata Institute, ed. P. V. Ramana Murthy & T. C. Weekes, 97 (1982).
Grindlay, J. E. et al., Ap. J., 197, L9 (1975).
Li, T. P., & Ma, Y. Q., Ap. J., 272, 317 (1983).
Roberts, M. D., et al., astro-ph/9902008, (1999).
Stecker, F. W., de Jager, O. C., & Salamon, M. H., Ap. J., 473, L75 (1996).
Stecker, F. W., astro-ph/9812286, (1998).

SEARCH FOR INTERACTIONS BETWEEN EJECTIONS OF GRS 1915+105 AND ITS ENVIRONMENT

S. Chaty[1,2], L.F. Rodríguez[3] & I.F. Mirabel[2]

[1] Open University, Milton Keynes, UK; [2] Service d'Astrophysique, CEA Saclay, France; [3] Instituto de Astronomía, UNAM, México

ABSTRACT

To unravel the effect of likely interactions between the energetic ejections of the galactic superluminal source GRS $1915 + 105$ and its surrounding interstellar medium, we observed its environment. Two IRAS sources are symmetrically placed with respect to GRS $1915 + 105$, and are aligned with the sub-arcsec ejections of this source. We analyzed these two sources IRAS $19124 + 1106$ and IRAS $19132 + 1035$ through near-infrared, millimeter and centimeter wavelengths. The evidence for these regions being interaction zones seems inconclusive.

KEYWORDS: Stars: individual: GRS 1915+105 – HII regions – ISM: individual objects: IRAS 19124+1106, IRAS 19132+1035 – ISM: jets and outflows – X-rays: stars

1. INTRODUCTION

The first known galactic superluminal source GRS $1915 + 105$ is a highly energetic and relativistically ejecting source. Consequently one wonders if there is an observable interaction when the frequently ejected plasma clouds collide at relativistic velocities with the interstellar medium, or heat molecular clouds surrounding this source. Indeed, the ensemble of ejections of such a microquasar must have an effect on its environment, as it is the case for the well-known ejecting source SS 433. Therefore, we undertook a comprehensive study of the environment of GRS $1915 + 105$ at near-infrared, mid-infrared, radio centimeter and millimeter wavelengths. Here some of the radio observations are described and discussed; the reader can refer to Chaty et al. (2000, hereafter C00) for a complete description of the study.

2. THE OBSERVATIONS: TWO AXISYMMETRIC SOURCES

The region surrounding GRS $1915 + 105$ was inspected and described by Rodríguez and Mirabel (1998, hereafter RM98). This search was performed at $\lambda = 20$ cm, with the VLA (Very Large Array) of NRAO[1], in C-configuration, giving a resolution of $15''$. The resulting map is shown in Figure 1. They discovered that there were

---

[1] The National Radio Astronomy Observatory is operated by Associated Universities, Inc., under cooperative agreement with the USA National Science Foundation

Source	J2000.0 coord.	gal. coord.
GRS 1915+105	$\alpha = 19^h 15^m 11\overset{s}{.}545$	$l^{II} = 45\overset{\circ}{.}40$
	$\delta = 10°56'44\overset{''}{.}80$	$b^{II} = -0\overset{\circ}{.}29$
IRAS 19124+1106	$\alpha = 19^h 14^m 45\overset{s}{.}77$	$l^{II} = 45\overset{\circ}{.}54$
	$\delta = 11°12'06\overset{''}{.}4$	$b^{II} = -0\overset{\circ}{.}007$
IRAS 19132+1035	$\alpha = 19^h 15^m 39\overset{s}{.}13$	$l^{II} = 45\overset{\circ}{.}19$
	$\delta = 10°41'17\overset{''}{.}1$	$b^{II} = -0\overset{\circ}{.}44$

TABLE 1. Positions of GRS 1915+105, IRAS 19124+1106 and IRAS 19132+1035. These coordinates are the positions of peak signal obtained with the VLA-C 20 cm observations.

two axisymmetrically placed continuum radio sources, each located at 17' from GRS 1915 + 105, and coincident with IRAS sources. Their positions are given in Table 1. Furthermore, the position angle of these sources is 157±1° from GRS 1915+ 105, very similar to the one of the well studied sub-arcsec radio-ejections from GRS 1915 + 105 (~ 150°). In order to interpret the radio data, we remind here that the angle between the ejections and the line of sight towards GRS 1915+105 is 70°, that the South component is approaching us, and the North component is receding (Mirabel and Rodríguez, 1994; Fender et al., 1999).

Although these two sources could be a chance alignment, the striking point-symmetric position of these two clouds suggests that they result from an association with the high-energy source GRS 1915 + 105.

2.1. Centimeter wavelength observations

High-resolution maps of the two continuum radio sources have been obtained with the VLA (RM98). These maps are shown in the Figure 2. Concerning the North lobe, the centimeter map shows that it resembles to a common cometary H II region, but it also shows a shockwave structure to the South, e.g. to the direction of GRS 1915 + 105. For the South lobe, the centimeter map shows to the northwest a non-thermal jet, pointing along the direction of GRS 1915 + 105. The flux densities of this jet are $\sim \leq 1$, 2 and 5 mJy respectively at 2, 6 and 20 cm, showing a spectral index of -0.8. Furthermore, the South lobe shows a sharp edge to the South, which could be either a bow shock, or an ionization front in the H II region. The following discussion emphasizes these two striking features of the South lobe.

2.2. Millimeter wavelength observations

We used the IRAM (Institut de Radio Astronomie Millimétrique) 30-m radio telescope, located on Pico Veleta, near Granada, Spain. The details of observations are described in C00, and the results for IRAS 19132+1035 are shown in Figure 3. The OFF position (position switching) was chosen at $(\alpha = -500'', \delta = -1200'')$ from

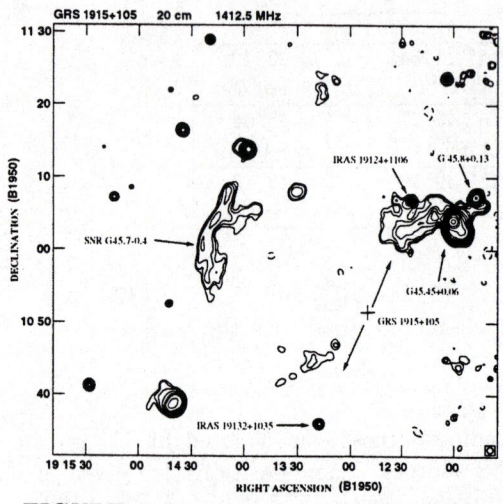

FIGURE 1. Map of the surroundings of GRS 1915 + 105 (VLA-C, $\lambda = 20$ cm, half power contour of the beam shown in the bottom right corner). The arrows around GRS 1915 + 105 indicate the position angle of the sub-arcsec relativistic ejecta.

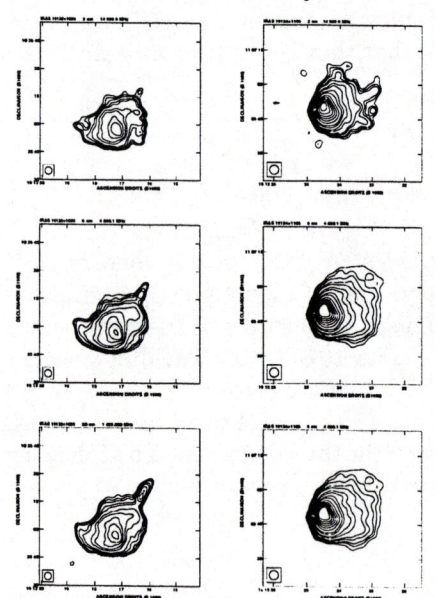

FIGURE 2. Maps of the two continuum radio sources IRAS 19132 + 1035 (left) and IRAS 19124 + 1106 (right), (VLA-D $\lambda = $ 2 cm (top), C 6 cm (middle) and B 20 cm (bottom), half power contour of the beam shown in the bottom right corner.

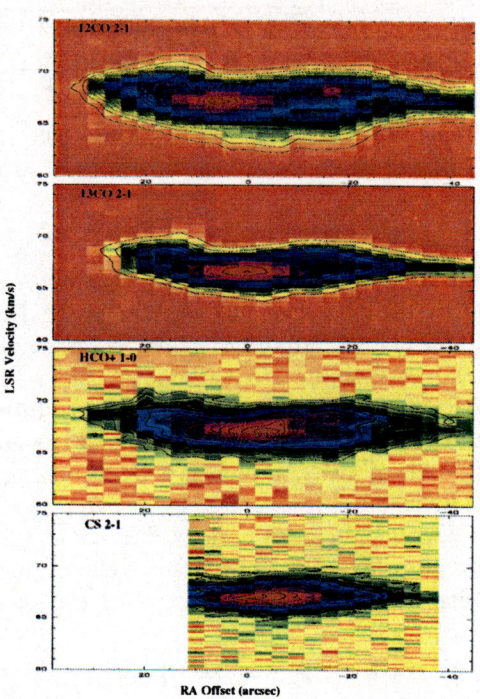

FIGURE 3. Observations of IRAS 19132+1035. Offsets are relative to the position of maximum radio emission observed at the VLA. The transitions are, from top to bottom of Figure ^{12}CO, ^{13}CO, H^{13}CO$^+$ and CS. The black contours are antennae iso-temperature, equal respectively from top to bottom: ^{12}CO: $T_A^* = -1$, and from 1 to $20K$ separated by an interval of $1K$; ^{13}CO: $T_A^* = -1$, and from 1 to 11 K separated by an interval of 1 K; H^{13}CO$^+$: $T_A^* = -1$, and from 0.2 to 2.1K separated by an interval of 0.1K; CS: $T_A^* = -1$, and from 0.2 to 2.2K separated by an interval of 0.2K. The counter-jet seen in the VLA centimeter continuum is located at $\sim -18''$.

GRS 1915 + 105. The main results are that i) the density profile of the cloud exhibits an asymmetric velocity distribution, ii) the maximum of the profile is closer to the counter-jet for high density tracers (compare e.g. ^{12}CO 2-1 to the CS 2-1 transition), and iii) there are two maxima in the ^{12}CO 2-1 transition and only one in the others. The other main result is that we detected a 4-σ SiO 2-1 line, localized on the position of the counter-jet.

All these facts could indicate the presence of an interaction, although they do not constitute a clear proof thereof. Is there an association, or the two H II regions are point-symmetric by chance?

3. DISCUSSION AND CONCLUSION

There is a possibility that these two IRAS sources received energy from GRS 1915 + 105 through shocks initiated by plasma clouds ejected by GRS 1915 + 105 and colliding with H II regions, creating the non-thermal jet seen in the South lobe. There is also a possibility that the relativistic ejecta have induced star formation, and this could have created the non-thermal jet as a Herbig-Haro-like feature. However, we consider this last possibility as unlikely because of the timescale of the different phenomena.

The other possibility is that these two IRAS sources have nothing to do with GRS 1915 + 105: the alignment could be a background coincidence. It is worthwhile to remember that there are two point-symmetric sources, and furthermore the IRAS fluxes and the molecular lines show that the two IRAS sources are in our Galaxy (RM98). This decreases the probability of a background coincidence.

Although our observations spanned in a large range of wavelengths (C00), and there are some striking facts, we can not clearly prove any association between GRS 1915 + 105 and the two IRAS sources.

ACKNOWLEDGEMENTS

S.C. acknowledges support from grant F/00-180/A from the Leverhulme Trust, and is grateful to C.A. Haswell for improving the language of the manuscript.

REFERENCES

Chaty, S., Rodríguez, L.F., Mirabel I.F. et al., 2000, A&A submitted
Fender, R. et al., 1999, MNRAS, 304, 865
Mirabel I.F. and Rodríguez, L.F., 1994, Nat, 371, 46
Rodríguez, L.F. and Mirabel I.F., 1998, A&A, 340, L47

EUVE/ASCA/RXTE OBSERVATIONS OF NGC 5548

James Chiang[1], Chris Reynolds[1,2], Omer Blaes[3], Mike Nowak[1], Norm Murray[4], Greg Madejski[5], Herman Marshall[6], and Pawel Magdziarz[7]

1) JILA 2) Hubble Fellow 3) Dept. of Physics, UCSB 4) CITA 5) NASA/GSFC 6) MIT
7) Deceased 1998 August

KEYWORDS: galaxies: individual (NGC 5548) — galaxies: Seyfert — X-rays: galaxies

1. INTERBAND VARIABILITY AND THE DISK/CORONA GEOMETRY

Reverberation mapping observations of Seyfert Broad Line Regions revealed unexpected highly correlated variability between the optical and UV continua. An early campaign for NGC 5548 showed that essentially no lag exists between continuum emission at 4870Å relative to emission at 1350Å within the 2-day temporal resolution of the observations (Peterson et al. 1991). If the observed optical/UV flux is produced by a multitemperature accretion disk, then the characteristic radii from which emission at these two wavelengths are produced will be quite different, and the degree of correlated variability should limited by the sound speed in the accretion flow. However, the upper limits on the delay for NGC 5548, as well for numerous other Seyferts, including NGC 4151, Fairall 9, NGC 3783, and NGC 7469, are several orders of magnitude shorter than the expected sound crossing times (see Peterson, et al. 1998 and references therein). Later *EUVE/HST* observations of NGC 5548 showed that this simultaneity also extends to the EUV ($\sim 0.1\,\text{keV}$) energies (Marshall et al. 1997). This led to the suggestion that the broad band optical/UV/EUV emission is produced in the disk by reprocessing of higher energy emission so that signal propagation would only be limited by light travel times.

Recent UV/X-ray monitoring observations of NGC 7469 (Nandra et al. 1998) reveal a complex relationship between the variability in the UV and the X-rays. The light curves in both energy bands exhibit factor of ~ 2 modulations on 10 day time scales with the UV maxima appearing to lead corresponding peaks in the X-rays by about 4 days while the troughs in the two light curves are nearly simultaneous. Nandra et al. suggest that this behavior could be due to several reprocessing regions which interact at different time scales depending on the flux state.

The data from our *EUVE/ASCA/RXTE* observations of NGC 5548 (Chiang et al. 1999) exhibit correlated variability which is also in conflict with the basic X-ray reprocessing hypothesis, but which is also much easier to interpret. Figure 1 shows the light curves from the various instruments for the longest of our monitoring observations. These data clearly show in all three energy bands a prominent step at about 20 ks in the figure, as well as subsequent modulations on ~ 50 ks time scales.

We have calculated the cross-correlation functions of these data using the z-transformed discrete correlation function of Alexander (1997). These data show that the variations in the EUV light curves precede similar variations in the higher energy X-rays implying that **the EUV emission is not produced by reprocessing.** Specifically, we find that the modulations of the soft, 0.5–1 keV X-rays observed by *ASCA*-SIS lag those of the EUV (at 0.16 keV) by 13 ± 6 ks (90% C.L.) and those of the 2–20 keV *RXTE*-PCA data lag the EUV by 34^{+11}_{-10} ks.

If we assume that the EUV light curve is representative of the seed photons which are Compton up-scattered by a hot corona to produce the hard X-ray continuum, then we can estimate of the size of the scattering region based upon the observed delays. With each scattering, a photon picks up a fractional energy $\Delta E/E \simeq 4 k_b T / m_e c^2$ where $k_b T \simeq 50$ keV is the temperature of the electrons in the corona. This value has been measured from spectral fitting of the high energy roll-over seen at energies $\gtrsim 50$ keV (Magdziarz et al. 1998). The final energy of a Comptonized seed photon will be $E \simeq E_0 (1 + 4 k_b T / m_e c^2)^n$ where E_0 is the initial seed photon energy and n is the number of scatterings. The time delay of this photon with respect to the seed photon source will be roughly proportional to the number of scatterings, $t \simeq n t_0$. Here $t_0 \sim l_T / c$ where l_T is the mean free path for Thomson scattering or the size of the corona if it is optically thin. Using the effective photon energies of 0.78 keV for the 0.5–1 keV *ASCA*-SIS data and 5.4 keV for the *RXTE*-PCA data, we solve for the characteristic time between scatterings and find $t_0 \simeq 4$ ks. For a homogeneous corona of Thomson depth $\tau \sim 2$, this implies a coronal size of $\sim 2 \times 10^{14}$ cm which is about $10 \, r_g$ for a $10^8 \, M_\odot$ black hole.

2. POWER DENSITY SPECTRA

In Fig. 1, we plot the power density spectra we have obtained using data from our pointed *RXTE*-PCA and *EUVE* observations and from archival *RXTE*-ASM data. Over frequencies $(3{-}100) \times 10^{-6}$ Hz, the *EUVE* data has a rms variability of $18 \pm 1.4\%$ while the *RXTE*-PCA data has a rms variability of $7.4 \pm 0.6\%$. The *RXTE*-ASM PDS shows a break from a nearly flat slope to a $\sim f^{-1}$ slope at 6×10^{-8} s corresponding to a 200 day time scale. This is similar to the 1 month time scale found for a similar break in the PDS of NGC 3516 (Edelson & Nandra 1999).

The combined PDS from the PCA and ASM looks remarkably similar to power spectra found over much shorter time scales for Galactic black hole candidates (GBHCs; e.g., Nowak et al. 1999). It is tempting to ascribe the various breaks seen in these power spectra to some time scale associated with the accretion process such as the viscous or thermal time scales. All such time scales are proportional to the central mass and imply a value of $\mathcal{O}(10^7) \, M_\odot$ for the central object in NGC 5548. This is very similar to values obtained from Broad Line Region reverberation mapping estimates (e.g., Peterson & Wandel 1999).

Using archival *RXTE*-PCA data, we have also constructed a PDS for MCG−6-

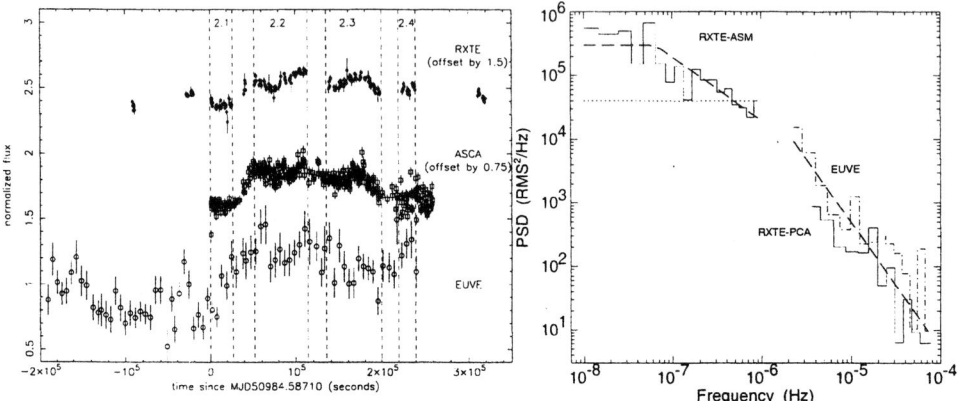

FIGURE 1. Left: *EUVE*, *ASCA*-SIS, and *RXTE*-PCA normalized light curves from our multiwavelength campaign. Right: Power Density Spectra constructed from *RXTE*-ASM, *RXTE*-PCA, and *EUVE* data for NGC 5548.

30-15 and find a knee at $\sim 10^{-5}$ Hz (Nowak & Chiang 1999). Identifying this with the lowest frequency breaks seen in the PDS of GBHCs, we infer a mass of $\mathcal{O}(10^6)\, M_\odot$. We note, however, that the breaks in the PDS of GBHCs can vary by as much as an order of magnitude for a given source. Furthermore, the various relevant time scales can also depend on effective values of the viscosity parameter as well as the Eddington ratio, both of which may vary substantially from source to source.

3. X-RAY SPECTRAL VARIABILITY AND CORRELATIONS

We divided our monitoring observations of NGC 5548 into 10 phases of 30–50 ks in order to examine X-ray spectral variability as a function of flux state. We find that the well-known correlation of spectral index with flux in Seyferts is borne out (Chiang et al. 1999). However, the behavior of the reprocessing signatures, the iron Kα line and the relative reflection normalization, are quite different than what has been seen for this object and Seyfert galaxies in general. Firstly, the iron line flux is consistent with being constant. This would be expected if the cold reprocessor were very distant from the source of the hard continuum, so that the iron line would not be seen to change even though the hard continuum varied by a factor of two over our observations. Our fits to the iron line shape, however, show that it is significantly redshifted in the rest frame of the AGN and are consistent with the bulk of the line emission being produced within $\sim 30\, r_g$ of the central source. This corresponds a response time of only ~ 20 ks for a $10^8\, M_\odot$ black hole.

In addition to the lack of iron line response to the continuum, the behavior of the relative reflection fraction normalization is also puzzling. Since the line equivalent width and reflection fraction, $\mathcal{R}$, are both *relative* quantities with respect to the underlying continuum and since both components are expected to be produced

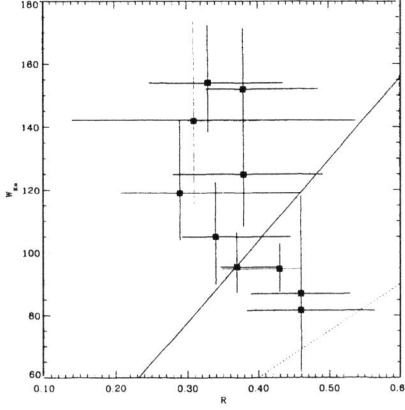

FIGURE 2. Iron line equivalent width $W_{K\alpha}$ versus the relative Compton reflection normalization $\mathcal{R}$. The lines show the expected proportionality for solar abundance (dotted) and an iron overabundance of $\simeq 1.7$ (solid).

in the same material, these two components should be proportional to one another. However, Fig. 2 shows that this is not the case. In contrast to the expected proportionality, **the iron line equivalent width and relative reflection normalization are anti-correlated.** The dotted line is the expected relationship for a cold reflector with solar abundances (George & Fabian 1991). The solid line is the best-fit proportionality relationship and corresponds to an iron overabundance of about 1.7 relative to solar. It is in conflict with the data at the 90% level. Other $RXTE$-PCA observations have revealed this anti-correlation of $W_{K\alpha}$ and $\mathcal{R}$ to be a prevalent feature of Seyfert galaxies. A similar anti-correlation is seen for MCG−6-30-15 (Lee et al. 1999) and the type 1 Seyfert NGC 3227 (Reynolds & Krishnamurthi 1999).

REFERENCES

Alexander, T. 1997, in Astronomical Time Series, ed. D. Maoz et al. (Netherlands: Kluwer Academic Publishers), 163

Chiang, J., et al. 1999, ApJ, in press (astro-ph/9907114)

Edelson, R. A., & Nandra, K., 1999, ApJ, 514, 682

George, I. A., & Fabian, A. C. 1991, MNRAS, 249, 352

Lee, J. C., et al. 1999, MNRAS, submitted (astro-ph/9909239)

Magdziarz, P., et al. 1998, MNRAS, 301, 179

Marshall, H. L., et al. 1997, ApJ, 479, 222

Nandra, K., et al. 1998, ApJ, 505, 594

Nowak, M. A., Wilms, J., & Dove, J. B. 1999, ApJ, 517, 355

Nowak, M. A., & Chiang, J. 1999, ApJL, submitted (astro-ph/9906371)

Peterson, B. M., et al. 1991, ApJ, 368, 119

Peterson, B. M., et al. 1998 PASP, 110, 660

Peterson, B. M., & Wandel, A. 1999, ApJL, 521, 95

Reynolds, C. S., & Krishnamurthi, A. 1999, in preparation

Zdziarski, A. A., Lubiński, P., & Smith, D. A. 1999, MNRAS, 305, 231

SUPER-AGILE - THE X-RAY DETECTOR FOR THE GAMMA-RAY MISSION AGILE

E. Costa[1], L. Barbanera[1], M. Feroci[1], M. Frutti[1], I. Lapshov[1], B. Martino[1], M. Mastropietro[1], E. Morelli[1], M. Rapisarda[1], A. Rubini[1], P. Soffitta[1], M. Tavani[2], S. Mereghetti[2], S. Vercellone[2], P. Caraveo[2], F. Perotti[2], G. Barbiellini[3], G. Budini[3], F. Longo[3], M. Prest[3], E. Vallazza[3], A. Morselli[4], P. Picozza[4], V. Cocco[4], C. Pittori[4], G. Di Cocco[5], C. Labanti[5]

1) Istituto di Astrofisica Spaziale - CNR - Roma
2) Istituto Fisica Cosmica "G. Occhialini" - CNR - Milano
3) Sezione Trieste - INFN - Trieste
4) Univ. Roma Tor Vergata - Roma
5) Istituto Tecnologie e Studio Radiazioni Extraterrestri - CNR - Bologna

ABSTRACT The gamma-ray observatory AGILE, the first ASI Small Mission, is planned to operate in 2002-2005. We present here Super-AGILE, an X-ray detector added on top of the gamma-ray tracker. Super-AGILE will have a large field of view, providing hard X-ray imaging and moderate spectroscopy together with the gamma-ray detector. Super-AGILE is composed by Si-microstrip detectors, equipped with a low-noise electronics allowing a sensitive range of 10-40 keV, and coupled with a set of mutually orthogonal one-dimensional coded masks. A bi-dimensional source location capability is obtained by dividing the total 1444 cm^2 geometric area of the detectors in two orthogonal directions.

KEYWORDS: instrumentation; x-ray astronomy.

1. INTRODUCTION

The mission AGILE (Astro-rivelatore Gamma ad Immagini LEggero) is one of the two space missions, selected by the Italian Space Agency (ASI) in the frame of the Small Satellite Program for phase B. AGILE will be operative in the year 2002-2005 as an observatory for gamma-ray (30 MeV - 50 GeV) and simultaneously for hard X-ray (10-40 keV). It will be open to the Italian and international scientific communities who will benefit of its good angular resolution, high time resolution and moderate spectroscopic capability. The payload of AGILE consists of an integrated gamma-rays and hard X-rays detecting system.

The hard x-ray detecting system Super-AGILE (SA) made of four thin silicon micro-strip detectors is located between the top tungsten converter and the top anti-coincidence system (ACS) of the AGILE gamma-ray instrument .

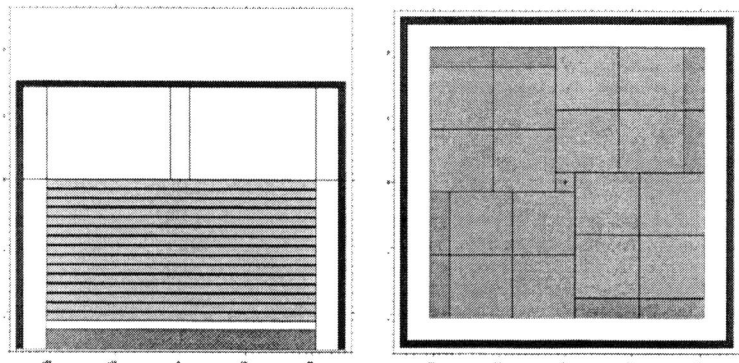

FIGURE 1. Baseline Super-AGILE design. *Left*: side view, showing the housing of Super-AGILE on top of the tracker, inside the anticoincidence shields; *Right*: view of the detector plane, showing the four detectors, composed by 4 Si microstrip tiles each, and the housing on the same plane for the front-end electronics. Both view's dimensions are 60cm×60cm.

Super-AGILE provides the mission with its burst monitor capability being able to locate hard X-ray transients with a position accuracy better than 5' (2' for sources above 100 mCrab) and with a field of view of 107°×68° (Full Width Zero Response - FWZR). Steady sources will be studied with an on-axis sensitivity better than 10 mCrab, and better than 20 mCrab in a field of view of 40°×40°, in one day of observation.

2. THE CONCEPT

The main elements of SA are designed to reach good scientific performances which complement and integrate those of the gamma-ray instrument. The mass and the materials employed here are such to avoid a reduction of efficiency and an increased background for the gamma ray instrument, with a fairly limited load in the telemetry. The design has a negligible impact on them. In Fig. 1 we show the schematic of the experiment, as it is currently used for the performances simulation of SA, and includes parts of the payload of AGILE as well.

As can be seen the ACS system completely surrounds Super-AGILE. This configuration is safe with respect to gamma-rays being possibly produced locally by particle interaction with the SA structure. The distance between the silicon plane of SA and the mask plane is 14 cm.

3. THE MASK AND THE COLLIMATOR

In our baseline design the key part of the SA encoding position system is a linear array mask with 751 element each one having a linear dimension of 242 μm. The

TABLE 1. Main characteristics of the Super-AGILE design

Energy Range	10-40 keV
Geometric Area	1444 cm^2
Effective Area	312 cm^2 (on-axis, 13 keV)
Energy Resolution	3 keV (FWHM)
Angular Resolution	6 arcmin
Source Location Accuracy	1.5 arcmin (radius)
Field of View	107°×68° FWZR
Timing Accuracy	10 - 20 μs

overall transparency is 50%. Since every detector, with the same field and orientation, is redunded twice, we are also envisaging a design with the masks of these identical detectors complementary (i.e. with the white exchanged with blacks in the code). This will afford a first order compensation of some systematic effect due to the presence of strong sources in the field. Change in the mask code is still possible but will not significantly impact in the expected performances. The mask will be made by a high Z material (baseline Au) deposited over a mechanical support integrated to the collimator. The baseline mechanical support is a thin carbon fiber sheet. The minimum opacity of the black strips ($<$ 10% to X-rays of energy $<$40 keV) and transparency of the white strips ($>$ 95% to X-rays of energy $>$8 keV) will be achieved with 100 μm Au on 500 μm carbon fiber.

The trade-off collimator design (obtained taking into account sensitivity, telemetry load and coverage of the AGILE field of view) provides a field of view of 107°×68° (FWZR) well matched with AGILE field of view. With this field of view the expected background counting rate is lower than 750 c/s keeping the telemetry load within the allocated figure of 20kb/s.

4. THE DETECTION UNITS

Super-AGILE exploits the technical implementations for AGILE. The SA detectors are identical to those used for the AGILE's Silicon Tracker, except for the fact that for the read-out low noise preamplifiers are used, the XA1.3 by IDE AS company, optimized for reading Si microstrips as X-ray detectors. The entire SA is composed of 16 Si tiles, organized in 4 detectors of 4 tiles each, with 2 detectors having the strips oriented 90° with respect to the other 2 detectors (along AGILE's X and Y axis respectively). The detectors are equipped with a low noise electronics so that signals corresponding to electron energy losses above 10 keV can be well separated from noise. The electronics is capable to self-trigger on individual events above a threshold.

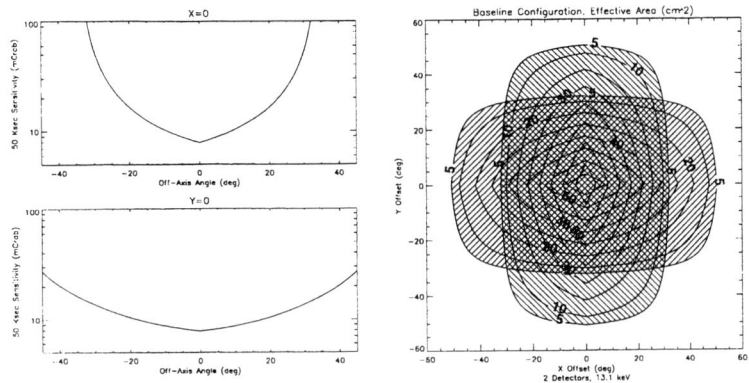

FIGURE 2. *Left*: Super-AGILE sensitivity (5σ in 50 ks, Crab-like spectrum) curves: for one detector, in mCrab units, as a function of the off-axis angle on the two orthogonal directions. *Right*: Effective area and field of view map: for two crossed detectors at 13.1 keV, numbers indicate effective area in cm^2 for one detector; contours are 5, 10, 20, 30, 40, 50, 60, 70 cm^2 from the edge to the center. Maximum effective area for one detector is 78 cm^2.

5. SUPER-AGILE PERFORMANCE

The performance of SA have been studied both by means of Monte Carlo simulation based on the Monte Carlo Neutron Photon code (MCNP) and by analytical calculations. The expected sensitivity of one single SA detector to a Crab-like point source observed for 50 ks, as a function of the off-axis position in two orthogonal directions is shown in Fig. 2. On the same figure we present the instrument's effective area over its field of view at 13.1 keV.

The combination of a rather wide field of view together with moderately high sensitivity will enable the study of a big variety of cosmic x-ray sources including gamma-ray bursts, persistent and transient Galactic x-ray sources, as well as a number of the brightest extragalactic sources. We have estimated an expected gamma-ray burst detection rate to be around 1.5 burst/month. For brightest events the localization accuracy with SA may be as good as 1.5 arc. minutes.

6. CONCLUSIONS

The concept and preliminary design of the Super-AGILE experiment show that this instrument can reach interesting astrophysical capabilities by itself, and it will have a great scientific impact when combined with the main gamma-ray detector for the study of high energy cosmic sources. The potential degradation of the performance of the main instrument due to the addition of Super-AGILE in its field of view is expected to be negligible.

NEW EXTREME SYNCHROTRON BL LAC OBJECTS

L. Costamante[1,2], G Ghisellini[2], P. Giommi[3], G. Tagliaferri[2], A. Celotti[4], M. Chiaberge[4], L. Chiappetti[5], G. Fossati[6], L. Maraschi[2], F. Tavecchio[2], A. Treves[7], A. Wolter[2]

1) Università degli Studi di Milano, Milano, Italy; 2) Osservatorio Astronomico di Brera, Milano, Italy; 3) BeppoSAX Science Data Center, ASI, Roma, Italy; 4) S.I.S.S.A., Trieste, Italy; 5) IFCTR/CNR, Milano, Italy; 6) CASS/UCSD, La Jolla, California, USA; 7) Università dell'Insubria, Como, Italy

ABSTRACT We report on the *Beppo*SAX observations of four "extreme" BL Lacs, selected to have high synchrotron peak frequencies. All have been detected also in the PDS band. For 1ES 0120+340, PKS 0548–322 and H 2356–309 the spectrum is well fitted by a convex broken power-law, thus locating the synchrotron peak around 1–4 keV. 1ES 1426+428 presents a flat energy spectral index ($\alpha_x = 0.92$) up to ~100 keV, thus constraining the synchrotron peak to lie near or above that value. For their extreme properties, all sources could be strong TeV emitters.

KEYWORDS: BL Lacertae objects: individual: 1ES 0120+340, PKS 0548-322, 1ES 1426+428, H 2356-309 — X–rays: general — TeV: general

1. INTRODUCTION

BL Lac objects are usually divided in two main classes, on the basis of their overall Spectral Energy Distribution (SED): LBL or HBL (low or high energy peaked BL Lacs), according as the peak of the synchrotron emission (in a νF_ν representation) is in the IR–optical or EUV–soft-X band, respectiveley. In the X-ray band this usually translates in a spectral index which is steep ($\alpha_x > 1$) for HBLs (corresponding to the tail of the synchrotron emission) and flat ($\alpha_x < 1$) for LBLs (corresponding to the upcoming of the Inverse Compton emission). In 1997, the *Beppo*SAX observations of Mkn 501 (Pian et al. 1997) and 1ES 2344+514 (Giommi et al. 1997) revealed that, at least in a flaring state, the peak of the synchrotron emission can actually reach very high energies, around 100 keV, with a consequently flat synchrotron X-ray spectral index. In order to find and study other sources with such "extreme" properties, we have selected several candidates from the Einstein Slew Survey and the Rosat All Sky Survey Bright Sources Catalogue (RASSBSC). The selection criteria were based on properties suggesting a high ν_{peak}:
a) very high F_x/F_{radio} ratio ($> 3 \times 10^{-10}$ erg cm^{-2} s^{-1} / Jy, at [0.1-2.4] keV and

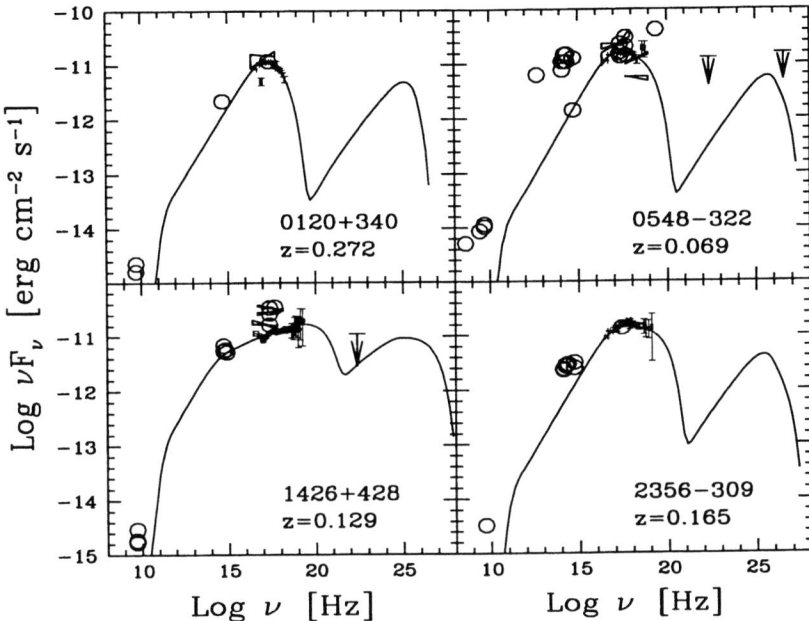

FIGURE 1. The SEDs of the 4 BL Lacs, made with *BeppoSAX* and literature data, together with a pure homogeneous SSC model (further details in Costamante et al., in prep.)

5 GHz respectively); b) flat X-ray spectrum (when available), connecting smoothly with the flux at lower frequencies; c) appropriate values of α_{ro}, α_{ox} and α_{rx} (Padovani & Giommi 1995). A high X-ray flux ($> 10^{-11}$ erg cm^{-2} s^{-1}) in the 2–10 keV band was also requested, to achieve a good detection in the PDS instrument.

We used the *BeppoSAX* satellite, whose wide X-ray energy range (0.1–200 keV) is ideal to constrain the synchrotron peak. Four objects have been observed, between June 1998 and April 1999: 1ES 0120+340, PKS 0548–322, 1ES 1426+428 and H 2356–309. In Fig. 1 and Table 1 only the best fit results are reported (for a complete discussion, see Costamante et al., in preparation). LECS, MECS and PDS data have been reduced and analysed according to the SDC Cookbook instructions, using the latest calibration matrices available. Standard extraction radii of 4' and 8' for MECS and LECS were used, except for the PKS 0548–322 observation of 20/2/99: in this case a 6' radius for the LECS has been used, due to the presence of a contaminating source in the field of view (identified as the star GSC_07061_01558 in the Guide Star Catalog, probably flaring).

The PDS instrument doesn't have imaging capabilities, and its f.o.v. (radius $\sim 45'$) is larger than LECS and MECS ($\sim 28'$ for the MECS). Therefore there is the possibility for PDS spectra to be contaminated by hard serendipitous sources in the f.o.v, not visible in the MECS images. Analyzing PDS data, we have taken this into account, also checking in the NED database for potentially contaminating sources.

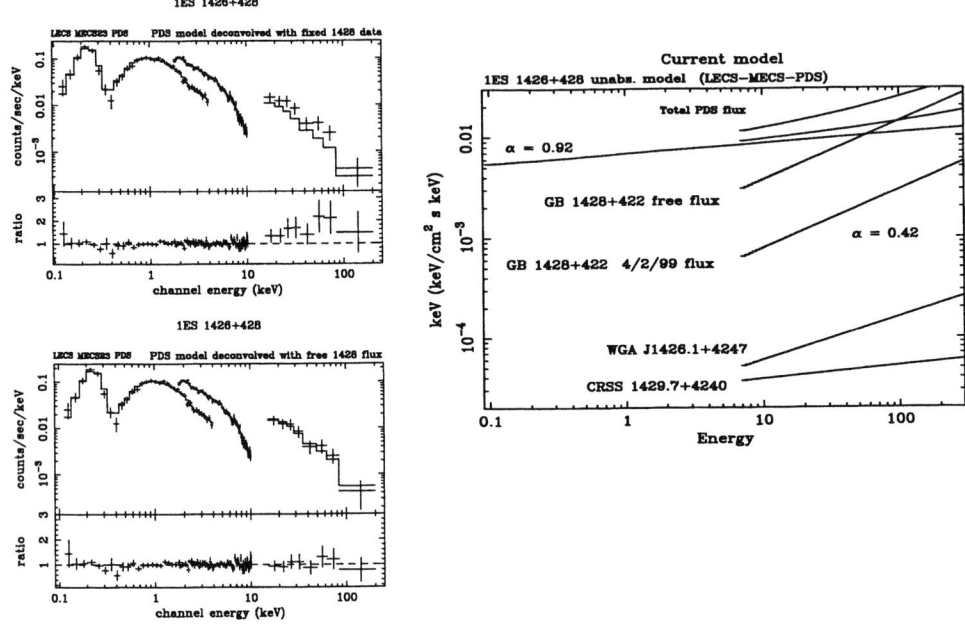

FIGURE 2. Left panel: LECS+MECS single powerlaw fits to 1ES 1426+428 data, with two different levels for GB 1428+422 flux. Right panel: the model used for PDS data.

2. 1ES 1426+428

At 41' from this source, thus in the PDS f.o.v., there is the quasar GB 1428+422 (Fabian et al 1998). To account for its contribution, we have added a component to the PDS model, based on the GB 1428+422 data from the *Beppo*SAX observation of 4/2/99 ($\alpha = 0.42$, $F_{1keV} = 0.30$ μJy; Celotti & Iwasawa, priv. comm.). We have also checked in the NED and WGACAT databases for other potentially contaminating objects: we added the contributions of the two most important objects (WGA J1426.1+4247 and CRSS 1429.7+4240, Fig 2 right panel), according to the fluxes and spectral indices extrapolated from the ROSAT band (when data were not available in literature, we used galactic N_H and a HR–α conversion by Giommi, priv. comm.). Summing all components, the different off-axis response of the instrument has been taken into account. The PDS/MECS normalization has been fixed at 0.9. With this model (Fig. 2 right panel), adding the PDS data to the LECS+MECS fit yields a $\chi_r^2 = 1.06$, with the PDS points still slightly above the model (Fig.2 left upper panel). A better χ_r^2 (0.95) is obtained with GB1428+422 flux as a free parameter: in this case the resulting flux is $F_{1keV} = 1.44 \pm 0.54$, a factor more than 4 higher during this observation than the week before. *Anyway, in both cases, the spectrum of 1ES 1426+428 remains flatter than unity up to 100 keV.*

TABLE 1. LECS+MECS best fits parameters

Source, date, model	N_H 10^{20} cm^{-2}	α_1	E_{break} keV	α_2	F_{1keV} μJy	F_{2-10} ergs/cm^{2}s	χ^2_r/d.o.f.
1ES 0120+340 2/2/99, BP	5.2 *gal*.	$0.82^{-0.96}_{+0.26}$	$1.4^{-0.7}_{+1.2}$	$1.32^{-0.08}_{+0.08}$	$4.5^{-0.6}_{+2.1}$	1.3×10^{-11}	0.92/93
PKS 0548-322 20/2/99, BP	$4.2^{-0.9}_{+1.1}$	$0.91^{-0.16}_{+0.10}$	$4.4^{-2.2}_{+1.8}$	$1.38^{-0.31}_{+0.59}$	$5.7^{-0.5}_{+0.5}$	2.3×10^{-11}	0.95/82
1ES 1426+428 8/2/99, SP	$1.5^{-0.3}_{+0.4}$	$0.92^{-0.04}_{+0.04}$	—	—	$4.6^{-0.2}_{+0.2}$	2.0×10^{-11}	1.00/89
H 2356-309 21/6/98, BP	1.3 *gal*.	$0.78^{-0.09}_{+0.06}$	$1.8^{-0.6}_{+0.6}$	$1.10^{-0.05}_{+0.05}$	$6.2^{-0.5}_{+0.5}$	2.5×10^{-11}	0.94/35

SP: Single Powerlaw　　BP: Broken Powerlaw　　Errors at 90% conf. level for 2 par. of interest

3. RESULTS

The main results for all sources are presented in Table 1. All have been detected in the PDS band. For three of them the spectrum is best fitted with a convex broken power-law: this locates the peak of the synchrotron emission in the X-ray band, around 1-4 keV, thus confirming the "extreme" nature of these sources. The spectrum of 1ES 1426+428 is instead well fitted by a single powerlaw, with a flat spectral index ($\alpha = 0.92$) up to 100 keV. **This constrains the synchrotron peak to lie near or above 100 keV.** Such high values of the synchrotron peak frequencies, flagging the presence of high relativistic electrons, make these sources good candidates for TeV emission through the Inverse Compton mechanism.

ACKNOWLEDGEMENTS

This research has made use of the NASA/IPAC Extragalactic Database (NED) which is operated by the Jet Propulsion Laboratory, California Institute of Technology, under contract with the National Aeronautics and Space Administration. We thank the *Beppo*SAX Science Data Center for their support in the data analysis. This research is financially supported by the Italian Space Agency.
L.C. thanks the Cariplo Foundation and the Italian MURST for support.

REFERENCES

Fabian A.C., Iwasawa K. et al., 1998, MNRAS 295L. 25F
Giommi P., Padovani P., Perlman E., Nucl. Physics B (Proc. Suppl.), vol. 69, p.407
Padovani P. & Giommi P., 1995, ApJ 444, 567
Pian E. et al. 1998, ApJ, 492, L17

THE COMPLEX AND VARIABLE ABSORPTION OF NGC 3516 OBSERVED BY BEPPOSAX

E. Costantini[1], C. Salvini[2], A. Comastri[2], A. Fruscione[1], S. Mathur[3], F. Nicastro[1], G.M. Stirpe[2], B. Wilkes[1]

[1] *Harvard-Smithsonian Center for Astrophysics, 60 Garden St. Cambridge MA 02138 USA*
[2] *Osservatorio Astronomico di Bologna, V. Ranzani, 1, 40127, Bologna, Italy*
[3] *The Ohio State University, 14 West 18th Ave., Columbus OH 43210, USA*

ABSTRACT   We present 2 BeppoSAX (0.1-150 keV) observations of the Seyfert 1 galaxy NGC 3516, taken 4 months apart (8-11-1996, 12-03-1997). We find that during the 1996 observation the nuclear X-ray continuum was absorbed by an unusually large amount of cold gas clearly associated with the AGN environment. Unabsorbed thermal emission is also seen below 3 keV during this observation. Finally, the 0.1-10 keV spectra of both the observations show the presence of a complex system of ionized absorbing matter which we parameterize with two "warm absorber" with quite different degrees of ionization and column densities.

KEYWORDS: AGN, X-ray spectroscopy, NGC 3516, BeppoSAX

1. INTRODUCTION

NGC 3516 is a bright ($M_B = -20.4$, de Vaucouleurs et al. 1991), nearby ($z = 0.009$) Seyfert 1 Galaxy known to host a multizones system of highly ionized and variable absorbers abscuring both the UV and X-ray nuclear radiation. The X-ray component has been directly observed by ROSAT (Mathur et al., 1997) and ASCA (Kriss et al., 1996) through the detection of two strong absorption features at the energies of the OVII and OVIII K edges. The column densities and ionization degrees of this absorber, as measured in the ROSAT and ASCA spectra using single zone photoionization models, are similar. Mathur et al. (1997) report a column density of $N_H = 7 \times 10^{21}$ cm^{-2}. Absorption by large amount of cold gas obscuring the line of sight has, instead, never been unambiguosly observed in NGC 3516. The *Ginga* data were in fact consistent with obscuration by either cold or ionized gas (Kolman et al., 1993, Nandra & Pounds, 1994), mainly due to the limited low energy coverage of *Ginga*.

Here we present the data of 2 BeppoSAX observations of NGC 3516, which clearly show, for the first time, the presence of variable cold absorption in the nuclear environment of this source. We screened and reduced the data following the standard criteria indicated in Fiore, Guainazzi & Grandi (1999).

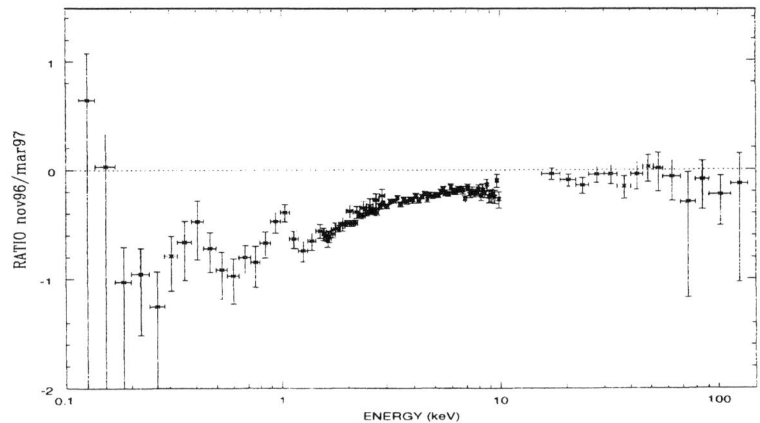

FIGURE 1. Ratio between the 1996 and 1997 BeppoSAX spectra of NGC 3516

2. DATA ANALYSIS

2.1. Model Independent Evidence for Spectral Variability

NGC 3516 underwent a strong spectral variability between the two BeppoSAX observations. This is shown in Figure 1, where we plot the ratio between the 1996 November and the 1997 March spectra. This clearly shows a flattening below 3 keV of the 1996 November spectrum of NGC 3516, compared to the 1997 March spetrum. We interpret this low energy spectral variability as due to a significant variation of the column density of cold material absorbing the nuclear radiation along the line of sight.

2.2. Spectral Fitting

The 1997 March BeppoSAX spectrum of NGC 3516 is well fitted by a nuclear continuum consisting of a 0.1-150 keV power law plus reflection by cold material. A gaussian emission line at ∼6.4 keV is also required by the data. In addition to these components our best fitting model includes: an ionized, single-zone, absorber and either an absorption edge (Model A), or a second ionized absorber with higher degree of ionization and column density (Model B). To model the 1996 November spectrum we replaced the lower ionization absorber with absorption by cold gas, largely exceeding the Galactic value (as a warm absorber model ionization degree is consistent with zero) and added a thermal emission component to account for the observed excess of counts below ∼2 keV. Strong emission from the Fe L complex has been detected around 1 keV. The results of spectral analysis are shown in Table 1.

97	[1]$\log N_H^{ion}$	log U	E_{edge}	τ_{edge}	[2]$\log N_H^{ion}$	log U	χ^2/dof
A	22.01±0.1	$0.85^{+0.12}_{-0.16}$	$7.80^{+0.23}_{-0.21}$	0.19±0.06			97/96
B	22.09±0.1	$0.76^{+0.13}_{-0.17}$			$23.25^{+0.05}_{-0.17}$	$2.35^{+0.12}_{-0.06}$	85/96
96	[3]$\log N_H^{cold}$	[4]kT	E_{edge}	τ_{edge}	[2]$\log N_H^{ion}$	log U	χ^2/dof
C	22.41±0.08	$1.78^{+0.91}_{-0.39}$	$7.93^{+0.30}_{-0.26}$	$0.22^{+0.09}_{-0.07}$			101/94
D	22.41±0.08	$1.78^{+0.91}_{-0.39}$			$23.34^{+0.21}_{-0.23}$	$2.34^{+0.09}_{-0.16}$	96/94

TABLE 1. [1]: ionized absorber responsible for the OVII, OVIII edges (1997). [2]: highly ionized absorber responsible for the ~8 keV edge. [3]: cold absorber responsible for the spectral flattening during the 1996 observation. [4]: thermal emission from optically thin plasma. Errors are quoted at a 90 % confidence level for 1 interesting parameter (i.e. $\Delta\chi^2 = 2.71$).

3. DISCUSSION

3.1. The Variable Cold Absorber

The 1996 November BeppoSAX spectrum of NGC 3516 is absorbed by an equivalent hydrogen column density of cold matter of $N_H^{Nuclear} = 2.6 \times 10^{22} cm^{-2}$, exceeding the Galactic value ($N_H^{Gal} = 3.4 \times 10^{20} cm^{-2}$, Dickey & Lockman 1990) by almost 2 order of magnitudes. The 1996 November LECS (0.1-3 keV) and MECS (3-10 keV) lightcurves show a similar pattern, and the amplitude variations are smaller than 30%, strongly suggesting no changes in the degree of obscuration during the entire duration of the observation (~ 20 hours). The absorption component is instead not required by the 1997 March data (4 months later). Considering these two time values as lower and upper limits respectively, we can estimate the linear size of a putative cloud of gas crossing the line of sight. Assuming a transverse velocity of 4000 km s^{-1} (the Hβ FWHM in NGC 3516, Wanders et al., 1993), we obtain: 3×10^{13} cm $\lesssim l_{cloud} \lesssim 4 \times 10^{15}$ cm. This in turn, along with our estimate of $N_H \sim 2.6 \times 10^{22}$ cm^{-2}, permits to constrain the range of values of the number density of this gas: 6×10^6 cm$^{-3} \lesssim n_H \lesssim 10^9$ cm^{-3}. We note that this interval would shift through higher values if the gas were mildly ionized (as it is the case for the broad emission line clouds, BELCs), since in this case we would be underestimating the equivalent N$_H$ obtained fitting the data with a model for neutral absorber. This range of densities is of the order of those expected for the BELCs (Netzer, 1991), confirming that a BELC crossing the line of sight could be responsible for the obscuration of the nucleus of NGC3516 during the 1996 BeppoSAX observation. Alternatively, this long term spectral variability could also be interpreted in terms of a dramatic change in the absorbing gas ionization degree.

3.2. The Complex Warm Absorbing System

Table 1 shows that a good fit to the BeppoSAX spectra of NGC 3516 during both observations can be obtained modelling the deficit of counts around 7.8-8 keV with an absorption edge (Model A and C). The energy and the optical depth of this feature are consistent with absorption by highly ionized gas with FeXX-XXIII dominating, and column density of log $N_H = 23.2$. Replacing the edge with a physical photoionization model (built with CLOUDY, vs. 90.04, Ferland, 1996), does indeed produce a very good fit to the data (model B and D), but modifies considerably the best fit continuum parameters, yielding a steeper X-ray power law ($\Delta\Gamma \sim 0.31\pm 0.21$) and a higher degree of reflection ($\Delta R \sim 0.64\pm 0.44$). This is due essentially to the Iron L photoelectric absorption between 1 and 2 keV, which contributes to flatten the apparent continuum up to 10 keV.

A detailed testing of the models applied and their consequences and implications is described in Costantini et al. 2000.

4. CONCLUSIONS

- We detected a marked variation of the amount of cold gas obscuring the line of sight to NGC 3516, between the two BeppoSAX observations. We propose, as a possible explanation, that a Broad Emission Line Cloud crossed the line of sight during the 1996 November observation, so obscuring the direct view to the nucleus.
- Emission by a optically thin plasma has been detected during the 1996 observation with a $L_{[0.5-4.5keV]} \sim 6 \times 10^{41}$ erg/s, just fairly consistent with emission from the host galaxy (David et al. 1992).
- The spectra of both the observations of NGC 3516 show narrow absorption features at 7.8-8 keV, which may be produced by a high column density, highly ionized gas, absorbing the X ray continuum.

ACKNOWLEDGEMENTS

We acknowledge partial support of the Italian Space Agency under the contract ASI-ARS-98-119 and of the Italian Ministry for University and Research (MURST) under grant Cofin-98-02-32 and the support of NASA grant NAG5-3545.

REFERENCES

Netzer H., in Blandford, R.D., Netzer, H, Woltjer, L., O'Brien, P., 1991, Obs, 111, 328;
de Vacouleurs, G. et al., 1991, S& T, 82, 621;
Costantini, E. et al. 2000, ApJ, in press (astro-ph 0007158).
David, L.P.,Jones, C. & Forman, W., 1992, ApJ, 388, 82;
Dickey, J.M. & Lockman, F.J., 1990, ARA& A, 28, 215;
Ferland, G.J. et al. , 1996, AAS, 188, 320;
Fiore, F., Guainazzi, M. & Grandi, P., 1999, "handbook for NFI spectral analysis";
Kolman, M. et al., 1993, ApJ, 402, 51;
Kriss, G.A. et al. 1996, ApJ, 467, 622 (1996);
Mathur, S., Wilkes, B. & Aldcroft, T., 1997, ApJ, 478, 182;
Nandra, K. & Pounds, K.A., 1994, MNRAS, 268, 405;
Wanders, I. & Horne, K., 1993, A& A, 269, 39;

ERRATIC VARIABILITY OF LMC X-1 IN THE 1.5-10 KEV RANGE. SPORADIC PRESENCE OF A QPO IN A BLACK HOLE CANDIDATE.

D. Dal Fiume [1], F. Haardt[2], M. R. Galli[2], A. Treves[2], L. Chiappetti[3]

1) Istituto TESRE/CNR, via Gobetti 101, 40129 Bologna, Italy
2) Dipartimento di Fisica, Università dell'Insubria, via Lucini 3, 22100 Como, Italy
3) IFCTR/CNR, via Bassini 15, 20133 Milano, Italy
On behalf of a larger collaboration

ABSTRACT

We observed LMC X-1 on October 5-6 1997 with *BeppoSAX*. The analysis of erratic variability yielded a power–law red noise with frequency index $\alpha=0.86\pm0.05$. Marginal evidence of a QPO centered at 0.075±0.007 Hz, with a power corresponding to 5±2.2% was found. These results are in good agreement with those from *GINGA* 1987 measurements (Ebisawa et al. 1989). Recently Schmidtke et al. (1999) failed to detect QPOs in 9 observations performed in 1996 with *RXTE*. The presence of QPOs in the power spectrum of LMC X-1 appears sporadic.

KEYWORDS: Stars: individual: LMC X-1; Stars: binaries; X-rays: stars

1. INTRODUCTION

LMC X-1 is, together with Cyg X-1 and LMC X-3, a persistent X-ray binary where the presence of a black hole is established by the accurate measurement of the mass function. After the identification of the optical counterpart (Cowley et al. 1995), an orbital period of 4.2 d was measured. The mass function $f(M) = 0.14 M_\odot$ implies a mass of the compact star $M_X \simeq 6 M_\odot$, indicative of a black hole (e.g. Hutchings et al 1983).

The spectrum of LMC X-1 was studied in detail with *GINGA* (Ebisawa et al 1993) and was modeled using a superposition of a multicolour disk black body (DBB) and a power–law high energy tail (PL). Ebisawa, Mitsuda and Inoue (1989) detected in the 1.2-15.7 keV count rate from *GINGA* the presence of quasi–periodic oscillations (QPOs) with a centroid frequency of 0.0751±0.0009 Hz and an amplitude of 2.9±0.2%. In addition a red noise continuum component $P(f) \propto f^{-\alpha}$ appeared in the Power Spectral Density (PSD) with a slope $\alpha = 0.81 \pm 0.12$. The recent *RXTE* observations of Schmidtke, Ponder and Cowley (1999) failed to detect QPOs in nine observations from February 1996 to October 1996, while the prominent red noise was confirmed. The 3σ upper limits for QPOs were between 0.2 and

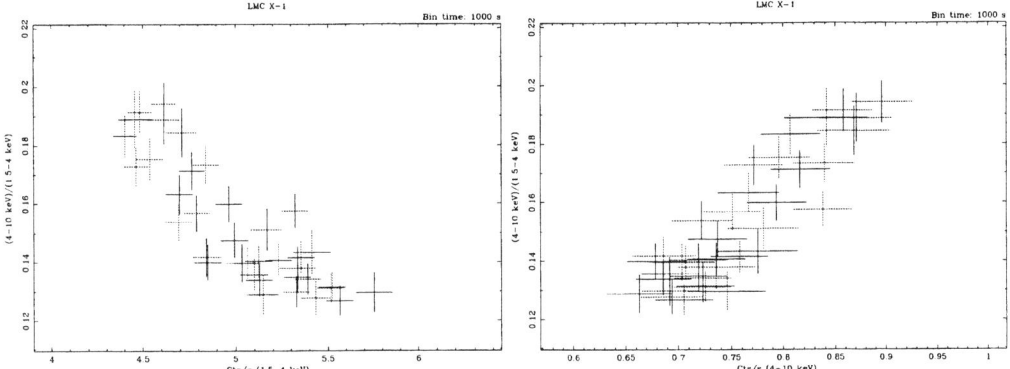

FIGURE 1. *Left panel:* hardness ratio versus 1.5–4 keV count rate. *Right panel:* hardness ratio versus 4–10 keV count rate (see text).

1.8%.

We report on temporal variability of LMC X–1 measured during a *BeppoSAX* observation performed on October 5/6, 1997. Preliminary results on the spectral analysis are reported in Treves et al. 1999. A complete presentation will be given in a forthcoming paper (Haardt et al. 1999 in preparation).

2. DATA ANALYSIS AND RESULTS

The MECS source count rate shows the known erratic variability of this source. In Figure 1 the colour versus intensity diagrams for two MECS energy bands (1.5–4 and 4–10 keV), binned on 1000s, show evidence of correlation. The spectral hardness correlates positively with the count rate in 4–10 keV and anti–correlates with the count rate in 1.5–4 keV. This differs from the behaviour observed by Schmidtke et al. (1999), where no correlation between colour and "soft" count rate was detected. A natural explanation of our observation is that the source spectrum is pivoting. Assuming a two–component spectral model this implies that the disk blackbody is anticorrelated with the power law component. We analyzed the temporal variations in the X–ray flux from LMC X–1 by calculating a Power Spectral Density (PSD) on the 1.5–10 keV count rate. 16 uninterrupted time intervals were selected, with a typical duration of 3000s. One PSD for each of these data run was calculated and the resulting spectra were averaged. The logarithmically rebinned averaged spectrum in coarse frequency bins is shown in Figure 2, left panel. The PSD is normalized to have its expected value for a purely white noise equal to 2 (Leahy et al. 1983), this holding when instrumental and dead time effects are negligible. In Figure 2 a fit with a power law plus constant is also shown. The value of the observed slope α is completely consistent with those obtained in previous measurements (Ebisawa et al 1989, Schmidtke et al. 1999).

In order to search for the presence of QPOs, we analysed the PSD using a linear rebinning, with an optimal choice to detect the QPOs at the expected frequency, if present. A visual inspection of this linearly rebinned PSD shows a possible power excess at a frequency compatible with the measure of Ebisawa et al. (near 7×10^{-2}

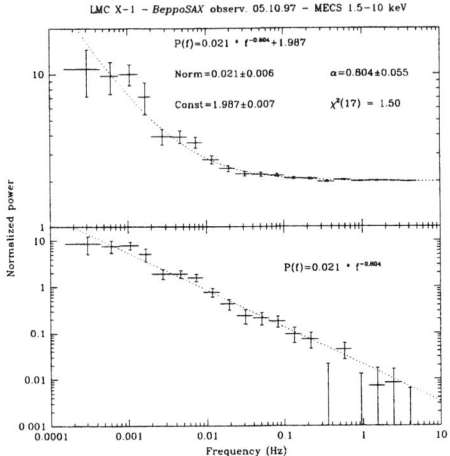
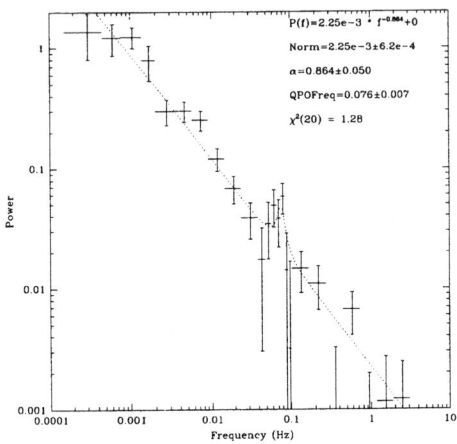

FIGURE 2. *Left panel:* Power Spectral Density of LMC X-1 in 1.5-10 keV from merged MECS2 and MECS3 events. A fit with a power law plus constant is also shown. Errors are 68% single parameter confidence level. *Right panel:* A finer frequency binning near the *GINGA* QPO frequence shows the presence of a marginal excess. A fit with a power law plus a Lorentzian (to model the QPO peak) is also shown.

Hz). The resulting broad-band PSD with fine binning in the vicinity of the expected QPO frequency is shown in Figure 2, right panel. The total significance of the excess power is rather low. It can be quantified adding a component to the spectral modeling and performing an F-test. We therefore compared a fit using a simple power law with a fit using a power law plus a Lorentzian. In the first case the minimum χ^2 is 31.1 for 23 degrees of freedom and in the second case the minimum χ^2 is 25.8 for 20 degrees of freedom. The F-test gives a probability of chance improvement adding the Lorentzian of 0.32. The total excess above the power law fit is 5±2.2% rms. This is close to the measured power by Ebisawa et al. (1989), but much higher than the observed upper limit by RXTE (~0.8%) and BBXRT (Schlegel et al. 1994 - ~1.6%).

3. DISCUSSION AND CONCLUSIONS

Low frequency QPO have been observed in a number of black hole candidates and in general their intensity is strongly variable (see van der Klis (1995) for a review and Nowak et al. (1999) at this conference). In these regards the case of LMC X-1 is not exceptional. The time lag between the QPO detection with Ginga and with BeppoSAX was 10 years, and the QPO frequency was reproduced within the uncertainties. In the same time interval negative searches were performed RXTE

TABLE 1: DISK LUMINOSITY

Experiment	L_{Disk} (erg/s)	QPO?
Ginga	$1.2 \times 10^{38}/\cos i$	Y
BBXRT	$1.0 \times 10^{38}/\cos i$	N
RXTE	$1.2 \times 10^{38}/\cos i$	N
BeppoSAX	$9.5 \times 10^{37}/\cos i$	Y

and BBXRT.

In order to understand the origin of the QPO and constrain physical models, it would be of interest to correlate the appearence of the QPOs with a spectral property, and an obvious parameter could be the X-ray intensity. This is summarized in Table 1. The presence of the QPO thererefore does not directly correlate with the intensity. However, as it was noted above, the X-ray spectral shape is rather complex. In particular it could be of interest to correlate the QPO presence with the power law component. Unfortunately present data are affected by too large uncertainties.

REFERENCES

Cowley, A. P. et al. 1995 *PASP*, **107**, 145
Ebisawa, K., Mitsuda, K., Inoue, H. 1989 *PASJ*, **41**, 519
Haardt, F., et al. 1999 in preparation
Hutchings, J. B. et al. 1983 *ApJL*, **275**, L43
Leahy, D. A. et al. 1983 *ApJ*, **266**, 160
Nowak, M. A. et al. 1999 *these proceedings*
Schlegel, E. M. et al. 1994 *ApJ*, **422**, 243
Schmidtke, P. C., Ponder, A. L., Cowley, A. P. 1999 *ApJ*, **117**, 1292
Treves, A. et al. 1999 *Adv. Space Res.*, Proc. of Nagoya E1.1 COSPAR Symp., in press
van der Klis, M. 1995 in *X–Ray Binaries*, W. H. G. Lewin, J. van Paradijs and E. P. J. van den Heuvel eds., Cambridge University Press.

THE PARKES MULTIBEAM PULSAR SURVEY: PRELIMINARY RESULTS

N. D'Amico[1], A.G. Lyne[2], R.N. Manchester[3], F.M. Camilo[2], V.M. Kaspi[4], J. Bell[3], I.H. Stairs[2], F. Crawford[4], D. Morris[2], A. Possenti[1]

1) Osservatorio Astronomico di Bologna, via Ranzani 1, 40127 Bologna,
and Istituto di Radioastronomia del CNR, via Gobetti 101, 40129 Bologna, Italy
2) University of Manchester, Jodrell Bank Observatory,
Macclesfield, Cheshire, SK11 9DL, UK
3) Australia Telescope National Facility, CSIRO, PO Box 76,
Epping NSW, 1710, Australia
4) Massachusetts Institute of Technology, Center for Space Research, 70 Vassar Street, Cambridge, MA 02139, USA

ABSTRACT

A high-frequency survey of the Galactic plane for radio pulsars is in progress, using the multibeam receiver on the 64-m Parkes radiotelescope. We describe the survey motivations, the observing plan and the inital results. The survey is discovering many pulsars, more than 500 so far. Eight of the new pulsars are binary, one with a massive companion. At least eight are young, with characteristic ages of less than 100 kyr. Two of these (Kaspi et al, this Conference) have surface dipole magnetic field strengths greater than any other known radio pulsar.

KEYWORDS: methods: observational; pulsars: general; pulsars: searches.

1. INTRODUCTION

Young pulsars are relatively rare objects in the pulsar population because they evolve rapidly, so on average their distance is relatively high. They are usually found at low Galactic latitudes, close to their places of birth, where their detection is limited by the high background temperature and by the broadening of pulses due to dispersion and interstellar scattering. On the other hand, they are interesting objects for many reasons: they are likely to be γ–ray sources; they exibhit rotational *glitches*, which are of interest in the understanding of the interior structure of neutron stars; they are likely to be associated with supernova remnants.

High frequency ($\nu \simeq 1400$ MHz) surveys (Clifton et al, 1992; Johnston et al 1992) and searches (Kaspi et al, 1996; Manchester, D'Amico & Tuohy, 1985) for young, low latitude distant pulsars proved to be successful, because the contribution of the Galactic synchrotron radiation to the radiotelescope system temperature is highly reduced, because the effect of dispersion is more easlily removed, and

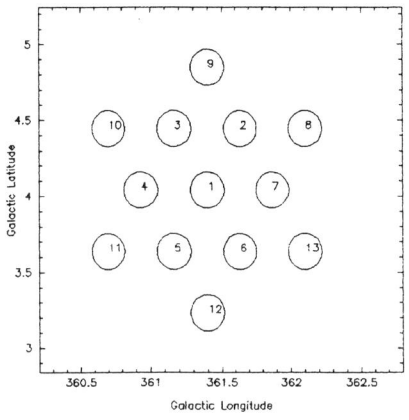

FIGURE 1. Beam pattern of the multibeam receiver at Parkes.

because the broadening of pulses due to interstellar scattering varies with frequency aproximately as $\nu^{-4.4}$. Triggered by the above motivations, we are undertaking a new survey for pulsars along the Galactic plane at 1.4 Ghz, using the 13-element multibeam receiver recently installed on the 64-m Parkes radiotelescope. In this paper we present the experiment configuration, the survey plan and the preliminary results of about 50% of the survey.

2. THE MULTIBEAM SURVEY

Each beam of the multibeam receiver system at Parkes is approximately $0.23°$ wide and the beams centres are spaced 2 beamwidth apart (see Fig. 1). The survey pointings are interleaved to give complete sky coverage on a hexagonal grid containing a total of 2670 pointings of 13 beams each. The parameters of the present experiment and those of two previous high frequency surveys of the Galactic plane are summarized in Table 1. Thanks to the long integration time adopted (35-min) and the high sensitivity of the new receiver system, the present survey has a sensitivity 7 times better than previous surveys. Fig 2. shows the theoretical sensitivity as a function of the pulsar period and dispersion measure.

3. RESULTS AND DISCUSSION

So far we have observed about 1600 pointings, 90% of which are analysed, corresponding to about 50% of the total survey region. The data reduction system is similar to that used in the Parkes low frequency survey (Manchester et al, 1996), and is carried out on a network of workstations. Because of the relatively long integrations adopted, we complement the standard search analysis with "acceleration search" to take into account possible binary motions. To date we have discovered

TABLE 1. Three 20 cm pulsar surveys

	Jodrell Bank	Parkes	Parkes
Latitude range, $|b|$	$< 1°$	$< 4°$	$< 5°$
Longitude range, l	$-5°\ldots 100°$	$-90°\ldots 20°$	$-100°\ldots 50°$
Center frequency (MHz)	1400	1520	1374
Number of beams	1	1	13
Integration time (min)	10	2.5	35
Sample interval (ms)	2.0	1.2	0.25
Bandwidth (MHz)	$2 \times 8 \times 5$	$2 \times 64 \times 5$	$2 \times 96 \times 3$
$S_{\rm sys}$ (Jy)	60	70	36
$S_{\rm min}$ (mJy)	1.2	1.0	0.15
Pulsars found/detected	40/61	46/100	513+/703+
Reference	Clifton et al.	Johnston et al.	this work

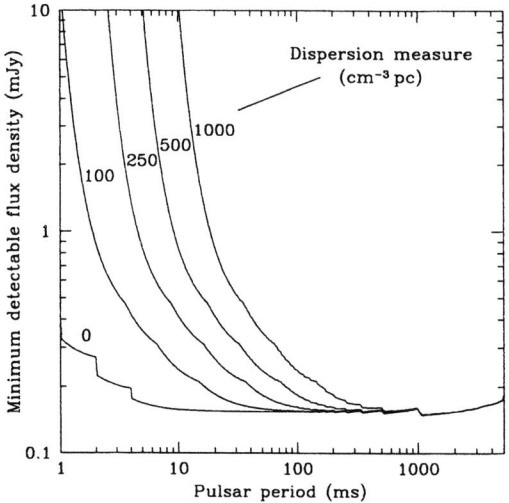

FIGURE 2. Theoretical sensitivity of the Parkes multibeam pulsar survey as a function of period and DM for a pulsar on the centre beam.

TABLE 2. New binary pulsars

PSR J	P (ms)	τ_c (10^6 y)	Distance (kpc)	P_b (d)	Ecc.	Min. M_c ($M_\odot$)
J1232−6501	88.28	1400	10.0	1.863	0.00	0.15
J1904+04	71.09	–	4.0	15.750	0.04	0.23
J1810−2005	32.82	4000	4.0	15.012	0.00	0.29
J1453−58	45.25	–	3.3	12.422	0.00	0.88
J1435−60	9.35	–	3.2	1.355	0.00	0.90
J1811−1736	104.18	950	5.9	18.779	0.83	0.87
J1141−6545	393.90	1.45	3.2	0.198	0.17	1.01
J1740−3052	570.31	0.36	10.8	231.039	0.58	11.07

513 new pulsars, and have detected 190 known pulsars. Accounting for the fact that so far we searched the regions closest to the Galactic plane, we believe that the number of new discoveries for the entire survey should be somewhat over 800.

Timing observations of the newly discovered pulsars are carried out at Jodrell Bank and Parkes. Observations are made at intervals of 4 – 8 weeks, or more closely spaced when pulse-counting statistics need to be resolved. Full timing solutions have been obtained so far for 80 pulsars. At least eight of the new discoveries are young pulsars ($\tau_c < 10^5$ years). Two radio pulsars with the highest known surface magnetic field have been discovered (Kaspi et al, these proceedings). One of these objects, PSR J1119-6127 is very young, with as characteristic age, $\tau_c = 1600$ years. For this pulsar we also measured a braking index $n=3.0\pm0.1$.

So far, eight of the newly discovered pulsars proved to be members of binary systems, including a pulsar (PSR J1811-1736) in a highly eccentric binary system (Lyne et al 2000) and a pulsar (PSR J1740-3052) with a very massive companion ($\simeq 11$ $M_\odot$). The basic parameters of the binary pulsars are shown in Table 2.

REFERENCES

Clifton, T.R., Lyne, A.G., Jones, A.W., McKenna, J., Ashworth, M. 1992, MNRAS, 254, 177

Johnston, S., Lyne, A.G., Manchester, R.N., Kniffen, D.A, D'Amico, N., Lim, J., Ashworth, M. 1992, MNRAS, 255, 401

Kaspi, V.M., Manchester, R.N., Johnston, S., Lyne, A.G., D'Amico, N. 1996, AJ, 111, 2028

Lyne, A.G. et al 2000, MNRAS, 312, 698

Manchester R.N., D'Amico, N., Tuohy, I.R. 1985, MNRAS, 212, 975

Manchester R.N., et al 1996, MNRAS, 279, 1235

THE ASCA HARD SERENDIPITOUS SURVEY (HSS): A PROGRESS UPDATE

R. Della Ceca[1], V. Braito[2], I. Cagnoni[3], T. Maccacaro[1]

1) Osservatorio Astronomico di Brera, Milan, Italy; 2) Università degli Studi di Milano, Milan, Italy; 3) International School for Advanced Studies, SISSA, Trieste, Italy.

ABSTRACT

We present here a status update on the ASCA Hard Serendipitous Survey (HSS), a survey program conducted in the 2-10 keV energy band. In particular we discuss the number-flux relationship, the 2-10 keV spectral properties of the sources, and of the spectroscopically identified objects.

KEYWORDS: Galaxies: active — diffuse radiation — surveys — X-ray: galaxies — X-ray: general

1. INTRODUCTION

At the *Osservatorio Astronomico di Brera* we started a few years ago the ASCA Hard Serendipitous Survey (HSS): a systematic search for sources in the $2-10$ keV energy band, using data from the GIS2 instrument onboard the ASCA satellite. The specific aims of this project are: a) to extend to faint fluxes the census of the X-ray sources shining in the hard X-ray sky, b) to evaluate the contribution to the Cosmic X-ray Background (CXB) from the different classes of X-ray sources, and c) to test the Unification Model for AGNs.

This effort has lead to a pilot sample of 60 sources that has been used to extend the number-counts relationship down to a flux limit of $\sim 7 \times 10^{-14}$ erg cm^{-2} s^{-1} (the faintest detectable flux) resolving *directly* about 30% of the $(2-10$ keV) Cosmic X-ray Background (CXB), and to investigate their X-ray spectral properties (Cagnoni, Della Ceca and Maccacaro, 1998; Della Ceca et al., 1999).

Recently the ASCA HSS has been extended: we discuss here this extension and the main results obtained so far.

2. THE ASCA HSS SAMPLE

The data considered for the extension of the ASCA HSS were extracted from the public archive of 1629 ASCA fields (as of December 18, 1997). The field(s) selection criteria, the data preparation and analysis, the source detection and selection, and the computation of the sky coverage are described in detail in Cagnoni, Della Ceca and Maccacaro (1998) and Della Ceca et al. (1999).

The 300 GIS2 images adequate for this project have been searched for sources with a signal-to-noise (S/N) ratio greater than 4.0 (a more restrictive criterion than that adopted in Cagnoni et al., (1998) where a S/N $\geq$ 3.5 was used). A sample of 189 serendipitous sources with fluxes in the range $\sim 1 \times 10^{-13} - \sim 7 \times 10^{-12}$ erg cm^{-2} s^{-1}, found over a total area of sky of ~ 71 deg^2, has been defined. Full details on this sample will be reported in Della Ceca at al., (2000).

3. THE 2–10 KEV logN(>S)–logS

In Figure 1 we show a parametric (solid line) and a non parametric (solid histogram) representation of the number-flux relationship obtained using the new ASCA HSS sample of 189 sources.

Also shown in Figure 1 (cross at $\sim 3 \times 10^{-11}$ erg cm^{-2} s^{-1}) is the surface density of the extragalactic population in the Piccinotti et al. (1982) HEAO 1 A-2 sample (as corrected by Comastri et al., 1995) and the surface density of X-ray sources as determined by Kondo (1991) using a small sample of 11 sources extracted from the Ginga High Galactic Latitude survey (filled triangle at $\sim 8 \times 10^{-12}$ erg cm^{-2} s^{-1}). The surface densities represented by the filled dots at $\sim 1.2 \times 10^{-13}$, $\sim 1.8 \times 10^{-13}$, and $\sim 3.0 \times 10^{-13}$ erg cm^{-2} s^{-1} are the results from the ASCA Large Sky Survey (Ueda et al., 1999); the filled dot at $\sim 5 \times 10^{-14}$ erg cm^{-2} s^{-1} has been obtained by Georgantopoulos et al. (1997) using 3 deep ASCA GIS observations; the filled dot at $\sim 4.0 \times 10^{-14}$ erg cm^{-2} s^{-1} has been obtained from Inoue et al. (1996) using data from a deep ASCA observation. Finally, the filled square at $\sim 5.0 \times 10^{-14}$ erg cm^{-2} s^{-1} has been obtained by Giommi et al. (1998) using data from the BeppoSAX deep surveys. As it can be seen, our determination of the number-flux relationship is in very good agreement with those obtained from other survey programs. A good agreement is also found with the fluctuations analysis of GINGA (Butcher et al., 1997) and ASCA SIS (Gendreau, Barcons and Fabian, 1998) data.

The logN(>S)–logS can be described by a power law model N(>S) = $K \times S^{-\alpha}$ with best fit value for the slope of $\alpha = 1.63 \pm 0.09$; the dotted lines represent the $\pm 68\%$ confidence interval on the slope. The normalization K is determined by rescaling the model to the actual number of objects in the sample and, in the case of the "best" fit model, is $K = 9.65 \times 10^{-21}$ deg^{-2}. At the flux limit of the survey ($\sim 7 \times 10^{-14}$ erg cm^{-2} s^{-1}) the total emissivity of the resolved objects is ~ 10 keV cm^{-2} s^{-1} sr^{-1}, i.e. about 30% of the 2–10 keV CXB. A flattening of the number-flux relationship, within a factor of 10 from our flux limit, is expected in order to avoid saturation.

4. THE 2–10 KEV SPECTRAL PROPERTIES OF THE SOURCES

To investigate the spectral properties of the sources in the 2 – 10 keV energy range we defined the Hardness Ratio, $HR2 = \frac{H-M}{H+M}$, where M and H are the observed (GIS2 + GIS3) net counts in the 2–4 keV and 4–10 keV energy band respectively (see Della Ceca et al.,1999 for details). In Figure 2a, for all sources, we plot the

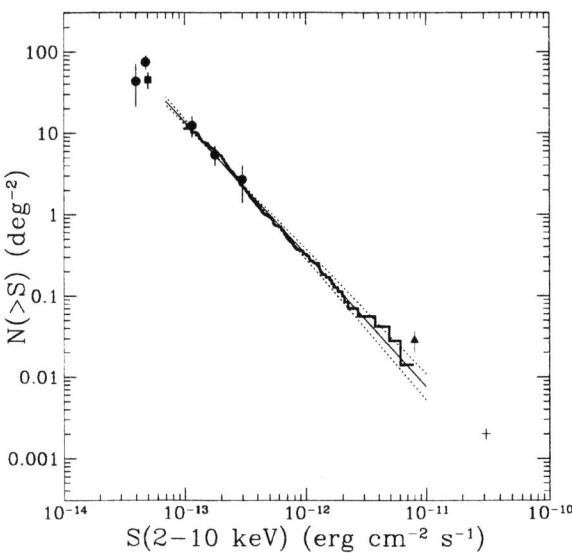

FIGURE 1. The 2−10 keV logN(>S)−logS. See section 3 for details.

HR2 value versus the GIS2 count rate; we have also reported the flux scale obtained assuming a count rate to flux conversion factor appropriate for a power law model with $\alpha_E \sim 0.6$, the median spectral energy index of the sample. The HR2 values are then compared with those expected from a non absorbed power-law model with α_E ranging from −1.0 to 2.0. It is worth noting the presence of many sources which seem to be characterized by a very flat 2−10 keV spectrum with $\alpha_E \leq 0.4$ and of a number of sources with "inverted" spectra (i.e. $\alpha_E \leq 0.0$).

A flattening of the mean spectrum of the sources with decreasing count rate is clearly evident. If we divide the sample into two subsamples (the bright sample is defined by the 60 sources with a count rate $\geq 4.3 \times 10^{-3}$ cts s^{-1}, while the faint sample is defined by the remaining 129 sources), then the fraction of sources with $\alpha_E \leq 0.4$ ($\alpha_E \leq 0.0$) is $15 \pm 5\%$ ($8 \pm 4\%$) in the bright sample and becomes $43 \pm 7\%$ ($18 \pm 4\%$) in the faint sample. These objects with very flat spectra could represent a new population of very hard sources or, alternatively, a population of very absorbed sources as expected from the CXB synthesis models based on the AGN Unification Scheme.

5. THE SPECTROSCOPICALLY IDENTIFIED SAMPLE

Up to now 47 sources have been spectroscopically identified using optical database (NED) and our observations. The optical breakdown is the following: 1 star, 5 cluster of galaxies, 5 BL Lac objects, 33 Broad Line Type 1 AGNs and 3 Narrow Line Type 2 AGNs. However we stress that this small sample of identified objects is probably not representative of the whole population since Type 1 AGNs are much better represented in literature than Type 2 AGNs.

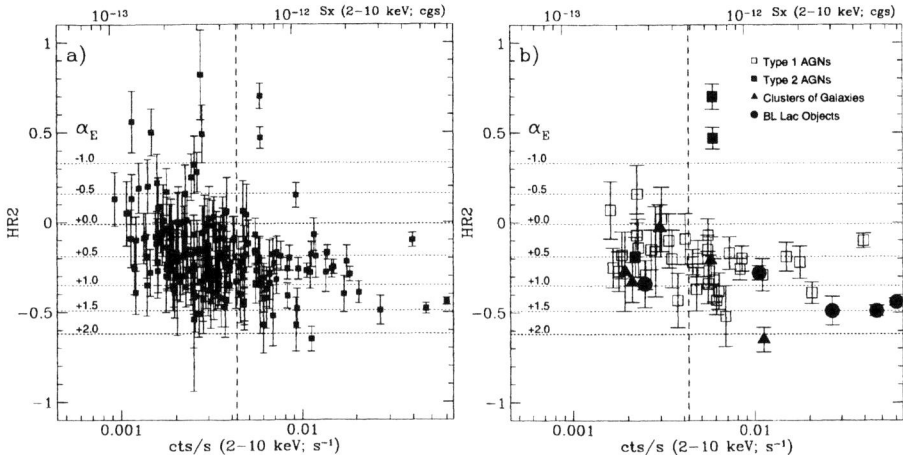

FIGURE 2. HR2 vs. count rate. Panel a: the complete ASCA HSS sample; Panel b: the identified objects

In Figure 2b we plot the HR2 value versus the GIS2 count rate for this subsample of identified objects. We note that 2 of the 3 objects classified as Type 2 AGNs have an inverted X-ray spectrum in the 2-10 keV band, and that some of the broad line Type 1 AGNs seem to have a very flat ($\alpha_E \leq 0.5$) spectrum. A similar result has been also found in the HELLAS survey (Fiore et al., these proceedings) and in the ASCA sample of Akiyama et al. (2000).

ACKNOWLEDGEMENTS

This work received partial financial support from the Italian Ministry for University Research (MURST) under grant Cofin98-02-32 and from the Fondazione CARIPLO.

REFERENCES

Akiyama, M., et al., 2000, Ap.J., 532, 700.
Butcher, J.A., et al., 1997, MNRAS, 291, 437.
Cagnoni, I., Della Ceca, R., and Maccacaro, T., 1998, Ap.J., 493, 54.
Comastri, A., Setti, G., Zamorani, G., and Hasinger, G., 1995, A&A, 296, 1.
Della Ceca, R., Castelli, G., Braito, V., Cagnoni, I., and Maccacaro, T., 1999, Ap.J., 524, 674.
Della Ceca, R., et al., 2000, in preparation.
Gendreau, K.C., Barcons, X., and Fabian, A.C., 1998, MNRAS, 297, 41.
Georgantopoulos, I., et al., 1997, MNRAS, 291, 203.
Giommi, P., et al., 1998, Nuclear Physics B (Proc. Suppl.), 69/1-3, 591.
Inoue, H., Kii, T., Ogasaka, Y., Takahashi, T., and Ueda, Y., 1996, MPE REP. 263, 323.
Kondo, H., 1991, Ph.D. Thesis Univ. of Tokyo.
Piccinotti, G., et al., 1982, Ap.J., 253, 485.
Ueda, Y., et al., 1999, Ap.J., 518, 656.

BEPPOSAX OBSERVATIONS OF ASYNCHRONOUS MAGNETIC CATACLYSMIC VARIABLES

D. de Martino [1], G. Matt[2], T. Belloni[3], K. Beuermann[4], B.T. Gänsicke[4], F. Haberl[5], M. Mouchet[6], K. Mukai [7], J.M. Bonnet-Bidaud[8]

1) Osservatorio Astronomico di Capodimonte, Napoli, Italy 2) Dipartimento di Fisica, Universitá Roma III, Roma, Italy 3) Osservatorio Astronomico di Brera, Italy 4) Universitäts Sternwarte Göttingen, Germany 5) Max Planck Institut für Extraterrestrische Physik Garching, Germany 6) DAEC Observatoire de Paris Meudon, France 7) Laboratory for High Energy Astropysics NASA/GSFC, USA 8) Service d'Astrophysique Saclay Gif-Sur-Yvette, France

ABSTRACT

We present a preliminary analysis of BeppoSAX observations of two Intermediate Polars, which show different temporal and spectral properties, characteristics of different accretion modes.

KEYWORDS: X-rays; cataclysmic variables; accretion; accretion disks

1. INTRODUCTION

Intermediate Polars (IPs) are a subclass of magnetic Cataclysmic Variables (mCVs) containing a weakly magnetized (B$\lesssim$ 10 MG) and asynchronously rotating (P_{rot} < P_{orb}) white dwarf (WD) accreting from a late type Roche lobe overflowing secondary star. The magnetic field influences the details of the flow down to the WD poles, where a stand-off shock is formed, below which hard X-rays and cyclotron radiation are emitted. In the high field synchronous mCVs (Polars), cyclotron radiation is an important cooling mechanism, whilst thermal Bremsstrahlung dominates in low field systems. IPs are generally hard (kT $\sim$ 5 – 30 keV) highly absorbed (up to 10^{23} cm^{-2}) X-ray sources, while Polars show a substantial blackbody soft X-ray/EUV emission, produced by the absorption of cyclotron and hard X-ray radiations and/or by the infall of dense blobs deep into the WD atmosphere (Beuermann 1998). However, there are a few exceptional IPs (RE 0751+14 (Mason et al. 1992), RX J0558+53 (Haberl et al. 1994; Haberl & Motch 1995), RX J0512–32 (Burwitz et al. 1996) and RX J0806+15 (Burwitz & Reinsch, these proceedings)), showing a soft X-ray component, and thus believed to be progenitors of Polars.

In Polars material is directly channelled towards the magnetic poles (accretion column), while in IPs matter is generally accreted via a partial disk in an arc-shaped curtain. Disk-less accretion can be expected in high magnetic field IPs. Disk-fed systems are generally identified by the dominance of the WD spin (ω) pulsation

(Norton et al. 1996; Hellier 1999) as material in the disk loses memory of the orbital motion. The additional presence of periodicities, at the beat (or synodic) $\omega - \Omega$ and orbital (Ω) frequencies are an indication of disk-overflow accretion. Amplitude variations of pulsations on time scales of years (Norton et al. 1996; de Martino et al. 1999) indicate changes in the proportion of disk-overflow and disk accretion.

In the framework of a core program on mCVs with BeppoSAX, here we present some results on two systems which are representative of different accretion modes.

2. THE OBSERVATIONS

BeppoSAX (Boella et al. 1997) observations of RX J1712–24 were carried out in July 1998 for a total of 83 ks, 40 ks and 32 ks in the MECS (1.3-10 keV), PDS (15-300 keV) and LECS (0.1-10 keV) detectors respectively. RX J0028+59 was observed in August 1998 for a total of 86 ks, 40 ks and 34 ks in the same detectors respectively.

3. THE PULSATION CHARACTERISTICS

Fig. 1 shows the raw MECS power spectra together with those obtained with the CLEANED algorithm, which removes the effects of windowing in the data.

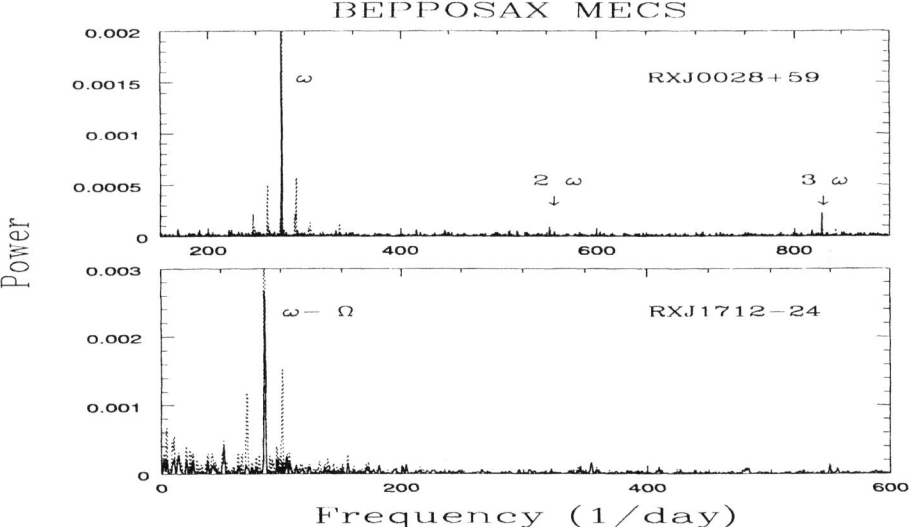

FIGURE 1. The raw MECS power spectra (dotted line) together with the CLEANED power spectra (solid line) of the IPs RX J1712–24, and RX J0028+59.

RX J1712–24 is a hard X-ray IP, but characterized by optical polarized radiation (Buckley et al. 1995; 1997). The BeppoSAX data show the dominance of the 1003 s beat pulsation, confirming previous ROSAT and XTE results, with no significant power at the 927 s spin period. It is hence an example of a stream-fed IP, where accretion occurs onto both poles each half of a beat cycle. The 1003 s modulation is single peaked with a pulsed fraction of ∼ 30% in the 0.5-2 keV band, similar to that observed in the ROSAT data. The pulsation is quite sinusoidal but structured below 2 keV (Fig. 2). A decrease in the pulsed fractions moving to higher energies is observed. This is typical of IPs, indicating absorption effects. Both LECS and MECS hardness ratios show hardening at beat maximum. This behaviour is different from other IPs, but similar to Polars. If an accretion column is present, which is likely to occur in this stream-fed IP, then the maximum is expected when the column points towards the observer, when the absorption is larger. A phase resolved spectral analysis (in progress) will provide further constraints to this scenario.

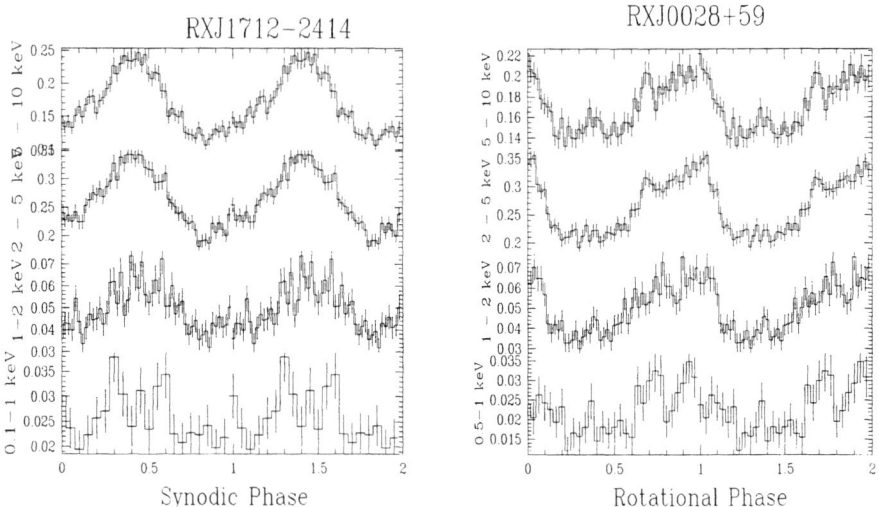

FIGURE 2. Modulations at different energies. Ordinates are counts s^{-1}.

RX J0028+59 shows a strong pulsation at the WD spin (313 s) as well as the presence of a weak signal at its second harmonic. The detection of the first harmonic in ROSAT data (Norton et al. 1999) is only marginal. The absence of orbital and beat modulations is a strong indication that this system is a disk accretor. The presence of X-ray flux at rotational minimum (Fig. 2) indicates that accretion occurs onto both poles. Although the ROSAT spin pulse is double-peaked (Norton et al. 1999; these proceedings), its modulation, especially at higher energies, is single-humped. A dip-structure centered on the pulse maximum is instead observed which increases in depth at low energies and hence mimics a double-peaked light curve in soft X-rays. It however does not reach the pulse bottom level as instead observed in YY Dra, an IP dominated by the first spin harmonic (Norton et al.

Pulse	kT keV	R^a	EW^b eV	C_F^c	N_H^d $10^{22}\,\mathrm{cm}^{-2}$
Max.	27	$0.20^{+0.93}_{-0.20}$	133^{+41}_{-59}	$0.22^{+0.11}_{-0.09}$	$3.73^{+5.87}_{-1.93}$
Min.	27	$0.75^{+1.07}_{-0.75}$	272^{+71}_{-62}	$0.37^{+0.09}_{-0.08}$	$3.74^{+1.99}_{-1.52}$

a: Relative normalization of the reflection component, representing the solid angle subtended by cold matter in units of 2π, for a viewing angle of $60°$. **b**: Equivalent width of the 6.4 keV fluorescent iron line. **c**: Covering fraction of partial absorber. **d**: Column density of the partial absorber.

TABLE 1. Spectral parameters as obtained from fits to the maximum and minimum pulse spectra of RX J0028+59.

1999). Furthermore, the hardness ratios in both MECS and LECS show a hardening at rotational minimum, indicating absorption is more efficient when viewing along the accretion curtain. Fits to the maximum and minimum combined LECS and MECS spectra have been performed using a composite model consisting of an optically thin isothermal plasma at kT = 27 keV, derived from the grand-average combined LECS, MECS and PDS spectra, with a partial covering and a reflection components. A gaussian line at 6.4 keV accounting for the presence of a cold iron line expected to go along with the Compton reflection continuum (Matt et al. 1991) was also included (Table 1). At spin maximum, we find a lower covering fraction and a lower reflection, consistent with the accretion curtain pointing away from the observer and when the projected area of the WD is smaller. As expected the phase changes in the reflection are also observed in the fluorescence K_α line.

REFERENCES

Beuermann K., 1998, in *High Energy Astronomy and Astrophysics*, P.C. Agrawal and P.R. Visvanathan eds., India Univ. Press, p. 100.
Boella G., et al., 1997, A&AS 122, 299.
Buckley D.A.H., et al., 1995, MNRAS 275, 1028.
Buckley D.A.H., et al., 1997, MNRAS 1997, 287 117.
Burwitz V., et al., 1996, A&A 310, L25.
de Martino D., et al., 1999, A&A 350, 517.
Haberl F. et al., 1994, A&A 291, 171.
Haberl F., Motch C., 1995, A&A 297, L37.
Hellier C., 1999, ASP Conf. Ser. 157, p.1.
Mason K., et al., 1992, MNRAS 258, 749.
Matt G., Perola G.C., Piro L., 1991, A&A 247, 25.
Norton A.J., Beardmore A.P., Taylor P., 1996, MNRAS 280 937.
Norton A.J., et al., 1999, A&A 347, 203.

THE BEPPOSAX LONG LOOKS AT THE SEYFERT 1 GALAXIES NGC 5548 AND NGC 3783

A. De Rosa[1], L. Piro[1], F. Nicastro[1,2,4], P. Grandi[1], M. Dadina[1,3], F. Fiore[3,4], F.Haardt[5], J. Kaastra[6], L.Maraschi[7], G. Matt[8], T. Mineo[9], G.C.Perola[8], P.O.Petrucci[7], A. Treves[5].

1)Istituto di Astrofisica Spaziale (IAS) CNR-Roma, 2)Harvard-Smithsonian Center for Astrophysics, 3)BeppoSAX Science Operation Center, 4)Osservatorio Astronomico di Roma, 5)Dipartimento di Scienze, Università dell'Insubria/Polo di Como, 6)Space Research Organization Netherlands (SRON), 7)Osservatorio Astronomico di Brera, 8)Dipartimento di Fisica "E. Amaldi", Università degli Studi "Roma Tre", 9)Istituto di Fisica Cosmica ed Applicazioni dell'Informatica CNR.

ABSTRACT

BeppoSAX has observed two Seyfert 1 galaxies, namely NGC 5548 and NGC 3783, for ~ 8 days and ~ 5 days, respectively. The long exposures have provided very high quality spectra. We were able to detect the high energy cut-off in these two sources for the first time. In addition, we observed an emission feature at 0.57 keV, probably due to the OVII Kα transition. Flux and spectral variations on time scales of hours and days were observed. A time resolved spectral analysis indicates that this spectral variability is due to a change of the intrinsic power law slope.

KEYWORDS: X-ray astronomy; Galaxies: Seyfert-individual (NGC 5548, NGC 3783); Methods: Data analysis

1. INTRODUCTION

NGC 5548 and NGC 3783 are very well known, low redshift ($z_{5548} = 0.017$, $z_{3783} = 0.009$), Seyfert 1 galaxies. The extensively studied X-ray spectra of these objects are characterized by narrow and broad features: a soft absorption due to ionized matter obscuring the line of sight to the source, a Compton reflection hump at $\sim$20-30 keV, and the iron emission line at 6.4 keV, due to reflection off optically thick cold matter (Nandra et al. 1993; Turner et al. 1993; Nandra & Pounds 1994; George et al. 1998). These sources are bright in the X-rays ($F_{2-10} > 2 \times 10^{-11}$ erg cm^{-2} s^{-1}) and exhibit large flux variations by a factor of about two on time scales of days. They are therefore ideal for spectral variability studies. Two BeppoSAX long looks were performed ($\sim$8 days for NGC5548 and $\sim$ 5 days for NGC 3783), within the "Broad band (0.2-200 keV) spectral variability in bright Seyfert 1 galaxies" Core Program. We present here the spectral and temporal analysis of these long looks.

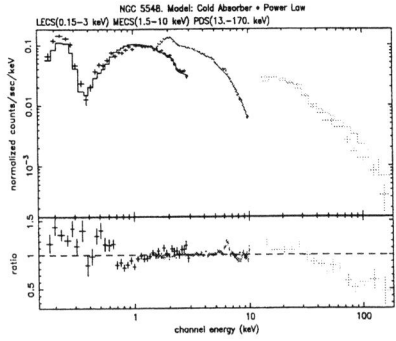 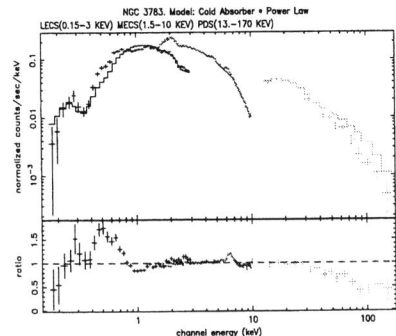

FIGURE 1. LECS, MECS and PDS data and data/model ratio when the continuum is fitted with a power law absorbed by the Galactic hydrogen column density.

2. THE TOTAL SPECTRUM

We first studied the total integrated spectra (t_{exp}^{5548}=314 ks, t_{exp}^{3783}=170 ks in the MECS instrument). We resolved different spectral components (Figure 1): a power law with exponential cut-off, a warm absorber, a Compton reflection hump, an iron emission line and an emission feature at 0.57 keV. The wide energy band of BeppoSAX coupled with the long exposure has allowed us to constrain with unprecedented precision all the spectral parameters (Table 1, the quoted errors refer to the confidence level of 90% for 1 interesting parameter). In addition, we revealed for the first time the high energy cut-off and a spectral feature in the soft energy band. Both thermal and non-thermal models predict the high energy cut-off (Svensson 1994, Haardt et al. 1997). The spectral feature at $\sim$ 0.6 keV in the case of NGC 5548 is probably an emission line arising from the warm absorber (Nicastro et al. 1999). In analogy with this previous work, we have fitted the excess at $\sim$0.6 keV in NGC 3783 with an emission line. A more detailed analysis of this spectral component is still in progress.

3. VARIABILITY STUDIES

3.1. Lightcurves

The lightcurves of the LECS (0.2-3 keV), MECS (2-10 keV) and PDS (13-200 keV) data for NGC 5548 and NGC 3783 are plotted in Figure 2. Both sources showed a strongly energy dependent flux variability. NGC 5548 exhibits a flux variation of 15% from 2-10 keV and $\sim$30% from 0.2-3 keV, on time scales of order of $\sim$2 days, while the 2-10 keV and 0.2-3 keV flux variations of NGC 3783 were respectively $\sim$40% and $\sim$50% on time scales of order a few hours. No any time lags between variations in the soft (0.2-3 keV) and hard (2-10 keV) energy bands were detected by a cross-correlation analysis.

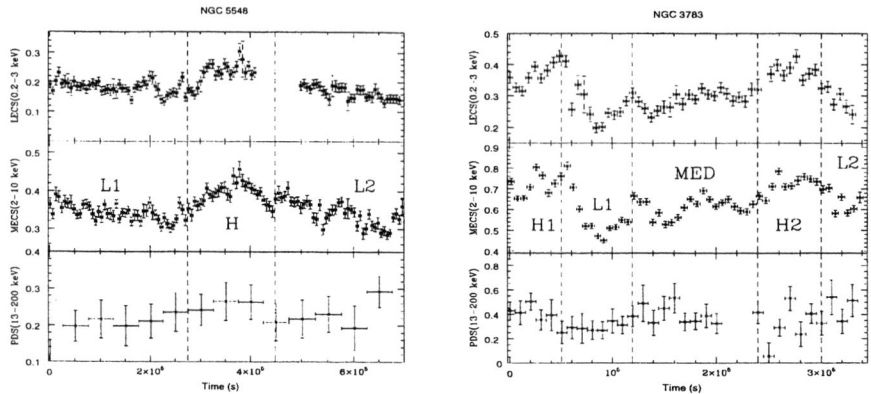

FIGURE 2. Lightcurves of the LECS, MECS and PDS data for NGC 5548 and NGC 3783. MECS and LECS data are grouped in bins of 6 ks, while PDS data are grouped in bins of 50 ks (NGC 5548) and 10 ks (NGC 3783).

3.2. Time resolved spectral analysis

We extracted separate spectra for high and low flux level states, as shown in Figure 2. In both sources the spectral slope gets steeper at higher fluxes (Table 2 and Figure 3). The spectral index is the only intrinsic parameter that clearly changes with the luminosity of the objects. We found marginal evidence of a shift of E_c with L_X in the case of NGC 5548 (Figure 3). No variations of the warm absorber are required to explain the spectral differences between the different states (Table 2).

4. SUMMARY

A Compton reflection component and the iron line are detected in both sources. We confirm the presence of a warm absorber in NGC 5548 and NGC 3783. An emission feature at ∼0.6 keV is found in both sources. The BeppoSAX broad band and the long look strategy adopted for the observations, has allowed us to determine the high energy cut-off in NGC 5548 and NGC 3783 for the first time (Figure 3). Our data do not allow us to distinguish between two different theoretical models (thermal and non-thermal) wich predict this spectral feature. We detect strong and energy dependent spectral variability. This behaviour can be accounted for by a change of intrinsic spectral slope, fully consistent with the α versus F_X relationship observed in NGC 4151 (Piro et al. 1999), and predicted by thermal emission models. The limited statistics of the separate states provide only a marginal evidence for variations of E_C in NGC 5548. This information would provide an important constraint on the radiative processes. The case of the thermal model where the continuum component is more complex than a cut-off power law, is discussed in Petrucci et al. (1999).

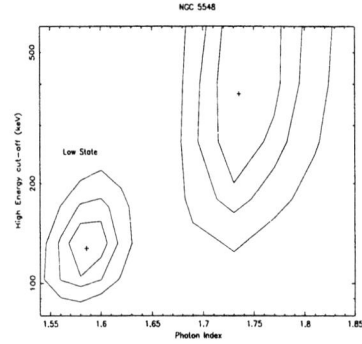
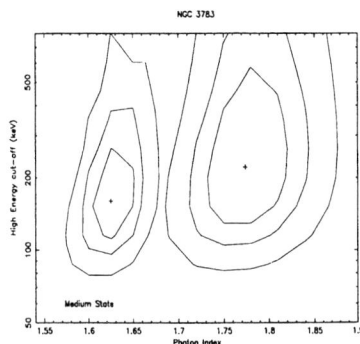

FIGURE 3. 1,2 and 3σ confidence levels E_c vs Γ for low and high flux level states.

TABLE 1. Total spectra: best fit parameters.

	Log(nH)	Log(U)	Γ	E_c (keV)	°Norm.	E_{Fe} (keV)
NGC 5548	21.55±0.08	$0.82^{+0.07}_{-0.30}$	$1.62^{+0.03}_{-0.02}$	154^{+47}_{-31}	7.8±0.02	6.31±0.05
NGC 3783	21.94±0.21	$1.39^{+0.20}_{-0.28}$	1.70±0.05	164^{+64}_{-43}	16 ± 0.1	6.40±0.04

	σ_{Fe} (keV)	EW_{Fe} (eV)	R	E_2 (keV)	σ_2 (keV)	$\chi^2(dof)$
NGC 5548	< 0.2	110±28	0.44±0.14	$0.56^{+0.04}_{-0.03}$	< 0.09	1.13/105
NGC 3783	0.34 ± 0.10	191±44	0.37±0.17	0.52 ± 0.09	0.14±0.06	1.11/104

° In 10^{-3} photons cm^{-2} s^{-1} Kev^{-1}.

TABLE 2. Spectral Variability: best fit parameters.

	Spectrum	Log(U)	Γ	E_c(keV)	°Norm.	EW_{Fe}(eV)	R	$\chi^2(dof)$
NGC 5548	H	$0.70^{+0.23}_{-0.30}$	1.74±0.03	> 200	$9.9^{+0.5}_{-0.4}$	110±40	$0.56^{+0.26}_{-0.21}$	0.94/107
	L (L1+L2)	$0.74^{+0.24}_{-0.26}$	1.59±0.02	128^{+39}_{-26}	7.12±0.22	123±34	$0.41^{+0.15}_{-0.12}$	0.95/105
NGC3783	H1	$1.44^{+0.51}_{-0.06}$	1.77±0.08	222^{+540}_{-100}	19.8 ± 1.3	169±68	$0.57^{+0.22}_{-0.33}$	0.94/92
	Med	$0.89^{+0.45}_{-0.10}$	1.62±0.05	160^{+290}_{-80}	13.6 ± 0.5	217±68	$0.51^{+0.08}_{-0.16}$	0.85/91

° In 10^{-3} photons cm^{-2} s^{-1} Kev^{-1}.

REFERENCES

George I.M., Turner T.J., Netzer H. et al. 1998, ApJS, 114, 73G.

Haardt F., Maraschi L., Ghisellini G. 1997, ApJ, 476, 620.

Nandra K., Fabian A.C., George I.M. 1993, MNRAS, 260, 504.

Nandra K., Pounds K.A. 1994 MNRAS, 268, 405.

Nicastro F., Piro L., De Rosa A. et al. 1999, ApJ, Submitted.

Petrucci P.O., Haardt F., Maraschi L. et al. 1999, in preparation.

Piro L., De Rosa A., Dadina M. et al. 1999, astro-ph/9908360.

Svensson R. 1994, ApJ, 92, 585.

Turner T.J., Nandra K., George I.M. et al. 1993, ApJ, 419, 127.

DISCOVERY OF HARD X-RAY EMISSION FROM TYPE II BURSTS OF THE RAPID BURSTER

F. Frontera[1,2], N. Masetti[1], M. Orlandini[1], L. Amati[1], E. Palazzi[1], D. Dal Fiume[1], S. Del Sordo[3], G. Cusumano[3], A.N. Parmar[4], G. Pareschi[5], I. Lapidus[6] and L. Stella[7]

[1] *Istituto Te.S.R.E., CNR, via Gobetti 101, I-40129 Bologna, Italy*
[2] *Dipartimento di Fisica, Università di Ferrara, via Paradiso 12, I-44100 Ferrara, Italy*
[3] *I.F.C.A.I., CNR, via Ugo La Malfa 153, I-90146 Palermo, Italy*
[4] *SSD, ESA/ESTEC, Postbus 299, 2200 AG Noordwijk, The Netherlands*
[5] *Osservatorio Astronomico di Brera, Via Bianchi 46, I-23807 Merate, Italy*
[6] *McKinsey & Co. Inc., London, UK*
[7] *Osservatorio Astronomico di Roma, via Frascati 33, I-00040 Monteporzio Catone, Italy*

ABSTRACT   We report on results of *BeppoSAX* Target Of Opportunity (TOO) observations of the source MXB 1730-335, also called the Rapid Burster (RB), made during its outburst of February–March 1998. We monitored the evolution of the spectral properties of the RB from the outburst decay to quiescence. During the first TOO, the X-ray light curve of the RB showed many Type II bursts and its broadband (1-100 keV) spectrum was acceptably fit with a two blackbody plus power law model. Moreover, to our knowledge, this is the first time that this source is detected beyond 30 keV.

KEYWORDS: Stars: individual: MXB 1730-335, stars: neutron, X-rays: stars, X-rays: bursts

1. OBSERVATIONS

Four Target Of Opportunity (TOO) observations were performed with *BeppoSAX* (Boella et al. 1997a) on the Rapid Burster (=MXB 1730–335; hereafter RB) during the activity state which started on January 28, 1998 (Fox et al. 1998). These TOOs spanned over one month (from February 18 to March 18) and caught the object in four different snapshots, from the post–maximum decay to the quiescent state. Figure 1, left panel, shows the ASM light curve of the *Rossi-XTE* satellite with superimposed the times of the four *BeppoSAX* observations. Here we report on RB data from three of the four instruments mounted on *BeppoSAX*: LECS (0.1-10 keV; Parmar et al. 1997), MECS (1.5-10 keV; Boella et al. 1997b, and PDS (15-300 keV; Frontera et al. 1997). For the PDS the default rocking collimator law was modified by offsetting the RB by 40′ from the center of the field of view in order to reduce as much as possible the contamination from a nearby variable X–ray source, GX 354-0 (=4U 1728-34), located at about 30′ from the RB. Unfortunately, due to

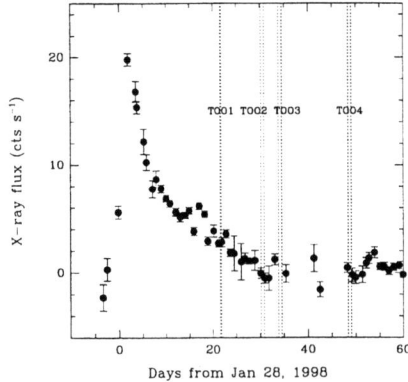
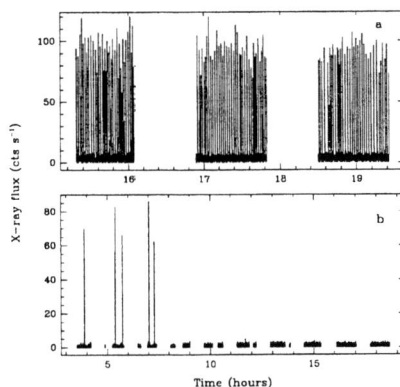

FIGURE 1. *(Left panel)* *XTE* ASM light curve of the February-March 1998 activity phase of the RB. The vertical dashed lines indicate the four *BeppoSAX* TOOs. *(Right panel)* (a) MECS light curve of TOO1. Times are expressed in hours from 0 UT of February 18, 1998; (b) MECS light curve of TOO2. Times are expressed in hours from 0 UT of February 27, 1998.

failure of the rocking law setup program, the collimator did not move as requested during TOO2 and TOO3; so, we have only LECS and MECS data for these two observations.

In this paper we report on preliminary results of these observations. Definitive results along with their implications will be the subject of another paper (Masetti et al. 2000). In the following, for the luminosity estimates we will assume that the RB lies at a distance $d = 8$ kpc (Ortolani et al. 1996).

2. SPECTRAL ANALYSIS AND TEMPORAL EVOLUTION OF THE RB

During TOO1, the RB was in a strong state of bursting activity. The 2-10 keV light curve obtained with MECS (see Fig. 1, right panel, part *a*) showed 113 Type II X-ray bursts during 9457 seconds of good observational data. Evidence of Type II bursts was also observed in the 0.1-2 keV data obtained with LECS. We divided the MECS TOO1 data into two subsets: persistent emission (PE; below 5 counts s^{-1}) and bursting emission (BE; above 5 counts s^{-1}).

The MECS PE and BE spectra could be well fit with a photoelectrically absorbed two-component blackbody (2BB); these BB components may originate from the neutron star (NS) surface, a boundary layer between the NS and the inner edge of the accretion disk, or the inner region of the disk itself. The same model was used for the RB by Guerriero et al. (1998) who found values consistent with ours. In Table 1 we report the best-fit parameters along with their 90% confidence errors. The temperature values of the two BB components were slightly higher during the PE

TABLE 1. Best-fit parameters for the TOO spectra. The reported uncertainties are given at 90% confidence level for a single parameter of interest.

Model	TOO1 1-10 keV BE	TOO1 1-10 keV PE	TOO1 1-100 keV BE − PE	TOO2 0.5-10 keV (mainly PE)	TOO3 0.5-10 keV PE
χ^2_ν (dof)	1.17 (395)	1.07 (203)	1.17 (223)	0.97 (113)	1.07 (55)
Wabs(2BB):					
N_H ($\times 10^{22}$ cm^{-2})	3.5 ± 0.5	1.5 ± 0.3	$4.9^{+1.7}_{-1.1}$	1.6 ± 0.3	$1.1^{+0.4}_{-0.3}$
kT_1 (keV)	0.43 ± 0.04	0.63 ± 0.06	$0.32^{+0.23}_{-0.12}$	0.65 ± 0.07	0.64 ± 0.09
$L_1^{BB(a)}$	15^{+6}_{-5}	$0.85^{+0.08}_{-0.06}$	10^{+22}_{-10}	$0.32^{+0.04}_{-0.03}$	0.06 ± 0.01
kT_2 (keV)	1.64 ± 0.03	$1.72^{+0.19}_{-0.17}$	1.67 ± 0.04	$1.78^{+0.10}_{-0.08}$	$2.1^{+0.4}_{-0.2}$
$L_2^{BB(a)}$	47 ± 0.8	$0.74^{+0.06}_{-0.05}$	40 ± 2	0.76 ± 0.03	$0.106^{+0.011}_{-0.008}$
+ power law:					
Γ			$3.1^{+0.3}_{-0.4}$		
$K^{(b)}$			$3.3^{+4.5}_{-2.5}$		
+ Fe emission line:					
E_l (keV)			6.5 ± 0.2		
EW (eV)			100^{+80}_{-60}		
FWHM (keV)			$0.4^{+0.3}_{-0.4}$		
$I_l^{(c)}$			7^{+5}_{-4}		

(a): in units of 10^{36} erg s^{-1}
(b): in units of photons keV^{-1} cm^{-2} s^{-1} at 1 keV
(c): in units of 10^{-3} photons cm^{-2} s^{-1}

than during the BE, while their luminosities were much higher (by a factor 20 for the cooler BB and 60 for the hotter BB) during the BE than during the PE. We also remark that during the BE the hotter BB component was brighter (by more than a factor 3) than the cooler BB, while during the PE they had similar luminosities. This implies that the BE influences more the higher temperature component than the other one. If the BE is due to spasmodic accretion onto the compact object, the higher temperature component should be the one coming from the NS surface.

The source was also visible in the hard X–ray (15-100 keV) energy range. However the statistics of the PDS light curve was much lower and did not allow distinguishing the Type II bursts. In order to construct the BE spectrum we used the time intervals in which the bursts were observed with MECS. Also, we could not derive the correct 15-100 keV flux and spectrum of both BE and PE given the residual source contamination by GX 354-0. Thus, in order to overcome this problem, we used as background level for the 1-100 keV BE spectrum the total count rate level measured during the PE time intervals. The combined LECS+MECS+PDS PE-subtracted bursting spectrum, shown in Fig. 2, was no longer fit with a 2BB model. By adding a power law component we obtained an acceptable fit (see Table 1). The further addition of a Fe K emission line at 6.5 keV slightly improved the fit, with parameter values found for this line in general agreement with the findings

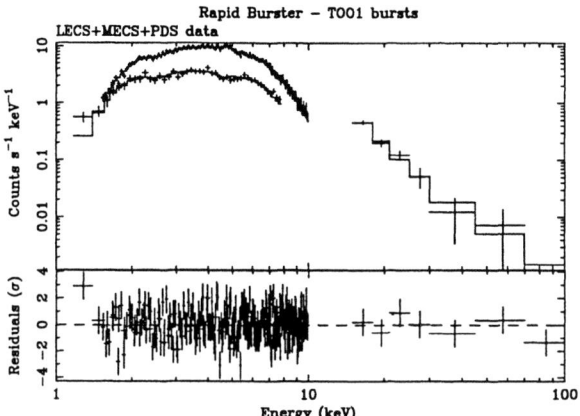

FIGURE 2. TOO1 LECS+MECS+PDS X-ray spectrum of the total PE-subtracted BE. The fit corresponds to an absorbed two-component blackbody plus power-law model plus an iron line at 6.5 keV.

by Stella et al. (1988) for Type II bursts.

During TOO2 the RB drastically reduced its bursting activity, and the bursts were concentrated at the beginning of this TOO (Fig. 1, right panel, part b). Also, the emission intensity level decreased. The best-fit model was a photoelectrically absorbed 2BB model (Table 1). No evidence of a Fe emission line was present.

During TOO3 the object further reduced its bursting activity, and no Type II bursts were seen throughout the observation. The best-fit model spectrum was still an absorbed 2BB (Table 1). As in the case of TOO2, no iron emission line at 6.5 keV was found.

The RB was instead no longer visible in MECS/LECS images during TOO4. Stray light from GX354-0 prevented us to get a deep observation of the source. The 3σ upper limit to the RB X-ray emission was 1.5×10^{-12} erg cm^{-2} s^{-1} in the 2-10 keV energy band.

REFERENCES

Boella G., Butler R.C., Perola G.C. et al., 1997a, A&AS 122, 299
Boella G., Chiappetti L., Conti G. et al., 1997b, A&AS 122, 327
Fox D., Guerriero R., Lewin W.H.G., 1998, ATEL n. 9
Frontera F. Costa E., Dal Fiume D. et al., 1997, A&AS 122, 357
Guerriero R., Fox D., Kommers J. et al., 1999, MNRAS 307, 179
Masetti N., Frontera F., et al., 2000, A&A, submitted
Ortolani S., Bica E., Barbuy B., 1996, A&A 306, 134
Parmar A., Martin D.D.E., Bavdaz M. et al., 1997, A&AS 122, 309
Stella L., Haberl F., Lewin W.H.G. et al., 1988, ApJ 324, 379

X-RAY BEAMING IN THE HIGH MAGNETIC FIELD PULSAR GX 1+4

D.K. Galloway [1,2,3] and K. Wu [2]

[1] *School of Mathematics and Physics, University of Tasmania, GPO Box 252-21, Hobart, Tasmania 7001, Australia*
[2] *RCfTA, University of Sydney, Camperdown, NSW 2006, Australia*
[3] *present address: Center for Space Research, MIT, 77 Massachusetts Avenue, Cambridge MA 01239-4307*

ABSTRACT Pulse profiles from X-ray pulsars often exhibit strong energy dependence and both periodic and aperiodic variations with time. The great variety of profiles observed in various sources, and even from individual sources, makes it difficult to separate the numerous factors influencing the phase-dependence of the X-ray emission. These factors include the system geometry and particularly the photon energy and angle dependence of emission about the neutron star poles.

Comptonisation may play an important role in determining beam patterns and hence pulse profiles in X-ray pulsars. A Monte Carlo simulation is used to investigate the beaming due to Comptonisation in a simple accretion column geometry. We apply the model to the extremely variable pulse profiles of the high-magnetic field pulsar GX 1+4.

KEYWORDS: scattering – X-rays: stars – pulsars: general – pulsars: individual (GX 1+4)

1. INTRODUCTION

To date essentially all the approaches which have been used to model the emission region in X-ray pulsars have limitations. Past efforts have typically adopted a geometry suitable for a particular accretion rate ($\dot{M}$) regime and then predict the emission properties by a range of techniques. Radiative transfer calculations (e.g. Burnard, Arons & Klein 1991) are necessarily restricted to symmetric, homogeneous emission regions, where in reality the accretion column may be hollow and even incomplete (an 'accretion curtain'). The geometric fitting approach, where a beam pattern is assumed and then the geometry is varied (e.g. Leahy 1991) cannot reproduce sharper features observed in several sources. Neither method can generate asymmetric pulse profiles without resorting to an off-center magnetic axis, for which there is no other observational evidence. Recent observations of the X-ray pulsar GX 1+4 suggest a rather different scenario.

The X-ray continuum spectrum of GX 1+4 is rather flat (with photon index ≈ 1.0) up to a cutoff around 10-20 keV, above which the decay is steeper; it is one of the hardest known amongst the X-ray pulsars. Analysis of recent Rossi X-ray Timing Explorer (*RXTE*) data shows that the spectrum is generally consistent with those predicted by unsaturated Comptonisation models (e.g. Galloway et al. 2000). Pulse profiles are extremely variable and typically asymmetric, often with a sharp dip forming the primary minimum (Greenhill, Galloway & Storey 1998). During a low flux episode in July 1996, the pulse profiles were found to shift in asymmetry from 'leading-edge bright' (with the maximum closely following the sharp primary minimum) to 'trailing-edge bright'. The entire observation which captured the change spanned only 34 hours, and occurred just 10 days before a short-lived transition from rather constant spin-down to spin-up and back again (Giles et al. 2000). We propose a model which seeks to explain the sharp primary minima seen in this and other sources (A 0535+262, Cemeljic & Bulik 1998; and RX J0812.4-3114, Reig & Roche 1999) and ultimately the change in the pulse profiles.

2. MODEL DESCRIPTION AND PRELIMINARY RESULTS

A Monte-Carlo code is used to generate the spectra and pulse profiles emitted by two semi-infinite homogeneous cyclindrical accretion columns of radius R_C, diametrically located on the surface of a 'canonical' neutron star of radius $R_* = 10$ km (Figure 1a).

The algorithms of Pozdnyakov, Sobol' & Syunyaev (1983) are used to draw the photon energy and direction, electron energies, and to calculate the fully relativistic (non-magnetic) cross-section for Compton scattering. Outside the accretion column, the redshift and bending of photon trajectories by the neutron star's gravity is calculated by assuming a Schwarzschild metric. We simulate a single column for both poles of the star, and generate pulse profiles for a range of geometries simultaneously.

The beam pattern and pulse profiles over a range of geometries are shown in Figure 1 b) and c). We note that when ($|i - \beta| \lesssim 45°$) the emission exhibits a strong modulation at the stars rotation period, with the primary minimum corresponding to the closest passage of the line of sight with one of the magnetic polar axes. As i and β increase the primary minimum becomes progressively narrower. When $i \approx \beta \gtrsim 50°$, a secondary minimum (from the passage of the second axis through the line of sight) is observed. The emission is beamed at an angle $> 90°$ with respect to the column axis; this corresponds to a 'fan' type beam. Emission at smaller angles is supressed as a consequence of the decreased escape probability for photons propagating along the column axis. That emission is beamed at $> 90°$ is a consequence of the gravitational light bending; this is also affected by the size of the column R_C.

Our simulations indicate that the mean spectra also depend strongly on the density of the column and the viewing geometry.

3. DISCUSSION AND APPLICATION TO X-RAY PULSARS

Since we assume a constant infall velocity v_C and neglect effects due to radiation pressure on the infalling electrons, the results described are only applicable to sys-

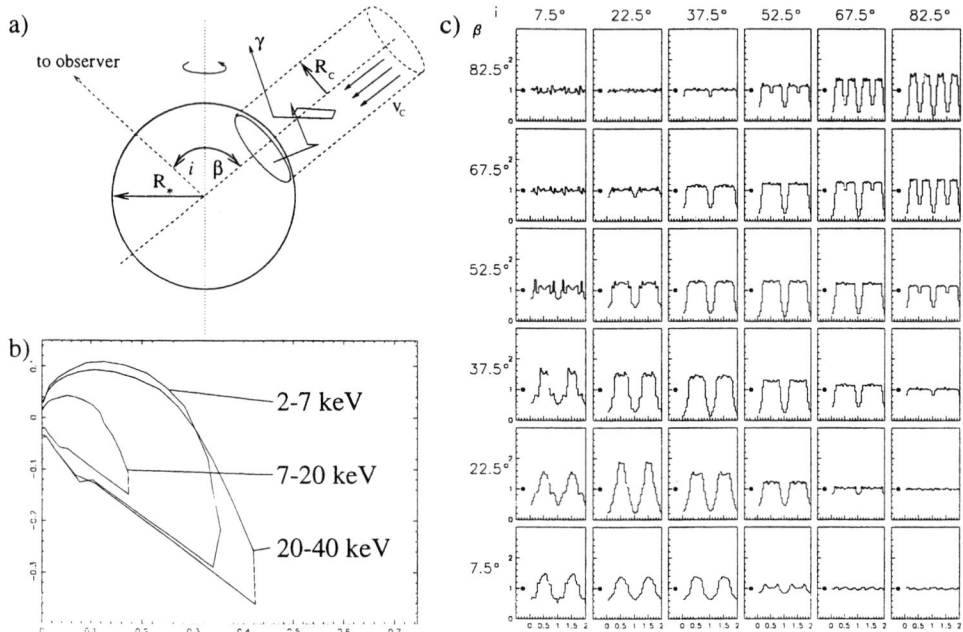

FIGURE 1. a) Model geometry (only one column shown for clarity). The magnetic axis of the star is aligned relative to the rotational axis by an angle β, with i the inclination angle of the system with respect to the observer. Photons are emitted isotropically from each (circular) polar cap, and are then Compton scattered by the column plasma before escaping towards the observer. The plasma is flowing towards the pole with speed v_C. The temperature of the polar cap T_0, temperature T_e and optical depth τ of the column plasma, and column diameter R_C are all free parameters in the model. Note that $\tau = \sigma_T R_C N_e$ where N_e is the electron density in the column; that is, τ corresponds to photon paths from the centre of the column radially outwards. Clearly the optical depth experienced by individual photons will depend on the trajectory, and in particular with the angle relative to the column axis.
b) X-ray beam pattern within several energy bands. The model parameters are $kT_0 = 1$ keV, $kT_e = 8$ keV, $\tau = 3$, $R_C = 1$ km, and $v_C = 0.5c$; these conditions are such as to approximately correspond to those expected for a low-luminosity X-ray pulsar. The origin corresponds to the base of the accretion column, which is aligned along the (positive) y-axis. The normalisation is arbitrary.
c) Predicted pulse profiles using the same model parameters over a range of geometries. Each profile shown corresponds to a particular choice of i and β, which vary between 7.5° and 82.5° along the x- and y-axes respectively. Profiles are normalised to the mean and plotted over two cycles; a typical error bar is shown at the left of each panel.

tems with low $\dot{M}$. Previous low-$\dot{M}$ models predict a 'pencil' rather than 'fan' beam, with emission reaching a maximum at small angles relative to the accretion column. This is probably because of the assumed 'slab' or 'mound' shaped emission region. Interactions between photons and the inflowing material in the accretion column, which is neglected by these models, is crucial for the formation of sharp primary minima in the pulse profiles as observed in GX 1+4 and several X-ray pulsars. The persistence of the sharp feature in GX 1+4 as X-ray flux drops almost to zero points to the continued importance of this effect, even at extremely low $\dot{M}$ (Giles et al. 2000).

A significant approximation is the use of the nonmagnetic Compton scattering cross-section. For GX 1+4 - with an estimated magnetic field strength of $2-3 \times 10^{13}$ G (e.g. Cui et al. 1997) - deviations from the nonmagnetic cross section will be significant within typical observational bands for X-ray astronomy. However we suspect that magnetic effects may only play a minor role in shaping the pulse profile, principally narrowing the primary minimum and possibly giving rise to the local maxima ('shoulders') immediately prior to and following the minimum (Giles et al. 2000).

Finally we note that the model-predicted pulse profiles are in general quite symmetric. A possible cause of asymmetry in the observed profiles is a variation in density across the accretion column, which could potentially develop in the region where the disc plasma becomes entrained onto the magnetic field lines and persist to the neutron star surface. This effect further suggests a mechanism for the rapid changes in profile asymmetry observed in GX 1+4 (Giles et al. 2000); that is, the sense of asymmetry in the column changes and consequently so does the pulse profile. The detailed structure of the entrainment region is rather poorly understood, and we feel it is not possible to rule out such a phenomenon.

We have described a model with homogeneous, axisymmetric, cylindrical emission regions. To explain other qualitative features of observed pulse profiles, our model needs to be modified to take into account effects due to inhomogeneities and more complicated geometry of the emission regions.

REFERENCES

Burnard D.J., Arons J., Klein R.I. 1991, ApJ 367, 575

Cemeljic M., Bulik T. 1998 AcA 48, 65

Cui W. 1997 ApJ 482, L163

Galloway D.K., Giles A.B., Greenhill J.G., Storey M.C. 2000 MNRAS 311, 755

Giles A.B., Galloway D.K., Greenhill J.G., Storey M.C., Wilson C.A. 2000 ApJ 529, 447

Greenhill J.G., Galloway D.K., Storey M.C. 1998 PASA 15, 2, 254

Leahy D.A. 1991 MNRAS 251, 203

Pozdnyakov L.A., Sobol' I.M., Syunyaev R.A. 1983 Astrophys. Space Phys. Rev. 2, 189

Reig P., Roche P. 1999, MNRAS 306, 95

VARIABLE FE-Kα LINE PROFILES IN SY 1 GALAXIES

J. Gelbord [1], K.A. Weaver [1,2]

1) Department of Physics & Astronomy, Johns Hopkins University, Baltimore, MD, USA
2) also at Goddard Space Flight Center, Greenbelt, MD, USA

ABSTRACT

We discuss Fe Kα line profile variability in an *ASCA* sample of Seyfert 1 galaxies. In particular, we examine the line core energies defined by fitting narrow Gaussians. Such an empirical analysis is simplistic but avoids modeling uncertainties. We find significant Fe flux and/or energy variability in ~60% of the sample on time scales ranging from years to hours. For example, the peak line energy in Mrk 279 changes from 6.3 to 6.5 keV in $\sim 10^4$ s, suggesting an emission region no more than a few light-hours across.

KEYWORDS: galaxies: Seyfert; line: profiles; X-rays: galaxies.

1. INTRODUCTION

According to the canonical model of Seyfert nuclei, Fe Kα emission can be produced by high energy photons reprocessed by iron in a number of possible regions ranging from the cold molecular torus to the relativistic inner accretion disk. Each region will imprint a distinct signature upon the emission feature it produces. The inner accretion disk approaching the innermost stable orbit around the black hole would produce a relativistically broadened line, with a strong gravitational redshift and possible rapid variability ($\sim 10^3$s time scales) which should not measurably lag continuum fluctuations (George & Fabian 1991). Radiation reprocessed farther out in the accretion disk will exhibit less drastic Doppler broadening, with less rapid variability ($\sim 10^{4-6}$s) which can lag the continuum considerably. Emission lines from the inner surface of the torus are expected to be narrow features which should not vary on time scales less than $\sim 10^{7-8}$s (Ghisellini *et al.* 1994, Krolik *et al.* 1994). By examining the properties of the emission features, we can learn about the reprocessing regions, and thereby impose constraints upon our models.

Some important results have come from studies of the Fe Kα profiles of a few Seyferts, notably MCG –6-30-15 (Iwasawa *et al.* 1996, Iwasawa *et al.* 1999), NGC 3516 (Nandra *et al.* 1999) and NGC 5548 (Chiang *et al.* 1999). However, there are substantial difficulties in measuring the line properties. The limited effective area of the current X-ray telescopes makes it impossible to measure any but the brightest Seyferts with a time resolution better than ~1 ks. The broad wings have been particularly problematic; until now, the best available spectral

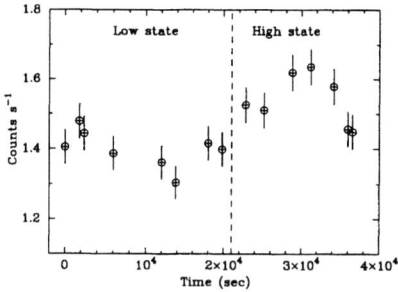 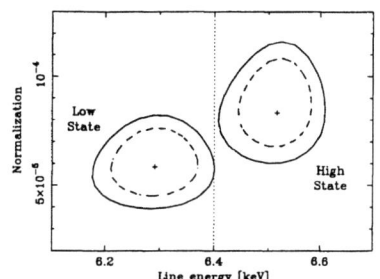

Figure 1: ASCA SIS0 light curve of Mrk 279, showing the transition from the "Low" to the "High" state.

Figure 2: Gaussian line energy vs. normalization for the Fe Kα line in the low (left) and high (right) flux states. The contours are 90% and 68% confidence.

resolution has been from *ASCA*, but the signal to noise ratio in the broad wings is generally low. The measured broad profiles therefore can be substantially model dependent and are strongly affected by the method of continuum subtraction. The S/N can be improved by averaging together multiple observations of a galaxy, or even multiple objects (e.g. Nandra *et al.* 1997), but such studies lose all variability information, and are limited in what they can tell us about the distribution of properties across the sample. Even when good data is available for a single galaxy, their interpretation can be complicated if there are multiple reprocessing regions (Weaver & Reynolds 1998). Furthermore, the assumptions of reverberation analysis probably fail for regions near the inner disk radii because the continuum source may no longer be compact compared to the reprocessor and may not even be localized but rather spread out in patchy flares. These combined difficulties create a danger of overinterpreting the data.

2. PROCEDURE

2.1. Developed for Mrk 279 ...

When reducing the data from *ASCA*'s 1994 observation of Mrk 279, we found the continuum flux increased by ~20% within 15 ks, neatly halving the observation into low and high states (fig. 1). Comparing the spectra in these two states, we saw no change in the continuum spectrum, but the Fe emission feature did change. However, modeling the entire broadened profile is particularly difficult when the data is subdivided into time bins, and both the line width and normalization depend upon our continuum model. In order to make the most robust quantitative measurement of this line, we fit an empirical model using an unresolved narrow Gaussian for the emission feature, replacing the poorly constrained reflection component with a second power law to represent the hard excess.

The resulting simplistic model underestimates the line variability because it is insensitive to fluctuations in the broad wings, which are expected to be more variable than the line cores. In addition, it hides any information about the physical origin of the variation (e.g. if the line fit shifts redward, this model cannot discriminate between a flare in the red wing or a disappearance of the blue wing). What this model does give us is a robust measurement of the line core energy minimizing the model dependence, and a lower limit to the variability intrinsic to the source.

By using this model, we find that the peak energy of the Fe Kα emission changes (>95% confidence) from 6.3 to 6.5 keV (fig. 2), with no measurable lag behind the continuum variation. 6.5 keV is strictly inconsistent with emission from neutral iron unless Doppler effects are involved. This rapid change tracking the continuum increase suggests a reprocessing region in close proximity to the continuum source.

2.2. ...and applied to a broader sample

With this success in hand, we realized that our method provides a powerful means of looking for Fe line variability without being limited to just the brightest Seyfert 1s. Better yet, we could use the large pool of data available in the *ASCA* archive to see whether the dominant reprocessing region varies from one Seyfert to another.

We defined our sample to be the 15 Seyfert 1 galaxies with multiple observations available in the *ASCA* archive (table 1), and we measured the time averaged data from each pointing (2.5-185 ks integrations). This approach underestimates variability because it is insensitive to fluctuations on time scales shorter than the observation[1], and limited sampling can hide some intrinsic variability.

3. RESULTS

We find iron line fluctuations through much of the sample. Our results are summarized in table 1. Among the 15 Seyferts examined, we find variability in the Fe line core energy at >90% confidence in 6 of them (with confidence >68% we also find 6), in the line flux of 3 (9) sources, and in the equivalent width of 7 (11) galaxies. Note that the bias due to limited sampling is immediately clear: the six galaxies with variable line energies were observed an average of 6.33 times while the other nine were observed 3.67 times each, and the four galaxies without significant equivalent width fluctuations are the only galaxies with just two observations each.

There is strong evidence that the Fe Kα line in some Seyferts originates in the inner regions of the accretion disk. Mrk 509, Fairall 9 and NGC 3783 all change their line energies (>90% confidence) on the shortest measured scales (3-4 days). The Mrk 279 data (§2.1) suggest even more rapid variability may be present. Across the sample, we find extreme peak line energies which lie either below 6.3 keV (six galaxies) or above 6.5 keV (three galaxies), and which exclude the energy of neutral Fe Kα (6.40 keV) with >90% confidence. This suggests a systemic change in ionization state, bulk motions, or a region in the depths of the central potential well,

---

[1] Averaging the whole Mrk 279 observation yields a line energy of $6.48^{+.10}_{-.12}$ keV (which is consistent with neutral Fe Kα), hiding the transition between the low and high states whose energies lie on opposite sides of 6.40 keV, each excluding it with 90% confidence (fig. 2).

Galaxy	z	log $\overline{L_{2-10}}$	N	E_{Fe}	F_{Fe}	EW
Mrk 509	.0344	44.11	11	6.26–6.96	2.6–6.7	*43–113*
MCG −2-58-22	.0467	44.10	3	—	1.8*–2.2	40*–81
Fairall 9	.0470	44.02	8	6.20–6.65	2.7–4.5	*84–171*
IC 4329a	.0161	43.72	5	—	—	58–85
Mrk 841	.0364	43.59	3	—	1.7*–2.8	*86*–170*
NGC 5548	.0168	43.55	8	—	2.6*–6.2	*29*–81*
MCG +8-11-11	.0200	43.30	2	—	—	—
NGC 7469	.0162	43.24	4	6.27–6.61	2.2–4.5	47–113
NGC 3783	.0085	43.01	6	6.24–6.39	7.5–10.8	*88–164*
NGC 3516	.0087	42.94	4	6.33–6.40	6.0–11.4	*94–149*
NGC 4593	.0090	42.85	2	—	—	—
MCG −6-30-15	.0077	42.70	5	6.27–6.42	3.5–4.8	*58–121*
NGC 4151	.0033	42.68	7	—	—	*72–109*
NGC 3227	.0038	41.97	2	—	—	—
NGC 4051	.0024	41.47	2	—	—	—

TABLE 1. Sample, with ranges of best fit values when variable with >68% confidence (>90% *when italicized*). When marked with an asterisk (*), the low end of the range of values is defined by the 68% (*or 90%*) upper limit on a non-detection. Columns are galaxy name, redshift, log of the average 2-10 keV luminosity (in erg/s), number of observations compared, Fe line energy (keV), line flux (10^{-13} erg/s/cm^2), and equivalent width (eV).

all of which point towards a reprocessor at the smallest physical scales[2].

On the other hand, some galaxies (notably NGC 4151; see Weaver, this volume) have constant line peak energies and fluxes, and equivalent widths which scale inversely with the continuum. The line fluxes in four other galaxies remain constant while the continuum changes. Hence, the dominant reprocessors in these Seyferts appear to lie at great distances ($\gtrsim 1$ pc) from the continuum source.

REFERENCES

George, I., Fabian, A. 1991, MNRAS, 249, 352
Ghisellini, G., Haardt, F., Matt, G. 1994, MNRAS, 267, 743
Iwasawa, K. *et al.* 1996, MNRAS, 282, 1038
Iwasawa, K., Fabian, A., Young, A., Inoue, H., Matsumoto, C. 1999, MNRAS, 306, L19
Krolik, J., Madau, P., Życki, P. 1994, ApJ, 420, L57
Nandra, K., George, I., Mushotzky, R., Turner, T., Yaqoob, T. 1997, ApJ, 477, 602
Nandra, K., George, I., Mushotzky, R., Turner, T., Yaqoob, T. 1999, ApJ, 523, L17
Weaver, K., Gelbord, J. Yaqoob, T. 2000, ApJ *in press*
Weaver, K., Reynolds, C. 1998, ApJ, 503, L39

---

[2] An alternative to the inner disk is reprocessed radiation dominated by (no more than a few) more distant disk regions near localized flares, with light crossing times of a few days or less.

AGN X–RAY ABSORPTION VS. OPTICAL CLASSIFICATION: HINTS FROM THE XRB MODELS

R. Gilli [1], G. Risaliti [1] and M. Salvati [2]

1) Dipartimento di Astronomia e Scienza dello Spazio, Università di Firenze, Largo E. Fermi 5, I–50125 Firenze, Italy
2) Osservatorio Astrofisico di Arcetri, Largo E. Fermi 5, I–50125 Firenze, Italy

ABSTRACT   We present a synthesis model of the cosmic X-ray background (XRB) with Active Galactic Nuclei (AGNs) where the evolution of absorbed AGNs is assumed to be faster than that of unabsorbed AGNs and to stop at a lower redshift. The model provides a good description of the XRB spectrum over the broadband 1–100 keV and is in agreement with the AGN counts in the three different bands 0.5–2 keV, 2–10 keV and 5–10 keV, covered by recent X-ray surveys.

We also find a relation between optical classification and X-ray absorption of AGNs, which reproduces the fractions of the different optical types observed in the X-ray surveys. As a result of our investigation, among the AGNs optically classified as type 1, the fraction of objects with X-ray absorbed spectra seems to increase with the redshift.

KEYWORDS: X-rays:galaxies; Galaxies:Seyfert; cosmology:diffuse radiation

1. INTRODUCTION

Most of, if not all, the intensity of the XRB above ~ 1 keV is thought to be produced by AGNs. About 70-80% of the 0.5–2 keV XRB intensity has been resolved by ROSAT into discrete sources (Hasinger et al. 1998), mostly identified with AGNs (Schmidt et al. 1998, hereafter Sc98). In the 5–10 keV band about 30% of the XRB has been resolved by the BeppoSAX HELLAS survey (Giommi et al. 1998) into single sources, mainly identified with AGNs (Fiore et al. 1999; hereafter Fi99).

The most popular synthesis models of the XRB are based on the unification schemes for AGNs, where the orientation of a molecular torus surrounding the nucleus with respect to the line of sight determines the obscuration and then the classification of the source. Indeed, it has been shown that the broad band 3–100 keV spectrum of the XRB and the source counts in the 0.5–2 keV and 2–10 keV energy bands could successfully be reproduced by an appropriate mix of unabsorbed and absorbed (obscured) AGNs (e.g. Comastri et al. 1995). The shape of the X-ray luminosity function (XLF) and the cosmological evolution of absorbed objects are unknown and have been usually assumed to be identical to those derived for unabsorbed AGNs. Also, the number ratio R between absorbed and unabsorbed AGNs and the distribution of the absorbing column densities N_H are free parameters

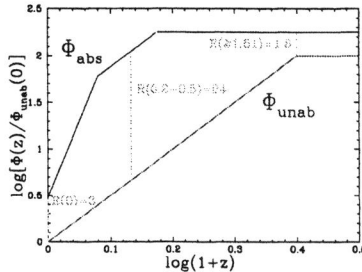

FIGURE 1. The evolution of the XLF for absorbed and unabsorbed AGNs. The absorbed to unabsorbed AGN ratio at different redshifts $R(z)$ is also indicated.

of standard models and have been assumed to be independent of redshift and of the intrinsic source luminosity.

By using a number of recent results about the local N_H distribution of absorbed AGNs (Risaliti et al. 1999), the local value of R (Maiolino & Rieke 1995)[1], the 5-10 keV source counts (Fiore et al. 2000) and the AGN XLF and evolution (Miyaji et al. 1998, who found a Luminosity Dependent Density Evolution of AGNs), Gilli et al. (1999) showed that the XRB spectrum and the hard X-ray counts cannot be reproduced simultaneously. Then, additional sources with hard X-ray spectrum seem to be required at intermediate or high redshift. We propose a model where absorbed AGNs evolve faster than unabsorbed AGNs and stop their evolution at a lower redshift. Also, by considering the optical identifications of the AGNs found in the deep X-ray surveys, we verify if the local relation between optical and X-ray type is still tenable at high redshift.

Throughout this work we will assume $q_0 = 0.5$ and $H_0 = 50$ km s^{-1} Mpc^{-1}.

2. THE MODEL

The AGN X-ray spectra have been divided into 6 classes (unabsorbed and absorbed with logN_H=21.5,22.5,23.5,24.5,25.5) and calculated following Gilli et al. (1999). The N_H distribution for absorbed objects is fixed to the local one derived by Risaliti et al. (1999). We assume for absorbed AGNs the XLF and evolution of Miyaji et al. (1998), where we have reduced the evolution exponent by 8% and the XLF normalization by 15%, respectively, in order to correct for the fraction of absorbed AGNs which are likely to appear in the ROSAT data because of the soft excesses or, for sources at high redshift, because of the K-correction. We consider this as a reasonable correction, which is confirmed by the agreement between the total AGN counts observed by ROSAT and those predicted by our model (see Fig. 2). For

---
[1] Since locally most of type 1 AGNs (Seyfert 1+1.2+1.5) are unabsorbed in the X-rays (Schartel et al. 1997), we assume that the type 2 to type 1 ratio found by Maiolino & Rieke corresponds to the local ratio of X-ray absorbed to unabsorbed AGNs.

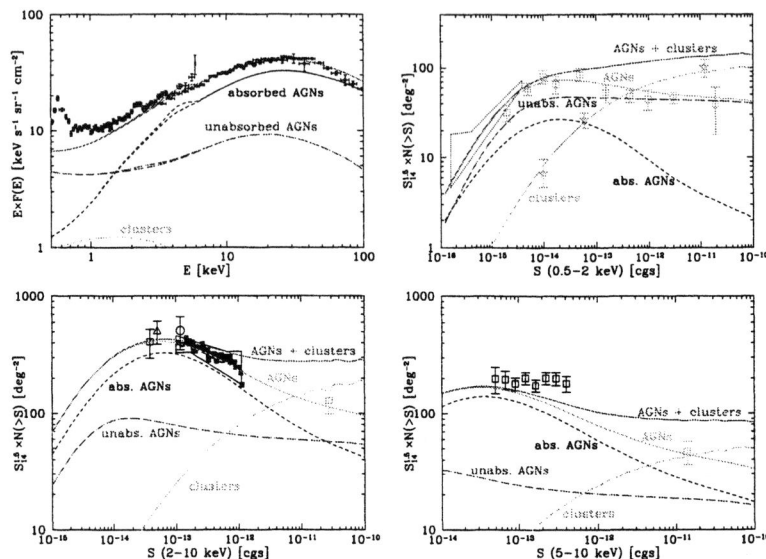

FIGURE 2. The fit to the XRB spectrum and the comparison with the source counts in the bands 0.5–2 keV, 2–10 keV, 5–10 keV. Datapoints are the same as in Figs. 2,3,4,5 of Gilli et al. (1999) except for the HELLAS data (Fiore et al. 2000).

absorbed AGNs we assume: $\phi_{abs}(z, L_x) = \phi_{unab}(z, L_x) \times R(z)$. Since high luminosity objects with a hard spectrum in the X-rays have been found both in the ASCA and BeppoSAX surveys, we assume that $R(z)$ does not drop with increasing luminosity.

The different evolutions of absorbed and unabsorbed AGNs are shown in Fig. 1. It is noted that $R(z)$ rapidly increases from 3 to 24 in the range $0 < z < 0.2$, is constant in the range $0.2 < z < 0.5$ and then decreases to 1.8 for $z \geq z_{cut}$. The very fast density evolution of absorbed AGNs below $z = 0.2$ (evolution exponent ≥ 10) is generally not observed for other classes of sources. However, it is not inconsistent with the evolution of Ultra Luminous IRAS Galaxies (Kim & Sanders, 1998). As shown in Fig. 2, the model provides a good fit both to the XRB spectrum and to the source counts in the three bands 0.5–2 keV, 2–10 keV and 5–10 keV. Also, we have verified that the model is in good agreement with the AGN N_H distribution in the Piccinotti et al. sample (Schartel et al. 1997) and in the Fi99 sample. Then, we have tried to reproduce the fractions of the different AGN optical types in the ROSAT Deep Survey (RDS; Sc98) and in the Fi99 sample. On the basis of their line properties, we have considered as type 1 AGNs the ID classes "a,b" for the RDS and "B" for HELLAS, and as type 2 AGNs the ID classes "c,d,e" for the RDS and "R,1.8,1.9" for HELLAS (see Sc98 and Fi99). Then, the ratios between optical type 2 and type 1 AGNs in the RDS and in the HELLAS sample are 0.39 and 1.41, respectively. In order to match this ratios, we have assumed that, for each bin of

TABLE 1. The assumed fraction of type 1 AGNs in the different bins of $N_{\rm H}$.

$\log N_{\rm H}$	< 21	21.5	22.5	23.5	24.5	> 25
type 1 %	1	0.35	0.30	0.25	0	0

$N_{\rm H}$ in the model, the fraction of optical type 1 AGNs decreases with increasing $N_{\rm H}$ according to Tab. 1. The model ratios between optical type 1 and type 2 in the RDS and in the HELLAS sample are 0.34 and 1.51, respectively. As a result of our choice, the redshift distribution of type 1 and type 2 AGNs in the RDS and HELLAS survey are also reproduced. Tab. 1 implies that, among type 1 AGNs, a significant fraction of objects is absorbed with $N_{\rm H} > 10^{22}$ cm^{-2}. Since locally the percentage of absorbed type 1 AGNs with $N_{\rm H} > 10^{22}$ cm^{-2} has been observed to be negligible, a redshift (or a luminosity) effect should be invoked. Some examples of type 1 AGN with $N_{\rm H} > 10^{22}$ cm^{-2} at $z > 1$ are already known (Fi99, Akiyama et al. 2000) The physical reason for the disagreement between optical and X-ray spectral classification is being investigated, and could involve a dust-to-gas ratio or dust properties different from Galactic. In any case, this possibility would solve the puzzle of the "missing" QSO2s.

REFERENCES

Akiyama M., et al., 2000, ApJ 532, 700
Comastri A., Setti G., Zamorani G., Hasinger G., 1995, A&A 296, 1
Fiore F. et al., 1999, MNRAS 306, L55 (Fi99)
Fiore F. et al., 2000, MNRAS, submitted
Gilli R., Risaliti G., Salvati M., 1999, A&A 347, 424
Giommi P., et al., 1998, Nuclear Physics B Proceedings Supplements 69/1-3, 591
Kim D.-C., Sanders D.B., 1998, ApJS 119, 41
Maiolino R., Rieke G.H., 1995, ApJ 454, 95
Miyaji T., Hasinger G., Schmidt M., 1998, in "Highlights in X-ray astronomy", astro-ph/9809398
Hasinger G. et al., 1998, A&A 329, 482
Piccinotti G., Mushotzky R.F., Boldt E.A., et al., 1982, ApJ 253, 485
Risaliti G., Maiolino R., Salvati M., 1999, ApJ 522, 157
Schartel N., Schmidt M., Fink H.H., Hasinger G., Trümper J., 1997, A&A 320, 696
Schmidt M. et al., 1998, A&A 329, 495 (Sc98)

IBIS/INTEGRAL GALACTIC PLANE OBSERVATIONS: X-RAY NOVAE

P. Goldoni [1], A. Goldwurm [1], P. Laurent [1], F. Lebrun [1]

1) CEA/SAp, Saclay

ABSTRACT

The INTEGRAL satellite, with the hard X-ray/soft gamma ray telescope IBIS onboard, will be launched in 2002. The core program observations include the Galactic Plane Survey and the Galactic Central Radian Deep Exposure. They will respectively provide a weekly coverage of a significant part of the Galactic plane and a deep exposure of the central region of the Galaxy. Both will lead to the discovery of new transient sources, X-ray Novae being the best example. We discuss the possible strategies and results of IBIS/INTEGRAL detection and follow-up observations of X-ray Novae.

KEYWORDS:

1. INTRODUCTION

The Imager on Board Integral Satellite (IBIS) is the imaging instrument of the INTEGRAL satellite, the hard-X/soft-gamma ray ESA mission to be launched in 2001. It provides diagnostic capabilities of fine imaging (12' FWHM), source identification and spectral sensitivity to both continuum and broad lines over a broad (15 keV–10 MeV) energy range. It has a continuum sensitivity of $2\ 10^{-7}$ ph cm^{-2} s^{-1} at 1 MeV for a 10^6 seconds observation and a spectral resolution better than 7 % @ 100 keV and of 8 % @ 1 MeV. The imaging capabilities of the IBIS are characterized by the coupling of its source discrimination capability with a very wide field of view (FOV), namely $9° \times 9°$ fully coded, $29° \times 29°$ partially coded FOV.

The INTEGRAL observing program is divided in a core program reserved to instrument teams and collaborators and an open observing program. The core program is mainly formed by two kind of observations: the Galactic Central Radian Deep Exposure and the Galactic Plane survey (GPS). The first one will consist of deep pointings of the central region of the Galaxy that will allow to probe this region with unprecedented accuracy. The GPS instead is composed of a series of separate short pointings arranged to perform a scan of a section of the galactic plane every week.

The GPS will monitor the galactic plane to search for galactic transient sources (especially X-ray Novae) during their outbursts. The discovery and arcminute positioning of a new transient source may trigger INTEGRAL TOO observations, depending on scheduling constraints and will generate an alert to the astronomical community enabling ground and space observatories to start observations at different wavelengths. If a TOO observations is not triggered, the GPS will anyway follow the transient during its outburst with an observation every week, if the pointing constraints allow it.

We present simulations of IBIS Galactic plane survey observations. They show the capabilities of this instrument in discriminating between different sources while at the same time monitoring a huge FOV. It is envisaged that a proper exploitation of both the FOV dimension and the source localization capability of IBIS will be a key factor in maximizing its scientific output.

2. THE IBIS TELESCOPE

The IBIS detection system is composed of two planes, an upper layer made of 16384 squared CdTe pixels (ISGRI) and a lower layer made of 4096 CsI scintillation bars (PICsIT). This system enables high sensitivity continuum spectroscopy ($E/\Delta E > 10$) and a wide spectral range (15 keV – 10 MeV).

The simulation we performed are for the moment limited to the ISGRI upper layer (Lebrun et al. 96). The ISGRI pixels are 4×4 mm^2, 2 mm thick crystals of Cadmium Telluride, a semiconductor operating at ambient temperature, providing a spectral resolution of about 8% at room temperature. The $128 \times 128 = 16384$ pixels are arranged in 8 modules separated by dead zones 2 pixel wide needed by the mechanical structures of the detector plane. The sensitivity loss caused by dead zones is not large, however the absence of sensitive elements in the detector plane must be properly taken into account during deconvolution procedures.

3. GPS PARAMETERS AND SIMULATIONS

One GPS scan will involve 18 to 57 (depending on visibility and satellite constraints) 1050 sec pointings along the Galactic Plane on a saw-tooth pattern with a 21° inclination. The angular distance between pointings along the scan path is 6°, the extreme of latitude pointing are at b=± 6.45°. This effectively results in a coverage of a region extending 100° to 320° in Galactic Longitude and 40° in Galactic Latitude. Every week the scan is repeated with a 27° shift in Galactic Longitude. The effective IBIS exposure for each sky point along the path is about 3250 seconds, which, on actual estimates translates in a 5σ sensitivity of ~30 mCrab in the 50-150 keV band. This value is valid in the fully coded field of view of the instrument and decreases to zero at the images limits. This is vastly better than the sensitivity level of the BATSE occultation analysis on similar timescales, assuring the monitoring of hard X-ray sources activity at a level never reached by past all-sky monitors.

We simulated a GPS scan containing the central regions of the Galaxy in the 50-150 keV energy band. We selected a sample of hard X-ray galactic sources extracted from the Heasarc X-ray binaries catalog (Van Paradijs et al. 1995) and containing the sources monitored by BATSE in its occultation program (http://cossc.gsfc.nasa.gov/-

TABLE 1. Sources, their simulated 50-150 keV flux and their detection likelihood. The detection likelihood depends both on source strength and on its effective exposure which is in general lowest for sources far from the Galactic Plane.

GRS 1716 − 249	100 mCrab	16σ	Bright Nova	1 Crab	170σ
SLX 1735 − 269	90 mCrab	16σ	4U 1630 − 472	100 mCrab	10σ
GX 1 + 4	40 mCrab	7σ	GX 339 − 4	200 mCrab	22σ
GRS 1758 − 258	60 mCrab	13σ	OAO 1657 − 415	100 mCrab	13σ
GS 1826 − 38	30 mCrab	2.6σ	GRO J1655 − 40	200 mCrab	27σ
GS 1845 − 03	100 mCrab	21σ	4U 1700 − 377	250 mCrab	38σ
A 1845 − 024	40 mCrab	8σ	GX 354 − 00	50 mCrab	8σ
GRS 1915 + 105	300 mCrab	51σ	Faint Nova	30 mCrab	6σ
QSO 0241 + 622	100 mCrab	20σ	1E1740.7 − 2942	100 mCrab	18σ
LSI + 61 303	20 mCrab	2.5σ	A 1742 − 294	20 mCrab	2.5σ
4U0115 + 63	50 mCrab	8σ	Terzan 2	30 mCrab	5σ

cossc/batse/ hilev/occ.html) and transient and persistent sources detected by the SIGMA hard X-ray/soft γ-ray telescope in the Galactic plane (see e.g. Goldwurm et al. 1994). The sample was selected in order to give a representative example of GPS results and possibilities but we consider that several other sources could be detected in an average scan, especially at lower energies. The simulated fluxes are representative of the average behaviour of these sources. To this sample we added two serendipitous transient events to check the sensitivity of our procedures to these occurrences. A bright (1 Crab) one to test the localization capability of our system and a faint (30 mCrab) one in a crowded region to test the source discrimination capability.

The simulation parameters and the deconvolution results are shown in Table 1. The GPS sensitivity is maximal (RMS noise $\sim$ 6 mCrab in images' centers) on the galactic plane and decreases towards the images' limits. From the results sketched in Table 1 it is clear that hard X-ray sources with a flux $\geq$ 30 mCrab can be detected if they are in the region of maximum sensitivity near the Galactic Plane. An image of a 30° x 25° region of this simulation is shown in Goldwurm et al. 1999.

We here focus on IBIS capability of positioning a new source in the two cases we simulated. A 1 Crab source can be positioned by IBIS with a 90 % error box of 10" (Ubertini et al. 1999) radius. However some systematic effects which can be partially corrected by in-flight calibration are expected to limit the accuracy to $\sim$ 20" This gives a reasonable possibility of promptly identifiying the optical/radio counterpart from the ground directly from this position even in the crowded regions

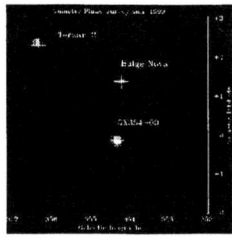

FIGURE 1. Contour, color coded image of the region containing the faint X-ray Nova, levels start from 1.5 σ, spaced by 1 σ.

of the Galactic Bulge.

The 30 mCrab transient is clearly detected in a crowded region (see Fig.1). We remark that present-day hard X-ray monitoring lacks the capability of identifying such a weak transient source with sources at a distance of $\sim$ 1 degree. A source detected at the 6σ level is expected to have a two-parameters 90 % error box of 3'. This value is comparable to the one obtained by RXTE/PCA in a short pointed observation (Remillard et al. 1997) of a stronger (200 mCrab) transient. However this error box is too big to allow a direct identification from the ground. An optical/radio search of the counterpart may however be pursued from the ground trying to identify highly variable objects in the error box.

4. CONCLUSIONS

We have presented simulations of IBIS/INTEGRAL GPS observations, the main results are the following:

1) Sources as faint as $\sim$ 30 mCrab will be detected at a $\sim$ 3 σ level allowing an unprecedented monitoring in these energy bands.

2) $\sim$ 1 Crab transient sources will be localised within few minutes with a 20" 90% error box.

3) New <50 mCrab transients, possibly more distant, not detected by present-day instruments will be detected and localised.

REFERENCES

Chen W., Shrader C.R. & Livio M., 1997, ApJ 491, 312

Goldwurm A. et al., 1994 Nature 371, 589

Goldwurm A. et al., 1999, to appear in Proceedings of the 5th Compton Symposium

Lebrun F. et al., 1996, SPIE 2806, 258

Remillard R. et al., 1997, IAU Circ. 6710

Ubertini P. et al., 1999, 3rd INTEGRAL Workshop Proc., Ap. Lett. & Comm. 529, 39(II), 331

van Paradijs J. in "X-ray Binaries", eds. W.H.G. Lewin, J.v.Paradijs and E.P.J. van den Heuvel, Cambridge University Press, 1995, P536

RXTE STUDIES OF THE STARBURST GALAXIES M82 AND NGC253

D. E. Gruber[1], Yoel Rephaeli[1,2]

1) Center for Astrophysics and Space Sciences, UCSD, La Jolla CA, 92093-0424, USA
2) School of Physics and Astronomy, Tel Aviv University, Tel Aviv, Israel

ABSTRACT   The two nearby starburst (SB) galaxies M82 and NGC253 were observed for ~ 100 ksec over a 10-month period in 1997. While the NGC253 flux did not vary significantly, the M82 flux jumped by a factor ~ 2 in July to a total luminosity of 4×10^{40} erg-s^{-1}. The earlier emission from M82 has a thermal shape with kT $\simeq$ 6.63$\pm$0.08. In the high flux phase, the additional emission requires either an absorbed second thermal component, or absorbed power law component, with the former providing a better fit. But given the unphysically weak Fe K emission if thermal, nonthermal power law emission is more likely. M82 could therefore be a combined SB/LLAGN, with the dominant emission powered by the SB activity. The best-fit emission from NGC253 if thermal has kT=4.2$\pm$0.4 keV or index 2.6$\pm$0.1 if power law.

KEYWORDS: galaxies; starbursts; AGN; spectra; x-rays; abundances.

1. INTRODUCTION

Starburst galaxies (SBGs) must also contain the end products of their abundant population of massive young stars, so that they should exhibit copious emission of X-rays from supernovas, supernova remnants, x-ray binaries, hot winds driven by SN shocks, and as secondary radiation resulting from interactions of particles accelerated in shocks. A SBG may also have an active, compact nucleus. Determination of the X-ray properties of SBG may help in classification and the assessment of the significance of the starburst phase in galactic evolution. Considering the obscuration of optical emission and re-processing of IR emission, X-rays allow a more penetrating view of the inner regions of SBGs.

M82 and NGC253 are the closest nearby SBGs. Their X-radiation has been observed many times (e.g., Fabbiano 1988, Tsuru et al. 1990, Ohashi et al. 1990, Boller et al. 1992, Matsumoto & Tsuru 1999). The X-ray emission is substantial, above 10^{40} ergs/s. Both thermal and power-law spectra have been claimed. At low X-ray energies, emission from these galaxies extends well beyond their optical disks, so starburst activity is likely to be present also outside the central nuclear region ($\leq$1 kpc). Monitoring as well as better spectral measurements can separate thermal and non-thermal components, possibly permitting comparison of the starburst-powered versus nuclear-powered emissions. We have performed a moni-

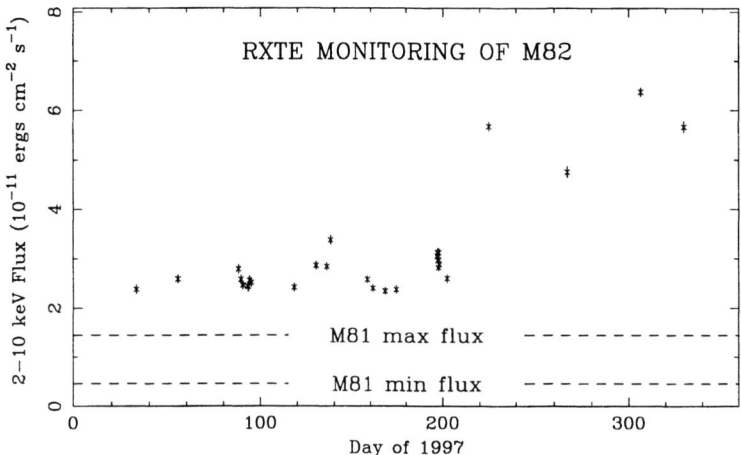

FIGURE 1. Light-curve of the 1997 RXTE observations of M82. Limits to the contribution from M81 are indicated.

toring program of these two galaxies with RXTE between February and November 1997. These were organized in four sets of seven or eight one-orbit samples spanning intervals of one day, one week, eight weeks, and 42 weeks. Total time was 100 ksec per source.

2. TEMPORAL ANALYSIS

Light curves of 2-10 keV flux were constructed for M82 and NGC253. NGC253 was quite faint for RXTE, and the light curve excludes only large factor-of-two variability. The light curve of M82 (Figure 1) on the other hand, shows immediate evidence for variability on a variety of time scales. although it strongly appears to be dominated by a single event which doubles the counting rate late in the monitoring period. A BeppoSAX observation (Cappi et al. 1999) weeks after the last RXTE sample shows a return to the flux level of the earlier RXTE samples. The variable Seyfert galaxy M81 was near the edge of the beam at 38% response. Figure 1 shows limits to the attenuated M81 flux seen by RXTE, based on roughly contemporary monitoring of M81 (Pellegrini et al. 2000). The observed variability is consistent with a red noise process of index 1.0, but may come in part from M81. Obviously, the large change after Day 220 can be little affected by M81.

3. SPECTRAL ANALYSIS

PCA data below 3.5 keV were ignored due to unreliability of the response matrix. Because of this, PCA was insensitive to the second, low-temperature thermal component with $kT = 0.7$ of Cappi et al. (1999). PCA data above 25 keV and HEXTE data above 50 keV were felt to add only noise in the fits and were not used.

NGC 253, observed at an average flux level of 2.6×10^{-12} erg cm^{-2} s^{-1} in the 2–10 keV band was significantly detected by the PCA but not HEXTE. Data from all thirty ∼3000 s samples were coadded, as well as the background estimates, to produce a net spectrum at about 30σ significance, strong enough for simple spectral modeling. We tried power-law and Raymond-Smith models. With the power-law we added an iron K line, whose energy floated to 6.7 keV. In both models the detection of the iron line is only marginally significant (2σ). With the Raymond-Smith model, the abundance was 0.16±0.08 relative to solar. The best-fit temperature was 4.2±0.4 keV, and the index for the power law was 2.60±0.10. Both the thermal and power-law forms fit better than could be expected, with reduced χ^2 values of 0.70 and 0.77, respectively, therefore there can be no formal preference.

Spectral analysis of M82 was complicated by variability and by the presence of M81 in the field of view. Irrespective of origin, the last four one-orbit samples had an average flux almost double the earlier average of 2.9×10^{-11} erg cm^{-2} s^{-1} (2×10^{40} erg-s^{-1}). We analyzed spectra summed over the 26 (low) earlier observations, the last 4 (high) observations, and also obtained a difference (outburst) spectrum between the high and low rate spectra. All spectra show an overwhelming preference for a bremsstrahlung shape over an unabsorbed power law, the reduced chi squares with the power law form lying in the range 3–4.

To account for the known X-ray flux of M81, we have assumed a mean 2–10 keV power law flux of 14×10^{-12} erg cm^{-2} s^{-1}, and photon index, $\alpha = 1.85$, as measured in 1997 by ASCA (Ishisaki et al. 1996). The fits were repeated in the presence of this power law component which, with the 38% transmission, amounts to ∼25% of the total counting rate in the low samples, and only 10% for the high average.

The best-fit model for the net M82 emission in the 'quiescent', low-brightness phase is thermal, with kT = 6.63±0.08 keV, and Fe abundance of 0.14±0.01 in solar units. This model is then our most probable representation of the emission from M82 in its 'normal' state. The above value of the temperature is somewhat higher than the best-fit value deduced from the BeppoSAX measurements of Cappi et al. (1999); their value for the iron abundance, 0.07±0.02 is particularly low, even when compared with the already low value deduced here. One possible way to obtain higher abundances is discussed in the next section.

For the outburst emission the best-fit temperature of 5.95±0.16 keV is modestly but significantly lower than for the earlier emission. Whether the outburst is assumed to have a thermal or power law form, the addition of an absorbing column results in a significant improvement in χ^2, dramatically so for the power law form. The best-fit column in the latter case is N$_H$ = 1.2×10^{23} cm^{-2} and the best-fit index is steep: 3.9±0.06. The abundance for the absorbed thermal fit is extremely low, 0.02±0.02. This unphysically low value argues against a thermal form. It also excludes origin from a super- luminous binary stellar system: the limit of 15 eV to equivalent width for the K line of iron is a factor of 20 below equivalent widths observed in binary sources in the Galaxy. The combination of heavy absorption and power law emission during the outburst strongly point to a buried single source, probably in the central region of M82. Such a large column is unlikely in M81.

4. INTERPRETATION AND SUMMARY

The emission from NGC253 and the early emission from M82 are consistent with origin in intense starburst activity. Such activity is clearly manifested by thermal emission and possibly also by appreciable nonthermal emission, as is reflected by our spectral fits. But the RXTE results do not permit quantitative estimates of multiple components.

A large temporal change of galactic-scale emission, such as observed here with M82 ($\Delta L = 2 \times 10^{40}$ erg-s^{-1}), is commonly attributed to accretion onto a massive central black hole, and is expected to have a power-law spectrum with very significant photoelectric absorption, such as is seen in this RXTE spectral work. The earlier emission is at most weakly varying, and in view of its thermal spectrum and previous evidence for its spatial extent, it may be dominated by starburst activity. M82 is therefore primarily a starburst galaxy with a low-luminosity active nucleus.

With ASCA data Ptak & Griffiths (1999) concluded that the M82 flux varied by a factor ~4 between 1993 and 1999, and that the emission originates from within a central 10 arcsec region. If due to accretion onto a massive blackhole, the implied mass is $M \geq 460\,M_\odot$. Matsumoto and Tsuru (1999) also give evidence for AGN activity in M82.

The low iron abundances we deduce here are puzzling given the intense star formation activity with consequent enrichment. The abundances can be higher, however, if more of the emission is nonthermal. For example, if the 2-10 keV flux in the power law component is taken to be 25% that of the thermal component, then the iron abundance in M82 increases from its best-fit value of 0.14 to 0.23 solar.

In summary, thermal spectra with Fe lines were confirmed in the steady emission from both objects. M82 (or, much less likely, M81) had a big outburst in July 1997 which doubled the flux. The outburst spectrum is most consistent with absorbed power-law emission. Therefore the outburst source could be a hidden AGN in M82.

ACKNOWLEDGEMENTS

This work was supported by NASA contract NASA UCSD-876359 and NASA grant NAG5-4623.

REFERENCES

Boller, Th. *et al.* 1992, A&A 261, 57
Cappi, M., *et al.* 1999, A&A, 350, 777
Fabbiano, G., 1988, ApJ 330, 672
Ishisaki, Y., *et al.* 1996, PASJ, 48, 237
Matsumoto, H., & Tsuru, T.G., 1999, PASJ, 51, 321
Ohashi, T. *et al.*, 1990, ApJ 365, 180
Pellegrini, S., & 2000, A&A 536, 153
Ptak, A., & Griffiths, R. 1999, ApJ 517, L85
Tsuru, T. *et al.*, 1990, Publ. Astron. Soc. Japan, 42, L75

COMPTONIZATION IN X-RAY BRIGHT NEUTRON STAR GLOBULAR CLUSTER SYSTEMS

M.Guainazzi[1], A.N.Parmar[2], T.Oosterbroek[2]

1) XMM SOC, VILSPA, ESA, Apartado 50727, E-28080 Madrid, Spain; 2) Astrophysics Division, Space Science Department of ESA, Postbus 299, NL-2200 AG Noordwijk, The Netherlands

ABSTRACT A BeppoSAX survey of bright Neutron Star (NS) systems in globular clusters, together with the results obtained on Galactic LMXRB, suggests that a two-component model is a reasonable description of the 0.1–200 keV spectra. At energies lower than a few keV, a thermal component dominates, either due to the direct view of the NS surface, to the emission of the boundary layer, or to a multi-temperature blackbody originating in the accretion disk. At higher energies, a power-law with an high-energy cutoff is the likely signature of thermal Comptonization of soft X-ray photons, probably produced by the same thermal mechanisms as above. The plasma optical depth (electron temperature) correlates (anticorrelates) with the X-ray luminosity, suggesting that the properties of the Comptonizing plasma are driven by the X-ray energy output.

KEYWORDS: Stars: neutron – *(Galaxy:)* globular clusters: general – Accretion, accretion disks

BeppoSAX (Boella et al. 1997) is carrying out a systematic survey of X-ray bright globular cluster sources. More than 90% of them (11 out of 12) are binary Neutron Star (NS) systems. The sources discussed in this paper are listed in Tab.1.

The "burster box" (see Fig.1) was invented by Barret et al. (1996) to discriminate between NS and Black Hole binaries (BHB). With the BeppoSAX observations (crosses) we can for the first time fill this box with simultaneous measures in the 1-20 and 20-200 keV energy bands (see also Barret et al. 1999). Most of the NS sources align within a strip corresponding to simple power-laws with photon index (Γ) between 1.5 and 2.0. However, a few exceptions exist. Is it possible to encompass also these exceptions in the same physical framework?

In several NS binaries, BeppoSAX measured high-energy cutoffs (see Fig.2 for two illustrative cases). Although there is no clear trend with the X-ray luminosity (proportional to the data point size in Fig.1), it is worthwhile to notice that the only source with $E_{cutoff} < 10$ keV is that with $\log(L_X) \sim 38$ (X1820-308).

In our sample, a model consisting of a disk blackbody and thermal Comptonization in a spherical geometry provides a good fit of the broadband BeppoSAX spectrum in all cases (see Fig.2). Some best-fit parameters are listed in Tab.1. The underlying scenario is a distribution of soft photons, originating in the accretion

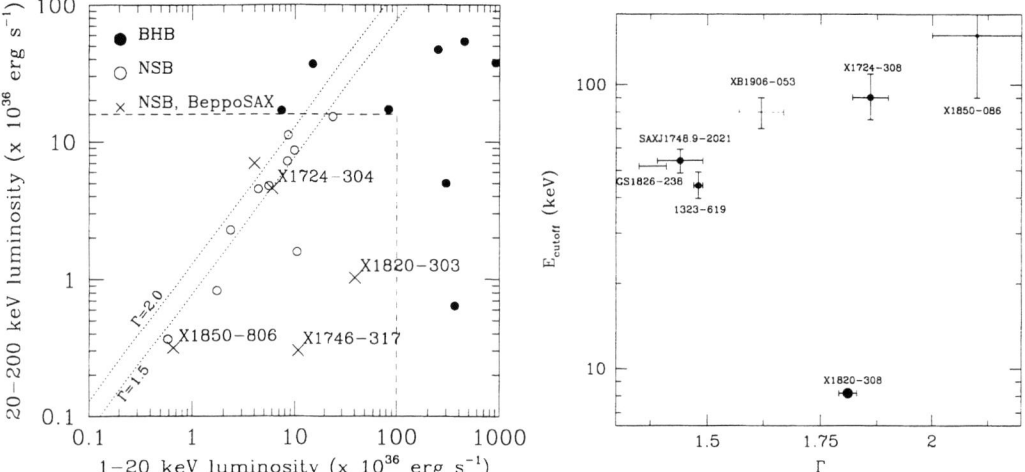

FIGURE 1. *Left panel*: 1–20 keV versus 20–200 keV luminosity for black hole (*filled circles*) and neutron star (*empty circles*) binary systems. The globular cluster sources observed by BeppoSAX are plotted with *crosses*; *Right panel*: cutoff energies versus spectral indices when a model constituted by a disk blackbody and a cut-offed power-law is applied to the LMXRB observed by BeppoSAX. All measures are from our analysis, except: XB1916-053 (Church et al. 1999), 4U1323-69 (Bałucińska-Church M. et al. 1999), GS1826-238 (in't Zand et al. 1999b), SAXJ1748.9-2021 (in 't Zand et al. 1999a)

disk, which is Comptonized in a region closer to the NS surface, *e.g.* the boundary layer. A model where the thermal emission originates from a single-temperature blackbody (*e.g.* the surface of the NS), and the Comptonization occurs in a disk corona above the disk yields comparably good χ_ν^2, and is discussed in Guainazzi et al. (1998). The properties of the Comptonization do not substantially differ in the two scenarios. The ratio of the thermal versus the Comptonization X-ray fluxes is 0.2–0.5. The ratio of the thermal component versus input Comptonization temperatures is on the average 0.7 (if one does not consider X1746-371, see below), and is consistent with the value of 1 in two out of three cases within the statistical uncertainties. In X1746-371 (Parmar et al. 1999), the Comptonization is strongly absorbed. This is a "dipping" source, thus likely to be observed at high inclination. This supports the idea that the Comptonization occurs in a region smaller than that, where the thermal component originates [see, however, a different point of view in Church et al. (1999)].

The Comptonizing plasma electron temperature and optical depth correlate cleanly with luminosity (see Fig. 3). In the above scenario (thermal photons from

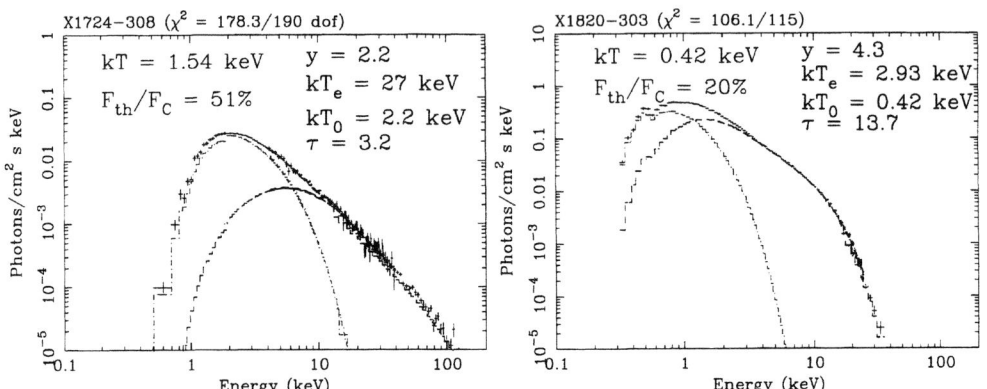

FIGURE 2. X-ray spectral energy distributions as measured by BeppoSAX (*crosses*) compared with the best fit models for X1724-308 ($L_{0.1-200 \text{ keV}} \sim 10^{37}$ erg s^{-1}, *left*), and X1820-303 ($L_{0.1-200 \text{ keV}} \simeq 8 \times 10^{37}$ erg s^{-1}, *right*). In both cases, the model is constituted by Comptonization in spherical geometry (*high energy component*) and a disk blackbody component (*low energy component*).

	kT_e	τ (keV)	kT_0 (keV)	kT_{disk} (keV)	F_{disk}/F_{Compt}	χ^2_ν
X1724-308 (Terzan 2)	27^{+11}_{-4}	$3.2^{+0.6}_{-0.8}$	$2.2^{+0.3}_{-0.2}$	$1.54^{+0.11}_{-0.08}$	0.5	0.94
X1746-376 (NGC6441)	< 110	< 3.7	0.43 ± 0.03	2.79 ± 0.04	6.9	1.08
X1820-303 (NGC6624)	2.93 ± 0.06	$13.7^{+0.5}_{-0.4}$	$0.46^{+0.08}_{-0.06}$	0.42 ± 0.06	0.2	0.92
X1850-086 (NGC6712)	110^{+240}_{-50}	$0.7^{+0.2}_{-0.4}$	$0.15^{+0.15}_{-0.12}$	$0.64^{+0.06}_{-0.04}$	0.3	1.02

TABLE 1. Best-fit parameters when the two components (disk blackbody +Comptonization) model is applied to the BeppoSAX observations of X-ray bright globular cluster (in brackets) sources. From left to right: Comptonizing plasma electron temperature and optical depth, input Wien distribution temperature, disk blackbody temperature, ratio between the disk blackbody and Comptonized fluxes, reduced chi-squared

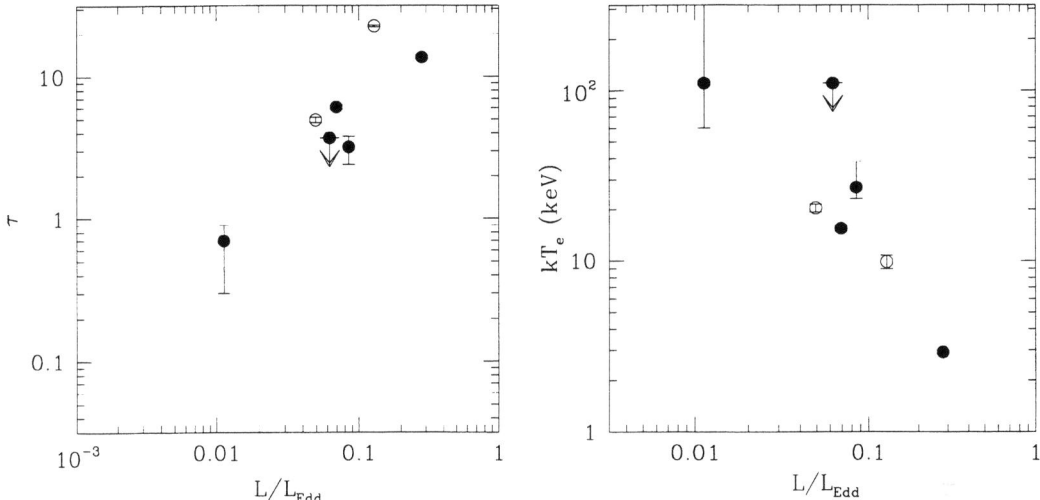

FIGURE 3. Optical depth (*left panel*) and Comptonizing plasma electron temperature (*right panel*) versus luminosity correlation in our sample, plus SAXJ1748.9-2021 (in' t Zand et al. 1999a; *filled circles*. The Galactic LMXRB X1822-371 (Parmar et al. 2000) and GC1826-238 (in't Zand et al. 1999b) are also displayed with *empty circles*

the disk; Comptonization in the boundary layer), the correlation may be qualitatively explained. If the X-ray luminosity of the boundary layer is proportional to the accretion rate (King 1995), the higher $\dot{m}$, the higher is the plasma influx beyond the boundary layer radius. Hence the Comptonizing plasma optical depth is higher and the Compton cooling is more efficient, yielding a lower Comptonizing electron temperature.

REFERENCES

Bałucińska-Church M., et al., 1999, A&A, 349, 495
Barret D., et al., 1996, ApJ, 473, 963
Barret D., et al., 2000, ApJ, 533, 329
Boella G., et al., 1997, A&AS, 122, 299
Church M., et al. 1999, AA&, 338, 556
Guainazzi M., et al., 1998, A&A, 339, 802
King A., 1995, in "X-ray binaries", Lewin G., van Paradijs J., van den Heuvel E.P.J. eds., 439
in' t Zand J., et al., 1999a, A&A, 345, 100
in 't Zand J., et al., 1999b, A&A, 347, 891
Parmar A.N., et al. 1999, A&A, 351, 225
Parmar A.N., et al., 2000, A&A, 356, 175

TRANSIENT TAXONOMY

C.A. Haswell[1], A.R. King[2]

1) Department of Physics, The Open University, Walton Hall, Milton Keynes MK7 6AA, E–mail: C.A.Haswell@open.ac.uk
2) Astronomy Group, University of Leicester, Leicester, LE1 7RH, E–mail: ark@star.le.ac.uk

ABSTRACT   We show that low–mass X–ray binary analogues of all the subtypes of cataclysmic variables exist. In particular neutron–star systems such as EXO 0748-68 or X2129+470, which abruptly switch on or off, are similar to the Z Cam systems. Their mass transfer rates lie close to the boundary between stable and unstable disc accretion, and they are carried across it by variations in these rates, perhaps resulting from spots on the companion star. More complex variability is possible if disc irradiation is only partially effective. This may occur in X1705–440.

KEYWORDS: accretion discs; black hole physics; stars: neutron; stars: cataclysmic variables.

1. INTRODUCTION

Low–mass X–ray binaries (LMXBs) and cataclysmic variables (CVs) are close binaries in which a compact star accretes from a Roche–lobe-filling low–mass companion. The two types are distinguished by the nature of the compact object: in LMXBs this is a neutron star or black hole, while in CVs the accretor is a white dwarf. Both LMXBs and CVs are powered almost entirely by accretion.

Comparison of these two classes has been extremely fruitful, as any differences must be ascribed to the nature of the compact object, which has in some way to communicate itself to an otherwise identical accretion flow. Here we pursue this comparison further by considering the long–term variability of the two classes. We restrict attention to the orbital period range $P \lesssim 10$ hr inhabited by most CVs, and consider only cases where the accretor can be regarded as non–magnetic.

Variability is the most striking feature of CVs. The very large group known as dwarf novae (DN) show outbursts lasting typically a few days and recurring on timescales of weeks. The remaining non–magnetic CVs, the novalikes (NL), are essentially steady on these timescales. The LMXB analogues of these two broad classes are the soft X–ray transients (SXT) and persistent LMXBs respectively. In both CV and LMXB cases the differing behaviour results from the stability or otherwise of the accretion discs, DN and SXTs having unstable discs, NL and persistent LMXBs having stable discs. For given system parameters, unstable discs appear for mass transfer rates below a critical value $\dot M_{\rm crit}$. However both the occurrence (i.e. the value of $\dot M_{\rm crit}$; van Paradijs, 1996; King, Kolb & Burderi, 1996) and the nature

(King & Ritter, 1998) of SXT outbursts are strongly affected by the fact that the central X–ray source is known to irradiate the disc. This can keep hydrogen ionized in the outer part of discs where viscous heating is too weak to do this, and thus makes LMXB discs more stable than CV discs. The same effect prolongs SXT outbursts, which are forced to continue until the central source has weakened to the point that it can no longer keep the disc in the hot state.

The CV/LMXB comparison largely supports the disc instability picture, suitably modified by irradiation. However there are some difficulties. First, there are some short–period SXTs which show variability apparently unlike any known CV. In particular at least two short–period transients, EXO 0748-68 and GS 1826-24 seem essentially to have simply 'turned on', i.e. they were undetected for the first $\sim$ 30 yr of X–ray astronomy, but have remained 'on' ever since their discovery. In compensation, several short–period transients, e.g. X2129+470, X1658-298, have also been observed to turn off during the same time.

Second, CV variability is more complex than the simple division into DN and NL described above, as both groups have conspicuous subgroups. A subset of DN, the Z Cam (ZC) systems, show standstills, in which the luminosity remains constant at a level close to, but slightly below, the peak of the outburst. Standstills can last for weeks or even years. We will show that these apparently discrepant modes of variability can be understood in terms of the disc instability picture. Moreover, this interpretation supports the current picture of the differing evolution of black–hole and neutron–star LMXBs, and the relation to CV evolution. We describe this first.

2. CV AND LMXB EVOLUTION

Mass transfer in short–period CVs and LMXBs (i.e. in the range 3 hr$\lesssim P \lesssim$ 10 hr) must be driven by angular momentum losses. The latter are usually assumed to involve a combination of gravitational radiation and magnetic stellar wind braking (see e.g. King, 1988 for a review). In CVs the secondary star is unevolved; it turns out that angular momentum losses drive average mass transfer rates $\dot M$ which are fairly close to the critical value $\dot M_{\rm crit}$(WD) throughout the relevant period range (e.g. Kolb, 1996). Thus dwarf novae and novalikes appear throughout this range, the Z Cam systems being precisely those systems with $\dot M$ very close to $\dot M_{\rm crit}$(WD). We can now classify LMXBs into 3 groups.

• Persistent LMXBs are the neutron–star analogues of CVs, with unevolved secondaries and similar mass transfer rates $\dot M$. Irradiation means that the critical rate $\dot M_{\rm crit}$(NS) for stability of disc accretion on to a neutron star is far lower than the typical $\dot M$, hence the persistence of these systems.

• Neutron–star SXTs must have much lower transfer rates $\dot M < \dot M_{\rm crit}$(NS) than persistent LMXBs, and must probably have nuclear–evolved secondaries despite their short periods (King, Kolb & Burderi, 1996). Formation constraints (King & Kolb, 1997; Kalogera, Kolb & King, 1998) make it plausible that a significant fraction of neutron–star LMXBs are of this type.

• Although one cannot rule out totally the possibility that some persistent

LMXBs have black–hole primaries, a large fraction of black–hole LMXBs are certainly transient. The fundamental black–hole property – the lack of a hard stellar surface – means that irradiation is generically weaker in black–hole LMXBs The critical rate $\dot M_{\rm crit}({\rm BH})$ for disc stability is therefore larger than $\dot M_{\rm crit}({\rm NS})$. At the same time, the extra inertia caused by the greater binary mass means that $\dot M$ is lower than for neutron–star systems in the magnetic braking regime 3 hr $\lesssim P \lesssim$ 10 hr. The result is that almost all black–hole LMXBs are SXTs (King, Kolb & Szuszkiewicz, 1997), i.e. $\dot M < \dot M_{\rm crit}({\rm BH})$. This is despite the fact that the formation constraints for black–hole binaries are considerably weaker than for neutron–star systems, so that most short–period black–hole LMXBs probably have unevolved secondaries.

3. MASS TRANSFER VARIATIONS

The general picture summarized in the previous section allows us to see what will happen if the mass transfer rate varies around its mean value in the various types of systems. Such variations might for example occur in response to the passage of starspots of various sizes across the region near the inner Lagrange point; however our considerations here are largely independent of the precise cause of the variations.

Persistent LMXBs, and black–hole SXTs have transfer rates respectively so far above or below the critical values $\dot M_{\rm crit}({\rm NS})$, $\dot M_{\rm crit}({\rm BH})$ that only an extremely large and prolonged decrease (increase) of $\dot M$ could change the stability properties of the disc. These systems are therefore the analogues of novalike (UX UMa) and dwarf nova (U Gem) CVs respectively.

The most interesting group are the neutron–star SXTs, where $\dot M$ is close to the critical rate $\dot M_{\rm crit}({\rm NS})$. Here changes in $\dot M$ could cause Z Cam–like standstills. However, since a standstill can only occur after an outburst has triggered the transition to the hot disc state, standstills must be separated by the usual very long quiescent intervals. If the standstills are themselves also very prolonged, the long–term X-ray behaviour of these systems will consist essentially of 'off' and 'on' states, with only short transitions ('outbursts' and 'decays') between them. This on–off behaviour is of course just what observed for several of these systems. Accordingly we identify the on–off transients as the LMXB analogues of the Z Cam systems. Further support for this idea comes from the fact that the amount of mass accreted by the neutron star during the 'on' states of systems like EXO 0748-678 is far too large to have been stored in a disc, even one very close to the critical density at all radii. The accreted mass can only have come from the companion star, implying steady and stable disc accretion during the 'on' state.

A possible difference between on–off systems and Z Cams is the length of the 'on' states compared with Z Cam standstills. The latter seem rarely to exceed 5 – 7 yr, with very short outburst intervals, while EXO 0748-678 was 'on' for at least 10 years after its discovery in 1985. If real, this difference might suggest that starspots are rarer in LMXB companions than in those in CVs. A systematic difference would hardly be suprising in view of their rather different evolutionary states (see Section 2 above), but might also occur because the side of the companion facing the neutron

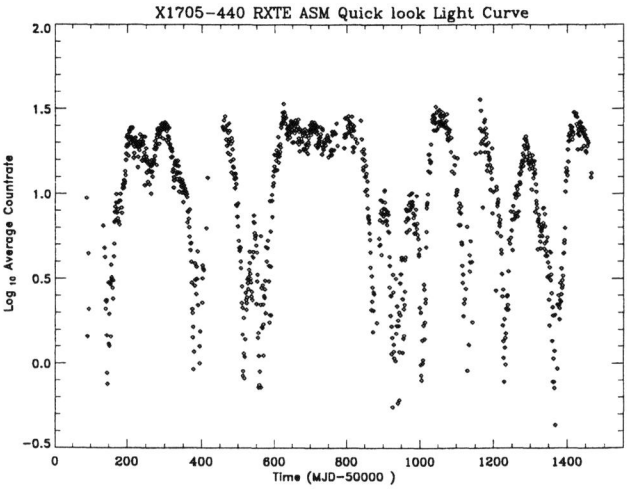

FIGURE 1. The X-ray light curve of X1705-440.

star is heavily irradiated by X-rays, possibly inhibiting starspot activity.

Finally we note that some LMXBs do not fit into the categories described above. In particular X1705-440 shows almost continuous X-ray activity (Fig. 1). This system cannot be described by the simple irradiated-disc model for transient outbursts proposed by King & Ritter (1998), which in general is reasonably successful. X1705-440 does not obey the expected relation between peak brightness and the e-folding timescale of an outburst. Such behaviour may well indicate that the inner accretion disc puffs up and shadows the outer parts. Detailed calculations of irradiated discs (e.g. Cannizzo et al., 1995; Dubus et al., 1999) actually predict this, in contradiction with the fact that most LMXB discs are observed to be irradiated.

CAH and ARK gratefully acknowledge support from the Leverhulme Trust, and ARK a PPARC Senior Fellowship.

REFERENCES

Cannizzo, J.K., Chen, W., Livio, M., 1995, ApJ, 454, 880
Dubus, G., Lasota, J.-P., Hameury, J.-M., Charles, P.A., 1999, MNRAS 303 139
Kalogera, V., Kolb, U., King, A.R., 1998, ApJ, 504, 967
King, A.R., 1988, QJRAS 29, 1
King, A.R., Kolb, U., 1997, ApJ, 481, 918
King, A.R., Kolb, U., Burderi, L., 1996, ApJ, 464
King, A.R., Kolb. U., Szuszkiewicz, E., 1997, ApJ, 488
King, A.R., Ritter, H., 1998, MNRAS 293, L42
Kolb, U. 1996, in Cataclysmic Variables and Related Objects, (Kluwer Academic Publishers), 433
van Paradijs, J., 1996, ApJ, 464, L139

THE VSOP MISSION AND THE POSSIBLE FUTURE

H. Hirabayashi [1], Y. Murata [1], P.G. Edwards [1], H. Kobayashi [2]

1) ISAS, Yoshinodai 3–1–1, Sagamihara, Kanagawa 229–8510, Japan
2) NAO, Ohsawa 2–21–1, Mitaka, Tokyo 181-8588, Japan

ABSTRACT The first space VLBI (Very Long Baseline Interferometry) satellite, HALCA, was launched in February 1997 and opened the new field of Space VLBI. HALCA has been operated for science observations at 1.6 and 5 GHz for the VLBI Space Observatory Programme (VSOP) in cooperation with many organizations and radio telescopes around the world. Many of the >100 MeV gamma ray sources detected by EGRET are targets for VSOP observations. The connections between space VLBI and high-energy X-ray and gamma-ray astronomy are important, and possible future plans after VSOP are described.

KEYWORDS: techniques: interferometric; galaxies: active; galaxies: jets; quasars: general

1. INTRODUCTION

HALCA was launched on 12 February 1997. The present observing orbit has an apogee height of 21,400 km, perigee height of 560 km, inclination angle of 31°, and orbital period of 6.3 hours. HALCA's radio telescope is an 8 m Cassegrain type deployable antenna composed of a mesh-surface main reflector.

HALCA observes in the bands 1.60–1.73 GHz and 4.7–5.0 GHz (Hirosawa and Hirabayashi, 1997). In mid-June 1997, the first images were generated from observations of the quasars 1519−273 and 1156+295 at 1.6 GHz. Images at 5 GHz were first generated in July 1997, and an angular resolution of about 0.3 milli-arcseconds (mas) was achieved.

For VLBI observations HALCA needs special real-time telemetry support. A reference signal, derived from the hydrogen-maser-based frequency standard at the tracking station, is uplinked to HALCA at 15.3 GHz, and HALCA downlinks the astronomical data to the tracking station at 14.2 GHz (at 128 Mbps) in real time for recording. Five 15 GHz-band tracking stations are employed: at Usuda (Japan), operated by ISAS; Green Bank (USA), operated by NRAO; and Tidbinbilla (Australia), Goldstone (USA), and Madrid (Spain), operated by NASA/DSN. The two-way link to the tracking station also enables orbit determination data to be gathered.

The 8 metre diameter radio-telescope makes possible the first dedicated space-VLBI observations on baselines up to 30,000 km. Due to the Earth's oblateness the longitude of the ascending node and argument of perigee precess on ∼1 year timescales, and so good imaging capability over the whole sky can be obtained during the life of the mission.

Participating VLBI networks include the Very Long Baseline Array (VLBA) and the European VLBI Network (EVN), Asia-Pacific Telescope (APT), DSN 70 m antennas, and individual telescopes. VSOP observations are correlated at the 10-station VSOP correlator in Mitaka, the 20-station VLBA correlator in Socorro, or the 6-station S2 correlator in Penticton.

2. VSOP MISSION SCIENCE

Observations are classified into two main categories; the General Observing Time Program and the mission-led Survey Program. The VSOP Survey Program is observing a large, statistically unbiased sample of flat-spectrum AGN to produce the highest resolution 5 GHz survey ever carried out (Fomalont et al. 1999). The VSOP World Wide Web site (http://www.vsop.isas.ac.jp/) contains details about Announcements of Opportunity as well as much other information about the mission, including images obtained from VSOP observations.

VSOP observations have been made of radio galaxies, OH masers and pulsars, but most observations are made of quasars and BL Lacs. Initial scientific results from the VSOP mission were presented in Hirabayashi et al. (1998). Data on the compact core of gamma-ray loud quasar 1156+295, the superluminal quasar 0212+735, and the quasars 1548+056 and 0014+813 are shown and discussed.

3. SOME HIGHLIGHTS OF VSOP RESULTS

In this section some highlights of VSOP observations related to high-energy phenomena are presented.

3C279, the first "superluminal" quasar, was observed with HALCA and the 10-element Very Long Baseline Array (VLBA) at 5 GHz. The resulting image, from Hirabayashi et al. (1999) reveals a bright core and a jet component lying at a position angle of $-115°$ at a distance of ~ 3 milli-arcseconds. The leading edge of the second component is much sharper than the trailing edge, suggesting that this may be the site of a shock front. An extension to the core at a position angle of approximately $-130°$ may be associated with the emergence of a new component from the core, seen in 22 GHz ground-based VLBI observations, which may possibly be associated with the gamma-ray flare of early 1996 (Unwin et al., 1998).

A preliminary estimate of the brightness temperature for the core of 3C279 in the source rest frame is several times 10^{12} K, consistent with the Doppler factors of ~ 6 inferred from 22 GHz ground-based VLBI and from the simultaneous gamma-ray and X-ray flare seen in early 1996 (Wehrle et al., 1998).

The core of PKS 1921-293 was found to be no bigger than 1.3 light-years, giving a brightness temperature of 5×10^{12} K, one of the highest ever directly measured, and implies a Doppler boosting factor of 10 for the material in the jet. Combining this with the reported jet component speed implies the jet is traveling at an angle of less than 5 degrees to the line of sight, and at a speed of $0.99c$ (Edwards et al., 1999; Shen, et al., 1999).

Piner et al. (1999) showed that the motions of components in the radio jet of the TeV gamma-ray source Mrk 421 are subluminal, implying that there has been rapid jet deceleration between the regions where the TeV photons and radio photons are being emitted.

VSOP polarization observations provide a unique opportunity to better resolve spatial variations in magnetic field structures and have revealed a twisted jet in the AGN 1803+784, with the magnetic field remaining perpendicular to the jet direction all along the bent structure (Gabuzda et al., 1999).

The VSOP image of M87 at 1.6GHz shows a helical structure (Reid et al. 1999). It is worth mentioning that the VSOP beam at 5 GHz is 300 times the Schwarzschild radius of suspected super-massive black hole revealed by the HST spectral observation.

Multi-epoch observations over the lifetime of the VSOP mission for the superluminal quasar 1928+738 and other sources enable the conflicting theoretical models to be tested (Murphy et al., 1999).

Imaging of high-redshift quasars at the VSOP frequencies of 1.6 and 5 GHz enables their milliarcsecond radio structures to be studied at emitted frequencies of 6.5–8.0 and 20–27 GHz respectively. Highlights of the VSOP observations of high-z quasars to date include the detection of extremely violent bending in the quasar 1351-017 ($z=3.71$) and the discovery of the most distant rich core-jet structure in the quasar 2215+020 at $z=3.55$ (Gurvits et al., 1999).

VSOP observations of the VLBI Pearson-Readhead sample of sources enable the brightness temperature distribution of bright core-dominated radio sources to be determined. Results obtained so far for 23 sources indicate that the distribution extends well beyond the Inverse Compton limit of 10^{12} K (Preston et al., 1999). Also, there appear to be no differences between the brightness temperature distribution of gamma-ray loud and gamma-ray quiet AGN, contrary to previous claims.

4. THE POSSIBLE FUTURE OF SPACE VLBI

VSOP's high angular resolution imaging capability and science are sensitivity-limited. Sources in such classes as gravitationally lensed objects, galactic superluminal or sub-luminal sources, flare stars, late-type stars, continuum AGN cores with rotating H_2O maser disks, etc. are generally outside of the VSOP sensitivity horizon. H_2O maser and continuum observations at 22 GHz were planned as one of the most important goals for the VSOP mission in terms of AGN and maser physics, but have been unable to be carried out.

A Japanese successor to VSOP, currently dubbed VSOP-2, is considering as a strawman model a 10 m class antenna in a slightly higher orbit than HALCA, with a range of frequencies possibly up to 43 GHz as a modest near-future mission (Hirabayashi, 1999). This model has 20 micro-arcsecond resolution at 43 GHz, and the sensitivity increase over the VSOP mission is one order of magnitude, allowing a wider scientific area and a larger number of observable sources. The desire to go to 86 GHz band to probe the central engine is natural because the AGN core can be optically thin and also we gain in angular resolution. A two spacecraft mission is also attractive for many reasons. We have better image quality, better angular resolution, good coverage of observable sky. Also, a space-space interferometer could give us a better coherence time — hence a longer integration time to give us better sensitivity, because of the lack of tropospheric disturbance.

The angular resolution with a HALCA-type orbit at 43 GHz and 86 GHz is 20 and 10 micro-arcseconds, respectively. This is only 10 and 5 times that of the suspected Schwarzschild radius of M87. ASCA's discovery of MCG–6-30-15 type phenomena seen by Fe-line spectroscopy is taking place in the region only several times that of the event horizon (Tanaka et al., 1995). High energy physics and low energy physics (radio) will become much more related in the next decades. As an example of this friendship, VSOP has made an imaging observation of PKS 0637−752 in August 1999, coordinated with the calibration observation of Chandra.

ACKNOWLEDGEMENTS

The VSOP Project is led by the Institute of Space and Astronautical Science, with significant contributions from National Astronomical Observatory of Japan, the Jet Propulsion Laboratory, the U.S. National Radio Astronomy Observatory, the Canadian Dominion Radio Astrophysical Observatory, the Australia Telescope National Facility, the European VLBI Network, the Joint Institute for VLBI in Europe, and the directors and staff of many of the world's radio observatories.

REFERENCES

Edwards, P.G. et al., 1999, in preparation
Fomalont, E.B. et al. 1999, Adv. Sp. Res. (in press)
Gabuzda, D., New Astronomy Reviews, in press (1999)
Gurvits, L.I. et al. 1999, Adv. Sp. Res. (in press)
Hirabayashi, H., Hirosawa, H., Kobayashi, H., Murata, Y., et al. 1998, Science, 281, 1825
Hirabayashi, H., Edwards, P.G., Wehrle, A.E., . et al. 1999, Adv. Sp. Res. (in press)
Hirabayashi, H. 1999, Adv. Sp. Res. (in press)
Hirosawa, H., and Hirabayashi, H. 1997, Advances in the Astronautical Sciences, 96, 229
Murphy, D.W. et al, 1999, Adv. Sp. Res. (in press)
Piner, B.G. et al., 1999, ApJ (in press)
Preston, R.A. et al., 1999, Adv. Sp. Res. (in press)
Reid, M. et al., 1999, in preparation
Shen, Z.-Q. et al., 1999, PASJ 51, 513
Tanaka, Y. et al., 1995, Nature, 375, 659
Unwin, S.C., Wehrle, A.E, et al., 1998, IAU Colloquium 164, ASP Conf. Series 144, 69
Wehrle, A.E., Pian, E. Urry, C.M. Maraschi, L., et al., 1998, ApJ, 497, 178

DIGITIZED ASTRONOMICAL PLATES: OPTICAL DATA FOR X-RAY ASTRONOMY

R. Hudec

Astronomical Institute, CZ-251 65 Ondrejov, Czech Republic

ABSTRACT   The worlds astronomical plate archives count nearly 3 million plates, some of them of very high quality with limiting magnitudes up to 22-23. The recent digitization of astronomical plates enables their wide use as data source in many areas of astronomy and astrophysics. I review the worlds major plate collections and discuss their impact on various regions of X-ray astronomy and astrophysics including long-term studies and monitoring of flares and active states of both galactic and extragalactic X-ray sources.

KEYWORDS: Sky surveys, sky monitoring, archival plates

1.   INTRODUCTION - PLATE ARCHIVES

The total number of plates in astronomical plate archives amounts to 3 millions. Some of the major plate archives are listed below:
   (1) Harvard HCO USA 600 000 plates, (2) Sonneberg Germany 270 000 plates, (3) Ondrejov Czech Republic 110 000 plates, (4) RGO Cambridge, UK 80 000 plates, (5) Dr Remeis Sternwarte Bamberg Germany 40 000 plates, (6) Leiden Observatory Netherlands 37 000 plates, (7) ROE UKSTU Edinburgh, UK 17 000 plates, (8) TLS Tautenburg, Germany 9 300 plates, (9) Klet Observatory, Czech Republic 10 000 plates, (10) Bologna University Observatory, Bologna, Italy 20 000 plates, (11) Budapest Konkoly Observatory Hungary, 12 000 plates, (12) Hamburg Bergedorf Observatory, Germany 35 000 plates incl. spectral plates with 18 mag limit, (13) Heidelberg-Koenigstuhl Observatory, Germany 10 000 plates, (14) Odessa Observatory, Ukraine 80 000 plates variable star patrol, (15) Dushanbe Observatory 40 000 meteor patrol plates, (16) Sternberg Astronomical Institute Moscow, Russia 41 000 plates, (17) Asiago Observatory Italy, 35 000 plates, (18) Tatranska Lomnica Observatory, Slovak Republic 12 000 plates, (19) Konkoly Observatory Budapest, Hungary, 12 000 plates, (20) Bucharest Observatory, Romania 12 000 plates, (21) Palomar Observatory USA 40 000 plates, (22) Pulkovo Observatory, Russia 52 000 plates, (23) Swathmore Observatory, Pennsylvania USA 100 000 plates, (24) Tonantzintla Observatory, Mexico 10 000 plates, (25) Mt Kanobili Observatory, Georgia 60 000 plates, (26) Uccle Observatory Belgium 15 000 plates, (27) Byurakan, Armenia 40 000 plates, (28) Charlottesville, Virginia USA 165 000 plates, (29) Krym Observatory, Ukraine 10 000 plates, (30) ESO Garching Germany 25 000 plates, (31)

Hoher List, Germany 11 000 plates, (32) Kitt Peak, AZ USA 17 000 plates, (33) Flagstaff AZ USA 12 000 plates, and many amateur collections, some of them quite numerous.

2. SUMMARY - PLATE ARCHIVES

There are numerous sky plate archives across the world, counting almost 3 million plates. The quality of particular plate series as well as particular plate archives is various, ranging from very poor up to very high quality plates. Particular project usually requires particular plate archives. The recent use of sky plate collections is very limited. There is lack of funding and lack of modern equipment, as well as lack of users. The plate catalogues are available as files only in a few cases and the same is valid for searching software for plate catalogues. There is in most cases no remote access and no access by computers. This situation however changes recently with extended use of scanners and digitization of plates. This enables wide use of data from archival plates in many areas of astronomy and astrophysics.

3. HIGH ENERGY SOURCES AND SKY PATROL PLATES

Many of the high energy sources exhibit optical emission and hence can be also observed and investigated from the ground. These investigations are far less expensive than satellite measurements, and despite of this, can also provide valuable data. The multispectral approach has proved to be crucial in understanding of physical processes in observed objects.

Although the large optical telescopes with CCDs and other advanced detectors can provide precise astrometry, photometry and spectroscopy of these sources, they are unable to provide long-time monitoring with good sampling. The real use of robotic telescopes is still very limited. However, the long-term monitoring of these sources may be crucial for their understanding since they often shows long-term evolution, different activity states, brightness variations and/or flaring.

This can be easily provided by sky patrols. The combination of data from several large plate collections can yield up to few tens of thousands of monitoring hours - one full life of an astronomer would be required to obtain the same observing every clear night by a CCD telescope.

4. GALACTIC SOURCES

Many of the galactic X-ray sources are related to variable stars. Some of them have been discovered as variable stars long time before their X-ray detection (e.g. HZ Her), another have been identified with variable stars after their X-ray detection. Many of variable optical counterparts of X-ray sources have been found and/or in details studied in Sonneberg, mostly using archival plates. Examples of objects and object categories where long term monitoring may provide crucial data: (1) X-ray binaries. Binary objects powered by either accretion or stellar wind consisting a stellar and a degenerated (WD, NS, BH) companion. Example: Her X-1 - HZ Her

(2) Low mass X-ray binaries, polars and related objects. Example: AM Her, TT Ari, AO Psc (3) Stars with coronal emission. Example: dMe star binaries with enhanced activity and X-ray emission (4) Flare stars (5) X-ray transients-novae (6) Be stars with X-ray emission. Example: X Per (7) Soft X-ray transients.

5. EXTRAGALACTIC SOURCES

5.1. AGNs-QSOs-blazars

Most of the faint X-ray sources are AGNs and related objects. Analogously, most of the gamma ray sources seen by the EGRET experiment onboard the US COMPTON GRO satellite, are AGNs-blazars. The evolution as well as the light curves of these objects can be very complex and violently variable. Most of QSOs are optically variable - about 97% of QSOs brighter than mag B 22.5 are variable (Trevese and Kron, 1994). This can be used as one of criteria for identification of X-ray sources. From optically variable objects of this class, only a negligible fraction of (mostly bright) objects has been investigated in detail, and good sampling is available only for a few dozens of them.

5.2. Supernovae

Supernovae are sources of optical, radio, X-ray and gamma ray emission. Multispectral observations are important to understand the related physics. Monitoring and searches in optical light are less expensive and easier than HE searches from satellites. Many of especially fainter SNe are poorly investigated with poor or even no sampling. Example: SN1998bw, source of X-rays and probably also of a gamma ray burst (Woosley et al. 1999).

5.3. Gamma ray bursts

The recent confirmation of both flaring as well as fading optical emission of Gamma Ray Bursts (GRBs) strongly supports the importance of sky patrol plates and other survey projects. What is confirmed: (1) Optical fading emission detected between $\sim$ hrs and $\sim$weeks or months after GRBs, magnitudes 18-23 or fainter. Afterglows of GRBs - Optical Afterglows, OAs. (2) Optical flaring emission detected $\sim$ 1 min after GRB (GRB990123), magnitude $\sim$ 9 (one observed case), duration $\sim$ minutes. Direct optical emission of GRB - Optical Transients, OTs. (3) One peculiar supernova coinciding in time and position with a GRB: SN1998bw, peak magnitude $\sim$ 13.6. (4) All above listed types of optical counterparts of GRBs can be detected by suitable sky survey telescopes/archival plates. (5) This opens a new observing window for GRBs in optical light, independently from satellites.

The rate of Optical Transients (OTs) and Optical Afterglows (OAs) detected this way may significantly (by a factor of 10 to 1000) exceeds the observed rate of GRBs due to very different beaming in gamma rays and in optical light predicted by theory. The OTs and OAs related to GRBs can be found on astronomical archival plates of high quality (Hudec 1999). High level of automation (digitization of plates, sky surveys, ASPA ...) as well as sophisticated recognition and elimination of background events (faults) is required.

6. IDENTIFICATION OF HE SOURCES

The recent HE satellites, especially those observing in X-rays, provide a large number of detected sources. The identification of them is an important but not easy task. ROSAT catalogues list more than 200 000 X-ray sources detected and 2000 sources with variability factor more than 3. A large fraction of faint X-ray sources is variable. The use of archival sky patrol plates provides an important tool for identification of especially variable X-ray sources since they provide large coverage in time and, in some cases, color information. So can archival plates be used to look for peculiar (mostly variable) optical objects at the positions of X-ray sources. Some collections (Hamburg, Sonneberg, Edinburgh, ...) include also spectral plates taken by objective prisma. This can also provide important tool for objects identification and classification.

7. CONCLUSION

Archival plates represent a valuable tool in HE astrophysics especially by: (1) Providing extended monitoring intervals with good sampling, allowing long-term evolution and changes to be studied, as well as detecting flares and other brightness variations (2) Providing extended database for identification and classification of sources (3) Providing database for monitoring the activity of selected HE sources to provide input for satellite observations (ToO - targets of opportunity).

Suggestions for the future: (1) Rapid evaluation of the plates obtained in still running patrol systems (within one day or so) - monitoring of activity of HE sources and providing inputs for satellites - ToO (2) Extended digitization of plates - automated search for HE counterparts as well as their classification and detailed study (3) Next generation sky patrol experiments with high quality imaging, deep magnitudes and excellent sampling (ROTSE, ASPA, BOOTES, USNO ...).

ACKNOWLEDGEMENTS

We acknowledge the support provided by the Grant Agency of the Czech Republic, grant 205/99/0145. The investigations of the ROE plates, UK, have been supported by the British Council in Prague, Academic link with the University of Westminster.

REFERENCES

Hudec R. 1999, A&ASS 138, 593.
Trevese D., Kron R. G., 1994, in Multi-Wavelength Continuum Emission of AGNs, Courvoisier T. and Blecha A. (Eds.), 412.
Woosley S. E., Eastman R. G., Schmidt B.P. 1999, ApJ 516, 788.

X-RAY ASTRONOMY 2000: WIDE FIELD X-RAY MONITORING WITH LOBSTER-EYE TELESCOPES

A. Inneman [1], R. Hudec [2], L. Pina [3] and P. Gorenstein [4]

1) Faculty of Mechanical Engineering, Czech Technical University Prague, Czech Republic 2) Astronomical Institute, CZ-251 65 Ondrejov, Czech Republic 3) Faculty of Nuclear Engineering, Czech Technical University Prague, Czech Republic 4) SAO, Cambridge MA, USA

ABSTRACT

The recently available first prototypes of innovative very wide field X-ray telescopes of Lobster-Eye type confirm the feasibility to develop such flight instruments in a near future. These devices are expected to allow very wide field (more than 1000 square degrees) monitoring of the sky in X-rays (up to 10 keV and perhaps even more) with faint limits. We will discuss the recent status of the development of very wide field X-ray telescopes as well as related scientific questions including expected major contributions such as monitoring and study of X-ray afterglows of Gamma Ray Bursts.

KEYWORDS: X-rays: X-ray optics, X-ray telescopes

1. WHY WIDE FIELD TELESCOPES?

The X-ray sky is violently variable, with a rich variety of galactic and extragalactic sources, escecially at low luminostities. There is an increasing need for high sensitivity X-ray monitoring to further study the physics and the evolution of these sources.

The lobster-eye (LE) geometry X-ray optics offer an excellent opportunity to achieve very wide fields of view while the widely used classical Wolter grazing incidence mirrors are limited to roughly 1 deg FOV. Further, the wide field X-ray telescopes with imaging optics are expected to represent an important tool in future space astronomy projects, especially those for deep monitoring and surveys in X-rays over a wide energy range including hard X- rays perhaps up to 100 keV.

The wide field X-ray optics has been suggested in 70ies by Schmidt (orthogonal stacks of reflectors) and by Angel (array of square cells) but has not been constructed yet. Up to 180 deg FOV may be achieved. The possible solution is offered by the replication technology as confirmed by first prototypes presented here for both Schmidt as well as Angel arrangements (for more details see Inneman et al. 1999).

2. SCHMIDT OBJECTIVES

One dimensional lobster-eye geometry was originally suggested by Schmidt (Schmidt 1975). The device consists of a set of flat reflecting surfaces. The plane reflectors are arranged in an uniform radial pattern around the perimeter of a cylinder of radius R. X-rays from a given direction are focussed to a line on the surface of a cylinder of radius R/2. Focussing is not perfect and the image size is finite. Angular resolution of the order of one tenth of a degree or better may be achieved. The 1D dimensional focusing device offers a wide field of view, up to maximum of 180 degrees with the coded aperture. Two such systems in sequence, with orthogonal stacks of reflectors, form a double-focusing device.

3. ANGEL OBJECTIVES

The idea of two dimensional lobster-eye type wide-field X-ray optics was first mentioned by Angel (Angel 1979). The full lobster-eye optical grazing incidence X-ray objective consists of numerous tiny square cells located on the sphere and is similar to the reflective eyes of macruran crustaceans such as lobsters. The field of view can be made as large as desired. It is possible to achieve good efficiency for photon energies up to 10 keV and another extension is possible if additional coatings would be applied. Spatial resolution of a few seconds of arc over the full field is possible, in principle, if very small reflecting cells can be fabricated.

This idea was however never been further developed because of difficulties with production of numerous polished square cells of very small size (with aperture/length ratios of 30 or more i.e. about 1 x 1 mm or smaller at lengths of order of tens of mm).

4. LOBSTER–EYE TELESCOPE PROTOTYPES

First Lobster-eye X-ray telescope prototypes have been finished. The prototype of the Schmidt geometry represents one module and consists of two perpendicular arrays of double-sided X-ray reflecting flats (36 and 42 double- sided flats 100 x 80 mm each). The flats are 0.3 mm thick and gold-coated. The microrougness is below 1 nm. The focal distance is 400 mm from the midplane. The FOV of one module is about 6.5 degrees. More such modules may create an array with substantially larger FOV. The optical and X-ray tests indicate performance close to those calculated and expected (e.g. by ray tracing).

For the Angel geometry, numerous square cells of very small size (about 1 x 1 mm or less at lengths of order of tens of mm, i.e. with the size/length ratio of 30 and more) are to be produced. This demand can be also solved by modified innovative replication technology. First test modules with LE Angel cells have been succesfully produced. First test module has 47 cells 2.5 x 2.5 mm, 120 mm long (i.e. size/length ratio of almost 50), surface microroughness 0.8 nm, f = 1.3 m. Second test module with 5 x 5 cells is finished recently. Third test module with 96 x 96 i.e. 9216 cells is in development. The surface microroughness of the replicated reflecting surfaces is better than 1 nm.

The first prototypes of lobster eye X-ray lenses of both the Schmidt as well as Angel geometries demonstrate the feasibility that the wide field telescopes may be constructed in the future based on this type of reflective X-ray optics (Inneman et al. 1999).

5. LE TELESCOPES AND GRB X-RAY AFTERGLOWS

The LE X-ray telescopes are extremely important since the discovery of X-ray afterglows of Gamma Ray Burst (GRBs) sources in 1997. The expected rate of GRBs is 1 per day, however the theoretical prediction assumes larger beaming angle in X-rays if compared with gamma rays, hence the actual rate of X-ray afterglows is expected to be substantially larger (nearly 10 x or even more) than the rate of GRBs, hence about 10 X-ray afterglows are expected daily. The recent discovery of X-ray flashes resembling X-ray afterglows of GRBs (but without detectable GRBs, Heise et al., 2000) seems to confirm these expectations. The sensitivity of LE telescopes is sufficient enough to detect the recently discovered X-ray GRB afterglows. The localization accuracy of the LE telescopes is of order of 1 arcmin, substantially exceeding the recent localization accuracy of most gamma ray instruments (2 deg and more). The LE telescopes are expected to provide a substantial contribution to the science and statistics of GRBs.

6. MORE SCIENCE WITH LE TELESCOPES

The additional science of LE X-ray telescopes includes supernova explosions, high energy binary sources, AGNs, blazars, X-ray novae, X-ray flares on stars, X-ray transients etc. The use of LE telescopes will allow these objects to be detected and studied by sky patrol monitoring.

7. FUTURE STEPS TOWARD THE REAL LOBSTER EYE TELESCOPE

For both Schmidt and Angel arrangements, the application of additional and/or alternative coatings such as multilayers should be exploited to extent the energy range to higher energies. Further developments are also necessary improving the surface quality (microroughness, slope errors) to further increase the reflectivity and the angular resolution. The construction of larger or multiple modules to achieve a larger FOV of order of at least 30 deg or more should be exploited. For Angel objective, the next development should reduce the cell apertures and to enhance the length/aperture ratios to achieve a better angular resolution.

8. CONCLUSION

The use of very wide field X-ray imaging system could be without doubts very valuable for many areas of X-ray and gamma-ray astrophysics. Results of analyses and simulations of lobster-eye X-ray telescopes have indicated that they will be able to monitor the X-ray sky at an unprecedented level of sensitivity, an order of magnitude better than any previous X-ray all-sky monitor. Limits as faint as 10^{-12} erg cm$^{-2}s^{-1}$ for daily observation in soft X-ray range are expected to be achieved, allowing monitoring of all classes of X- ray sources, not only X-ray binaries, but also fainter classes such as AGNs, coronal sources, cataclysmic variables, as well as fast X-ray transients including gamma–ray bursts and the nearby type II supernovae. For pointed observations, limits better than 10^{-14} erg sec$^{-1}cm^{-2}$ (0.5 to 3 keV) could be obtained, sufficient enough to detect X-ray afterglows of GRBs.

The production of corresponding optical elements can be reasonably achieved by methods of electroforming and composite replication as an alternative to other methods. The first prototypes of both Schmidt as well as Angel arrangements have been produced successfully for the first time, demonstrating the possibility to construct these lenses by innovative but feasible technologies. This makes the proposals for space projects with very wide field lobster eye optics possible.

ACKNOWLEDGEMENTS

The development of double-sided reflecting X-ray foils was supported by a grant within the US-Czech Science and Technology program, No. 930 37. The development of the Angel objective is supported by the grant provided by the Grant Agency of the Czech Republic No. 106/97/1223. The design and development of innovative X-ray telescopes for future space projects is supported by the grant provided by the Grant Agency of the Czech Republic, No. 105/99/1546. We also acknowledge the support provided by the UK X-ray astronomy group at the University of Leicester in X-ray tests of the LE telescope Schmidt prototype.

REFERENCES

Angel, J. R. P. 1979, ApJ 233, 364.
Heise, J. et al. 2000, in Proc. 5th Huntsville GRB Symposium, AIP Conf. Proc., in press.
Inneman, A., Hudec, R., Pina, L. and Gorenstein, P. 1999, SPIE Proc. 3766 (X–ray Optics, Instruments and Missions II, Eds. R. B. Hoover and A. B. Walker), in press.
Schmidt, W. H. K. 1975, Nucl. Instr. and Methods 127, 285.

SPECTRAL EVOLUTION DURING DIPPING IN X 1624-490 FROM THE BEPPOSAX OBSERVATION

P. J. Humphrey[1], M. J. Church[1], M. Bałucińska-Church[1] and A. N. Parmar[2]

1) School of Physics and Astronomy, University of Birmingham, Birmingham B15 2TT, UK
2) Astrophysics Division, Space Science Department of ESA, ESTEC, Postbus 299, NL-2200 AG Noordwijk, The Netherlands

ABSTRACT   We report analysis results for the BeppoSAX observation of the unusual LMXB dipping source X 1624-490 revealing the nature of spectral evolution during dipping. The broad-band spectrum (1 – 100 keV) is well-fitted by the two-component model consisting of blackbody emission from the neutron star with $kT = 1.31 \pm 0.07$ keV plus Comptonized emission from an extended ADC with photon index $2.0^{+0.5}_{-0.7}$ and cut-off energy ~12 keV. Intensity-selected dip spectra are well described assuming that the emission components are progressively covered during dipping. It now appears that Progressive Covering of this two emission component model can explain all of the dipping LMXB.

KEYWORDS: X rays: stars – stars: neutron – binaries: close – accretion: accretion discs.

1. INTRODUCTION

X 1624-490 is arguably the most interesting member of the dipping LMXB sources, having the highest luminosity approaching 10^{38} erg s^{-1}, and the longest orbital period of 21 hr. It does not exhibit bursting like most of the dipping sources, but instead often has strong, hard flaring equivalent to increases in the count rate above 5 keV of 30% or more. The non-dip, non-flaring *Exosat* ME spectrum could be well-represented by an emission model consisting of a point-like blackbody plus an extended Comptonized term (Church & Bałucińska-Church 1995). This blackbody was much stronger than in other dipping sources, dominating the shape of the source spectrum, and contributing 72% to the 1.2 – 4.7 keV count rate, and 79% in the 4.7 – 9.8 keV band. These values were in good agreement with the percentage depths of dipping in these bands showing that deep dipping consisted primarily of complete absorption of the point-like blackbody but relatively little absorption of the extended Comptonized emission.

Spectral analysis in this source is complicated as flaring may often coincide with dipping so that the two effects are difficult to separate. In the *Exosat* data, only two short spectra could be extracted essentially free from flaring: one non-dip and one deep dip, so a systematic study of spectral evolution in dipping was not possible.

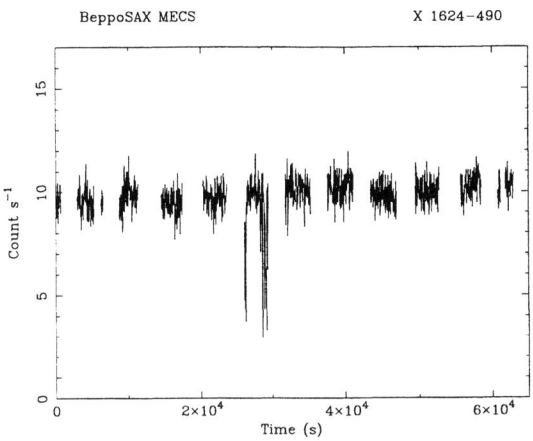

FIGURE 1. The MECS lightcurve in the band 1.8 – 10.0 keV with 32s binning

In more recent observations of X 1624-490 using *Rossi-XTE*, very strong flaring took place and this has allowed a detailed study of the flaring process to be made (Bałucińska-Church et al. 2000a).

Although the dipping LMXB sources have very disparate behaviour, for example, exhibiting hardening, softening or energy independence during dipping, they may all be described by the same model consisting of the above two-component emission model combined with the "Progressive Covering" explanation of dipping (Church et al. 1997). Several dipping sources clearly have an unabsorbed part to their dip spectra below 5 keV, including XB 1916-053, XBT 0748-676 and XB 1254-690, previously treated by "Absorbed + Unabsorbed" modelling (Parmar et al. 1986). It has now been shown that this behaviour is simply explained by a large, dense absorber moving progressively across the extended source region so that at any stage the uncovered emission is not absorbed. In this work, we present a detailed examination of spectral evolution during dipping using high quality data from the *BeppoSAX* observation in the very broad band 0.1 – 300 keV which strongly constrains spectral models.

2. RESULTS

The observation was made on 1998, August 11, starting at 00:47 UT and lasting 17.5 hr. Data from the Narrow Field Instruments (NFI) were screened appropriately and analysed using the SAXDAS data analysis package. Fig. 1 shows the MECS light curve in the band 1.8 – 10.0 keV in which strong dipping is evident, which closer inspection reveals is highly variable demonstrating the blobbiness of the absorber. Careful examination of the light curve in the band 5.0 – 10.0 keV revealed little evidence for flaring, except for a single weak flare at ~33 ks. A broadband spectrum of non-dip, non-flaring emission was extracted by selecting MECS data in the intensity band 9.7 – 10.3 count s^{-1} and removing data from the dipping period. LECS data were selected similarly and HPGSPC and PDS data selected via timing filters equivalent to the MECS selection. A range of spectral models was applied to the 4 NFI spectra fitted simultaneously including all simple models which were quite incapable of fitting the broadband spectrum. The best fit model consisted of a blackbody with kT = 1.31 ± 0.07 keV plus a cut-off power law with photon index $\Gamma = 2.0^{+0.5}_{-0.7}$ and cut-off energy 12^{+14}_{-5} keV and $N_H = 8.6^{+0.8}_{-1.2} \times 10^{22}$ atom cm^{-2} as shown in Fig. 2.

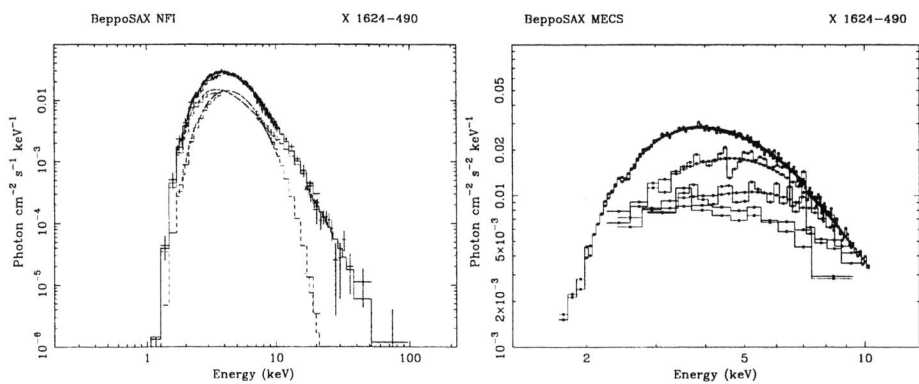

FIGURE 2. (left) The broadband non-dip, nonflaring spectrum in the NFI. The best-fit total model, the blackbody and the cut-off power law are shown. (right) Best fits to the MECS non-dip and dip spectra

TABLE 1. Spectral fitting results for MECS non-dip data and 3 dip levels. Total column densities include the Galactic contribution (see text)

intensity (count s^{-1})	$N_H^{BB,total}$	$N_H^{CPL,total}$	f	χ^2
non-dip	8.6	8.6	0.	139/157
6.0 – 8.0	24 ± 8	24 ± 14	0.23 ± 0.13	37/32
4.0 – 6.0	56 ± 19	38 ± 20	$0.374^{+0.153}_{-0.074}$	16/21
2.9 – 4.0	> 600	34 ± 27	$0.457^{+0.209}_{-0.107}$	8/7

MECS dip spectra were selected in 3 appropriate intensity bands. An acceptable model requires the emission parameters kT, and power law index Γ etc., not to change, and so the non-dip fit was applied to the dip spectra with various absorption models. Simple (one-term) absorption could not fit the data but a good fit was obtained with the Progressive Covering model which may be expressed as $e^{-\sigma_{MM}N_H^{Gal}}(I_{BB}e^{-\sigma_{MM}N_H^{BB}} + I_{CPL}(f\ e^{-\sigma_{MM}N_H^{CPL}} + (1-f)))$, where σ_{MM} is the photoelectric absorption cross-section, N_H^{Gal} is Galactic absorption, f is the progressive covering factor allowing the overlap between absorber and source to increase progressively and I_{BB} and I_{CPL} are the blackbody and cut-off powerlaw intensities. The fitting results are shown in Table 1.

3. DISCUSSION

Spectral fitting in the very broad band of the *BeppoSAX* NFI strongly constrains models, and our results show that the best-fit to the non-dip spectrum is given by the two-component model discussed above. The 1 – 30 keV non-dip, non-flaring luminosity of X 1624-490 is 7.3×10^{37} erg s^{-1} for a distance of 15 kpc (Christian & Swank 1997). The MECS dip spectra are well modelled by Progressive Covering of the two-component model. Dipping is dominated by absorption of the strong blackbody component. A substantial fraction of the Comptonized emission remains in deep dipping and the covering fraction apparently rises to only 0.43. In a more complete treatment including the effects of dust scattering, the emission parameters are unchanged, however the covering fraction actually rises to 0.82 (Bałucińska-Church et al. 2000b).

It is now apparent that *all* LMXB dipping sources with absorbed and unabsorbed parts of the spectrum can be described by Progressive Covering of the two-component model (see Church et al. 1998a,b). Recent *ASCA* results on the sources XB 1746-371 and XB 1254-690 (Bałucińska-Church et al. 2000c) show that these two sources are also well described by the emission model of Church & Bałucińska-Church (1995), so that now *all* members of the dipping class are well fitted by this model. The model has been extended to Atoll and Z-track sources, and a recent *ASCA* survey of the spectra of these sources shows that the model fits well in all cases tested (Church & Bałucińska-Church 2000).

REFERENCES

Christian D.J. and Swank J.H., 1997, ApJS 109, 177

Bałucińska-Church M., Barnard R., Church M.J. and Smale A.P., 2000a, ApJ In Preparation

Bałucińska-Church M., Humphrey P.J., Church M.J. and Parmar A.N., 2000b, A&A In Press.

Bałucińska-Church M., Church M.J. and Dotani T., 2000c, A&A In Preparation

Church M.J. and Bałucińska-Church M., 1995, A&A 300, 441

Church M.J., Dotani T., Bałucińska-Church M., Mitsuda K., Takahashi T., Inoue H. and Yoshida K., 1997, ApJ 491, 388

Church M.J., Bałucińska-Church M., Dotani T. and Asai K., 1998a, ApJ 504, 516

Church M.J., Parmar A.N, Bałucińska-Church M., Oosterbroek T., Dal Fiume D. and Orlandini M., 1998b, A&A 338, 556

Church M.J. and Bałucińska-Church M., 2000, ApJ Submitted

Parmar A.N., White N.E., Giommi P. and Gottwald M., 1986, ApJ 308, 199

OPTICAL OBSERVATIONS OF SUPERSOFT SOURCE 0925-47

J.B.Hutchings [1], A.P.Cowley [2], D.Crampton [1], P.Schmidtke [2]

1) Dominion Astrophysical Observatory, Herzberg Institute of Astrophysics, NRC of Canada 2) Dept of Physics and Astronomy, Arizona State University

ABSTRACT

We present an orbital analysis of the system from new optical observations.

KEYWORDS: stars:binaries:fundamental parameters, x-rays: stars

1. INTRODUCTION

The supersoft X-ray system 0925-47 was observed at CTIO in Chile in March 1999. Spectroscopic observations were obtained with the 4-m telescope over 4 consecutive nights and photometry (in V, B, and R) was obtained using the 0.9-m telescope over 8 consecutive nights. We also have photometry obtained during 5 consecutive nights in February 1998.

It has been shown by Motch (1996) that the period for this system is near 3.8 days, or some shorter alias of that period. Analysis of both the radial velocities and the light curve together rule out any of the possible shorter periods found from photometry alone. The best period using our photometric data from both 1998 and 1999 and fitting it to Motch's 1992 photometry gives P = 3.83 days. The best period indicated by the 1999 velocities alone is 3.87 days.

2. SPECTRUM AND LIGHTCURRVE

The strength of the interstellar absorptions and the very red color of the system (B-V$\sim$2) shows that the system lies behind considerable interstellar material. The unreddened SSS disk systems in the Magellanic Clouds have B-V$\sim$0.

Figure 1 shows a phased plot of the light curve, the velocity curves of He II (4686Å) and Hα (6562Å), and the equivalent width of He II. He II shows a somewhat larger velocity amplitude than Hα, although it also shows a greater scatter since it is in a weaker part of the spectrum (due to the reddening mentioned above). The emission line velocity phase differences between Hα and He II are significant at the 3σ level. The velocity minima slightly precede the light maximum, although the exact shifts are indeterminate since our light curve coverage is not complete, and the curve is non-sinusoidal. It also appears that the equivalent widths vary in phase

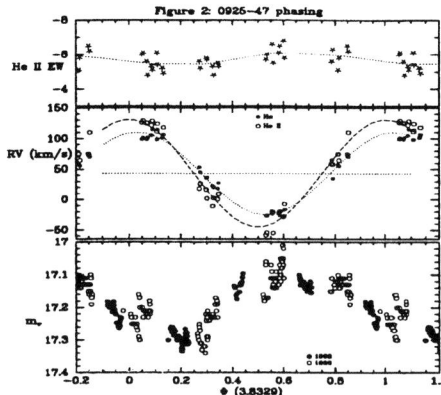

FIGURE 1. Binary phased measurements of 0925-47.

with the light curve. This implies that the line emission is actually modulated more strongly than the light curve, with roughly the same phasing.

The Hα profiles at velocity minimum (near maximum light) are asymmetrical with a red wing, and the overall emission flux is higher. Thus, the asymmetry may be due to an extra component being present near this phase (~0.5).

Motch (1998) reported an observation of twin jets seen as satellite Hα emission lines on one night. Our data do not show this, but we do see emission at 6678Å, one of the positions observed by Motch. While this might be He I, there is no other evidence for He I.

3. MASSES AND LUMINOSITY

Figure 2 shows a mass diagram, based on adopting the He II velocity amplitude as representing the orbital motion of the compact star. Since X-ray eclipses are not seen, and the optical light curve has a small amplitude (with no known evidence of eclipses), the inclination must be less than ~65° and perhaps as low as ~50°. If the compact star is a white dwarf the mass would lie in the range 0.5 - 1.4 and if it is a neutron star it could be as large as about $1.7 M_\odot$. The box of preferred masses is enclosed by these limits. This implies that the mass of the donor star is in the range of 1 - $1.6 M_\odot$ for a WD or $2 M_\odot$ for a NS. The Roche lobe for these masses is in the range 4 to 7 $R_\odot$. Thus, to fill its Roche lobe and allow mass transfer, the donor star must be an evolved (giant) star.

A giant star of mass 1 - 2 has absolute magnitude between 1.5 and 2.0. We know the system is highly reddened. If the true color is that of the disk is B-V~0, then E_{B-V} ~2.1 (Motch 1994) and A_v ~7. The unreddene d magnitude of the system would then be ~10. For the donor star not to be seen in the composite spectrum it would have to be at least 2 magnitudes fainter, or fainter than 12. This implies a distance modulus of about 10 or more, and hence an X-ray luminosity of Lx ~5.10^{35}

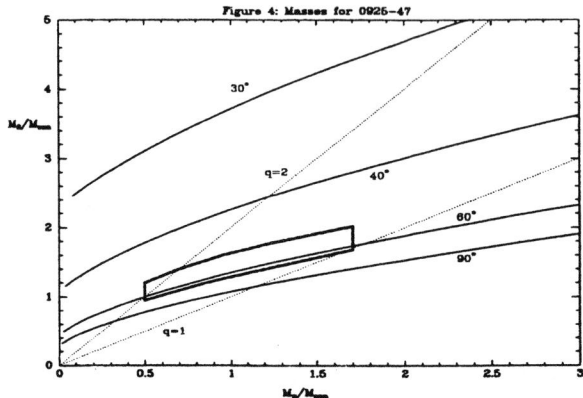

FIGURE 2. Mass diagram for 0925-47. Box shows most likely values.

erg/s, which is lower than the other SSS discussed by Cowley et al. To have the luminosity of the MC sources, the distance would have to be 10 Kpc or more.

The non-visibility of the donor star spectrum implies that the disk absolute magnitude $M_V \leq 0$. This is comparable with other supersoft sources with better-known distances and reddening (see Cowley et al. 1998). It also eliminates lower inclinations for the system since that would imply an even higher donor mass and hence a brighter magnitude for that star making it visible in the spectrum.

It is worth considering the implications if the He II velocities are not fully orbital. The phase shift and amplitude difference from the Hα velocities would be hard to explain if the orbital velocity and phasing were very different from what we have adopted. We also note that for the donor star to have a main-sequence radius, the orbital velocity would have to be much lower - about 25 km/s or less.

4. MODEL

Figure 3 shows a possible model for the system. Motch has discussed the simplest explanation of the light and velocity phasing as heating of the primary, as seen classically in Her X-1. In 0925-47, however, the phase and velocity differences between lines are not consistent with this, so we propose an alternative model. We have adopted the He II velocity curve as showing the motion of the compact star. We assume that the light curve is modulated by a thickening of the disk by cooler (dark) material where the mass-transfer stream impacts it. Such a model has been described by Meyer-Hofmeister et al (1997).

This model is consistent with the data, as follows.

1. The radial velocity curves for Hα and He II have the same V_0, indicating that they are predominantly orbital: there is no outflow, absorption, or asymmetry that would change the fitted V_0 values.

2. The differences in the semiamplitude K and phasing may be explained if there

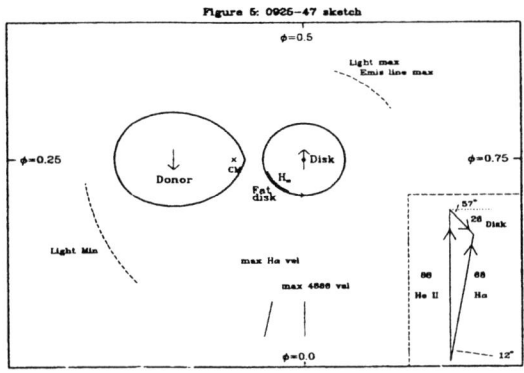

FIGURE 3. Orbital plane sketch of model for 0925-47

is additional Hα seen when the disk thickening is on the far side, and this emission has some velocity appropriate to the outer disk where it arises. There is little of this extra emission seen in He II because it is predominantly formed in the inner (hotter) part of the disk. There does not appear to be a hot-spot on the disk, as in some systems. Instead we propose a cool spot perhaps due to build-up of density and low streaming velocities.

3. The difference between the He II and Hα velocities amounts to a streaming motion of 24 km/s at an angle of 24° below the line of centers of the binary.

4. Both Hα and He II 4686Å show maximum scatter near light maximum, when the inside of the proposed thick disk is seen. There is also considerable scatter in the photometry at this time as well as an asymmetrical Hα profile. We suggest this is caused by the presence of an additional weaker component from the disk edge contribution. This asymmetry is not seen in the He II profiles.

5. Although we have not observed all phases, the equivalent widths appear to vary with the light curve. This implies that the line emission is actually modulated more strongly than the light curve, but with roughly the same phasing. The reasons the He II line shows more variation in equivalent width than Hα, could be because of the obscuration of the central disk by the thick disk edge. Hα would suffer less from this as it is formed in the outer cooler parts of the disk which are less obscured. However, the extra emission inside the thick disk edge would enhance the Hα variation.

REFERENCES

Cowley, A.P., Schmidtke, P.C., Crampton, D., & Hutchings, J.B. 1998, ApJ, 504, 854
Meyer-Hofmeister, E., Schandl, S., & Meyer, F. 1997, A&A, 321, 245
Motch, C. 1998, A&A (astro-ph 9807226)
Motch, C. 1996, Supersoft X-ray Sources (Springer: New York) ed. J. Greiner, 472

X-RAY EMISSION FROM SUPERNOVAE

S. Immler[1], W. Pietsch[1]

1) Max-Planck-Institut für extraterrestrische Physik, Postfach 1603, D–85740 Garching, Germany

ABSTRACT ROSAT observations of young supernovae (SNe) in nearby galaxies are presented. The observations led to the detection of X-ray emission from the type II SNe 1979C in M100, 1951H and 1970G in M101. In contrast to radio observations, subsequent monitoring with ROSAT shows a steep decline in the X-ray count rate for SN 1979C over years. We also found evidence for the first detection of X-ray emission from a type I SN (SN 1994I in M51). Mass-loss rates of the progenitors and densities of the circumstellar matter are derived and discussed in the context of the Chevalier 'mini-shell' model.

KEYWORDS: supernovae: individual: SN 1994I, SN 1951H, SN 1970G, SN 1979C; X-rays: general

1. SN 1979C IN M100

In a 43 ks ROSAT HRI X-ray observation of the spiral galaxy NGC 4321 (M100) X-ray emission from the supernova SN 1979C was discovered, sixteen years after its outburst, with an (0.1–2.4 keV) X-ray luminosity $L_x = 1.2 \times 10^{39}$ erg s^{-1} (Immler, Pietsch & Aschenbach 1998a; cf. Fig. 1). For three earlier *Einstein* observations, taken on days 64, 239 and 454 after the outburst, we derived 3σ upper limits of 1.8×10^{40} erg s^{-1}, 7.6×10^{39} erg s^{-1} and 6.9×10^{39} erg s^{-1}, respectively.

Applying the Chevalier 'mini-shell' model to the luminosity of SN 1979C observed 16 years after the outburst (cf. Chevalier 1984; Chevalier & Fransson 1994), we find a mass-loss rate of the progenitor of $\dot{M} = 1 \times 10^{-4} M_\odot$ yr^{-1}. The same mass-loss rate was derived from radio observations of this SN (Weiler et al. 1993). SN 1979C was re-observed with the ROSAT HRI in January 1998 (25 ks) and with the PSPC detector during the ROSAT 'last light' campaign in December 1998 (10 ks), and soft X-rays (0.5 keV) were recorded (Immler 2000). Whereas the two HRI observations on days 5 900 and 6 800 after the outburst give a best fit X-ray rate of decline of $L \propto t^{-4}$, the rate of decline including the PSPC observation (day 7 100) is $t^{-7.5}$. The results are significantly different from the slow rate of decline ($t^{-0.7}$) observed in the radio regime over a period of more than 10 years (Weiler at al. 1991).

It has been proposed that X-rays from type II SNe are created by the interaction of gas ejected by the supernova with circumstellar medium, which is likely to be the wind blown material by the progenitor star prior to the explosion (cf., Chevalier

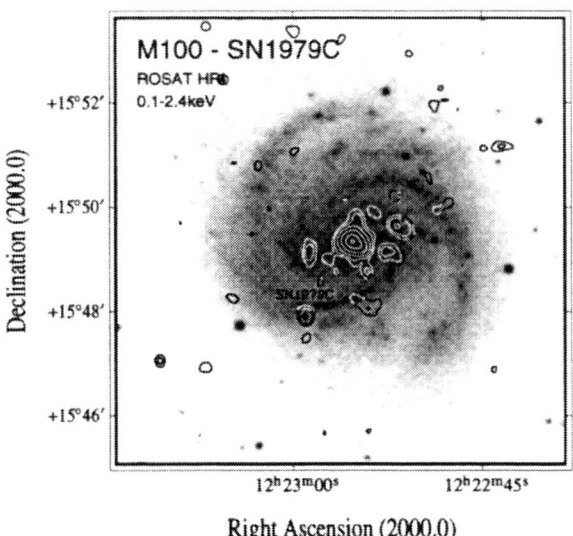

FIGURE 1. ROSAT HRI band X-ray contours (12″ FWHM) of M100 and SN 1979C, superposed on an digitized POSS plate. Contour levels are $3, 5, 10, 20$ and 40σ above the mean background rate (4.2×10^{-3} cts s^{-1} arcmin^{-2}, $1\sigma \hat{=} 1.2 \times 10^{-3}$ cts s^{-1} arcmin^{-2}).

1984; Chevalier & Fransson 1994). In this 'mini-shell' model both radio and X-ray luminosity depend on the shocked matter density, which is proportional to the mass loss rate $\dot{M}$ of the progenitor star divided by the wind velocity v_w, assuming a constant v_w. The supernova shock front radius increases with time as t^m with $m = (n-3)/(n-2)$ (Chevalier & Fransson 1994). The change in radius with time is reflected in the radio and X-ray light curves, and Weiler et al. (1991) have used the 10 year radio light curve of SN 1979C to determine $\dot{M}/v_w = 12$, with $\dot{M}$ in units of 10^{-5} $M_\odot$ yr^{-1}, and v_w, measured in 10 km/s. The index m has been determined to be very close to unity, which means that n is large, and the best fit to the radio light curve by Weiler et al. gives n = 79 with a lower limit of n > 22. We have used these values to determine the X-ray luminosity expected from the circumstellar interaction model, and it turns out that the ROSAT measurements give a luminosity at least a factor of 30 lower than the model predicts.

2. SN 1994I IN M51

X-ray observations of the galaxy NGC 5194 (M51) with the HRI onboard ROSAT were analysed to search for X-ray emission from the type Ic SN 1994I (Immler, Pietsch & Aschenbach 1998b).

Since SN 1994I is only $\sim 18''$ offset from the nucleus of NGC 5194, the contri-

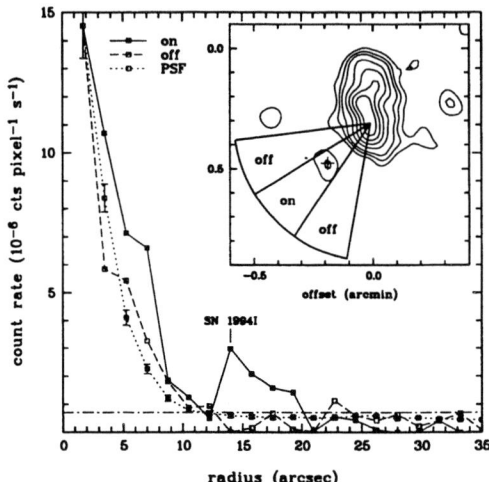

FIGURE 2. Radial surface brightness profiles of the X-ray emission for three sectors in the 6.4 ks ROSAT HRI observation, centred on the nucleus of M51. Contours of the inset image (4″.7 FWHM) are 1,1.5,2,3,4,5 and 6 in units of 1 count per detection cell (5.6×10^{-1} cts s^{-1} arcmin^{-2}). The position of the nucleus of M51 is at the origin of the sectors, the radio position of the SN 1994I is indicated by a cross.

bution of emission from the extended bulge region to the emission of the source at the position of SN 1994I was estimated by constructing surface brightness profiles, centred on the X-ray peak emission of the galaxy. Three sectors were extracted, with radially binning of 1″.75 (cf. Fig. 2).

An X-ray source with a (0.1–2.4 keV) luminosity of 1.6×10^{38} erg s^{-1} is found at the position of SN 1994I, 79–85 days after the explosion. We believe this to be strong evidence for the first detection of X-ray emission from a type I supernova. Assuming the emission arises from shocked circumstellar gas, deposited by the progenitor through non-conservative mass-transfer to a companion, we estimate a constant gas density of $\rho = 2 \times 10^5$ cm^{-3} $v_{16\,500}^{2/3}$ and a total mass of X-ray luminous gas of $M = 1 \times 10^{-3} M_\odot$ inside a sphere of radius 1.2×10^{16} cm. If the emission arises from the shocked stellar wind of the progenitor, heated by the outgoing wave, we derive a mass-loss rate prior to the outburst of $\dot{M} = 3.6 \times 10^{-6}$ $M_\odot$ yr^{-1}. This mass-loss rate is similar to that of other type I SNe, inferred from radio observations. (cf., e.g. SN 1983N, SN 1984L: $2 \times 10^{-6} M_\odot$ yr^{-1}, SN 1990B: $3 \times 10^{-6} M_\odot$ yr^{-1}, Weiler et al. 1993), whereas the mass-loss rate of type II SNe is more than an order of magnitude higher (cf., e.g. SN 1979C: $1 \times 10^{-4} M_\odot$ yr^{-1}, SN 1980K: $3 \times 10^{-5} M_\odot$ yr^{-1}, SN 1996J: $2 \times 10^{-4} M_\odot$ yr^{-1}, Weiler et al. 1993; SN 1988Z: $7 \times 10^{-5} M_\odot$ yr^{-1}, van Dyk et al. 1993).

3. SN 1951H AND SN 1970G IN M101

We have searched for X-ray emission from the type II supernovae SN 1951H and SN 1970G, located in giant H II regions within the face-on spiral galaxy M101 (NGC 5457) using ultra-deep ROSAT HRI (229 ks) and PSPC (34 ks) observations (Immler 2000; Wang, Immler & Pietsch 1999). At the SNe positions, X-ray sources are both detected with the ROSAT HRI and PSPC.

Using the HRI luminosity observed for SN 1970G on day 9 300 after the outburst ($L_x = 4.5 \times 10^{37}$ erg s^{-1} for a 0.5 keV thermal bremsstrahlung spectrum), we derive a pre-outburst mass-loss rate of $\dot{M} = 2.0 \times 10^{-5}$ $M_\odot$ yr^{-1}. This is the same mass-loss rate as derived in the radio regime (Weiler 1993). The hardness-ratios from the PSPC observation (HR1 = 1.73 ± 0.87, HR2 = -0.23 ± 0.24; Wang, Immler & Pietsch 1999) indicate that the emission is due to highly-absorbed, relatively low-temperature plasma located within the extended H II emission complex NGC 5455.

The mass-loss rate of the progenitor of SN 1951H inferred from the HRI luminosity ($L_x = 2.0 \times 10^{38}$ erg s^{-1}) on day 16 400 after the outburst is $\dot{M} = 5.5 \times 10^{-5}$ $M_\odot$ yr^{-1}, rather typical for type II SNe. The PSPC spectrum of the X-ray source shows that the emission is very soft (best fit temperature 0.5 keV for a thermal bremsstrahlung spectrum and relatively unabsorbed ($N_H = 4 \times 10^{20}$ cm^{-2}), possibly contaminated with emission from hot gas associated with the extended H II complex NGC 5462 and the H I hole coinciding with this position.

4. CONCLUSIONS

The observations presented in this paper demonstrate that X-ray observations are an important diagnostical tool to derive physical parameters (e.g. temperature, gas density and mass) of both the SN ejecta and the circumstellar medium, deposited by the progenitor at the late stage of stellar evolution before explosion.

REFERENCES

van Dyk, S.D., Weiler, K.W., Sramek, R.A., et al., 1993, ApJ 419, L69

Chevalier, R.A., 1984, ApJ 285, L63

Chevalier, R.A., Fransson, C., 1994, ApJ 420, 268

Immler, S., Pietsch, W., Aschenbach, B., 1998a, A&A 331, 601

Immler, S., Pietsch, W., Aschenbach, B., 1998b, A&A 336, L1

Immler, S., 2000, PhD Thesis, University of Munich

Wang, Q.D., Immler, S., Pietsch, W., 1999, ApJ 523, 121

Weiler, K.W., van Dyk, S.D., Discenna, J.L., et al., 1991, ApJ 380, 161

Weiler, K.W., van Dyk, S.D., Panagia, N., et al., 1993, Radio Supernovae & Massive Stellar Winds. In: Massive Stars: Their Lives in the Interstellar Medium, ASP Conf. Series, vol. 35, 436

DETECTION OF SUPERNOVA REMNANT AND BLACK HOLE CANDIDATES IN M83 WITH ROSAT

S. Immler[1], M. Ehle[1], W. Pietsch[1], A. Vogler[2]

1) Max-Planck-Institut für extraterrestrische Physik, Postfach 1603, D–85740 Garching, Germany
2) CEA/Saclay, DAPNIA, Service d'Astrophysique, L'Orme des Merisiers, Bât. 709, F–91191 Gif-sur-Yvette, France

ABSTRACT   Within the D_{25} ellipse of M83, 21 X-ray sources are detected with the ROSAT HRI. A variable super-Eddington (3.8×10^{39} erg s^{-1}) X-ray source is found to coincide with a faint, extended optical counterpart. Based on the multiwavelength characteristics, the source most likely represents a massive (~ 30 M$_\odot$) accreting black hole binary, located in a compact HII region or in a globular cluster in M83. We also detect two luminous (3.7 and 6.7×10^{38} erg s^{-1}) previously unknown supernova remnant candidates, located in extended Hα emission complexes and coinciding with compact 6 cm and 20 cm radio sources.

KEYWORDS: galaxies: individual: NGC 5236 (M83); X-rays: general; X-rays: galaxies

1. GENERAL X-RAY MORPHOLOGY OF M83

Based on high-resolution ROSAT HRI (48 ks) and PSPC (25 ks) observations, spatial, spectral and timing characteristics of X-ray sources in the M83 field were studied. The main results and conclusions are as follows (Immler et al. 1999):

Significant X-ray emission is detected from M83 (distance 8.9 Mpc) over an approximately circular region of radius ~ 7 kpc from the galaxy's nucleus ($\sim 60\%$ of the D_{25} ellipse of M83; cf. Fig. 1). The total luminosity of the galaxy in the (0.1–2.4 keV) ROSAT band is $L_x = 4.1 \times 10^{40}$ erg s^{-1}.

Within the HRI field-of-view centred on M83, 37 X-ray sources are detected with a likelihood of > 8 (3.6σ significance). Eight sources outside the galaxy are identified with interlopers (AGN, stars, etc.). 21 sources are assumed to be associated with M83, with luminosities in the range $(1.3 - 106) \times 10^{38}$ erg s^{-1}. Timing analysis of the HRI observations and comparisons with previous ROSAT PSPC (Ehle, Pietsch & Beck 1995) and Einstein HRI and IPC observations (Fabbiano, Trinchieri & Macdonald 1984) show that only the X-ray source at the nucleus of M83 and a superluminous X-ray source at the edge of the south-eastern spiral arm are variable.

The detected point sources account for 50% (2.0×10^{40} erg s^{-1}) of the total X-ray emission observed inside the D_{25} ellipse of M83. A substantial fraction of the residual emission likely arises in the hot ISM at a characteristic temperature

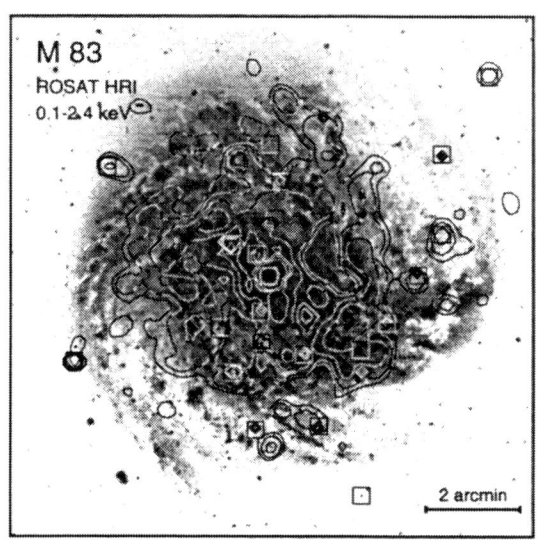

FIGURE 1. Contour map of the ROSAT HRI image of M83 (0.1–2.4 keV), smoothed with an adaptive filter (FWHM $\leq 170''$) and superposed on an AAO optical image.

of some 10^6 K. Within the bulge region (< 1 kpc radius from the centre of the galaxy), and similarly to the total galactic emission, half of the extended emission (1.0×10^{40} erg s^{-1}) can be resolved into X-ray point sources.

Four sources inside the D_{25} ellipse of M83 have fluxes greater than $\sim 5 \times 10^{-14}$ erg cm^{-2} s^{-1} in the 0.1–2.4 keV band. If associated with M83, each source has a luminosity of $\geq 5 \times 10^{38}$ erg s^{-1}, clearly exceeding the Eddington limit for a $1.4 M_\odot$ accreting neutron star.

2. DETECTION OF A BLACK HOLE CANDIDATE

Apart from the nuclear source, H30 (cf. Fig. 2) is the brightest X-ray source in M83 (3.8×10^{39} erg s^{-1}), clearly exceeding the Eddington limit for an $\sim 1.4 M_\odot$ accreting neutron star. Timing analysis shows that H30 is variable during the HRI period of observation by a factor ≥ 2. The source coincides with a faint, extended optical object found on deep AAO images (offset $1\rlap.''6$). The optical counterpart has approximately the same magnitude in all bands ($\sim 19^m$), leading to the assumption that the optical emission is due to a compact HII region or a globular cluster at the fringe of the south-eastern spiral arm. The object cannot be a foreground star, given the optical extent and that the X-ray-to-visual flux ratio $\log(f_x/f_v)$ is $+0.58$. The X-ray-to-visual flux ratio is also different from galaxies, which cover the range from -0.5 to -2.0 (Maccacaro et al. 1988). AGN essentially all have $B - R < 1$ (blue objects), while the counterpart to H30 shows no difference in colour. Also, the morphology is rather untypical for AGN, which usually have bright and compact

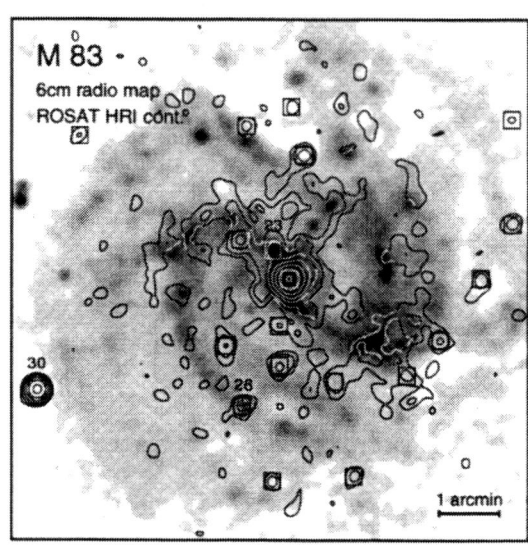

FIGURE 2. Overlay of ROSAT HRI contours (12″ FWHM) on a 6 cm radio continuum map of M83. X-ray contour levels are $3, 5, 10, 20, 40, 80$ and 160σ above the mean background rate (7.5×10^{-3} cts s^{-1} arcmin^{-2}). HRI sources are enclosed by boxes, SNRs (H23, H26) and black hole (H30) candidates are marked.

nuclei. The chance probability of a background AGN at the given X-ray flux of H30 (4.1×10^{-13} erg cm^{-2} s^{-1}) within the $34' \times 34'$ HRI field is $\sim 10^{-3}$ (using results from the ROSAT Medium Sensitivity Survey; Hasinger, Schmidt & Trümper 1991). The two spectral indices 'optical-to-X-ray' and 'radio-to-optical' are $\alpha_{OX} = 1$ and $\alpha_{RO} < 0$. All radio-loud objects (BL Lac's, AGN, quasars) can be ruled out based on their large α_{RO} values of > 0.3 and the absence of radio emission for H30. Also, 'normal' galaxies and ellipticals essentially have α_{OX} values exceeding 1.5. Even the region in the α_{RO}–α_{OX} diagram covered by radio-quiet AGN and Seyferts ($\alpha_{OX} > 1.2$) is different from the spectral indices derived for H30. Spectral models fitted to the PSPC data give reasonably good results with χ^2/ν values of the order of ~ 1. For a thermal bremsstrahlung fit, a temperature of (2.30 ± 1.30) keV, fixing the total absorbing column to the best fit value of 5.46×10^{20} cm^{-2}, is obtained.

From the above discussion, we conclude that H30 is one of the best candidates for a mass accreting black hole found in a nearby galaxy, with a mass of the compact object of $\sim 30 M_\odot$.

3. DETECTION OF TWO X-RAY LUMINOUS SNR CANDIDATES

The X-ray source H23 (8.1×10^{38} erg s^{-1}) shows no variability and coincides with a radio source and an H II region located $\sim 30''$ north-east of the nucleus (cf. Fig. 2). Comparison with radio maps of Cowan et al. (1985; 1994) shows that the radio source is detected both at 6 cm and at 20 cm. Spectroscopic observations of the H II region coinciding with H23 (source I in Dufour et al. 1980), show an emission line ratio of [S II]/Hβ = 0.21. If this is scaled to the emission line intensity ratio $I(H\alpha)/I(H\beta)$ = 3.03, as assumed by Dufour et al., a high emission line ratio of [S II]/Hα = 0.63 is derived. The detection of radio emission from this region further strengthens the assumption that the X-ray emission is due to a single luminous SNR or a collection thereof.

H26 also coincides with a giant H II region (source 120 in Rumstay & Kaufman 1983; $\log S_\alpha$ = −12.149). However, the extent of the H II complex clearly exceeds that of the point-like X-ray source. The observed anti-coincidence of Hα and Hβ with H I emission (Tilanus & Allen 1993) supports the assumption of an X-ray luminous SNR, since the hydrogen is expected to be ionized at the position of SNRs.

The radio counterpart of H26 is detected both at 6 cm and 20 cm (source 8 in Cowan, Roberts & Branch 1994; $S_{6cm,20cm}$ ∼ 0.7 mJy). Studies of the eastern spiral arm of M83 in the optical and radio regime (Deutsch & Allen 1993) have demonstrated that the emission in the neighbourhood of giant H II regions is predominantly non-thermal and mainly due to enhanced production of high-energy radiation (hard X-rays and cosmic rays) from historical supernovae. The observation of a bright CO emission complex coinciding with that region, excited by the enhanced production of cosmic rays from a SNR, points also towards the existence of an X-ray luminous SNR (Wiklind et al. 1990; Rand, Steven & Higdon 1999).

ACKNOWLEDGEMENTS

We are indebted to R.J. Allen (STScI), R. Beck (MPIfR) and S. Sukumar (UC Berkeley) for providing the 6 cm radio continuum map of M83 prior to publication and D. Malin (AAO) for the optical image.

REFERENCES

Cowan, J.J., Branch, D., 1985, ApJ 293, 400
Cowan, J.J., Robert, D.A., Branch, D., 1994, ApJ 434, 128
Deutsch, E.W., Allen, R.J., 1993, AJ 106, 1812
Dufour, R.J., Talbot, R.J. Jr., Jensen, E.B., et al., 1980, ApJ 236, 119
Ehle, M., Pietsch, W., Beck, R., 1995, A&A 295, 289
Fabbiano, G., Trinchieri, G., Macdonald, A., 1984, ApJ 284, 64
Hasinger, G., Schmidt, M., Trümper, J., 1991, A&A 246, L2
Immler, S., Vogler, A., Ehle, M., Pietsch, W., 1999, A&A 352, 415
Maccacaro, T., Gioia, I.M., Wolter, A., et al., 1996, ApJ 464, 829
Rand, R.J., Steven, D.L., Higdon, J.L. 1999, ApJ 513, 720
Rumstay, K.S., Kaufman, M., 1983, ApJ 274, 611
Sukumar, S., Allen, R.J., Beck, R., et al., 1999, in prep.
Tilanus, R.P.J., Allen, R.J., 1993, A&A 274, 707
Wiklind, T., Rydbeck, G., Hjalmarson, A., et al., 1990, A&A 232, L11

A SYSTEMATIC SEARCH FOR NEW X-RAY PULSATORS IN PUBLIC ROSAT HRI AND BEPPOSAX SMC FIELDS

G.L. Israel[1], S. Campana[2], S. Covino[2], D. Dal Fiume[3], D. Lazzati[2], T. Oosterbroek[4], M. Orlandini[3], M.R. Panzera[2], A.N. Parmar[4], D. Ricci[5], G. Tagliaferri[2] and L. Stella[1]

1) Osservatorio Astronomico di Roma, Monteporzio Catone, Italy, 2) Osservatorio Astronomico di Brera, Merate, Italy, 3) TeSRE–C.N.R., Bologna, Italy, 4) ESA Astrophysics Division, SSD–ESTEC, Noordwijk, The Netherlands, 5) SAX SDC, ASI, Roma, Italy

ABSTRACT We present the first results obtained from a systematic search of periodicities carried out over two samples of X-ray sources: (i) those detected in the public ROSAT HRI fields, and (ii) a selected number of candidate High Mass X-ray Binaries in the Small Magellanic Cloud observed by BeppoSAX.

KEYWORDS: binaries: general — pulsars: individual: RX J0806.3+1527, 1E 0101.5−7225 — novae, cataclysmic variables — pulsar: general — star: rotation — X-rays: stars

1. INTRODUCTION

Several classes of X-ray sources are either known or expected to show coherent periodicities in their fluxes: these include X-ray binaries, cataclysmic variables, isolated neutron stars accreting from the interstellar medium or molecular cloud, and rotation powered and/or cooling neutron stars.

Based on the results already obtained during a timing survey of the ROSAT Position Sensitive Proportional Counter (PSPC; 0.1–2.4 keV; see Israel et al. 1998a and reference therein) data, we performed a similar analysis on the public fields of the ROSAT High Resolution Imager (HRI). Sources were detected by means of a wavelet-based detection technique, the Brera Multiscale Wavelet (BMW) algorithm (Lazzati et al. 1999; Campana et al. 1999). From the analysis of the ROSAT HRI dataset a catalog of ~ 20000 sources with significance $\geq 4.5\sigma$ has been obtained (the BMW-HRI catalog; Panzera et al. 2000), 4000 of which have more than ~ 200 photons, which we set as the minimum number required to carry out a meaningful search for periodic signals (see Israel et al. 1998a). These light curves were analysed in a systematic way by using a power spectrum-based algorithm described in Israel & Stella (1996).

BeppoSAX carried out 10 pointings in the direction of the SMC between 1998–1999. We monitored a number of candidate High Mass X-ray Binaries selected by means of long-term variability between Einstein and ROSAT observations and/or whithin the ROSAT pointings. Also in this case we analysed the light curves of the detected sources in a systematic way.

2. 1BMW J080622.8+152732 = RX J0806.3+1527

This source was detected at a level of 0.082 ± 0.004 and 0.052 ± 0.003 cts s^{-1} in 1994 Oct. 23 and 1995 Apr. 15, respectively. In the longest observation (1995) the source was detected at R.A.=$08^h06^m22^s.84$ and Dec.=$+15°27'31''.5$ (equinox 2000; 10" uncertainty radius).

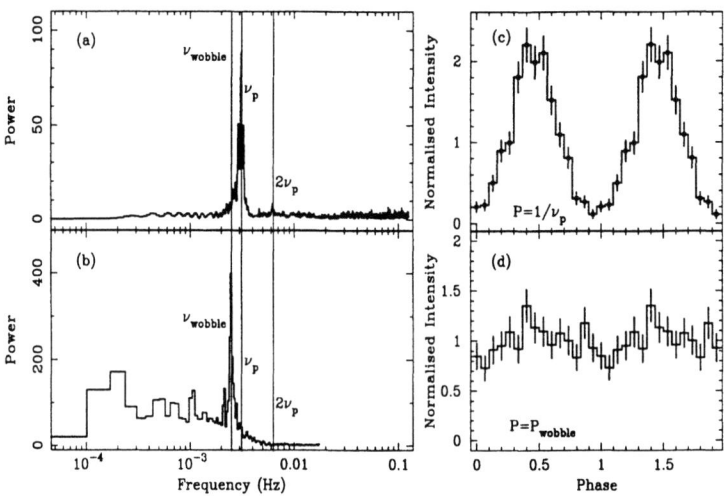

FIGURE 1. The average power spectra obtained for (a) the RX J0806.3+1527, and (b) the pointing R.A. housekeeping data. The frequency of the wobble (ν_{wobble}) and the 321 s signal (ν_p) are marked. The RX J0806.3+1527 light curves folded at the best period and at the wobble period are also shown (panel c and d).

The photon arrival times were corrected to the barycentre of the solar system and background subtracted light curves accumulated in 1 s bins. An average power spectrum was calculated for the two observations in order to achieve a better statistics. A highly significant peak ($\sim 9\sigma$) was found at a frequency of 0.0031 Hz. We derived an accurate period measurement by means of a Rayleigh periodogram which gives a best period of of 321.25 ± 0.25 s ($|\dot{P}| < 2 \times 10^{-8}$ s s^{-1}; see Fig. 1). The modulation is fairly sinusoidal, with a pulse fraction (semiamplitude of modulation divided the mean source count rate) of $\sim 90\%$.

RX J0806.3+1527 lies away from the Galactic plane at a Galactic latitude of $b_{II}=23°.96$ (and $l_{II}=207°.37$). Correspondingly we expect that the source is unlikely more distant than ~ 750 pc. The most likely interpretation is that RX J0806.3+1527 is a cataclysmic variable. However we can not exclude that this source is an isolated neutron star even if more unlikely (Israel et al. 1999). Further observations at X-ray and optical wavelengths are needed to firmly assess the nature of this new X-ray pulsator.

3. SAX J0103.2–7209 = 2E 0101.5–7225

The Small Magellanic Cloud field including the position of 2E 0101.5–7225 was observed by BeppoSAX during 1998 July 27–28. A bright X-ray source at a position of R.A. = $01^h03^m13^s$, Dec. = $-72°09'16''$ (equinox 2000; 30'' of uncertainty radius, 90% confidence level) was found in the MECS at a count rate of $\sim 3.7 \times 10^{-2}$ cts s^{-1} (1–10 keV).

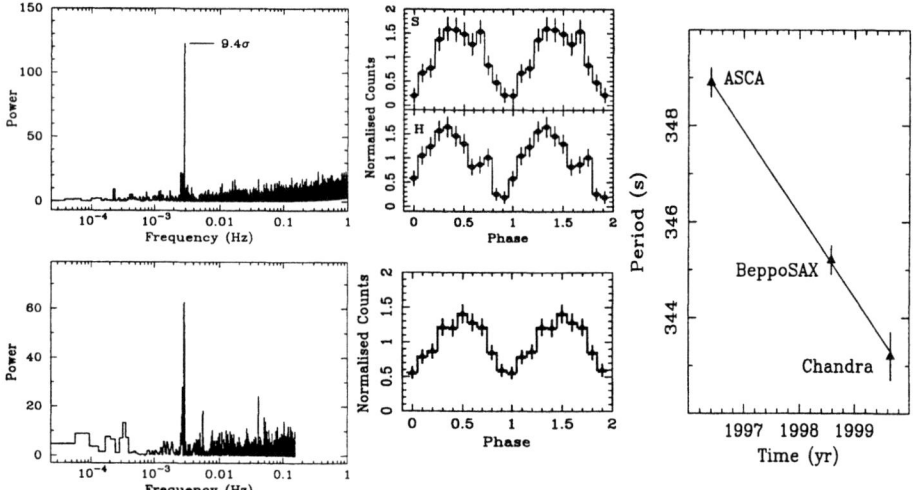

FIGURE 2. 1–10 keV BeppoSAX (upper left panel) and Chandra (lower left panel) power spectra of SAX J0103.2–7209. The corresponding folded light curves at the best periods are also shown in the central panels. The period history of SAX J0103.2–7209 is shown on the right.

The photon arrival times were corrected to the barycentre of the solar system and background subtracted light curves accumulated in 0.5 s bins. A single power spectrum was calculated over the whole observation. A highly significant peak ($\sim 9.4\sigma$ based on the fundamental only; see upper left panel of Fig. 2) was found at a frequency of 0.0029 Hz. A better evaluation of the period was determined by a phase fitting technique which gives a best period of 345.2 ± 0.3 s (90% confidence level). The pulse shape profile is nearly sinusoidal, while the pulsed fraction is $\sim 45\%$ (Fig. 2; central panels). The BeppoSAX X-ray spectrum is well fit by an absorbed power-law model with photon index ~ 1.0. The unabsorbed luminosity in the 2–10 keV energy range is $\sim 1.2 \times 10^{36}$ erg s^{-1} (for a distance of 62 kpc).

The position of SAX J0103.2–7209 was found to be consistent with that of the

Einstein source 2E 0101.5−7225 and the ROSAT source RX J0103.2−7209. The latter source was detected at a nearly constant flux level in all the Einstein, ROSAT and ASCA pointings which surveyed the SMC (see Hughes & Smith 1994). After the discovery of 345 s pulsations in SAX J0103.2−7209 (Israel et al. 1998b), Yokogawa & Koyama (1998) found a signal at a period of 348.9 s in a 1996 ASCA dataset.

The field including the position of SAX J0103.2−7209 was also observed by the Chandra satellite on 1999 August 23. After extracting a 3.24 s binned light curve around the position of the source, we carried out a power spectrum and performed a search of significant peaks in a period interval around that detected by BeppoSAX. A peak was detected at a significance level of 6.4σ, corresponding to a period of 343.5 ± 0.5 s. This result implyies that the pulsar is continuing to spin–up at a constant rate of $-1.7\,{\rm s\,yr^{-1}}$ since 1996. The detection of the pulsations clearly associates the Chandra source with that of BeppoSAX and of ROSAT HRI. Moreover the Chandra coordinates (R.A. = $01^h03^m14\overset{s}{.}06$, Dec. = $-72°09'15\overset{''}{.}25$; equinox 2000; uncertainty radius of $\leq 8''$) differ of less than $2''$ from the proposed optical counterpart, a O9–B1 III–Ve spectral–type star (Hughes & Smith 1994), making the association very likely.

The nearly absence (within a factor of 5–10) of long–term variability over an interval of ~ 30 years strongly point towards a *persistent* X–ray pulsar, the second one in the Magellanic Clouds after the discovery of SMC X–1. Moreover the association of the companion with a Be spectral–type star makes 2E 0101.5−7225 the first example of persistent Be/X–ray binary system in the MCs. The relatively high value inferred for the magnetic field (4–12×10^{12} Gauss) and for the period derivative point towards a rather young object. Finally, the period derivative corresponds to a secular spin–up time–scale of 200 yr which is the shortest of any known X–ray Pulsar in a High Mass X–ray binary.

ACKNOWLEDGEMENTS

We thank the Chandra Science Data Center at CfA for a prompt release of the data.

REFERENCES

Campana S., et al. 1999, ApJ, 524, 423
Hughes, J.P. & Smith, R.C. 1994, PASP, 107 (4), 1363
Israel G.L., Stella L. 1996, ApJ, 468, 369
Israel G.L., et al. 1998a, in "The Many Faces of Neutron Stars" NATO–ASI Series, Vol. 515, Kluwer Academic Publishers, p. 411
Israel G.L., et al. 1998b, IAU Circ. 6999
Israel G.L., et al. 1999, A&A, 349, L1
Leahy D.A., Elsner R.F., Weisskoff, M.C. 1983, ApJ, 272, 256
Lazzati D., et al. 1999, ApJ, 524, 414
Panzera M.R. et al., 2000, in preparation
Yokogawa, J. & Koyama, K. 1998, IAU Circ. 7009

THE RELATIVISTIC ASTROPHYSICS EXPLORER: A NEW MISSION FOR X-RAY TIMING

P. Kaaret[1], J. Grindlay[1], F.K. Lamb[2], E.H. Morgan[3], J.H. Swank[4], W. Zhang[4]

1) Harvard-Smithsonian Center for Astrophysics
2) University of Illinois
3) Massachusetts Institute of Technology
4) NASA/Goddard Space Flight Center

ABSTRACT

The great success of the Rossi X-Ray Timing Explorer (RXTE) has given us a new probe to study strong gravitational fields and to measure the physical properties of black holes and neutron stars. Here, we describe a "next-generation" x-ray timing mission, the Relativistic Astrophysics Explorer (RAE), designed to fit within the envelope of a "medium-sized" mission. The main instruments will be a narrow-field x-ray detector array with an area of at least 60,000 cm^2 equal to ten times that of RXTE, and a wide-field x-ray monitor with good sensitivity and few arcminute position resolution. We describe the design of the instruments and the science which will be possible with a factor of ten increase in collecting area.

KEYWORDS: instrumentation: detectors, X-rays: general

1. SCIENTIFIC MOTIVATION FOR A NEW X-RAY TIMING MISSION

The Rossi X-Ray Timing Explorer (RXTE) has made substantial and unique contributions to the study of the behavior of matter in strong gravitational fields near accreting compact objects, the formation of relativistic jets, the emission mechanisms of active galactic nuclei, the evolution of neutron stars in binaries, the x-ray emission regions in cataclysmic variables, and many other aspects of high-energy astrophysics (for a review see Bradt 1999). The key feature of RXTE is a large effective area x-ray detector coupled with a high telemetry bandwidth. The prowess of RXTE for fast timing opened a new "discovery space" in rapid x-ray variability, allowing timing studies at the dynamical time scales of the innermost orbits around stellar mass compact objects, and leading to the discovery of millisecond quasiperiodic oscillations from accreting neutron stars and black holes. The large x-ray detector area also made possible many other advances such as the discovery of coherent millisecond pulsations from an accreting neutron star, and the study of rapid spectral variations, such as the ~ 200 s cycles in the microquasar GRS 1915+105 related to ejection of the inner regions of the accretion disk.

The great success of the Rossi X-Ray Timing Explorer (RXTE) is a strong indication that further progress in x-ray timing will lead to new scientific advances. Here, we describe a next generation x-ray timing mission which would offer an order of magnitude increase in x-ray timing capabilities via an x-ray detector with a geometric area of at least 60,000 cm^2, equal to ten times that of RXTE. The most important advances made with this order of magnitude increase in collecting area are likely to be true discoveries and thus cannot be anticipated. However, an order of magnitude increase in area would benefit many scientific investigations. Here, we describe three particular examples.

Fast quasiperiodic oscillations from black hole candidates (BHCs) have been discovered in three systems with frequencies of 67-300 Hz (Remillard et al. 1999). The fast QPOs from BHCs are rather weak (rms amplitudes near 1%) and difficult to study in detail with RXTE. A number of models of the QPOs have been proposed, all of which involve strong-field general relativistic effects, but distinguishing amongst the various models will be difficult with the RXTE data. The increase in the photon statistics with RAE would make possible much more accurate measurements of the QPO parameters and their variations with time or correlations with spectral or other timing parameters. This may lead to a unique identification of the QPO generation mechanism. Understanding these QPOs would provide a unique probe of strong-field gravity.

Millisecond oscillations in x-ray bursts have been discovered from a number of neutron stars. The oscillations have periods in the range 1.7-3 ms and are interpreted as due to inhomogeneous nuclear burning of matter initially located on the neutron star surface. The burst oscillations provide a means to constrain the neutron star mass-radius relation. Currently, the best constraint comes from a deep modulation (75% ± 17%) seen in the initial 62.5 ms of one burst (Strohmayer et al. 1998). RAE would detect roughly 1000 counts in each oscillation cycle near the peak of a typical bright burst. This would permit detailed examination of individual oscillation cycles and allow accurate measurement of the modulation amplitude in the first few oscillation cycles. Both our understanding of the burst oscillations and constraints on the neutron star mass-radius relation would improve.

Eclipse mapping of the accreting magnetic white dwarf XY Arietis showed that the x-ray flux emerges from eclipse egress in < 2 s (Hellier 1997). For the previous 15 years, the fraction, f, of the white dwarf surface involved in x-ray emission had been debated with values ranging from 0.001 to 0.3. Hellier's result, obtained by combining 20 RXTE observations, shows that $f < 0.002$. Using RAE, an accurate estimate could be made of the emitting region location on each egress which would allow direct mapping of movement of the emitting region. Similar mapping can also be done in neutron star and black hole binaries. The best constraints currently available on the size of the x-ray emitting regions in black hole systems come from x-ray dips (e.g. Tomsick et al. 1997). RAE would lead to significant advances in mapping x-ray emission from many different x-ray sources.

2. MISSION OVERVIEW

The Relativistic Astrophysics Explorer (RAE) will consist of two scientific instruments: a large area x-ray detector and a wide-field x-ray monitor. RAE will be designed to have telemetry sufficient to transmit the large event rate and flexible operations with multiple repointings each day to permit study of transient sources and rare states of known sources.

The goal for the large area x-ray detector is to provide an order of magnitude increase in x-ray timing capabilities relative to RXTE. The design goals are: a useful detector area of at least 6 m^2, sensitivity from 2 keV to 30 keV, absolute timing better than 10 μs, minimal dead time effects for sources 10 times as bright as the Crab nebula, an energy resolution of 1.2 keV (preferably 300 eV) at 6 keV, no imaging, and a field of view of 1° or smaller.

All-sky x-ray monitoring is needed for several reasons. First, the x-ray monitor provides continual long-term light curves. As many x-ray sources are highly variable, knowledge of the long term behavior in important in understanding the physical nature of the sources and in placing pointed observations in the context of the source state. Second, an x-ray monitor provides a means to trigger pointed observations when a selected source reaches a state of particular interest. Finally, an x-ray monitor allows discovery of new sources or new, unpredicted, outbursts of known sources. Many of the sources of interest are transients with unknown or irregular recurrence intervals. An x-ray monitor is essential to detect transient events. The design goal for the x-ray monitor is a sensitivity of several mCrab for daily observations, sufficient to monitor a large sample of AGNs ($\sim$ 40) on a daily basis.

3. LARGE AREA X-RAY DETECTOR ARRAY

An effective area of 6 m^2 will require a total geometric detector area near 10 m^2. A detector with a cross sectional area of 10 m^2 and a thickness of 0.75 m^2 fits within the 3 m diameter fairing of a two-stage Delta II. Thus, a 6 m^2 detector can be accommodated in a "medium-sized" mission without a deployment mechanism.

The x-ray detector must have low mass per unit effective area, reasonable cost, highly reliable and stable operation, efficient rejection of particle backgrounds, and good energy resolution. After extensive review of the available detector technologies, we have selected silicon detectors as the most promising candidate for large format x-ray astronomy detectors. A 2 mm thick silicon detector provides 40% efficiency up to 30 keV at a mass of 0.5 gm/cm^2; this compares favorably to PCA on XTE at a mass of 90 gm/cm^2. Silicon is widely used and can be obtained at low cost due to large economies of scale; 10 m^2 of commercially available silicon strip detectors (see below) can be procured for less than US$4M. Silicon has a low ionization potential which leads to good energy resolution and allows silicon detectors to be operated without internal amplification. While the lack of internal amplification mandates the use of low capacitance detectors and low noise electronics for good performance, it also eliminates the need for high voltage and facilitates reliable and stable operation.

Silicon detectors can be configured in different geometries including PIN diodes, silicon strip detectors, and silicon drift chambers. Silicon strip detectors (SSDs) are widely used in particle physics. SSDs offer one-dimensional imaging which may provide a means of effective discrimination against particle backgrounds. However, SSDs employ charge collection strips which run the length of a wafer and, thus, have relatively high capacitance which leads to high electronic noise and poor energy resolution. Given mission constraints, we estimate that the best energy resolution achievable with SSDs will be 1–2 keV at 6 keV (see Costa et al. these proceedings).

Silicon drift chambers (SDCs) (Gatti & Rehak 1984) have an internal electric field arranged so that electrons, produced by interaction of radiation within the silicon, drift toward a single charge collection point. SDCs have a great advantage in that the readout electrode can be made very small even if the detection area is large; thus, the readout capacitance is small, which leads to low readout noise and good energy resolution. Cylindrical SDCs, in which a single detector electrode is used to collect charge from a cylindrical drift region, are particularly well-optimized for x-ray spectroscopy (Rehak et al. 1985; Lechner et al. 1996). Excellent performance has been demonstrated from cylindrical SDCs in the laboratory (Gauthier et al. 1994; Fiorini et al. 1997) and in the field (Longoni et al. 1998). Operating at $-15°$ C, a resolution of 155 eV (FWHM) at the Mn-$K\alpha$ line has been obtained (Fiorini et al. 1997). The main question for x-ray astronomy is whether effective particle background rejection can be achieved. We currently are engaged in a technology development program to study application of SDCs to x-ray astronomy and to develop effective techniques for charged particle background rejection.

4. CONCLUSION

We have presented a conceptual design for a next generation x-ray timing mission and identified key technologies which must be developed. The outstanding results from the Rossi X-Ray Timing Explorer are strong motivation for a new mission with an order of magnitude increase in x-ray timing capabilities.

REFERENCES

Bradt, H.V. 1999, astro-ph/9901174
Fiorini, C. et al. 1997, Rev. Sci. Instrum. 68, 2461
Gatti, E. & Rehak, P. 1984, NIM, A255, 608
Gauthier, C. et al. 1994, NIM, A349, 258
Hellier, C. 1997, MNRAS, 291, 71
Lechner et al. 1996, NIM, A377, 346
Longoni, A. et al. 1998, NIM, A409, 407
Rehak, P. et al. 1996, NIM, A235, 224
Remillard et al. 1999, ApJ, 517, L127
Strohmayer, T.E. et al. 1998, ApJ, 498, L135
Tomsick, J.A., Lapshov, I., & Kaaret, P. 1998, ApJ, 494, 747

LUMINOUS SUPERSOFT X-RAY EMISSION FROM THE RECURRENT NOVA U SCORPII

P. Kahabka

Astronomical Institute, University of Amsterdam, Kruislaan 403, 1098 SJ Amsterdam, Netherlands

ABSTRACT BeppoSAX detected luminous 0.2–2.0 keV supersoft X-ray emission from the recurrent nova U Sco ~19-20 days after the peak of the optical outburst in February 1999. The spectral fit requires a very hot $\sim 9\,10^5$ K and luminous supersoft X-ray component and in addition a wind component. In addition applying the model of an irradiated accretion disk and subgiant donor star we can reproduce the UV and optical continua observed 12, 33, and 263 days after the peak of the previous (1979/80) outburst assuming that steady nuclear burning proceeded for ~3 weeks and the cooling time for the white dwarf envelope lasted another ~3 weeks.

KEYWORDS: accretion, accretion disks; binaries: close; stars: evolution; stars: mass-loss; novae, cataclysmic variables; X-rays: stars

1. SUPERSOFT X-RAY EMISSION DETECTED WITH BEPPOSAX

Recurrent novae are a small and diverse subclass of cataclysmic variables, which show multiple outbursts resembling those of classical novae, though of lesser magnitude (see Webbink et al. 1987; Sekiguchi 1995). U Scorpii (U Sco) is one of the best known members of this class. Recently (in February 1999) an optical outburst has been observed from this source.

During a target of opportunity observation performed with the BeppoSAX X-ray satellite ~20 days after the outburst U Sco has been detected in a supersoft X-ray phase (cf. Kahabka, Hartmann, Parmar & Negueruela 1999). This observation for the first time confirmed the theoretical predictions that recurrent novae have a supersoft X-ray phase (Yungelson et al. 1996; Kato 1996). Helium enhanced non-LTE white dwarf atmosphere model spectra, with a high N/C ratio (using abundances derived from the ejecta) were required to fit the BeppoSAX X-ray spectrum. This is considered to be evidence that the outburst of U Sco was triggered by a thermonuclear runaway and that the CNO cycle was active. From the temperature of the optically thick supersoft component of $\sim 9 \times 10^5$ K, the white dwarf is constrained to be very massive (> 1.2 M$_\odot$) and consistent with having a mass close to the Chandrasekhar limit.

Besides the supersoft emission an additional optically thin component was observed and explained as emission from a strongly shocked wind from the white

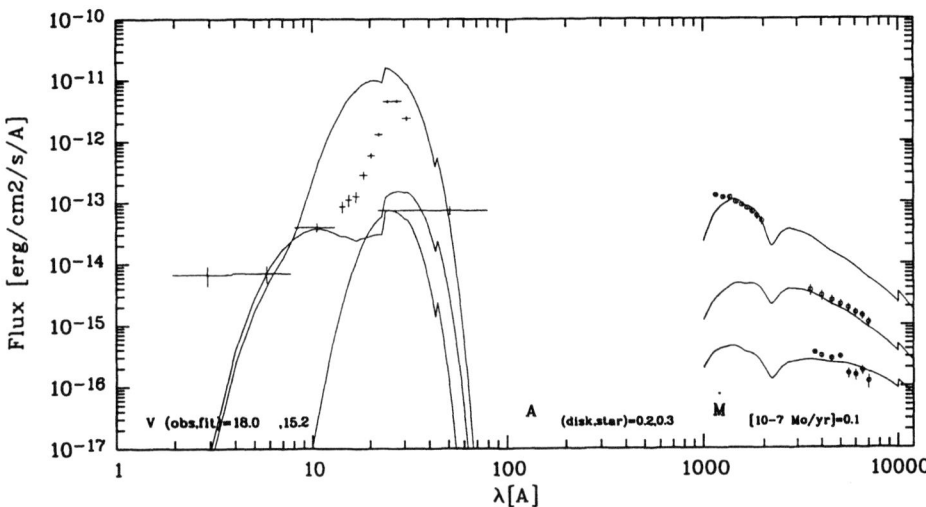

FIGURE 1. Calculated multiwavelength spectra of U Sco assuming irradiation of an accretion disk and a near Roche lobe filling subgiant star by the steady nuclear burning white dwarf and fitted to the UV and optical spectra observed on days 12, 33 and 263 after the peak of the 1979/80 outburst. The 1979/80 data are taken from Barlow et al. (1981) and Williams et al. (1981). To reproduce the optical and UV spectra a viscous disk with additional X-ray heating has been used. Also shown is the deconvolved BeppoSAX X-ray spectrum observed on days 19-20 after the 1999 outburst (Kahabka, Hartmann, Parmar & Negueruela, 1999).

dwarf with a mass loss rate of $\dot{M} = (2.4 - 6.9) \times 10^{-8}$ $M_\odot$ yr^{-1} (d/kpc). Such a component is also consistent with the theoretical predictions for the winds from a white dwarf with a mass just below the Chandrasekhar mass (Hachisu et al. 1999). According to the calculations of the latter authors U Sco emerged from an optically thick wind phase when the BeppoSAX observation was being performed.

2. CONSTRAINTS DERIVED FROM THE UV AND OPTICAL CONTINUA

We calculated the X-ray to IR spectra for an irradiated accretion disk and donor (subgiant) star in the U Sco binary system and reproduced the UV and optical spectra observed 12, 33, and 263 days after the (1979/80) outburst (Barlow et al. 1981; Williams et al. 1981). We in addition considered viscous heating of the accretion disk assuming an accretion rate $\dot{M} = 10^{-8}$ $M_\odot$ yr^{-1}.

We used in our calculation for the disk and donor star an albedo A~0.1-0.3 (cf. Table 1) and we used an inclination for the system i = 70°. We in addition assumed a distance of 14 kpc. We could reproduce the observed UV and optical continua (cf. Figure 1) with temperatures and luminosities for the supersoft component which

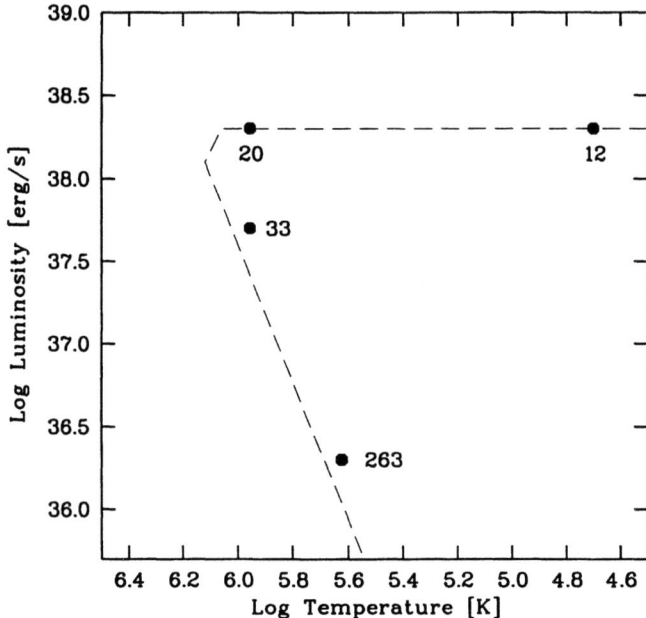

FIGURE 2. Hertzsprung-Russell diagram of a Chandrasekhar mass white dwarf (plateau and cooling track, from Iben 1982) together with the data derived from the multiwavelength (X-ray, UV and optical) fit to the observations on days 12, 20, 33, and 263 after the 1979 and 1999 outburst, respectively. The observations can be explained due to X-ray heating with additional viscous heating in a disk with a $\dot{M} = 10^{-8}$ $M_\odot$ yr^{-1}.

are in agreement with an almost Chandrasekhar mass white dwarf evolving from the plateau towards the cooling track in the Hertzsprung-Russell diagram (cf. Iben 1982). The time spent by the steadily nuclear burning white dwarf on the plateau is found to be ~3 weeks and the cooling time continues for another ~3 weeks. These numbers are in agreement with the numbers predicted from the calculations of Kato (1999). The change in visual magnitude derived in our calculation is V~14-18 and is consistent with the observed change in visual magnitude.

3. U SCO: A TYPE IA SUPERNOVA PROGENITOR ?

From optical observations of the 1999 outburst follows that a negligible amount of mass was ejected during the outburst (Munari et al. 1999). Together with the

TABLE 1. Parameters used in the multiwavelength calculations for U Sco (donor 1.0 $M_\odot$ subgiant filling 0.9 of Roche lobe, degenerate star 1.38 $M_\odot$). Sets of parameters are given to reproduce observations during the 1979/80 and 1999 outbursts.

Parameter	Value			
	1979 (day 12)	1999 (day 20)	1979 (day 33)	1980 (day 263)
		supersoft		
T_{WD} [K]	$5.0\ 10^4$	$9.0\ 10^5$	$9.0\ 10^5$	$4.2\ 10^5$
L_{WD} [erg/s]	$2.0\ 10^{38}$	$2.0\ 10^{38}$	$5.0\ 10^{37}$	$2.0\ 10^{36}$
		wind		
T_{WD} [K]	$2.3\ 10^6$	$2.3\ 10^6$	$2.3\ 10^6$	$2.3\ 10^6$
L_{WD} [erg/s]	$2.5\ 10^{34}$	$1.5\ 10^{34}$	$1.5\ 10^{34}$	$1.0\ 10^{30}$
Albedo Disk	0.1	0.3	0.3	0.3
Albedo Star	0.2	0.3	0.2	0.2
N_H [cm^{-2}]	$2.0\ 10^{20}$	$2.0\ 10^{20}$	$1.4\ 10^{20}$	$1.4\ 10^{20}$
V magnitude	13.4	15.1	15.9	17.9

fact that the white dwarf is very massive and underwent a phase of steady nuclear burning for ∼3 weeks gives the perspective that in U Sco for the first time a candidate binary system has been found in which the white dwarf can grow towards the Chandrasekhar mass limit. U Sco therefore can be considered to be a realistic progenitor for a type Ia supernova (SN Ia).

The fact that the system belongs to the galactic halo population may indicate that SNe Ia may well also occur in the galactic halo.

REFERENCES

Barlow M.J., Brodie J.P., Brunt C.C., et al., 1981, MNRAS 195, 61
Hachisu I., Kato M., Nomoto K., 1999, ApJ 519, 314
Iben I.Jr., 1982, ApJ 259, 244
Kahabka P., Hartmann H.W., Parmar A.N., & Negueruela I., 1999, A&A 347, L43
Kato M., 1996, in Greiner, "Supersoft X-ray sources", LNP 472, Springer, p. 15
Kato M., 1999, PASJ 51, 525
Munari U., et al., 1999, 1999, A&A 347, L39
Sekiguchi K., 1995, Ap&SS 230, 75
Webbink R.F., Livio M., Truran J.W., Orio M., 1987, ApJ 314, 653
Williams R.E., Sparks W.M., Gallagher J.S., et al., 1981, ApJ 251, 221
Yungelson L., Livio M., Truran J.W., et al., 1996, ApJ 466, 890

AN OPTICAL AND X-RAY STUDY OF THE PECULIAR NARROW-LINE QUASAR QSO 0117-2837

Stefanie Komossa[1], D. Grupe[1], V. Burwitz[1], M. Janek[2]

[1] Max-Planck-Institut für extraterrestrische Physik, Giessenbachstr., 85748 Garching, Germany; skomossa@xray.mpe.mpg.de, [2] 3063 Obernkirchen, Germany

ABSTRACT   We present an optical and X-ray study of the quasar QSO 0117-2837. It exhibits an interesting combination of optical and X-ray properties: Despite its extremely steep observed X-ray spectrum ($\Gamma_x \simeq -4$ when fit by a simple powerlaw), its Balmer lines are fairly broad. A two-component Gaussian fit to Hβ yields FWHM$_{H\beta, \rm broad} \simeq 4000$ km/s, and places QSO 0117-2837 in the 'zone of avoidance' in the FWHM$_{H\beta}$-Γ_x diagram. A time variability analysis shows that QSO 0117-2837 is another case of Narrow-line Sy 1 galaxy (NLSy1 hereafter) which does *not* show any X-ray variability during the observation. The results are discussed in view of the NLSy1 character of QSO 0117-2837.

KEYWORDS: Galaxies: active, quasars: individual: QSO 0117-2837, quasars: emission lines, X-rays: galaxies

1. INTRODUCTION

QSO 0117-2837 (1E 0117.2-2837) was discovered as an X-ray source by *Einstein* and is at a redshift of $z=0.347$ (Stocke et al. 1991). It is serendipituously located in a *ROSAT* PSPC pointing. Its X-ray spectrum is extremely steep, as was briefly noted by Schwartz et al. (1993). We present here the first detailed analysis of the *ROSAT* observations of this AGN (see Komossa & Fink 1997b for first results, and Komossa & Meerschweinchen 2000 for the complete ones), and a discussion of a high-quality optical spectrum.

2. OPTICAL PROPERTIES

We have obtained an optical spectrum of QSO 0117-2837 with the ESO 1.52 m telescope at LaSilla in September 1995. The spectrum covers the wavelength range 3900-7800Å with a resolution of 6Å. The optical spectrum reveals several signs of a NLSy1 galaxy (we do not distinguish between NL Seyferts and NL quasars, here): weak [OIII]λ5007 emission and strong FeII complexes (Fig. 1a). After subtraction of the FeII spectrum (see Grupe et al. 1999 for details) we derive FWHM$_{H\beta}$=2100±100 km/s, FWHM$_{\rm [OIII]}$=820±150 km/s (based upon single-component Gaussian fits to the emission lines), and [OIII]/Hβ=0.056. Alternatively, Hβ was fit with a two-component Gaussian in which case we obtain FWHM$_{H\beta,broad}$ $\simeq$ 4000 km/s and FWHM$_{H\beta,narrow}$ $\simeq$ 1100 km/s. Our spectrum also covers the important region around [OII]λ3727, which provides an important discriminant between different models to account for the excitation of the NLR in NLSy1s (Komossa & Janek 1999). [OII] is not detected, implying [OII] to be much weaker than [OIII].

3. X-RAY PROPERTIES

The *ROSAT* PSPC observation of QSO 0117-2837 was performed in Dec. 1991 with an exposure time of 4.5 ksec. The source is detected with a countrate of 0.44 cts/s (see Komossa & Meerschweinchen 2000 for further details on X-ray data reductions carried out).

Spectral analysis. The *ROSAT* X-ray spectrum of QSO 0117-2837 is extremely steep. Three models provide a successful spectral fit; we discuss each in turn:

(i) When the X-ray spectrum is fit by a single powerlaw with Galactic cold absorption of $N_{\rm Gal} = 1.65\,10^{20}$ cm^{-2}, we derive a photon index $\Gamma_{\rm x} \simeq -3.6 \pm 0.1$ (-4.3 ± 0.4, if $N_{\rm H}$ is treated as free parameter). The overall quality of the fit is good ($\chi^2_{\rm red} = 0.8$), but slight systematic residuals remain (Fig. 1b).

(ii) A successful alternative description is a *warm-absorbed* flat powerlaw of fixed canonical index $\Gamma_{\rm x} = -1.9$. The warm absorber models are based on Ferland's (1993) code *Cloudy*, and the model assumptions and calculations are described in more detail in Komossa & Fink (e.g., 1997a). We find a very large column density of the warm absorber, and the contribution of emission and reflection is no longer negligible; there is also some contribution to Fe Kα. For the pure absorption model, the best-fit values for ionization parameter and warm column density are $\log U \simeq 0.8$, and $\log N_{\rm w} \simeq 23.6$ ($N_{\rm H}$ is now consistent with the Galactic value when treated as free parameter), with $\chi^2_{\rm red}$=0.7. Including the contribution of emission and reflection for 50% covering of the warm material as calculated with *Cloudy* gives $\log N_{\rm w} \simeq 23.8$ ($\chi^2_{\rm red} = 0.65$). Several strong EUV emission lines are predicted to arise from the warm material (e.g., FeXXIλ2304/H$\beta_{\rm wa}$ = 10, NeVIIIλ774/H$\beta_{\rm wa}$ = 9, and FeXXIIλ846/H$\beta_{\rm wa}$ = 113). No absorption from CIV and NV is expected to show up in the UV. Both elements are more highly ionized.

(iii) Thirdly, the spectrum can be fit with a flat powerlaw ($\Gamma_{\rm x}$ fixed to -1.9) plus soft excess which was parameterized as black body, or the standard accretion-disk emission model available in EXSAS. We find $kT_{\rm bb} \simeq 0.10$ keV for the black body description ($\chi^2_{\rm red}$=0.7). Using the accretion disk description, and fixing the black hole mass to $M_{\rm BH} = 0.6\,10^4 M_\odot$ yields $\frac{\dot{M}}{M_{\rm edd}} = 0.6$.

Temporal analysis. An analysis of the temporal variability reveals *constant* source flux within the 1σ error during the observation. This makes QSO 0117-2837 another example of a NLSy1 with constant X-ray flux during the time-interval of observation, showing that *not all* NLSy1s are characterized by permanent rapid variability (see also the discussion and examples in Wisotzki & Bade 1997).

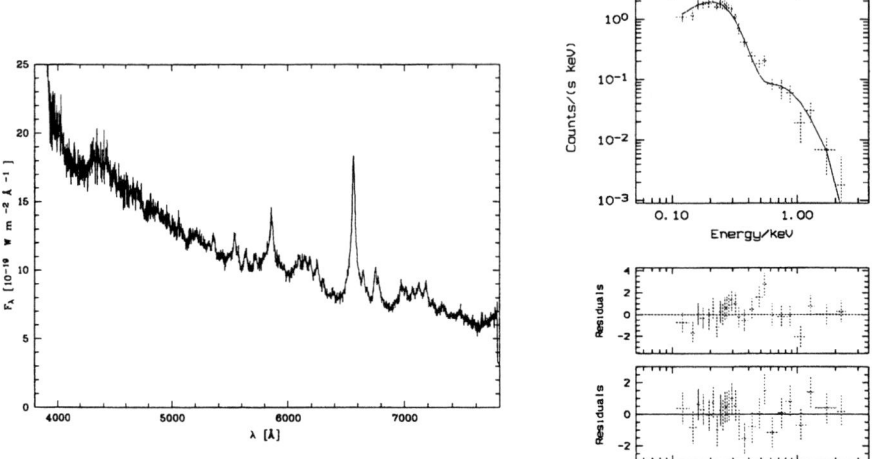

FIGURE 1. (**a**, left) Optical spectrum of QSO 0117-2837. (**b**) *ROSAT* X-ray spectrum of QSO 0117-2837. The first panel shows the observed X-ray spectrum (crosses) and best-fit powerlaw (solid line). The second panels shows the residuals for this model description whereas the third panel displays the residuals after fitting a warm absorber model.

4. DISCUSSION

Given the rather large width of Hβ, the X-ray spectrum of QSO 0117-2837 is exceptionally steep; among the steepest observed in NLSy1s. In fact, whereas there is a very large scatter in the X-ray spectral steepness of NLSy1s, with several as flat as 'normal' Seyferts (e.g., Xu et al. 1999), broad line objects always tend to show flat X-ray spectra (e.g., Boller et al. 1996, Grupe et al. 1999); the corresponding region in the FWHM$_{H\beta}$–Γ_x diagram is occasionally referred to as 'zone of avoidance'. QSO 0117-2837 appears to be an important transition object: Depending on which representation of the line profile is chosen – a one-component Gaussian with FWHM$_{H\beta}$=2100 km/s, or a two component Gaussian which gives a better fit and FWHM$_{H\beta,\text{broad}}$=4000 km/s – QSO 0117-2837 is placed at the border of the 'zone of avoidance', or inside it, respectively.

The origin of the steep observed X-ray spectra of NLSy1 galaxies and the relation to their optical properties is still not well understood. In particular, it is interesting to point out the following: Whereas among the early suggestions to explain the X-ray spectral steepness of NLSy1s was the presence of a strong soft excess, in analogy to 'normal' Seyferts and quasars that were believed to show a soft excess, there is now growing evidence that many soft excesses in Seyferts were in fact mimicked by the presence of warm absorbers. On the other hand, the soft excesses in NLSy1s, although originally inferred in analogy to Seyferts, turn out to be real, as judged by recent *ASCA* and *SAX* observations (e.g., Vaughan et al. 1999). Often, it turns out, several components are simultaneously present that contribute to the spectral steepness in NLSy1s: A steeper-than-usual powerlaw, a soft excess *and a* warm absorber pointing to the spectral complexity in these objects (see, e.g., the discussion in Komossa & Fink 1997a,b).

In the case of QSO 0117-2837 the limited *ROSAT* spectral resolution does not allow us to distinguish between models (i)-(iii) of Sect. 3. It still allows to determine the maximal possible contribution of each component (i)-(iii). In fact, the rather large inferred column density N_w of the warm absorber (model ii) suggests that further mechanisms contribute to, or dominate, the X-ray steepness. Given the very steep rise towards the blue of QSO 0117-2837's optical spectrum (with $\alpha_{opt,x}=0.0$, Grupe et al. 1998), it is tempting to speculate that a giant soft-excess dominates the optical-to-X-ray spectrum. We strongly caution, though, that simultaneous optical-X-ray variability studies in other Seyferts and NLSy1s (e.g., Done et al. 1995) do *not* favor a common origin of X-ray and optical components. Secondly, one giant optical-to-X-ray bump seems to be inconsistent with the finding of Rodriguez-Pascual et al. (1997) that NLSy1s tend to be underluminous in the UV. The possibility of an *indirect* relation between optical and X-ray component remains.

Given the abundant presence of warm absorbers in 'normal Seyferts', and the additional trend that warm absorbers are more abundant in FeII-strong objects (Wang et al. 1996), combined with QSO 0117-2837's large $FWHM_{H\beta}$ and steep X-ray spectrum, it is likely that both, a soft excess and a warm absorber contribute to its spectral steepness in the *ROSAT* band. Its peculiar optical–X-ray properties make the quasar a good target for (a) follow-up X-ray spectral observations with, e.g., *XMM*, and (b) high-resolution optical studies of the $H\beta$ profile.

This and related papers can be retrieved from our webpage at
http://www.xray.mpe.mpg.de/~skomossa/

REFERENCES

Boller T., Brandt W.N., Fink H.H., 1996, A&A 305, 53
Done C. Pounds K.A., Nandra K., Fabian A.C., 1995, MNRAS 275, 41
Ferland G.J., 1993, University of Kentucky, Physics Department, Internal Report
Grupe D., Beuermann K., Thomas H.-C., Mannheim K., Fink H., 1998, A&A 330, 25
Grupe D., Beuermann K., Mannheim K., Thomas H.-C., 1999, A&A 350, 805
Komossa S., Fink H., 1997a, A&A 322, 719
Komossa S., Fink H., 1997b, in Lecture Notes in Physics 487, 250 (astro-ph/9612185)
Komossa S., Meerschweinchen J., 2000, A&A 354, 411
Komossa S., Janek M., 1999, in 'Heating and Acceleration in the Universe', Astron. Nachr. 320, 310 (astro-ph/9907373)
Rodriguez-Pascual P.M., Mas-Hesse J.M., Santos-Lleo M., 1997, A&A 327, 72
Schwartz D.A., Zhao P., Remillard R., 1993, BAAS 25, 811
Stocke J.T., Morris S.L., Gioia I.M., et al., 1991, ApJS 76, 813
Vaughan S., Reeves J., Warwick R., Edelson R., 1999, MNRAS in press
Wang T., Brinkmann W., Bergeron J., 1996, A&A 309, 81
Wisotzki L., Bade N., 1997, A&A 320, 395
Xu D.W., Wei J.Y., Hu J.Y., 1999, ApJ 517, 622

SIMULTANEOUS X-RAY/OPTICAL BURST FROM GS 1826-24

Albert Kong[1], Lee Homer[1], Erik Kuulkers[2,3], Phil Charles[1] and Alan Smale[4]

1) NAPL, University of Oxford, Oxford OX1 3RH, U.K.
2) SRON, Sorbonnelaan 2, 3584 CA Utrecht, The Netherlands
3) Utrecht University, P.O. Box 80000, 3507 TA Utrecht, The Netherlands
4) LHEA, Code 660.2, NASA/GSFC, Greenbelt, MD 20771, USA

ABSTRACT   We report results from the first simultaneous X-ray ($RXTE$) and optical (SAAO) observations of the low-mass X-ray binary GS 1826-24 in June 1998. A type-I burst was detected in both X-ray and optical wavelengths. A $\sim$ 3 s optical delay with respect to the X-ray burst is observed and we infer that this is related to the X-ray reprocessing in the accretion disc. The delay provides additional support for the recently proposed orbital period of $\sim$ 2 hr.

KEYWORDS: accretion – binaries: close – stars: individual (GS 1826-24) – X-rays: bursts

1. INTRODUCTION

GS 1826-24 is a bursting X-ray source which was discovered by Makino et al. (1988) with *Ginga*. Recently, more than 70 X-ray bursts were recorded by *BeppoSAX* (Ubertini et al. 1999) and *ASCA* (Kong et al. 2000) and two optical bursts were detected by Homer et al. (1998). All these results confirm the presence of a neutron star accretor. We observed GS 1826-24 with the Proportional Counter Array (PCA) instrument on *RXTE* (Bradt et al. 1993) between 1998 June 23 and July 29. In order to maximize our timing and spectral resolution, we used a 125μs time resolution, 64 spectral energy channel mode over 2-60 keV in addition to the standard configuration. Optical observations of GS 1826-24 were made using the 1.9m telescope at South African Astronomical Observatory (SAAO), with the UCT-CCD fast photometer (O'Donoghue 1995) from 1998 June 23 to 26. The conditions were generally good with typical seeing $\sim$ 1.5-2.5 arcsec and the timing resolution for the source was 5 s.

2. ANALYSIS AND RESULTS

During simultaneous X-ray/optical observations on 24 June 1998, both *RXTE* PCA and the SAAO 1.9m + UCT CCD detected a burst (see Figure 1). The burst lasted for $\sim$ 150 s and the time profiles of the burst in both X-ray and optical are of the

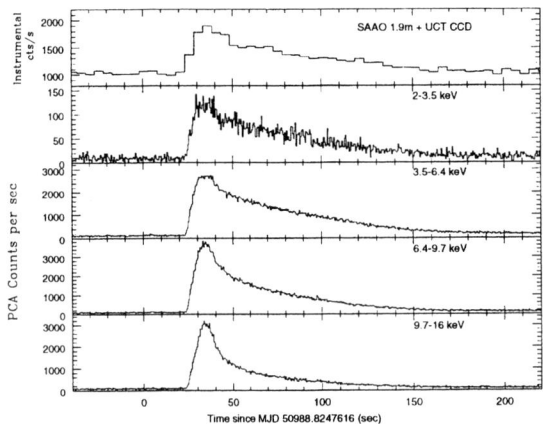

FIGURE 1. The optical (SAAO) and X-ray (*RXTE*) burst profiles in various energy bands. The timing resolution is 5 s (optical) and 0.5 s (*RXTE*/PCA). The decay times strongly depend on photon energy with decays being shorter at higher energies.

fast-rise exponential-decay form. The optical and X-ray bursts started almost at the same time but a delay is present between the peaks. The optical burst resembles the low energy (2–3.5 and 3.5–6.4 keV) X-ray light curves in which they all have a flat peak and a shoulder during the decay phase. At higher energies (> 6.4 keV), the peak is much sharper and the decay is faster during the initial decay phase. Detailed X-ray spectral analysis of the burst is presented by Kong et al. (2000).

Delay between X-ray and optical burst has been seen in some bursting sources (e.g. Lewin et al. 1993). In order to search for delay in our work, we performed: (i) cross-correlation analysis, (ii) modelled the optical burst by convolving the X-ray light curve with a Gaussian transfer function. In using a cross-correlation method we are essentially limited to the 5 s time resolution of the optical data (the PCA data have a much higher time resolution). Moreover, the delay appears to vary, from almost nothing at the start of the burst to a few seconds at the peak. This suggests that the delay might be a function of flux. Therefore a cross correlation analysis cannot provide a full picture of the delay between the X-ray and optical fluxes and instead we model the optical burst by convolving the X-ray light curve with a transfer function.

In order to model the time delay between the optical and X-ray bursts, we convolve a Gaussian transfer function with the X-ray light curve and use χ^2 fitting to model the optical light curve. The same method was used by Hynes et al. (1998) to model the *HST* light curve of GRO J1655−40 from the *RXTE* light curve. The Gaussian transfer function is given by:

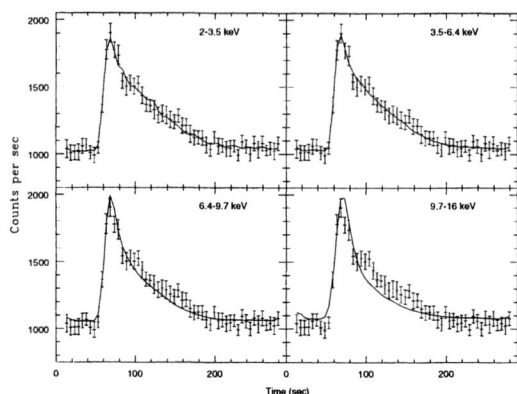

FIGURE 2. Best-fitting predicted light curves using a Gaussian transfer function on the four energy bands. The resulting curves are superimposed on the optical data points.

$$\psi(\tau) = \frac{\Psi}{\sqrt{2\pi}\Delta\tau} e^{-\frac{1}{2}(\frac{\tau-\tau_0}{\Delta\tau})^2} \qquad (1)$$

where τ_0 is the mean time delay and $\Delta\tau$ is the dispersion or 'smearing' which is a measure of the width of the Gaussian. Ψ is the strength of the response.

We performed a series of convolutions of the transfer function with the lightcurves from the four energy bands, varying both τ_0 and $\Delta\tau$ independently. Figure 2 shows the best fit predicted light curves from each convolution superimposed on the optical light curve. The principal features of the optical burst profile are reproduced well in the predicted light curves from the 2–3.5 and 3.5–6.4 keV energy bands. The fits are good ($\chi^2_\nu < 1$ for 51 d.o.f.) for the two lower energy bands but not for the higher energy ones. The mean delay and dispersion between the optical and lower X-ray energy bands are both ~ 3 s. Note that the delay should be interpreted with some caution, as the optical timing resolution is only 5 s. The strength of the response, Ψ, is almost the same for the two lower X-ray energy bands, at a value of ~ 0.0137, while it is 2–3 times smaller for the higher X-ray energy band. This is an indication that a greater proportion of the reprocessing occurs at lower energies.

3. DISCUSSION

We for the first time detected a simultaneous X-ray/optical burst from GS 1826–24. On the basis of its spectral property and time profile, the burst has a cooling trend during its decay. The time profile show fast rise times of 7–9 s and long decay times ranging from 20–60 s depending on the energy band. We therefore interpret the three bursts detected from GS 1826–24 as type I bursts (Hoffmann et al. 1978). The relatively long rise time ($\sim 7-9$ s), as compared to bursts in other systems, indicates that the burst front may have enough time to spread over the whole neutron star surface during the rise to burst and suggests that the burning is homogeneous over the surface of the neutron star. We note that X-ray burst rise times in other sources have been observed to be smaller than ~6 s (see e.g. Sztajno et al. 1986; Lewin et al. 1987), making this system rather unique. The burst rise times derived by in't Zand et al. (1999) and Kong et al. (2000) range from 5–9 s which are also consistent with our observations. This relatively long burst also resembles the theoretical results of X-ray bursts driven by rapid proton capture process, or rp-process (see Bildsten 1998; Schatz et al. 1998). The long time energy release from this process predicts the bursts with low α^1 values and long tails, as is observed (see Lewin et al. 1983; Kong et al. 2000).

Based on our observed mean delay of 3 ± 1 s for the optical burst with respect to low energy X-rays, we can then constrain the orbital period of the system. By Kepler's law, the light travel time of 2–4 s corresponds to an orbital period of 1.6–5.5 hr if we assume a $1.4 M_\odot$ neutron star and a companion star mass of 0.1–$1.1 M_\odot$ (i.e. for a low-mass main sequence star and stable mass transfer). Hence, this range of periods provides support for the 2.1 ± 0.1 hr orbital period proposed by Homer et al. (1998). Lastly, from only one simultaneous optical/X-ray burst, we cannot draw a firm conclusion as to whether the optical burst is due to reprocessing in the disk or on the surface of the companion star. However, given that the source is a low inclination system ($< 70°$; Homer et al. 1998) and the ratio of smearing to delay is ~ 1, the reprocessing is expected to be dominated by the accretion disc. An extensive simultaneous campaign has been performed and such results will shed light on the accretion geometry of the system.

ACKNOWLEDGEMENTS

We are grateful to Darragh O'Donoghue (SAAO) for his advice on the the use of the UCT-CCD. We also thank Lars Bildsten for valuable comments.

REFERENCES

Bildsten, L., 1998, in The Many Faces of Neutron Stars, ed. A. Alpar, L. Buccheri, & J. van Paradijs, (Dordrecht: Kluwer)
Bradt, H.V., Rothschild, R.E. & Swank, J.H., 1993, A&AS, 97, 355
Hoffman, J.A., Marshall, H., J.A. & Lewin, W.H.G., 1978, Nature, 271, 630
Homer, L., Charles, P.A. & O'Donoghue, D., 1998, MNRAS, 298, 497
Hynes, R.I., O'Brien, K., Horne, K., Chen, W. & Haswell, C.A., 1998, MNRAS, 299, L37
in't Zand, et al., 1999, A&A, 347, 891
Kong A.K.H., et al., 2000, MNRAS, 311, 405
Lewin, W.H.G., et al., 1987, ApJ, 319, 893
Lewin, W.H.G., et al., 1993, Sp. Sci. Rev., 62, 223
Makino, F., et al., 1988, IAUC 4653
O'Donoghue, D, 1995, Baltic Astronomy, 4, 519
Schatz, H., et al., 1998, Physics Reports, 294, 167
Sztajno, M., et al., 1986, MNRAS, 222, 499
Ubertini, P., et al., 1999, ApJ, 514, L27

---

[1] the ratio between gravitational and nuclear energy per gram accreted matter.

AN X-RAY/TEV GAMMA-RAY STUDY OF MKN 501 DURING ITS EXTRAORDINARY OUTBURST OF 1997

H. Krawczynski*, P.S. Coppi[†], T. Maccarone[†], F.A. Aharonian*

*Max Planck Institut für Kernphysik, Heidelberg, Germany
[†]Yale University, New Haven, CT 06520-8101

ABSTRACT For more than 6 months in 1997, the BL Lac object Mkn 501 has been in an exceptionally bright state, both in the X-ray band and in the Very High Energy (VHE) band. In this paper we present a multiwavelength study of Mkn 501 during this extraordinary outburst based on HEGRA VHE and RXTE X-ray data for more than 20 days with detailed spectral information in both bands. We interpret the results in the framework of a Synchrotron Self Compton models and study the implications for estimates of the intergalactic extinction due to pair production processes of the VHE photons on the Diffuse Extragalactic Background Radiation (DEBRA).

KEYWORDS: BL Lacertae objects: individual: Mkn 501, Gamma rays: observations, X-rays: galaxies

1. INTRODUCTION

In the northern hemisphere, two BL Lac objects have firmly been established as emitters of TeV γ-rays: Mkn 421 ($z=0.031$) and Mkn 501 ($z=0.034$). The nonthermal X-ray and TeV continuum emission from these objects is widely believed to originate in a relativistic jet due to a population of high energy electrons, producing synchrotron radiation at longer wavelengths and Inverse Compton (IC) photons at shorter wavelengths. The origin of the IC seed photons has not yet been established. In so-called "Synchrotron Self Compton" (SSC) models the target photon population is dominated by low energy synchrotron photons. In "External Compton" models the seed photons originate outside the emission volume and are, e.g. radiation from the nuclear continuum scattered or reprocessed in the broad-line regions or accretion disc photons. The study of these extragalactic objects does not only provide information about the nonthermal processes in the jet but also about the extragalactic extinction of the TeV γ-rays due to pair production processes on DEBRA photons (Gould & Schreder 1966; Stecker et al. 1992). Since the extragalactic extinction is expected to rapidly increase with γ-ray energy, the TeV energy spectra of extragalactic objects contain information about the DEBRA intensity. The TeV observations constrain the 1 μm to 50 μm intensity which is of utmost interest since it depends on the star formation history and on the cosmological parameters (Primack et al. 1999). Direct measurements in this wavelength region are extremely difficult due to instrumental and galactic contamination effects.

In 1997 the BL Lac object Mkn 501 showed at TeV and X-ray energies a phase of very high activity and dramatic flux variability (see e.g., Aharonian et al. 1999a-b, called *A99a-b* in the following; Pian et al. 1998, and references therein). The observations with the stereoscopic Cherenkov telescope system of HEGRA made it

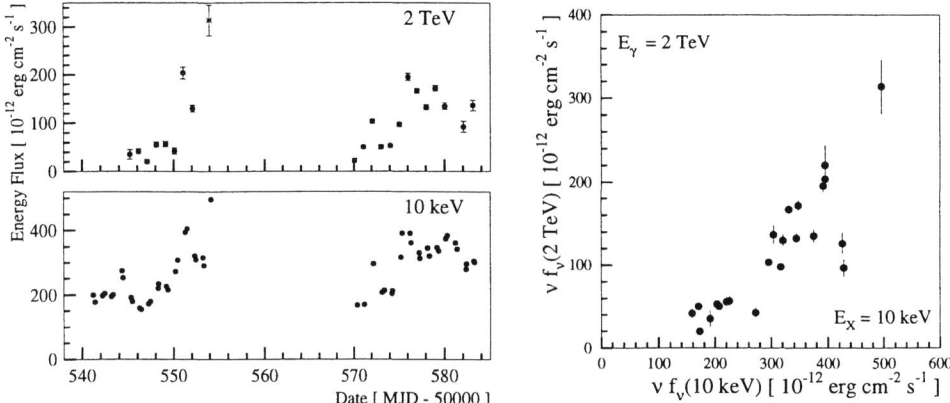

FIGURE 1. The left panel shows the Mkn 501 TeV γ-ray and X-ray lightcurve for the April and May 1997 observations. The solid TeV points are HEGRA observations taken from A99a; for April 16th (MJD = 50554) the flux level was estimated from CAT observations (Djannati-Atai et al. 1999). The right panel shows the correlation of TeV and keV fluxes for all TeV/X-ray observations with less than 6 h time difference (April,May, and July, 1997 data).

possible to measure the TeV energy spectrum deeply into the regime of an exponential cutoff. Over the broad energy range from 500 GeV to 24 TeV the energy spectrum could be described by a power law model with an exponential cutoff: $dN/dE \propto (E/1 \text{ TeV})^{-1.9} \exp(-E/6.2 \text{ TeV})$ (A99b).

In this paper, we present a large sample of simultaneous and nearly simultaneous TeV/X-ray observations of the BL Lac object Mkn 501 during its spectacular outburst in 1997. Based on the TeV/X-ray characteristics of the source we will discuss the question whether the exponential cutoff of the VHE spectrum results from intrinsic source properties or from extragalactic extinction. Details on the analysis and interpretation of the data can be found in (Krawczynski et al. 2000).

2. RXTE AND HEGRA OBSERVATIONS AND RESULTS

We report on public RXTE (Bradt et al. 1993) observations made between April 3rd, 1997 to July 14th, 1997 (two per night in April and May). We only use data from the Proportional Counter Array with good statistical information from 3 keV to 25 keV. We use the published TeV data from the HEGRA Cherenkov telescope system given in (A99a-b).

The most important result, the joint TeV and X-ray lightcurve is shown in Fig. 1, left side (data of April and May, 1997). The flux was strongly variable on a time scale of less than 1/2 day at TeV and less than 1 day at X-ray energies. While the TeV energy flux at 2 TeV varied by a factor of 10, the X-ray flux at 10 keV changed only by a factor of 3. Several well correlated strong flares can clearly be recognized in both energy bands. The TeV and X-ray fluxes measured with a time delay of less than 6 h are tightly correlated (Fig. 1, left and right side). An analysis based on

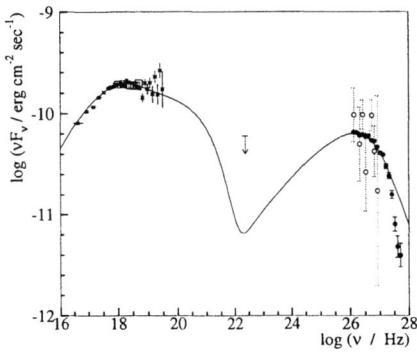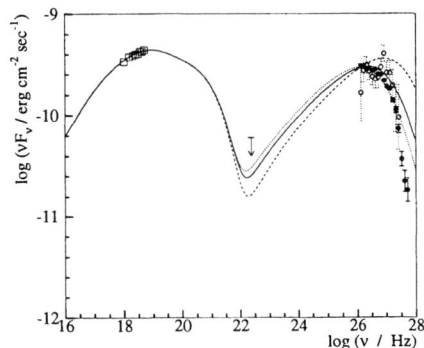

FIGURE 2. SSC model fits to the X-ray and TeV data of April 7th (left side) and April 13th (right side): BeppoSAX (asterisks) from (Pian et al. 1998), RXTE (squares), HEGRA (open circles), and the HEGRA 1997 time-averaged Mkn 501 spectrum scaled according to the detection rate of this day (full circles). The solid line shows a SSC model with Doppler factor $\delta_j = 25$, magnetic field $B = 0.037$, and a radius of the emission volume of $R = 1.5 \cdot 10^{16}$ cm (left side) and $R = 1.1 \cdot 10^{16}$ cm (right side). The dashed and dotted lines show the uncertainty in the shape of the predicted VHE spectrum arising from changing the values of 0.1 and δ_j and B (dashed line: $\delta_j = 100$, $B = 0.012$ G, dotted line: $\delta_j = 25$, $B = 0.12$ G). The 2σ upper limit at $2.4 \cdot 10^{22}$ Hz has been derived from EGRET observations (April 9th – April 15th, 1997) under the assumption of a constant emission level (Catanese et al. 1997).

the Discrete Correlation Function (DCF) (Edelson & Krolik 1988) shows that the time lag between hard (25 keV) and soft (3 keV) X-rays is smaller than $\sim 1/2$ day. Furthermore, the DCF analysis shows that the VHE (2 TeV) and X-ray fluxes vary simultaneously on a time scale of half a day or shorter. The data do not allow us to conclude definitively whether X-ray and TeV photon indices are correlated. The X-ray photon index (β from $dN/dE \propto E^\beta$) varies from -1.7 to -2.1 with typical rates of 0.01/h. In drastic contrast to the TeV flux variability by factors of 10, the TeV spectrum is surprisingly stable and only marginally significant deviations of the 1-5 TeV photon index from the 1997 mean value of -2.25 have been reported (A99a). Importantly, the accuracy of the diurnal TeV measurements – typically between 0.1 and 0.3 in the 1-5 TeV photon index – does not allow us to exclude a variability of the TeV photon index comparable to the modest changes of the X-ray photon index.

3. DISCUSSION

We have presented the analysis of a large data base of X-ray observations of Mkn 501 taken simultaneously or nearly simultaneously with TeV observations. The observations of several strong flares in both energy bands allowed us to study in unprecedented depth the X-ray/TeV γ-ray correlation properties. In the following we will show SSC model fits of the data and discuss the implications on the extragalactic extinction of the TeV radiation.

In Figure 2 the RXTE, BeppoSAX, and HEGRA data of April 7th and April 13th, 1997 as well as fits of a one zone homogeneous SSC model (Coppi 1992) are shown. Actually only the low frequency synchrotron component ($< 10^{22}$ Hz) is fitted, the spectral shape of the high frequency IC component is a prediction of the model. The radius of the spherical emission volume has been chosen so that the TeV-flux at 500 GeV (computed neglecting extragalactic extinction) matches the measured flux. Any discrepancy between the predicted TeV spectrum and the measured one above 500 GeV can be interpreted as evidence for a change of the extragalactic extinction with γ-ray energy. The model parameters have been chosen to satisfy the constraints from the X-ray and TeV variability and correlation properties discussed in detail by Krawczynski et al. (2000). Note that the shape of the synchrotron spectrum outside the frequency range constrained by the RXTE and BeppoSAX observations has been chosen as to be consistent with (i) the OSSE measurements during April 1997 (Catanese et al. 1997), and (ii) a simple scenario in which the spectral break observed around 1 keV is caused by synchrotron cooling of a power law spectrum of electrons. Fig. 2 clearly shows that the model predicts an emitted VHE spectrum which is substantially curved. Although the Doppler factor δ_j and the magnetic field B are not tightly constrained the corresponding uncertainties for the prediction of the emitted VHE spectra are small (Fig. 2, right side, dashed and dotted lines). For April 13th, the value -1.83 of the X-ray photon index approximately equals the mean index during the 1997 RXTE observations of this source. Taking into account that the HEGRA VHE spectra did not vary during 1997 within statistical errors, we think it is justified to compare the prediction of the emitted VHE spectrum of April 13th with the observed VHE spectrum as averaged over the whole 1997 observation period. This comparison suggests that the extragalactic extinction substantially increases for γ-ray energies above $\simeq 5$ TeV. Notably, the observed curvature of the TeV spectrum seems to arise from intrinsic source properties *and* extragalactic extinction.

ACKNOWLEDGEMENTS

The analysis of the RXTE data has been possible due to the High Energy Astrophysics Science Archive Research Center at NASA/Goddard Space Flight Center.

REFERENCES

Aharonian F.A., Akhperjanian A.G., Barrio J.A., et al., 1999a, A&A 342, 69 (*A99a*)
Aharonian F.A., Akhperjanian A.G., Barrio J.A., et al., 1999b, A&A 349, 11 (*A99b*)
Bradt H.V., Rothschild R.E., Swank J.H., A&AS 1993, 97, 355
Catanese M., Bradbury S.M., Breslin A.C., et al., 1997, ApJ 487, L143
Coppi P.S., 1992, MNRAS 258, 657
Djannati-Atai A., Piron F., Barrau A., et al. 1999, A&A 350, 17
Edelson R.A., Krolik J.H., 1988, ApJ 333, 646
Gould J., Schréder G., 1966, Phys. Rev. Lett. 16, 252
Krawczynski H., Coppi P.S, Maccarone T., Aharonian F.A. 2000, A&A 353, 97.
Pian E., Vacanti G., Tagliaferri G., et al., 1998, ApJ 492, L17
Primack J.R., Bullock J.S., Somerville R.S., Macminn D., 1999, Astropart. Phys. 11, 93
Stecker F.W., De Jager O.C., Salamon M.H.,1992, ApJ. 390, L49

WIND ACCRETION IN HMXRB

I. Kreykenbohm[1], J. Wilms[1], P. Kretschmar[1,3], R. Staubert[1],
R. E. Rothschild[2], W. A. Heindl[2], D. E. Gruber[2]

1) Institut für Astronomie und Astrophysik – Astronomie, Univ. of Tübingen, Germany
2) CASS, University of California, San Diego, La Jolla, CA 92093, U.S.A.
3) INTEGRAL Science Data Center, ISDC, 1290 Versoix, Switzerland

ABSTRACT We present light curves and spectra of a ∼35 ksec $RXTE$ observation of Vela X-1 and a ∼300 ksec observation of Cen X-3.

Both light curves show strong absorption which increases from no absorption at low orbital phases to large values at high orbital phases; in the case of Vela X-1 by a factor of almost 100. Superimposed are absorption dips which can be observed at any orbital phase, though mostly at high orbital phases. We study an unusually strong dip in Vela X-1 during which the flux is reduced by ∼90 % and almost no pulses are detectable. The increase of N_H and the scattered component show the in- and egress of material in the line of sight. A similar behaviour is observed in a pre-eclipse dip in Cen X-3.

KEYWORDS: X-rays: stars – stars: individual: Vela X-1 – stars: individual: Cen X-3

1. VELA X-1

Vela X-1 is an eclipsing High Mass X-ray Binary with an orbital period of 8.94 days (van Kerkwijk et al., 1995) and a spin period of ∼283 seconds . The optical companion is the massive B0.5 star HD 77581 emanating an intense stellar wind with $4 \cdot 10^{-6}$ M⊙/yr (Nagase et al., 1986).

The X-ray light curves of Vela X-1 show the typical behaviour of violent wind accretion: the X-ray luminosity varies on all timescales: periods of relative X-ray quiescence can be followed a few minutes later by an X-ray outburst (Haberl, 1994). Luminosity changes of more than a factor of 10 are not uncommon.

Previous observations of Vela X-1 have shown that its X-ray spectrum is complex and very difficult to describe (Kreykenbohm et al., 1999a; Orlandini et al., 1998). We used a partial covering model consisting of two NPEX (Negative Positive EXponential, Mihara, 1995) components: one is absorbed while the other one is scattered; the photon indices and the temperature of the NPEX components are coupled. This model had to be further modified by an additive iron fluorescence line at 6.4 keV and two absorption lines (see Kreykenbohm et al., 1999b, for details).

We observed Vela X-1 with the Rossi X-ray Timing Explorer ($RXTE$) from 1998 January 21 to 1998 January 22 continously for more than 20 hours (see Fig. 1) re-

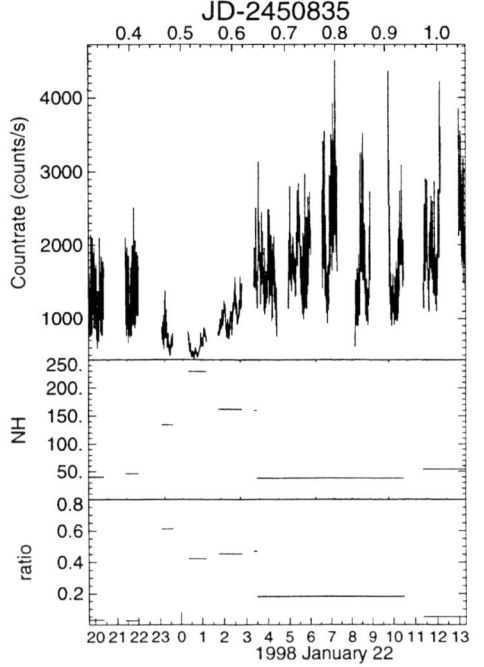
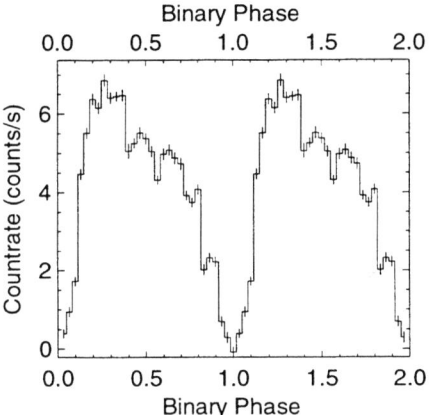

FIGURE 1. Light Curve and some spectral parameters of our *RXTE* observation of Vela X-1.

FIGURE 2. Orbital profile of the *ASM* light curve, folded with the orbital period of Vela X-1 of 8.94 days.

sulting in ∼33 ksec effective observing time. Although we scheduled our observation at orbital phase ∼0.3 where one expects the highest X-ray flux (see Fig. 2) and only minimal absorption, we encountered an extraordinary long and deep extended low which covers three *RXTE* orbits (see Fig. 1). Similar – still unexplained – off-states of Vela X-1 have first been reported by Inoue et al. (1984) and were also seen during our previous *RXTE* observation (Kreykenbohm at al., 1999a).

During the first two *RXTE* orbits, Vela X-1 was in a "normal" state. In the next orbits, the X-ray luminosity has dropped to a fraction of its previous value. Note the increase of N_H and the change of the ratio between the absorbed and the scattered component during the extended low. At the same time, the photon indices α_1 and α_2 of the NPEX model are not correlated with the count rate or N_H. Preliminary analysis of pulse profiles from this extended low show a very small pulsed fraction, while the pulses are strong before and after the low state. This agrees with the increase of the scattered component by a factor of almost 5. For a more detailed discussion of this subject, see Kretschmar et al. (1999). After the extended low the X-ray emission is very variable. Vela X-1 remains in this state till the end of our observation, 7 *RXTE* orbits later.

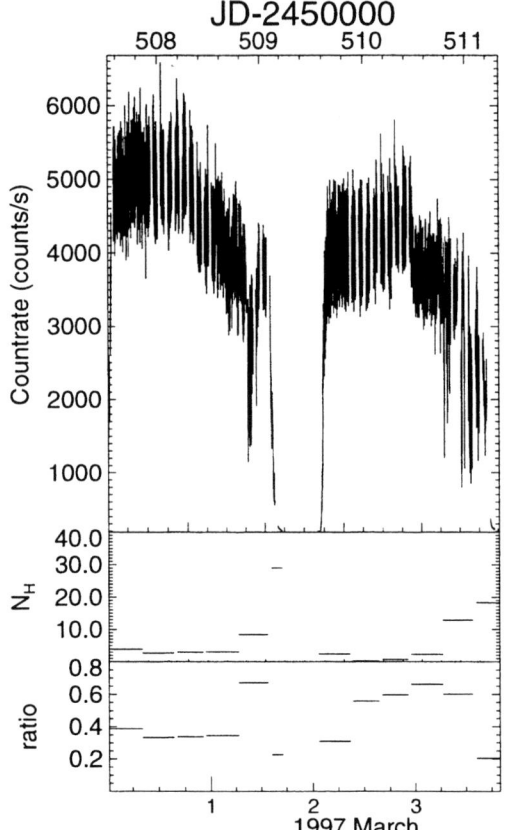

FIGURE 3. The upper panel shows the X-ray light curve of Cen X-3 as seen by RXTE from 1997 February 28 to 1997 March 4 covering two binary orbits. The two lower panels show the equivalent H column and the ratio between the scattered and the absorbed component.

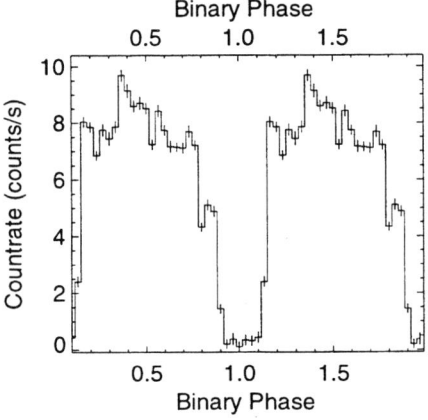

FIGURE 4. Orbital profile of the ASM light curve, folded with the orbital period of Cen X-3 of 2.087 days. Note the overall difference to the orbital profile of Vela X-1 (see Fig. 2). This indicates that the absorbing features and the accretion mechanism in the Cen X-3 system are different from those present in the Vela X-1 system. Note the decrease and afterwards the small increase of the X-ray flux just before the eclipse (see also orbit 1 in Fig. 3).

The extended low might be the result of very strong absorption due to a dense clump of material passing through the line of sight very near the neutron star or a "choke" in the accretion process. The first scenario has the advantage that it explains the observed value of N_H and the higher scattered component (see Fig. 1).

The decrease of the *ASM* count rate over the orbit shows the increase of the absorption with orbital phase (Haberl & White, 1990) in greater detail than previous observations since the database of the *ASM* covers now $\sim$3.5 years. Superimposed on this general trend are three dips. A possible interpretation is that these dips are caused when the edges of the accretion wake passes through the line of sight.

2. CENTAURUS X-3

Cen X-3 is an eclipsing High Mass X-ray Binary with an orbital period of 2.087 days and a spin period of $\sim$4.83 seconds (Nagase, 1989). The optical companion is the massive O6-9II-III star V 779 Cen (Krzeminski star) with an intense stellar wind of $\sim 2 \cdot 10^{-6} M_\odot$/yr (Blondin, 1994). This wind is probably excited by X-rays from the neutron star (Day & Stevens, 1993), suggesting that this wind is responsible for the mass transfer from the optical companion to the neutron star rather than Roche lobe overflow. The strongly varying X-ray light curve (see Fig. 3) supports this.

We used the same spectral model as for Vela X-1 to fit the data without the inclusion of absorption lines (see Fig. 3). Note the increase of N_H during the pre-eclipse dip in orbit 1 and at the end of orbit 2, and compare with the extended low in Vela X-1 (Fig. 1). Note also the increase of the X-ray luminosity just before the eclipse in orbit 1 and in the *ASM* orbital profile (Fig. 4). This dip could be well explained by the accretion stream passing through the line of sight.

REFERENCES

Blondin J.M., 1994, Astrophys. J. 435, 756

Day C.S.R., Stevens I.R., 1993, Astrophys. J. 403, 322

Haberl F., 1994, Astron. Astrophys. 288, 791

Haberl F., White N.E., 1990, Astrophys. J. 361, 225

Inoue H., Ogawara Y., Ohashi T., Waki I., 1984, Publ. Astron. Soc. Jpn. 36, 709

Kretschmar P., Kreykenbohm I., Wilms J., et al., 1999, In: Proc. of 5th Compton Symposium (eds. M. McConnell, J. Ryan), AIP Conf. Proc. 510

Kreykenbohm I., Kretschmar P., Wilms J., et al., 1999a, Astron. Astrophys. 341, 141

Kreykenbohm I., Kretschmar P., Wilms J., et al., 1999b, In: Proc. of X-ray Astronomy '99.

Mihara T., 1995, Ph.D. thesis, RIKEN, Tokio

Nagase F., 1989, Publ. Astron. Soc. Jpn. 41, 1

Nagase F., Hayakawa S., Sato N., et al., 1986, Publ. Astron. Soc. Jpn. 38, 547

Orlandini M., Fiume D.D., Frontera F., et al., 1998, Astron. Astrophys. 332, 121

van Kerkwijk M.H., van Paradijs J., Zuiderwijk E.J., et al., 1995, Astron. Astrophys. 303, 483

TWO CYCLOTRON LINES IN VELA X-1?

I. Kreykenbohm[1], P. Kretschmar[1,3], J. Wilms[1], R. Staubert[1],
W. A. Heindl[2], D. E. Gruber[2], R. E. Rothschild[2]

1) Institut für Astronomie und Astrophysik – Astronomie, Tübingen, Germany
2) CASS, University of California, San Diego, La Jolla, CA 92093, U.S.A.
3) INTEGRAL Science Data Center, ISDC, 1290 Versoix, Switzerland

ABSTRACT We present the spectral analysis of our 35 ksec continuous $RXTE$ observation of the High Mass X-ray Binary Vela X-1. The spectral models used normally to describe pulsar spectra like a power law with an exponential cutoff or the NPEX model (Negative Positive EXponential), do not work for our data. Our successful model consists of two NPEX components: one is heavily absorbed, while the other is due to reflection. But we still find large deviations around ~ 21 keV and ~ 50 keV. We interpret the feature at ~ 55 keV as a cyclotron absorption line while the lower deviation might be due to instrument calibration problems around 25 keV in the PCA.

KEYWORDS: X-rays: stars – stars: neutron-stars: pulsars: individual: Vela X-1

1. INTRODUCTION

Vela X-1 is an eclipsing High Mass X-ray Binary with an orbital period of 8.94 days and a spin period of ~ 283 seconds. The optical companion is the massive B0.5 star HD 77581.

Previous observations of Vela X-1 have shown that its X-ray spectrum and its temporal behaviour are complex and very difficult to describe. The spectrum is usually modeled by a power law with an exponential cutoff (White et al., 1983) or the Negative Positive EXponential (NPEX) model (Mihara, 1995). The spectrum is further modified by strongly varying photoelectric absorption, an iron fluorescence line, and occasionally an iron edge at ~ 7 keV. At higher energies, cyclotron absorption lines at ~ 55 keV have been reported from observations with $HEXE$ (Kendziorra et al., 1992), while Makishima et al. (1992) and Choi et al. (1996) reported an absorption feature at ~ 25 keV to 32 keV from $Ginga$ data. Later observations (Kreykenbohm et al., 1999; Kretschmar et al., 1996) and more detailed analysis of older data supported the existence of both lines (Mihara, 1995).

We observed Vela X-1 with the Rossi X-ray Timing Explorer ($RXTE$) from 1998 January 21 to 1998 January 22 continously for more than 20 hours resulting in ~ 35 ksec effective observing time. The observation is divided in three parts (Tab. 1).

TABLE 1. Date and lifetime of the *RXTE* observations of Vela X-1 taken in 1998 January.

obs	starttime [MJD]	stoptime [MJD]	PCA sec	HEXTE-A sec	HEXTE-B sec
000	50834.82	50835.15	14200	4560	4880
001	50835.15	50835.44	12100	4390	4780
00	50835.47	50835.55	7200	1300	1390

2. SPECTRAL ANALYSIS

We first tried to fit our combined *PCA* / *HEXTE* spectra with the models usually applied to X-ray pulsar spectra. The canonical models consisting of a power law modified by some sort of high energy cutoff were either not able to describe the data or the fit resulted in unphysical parameters like negative photon indices or in the inclusion of unphysical components like a blackbody with $kT \approx 3\,\mathrm{keV}$. We then tried newer models such as the NPEX model (Mihara, 1995), but they also did not work.

We finally used a partial covering model which consists of a (heavily) absorbed NPEX component and a scattered (unabsorbed) NPEX component (using the same photon indices and temperature for both components). This model had to be further modified by an additive iron line at 6.4 keV.

$$f(E) = \left(N_1 + e^{-\sigma N_\mathrm{H}} \cdot N_2\right) \cdot \left(E^{\alpha_1} + N \cdot E^{-\alpha_2}\right) \cdot e^{-E/kT} + \mathrm{Gauss} \qquad (1)$$

This model enabled us to fit the data. Large deviations occur around $\sim 20\,\mathrm{keV}$ and $\sim 55\,\mathrm{keV}$ (see Fig. 1b). As the NPEX model has a smooth turnover – no artificial residuals which might be interpreted as a cyclotron line (Kretschmar et al., 1997b) are introduced as in the case of the High Energy cutoff – and as it can model almost any continuum including an additional spectral break (Kretschmar et al., 1997a), these two deviations can be interpreted as a cyclotron absorption line and its first harmonic. As indicated by our previous observations the two features are not coupled by the canonical value of 2 (Kreykenbohm et al., 1999); therefore we included the coupling factor as a free parameter. We coupled the line energy and the width with the same parameter (Harding, priv. comm.).

The inclusion of two Lorentzian absorption lines resulted in accteptable fits: see Fig. 1a. The parameters obtained with this model are given in Tab. 2.

3. DISCUSSION

We have shown that our *RXTE* data of Vela X-1 can be fit by a model consisting of two NPEX components: an absorbed and a scattered part. At higher energies two large deviations around $\sim 20\,\mathrm{keV}$ and $\sim 55\,\mathrm{keV}$ are present.

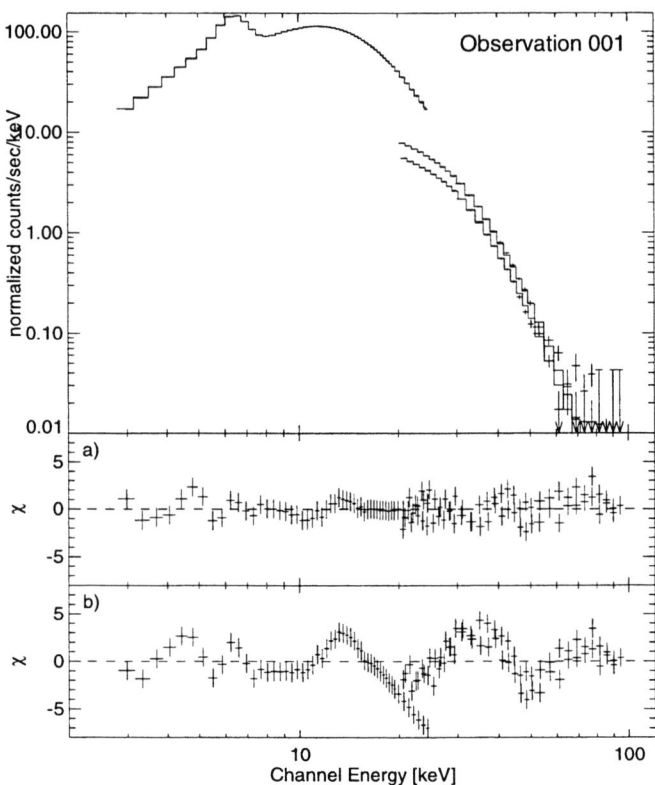

FIGURE 1. Fits to observation 001 with (panel a) and without (panel b) the inclusion of two cyclotron absorption lines. The two deviations at ~21 keV and ~55 keV are very apparent in panel b). While we are quite sure that the deviation around ~55 keV is a cyclotron absorption line, the nature of the feature at 21 keV is less clear. The line like residuals might result from the uncertainty of the *PCA* response matrix and general calibration problems above ~20 keV in the *PCA*.

While it is possible to obtain an acceptable fit after the inclusion of two cyclotron absorption lines (see Fig. 1a and Tab. 2), we have some reservations against this interpretation. The line at ~21 keV coincides closely with large uncertainties in the *PCA* response matrix due to the Xenon K-edge (Wilms et al., 1999). These uncertainties might be responsible for the behaviour of the residuals in the *PCA* around 20 keV. This could result in an absorption line like feature. If we ignore the *PCA* data above ~14 keV we detect only one absorption line at ~51 keV in the *HEXTE* data with a width of ~4 keV and a depth of ~0.6. This line shape is much more consistent with the theoretical models of Araya & Harding (1999). Additionally, a line at ~21 keV requires sometimes a coupling factor greater than

TABLE 2. Fit parameters for the observations 001 and 00. The uncertainties are on a 90 % confidence level. Norm1/Norm2 is the ratio between the absorbed (Norm1) and the scattered (Norm2) component.

obs	cts	N_H	α_1	α_2	kT	ratio
001	1076.	$39.^{+2.}_{-2.}$	$0.97^{+0.05}_{-0.05}$	$-2.58^{+0.07}_{-0.02}$	$5.1^{+0.0}_{-0.0}$	6.
00	1198.	$54.^{+2.}_{-2.}$	$1.53^{+0.05}_{-0.07}$	$-2.40^{+0.03}_{-0.16}$	$5.1^{+0.0}_{-0.0}$	20.

obs	E_{cyc}	Depth 0	Depth 1	width	factor	χ^2_{red} (DOF)
001	$21.3^{+0.3}_{-0.4}$	$0.19^{+0.03}_{-0.02}$	$0.37^{+0.16}_{-0.09}$	$6.0^{+1.2}_{-1.3}$	$2.39^{+0.17}_{-0.13}$	1.3 (103)
00	$21.2^{+0.4}_{-0.4}$	$0.17^{+0.00}_{-0.00}$	$0.47^{+1.36}_{-0.24}$	$4.6^{+0.7}_{-0.7}$	$2.75^{+0.75}_{-0.37}$	1.3 (103)

2.5 (see Tab. 2) which is extremely difficult to explain. Note that the *Ginga* LAC observations of Makishima et al. (1992), Mihara (1995) and Choi et al. (1996) might also have been affected by the response uncertainty close to the Xenon K-edge at ~30 keV.

We therefore interpret only the deviation around ~55 keV as a cyclotron absorption line. This interpretation is backed by the observation of only one line at ~57 keV by *BeppoSAX* (Orlandini et al., 1998). However, the nature of the 21 keV feature is still unclear.

Since previous (*HEXE*) and newer (*BeppoSAX*, *RXTE*) observations have consistently seen a line around ~55 keV, we consider this cyclotron absorption line as well established.

REFERENCES

Araya R.A., Harding A.K., 1999, Astrophys. J. 517, 334
Choi C.S., Dotani T., Day C.S.R., Nagase F., 1996, Astrophys. J. 471, 447
Kendziorra E., Mony B., Kretschmar P., et al., 1992, In: Tanaka Y., Koyama K., (eds.) 1992, Frontiers of X-Ray Astronomy, Frontiers Science Series 2, p.51
Kretschmar P., Kreykenbohm I., Staubert R., et al., 1997a, In: Dermer C.D., Strickman M.S., Kurfess J.D. (eds.) Proc. 4th Compton Symp.. AIP Conf. Proc. 410, AIP, Woodbury, p.788
Kretschmar P., Kreykenbohm I., Wilms J., et al., 1997b, In: Winkler C., Courvoisier T.J.L., Durouchoux P. (eds.) The Transparent Universe. ESA SP 382, Noordwijk, p.141
Kretschmar P., Pan H.C., Maisack E.K.M., et al., 1996, Astron. Astrophys.
Kreykenbohm I., Kretschmar P., Wilms J., et al., 1999, Astron. Astrophys. 341, 141
Makishima K., Mihara T., Nagase F., Murakami T., 1992, In: Tanaka Y., Koyama K., (eds.) 1992, Frontiers of X-Ray Astronomy, Frontiers Science Series 2, p.23
Mihara T., 1995, Ph.D. thesis, RIKEN, Tokio
Orlandini M., Fiume D.D., Frontera F., et al., 1998, Astron. Astrophys. 332, 121
White N.E., Swank J.H., Holt S.S., 1983, Astrophys. J. 270, 711
Wilms J., Nowak M.A., Dove J.B., et al., 1999, Astrophys. J. 522, 460

HER X-1 X-RAY TURN-ON MONITORED BY RXTE

M. Kuster [1], J. Wilms [1], R. Staubert [1], D. E. Gruber [2], R. E. Rothschild [2], W. A. Heindl [2]

1) Institut für Astronomie und Astrophysik – Astronomie, Waldhäuser Str. 64, D-72076 Tübingen, Germany
2) CASS, University of California at San Diego, La Jolla, CA 92093, U.S.A.

ABSTRACT   We report on a two day long continuous monitoring of the X-ray binary Her X-1 by the Rossi X-ray Timing Explorer (RXTE). The observation resulted in a complete and highly time-resolved coverage of an X-ray Turn-On in the energy range from 3 to 100 keV. The increase in flux from the off-state to the maximum level in the main on-state took almost two days. During these days a pre-eclipse dip, an anomalous dip, and an eclipse occured. The spectral modeling of the Turn-On data showed that $N_{\rm H}$ is increasing during the early phases of the Turn-On. This is counterintuitive to a pure absorption model. In this paper we present a physically more realistic interpretation with a scattering model that is able to reproduce the measured data.

KEYWORDS: stars: neutron — stars: individual (Her X-1)

1. INTRODUCTION

The 35 d cycle of Her X-1 (Giacconi et al., 1973) is one of the best evidences for the presence of inclined, precessing and warped accretion disks in X-ray binary systems. The warping is caused either by radiation driven accretion disk winds (Schandl & Mayer, 1994), or by radiation pressure (Maloney & Begelman, 1997). Due to the low inclination of the binary orbit, the warped disk blocks the line of sight to the central neutron star during about 60% of the 35 d cycle. The ~12 d long "main on" state begins when the outer accretion disk rim frees the line of sight to the neutron star, the end of the main on happens when the inner edge of the disk starts to cover the neutron star again. While the end of the main on has been the subject of a previous *Ginga* observation (Deeter et al., 1998), observational data on the spectral evolution during the start of the main on has been rare.

In 1997 September we observed a Turn-On of the 35 day cycle of Her X-1 with a two day continuous observation by the Rossi X-ray Timing Explorer (RXTE). In this paper we present results from the spectral and temporal analysis of this observation. Results from the preliminary data analysis have been presented earlier (Kuster et al., 1999). Fig. 1 displays the RXTE PCA light curve. During the Turn-On an anomalous dip, a pre-eclipse dip, and an eclipse occurred. For each of the numbered spacecraft orbits in Fig. 1 we extracted PCA spectra and light curves for both pointed instruments on RXTE. In Sect. 2 we discuss the evolution of the spectral parameters during the Turn-On. In Sect. 3 we describe the development of the energy dependent pulse profile.

2. EVOLUTION OF SPECTRAL PARAMETERS

The observed spectrum is modeled as a partial covering model, i.e. as the sum of an absorbed spectral component and a second unabsorbed component. We modeled both spectra using an exponentially cutoff power-law with a cyclotron scattering feature at 39 keV. With the exception of their normalization, the parameters in the unabsorbed and in the absorbed spectral component were identical. To this model we add a Gaussian emission line, fixed at 6.4 keV. Our modeling showed that, for times of large N_H, using the standard Morrison & McCammon (1983) absorptivity (the wabs model of XSPEC) does result in unphysical residuals. This is caused by weak absorption edges which are not taken into account in the wabs model. Using the cross sections of Bałucińska-Curch & McCammon (1992) removes this problem (the phabs model of XSPEC). We note, however, that the resulting N_H values are smaller than those inferred from the wabs model.

In Fig. 2 we display the evolution of the spectral parameters over the Turn-On. We omitted the data from the pre-eclipse dip (caused by small absorbing blobs from the accretion stream; Stelzer et al., 1999) and the anomalous dip (caused by the thickening of the accretion disk rim following the impact of the accretion stream onto the disk; Schandl, 1996), since our spectral model is not applicable during these periods. During the end of the Turn-On, it was necessary to add an additional black body component with a temperature of ~ 0.2 keV to the model. This latter component might be due to the accretion disk radiation itself. Such a component is also seen by Parmar et al. (this volume) in *Beppo SAX* observations of Her X-1.

When modeling the spectra with a partial covering model, N_H *is smaller* during the first few RXTE orbits and then increases. This is counterintuitive to a model, in which the gradual increase in flux is attributed to a gradual decrease of absorption. Interpreting the data with a physically more realistic scattering model indicates that during the early stages of the Turn-On (observations 00 to 07 of Fig. 2), *only the scattered component is seen in absorption*, whereas the direct line of sight to the neutron star is blocked. Later in the Turn-On, around observation 08, unscattered and strongly absorbed photons start to dominate the spectra. After this observation, the line of sight to the neutron star is freed by the accretion disk rim. Consequently, N_H decreases. After the anomalous dip, the observed flux is dominated by the absorbed spectral component.

3. EVOLUTION OF THE PULSE PROFILE

The same effect is also seen in the evolution of the pulse profile (see Fig. 3). Contrary to the end of the main on, the pulse profile does not change in shape during the Turn-On (Kuster et al., 1999). This is a result of the fact that the outer accretion

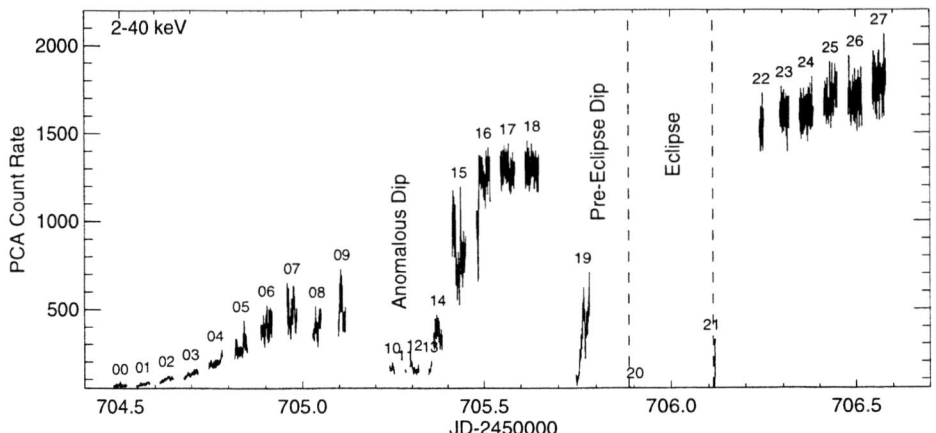

FIGURE 1. RXTE PCA count rate from 2 to 40 keV. The numbers identify the individual RXTE orbits analyzed and correspond to the notation in Fig. 3.

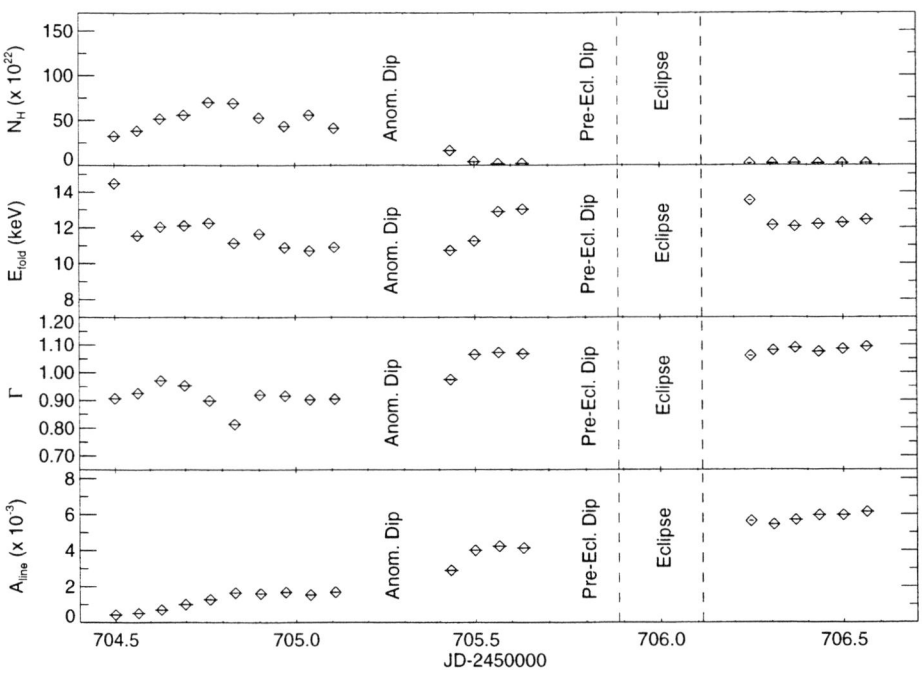

FIGURE 2. Top to bottom: total effective $N_{\rm H}$, $E_{\rm fold}$, Γ, and $A_{\rm line}$ of the iron line. The values are determined by fitting spectra for each spacecraft orbit with a partial covering model.

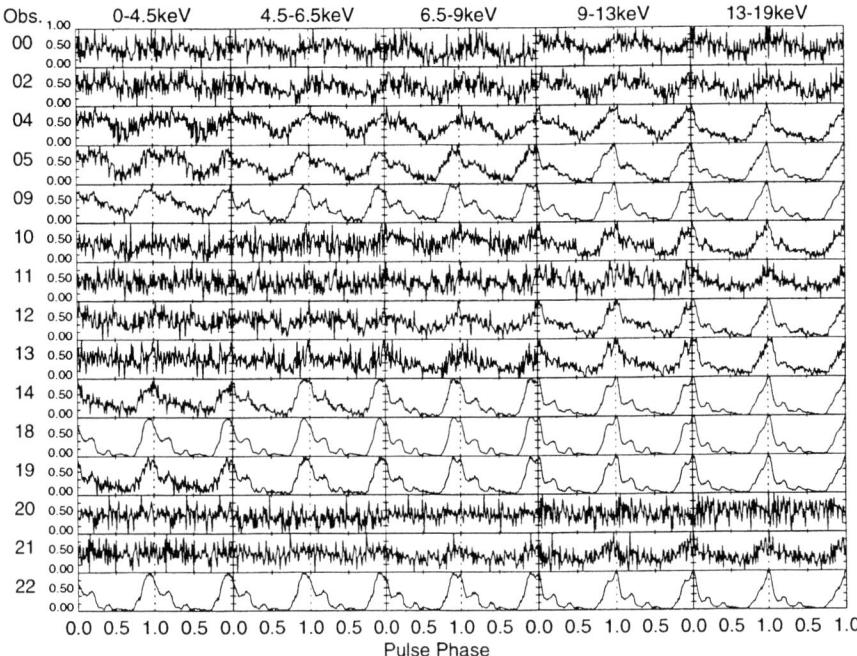

FIGURE 3. Energy dependent evolution of the pulse profile over the Turn-On. Each pulse profile was determined from one RXTE orbit over the Turn-On. All profiles are normalized to unity, after subtraction of the none pulsed flux. Pulse phase 0 was arbitrarily defined by the main pulse of the 13 – 19 keV band. The large noise present in observations 00 – 04 and 10 – 13 is due to the strong photoelectric absorption.

disk rim opens up the line of sight to the neutron star. The energy dependence of the shape of the pulse profile is in agreement with previous observations. Note the difference between the off-state absorption in, e.g., observation 02, where all energies appear to be influenced, and the pure photoelectric absorption in observations 10 – 14.

REFERENCES

Bałucińska-Church, M., McCammon, D., 1992, Astrophys. J., 699, 400
Deeter, J.E., Scott, D. M., et al., 1998, Astrophys. J., 502, 802
Giacconi, R., Gursky, et al., 1973, Astrophys. J., 184, 227
Kuster, M., Wilms, J., et al., 1999, Astrophys. Lett. Commun., 161, 38
Maloney, P., & Begelman, M. C., 1997, Astrophys. J., 491, L43
Morrison, R., & McCammon, D., 1983, Astrophys. J., 270, 119
Parmar, A. N., this volume
Schandl, S., 1996, Astron. Astrophys., 307, 95
Schandl, S., & Meyer, F., 1994, Astron. Astrophys., 289, 149
Stelzer, B., et al., 1999, Astron. Astrophys., 342, 736

RXTE OBSERVATIONS OF SEYFERT GALAXIES: EVIDENCE FOR REFLECTION FROM DISK AND TORUS

G. Lamer, P. Uttley, and I.M. McHardy

Department of Physics&Astronomy, University of Southampton, United Kingdom

ABSTRACT

We have monitored the Seyfert galaxies NGC 4051 and NGC 5506 with RXTE for the last 3 years. During this time, we have witnessed large variations in the X-ray flux of both sources, coupled to spectral variability. In both objects the 6.4 keV iron fluorescence line and the reflection hump are relatively stronger during fainter states of the X-ray source. These variations can be simply explained if there are not one, but two reflection components contributing to the X-ray spectrum of these sources: a reflection component which scales linearly with the primary continuum flux (as expected from the accretion disk), and a constant component due to reflection off distant cold matter.

KEYWORDS: Galaxies: Seyfert; X-rays: galaxies; Galaxies:individual:NGC 5506; Galaxies:individual:NGC 4051

1. INTRODUCTION

The 2-30 keV X-ray spectra of Seyfert galaxies are dominated by two components: a power law component that is believed to originate from the primary source of hard X-rays near the central black hole and a reflected component produced by Compton scattering of the primary radiation by neutral material (eg Pounds et al. 1990). The principal spectral signatures of the reflected component are a broad hump peaking at ~ 30 keV and iron K_α fluorescence emission at 6.4 keV. The reflection is believed to occur in the accretion disk, as the associated fluorescence line is often broadened by relativistic motion and gravitationally redshifted (Tanaka et al. 1995, Nandra et al. 1997, Reynolds et al. 1997). However, Ghisellini et al. (1994) showed that the molecular torus required by unified models of Seyfert galaxies is also likely to modify the X-ray spectra of Seyfert 1 as well as Seyfert 2 galaxies by reprocessing the primary radiation. Any reflection component originating from the torus will lag the variability of the primary X-ray source by months or years and hence will manifest itself as a virtually constant reflection component.

We present evidence from long term X-ray monitoring of the Seyfert galaxies NGC 4051 and NGC 5506, that non-variable reflection components are present in these sources.

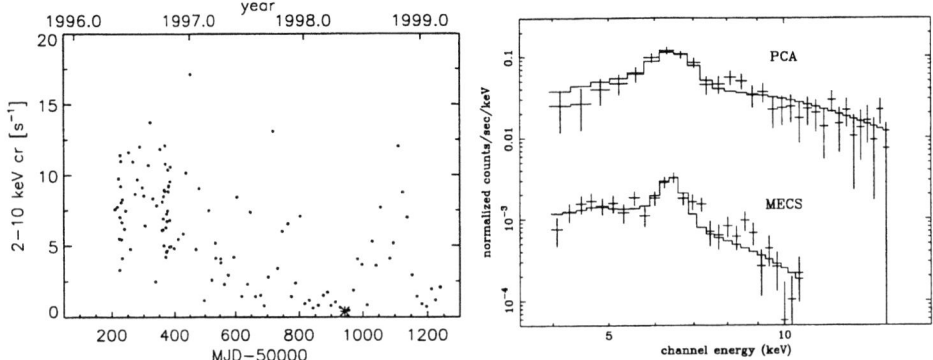

FIGURE 1. a.: (*left*) *RXTE* PCA lightcurve of NGC 4051 during the years 1996 - 1999. The asterisk marks the epoch and PCA countrate at the time of the simultaneous *BeppoSAX* and *RXTE* long look observations in May 1998 at the end of a $\sim$ 150 day low state. b.: (*right*) *RXTE* PCA and *BeppoSAX* MECS spectra during the May 1998 long look observation. The data are fitted by a pure reflection model with no contribution from the primary X-ray source (Uttley et al. 1999).

2. NGC 4051

The Seyfert 1 galaxy NGC 4051 has been observed by *BeppoSAX* and *RXTE* during an extreme low state in May 1988. Guainazzi et al. (1998) showed that the X-ray spectrum revealed a "bare" reflection hump and very strong iron fluorescence emission with $EW \sim 1000$ eV (Figure 1a). These features require an incident spectrum with much higher flux than observed in May 1998. Our long term *RXTE* PCA lightcurve (Figure 1b) shows, that the primary X-ray source had been switched off for ~ 150 days before the observations, giving a measure for the minimum time delay of the reflected radiation (Uttley et al. 1999). This places the reflecting material at a distance $> 10^{17}$cm from the primary X-ray source.

3. NGC 5506

We have observed the Seyfert 2 galaxy NGC 5506 with *RXTE* since May 1996 as part of a long term monitoring campaign and more intensely during June/July 1997.

When we attempt to fit the *RXTE* PCA spectra with a single power law spectrum, the iron K_α iron fluorescence line at 6.4 keV and a broad reflection hump are clearly visible in the fit residuals. We therefore fit the X-reay spectra with the XSPEC PEXRAV reflection model (Magdziarz & Zdziarski 1995) plus a gaussian emission line. The results of the model fits to the June 1997 spectrum are listed in Table 1.

We have accumulated count rate selected spectra in 4 different count rate bands for the June 1997 observations and in 6 count rate bands for the long term monitoring. We fit these spectra with the PEXRAV reflection model (Magdziarz & Zdziarski 1995) and a gaussian emission line. In order to minimise the errors of the interesting

TABLE 1. Spectral fits for June 1997

model	Γ	line EW [eV]	line E_{obs} [keV]	line σ [keV]	R_{refl}	χ^2 (dof)
pwl	2.00±.005	-	-	-	-	3089 (48)
pwl+gauss	1.85±.007	248±15	6.23±.01	0.41±.22	-	726 (45)
pexrav	2.19±.008	-	-	-	1.77±.06	1781 (48)
pexrav+gauss	2.13±.014	149±27	6.37±.02	0.30±0.08	1.20±.08	52.7 (45)

parameters we have fixed the parameters that are unlikely to vary. The absorbing column density was set to $3.6 \cdot 10^{22}$ as derived from *BeppoSAX* spectra (Perola et al. 1998). The high energy cuttoff parameter of the PEXRAV model has been set to 300 keV and the inclination angle of the reflecting slab was assumed to be 40°, the best fit value from *ASCA* spectroscopy (Wang et al. 1999). The source frame energy of the iron fluorescence line and its width were set to $E = 6.4$ keV and $\sigma = 0.35$ keV as these values are compatible with all our *RXTE* PCA spectra.

The relations of the intrinsic photon index Γ, the reflected fraction R and the iron line flux with the count rate are plotted in Figure 3. The spectra derived from the monitoring observations show a strong correlation of Γ and count rate, whereas during the June 1997 observations the intrinsic photon index was constant at Γ=2.13. We find a strong anti-correlation of the Fe K_α equivalent width and count rate. The values of the reflected fraction R tentatively show a similar trend with count rate. This is indicative of a constant reflection component which contributes relatively stronger during epochs of low luminosity. However, it is clear from the positive correlation of the absolute line flux and total count rate, that a variable reprocessed component is also present.

We have modeled the countrate dependency of the reflected fraction R, the absolute flux $F_{K\alpha}$ and the equivalent width EW of the iron line by simulating PCA spectra using PEXRAV+GAUSS models (see Lamer et al., 2000). The simulated spectra combine two reprocessed components: one constant reflection component based on the mean value of the primary X-ray flux and one component that is varying proportionally to the primary X-ray flux. Both reflection components are accompanied by iron K_α fluorescence emission. The ratios of reflected continuum flux and fluorescence flux are assumed to be the same in both components. We then fitted the simulated spectra with single PEXRAV+GAUSS models. The lines in Figure 3 indicate the best fitting simulation with a reflective fraction for the variable component $R_{var}=0.7$ and $R_{const} = 0.5$ for the constant component. The contributions to the iron line equivalent width are 100 eV from the variable component and 75 eV from the constant line component when the X-ray source is at its average flux level.

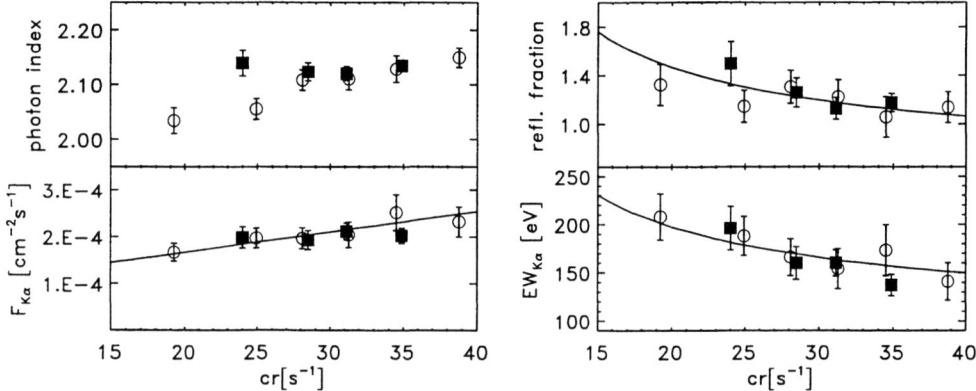

FIGURE 2. PEXRAV+GAUSS fit parameters of count rate selected PCA spectra of NGC 5506. Filled squares: June 1997 observations. Open circles: Monitoring observations. The lines indicate the values of reflected fraction and line fluxes expected from a self-consistent model with two reflectors: One close to the primary X-ray source and a more distant reflector causing a non-variable reflected component.

4. CONCLUSIONS

The 1998 low state of NGC 4051 enabled the first direct proof for X-ray reflection from matter located at a large distance from the nuclear X-ray source of a Seyfert galaxy. The minimum distance of 10^{17} cm makes a molecular torus the most likely reflector. In NGC 5506 the X-ray spectral variability revealed two reflection components. The variable reflection component is probably associated with the accretion disk while the torus is the likely origin of the constant reprocessed radiation.

REFERENCES

Ghisellini G., Haardt F., and Matt G., 1994, MNRAS, 267, 743
Guainazzi M., Nicastri F., Fiore F., et al., 1998, MNRAS 301, L1
Lamer G., Uttley P., and McHardy I.M., MNRAS, in press
Magdziarz P., and Zdziarski A., 1995, MNRAS, 273, 837
Perola G.C., 1998, Nuclear Physics B (Proc. Suppl.), 69/1-3, 477
Pounds, K.A., Nandra, K., Steward, G.C., George, I.M., and Fabian, A.C., 1990, Nature, 344, 132
Tanaka, Y., Nandra, K., Fabian, A.C, et al., 1995, Nature, 375, 659
Uttley P., McHardy I.M., Papadakis I.E., Guainazzi M., Fruscione A., 1999, MNRAS, 307, L6
Wang T., Mihara T., Otani C., Matsuoka M., and Awaki H., 1999, ApJ, 515, 567

GINGA OBSERVATIONS OF THE SHORT HIGH STATE IN HER X-1

D. Leahy

Department of Physics and Astronomy, University of Calgary, Alberta, Canada, T2N 1N4

ABSTRACT   The Short High state in Her X-1 has only been observed once so far with good coverage and good sensitivity. These observations were made with the GINGA LAC in 1989, and have not been analyzed in any detail previously. Here are reported results from analyses of the light curves and spectra from this data set.

KEYWORDS: stars: neutron — X-rays: stars — binaries: eclipsing — stars: individual: Her X-1

1. INTRODUCTION

The X-ray binary system Hercules X-1 is well reviewed by Scott (1993). The properties of the 35 day cycle are recently discussed by Scott & Leahy (1999) and Shakura et al (1998). Deeter et al (1998) discuss the X-ray pulse profile. Recent X-ray spectra of Her X-1 are studied in Oosterbroek et al. (1998), Dal Fiume et al (1998) and Choi et al (1997). Leahy & Scott (1998) give an updated set of binary parameters. Ultraviolet spectra of Her X-1 are presented by Boroson et al. (1997) and Vrtilek and Cheng (1996). Still et al (1997) discuss optical signatures of reprocessing on the companion and accretion disk.

The shape of the 35 day cycle and durations of the different states has only recently been well determined (Scott & Leahy 1999). The Main High state has been well studied, however the Short High state has not. Yet the properties of the Short High state are sensitive to the physical state of the accretion disk causing obscuration and reprocessing of the radiation from the neutron star. Only one set of comprehensive set of observations of the Short High state exists to date, those made by GINGA in 1989. These observations have not been analyzed in any detail yet. Here is reported results from analysis of these observations of the Short High state of Her X-1 by GINGA.

2. OBSERVATIONS

Hercules X-1 was observed with GINGA during the period May 16-20, 1989 (MJD = JD-2400000.5 =47662 to 47666). The observations were made with the GINGA LAC detectors with modes: MPC1, MPC2 and MPC3. MPC1 and MPC2 modes have 48 spectral channels and MPC3 mode has 12 spectral channels. The analysis here uses all three modes. Four energy bands were defined: Band 1 (2-4.6 keV); Band 2 (4.6-9.2 keV); Band 3 (9.2-13.8 keV) and Band 4 (13.8-36.8 keV). Background subtraction was performed for all of the data prior to lightcurve and spectral analysis.

3. LIGHT CURVE

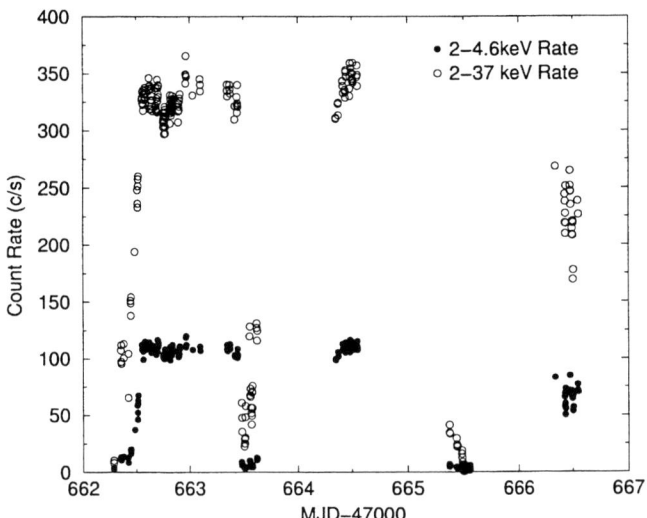

FIGURE 1. The GINGA light curve of the 1989 Short High observation campaign of Her X-1. The count rate is given for total (2-37 keV) and Band 1 (2-4.6 keV) energy bands.

The GINGA lightcurve of the May 1989 Short High State is given in Figure 1, for band 1 and for the total energy range. All of the GINGA LAC observations are included. The turn on shows the fastest rise in the lowest energy band, with the rise being more gradual as one considers higher energy bands. The energy dependence is consistent with reduced absorption being the cause of the turn on. The shape and timescale of the Short High turn on are similar to those for Main High turn on, consistent with the idea that both are caused by emergence of the neutron star from behind the cold outer edge of the accretion disk (Scott, Wilson and Leahy, in preparation). The count rates and spectrum are fairly steady during the early part of the bright phase of the Short High phase (on MJD47662) and the middle part (on MJD47664).

A major dip episode occurs near MJD 47663.5. The 2-4.6 keV band decreases the most and the 13.8-36.8 keV band the least consistent with cold matter absorption. Ingress into eclipse occurs near MJD 47665.45. Eclipse center is at MJD 47665.60061. The decline of the Short High state is observed on MJD 47666. The count rates are lower and fluctuate more than in the bright phase of the Short High, but the ratios of the different bands are similar, consistent with electron scattering causing the intensity fluctuations.

4. SPECTRAL ANALYSIS

The data set was divided into a number of sections, which were processed into a set of spectra, some from 48-channel MPC1 and MPC2 modes, and others from 12-channel MPC3 mode. These were fitted with a model consisting of an absorbed continuum component and an unabsorbed component plus an iron emission line. This model was the only simple model which could adequately represent all of the spectra. Both continuum components of the model were a power law times a cyclotron absorption line. Several parameters were found to be consistent with a constant for all of the spectra and thus were fixed (the cyclotron energy, width and depth parameters, the iron line energy and width, and the power law index). This left as free parameters: unabsorbed continuum normalization, absorbed continuum normalization, column density, and iron line intensity. The spectral fits resulted in measurement of the free parameters as a function of time throughout the Short High observation.

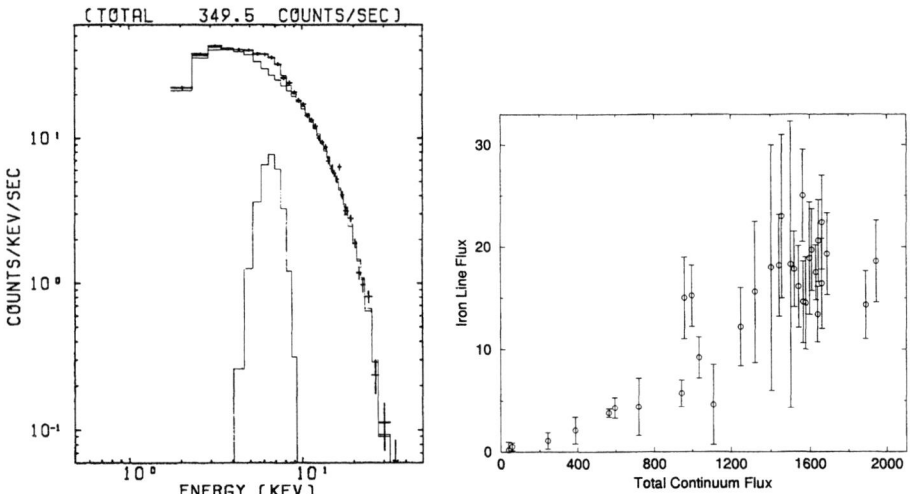

FIGURE 2. a. (left panel) GINGA Spectrum from MJD47663 and model spectral fit. b. (right panel) Iron line flux vs. total (absorbed plus unabsorbed) continuum flux.

A 48-channel MPC2 mode spectrum (crosses) and its best fit model (histograms, with unabsorbed continuum, iron line and total shown separately) are shown in Fig. 2a. In this case the fit did not required an absorbed continuum component, but for many ($\sim 2/3$) of the spectra the absorbed continuum component was required.

5. RESULTS AND DISCUSSION

During turn on, the dip episode and the eclipse ingress the spectra are dominated by the absorbed component. However all other spectra are dominated by the unabsorbed component, and for most of the bright phase of the Short High (MJD47662.55 to MJD47665) the absorbed component is either very small or consistent with zero. For the late Short High (during MJD 47666), the absorbed component is 10% to 25% of the unabsorbed component. The column density, for the spectral fits where an absorbed component is required, is nearly constant.

The absorbed intensity is correlated with the unabsorbed intensity: as the unabsorbed intensity increases, the absorbed intensity at first rapidly increases then slowly decreases back to zero. This can be explained by an extended source which is uncovered by an absorber which has an edge region with column density of $\sim 10^{24} cm^2$ and a main region that is optically thick. The iron line flux shows a very good linear correlation with total intensity (absorbed plus unabsorbed): see Fig.2b. This is good evidence that the fluorescent region is the same size as the extended source region. For the turn-on, the column density is constant but the absorbed normalization increases then decreases while the unabsorbed normalization monatonically increases. This behaviour is consistent with an extended source which is gradually uncovered by an absorber with constant column density. The dip episode at MJD47663.5 can be explained by the same mechanism. The eclipse ingress on MJD47665 shows similar behaviour, consistent with a source that is extended and partially covered before eclipse. In summary, several lines of evidence point towards there being an extended source in Short High. This source is likely the inner accretion disk scattering X-rays from the neutron star and emitting flourescent iron line photons.

ACKNOWLEDGEMENTS

This work was supported in part by the Natural Sciences and Engineering Research Council of Canada.

REFERENCES

Boroson, B., Blair, W., Davidsen, A., et al., 1997, ApJ, 491, 903
Choi, C., Seon, K., Dotani, T. Nagase, F., 1997, ApJ, 476, 81
Dal Fiume, D., et al. 1998, A&A, 329, L41
Deeter,J., Scott, D., Boynton,P., et al., 1998, ApJ, 502, 802
Leahy, D., Scott D., 1998, ApJ, 503. L63
Oosterbroek, T., Parmar, A., Martin, D., Lammers, U. 1998 A&A 329, L41
Scott D. M., 1993, PhD Thesis, University of Washington
Scott D., Leahy, D. 1999, ApJ, 510, 974
Shakura, N., Ketsaris, N., Prokhorov, M., Postnov, K. 1998, MNRAS, 300, 992
Still, M., Quaintrell, H., Roche, P., Reynolds, A., 1997, MNRAS, 292, 52
Vrtilek, S., Cheng, F. 1996, ApJ, 465, 915

TWO PHOTON BREMMSTRAHLUNG IN STRONG MAGNETIC FIELDS

D. Leahy & L. Semionova

Department of Physics and Astronomy, University of Calgary, Alberta, Canada, T2N 1N4 & Department of Physics, Universidad National, Heredia, Costa Rica

ABSTRACT

The results are given for the second order calculation for an electron in a strong magnetic field which emits two photons. The electron transition can be between any Landau level and any lower Landau level. We use the Sokolov electron wavefunctions so the calculation is valid for any magnetic field. The results are of particular interest for processes in magnetars, for which the magnetic field is near or can exceed the critical value.

KEYWORDS: radiation mechanisms: non-thermal — magnetic fields — polarization

1. INTRODUCTION

We analyse the emission of two photons by an electron in a strong magnetic field, a process which is of second order. This process is also called two-photon emission, second order synchrotron radiation or second order Bremsstrahlung. The calculations here are valid for any value of magnetic field. The critical value of the magnetic field is defined as $B_{cr} = \frac{m^2 c^3}{e\hbar} = 4.414 \times 10^{13}$ Gauss.

It is possible that this process plays an important role in understanding the radiation mechanism in pulsars and in γ-ray burst events. With the observational discovery of magnetars (Vasisht & Gotthelf 1997), which are pulsars with magnetic fields in the range $\sim 10^{14} - 10^{15} Gauss$, calculations which are valid for such high fields are of great interest. Previous work related to photon emission by electrons in strong magnetic fields includes the following. The first order emission process is discussed by Harding & Preece (1987), Latal (1986), Semionova (1983), Herold et al (1982), White (1978) and White (1974). Bussard et al (1986) calculate first and second order Compton scattering, in which there is one photon in both the initial and final states. They use an S-matrix formalism which is applicable to arbitrarily strong fields, but the photon normal modes they use to include the effects of vacuum polarization and plasma apply only in the weak field limit ($B << B_{cr}$). They also use the Johnson-Lippmann electron wavefunctions (Johnson & Lippmann 1949) which makes their calculation valid only in weak fields, as shown by Graziani (1993). Graziani (1993) also has given the correct treatment of the resonant line width for strong field cyclotron scattering using the appropriate electron wavefunctions, which are those of Sokolov & Ternov (1968). Alexander & Meszaros (1991) use the results of Bussard et al (1986) to calculate opacities and source functions, thus are also valid only in the weak field limit.

Here we calculate the second order emission process. The calculation here utilizes the electron wavefunctions of Sokolov & Ternov (1968), and is valid for strong fields. We also treat electron spin correctly when the parallel momentum of the electron is non-zero.

2. CALCULATION OF THE TRANSITION RATE

The details of the calculation are lengthy. They are to be published in Phys. Rev. D, and are not given here. We use the S-matrix formalism and the electron wavefunctions of Sokolov, which are also used by Herold (1979). These wavefunctions correctly describe the spin of the electron in the presence of a magnetic field. We also explicitly keep track of photon polarization, which can be described by the polarization unit vectors for linear polarization or for circular polarization. The four possible options for the spins of the incoming and outgoing electrons are: (up, up), (up, down), (down, up) and (down, down). The transition probability into the states $(N' \to N)$ is evaluated from the S-matrix, with a summation over the quantities not observed in the final state, yielding the expression for the transition rate from the initial state (to be given in the Phys. Rev. D paper).

3. RESULTS

The formula for the transition rate shows that the total cross section depends on the frequency, angle and polarization of both photons emitted, as well as on the field strength. Photon scattering shows resonances, where the denominator of the expression for the transition rate vanishes. The resonance energies may be smaller or larger than $\omega_B = eB/(mc)$. In our results for the resonance frequency, we may take $\omega' = 0$, which corresponds to the process in which only one photon is emitted. This agrees with previous calculations for bremsstrahlung of the first order (Semionova & Paez 1992).

Figure 1 presents the dependence of the resonance frequency ω_{res}/m (i.e. in units of electron mass energy) on the angle of emission of the photon, θ, for $B = B_{cr}$. The direction of the magnetic field is the horizontal axis in the figure; ω_{res} is independent of azimuthal angle. The different curves are for values of $p_z'/m = 0, 1, 2, 3, 4, 5, 6$ (i.e. p_z' from zero to $6mc = 3.07 MeV/c$ in more common units). The curves have monotonically increasing resonance frequency in the $\theta = 0$ direction as p_z' increases, and monotonically decreasing resonance frequency in the $\theta = 180°$ direction. So, an electron moving in the x-y plane ($p_z' = 0$) has an approximately constant resonance frequency, almost independent of the emission angle of the photon. With an increase in p_z', a maximum value for ω_{res} is found when the emission of the photon ω_{res} is in the direction of the magnetic field. A minimum in the resonance energy is found when the photon is emitted at $180°$ to the magnetic field. The resonance energy is

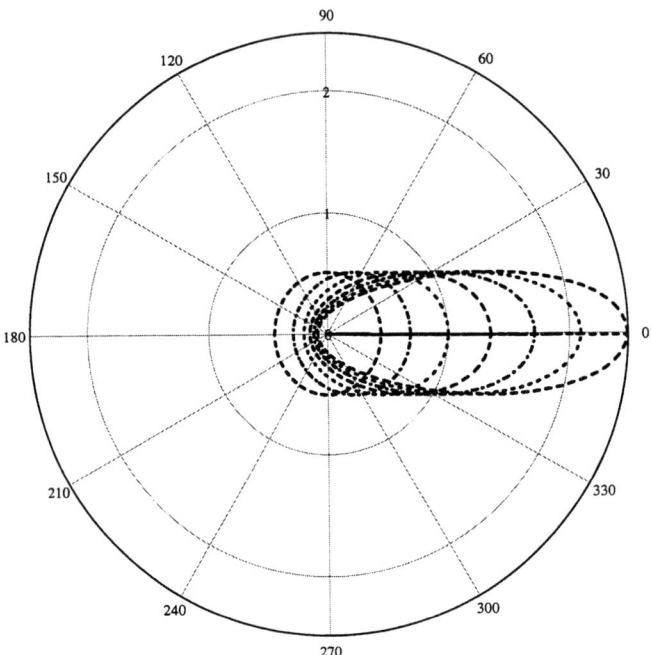

FIGURE 1. The dependence of the resonance frequency on the angle of emission of the photon, θ, for $p_z'/m = 0, 1, 2, 3, 4, 5, 6$ and for $B = B_{cr}$.

seen to be strongly dependent on both the photon emission angle and on p_z'.

We now present some results of evaluation of the transition rates. In practise the by-far most common final state is the ground state ($N = 0$), which has electron in the spin down state. The most common (highest rate) transitions have the electron in initial state $N' = 2$, spin-down, and have intermediate state $n = 1$. We present the differential transition rate, $dW/d\Omega$, summed over photon polarization, as a function of the magnetic field strength. The two photons are emitted at specific angles to the magnetic field and for the figures, the azimuthal angles of the photons are taken to be equal ($\phi' = \phi$). Figure 2 presents the results for the case of one photon emitted with energy 58 keV and emitted at 5° to the field (with Γ denoting $dW/d\Omega$ in the figure labels). The other photon has energy determined by energy conservation. We present transition rates for three cases of emission angle with respect to the magnetic field, 5° (solid line), 15° (dotted line) or 60° (dashed line). The incoming electron parallel momentum is $p_z' = 0$. The transition rate is peaked near $0.12 B_{cr}$, which is expected, since that is the field strength for which the resonant energy is 58 keV. The transition rate is largest for the second photon emitted at the largest angle.

We have also calculated the differential transition rates, $dW/d\Omega$, for different photon polarizations and the angular dependences of the transition rates. Details are to be given in the Phys. Rev. D paper.

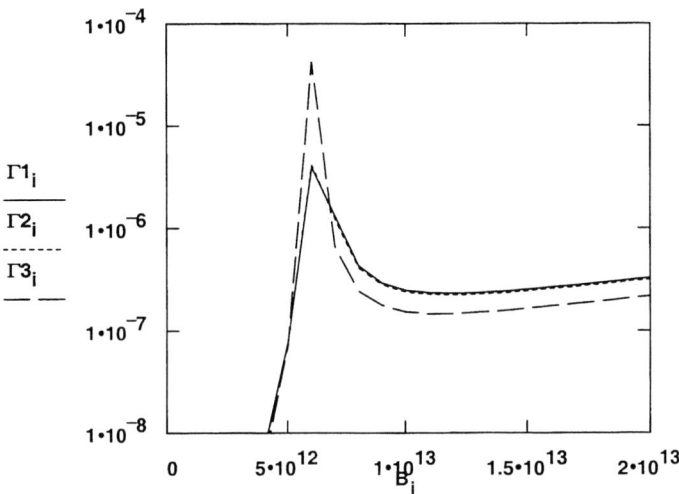

FIGURE 2. The dependence of the transition rate per steradian on magnetic field, for an electron with $p'_z = 0$, with one photon emitted at 58 keV and at angle $\theta = 5°$ to the field, and for the second photon emitted at $\theta = 5°$ (solid line), $\theta = 15°$ (dotted line), and $\theta = 60°$ (dashed line). The rate is summed over polarization of the photons.

ACKNOWLEDGEMENTS

This work was supported in part by the Natural Sciences and Engineering Research Council of Canada.

REFERENCES

Alexander, S. Meszaros, P. 1991, Astrophysical Journal, 372, 554
Bussard, R. Alexander, S. Meszaros, P. 1986, Physical Review D, 34, 440
Graziani, C. 1993, Astrophysical Journal, 412, 351
Harding, A. Preece, R. 1987, Astrophysical Journal, 319, 939
Herold, H. 1979, Physical Review D, 19, 2868
Herold, H., Ruder, H., Wunner, G. 1982, Astronomy and Astrophysics, 115, 90
Johnson, M. Lippmann, B. 1949, Physical Review D, 76, 828
Latal, H. 1986, Astrophysical Journal, 309, 372
Semionova, L., M.Sc. Thesis, University of Costa Rica (1983)
Semionova, L. Paez, J. 1992, Science and Technology, UCR, 16, 79
Sokolov, A. Ternov, I., Synchrotron Radiation (Berlin: Akademie 1968)
Vasisht, G. & Gotthelf, E. 1997, ApJ, 486, L129
White, D. 1978, Physical Review D, 18, 2166
White, D. 1974, Physical Review D, 9, 868

A SERIES OF ECLIPSES OF HER X-1 OBSERVED WITH RXTE

D. Leahy & D. M. Scott

Department of Physics and Astronomy, University of Calgary, Alberta, Canada, T2N 1N4 & Space Science Laboratory ES-84, NASA/Marshall Space Flight Center, Huntsville, AL 35812

ABSTRACT One of the main goals of an RXTE AO2 observing campaign was to observe a series of eclipse ingresses and egresses in Her X-1. Nine ingresses and egresses were observed during AO2. They fall into two categories: rapid ingresses and egresses, which are due to extinction of a point-like x-ray source by the extended atmosphere at the limb of HZ Her; and extended ingresses and egresses, which are not well understood.

KEYWORDS: stars: neutron — X-rays: stars — binaries: eclipsing — stars: individual: Her X-1

1. INTRODUCTION

Hercules X-1 is one of the brighter, and most studied, of the x-ray binary pulsars. It exhibits a wealth of phenomena, including pulsations at 1.23 seconds, eclipses at the orbital period of 1.7 day, and a 35 day cycle in the x-ray intensity. Her X-1 is reviewed by Scott (1993). Recent discussions of the properties of the 35 day cycle are given by Scott & Leahy (1999) and Shakura et al. (1998). The x-ray pulse profile evolution is discussed in Deeter et al. (1998). Recent x-ray spectra of Her X-1 are given by Oosterbroek et al. (1998) and Dal Fiume et al. (1998) (from BeppoSAX) and Choi et al. (1997) (from ASCA). An updated set of binary parameters is given by Leahy & Scott (1998). Analysis of ultraviolet spectra of Her X-1 are presented by Boroson et al. (1997), and Vrtilek and Cheng (1996). Optical signatures of reprocessing on the companion and accretion disk are discussed by Still et al. (1997).

The origin of the 35-day cycle is of great interest, and is attributed to be due to changing obscuration by a tilted precessing accretion disk. The shape of the 35 day cycle and durations of the different states has only recently been well determined (Scott & Leahy 1999): the Main High state covers 35 day phases 0-0.31, and the Short High covers approximately phases 0.56-0.76. Eclipse observations have the promise to reveal much about the geometry of the system. Eclipse ingress and egress timing restricts tightly the binary parameters (Leahy & Scott 1998). The atmosphere of the companion has been probed by the ingress profile (Leahy & Yoshida 1995 and references therein). Extended ingresses and egresses give evidence that the x-ray source is extended at certain times during the 35-day cycle (Leahy 1995b) instead of point-like as during the main high state. However the set of eclipse ingress and egress observations to date has been very limited.

An RXTE observation campaign on Her X-1 was initiated with the express purpose of observing as many eclipse ingresses and egresses as possible during the main high and short high states. Here we report results from an analysis of these observations.

2. OBSERVATIONS

Her X-1 was observed with RXTE during Sept-Oct 1997, with 5 pointings during the short high on MJD50340 - MJD50346, 12 pointings during the following main high on MJD50356 - MJD50366, and 10 pointings during the following short high. 2 ingresses and 3 egresses were observed during the two short highs, and 1 ingress and 3 egresses were observed during the main high.

Background subtraction was performed for all of the data using version 4.0 of PCABACKEST. The times were barycentered and corrected for the orbit of the pulsar. Orbital phase was determined using the ephemeris of Wilson, Scott & Finger (1997) and 35-day phase using Scott & Leahy (1999). Eclipse center is at MJD 47665.60061 (accurate to .00001).

3. LIGHT CURVES

The light curves in several energy bands, including Band 1 (defined as count rate in 0-4.6 keV), Band 2 (4.6-9.4 keV), Band 3 (9.4 to 13.8 keV), and Total (0-90 keV), were determined for for the 3 ingresses and 6 egresses. The softness ratio, Band 1 divided by Band 3 , is a good indicator of low energy absorption associated with eclipse ingresses and egresses (e.g. see Leahy & Yoshida 1995) and marks the times when the x-ray point source disappears behind the limb of HZ Her. Figure 1 shows all of the observed egresses plotted as a function of orbital phase. The egresses are labeled by MJD-50000 and plotted with the symbols shown in the figure legend.

4. DISCUSSION

Several topics can be studied with the ingress and egress light curves, such as the atmospheric structure of HZ Her, or the extent of a scattering region around the pulsar (e.g. see Leahy & Yoshida 1995 and Leahy 1995). Here we discuss a new asymmetry we have detected comparing ingresses and egresses and also a broadening of egresses in Short High compared to Main High, which was also detected previously in GINGA data (Leahy 1995).

Six of the nine eclipse ingresses and egresses have good enough data to determine accurately when the x-ray source was completely covered by the companion (i.e. second contact for ingress or third contact for egress). The mean ingress (second contact) is farther from orbital phase 0.0 than the mean egress (third contact) by a significant amount: $0.0009(\pm.0002)$.

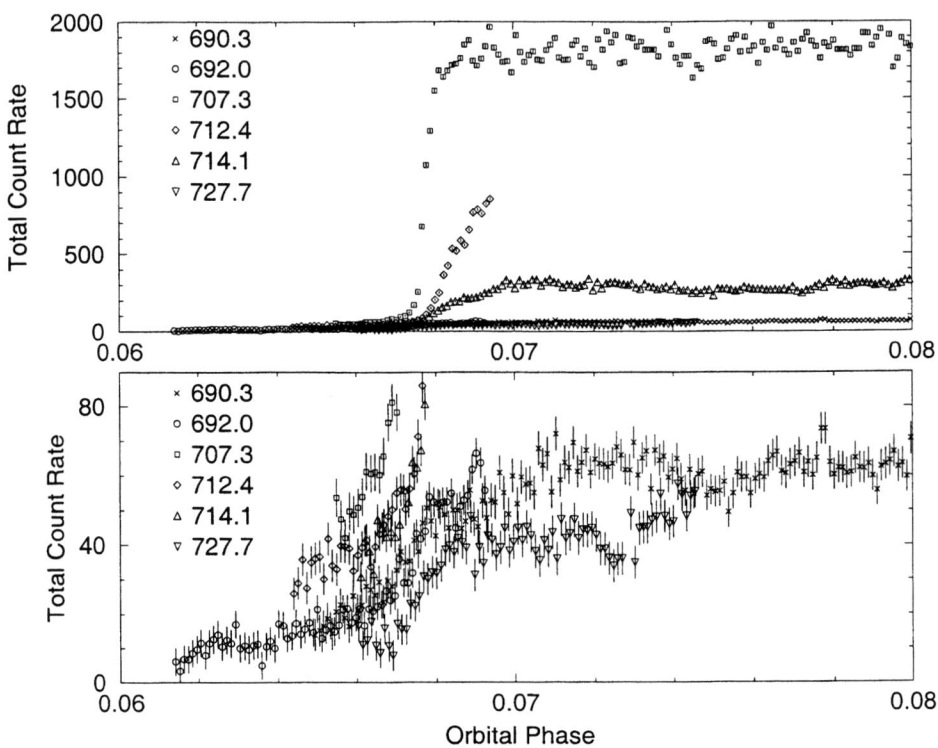

FIGURE 1. The set of six egresses observed during the RXTE campaign on Her X-1 during Sept.-Oct. 1997. The lower panel has the same light curves as the upper panel but displayed with an expanded count rate scale.

This implies that the leading limb of HZ Her is farther from the HZ Her center of mass than the trailing limb by $\sim 37000 km$. This is about 1.4% of the radius of HZ Her. Could this be due to heating of one side of HZ Her due to x-ray illumination on one limb and shadowing on the other limb? This is not plausible, since the heated atmosphere of HZ Her has a measured scale height of $5 \times 10^8 cm$ (Leahy & Yoshida 1995), consistent with $T/\mu \simeq 25000 K$ caused by x-ray heating. Thus the difference in leading and trailing limbs of HZ Her from the center of mass is ~ 10 times the (heated) scale height. However, the difference could be due to x-ray heating induced mass flows from the surface of HZ Her, which should extend to great distances from the surface of HZ Her. Such mass flows have been discussed by Shakura et al. (1999) as likely responsible for the well observed pre-eclipse dips in Her X-1.

The x-ray illumination of HZ Her is modulated at the 35-day period due to the variation of the shadow of the accretion disk on HZ Her. Thus for a cause due to mass flows, and for any other cause which is also related to the x-ray heating, the difference between eclipse ingress and egress times should be a function of 35-day phase. This will be testable with more good observations of ingresses and egresses, such as those obtainable with RXTE.

The orbital phases of start and of end of ingress and egress have been examined as a function of 35 day phase. The egresses are most interesting, since we have data for six of them. For Short High, both totality and end of egress are far away from the nominal eclipse phase. The start of egress is earlier than expected and the end of egress is later than expected, and by an equal time interval. This is in agreement with the results reported from the GINGA eclipse observations (Leahy 1995). This symmetric broadening of egress during Short High implies an extended source. This same conclusion was given by Leahy (1995). The alternative that HZ Her is changing in size by a large amount is not feasible. The symmetric source in Short High is likely due to scattering of x-rays by the accretion disk, with the point source mostly hidden, also by the disk. Further studies of Her X-1 utilizing data with a larger set of ingresses and egresses will be valuable in extracting more information on the variability of the scattering region in the disk as a function of 35 day phase.

ACKNOWLEDGEMENTS

This work was supported in part by the Natural Sciences and Engineering Research Council of Canada.

REFERENCES

Boroson, B., Blair, W., Davidsen, A., Vrtilek, S., Raymond, J., Long, K., McCray, R., 1997, ApJ, 491, 903

Choi, C., Seon, K., Dotani, T. Nagase, F., 1997, ApJ, 476, L81

Dal Fiume, D., et al., 1998, A&A, 329, L41

Deeter,J., Scott, D., Boynton,P., Miyamoto,S., Kitamoto,S., Takahama, S., Nagase,F., 1998, ApJ, 502, 802

Leahy, D.A., 1995, ApJ, 450, 339

Leahy, D.A. & Yoshida, A., 1995, MNRAS, 276, 607

Leahy, D., Scott D., 1998, ApJ, 503, L63

Oosterbroek, T., Parmar, A., Martin, D., Lammers, U., 1998, A&A, 329, L41

Scott D. M., 1993, PhD Thesis, University of Washington

Scott D., Leahy, D., 1999, ApJ, 510, 974

Shakura, N., Ketsaris, N., Prokhorov, M., Postnov, K., 1998, MNRAS, 300, 992

Shakura, N., Prokhorov, M., Postnov, K., Ketsaris, N., 1999, A&A, 348, 917

Still, M., Quaintrell, H., Roche, P., Reynolds, A., 1997, MNRAS, 292, 52

Vrtilek, S., Cheng, F., 1996, ApJ, 465, 915

Wilson, Scott & Finger, 1997, "Proceedings of the Fourth Compton Symposium", ed. C. D. Dermer, M. S. Strickman & J. D. Kurfess, (AIP Conference Proceedings 410), 739

THE OPTICALLY BRIGHT REX SAMPLE

D. Lentini[1], R. Della Ceca[1], T. Maccacaro[1], A. Wolter[1], A. Caccianiga[2], I.M. Gioia[3,4]

1) Osservatorio Astronomico di Brera, Milano Italy; 2) Observatorio Astronomico de Lisboa, Lisboa Portugal; 3) Istituto di Radioastronomia del CNR, Bologna Italy; 4) Institute for Astronomy, Honolulu, HI USA.

ABSTRACT We describe the selection technique adopted to extract from the REX sample (Radio Emitting X-ray sources) a well-defined subsample of 109 Optically Bright REX (OB-REX). Here we present the optical, X-ray and radio properties of the OB-REX sources and the results of the selection effects analysis, obtained by use of properly selected comparison samples.

KEYWORDS: galaxies: general — galaxies: active — radio continuum: galaxies

1. SELECTION OF THE OB-REX SAMPLE

At present almost 500 of the 1600 REX (see Maccacaro et al. these proceedings, Caccianiga et al. 1999) have been spectroscopically identified. Limiting ourselves to the optically bright REX, we have selected a well defined and fully identified subsample, with known optical counterparts and spectral properties. To associate an homogeneous magnitude to the optical counterparts of the REX sources, we have used the *Lyon/Meudon Extragalactic Database* (LEDA, Paturel et al. 1997) which provides magnitudes converted to a standard system, the B_T magnitude (Paturel et al. 1994) for galaxies down to a completeness limit of $B_T = 15.5$. Through a positional cross-correlation between the 1600 REX and the LEDA galaxies, we have extracted 109 REX sources with apparent B_T magnitude brighter than 15.5: the OB-REX subsample. The OB-REX sources are thus selected simultaneously in three bands with the following flux limits: $F_{X[0.5-2keV]} \gtrsim 5 \times 10^{-14} erg\ s^{-1} cm^{-2}$, $F_{R[1.4GHz]} \geq 5\ mJy$, and $B_T \leq 15.5$ and are all associated with known optical counterparts.

2. PROPERTIES OF THE OB-REX SAMPLE

Due to the bright magnitude limit, the OB-REX sample is a local subsample ($<z>\sim$ 0.02) of the REX sample. It is all made up by objects morphologically classified as galaxies: 34% ellipticals, 53% spirals and 3% irregulars (for the remaining 10%, the morphological type is not known). More than 25% of the OB-REX sources are AGN, of which 6% are LINER and 19% are Seyfert galaxies, with a Sy2 to Sy1 ratio of ~ 5.

The absolute magnitude distribution of the OB-REX galaxies is peaked at $M_{B_T} = -22$ ($H_0 = 50\ km\ s^{-1} Mpc^{-1}, q_0 = 0$), with a shift of 1 – 1.5 magnitude towards brighter magnitudes compared to the typical B_T absolute magnitude expected for normal galaxies (Marzke et al. 1994). The radio and X-ray luminosity distributions span a wide range, covering 6 – 7 orders of magnitude, from 10^{26} to $10^{33} erg\ s^{-1} Hz^{-1}$ at 1400 MHz and from 10^{38} to $10^{45} erg\ s^{-1}$ in the 0.5 – 2 keV band respectively, suggesting the presence of different populations contributing to the radio and X-ray emission. In particular, the high radio luminosity tail ($L_R > 10^{30} erg\ s^{-1} Hz^{-1}$) is dominated by elliptical galaxies with structured radio morphology and high radio–optical flux ratio $F_R/F_O > 1$ (i.e. by radio-galaxies). They are also characterized by high X-ray luminosities (10^{41} – $10^{45} erg\ s^{-1}$) and constitute 30% of the OB-REX sample. The fact that most of these galaxies are in clusters suggests that the cluster itself contributes to the high X-ray luminosity observed, as confirmed by the observed extended X-ray emission of almost all of these sources (X-ray extension ≥ 1.4 ROSAT PSPC PSF).

3. SELECTION OF THE COMPARISON SAMPLES

From the analysis of the OB-REX properties, it is clear that they are strongly affected by the three flux limits adopted in the optical, X-ray and radio bands. In order to understand the selection effects present in the OB-REX sample, we have defined three comparison samples of optically bright galaxies, lifting in turn one or both the X-ray and radio emission requirement : i.e. an X-ray–optical, a radio–optical and an optical comparison sample. To guarantee a correct and significant comparison, we have extracted these samples from the *same catalogues* and using the *same flux limits* adopted for the OB-REX sample.

4. SELECTION EFFECTS

The main results of the selection effects analysis are the following:

– *Optical band selection*: as expected, the bright flux limit in the optical band ($B_T = 15.5$) leads to the selection of local samples ($<z> \sim 0.02$). The redshift distributions of the three comparison samples are in fact consistent with the OB-REX sample distribution.

– *X-ray band selection*: to understand the selection effects introduced by the X-ray selection, we compared the OB-REX sample with the radio–optical comparison sample. From the radio luminosity and the radio–optical flux ratio distributions, shown in Fig. 1 and Fig. 2, it is evident that the population of radio-galaxies ($L_R > 10^{30} erg\ s^{-1} Hz^{-1}$, $F_R/F_O > 1$) present in the OB-REX sample is almost completely missing in the radio–optical sample. Thus the X-ray selection in a "local" sample down to a flux limit of $\sim 5 \times 10^{-14} erg\ s^{-1} cm^{-2}$ favors the selection of high X-ray luminosity radio-galaxies in clusters. These radio-galaxies would otherwise be underrepresented (in percentage) in a sample selected in the radio band alone.

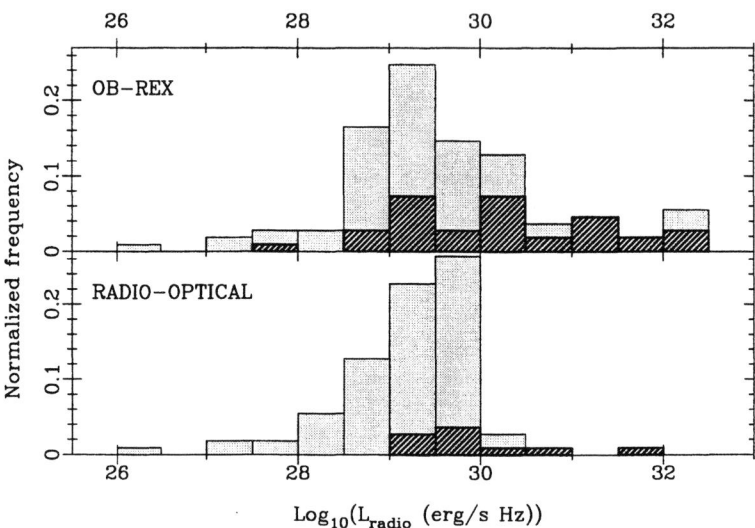

FIGURE 1. Radio luminosity distributions for the OB-REX and the radio–optical samples (full: total sample, dashed: elliptical galaxies contribution).

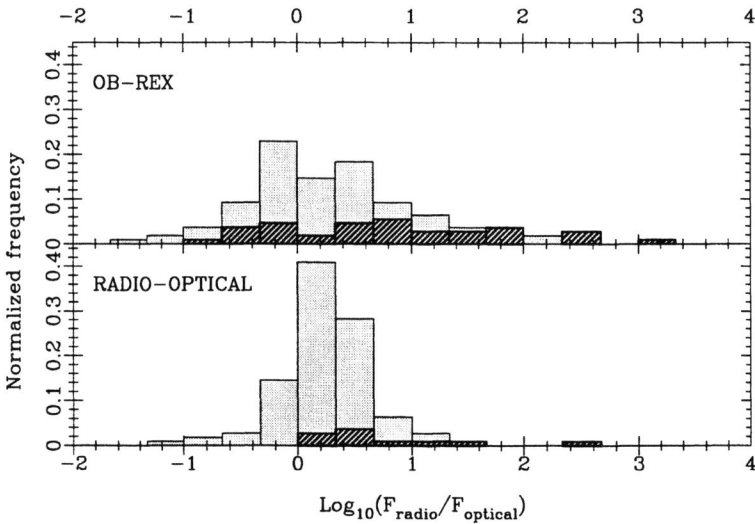

FIGURE 2. Radio–optical flux ratio distributions for the OB-REX and the radio–optical samples (full: total sample, dashed: elliptical galaxies contribution).

– *Radio band selection*: the fact that the OB-REX properties (X-ray luminosity, X-ray to optical flux ratio, percentage of X-ray extended sources) do not differ significantly from those of the X-ray–optical sample suggests that the radio band selection at a flux limit of 5 mJy does not affect much the X-ray properties of a local sample.

– *X-ray and radio band selection*: the simultaneous X-ray and radio band selection introduces two main selection effects in the OB-REX sample. The first selection effect is related to the optical luminosity. In fact, the OB-REX sources are more luminous, by 1 – 1.5 absolute B_T magnitude, than the sources in any of the three comparison samples. The second selection effect concerns the Sy2 to Sy1 ratio which is higher in the OB-REX sample than in the X-ray–optical sample ($\sim$ 5 vs. $\sim$ 1.5). The X-ray band selection makes Sy2 galaxies hardly detectable above the mean redshift of the samples ($<z>= 0.02$), since the X-ray flux limit corresponds, at this redshift, to a luminosity of $L_X = 10^{41} erg\ s^{-1}$, that is the typical X-ray luminosity of Sy2. Sy1 galaxies are instead still detectable since they are characterized by a higher X-ray luminosity ($L_X \sim 10^{43} erg\ s^{-1}$). Therefore, Sy1 galaxies have on average a higher redshift than Sy2 galaxies because of the X-ray selection. On the other hand the radio flux limit of 5 mJy corresponds, for $z = 0.02$, to a radio luminosity of $10^{29} erg\ s^{-1} Hz^{-1}$, a typical value for both Sy1 and Sy2 galaxies. As a consequence, Sy1 galaxies, that are at larger redshift, are lost preferentially because of the radio flux limit, while Sy2 galaxies, which are already confined to $z \leq 0.02$ by the X-ray selection, are not lost, making their percentage higher, in the OB-REX sample, with respect to Sy1 galaxies.

5. CONCLUSIONS

The comparison of the OB-REX sample with properly selected comparison samples turned out to be a very powerful tool to understand in detail the selection effects introduced in flux-limited samples by the single band selection.

ACKNOWLEDGEMENTS

This work has received partial financial support by the Italian Space Agency (ASI).

REFERENCES

Caccianiga, A. *et al.* 1999, ApJ 513, 51C.
Marzke, R. D. *et al.* 1994 ApJ 428, 43
Paturel, G. *et al.* 1994, A&A 286, 76.
Paturel, G. *et al.* 1997, A&AS 124, 109.

THE REX SURVEY: THE CATALOG

T. Maccacaro[1], R. Della Ceca[1], A. Wolter[1], A. Caccianiga[2], and I.M. Gioia[3,4]

1) Osservatorio Astronomico di Brera, Milano, Italy 2) Observatório Astronómico de Lisboa, Lisboa, Portugal 3) Institute for Astronomy, Hawaii, USA 4) Istituto di Radioastronomia del CNR, Bologna, Italy

ABSTRACT We recall the basic properties of the REX survey and present a status update.

KEYWORDS: X-ray sources: general—radio sources: general—surveys

1. INTRODUCTION

The purpose of the REX survey is to create a large and statistically well defined sample of Radio Emitting X-ray sources. This survey makes use of the best X-ray and radio data currently available over a large fraction of the sky: the archive of the ROSAT PSPC observations and the NRAO VLA Sky Survey (NVSS, Condon et al., 1998). The scientific goals, strategy and first results of the REX project can be found in Caccianiga et al. (1999). In the interest of clarity here we recall the basic properties of the survey and present a status update.

2. X-RAY DATA

From the public archive of PSPC observations, 1202 PSPC images have been selected according to the following criteria:
- Declination north of $-40°$
- High galactic latitude ($|b| > 10°$)
- Exposure time > 1000s
- On suitable target (i.e. excluding M31, stellar and galaxy clusters, dark clouds, etc.)
- Non overlapping to each other

Their distribution in the sky is shown in Figure 1. The total area of sky covered by the REX survey is ~ 2700 square degrees. This is ~ 16 times better than the EMSS (Gioia et al., 1990) at 10^{-13} erg cm^{-2} s^{-1} and ~ 3.5 times better at 10^{-12} erg cm^{-2} s^{-1}

X-ray sources have been searched in the 0.5-2.0 keV band with a wavelet based detection algorithm (developed by Damiani et al. 1997) and accepted in the sample

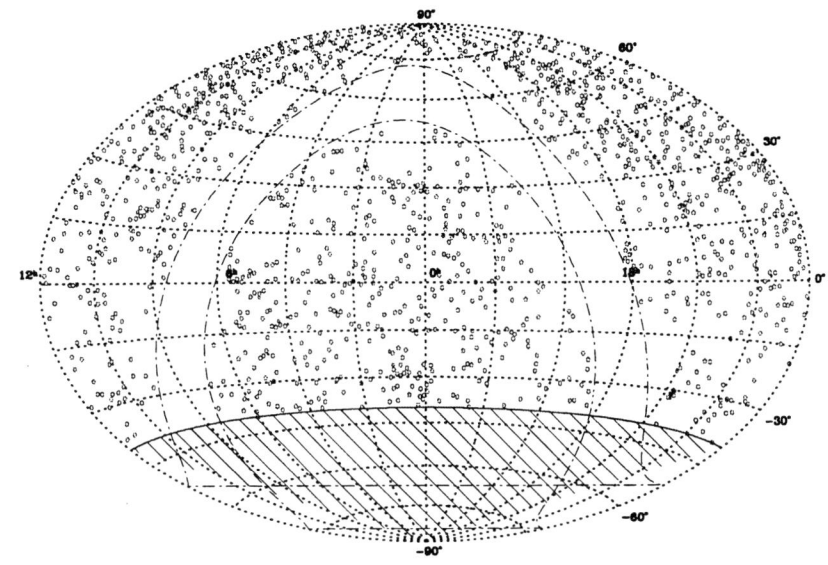

FIGURE 1. Sky distribution of the 1202 PSPC field used (celestial coordinates)

if the significance of the detection equals or exceeds 6.0σ. The outermost part of the PSPC ($r > 47'$) has not been used because of the significant degradation of the data quality. Also the innermost part of the image, if containing the target of the observation, has been ignored. Usually, this corresponds to a circle of radius $r = 4'$, but a larger area is masked out if the target is particularly bright or extended. Targets and target related sources have been removed from the source sample.

3. RADIO DATA

Data from the NRAO VLA Sky Survey (Condon et al. 1998) have been used. The NVSS covers the whole sky north of $-40°$ declination at 1400 MHz. The survey has been completed recently and only one of the few "holes" remaining in the data overlaps with a PSPC image.

A cross correlation between the X-ray catalog and the NVSS catalog has been performed, considering only radio sources with flux larger or equal to 5 mJy ($\sim 10\times$ r.m.s.). Since the catalog derived from the NVSS is not a catalog of radio sources but a catalog of components, particular attention has been paid to the way a coincidence is defined, so as to retain also highly structured sources (e.g. double radio sources, see Figure 2). In particular, if only one radio component is present in the vicinity (2.5′) of the X-ray source, a REX is defined when the radio position is consistent with the X-ray position (combining the two 90% error circles). If two or more radio components are present in a circle of 5′ diameter centered on the X-ray position, a T-REX (Temporary REX) is defined. Visual inspection of optical and radio data is then required to decide whether the T-REX is a bona-fide REX or a spurious association.

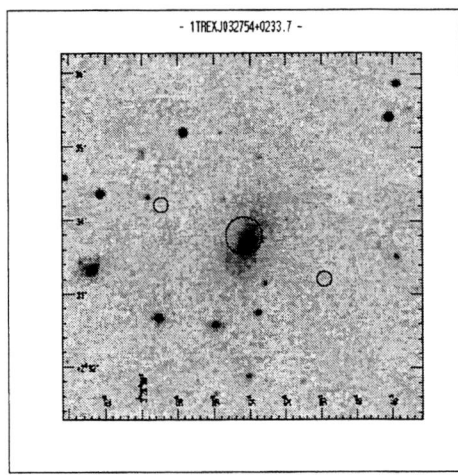
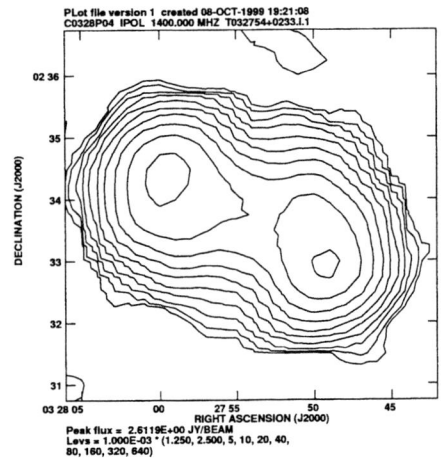

FIGURE 2: Example of a T-REX that would have been missed from a "blind" positional cross-correlation between X-ray and radio sources. The larger circle represents the X-ray error circle, the smaller circles represent the position of the two radio components found in the NVSS (*left panel*: optical image; *right panel*: radio isointensity contours).

4. RESULTS

The cross correlation has yielded: 1163 REX and 977 T-REX. Inspection of the optical and radio images for the 977 T-REX has allowed us to classify 488 T-REX as REX and 288 as spurious associations. Further data are needed to classify the remaining 201 T-REX.

Therefore, the REX survey currently consists of 1651 REX and 201 T-REX.

A source catalog has been constructed and it is undergoing now a quality control. It will be made available to the astronomical community as soon as possible, together with all the optical and radio images and other relevant information on the individual sources. A web site is being developed (*http://www.brera.mi.astro.it/∼ rex*) that will be the repository of all the relevant information on the REX survey.

An optical identification program has been initiated a few years ago and it is still under way, together with a recurrent search of the NED database and literature.

TABLE 1. Breakdown of optical identifications

Class	#	%
AGN	232	48.3
BL Lac objects	72	15.0
Galaxies	176	36.7
Stars	3	< 1
TOTAL	483	–

So far almost 500 sources have been identified either trough our own spectroscopic observations or from the literature (see Table 1).

The identification program prioritizes sources as a function of their X-ray flux so as to allow the definition of subsamples of increasing size with virtually complete optical identification. Currently, the REX sample has an identification rate of 90% at a flux level of $5. \times 10^{-13}$ erg cm^{-2} s^{-1}. Preliminary statistical results derived from this sample are discussed in Caccianiga et al. (these proceedings). A fully identified subsample defined by the REX sources with optical counterparts of apparent magnitude $B_T \leq 15.5$ has also been extracted from the REX survey. Its properties are discussed in Lentini et al. (these proceedings).

ACKNOWLEDGEMENTS

This work has received partial financial support from the Italian Space Agency (ASI) and from the Italian Ministry for University and Research (MURST) under grant Cofin98-02-32.

REFERENCES

Caccianiga, A., Maccacaro, T., Wolter, A., Della Ceca, R., and Gioia, I.M., 1999, ApJ, 513, 51.
Condon, J.J., Cotton, W.D., Greisen, E.W., Yin, Q.F., Perley, R.A., Taylor, G.B., and Broderick, J.J., 1998, AJ, 115, 1693.
Damiani, F., Maggio, a., Micela, G., and Sciortino, S., 1997, ApJ, 483, 350.
Gioia, I.M., Maccacaro, T., Morris, S.L., et al., ApJS, 72, 567, 1990.

MEASUREMENTS OF FLUCTUATIONS IN THE HARD X-RAY BACKGROUND WITH RXTE

Daniel R. MacDonald [1], Duane E. Gruber [2], Elihu A. Boldt [3]

1) Jet Propulsion Lab
2) Center for Astrophysics and Space Sciences, UCSD
3) Laboratory of High Energy Astrophysics, GSFC

ABSTRACT

During normal observations, the High Energy X-ray Timing Experiment instrument (HEXTE) on the Rossi X-ray Timing Explorer (RXTE) Spacecraft makes nearly simultaneous observations of two pairs of background fields separated by 3.0 degrees. Differences in fluxes from these pairs of background fields can be used to detect a source population much deeper than can be studied from individual detections.

We have measured the fluctuations in the X-ray background (XRB) from 15-40 keV using differences between adjacent HEXTE background fields. The fluctuations as a fraction of the XRB flux are : 0.092 ± 0.014 % from 15 to 20 keV, 0.11 ± 0.02 % from 20 to 25 keV and 0.23 ± 0.08 % from 34 to 41 keV for a 1.1 square degree field of view. The measured fractional fluctuations increase with energy, which implies that the effective number of contributing sources is decreasing with energy. The fluctuation level and energy dependence are consistent with HEAO-1 results.

KEYWORDS: Background Fields; X-rays; Fluctuations; AGN

1. INTRODUCTION

The origin of the X-ray diffuse background continues to be a challenging problem. The background intensity is best explained as the sum of emission from point sources (e. g. Almaini and Fabian 1997; Gilli, Risalti and Salvati, 1999). Near 1 keV, source counts from deep x-ray imaged fields have recently accounted for the entire XRB flux (Mushotzky et al. 2000). This result is encouraging, but another important challenge to the superposition-of-sources origin appears at higher energies, particularly in the interval 10-40 keV, where the humped background spectrum (Gruber 1999) is completely unlike that of any ensemble of directly observed extragalactic sources. Subtraction of the contribution due to known sources only enhances the discrepancy.

In recent years, however, steady advances toward resolution of this "spectral paradox" have been made both in theory and observation: (Madau, Ghisellini and Fabian 1994) decompose the XRB into a sum of relatively unabsorbed AGN and highly absorbed objects.This simple model is consistent with unified theories of AGN in which Seyfert 2 galaxies contain a generic AGN engine seen through a molecular torus, while Seyfert 1's are seen through little absorption. This basic model has since been refined (e.g. Comastri 1995; Gilli, Risalti and Salvati, 1999.) Recent results from the BeppoSAX satellite have provided high energy spectra of both Seyfert 1 and Seyfert 2 galaxies (e.g. Matt et al. 2000, Nicastro et al. 2000 and Perola et al. 2000).

All models predict that the observed X-ray background log N-log S at high energies has a relatively higher normalization than at lower energies, as required by a comparison of the HEAO-1 A2/GINGA fluctuation analysis in the 2-10 keV (e.g. Carrera et al. 1995) with the Einstein/Rosat soft X-ray logN-logS and tentative results from the HEAO-1 A4 data (Boldt, reported by Gruber 1992). In this scenario the spectrum of faint, hard X-ray selected objects should show the presence of either very strong reflection or of strong absorption.

In spite of the recent progress sketched above, there are still numerous points of difficulty (e.g. Johnson et al. 1994) in relating individual and collective source measurements with even the overall background properties. The steep OSSE spectra of AGN (Zdziarski et al. 1995) indicate a break in the range, 40-100 keV, which can be sensitively measured by HEXTE with long exposures. However a re-analysis of the OSSE data (Madejski et al 1995) concludes that such a break is not a necessary consequence of the OSSE data if one includes the effects of reflection. Zdziarski et al. (1995) discuss whether present epoch AGN's with such a break can be used as a satisfactory basis for a spectral model of the background.

2. METHOD

The HEXTE instrument consists of two independent clusters, each with four detectors, which switch the 1 deg FWHM beams on two perpendicular axes to collect background data. This beamswitching, 1.5° on either side of the on-source positions, produces nearly simultaneous measurements of four background fields, two for cluster 1 and two for cluster 2. The flux difference between each pair is a measurement of the variation in the background sky at a 3° separation. For individual observations, the fluctuations in the XRB are below the sensitivity of the HEXTE instrument for any reasonable observation length. Although too small to be observed individually, combining observations of different sky regions allow us to measure average properties of the sky fluctuations. To do this, we created a database of 36 Ms of observations consisting of all the long HEXTE pointings which are outside of the galactic disk and bulge.

The measurement of the fluctuations is a comparison of the data set to a series of models. These models are created by taking the errors from the counting statistics in the data set and adding in quadrature an extra width which represents Gaussian sky fluctuations

Internal instrument background subtraction is the most important single problem to overcome in performing this observation. The expected signal is only a few percent of the instrumental background (MacDonald, 1999) and any systematic effects that are comparable to the counting statistics in an observation could be confused with an XRB fluctuation.

Since the XRB signal is so small, the spectra must be binned in wide energy bands to see any signal from the XRB. The energy bands, 15 – 20, 20 – 25, 34 – 41, 80 – 132 and 132 – 261 keV are chosen to avoid the large line features at 30 and 66 keV in the HEXTE instrument background (Gruber, 1996). Using differences between off-source pointings of the same detectors subtracts out any instrumental background that varies slowly compared to the 16 second timescale of the beam-switching.

The fundamental source of uncertainty in the data is in the counting statistics of the photons in the various different observations. This has been accounted for explicitly in the model, so the flux we measure is in addition to the counting statistics. The extra RMS flux so determined can contain both sky fluctuations and systematic errors.

The higher energy bands in the HEXTE instrument provide a means of measuring the effects of the systematics. The sky fluctuation level predicted in the highest two bands is more than an order of magnitude less than the measured additional RMS fluctuation (MacDonald, 1999). The signal in these energy bands is then used as a direct monitor of the level of the systematics at lower energies.

Using the upper energy bands as a measurement of the systematic noise, one makes an assumption about the energy dependence of the processes creating the systematic noise, since it is an average over different energy bands. At first glance, the agreement of the two measurements of the systematics in each test appear to indicate that the systematics are a constant fraction of the instrument background over the upper $\frac{3}{4}$ of the instruments energy band. This is equivalent to assuming that the systematics are due to random errors in the livetime, since errors in the livetime produce incomplete subtraction of the instrument background. The three lower corrected HEXTE bands are shown as the square datapoints in figure 1.

3. RESULTS

The increase in the fractional fluctuations of the three HEXTE datapoints is suggestive of a trend, but not significant by themselves. To make a significant statement on the spectrum of the fluctuations, the HEXTE results are compared to the results from the HEAO-1 A4 experiment (Gruber 1992) in figure 1. The data have been scaled to a hypothetical instrument with a 1 sr FWHM.

The combined results show a 3 σ significant increase in the fractional fluctuations of the XRB with energy from 15 to 80 keV with a power law index of 0.28 ± 0.09. These results require a *decrease* in the density of sources at higher energies.

REFERENCES

Almaini, O. and Fabian, A. C., MNRAS, 288, L19, 1997.

Carrera, F.J., et al. MNRAS, 275, 22, 1995.

Comastri, G., et al. A&A, 296, 1, 1995.

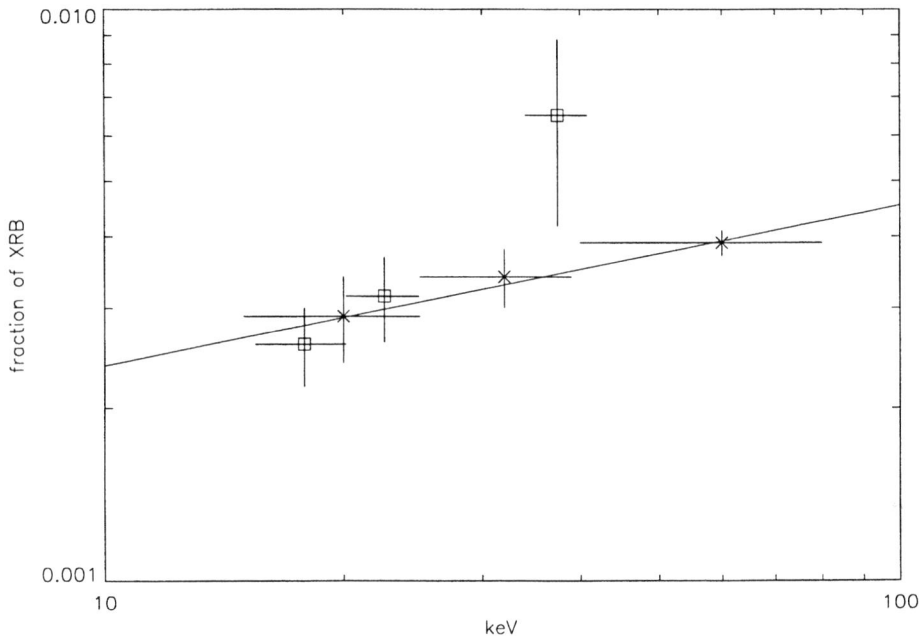

FIGURE 1. HEXTE and A4 fluctuation measurements as a fraction of the XRB have been normalized to 1 sr FWHM instrument. The spectrum of the fluctuations is harder than the XRB spectrum with a best fit relative power law index of 0.28 ± 0.09. The HEXTE datapoints are squares and the A4 datapoints are crosses.

Gilli, R., Risaliti, G. and Salvati, M., 1999, A&A, in Press.
Gruber, D. E., 1992, in The X-ray Background, Cambridge university press, 44–53.
Gruber, D. E., et al, 1996, A&A Sup., 120, 641.
Gruber, D. E., et al, 1999, Ap. J., 520, 124.
Johnson, W. N. et al., in The Second Compton Symposium, p. 515, 1994.
Lightman, A. and White, T., Ap. J., 335, 57L, 1988.
MacDonald, D. R., 1999, dissertation, UCR.
Madau, P., Ghisellini, G., and Fabian, A. C., MNRAS, 270, L17, 1994.
Madejski, G.M., et al., Ap. J., 438, 672, 1995.
Matt, G., et al. 2000, submitted to MNRAS.
Mushotzky R. F., et al., Nature, 404, 459.
Nicastro, F., et al., 2000, preprint.
Perola, G. C. , et al., 2000, A&A, in Press.
Zdziarski, et al., Ap. J., 438, L63, 1995.

THE RADIO TO HARD X-RAY SPECTRAL ENERGY DISTRIBUTION OF BEPPOSAX EMISSION LINE AGN

A. Malizia [1,2] and F. Fiore [1,3]

1) BeppoSAX Science Data Center, Via Corcolle 19, Roma, Italy
2) Southampton University, SO17 1BJ, England
3) Osservatorio Astronomico di Roma, Via dell'Osservatorio, Monteporzio Catone, Italy

ABSTRACT We have studied the correlation between the hard X-ray (10 keV) luminosity, the radio (5GHz) luminosity and the [OIII]5007Å luminosity in an heterogeneous sample of emission line AGN observed by BeppoSAX. We find strong correlations between all these luminosities. We also find that the [OIII]5007 luminosity is correlated with the ratio between the 10 keV and the 5 GHz luminosity at the 98% confidence level. This suggests that the [OIII] emission is probably not related only to the UV ionizing radiation from the AGN, and therefore that the [OIII] emission is not a perfect indicator of the isotropic UV radiation field.

KEYWORDS: Active Galactic Nuclei; X-Rays, Radio, [OIII]line emission.

1. SAMPLE AND OBSERVATIONS

The correlations between the radio and optical properties of Seyfert galaxies were first explored in 1978 by de Bruyn and Wilson. They found a strong correlation between the 21 cm luminosity and the [OIII]5007Å narrow line luminosity. This correlation holds both correcting and not correcting for the reddening, as estimated through the Balmer decrement. A similar result has been found later for radio loud AGN by e.g. Jackson and Rawling (1997) and Tadhunter et al. (1998) who considered the total core+lobes radio luminosity. The study of these correlations can provide fundamental informations about the physics of the energy-generating mechanisms in the radio loud AGN. In fact the emission line regions may be energized by EUV photons emitted in the central region where the radio jets have their origin. We consider here an heterogeneous sample of bright emission line AGN (44), all showing strong optical and UV emission lines, witnessing a large O-UV radiation field presumably due to accretion onto a central black hole. The sample includes: Seyfert 1 galaxies; Seyfert 2 galaxies showing the direct component at least at high energies (N_H < a few 10^{24} cm^{-2}, therefore excluding NGC1068); radio-quiet quasar; radio galaxies; Flat and Steep Spectrum Radio Quasars present in the *Beppo*SAX public archive (26) as Aug 1999. The *BeppoSAX* broad 0.1-200 keV band allow to adequately constrain the continuum spectral shape even in presence of large absorption and/or strong emission lines and/or additional continuum components such as

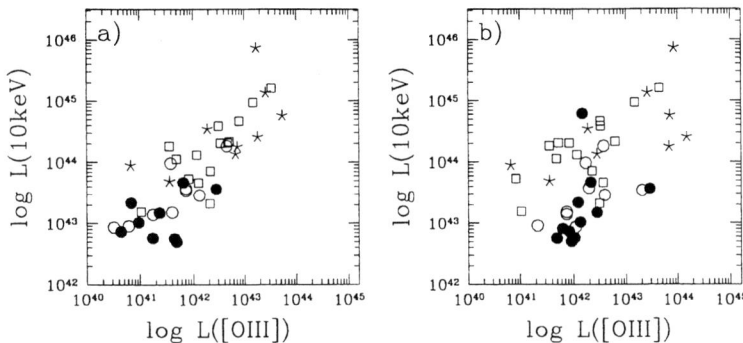

FIGURE 1. logL(10keV) vs logL([OIII]). a) [OIII] luminosity NOT corrected for the Balmer decrement; b) [OIII] luminosity corrected for the Balmer decrement. The luminosities are in units of $erg\ s^{-1}$. Filled and empty symbols are respectively absorbed and unabsorbed AGN; stars identify radio-loud galaxies and radio-loud quasars; circles identify radio-quiet AGN at redshift < 0.05 and squares are radio quiet AGN at higher redshift ($0.05 < z < 0.4$).

Compton reflection and soft X-ray excesses, providing in turn a reliable estimate of the high energy flux and luminosity. In addition, X-ray fainter PG quasar at z<0.4 observed by the BeppoSAX and ASCA imaging instruments up to 10 keV are also considered (18 quasars). The inclusion of these objects allows to expand the dynamic range of luminosities, favoring the discovery of correlations.

At radio wavelengths, the 5 GHz, small beam (VLA/ATCA) observations have been considered to isolate as much as possible the nuclear emission against the extended lobes in both radio-loud and radio-quiet AGN. Measurements have been taken from literature and in particular from Morganti et al. (1999) and references therein for the Seyfert galaxies, Morganti et al. (1997) for radio sources and Kellermann et al. 1994 for the PG quasars.

The [OIII]5007Å luminosities have been collected form the literature as well as the Balmer decrements (Bassani et al. 1998 and references therein, Mulchaey et al. 1994, Tadhunter et al. 1998, Miller at al. 1992). Balmer decrement measurements are necessary to correct for the extinction. They are not available for two radio galaxies (HER A and PictorA) and for 5 of the 18 PG quasars (PG1216, PG1415, PG1444, PG1543, PG1626); in this case we have assumed the value of 3, expected from atomic physics.

2. CORRELATIONS

Figures 1 and 2 show the correlations between X-ray (10 keV), radio-core (5GHz) and [OIII] 5007Å luminosities. In each figure the left panel refers to [OIII] lumi-

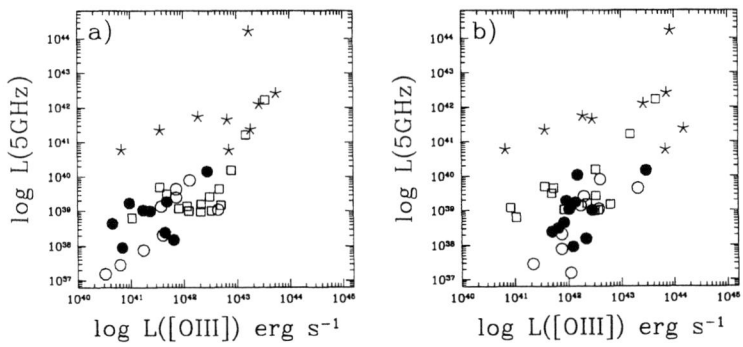

FIGURE 2. logL(5GHz) vs logL([OIII]). Symbols, units and OIII correction as in figure 1.

TABLE 1. Correlation Statistics

Quantity	NOT Corrected			Corrected		
	N	R_s	Prob	N	R_s	Prob
L_{10} vs OIII	44	0.722	>99.99%	44	0.503	99.94%
L_{10} vs OIII (without RL)	35	0.677	>99.99%	35	0.406	98.45%
R vs OIII	44	0.638	>99.99%	44	0.507	99.96%
R vs OIII (without RL)	35	0.569	99.96%	35	0.477	99.63%
OIII vs R/L_{10}	44	0.372	98.72%	44	0.358	98.30%
OIII vs R/L_{10} (without RL)	35	0.152	61.79%	35	0.274	88.96%

nosities not corrected for the Balmer decrement and the right panel to corrected [OIII] luminosities. Table 1 gives the Spearman rank correlation coefficients and the associated probabilities. The correlations are quite strong in all cases, although we note that the correlation coefficients are slightly higher when the [OIII] luminosities are not corrected for the Balmer decrement. The correlations remain strong also when the radio-Loud AGN are excluded from the sample (see Table 1).

Figure 3 shows the correlation between the [OIII] luminosity and the ratio between the radio and X-ray luminosities at the 98-99 confidence level.

3. DISCUSSION

Using an heterogeneous sample of emission line AGN observed by BeppoSAX and ASCA and collecting the OIII and the 5GHz (core component) measurements from the literature, we have found strong correlation (Prob>99%) between the hard X-

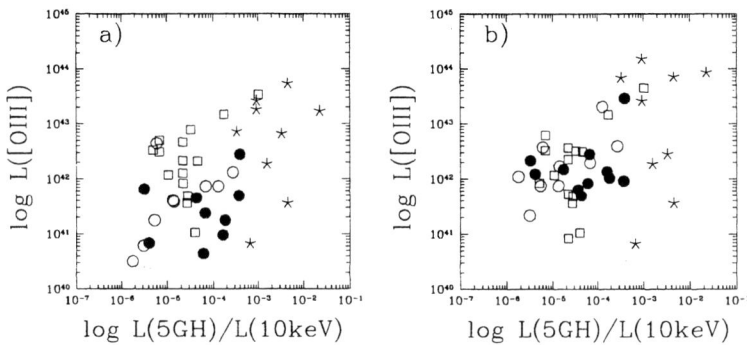

FIGURE 3. logL([OIII] *vs* log L(5GHz)/L(10keV). Symbols, units and OIII correction as in figure 1.

ray (10 keV) and the [OIII] and radio luminosities. This result is in agreement with studies performed so far for Seyfert galaxies and radio-loud AGN (e.g. de Bruyn and Wilson 1978 Jackson and Rawling 1997, Mulchaey et al. (1994)). We also find a correlation between the corrected [OIII] luminosity and the ratio between the radio and X-ray luminosity. Our results suggest that the [OIII] emission is not a perfect indicator of the isotropic UV radiation field. One possible explanation of the stronger [OIII] emission in objects with higher core radio emission, is that the radio jets, often present in radio-quiet objects too (Tadhunter et al. 1998), help in confining the Narrow Emission Line clouds and in compressing the NEL gas to the densities at which forbidden line emission, probably excited by the UV ionizing continuum, starts to be efficient (Capetti et al. 1996, Tadhunter et al. 1998).

REFERENCES

Bassani L., Dadina M.,Maiolino R., et al. 1999, ApJ, 121, 473
Capetti A., Axon D. J., Macchetto F., Sparks W. B. and Boksenberg A. 1996, ApJ, 469, 554
de Bruyn A. G. and Wilson A. S., 1978, A&A 64, 433
Jackson N. and Rawlings S., 1997, MNRAS, 286, 241
Kellermann, K. I., Sramek, R. A., Schmidt, R. A. et al. 1994, AJ, 108, 1163
Miller P., Rawlings S., Sauders R., Eales S., 1992, MNRAS, 254, 93
Morganti R., Oosterloo T. A. et al. 1997, MNRAS 284, 541
Morganti R., Tsvetanov Z. I, Gallimore J., Allen H. G., 1999, A&AS, 137, 457
Mulchaey J. S., Koratkar A., Ward M., et al. 1994, ApJ 463, 586
Tudhunter C. N., Morganti R., Robinson A., et al. 1998, MNRAS, 298, 1035

TIME DEPENDENT COMPTONISATION

J. Malzac[1], E. Jourdain[1]

CESR (CNRS/UPS), 9 av. du Colonel Roche, BP 4346, 31028 Toulouse Cedex 4, France

ABSTRACT

We investigate the radiative response of an accretion disk corona system to short soft photons flares originating from the disk. The coronal response to the soft photon shots produces a strongly non-linear Comptonised radiation output, with complex correlation/anti-correlations between energy bands. This behaviour strongly differs from those found with usual linear calculations. These results show that the feedback effects from the corona are important and should be taken into account by any model invoking the soft photon flux variability as a source for the hard X-ray variability of X-ray binaries or Seyfert galaxies.

KEYWORDS: Accretion, accretion disks – Radiation mechanism: non-thermal – Methods: numerical – binaries: general – Galaxies: Seyfert

1. THE SLAB CORONA MODEL

We consider a classical slab corona coupled with a cold accretion disk (Haardt & Maraschi 1993). The corona is composed of electrons (associated with ions) with a fixed Thomson optical depth τ_s. This corona is assumed to be uniformly heated by an unspecified process, which is quantified using the usual local dissipation parameter l_c. The disk emit black-body radiation with a fixed colour temperature, $kT_{\rm bb} = 0.1$ keV, both by reprocessing of the hard Comptonised radiation and by intrinsic dissipation processes. The internal dissipation is parametrised by l_d which is a simple function of time.

We then compute the coronal temperature kT_e and optical depth τ_T evolution, as well as the system light curves. The calculations are performed using a non-linear Monte-Carlo code developed according to the method proposed by Stern et al. (1995). In our computation we take self-consistently into account the energy and pair balance, reprocessing and reflection on the disk.

Hereafter, time is given in H/c units, where H is the height of the slab corona.

2. RESULTS FROM SIMULATIONS

The left panel in Fig. 1 presents the evolution of the coronal physical parameters. The system is initially in a steady state where the disk internal dissipation parameter

$l_{d0} = 1$ and $l_c = 100$. The dissipation parameter l_d is then increased by a factor of 100 during $\Delta t = 1\ H/c$. The Compton cooling of the plasma is quasi-instantaneous and temperature drops by $\sim 15\%$. The pair production rate thus decreases, leading to a lower optical depth. After the perturbation the optical depth increases, but slower than the lepton kinetic energy. Thus the temperature increases slowly and reaches a maximum higher than the initial equilibrium temperature. This arises from the pair production time being longer than the heating time. After the event the system relaxes toward equilibrium.

The associated light curves are also shown (Fig. 1, right). The soft disk blackbody luminosity in the lower energy band (E<2 keV), increases strongly by a factor of ~ 2 around $1\ H/c$ after the beginning of the flare. This delay is due to the corona light crossing time. The flux in the highest energy bands (20-200 and 200-2000 keV) decreases due to the temperature drop. On the other hand, the flux in the intermediary band (2-20 keV) increases slightly.

The overall Comptonised radiation flux is constant, since the dissipation in the corona is kept constant. *If l_c is constant a modification of the seed photon flux does not change the integrated luminosity in the 1-2000 keV band.* The temperature adjusts very quickly to maintain constant Compton losses. The spectrum thus appears to have a pivot located in the 2-200 keV energy range.

Note however that the variations in the energy range which is usually used in observations (2-50 keV) are very weak (at most 20 %), just slightly larger than the statistical fluctuations and far lower than what is expected in linear calculations. Higher amplitude fluctuations can be obtained for a larger perturbation amplitude. For example, Fig. 2 displays the temperature and optical depth evolution for $l_d = 1000$ between $t = 2$ and $t = 3$, instead of $l_d = 100$ in the previous example, the other parameters being unchanged. As the cooling effect is now stronger the temperature drops by $\sim 80\%$. The spectral evolution is important, the light curves, shown in Fig. 4, display significant fluctuations in the four energy ranges. Important fluctuations can also be obtained by increasing the shot characteristic time scale. Longer durations indeed enable to reach the low temperatures obtained at equilibrium for large l_d.

By varying the shot time scale and amplitude we can thus get complicated correlation and anti correlation between energy bands.

3. DISCUSSION

Current models which invoke the intrinsic seed photon variability as the source of the hard Comptonised radiation variability in X-ray binaries do not take into account the response of the corona (see e.g. Kazanas et al. 1997, Hua et al.1997). The light curves and PDS are computed using linear Monte-Carlo codes (e.g. Pozdniakov et al. 1983). The coronal characteristics (kT_e and τ) are fixed. The high energy flux thus varies linearly with the soft photon flux changes. Actually, the calculations are made as if the heating rate in the Comptonising medium was changed according to the soft photon flux to compensate exactly the Compton losses. We do not know

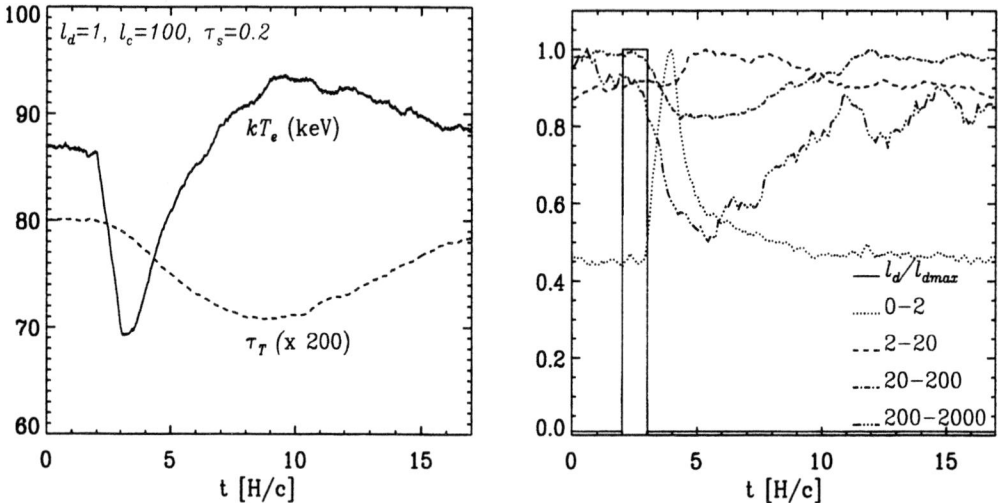

FIGURE 1. (Left) Evolution of the mean coronal temperature (solid line) and optical depth (dashed line) as a response to a flare in the disk. The system is initially at equilibrium with $\tau_s = 0.2$, $l_c = 100$, $l_d = 1$. Between t=2 and t=3, $l_d = 100$. (Right) Light curves associated with the disk flare. The profile of the disk dissipation parameter l_d is shown in solid line. Dotted, dashed, dot-dash and dot-dot dash line are respectively the light curves in the 0-2, 2-20, 20-200 and 200-2000 keV bands. All curves are normalised to their maximum.

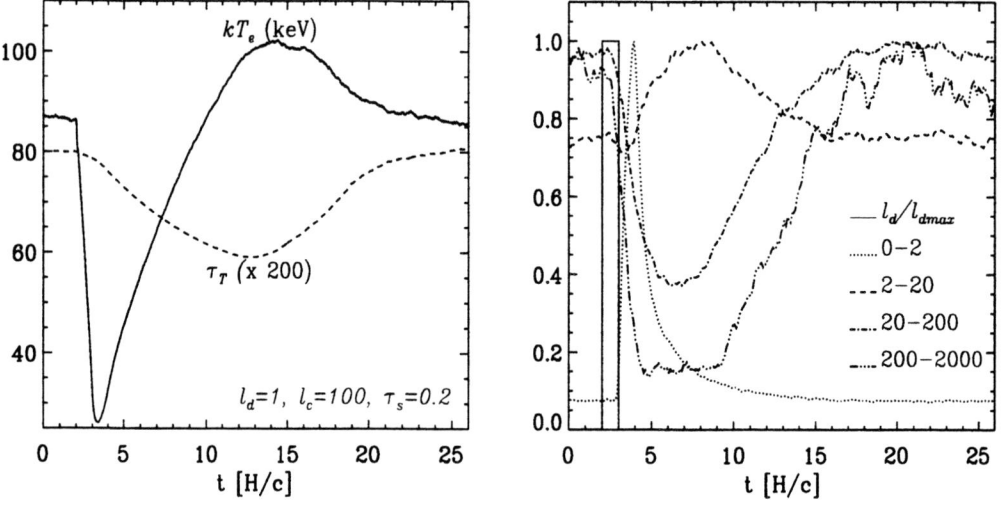

FIGURE 2. Same as Fig. 1 with $l_d = 1000$ between $t = 2$ and $t = 3$.

how the disk is physically coupled with the corona, there may be some correlation between l_c and l_d. It seems however very unlikely that they adjust so perfectly to keep the temperature constant. Our results show that when taken into acount, these coupling effects lead to very different results with a high energy variability wich is very weak compare to the input variability. Indeed, if l_c is assumed to be constant, the Comptonised radiation varies only due to spectral evolution with a constant integrated luminosity.

The amplitude of fluctuations induced by soft photon shots, increases with energy. Indeed the higher energy part of the spectrum is most sensible to temperature fluctuations. This seems to be inconsistent with observations of X-ray binaries showing that the RMS variability is nearly independent of energy (e.g. Nowak et al. 1999). On the other hand, if the shots have an amplitude and time scale long enough to make the corona very cool, the spectral pivot point may shift toward the soft X-ray energies, leading to an anticorrelation between the disk and Comptonised radiation. Such a mechanism might explain the strange anticorrelation between UV an X-rays observed in the Seyfert 1 galaxy NGC7469 by Nandra et al. (1998).

Another complication that we do not consider here is that there may be a rapid succession of shots in the disk. If these shots are close enough in time and space, the corona has no time to relax toward equilibrium between each event. The resulting light curves will thus also depend on the details of the temporal shot distribution.

However that may be, any model where variability is driven by changes in the soft photon flux should also specify a coronal heating and account for the corona response.

REFERENCES

Haardt, F.,Maraschi, L., 1993, ApJ, 413, 501

Hua, X. M., Kazanas, D., Titarchuk, L., 1997, ApJ, 482, L57

Hua, X. M. and Titarchuk, L., 1996, ApJ, 469, 280+

Kazanas, D., Hua, X. M., Titarchuk, L., 1997, ApJ, 480, 735+

Nandra, K., Clavel, J., Edelson, R. A., George, I. M., Malkan, M. A., Mushotzky, R. F., Peterson, B. M., and Turner, T. J., 1998, ApJ, 505, 594

Nowak, M. A., Vaughan, B. A., Wilms, J. R., Dove, J. B., Begelman, M. C., 1999, ApJ, 510, 874

Pozdniakov, L. A., Sobol, I. M., Sunyaev, R. A., 1983, Sov. Ast. Space Sci. Rev., 2, 189

Stern, B. E., Begelman, M. C., Sikora, M., Svensson, R., 1995, MNRAS, 272, 291

SPECTRAL FEATURES FROM X-RAY ILLUMINATED ACCRETION DISCS AS DIAGNOSTICS OF THE BLACK HOLE ANGULAR MOMENTUM

Andrea Martocchia [1], Giorgio Matt [2] and Vladimir Karas [3,1]

1) SISSA-ISAS, Via Beirut 2/4, I-34014 Trieste (Italy); E-mail: martok@sissa.it; 2) Dipartimento di Fisica, Universita' degli Studi "ROMA TRE", Via della Vasca Navale 84, I-00146 Roma (Italy); 3) Astronomical Institute of the Charles University, Faculty of Mathematics and Physics, V Holesovickach 2, CZ-180 00 Praha (Czech Republic).

ABSTRACT

Detailed calculations of the relativistic effects on both the reflection continuum and the iron line from accretion discs around rotating black holes are performed using a method which takes into account the relativistic transfer of both illuminating and reprocessed photons. These may be used as a diagnostics of the black hole spin when data from high throughput X-ray missions, like Constellation-X, will be available.

KEYWORDS: Accretion, accretion discs; Relativity; Line: formation; Galaxies: Active; X-rays: galaxies

1. INTRODUCTION

X-rays from Active Galactic Nuclei (AGN) and Black Hole Candidates ($BHCs$) can exhibit the imprints of black hole gravitational fields, both in the *line profiles* - particularly the iron Kα feature at $E = 6.4$ keV - and in the *continuum* - direct as well as Compton-reflected, i.e. from about 1 up to several hundreds keV. Such imprints have already been observed by ASCA and BeppoSAX (e.g. Tanaka et al. 1995; Nandra et al. 1997; Guainazzi et al. 1999). Even before the ASCA discovery of a relativistic iron line profile in the spectrum of the Seyfert 1 galaxy MGC-6-30-15 (Tanaka et al. 1995), many authors modelled general relativistic effects on line profiles under various physical and geometrical assumptions, also considering spinning black holes, i.e. Kerr metric, surrounded by an accretion disc (for references see Martocchia, Karas & Matt (1999) - in the following: MKM99).

Among the various approximations usually made in calculating iron line profiles two appear to be critical. *First*, the line profile is calculated and shown without regard to the effects on the reflection continuum, which is produced along with the iron features by illumination of the disc from a primary X-ray source. *Second*, a simple power law parametrization of the *disc emissivity* $\epsilon(r)$ is usually adopted,

Source height	A	B	C	D
$h/m = 2$	0.02268	38.38706	0.000122	6.14786
$h/m = 3$	0.00146	24.47516	0.000059	0.00006
$h/m = 4$	0.00324	30.85495	0.000039	4.70109
$h/m = 5$	0.00640	35.36543	0.000025	4.49187
$h/m = 6$	0.00968	34.27642	0.000013	3.75576
$h/m = 8$	0.27139×10^{-4}	8.74175	0.01662×10^{-4}	1.74206
$h/m = 10$	0.99815×10^{-5}	6.05660	0.04793×10^{-5}	1.14952
$h/m = 12$	0.42940×10^{-5}	4.42939	0.01590×10^{-5}	0.81154
$h/m = 15$	0.27182×10^{-5}	4.64923	0.01451×10^{-5}	0.89655
$h/m = 20$	0.98182×10^{-6}	3.66965	0.05796×10^{-6}	0.72744
$h/m = 100$	0.59002×10^{-5}	1.05533	-0.59023×10^{-5}	1.05622
$h/m = 5_{\rm CL}$	0.10084×10^{-4}	0.29168	-0.09948×10^{-4}	0.28796

TABLE 1. Coefficients of the fitting laws for the emissivities obtained by Martocchia & Matt (1996). All results are for a maximally rotating black hole, except for the row $h/m = 5_{\rm CL}$ which refers to a purely Euclidean case.

while the actual emissivity, which depends on the geometry of the illuminating matter, is in general more complex.

By making use of a fully relativistic code in Kerr metric (see MKM99 for details), we calculated the effects expected on both the reflection continuum *and* the line profile in a geometrically thin, optically thick scenario. We used emissivity laws appropriate when the disc is illuminated by a X-ray primary source, which is here assumed for simplicity to be point–like and located at a height h on the system symmetry axis. This allows us to estimate the effects resulting from various degrees of anisotropy of the illumination, just by varying h (Martocchia & Matt, 1996). Of course, extensions to more refined accretion theories and assumptions are desirable and will be performed in the future. We are also currently developing a code suitable for spectral fitting of real data.

2. PROPERTIES OF SPECTRAL FEATURES FROM AN ILLUMINATED DISC IN KERR METRIC

Monte Carlo simulations are used to calculate the photon transfer within disc matter. Our code computes such reflection using standard assumptions on the disc structure and ionization (Matt et al. 1991).

The radial dependence of the emissivity accounts for the local illuminating flux and incident angle; we used fully relativistic calculations of the primary flux impinging onto the disc (Martocchia & Matt 1996, 1997). The resulting emissivity law, which depends on h (it steepens when h decreases because of the enhanced anisotropy of the primary emission; see MKM99), can be approximated through polynomials of the form $\epsilon(r) = Ar^{-B} + Cr^{-D}$.

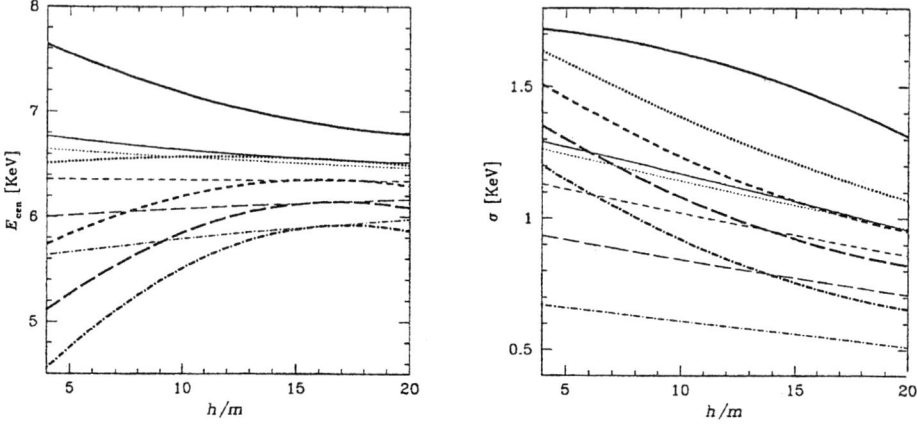

FIGURE 1. Left: Line centroid energy vs. h/m for $cos(\theta_{obs}) = 0.1, 0.3, 0.5, 0.7$ and 0.9 (from top to bottom), both for a static (thin line) and a maximally rotating (thick line) black hole. Right: Geometrical width of the spectral line as a function of h/m.

In Table 1 the best-fit coefficients for the above formula are listed. At this point it is possible to compute the spectra which are seen from a distant observer. In MKM99 we show iron line profiles with the underlying Compton-reflected continuum. With decreasing h, the effect of light bending is enhanced (Martocchia & Matt 1996) and the fraction of (primary) X-ray photons impinging onto the disc increases; this causes a substantial increase of the reflected component with respect to the direct one.

We also considered integral quantities: the line equivalent width (EW), centroid energy (E_c) and geometrical width (σ; for the definitions, as well as previous results in the static case, see Matt et al. 1992). The EW behaviour with varying h is very interesting to investigate, as discussed in Martocchia & Matt (1996), because very large values are predicted in extreme cases. This may apply to MCG-6-30-15 in some of its spectral states (Iwasawa et al. 1996).

It is possible to approximate the observable quantities $E_c(h)$, $\sigma(h)$ and $EW(h)$ using quadratic polynomials of the form $a_0 + a_1 h/m + a_2(h/m)^2$. The corresponding coefficients of the least-square fitting are provided in Martocchia, Karas & Matt (1999). They approximate h-dependences in the range between $4m$ and $20m$ with good accuracy (see Figures). For the whole interval, up to $h = 100m$, we used more precise spline fits; the corresponding MATLAB script, which can also be used for the numerical inversion to obtain the parameters of the model in terms of the three observables, is available from the authors upon request.

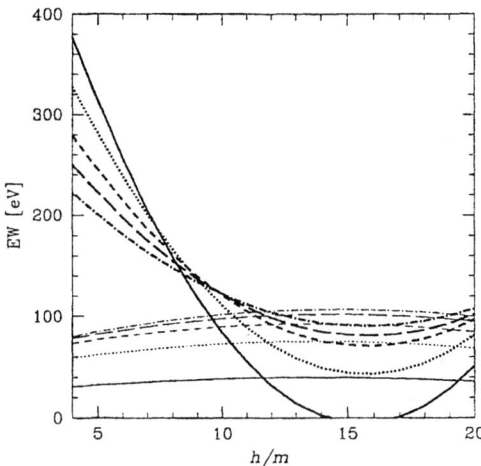

FIGURE 2. Equivalent width dependence on h/m. Line conventions are the same as in the previous figures.

3. CONCLUSIONS

The results presented here have been obtained with fully relativistic codes which enable fast computations of spectral features produced in the vicinity of rotating black holes. In particular, the development of a fitting code, making use of a big atlas of geodesics (Karas, Lanza & Vokrouhlický 1995), is in progress.

Our software package, which can be used for different assumptions on the accretion and primary source geometry and physics, will hopefully be useful for the near future high-sensitivity X-ray observatories, like XMM, as far as the iron line is concerned. We have to wait for mission like *Constellation-X*, with its very large sensitivity and broad band coverage, in order to simultaneously examine both the iron line and the reflected continuum in the desired detail.

REFERENCES

Guainazzi M., et al. (1999), A&A 341, L27
Iwasawa K. et al. (1996), MNRAS 282, 1038
Karas V., Lanza A. & Vokrouhlický D. (1995), ApJ 440, 108
Martocchia A., Karas V. & Matt G. (1999), MNRAS, in press
Martocchia A. & Matt G. (1996), MNRAS 282, L53
Martocchia A. & Matt G. (1997), in: "Proc. 12th Italian Conference on Gen. Rel. & Grav. Phys.", World Scient. Pub. Co., Singapore
Matt G., Perola G. C., Piro L. & Stella L. (1992), A&A 257, 63; Erratum in A&A 263, 453
Nandra K. et al. (1997), ApJ 477, 602
Tanaka Y. et al. (1995), NATURE 375, 659

EPIC/XMM CALIBRATIONS

P. Marty[1], J.-P. Bernard[1]

1) Institut d'Astrophysique Spatiale (IAS), Campus Université Paris-Sud, bat. 121, 91405 Orsay cedex, France - marty@ias.fr

ABSTRACT   The European Space Agency's (ESA) X-rays Multi Mirrors (XMM) observatory satellite is about to be launched. Its three CCD cameras (EPIC instrument) and two CCD spectroscopes (RGS instrument) will allow high troughput and high resolution spectro-imaging.

Among the various goals of this mission, measuring the X-rays coming from clusters of galaxies (bremsstrahlung light) will constrain our large scale matter distribution models. Inside a cluster, X-rays brightness variations will leave clues regarding matter dynamics.

Given clusters megaparsecs large and up to one gigaparsec away, one obviously needs to resolve such objects at far better than 6 arcmins. EPIC instrument has been designed to reach a few arcsecs. Anyway, one has to care for good flat field and quantum efficiency (QE) calibrations.

In this paper, we will briefly describe the Orsay synchrotron facility were such calibrations were undergone in 1998, the latest results concerning EPIC QE measurements, and then summarize our involvment in clusters of galaxies and X-rays background observations.

KEYWORDS: XMM; EPIC; calibrations; synchrotron; clusters of galaxies; X-rays background.

1. EPIC CALIBRATIONS AT ORSAY FACILITY

1.1. The Orsay synchrotron facility

The calibration of EPIC CCD cameras took place in the IAS calibration facility in 1998. In order to cover the 0.1 to 15.0 keV energy range, we used the synchrotron light produced by the two positrons storage rings of the Laboratoire pour l'Utilisation du Rayonnement Electromagnétique (LURE): SACO & DCI (table 1).

Two whole beam lines were build through 1994 to 1997, each starting at one different ring (fig. 1), including a monochromating device to select required energies (150 to 1500 eV on SACO; 1.5 to 15.0 keV on DCI) with high accuracy (better than 1 eV) and cut-off mirrors to remove reflected harmonics.

The camera to be calibrated is located into a 23 m^3 vaccuum chamber (called Jupiter tank) and can be moved toward the requested beam lines by a Mechanical and Optical Ground Support Equipment (MOGSE).

Since it was too difficult to drive from the ring a perfectly homogeneous beam wider enough to illuminate one EPIC camera (75 by 75 mm) at a time, a resizing slit has been plunged in front of the camera in the Jupiter tank in order to select

TABLE 1. Storage rings description.

	SACO	DCI
Positrons energy	0.80 GeV	1.85 GeV
Maximum current	400 mA (24-paquets)	330 mA (4-paquets)
Synchrotron peak e.	0.68 keV	3.62 keV
Positrons life time	9 hours	200 hours
Positrons period	240 ns	315 ns

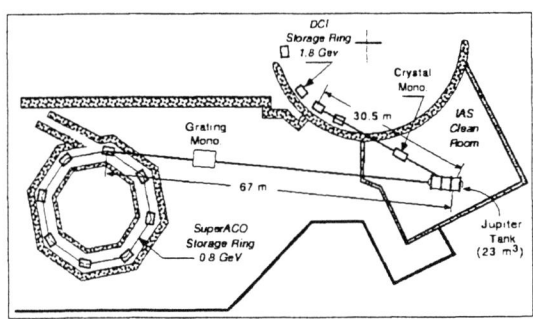

FIGURE 1. Orsay synchrotron facility map.

an horizontal slice (75 by 1 mm) of the beam. The camera was then scanned up and down behind the slit and read out between each cycle. Both slit position and camera movement could be actuated again by the MOGSE.

All instruments can be continuously driven by a network of computers operating a specially designed software under Wind*ws NT. This Extended Instrument Computer Control (EICC) software uses network and multitasking capabilities, allowing a complete remote control by a single operator. The EICC also integrates some basic file saving and database updating features.

1.2. Detectors and Results

An absolute monitoring of the beam profile behind the slit was needed to analyse the horizontal response of the camera (the vertical response depended only on its up-and-down movement speed). Two "home made" detectors were thus mounted on a scanning plate between the slit and the camera:
* a solid state semiconductor, Si(Li), detector for 0.3 to 15 keV energy range;
* a gaz proportional counter (GPC) for energies below 300 eV and around the Silicium absorbtion edge (around 600 eV).

Furthermore, since the camera exposures lasted for tens of minutes, it was necessary to monitor the beam variations without screening the camera field of view.

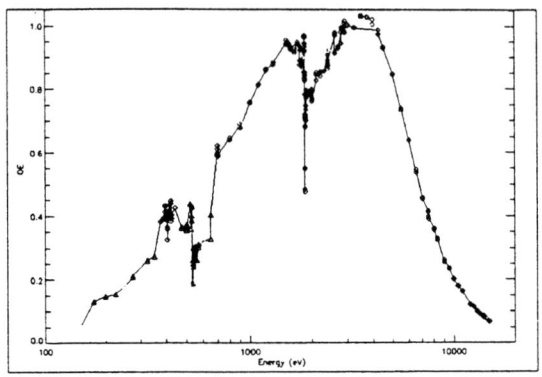

FIGURE 2. EPIC MOS QE (as of oct, 1. 1999).

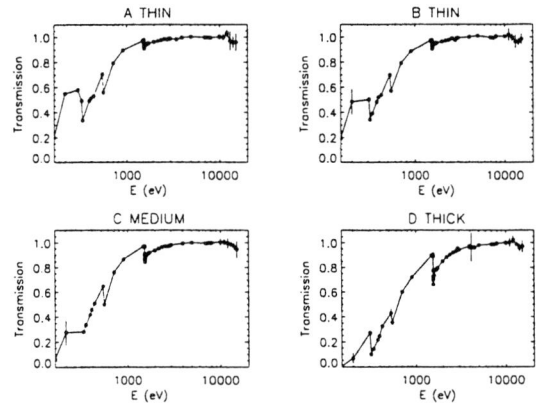

FIGURE 3. EPIC MOS Filter Transmission (as of oct, 1. 1999).

Hence, a network connection to the LURE allowed to acquire on a regular time basis the synchrotron current inside the rings (to monitor the positrons decay); a diode mounted on the monochromator in front of the second order reflection allowed to see the variations due to mirrors (thermalization, vibrations, etc.).

Typically three types of calibration runs were performed:

QE measurement runs (fig. 2) consisting of camera read out with filter wheel in OPEN position interleaved with absolute measurements of the beam (typically: OPEN, absolute, OPEN);

Filter transmission runs (fig. 3) consisting of successive camera read out with filter wheel alternating OPEN and filter positions (typically: OPEN, THINA, OPEN, THINB, OPEN, MEDIUM, OPEN, THICK);

A combination of the above, at energies where both absolute QE and filter transmission measurements were required.

2. CONCLUSIONS AND PERSPECTIVES

The grating monochromator (SACO) and the double crystals monochromator (DCI) can be both set and tuned separately and without the need of the Jupiter tank. This optimizes the calibration needs and allows taking into account the schedule and running time of each storage ring. Using great length lines, it has been achieved monochromatic beams with known profile along a 75 mm width to match the EPIC focal plane, while minimizing harmonics contamination and monitoring scattered light. Each main element of the whole Calibration Facility can be monitored and tuned from a remote computer control network (EICC), also designed on purpose. Now that the EPIC calibrations are over, all these equipments (beamlines, EICC...) have been reoriented toward a broader scientific program and open to other users.

Among the XMM Guaranted Time targets, an X-rays deep sky survey and some clusters for which Sunyaev-Zel'dovich Effect (SZ) submillimetric data have been provided by PRONAOS or DIABOLO experiments. In addition to characterizing temperature, density and morphology using X-rays alone, matching correlated X-rays and SZ data sets will allow making assumptions about cosmological parameters without having to estimate the cluster distance. For instance, the cluster proper motion could be retrieved by measuring its SZ effect at two different wavelengths; the Hubble Constant by comparing a cluster SZ effect to its X-rays brightness. But the accuracy of all this strongly depends on the X-rays camera spatial resolution.

Regarding the X-rays Background puzzle, we have submitted an XMM AO1 proposal where we suggest to correlate the structure of the Far-Infrared (FIR) and X-rays background emissions in the FIRBACK southern field. It was surveyed by ISO at $170\,\mu m$ in order to estimate the fraction of the FIR background powered by Active Galactic Nuclei (AGNs). This correlation is expected to decrease with decreasing energy as dust extinction increases at lower X energies. This will give the answer to the crucial cosmological question of knowing what fraction of the energy radiated by galaxies over the history of the Universe comes from AGNs.

ACKNOWLEDGEMENTS

Many thanks to the IAS Calibration Facility crew members, LURE staff (congratulations for the SOLEIL project...), Leicester MOS and MPE PN teams, Bologna ITeSRE and Saclay CEA colleagues !

REFERENCES

(EPIC) Bernard, J.P., Marty, P., 1999, EPIC/XMM Calibrations product release v1.0
(EPIC) Pigot, C., et al, 1999, SPIE Proceedings (july 1999)
(Clusters) Holzapfel, Arnaud, M., et al, 1997, ApJ (may, 10. 1997)
(Clusters) Desert, F.-X., et al, 1999, New Astronomy (ed. Elsevier)
(XRB) Comastri, A., et al, 1999, astro-ph/9902060
(XRB) Puget, J.-L., et al, 1999, A&A, 245, 29

BEPPOSAX SPECTRA OF FIVE LOW MASS X-RAY BINARIES

Nicola Masetti, Elena Pian, Filippo Frontera*, Eliana Palazzi, Lorenzo Amati, Mauro Orlandini and Daniele Dal Fiume

Istituto Te.S.R.E., CNR, via Gobetti 101, I-40129 Bologna, Italy
*also Dipartimento di Fisica, Università di Ferrara, Italy

ABSTRACT   Observations of five Low Mass X-ray Binaries (LMXBs) were performed with BeppoSAX in order to investigate emission processes in LMXBs in a wide X-ray energy range and understand the crucial spectral characteristics of compact stellar systems harbouring neutron stars. We find that two thermal components are present in the spectra of both faint and bright LMXBs and that the high energy spectra of the observed LMXBs are systematically softer than those of black hole candidates. We also detect an iron emission line around 6.5 keV in bright LMXBs.

KEYWORDS: X-rays: stars — Stars: neutron — Accretion, accretion disks

1. INTRODUCTION

Low Mass X-ray Binaries (LMXBs) are the most numerous class of galactic X-ray binaries. They include many persistent low magnetic field neutron stars, whose emission properties have been extensively studied in the low X-ray energy band (≤ 20 keV; see e.g. van der Klis 1994). The high sensitivity of the Narrow Field Instruments (NFI) onboard the satellite *BeppoSAX* (Boella et al. 1997a) allows the study of the spectral behavior of LMXBs on a wide range of energies (0.1-200 keV).

We exploited the capabilities of the NFI in order to: (i) test the existence of thermal components in LMXBs in the 0.1-10 keV band; (ii) compare the high-energy (>20 keV) spectra of LMXBs with those of black hole candidates.

Here we present the results of *BeppoSAX* observations for five LMXBs: two faint ($F_{2-10 \text{ keV}} \sim 20$ mCrab) sources, X1543-624 and X1556-605, and three bright ($F_{2-10 \text{ keV}} \sim 500$ mCrab) objects, X1642-455, X1744-265 and X1813-140. A more complete presentation of the results will be given elsewhere (Masetti et al. 2000).

2. THE *BEPPOSAX* OBSERVATIONS

As part of a monitoring campaign on LMXBs, *BeppoSAX* NFI observations were carried out between 1997 Feb 21 and 1999 Mar 1. The five LMXBs listed above have been imaged and observed with the LECS (0.1–10 keV; Parmar et al. 1997), the MECS (1.5–10 keV; Boella et al. 1997b), and the PDS (15–300 keV; Frontera et al. 1997); see the journal of observations in Table 1. For the PDS instrument the standard rocking collimator law has been modified during the observation of X1744-265 in order to reduce as much as possible the background contamination from nearby variable X–ray sources.

The joint LECS+MECS+PDS spectra (0.1-200 keV) were fitted using *XSPEC*, v. 10.0. In Table 2 are reported the best-fit parameters of the adopted spectral models. An absorption hydrogen column density N_H has been applied to every model we used.

3. RESULTS

The analysis of the *BeppoSAX* data yielded the following preliminary results: [i] all the observed sources show a hard X-ray tail which extends above 20 keV up to ~100 keV. [ii] The X–ray spectra of faint LMXBs (Fig. 1) are fitted with 2 thermal components, i.e. blackbody (BB, of temperature T_1) plus disk-blackbody (DISKBB, with T_{in} as the inner disk temperature) model for X1543-624 and two-blackbody (2BB, with temperatures T_1 and T_2) model for X1556-605 (with the high temperature component $kT_1 \sim 1.5$ keV and the lower one, kT_2 or kT_{in}, ~0.5 keV in both cases). A cutoff power law (CPL) of index α and cutoff energy E_c must be added in both cases to fit the hard X–ray tail. Alternatively, one can fit the spectra of these sources with a single BB (or DISKBB) plus Sunyaev-Titarchuk Comptonization model (COMPST; Sunyaev & Titarchuk 1980) in which T_e is the plasma temperature and τ is the plasma opacity. The present analysis does not allow us to choose either model: the BB, or DISKBB, plus COMPST is better constrained but the two-temperature model plus CPL gives a better fit of the high-energy data. [iii] The X–ray spectra of bright LMXBs are fitted with a DISKBB+COMPST with typical temperatures $kT_{in} \sim 2$ keV and $kT_e \sim 4$ keV (Figs. 2 and 3), although the BB+CPL model gives a nearly equivalent fit. However, a power law (PL) must be added in the first pointing spectrum of X1813-140 to achieve a reasonable fit (Fig. 3, left panel). One can note that the DISKBB+COMPST does not adequately fit the high-energy data of X1744-265 (Fig. 2, right panel); this suggests that a more complex model should be applied. [iv] An iron emission line around 6.5 keV is seen in bright X-ray LMXBs.

We also note that the reduced χ^2 values are in all cases somewhat larger than 1. This is usually found when fitting high signal-to-noise X–ray spectra over very wide energy ranges.

REFERENCES

Boella G., Butler R.C., Perola G.C. et al., 1997a, A&AS 122, 299
Boella G., Chiappetti L., Conti G. et al., 1997b, A&AS 122, 327
Frontera F. Costa E., Dal Fiume D. et al., 1997, A&AS 122, 357
Masetti N. et al., 2000, in preparation
Parmar A.N., Martin D.D.E, Bavdaz M. et al., 1997, A&AS 122, 309
Sunyaev R.A., Titarchuk L., 1980, A&A 86, 121
van der Klis M., 1994, ApJS 92, 511

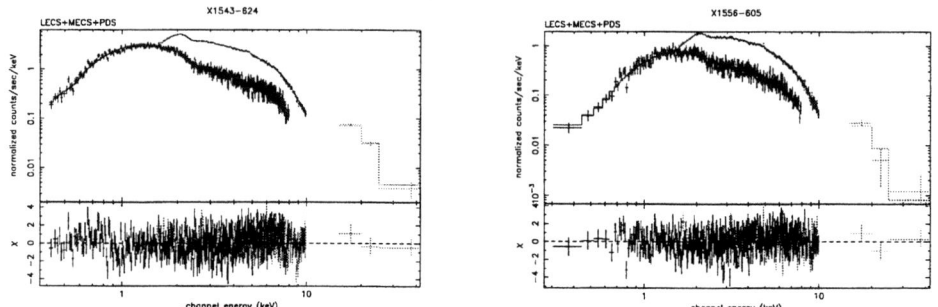

FIGURE 1. *(Left panel)* Broad band X–ray spectrum of X1543-624 from the first *BeppoSAX* observation. The fitting curve represents a BB+DISKBB+CPL model. *(Right panel)* Broad band X–ray spectrum of X1556-605 from the first *BeppoSAX* observation. The fitting curve represents a 2BB+CPL model.

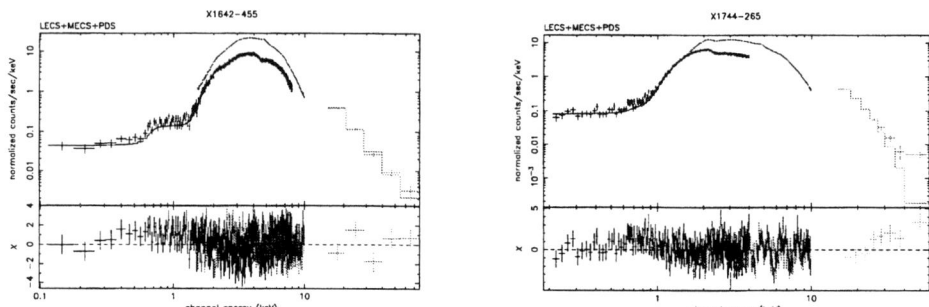

FIGURE 2. *(Left panel)* Broad band X–ray spectrum of X1642-455 from the first *BeppoSAX* observation. *(Right panel)* Broad band X–ray spectrum of X1744-265 from the single *BeppoSAX* observation. The fitting curves superimposed on the data in both panels represent a DISKBB+COMPST+Fe line model.

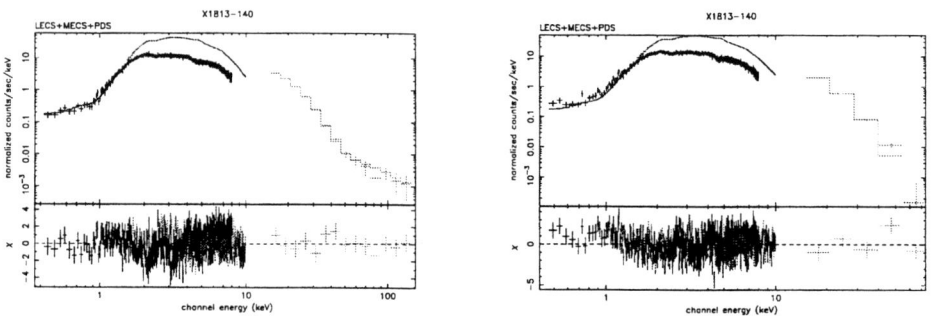

FIGURE 3. Broad band X–ray spectrum of X1813-140 from the first *(left panel)* and from the second *(right panel) BeppoSAX* observation. The fitting curves superimposed on the data in both panels represent DISKBB+COMPST+Fe line model, with the further addition of a PL to the left panel data.

TABLE 1. Log of the nine NFI observations presented in this paper.

Source	Start day	Start time (UT)	Duration (sec)	MECS obs. time (sec)
X1543-624	1997 Feb 21	09:38:53	28883	17938
	1997 Apr 1	22:40:51	26136	17783
X1556-605	1997 Mar 10	17:10:56	24528	16575
	1997 Apr 3	18:09:35	27921	18253
X1642-655	1997 Sep 4	12:16:46	27563	14778
	1997 Oct 2	20:33:35	27823	12141
X1744-265	1999 Feb 28	11:06:15	89897	44809
X1813-140	1997 Apr 3	00:28:15	19934	10688
	1997 Apr 21	02:29:37	13495	6482

TABLE 2. Best-fit spectral parameters of the observed LMXBs. Quoted errors are at 1-σ confidence level.

Parameter	X1543-624 obs. 1	obs. 2*	X1556-605 obs. 1	obs. 2	X1642-455 obs. 1	obs. 2	X1744-246 obs. 1	X1813-140 obs. 1	obs. 2
χ_ν^2 (dof)	1.32 (826)	1.23 (731)	1.11 (628)	1.06 (625)	1.35 (846)	1.45 (813)	1.46 (517)	1.53 (861)	1.28 (849)
N_H^a	0.176±0.007	0.206±0.008	0.22±0.14	0.19±0.14	7.23±0.13	7.09±0.11	2.08±0.03	2.78±0.03	2.75±0.04
kT_1 (keV)	1.55±0.03	1.58±0.01	1.41±0.14	1.31±0.17	—	—	—	—	—
kT_2 (keV)	—	—	0.52±0.03	0.52±0.05	—	—	—	—	—
kT_{in}^b (keV)	0.66±0.03	0.655±0.007	—	—	1.811±0.007	1.91±0.02	2.14±0.05	2.60±0.11	2.42±0.05
kT_e (keV)	—	—	—	—	13.8±4.0	4.5±0.3	3.2±0.2	3.76±0.12	3.83±0.12
τ	—	—	—	—	2.7±0.6	6.2±0.4	9.6±0.8	10.5±0.5	9.7±0.4
α	~0^c	—	~0^c	~0^c	—	—	—	2.2±0.6	—
E_c (keV)	~5^c	—	~4^c	~4^c	—	—	—	—	—
+ Fe em. line									
E_{Fe} (keV)	—	—	—	—	6.83±0.03	6.76±0.03	6.33±0.06	6.75±0.03	6.80±0.04
σ_{Fe} (keV)	—	—	—	—	0.2^c	0.1^c	0.86±0.06	0.2^c	0.2^c
EW (eV)	—	—	—	—	46±6	32±5	169±18	39±4	28±6
$F_{2-10\,keV}^d$	8.20±0.02	7.26±0.02	3.43±0.01	3.61±0.01	138.8±0.1	164.4±0.1	50.39±0.03	149.0±0.1	157.0±0.1

* LECS+MECS only
[a] In units of 10^{22} cm^{-2}
[b] Temperature of the inner disk radius if a DiskBB model is used
[c] Not well constrained
[d] Unabsorbed flux in units of 10^{-10} erg cm^{-2} s^{-1}

X-RAY EVIDENCE OF AN AGN IN M82

H. Matsumoto [1], T.G. Tsuru [2]

1) CSR/MIT, 77 Massachusetts Avenue, Cambridge, MA 02139-4307, USA
2) Department of Physics, Kyoto University, Kyoto 606-8502, Japan

ABSTRACT

M82, which is the most famous starburst galaxy, shows hard X-ray emission. However, the origin of the hard component is unclear. Therefore, we conducted a monitoring observation of M82 with ASCA in 1996. A significant time variability of the hard component was found between 3×10^{40} erg/s and 1×10^{41} erg/s on various time scales from 10 ks to a month. This strongly suggests that a low-luminosity AGN (LLAGN) exists in M82. A broad iron line emission was found, which is similar to other LLAGNs.

KEYWORDS: galaxies:active — galaxies:starburst — galaxies:individual (M82)

1. INTRODUCTION

Since M82 is thought to be an archetypical starburst galaxy, many X-ray observations of M82 have been made. The first observation of M82 with ASCA was conducted in 1993. Tsuru et al. (1997) analyzed the ASCA spectrum, and found that it consists of three components: soft, medium, and hard. Since the soft and medium components show emission lines from various elements and their X-ray images are extended compared with the ASCA point spread function (PSF), they could be due to a galactic wind driven by stellar winds from massive stars and supernovae.

The ASCA spectrum of the hard component can be described by either a power-law model (photon index is $\Gamma \sim 1.7$) or a thermal plasma model (temperature is $kT \sim 14$ keV). Tsuru et al. (1997) compared the ASCA flux in the 2 – 10 keV band with those of Ginga (Tsuru 1992) and EXOSAT (Schaaf et al. 1989), and found a time variability. Furthermore, the spatial extent of the hard component is consistent with the ASCA PSF. These may suggest that the origin of the hard component is a low-luminosity AGN (LLAGN) of M82. However, since Ginga and EXOSAT are non-imaging detectors, possible contamination from other hard sources is not excluded. Furthermore, the Ginga spectrum can be fitted with a thermal model, but cannot be fitted with a power-law model (Tsuru 1992), which is different from the typical X-ray spectra of AGNs. The same conclusion was suggested by Cappi et al. (1999) using the BeppoSAX data. Thus, the origin of the hard component is still a matter of debate.

The first key to revealing the origin of the hard component is to clarify whether or not it shows a time variability. The second key is to detect the iron K-line emission and to determine its central energy. The central energy of the line can be direct evidence. For these purposes, we made a monitoring observation (9 times) of M82 with ASCA in 1996. See Matsumoto and Tsuru (1999) and Ptak and Griffiths (1999) for more details about the observation and our results.

2. LIGHT CURVE

First, we studied the light curve of each observation. We found that the data on April 24, 1996, show short-term variability only in the hard-energy band, which is shown in figure 1. The time scale is about 10^4 s. We found no short-term variability in the other observations.

Next, we compared the average counting rate of each observation, which is shown in figure 2. We can see that only the light curves in the high-energy band show time variability.

These light curves strongly suggest that only the hard component of M82 varies with time.

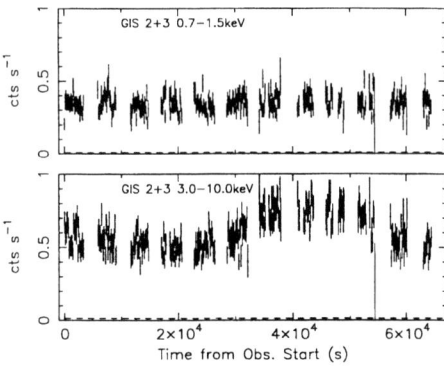
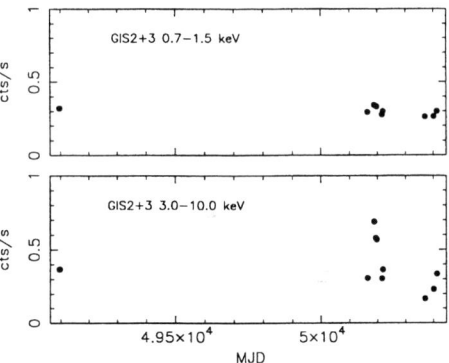

FIGURE 1. Light curves of the GIS2+3 data during the observation on April 24, 1996. The background levels are indicated by the dashed lines.

FIGURE 2. Average counting rates of the GIS 2+3 data.

3. SPECTRAL ANALYSIS

Then we fitted the ASCA spectrum of each observation with the three-temperature thermal plasma model proposed by Tsuru et al. (1997). For the soft and medium components, we fixed their spectral shapes to the best-fit results of Tsuru et al. (1997). Therefore, only their normalizations are free parameters. For the hard component, the free parameters are its absorption column density, temperature, metallicity, and normalization. This three-temperature model can fit the spectra of all observations quite well. The X-ray luminosity of each component in the 0.5 – 10 keV band is shown in figure 3. It is clear that only the hard component shows time variability. The temperature of the hard component also changed. However, we found no correlations between the temperature and luminosity of the hard component.

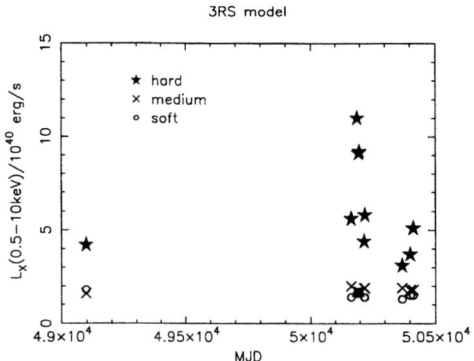

FIGURE 3. Time variability of each component in the three-temperature model. The stars, crosses, and circles show the hard, medium, and soft components, respectively.

We subtracted the spectrum of the lowest state (Apr. 15, 1996) from that of the highest state (Oct. 14, 1996), which is shown in figure 4. This spectrum can be fitted by a heavily absorbed ($N_H \sim 10^{22}$ cm^{-2}) thermal plasma model. This strong absorption suggests that the variable source is embedded in the center of M82.

Finally, we examined the iron line feature. Because the statistics of each observation were rather limited, we combined the spectra of all observations above the 4 keV band. The spectrum can be fitted by a thermal bremsstrahlung continuum plus a Gaussian line model, which is shown in figure 5. The significance of the detection of the Gaussian line is larger than 99%. The center energy, sigma, and equivalent width of the line are 6.56 ± 0.14 keV, 0.30 ± 0.18 keV, and 121 ± 60 eV, respectively (errors are at 90% confidence level). The broad iron line feature is similar to other LLAGNs such as M81 (Ishisaki et al. 1996).

4. CONCLUSION

We found that the hard component of M82 shows time variability between 3×10^{40} erg/s and 1×10^{41} erg/s at various time scales from 10 ks to one month. The variable source shows a heavily absorbed feature with a column density of 10^{22} cm^{-2}, which means the variable source is embedded in the center of M82. Furthermore, the X-ray images of the hard component are consistent with a point source. All these suggest that there is a LLAGN in M82. Other possibilities, such as a collection of X-ray binaries, a super-Eddington source, a young supernova remnant, and inverse

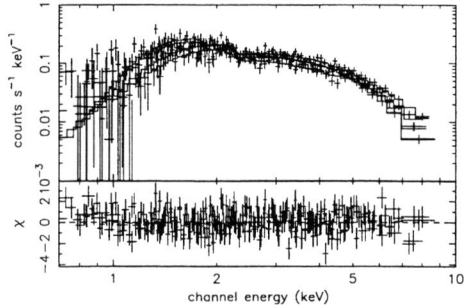 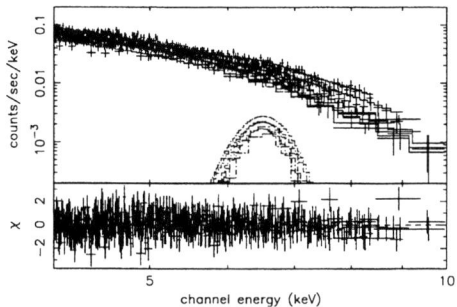

FIGURE 4. Residual SIS and GIS spectra obtained by subtracting the lowest state (Apr. 15, 1996) from the highest state (Oct. 14, 1996). The lines show the best-fit heavily absorbed thermal plasma model.

FIGURE 5. Composite SIS and GIS spectra above in the 4 – 10 keV energy band. The lines show the best-fitting thermal bremsstrahlung plus Gaussian line model.

Compton scattering of IR photons from relativistic electrons which produce radio emission, cannot explain the variability (Matsumoto and Tsuru 1999). We also found a broad iron K line, which is similar to other LLAGNs. The time variability on a time scale of 1×10^4 s implies the size of the emitting region to be smaller than 3×10^{14} cm. Assuming that the origin of the hard component is the LLAGN and the X-rays are mainly emitted from a region as large as 6 times its Schwarzschild radius, the mass of the central object is estimated to be less than $2 \times 10^8 M_\odot$. The mass must be larger than $1 \times 10^3 M_\odot$, so that the peak luminosity does not exceed its Eddington luminosity.

ACKNOWLEDGEMENTS

We thank the ASCA team members for their support. We are also grateful to Miss Deborah Gage for careful review of the manuscript. HM is supported by the JSPS Postdoctoral Fellowships for Research Abroad.

REFERENCES

Cappi, M. et al. 1998, A&A, 350, 777
Ishisaki, Y. et al. 1996, PASJ, 48, 237, 1996
Matsumoto, H., Tsuru, T. G. 1999, PASJ, 51, 321
Ptak, A., Griffiths R. 1999, ApJ, 517, L85
Schaaf, R., Pietsch, W., Biermann, P. L., Kronberg, P. P., Schmutzler, T. 1989, ApJ, 336, 722
Tsuru, T. G. 1992, PhD Thesis, The University of Tokyo
Tsuru, T. G., Awaki, H., Koyama, K., Ptak, A. 1997, PASJ, 49, 619

RESOLVING THE 10-40 KEV COSMIC X-RAY BACK-GROUND WITH CONSTELLATION-X

Giorgio Matt, Fulvio Pompilio & Fabio La Franca

Dipartimento di Fisica, Università degli Studi Roma Tre, via della Vasca Navale 84, I–00146 Roma, Italy

ABSTRACT

The energy density of the Cosmic X-ray background (XRB) peaks around 30 keV (see Figure 1), an energy not yet probed by focussing imaging instruments. The first hard X-ray telescope due to fly on a space mission will be that on board Constellation-X. The imaging capability, besides providing an improvement of several orders of magnitude in sensitivity over current passively collimated detectors, will permit for the first time to resolve a fraction of the XRB at this most crucial energy. Synthesis models of the XRB based on obscured AGN predict that at least 40% of the 10-40 keV XRB will be resolved by Constellation-X.

KEYWORDS: X-rays: galaxies; galaxies: nuclei

1. THE CONSTELLATION X-RAY MISSION

The Constellation X-ray mission is a high throughput X-ray facility emphasizing observations at high spectral resolution (E/Δ=300–3000) while covering a broad energy bandpass (0.25–40 keV). Constellation-X will provide a factor of nearly 100 increase in sensitivity over current high resolution X-ray spectroscopy missions. The large collecting area is achieved with a design utilizing several mirror modules, each with its own detector system. Each (or a few) science unit will fly on a separate spacecraft.

Two telescopes will be on–board: the low–energy Spectroscopy X-ray telescope (SXT), operating simultaneously with a 2 eV resolution calorimeter and a set of reflection gratings; a high–energy system (HXT), that will be the first focusing telescope system operating at several tens of keV, where the energy density of the XRB peaks (Figure 1).

The Baseline Mission Characteristics are:

- Effective Area: 15000 (6000, 1500) cm^2 at 1 (6.4, 40) keV

- Angular resolution: 15" HPD from 0.25 to 10 keV; 1' HPD at 40 keV

- Band Pass: 0.25 to 40 keV

More information can be found at: HTTP://CONSTELLATION.GSFC.NASA.GOV/

2. THE XRB SYNTHESIS MODEL

To evaluate the fraction of hard XRB resolved by the HXT onboard Constellation-X, we first developed a synthesis model based on the standard assumption that the XRB is mostly made by a combination of type 1 and 2 AGN (Setti & Woltjer 1989; Comastri et al., 1995, and references therein). Details on the model can be found in Pompilio, La Franca & Matt (1999 and this volume). Here we summarize the main features of the model.

1. AGN spectra

    (a) type 1 (AGN1) spectrum:
    - power law ($\alpha = 0.9$) + exponential cut-off ($E_c = 400$ keV);
    - Compton reflection component (accretion disk, $\theta_{obs} \sim 60°$);

    (b) type 2 (AGN2) spectrum (Matt, Pompilio & La Franca, 1999):
    - primary AGN1 spectrum obscured by cold matter:
      $10^{21} \leq N_H \leq 10^{25} cm^{-2}$, $\frac{dN(log N_H)}{d(log N_H)} \propto log N_H$;
    - Compton scattering within the absorbing matter fully included.

2. Cosmological evolution

    (a) PLE ($\Phi^*(z=0) = 1.45 \times 10^{-6} Mpc^{-3} (10^{44} erg\ s^{-1})^{-1}$);

    (b) power law evolution for the break-luminosity:
    $L^*(z) \propto (1+z)^k$ up to $z_{max} = 1.73$, with
    $L^*(z=0) = 3.9 \times 10^{43} erg\ s^{-1}$ and $k = 2.9$ (model H of Boyle et al., 1994);

    (c) the redshift integration is performed up to $z_d = 4.5$.

3. PREDICTIONS

We are now able to predict the fraction of the Cosmic XRB which can be resolved by Constellation-X in the 10–40 keV energy range. Our estimate is based on the baseline spatial resolution (1' HPD, which appears to be rather conservative given the large efforts of the hardware teams towards a better resolution). Any improvement in this resolution will of course increase the fraction of XRB resolved.

The predictions are:

1. Number densities of type 1 and type 2 AGN detected in the 10-40 keV band down to a flux limit of $10^{-14}\ erg\ cm^{-2} s^{-1}$ (reachable with an exposure time of a few thousand seconds, Harrison et al. 1999):

    - AGN1 $\longrightarrow$ 53.3 deg^{-2};

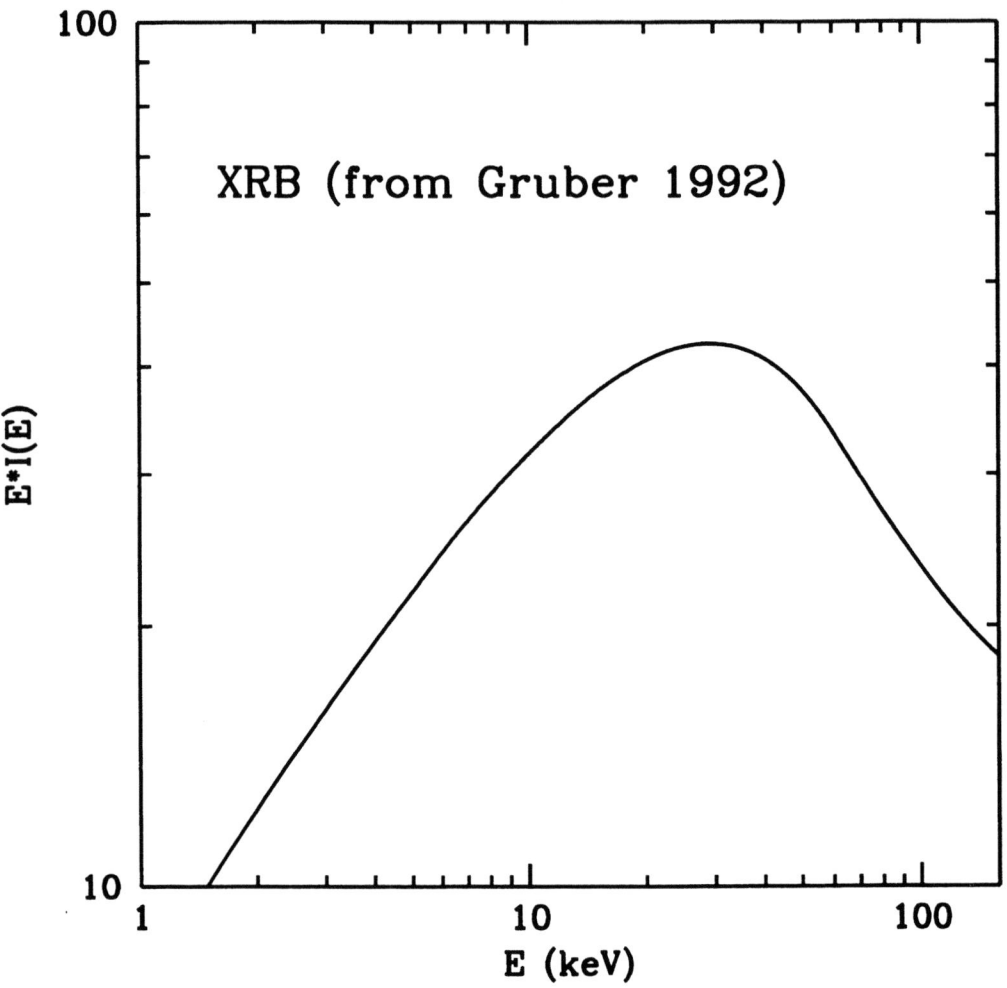

FIGURE 1. The spectrum of the Cosmic X-ray Background after Gruber (1992). The νF_ν representation shows that the energy density peaks at 30 keV.

- AGN2 $\longrightarrow$ 123.4 deg^{-2};

$\Rightarrow \sim 70\%$ are absorbed sources. (Note that these numbers implied a density of about a source per 20 beams, which should ensure a tolerable level of confusion. If the final spatial resolution will be better than 1', as it appears likely, confusion problems will be of course less severe).

2. Integrated XRB spectrum in the 10-40 keV band:
   - measured value $\longrightarrow I(10-40\ keV) \simeq 9.4 \times 10^{-8}\ erg\ cm^{-2}s^{-1}sr^{-1}$ (HEAO-1 A2 data, Marshall et al. 1980);
   - fraction of XRB down to a flux of $10^{-14}\ erg\ cm^{-2}s^{-1}$
   $\longrightarrow I_{Const-X}(10-40\ keV) \simeq 4.0 \times 10^{-8}\ erg\ cm^{-2}s^{-1}sr^{-1}$;

$\Rightarrow \sim$ **40% of the 10–40 keV XRB will be resolved by Constellation-X.**

ACKNOWLEDGEMENTS

This work is part of the activities of the Constellation–X Facility Science Team. The authors acknowledge financial support from the Italian Space Agency.

REFERENCES

Boyle B.J., Shanks T., Georgantopoulos I., Stewart G.C., Griffiths R.E., 1994, MNRAS, 271, 639
Comastri A., Setti G., Zamorani G., Hasinger G., 1995, A&A, 196, 1
Gruber D.E., 1992, in "The X-ray Background", X. Barcons and A.C Fabian eds., Cambridge University Press, p.45
Harrison F., et al., 1999, SPIE Proc., 3765, 104
Marshall F.E., et al., 1980, ApJ, 235, 4
Matt G., Pompilio F., La Franca F., 1999, New Astron., 4, 191
Pompilio F., La Franca F., Matt G., 1999, A&A, in press (astro-ph/9909390)
Setti G., Woltjer L., 1989, A&A, 224, 21

CORRELATED VARIABILITY IN LMC X-2

K.E. McGowan [1], P.A. Charles [1], D. O'Donoghue [2], A.P. Smale [3]

1) Nuclear and Astrophysics Lab., Keble Road, Oxford OX1 3RH
2) SAAO, PO Box 9, Observatory 7935, Cape Town, South Africa
3) LHEA, Code 662, NASA/GSFC, Greenbelt, MD 20771

ABSTRACT   We report simultaneous X-ray and optical photometry of LMC X-2, to search for correlated variability on timescales from seconds to hours. Our results show evidence for a lag between the X-ray and optical variations of ∼6-16 s. Given the likely binary size this suggests that the X-ray reprocessing time is not instantaneous, but has a finite time added, possibly due to the (poorly understood) interaction with the disc's "atmosphere". These data are compared with similar archival studies of Sco X-1.

KEYWORDS: binaries: close - stars: individual: LMC X-2

1. INTRODUCTION

LMC X-2 is the most luminous low mass X-ray binary (LMXB) known and shows substantial variability with $L_x \sim 0.6$ to $3 \times 10^{38}\,\mathrm{erg\,s^{-1}}$ (Markert & Clark 1975; Long et al. 1981). From a precise X-ray location (Johnston et al. 1979) LMC X-2 was optically identified as a faint, V∼18.8, blue star (Pakull 1978; Pakull & Swings 1979).

In spite of a number of studies, the period of LMC X-2 remains uncertain. Callanan et al. (1990) found a periodicity of 8.15 h, Crampton et al. (1990) suggested a much longer period of ∼12.5 d. No correlation between simultaneous optical and X-ray data has previously been found (Bonnet-Bidaud et al. 1989).

2. OBSERVATIONS

White-light high-speed photometry (2 s integrations) of the optical counterpart of LMC X-2 were performed using the UCT-CCD fast photometer (O'Donoghue 1995) on the SAAO 1.9 m. Simultaneous *RXTE* observations of LMC X-2 were obtained with the Photon Counting Array (PCA), and the data were extracted to create a lightcurve with 1 s resolution.

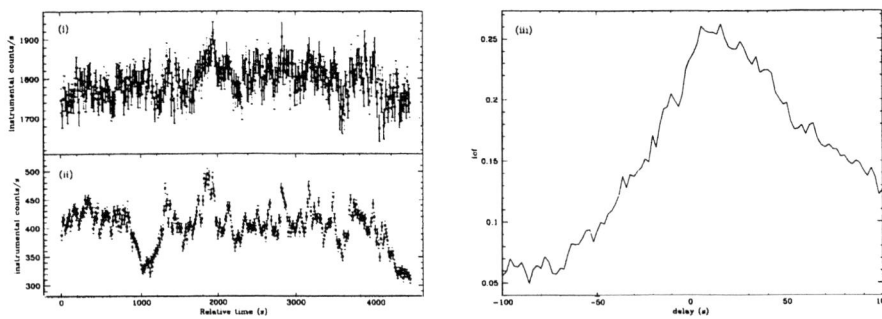

FIGURE 1. Simultaneous optical (i) and X-ray (ii) observations of LMC X–2 (both binned on 16 s). Sign convention for the ICF (iii) is such that positive lags correspond to X-rays leading the optical.

3. RESULTS

The optical and X-ray data were cross-correlated using a version of the Interpolation Correlation Function, ICF (Gaskell and Peterson 1987; Hynes et al. 1998), aswell as standard cross-correlation routines; both methods agreed well. For the one simultaneous run where LMC X–2 was in an X-ray high state (Figure 1), a double peaked ICF was produced. A broad range of significant delays were found, the two highest peaks correspond to lags of $\sim$6 s and $\sim$16 s, both with $\sim$12 σ-confidence. The sign convention of the ICF indicated that the optical lagged the X-rays as expected. In all other runs the source was in a low state and no significant peaks in the ICFs were found.

Monte-Carlo simulations were performed to create optical and X-ray light curves with the same mean and standard deviation as in each dataset. These were cross-correlated and statistics computed on the spread of delay values, giving a lag of $12.4^{+5.9}_{-5.9}$ s.

4. COMPARISON WITH SCO X–1

It is instructive to compare this behaviour with that of the brightest LMXB, Sco X–1. As we found with LMC X–2, Ilovaisky et al. (1980) detected correlated optical and X-ray variability only when Sco X–1 was in a high state (Figure 2). They also found the optical was delayed with respect to the X-rays, which they attributed to X-ray reprocessing.

We performed cross-correlations of the Ilovaisky et al. Sco X–1 data and found delays for the first two datasets of $\sim 1.7^{+0.9}_{-0.9}$ s and $17.6^{+5.4}_{-5.4}$ s (Figure 2). For the third dataset no significant delay was found, but we note the decline in the X-rays compared to the constant optical. The lack of a delay confirms that correlated variability is only seen when the X-rays are in a high flaring state. Petro et al. (1981) also found correlated variability in Sco X–1 (Figure 3). Again we cross-correlated

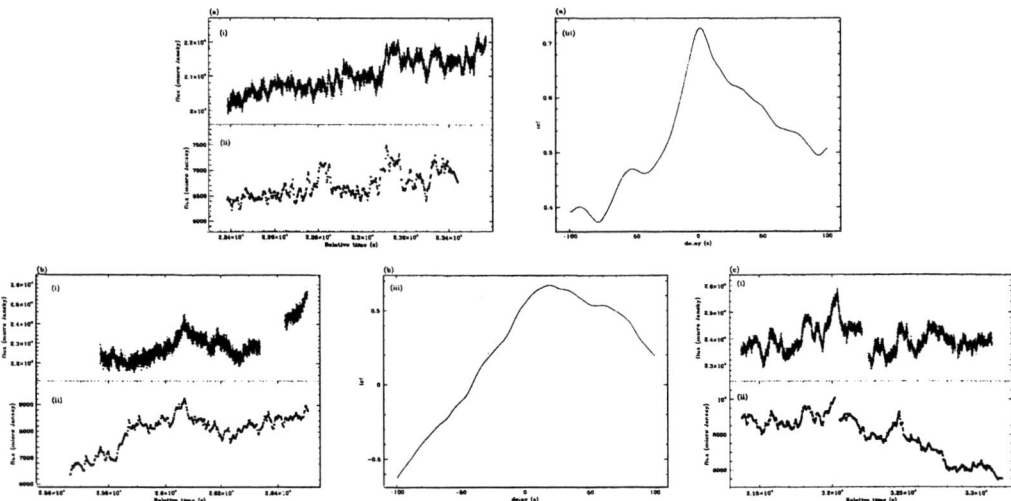

FIGURE 2. Three sets (a-c) of simultaneous optical (i) and SAS-3 X-ray (ii) observations of Sco X-1 in an X-ray high state taken March 1977 (Ilovaisky et al. 1980). Note that in (c) the source declines into a low state. The ICF's are shown in (iii). All light curves are binned on 2 s resolution.

the two simultaneous high state datasets, finding delays of $5.9^{+}_{-}2.5$ s and $5.7^{+}_{-}2.8$ s (Figure 3).

As in LMC X-2, Sco X-1 exhibits an increase in correlation between optical and X-rays as it becomes more active. Also, as for LMC X-2, the time delays for Sco X-1 are in general longer than the light travel time across the disc (for Sco X-1 ~4-5 s) given its orbital period of 18.9 h (LaSala & Thorstensen 1985).

5. DISCUSSION

The variability of LMC X-2 in the optical is delayed with respect to that in X-rays indicating that the lag is due to X-ray reprocessing in material at some distance from the source. Potential candidates as reprocessing regions include the outer areas of the accretion disc and the heated face of the secondary star. However, to identify such regions with our observed time delay requires a detailed knowledge of the binary system parameters. These same *RXTE* observations have provided a new confirmation of the 8.2 h orbital period in the X-ray data (Smale & Kuulkers 1999). Assuming that 8.2 h is indeed the orbital period, then the compact object could either be : (a) a neutron star, assumed mass $1.4M_\odot$ in which case the tidal disc radius will be ~$1R_\odot$ (equiv. 2-3 s light travel time) and the secondary at ~$1.3R_\odot$ (~5-6 s), which is difficult to reconcile with our observed delays, unless there is an additional component due to the actual reprocessing physics within the disc or stellar atmosphere; or (b) a $10M_\odot$ black hole with disc radius ~$2R_\odot$ (~4-5 s) and

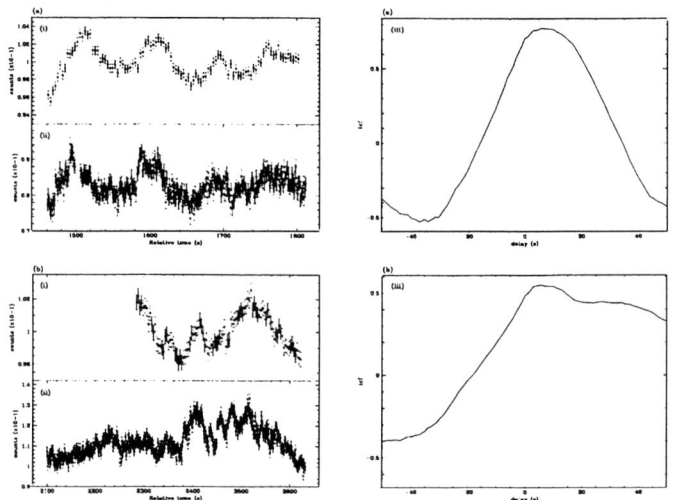

FIGURE 3. Two sets (a,b) of simultaneous optical (i) and X-ray (ii) observations of Sco X–1 in an X-ray high state taken March 1979 (Petro et al. 1981). The ICF's are shown in (iii).

separation of secondary $\sim 2.5 R_\odot$ ($\sim$10–11 s). [Note, however, that if the period were to be 12.5 d, then these delays would be (a) 25-28 s and 60-65 s (b) 47-54 s and 117-124 s]. Hence the delay found is longer than even the distance to the secondary (assuming an 8.2 h period). In Sco X–1 the delays found are longer than allowed by the disc size but within the orbit of the secondary.

REFERENCES

Bonnet-Bidaud, J.M. et al. 1989, A&A, 213, 97
Callanan, P.J. et al. 1990, A&A, 240, 346
Crampton, D. et al. 1990, ApJ, 355, 496
Gaskell, C.M., Peterson, B.M. 1987, ApJS, 65, 1
Hynes, R.I. et al. 1998, MNRAS, 299, L37
Ilovaisky, S.A. et al. 1980, MNRAS, 191, 81
Johnston, M.D., Bradt, H.V., Doxsey, R.E. 1979, ApJ, 233, 514
LaSala, J., Thorstensen, J.R. 1985, AJ, 90, 2077
Long, K.S., Helfand, D.J., Grabelsky, D.A. 1981, ApJ, 248, 925
Markert, T.H. and Clark, G.W. 1975, ApJ, 196, L55
O'Donoghue, D. 1995, Baltic Astron., 4, 519
Pakull, M.W. 1978, IAU Circ 3313
Pakull, M.W., Swings, J.P. 1979, IAU Circ 3318
Petro, L.D. et al. 1981, ApJ, 251, L7
Smale, A., Kuulkers, E. 1999, ApJ, in press

THE MULTICOLOUR DISC MODEL PARAMETER FOR BLACK HOLE CANDIDATES: CAN WE TRUST THE INNER DISC MEASUREMENTS ?

A. Merloni[1], A.C. Fabian[1] & R.R. Ross[2]

1) Institute of Astronomy, Madingley Road, Cambridge, CB3 0HA
2) Physics Department, College of the Holy Cross, Worcester, MA 01610, USA

ABSTRACT We present a critical analysis of the usual interpretation of the multicolour disc model parameters for black hole candidates in terms of the inner radius and temperature of the accretion disc. Using a self-consistent model for the radiative transfer and the vertical temperature structure in a Shakura-Sunyaev disc, we simulate the observed spectra, taking into account doppler blurring and gravitational redshift, and fit them with multicolour models. We show not only that such a model systematically underestimates the value of the inner disc radius, but that when the accretion rate and/or the energy dissipated in the corona are allowed to change the inner edge of the disc, as inferred from the multicolour model, appears to move even when it is in fact fixed at the innermost stable orbit.

KEYWORDS: accretion, accretion discs - black hole physics - radiative transfer - X-rays: stars

1. INTRODUCTION

Black Hole Candidates in their Soft/High state exhibit a strong thermal component with characteristic temperatures $kT \leq 1$ keV. This ultra-soft thermal component is usually interpreted as the product of the emission from an optically thick accretion disk, and is often considered one of the best evidences that the observed object is indeed a black hole.

The value of the inner radius of the disc is determined by the innermost stable orbit (ISO) of a particle around a black hole. The actual value of the radius of the ISO, in turn, depends on the mass and on the specific angular momentum of the black hole only. Therefore, the knowledge of its value could in principle lead us to the exact determination of the compact object's parameters and to test General Relativity in the strong field regime.

The multicolour disc (MCD) model (Mitsuda et al. (1984)) can be regarded as the zeroth order approximation in describing emission from an accretion disc and for that reason has been intensively used in interpreting observed spectra over the last 15 years. It assumes geometrically thin and optically thick disc (Shakura & Sunyaev (1973)); no treatment of vertical structure; no effect of Comptonization on emergent spectrum; consequently the local emission is Planckian, with a temperature profile $T(r) \propto r^{-3/4}$.

The MCD fitting model has two free parameters: a temperature T_{col} [keV] and a normalization factor $n = (R_{col}/1\text{km})^2 \cos(i)/(D/10\text{kpc})^2$. The parameter R_{col} is *neither* the inner radius, *nor* the radius at which the temperature of the disc is the highest; in fact the assumption behind using the MCD model is that the Comptonized emergent spectrum can be approximated by a diluted blackbody spectrum with a colour temperature (T_{col}) higher than the effective temperature (T_{eff}). The ratio $f_{col} = T_{col}/T_{eff}$ (the so called *spectral hardening factor*, (see Shimura & Takahara (1994)), is then assumed to be constant throughout the disc and for varying physical parameters. Thus the actual inner edge of the disc would be given by

$$R_{in} = \eta R_{eff} = \eta g(i) R_{col}(T_{col}/T_{eff})^2, \tag{1}$$

where η is the ratio of the inner radius of the disc to the radius at which the emissivity actually peaks, typically $\eta \simeq 0.6 - 0.7$, while $g(i)$ takes into account Doppler blurring and General Relativistic corrections and is of the order $g \simeq 0.7-0.8$ (Ebisawa et al. (1994)).

2. A 'TEST' FOR THE MCD MODEL

We present here a more realistic disc model (Ross & Fabian (1996)) assuming Shakura & Sunyaev "standard" solution for geometrically thin discs in the r direction, but we also allow a fraction f of the accretion power to be dissipated in a hot corona above the disc (Svensson & Zdziarski (1994)). We consider a $10 M_\odot$ Schwarzschild Black Hole and a fixed inner boundary of the disc at the innermost stable orbit $R_{in} = 6GM/c^2$. Then we implement the self-consistent (numerical) treatment of radiative transfer and of vertical temperature structure using Fokker-Planck/diffusion equations in plane parallel geometry: it gives $T(z)$ by balancing heating rate due to dynamic heating, Compton scattering and free-free absorption with the cooling rate due to inverse Compton scattering and free-free emission (Ross et al. (1978)).

We use dimensionless parameters $m = \frac{M}{M_\odot}$, $\dot{m} = \frac{\dot{M}}{\dot{M}_{Edd}}$, $r = \frac{R}{R_S}$, where $\dot{M}_{Edd} = 3.1 \times 10^{-8} m\, M_\odot\, \text{yr}^{-1}$ is the Eddington accretion rate for a disk efficiency $\epsilon = 0.083$. X is the mass fraction of hydrogen and will be fixed to 0.71. The radial structure of the disc is summarized by the values of the half-thickness (in units of the Schwarzschild radius) h and the uniform gas density ρ, for the radiation and gas pressure dominated regions respectively:

$$h_r = \frac{H}{R_S} = 5.3(1+X)[\dot{m}J(r)](1-f) \tag{2}$$

$$\rho_r = 3.4 \times 10^{-7} \alpha^{-1} m^{-1} r^{3/2}[\dot{m}J(r)]^{-2}(1-f)^{-3}\, \text{g cm}^{-3}, \tag{3}$$

$$h_g = 3.4 \times 10^{-2}(1+X)^{1/10}\alpha^{-1/10}m^{-1/10}r^{21/20}[\dot{m}J(r)]^{1/5}(1-f)^{1/10} \tag{4}$$

$$\rho_g = 4(1+X)^{-3/10}\alpha^{-7/10}m^{-7/10}r^{-33/20}[\dot{m}J(r)]^{2/5}(1-f)^{-3/10}\text{g cm}^{-3}, \quad (5)$$

with $J(r) = (1 - \sqrt{3/r})$.

Let us now describe the effect of Comptonization on the emergent spectra. We define the *effective optical depth* for absorption: $\tau_*(\nu) = \sqrt{3\tau_{\text{ff}}(\nu)(\tau_T + \tau_{\text{ff}}(\nu))}$, where τ_T is the Thomson depth below the surface, and $\tau_{\text{ff}}(\nu)$ is the optical depth due to free-free absorption. In the inner region of the disc, for intermediate values of the accretion rate (and $f < 0.5$) the thermalization depth (where $\tau_*(\nu) = 1$) for high energy photons can be reached at very high Thomson depths, where the temperature is considerably higher than near the surface. Compton scattering in the outer layers of the disc downscatters these photons to lower energies while low energy photons are upscattered via inverse Compton scattering. The competition between these two effects produces a hard Wien-law tail in the emergent spectra (Ross & Fabian (1996)).

Interestingly, anyway, if we decrease significantly the flux emerging from the disc itself, either by lowering the accretion rate ($\dot{m}$) or increasing the fraction (f) of the accretion power dissipated in the corona, the disc becomes much denser and cooler (see eq. 4), both in its outermost portions and also, for very low fluxes, in the very innermost ones. In those cases Comptonization is not complete or 'saturated' because

$$y = \frac{4kT}{m_e c^2}\tau_T^2 < 1 \quad (6)$$

at the thermalization depth. In these regions we treat Compton scattering as *coherent* (see discussion in Ross & Fabian (1996)), dropping the Fokker-Planck term in the radiative transfer equation.

3. RESULTS AND DISCUSSION

We have simulated disc spectra in nine different physical situations, keeping the viscosity parameter $\alpha = 0.1$, $\eta g(i) = 0.5$, and varying the accretion rate and the amount of power dissipated in the corona (see Table 1). By fitting the simulated spectra with the MCD model we are able to show that the model systematically underestimates the value of the inner radius of the accretion disc for a black hole candidate. We also have shown that the spectral hardening factor f_{col}, which is needed to correct the results of the fits, is not, as is usually assumed, constant when the accretion rate and/or the fractional coronal activity change.

Recent observations (see e.g. Sobczak et al. (1999), Muno et al. (1999)) of galactic black hole candidates seem to point towards an extreme dependence of the observed *colour* radius on f.

In both these cases the inferred inner disc radii can shrink by more than a factor of four when the flux in the power law component dominates the blackbody one.

This behaviour is exactly what our results predict we should expect from the MCD model when we increase f and/or reduce the accretion rate. Thus, every time the observations imply that the coronal activity is dominant, the multicolour fits

TABLE 1. Physical parameters of the simulations and resulting MCD fit parameters in the range 2-20 keV. All the models refer to a $10 M_\odot$ black hole with inclination angle and distance set to the values of the black hole candidate GRS 1915+105, namely $i = 70°$ and $D = 12.5$ kpc. The actual value of the inner disc is $88.6 km$.

SET	$\dot m$	f	R_{col}[km]	f_{col}
S1	0.3	0.0	54.5	1.80
S2	0.2	0.0	56.9	1.76
S3	0.1	0.0	53.3	1.82
S4	0.2	0.5	52.1	1.84
S7	0.1	0.4	46.5	1.95
S8	0.05	0.0	41.1	2.08
S9	0.1	0.5	41.5	2.07
S10	0.05	0.5	39.0	2.12
S11	0.05	0.8	24.5	2.68

should be corrected with a *varying* hardening factor $1.7 < f_{col} < 3$ in order to recover the actual value of the inner disc radius.

Also, recent observations of a family of ultraluminous compact X-ray sources in nearby spiral galaxies (Makishima et al. (1999)) seem to stress the shortcomings of the MCD model we predict. For such a family of black hole candidates the multicolour model gives too small colour radii with respect to the masses required to produce the observed luminosity. That would imply an inner radius smaller than the ISO for most of those objects. These problem could be solved either introducing black hole rotation, or, again rescaling with a much higher hardening factor, as our study would suggest.

ACKNOWLEDGEMENTS

AM thanks the TMR network "Accretion onto black holes, compact stars and protostars" funded by the E.C. (contract number ERBFMRX-CT98-0195) for support.

REFERENCES

Ebisawa K. et al., 1994, PASJ, 46, 375

Makishima et al., 1999, submitted to ApJ.

Mitsuda K. et al., 1984, PASJ, 36, 741.

Muno M.P., Morgan E.H. & Remillard R.A., astro-ph/9904087.

Ross R.R. & Fabian A.C., 1996, MNRAS, 281, 637.

Ross R.R., Weaver R., McCray R., 1978, ApJ, 219, 292.

Shimura T. & Takahara F., 1995, ApJ, 445, 780.

Shakura N.I., Sunyaev R.A., 1973, A&A, 24, 337.

Sobczak G.J. et al., astro-ph/9809195 v2.

Svensson R. & Zdziarski A.A., 1994, ApJ, 436, 599.

SPATIAL DISTRIBUTION OF SPECTRAL CHARACTERISTICS OF THE SUPERNOVA REMNANT CAS A

T. Mineo[1], M.C. Maccarone[1], A. Preite-Martinez[2], J. Vink[3], J.S. Kaastra[4]

1) CNR-IFCAI, Palermo, Italy
2) CNR-IAS, Rome, Italy
3) Astrophysikalisches Institut Potsdam, Germany
4) SRON Laboratory for Space Research, Utrecht, NL

ABSTRACT   We present preliminary results about the spatially resolved spectroscopy of the supernova remnant Cas A in the energy band 3-10 keV. Data refer to observations performed with the MECS instrument on board BeppoSAX. Images of Cas A have been accumulated in very narrow energy intervals, and then deconvolved with Lucy's method. The spectral analysis of the deconvolved images allowed us to derive the spatial distribution of the relevant spectral parameters of the source with a resolution of about 30 arcsec.

KEYWORDS: ISM: supernova remnants - ISM: individual objects: Cas A - X-rays: ISM - Methods: data analysis - Techniques: spectroscopy

1. INTRODUCTION

Cas A, the youngest known supernova remnant in our galaxy, is an object continuosly investigated at several wavelengths. In X-rays (but also in optical) Cas A shows a heterogeneous morphology, not spherically symmetric with two prominent emission regions in the north-west and south-east (Holt et al. 1994, Vink et al. 1996 and references therein). In a first paper (Favata et al., 1997) we presented the broad-band spectrum of Cas A as seen by all four co-aligned BeppoSAX instruments (LECS, MECS, HPGSPC, PDS). In a subsequent paper (Maccarone et al., 1998) we presented BeppoSAX spatially resolved observations of Cas A, mapping for the first time the Fe line intensity with arcmin angular resolution. Recently (Vink et al., 1999), we obtained a more detailed morphological description of the emission regions, sharpening Cas A image data via a suitable deconvolution method to take into account the MECS instrumental PSF. The deconvolution, applied to a restricted number of narrow energy bands, confirmed that Cas A has different morphologies in the continuum and line bands. In this paper we present the first spatially resolved spectroscopy of Cas A at full energy resolution in the MECS energy band, limiting the discussion to the high energy range 3-10 keV.

Cas A was observed four times in 1996 and once in 1997, for a total of 128 Ksec of net exposure time. The Cas A space-energy cube (X, Y, E) was generated

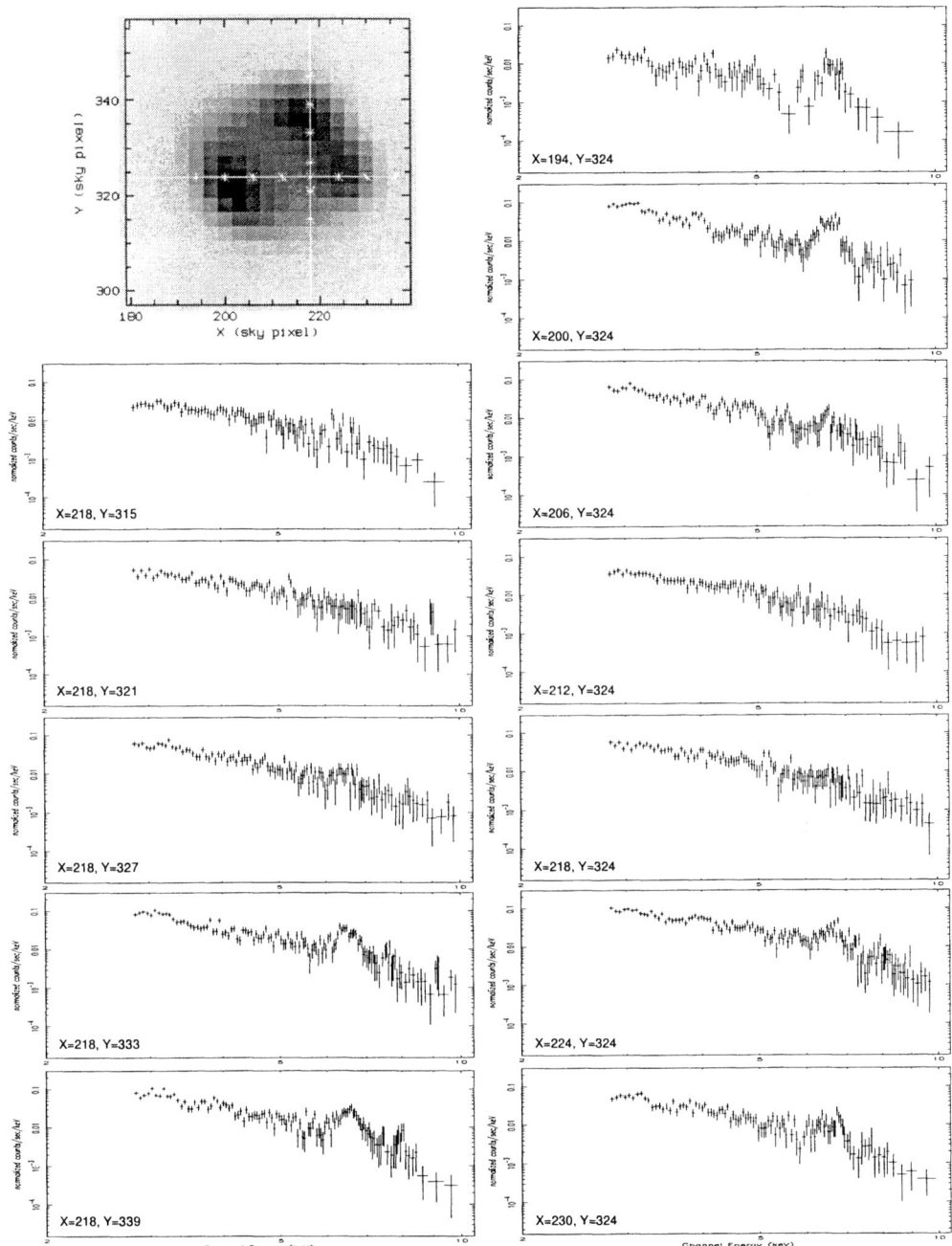

FIGURE 1. Spectra extracted in the selected locations indicated in the top left panel, where the map of total counts is shown.

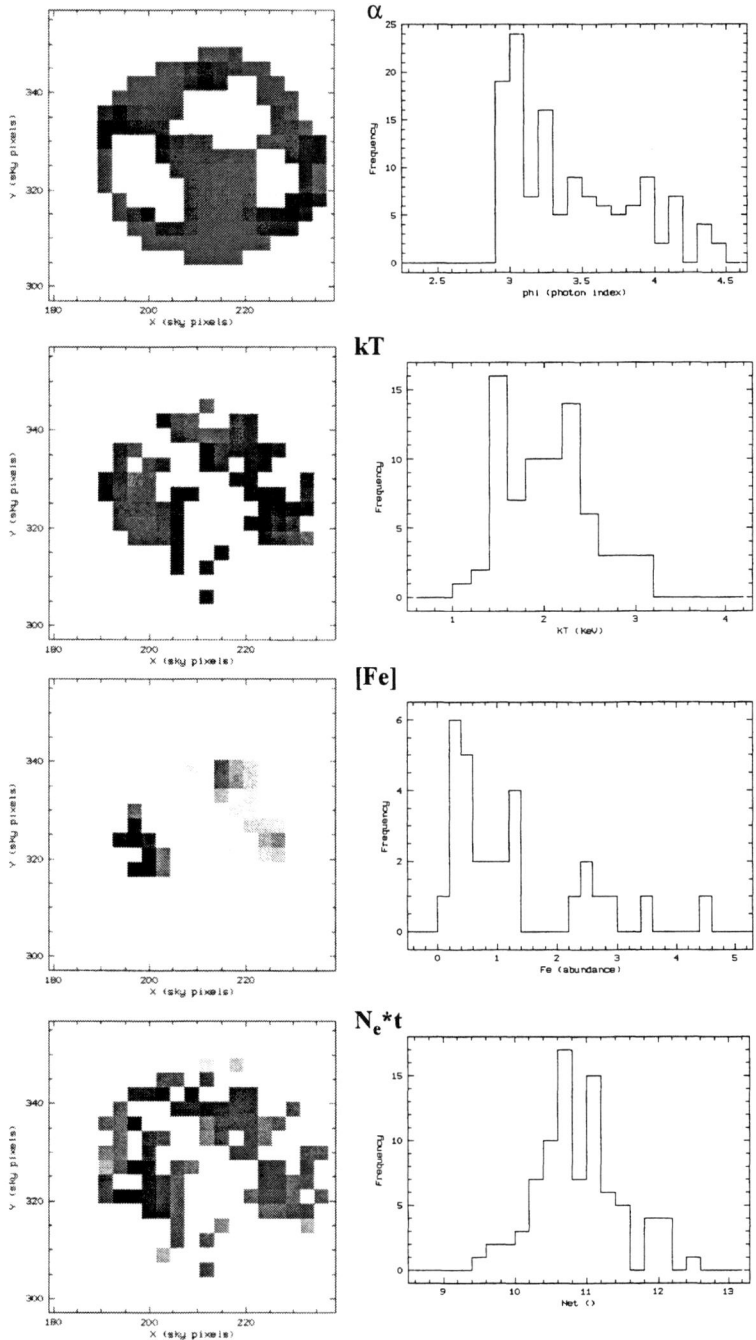

FIGURE 2. Spatial (left panels) and frequency (right panels) distributions of the derived parameters α, kT(keV), Iron abundance, and $\mathrm{Log}(N_e*t)$.

(i) selecting data from each observations, (ii) converting data in X,Y images (sky pixels), (iii) merging and rebinning images to improve statistics, (iv) generating the appropriate PSF, and then (v) deconvolving each monochromatic rebinned image. Fig.1 shows a set of spectra extracted at selected locations in the remnant.

2. DISCUSSION OF THE RESULTS

As shown in Fig.2, the remnant is clearly divided in two zones: a region best fitted by a NEI model, coincident with the broken shell, and a diffuse region of featureless emission, best fitted by a power law model. This last region extends around and inside the broken shell of predominantly thermal emission.

The SE region of the shell is cooler than the NW rim. kT ranges from 1 to 3.2 keV, with two peaks at 1.6 and 2.3 keV. Iron abundance is essentially solar in the W-NW region of the shell. In the SE, [Fe] can be as high as 4.5. The SE rim is closer to ionisation equilibrium than the rest of the shell.

The spatial distribution of the power law index α is rather peculiar. Index α ranges from 2.9 up to 4.5. The distribution of values peaks around 3, with a very asymmetric distribution. The peak value $\alpha=3$ corresponds to the index of the power law component found in the fit of the global spectrum of Cas A (Favata et al., 1997). The spatial distribution of α can explain the distribution of values at $\alpha > 3$: α is around peak value far from the thermal shell, while it is higher if we get close to the shell. In this case there is probably also a thermal component present, steepening the spectrum.

Our spatially resolved spectroscopy of Cas A showed the following results:

1. The remnant can be divided in two distinct regions: a broken shell were line emission is very strong and a diffuse region all around and inside the shell, best fitted with a power law model. Our interpretation is that the featureless emission can be synchrotron radiation from shock-accelerated electrons, then originating from a thin shell at the position of the shock. In the central region of the remnant, the non thermal emission could be related to the central object recently discovered by CHANDRA (Tananbaum, 1999).

2. The sub-regions of the broken shell of predominantly thermal emission show different characteristics. The SE rim is hotter and closer to ionisation equilibrium (higher values of N_e*t), while the NW is colder and farther from equilibrium (lower values of N_e*t). Iron abundances are almost solar in the SE, higher in the NW.

REFERENCES

Favata, F., Vink, J., Dal Fiume, D., et al., 1997, A&A, 324, L49

Holt, S.S., Gotthelf, E.V., Tsunemi, H., et al., 1994, PASJ, 46, L151

Maccarone, M.C., Grandi, P., Mineo, T., et al., 1998, Nuclear Physics B Proc. Suppl. 69/1-3, 74

Tananbaum, H., 1999, IAUC 7246

Vink, J., Kaastra, J.S., Bleeker, J.A.M., 1996, A&A Letters, 307, L41

Vink, J., Maccarone, M.C., Kaastra, J.S., et al., 1999, A&A Suppl., 344, 289

VARIABILITY OF AN IRON LINE PROFILE AND A POWER-LAW CONTINUUM

K. Misaki[1,3], H. Kunieda[1], Y. Terashima[2]

1) High Energy Astrophysics Division, Institute of Space and Astronautical Science, Yoshinodai 3-1-1, Sagamihara, Kanagawa 229-8510, Japan   (e-mail address : misaki@astro.isas.ac.jp)
2) Laboratory for High Energy Astrophysics, NASA Goddard Space Flight Center, Code 662, Greenbelt, MD20771, USA
3) Department of Physics, Nagoya University, Furo-cho, Chikusa, Nagoya 464-8602, Japan

ABSTRACT   The iron lines from Seyfert 1 galaxies exhibit broad red-wings in general, which are attributed to the relativistic accretion disk close to the black hole. The line profile provides us with the information of the geometry of the accretion disk and the central source. We examine a time behavior of the iron line in accordance with a spectral variability of the power-law continuum of bright Seyfert galaxy Mrk 841 observed with $ASCA$. The emission mechanism and the structure around a black hole are discussed to accommodate both the spectral change of the continuum and the line profile change.

KEYWORDS: galaxies: individual(Mrk 841); galaxies: Seyfert; X-rays: galaxies

1. INTRODUCTION

$ASCA$ observations of Seyfert 1 galaxies have revealed the iron Kα fluorescent line being extremely broad and skewed to the low energy end (e.g. Tanaka et al. 1995; Nandra et al. 1997). The most plausible explanation of the skewed nature of the iron line profiles is the combined effect of relativistic Doppler-shift and gravitational redshift in the deep gravitational potential of the central massive black holes (Fabian et al. 1989, Fabian et al. 1995). The line profile gives us information about physical conditions in the vicinity of the black hole. On the X-ray continuum, the general tendency has been known that Seyferts show the softer spectrum when they are brighter. However, it remains unclear whether the change of the spectral indices of the X-ray continuum is an intrinsic change or due to the additional components such as Compton reflection, soft excess, warm absorber.

We report here spectral changes of the luminous Seyfert galaxy Mrk 841 which has shown a violent spectral variability in the previous X-ray observations (George et al. 1993; Nandra et al. 1995). Mrk 841 was observed twice with $ASCA$. The first observation was carried out in 1993 and the second was 6 months later (in 1994).

2. RESULTS

2.1. X-ray Continuum Component

Fitted With the simple power-law model, the photon index is found to be 1.87±0.2 and 1.65±0.2 in the first (1993) and second (1994) observations. The X-ray luminosity in the 2 – 10 keV band is 6.8×10^{43} ergs s^{-1} and 6.2×10^{43} ergs s^{-1} respectively. The simple power-law model failed to fit the observed spectra, because significant systematic excess is seen in the low energy band. Therefore, we introduced an additional thermal component to fit the soft excess component. After addition of the soft excess component, the photon index is reduced by $\Delta\Gamma \sim 0.10$ and 0.07 in the 1993 and 1994 observation, respectively. The Compton reflection is also introduced in order to examine whether the change of spectral slope is intrinsic or not, although it is difficult to constrain the reflection hump with the limited bandpass of ASCA. The reflection model makes the spectral slope steeper by $\Delta\Gamma \sim 0.08$ in both the cases.

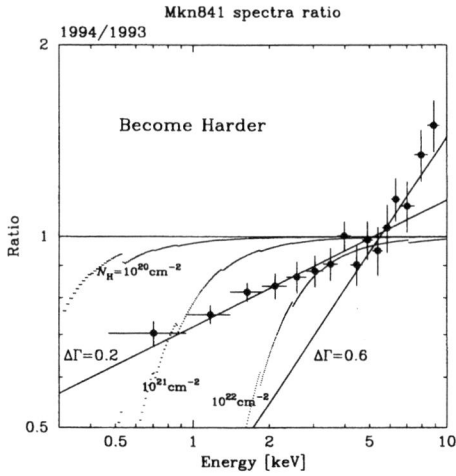

FIGURE 1. Spectra ratio of Mrk 841 (1994/1993)

We cannot find a plausible model to explain the observed change without the intrinsic change of spectral slope. The spectral ratio of the two observations (1994/1993) is shown in Fig. 1. The positive slope shows the spectral hardening between the two observations. It can be attributed neither to the increase of the Compton reflection nor to the increase of absorption. Any models suggest the power-law continuum is intrinsically flatter during the second observation than the first one.

2.2. Iron line profile

In Fig. 2, the ratio of data to the power-law continuum model are plotted for the two observations, highlighting the iron Kα line profile. The best fit parameters of iron Kα emission obtained from the first observation are summarized in Table 1.

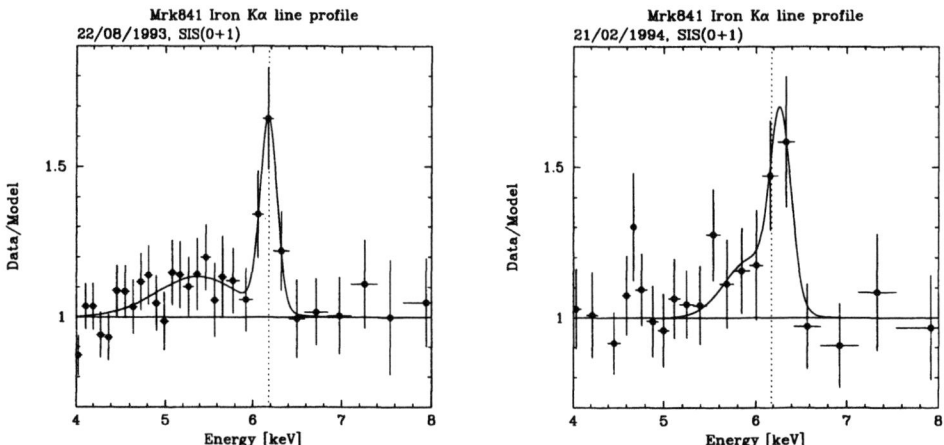

FIGURE 2. Iron line profile of Mrk 841 (left: 1993, right: 1994)

TABLE 1. Mrk 841 iron Kα line parameters (1993 observation)

double Gaussian		energy [keV]	width [eV]	E.W. [eV]	
narrow		$6.37^{+0.06}_{-0.04}$	10 (fixed)	135^{+70}_{-50}	
broad		$5.45^{+0.93}_{-0.65}$	460^{+5500}_{-260}	81^{+195}_{-60}	
Diskline	energy [keV]	inclination	R_{in}	R_{out}	q^a
	$6.36^{+0.12}_{-0.10}$	$12°^{+15}_{-12}$	$3R_s$	$600R_s{}^{+\infty}_{-400}$	-2 (fixed)

^{a} emissivity parameter : line emissivity is assumed to vary as $\propto R^q$

During 1994 observation, the red-wing becomes less than the first observation. On the other hand, the narrow component is still clearly seen. We tried to apply the diskline model as is the first observation. However, the data quality is too poor to determine many parameters independently. Therefore, the most parameters of diskline are fixed at the same value found in the first observation, while one free parameter is picked out to represent the observed line profile by the following two models. The inner radius of the line emitting region is found to be pushed out to 8 R_s with the other parameters fixed. Alternatively, when the emitting region is fixed to $3R_s - 600R_s$, emissivity parameter q of -0.5 can explain the observed profile. Although it is impossible to distinguish these alternatives, both models suggest more flux of the iron line arises from the outer disk.

3. DISCUSSION

We consider possible scenarios to explain the spectral variability found in the two observations. The first simple idea to explain the change of the iron Kα profile is a disappearance of the inner part of the accretion disk, which is responsible to the red part of the broad iron line. Ionization effect or ADAF is considered to be possible way to evaporate the inner disk, however, both are not the case in the change of Mrk 841 from 1993 to 1994. Because the continuum flux is almost same level, it doesn't show a drastic change between two observations. Therefore we conclude a change of the disk can not explain the present phenomena. The next approach is a change of illumination of the disk. If the illuminating source is assumed to be a point source on the axis of the disk, the flux distribution is simply given as: $F_X(r) = hL_X/4\pi(r^2 + h^2)^{3/2}$ where r is the radius and h is the height of an illuminating source from the disk. When h is much larger than r, a value of q approaches 0. The observed change of the line profile can be explained by the change of q from -2 to -0.5. This suggests the increase of the height of the continuum source.

The idea with the change of the source height can be related to the change of the spectral slope in the following scenario. When the soft seed photon field is strong, the high energy electrons should be frequently scattered by the dense soft photon flux close to the accretion disk. In this case, the high energy corona should be cooled resulting in a softer Comptonized spectrum. On the contrary, when the seed photon flux decreases, high energy electrons keep its energy (less cooled) and can diverse in larger volume and may have taller distribution above the disk. Then a hard power-law slope and iron line with faint red-wing should be observed.

Another possibility is a multiple hot spots model, which are not on axis but distributed on the disk. The distribution change of variable hot spots can be inferred from the profile change. For example, many flares happening at outer radii will create a line profile dominated by a narrow component. Iwasawa et al. (1999) suggested this scenario from the variation of the broad iron line profile of MCG $-6-30-15$. In the model of multiple hot spots on the accretion disk, what mechanism causes the spectral slope change? We hope for future variability data with better sensitivity to reveal the emission mechanisms of AGN.

ACKNOWLEDGEMENTS

The authors are grateful to all the *ASCA* team members. K. M. and Y. T. thank Japan Society for the Promotion of Science for support. I express special thanks to Kazushi Iwasawa. He always encourages me and gives us many useful suggestions. We thank B. Paul for his careful reading to make this paper easily readable.

REFERENCES

Fabian, A.C., Rees, M.J., Stella, L., & White, N.E. 1989, MNRAS, 238, 729
Fabian, A.C., et al. 1995, MNRAS, 277, L11
George, I.M., et al. 1993, MNRAS, 260, 111
Iwasawa, K., Fabian, A.C., Young, A.J., Inoue, H., & Matsumoto, C. 1999, MNRAS, 306, L19
Nandra, K., et al. 1995, MNRAS, 273, 85
Nandra, K., George, I.M., Mushotzky, R.F., Turner, T.J., & Yaqoob, T. 1997, ApJ, 477, 602
Tanaka, Y., et al. 1995, Nature, 375, 659

WHY DO BROAD LINE AGN SHOW UP AMONG FAINT HARD X-RAY SOURCES?

J.P.D. Mittaz [1], M.J. Page [1], F.J. Carrera [1,2]

1) Mullard Space Science Laboratory, Holmbury St. Mary, Nr. Dorking, Surrey, RH5 6NT, UK
2) Instituto de Física de Cantabria, Consejo Superior de Investigaciones Científicas - Universidad de Cantabria, Santander, Spain

ABSTRACT   We have assembled a sample of ROSAT sources that are harder than the X-ray Background (XRB). By studying a hard X-ray spectrum population we are trying to discover the nature of the as yet unidentified source population needed to resolve the XRB spectral paradox. Current theory would suggest that narrow line QSOs (so called QSO2 objects) should be important, but we find that broad line AGN predominate. Here we discuss the nature of the broad line AGN and show that the absorption responsible for hardening the X-ray spectrum has a variety of optical signatures.

KEYWORDS: X-rays: galaxies, quasars: absorption lines, quasars: emission lines

1. INTRODUCTION

The exact nature of the source population that produces the XRB is still hotly disputed. It is known that the background must be the integrated emission of many discrete extragalactic sources and surveys of the soft X-ray sky with *Rosat* have succeeded in resolving ~ 40% of the 1-2 keV XRB into individual sources (e.g. Hasinger et al., 1993), the majority of which are extra-galactic. However, right down to the faintest flux at which *Rosat* sources can be resolved, the integrated source spectrum is still softer than that of the background (Hasinger et al., 1993, Mittaz et al., 1999). This problem is known as the 'spectral paradox'.

Since the average spectrum of the resolved sources is too soft to make the XRB, there must be a population of much harder sources at faint fluxes. We have performed a survey to find just these objects using archival *Rosat* data and have built up a sample of 147 objects whose spectral slope is harder than the XRB (see Page et al. this conference). We have a total of 62 identified objects of which 28 have been classified as broad line AGN. It is these AGN which are the subject of this paper.

2. THE BROAD LINE AGN SAMPLE

We find a whole range of broad line AGN types with hard X–ray spectra. However, we do not find any of the so called narrow line QSOs (the QSO equivalents of Seyfert 2 galaxies) that might be expected - only broad line QSOs are seen. Generally, X–ray selected broad line AGN have soft X–ray spectra, but our objects are spectrally hard implying that absorption is important. This is borne out in the one case where there exists sufficient signal in the X–ray data to fit the hydrogen absorbing column ($\sim 10^{22}$ atoms cm^{-2} see figure 16 in Mittaz et al. 1999). Further, many of our broad line AGN show evidence for absorption in their optical spectra. We find that the nature of this absorption varies from object to object, and sometimes more than one process is visible in the same object. Broadly speaking, the absorption can be categorised into absorption intrinsic to the source or to intervening absorption by material along our line of sight.

3. INTRINSIC ABSORPTION

3.1. Narrow associated absorption lines

A number of our AGN show strong narrow absorption features associated with either broad Mg II, C IV or Ly-α emission lines, some showing multiple velocity components (e.g. Figure 1a). In one case we see large amounts of Fe absorption (Figure 1b); there is only one other object known, the BALQSO QSO0059-2735 (Hazard et al. 1987), which shows similar extreme Fe absorption. In all cases the presence of strong narrow absorption suggests significant absorbing material to our line of sight associated with the AGN. Since these sources are radio-quiet, strong (EW > 1Å) narrow line absorption is relatively uncommon (e.g. Foltz et al., 1986) and it is highly likely that the hard X–ray spectra are related to the presence of these narrow absorption lines.

3.2. Red spectrum AGN

A number of AGN show very red optical spectra and also show large B-K colours (see Carrera et al. this conference). Such 'red' quasars are generally considered to be reddened by dust (e.g. Elvis et al. 1994) and are also X–ray weak (Ledden & O'Dell 1983). While dust obscuration is clearly important in these systems, the fact that we find such objects in our hard X–ray sample suggests an absorber where both gas (X–ray) and dust (optical) obscuration is present.

3.3. Intermediate type AGN

We have also found a number of intermediate class AGN. Two of our objects are Seyfert 1.8s where the broad component of the permitted lines is suppressed relative to the narrow line emission. This is thought to be due to obscuration of the broad line region by dust. In our sample we have also found two QSOs which show line profiles analogous to those of intermediate type Seyferts. In the one case where we

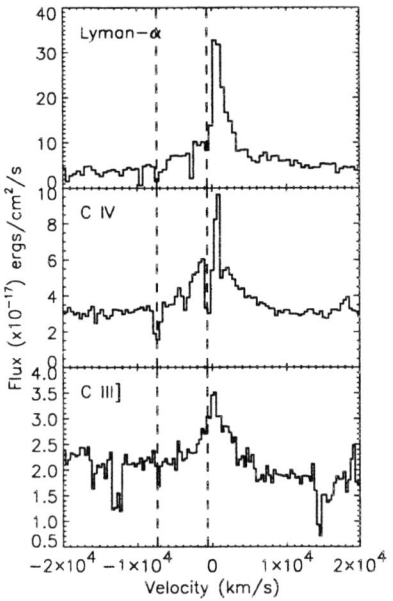
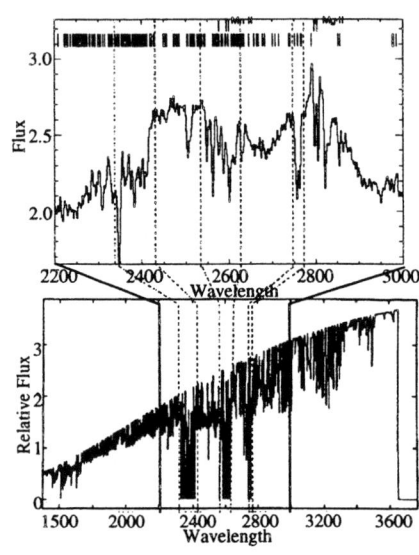

FIGURE 1. a) The left panel shows an object with strong narrow absorption lines with multiple velocities of 650 and 7400 km/s shown by the dashed lines.
b) The right panel shows an object with extreme Fe absorption together with a theoretical Fe line absorption spectrum from Wills, Netzer & Wills (1985). There is a clear match between the observed and theoretical absorption features.

have observed both CIV and MgII we also see an apparent difference between the ratio of broad and narrow components in the two lines, with the broad component of CIV being more suppressed (Figure 2a). This provides clues to the location of the obscuring material.

4. INTERVENING SYSTEMS

Three of our sources show strong metal absorption systems with several Å equivalent width in MgII (see Figure 2b). Absorption line systems of this strength at $z < 1$ are rare (e.g. Boissé et al. 1992), so the probability of finding such strong systems in 3 out of 28 hard AGN is small. Excluding the low redshift AGN and those objects with poor S/N we estimate that the probability of finding 3 systems is $\sim 1\%$. It is therefore likely that the intervening systems are related to the observed hard X-ray spectrum.

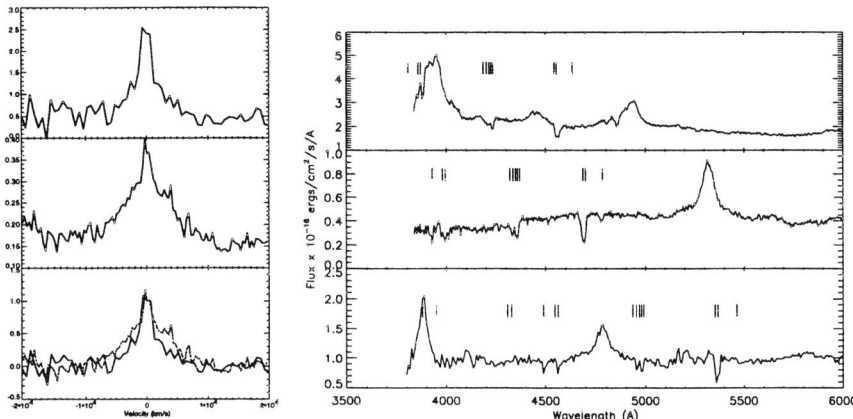

FIGURE 2. a)The left hand figure shows an object with both broad and narrow components to CIV (top panel) and MgII (middle panel), the QSO equivalent of a Seyfert 1.8 object. The bottom panel shows a comparison of the two profiles.
b) The right hand figure shows three objects showing intervening absorption from metal line systems along our line of sight. The position of known interstellar absorption lines is shown at redshifts 0.625, 0.673 and 0.920 respectively.

5. CONCLUSIONS

Absorption is likely to explain the hard X–ray spectra of our broad line AGN but the absorption manifests itself in different ways as shown in the optical spectra. The optical signatures range from narrow absorption lines to broad absorption lines (there is one BAL QSO in our sample), Seyfert 1.8 and QSO 1.8 behaviour, extremely red optical continua and intervening absorption lines. Therefore the absorption responsible for hardening the X–ray spectra of broad line AGN does not always originate in the same place or under the same conditions, and there are a number of different scenarios which give rise to hard X–ray spectra in broad line AGN.

REFERENCES

Boissé, P. et al., 1992, A&A, 262, 401
Elvis, M. et al., 1994, ApJ, 425, 103
Foltz, C.B, et al., 1986, ApJ, 307, 504
Hasinger, G. et al., 1993, A&A, 271, 1
Hazard, C. et al., 1987, ApJ, 323, 263
Ledden, J.E. & O'Dell, S.L., 1983, ApJ, 270, 434
Mittaz, J.P.D. et al., 1999, MNRAS, 308, 233
Wills, B.J., Netzer, H. & Wills, D., 1985, ApJ, 288, 94

HYSTERETIC BEHAVIOR AND COHERENCE OF THE BLACK HOLE CANDIDATE X-RAY BINARIES

Sigenori Miyamoto[1], Shunji Kitamoto[2]

[1] *Osaka University of Health and Sport Science, Noda 1558-1, Kumatori, Osaka, 590-0496, Japan*
[2] *Osaka University, Graduate School of Science, Machikaneyama 1-1, Toyonaka, Osaka, 560-0043, Japan*

ABSTRACT

It is shown that transitions between the hard/low state and the soft/high state are determined not only by the rate of the accreting matter onto the black hole but also by the history of the matter accretion, i.e. the transitions show the hysteresis just as the magnetic materials do. We found also that in the soft/high state of GS1124-683, coherence of time variability is small between different energy X-rays and that the coherence depends on the fraction of the soft power law component. This is quite different from what is observed in the hard/low state of Cyg X-1 and GX339-4, where the coherence between different energy X-rays is near unity over a wide range of frequencies.

KEYWORDS: black hole candidate, X-ray astronomy, time variation, coherence, hysteresis, state

1. HYSTERETIC BEHAVIOR

It is known that black hole candidate X-ray binaries (BHC-XBs) have three main states: the hard/low state, the soft/high state and the quiescent/off state (e.g. Tanaka & Shibazaki 1996). But it is not yet known whether the transitions between these states are determined only by the rate of the accreting matter onto the black hole, or they are controlled also by the history of the matter accretion. In the latter case, hysteretic behavior should be observed in the transitions between these states. For instance, in an X-ray nova, the transition from the hard/low state to the soft/high state during its flare-up phase should take place at a much larger X-ray luminosity than in the opposite transition during its decay phase. We examined data obtained with the Ginga ASM (Kitamoto et al. 1992) and the LAC on two X-ray novae: GS1124-683 (Miyamoto et al. 1993,) and GX339-4, and confirmed the large hysteretic behavior in these BHC-XBs, as suggested by Miyamoto et al. (1995). The results are shown in Fig.1 and Fig. 2.

The X-ray energy spectrum obtained with the Ginga ASM on Jan. 8, 1991 was hard and was fit with a power law with a photon index $=1.73 \pm 0.03$ (this observation caught the source in the hard/low state), while the spectrum of Jan. 10 showed a power law photon index $= 2.23 \pm 0.03$, indicating that the source was

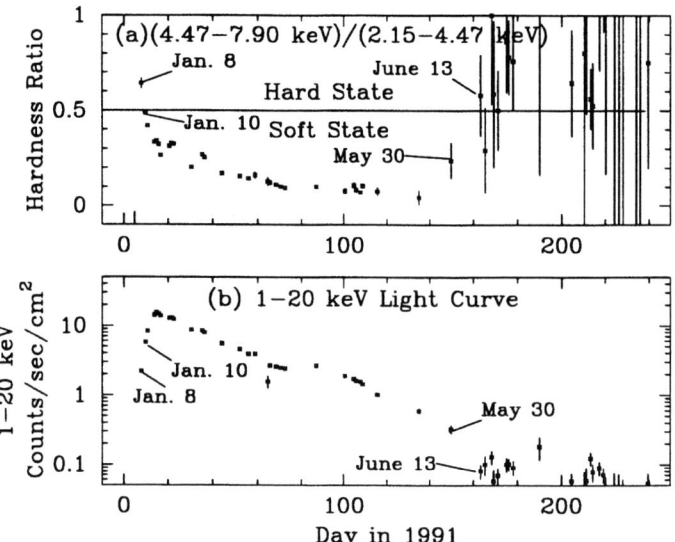

FIGURE 1. a) The evolution of the hardness ratios (4.47-7.90 keV / 2.15-4.47 keV) observed with the Ginga ASM. b) The X-ray light curve of the X-rays from GS1124-683 observed with the Ginga ASM.

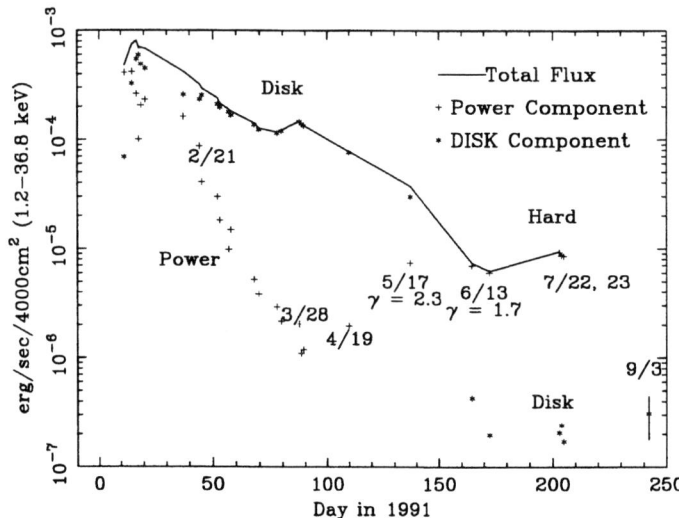

FIGURE 2. The light curves of the two X-ray components observed with the Ginga LAC (Miyamoto et al. 1993).

in the soft/high state. Thus the transition from the hard/low state to the soft/high state during the flare-up phase took place between Jan. 8 and Jan. 10 in 1991. Fig.1-a shows that this transition took place with a hardness ratio (4.47-7.90 keV / 2.15-4.47 keV) of about 0.5 in the data obtained with the Ginga ASM.

During the decay phase, the hardness ratio crossed the 0.5 value again in the opposite direction between May 30 and June 13 in 1991 as shown in Fig.1 and Fig.2. This transition from the soft/high state to the hard/low state was also observed with the Ginga LAC and was reported by Miyamoto et al. (1993) and Ebisawa et al. (1994). We can compare the X-ray luminosity at these two transitions, because we observed the X-ray fluxes in the hard/low state, i.e. in the same and simple energy spectrum state. Thus during the decay phase, the opposite transition took place with about an one order of magnitude lower X-ray luminosity than in the case of the transition during the flare-up phase. The same hysteretic behavior was confirmed similarly in GX339-4 observed with the Ginga ASM (Miyamoto & Kitamoto, 2000).

As the difference of the X-ray luminosity between the transition during the flare-up phase and the opposite transition during the decay phase is quite large, the matter accretion rates in these transitions should be quite different. Thus we concluded that the transitions between these two states are not only determined by the rate of the accreting matter onto the black hole, but also controlled by the history of the matter accretion. This is a confirmation of the large hysteretic behavior suggested by Miyamoto et al. (1995).

2. COHERENCE OF X-RAYS IN THE SOFT/HIGH STATE

In the hard/low state of BHC-XBs, the X-ray spectrum has mainly one component, i.e. the hard power law component (with a photon index = 1.5-1.8), and the coherence of time variation between different energy X-rays is near unity over a wide range of frequencies (Nowak et al. 1999a; 1999b). Thus, low and high energy X-rays seem to be produced in the same dynamic process during this state. What about the behavior in the soft/high state, where X-rays consist of two components, the disk blackbody component and the soft power law component (with the photon index = 2.2-2.7) ?

We calculated the coherence between different energy X-rays from GS1124-683 in the soft/high state observed with the Ginga LAC (Miyamoto et al. 1993), and found the following: a) The coherence of time variability is small (about 0.03) between different energy X-rays of the disk blackbody component (see the time interval between 70-90 days in 1991 in Fig.2 and Fig.3). b) The coherence between the power law component and the disk blackbody component is small (see Fig. 4).
c) The coherence between X-rays of the soft power law component is much larger than that of the disk blackbody component in the soft/high state (see Fig.2, Fig.3 and Fig.4). d) The value on April 19 in Fig.4 shows that the coherence increases before the soft power law component begins to increase.

Point a) suggests that X-rays of different energy are produced in different parts of the disk, and variability of a part of the accretion disk is independent of other

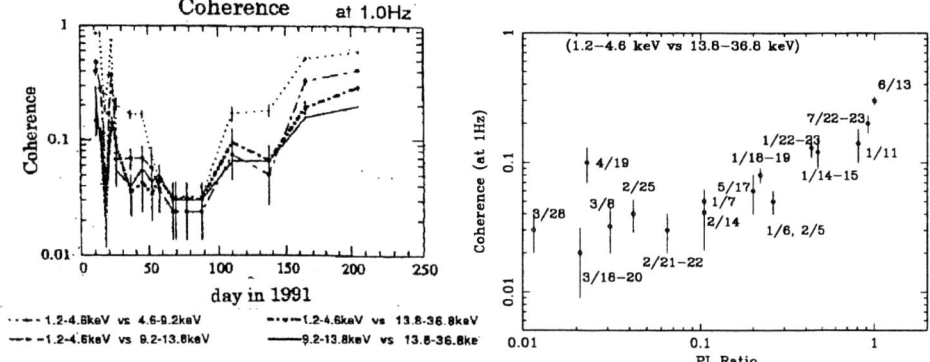

FIGURE 3. The evolution of the coherence between various energy X-rays at 1.0 Hz, observed with the Ginga-LAC.

FIGURE 4. Relation of the coherence of the X-rays (1.2-4.6 keV vs. 13.8-36.8 keV) at 1Hz to the ratio of the power law component to total X-rays.

parts of the disk. If Compton scattering is the production process of the soft power law component, and the seed photons are from the disk, b) and c) suggest either that the seed photons could be produced in the inner edge of the accretion disk, where variability is much larger than in the outer part of the disk, or that the Compton cloud could have large variability and this destroys the coherence between the two components. Point d) suggests that the Compton cloud or the inner edge of the disk seems to increase its variability before the increase of the soft power law component.

Thus, in the soft/high state of GS1124-683, the coherence of time variability is small between different energy X-rays and it depends on the fraction of the soft power law component. This is quite different from what is observed in the hard/low state of Cyg X-1 and GX339-4 (Nowak et al. 1999a; 1999b). These should reflect differences between the X-ray production mechanisms in these two states, which is described by Miyamoto et al.(1994).

REFERENCES

Ebisawa, K., et al. 1994, PASJ 46, 375
Kitamoto, S., et al. 1992, ApJ, 394, 609
Miyamoto, S., et al. 1993, ApJ Letters, 403 L39
Miyamoto, S., et.al. 1994, ApJ , 435, 398
Miyamoto, S., et.al. 1995, ApJ Letters, 442, L13
Miyamoto, S., & Kitamoto, S., 2000, in preparation
Nowak, M. A., et al. 1999a, ApJ, 510,874
Nowak, M. A., et al. 1999b, ApJ, 517, 355
Tanaka, Y., & Shibazaki, N. 1996, ARA&A, 34, 607

AN EXTENDED MULTI-ZONE MODEL FOR THE MCG−6-30-15 WARM ABSORBER

R. Morales[1], A.C. Fabian[1] and C.S. Reynolds[2]

1. Institute of Astronomy, Madingley Road, Cambridge CB3 0HA
2. JILA, University of Colorado, Campus Box 440, Boulder, CO 80309-0440 USA

ABSTRACT   The variable warm absorber seen with $ASCA$ in the X-ray spectrum of MCG−6-30-15 shows complex time behaviour in which the optical depth of OVIII anticorrelates with the flux whereas that of OVII is unchanging. The explanation in terms of a two zone absorber has since been challenged by $BeppoSAX$ observations. These present a more complicated behaviour for the OVIII edge. We demonstrate here that the presence of a third, intermediate, zone can explain all the observations. In practice, warm absorbers are likely to be extended, multi-zone regions of which only part causes directly observable absorption edges at any given time.

KEYWORDS: galaxies: active − galaxies: individual: MCG−6-30-15 − galaxies: Seyfert − X-rays: galaxies.

1. INTRODUCTION

To explain the behaviour of the OVII and OVIII edges, Otani et al. (1996) adopted a multizone model in which the OVII and OVIII edges originate from spatially distinct regions. Orr et al. (1997) found during their MCG−6-30-15 $BeppoSAX$ observation that the optical depth for OVIII, $\tau(OVIII)$, exhibited significant variability. The authors claimed that its large value during epoch 1^1 (1.7 ± 0.5, 1σ uncertainty) was inconsistent with the values at all other epochs.

In this poster we present a simple photoionization model that accounts for the experimental results of both $ASCA$ and $BeppoSAX$ observations.

2. TIME DEPENDENT PHOTOIONIZATION CODE

The state of the inner absorber has been modelled using a time dependent photoionization code for oxygen [7]. The model has been compared against the photoionization code Cloudy [1] (see Morales et al. (2000) for a detailed description of the code and the comparison).

---

[1]The epochs in the $BeppoSAX$ observation are chronologically numerated (i.e. number 1 corresponds to the first epoch of the observation, etc.).

3. APPLICATION TO THE MCG−6-30-15 WARM ABSORBER

The model we propose to explain both *ASCA* and *BeppoSAX* observations incorporates a new zone for the warm absorber. Let warm absorber 1 ≡ WA1 be the inner warm absorber. The outer warm absorber will be warm absorber 2 ≡ WA2. The WA3 radius and density will have values between those of WA1 and WA2.

Assuming then the presence of WA3 and also an epoch of low luminosity previous to the *BeppoSAX* observation, the expected WA3 behaviour would be[2]:

i) when $L \approx 10^{42}$ erg s^{-1}, the ionization parameter $\xi \approx 50$ erg cm s^{-1}, giving a high value for f_{O8}.

ii) when $L \approx (1,5) \times 10^{43}$ erg s^{-1}, then $\xi \approx 500, 2500$ erg cm s^{-1}. For these high values of ξ, oxygen is practically fully stripped and therefore there is a very small contribution to the optical depth for OVIII.

The range of parameters investigated for WA3 is $R = (2,8) \times 10^{17}$ cm, $n = 5 \times 10^5, 10^7$ cm^{-3} and ΔR in the interval that gives a column density for WA3 approximately equal to 4×10^{22} cm^{-3}. An example of the results obtained is presented in figure 1, where the general good agreement is also extended to point 1.

4. CONCLUSIONS

The explanation we offer for the time variability of the MCG−6-30-15 warm absorber during both *ASCA* and *BeppoSAX* observations invokes a very simple photoionization model together with the presence of a multi-zone warm absorber. This would be constituted by a continuum of clouds at different radii and different densities, such that only some of them contribute to the total optical depth for OVIII depending on the value of the luminosity.

ACKNOWLEDGEMENTS

This work has been supported by PPARC and Trinity College (R.M.) and by the Royal Society (A.C.F.). C.S.R. thanks support from Hubble Fellowship grant HF-01113.01-98A. This grant was awarded by the Space Telescope Institute, which is operated by the Association of Universities for Research in Astronomy, Inc., for NASA under contract NAS 5-26555. C.S.R. also thanks support from NASA under LTSA grant NAG5-6337.

References

[1] Ferland G. F., 1996, Hazy, a Brief Introduction To Cloudy. U. of Kentucky, Dept. of Physics and Astronomy Internal Report

[2] Grevesse N. & Anders E., 1989, in Waddington C.J., ed, Cosmic Abundances of Matter, AIP Conference Proceedings 183. AIP, New York

---

[2] All luminosities quoted, unless otherwise stated, are in this spectral range.

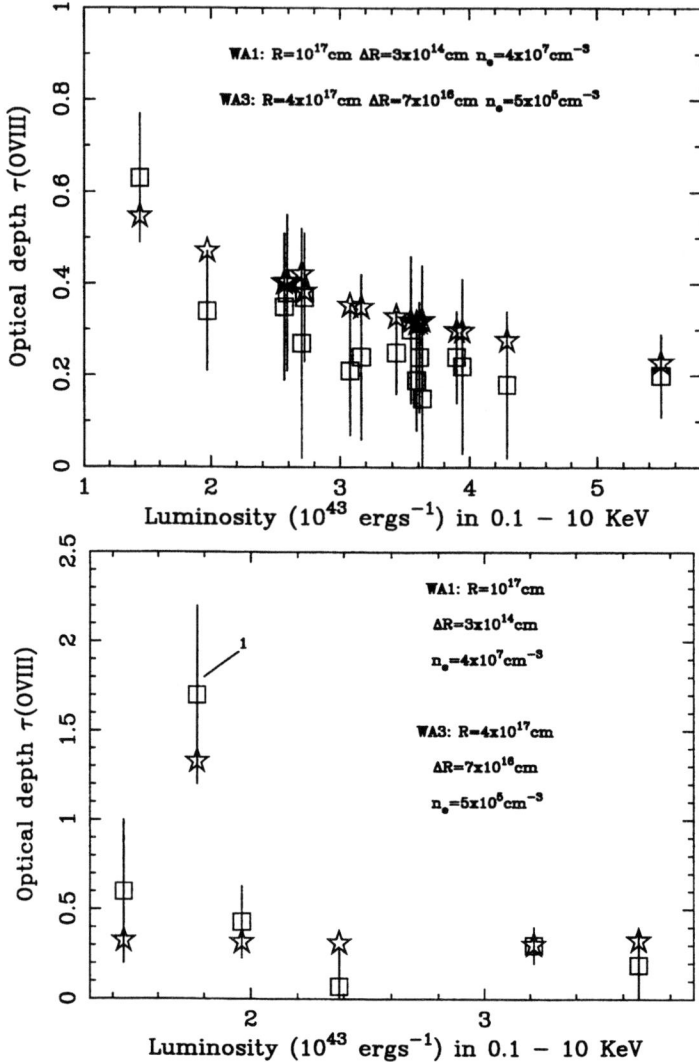

FIGURE 1. Comparison *ASCA* data (square)-top panel and *BeppoSAX* data (square)-bottom panel with the model computations (star) for the following warm absorber parameters: WA1: distance to the ionizing source, $R = 1.0 \times 10^{17}$ cm, line-of-sight distance through the warm absorber, $\Delta R = 3.0 \times 10^{14}$ cm, and electron density $n_e = 4.0 \times 10^7$ cm^{-3}. WA3: $R = 4.0 \times 10^{17}$ cm, $\Delta R = 7.0 \times 10^{16}$ cm, and $n_e = 5.0 \times 10^5$ cm^{-3}.

[3] Morales R., Fabian A.C. and Reynolds C.S., MNRAS, 2000, 315, 149M

[4] Osterbrock D., 1989, Kelly A., ed, Astrophysics of Gaseous Nebulae and Active Galactic Nuclei. Univ. Science Books, Mill Valley, Ch. 11

[5] Orr A. et al., 1997, A&A, 324, L77

[6] Otani C. et al., 1996, PASJ, 48, 211

[7] Reynolds C. S., 1996, Ph.D. thesis, University of Cambridge

DUST SCATTERED X-RAY HALO OF 4U 1538-52 OBSERVED WITH ASCA

F. Nagase [1], T. Dotani [1], T. Endo [1], H. Ozawa [1], S. Uno [2], T. Kotani [3], T. Mihara [4],

1) The Institute of Space and Astronautical Science, 3-1-1 Yoshinodai Sagamihara, Kanagawa 229-8510, Japan
2) Faculty of Social and Information Science, Nihon Fukushi University, 26-2 Higashihaemi-cho, Handa, Aichi 475-0012, Japan
3) Laboratory of High-Energy Astrophysics, NASA/GSFC, Greenbelt MD 20771, USA
4) The Institute of Physical and Chemical Research, 2-1 Hirosawa, Wako, Saitama 351-0198, Japan

ABSTRACT   The eclipsing X-ray binary pulsar 4U 1538−522 was observed with ASCA in March 1994 throughout the eclipse transition. Evidence of extended X-ray halo due to scattering by interstellar dust grains was seen in both the X-ray image and spectrum obtained during the eclipse phase. The radial profiles obtained in 1-2 keV and 2-5 keV bands from 4U1538−52 are consistent with the previous results obtained with ROSAT and ASCA from other Galactic sources. The 1-10 keV spectrum during out-of-eclipse phase can be fitted by a simple power law model of photon index $\Gamma = 1.2$ with relatively large interstellar absorption column density. On the contrary, a steep continuum spectrum with a photon index of $\Gamma = 2.6$ was observed during the eclipse. Comparing the two spectra, we can estimate the column density of the dust grains toward the source, which is substantially less than the density of the astronomical interstellar grains. This support the idea that interstellar grains are "fluffy" aggregates of smaller solid particles.

KEYWORDS: eclipsing binaries; interstellar dust; X-ray pulsars; X-ray sources.

1. INTRODUCTION

The X-ray source, 4U 1538−52 is an eclipsing X-ray binary pulsar accreting mass from the massive companion star, QV Nor via stellar wind, with a binary period 3.73 d and a long pulse period of 530 s (e.g., Bildsten et al. 1997). From analysis of the Ginga X-ray spectra throughout an eclipse transition from ingress to egress, Clark et al. (1994) found the existence of a soft component that is superimposed on the power law spectrum at energies below 4.5 keV and they interpreted this component to be due to the scattering by interstellar dust grains. This X-ray binary pulsar, 4U 1538−52/QV Nor is a system similar to the X-ray pulsar Cen X-3 and the ASCA observation of Cen X-3 revealed dust scattering halo in the soft X-ray image observed during the eclipse phase (Woo et al. 1994). Motivated by the similarity of

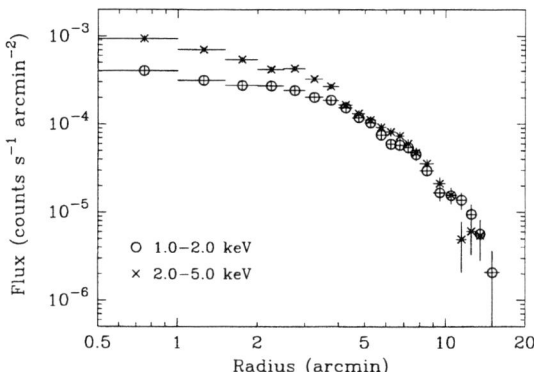

FIGURE 1. Radial profiles of the GIS images in the 1-2 keV and 2-5 keV bands observed with ASCA during the eclipse phase of 4U 1538−52. Background and contamination from the point source are subtracted.

the 4U 1538−52/QV Nor system to the Cen X-3 system, we observed 4U 1538−52 with ASCA (Tanaka et al. 1994) throughout the eclipse transition in March 1994 to search soft X-ray halo that might be produced by the interstellar dust-grain scattering.

2. IMAGE ANALYSIS

The image obtained with the ASCA SISs (Solid-state Imaging Spectrometer, i.e., CCD cameras) in the low energy band during an eclipse phase of 4U 1538−52 clearly shows the evidence of an extended halo. It was confirmed that the feature of halo extends over a radius of about 10 arcminutes, using the image of the GISs (Gas Imaging Spectrometer) which have a larger field of view of about 40 arcminutes diameter than that of SIS. The radial profiles of X-ray surface brightness derived from the GIS image taken during the eclipse phase of 4U 1538−5 show excess over the point spread function (PSF) for a point source in the energy ranges 1-2 keV and 2-5 keV.

Taking the results of spectral analysis together into account, we obtain the radial profile of dust scattering component after subtracting the residual contribution of the point source, contribution from a contamination source and the background. The resulting radial profiles of dust scattered component are shown in Figure 1 for 1-2 keV and 2-5 keV bands.

The radial profiles of the dust scattered halo observed from 4U 1538−52 in the 1-2 kev and 2-5 keV bands are relatively flat at small angular radius with turn over around 3-5 arcminutes. This feature is consistent with the previous ROSAT and ASCA observations for other Galactic sources (Predehl et al. 1991; Woo et al. 1994; Predehl & Schmitt 1995). In this analysis we for the first time derived the halo radial profile in the high energy band of 2-5 keV.

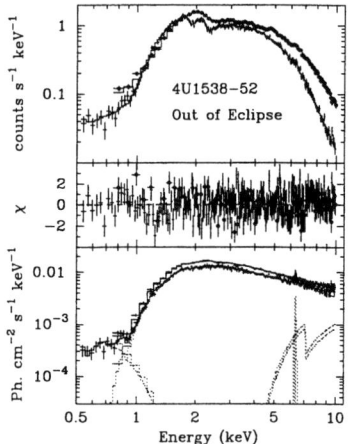
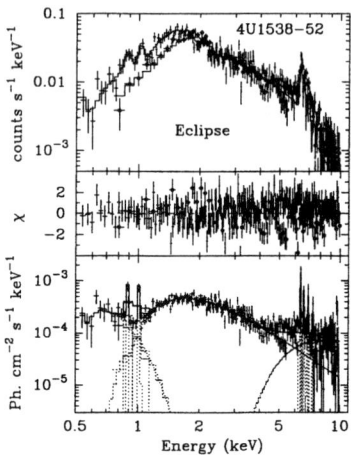

FIGURE 2. LEFT: Simultaneous fit to the GIS and SIS spectra observed from 4U 1538−52 during the out-of-eclipse phase: (top) observed counts spectra and the best fit model, (middle) residuals from the best-fit model, and (bottom) an unfolded spectrum derived from the best-fit parameters. RIGHT: Same as the left figures for the 4U 1538−52 eclipse phase spectrum.

3. SPECTRAL ANALYSIS

The energy spectrum of 4U 1538−52 is known by the feature of a hard power law with large soft X-ray absorption. The spectral feature changes gradually during the transition to eclipse along the ingress and egress, mainly due to soft X-ray absorption. Finally when the neutron star is totally eclipsed by the companion star, the direct beam from the neutron star surface disappears completely. During the eclipse of the neutron star, however, finite residual X-rays are visible with intensity of a few % of the direct beam because of scattering by the stellar wind in the binary system and interstellar dust grains. Spectra observed with ASCA during out-of-eclipse and during eclipse are shown in Figure 2 together with the best fit models and inferred incident spectra that are unfolded from the best-fit parameters.

The out-of eclipse spectrum in the left panel of Figure 2 shows that the direct pulsating component has a flat power law spectrum with soft X-ray absorption of $N_H = 1.7 \times 10^{22}$ cm^{-2}. The eclipse spectrum in the right panel indicates that the extended component seen in the 1-5 keV image corresponds to the steep power law spectrum with the soft X-ray absorption similar to the direct pulsating component. From the detailed analysis (see Nagase et al. 1999 for further details), the spectra of the direct beam and the dust scattered halo and their ratio in the 1-5 keV range are derived as follow:

$$F_{\text{direct}}(E) = (7.58 \pm 0.21) \times 10^{-2} \times E^{-(1.25 \pm 0.02)},$$

$$F_{\text{dust}}(E) = (0.52 \pm 0.15) \times 10^{-2} \times E^{-(2.6 \pm 0.2)},$$

$$R = \frac{F_{\text{dust}}}{F_{\text{direct}}} = (6.9 \pm 2.0) \times 10^{-2} \times E^{-(1.4 \pm 0.2)}.$$

This ratio shows an energy dependent dust scattering cross section integrated over the size of dust grains and the scattering angle. We compared this result with calculated energy dependence of dust scattering adopting the Rayleigh-Gans approximation (e.g., Mathis & Lee 1991; Smith & Dwek 1998). From this comparison, we obtained the amount of dust to be $M_{\text{dust}} = (1.1 \pm 0.4) \times 10^{-5}$ g cm^{-2} assuming that the dust grains are made only from silicate. This value is substantially smaller than the dust column density $M_{\text{dust}} = 4 \times 10^{-5}$ g cm^{-2} derived from the IRAS data on the assumption of $K = 40$ cm^2 g^{-1} (Hauser et al. 1984).

4. CONCLUSIONS

Results from present analysis of 4U 1538−52 data observed with ASCA are summarized as follow:
(1) We obtained radial profiles of dust scattered halo of 4U 1538−52 in both the low (1-2 keV) and medium (2-5 keV) energy bands.
(2) We derived the energy dependence of the dust scattering in the energy range of 1-5 keV.
(3) The column density of the dust grains derived from the present analysis is substantially less than the density estimated from the IRAS infrared observation.
(4) This supports the idea that interstellar grains are "fluffy" aggregates of smaller solid particles, as proposed by Mathis & Whiffen (1989).

REFERENCES

Bildsten, L., Chakrabarty, D., Chiu, J., et al. 1997, ApJS, 113, 367
Clark, G. W., Woo, J. W., Nagase, F. 1994, ApJ, 422, 336
Hauser, M. G. et al. 1984, ApJ, 285, 74
Mathis, J. S., Lee, C. -W. 1991, ApJ 376, 490
Mathis, J. S., Whiffen, G. 1989, ApJ, 341, 808
Nagase, F., et al. 2000, in preparation
Predehl, P., Bräuninger, H., Burkert, W., Schmitt, J. H. M. M. 1991, AAp, 246, L40
Predehl, P. Schmitt, J. H. M. M. 1995, AAp, 293, 889
Smith, R.K., Dwek, E. 1998, ApJ, 503, 831
Tanaka, Y., Inoue, H., Holt, S. S. 1994, PASJ, 46, L37
Woo, J. W., Clark, G. W., Day, C. S. R., Nagase, F., Takeshima, T. 1994, ApJ, 436, L5

SHORT-TERM SPECTRAL VARIATIONS DURING X-RAY FLARES IN BLACK HOLE CANDIDATES AND AGNS

Hitoshi Negoro

RIKEN, Cosmic Radiation Lab., 2-1 Hirosawa, Wako 351-0198, Japan

ABSTRACT

It has been pointed out that stellar black hole candidates (BHCs) and AGNs, especially Seyfert galaxies, have similar properties in X-ray time variations and energy spectra. We, however, have no crucial evidence that physical processes of these properties are the same. Here, is presented another common feature; spectral hardening during X-ray flares in these objects. Ginga data have revealed that X-ray "shots" in BHCs in the hard state have softer energy spectra then average spectra, and the spectra suddenly harden after the peak intensities of the shots. Using ASCA data of two Seyfert galaxies, MCG-6-30-15 and NGC 7314, it has been also found that spectra during bright "flares" are soft and become harder as the flares progress. The time variations in BHCs can be well explained by density fluctuations on the optical thin branch of an ADAF. The obtained similarity in the spectral evolution during the X-ray shots/flares and luminosity of these AGNs imply that the time variations in AGNs come from density fluctuations on the optical thick branch of an ADAF.

KEYWORDS: black hole candidates, Seyfert galaxies, time variations, Ginga, ASCA

1. INTRODUCTION

Stellar black hole candidates (BHCs) such as Cyg X-1 and AGNs, especially Seyfert galaxies, have the following common features:

- Rapid and chaotic time variations, characterized by a power-law PSD with a few knees (Miyamoto et al. 1991/ McHardy 1989).
- Energy spectra described by a power-law with a cutoff around 50-100 keV (Gierliński et al. 1997/ Zdziarski et al. 1995).
- Excess components from the power-law energy spectra below 2 keV (Balusińska et al. 1991/Turner et al. 1989).
- Soft energy spectra during X-ray flares (Negoro et al. 1994/ Matsuoka et al. 1990).

ASCA observations, however, show that only Seyfert galaxies exhibit relativistically skewed iron-lines with the centroid energy of 6.4 keV and an equivalent width

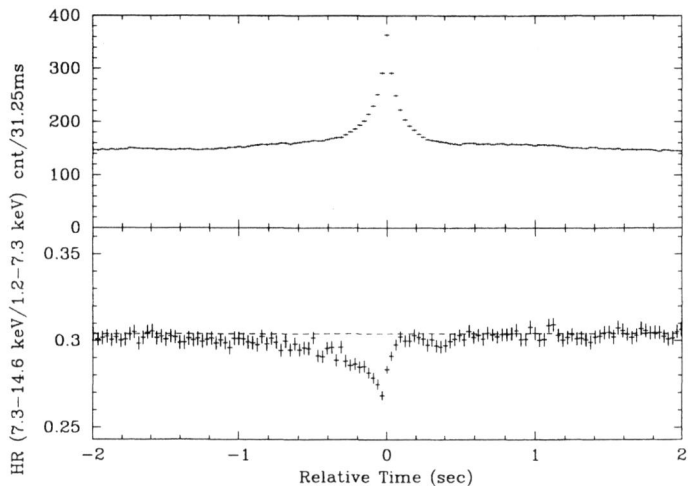

FIGURE 1. An average shot profile of Cyg X-1 in 1990 and hardness ratios. The dashed line in the lower panel indicates the average hardness ratio of all the X-rays (Negoro 1995).

of more than 100 eV, implying that a cold disk extends to the innermost radius (c.f., Matt et al. 1993). These naturally give rise to a question: are X-ray radiation processes and the origin of the time variations of these objects the same ? Here, another crucial similarity in spectral evolution during X-ray flares will be shown.

2. SPECTRAL CHANGES DURING FLARES

2.1. Stellar Black Hole Candidates

A superposition technique applied to Ginga data has revealed that shots of the BHCs in the hard state commonly have soft energy spectra, and that the spectra suddenly harden at the peak intensities of the shots, followed by complex changes (Fig. 1; Negoro et al. 1994; Negoro 1998). Time scales of the shots are about 0.1–0.2 s (FWHM), but the shots last more than 2-3 sec before and after the peaks. In RXTE data, similar properties have been found not only in a hard state but also in other states (Focke 1998; Feng et al. 1999).

2.2. Fourier Analyses

These properties in the superposed shots are consistent with structures in PSDs and phase lags below a few Hz (Negoro 1995). These results completely exclude a number of theoretical models based on a simple Comptonization process (Negoro 1995; also see Nowak 1999). The structures seen in PSDs and phase lags are mixtures of at least two components, two radiative processes (Negoro 1995).

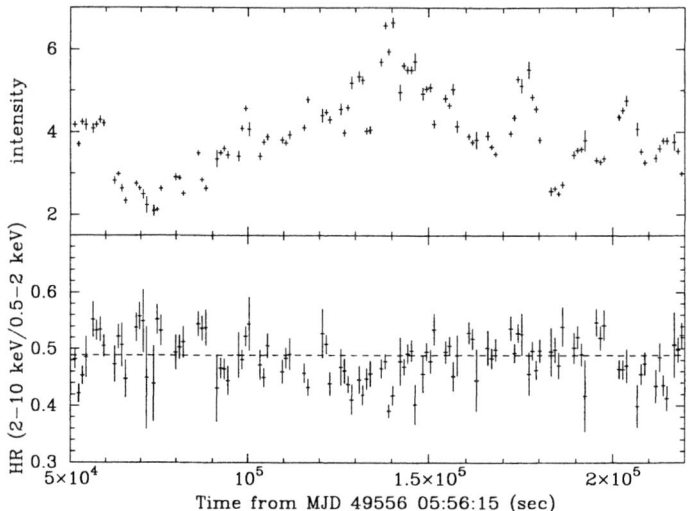

FIGURE 2. Light curve and hardness ratios of MCG-6-30-15. The dashed line in the lower panel indicates the average ratio.

2.3. Active Galactic Nuclei

ASCA observed a bright Seyfert galaxy MCG-6-30-15 in 1994, and detected a bright flare lasting for about 50 ksec (Iwasawa et al. 1996). An energy spectrum during the flare was confirmed to be softer than the average spectrum of all the data as previously shown. Furthermore, we have found that the spectrum becomes harder as the flare progresses (Fig. 2).

A bright flare was also observed in NGC 7314 (Yaqoob et al. 1996). Preliminary results showed that a tendency similar to MCG-6-30-15 was recognized. The spectrum is the softest in the rise phase, and becomes harder. The hardest spectrum is obtained after the flare. Due to poor count statistics, a spectral change near the peak intensity is not clear. Detailed results about these Seyfert galaxies will be reported elsewhere (Negoro et al. 2000).

3. DISCUSSION

The mass of a central black-hole, the origin of the time variations and an iron-line profile are closely related. If the iron-line profiles really reflect the relativistic effect as observed with ASCA, an optically thick disk would extend to the innermost radius. The observed iron-lines with the centroid energy of 6.4 keV further constrain physical environment of the accretion disk.

If an external X-ray source illuminates the disk as usually assumed, luminosity should be below 1/10 of the Eddinton luminosity (e.g., Matt et al. 1993). In this case, the mass of black holes in these Seyferts can be estimated to be more than

10^6 $M_\odot$ from observed X-ray luminosities of $\sim 10^{43}$ erg/s. If the flares observed originate from hot spots rotating at small (< 20 r_g) radii, the black holes would be further massive, say 10^8 $M_\odot$ (Iwasawa et al. 1996). In this case, however, the spectrum during the flares is expected to be almost time-symmetric if the original spectrum does not change, and spectral hardening such as observed is not expected.

As shown before, the spectral hardening observed in the two Seyferts is similar to that in X-ray shots of BHCs, suggesting that the bright flares have the same origin with the X-ray shots, aperiodic mass accretion (Negoro 1995; Manmoto et al. 1996). The time scale of the shots, $\tau_{shot} = 0.1\text{--}0.2$ sec, is more than 100 times longer than the dynamical time scale in an accretion disk of a BH with 10 $M_\odot$. Thus, if the origin is the same, the black-hole mass of these Seyferts can be estimated to be $\sim (M_{BHC} \times (\tau_{flare}/\tau_{shot}) =) 10^6 M_\odot$.

Such small mass and the luminosities indicate $L/L_{Edd} \sim 0.1\text{--}1$, where the disks are likely to be on the optically thick branch of an ADAF (e.g., Abramowicz et al. 1995). This may conflict with the above discussion about the 6.4 keV iron-lines. One possibility to avoid this conflict is that narrow cores of the observed iron-lines come from outer regions of accretion disks.

ACKNOWLEDGEMENTS

This work was supported by the special postdoctoral researchers program of RIKEN.

REFERENCES

Abramowicz, M. et al. 1995, ApJ, 438, L37

Balucińska, M. & Hasinger, G. 1991, A&A, 241, 439

Fabian A., Rees, M., Stella, L. & White, N. 1989, MNRAS, 238, 729

Feng, Y, Li, T. & Chen, l. 1999, ApJ, 514, 373

Focke, W., 1998, Doctral thesis Univ. of Maryland; LHEA, GSFC/NASA

Gierliński, M., et al. 1997, MNRAS, 288, 958

Iwasawa, K. et al. 1996, MNRAS, 282, 1038

Manmoto, T., Takeuchi, M., Mineshige, S., Matsumoto, R., and Negoro. H. 1996, ApJ, 464, L135

Matsuoka, M., Piro, L., Yamauchi, M. & Murakami, Y. 1990, ApJ, 361, 440

Matt, G., Fabian, A. & Ross, R. 1993, MNRAS, 262, 179

McHardy, I., 1989, in Proc. 23rd ESLAB Symposium, ed. J. Hunt & B. Battrick, Vol. 1, 1111

Miyamoto, S., Kitamoto, S., Iga, S., Negoro, H. & Terada, K., 1991, ApJ, 383, 784

Negoro, H., Miyamoto, S. & Kitamoto, S. 1994, ApJ, 423, L127

Negoro, H., 1995, Doctral thesis Osaka Univ.; ISAS Research Note, 616

Negoro, H., 1998, Nuclear Physiscs B (Proc Suppl.), 69/1-3, 344

Negoro, H. et al, 2000, (in preparation)

Nowak, M. 1999, (in this voluume)

Zdziarski, A., et al. 1995, ApJ, 438, L63

Turner, T.& Pounds, K. 1989, MNRAS, 240, 833

Yaqoob, T., Serlenmitsos, P., Turner, T., George, I. & Nandra, K., 1996, ApJ, 470, 27

HIGH RESOLUTION OBSERVATIONS OF X-RAY ABSORBERS/EMITTERS

Fabrizio Nicastro[1,2,3], Martin Elvis[1], Fabrizio Fiore[2], Giorgio Matt[4] and Sandra Savaglio[2]

[1] *Harvard-Smithsonian Center for Astrophysics, 60 Garden st. Cambridge MA. 02138 USA.* [2] *Osservatorio Astronomico di Roma, via Osservatorio, Monteporzio-Catone (RM), I00040 Italy.* [3] *Istituto di Astrofisica Spaziale - CNR, Via del Fosso del Cavaliere, Roma, I-00133 Italy.* [4] *Department of Physics, Universitá degli Studi Roma Tre Via della Vasca Navale 84, Roma I-00146, Italy*

ABSTRACT We present photoionization and collisional ionization models, and their application to three important fields: (a) the Warm Absorbers/Emitters in type 1 AGN, (b) the Warm Reflectors in Type 2 AGN, and (c) X-ray absorption of background quasars by intergalactic gas. A number of cases are investigated, and the dependences of the main parameters explored.

KEYWORDS: AGN, Ionized Absorber, Ionized Emitter

1. INTRODUCTION

Two, apparently distinct, main components are resolved in low resolution X-rays spectra of AGN: (a) the so called "warm absorber" (and possibly emitter), found in half of the Seyfert 1 galaxies observed by ASCA (Reynolds, 1997; George et al., 1998), and (b) the ionized reflector, seen in a number of Seyfert 2 galaxies observed by ASCA and BeppoSAX (i.e. Turner et al., 1997; Comastri et al., 1998; Guainazzi et al., 1999).

Ionized matter, potentially absorbing background quasar radiation, is also expected to be present (maybe under the form of filaments) in the intergalactic space, and its presence has relevant cosmological consequences (Hellsten, Gnedin & Miralda-Escudé, 1998). This matter is expected to produce strong resonant absorption lines in the X-ray band (i.e. "X-ray Forest"), which can be used as powerful diagnostics of the ionization state, and the temperature of this diffuse gas.

2. THE MODELS

Our models for photoionized and collisionally ionized gas (for a detailed presentation see Nicastro, Fiore & Matt, 1999: NFM99; Nicastro, Fiore, Matt & Elvis, in preparation: NFME99), include all the strongest (oscillator strength > 0.1) absorption lines as well as permitted, intercombination and forbidden emission lines in the

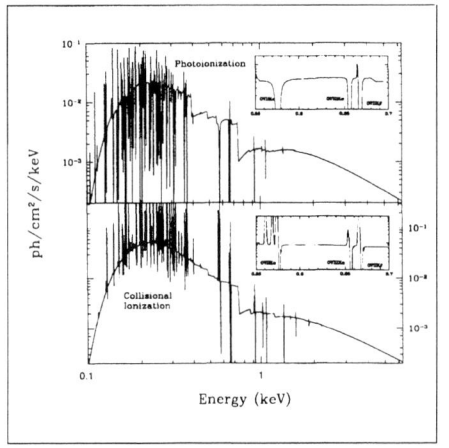 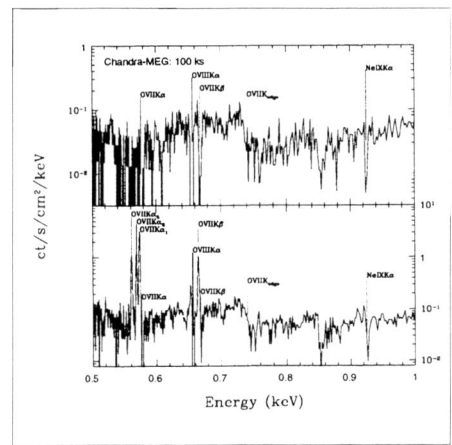

FIGURE 1. Left panel: pure photoionization (upper panel) and mix photoionization/collisional ionization (lower panel) models for Warm Absorber/Emitters in type 1 AGN. Right panel: 100 ks Chandra-MEG simulations of the models in the left panel.

50 eV to 10 keV band. The ionization structure of the gas is computed by using CLOUDY (vs. 90.04, Ferland, 1996). Resonant absorption is included as in NFM99, while the emission contribution is that predicted by CLOUDY. The lines profile (both in emission and absorption) is the correct voigt profile (NFM99). The geometrical configuration of the absorbing/emitting clouds is properly accounted for by weighting the relative (absorption versus emission) intensity with the covering factor f_c as seen by the central source (for details see NFME99).

3. WARM ABSORBERS/EMITTERS IN TYPE 1 AGN

Figure 1a (left panel) shows two spectra reprocessed by: (1) photoionized outflowing and turbulent gas, with $v_{out} = 1,000$ km s^{-1}, $v_{turb} = 500$ km s^{-1}, $f_c = 0.5$, log N_H = 22 (in cm^{-2}), log n_H = 10 (in cm^{-3}), log U = 0.5, and equilibrium temperature of $T = 4.5 \times 10^4$ K (upper panel); (2) gas with the same dynamical/geometrical parameters and densities, but with log U = -0.2 and $T = 3.2 \times 10^6$ K (lower panel). In the latter case the gas is not in photoionization equilibrium: the temperature is kept higher by an external source of heating (as suggested in the case of the "truly-warm" absorber of NGC 5548: Nicastro et al., 1999). The value of U in the non-equilibrium case (lower panel) has been chosen to give OVII-OVIII relative abundances similar to the pure photoionization case. We note that the emissivity of the gas is strongly enhanced in the non-equilibrium case, and the OVIIKα triplet is now clearly visible. The right panel (Fig. 1b) shows 100 ks Chandra-MEG simulations of the models of Fig. 1a. The 2-10 keV source flux is of 1 mCrab = 2×10^{-11} erg s^{-1} cm^{-2}. The

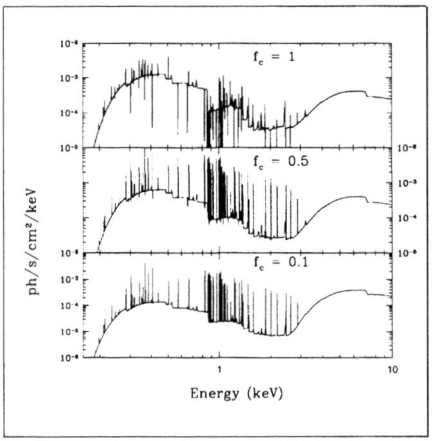 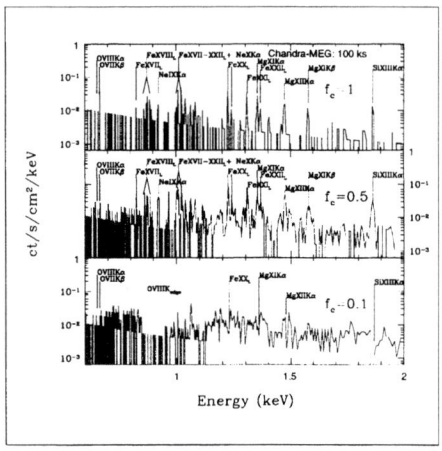

FIGURE 2. Left panel: Photoionization models for "warm-reflectors" in Seyfert 2s. Right panel: 100 ks Chandra-MEG simulations of the models in the left panel.

Chandra-MEG, clearly resolves most of the 0.5-1 keV absorption and emission lines by highly ionized Oxygen and Neon, and allows one to measure their relative intensity and width. The OVIIKα triplet is clearly resolved in the spectrum reprocessed by gas with $T = 3.2 \times 10^6$ K.

4. WARM REFLECTORS IN TYPE 2 AGN

Figure 2a (left panel) shows the same photoionzation model as in the upper panel of Figure 1a, except the direct nuclear continuum is now obscured by a column of neutral gas of $N_H^{Cold} = 3 \times 10^{23}$ cm^{-2} (Seyfert 2-like). Only the *Warm-Reflector* is visible in this case. The three cases correspond to three different values of the covering factor of the warm medium as seen by the central (obscured) source. Figure 2b (right panel) shows 100 ks Chandra-MEG simulations of the models in Fig. 2a. The 2-10 keV source flux is of 1 mCrab. The Chandra-MEG resolves most of the predicted 0.5-2 keV emission lines by O, Ne, Mg, Si and Fe highly ionized.

5. ABSORPTION LINE DIAGNOSTICS OF HOT INTERGALACTIC PLASMA

Figure 3a (left panel) shows two spectra of a bright background quasar transmitted by diffuse hot gas with log $N_H = 21$ (in cm^{-2}), log $n_H = -3$ (in cm^{-3}), and two different temperatures: log $T = 6.5$ (upper panel), and log $T = 7$ (lower panel). Highly ionized oxygen produces the strongest absorption features from the $10^{6.5}$ K plasma, while L absorption lines by FeXVII-XVIII are imprinted on spectra transmitted by the hotter gas. Figure 3b (right panel) shows 100 ks Chandra-MEG simulations of the models of Figure 3a. The 2-10 keV flux of the background quasar is of 1 mCrab. The Chandra-MEG resolves most of the predicted 0.5-1 keV absorption

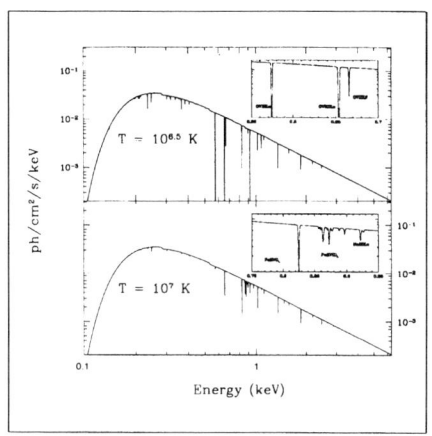 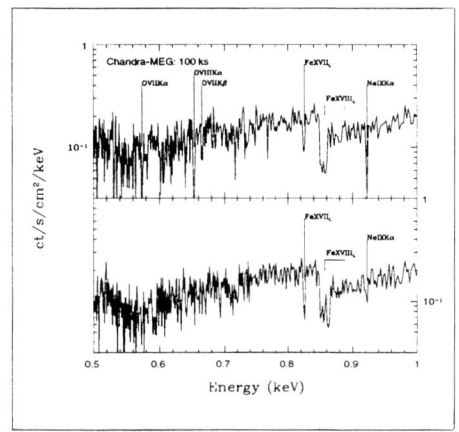

FIGURE 3. Left panel: models for absorption of a bright background quasar from hot intracluster gas. Right panel: 100 ks Chandra-MEG simulations of the models in the left panel.

lines, and clearly allows one to distinguish the two considered cases: the intensities of OVII-VIII Kα, β and NeIX Kα absorption lines are a very powerful diagnostics.

6. CONCLUSIONS

We have presented our photoionization and collisional ionization models, and briefly discussed three different applications to as many important astronomical fields. Simulations with the Chandra-MEG of each of these cases have been shown.

ACKNOWLEDGEMENTS

F.N. thanks H. Netzer and I.M. George for the useful discussions during the meeting. This work has been partly supported by the NASA grant NAG5-2476.

REFERENCES

Comastri A. et al., 1998, MNRAS, 259, 443.
Ferland G.J., 1996, CLOUDY, Version 90.04
George I.M. et al., 1998, ApJS, 114, 73.
Guainazzi M. et al., 1999, MNRAS in press (astro-ph/9905261).
Hellsten U., Gnedin N.Y. & Miralda-Escudé J., 1998, ApJ, 509, 56.
Nicastro F., Fiore F. & Matt G., 1999, ApJ, 517, 108.
Nicastro et al., 1999, ApJ, submitted.
Nicastro F., Fiore F., Matt G. & Elvis M., 1999, in preparation.
Reynolds C.S., 1997, MNRAS, 286, 513.
Turner T.J., George I.M., Nandra K. and Mushotzky R.F., 1997, ApJS, 113, 23.

INTERMEDIATE POLAR SPIN PULSE PROFILES

A.J. Norton[1]

1) Department of Physics and Astronomy, The Open University, Walton Hall, Milton Keynes MK7 6AA, U.K.

ABSTRACT The X-ray spin pulse profiles of fast rotating intermediate polars ($P_{\rm spin} < 700$ s) are double peaked, whilst those of slower rotating systems are not. We believe this is due to the lower magnetic field strength in the short spin period systems, so the radius at which material is captured by the field lines is relatively small. Consequently the footprints of the disc-fed accretion curtains on the white dwarf surface are large. The optical depths to X-ray emission within the accretion curtains are such that the emission from the two poles conspires to produce double-peaked X-ray pulse profiles. We emphasise, however, that a double-peaked pulse profile is *not* a unique indicator of two-pole accretion. Indeed, two-pole accretion onto smaller regions of the white dwarf surface may be considered the 'normal' mode of behaviour in a disc-fed intermediate polar with a longer white dwarf spin period, resulting in a single-peaked pulse profile.

KEYWORDS: stars: novae, cataclysmic variables – X-rays: stars

1. INTRODUCTION

Intermediate polars (IPs) are semi-detached interacting binaries in which a magnetic white dwarf accretes material from a Roche-lobe filling, usually late-type, main sequence companion star. The accretion flow from the secondary proceeds towards the white dwarf either through an accretion disc, an accretion stream, or some combination of both (known as disc overflow accretion), until it reaches the magnetospheric radius. Here the material attaches to the magnetic field lines and follows them towards the magnetic poles of the white dwarf. The infalling material takes the form of arc-shaped accretion curtains, standing above the white dwarf surface. At some distance from this surface, the accretion flow undergoes a strong shock, below which material settles onto the white dwarf, releasing X-rays as it cools by thermal bremsstrahlung processes. Since the magnetic axis is offset from the spin axis of the white dwarf, this gives rise to the defining characteristic of the class, namely X-ray emission pulsed at the white dwarf spin period.

2. TWO-POLE DISC-FED ACCRETION

Two-pole disc-fed accretion is the 'normal' mode of behaviour in IPs as material attaches to the field lines from the inner edge of the accretion disc, yet both single-peaked and double-peaked pulse profiles are seen. Hence, it is important that the paradigm which states 'single-peaked profile equals one-pole accretion; double-peaked profile equals two-pole accretion' is put to rest for intermediate polars: double-peaked pulse profiles *are not* a unique indicator of two pole accretion.

2.1. Fast rotators

IPs with 'fast rotating' white dwarfs ($P_{spin} < 700$s) exhibit double peaked pulse profiles: namely AE Aqr (33 s), DQ Her (142 s), XY Ari (206 s), V709 Cas (313 s), GK Per (351 s), YY Dra (529 s), V405 Aur (545 s) – see Figure 1 for examples. If they are in a state of equilibrium rotation (i.e. the accretion disc is disrupted at the radius where the Keplerian period of the disc is equal to the rotation period of the white dwarf), then the magnetic moment $\mu \propto P_{spin}^{7/6}$. As shown in Figure 2a, if the white dwarf has a relatively *weak* magnetic field, then material threads onto the field lines relatively *close* to the white dwarf. This results in *large* emission regions, whose 'vertical' optical depth (along the magnetic field lines) is *less* than their 'horizontal' optical depth (parallel to the white dwarf surface). In this case minimum attenuation of the X-ray flux occurs when the emission region is seen from above. When the upper pole points towards the observer, maximum flux is seen from it, so giving the first peak in the pulse profile (viewpoint A). The contribution to the modulation from the lower pole is *in anti-phase* with that from the upper pole since, when the upper pole is pointing towards the observer, the lower pole will generally be occulted, and when the upper pole is pointing away from the observer, the lower pole is at its most visible, so giving a second peak in the pulse profile (viewpoint B). A double-peaked X-ray pulse profile is therefore seen, as shown in Figure 2b.

2.2. Slow rotators

By contrast, slow rotators mostly show single-peaked X-ray pulse profiles, as shown in Figure 3. With a longer spin period, the magnetic moment must be larger and with a relatively *strong* magnetic field, the accreting material attaches to the field lines whilst still quite *distant* from the white dwarf. This results in *small* emission regions, whose 'vertical' optical depth (along the magnetic field lines) is *greater* than their 'horizontal' optical depth (parallel to the white dwarf surface), as shown in Figure 4a. In this case minimum attenuation of the X-ray flux occurs when the emission region is seen from the side. Hence, when the 'upper pole' points away from the observer, then pulse maximum is seen (viewpoint B), and when it points towards the observer, pulse minimum is seen (viewpoint A). The contribution to the modulation from the lower pole is *in phase* with that from the upper pole, since when the upper pole is pointing towards the observer, the lower pole will generally be occulted. Conversely, when the upper pole is pointing away from the observer, the lower pole is viewed essentially from the side too. A single-peaked X-ray pulse profile is therefore seen, as shown in Figure 4b.

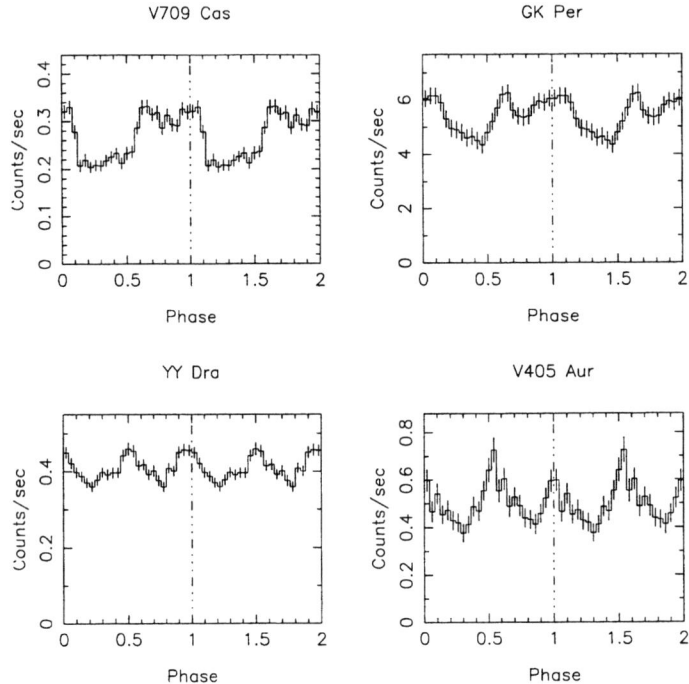

FIGURE 1. Soft X-ray spin pulse profiles of some example fast rotators – all are double-peaked.

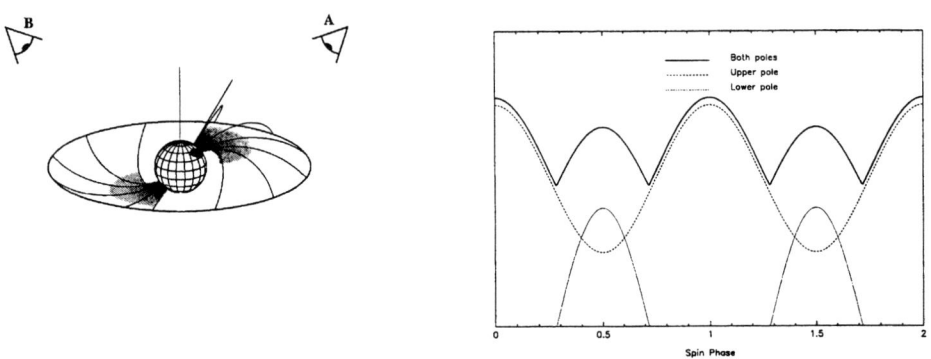

FIGURE 2. (a) Accretion in fast rotators. (b) A schematic pulse profile resulting from the situation in Figure 2a. Viewpoint A corresponds to phase 0.0 and viewpoint B corresponds to phase 0.5.

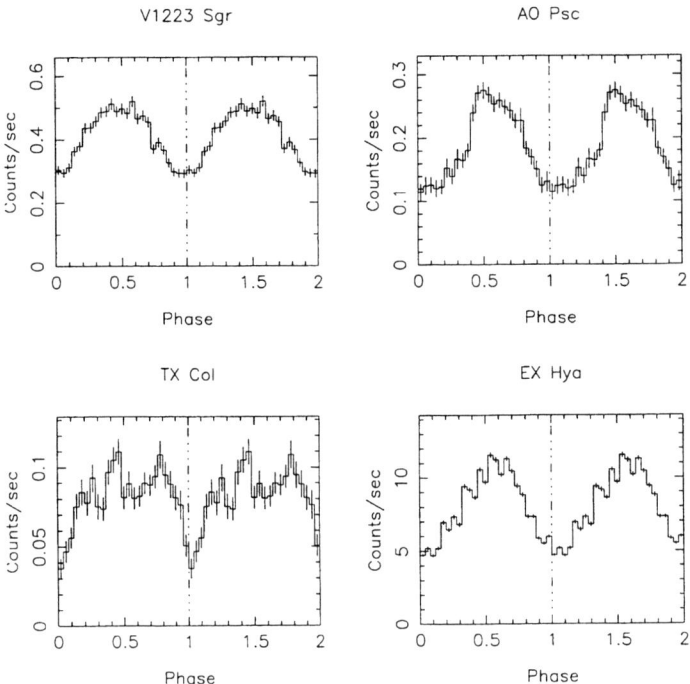

FIGURE 3. Soft X-ray spin pulse profiles of some example slow rotators – all are single-peaked.

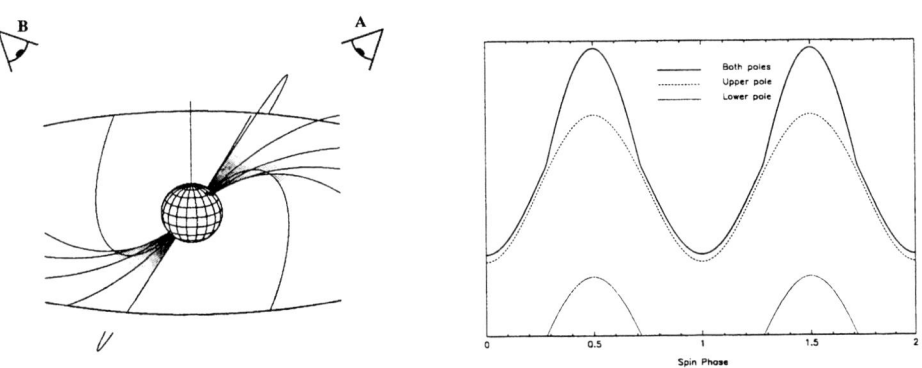

FIGURE 4. (a) Accretion in slow rotators. (b) A schematic pulse profile resulting from the situation in Figure 4a. Viewpoint A corresponds to phase 0.0 and viewpoint B corresponds to phase 0.5.

RESONANT TRUNCATION OF BE DISKS AND TYPE I OUTBURSTS IN BE/X-RAY BINARIES

A.T. Okazaki [1], I. Negueruela [2]

1) Faculty of Engineering, Hokkai-Gakuen University, Sapporo 062-8605, Japan
2) SAX SDC, ASI, c/o Nuova Telespazio, I00131 Rome, Italy

ABSTRACT We study the resonant interaction of neutron stars with disks around Be stars in Be/X-ray binaries, assuming that Be disks are formed by viscous decretion. Comparing the viscous and resonant torques exerted on the gas, we find that the Be disk is truncated at a radius which depends on the orbital parameters and viscosity. Based on this result, we propose two posssible scenarios for Type I X-ray outbursts in Be/X-ray binaries.

KEYWORDS: binaries: general – circumstellar matter – stars: emission line, Be – stars: neutron – X-rays: bursts

1. INTRODUCTION

Be/X-ray binaries form a major subgroup of high mass X-ray binaries. These are systems consisting of a Be star earlier than B2 and a compact star, presumably a neutron star. The orbit is generally wide and eccentric.

Be/X-ray binaries exhibit three types of X-ray activity:

- persistent low-luminosity X-ray emission ($L_X \lesssim 10^{34}\,\mathrm{erg\,s^{-1}}$),

- periodical (Type I) X-ray outbursts, coinciding with periastron passage ($L_X \approx 10^{36-37}\,\mathrm{erg\,s^{-1}}$),

- giant (Type II) X-ray outbursts ($L_X \gtrsim 10^{37}\,\mathrm{erg\,s^{-1}}$), which show no orbital modulation

(Stella et al. 1986; see also Negueruela et al. 1998), which suggest a complicated interaction between the Be-star envelope and the neutron star.

2. DECRETION DISKS AROUND BE STARS

A Be star has two-component envelope, a polar wind and an equatorial disk. The former consists of a low-density, fast outflow emitting UV radiation, while the latter consists of a high density plasma rotating at near Keplerian speed. Optical emission lines and the infrared excess arise from the equatorial disk.

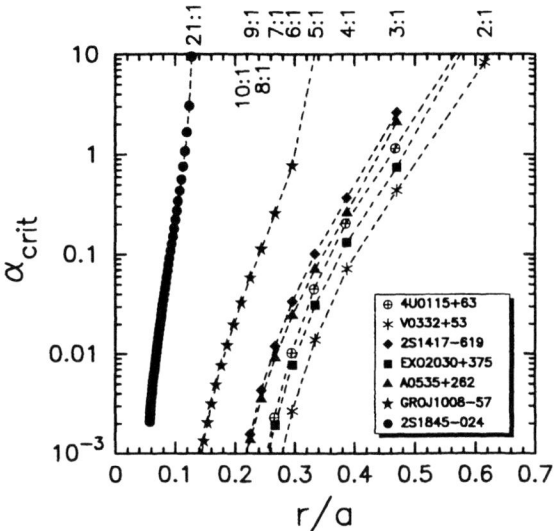

FIGURE 1. $\alpha_{\rm crit}$ at the $n:1$ resonance radii for some Be/X-ray binaries. Numbers on top indicate the locations of the $n:1$ commensurabilities of disk and binary orbital periods. The disk temperature is assumed to be 80% of the effective temperature of the Be star.

Although there is no widely-accepted model for Be disks, the viscous decretion disk model proposed by Lee et al. (1991) seems promising (see Porter 1999). In this model, the matter supplied from the equatorial surface of the star drifts outward by viscosity and forms the disk. Basic equations for viscous decretion are the same as those for viscous accretion, except that the sign of the mass flow rate is opposite. Thus, viscous decretion produces a geometrically thin, near Keplerian disk around a Be star. The outflow in the disk is very subsonic at least within several tens of stellar radii (Okazaki 2000). In addition, such a disk is unstable to the very-low frequency, $m=1$ oscillation mode (Negueruela and Okazaki 2000). These features agree well with the observed characteristics of Be disks.

3. RESONANT TRUNCATION OF BE DISKS

From the direct comparison of the viscous torque (which is providing angular momentum to the material) and the tidal torque (which is taking angular momentum from the material) at a given resonance radius, we have the criterion for resonant truncation as

$$\alpha \leq \alpha_{\rm crit} \equiv \frac{|T_{\rm reson}|}{3\pi \Sigma \Omega^2 r^2 H^2} \qquad (1)$$

(Artymowicz & Lubow 1994), where α is Shakura-Sunyaev's viscosity parameter, H the scale-height of the disk, Σ the disk surface density, Ω the angular frequency of

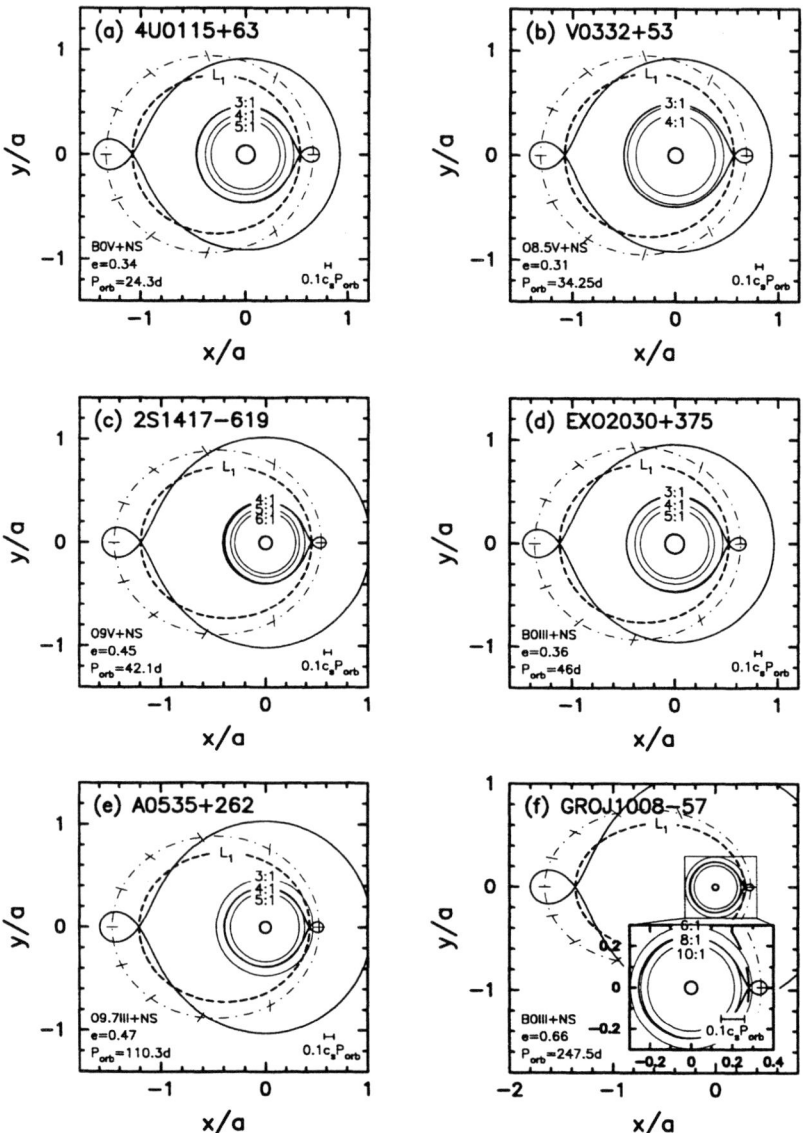

FIGURE 2. Truncated disk radii and orbital models for some Be/X-ray binaries. The dash-dotted line represents the orbit of the neutron star. The dashed line represents the L_1 point around the orbit. The thick solid lines denote the effective Roche lobes of the two stars at apastron and periastron. The thin solid lines denote the radii of the resonantly-truncated Be disk. They are for $\alpha = 0.3$, 0.1, and 0.03 from outside. Also shown is an upper limit of the distance over which the disk outer radius can expand during one orbital period.

disk rotation, and $T_{\rm reson}$ the resonant torque by the neutron star. [See Negueruela and Okazaki (2000) for details about how to compute the resonant torque.]

In Figure 1, we plot $\alpha_{\rm crit}$ at the $n:1$ resonance radii for seven Be/X-ray binary systems. The resonant torques at the $n:1$ radii are stronger than those at radii with other period commensurabilities located nearby. Figure 1 shows that the Be disk in Be/X-ray binaries should be truncated at a radius which depends on the orbital parameters and viscosity.

In Figure 2, we show the truncated disk radii and orbital models for six systems among the seven shown in Figure 1. We also show the distance scale corresponding to $0.1 c_s P_{\rm orb}$ in each panel, where c_s and $P_{\rm orb}$ are the sound speed and the orbital period, respectively. This scale should be taken as an upper limit of the distance over which the disk outer radius can expand during one orbital period, because the outflow velocity in Be disks is certainly much smaller than a few $\rm km\,s^{-1}$.

From Figure 2, we observe that the neutron star never passes through the disk at periastron, as has been described by many authors; the Be disk cannot be that large because of the resonant truncation.

4. SCENARIOS FOR TYPE I OUTBURSTS

Although the viscous decretion is a very slow process, in systems with wide orbits like A 0535+262, GROJ 1008-57, and 2S 1845-024, the outer radius of the resonatly-truncated Be disk can expand significantly during the neutron star orbits far from periastron. Such systems will exhibit regular periodic (Type I) outbursts around periastron, if the viscosity is large enough ($\alpha \gtrsim 0.1$) to allow the disk outer radius to expand beyond the L_1 point. Systems with close orbits can also show regular Type I outbursts, if the viscosity is so large ($\alpha \lesssim 1$) that the truncated disk almost fills the effective Roche lobe at periastron.

For other systems such as wide orbit systems with small viscosity ($\alpha \ll 0.1$) or close orbit systems without very high viscosity ($\alpha \lesssim 0.1$), we expect no Type I outbursts as long as the disk is axisymmetric. Be disks are, however, known to often suffer global $m = 1$ perturbations and become elongated. If the $m = 1$ perturbation makes the disk elongated toward periastron by chance, the Roche lobe overflow can occur around periastron passage. In such systems, we will observe transient and/or irregular Type I outbursts.

REFERENCES

Artymowicz, P., Lubow, S.H. 1994, ApJ 421, 651
Lee, U., Saio, H., Osaki, Y. 1991, MNRAS 250, 432
Negueruela, I., Okazaki, A.T. 2000, A&A, accepted
Negueruela, I., Reig, P., Coe, M. J., Fabregat, J. 1998, A&A 336, 251
Okazaki, A.T. 2000, PASJ, submitted
Porter, J. 1999, A&A, 348, 512
Stella, L., White, N. E., Rosner, R. 1986, ApJ 208, 669

TIME LAGS IN LOW MASS X-RAY BINARIES

Jean-François Olive & Didier Barret

Centre d'Etude Spatiale des Rayonnements, CNRS-UPS, 9 Av. du Colonel Roche, 31028 Toulouse, Cedex 04, France

ABSTRACT Using RXTE/PCA data, we have studied the time lag (TL) properties of a sample of four accreting neutron stars (NSs), namely 1E1724-3045, GS1826-238, 4U1705-44 and 4U1728-34. The aim of the study is to identify the spectral and timing state(s) in which TLs are detected. Along this work, we have discovered TLs between the 7-40 keV hard and 2-7 keV soft photons from 4U1728-34 with amplitudes similar to those seen in 4U1705-44 (i.e. ~ 2 ms at 5 Hz). We show that the TLs are only seen in the low states of those sources, but that within the so-called "island" spectral state, some sources display TLs whereas some do not. On the other hand, we have found that TLs are detected when the associated Power Density Spectrum (PDS) shows excess power at high frequencies (above ~ 1 Hz).

KEYWORDS: Stars: individual: 1E1724-3045, GS1826-238, 4U1705-44, 4U1728-34. - X-rays : stars - stars : neutron - stars : binaries.

1. INTRODUCTION

TLs have been reported so far from both accreting black holes (BHs, Cyg X-1, GX339-4) and NSs (4U0614+09, 4U1705-44, Ford et al. 1999). The amplitude of these TLs is typically 20 ms at 1 Hz for NSs and BHs (Ford et al. 1999). The origin of these TLs is currently under debate. Uniform comptonization models predicting constant TLs are however ruled out by the data. Comptonization in a non-uniform medium might account for the observed TLs (Kazanas et al. 1997), but in this case, their magnitude implies that the Comptonizing cloud extends to very large radii, which in turn poses an energy problem. Recently, Poutanen and Fabian (1999) have proposed a "magnetic flare" model that could reproduce the observed PDS and TL magnitude without requiring a large size region.

Using proprietary and archival RXTE/PCA data, we have initiated a systematic study of TLs from NSs, with the aim of determining whether TLs were associated with a singular spectral and/or timing state. We characterize the spectral state of a source using color-color diagrams (CCDs). For the timing state, we compute the PDS over a broad X-ray energy band (2-40 keV). To start with, we have selected two NSs that are always in a low state (1E1724-3045, GS1826-238, Barret et al. 2000), and two that are more variable; undergoing occasionally low states (4U1705-44, 4U1728-34). For the latter sources, we have sampled both their low and high

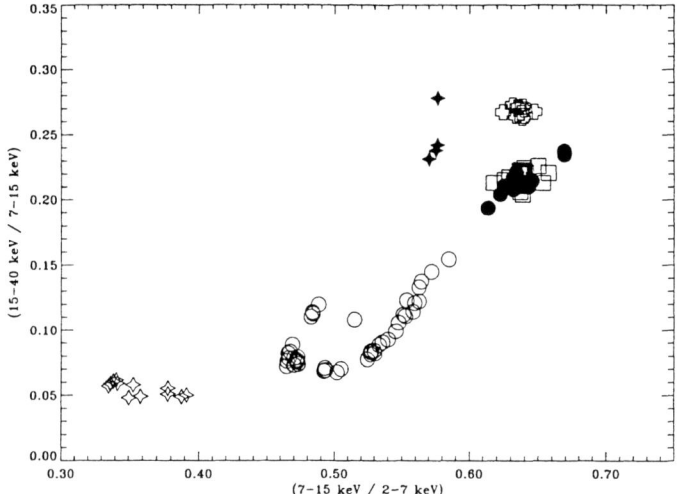

FIGURE 1. Color-Color diagram for 4U1705-44 (stars opened and filled), 4U1728-34 (circles opened and filled), 1E1724-3045 (squares), and GS1826-238 (crosses). Filled symbols correspond to time periods during which TLs were detected. The integration time of each data point is $\sim$ 3000 seconds. For 4U1728-34 and 4U1705-44, both high and low states have been sampled. TLs are only detected in their low states (up right in the figure).

states. Their low state PDS are typical of "island" state PDS, being characterized by a flat top below ν_{Break} and a broad QPO-like feature in the declining part of the PDS above ν_{Break} (at ν_{QPO}). TLs were computed using the techniques described in Nowak et al. (1999a) between the 7-40 keV hard and 2-7 keV soft photons.

2. TIME LAGS VERSUS SPECTRAL/TIMING STATE

Figure 1 shows the CCD of the four sources. Fig. 2 shows the corresponding PDS (top) and TL spectra (bottom). TLs were detected only from 4U1728-34 and 4U1705-44, and in their lowest/hardest intensity states (namely their "island" state). This is the first report of TLs from 4U1728-34. No TLs were detected in their high states, with upper limits of 0.1-0.01 seconds between 1 and 10 Hz; i.e. a factor of 10 larger than the values detected during their low states. No TLs were detected from the two steady low state sources, and the upper limits we derived are lower than the observed values for 4U1728-34 and 4U1705-44, indicating that the non detection of similar TLs is not due to a lack of sensitivity.

Looking at Fig. 1 and 2, a few points can be drawn. First, TLs are not associated with a singular spectral state; 4U1728-34 and 1E1724-3045 occupy the same region of the CCD, and only the former shows TLs. Second, although the overall shape

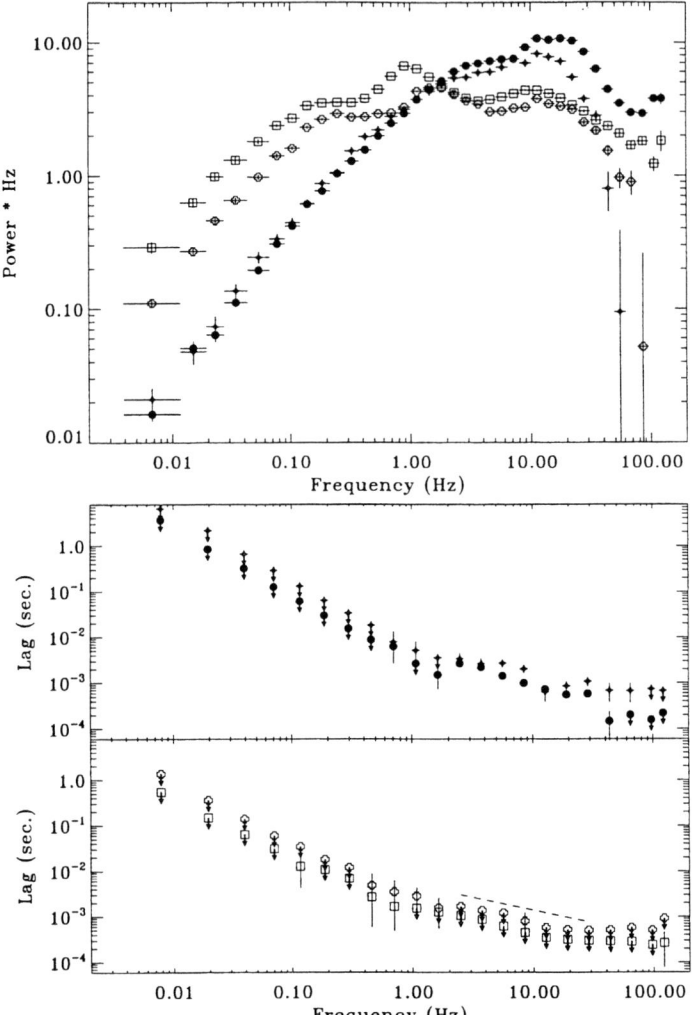

FIGURE 2. *Top panel:* Power Density Spectra, *Bottom panel:* TL spectra for 4U1728-34 and 4U1705-44 (top) and upper limits on TLs for GS1826-238 and 1E1724-3045 (bottom). For indication, the magnitude of the TLs detected from 4U1728-34 and 4U1705-44 is plotted with a dashed line.

of their PDS is broadly similar, there is one noticeable difference that shows up very clearly in the $\nu F \nu$ representation of the PDSs; that TLs are associated with a timing state in which the whole PDS is shifted towards high frequencies (ν_{Break} and ν_{QPO} are a factor of 10 larger for 4U1728-34 and 4U1705-44 than for 1E1724-3045 and GS1826-238). Third, when ν_{Break} and ν_{QPO} are high, TLs are significantly detected at frequencies between ν_{Break} and up to or slightly above ν_{QPO}. Fourth, although of lower significance than the effect observed in the two BHs Cyg X-1 and

GX339-4 (Nowak et al. 1999a,b), there is an indication that the TL decreases with frequency, especially for 4U1728-34. Finally, TLs do not depend upon the intensity of the aperiodic variability, as the four PDSs of Fig 2. have comparable integrated RMS (ranging from 17 to 25% in the 2-40 keV band).

3. CONCLUSIONS

We have discovered TLs in the low state of 4U1728-34, and confirmed the previous detection of TLs from 4U1705-44 (Ford et al. 1999). In our attempt to associate the presence of TLs with a singular spectral state, we have shown that for the same spectral state (same position in the CCD) some sources do have TLs whereas some do not. The presence of TLs must thus be tracked somewhere else.

Despite having a limited sample, we have found that TLs are detected when the characteristic low state PDS shows excess power at high frequencies (with a $\nu_{Break} \sim 1$ Hz instead of $\sim 0.1 - 0.2$ Hz). Recently, Pottschmidt et al. (2000) showed that the shot relaxation time in the hard state of Cyg X-1 (scaling as ν_{Break}^{-1}) anticorrelates with the TL amplitude. We cannot test the presence of this effect within the data set used here. However, if the same anticorrelation applies to NSs, it could provide an explanation for the non detection of smaller TLs from the sources with the lowest ν_{Break}. If ν_{Break} and ν_{QPO} are somehow related to the position of the inner disk radius within the corona, as possibly suggested by recent observations (Revnivtsev et al. 2000), the closer the disk gets to the central object (i.e the deeper inside the corona), and the larger the TLs are. In this picture, it is interesting to note that this subtle change in the accretion geometry is not reflected in the CCD, implying similar parameters for the Comptonizing cloud and input seed photon energy for all sources.

To conclude, the above study, although not addressing the origin of the TL itself has provided some hints about the conditions under which they are produced. Further work is needed to determine whether TLs are indeed always associated with the peculiar timing state described in this paper. Confirming this result would be extremely valuable for constraining any theoretical models attempting to reproduce the TL properties of these systems.

We thank E. Ford and A. Zdziarski for helpful comments on this paper.

REFERENCES

Barret, D. et al., 2000, ApJ, 533, 329
Ford, E. C. et al., 1999, ApJL, 512, L31
Kazanas, D., Hua, X. & Titarchuk, L., 1997, ApJ, 480, 280
Nowak, M. et al., 1999a, ApJ, 510, 874
Nowak, M. et al., 1999b, ApJ, 517, 355
Poutanen, J. & Fabian, A., 1999, MNRAS, 306, L31
Pottschmidt K. et al., 2000, A&A, 357, L17
Revnivtsev, M., Gilfanov, M., Churazov, E., 2000, A&A submitted, (astro-ph/9910423)

THE X-RAY SPECTRA OF AGN ACCRETION DISK CORONAE: FIRST RESULTS

A. Orr [1,2], G. Torricelli-Ciamponi [3,4], P. Pietrini [5]

1) IAAT, Tübingen, Germany; 2) Astrophysics Division, ESA/ESTEC, Noordwijk, The Netherlands; 3) Integral Science Data Center, Versoix, Switzerland; 4) Osservatorio Astrofisico di Arcetri, Firenze, Italy; 5) Dipartimento di Astronomia e Scienza dello Spazio, Università di Firenze, Italy

ABSTRACT Hot accretion disk coronae may account for substantial X-ray flux in AGN. We present preliminary results from a novel coronal emission model which builds upon the knowledge we have of solar-stellar coronae. The model considers a population of accelerated electrons (such as produced at the onset of a flare) injected at the top of an accretion disk magnetic loop. The inverse Compton and bremsstrahlung emission for the electron population is computed and compared with AGN X-ray observations. The constraints on model parameters are discussed.

1. Introduction

In these last years the presence of hot, tenuous, optically thin plasma, a so called *corona*, lying above the central part of AGN accretion disks has been invoked in order to account for the observed X-ray emission (see Di Matteo (1998) and references therein). The analogy between stellar and AGN coronae comes from the idea that differential rotation in accretion disks may act in the same way as photospheric motions do on stellar surfaces. Magnetic loops, originated by buoyancy of magnetic field lines of force, would be stretched by these motions and then, through reconnection, release energy in the form of accelerated particles. In stellar physics it is assumed that the signature of the energy conversion from the kinetic energy of the photosphere to that of the accelerated particles (through magnetic storage) is a flare. Once it is accepted that for AGN coronae the same picture as that for stellar coronae may hold, it seems important to test whether the emission can be directly derived from the accelerated electron distribution for the AGN case as well. This is the aim of the present paper.

2. The model

In our model (described in detail by Torricelli et al., in preparation) we suppose that a population of accelerated electrons (such as that produced at the onset of a flare) is injected at the top of an accretion disk magnetic loop. While electrons stream along magnetic lines of force in part trapped in the magnetic structure, their energy distribution evolves in time by collisional, bremsstrahlung and inverse

Compton losses. Supposing that a constant source of UV photons exists at a certain distance, R_{UV}, from the loop position, the inverse Compton and bremsstrahlung emission from the evolved electron population can be computed at different times. The results are presented in Fig. 1, where the time evolution of the emission is shown (left panel), as well as the contributions of different processes to the overall energy spectrum (right).

As is evident from Fig. 1, the typical decay time of a flare is shorter than the integration time for data acquisition in the X-ray domain (typically longer than 1000 s). Hence, in order to compare our model with AGN X-ray observations, which provide a temporal mean of the time dependent real emission, we compute the final spectrum as the average in time of the spectra shown in Fig. 1. If Δt is the time for a single flare to decrease by a factor of 10, we suppose that the observed emission is due to the temporal mean of a series of consecutive and identical flares exploding every Δt seconds.

In the present example the parameters for the UV component are set to the values observed for a given source, while the parameters of the initial electron distribution (energy interval and power-law index) are fixed to reasonable values. The free input parameters are then the following:
- R_{UV}, the distance between the reprocessor and the UV source.
- N_{tot}, the total number of accelerated electrons.
- n_{th}, the thermal plasma number density in the corona.

For this first test of the model we have chosen to fix the thermal plasma density according to Di Matteo (1998), $n_{th} = 3 \times 10^{11}$ cm^{-3}, and to look for the best fit of an AGN X-ray spectrum by varying R_{UV} and N_{tot}.

3. Fits to BeppoSAX data of 3C 273

The very extended energy range of BeppoSAX (Boella et al. 1997) between 0.1 and ~300 keV make it extremely well suited for comparisons with our model. We have generated a grid of our spectra using the XSPEC ATABLE format. The input parameters are set for each individual source. Preliminary fits were performed using BeppoSAX observations of the QSO 3C 273. We chose this QSO because it is bright in the X-rays and therefore well detected by all the BeppoSAX Narrow Field Instruments (NFI). We also chose an epoch when the X-ray spectrum of 3C 273 was essentially characterized by continuum emission (i.e. very weak Fe Kα line, no significant warm absorber). Because it is controversial whether the contribution of the jet to the core X-ray emission in 3C 273 is significant or not (Courvoisier 1998), we assume that it is negligible in the present study.

3C 273 was observed by BeppoSAX on 1998 June 24 5:59 UT to June 26 4:19 UT. The reduction of the NFI data followed the standard SAXDAS procedures and used the most recent version (V.1.5) of the SAXPIPE software. The September 1997 release of the BeppoSAX calibration was adopted.

Since no simultaneous IR to UV data for 3C 273 are available, the parameters describing the blue bump spectrum were obtained from the literature. Between 10^{-4} keV (12.4 μ) and 1.039×10^{-3} keV (1.2 μ) we assume a power-law with photon index

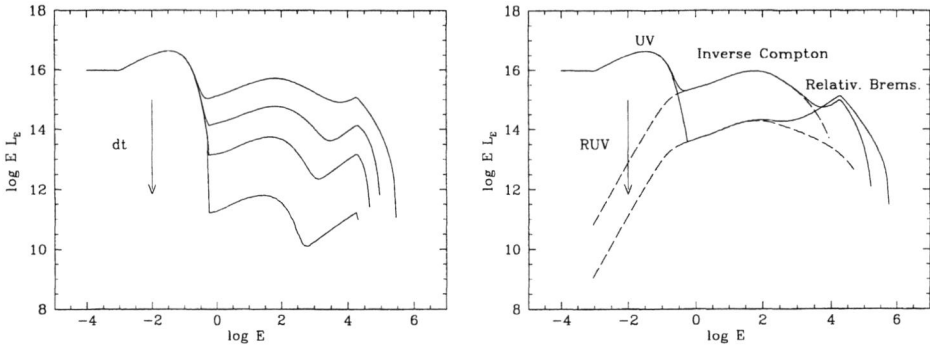

FIGURE 1. Left: Decrease in time of (E L_E) versus E, in units of 10^{30} erg/sec and keV, for a single flare. The model input parameters are set at values appropriate for 3C 273. The first (upper) curve is at $dt = 30$ s. The time increment is 100s. **Right:** Illustration of the effect of the parameter R_{UV}. The other input parameters are set at values appropriate for 3C 273. The solid line shows the three emission components in the case of $R_{UV} = 10 \times 10^{15}$ cm. The lower, dashed line shows the inverse Compton component for $R_{UV} = 85 \times 10^{15}$ cm. The UV component remains identical and the relativistic bremsstrahlung spectrum changes only slightly

$\Gamma_{UV1} = 2$ (Malkan & Sargent 1982). At higher energies we assume a power-law with $\Gamma_{UV2} = 1.39$ and an exponential cut-off at $E_{cut} = 0.056$ keV, as measured during a simultaneous IUE, ROSAT and Ginga campaign (December 1990; Walter et al. 1994). The blue bump flux normalization is based on optical data obtained at the Lowell Observatory in Spring 1991 (Lichti et al. 1995), when $f_\nu = 33.6 \pm 0.6$ mJy was measured at 5500 Å. Assuming $H_0 = 60$ km s^{-1}Mpc^{-1} and $q_0 = 0.5$, 3C 273 is at a distance of $\sim$820 Mpc and f_ν corresponds to a specific luminosity $L_E \sim 6.5 \times 10^{48}$ erg s^{-1} keV^{-1}.

A simple power-law fit to the BeppoSAX data between 0.1 and 200 keV gives a slope $\Gamma = 1.61 \pm 0.01$ (90% confidence for 1 parameter). Because of the presence of a strong soft excess the absorption, $N_H = (8.7 \pm 1.1) \times 10^{19}$ cm^{-2} is lower than the Galactic value $N_H(Gal) = 1.68 \times 10^{20}$ cm^{-2}. The fit statistics are $\chi^2_\nu = 1.22$ (172 degrees of freedom = d.o.f.).

We have fitted the same data with our model, allowing only two free parameters: R_{UV} and N_{tot}. The latter is basically the normalization of the spectrum. The fits show us that those two parameters alone do not allow an adequate fit over the *entire* energy range. As indicated in Fig. 1 (right), low values of R_{UV} lead to steep spectra above $\sim$100 keV, which are not required by our data. On the other hand, large values of R_{UV} improve the fit at high energies but produce a soft excess which is much too large. Although the 1991 optical flux used as input to our model is not simultaneous to the 1998 BeppoSAX observation of 3C 273, the contemporaneous 1 keV X-ray flux in 1991 was only $\sim$25% higher (Lichti et al. 1995) than in 1998.

Therefore the soft excesses of 1991 and 1998 are probably of comparable intensity.

A fit between 0.1–200 keV gives $\chi_\nu^2 = 1.76$ (172 d.o.f.), with $R_{UV} = (12^{+2.9}_{-0.4}) \times 10^{15}$ cm, $N_{tot} = (0.27 \pm 0.01) \times 10^{54}$ electrons and $N_H = (3.7 \pm 1.1) \times 10^{20}$ cm^{-2}. Fig. 2 shows the spectrum and the model fit. The deviation at large E is evident.

However, our model gives a reasonable first order fit to the X-ray spectrum of 3C 273 despite making exceedingly simplifying assumptions and only using two free model parameters. This suggests that our model, in which the X-ray spectrum is due to the nonthermal emission from accelerated electrons, is plausible. Our model has also the advantage of being based on variable emission, even though it is for the moment roughly modeled as periodic. Of course, the effects of other parameters, e.g. those describing the initial electron distribution, should be investigated. Other sources must also be studied, e.g. Sy 1 galaxies, and over a wider range of energy.

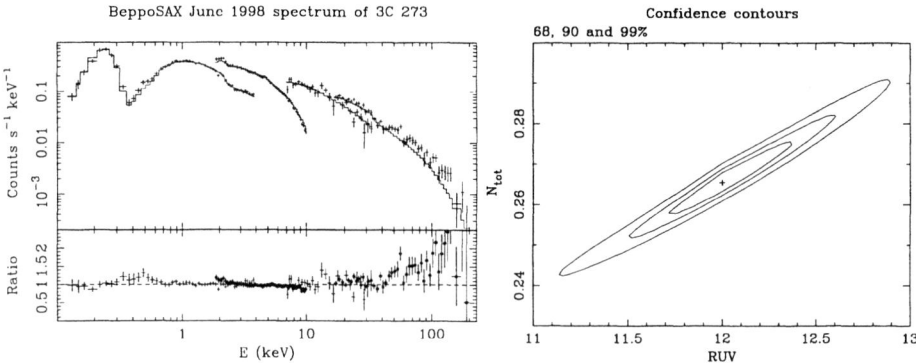

FIGURE 2. **Left**: Best fit to the X-ray spectrum of 3C 273 with our model and data-to-model ratio. The ratio data for the MECS and the PDS are indicated with dark symbols. **Right**: 3C 273 fit: contour plot for the two free parameters, R_{UV} and N_{tot}. N_H is frozen at the best-fit value. See the text for values and units.

Acknowledgements

AO thanks Arvind Parmar for a fruitful collaboration with the BeppoSAX/LECS team. AO also acknowledges a Post-Doctoral Fellowship of the Swiss National Science Foundation and a contract at the Astrophysics Division of ESA/ESTEC.

REFERENCES

Boella G., Butler R., Perola G., et al., 1997, A&AS 122, 299
Courvoisier T., 1998, A&A Rev. 9, 1
Di Matteo T., 1998, MNRAS 299, L15
Lichti G., Balonek T., Courvoisier T., Johnson N., et al., 1995, A&A 298, 711
Malkan M., Sargent W., 1982, ApJ 254, 22
Walter R., Orr A., Courvoisier T., Fink H., et al., 1994, A&A 285, 119

A SURVEY FOR HARD SPECTRUM ROSAT SOURCES

M.J. Page[1], J.P.D. Mittaz[1], F.J. Carrera[1,2]

1) Mullard Space Science Laboratory, University College London, Holmbury St Mary, Dorking, Surrey RH5 6NT, UK.
2) Instituto de Física de Cantabria (Consejo Superior de Investigaciones Científicas–Universidad de Cantabria), 39005 Santander, Spain.

ABSTRACT

We have performed a survey of 188 $Rosat$ fields looking for sources significantly harder than the X-ray background ($\alpha \sim 0.5$), with the aim of understanding the nature of sources that produce its remaining unresolved fraction. Our sample of 147 hard sources has a steeper-than-Euclidean log N - log S. We have spectroscopic identifications for 62 of the hard sources: 28 are broad line AGN, 12 are narrow emission line galaxies, and the remainder are galaxies without emission lines, clusters of galaxies, and Galactic stars. The properties of these sources are discussed. Contrary to currently favoured theories for the production of the X-ray background, we do not find a single example of a high redshift type 2 QSO.

KEYWORDS: QSOs: absorption; X-rays: galaxies; galaxies: active; cosmology: diffuse radiation

1. INTRODUCTION

The X-ray background (XRB) is now known to be primarily extragalactic in origin, but its spectrum is much harder ($\alpha \sim 0.5$ from 1-10 keV, Miyaji et al 1998) than that of a typical extragalactic source ($\alpha \sim 1$, Mittaz et al 1999). To produce the remaining XRB there must be a population of extragalactic hard ($\alpha < 0.5$) sources, some of which will be visible as faint, hard sources in the $Rosat$ energy range (0.1-2 keV). The majority of recent models for the X-ray background (eg Comastri et al 1995) postulate that these hard sources will be absorbed, narrow line 'type 2' AGN. To investigate the faint, hard source population, we have constructed a sample of secure hard spectrum $Rosat$ sources.

2. THE SAMPLE OF HARD SOURCES AND THEIR LOG N - LOG S

We have analysed 188 high Galactic latitude $Rosat$ PSPC observations. The central 20 arcminutes of each field was source-searched using the Starlink ASTERIX PSS routine. 3 colour spectra (channels 11-41, 52-90 and 91-201) were extracted for each source, and fitted with a power law model using the Cash statistic (Cash 1979, Mittaz et al 1999); more than 7000 sources were fitted in total. For our hard source

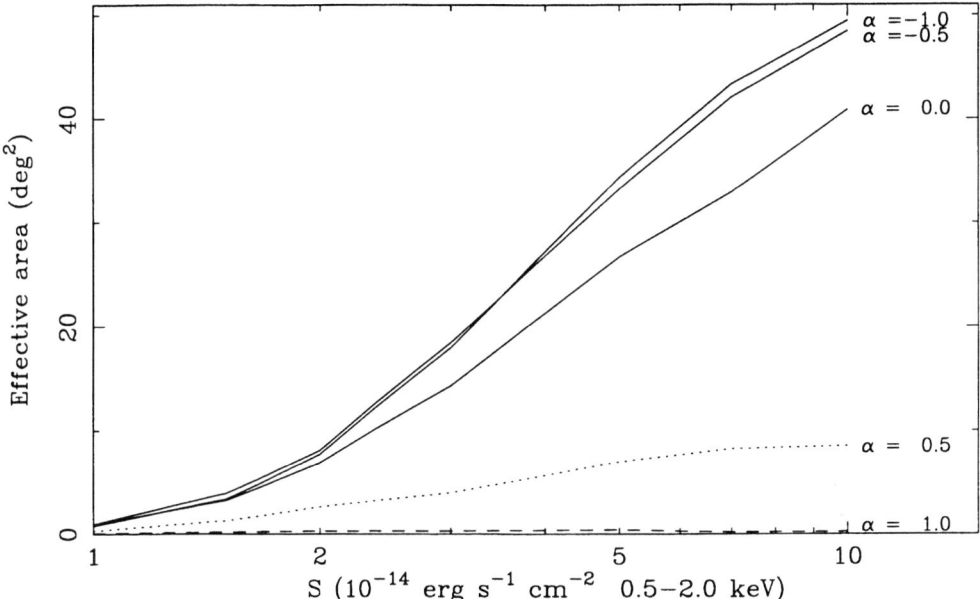

FIGURE 1. Effective area of the *Rosat* hard source survey for sources with different spectral slope α

sample, we selected only those serendipitous sources (i.e. sources unrelated to the observation targets) whose 1σ upper error bound was harder than $\alpha = 0.5$, i.e. only those sources that have a greater than $\sim 85\%$ probability of being harder than the XRB. The hard source sample contains 147 sources.

The sensitivity of our survey varies with source flux and spectrum, and from field to field. For this reason, the effective area of the survey has been computed from Monte Carlo simulations. Sources were simulated for each field, taking into account background, exposure, and Galactic N_H, and over a grid of flux and spectral slope values. The effective area of the survey as a function of flux to sources with different spectral slopes is shown in Figure 1. As intended, the survey has a very small effective area to normal sources ($\alpha \sim 1$, dashed line), compared to its effective area for hard sources ($\alpha < 0.5$, solid lines).

The integral log N - log S of the hard sources is shown in Figure 2. It was constructed by summing $1/A_{eff}(F)$ for all sources with flux $F > S$ where $A_{eff}(F)$ is the effective area at flux F. For comparison we show the log N - log S of 0.5 - 2 keV sources directly observed (Branduardi-Raymont et al 1994, stepped line) and inferred from fluctuations (Barcons et al 1994) in the UK deep survey. The hard sources have a steep log N - log S, which means they probably constitute a large fraction of the sources at faint fluxes and are therefore likely to be important contributors to the XRB.

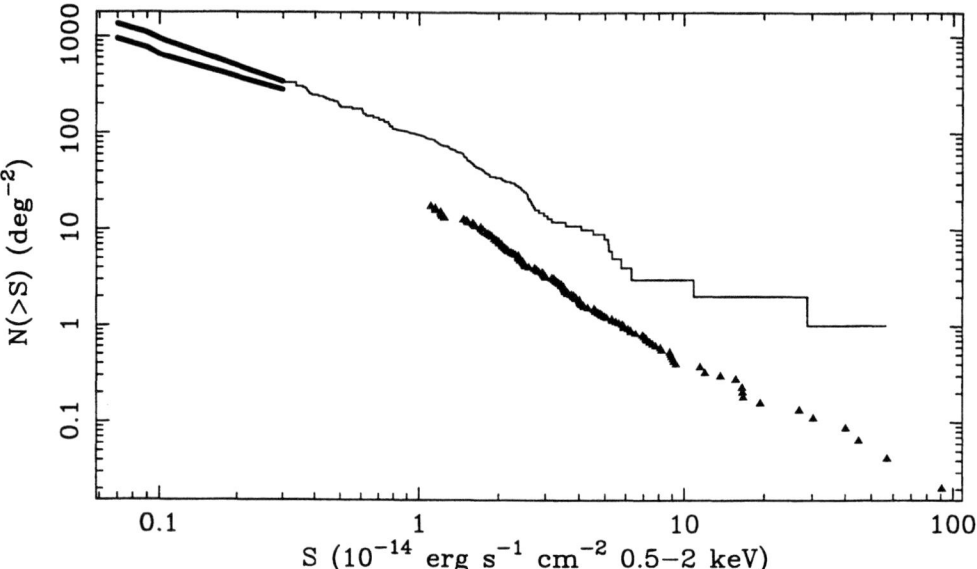

FIGURE 2. Log N - log S of the hard sources (triangles) compared to the total 0.5-2 keV source population (stepped line and bold trumpet, from Branduardi - Raymont et al 1994 and Barcons et al 1994 respectively).

3. WHAT ARE THE HARD SOURCES?

From optical spectroscopy (on the WHT and ESO 3.6m) and using existing catalogues we have identified 62 hard sources. Galaxy classifications are based entirely on their emission line properties: objects with a broad component to one or more emission lines have been classified as 'broad line AGN', those which show only narrow (< 1000 km/s) emission lines have been classed as 'NELGs', and galaxies in which we see no emission lines have been classified as 'normal galaxies'. The identifications are summarised in Table 1. Contrary to expectations, over half of the identified extragalactic ROSAT hard sources are broad line AGN, and these are found with a wide range of redshifts (up to $z = 2.8$). Some of the NELGs are Seyfert 2s, but the NELGs and normal galaxies are only found at $z < 0.6$, hence no high redshift, high luminosity, 'type 2' QSOs have been found.

The X-ray spectral slopes and fluxes of the different source types are shown in Figure 3. The galaxy clusters have different X-ray properties to all the other source types: they are brighter and are not as hard. This means that they are unlikely to constitute much of the faint, hard, XRB producing population. The broad line AGN, NELGs and normal galaxies are found with similar ranges of flux and spectral slope. As they are by far the most numerous source type, broad line AGN are the best candidates for the population of faint, hard sources required to solve the XRB spectral paradox.

TABLE 1. Optical identifications of ROSAT hard sources

Galactic stars	9
clusters of galaxies	8
normal galaxies (without emission lines)	4
narrow emission line galaxies (NELGs)	13
broad line AGN	28

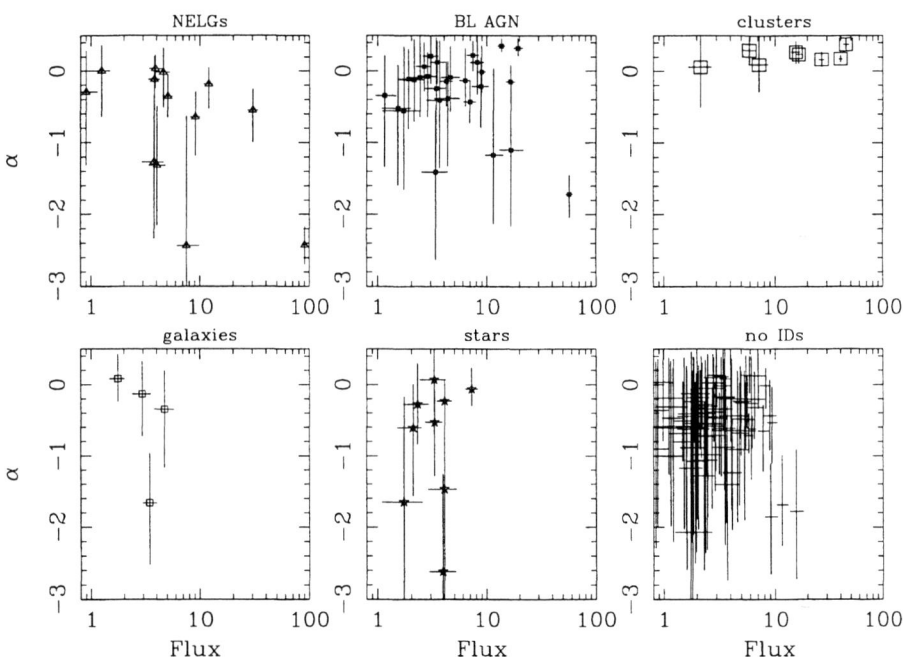

FIGURE 3. X-ray spectral slopes and fluxes of the different types of hard source.

REFERENCES

Branduardi-Raymont G., et al 1994, MNRAS, 270, 947
Barcons X., et al 1994, MNRAS, 268, 833
Cash J., 1979, ApJ, 228, 939
Comastri A., et al 1995, A&A, 296, 1
Mittaz J.P.D., et al 1999, MNRAS, 308, 233
Miyaji T., et al 1998, A&A 334, L13

AUTOMATIC SEARCH FOR PERIODIC SOURCES IN THE ROSAT DATABASE

S. Paltani[1,2], P. Bartholdi[2]

1) INTEGRAL Science Data Centre, 16 ch. d'Ecogia, 1290 Versoix, Switzerland
2) Geneva Observatory, 51 ch. des Maillettes, 1290 Sauverny, Switzerland

ABSTRACT   We have started a systematic search for periodic sources in the very rich archive of ROSAT PSPC observations. We have devised a completely automatic procedure, and we extend the search up to high frequencies (100 Hz), wich is made possible by the use of an array of 64 multiprocessor PCs. We present here the method, and the first results on about 6500 sources

KEYWORDS: Methods: statistical – (Stars:) pulsars: general – X-rays: general.

1. INTRODUCTION

One of the major breakthrough in X-ray astronomy that occurred in the course of the present decade is the generalization of X-ray telescopes equipped with grazing incidence mirrors. This made the number of detected X-ray sources increase dramatically; so dramatically in fact that it becomes quite impossible to analyze each source separately without automating the procedures on a computer.

In particular, the ROSAT satellite has detected tens of thousands of sources. Among those that have not yet been investigated are certainly some with interesting properties. We concentrate here on the detection of periodic sources in the ROSAT database. We present here a fully automated procedure that we apply on a large part of the ROSAT archival data. This method has been applied to about 6500 sources, and we present here the first results.

2. DATA AND SOURCE DETECTION

We use here the ROSAT PSPC archival data from the periods of observation 0 and 1. For each observation we apply the standard EXSAS spatial analysis on the standard events. The photons are then extracted, and their arrival times are barycentred using the coordinates found by EXSAS. In a unique ROSAT PSPC observation, it is not uncommon to observe more than 100 sources. For each source, we get a list of photons and a list of intervals during which the photons have been acquired. We have obtained about 6 500 sources.

3. RAYLEIGH'S TEST

The presence of periodicities is tested using Rayleigh's test (e.g. Fischer 1993). This test is generally used to determine whether a circular distribution is uniform or not. It can be applied directly to a photon list from of a potentially periodic source without need to build a light curve: If we receive N photons at times t_j, we calculate their phases φ_i ($0 \leq \varphi < 2\pi$) for a given test frequency f_0. We then calculate the complex sum:

$$R = \sum_{j=1}^{N} e^{i\varphi_j}$$

If N is sufficiently large (i.e., 30) and the parent population is uniform, the expected distribution of $Z = 2|R|^2/N^2$ is a χ^2 distribution with 2 degrees of freedom.

The maximum frequency at which we can detect periodicities is limited only by the accuracy on the time tag of individual photons, and not at all by their number, or by the intensity of the sources; this means that we can search for very high frequencies up tp 100 Hz that are unattainable by Fourier transform methods. Note that one can detect periodicities even when much less than one photon is received every period.

A period in the photon arrival time will produce a peak in the $Z(f)$ function, whose width decreases proportionally to the duration of the observation. We therefore search the frequency space from $f = 0$ up to $f = 100$ Hz with a step given by:

$$\Delta f = \frac{1}{5 \cdot (T_N - T_1)}$$

For a 1-day observation, $Z(f)$ must be calculated for over 40 million frequencies. Considering the huge number of sources, we have performed the calculation on the *Gravitor I* PC-farm, a network of 94 400-MHz Pentium II processors running Linux.

4. SPURIOUS PEAK DETECTION

Two major effects can create spuriously significant frequencies. We have implemented algorithms that remove automatically these frequencies.

4.1. Spectral window contamination

The photons are actually obtained only during a limited time span, defined "good time intervals" (GTIs). The spectral window power-spectrum is the Fourier power-spectrum of a function equal to 1 inside a GTI and 0 outside. It can be expressed analytically:

$$W(f) = \left| \sum_{j=1}^{N_{\text{GTI}}} \frac{i \cdot e^{2\pi i f t_j^s}}{2\pi f} - \frac{i \cdot e^{2\pi i f t_j^e}}{2\pi f} \right|$$

where N_{GTI} is the number of GTIs, $i \cdot i = -1$, and t_j^s and t_j^e define respectively the start and end time of the jth GTI. Thus, for each "significant" peak, we evaluate

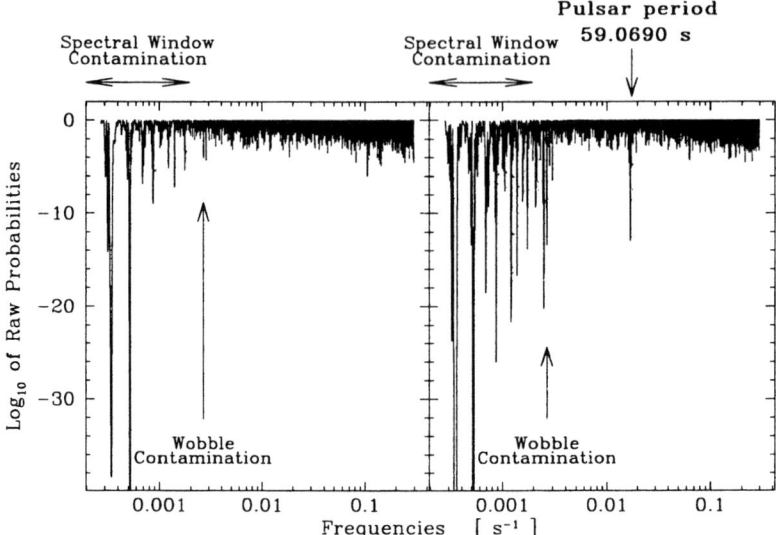

FIGURE 1. Probabilities that the parent distribution of the phases of the photons is uniform for an unknown source (left) and for 2E0053.2-7242 (right). The peaks due to the spectral window are clearly visible. The effect of the wobble is significantly present only in 2E0053.2-7242. The pulsar period emerges very clearly, although it is not, by far, the most prominent peak.

$W(f)$, and discard the peak if the contribution of $W(f)$ to $Z(f)$ is not negligible with respect to the expected value of $Z(f)$.

4.2. Mask wobble

Because of the shadowing effect of the PSPC window support structure, the spacecraft's optical axis moves back and forth with a period of about 400 s. This can cause a periodic modulation of the sources that are partly covered by this structure. As the exact frequency can change with time, and as the modulation is itself convoluted to the spectral window, in practice we discard all frequencies in the range (1/1000s)–(1/250s). This simplistic treatment should be improved in the future. The intensity of this contamination depends strongly on the position of the source on the detector, and is inexistent in some cases.

5. APPLICATION ON 2E0053.2-7242

We have first applied the automatic procedure to the ROSAT PSPC observation RP600195A00, in which Israel et al.(1998) have (re-)discovered the X-ray pulsar

2E0053.2-7242. The spatial analysis has extracted 49 sources. 30 GTIs are found in this observation. Figure 1 shows the probabilities that the parent distribution of the phases of the photons is uniform for an unknown source and for 2E0053.2-7242. The spectral window contamination is clearly visible at frequencies below $2 \cdot 10^{-3}$ s^{-1}, as is the effect of the wobble at a period of 402.181 s in 2E0053.2-7242.

In addition to these contaminations, one peak found at a period of 59.0690 s clearly stands out in the $Z(f)$ function of 2E0053.2-7242. This is perfectly in agreement with the result of Israel et al.(1998). The probability that a uniform parent population produces a given value of $Z(f)$ has to be "rescaled" to correct for the fact that the tested frequencies are not independent (Paltani & Bartholdi in preparation). We finally obtain that a probability of $10^{-7.84}$, which is highly significant. It is very important to notice that this peak has not been found after visual inspection of the $Z(f)$ functions, but that it has been singled out by the automatic processing as the *only valid candidate period* for the whole observation.

6. RESULTS ON THE FIRST ROSAT CD

We have applied the processing described above to all ROSAT observations whith observation ID ranging from RP000002N00 to RP170075N00. 30 sources have been detected by our processing. The source identification process is however still in progress. However this figure appears reasonable, considering that X-ray pulsars are privileged targets for X-ray satellites.

7. CONCLUSIONS

The algorithm that we have presented here has the advantage of considering directly the photons, and not a light curve, which makes possible to search for very short periods. We have demonstrated the feasibility of the automatic detection of periodic sources. Considering the number of sources that will be detected by the new X-ray satellites like Chandra and XMM, the automatic procedure that we have developed may ensure that the potentially numerous periodic sources will not remain undetected in the flood of serendipitous sources. While a treatment specific for each source might prove to be more sensitive, the sheer number of sources make this case-by-case analysis impossible.

ACKNOWLEDGEMENTS

This research has made use of data obtained through the High Energy Astrophysics Science Archive Research Center Online Service, provided by the NASA/GSFC. We also thank Daniel Pfenniger, who gave us access to the Gravitor I computer.

REFERENCES

Fischer N.I., 1993, Statistical Analysis of circular data. Cambridge University Press

Israel G.I., Campana S., Cusumano G. et al., 1998, A&A 334. L65

THE X-RAY SPECTRUM OF $[OIII]\lambda 5007$ SELECTED SEYFERT 2 GALAXIES.

A.Pappa [1], I.Georgantopoulos [2], G.C. Stewart [1], A.L. Zezas [1]

[1] *Department of Physics and Astronomy, The University of Leicester, Leicester LE1 7RH*
[2] *Astronomical Institute, Lofos Koufou, Palaia Penteli, 15326, Athens*

ABSTRACT   We present the results from an $ASCA$ spectral analysis of a sample of 9 Seyfert II galaxies, selected by their high $[OIII]\lambda 5007$ flux, taken from Ho et al. and De Grijp et al. Such a sample should be relatively free from selection effects such as intrinsic absorption and differences in viewing angle. Our major results are as follow: a) four of our objects are described by the standard model expected from a Seyfert 2 galaxy and b) five show no or small excess absorption. We investigated whether the latter subsample includes Compton thick objects using the $[OIII]\lambda 5007$ flux as an indicator. We find no Compton thick object other than the NGC1068. We discuss the implications of the lack of a universal description of the spectrum of the Seyfert 2 galaxies in conjunction with the models that try to reconstruct the spectrum of the X-ray Background.

KEYWORDS: surveys-galaxies:active-galaxies:general-X-rays:galaxies-X-rays: general

1. INTRODUCTION

In the standard unification model for Active Galactic Nuclei (AGN, e.g Antonucci et al. 1993) the difference between type I and type II Seyfert galaxies is attributed to orientation effects. According to the above model both Seyfert 1 and 2 nuclei have an accretion disk, a broad-line region- and a thick molecular torus. Objects observed within the opening angle of the torus are classified as Seyfert 1 objects whereas objects seen at angles intersecting the torus appear as Seyfert 2 sources. Above the nucleus (~ 1 pc), there is ionised gas (a warm scatterer) where the primary emission is scattered along our line of sight and in some cases starburst regions. According to the above scenario the X-ray spectrum of a Seyfert 2 galaxy should consist of: a) an absorbed (N_H of the order of a few $10^{23} cm^{-2}$) power-law with intrinsic index of ~ 1.9, b) a scattered power-law component produced by the elastic scattering of the primary emission by warm plasma, c) a "reflection" component which is produced by Compton scattering of the primary emission by the accretion disk and/or the torus d) iron lines both narrow and broad, produced by the accretion disk, torus and warm plasma, e) thermal emission from the warm scattering plasma, and f) thermal emission from starburst regions, if present.

TABLE 1. The Sy2 sample

Name	flux[a]	$[OIII]\lambda 5007 flux$[b]	$F_x/F_{[OIII]}$
NGC4388	3.74	19	5.08
NGC4507	1.58	89	56.33
IC5063	3.53	37	10
M51	2.28	1.5	0.66
NGC1068	158.4	3.6	0.023
NGC3079	1.2	0.9	1.3
NGC4698	1.1	0.02	46
NGC3147	1.6	0.09	18
IR18325-5926	14	7.52	1.86

[a] 2 − 10 keV flux in units of $10^{-12} \mathrm{erg\,sec^{-1}cm^{-2}}$ corrected for absorption.
[b] $[OIII]\lambda 5007$ flux in units of $10^{-12} \mathrm{erg\,sec^{-1}cm^{-2}}$ corrected for absorption.

Some objects (e.g. NGC1068 and Circinus, Matt et al. 1997, 1999 respectively) appear to be thick to Compton scattering. As a result in Compton thick Seyfert 2 galaxies, the direct component in the 2-10keV range is suppressed and we only observe the scattered (warm and/or cold) emission. Thus the Compton thick objects show low obscuration below 10keV. BeppoSAX observations with PDS revealed that a power-law emerges through a high column density ($> 10^{24} \mathrm{cm^{-2}}$) above 10 keV in such cases. The obscuration is large enough to completely block the direct emission below 10keV but small enough to allow transmission at higher energies.

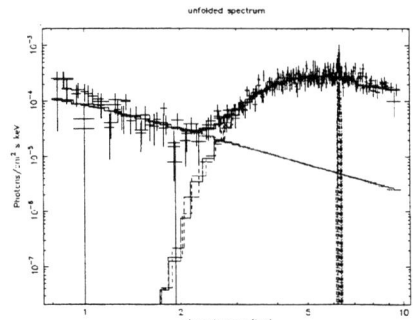
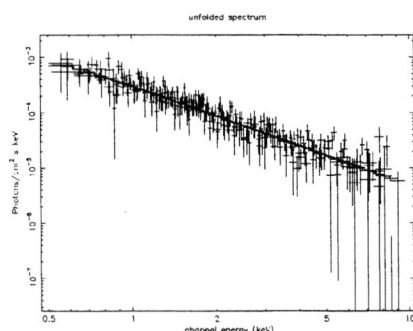

FIGURE 1. The unfolded model spectrum of IC5063 for the partial covering model (left) and for NGC4698 for a single power-law model (right).

2. DOES THE STANDARD MODEL APPLY TO ALL SY2S?

We study 9 Seyfert 2 galaxies, selected by their high $[OIII]\lambda 5007$ flux, taken from the Ho et al., (1997) and DeGrijp (1992) samples. Such a sample should be relatively free from the selection effects and biases that might appear in eg hard X-ray selected samples, such as intrinsic absorption or differences in viewing angle (Ueno et al. 1998). The objects are classified by their authors using emission line diagnostics. Data come from both our own proprietary observations and from the *ASCA* archive.

Our analysis shows that not all X-ray spectra can be described by one model type. We find that they fit roughly into two categories. One in which the spectra are consistent with the expectations of the unification models (for example IC5063, fig.1), where the objects are well represented by a partial covering model with columns of a few 10^{23} and covering fractions of typically 80% (NGC4507, NGC4388, M51, IC5063) and another, which includes the other 5 Sy2s of which NGC4698 is shown as an example in fig.1.

The objects included in the second category, require no intrinsic absorption - NGC1068, NGC3147, NGC4698, while IRAS18325-5926 is well fitted by an absorbed power-law ($N_H \sim 8 \times 10^{21} \mathrm{cm}^{-2}$) and NGC3079 is well described by a power-law and a Raymond-Smith component both absorbed by a column of $\sim 6 \times 10^{21} \mathrm{cm}^{-2}$. Objects with similar X-ray spectra have also been discussed by Ptak et al. (1996) and Bassani et al. (1999). The deficit of absorption could mean that these galaxies are Compton thick. We investigated that option by examining the 2-10keV to $[OIII]\lambda 5007$ fluxes ratio. All the galaxies but NGC1068, give ratios ~ 1 comparable to that of Seyfert 1s (table 1), thus ruling out the Compton thick possibility. This intriguing result could be explained either by an abnormal gas to dust ratio or absence of absorbing material. In the latter case the objects may have no broad line region so that their appearance as Sy2 is intrinsic and not due to absorption.

3. CONCLUSIONS

We analysed ASCA data for 9 Seyfert 2 galaxies. Our analysis shows that although half of our objects are well described by the standard Seyfert 2 model, a fraction of objects present no or low absorption. Investigation of their 2-10keV to $[OIII]\lambda 5007$ fluxes ratio indicates that only one of them may be a Compton thick object. The above result is rather in disagreement with previous work by Maiolino et al. (1998) and Risaliti et al. (1999), which suggests that half of the Seyfert 2 population consist of Compton thick galaxies. However, it should be pointed out that our work contains only a limited number of objects. It is evident that there is not a universal description of the spectrum of the Seyfert 2 galaxies. This result is important particularly in conjunction with the X-ray Background (XRB) models. Current models that reconstruct the cosmic XRB spectrum assume that it could be explained by the superposition of highly obscured objects at high redshifts e.g the analogues of the nearby hard X-ray selected Seyfert 2s. All these models are based on the unification schemes of active galaxies and in general assume a heavily absorbed power-law representing the spectral properties of the sources. Our results suggest that there may be some appreciable fraction of Compton thick objects and *unabsorbed* Seyfert 2s which has to be taken into account in the above models.

REFERENCES

Antonucci R., 1993, ARA&A, 31, 473
Bassani et al. 1999, ApJS, 121, 473
DeGrijp M.H.K. et al., 1994, A&AS, 96,389
Gendrau K.C., et al., 1995, PASJ,, 47, L5
Gilli R., Comastri A., Bruneti G., Setti G., 1999, astro-ph/9902256
Ho L.C., Fillipenco A.V., Sargent W.L.W., Peng C.Y., 1997, ApJS, 112, 315
Maiolino et al., 1998, $A\&A$, 338, 781
Matt G. et al., 1997, $A\&A$, 325, 13
Matt G. et al., 1999, $A\&A$, 341, 39
Ptak A., Yaqoob T., Serlemitsos P.J, Kunieda H., Terashima Y.,1996, 459, 542
Risaliti G., Maiolino R., Salvati M., 1999, ApJ, 522, 157
Ueno S., Law-Green S.D., Awaki H., Koyama K., 1998, AUS, 188, 342

THE 0.5-10KEV SPECTRA OF BROAD-LINE QUASARS AND THE X-RAY BACKGROUND

A.Pappa [1], G.C. Stewart [1], I.Georgantopoulos [2], R.E. Griffiths [3], B.J. Boyle [4], T. Shanks [5], O.Almaini [6],

[1] *Department of Physics and Astronomy, Leicester University, Leicester LE1 7RH*
[2] *Astronomical Institute, Lofos Koufou, Palaia Penteli, 15326, Athens*
[3] *Department of Physics, Carnegie Mellon University, Pittsburgh, PA15213, USA*
[4] *Anglo-Australian Observatory, PO Box 296, Epping, NSW 2121, Australia*
[5] *Physics Department, University of Durham, South Road, Durham DH1 3LE*
[6] *Institute of Astronomy, University of Edinburgh, Blackford Hill, Edinburgh, EH9 3HJ*

ABSTRACT We present an analysis of the $ROSAT$ and $ASCA$ spectra of 21 broad line quasars detected in six $ROSAT$ deep fields with the $ASCA$ GIS. The summed spectrum in the $ASCA$ band is well described by a power-law with $\Gamma = 1.56 \pm 0.2$, flatter that the average spectral index of bright quasars. The 0.5-8keV $ASCA$-$ROSAT$ spectrum is well fit by a power-law of $\Gamma = 1.7 \pm 0.2$ when a soft excess component is included. The analysis of the individual hardness ratios confirms the flattening of the spectral index towards fainter fluxes.

KEYWORDS: surveys-galaxies:active-galaxies:general-X-rays:galaxies-X-rays:general

1. INTRODUCTION

It has been almost 40 years since Giacconi et al. (1962) discovered the X-ray Background (XRB), the first cosmic background detected. Nevertheless its origin is still not fully understood. At a flux limit of $\sim 10^{-15} \mathrm{erg\, cm^{-2} s^{-1}}$ 70 − 80% of the 0.5-2 keV XRB is resolved into discrete sources and the majority of these are AGN (Schmidt et al. 1998). In the hard X-ray band (2-10keV) only $\sim 30\%$ of the XRB has been resolved into discrete sources with ASCA down to a flux limit of $\sim 5 \times 10^{-14} \mathrm{erg\, cm^{-2} s^{-1}}$ (Georgantopoulos et al. 1997, Ueda et al. 1998).

The spectrum of the X-ray background, however, is not consistent with the mean spectrum of X-ray bright AGN - the so called "spectral paradox". Over the 3-10 keV band the spectral index of the background is ~ 1.4 (Marshall et al. 1980) and a similar slope is found over the 0.5-2.0 keV $ROSAT$ band (Georgantopoulos et al. 1996). Recent results from BeppoSAX (Vecchi et al. 1999) indeed confirm that the X-ray background spectrum over the 0.5-10keV is well fit by a single power-law with a slope of $\sim 1.4 - 1.5$ in agreement with the slope determined over this energy range with $ASCA$ (Gendrau et al. 1995) although the $ASCA$ and $HEAO$-1 normalisations are only ∼ 0.7 those of *ROSAT* and BeppoSAX. The spectra of quasars, however, in the 2-10keV band have an average slope of ∼ 1.8 (Reeves et al. 1999) and in the softer *ROSAT* band are even steeper ($\Gamma \sim 2.5$, Schartel et al. 1996) although this may be explained in part by the presence of soft excesses.

Here, we derive the broad-band (0.5-8 keV) spectral properties of the 'typical' (ie high redshift, faint) hard X-ray selected QSOs in our survey. These contribute a large fraction (30%) of the XRB and therefore comparison of their spectrum with that of the XRB is expected to shed more light on the spectral paradox.

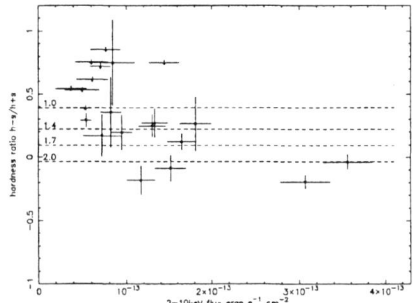 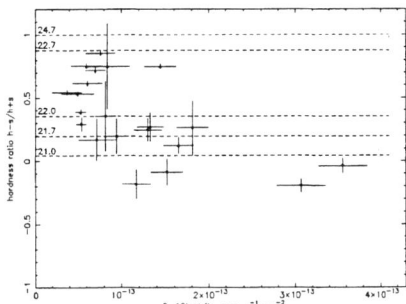

FIGURE 1. The HR versus flux for our broad-line quasars with (left) predicted HRs for various power-law indices and (right) for a power-law with $\Gamma = 1.9$ and a range of absorption labeled by the log at the column density. Arrows indicate upper limits.

2. THE SAMPLE

We have performed an *ASCA* follow-up (Georgantopoulos et al. 1997) of our deep *ROSAT* survey (Georgantopoulos et al. 1996) We have observed 6 fields and in out first, quick-look analysis, we detected 39 sources down to a flux limit of $S_{(2-10\text{KeV})} \sim 5\times 10^{-14}$ erg cm^{-2} s^{-1} in the $2-10$ keV band. We obtained optical identifications for the vast majority of our sources using the 3.9m AAT telescope. The majority of sources are broad-line QSOs. Their redshifts range from z=0.145 to z=1.952 with a mean of z∼ 1. There is also evidence for the presence of QSOs obscured in X-rays with some modest amount of optical obscuration in the optical. These objects have narrow lines in the optical range but they clearly present broad lines in their infrared spectra (eg Georgantopoulos et al. 1999). Here we restrict our analysis to the broad-line QSOs. Cross correlations of the broad line quasars with the NVSS list of radio sources, down to a flux limit of S=2.5mJy showed that 2 of the sources have a possible radio counterpart, but their X-ray properties do not differ from the rest of our QSOs.

3. X-RAY SPECTRUM

3.1. Individual source spectra

Since the photons of the individual QSOs are too few to give reliable spectra, clues for the properties of each QSO come from their hardness ratios (HR). Here we define the hardness ratio as h-s/h+s, where h and s are the total number of counts in the detection cells, in the 2-10 keV and 1-2 keV bands respectively. In fig 1. (left panel) the hardness ratio plot of each object versus the observed flux in the 2-10keV band is shown. In the case where there was no detection in the soft band (1-2 keV) we plot the corresponding 3σ upper limit. The hardness ratios for four power-law models assuming Galactic absorption are indicated as well. It is clear that the spectrum flattens towards faint fluxes in agreement with the results of Della Ceca et al. (1999). Most of the sources lie in the 1.4-1.7 spectral index range. In the right panel the hardness ratio data are compared with the expected spectra for a power-law model with $\Gamma = 1.9$ and varying absorption. Note that only a moderate absorption is needed ($\sim 10^{22} cm^{-2}$) to reproduce the flat spectra observed.

3.2. The integrated QSO spectrum

We first study the stacked spectrum of the QSOs in the 0.8-8keV *ASCA* band. We find that a single power law with $\Gamma = 1.56 \pm 0.18$ for $\chi^2 = 104.89/99$ dof with the hydrogen column density fixed to the Galactic value (in the range of $1.7 \times 10^{20} cm^{-2} - 1.9 \times 10^{20} cm^{-2}$) is a reasonable fit. For the *ROSAT* data over the 0.5-2.0 keV range we find $\Gamma = 2.3 \pm 0.1$ ($\chi^2 = 172/173$ dof) while over the full 0.1-2.0 keV range a single power-law gives a poor fit a poor fit ($\Gamma = 2.46$ with $\chi^2 = 708/303$ dof). The joint *ASCA* -*ROSAT* spectra over the 0.5-8.0keV energy range are described by a single power-law with $\Gamma = 2.2 \pm 0.1$ (N_H = galactic and $\chi^2 = 309.21/273$ dof), but this index is clearly discrepant with the *ASCA* data alone. While this may be due to a possible mis-calibration between *ROSAT* and *ASCA* (Iwasawa et al. , 1999) we have also tested the addition of a soft-excess black-body component with kT=0.1keV to the joint *ROSAT* -*ASCA* fit. In this case we obtain a good fit ($\chi^2 = 271.46/261$ dof) with $\Gamma = 1.71^{+0.19}_{-0.17}$. We emphasize that the above thermal model has no physical meaning as our data span a wide range of redshifts. However, it demonstrates the need for a spectral upturn at low energies (below $\sim$ 2keV).

4. CONCLUSIONS

We confirm previous results (Ueda et al. , 1999, Della Ceca et al. 1999) for the flattening of the spectral index towards fainter fluxes. The above study shows that the *ASCA* yields an effective spectral index of $\Gamma = 1.56 \pm 0.18$, for type-1 QSOs. This may represent either an intrinsic flattening of the QSO spectral index or the presence of moderate obscuration. If the latter case it would be difficult to reconcile the spectral upturn at soft energies with the presence of absorption. The effective spectral index derived above is lower than the canonical spectral index of bright

QSO and is more consistent with that of the XRB, reducing the need for a large number of type-2 QSOs (with narrow optical lines and heavily obscured at X-ray wavelengths) at high redshifts.

REFERENCES

Almaini O., Shanks T., Boyle B.J., Griffiths R.E., Roche N., Stewart G.C., Georgantopoulos I., 1996, MNRAS, 282, 295

Antonucci R., 1993, ARA&, 31, 473

Comastri A., Setti G., Zamorani G., Hasinger G., 1995, A&A,, 296, 1

Della Ceca R., et al., these proceedings

Gendrau K.C., et al. , 1995, PASJ,, 47, L5

Georgantopoulos I., Stewart G.C.,Shanks T., Boyle B.J., Griffiths R.E., 1996, MNRAS, 280, 276

Georgantopoulos I., et al. , 1997, MNRAS, 291, 203

Georgantopoulos I., Almaini O., Shanks T., Stewart G.C., Griffiths R.E., Boyle B.J., Gunn K.F., 1999, MNRAS, 305, 125

Giacconi R., et al. , 1962, Phys. Review Letters, 9, 439

Marshall F.E., et al. , 1980, ApJ, 235, 4

Parmar A.N., Guainazzi M., Oosterbroek T., Orr A., Favata F., Lumb D., 1999, astro-ph/9903109

Reeves J.N, 1999, PhD thesis, University of Leicester

Schartel N., et al. , 1996, MNRAS, 283,101

Schmidt M., et al. , 1998, A&A, 329, 495 B.J., Griffiths R.E., 1991, Natur, 353, 315

Ueda Y., Takahashi Y., Ohashi T., Makishima K., 1999, ApJ, 524, 11

Vecchi A., Molendi S., Guainazzi M., Fiore F., Parmar A.N., 1999, A&A, 349, 73

BEPPOSAX OBSERVATIONS OF THE HER X-1 SHORT-ON AND ANOMALOUS LOW-STATES

A.N. Parmar [1], T. Oosterbroek [1], D. Dal Fiume [2], M. Orlandini [2], A. Santangelo [3], S. Del Sordo [3], A. Segreto [3]

1) Astrophysics Division, Space Science Department of ESA, The Netherlands. 2) ITESRE, Bologna, Italy. 3) IFCAI, Palermo, Italy

ABSTRACT BeppoSAX recently observed the short-on and anomalous low-states of Her X-1. During the short-on state there are long intervals where the N_H is $>5 \times 10^{22}$ atom cm^{-2}. These intervals become longer and occur ~0.3 day earlier during each orbital cycle as the short-on state progresses. During intervals of high absorption the 0.1 keV blackbody is clearly detected. This implies the presence of separate scattering and absorbing components, or partial covering. Using BeppoSAX and RXTE data we discovered low-amplitude pulsations during the anomalous low-state of Her X-1 in 1999 July. The pulse period indicates that the Her X-1 underwent substantial spin-down close to the start of the anomalous low-state.

KEYWORDS: accretion, accretion disks, Stars: individual: Hercules X-1, Stars: neutron

1. INTRODUCTION

Her X-1 is an eclipsing X-ray pulsar with a pulse period of 1.24 s and an orbital period of 1.70 days. The source usually exhibits a 35 day X-ray intensity cycle consisting of a ~10 day duration main on-state and a fainter ~5 day short-on state. At other 35-day phases Her X-1 is still visible as a low-level X-ray source. In the warped disk model of Petterson (1977) and Schandl & Meyer (1994), the onset of the main-on state is caused by the edge of the accretion disk moving out of the line of sight to the neutron star and the end of the main-on state is caused by the corona crossing the line of sight. As predicted, the onset of the main-on state is associated with an increase in photoelectric absorption by cold material, while no such increase is seen during its more gradual decay (e.g., Parmar et al. 1980).

The 35-day cycle has been evident in RXTE ASM data with the main-on state being clearly detected every 35 days for >3 years (e.g., Scott & Leahy 1999). An exception occurred when the on-state expected around 1999 March 23 was not detected (Levine & Corbet 1999). Similar exceptions have been detected twice before. In 1983 June to August by EXOSAT (Parmar et al. 1985) and in 1993 August by ASCA (Vrtilek et al. 1994; Mihara & Soong 1994). Optical observations indicate that strong X-ray irradiation of the companion continues during these anomalous low-states.

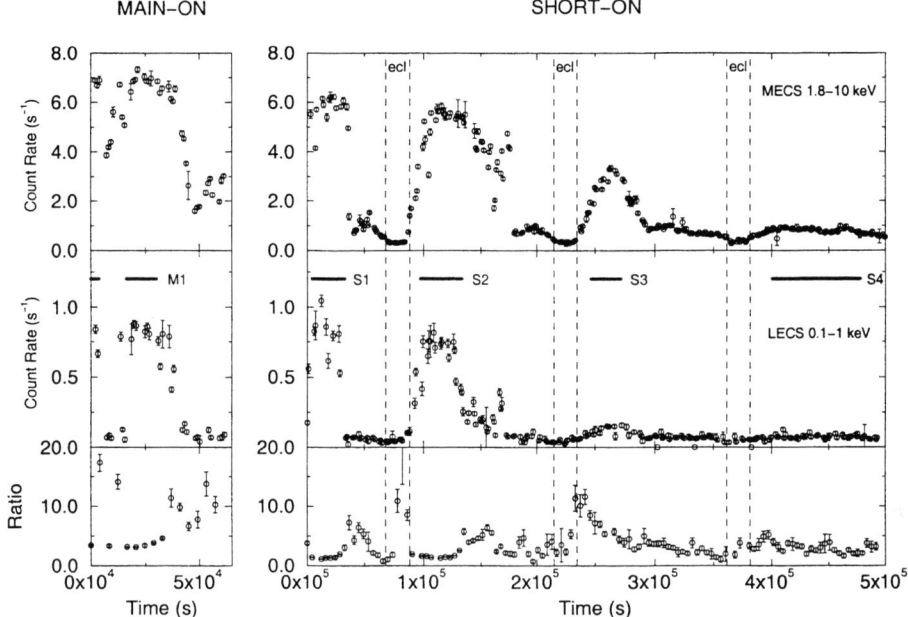

FIGURE 1. MECS 1.8–10 keV (upper panels) and LECS 0.1–1.0 keV (middle panels) lightcurves obtained during the main- (left) and short-on state (right) observations with a binning of 1024 s. The lower panels show the LECS hardness ratio

2. SHORT-ON STATE

As well as a short observation towards the end of the preceeding main-on state, Her X-1 was observed by BeppoSAX from 1998 July 9 20:38 until July 15 15:13 (UTC) during the short-on state. The 0.1–1.0 keV LECS and 1.8–10 keV MECS on-state lightcurves are shown in Fig. 1 together with the hardness ratios (LECS counts between 4–10 keV divided by those between 0.1–1.0 keV). A number of new and interesting features are evident. Soon after the start of the main-on state observation there is an interval of dipping activity (the modulation is much stronger in the LECS than the MECS, consistent with the known energy dependence of dipping), followed by a dip-free interval and a longer interval of deep dipping at the end of the observation. The short-on state observation covers parts of 4 orbital cycles and includes 3 eclipse intervals. Both the LECS and the MECS show a gradual reduction in count rate, with this effect being more pronounced in the LECS, such that the fourth orbital cycle appears to be absent, whereas a small modulation is still visible in the MECS. Superposed on this decay are the eclipses and what is normally taken to be dipping activity. However, this appears to be present for up to 20 hrs during each orbital cycle, whereas the main-on state dip duration is usually 5–10 hr. During each orbital cycle, the centroid of the emission occurs at successively earlier

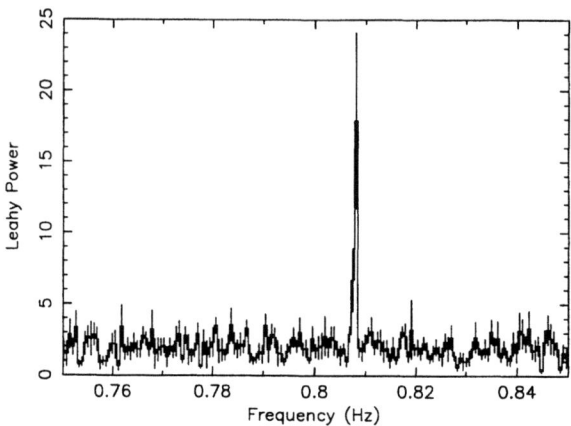

FIGURE 2. The power spectrum of Her X-1 obtained with the RXTE PCA during the anomalous low-state showing the clear detection of pulsations

orbital phases. This is confirmed by the hardness ratio plot which shows intervals of increased hardness (consistent with increased absorption) that occur progressively earlier in each of the first 3 (and possibly the fourth) orbital cycles. The peaks of the MECS centroids and the increases in LECS hardness ratio seen in Fig. 1 are separated by an average of ~1.4 days. This means that the intervals of strong absorption march back by ~7 hr each orbital cycle. This rapid marching back is in strong contrast to the main-on state dips which have a period only 0.5 hr less than the orbital one (Scott & Leahy 1999).

3. ANOMALOUS LOW-STATE

BeppoSAX observed Her X-1 on 1999 July 08 08:16 UT to July 10 06:20 (UTC), just after the expected time of turn-on to the main-on state three 35-day cycles after the non-detection by Levine & Corbet (1999). A period search in the range 1.23770–1.23776 s, predicted by extrapolating the spin-period history in Bildsten et al. (1997), was performed. Arrival times were converted to the solar barycenter and to the center of system mass using the ephemeris of Deeter et al. (1991). A peak at P_s=1.237747 s was found with a χ^2 of 42 for 16 dof. Taking into account the number of trials, and the other peaks neighboring this period, the chance-probability of detecting such a modulation is estimated to be <0.5%. The lightcurve folded over the best-period was fit with a sine curve to give a semi-amplitude of 2.1 ± 0.8%. To confirm this result public RXTE PCA data of Her X-1 obtained on 1999 April 26 were analyzed. A 2–60 keV lightcurve was extracted from the event data with a resolution of 0.05 s. A clear modulation is detected at >99.999% confidence with a period of 1.237754(6) s, confirming the BeppoSAX discovery (Fig. 2).

The pulsations discovered during the anomalous low-state have a semi-amplitude of 2.1±0.8%, consistent with the previous upper limits obtained by Mihara & Soong

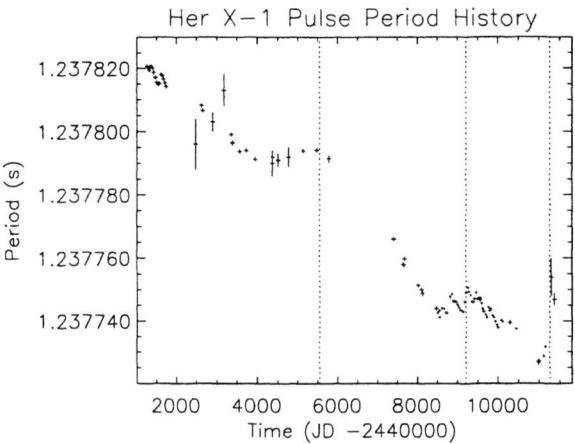

FIGURE 3. The pulse period history of Her X-1. The pulse periods determined here are the last two points. The approximate times of the 3 known anomalous low-states of Her X-1 are indicated with dotted lines

(1994) during the 1993 anomalous low-state (<1.5%) and Mihara et al. (1991) during the low-state (<2.4%). The pulse periods measured here indicate that Her X-1 underwent an interval of rapid spin-down at around the time of the start of the anomalous low-state (Fig. 3). A similar occurrence is evident in the case of the 1993 anomalous low-state (see Vrtilek et al. 1994), while the 1983 anomalous low-state appears to occur close to the end of an extended interval of spin-down. These measurements therefore strengthen the link between intervals of spin-down and the occurrence of anomalous low-states.

REFERENCES

Bildsten L., Chakrabarty D., Chiu J., et al., 1997, ApJS 113, 367
Deeter J.E., Scott D.M., Boynton P.E., et al., 1998, ApJ 502, 802
Mihara T., Ohasi T., Makishima K., et al., 1991, PASJ 43, 501
Mihara T., Soong Y., 1994, In: Makino F. (ed.) Proc. of New Horizon of X-ray astronomy. Universal Academy Press, Tokyo, p. 419
Levine A.M., Corbet R., 1999, IAU Circ. 7139
Parmar A.N., Sanford P.W., Fabian A.C., 1980, MNRAS 192, 311
Parmar A.N., Pietsch W., McKechnie S., et al., 1985, Nat 313, 119
Petterson J.A., 1977, ApJ 214, 550
Schandl S., Meyer F., 1994, A&A 289, 149
Scott D.M., Leahy D.A., 1999, ApJ 510, 974
Vrtilek S.D., Mihara T., Primini F.A., et al., 1994, ApJ 436, L9

XEUS - THE X-RAY EVOLVING UNIVERSE SPECTROSCOPY MISSION

A.N. Parmar [1], T. Peacock [1], M. Bavdaz [1], G. Hasinger [2], M. Arnaud [3], X. Barcons [4], D. Barret [5], A. Blanchard [6], H. Böhringer [7], M. Cappi [8], A. Comastri [9], T. Courvoisier [10], A.C. Fabian [11], R. Griffiths [12], P. Malaguti [8], K.O. Mason [13], T. Ohashi [14], F. Paerels [15], L. Piro [16], J. Schmitt [17], M. van der Klis [18], M. Ward [19]

1) SSD/ESA, 2) AIP Potsdam, 3) CEA Saclay, 4) IFCA (CSIC-UC) Santander, 5) CESR-CNRS/UPS, 6) Strasbourg, 7) MPE Garching, 8) TESRE Bologna, 9) Bologna Obs 10) Obs de Geneva, 11) IoA, Cambridge, 12) Carnegie Mellon U, 13) MSSL/UCL, 14) Tokyo Met. U, 15) SRON/Utrecht, 16) IAS/Roma, 17) U of Hamburg, 18) U of Amsterdam, 19) U of Leicester

ABSTRACT XEUS is under study by ESA as part of the Horizon 2000+ program to utilize the International Space Station (ISS) for astronomical applications. XEUS will be a long-term X-ray observatory with an initial mirror area of 6 m^2 at 1 keV that will be grown to 30 m^2 following a visit to the ISS. The 1 keV spatial resolution is expected to be 2-5$''$ HEW. XEUS will consist of separate detector and mirror spacecraft aligned by active control to provide a focal length of 50 m. A new detector spacecraft, complete with the next generation of instruments, will also be added after visiting the ISS. The limiting sensitivity will then be $\sim 4 \times 10^{-18}$ erg cm^{-2} s^{-1} - around 250 times better than XMM. The properties of a 350 eV (rest-frame) equivalent width Fe line from a 10^{44} erg s^{-1} AGN will be measurable out to $z = 10$, paving the way for detailed spectroscopic X-ray studies of some of the earliest known objects.

KEYWORDS: Instrumentation: detectors – Techniques: miscellaneous X-rays: general

1. INTRODUCTION

XEUS, the X-ray Evolving Universe Spectroscopy mission, is a potential follow-on mission to XMM and is being studied as part of the Horizon 2000+ program within the context of the International Space Station (ISS) utilization. The XEUS mission aims to place a long lived X-ray observatory in space with a sensitivity comparable to the next generation of ground and space based observatories such as ALMA and NGST (Fig. 1). By making full use of the facilities available at the ISS and by ensuring in the design a significant growth and evolution potential, the overall mission lifetime of XEUS could be >25 years.

The key characteristic of XEUS is the large aperture X-ray mirror. This will capitalize on the successful XMM mirror technology and the industrial foundations

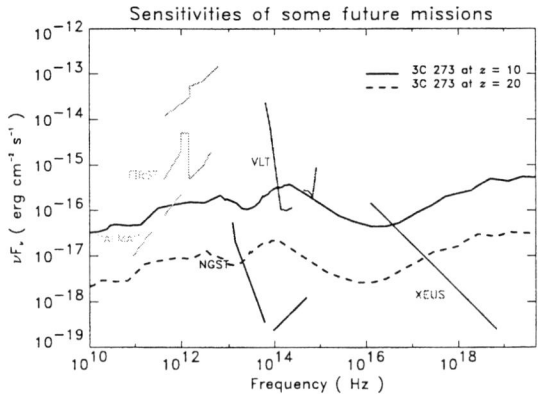

FIGURE 1. Comparison of the sensitivities of future missions in different wavebands. A horizontal line corresponds to equal power output per decade of frequency. For ALMA an 8 hr integration was assumed, for FIRST a 5σ detection in 1 hr, for NGST a 5σ detection in 10 ks, and for XEUS a 100 ks exposure

which have been already laid in Europe for this program. The XEUS mirror aperture of 10 m diameter will be divided into annuli with each annulus sub-divided into sectors. The basic mirror unit therefore consists of a set of heavily stacked thin mirror plates. This unit is known as a "mirror petal" and is a complete, free standing, calibrated part of the overall XEUS optics with a spatial resolution of 2–5" HEW and a broad energy range of 0.05–30 keV. Narrow and Wide field imagers will provide FOVs of 1' and 5', and energy resolutions of 1–2 eV and 50 eV at 1 keV.

2. MISSION PROFILE

XEUS will consist of separate detector (DSC) and mirror spacecraft (MSC) separated by 50 m and aligned by active control. The large aperture mirror cannot be deployed in a single launch. Instead, the "zero growth" XEUS (MSC1+DSC1) will be launched directly into a Fellow Traveler Orbit (FTO) to the ISS using an Ariane V or similar. The FTO is a low Earth orbit with an altitude of ~600 km and an inclination similar to the ISS. The mated pair will then decouple and DSC1 will take up station 50 m from the MSC1 and the zero growth astrophysics observation program will commence with an aperture of 6 m^2 at 1 keV.

After 4–5 years of observations, the XEUS spacecraft will re-mate and maneuver to the vicinity of the ISS. At the ISS the MSC1 will separate from DSC1 and then dock with the ISS. The DSC1, with its usefulness at an end, will undergo a controlled de-orbit. At the ISS the mirror area is expanded to 30 m^2 and MSC1 becomes MSC2. The extra mirror petals will have already been transported to the ISS using the STS or the European Automated Transfer Vehicle. Once the mirror growth and other checks are complete, MSC2 will leave the ISS and mate with the recently launched

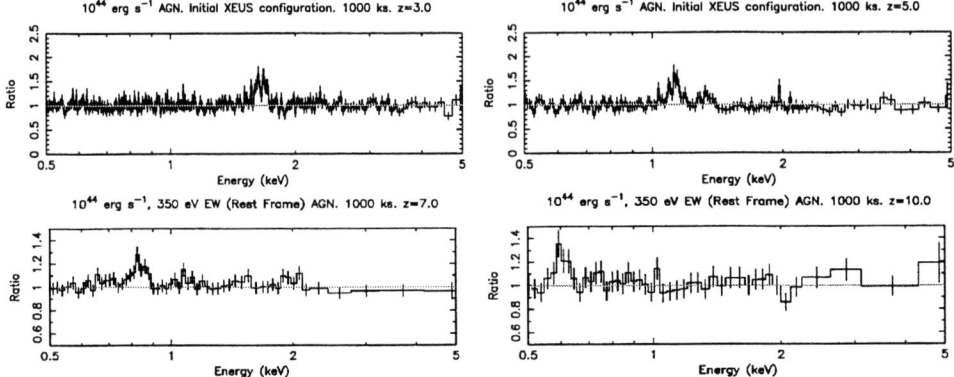

FIGURE 2. The residuals when the 350 eV EW Fe line normalization is set to zero for an AGN with a (rest-frame) luminosity of 10^{44} erg s^{-1} for $z = 3, 5, 7$, and 10 for the fully grown XEUS configuration

DSC2. Using the DSC2 propulsion system the pair will return to FTO and the fully grown XEUS will start its observing program.

3. SCIENCE GOALS

XEUS will study the evolution of the hot baryons in the Universe and in particular:

- Detect massive black holes in the earliest AGN and estimate their mass, spin and z through studies of relativistically broadened Fe-K lines and variability.

- Study the formation of the first gravitationally bound, dark matter dominated, systems ie. small groups of galaxies and trace their evolution into today's massive clusters.

- Study the evolution of metal synthesis down to the present epoch, using in particular, observations of the hot intra-cluster gas.

- Characterize the mass, temperature, density of the intergalactic medium, much of which may be in hot filamentary structures, using aborption line spectroscopy. High z luminous quasars and X-ray afterglows of gamma-ray bursts can be used as background sources.

3.1. Spectroscopy of Massive Blackholes

Currently, X-ray astronomy can only detect AGN to a z of ∼5. XEUS will be able to undertake *detailed* X-ray spectroscopy of much more distant AGN. Fig. 2 illustrates the results of a series of simulations of a "typical" AGN with a 2–10 keV rest-frame luminosity of 10^{44} erg s^{-1} at different red-shifts. An exposure time of 10^6 s was

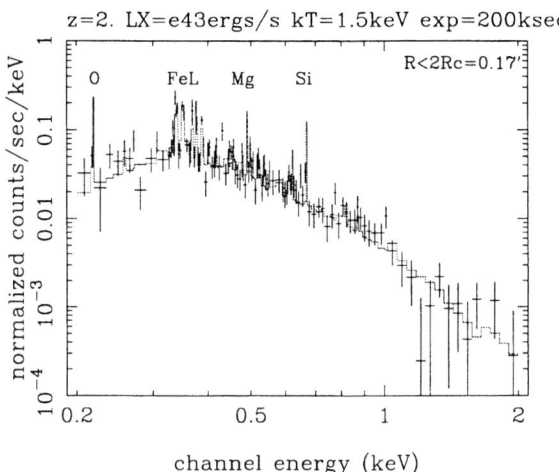

FIGURE 3. Simulated XEUS spectrum of a $z = 2$ galaxy group

assumed for the fully grown XEUS. Values for H_0 and q_0 of 50 km s^{-1} Mpc^{-1} and 0.5 together with an underlying $E^{-2.0}$ spectrum with a Galactic N_H of 10^{21} atom cm^{-2} and a local (red-shifted) N_H of 5×10^{21} atom cm^{-2} were assumed. A "double-horned" relativistically distorted and Doppler broadened Fe line at 6.4 keV with a rest-frame equivalent width of 350 eV was simulated. The other line parameters were taken to be as for MCG-6-30-15. Fig. 2 shows the residuals when the source is red-shifted to $z = 3, 5, 7$, and 10, demonstrating that such a line can be clearly detected even at $z = 10$.

3.2. Spectroscopy of Distant Galaxy Groups

To illustrate the potential of XEUS to study the formation of large scale structure Fig. 3 shows a simulation of a distant ($z = 2$) galaxy group. In standard cosmological models these groups are the first emerging massive objects, with masses of $\sim 10^{13} M_\odot$. The epoch of their first formation depends critically on the adopted cosmology, and is likely to be $z \sim 2$-5. Therefore the study of groups will provide a deep probe of the early Universe. These systems and their dark matter haloes are the smallest units by which to study the hot thermal intergalactic gas trapped in deep gravitational wells. Emission lines of O, Fe, Mg and Si are clearly evident. The temperature can be determined to better than $\pm 3\%$ and the Fe and O abundances to better than 10% and 20%, respectively.

ACKNOWLEDGEMENTS

We acknowledge the enthusiastic support of the ESA ISS and XMM project teams.

SPECTRAL EVOLUTION OF GRS 1915+105 DURING THE REGULAR FLARES

B. Paul[1,2] and A. R. Rao[2]

1) ISAS, 3-1-1 Yoshinodai, Sagamihara, Kanagawa 229-8510, Japan
2) TIFR, Homi Bhabha Road, Mumbai 400005, India

ABSTRACT

We have investigated spectral evolution of GRS 1915+105 during the regular flares by high time resolution spectroscopy. Twenty-four colour light-curves were converted into 1 s and 3 s spectra. A spectral model consisting of a disk black-body and a power-law, shows that the flux variations during the regular flares (in 2–20 keV range) can mainly be attributed to changes in the disk black-body emission, with a small increase in the power-law flux. The disk black-body temperature increases gradually with the flux, and is associated with a decrease in the size of the emission region. Two different types of intensity variations, that is spectral-state transitions and regular flares involving a disappearing disk are found to occur simultaneously.

KEYWORDS: accretion, accretion disks – stars: individual (GRS 1915+105) – X-rays: stars

1. INTRODUCTION

The transient Galactic black hole candidate GRS 1915+105 provides a very attractive site to test the concepts of accretion onto black holes. The source has shown a wide range of X-ray luminosity (presumably corresponding to a wide range of accretion rates) and temporal variability. The temporal structure of the regular flares observed in GRS 1915+105 consists of a slow rise ($\sim$ 10–20 s, figure 1) and a sharp decay which is very different from the type-I bursts seen in neutron star sources. The associated spectral changes observed with the Indian X-ray Astronomy Experiment (IXAE) and the RXTE indicate that the bursts are produced by a quasi-periodic disappearance of inner part of the accretion disk into the black hole through the event horizon (Belloni et al. 1997; Paul et al. 1998). Among many other types of variabilities (Morgan et al. 1997) are the slightly irregular and longer flares which show a dramatic change in the spectral and temporal characteristics of this source. The energy spectrum and power density spectrum of the quiescent and flare phases are found to be remarkably similar to the same obtained during the low and high states of the source. This indicates that the source makes rapid transition between the high and low states. In GRS 1915+105, co-existence of a standard thin disk and an advective hot disk is possible. Changes in the advective component result in the fast spectral state changes (Yadav et al. 1999, Rao et al. 2000).

2. OBSERVATIONS AND ANALYSIS

Based on the temporal characteristics, we have selected two observations made on 09/06/1997 and 19/02/1999 with RXTE. Regular flares with about 80 seconds duration are present in the 1997 observation. The 1999 observation shows alternate high and low states each lasting for about 1000 seconds and a gradual state transition in between. In the high state, large flares similar to the regular flares are observed whose intensity is found to decrease and the duration is found to increase with progress of the high state.

We have generated light curves in 24 bands with 1 s time resolution and combined them to make series of time resolved spectra. The response matrices of the PCA detector array for the two observations were appropriately binned to be used with these spectra. To improve the statistical significance of the 1996 observation, spectra from 25 flares were co-added by aligning the falling edge. For the 1999 observation, the spectra were co-added with 3 s time resolution to improve statistical significance. A spectral model consisting of a power-law, a disk black-body and absorption in the low energy band (with $N_H = 6 \times 10^{22}$) was used to fit the spectra.

In the regular flares, we found that the flux variation (in 2–20 keV range) can mainly be attributed to the change in the disk black-body emission. The flux of the black-body component of the spectrum increases by a large factor (4–6) with relatively small change in the power-law flux ($\sim 40\%$). The disk blackbody temperature was found to increase gradually with the flux, and the size of the emission region was found to decrease rapidly (figure 1, left). The evolution of the spectrum with respect to that in the quiescent period is shown on the right hand side of figure 1. Ratio of the time resolved spectra and the spectrum during the quiescent state are plotted here for two cycles. It can be seen that during the bursts, most dramatic changes take place in the 3-13 keV band when the black body temperature increases to ~ 2.5 keV and the emission is from a relatively small region.

The spectral parameters measured with 3 s time resolution during the 1999 observation show some very interesting characteristics. During the low state and the transition between the two states the black body component has a low temperature and constitutes a small fraction of the observed luminosity. During the state changes, only the power-law component changes in shape and intensity. But, during the rapid flares in the high state, both the spectral components show related variations. The disk black-body component shows an increase in temperature and a decrease in radius whereas the power-law component becomes softer with an increase in intensity. Variations of the spectral parameters are shown on the left hand side of figure 2. Also shown on the right hand side of figure 2 are the two components of the energy spectra for the low state, transition state, and the peak of one burst.

3. DISCUSSION

Using a two component model for the energy spectrum, we have found interesting patterns of variations in the spectral parameters in GRS 1915+105 during different types of repetitive intensity variations. These changes can be identified with the

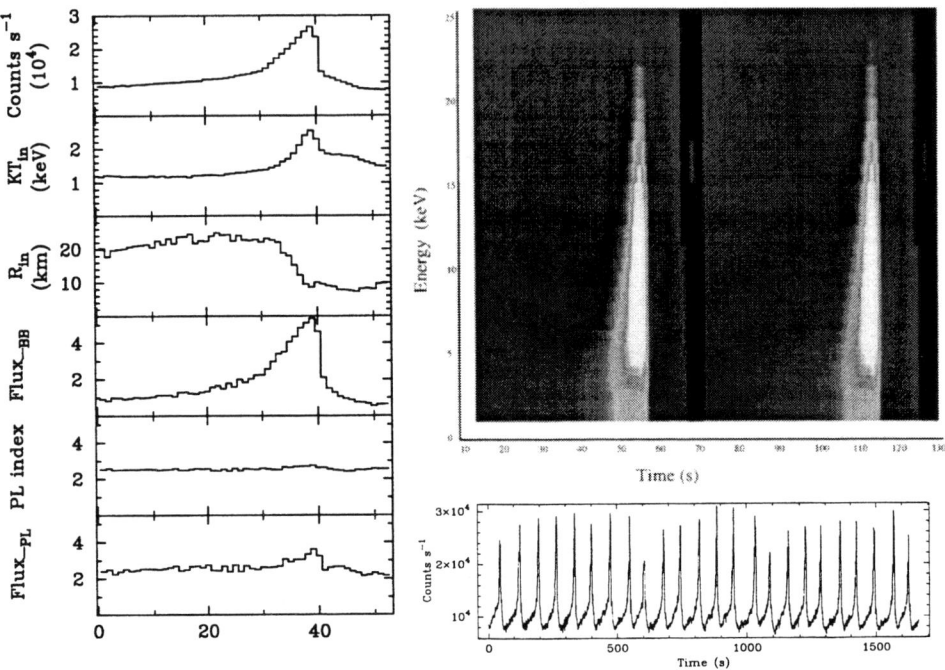

FIGURE 1. Evolution of the spectral parameters during the regular flares, derived from multi band light curve obtained with the PCA onboard RXTE are shown on the left. The flux values are in unit of 10^{-8} erg cm^{-2} s^{-1}. Evolution of the spectrum during the flares is shown on the right in grey scale. Ratio of the time resolved spectra and the spectrum during the quiescent period are used to create this diagram. The light curve is shown in the bottom-right corner.

changes in the accretion disk structure as the source goes through different long and short time scale variations.

During the regular flares which have slow rise and sharp decay, the power-law component shows little variation but the blackbody component shows increase in temperature and decrease in radius. This supports the scenario that the inner accretion disk periodically disappears into the event horizon giving rise to the flares and an oscillation in the outer disk provides matter to refill the inner disk in a 10–50 s time scale. There is a sharp discontinuity in the derived parameters near the flare maximum. This is probably due to the movement of accreted matter across some discontinuity like the event horizon or a shock region.

The spectral changes associated with the alternate low and high states and with the flares in the high states suggest that the long period variations are similar to the spectral state changes in black hole sources. The short flares have spectral

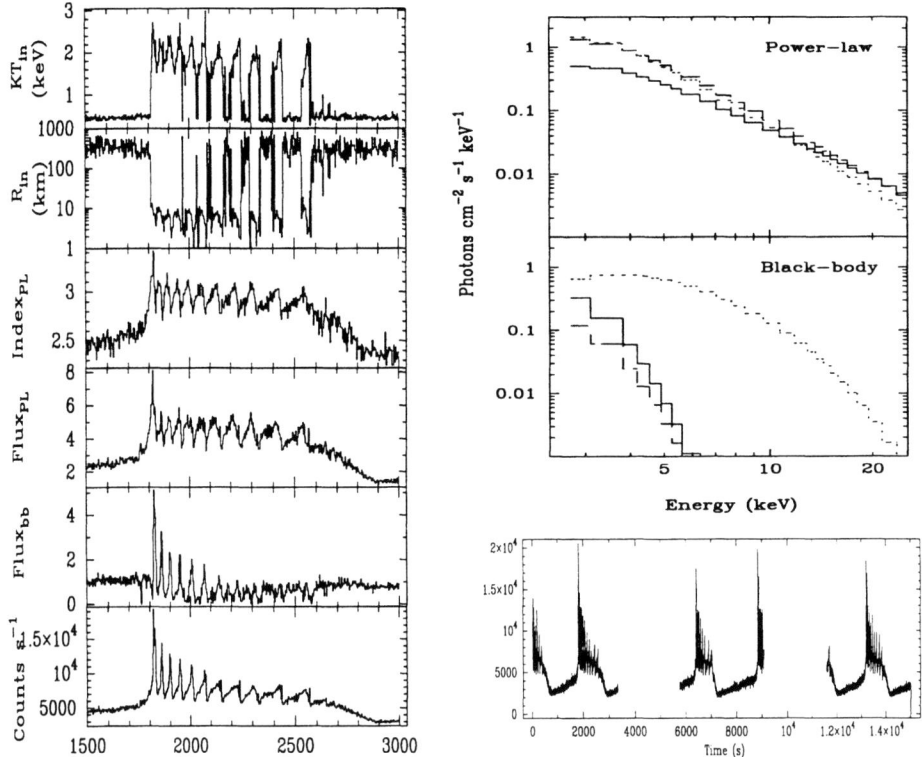

FIGURE 2. Evolution of the spectral parameters during the 1999 observation are shown on the left. The components of the energy spectra are shown for the low state, transition state, and the peak of one burst in solid, dashed and dotted lines respectively. The light curve is shown in the bottom-right corner.

characteristics which are very similar to the regular flares. The short and intense flares may arise due to the disappearance of the inner disk which takes place only in the high state of the source.

REFERENCES

Belloni, T., et al., 1997, ApJ, 488, L109
Morgan, E. H., et al., 1997, ApJ, 482, 993
Paul, B., et al., 1998, ApJ, 492, L63
Rao, A. R., et al., 2000, ApJ, 544 (in press)
Yadav, J. S., et al., 1999, ApJ, 517, 935

DETECTION OF A PULSATING SOFT COMPONENT IN THE X-RAY PULSAR XTE J0111.2–7317

B. Paul[1,5], J. Yokogawa[2], M. Ozaki[1], F. Nagase[1], D. Chakrabarty[3], and T. Takeshima[4]

1) ISAS, 3-1-1 Yoshinodai, Sagamihara, Kanagawa 229-8510, Japan
2) Department of Physics, Kyoto University, Sakyo-ku, Kyoto 606-8502, Japan
3) Dept. of physics and Center for Space Research, MIT, Cambridge, MA 02139, USA
4) LHEA, NASA, Goddard Space Flight Center, Greenbelt, MD 20771
5) TIFR, Homi Bhabha Road, Mumbai 400005, India

ABSTRACT The newly discovered transient X-ray pulsar XTE J0111.2–7317 was observed with *ASCA* in its bright phase. The pulsation was found to be nearly sinusoidal with some energy dependence (17%, 21% and 30% pulse fractions in 0.5–1.0, 1.0–2.0 and 2.0–10.0 keV bands respectively). In addition to a hard power-law component in the 2–10 keV band, the energy spectrum has a soft excess in the 0.5–2 keV band. Thermal models for the soft excess require very large emission zone for the soft component and can be ruled out based on the pulsations observed at low energies. The low energy excess can however be accommodated if the incident spectrum is modeled as an inversely broken power-law with line of sight absorption. We have also detected an iron K_α emission line with equivalent width of 50 ± 14 eV.

KEYWORDS: binaries: general — pulsars: individual (XTE J0111.2–7317) – X-rays: stars

1. INTRODUCTION

The transient X-ray pulsar XTE J0111.2–7317 was discovered with the Proportional Counter Array (PCA) of the *Rossi X-ray Timing Explorer (RXTE)* in 1998 November (Chakrabarty et al. 1998a) and was simultaneously detected in hard X-rays (Wilson & Finger 1998) with the Burst and Transient Source Experiment (BATSE) onboard the *Compton Gamma Ray Observatory (CGRO)*. Initial BATSE observations also found the pulsar to be spinning-up with a very short time scale of $\left(\frac{P}{\dot{P}}\right) \sim$ 20 yr. The pulsed X-ray luminosity in the hard X-ray band of BATSE is also high, and shows a positive correlation with the spin-up rate. Its association with the Small Magellanic Cloud indicates a large distance and very high luminosity, in excess of 10^{38} erg s^{-1} in the 2.0–10.0 keV band during the transient phase. A B type star with strong emission lines has been detected in the X-ray error box and this is likely to be the optical counterpart of this object (Israel et al. 1999). Following its discovery, a *Target of Opportunity* observation was made during 1998 November 18–19 with the two imaging spectroscopic instruments SIS and GIS onboard the *Advanced Satellite for Cosmology and Astrophysics (ASCA)* and the position of the X-ray source was determined precisely (Chakrabarty et al. 1998b). The *ASCA* observation of the source was made to study the pulsations and the soft X-ray spectrum extending upto 10 keV.

2. TIMING AND SPECTRAL ANALYSIS

To study the temporal properties, we used light curves from the two GIS detectors with 0.5 s time resolution. The arrival times of the photons were corrected to the solar system barycenter. Data from the two detectors were added and a power spectrum was generated. Pulsations were unambiguously detected with a period of 30.9497 ± 0.0004 s. The pulse profile is predominantly sinusoidal with indication of one small additional peak. To see the pulsations at low energy we have also generated pulse profiles in three different energy bands (0.5–1.0, 1.0–2.0 and 2.0–10.0 keV) from the barycentered SIS light curve. The normalized pulse profiles obtained from the SIS light curves in different energy bands are plotted in the left panel of figure 1. Slight difference in shape and larger modulation at higher energy are evident from these pulse profiles. A smaller pulse fraction at lower energy suggests spectral hardening during the pulse peak. The pulse fraction, defined as the ratio of the pulsed emission to the total emission was found to be 17%, 21% and 30% in the three energy bands of 0.5–1.0, 1.0–2.0 and 2.0–10.0 keV respectively. The pulse fraction is found to be nearly constant ($\sim$30%) in the entire 2.0–10.0 keV range.

The energy ranges chosen for spectral fitting are 0.55–10.0 keV for the SIS and 0.7–10.0 keV for the GIS. A simple power-law model with neutral absorber was found to be inadequate to represent the spectrum well, as residuals to the model fit showed the presence of a soft excess and an indication of an emission line around 6.4 keV. A black-body component and an emission line component were added to the model. These improved the fit considerably with power-law photon index 0.77 ± 0.02 and black-body temperature 0.148 ± 0.004 keV. If the soft component is assumed to be isotropic black-body emission from a spherical surface, the corresponding radius is estimated to be > 800 km, which is much larger than the neutron star surface area. The soft component could also be fitted with a thermal bremsstrahlung model of temperature 0.25 keV, but the emission measure is very high ($> 10^{62}$ cm^{-3}). In these spectral models in which a thermal component is assumed for the soft excess, the power-law contributes to only about 30% of the emission below 1.0 keV. Though the pulse fraction is smaller at lower energies, a pulsating nature of the soft component is evident. But, if the soft component is of thermal nature, as it requires a very large emission region the pulsations of this component remains unexplained. We therefore reject a thermal nature for the soft excess.

We found that the soft excess could be satisfactorily explained if we model the entire spectrum as an inversely broken power-law with iron K$_\alpha$ emission line at 6.4 keV and line of sight absorption. For details about the spectral analysis including pulse phase resolved spectroscopy please refer to Yokogawa et al. (1999).

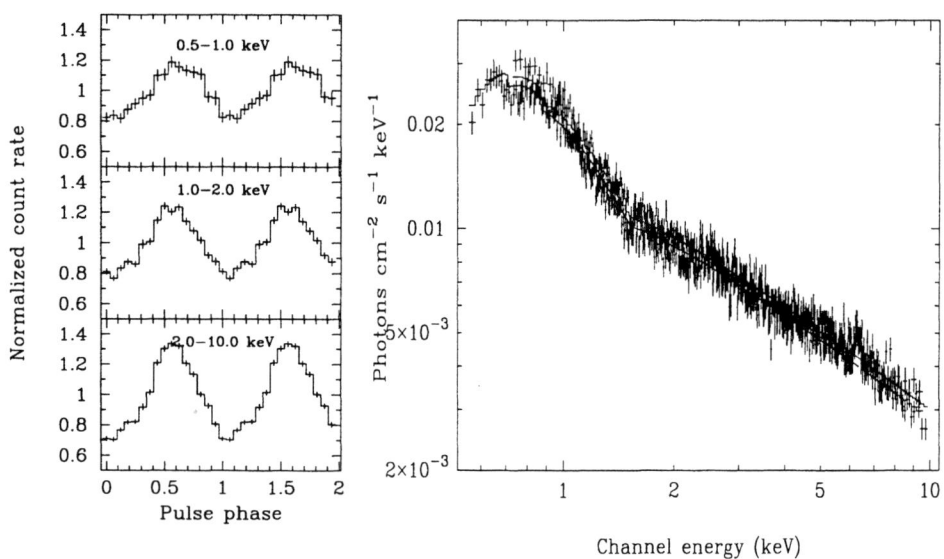

FIGURE 1. Background subtracted pulse profiles of XTE J0111.2-7317 in different energy bands obtained from the SIS light curves are shown on the left panel. The best fitted incident spectrum and the observed spectra (SIS and GIS) deconvolved through the detector response matrices are shown on the right.

The best fitted incident spectrum is shown on the right hand side of figure 1 along with the observed SIS and GIS spectra deconvolved through the detector response matrices. The observed X-ray flux in the 0.7–10.0 keV band is 3.6×10^{-10} erg s^{-1} cm^{-2}, which for an assumed distance of 65 kpc for this pulsar in the SMC and isotropic emission indicates an intrinsic luminosity of 1.8×10^{38} erg s^{-1}. The model spectrum described above has the following form

$$N(E) = IE^{-\Gamma_1}e^{-\sigma N_H} \text{ for } E < E_c \text{ and}$$

$$N(E) = IE_c^{(\Gamma_2-\Gamma_1)}E^{-\Gamma_2}e^{-\sigma N_H} \text{ for } E > E_c$$

where I is a normalization factor in units of photons cm^{-2} s^{-1} keV^{-1} at 1 keV.

The parameters for the inversely broken power-law model are :

$$N_H = 1.8 \pm 0.3 \times 10^{21} \text{ atoms cm}^{-2},$$
$$\Gamma_1 = 2.3 \pm 0.2,$$
$$E_c = 1.53 \pm 0.02 \text{ keV, and}$$
$$\Gamma_2 = 0.76 \pm 0.02.$$

The equivalent width of the iron emission line at 6.4 keV is 50 ± 14 eV.

3. DISCUSSION

The binary X-ray pulsars which are away from the galactic plane and therefore experience less interstellar absorption, show the presence of a soft component in the spectrum which is often described as a black-body and/or thermal bremsstrahlung emission (SMC X-1, Woo et al. 1995, Wojdowski et al. 1998; LMC X-4, Woo et al. 1996; 4U 1626-67, Orlandini et al. 1998). A thermal component for the soft part of the spectrum of 4U 1626-67 requires the size of emission region to be comparable to that of the neutron star, because the intrinsic luminosity of this source is of the order of 10^{35} erg s^{-1}. But for the pulsars in the Magellanic Clouds for which the distance is of the order of 60 kpc (and the luminosity is close to the Eddington value), a soft component dominating in the lower energy part of the spectrum will require an emission region which is a few orders of magnitude larger than the size of the neutron star. The bremsstrahlung component in LMC X-4, which is dominant in the intermediate energy range of 0.5-1.5 keV is also found to be pulsating (Woo et al. 1996). The pulsating nature of the soft component in XTE J0111.2-7317 and LMC X-4, as well as for SMC X-1 is difficult to explain if a thermal origin is assumed for the low energy part of the spectrum. A two component power-law with different absorptions can instead explain the pulsations at low energy. A soft power-law component may also be a common feature of the binary X-ray pulsars, which is difficult to observe because most of the sources are in the galactic plane and experience large interstellar absorption.

The exact nature of this system is unknown but a possible optical counterpart has been identified which is a Be-type star (Israel et al. 1999). The characteristics of this pulsar, like its transient nature, hard X-ray spectrum and high X-ray luminosity are analogous to the pulsars in binary systems with high mass companion. A pulse period of 31 s is also much larger than the period of X-ray pulsars in low mass binaries. The rapid spin-up property during the transient phase is similar to the X-ray pulsars with high mass type companions (2S 1417-624 etc., Bildsten et al. 1997). The present outburst is analogous to the giant outbursts seen in Be X-ray pulsars characterized by high luminosity and high spin-up rates.

REFERENCES

Bildsten, L., et al. 1997, ApJS, 113, 367
Chakrabarty, D., et al. 1998a, IAU Circ., No. 7048
Chakrabarty, D., et al. 1998b, IAU Circ., No. 7062
Israel, G. L., et al. 1999, IAU Circ., No. 7101
Orlandini, M., et al. 1998, ApJ, 500, L163
Wilson, C. A., & Finger, M. H. 1998, IAU Circ., No. 7048
Wojdowski, P., et al. 1998, ApJ, 502, 253
Woo, J. W., et al. 1995, ApJ, 445, 896
Woo, J. W., et al. 1996, ApJ, 467, 811
Yokogawa, J., et al. 1999, ApJ (submitted)

BEPPOSAX DETECTION OF HIGHLY IONIZED EMISSION AND ABSORPTION FEATURES IN M81

S. Pellegrini [1], M. Cappi [2], L. Bassani [2], G. Malaguti [2],
G. G. C. Palumbo [1], M. Persic [3]

1) Dipartimento di Astronomia, Università di Bologna, via Ranzani 1, I-40127 Bologna
2) Istituto TeSRE/CNR, via Gobetti 101, I-40129 Bologna
3) Osservatorio Astronomico di Trieste, via G.B. Tiepolo 11, I-34131 Trieste

ABSTRACT The LINER nucleus of the nearby spiral galaxy M81 was pointed by $BeppoSAX$, which caught it at the highest (2–10) keV flux level observed so far. The LECS, MECS and PDS data (extending over 0.1–100 keV) are used to investigate the similarities and differences between LINERs and AGNs. The continuum is well fitted by a power law of photon index $\Gamma \sim 1.84$, modified by little absorption due to cold material; signs of reflection from the optically thick material of an accretion disk are not observed. Instead, $BeppoSAX$ detects a 6.7 keV emission line (confirming an $ASCA$ result), and an absorption edge at ~ 8.6 keV. Both spectral features are consistent with being produced by iron at the same high ionization level, and possibly also with the same column density. We suggest that they come from transmission through a highly photoionized medium. The origin of the continuum emission is uncertain, since typical signatures of a standard accretion disk are absent, and the steep spectrum observed over (0.1–100) keV requires Compton scattering to dominate the X-ray emission even in a modeling via advection dominated accretion.

KEYWORDS: Galaxies: active; Galaxies: individual: M81; Galaxies: nuclei; Galaxies: Seyfert; X-rays: galaxies.

1. INTRODUCTION

M81 is a nearby spiral galaxy with a prominent bulge and well defined spiral arms, that has been well studied at all frequencies, from radio to optical and UV (e.g., Ho et al. 1997), to X-rays. At these frequencies it was first observed with $Einstein$, then by $GINGA$, $BBXRT$, and $ROSAT$ (see Pellegrini et al. 1999 for more details). $ASCA$ pointed M81 several times between 1993 and 1999 (Ishisaki et al. 1996, Iyomoto 1999; see Fig. 1). The average spectrum was well fitted by a power law continuum of photon index $\Gamma \sim 1.85$, absorbed by a column density of $N_H \sim 10^{21}$ cm^{-2}, plus a thermal component with a temperature of $kT \sim 0.6 - 0.8$ keV. An apparently broad or complex iron emission line centered at 6.6–6.9 keV, with an equivalent width of 170^{+60}_{-60} eV, was also detected.

On the basis of these observations, M81 turned out to be the closest galaxy to

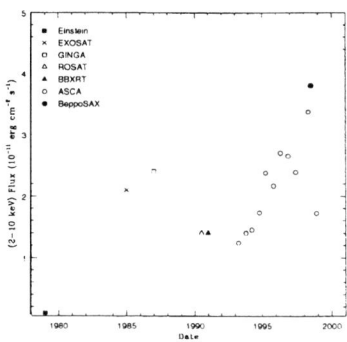

FIGURE 1. Long term variation of the (2-10) keV flux from the nucleus of M81. *ASCA* fluxes have been kindly provided by N. Iyomoto (1999), and that from *EXOSAT* data by P. Barr.

show the spectroscopic signatures of a LINER (Ho et al. 1997) and to be a good candidate for a low luminosity AGN (LLAGN). Dynamical studies also suggest the presence of a super-massive object at the galaxy nucleus, of mass $4 \times 10^6 M_\odot$ (Ho 1999). Here we report the results of an analysis of the nucleus of M81 over 0.1-100 keV, and discuss them in relation to the unsolved issue of sorting out the physical similarities and differences between LINERs and AGNs.

2. RESULTS

2.1. Variability

The long term variation of the M81 (2-10) keV flux, obtained by collecting observations from various observatories, is plotted in Fig. 1. During the *BeppoSAX* pointing the nucleus was at the highest level ever observed.

A short term variation is clearly detected by *BeppoSAX* (Fig. 2). The MECS (2-10) keV count rate varies by $\sim 30\%$ from peak to valley over roughly one day. The fit of the light curve with the sine function is significantly better than that with a constant value; the period turns out to be almost 2 days (43 hours). Future observations might reveal whether the periodicity is real. The *ASCA* short term light curve showed variability of the order of $\sim 20\%$ on a timescale of ≤ 1 day (Serlemitsos et al. 1996).

2.2. Spectral Analysis

A single power law component of photon index $\Gamma \sim 1.85$ can reproduce the data over the whole energy band (Fig. 3). As can be seen looking at the residuals, excess emission is present between 6-7 keV. In fact, the presence of an emission line is statistically significant at $> 99.9\%$ confidence level. The line is centered at $6.70^{+0.11}_{-0.11}$ keV, and its equivalent width is EW=104^{+29}_{-40} eV. The 6.4 keV value for the line center energy is excluded at $> 99\%$ confidence level.

There is another spectral region, located around 9 keV, that shows significant residuals when fitting with a simple power law. The fit improves by adding an absorption edge, that turns out to be located at $E = 8.6^{+0.4}_{-0.8}$ keV, with optical

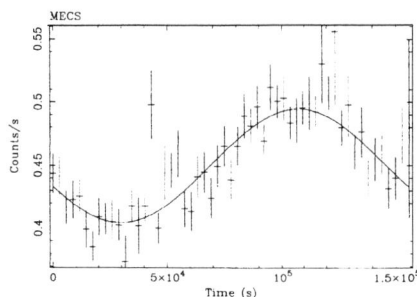

FIGURE 2. MECS light curve; counts are from 3′ radius, and 2-10 keV. The sine function that best fits the data is also plotted.

depth $\tau = 0.15^{+0.09}_{-0.08}$. The presence of one edge is statistically significant at 99% confidence level, and we cannot exclude that the 'hole' around 9 keV could be produced by more than one ionization stage.

3. DISCUSSION

3.1. Origin of the X-ray continuum

The power law component that well reproduces the (0.1-100) keV spectrum is usually related to the presence of an AGN, and the value of Γ is close to that found for Seyfert 1's (Nandra & Pounds 1994). Moreover, no absorption much in excess of the Galactic value is seen in the X-ray spectrum. All this makes M81 similar to Seyfert 1's. The observed bolometric luminosity of the nucleus of M81 corresponds to a low accretion rate. Adopting the central mass value of $4 \times 10^6 M_\odot$, the bolometric luminosity of the nucleus obtained by integrating the observed spectral energy distribution from the radio to 10 keV ($L = 2.1 \times 10^{41}$ erg s^{-1}, Ho 1999), and a radiative efficiency of 0.1, it turns out that $\dot{m} = \dot{M}/\dot{M}_{Edd} \sim 4 \times 10^{-4}$. A problem with this scenario is that we do not detect signs of reflection from optically thick cold material, usually found in Seyfert 1's, and attributed to the presence of an accretion disk (a 6.4 keV emission line with typical EW of 100-150 eV, and a broad bump peaking at 10-20 keV, Nandra & Pounds 1994). As a further comparison with the continuum of classical, more luminous AGNs, also the 'big blue bump', traditionally attributed to thermal radiation from an accretion disk, is absent in the spectral energy distribution observed for the nucleus of M81 (Ho 1999). So, we conclude that this LLAGN is not a simple extension of high luminosity ones.

An alternative possibility for the origin of the continuum emission could be the presence of an advection dominated accretion flow (ADAF, Narayan & Yi 1995), since the conditions for an ADAF are satisfied (low L_{bol}/L_{Edd}). This modeling could reproduce the slope of the continuum over (0.1-100) keV by assuming Compton scattering of synchrotron photons from the thermal electrons to produce the X-ray emission (Quataert et al., astro-ph/9909193).

3.2. Origin of the iron-K emission and absorption

The energy of the line indicates Kα emission from He-like iron (FeXXV); the absorption edge can be produced by ions from FeXVII to FeXXV, within the 90% con-

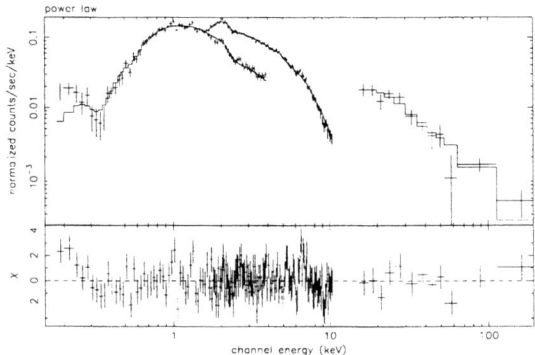

FIGURE 3. *BeppoSAX* LECS, MECS and PDS observed spectra of M81 (crosses), modeled with a power law of $\Gamma = 1.85$ (solid line). The residuals, in terms of σ's, between the data and the model are plotted below.

fidence interval, and its best fit energy corresponds to the FeXXIV K-edge (~ 8.6 keV; Makishima 1985). Both features therefore come from highly ionized material. Where is this highly ionized material located? In the case of reflection from a disk, its ionization state should be higher than in Seyfert 1's, because the iron line energy is higher (6.7 keV instead of 6.4 keV). But a reflection continuum from an ionized disk is absent in the *BeppoSAX* data; moreover, a fit with ionized reflection is of poor quality because it cannot reproduce a sharp absorption edge at 8.6 keV as observed. A physical problem is also that the observed accretion rate is too low for the required ionization level in the disk (Pellegrini et al. 1999).

We suggest that line and edge are produced by transmission through a warm highly photoionized medium, located close to the nucleus (Pellegrini et al. 1999). This points out an interesting similarity with the presence of a warm absorber in Seyfert 1's (Nandra & Pounds 1994).

ACKNOWLEDGEMENTS

We are grateful to G. Fabbiano, N. Iyomoto, H. Netzer, F. Nicastro and T. Di Matteo for discussions. ASI and MURST (contract CoFin98) are acknowledged for financial support.

REFERENCES

Ho, L.C. 1999, ApJ, 516, 672
Ho, L. C., Filippenko, A. V., Sargent, W. L. W., Peng, C. Y. 1997, ApJS, 112, 315
Ishisaki, Y., et al. 1996, PASP, 48, 237
Iyomoto, N. 1999, PhD Thesis, University of Tokyo
Makishima, K. 1985, in "The Physics of Accretion onto Compact Objects", eds. K.O. Mason, M.G. Watson and N.E. White. (Berlin: Springer-Verlag), p.249
Nandra, K., Pounds, K.A. 1994, MNRAS, 268, 405
Narayan, R., Yi, I. 1995, ApJ, 452, 710
Pellegrini, S., et al. 1999, submitted to A&A
Serlemitsos, P., Ptak, A., Yakoob, T. 1996, in The Physics of Liners, ASP Conf. Ser., eds. M. Eracleous, A. Koratkar, C. Leitherer, and L. Ho, (Baltimore: STScI), p.70

THE BEPPOSAX GRATIS SURVEY

M. Perri[1], P. Giommi[1], F. Fiore[1,2], M. Capalbi[1]

1) BeppoSAX Science Data Center, ASI, Via Corcolle 19, I-00131 Rome, Italy
2) Osservatorio Astronomico di Roma, Via Frascati 33, I-00044 Monteporzio, Italy

ABSTRACT We present the preliminary results of the BeppoSAX GRATIS Survey, an ongoing survey in the 2-10 keV energy band based on short exposures of "random" parts of the sky. At present GRATIS covers about 20 deg^2 of the high galactic latitude sky with a flux limiting sensitivity of 3-4 10^{-13} erg s^{-1} cm^{-2}. GRATIS source counts are in good agreement with those of ASCA and BeppoSAX deeper surveys and provide an unbiased estimation of the bright end of the 2-10 keV LogN-LogS relationship. A BeppoSAX follow-up observation of one of the GRATIS X-Ray sources shows evidence of a strongly absorbed spectrum, indicating that it could be a heavily absorbed (type 2) AGN, as predicted by population synthesis models for the X-Ray Background.

1. INTRODUCTION

It is now widely accepted that the cosmic X-Ray Background (XRB), discovered more than 35 years ago (Giacconi et al. 1962), is the result of the superposition of faint discrete sources. In the *soft* X-ray energy band (0.5-2 keV) the most sensitive survey performed to date is the ROSAT Deep Survey in the "Lockman Hole" (Hasinger et al. 1993, 1998). This survey has resolved most (70-80%) of the soft XRB, reaching a source density of $\sim$ 1000 deg^{-2} at a limiting flux of $\simeq 10^{-15}$ erg s^{-1} cm^{-2}. Spectroscopic identification of optical counterparts of these faint soft X-ray sources (Schmidt et al. 1998) have revealed that the large majority (> 80%) are AGNs, mostly with broad emission lines (QSOs and Seyfert 1 galaxies).

In contrast, our knowledge of the nature of the sources which produce the *hard* (2-50 keV) XRB, where the energy density of the XRB peaks, is quite poor compared to our understanding of the softer sky. Only in the last few years, thanks to the imaging instruments on board ASCA and *Beppo*SAX, the faint hard (up to 10 keV) X-ray sky has become accessible to deep surveys (Georgantopoulos et al. 1997; Cagnoni et al. 1998; Giommi et al. 1998; Ueda et al. 1999; Giommi et al. 1999; Fiore et al. 1999). At present, ASCA and *Beppo*SAX hard X-ray surveys have resolved about 30% of the 2-10 keV XRB into discrete sources and found a source density of 20-30 deg^{-2} at a flux limit of $\simeq 5 \ 10^{-14}$ erg s^{-1} cm^{-2}.

Despite these improvements, little is still known about the *nature* of these hard sources: optical identification programs have just started, but due to the large ASCA and, to a lower extent, *Beppo*SAX position uncertainties firm results are expected to be achieved in a relatively long time. However, first indications clearly suggest

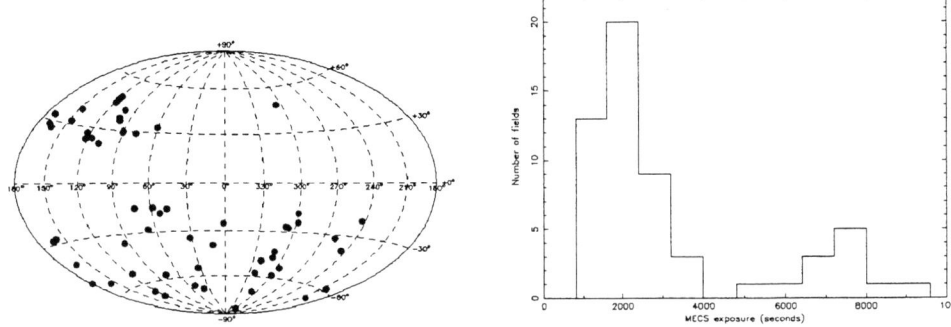

FIGURE 1. a) the initial set of 58 intermediate pointings of the GRATIS Survey plotted in galactic coordinates. b) MECS exposures distribution of this initial sample of intermediate pointings.

that most of these hard X-ray sources are AGNs (Ueda et al. 1999; Fiore et al. 1999). Moreover, none of the single class of known X-ray sources is characterized by an energy spectrum similar to that of the XRB (this problem is known as the *spectral paradox*): bright AGNs have a "canonical" 2-10 keV energy index $\alpha \simeq 0.7$, while the XRB have a significantly softer slope ($\alpha \simeq 0.4$).

This fact have been interpreted as evidence for a population of cosmic sources which, undersampled in soft X-ray surveys, dominate the hard X-ray sky. This population could include heavily obscured (type 2) AGNs, with an high intrinsic neutral hydrogen column densities ($N_H \sim 10^{23}$ cm^{-2}), as predicted from several theoretical models (Setti & Woltjer 1989; Madau, Ghisellini & Fabian 1994; Comastri et al. 1995). However these heavily cutoff sources may be still detectable in the soft band because of the emerging of either non-nuclear components, or of reflected, or partially transmitted nuclear X-rays (Giommi, Fiore & Perri 1999).

In this paper we present the preliminary results of the *Beppo*SAX GRATIS Survey (GRand Area Target acquisition Intermediate pointings Survey), a shallow survey in the 2-10 keV energy band based on *Beppo*SAX intermediate pointings.

2. THE BEPPOSAX GRATIS SURVEY

Following the failure of four of the six gyros on board *Beppo*SAX (Boella et al. 1997a) a new procedure for the spacecraft pointing that makes use of only one gyro has been implemented. This has the consequence that during slews from one target to the following the spacecraft remains pointed to an intermediate "blank" part of the sky with the MECS instruments (two identical X-Ray telescopes on board the *Beppo*SAX satellite operating in the 1.3-10.5 keV band; Boella et al. 1997b) producing useful scientific data not foreseen in the nominal mission.

We have started a systematic analysis of all MECS intermediate pointings and

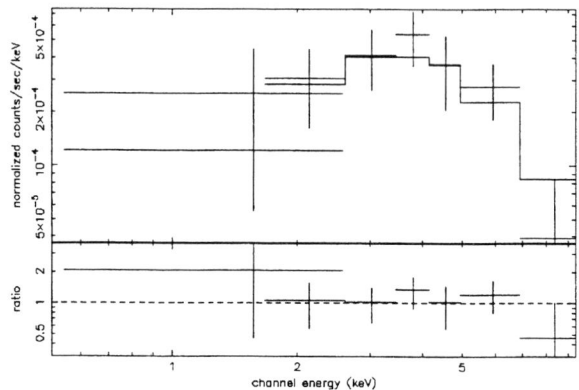

FIGURE 2. LECS + MECS 0.5-10 keV energy spectrum of 1SAXJ2335.4-5618 measured during the follow-up observation of June 1999. The best fit model assuming a power law spectrum with a fixed energy index $\alpha = 0.7$ plus an intrinsic absorption is also shown.

named this project the *Beppo*SAX GRATIS Survey. Figure 1a shows, in galactic coordinates, an initial set of 58 distinct pointings covering about 20 square degrees of the high ($|b| > 15$ deg) galactic latitude sky. Figure 1b shows the exposures distribution of this sample: typical exposures are $\sim$ 3000-4000 seconds setting the 2-10 keV flux sensitivity limit to about $3\ 10^{-13}$ erg cm^{-2} s^{-1}. All 58 MECS images have been analysed with the detect routine of the XIMAGE package (Giommi et al. 1991). Sources have been accepted according to the following criteria: 1) statistical probability that the source is a fluctuation of the local background $< 5\ 10^{-4}$ and 2) off-axis angle < 8.5 arcminutes or > 11 arcminutes to avoid problems connected to the window support structure. Count rates have been converted to 2-10 keV fluxes assuming a power law spectrum with energy index $\alpha = 0.6$ and corrected for the Galactic absorbing neutral hydrogen column density.

We detected 16 X-ray sources with 2-10 keV fluxes ranging from $3.6\ 10^{-13}$ erg cm^{-2} s^{-1} to $4.4\ 10^{-12}$ erg cm^{-2} s^{-1}. At GRATIS flux sensitivity limit we find $\sim$1 sources/deg^2, a value that is in good agreement with ASCA and *Beppo*SAX 2-10 keV LogN-LogS relationship (Ueda et al. 1999; Giommi et al. 1999).

3. BEPPOSAX FOLLOW-UP OBSERVATION OF 1SAXJ2335.4-5618

Cross-correlations with catalogs of cosmic sources have led to the identification of two of the GRATIS X-ray sources, one with a cluster of galaxies (Gioia & Luppino 1994) and another with a type 1.8 Seyfert galaxy (Pietsch et al. 1998), all the others are still unidentified. A *Beppo*SAX follow-up observation of one of these unidentified X-ray sources, 1SAXJ2335.4-5618, has been performed on June 1999. The measured 2-10 keV flux is $\simeq 2.5\ 10^{-13}$ erg cm^{-2} s^{-1}, about a factor 10 below the value observed during the discovery observation performed on June 1998. Figure 2 shows LECS+MECS 0.5-10 keV energy spectrum of 1SAXJ2335.4-5618 during the follow-up observation. The best fit model assuming a power law spectrum with a fixed energy index $\alpha = 0.7$ plus an intrinsic absorption is also shown. We find a best fit value of the amount of absorbing neutral hydrogen column densities $N_H = 2.89^{+3.50}_{-2.06}$ 10^{22} cm^{-2}, indicating a strongly absorbed spectrum.

4. CONCLUSIONS

We have presented the preliminary results of the *Beppo*SAX GRATIS Survey, an ongoing survey in the 2-10 keV energy band based on *Beppo*SAX intermediate pointings. Although its sensitivity limit is only moderate (S $\sim 3\ 10^{-13}$ erg s^{-1} cm^{-2}) the GRATIS survey has the important advantage of not suffering from incompleteness at fluxes comparable or higher than those of the targets, a bias that is hardly avoidable in serendipitous surveys. A *Beppo*SAX follow-up observation of 1SAXJ2335.4-5618, one of the GRATIS X-Ray sources, shows that it is a strongly variable object: the follow-up *Beppo*SAX pointing showed a factor of $\sim$10 variability compared with the discovery observation. Moreover the analysis of the available energy spectrum indicates that 1SAXJ2335.4-5618 shows a strongly absorbed spectrum ($N_H \sim 10^{22}$ cm^{-2}), and therefore it could be a heavily absorbed (type 2) AGN, as predicted by population synthesis models for the X-Ray Background.

REFERENCES

Boella, G. et al. 1997a, A&AS, 122, 299
Boella, G. et al. 1997b, A&AS, 122, 327
Cagnoni, I. et al. 1998, ApJ, 493, 54
Comastri, A., Setti, G., Zamorani, G. & Hasinger, G. 1995, A&A, 296, 1
Fiore, F., La Franca, F., Giommi, P. et al. 1999, MNRAS, 306, L55
Georgantopoulos, I. et al. 1997, MNRAS, 291, 203
Giacconi, R., Gursky, H., Paolini, F. & Rossi, B. 1962, Phys. Rev. Lett., 9, 439
Gioia, I.M. & Luppino, G.A. 1994, ApJS, 94, 583
Giommi, P. et al. 1998, Nuclear Physics B (proc. Suppl.) 69/1-3, 591
Giommi, P., Fiore, F. & Perri, M. 1999, Astrophys. Letters and Communications, Gordon & Breach, in press (astro-ph/9812305)
Giommi, P. et al. 1999, in preparation
Hasinger, G. et al. 1993, A&A, 275, 1
Hasinger, G. et al. 1998, A&A, 329, 482
Madau, P., Ghisellini, G. & Fabian, A. 1994, MNRAS, 270, L17
Mather, J.C. et al. 1994 ApJ, 429, 439
Pietsch, W. et al. 1998, A&A, 333, 48
Schmidt, M. et al. 1998, A&A, 329, 495
Setti, G., Woltjer, L. 1989, A&A, 224, L21
Ueda, Y. et al. 1999, ApJ, in press (astro-ph/9901101)

PAIR CREATION AT SHOCKS: A POSSIBLE ORIGIN OF THE HIGH ENERGY VARIABILITY OF AGNS

P.O. Petrucci [1], G. Henri [2], G. Pelletier [2]

1) Osservatorio Astronomico di Brera, via Brera 28, 20121 Milano, Italy, 2) Laboratoire d'Astrophysique, Observatoire de Grenoble, B.P 53X, F38041 Grenoble Cedex, France

ABSTRACT   We investigate the effect of pair creation, via high energy photon-photon interactions, on a shock structure. The high energy photons are produced via Inverse Compton process by particles accelerated by the shock itself. The increase of the associated pair pressure may thus be able to modify the plasma flow and we show that there exists a pair pressure upper-limit above which the shock even disappears. Such processes could be the possible origin of the AGNs high energy variability.

KEYWORDS: acceleration of particles; shock waves; galaxies: active

1. INTRODUCTION

It is well-known that particles can be accelerated to high energy by crossing a magnetized shock front many times and thus experiencing the so-called first order Fermi process (Bell, 1978; Blandford & Ostriker, 1978). The simple linear theory may be however insufficient to decribe a realistic situation because the accelerated particles may induce a strong non-linear back reaction on the shock itself. For instance, particles accelerated by the shock can be sufficiently energetic to boost, via Inverse Compton process (IC) for example, surrounding soft photons above the rest mass electron energy and thus to trigger the pair creation process. The increase of the associated pair pressure is thus able to disrupt the plasma flow and possibly, for too high pressure to smooth it completely. Reversely, significant changes of the flow velocity profile may modify the distribution function of the accelerated particles, modifying consequently the pair creation rate. The aim of this paper is to investigate, using simplifying assumptions, the effect of pair creation on the shock structure (see Petrucci et al., 2000, hereafter P00, for a complete discussion)

2. MAIN HYPOTHESES

- We assume a 1D geometry and we suppose the shock to be located in $x = 0$ This will be insured by imposing that the flow velocity possesses an inflection point in $x = 0$ that is (noting $u(x)$ the flow velocity): $\left.\frac{\partial^2 u}{\partial x^2}\right|_{x=0} = 0$ \quad (1)

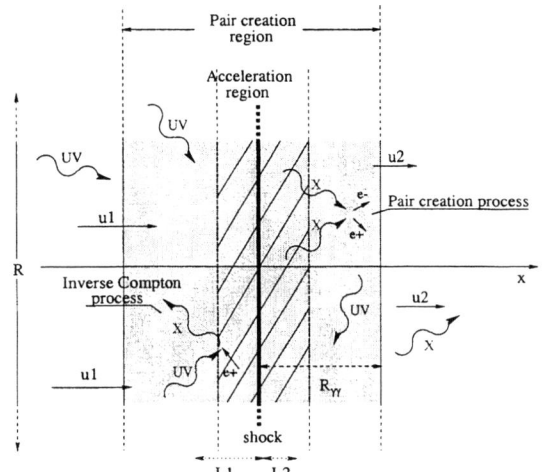

FIGURE 1. Schematic view of a shock. The shock discontinuity is represented by the vertical bold line. We have also indicated the different parameters defining the acceleration and pair creation region (cf. section 3). Particles are represented by straight arrows and photons by warped ones. Scales are not respected.

- We assume the shock region to be embedded in a isotropic external soft photon field and that particle cool by inverse Compton process on these soft photons, producing X-rays that may annihilate to produce pairs.

- We suppose the spatial diffusion coefficient D to be independent of x and of the energy of the particles.

- We assume the plasma pressure P_{tot} to be dominated by the pressure P_{rel} of the relativistic leptons population.

- Only particles having a Larmor radius comparable to the wavelength of the Alfven spectrum will undergo scatterings. Consequently there exists a lower Lorentz factor γ_{min} for a relativistic lepton to be accelerated in a shock. We thus suppose that pre-accelerator processes exist (like magnetic reconnection, whistler) to bring particles above γ_{min}. Consequently, since particles annihilate preferentially for $\gamma \simeq 1$, the annihilation process can be neglected.

3. GEOMETRY OF THE MODEL

3.1. The acceleration region

A particle of Lorentz factor γ will interact with the shock if it is located within about one diffusion length $L_{D,i}(\gamma) = D/u_i$ (i=1,2 for the up and downstream flow respectively) from the shock. On the other hand it will be cooled (via radiative processes) on a cooling length scale $L_{cool}(\gamma) \propto \gamma^{-1}$. There thus exists a Lorentz factor γ_c for which $L_{D,i}(\gamma_c) = L_{cool}(\gamma_c) = L_i$. We will define the acceleration region as the physical space $-L_1 \leq x \leq L_2$ (hatched region in cf. Fig. 1). Consequently, in this region the coolings are negligible for particles with $\gamma < \gamma_c$ (the majority).

3.2. The pair creation region

The high energy photon density will suffer from photon-photon depletion and geometrical dilution, the latter becoming important at a distant $x > R$ (R being of the order of the shock transverse size, cf. Fig. 1). We define the pair creation region as the region lying between $-R_{\gamma\gamma}$ and $+R_{\gamma\gamma}$, $R_{\gamma\gamma}$ being the minimum between R and the distance l_0 corresponding to a pair creation optical depth $\tau_{\gamma\gamma}$ of 1 (for a photon energy $\epsilon = m_e c^2$). In the case of AGNs (i.e. for high compactness of the order of a few hundreds) $R_{\gamma\gamma} > L_i$ for $i = 1, 2$ (cf. P00).

4. BASIC EQUATIONS

We suppose the existence, in the vicinity of the shock, of a magnetic field B, slightly perturbed by Alfven waves. The particles are thus scattered by these waves through pitch angle scattering (Jokipii, 1976; Lacombe, 1977) and gain energy through the well known first order Fermi process. We suppose the magnetic pertubations to have sufficiently small amplitudes to treat the problem in quasilinear theory using the Fokker-Planck formalism. From the evolution equation followed by the particle distribution function (coolings and particle annihilation being negligible in the acceleration region, cf. sections 2 and 3) and the momentum conservation equation, we can show that the flow velocity $u(x)$ must verify the following equation (cf. P00):

$$\frac{\partial \widetilde{u}}{\partial \widetilde{x}} = \frac{7}{6}(1 - \widetilde{u})\left(\frac{1}{7} - \widetilde{u}\right) + A\widetilde{x} + A\widetilde{R}_{\gamma\gamma} \qquad (2)$$

where $\widetilde{u} = u/u_1$, $\widetilde{x} = x/L_1$, $\widetilde{R}_{\gamma\gamma} = R_{\gamma\gamma}/L_1$ and $A = \frac{\dot{P}_\pm L_1}{\rho u_1^3}$ with $\dot{P}_\pm$ the pair pressure creation rate (supposed to be constant in the pair creation region and null outside).

5. SHOCK DISAPPEARANCE

5.1. Without hydrodynamical feedbacks

We firstly consider the feedback of the pair processes on the hydrodynamics of the shocked flow and not the reverse, i.e. the pair pressure creation rate A is considered as a constant free parameter. The shock still exists as long as Eq. (1) is verified. It appears that, for a given value of $R_{\gamma\gamma}$, there exists a limit value $\dot{P}_\pm^{lim}$ of $\dot{P}_\pm$ above which Eq. (1) has no consistent solutions anymore and the shock disappears. We show that $\dot{P}_\pm^{lim}$ is necessarily smaller than $\simeq 0.2$ meaning that at most 20% of the kinetic power of the upstream flow can be transformed in pairs. Above this value the shock cannot exist anymore.

5.2. Stationary states

If the pair pressure modifies the flow velocity profile, reversely a change of the flow velocity profile can modify the distribution function of the accelerated particles, modifying consequently the pair creation rate. Stationary states are then obtained by solving self-consistently for the particle distribution function $n(\gamma)$ and the flow velocity profile $u(x)$, linked through the pair pressure creation rate. We have reported on Fig. 2 the contour plots of the spectral index and high energy cut-off of the high energy spectra produced by IC by the accelerated particles for stationary

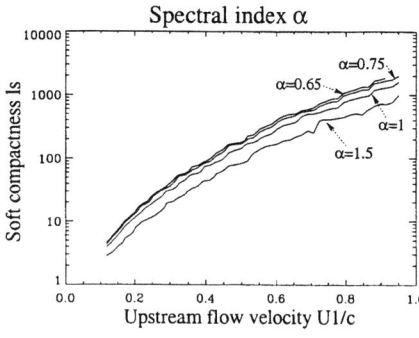 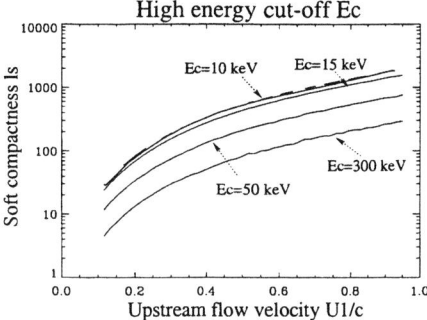

FIGURE 2. Contour plots in the $(l_s, u_1/c)$ space of the spectral index α and the high energy cut-off E_c of the high energy spectrum ($F_E \propto E^{-\alpha} \exp(-E/E_c)$) emitted by IC by the accelerated particles when the system is in stationary state. We assume a soft photon energy of 1 eV

states characterized by different values of the soft compactness and the upstream flow velocity. We also assume a soft photon energy of 1 eV.

6. VARIABILITY

As seen before, for pair pressure creation rate larger than $\dot{P}_{\pm}^{lim}$ the shock disappears. However, with the corresponding fall down of the acceleration processes, the rate of pair creation, and consequently the pair pressure creation rate, decreases and the shock may appear again, initiating pair processes until a new destruction of the shock. These cycles shock appearance/disappearance could thus be the origin of the high energy variability observed in AGNs (and perhaps in prompt γ-ray bursts where the compactness may be large enough).

7. CONCLUSION

We have presented the main results of the effect of pair creation, via high energy photon-photon interaction, on a shock structure taking also into account the hydrodynamical feedback on the pair process.

ACKNOWLEDGEMENTS

Work supported in part by the EC under contract number ERBFMRX-CT98-0195 (TMR network "Accretion onto black holes, compact stars and protostars")

REFERENCES

Bell, A. R. 1978, MNRAS, 182, 147
Blandford, R. & Ostriker, J. 1978, ApJL, 221, L29
Jokipii, J. R. 1976, ApJ, 208, 900
Lacombe, C. 1977, A&A, 54, 1
Petrucci et al., 2000, submitted (P00)

BEPPOSAX OBSERVATIONS OF MARKARIAN 501 IN JUNE 1999

E. Pian[1], L. Chiappetti[2], P. Giommi[4], F. Tavecchio[3], L. Maraschi[3], E. Palazzi[1], F. Aharonian[5], M. Catanese[6], A. Celotti[7], B. Degrange[8], A. Djannati-Atai[8], G. Fossati[9], G. Ghisellini[3], H. Krawczynski[5], C. M. Raiteri[10], R. M. Sambruna[11], D. Smith[12], G. Tagliaferri[3], G. Tosti[13], A. Treves[14], C. M. Urry[15], M. Villata[10]

[1] *ITESRE/CNR, Bologna, Italy,* [2] *IFCTR/CNR, Milan, Italy,* [3] *Astronomical Obs. of Brera, Milan, Italy,* [4] *SAX Science Data Center, Rome, Italy,* [5] *Max-Planck-Institut für Kernphysik, Heidelberg, Germany,* [6] *Dept. of Physics and Astronomy, Iowa State University, Iowa,* [7] *SISSA/ISAS, Trieste, Italy,* [8] *Astroparticle Group, PCC - College de France, Paris, France,* [9] *CASS/UCSD, La Jolla, California,* [10] *Astronomical Obs. of Torino, Pino Torinese, Italy,* [11] *Penn State University, Pennsylvania,* [12] *Centre d'Etudes Nucleaires de Bordeaux-Gradignan, France,* [13] *Astronomical Obs., Univ. of Perugia, Perugia, Italy,* [14] *Dept. of Physics, Univ. of Insubria, Como, Italy,* [15] *Space Telescope Science Institute, Baltimore, Maryland*

ABSTRACT We present the preliminary results of a long BeppoSAX observation of the BL Lac object Mkn501 carried out in June 1999. The source was fainter than found during the BeppoSAX pointings of 1997 and 1998, but is still detected with a good signal-to-noise ratio up to ∼40 keV. The X-ray spectrum in the energy range 0.1-40 keV, produced through synchrotron radiation, is steeper than in the previous years, it is clearly curved, and peaks (in νF_ν) at 0.5 keV. This energy is much lower than those at which the synchrotron component was found to peak in 1997 and 1998. Some intraday variability suggests that activity of the source on small time scales accompanies the large long time scales changes of brightness and spectrum.

KEYWORDS: BL Lacertae objects: individual (Mkn 501); X-rays: galaxies; radiation mechanisms: non-thermal

1. SCIENTIFIC GOAL OF THE PROGRAM

The radio-to-γ-ray spectra of blazars (νF_ν) are typically "double-humped", with the first peak commonly attributed to synchrotron radiation within a relativistic jet and the second to inverse Compton scattering of relativistic electrons off synchrotron or ambient soft photons (Ulrich, Maraschi & Urry 1997, and references therein). In X-ray bright BL Lacs (XBL) the synchrotron maximum generally occurs in or close to the X-ray band, and the inverse Compton emission peaks above the GeV spectral region extending in some cases to the TeV band, as observed so far for several sources by ground based Cherenkov telescopes. Among these is the nearby

($z = 0.034$) BL Lac object Mkn 501, one of the brightest blazars at UV, X- and γ-ray energies, and a typical XBL according to the numerous multiwavelength data taken prior to the BeppoSAX launch. Repeated observations with the satellites ASCA and XTE have detected large amplitude X-ray variability at different time scales, often correlated with strong activity in the TeV band (e.g., Kataoka et al. 1999; Sambruna et al. 2000).

BeppoSAX observations of Mkn 501 in April 1997, during an outburst, revealed a completely new behavior. In fact, the joint LECS, MECS and PDS spectra showed that at that epoch the synchrotron component peaked at 100 keV or higher energies (Fig. 1). Correspondingly the source was extremely bright in the TeV band and exhibited rapid flares (Catanese et al. 1997; Aharonian et al. 1999). The flux at 10 keV was an order of magnitude higher than the historical level, while around 1 keV the flux was only moderately brighter than usual. The X-ray spectrum hardens with increasing intensity, and the peak energy of the synchrotron component varies by more than two decades with respect to the quiescent state, a unique behavior in blazars, if compared with the variations of the peak energy exhibited by similar sources, never exceeding an order of magnitude with respect to quiescence (Fossati et al. 2000; Giommi et al. 1999). A discussion of the SED of Mkn 501 in the frame of the SSC model is reported in Tavecchio & Maraschi (this volume).

Further BeppoSAX observations in April-May 1998 showed that the synchrotron peak energy was located at ~ 20 keV (Fig. 1), indicating a decrease of an order of magnitude with respect to the previous year (Pian et al. 1999). The simultaneously measured TeV flux was also much lower than in 1997 (Krawczinsky, priv comm.) However, the fact that, despite the radiation losses, the synchrotron peak was still at such high energies one year after the huge outburst, clearly indicates the presence of very powerful, efficient and continuously active mechanisms of particle energization and acceleration in this source.

2. BEPPOSAX OBSERVATIONS: JUNE 1999 CAMPAIGN

We have carried out observations with BeppoSAX in June 1999, simultaneously with TeV Cherenkov telescopes and coordinated with XTE, to investigate the

- X-ray variability of the continuum on short (hours and sub-hour) time scales;
- the presence of possible time lags between soft and hard X-ray emission;
- the correlation between X-ray and gamma-ray flux and spectral index variations.

A single, 180 ks long BeppoSAX pointing has been performed between 1999 June 10, 23:01:18 UT and June 16, 02:11:08 UT, during one of the longest observing windows of the Cherenkov telescopes Whipple, HEGRA, CAT and CELESTE. The X-ray data have been cleaned and linearized at the BeppoSAX Science Data Center (SDC). Spectra and light curves have been extracted from the images with the

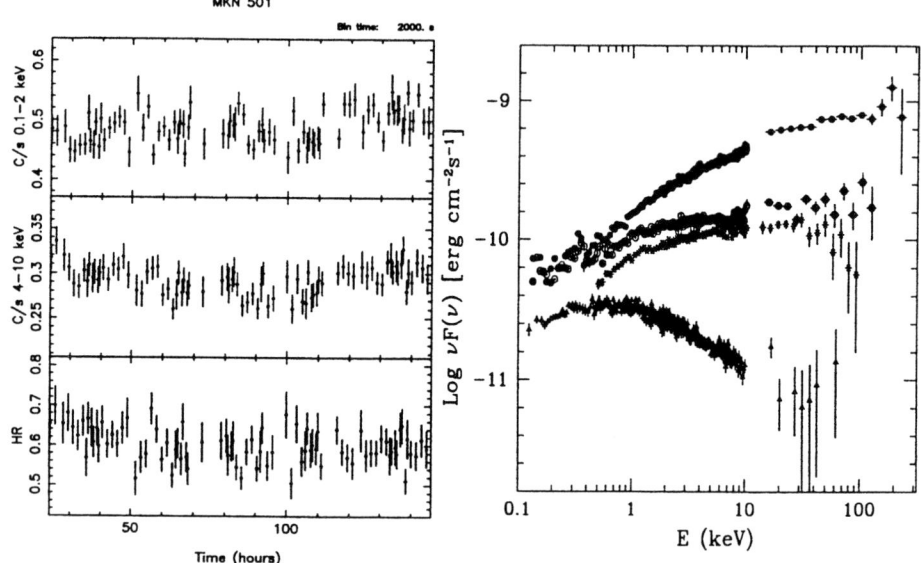

FIGURE 1. Left panel: Light curves (0.1-2 keV and 4-10 keV) and Hardness Ratio of Mkn 501 during the 1999 June observation. Right panel: History of the X-ray spectrum of Mkn 501. From top to bottom: 1997 Apr. 16 (filled circles), 1997 Apr. 7 (open circles), 1998 Apr. 28 (stars) and 1999 June (triangles).

standard XSELECT package. For the spectral analysis, we used the background files and response matrices distributed by the SDC.

3. RESULTS

The source has been clearly detected by the BeppoSAX LECS and MECS instruments, and by the PDS up to 40 keV. The flux exhibits variability of up to 20-30% in amplitude on time scales of 10-12 hours or more (Fig. 1).

The June 1999 flux level at 1 keV is similar to the one observed for Mkn 501 prior to 1997 ("historical" state), namely more than a factor of 2 fainter than detected by BeppoSAX in April-May 1998 and in 1997 April 7, and an order of magnitude fainter than observed by BeppoSAX during the outburst of 1997 April 16 (Fig. 1).

The 0.1-40 keV spectrum is steeper than found in 1997 and 1998 and is progressively steepening with energy. It is not well fitted ($\chi_r^2 \sim 5$) by a single power-law plus Galactic absorption ($N_H = 1.73 \times 10^{20}$ cm^{-2}), therefore two power-laws have been used to fit the data. The fitted energy break is around 1 keV and the spectral indices are $\alpha_1 = 0.89 \pm 0.03$ and $\alpha_2 = 1.44 \pm 0.02$ (errors are at 90% confidence level; $\chi_r^2 = 1.4$). Therefore, the X-ray spectrum is consistent with being produced with a unique emission component, which we identify with synchrotron radiation.

In a νf_ν representation, the X-ray spectrum appears to peak at the energy of ~ 1 keV, which can be identified with the maximum of the synchrotron component. The comparison with the BeppoSAX spectra of the previous years (see Fig. 1) indicates that in 1999 the synchrotron peak has shifted toward lower energies, following t ie trend already noted in 1997 and 1998, when the peak was observed at >100 keV and ~20 keV, respectively.

Optical and TeV coverage simultaneous with the present BeppoSAX campaign was limited due to bad weather. The optical flux level is similar to that usually observed for Mkn 501. At TeV energies, only marginal detections have been obtained on each night. The analysis of these data is still underway.

Our preliminary conclusions are that

- The energy at which the synchrotron component peaks can vary on long term by a large amount (more than a factor of ~ 200), in correspondence with long time scale large amplitude flux variations (2 orders of magnitude at 10 keV).

- This "shift" in energy takes place on much longer time scales (years) than the synchrotron cooling times at X-ray energies, estimated from multiwavelength energy distribution fitting (e.g. Tavecchio & Maraschi, this volume). This further indicates that electrons are continuously accelerated, while the high state is gradually turning to quiescence.

- The low TeV flux in June 1999 indicates that the TeV flux variations are probably correlated with the X-ray variations.

ACKNOWLEDGEMENTS

We thank the BeppoSAX SDC and Mission Planning Team for their support of this project.

REFERENCES

Aharonian F.A., Akhperjanian A.G., Barrio J.A., et al., 1999, A&A 342, 69

Catanese, M., et al. 1997; ApJ, 487, L143

Fossati, G., et al. 2000, ApJ in press (astro-ph/0005067)

Giommi, P., et al., 1998, Nucl. Phys. B Proc. Suppl., 69, 407

Kataoka, J., et al. 1999, ApJ, 514, 138

Pian, E., et al. 1998, ApJ, 492, L17

Pian, E., et al. 1999, in BL Lac Phenomenon, ASP Conf. Ser. 159, ed. L. Takalo (San Francisco: ASP), p. 180

Sambruna, R. M., et al. 2000, ApJ in press (astro-ph/0002215)

Ulrich, M.-H., Maraschi, L., & Urry, C. M. 1997, ARAA, 35, 445

ACCRETION DISK BOUNDARY LAYERS IN LOW-MASS X-RAY BINARIES

R. Popham [1], R. Sunyaev [1]

1) Max-Planck-Institut für Astrophysik, Karl-Schwarzschild-Str. 1, 85740 Garching, Germany

ABSTRACT   We present numerical models of the disk-star boundary layer region in accreting neutron stars. We examine the size and temperature of the region which radiates the boundary layer luminosity, and the spectrum it will produce. We study the dependence of the boundary layer characteristics on mass accretion rate, stellar rotation rate, and alpha. We discuss the implications of these results for understanding X-ray production in low-mass X-ray binaries.

KEYWORDS: accretion, accretion disks; X-rays: stars.

1. INTRODUCTION

Many LMXBs appear to contain neutron stars with magnetic fields so weak as to be dynamically unimportant, as evidenced by the lack of observed coherent pulsations. In these systems, a boundary layer should exist where the rapidly rotating disk reaches the stellar surface. Unless the neutron star is rotating near breakup, a large fraction of the accretion luminosity should originate in the boundary layer, and be radiated as X-rays.

X-ray production in LMXBs is poorly understood, and their X-ray spectra are usually fitted as the sum of two or more emission components. However, the temperatures, luminosities, sizes, and other characteristics of these components are selected only to fit the data, and may have little or no relation to physical models of accretion. We have self-consistently modeled the accretion flow in the boundary layer to provide a physical basis for understanding the spectra and timing behavior of LMXBs.

By calculating the gas dynamics and radiation in the boundary layer, we can directly determine the size, density, temperature, and optical depth of the region which radiates the boundary layer luminosity. We can also explore how the boundary layer changes with the mass accretion rate; these changes may be responsible for the spectral variations observed in atoll and Z sources. In particular, we can examine the effects of radiation pressure on the boundary layer as the accretion rate nears the Eddington limit. We can also see how the boundary layer disappears as the star spins up. Finally, we can see how sensitive the results are to the choice of α.

Due to space limitations, we can only briefly summarize our methods and results here. A more complete description of this work will appear shortly.

2. METHOD

We solve the slim disk equations for the structure of the disk and boundary layer. These resemble the standard thin disk equations, but they include a number of additional terms which make them applicable to disks which are rather thick, such as radial pressure gradient and acceleration, and radial energy transfer by radiation and advection. Most of the equations are the same as those described by Popham & Narayan (1995). Note that we do not include relativistic effects, which could lead to rapid infall of the accreting gas inside the last stable orbit (e.g. Kluźniak & Wilson 1991).

Together with these, we also solve a set of simple radiative transfer equations. These include absorption and scattering, and a simple treatment of Compton scattering based on the average photon energy, but they include no frequency dependence.

We assume that the flow is time-independent and axisymmetric, and make standard simple assumptions about vertical structure, so that the equations can be solved as a function of radius only. We apply boundary conditions at the stellar surface $R_* = 10$ km (most notably $\Omega = \Omega_*$) and in the disk at $R = 100R_*$.

We have found solutions over the ranges: $\dot{M} = 10^{-8} - 10^{-10}$ M$_\odot$ yr^{-1} ($\sim 0.01 - 1\ L_{Edd}$), $\alpha = 0.01 - 0.1$, and $\Omega_* = 0$ − breakup.

3. NOTABLE FEATURES OF THE SOLUTIONS

The boundary layer region is quite clearly visible in the solutions shown in Figure 1. The accreting gas falls inward quite rapidly when it reaches the boundary layer. This produces a dramatic decrease in the density and the absorptive optical depth. The gas can no longer radiate efficiently, and it heats up rapidly to temperatures on the order of 10^8 K. This further decreases the absorptive opacity ($\propto \rho T^{-7/2}$). The same effect has been seen in models for cataclysmic variable boundary layers (Narayan & Popham 1993), and was predicted for neutron star boundary layers by King & Lasota (1987). The gas is optically thick to scattering, but optically thin to absorption even when the increased path length due to scattering is accounted for ($\tau_* = (\tau_s \tau_a)^{1/2} < 1$).

Compton scattering is important in cooling the boundary layer - there is a large flux of softer photons into the hot optically thin zone from the adjacent optically thick zone where the material piles up on the stellar surface. These photons are heated to high energies as they cool the hot gas in the inner boundary layer.

4. DEPENDENCE ON $\dot{M}$, α, Ω_*

We find that the radial width of the boundary layer region increases rapidly with increasing $\dot{M}$ (Fig. 1). At $\dot{M} = 10^{-10}$ M$_\odot$ yr^{-1}, the boundary layer width is a few per-

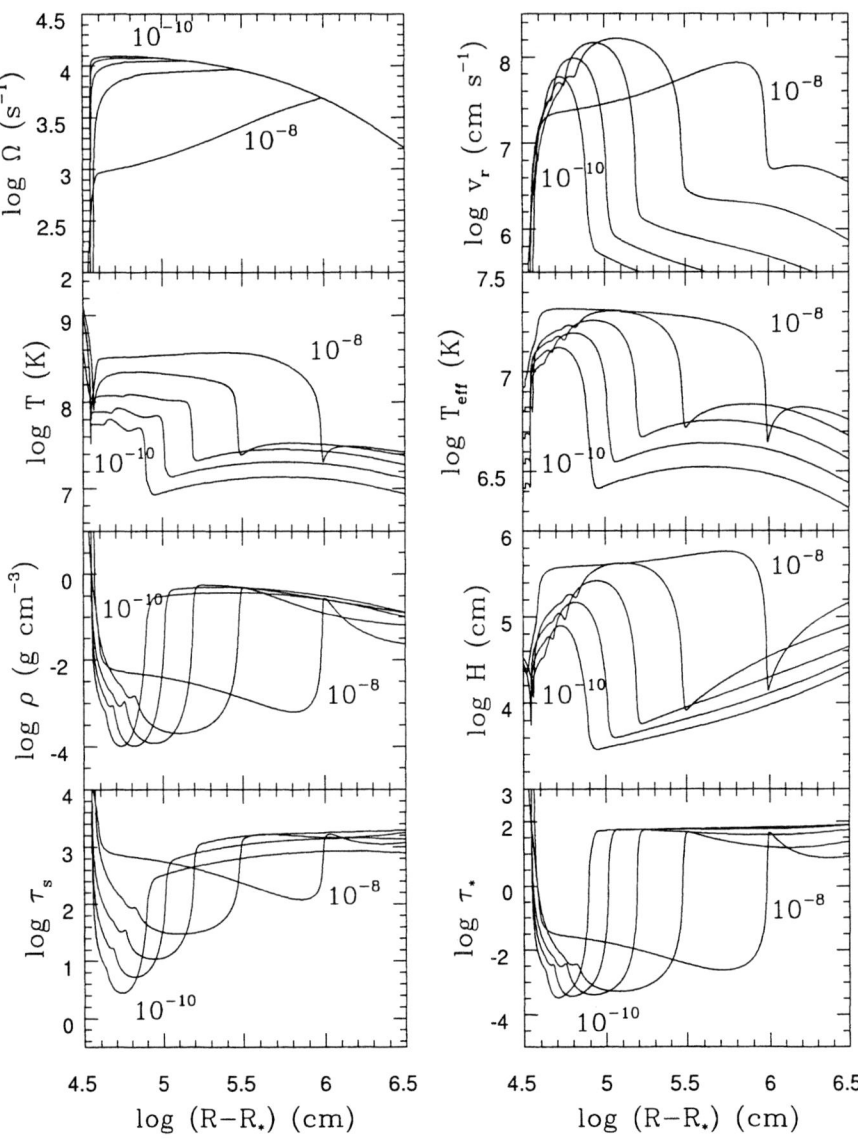

FIGURE 1. Boundary layer solutions for mass accretion rates 10^{-8}, $10^{-8.5}$, 10^{-9}, $10^{-9.5}$, 10^{-10} $M_\odot$ yr^{-1}, all with $\Omega_* = 0$, $R_* = 10^6$ cm, and $\alpha = 0.1$.

cent of the stellar radius, but this increases to $\sim 15\%$ of R_* at $\dot{M} = 10^{-9}$ M$_\odot$ yr^{-1}, and increases rapidly to $\sim R_*$ at $\dot{M} = 10^{-8}$ M$_\odot$ yr^{-1}. This rapid increase occurs because $\dot{M}$ is approaching the Eddington limit, and so radiation pressure can play a significant role in supporting the infalling gas against gravity, even at a substantial distance from the stellar surface.

The density, temperature, and optical depth of the boundary layer also increase with increasing $\dot{M}$ (Fig. 1). The temperature increases from about 5×10^7 K to 4×10^8 K as $\dot{M}$ goes from 10^{-10} to 10^{-8} M$_\odot$ yr^{-1}. The optical depth to scattering goes from a few to a few hundred over the same range in $\dot{M}$, but the optical depth to absorption remains very low, even when the longer path length due to scattering is taken into account.

As the rotation rate of the star Ω_* increases, the boundary layer becomes less luminous and cooler, and its radial width gets smaller. This is expected since less energy is dissipated in the boundary layer for a rotating star (by a factor $(1-f)^2$, where f is the star's rotation rate as a fraction of the breakup rotation speed). The viscosity parameter α has little effect on the size of the boundary layer, but solutions for lower α have higher density and optical depth and lower temperature. Unfortunately we do not have space to show these results here.

5. EXPECTED SPECTRA FROM THE BOUNDARY LAYER

The Compton y-parameter ($y = (4kT/m_e c^2) Max(\tau_s, \tau_s^2)$) ranges from $y < 1$ at $\dot{M} = 10^{-10}$ M$_\odot$ yr^{-1} to $y \sim 10^5$ at $\dot{M} = 10^{-8}$ M$_\odot$ yr^{-1}. Therefore, for accretion rates in the lower part of this range, the hot boundary layer region should have a power-law spectrum produced by unsaturated Comptonization, with a cutoff at $kT \sim 5 - 10$ keV.

At higher accretion rates, the Comptonization will be saturated due to the high scattering optical depth. This should produce a Wien spectrum which peaks at $\sim 3kT \sim$ tens of keV. This is hotter than observed spectra of luminous LMXBs, which are generally blackbody with $kT \sim 2$ keV. This suggests that the boundary layer may be optically thick, so that the spectrum is characterized by the effective temperature ($kT_{eff} \sim 2$ keV) rather than the midplane temperature ($kT > 10$ keV). Our $\dot{M} = 10^{-8}$ M$_\odot$ yr^{-1}, $\alpha = 0.01$ solution is becoming optically thick: the effective optical depth is $\tau_* > 1$.

The optically thick, scattering-opacity-dominated accretion disk will also contribute a multi-temperature modified blackbody component to the overall spectrum.

REFERENCES

King, A. R., Lasota, J. P. 1987, A&A, 185, 155
Kluźniak, W., Wilson, J. R. 1991, ApJ, 372, L87
Narayan, R., Popham, R. 1993, Nature, 362, 820
Popham, R., Narayan, R. 1995, ApJ, 442, 337

WEIGHING THE BLACK HOLE IN THE NARROW-LINE SEYFERT 1 GALAXY, RE J1034+396

E. M Puchnarewicz,[1] K. O. Mason[1], A. Siemiginowska[2], I. Cagnoni[2], A. Comastri[3], F. Fiore[4] and A. Fruscione[2]

[1] Mullard Space Science Laboratory, University College London, Holmbury St. Mary, Dorking, Surrey RH5 6NT, U.K.
[2] Harvard-Smithsonian Center for Astrophysics, 60 Garden Street, MS-4, Cambridge MA 02138, USA
[3] Osservatorio Astronomico di Bologna, via Ranzani 1, I-40127 Bologna, Italy
[4] Osservatorio Astronomico di Roma, via Frascati 33, I-00044 Monteporzio, Italy

ABSTRACT The mass of the black hole in the narrow-line Seyfert 1 (NLS1) galaxy RE J1034+396 is measured by fitting an accretion disc and power-law to quasi-simultaneous optical, UV and X-ray spectra. The fits favour accretion onto a low-mass black hole ($M \sim 5 \times 10^6$ $M_\odot$) at ~ 0.3-0.5 of the Eddington rate. They also prefer a disk viewed almost edge-on ($75°$ from the disk axis). The implication of high accretion rate onto a low mass black hole supports the model where NLS1s are the Seyfert-scale analogies of Galactic Black Hole Candidates (GBHCs).

KEYWORDS: accretion disks - galaxies: Seyfert - galaxies: individual (RE J1034+396) - galaxies: nuclei - X-rays: galaxies

1. INTRODUCTION

A link between the full width at half maximum (FWHM) of the $H\beta$ line and the steepness of the soft X-ray (~ 0.1-2 keV) spectrum has now been well established. It shows that the line-of-sight velocity of the Balmer line-emitting gas in the broad line region (BLR) is related to the observed strength of the soft X-ray component in active galactic nuclei (AGN). Boller, Brandt & Fink (1996) demonstrated that Nature discriminates against AGN with strong soft X-ray excesses and high velocity BLRs.

This trend was first reported in a sample of ultrasoft X-ray Seyferts and quasars by Puchnarewicz et al. (1992). It was subsequently confirmed by Boller et al. (1996) and in a sample of PG quasars by Laor et al. (1997). Thus it is observed across a wide range of luminosities and, by implication, black hole mass, and is one of the most fundamental observed properties of AGN.

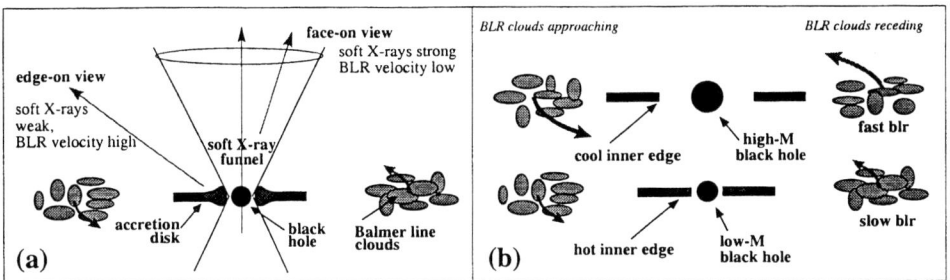

FIGURE 1. (a) The orientation-dependent model. (b) The mass-temperature model.

1.1. Orientation effects versus the low mass black hole

There are currently two popular hypotheses put forward to explain this correlation. The first is a consequence of observing systems at different angles of orientation. The model assumes that the outer BLR is disk-shaped and co-planar with an accretion disk (AD). The inner edge of the AD is puffed up by radiation pressure so that the soft X-rays are emitted anisotropically in a 'funnel' along the axis of the AD (see Figure 1a). Thus when viewed along the axis of the disk, the observed velocity of the BLR is low and the soft X-ray emission is high, and vice-versa.

The second model is based on a property of standard AD models where the temperature of the inner edge of the disk decreases as the black hole mass (M) increases. Also, as M increases, the gravitational potential at any given radius increases. Thus assuming the BLR forms at similar distances in all AGN and that the BLR clouds have Keplerian velocities, the BLR velocity would be low when the soft X-ray emission from the inner AD is high, and vice-versa (see Figure 1b).

2. DATA

We have begun a programme to test the hypothesis that the AD heating and slowing down of the BLR are due to the presence of a low-mass black hole, by fitting the the optical to X-ray spectra of Seyfert 1s and quasars.

2.1. The narrow-line Seyfert 1: RE J1034+396

As the 'prototype' for this programme, we have selected the narrow-line Seyfert 1 (NLS1) galaxy RE J1034+396 (Puchnarewicz et al. 1995, Pounds et al. 1995; Puchnarewicz et al. 1998). Using a combination of power-law and sophisticated AD models, we have fitted quasi-simultaneous optical, UV and X-ray data of this AGN, to place realistic constraints on the black hole mass and accretion rate, and on the inclination of the AD.

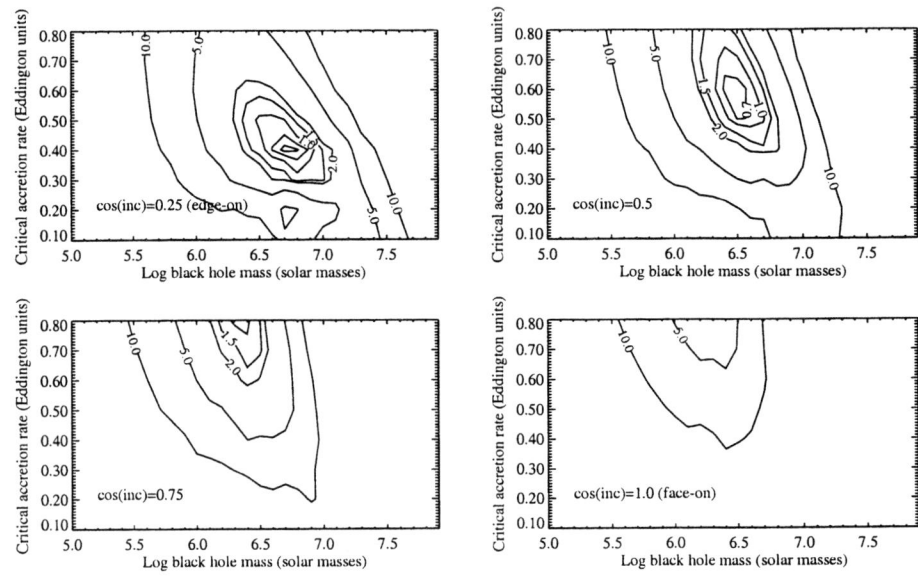

FIGURE 2. Reduced chi-squared (χ_ν^2) grids for the AD plus power-law models when compared to the observed data for RE J1034+396.

2.2. Optical to X-ray spectra

Three sets of data were used in the actual fitting. A deconvolved spectrum of the nucleus in the 3500 Å to 7200 Å range, was extracted from a spatially-resolved optical spectrum, taken using ISIS on the William Herschel Telescope. A UV spectrum, covering 1100 Å to 3300 Å was taken with the Faint Object Spectrograph on the Hubble Space Telescope. Finally, a Beppo-SAX observation, obtained with the LECS and MECS instruments and covering 0.1 keV to 10 keV, was used to measure the X-ray spectrum.

2.3. Models

A combination of two separate components were used to represent the data, a simple power-law underlying the optical to X-ray spectrum, and the AD model used by Siemiginowska & Elvis (1997). This AD model uses a sum of black bodies approximation to an AD, modified for the effects of electron scattering. Kerr or Schwarzschild models for the black hole may be used and the effects of a hot, Comptonizing corona around the disk may be included. Note that these are not 'fits' in the formal sense, since no parameters were allowed to be free. The power-law slope and normalization were fixed, having been derived from the best-fit to the optical and hard X-ray data. The AD parameters; inclination, black hole mass and accretion rate, were fixed at the values described in the plots.

3. RESULTS

Reduced chi-squared (χ^2_ν for 268 data points) grids are shown in Figure 2. The models which compare most favourably are found for a black hole mass, M=5×10^6 M$_\odot$ and an accretion rate of 0.4 $\dot{m}_{Edd}$ at almost edge-on inclinations, 60-75° from the axis of the disk (ie. at cos(inclination)=0.25-0.5). Plots of the best-fitting models are available on www.mssl.ucl.ac.uk/ emp/presentations/bologna/bologna.html at four inclinations (cos(inc)=0.25, 0.5, 0.75 and 1.0).

Over all inclinations, the grids show that relatively small black hole masses, between about 10^6 and 10^7 M$_\odot$, are preferred. The lowest χ^2_νs, at the nearly edge-on inclinations, are for masses between about 2×10^6 and 8×10^6 M$_\odot$.

The best-fitting models occur at the highest inclinations, ie. where the disk is preferentially *edge-on*. The best pole-on models require a very high accretion rate and still give a very poor fit to the data.

The preferred accretion rates fall significantly as the inclination of the disk increases. For an almost edge-on inclination (ie cos(inc)=0.25), accretion rates between $\sim$ 0.3 and 0.5 $\dot{m}_{Edd}$ provide the best fit. The anti-correlation of accretion rate with inclination is a physical effect predicted by the model. Relativistic effects boost photon energy preferentially along the plane of the disk, so that as the viewing angle becomes more pole-on, higher accretion rates are required to produce the observed energy output.

4. DISCUSSION

While the relationship between the soft X-ray spectrum and BLR velocity is now significant enough to warrant a conference on its own, firm evidence to support any of the models has been lacking. In this paper, we have, for the first time, explored the range of black hole mass, accretion rate and inclination which is favoured by high signal-to-noise, quasi-simultaneous spectra of the NLS1 RE J1034+396 using a sophisticated AD model plus a power-law. The results favour relatively high (but still sub-Eddington, $\dot{m}_{Edd}$ $\sim$ 0.3 $-$ 0.5) accretion onto a low-mass black hole, in support of the GBHC analogy (Pounds et al 1995). They also prefer a relatively edge-on disk, which in turn suggests that any molecular torus in RE J1034+396 is relatively small in height, that it lies relatively further out from the centre, or that it does *not* lie co-planar to the disk and is relatively face-on to us.

REFERENCES

Boller Th., Brandt W. N., Fink H., 1996, A&A, 305, 53
Laor A., Fiore F., Elvis M., Wilkes B. J., McDowell J. C., 1997, ApJ, 477, 93
Pounds K. A., Done C., Osborne J., 1995, MNRAS,
Puchnarewicz E. M. et al., 1992, MNRAS, 256, 589
Puchnarewicz E. M., Mason K. O., Siemiginowska A., Pounds K. A., 1995, MNRAS, 276, 20
Puchnarewicz E. M., Mason K. O., Siemiginowska A., 1998, MNRAS, 293, L52
Siemiginowska A., Elvis M., 1997, ApJ, 482, L9

SPECTRAL SIGNATURES OF REPROCESSING ON HERCULES X-1/HZ HERCULIS

H. Quaintrell[1,2], M. D. Still[3,4,5], S. D. Vrtilek[6], B. Boroson[3] and P. Roche[2,7]

1) Department of Physics & Astronomy, The Open University, Walton Hall, Milton Keynes MK7 6AA, UK
2) Astronomy Centre, CPES, University of Sussex, Falmer, Brighton BN1 9QJ, UK
3) NASA/Goddard Space Flight Center, Code 662, Greenbelt, MD 20772, USA
4) Universities Space Research Association, 7501 Forbes Blvd, Suite 206, Seabrook, MD 20706, USA
5) Physics & Astronomy, University of St. Andrews, North Haugh, St. Andrews, Fife KY16 9SS, UK
6) Harvard-Smithsonian Center for Astrophysics, Cambridge, MA 02138, USA
7) Department of Physics & Astronomy, University of Leicester, Leicester LE1 7RH, UK

ABSTRACT We use two optical and one ultraviolet spectrophotometric datasets of Her X-1/HZ Her, from the *Isaac Newton Telescope* and *Hubble Space Telescope* respectively, to construct Doppler maps (maps in velocity space) and light curves, which are then used as diagnostics with which to test the validity of two warped disk models published in the literature. We find that whilst the light curves may be explained with these models, the maps are not. The maps may be explained if the accretion stream is thick enough to cast a significant shadow over the heated face of the donor star, but in this case the modelled light curves do not match observations.

KEYWORDS: accretion, accretion disks; stars: individual (V* HZ Her, X Her X-1); X-rays: stars; stars: imaging; line: profiles

1. INTRODUCTION

Hercules X-1/HZ Herculis is an eclipsing X-ray binary system (Tananbaum et al., 1972) wherein a non-degenerate star (HZ Her) donates mass to a neutron star (Her X-1). The donor is assumed to be filling its critical Roche surface and approximately co-rotating with the binary. X-ray observations reveal a $1^{s}\!.24$ neutron star rotational period, $1^{d}\!.7$ orbital period and super-orbital 35^{d} period: the X-rays "turn-on" for $\sim 10^{d}$ (main-high state), "turn-off" for $\sim 10^{d}$, "turn-on" for $\sim 5^{d}$ (short-high state) and "turn-off" for $\sim 10^{d}$ (see e.g. Scott & Leahy, 1999). All three X-ray periodicities have been reflected in the optical data, indicating that X-rays are strongly irradiating the inner face of the donor (see e.g. Davidsen et al., 1972, Gerend & Boynton, 1976). Although the optical light curves vary over the 35^{d} cycle, the optical flux integrated over one orbital cycle stays approximately the same throughout the super-orbital cycle (Gerend & Boynton, 1976). This implies that the X-rays are not actually turning off twice every 35 days, but rather the X-ray emitting regions near the pulsar are periodically being obscured from our line of sight. The explanation for the X-ray "turn-on/off" phenomenon employs a warped accretion disk, precessing retrograde, periodically obscuring the neutron star (see e.g. Petterson, 1977). 35^d day optical modulation results from the disk casting an X-ray shadow over parts of the donor star.

The 35^d "on/off" cycle has been evident in X-ray observations since the discovery of the X-ray source, except for a number of anomalous X-ray low states (AXLS): < 0.8 year in 1983-1984, ~0.1 year in 1993 and for > 1 year from 1999 March 23 (Parmar et al., 1985, Vrtilek et al., 1994, Parmar et al., 1999, Vrtilek et al., 2000). During these AXLS the X-ray flux level is comparable with the levels observed during the regular low states in the 35^d cycle, but for weeks-months rather than days-weeks. Optical observations made during these AXLS detected the effects of X-ray heating of the companion, as usual, indicating the X-ray source to still be active but obscured from our line of sight (Delgado et al., 1983, Margon et al., 1999). One explanation for these AXLS is that the scale height of the disk increases, obscuring the X-ray source.

2. OBSERVATIONS

The optical spectrophotometric data analysed in this paper (see also Still et al., 1997 and Quaintrell et al., 2000a), are the result of two observing runs at the 2.5m *Isaac Newton Telescope* (INT) on La Palma (Canary Islands, Spain). They cover $\phi_{35} = 0.237 - 0.416$ (July 1995) and $0.643 - 0.823$ (June 1997), where $\phi_{35} = 0.0$ is the beginning of the main-high state. The spectral region covered ($\sim 4100 - 5100$Å) includes a variety of absorption and emission lines, e.g.: Hβ, MgIIλ4481, HeIλ4922 in absorption and NVλ4604, CIII/NIII $\lambda\lambda$4634–4650, HeIIλ4686 in emission.

The ultraviolet spectra of Her X-1/HZ Her (see also Boroson et al., 2000, Quaintrell et al., 2000b and Vrtilek et al., 2000) were obtained using the *Hubble Space Telescope* (HST) *Space Telescope Imaging Spectrograph* (STIS). The 38 HST orbits were split evenly between the short-high (1998 July observations) and the main-high (1999 July observations) states, although the system was in an AXLS for the 1999 observations. 34 HST orbits of observations used the E140M grating in TIME-TAG mode, giving a dispersion of 0.013 Å/pixel and a temporal resolution of 125μs. The remaining 4 orbits (1998 July) used the G140L grating and the ACCUM acquisition mode, giving better sensitivity during X-ray eclipse and a dispersion of 0.584 Å/pixel. A number of line features are evident in our HST spectra of HZ Her (see fig. 1): CIII λ1176, NV $\lambda\lambda$1239, 1243, OV λ1218, 1371, SiIV $\lambda\lambda$1394, 1403, NV] $\lambda\lambda$1487, CIV $\lambda\lambda$1548, 1551, HeII λ1640 in emission. All of these emission lines are seen for both the short-high state and the AXLS, at similar flux levels.

Sharp absorption features are blended into the OV, SiIV and CIV lines. The features present in the SiIV and CIV lines do not move with orbital phase, hence they must be interstellar in origin, whereas the absorption in the OV line does

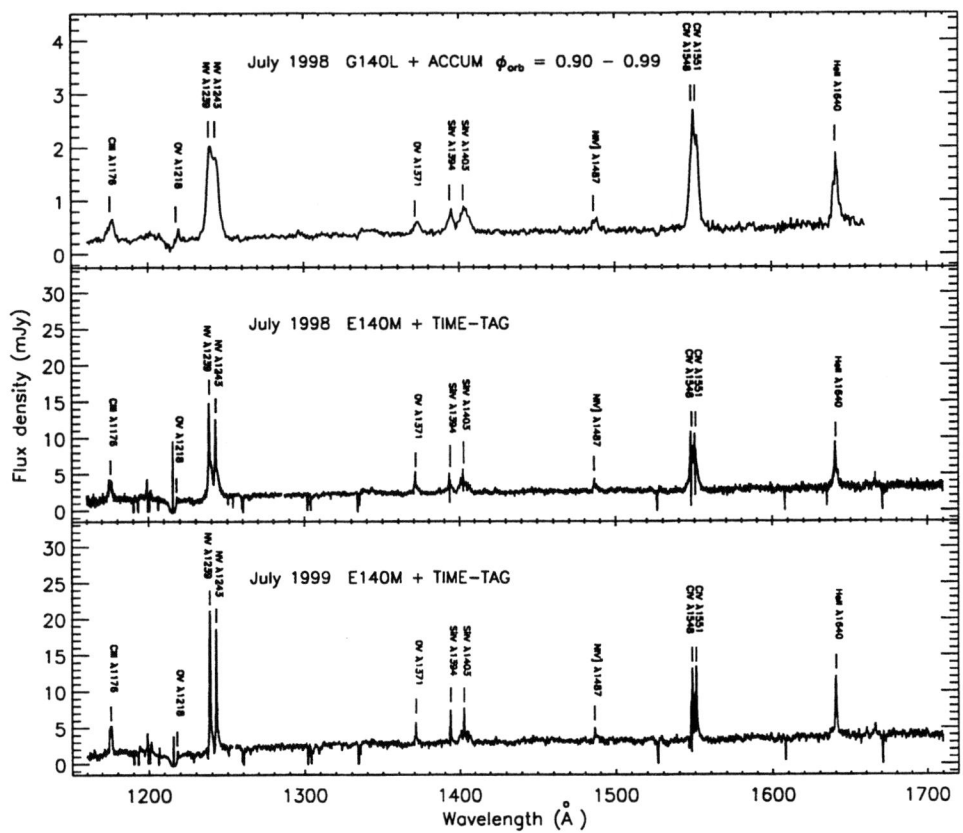

FIGURE 1. HST STIS ultraviolet spectra of Hercules X-1/HZ Herculis.

move, indicating that its origin is within the binary. P-Cygni absorption is found at ~ -400 km/s on the blue wings of the Nv $\lambda1239$, Ov $\lambda1371$, Silv $\lambda1394$ and Civ $\lambda1548$ emission lines for the AXLS. The orbital phases at which we see significant P-Cygni absorption in the AXLS spectra, are not sampled in the short-high data, hence we cannot tell whether this absorption is peculiar to the AXLS.

3. DOPPLER MAPS

A dataset of spectra can be transformed into a kinematic map, also known as a Doppler map or tomogram (Marsh & Horne, 1988). Although emission in a pixel of a Doppler map may correspond to more than one spatial site within the binary, certain structures within the binary have well defined positions on a Doppler map, e.g. the centres of mass and the Roche lobes of the donor and accretor. Trailed spectrograms and Doppler maps constructed using the 1997 INT and 1999 HST spectra are presented in fig. 2. and 3., respectively.

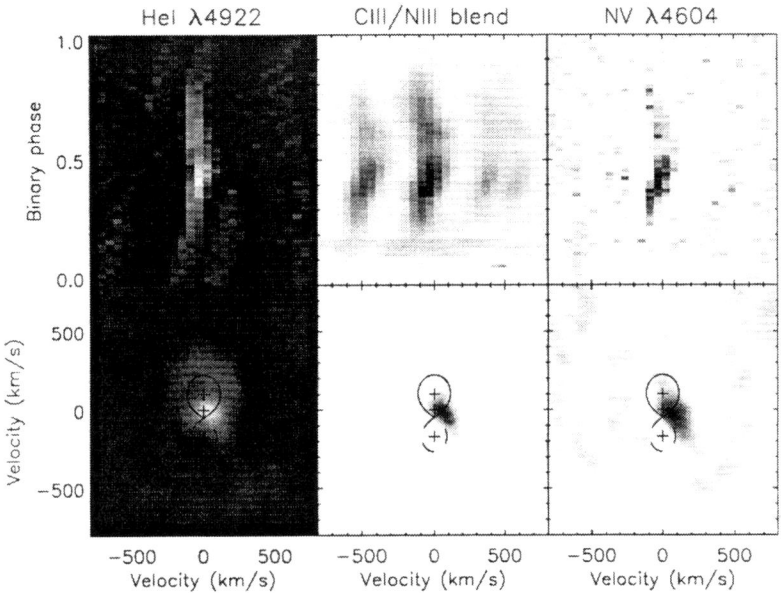

FIGURE 2. Trailed spectrograms (top panels) and Doppler maps (bottom panels) constructed using 1997 INT spectra. Overplotted on the Doppler maps are the Roche lobes of the donor star (solid line) and accretor (dashed line). The greyscales increase from minimum (white) to maximum flux (black).

Inspection of the maps shows that the line flux is consistent with an origin on the heated face of the donor star, HZ Her. The flux is not distributed evenly, it is biased towards the trailing side of the donor. The absorption feature in the Ov $\lambda1371$ line appears within the donor Roche lobe.

4. MODELLING

Our synthetic binary code (Still et al., 1997, Quaintrell, et al., 2000a) is applicable to a binary with a Roche lobe filling donor transfering material to a compact accretor via a ballistic accretion stream and Keplerian accretion disk. The surfaces of the lobe, stream and disk are constructed from small triangular tiles. The flux from a donor tile is calculated to be directly proportional to the X-ray flux incident upon it, assuming the accretor irradiates the system as a point source. Since there is no evidence from the Doppler maps for line emission from the stream or disk, we treat these as occultors only. An emission line spectrum is calculated by summing the contributions from all the visible tiles. Using these synthetic spectra we construct Doppler maps and light curves in an attempt to reproduce those of the CIII/NIII $\lambda\lambda4634\text{--}4650$ and Nv $\lambda4604$ emission lines and hence test the validity of two warped

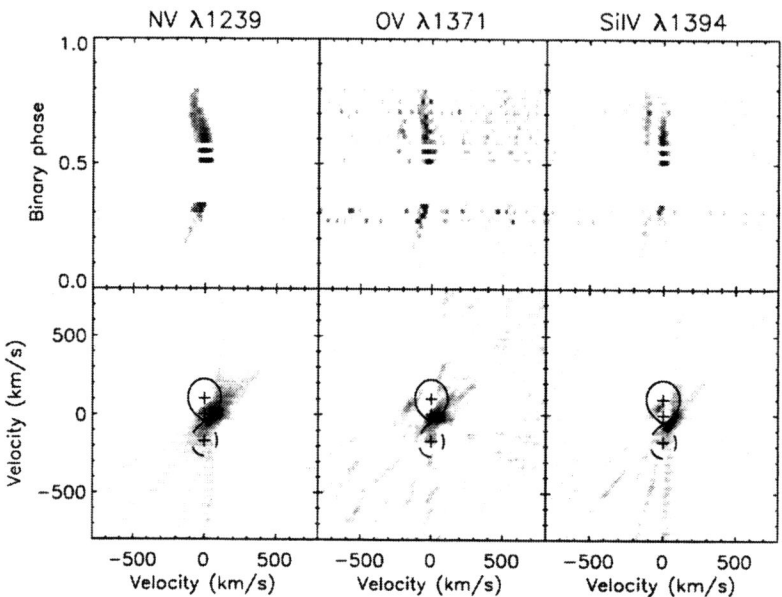

FIGURE 3. Trailed spectrograms (top panels) and Doppler maps (bottom panels) of the NV λ1239, OV λ1371 and SiIV λ1394 emission lines, constructed using 1999 HST spectra. Gaps in the spectrograms correspond to gaps in the data.

disk models: Schandl & Meyer (1994) and Wijers & Pringle (1999).

Schandl & Meyer (1994) assert that X-rays from the neutron star heat the surface of an optically thick, geometrically thin accretion disk, producing a hot corona. In the outer regions of the disk, $r > 10^{10.2}$ cm, the sound velocity of the gas exceeds the escape velocity and leaves the binary, exerting repulsive forces on the disk which leading to warping. The time taken for the wind to cross the the disk combined with the Keplerian motion of the disk results in a phase shift between the illumination of the disk and the resulting wind, which exerts a torque on the disk. They find that such a wind is capable of maintaining the warped shape of the disk, which can reproduce the main-high and short-high of the X-ray light curve.

Pringle (1996) investigated the effect of X-irradiation on the dynamics of an accretion disk. If an optically thick accretion disk is irradiated by a central source, then radiation received at a point on the disk must be reemitted normal to the disk surface at that point, producing a back reaction. If there is an uneven distribution of forces a torque will be exerted on the disk, twisting it. Wijers & Pringle (1999) modelled in some detail how this irradiation driven warping might be relevant to X-ray binaries, also taking into account self-shadowing, and were thus able to produce a warped disk in the shape of a prograde spiral. Once the disk has been warped, the parts of the disk visible to the neutron star become even more irradiated and

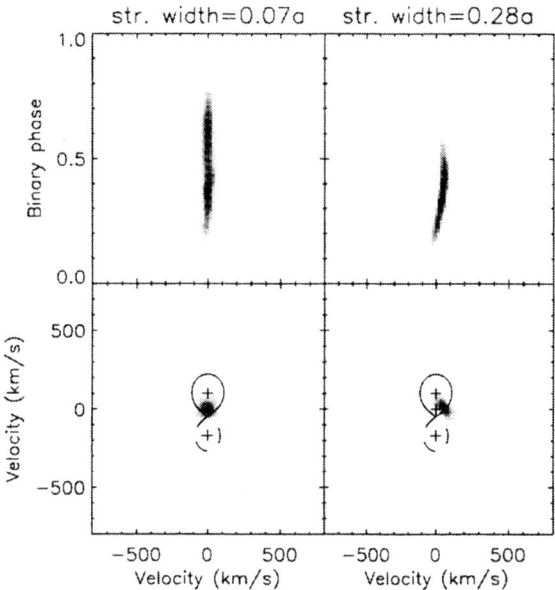

FIGURE 4. Trailed spectrograms (top panels) and Doppler maps (bottom panels) constructed from synthetic spectra, using the Wijers & Pringle 1999 disk model. Stream widths are given in terms of the binary separation, a.

the resulting back pressure causes the disk to precess. Given a prograde spiral disk shape, the radiation driven precession will then be retrograde. Subjecting such a disk to forced precession, which acts in the retrograde sense, tends to "unwind" the prograde spiral and the disk settles back down into the orbital plane. However, they compare the precession rates produced by tidal and radiative forces for a number of systems which exhibit a super-orbital period. They find that the distribution of super-orbital periods is better matched by radiative rather than tidal precession and subsequently disregard the effects of tidal precession.

One of the disk models above may explain the behaviour of Her X-1. Although the coronal wind model of Schandl & Meyer (1994) appears to explain the X-ray behaviour of Her X-1, Still et al. (1997) found that it could not reproduce the optical data. Also, Wijers & Pringle (1999) point out that the Schandl & Meyer (1994) model ignores self-shadowing and claim that they use an older and incorrect equation of motion for the disk. The radiative warping models of Pringle (1996) and Wijers & Pringle (1999) ignore tidal precession and the effect of changing the stream/disk impact point, which occurs as a consequence of the warped disk shape; as the donor orbits the disk the accretion stream will impact at different radii. The correct disk prescription must reproduce both the X-ray behaviour and the optical data.

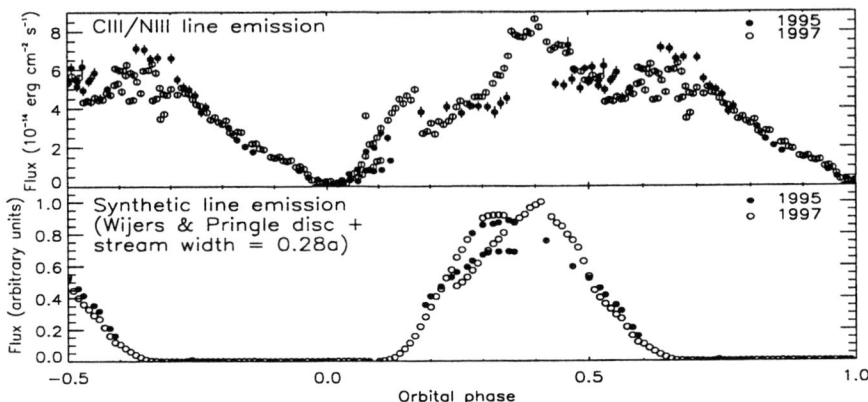

FIGURE 5. Light curves.

We find, using a moderate stream width, that whilst the flux on the synthetic maps (see e.g. fig. 4, left panels) is biased towards the accretor, there is no significant leading-trailing side asymmetry, for both disk models, contrary to observations. If the stream width is increased by a factor of 4, we may achieve the degree of asymmetry observed for some of the lines (see e.g. fig. 4, right panels). However, for such a thick stream the light curves become much more asymmetric than is observed (see e.g. fig. 5).

5. CONCLUSIONS

We find that the majority of the optical and ultraviolet line flux is consistent with an origin on the X-ray irradiated inner face of the donor star, HZ Her, rather than an accretion disk around the neutron star, Her X-1. At the disk precession phase that the 1999 HST observations were made we expected to see ultraviolet emission from the heavily irradiated inner regions of the accretion disk, however, when these observations were made the source was in an AXLS, which means the neutron star, and possibly these very hot inner parts of the accretion disk, were obscured.

From the optical and ultraviolet Doppler maps it is also evident that the heating over the inner face of HZ Her is not uniform, with line flux biased towards the trailing side of the donor, implying shadowing from an accretion stream and/or disk. If an extended accretion stream does cause the leading-trailing side asymmetry, then the asymmetry will persist throughout the $35^{\rm d}$ precession cycle. Otherwise the asymmetry must be related to the accretion disk, which precesses over the $35^{\rm d}$ cycle. The current data is unable to discriminate between the two because the two INT datasets are separated by $\Delta\phi_{35} \sim 0.4$. The majority of disk models are point symmetric between the top and bottom surfaces, consequently the illumination patterns for observations with $\Delta\phi_{35} = 0.5$ would be mirrored about the stellar equator, resulting in identical Doppler tomograms. The 1999 HST data were not modelled here as it is not clear what the $35^{\rm d}$ phases of the these observations are, also it is likely that the disk structure in the AXLS differs from what it is usually. Observations which sample a different part of the $35^{\rm d}$ cycle to those of our existing data will enable us to discriminate between disk-dominated and stream-dominated shadowing of HZ Her.

P-Cygni like absorption features are observed on the blue wings of the Nv λ1239, Ov λ1371, SiIV λ1394 and CIV λ1548 emission lines during the AXLS.

ACKNOWLEDGEMENTS

This paper is based on observations with the *Isaac Newton Telescope* and the *Hubble Space Telescope*. The *Isaac Newton Telescope* is operated on the island of La Palma by the Isaac Newton Group in the Spanish Observatorio del Roque de los Muchachos of the Instituto de Astrofisica de Canarias. The NASA/ESA *Hubble Space Telescope* observations were obtained at the Space Telescope Science Institute, which is operated by the Association of Universities for Research in Astronomy, Inc., under NASA contract GO-05874.01-94A. H.Q. is employed on PPARC grant L64621. S. D. V. and B. B. are supported in part by NASA (NAG 5-2532, NAGW-2685) and NSF (DGE-9350074). B. B. acknowledges an NRC postdoctoral associateship.

REFERENCES

Boroson, B., et al., 2000, ApJ, 529, 414
Davidsen A., Henry J. P., Middleditch J., Smith H. E., 1972, ApJ, 177, L97
Delgado, A. J., Schmidt, H. U., Thomas, H. C., 1983, A&A, 127, L15
Gerend D., Boynton P. E., 1976, ApJ, 209, 562
Margon, B., et al., 1999, IAU Circ. 7144
Marsh, T. M., Horne, K., 1988, MNRAS, 235, 269
Parmar, A. N., et al., 1999 A&A, 350, L5
Parmar, A. N., et al., 1985, Nat, 313, 119
Petterson, J. A., 1977, ApJ, 218, 783
Pringle, J. E., 1996, MNRAS, 281, 357
Quaintrell, Still, M. D., Roche, P. D., 2000a, MNRAS, submitted
Quaintrell, H., et al., 2000b, ApJ, in prep.
Schandl, S., Meyer, F., 1994, A&A, 289, 149
Scott D. M., Leahy D. A., 1999, ApJ, 510, 974
Still, M. D., Quaintrell, H., Roche, P. D., Reynolds, A. P., 1997, MNRAS, 292, 52
Tananbaum H., et al., 1972, ApJ, 174, L143
Vrtilek, S. D., et al., 1994, ApJ, 436, L9
Vrtilek, S. D., et al., 2000, ApJ, in prep.
Wijers, R. A. M. J., Pringle, J. E., 1999, MNRAS, 308, 207

X-RAYS FROM THE SEYFERT GALAXY IC 4329A, ITS NEIGHBOURS AND ITS GALAXY GROUP

A.M. Read [1], W. Pietsch [1]

1) Max-Planck-Institut für extraterrestrische Physik, Postfach 1603, D–85740 Garching, Germany

ABSTRACT   ROSAT PSPC and HRI observations of the nearby type 1 Seyfert galaxy IC 4329A field show many point sources. The brightest ($L_X = 6 \times 10^{43}$ erg s^{-1}) is associated with IC 4329A itself, having a single power-law spectrum ($\Gamma = 1.73$) and a spectral edge-like feature at 0.7 keV. The giant lenticular companion galaxy IC 4329 is also detected, as is shocked interaction-induced gas between the IC 4329A/IC 4329 pair. Residual, unresolved emission, extending for ~ 200 kpc, appears two-component, with a hard, circularly-distributed, smooth component, and a softer, clumpier component to the south-east. This hard component appears itself two-component, due to the 'wings' of IC 4329A, and to hot (~ 1.5 keV) galaxy group gas. The soft component may be a large ULIRG-like superwind, or a 'stripped wake' of intragroup gas.

KEYWORDS: Galaxies: individual: IC 4329A; Galaxies: interactions; Galaxies: intergalactic medium; Galaxies: Seyfert; X-rays: galaxies

1. GENERAL RESULTS

We have observed both the ROSAT HRI and PSPC data from fields centred on the edge-on, type 1 Seyfert galaxy IC 4329A and its nearby companion, the giant lenticular IC 4329. 17 and 22 sources are detected respectively in the full HRI and PSPC fields of view, the brightest being associated with the two central galaxies and a further source to the south-west. Many coincidences are seen in the two datasets and the nine most significant HRI sources all have equivalent PSPC counterparts. None of the sources detected show any significant temporal variability. Our findings with regard to the observed point-source and residual unresolved emission can be summarized as follows (Read & Pietsch 1998).

2. THE POINT SOURCES IN AND AROUND IC 4329A

An extremely bright ($L_X = 6 \times 10^{43}$ erg s^{-1}) source is detected associated with the central Seyfert IC 4329A (source P8; see Fig. 1). A power-law spectral model ($N_H = 2.8 \times 10^{21}$ cm^{-2}, photon index $\Gamma = 1.73$) with a spectral edge-like feature at 0.7 keV fits the IC 4329A data well, with a reduced χ^2 of 0.9.

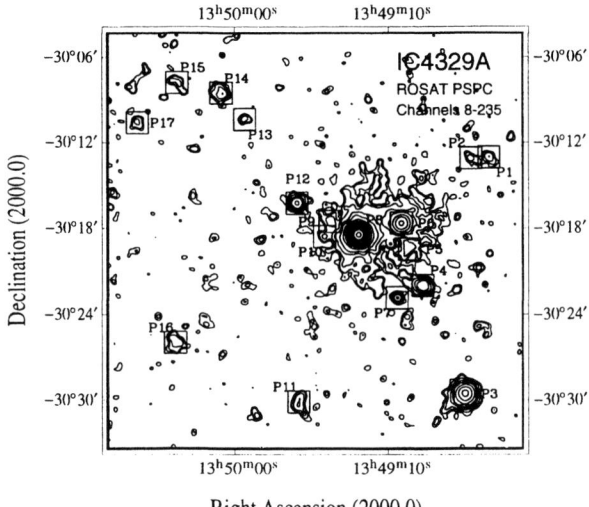

FIGURE 1. ROSAT PSPC map of the IC 4329A field in the broad (channels 8−235, corresponding approximately to 0.08−2.35 keV) band. The contour levels are at 2, 3, 5, 9, 15, 31, 63, 127, 255, 511, 1023, 2047 and 4095σ (σ being 1.25×10^{-3} cts s^{-1} arcmin^{-2}) above the background (2.81×10^{-3} cts s^{-1} arcmin^{-2}). Source positions are marked on the image.

Two other very bright sources are detected associated with the nearby giant lenticular IC 4329 (P6) ($L_X = 8\times10^{41}$ erg s^{-1}, $N_H = 2.7\times10^{20}$ cm^{-2}, $kT = 1.1$ keV), and with a likely quasar to the south-west (P3) ($L_X = 1\times10^{42}$ erg s^{-1}, $N_H = 3.3\times10^{20}$ cm^{-2}, $\Gamma = 2.4$).

Many other bright sources are detected both in the HRI and PSPC fields of view, including three point-like HRI sources (blended into two in the PSPC; P4 & P12), symmetrically positioned with respect to the disc of IC 4329A (see Fig. 1). Optical follow-up observations of these sources with the 2.2 m ESO/MPG telescope at La Silla, Chile, established that they are not associated with the central Seyfert, being merely foreground and background objects.

3. THE UNRESOLVED EMISSION AROUND IC 4329A

In addition to point source emission, unresolved residual emission is also detected, both in the HRI and in the PSPC, surrounding the IC 4329A/IC 4329 pair. This emission appears markedly two-component, comprising of a hard, smooth, circularly-distributed component, centred somewhere between IC 4329A and IC 4329 (Fig. 2 [left]), and a soft, irregular component, situated almost entirely to the south-east of the IC 4329A disc (Fig. 2 [right]).

The hard component of the residual emission appears itself to be made up of two components. One of these is purely the 'wings' of the extremely bright IC 4329A

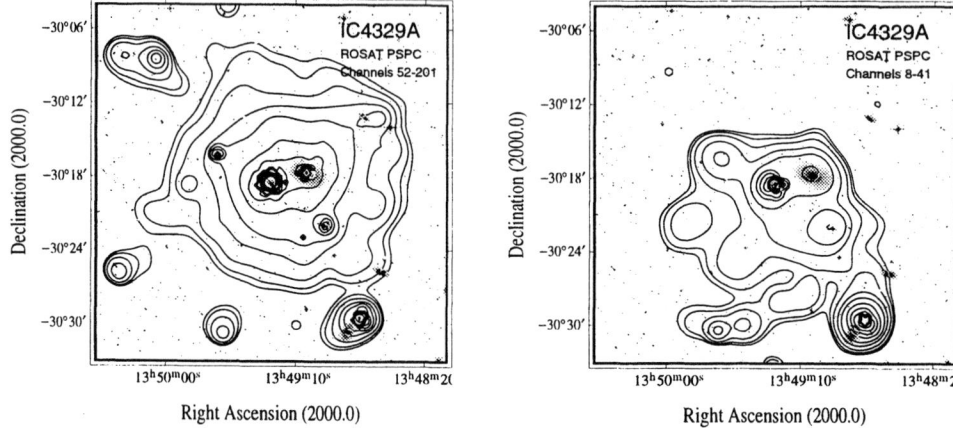

FIGURE 2. ROSAT PSPC maps of the IC 4329A field in the (left) hard (channels 52–201) and (right) soft (channels 8–41) bands, obtained using an adaptive filtering technique, and overlayed on optical images. The contour levels are at 2, 3, 5, 9, 15, 31, 63, 127, 255, 511, 1023 and 2047σ (σ being 4.3×10^{-5} (hard) and 2.3×10^{-5} (soft) cts s^{-1} arcmin^{-2}) above the background (6.4×10^{-4} (hard) and 7.1×10^{-4} (soft) cts s^{-1} arcmin^{-2}).

source, visible out to several arcminutes. The second component appears to be hot ($kT \sim 1.5$ keV) diffuse gas, with a luminosity of $\approx 5\times10^{41}$ erg s^{-1} and a mass of perhaps 2×10^{11} $M_\odot$.

The soft component of the residual emission appears to be absorbed merely by the Galactic column along the line of sight, and is very soft ($kT \approx 0.1$ keV). It has an X-ray luminosity of $L_X = 9\times10^{41}$ erg s^{-1}.

4. INTERACTIONS AND THE GALAXY GROUP

The properties of the hot ($kT \sim 1.5$ keV) diffuse component of the hard residual emission are very suggestive of it being due to hot gas within the galaxy group of which IC 4329A and IC 4329 are members. The high gas temperature agrees well with the group's high velocity dispersion, and though the luminosity ($\approx 5\times10^{41}$ erg s^{-1}) and estimated gas mass ($\approx 2\times10^{11}$ $M_\odot$) are lower (by factors of $\sim 2-4$) than would be expected for such a hot, high-velocity dispersion group (see *e.g.*, Ponman et al. 1996; Mulchaey et al. 1996), they are not unusually so.

The soft component of the residual emission, in terms of its temperature and one-sided nature, bears a good deal of resemblance to proposed starburst driven winds seen in some far-infrared ultraluminous systems (e.g. Arp 220, NGC 2623; Read & Ponman 1998), though it is very much brighter and larger. Another possibility is that the soft emission may be a 'wake' of stripped gas from the galaxy group, as

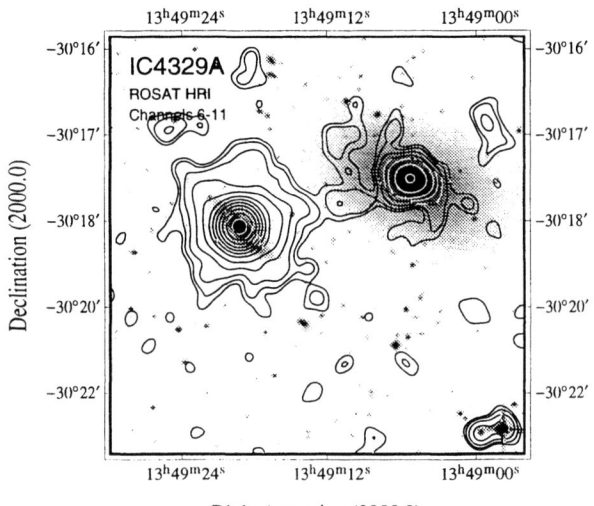

FIGURE 3. ROSAT HRI map of the IC 4329A field, obtained using an adaptive filtering technique and overlayed on a digitized sky survey image. Only channels 6−11 are used. The contour levels are at 2, 3, 5, 9, 15, 31, 63, 127, 255, 511, 1023 and 2047σ (σ being 7.1×10^{-4} cts s^{-1} arcmin^{-2}) above the background (2.1×10^{-3} cts s^{-1} arcmin^{-2}).

it moves through the surrounding A3574 cluster medium. This would result in a reduction in the X-ray luminosity and estimated gas mass of the group gas, as is observed.

It appears unambiguous, through evidence taken across the whole of the EM spectrum, that some sort of interaction between the IC 4329A/IC 4329 pair is taking place. Within the ROSAT HRI data, what appears to be a 'bridge' of emission connecting the two galaxies is seen (Fig. 3). This feature, similar to features seen in other systems (e.g. NGC 3395/6; Read & Ponman 1998, the *Antennae*; Read *et al.* 1995, HCG 92; Pietsch *et al.* 1997), is believed to be due to shocks resulting from the strong galaxy interaction.

REFERENCES

Mulchaey J.S., Wilson A.S., Tsvetanov Z., 1996, ApJS, 102, 309
Pietsch W., Trinchieri G., Arp H., Sulentic J.W., 1997, A&A, 322, 89
Ponman T.J., Bourner P.D.J., Ebeling H., Böhringer H., MNRAS, 283, 690
Read A.M., Ponman T.J., Wolstencroft R.D., 1995, MNRAS, 277, 297
Read A.M., Pietsch W., 1998, A&A, 336, 855
Read A.M., Ponman T.J., 1998, MNRAS, 297, 143

NATURE OF "PASSIVE" ELLIPTICAL GALAXIES

T.A. Rector [1] and J.T. Stocke [2]

1) National Optical Astronomy Observatories, 950 N. Cherry Ave., Tucson, AZ 85719 USA

2) Center for Astrophysics and Space Astronomy, University of Colorado, Campus Box 389, Boulder, CO 80309 USA

ABSTRACT

We present ROSAT HRI images of three candidate "passive" elliptical galaxies from the *Einstein* EMSS. Passive Es are the third most abundant identification type in deep ROSAT surveys (e.g., Griffiths et al. 1995) and could be major contributors to the XRB if they have hard X-ray spectra from an AGN. Deep-survey passive Es are too faint and too distant to address the question of their nature with the ROSAT HRI. Therefore, we have imaged three nearby, unusually X-ray luminous elliptical galaxies from the EMSS to determine whether they have weak AGN and are thus potential contributors to the XRB; or if they are diffuse, hot gas emitters, in which case they add significantly to the dark matter content of the Universe. No evidence is found for a significant AGN contribution to the X-ray flux in any of these objects, suggesting that passive ellipticals are not significant contributors to the hard XRB.

KEYWORDS:

1. INTRODUCTION

The three galaxies chosen for study were identified as early-type galaxies in poor clustering environments (Stocke et al. 1991). They are unusually luminous at X-ray energies (log L_x = 43–44 erg s^{-1} in the 0.3–3.5 keV band; e.g., Maccagni et al. 1987), having X-ray luminosities typical of radio galaxies which are ~100 times more luminous at radio wavelengths. Low-resolution (*Einstein* IPC and/or ROSAT PSPC) observations did not clearly resolve these objects, leaving the possibility that a significant fraction (50% or greater) of their X-ray flux could be from an embedded, low-luminosity AGN. Two of the three galaxies were detected at radio wavelengths, also suggesting the presence of an AGN. We also note that these galaxies are more X-ray luminous than the "isolated elliptical" galaxy NGC 1132 (Mulchaey & Zabludoff 1999).

2. OBSERVATIONS

2.1. MS 0116.3−0115

The unusual nature of this source was first presented by Maccagni et al. (1987). Our 29.2-ksec ROSAT HRI image reveals a diffuse structure (Figure 1) which extends well beyond the galaxy, indicating that the X-ray emission is not from the envelope of the galaxy but must be cluster emission. The maximum point source contribution at the location of the radio source is roughly 3% (Figure 2), putting very stringent limits on the presence of an embedded AGN.

2.2. MS 1204.1+2826 (NGC 4104)

Dahlem & Stuhrmann (1998) first presented a 1.8-ksec ROSAT HRI observation of this object. We confirm their discovery of strong, centrally-peaked X-ray emission centered on the galaxy. Our deeper, 15.1-ksec observation also detects a diffuse halo which extends well beyond the galaxy; this halo was also noticed in PSPC observations (Thiering & Dahlem 1999). Despite the strong central peak, the maximum potential contribution of a point source to the X-ray flux is only 15%.

2.3. MS 1306.7−0121

Like the other three objects, our 17.5-ksec ROSAT HRI image reveals diffuse emission which extends well beyond the galaxy. This cooling-flow cluster has also been detected via its Hα emission (Donahue, Stocke & Gioia 1992). The maximum potential contribution of a point source to the X-ray flux is 4%.

3. CONCLUSIONS

Prior low-resolution (Einstein IPC and/or ROSAT PSPC) observations did not clearly resolve these objects, leaving the possibility that a significant fraction of their X-ray flux could be from an AGN. However, in all three cases we detect only diffuse, extended emission from the poor cluster environments in which these ellipticals reside. No evidence is found for a significant AGN contribution, suggesting that passive ellipticals do not significantly contribute to the hard XRB.

REFERENCES

Dahlem, M. & Stuhrmann, N. 1998 A&A 332, 449.
Donahue, M., Stocke, J.T. & Gioia, I.M. 1992 ApJ 385, 49.
Gioia, I.M. et al. 1983 ApJ 271, 524.
Griffiths, R.E. et al. 1995 MNRAS 275, 77.
Maccagni, D. et al. 1987 ApJ 316, 132.
Mulchaey, J.S. & Zabludoff, A.I. 1999 ApJ 514, 133.
Stocke, J.T. et al. 1991 ApJS 76, 813.
Thiering, I. & Dahlem, M. 1999 A&A (submitted).

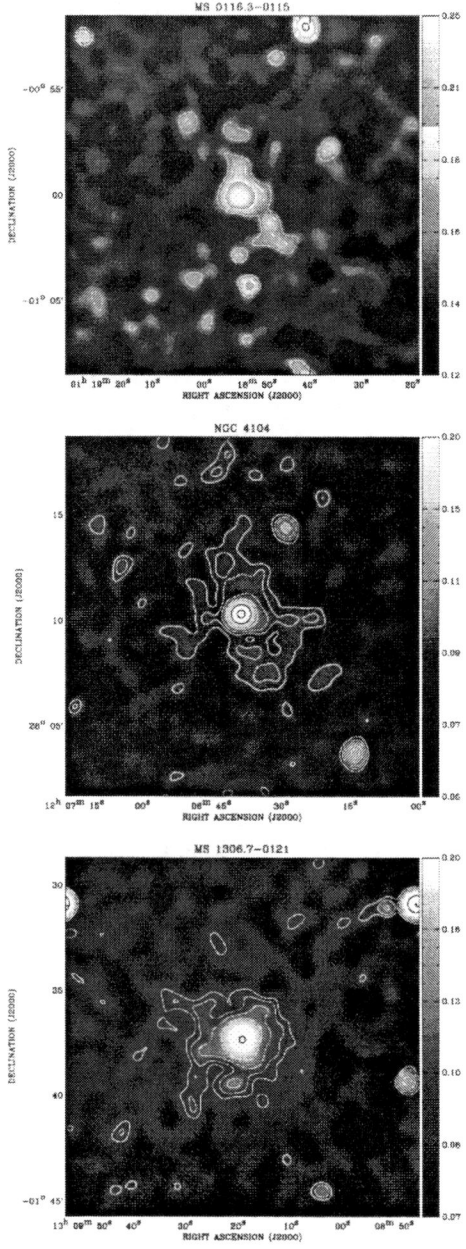

FIGURE 1. The ROSAT HRI images for each object. Each map is smoothed with a Gaussian kernel of FWHM = 10", roughly twice the HRI PSF, to enhance the extended flux.

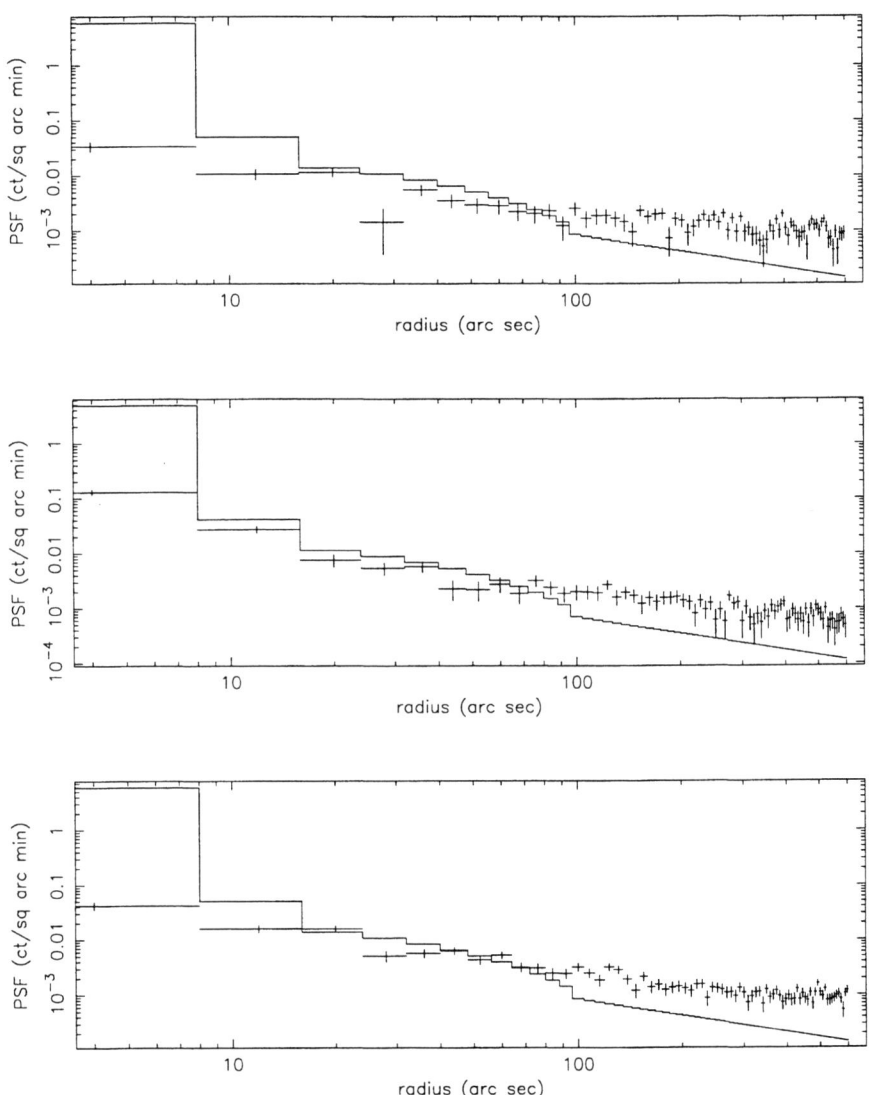

FIGURE 2. The binned radial profiles for (from top to bottom) MS 0116.3–0115, MS 1204.1+2826 and MS 1306.7–0121. Each source is plotted against the differential point spread function of the HRI (solid line). Of most interest is the fit to the central bin, which gives an estimate of what fraction of the flux could be from an embedded point source.

PDS 456: AN EXTREME ACCRETION RATE QUASAR?

J. Reeves[1], P. O'Brien[1], S. Vaughan[1], D. Law-Green[1], M. Ward[1], C. Simpson[2], K. Pounds[1], R. Edelson[1,3]

1) X-Ray Astronomy Group; Department of Physics and Astronomy; Leicester University; Leicester LE1 7RH; U.K.
2) Subaru Telescope, National Astronomical Observatory of Japan, 650 N. A'ohōkū Place, Hilo, HI 96720, U.S.A.
3) Department of Physics and Astronomy; University of California, Los Angeles; Los Angeles, CA 90095-1562; U.S.A.

ABSTRACT

X-ray and multi-wavelength observations of the most luminous known local ($z < 0.3$) AGN, the recently discovered radio-quiet quasar PDS 456, are presented. The spectral energy distribution shows that PDS 456 has a bolometric luminosity of 10^{47} erg/s, peaking in the UV. The X-ray spectrum obtained by $ASCA$ and $RXTE$ shows considerable complexity. The most striking feature observed is a deep, highly-ionised, iron K edge (8.7 keV, rest-frame), originating via reprocessing from highly ionised material, possibly the inner accretion disk. PDS 456 was found to be remarkably variable for its luminosity; in one flare the X-ray flux doubled in just ~ 15 ksec. If confirmed this would be an unprecedented event in a high-luminosity source, with a light-crossing time corresponding to $\sim 2R_S$. The implications are that either flaring occurs within the very central regions, or else that PDS 456 is a 'super-Eddington' or relativistically beamed system.

KEYWORDS: galaxies: active – quasars: individual: PDS 456 – X-rays: quasars

1. INTRODUCTION

PDS 456 is a bright, *radio-quiet* QSO (V=14) recently discovered by Torres et al. (1997). It lies fairly close to the Galactic plane ($\beta = 12$) and is seen through an extinction of $A_V = 1.5$ (or a column of $\sim 2 \times 10^{21}$ cm^{-2}). PDS 456 is at a similar redshift, at z=0.184, to 3C 273, but has a higher bolometric luminosity. The dereddened optical-IR spectrum of PDS 456 is compared in Figure 1a to that of 3C 273. They are remarkably similar in this bandpass. Overall, PDS 456 is the most luminous object in the local ($z < 0.3$) Universe ($M_V = -27$, $L_{BOL} \sim 10^{47}$ erg s^{-1}). Values of $H_0 = 50$ km s^{-1} Mpc^{-1} and $q_0 = 0.5$ have been assumed throughout this paper.

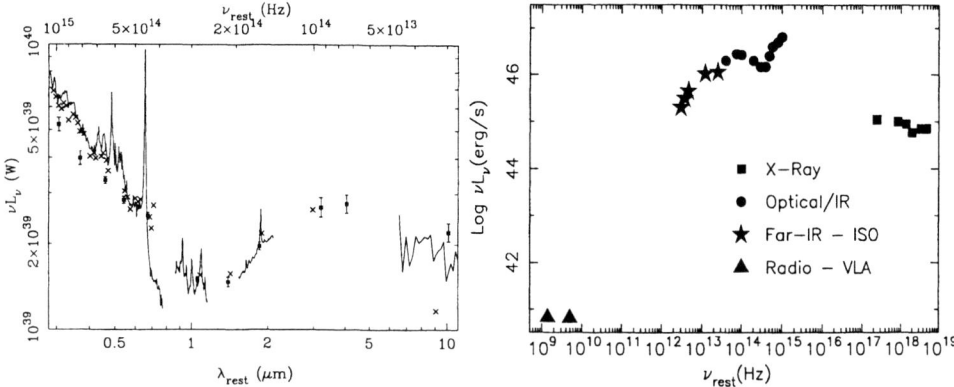

FIGURE 1. (a) Dereddened optical-to IR spectrum and photometry of PDS 456, plotted in the quasar rest-frame. The crosses represent the measurements of 3C 273 over the same waveband, but scaled up by a factor of 1.77. Notice the similarity in the emission between PDS 456 and 3C 273 over this waveband. (b) The radio to X-ray SED of PDS 456; the emission peaks in the blue/UV part of the spectrum. It is seen that the bolometric luminosity of PDS 456 approaches 10^{47} erg s^{-1}.

2. MULTI-WAVELENGTH PROPERTIES OF PDS 456

The spectral energy distribution (SED) of PDS 456 is shown in Figure 1b. In the optical and IR, PDS 456 shows strong H I emission lines, with moderately broad Hβ (FWHM = 3000 km s^{-1}) and strong optical Fe II emission (Simpson et al. 1999). VLA observations confirm that, unlike 3C 273, PDS 456 is radio-quiet (R_L=-0.7) and has little extended radio emission. Overall the SED is dominated by the optical/UV 'big blue bump'. The bolometric luminosity of PDS 456 is $\sim 10^{47}$ erg s^{-1}.

3. X-RAY OBSERVATIONS OF PDS 456

3.1. The X-ray Spectrum

As part of our campaign, we observed PDS 456 with ASCA on 7-8 March 1998 and with RXTE on 7-10 March 1998. The hard X-ray spectrum of PDS 456 obtained by both ASCA and RXTE shows complex features (see Reeves et al. 2000). The data/model residuals from a power-law fit ($\Gamma = 2.4$) to the RXTE data are shown in Figure 2a. Unsurprisingly a power-law gave an inadequate fit to the data in this band. An unusually deep and ionised Fe K edge is observed, with best-fit parameters of E=8.7±0.2 keV and $\tau = 0.75 \pm 0.15$ (errors at 90% confidence). The edge is detected in both the ASCA and RXTE data to >99.99% confidence. There is also some evidence in the X-ray spectrum for a broadened ($\sigma \sim 1$ keV) line at 6 keV; this line may originate from the inner disk, as hypothesised in Seyfert 1s (Tanaka et al. 1995). The best fit parameters for the line are E=6.1±0.5 keV, $\sigma = 1.2 \pm 0.7$ keV and EW=340^{+410}_{-200} eV. The line is detected at 99% confidence ($\Delta\chi^2 = 11$, for 3

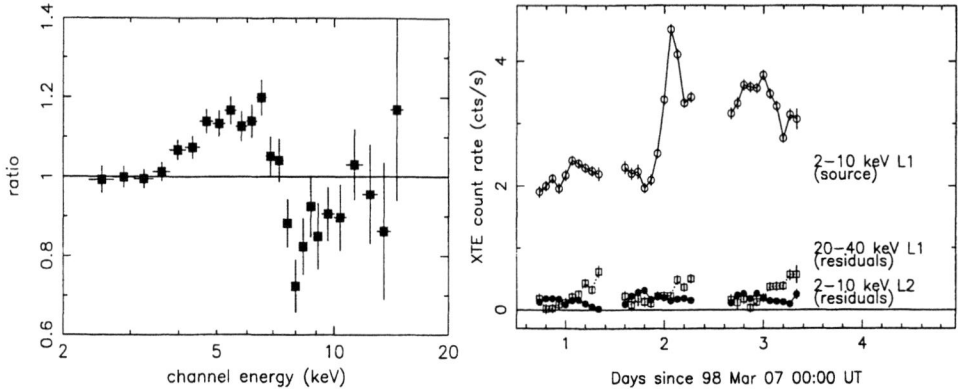

FIGURE 2. (a) The data/model ratio residuals from a simple ($\Gamma = 2.4$) power-law fit to the $RXTE$ data of PDS 456. The effect of the deep, ionised iron K edge is clearly seen in the residuals. (b) The 2-10 keV $RXTE$ lightcurve of PDS 456. The source appears to flare by a factor of 2.1 in 17 ksec, corresponding to a light-crossing size of $\sim 2R_S$ for a $10^9 M_\odot$ black hole.

parameters).

A model consisting of reflection off a highly ionised accretion disk provides the best-fit to the hard X-ray spectrum; with disk solid angle, R=$\Omega/2\pi = 1.0$, ionisation parameter, $\xi = 6000$ erg cm s^{-1} and $T_{disk} = 10^6$ K. The high-ionisation of the disk reflection component can match both the depth of the edge and its energy at 8.7 keV. Although the high ionisation appears inconsistent with the line energy (at 6.1 keV), the line constraints are poor as it is rather broad; in addition one may also expect some reddening close to the black hole. High-ionisation reflection features are predicted in some disk photoionisation models (e.g. Ross, Fabian & Young 1999, Nayakshin 2000), particularly at high accretion rates and when the primary X-ray emission is steep. An alternative hypothesis is a high ionisation warm absorber; however we note that the ionisation ($\xi = 10^4$, with $N_H = 5 \times 10^{23}$cm^{-2} and T=3×10^5 K) required to produce the edge is rather high (see Reynolds & Fabian 1995) and that such an absorber may not be thermally stable. Therefore we find the ionised reflection model more physically appealing.

The high ionisation of the reflector could imply a high accretion rate in PDS 456, particularly as $\xi \propto \dot{m}^3$ in a photoionised accretion disk (e.g. Matt, Fabian & Ross 1993). This interpretation is consistent with the other X-ray properties of PDS 456, namely a steep underlying continuum and rapid X-ray variability, both of which are commonplace in NLS1s (Boller et al. 1996). NLS1s are also thought to be accreting near the Eddington limit (e.g. Pounds, Done & Osborne 1995); indeed recent evidence has been found in one NLS1 (Ark 564) for a spectrum consistent with ionised disk reflection (Vaughan et al. 1999).

3.2. X-ray Variability

Both the ASCA and RXTE data were examined to search for X-ray variability. A strong hard X-ray flare is observed in the RXTE observation (figure 2b), well above any residual fluctuations in the detector background; the doubling time for the flare was ~ 15 ksec. We also calculated that the probability of finding another contaminating source of comparable brightness in the RXTE beam was low ($< 2\%$), although not totally excluded. Additionally there is no other X-ray source detected in the RASS (ROSAT All Sky Survey) to within a degree of PDS 456.

Therefore, if confirmed, this would be unprecedented behaviour in such a high-luminosity source. This suggests, from simple light-crossing arguments, a maximum size of $l = 4.5 \times 10^{12}$ m for the varying region. For a black hole of mass $10^9 M_\odot$ (corresponding to PDS 456, with $L_{BOL} = 10^{47}$ erg s^{-1}, at the Eddington limit), this implies that the X-ray flare occurs **within a region of less than 2 Schwarzschild radii** ($2R_S$). A smaller mass black hole would loosen this requirement somewhat, but would then imply a super-Eddington accretion rate. *Therefore one possible implication of the rapid variability is accretion near to or greater than* L_{Edd}. The variability also implies a (non-beamed) efficiency of converting matter to energy of $\sim 5\%$, close to the limit for a Schwarzschild black hole (see Fabian 1979). In any event, the flare must come from a very compact region (for instance, a small hot spot on the disk), presumably very close to the 'central engine' in which the luminosity is actually generated.

4. CONCLUSIONS

In conclusion PDS 456 is a remarkable object, showing clear features of a high ionisation reprocessor, one possible interpretation of which is through reflection off a highly ionised accretion disk. Overall the high ionisation spectral features, steep X-ray emission and the extreme rapid variability suggest that the super-massive black hole in PDS 456 could be running at an unusually high accretion rate.

REFERENCES

Boller, Th., Brandt, N., Fink., H.H., 1996, A&A., 1996, 305, 53

Fabian, A.C., 1979, Proc. R.Soc London., SerA., 336, 449

Matt, G., Fabian, A.C., Ross, R., 1993, MNRAS, 262, 179

Nayakshin, S., 2000, ApJ, 534, 718

Pounds, K., Done, C., Osborne, J., 1995, MNRAS, 277, L5

Reeves, J.N., et al., 2000, MNRAS, 312, L17

Ross, R., Fabian, A.C., Young, A., 1999, MNRAS, 306, 461

Reynolds, C., Fabian, A.C., 1995, MNRAS, 273, 1167

Simpson, C., Ward, M., O'Brien., P.T., Reeves, J., 1999, MNRAS, 303, L23

Tanaka, Y., et al., 1995, Nature, 275, 659

Torres, C.A.O., et al., 1997, ApJL, 488, 19

Vaughan, S., Pounds, K., Reeves., J., Warwick, R., Edelson, R., 1999, MNRAS, 308, L34

THE X-RAY PROPERTIES OF LUMINOUS INFRARED GALAXIES

G. Risaliti[1], R. Gilli[1], R. Maiolino[2], M. Salvati[2]

1) Dip. Astronomia, Università di Firenze, Italy; 2) Osservatorio di Arcetri, Firenze, Italy

ABSTRACT   We present a study of a sample of luminous infrared galaxies (LIGs, $L_{IR} > 10^{11} L_\odot$) observed in the hard (2-10 keV) X rays. The main results are: 1) an analysis of the correlation between X-ray and infrared emission reveals that most sources are powered both by AGN and starburst activity and that the AGNs in our sample are less absorbed in the infrared than in the X-rays; 2) the study of a subsample of sources observed in the 20-200 keV band indicates that most of the AGNs hosted by the LIGs are heavily obscured up to 100 keV and, therefore, their contribution to the X-ray background must be small.

KEYWORDS: Galaxies: infrared; Galaxies: X rays; X-ray background.

1. INTRODUCTION

The Luminous Infrared Galaxies (LIGs) are a class of objects characterized by high infrared luminosities ($L_{IR} > 10^{11}$ erg s^{-1}). It is widely accepted that in general both an AGN and a starburst can power LIGs, but the relative contribution of these two energy sources to the bolometric luminosity is still unclear.

X-ray observations in the hard (2-10 keV) band can be a powerful tool to unveil the AGN emission in the LIGs and to estimate its contribution to the total luminosity. It is at present impossible to analyze a representative LIGs sample in the 2-10 keV band, because the hard X-ray observations performed up to now are strongly biased in favor of AGN–dominated sources. Nevertheless, a significant number of IR-selected LIGs with X-ray observations is now available both in the literature and in public archives, and therefore a comparison between the X-ray properties of AGN–dominated and IR–selected LIGs is possible. We collected the data of all the luminous infrared galaxies observed so far in hard X-rays and studied their X-ray emission and the correlation between their X-ray and infrared properties.

2. X–RAY AND IR PROPERTIES OF THE SAMPLE

The X-ray properties of our sample of objects are very heterogeneous both in terms of brightness and spectral shape. Most of the objects optically classified as type 1 Sy or QSO are characterized by bright X-ray emission (relative to the IR luminosity) and their X-ray spectrum does not show indications for significant cold absorption.

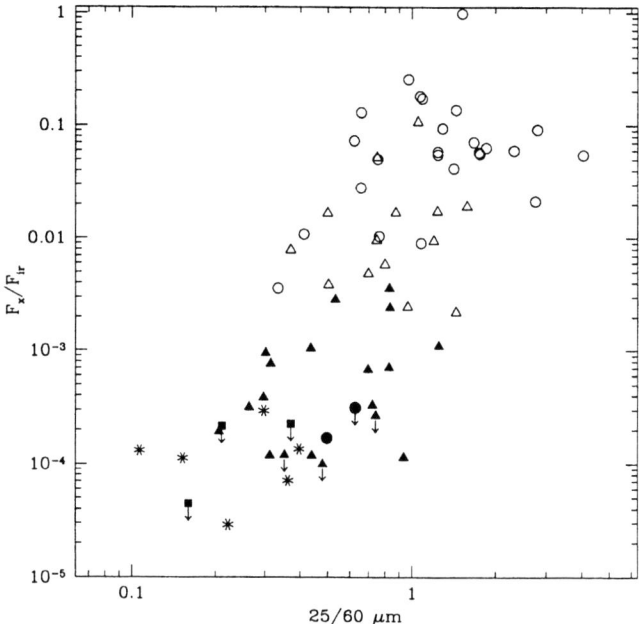

FIGURE 1. X/IR flux ratio versus infrared colour for the sources of our sample. Symbols: open circles: Seyfert 1s; open triangles: Compton thin Seyfert 2s; filled triangles: Compton thick Seyfert 2s; filled circles: broad absorption line quasars; stars: starbursts.

A significant fraction of the narrow line AGNs are also relatively bright in the X-rays and their spectrum is characterized by a photoleletric cutoff ascribed to Compton thin absorbing gas ($N_H < 10^{24} cm^{-2}$). The remaining objects optically classified as AGNs are very weak in the X-rays. This can be due to the intrinsic weakness of the AGN component or to an absorbing column density higher than $10^{24} cm^{-2}$ (Compton thick AGNs). Finally, a few objects are optically classified as starbursts and are all very weak in the X rays. In Fig. 1 we plot the X/IR flux ratio versus the infrared colour defined as $C_{IR} = 2 \times \frac{f_{25}}{f_{60}}$, where f_{25} and f_{60} are the flux densities at 25 μm and at 60μm. A clear correlation is apparent in Fig. 1: type 1 AGNs are preferentially in the high X/IR ratio and warm infrared colour part of the diagram. Moving towards lower 25/60μm ratios we find lower X/IR ratios and an increasing fraction of obscured AGNs at first, and of starbursts afterwards.

A simple model in agreement with this correlation is shown in Fig. 2: starting from a "pure Seyfert 1" point, at the top-right corner in the plot, the oblique rightmost (light) line gives the expected location of AGN–dominated objects, with an increasing X-ray absorbing column density moving towards the bottom. The

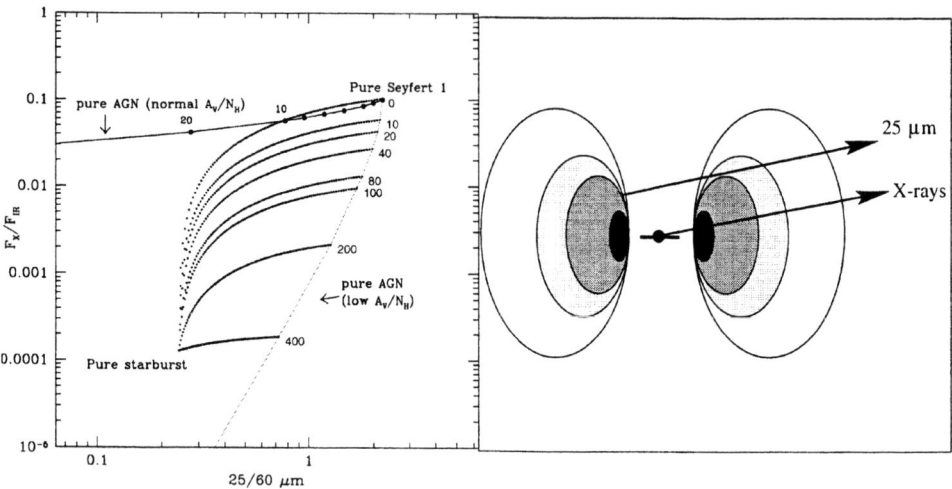

FIGURE 2. a) The results of the model described in the text (numbers refer to N_H in units 10^{22} cm^{-2}). b) A schematic view of the central region of an AGN: the X rays from the nucleus are heavily absorbed, while the 25μm emission, originating from the warm dust on the inner side of the torus, is much less obscured.

dotted (dark) curves give different degrees of mixing of starburst and AGN, with the AGN contribution lowering moving to the left.

In this model the absorption suffered by the IR radiation is only a small fraction of that inferred from the X-ray column density by assuming a standard dust-to-gas ratio. Indeed, we find that there is no way to explain the correlation if we assume that (1) the same amount of material obscures both the X-rays and the infrared and that (2) the dust-to-gas ratio is galactic. This is because if none out of the two hypothesis above are relaxed, the pure–AGN curve is almost horizontal, the IR colour decreasing too fast with respect to the X/IR flux ratio.

A simple explanation for the difference between the absorption in the IR and in the X-rays is shown in Fig. 2b: if the density in the circumnuclear torus decreases radially, the X-rays, emitted in the very central region, are much more absorbed than the 25μm radiation, emitted by the warm dust located along the inner face of the torus, far from the plane of the accretion disk.

Besides the optical classification, the simple picture depicted above is supported by additional pieces of evidence:

1) According to our scheme, the type 2 sources having the warmest IR colour allowed by their X/IR flux ratio are dominated by the AGN component, with low absorption on large scales. Therefore we expect to find broad polarized or near-IR broad lines more often in these objects than in colder ones, since in the latter the

starburst is more and more important, and the AGN lines could be diluted by the radiation contributed by the starburst, or suppressed by the large scale absorption associated with it. This expectation is well confirmed by the optical and near-IR data available in the literature.

2) The X-ray signature of the starburst contribution is a thermal component with typical kT values of 0.1-3 keV. Therefore, in the case of a strong starburst contribution the overall X-ray spectrum should be steeper, and the fit with a single powerlaw should give a photon index larger than in typical AGNs, e.g. $\Gamma > 2$. The observed data go exactly in this direction: the large majority of the IR-cold sources have steep (starburst-like) X-ray indices while IR-warm sources have flatter (AGN-like) indices.

3. THE 20-200 KEV EMISSION OF LIGS

Our sample includes a subsample of 13 sources optically classified as type 2 AGNs and observed by BeppoSAX up to 200 keV. 6 out of the 11 sources that are Compton–thick in the 2-10 keV range are completely absorbed also in the 10 to 200 keV band, therefore implying a column density $N_H > 10^{25}$ cm^{-2}. Only two of the 11 Compton thick sources have an excess in the 15-100 keV range, while for the remaining three the hard 15-100 keV X-ray emission is unconstrained. The shortage of objects with 10^{24}cm^{-2} $<N_H<10^{25}$ cm^{-2} already pointed out in a sample of optically selected Seyfert 2s (Risaliti, Maiolino & Salvati 1999), would have important consequences in the synthesis models of the X-ray background. If a large fraction of sources have $N_H > 10^{25}$ cm^{-2}, the integrated flux of the absorbed AGNs is dimmed and they are on average less detectable at a given flux. Therefore, the ratio between absorbed and unabsorbed AGNs should be increased with respect to the local value to account for the hard XRB intensity and hard X-ray source counts (Gilli, Risaliti & Salvati 1999).

ACKNOWLEDGEMENTS

The authors acknowledge the partial financial support from the Italian Space Agency (ASI) through the grant ARS–99–15 and from the Italian Ministry for University and Research (MURST) through the grant Cofin98-02-32.

REFERENCES

Genzel R., Lutz D., Sturm E., et al. 1998, ApJ 498, 579
Gilli R., Risaliti G., & Salvati M. 1999, A&A, 347, 424
Lutz D., Spoon H. W. W., Rigopoulou D., Moorwood A. F. M., & Genzel R. 1998, ApJ 505, L103
Risaliti G., Maiolino R., & Salvati M. 1999 ApJ 522, 157

THE ORBITAL LIGHT CURVE OF AQUILA X-1

E. L. Robinson, W. F. Welsh, & P. Young

Department of Astronomy, The University of Texas, Austin, TX 78731, USA

ABSTRACT

The R–band light curve of the X-ray nova/neutron-star binary Aql X-1 is dominated by ellipsoidal variations when it is at quiescence, although the ellipsoidal variations are severely distorted and have unequal maxima. The peak-to-peak amplitude of the variations is ≈ 0.25 mag. The orbital period measured from the ellipsoidal variations is consistent with the 18.95 hr period measured by Chevalier & Ilovaisky (1998); and the orbital inclination must be greater than $36°$, and probably lies between $36°$ and $55°$. During outbursts the light curve of Aql X-1 is dominated by the "reflection effect," that is, by heating of the side of the companion star facing the neutron star.

KEYWORDS: X-Ray binaries, Soft X-ray Transients, X-Ray Novae

1. INTRODUCTION

The X-ray nova Aql X-1 undergoes transient outbursts at intervals of somewhat less than one year. Type I X-ray bursts are present at both X-ray and optical wavelengths during the outbursts, demonstrating that the accreting star in the system is a neutron star, not a black hole (Koyama et al. 1981, Robinson & Young 1997).

Recent observations by Callanan et al. (1999) and Chevalier et al. (1999) have shown that the object previously thought to be the optical counterpart of Aql X-1 is, in fact, two stars separated by only 0.46″. Aql X-1 is the fainter of the pair, contributing only 12% of the combined V-band flux when it is at quiescence. The revised quiescent magnitude of Aql X-1 is $V = 21.6$ and it has a late K spectral type, which is presumed to be the spectral type of the companion to the neutron star.

During outbursts the light curve of Aql X-1 has a period of 18.95 hours (Chevalier & Ilovaisky, 1998). The quiescence light curve of Aql X-1 is exceptionally difficult to measure, however, because Aql X-1 is in a crowded field and because ground-based measurements do not generally resolve the close optical pair. Chevalier & Ilovaisky (1998) find a period of 18.95 hours for the quiescent light curve – the same period as for the eruption light curve – and identify this period as the orbital period. Shahbaz et al. (1998) find a period of 19.3 hours for the quiescent light curve. They interpret their 19.3-hour period as the orbital period and the 18.95-hour period during eruption as a reverse superhump period.

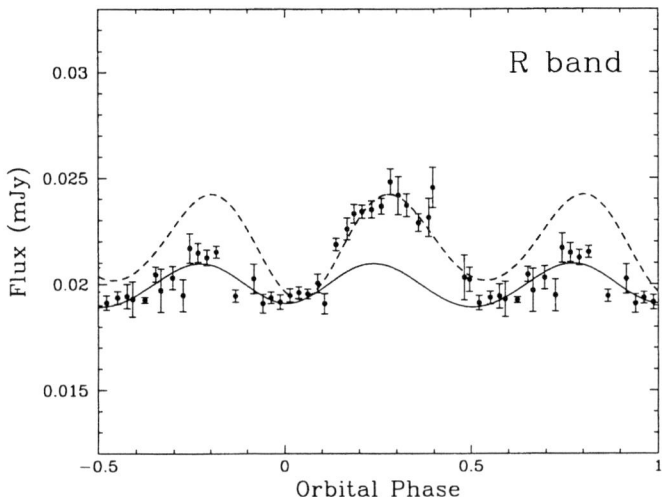

FIGURE 1. The light curve of Aql X-1 during quiescence, and two fits of a model for the ellipsoidal variations. The solid curve is fit to phases 0.5 – 1.0 (the smaller hump) and the dashed curve to phases 0.0 – 0.5 (the larger hump). The two fits require different inclinations, 36° and 55°, which probably bracket the true inclination.

2. THE ORBITAL PERIOD AND LIGHT CURVE AT QUIESCENCE

We obtained R–band CCD photometry of Aql X-1 on 1998 June 24 – 28 using the 2.1-m telescope at McDonald Observatory. Aql X-1 was in quiescence at the time; the X-ray light curve of the previous eruption peaked $\sim$ 85 days before our observations and had been in quiescence for 45 days.

Aql X-1 was not resolved from the field star 0.46" away in our data, so we measured the combined flux from the pair (stars $a + e$ in Chevalier et al. 1999). Since the $a+e$ pair is separated by only $\sim$ 2" from three other nearby field stars, the brightness of the pair cannot be measured reliably by simple aperture photometry of the CCD images as was done for previously published light curves. We measured the brightness of the $a+e$ pair by profile fitting and subtraction of surrounding stars using DAOPHOT with the full prescription outlined by Stetson for crowded-field photometry (Stetson 1987, 1990).

We derived the period of the quiescent light curve by fitting it with a sine curve plus its first harmonic, finding P = 18.71±0.06 hr. (The quoted error is the internal error, the systematic error is about 50% larger.) This period is consistent with the Chevalier et al. (1998) period and inconsistent with the Shahbaz et al. (1998) period. Thus, a single period, P = 18.95 hr (we use the Chevalier period henceforth), fits the light curve of Aql X-1 both in quiescence and during eruption. We identify this period as the orbital period.

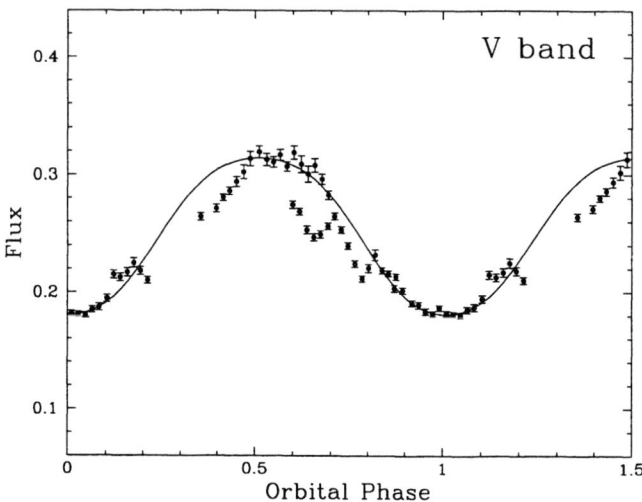

FIGURE 2. The V-band light curve of Aql X-1 during outburst measured by Garcia et al. (1999). The solid line is the synthetic light curve produced by an irradiated, late-K star in a binary with an orbital inclination of 40° and mass ratio $M_K/M_X = 1/3$. While the match to the data is not perfect, the reflection effect does account for the shape and the amplitude of the light curve.

Interpolating between the V- and I-band measurements of Chevalier et al. (1999) we find that star a contributes $\sim 83\%$ of the combined flux from the $a + e$ pair and have subtracted this contaminating flux from the R-band light curve. Figure 1 shows the resulting R-band light curve of Aql X-1 folded at 18.71 hours. The light curve clearly shows the double humps indicative of ellipsoidal variations, but one of the humps is much larger than the other, demonstrating that the light curve is severely distorted.

3. THE ELLIPSOIDAL VARIATIONS AND ORBITAL INCLINATION

The physical mechanism distorting the ellipsoidal variations is unknown, although there are two likely candidates, star spots on the K star and non-axisymmetries in the accretion disk around the neutron star. At visual wavelengths the dominant effect of both mechanisms is to increase the amplitude of one ellipsoidal hump and decrease the amplitude of the other. Thus, fitting an ellipsoidal light curve to the larger hump yields an overestimate the true amplitude of the ellipsoidal variations and an upper limit to the orbital inclination, while fitting an ellipsoidal light curve to the smaller hump yields an underestimate the true amplitude of the variations and a lower limit to the orbital inclination.

We fit the R-band light curve of Aql X-1 with synthetic light curves calculated from a new version of the light-curve program discussed by Zhang at al. (1986). We assumed that the late K star fills its Roche lobe and has a temperature of 4000 K, and adopted $M_K/M_X = 1/3$ for the mass ratio, $\beta = 0.08$ for the gravity darkening coefficient and $u = 0.8$ for the linear limb darkening coefficient (Claret 1998). The derived orbital inclinations, i, are relatively insensitive to any of these parameters. Changing the effective temperature to 5000 K, for example, changed the inclinations by only 2 – 3°.

If there is no other source of light in the R band from, for example, the accretion disk, fits of the models to the ellipsoidal variations yield orbital inclinations of 36° and 55° for the smaller and larger humps respectively. If half the light is coming from the disk, the inclination derived from the smaller hump increases to $i = 50°$, and no fit is possible for the larger hump. Since the Balmer lines in the spectrum of Aql X-1 are weak and since the spectral fits by Chevalier et al. (1999) leave little room for any other light source, there is probably little extra light in the system, and the lower inclinations are more likely to be correct. In any case, the orbital inclination of Aql X-1 must be greater than 36°.

4. HEATING AND THE ERUPTION LIGHT CURVE

During eruptions the orbital light curve of Aql X-1 becomes single peaked. Figure 2 shows the V-band light curve of Aql X-1 obtained by Garcia et al. (1999) during the 1996 July eruption. Adopting the orbital period of Chevalier et al. (1998) we find that the minimum of the V-band light curve is precisely in phase ($|\Delta\phi| < 0.012$) with the $\phi = 0$ minimum of our quiescent light curve obtained in 1998.

The solid line in Figure 2 is the synthetic light curve produced by an irradiated star with an undisturbed temperature of 4000 K filling its Roche lobe in a binary with $i = 40°$ and $M_K/M_X = 1/3$. The match to the data is hardly perfect, especially since Aql X-1 was flickering during the observations, but irradiation does account for both the shape and the amplitude of the light curve. The optical light curve of Aql X-1 was, therefore, dominated by heating of the K star during the eruption.

REFERENCES

Callanan, P., Filippenko, A. V., & Garcia, M. R. 1999, IAU Circ. 7086
Chevalier, C., & Ilovaisky, S. A. 1998, IAU Circ. 6806
Chevalier, C., Ilovaisky, S. A., Leisy, P., & Patat, F. 1999, A&A, 347, L51
Claret, A. 1998, A&A, 335, 647
Garcia, M. R., Callanan, P. J., McCarthy, J., Eriksen, K., & Hjellming, R. M. 1999, ApJ, 518, 422
Koyama, K., et al. 1981, ApJ, 247, L27
Robinson, E. L., & Young, P. 1997, ApJ, 491, L89
Shahbaz, T., Thorstensen, J. R., Charles, P. A., & Sherman, N. D. 1998, MNRAS, 296, 1004
Stetson, P. B. 1987, PASP, 99, 191
Stetson, P. B. 1990, PASP, 102, 932
Zhang, E.-H., Robinson, E. L., & Nather, R. E. 1986, ApJ, 305, 740

MAGNETIC FIELD LIMIT ON SGR 1900+14

R.E. Rothschild [1], D. Marsden[1,2], & R.E. Lingenfelter [1]

1) Center for Astrophysics and Space Sciences, University of California, San Diego, La Jolla, CA USA,
2) presently NAS/NRC Research Associate at Goddard Space Flight Center

ABSTRACT

We measured the period and spin-down rate for SGR 1900+14 during the quiescent period two years before the recent interval of renewed burst activity. We have shown that the spin-down age of SGR 1900+14 is consistent with a braking index of ~ 1 which is appropriate for wind torques and not magnetic dipole radiation. We have shown that a combination of dipole radiation, and wind luminosity, coupled with estimated ages and present spin parameters, imply that the magnetic field for SGR 1900+14 is less than 6×10^{13} G and that the efficiency for conversion of wind luminosity to x-ray luminosity is <2%.

KEYWORDS: gamma-rays: bursts, pulsars: individual (SGR 1900+14), magnetic fields

1. SPIN-DOWN HISTORY OF SGR 1900+14

The spin-down of SGR 1900+14 from 1966 September to 1999 April (Figure 1) is characterized by three intervals of time for which the spin-down rate was essentially constant within the interval. This characterization of the SGR 1900+14 spin-down is based upon either direct measurements of $\dot{P}$ as part of the period determination, or upon differences in measured spin periods between two different observations. The first interval begins with the RXTE observation in September of 1996 and ends with the ASCA observation at the beginning of May, 1998. The mean spin-down $\dot{P} \sim 6 \times 10^{-11}$ s/s (Marsden, Rothschild and Lingenfelter 1999a; Hurley et al. 1999; Woods et al. 1999a). The second interval begins with the onset of bursting on May 26, 1998 and continues until mid-September 1998. The mean spin-down during this time $\dot{P} \sim 13 \times 10^{-11}$ s/s (Kouveliotou et al. 1999; Marsden, Rothschild and Lingenfelter 1999a; Murakami et al. 1999). The third interval begins in mid-September 1998 and continues at least until March 30, 1999. The mean spin-down at that time $\dot{P} \sim 6 \times 10^{-11}$ s/s (Woods et al. 1999b).

Woods et al (1999b) have suggested that the data may be consistent with a discontinuous spin-down event during the second interval as a result of the Superburst, as opposed to a doubling of $\dot{P}$ during the entire second interval (Marsden, Rothschild and Lingenfelter 1999a). This appears to be at odds with the measurement of $\dot{P} = (11.0 \pm 1.7) \times 10^{-11}$ s/s in early June, 1998 (Kouveliotou et al. 1999), approxi-

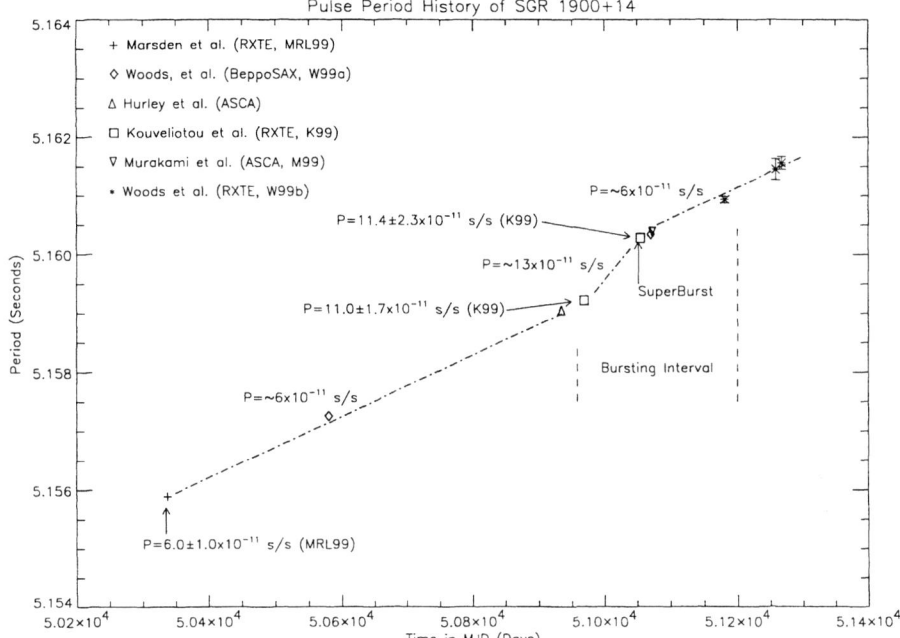

FIGURE 1. The pulse period history for SGR 1900+14 versus time. All of the published values of the pulse period are given along with the three measurements of $\dot{P}$ made as part of the period determination analysis. The mean spin-down in the three time intervals are also given.

mately 3 months before the Superburst and with the RXTE/ASCA determination of $\dot{P} \sim 10 \times 10^{-11}$ s/s just after the event (Murakami et al. 1999).

2. SPIN-DOWN IN SGR 1900+14 IS DUE TO A RELATIVISTIC WIND

Assuming that the spin-down torque is given by $\dot{\Omega} \propto \Omega^n$, the age of a pulsar with period P and spin-down $\dot{P}$ is given by

$$t_{age} = P/[(n-1)\dot{P}]$$

where the spin-down braking index, $n = 3$ for pure magnetic dipole radiation and $n \sim 1$ for wind torques. Inverting the age equation yields

$$n = 1 + (P/\dot{P})(t_{age})^{-1}.$$

Using parameters appropriate for SGR 1900+14, we find that

$$n = 1 + 0.27/(t_{age}/10^4 \text{yr}).$$

This indicates that the braking index for SGR 1900+14 must be ~ 1, and that the spin-down of SGR 1900+14 is dominated by torques due to the relativistic wind and not magnetic dipole radiation.

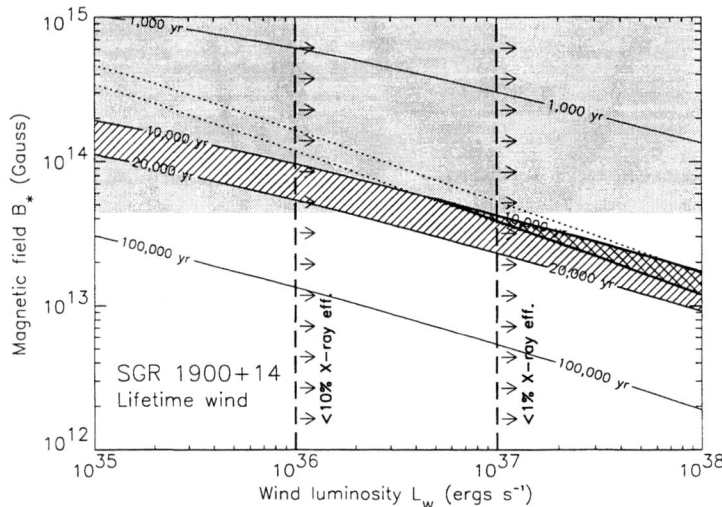

FIGURE 2. Age contours for SGRs 1900+14 for constant magnetic field and luminosity. The cross-hatched areas denote the allowed regions of parameter space given the constraints provided by the age of the associated supernova remnant (solid lines) and long term present-day spin-down rate (dotted lines). The vertical dashed lines denote the 10% and 1% efficiencies for producing the observed flux of 10^{35} ergs/s.

3. SPIN-DOWN TORQUES OF SGRS

In reality, there will be more than one torque spinning down the pulsar at any given time. The torque provided by the emission of a relativistic wind is (Thompson et al. 1999):

$$I_*\dot{\Omega}_w = -\Lambda(L_w/c^2)R_A^2\Omega$$

where I_* is the neutron star moment of inertia, L_w is the luminosity of the wind, $\Omega \equiv 2\pi/P$ is the spin frequency, $\dot{\Omega}_w$ is the spin-down rate due to the wind, and R_A is the Alfven radius. Λ is a constant equal to 2/3 for a magnetic dipole field aligned with the rotation axis. The Alfven radius is given by:

$$\frac{L_w}{4\pi R_A^2 c} = \frac{B_*^2(R_A)}{8\pi}$$

where B_* is the magnetic field of the neutron star. When the Alfven radius is inside the light cylinder radius ($R_A < R_{lc}$, where $R_{lc} = c/\Omega$),

$$I_*\dot{\Omega}_w = -\Lambda B_* R_*^3 \left(\frac{L_w}{2c^3}\right)^{1/2}\Omega$$

where R_* is the radius of the neutron star and dipole geometry is assumed. When the Alfven is outside the light cylinder radius, the torque is limited to

$$I_*\dot{\Omega}_w = -\Lambda L_w \Omega^{-1}$$

The transition frequency between these two wind spin-down regimes is

$$\Omega_{tr} = 8.572 \left(\frac{L_w}{10^{36} \text{ergs/s}}\right)^{1/4} \left(\frac{B_*}{10^{14} \text{G}}\right)^{-1/2} \text{ radians/s}.$$

The torque due to a rotating magnetic dipole is (Shapiro & Teukolsky 1983):
$$I_* \dot{\Omega}_{mdr} = -k \frac{B_*^2 R_*^6}{6c^3} \Omega^3$$
where k = 1 (Harding, Contopoulos, & Kazanas 1999).

Once the total spin-down torque is specified as a function of Ω, the age of the SGR can found by the integral of $d\Omega$ over the total torque divided by I_*, where the integration is performed from initial to present-day angular frequency.

4. MAGNETIC FIELD AND WIND LUMINOSITY LIMITS

Using the above model we explore a wide range of magnetic fields B_* and wind luminosity L_w, shown in Fig. 2. We see that the presently observed period of $P = 5.157$ s, the spindown rate of $\dot{P} = 6 \pm 1 \times 10^{-11}$ s/s (dotted lines) of SGR 1900+14, and the 10 to 20 Kyr range of ages (solid lines) of its associated supernova remnant G42.8+0.6 (Vaisht et al. 1994), tightly constrain the allowable magnetic field to $B_* < 6 \times 10^{13}$ G and wind luminosities $L_w > 5 \times 10^{36}$ erg/s. Compared to the quiescent 2-10 keV x-ray luminosity of $\sim 10^{35}$ erg/s (Murakami et al. 1999), this wind luminosity implies a $< 2 \%$ conversion efficiency of wind energy to x-rays in that band which is quite consistent with theoretical calculations (Tavani 1994, Harding 1995; Harding, Contopoulos & Kazanas 1999). The magnetic field limits are also quite consistent with the limiting values inferred for radio pulsars, but not with those expected for magnetars. Very similar limits are set by comparable analyses (Harding, Contopoulos, & Kazanas 1999; Marsden, Rothschild & Lingenfelter 1999b) of SGR 1806-20 and its supernova remnant G10.0-0.3.

REFERENCES

Harding, A.K. 1995, in High Velocity Neutron Stars and Gamma-ray Bursts, AIP Conf. Proc. 366, eds. R.E. Rothschild & R.E. Lingenfelter, (AIP Press: New York), 118.

Harding, A.K., Contopoulos, I., & Kazanas, D. 1999, Ap. J., 525, L125.

Hurley, K. et al. 1999, Ap. J., 510, L111.

Kouveliotou, C. et al. 1999, Ap. J., 510, L115.

Marsden, D., Rothschild, R.E., & Lingenfelter, R.E. 1999a, Ap. J., 520, L107.

Marsden, D., Lingenfelter, R.E., Rothschild, R.E., & Higdon, J.C. 1999b, Ap. J., submitted (astro-ph/9912207).

Murakami, T. et al. 1999, Ap. J., 510, L119.

Shapiro, S.L. & Teukolsky, S.A. 1983, Black Holes, White Dwarfs, and Neutron Stars, (John Wiley & Sons: New York).

Tavani, M. 1994, Ap. J., 431, L83.

Thompson, C. et al. 1999, Ap. J., in press (astro-ph/998086).

Vasisht, G., et al. 1994, Ap. J., 431, L35.

Woods, P.M. et al. 1999a, Ap. J., 518, L103.

Woods, P.M. et al. 1999b, Ap. J., 524, L55.

X-RAY IRON LINE VARIABILITY FOR THE MODEL OF AN ORBITING FLARE ABOVE A BLACK HOLE ACCRETION DISC

Mateusz Ruszkowski

Institute of Astronomy, University of Cambridge, Madingley Road, Cambridge CB3 OHA, UK

ABSTRACT The broad fluorescent iron line profiles detected in many AGNs contain information about the black hole spin and the accretion disc and may help to test general relativity in the strong gravity regime. We embark on the computation of the temporal response of the line to the illuminating flux. Previous studies concentrated on the calculation of reverberation signatures from static sources illuminating the disc. We focus on the more physically justified case of flares located above the accretion disc and corotating with it. We compute the time dependent iron line taking into account all general relativistic effects and show that it is of very complex nature. We suggest that the temporal behaviour of Fe line profiles seen in future data from XMM or Constellation-X may not necessarily imply a complicated model but could be explained for example in the framework of the orbiting flare model.

KEYWORDS: accretion, accretion discs - black hole physics - galaxies: Seyfert - X-rays: galaxies

1. INTRODUCTION

The fluorescent X-ray iron line may be produced by irradiation of the disc material by flares located above the accretion disc close to the supermassive black hole. As a result, the observed time-averaged spectra have distinctive, skewed, double-peaked profiles which reflect the Doppler and gravitational shifts in a strongly curved space-time (Fabian et al. 1989, Laor 1991). Since the spectral properties of the Fe Kα line are well known, the time-averaged iron line may be used as a probe of both the black hole and its near environment. This may permit the study of the geometry of the emission region and the search for the dragging of inertial frames.

1.1. Reverberation mapping

A promising approach for searching for the observational signatures of the spin and mass of the black hole and the geometry of emission region was first suggested by Fabian et al. (1989) and was later considered in any detail by Stella (1990). He computed the iron line variability from a disc around a Schwarzschild black hole assuming a pointlike variable X-ray source located in the geometric centre of the disc. For an observer at infinity any such variations of the primary X-ray source would be 'echoed' by different locations on the disc leading to time evolution of the line profile. Recently Reynolds et al. (1999) generalized these results by assuming arbitrary black hole spin and searched for observational signatures of this parameter but mainly focused on the case of a static, instantaneous, on-axis flare.

1.2. Orbiting flare model and iron line variability

The duration of bright flares can be of order $100m_7$ in geometric time units ($GM/c^3 \approx m_7 49\,s$ for a black hole mass $M = 10^7 m_7 M_\odot$) which is comparable to the orbital period close to the black hole. It is very likely that any variability associated with blobs above accretion discs comes from moving sources. Thus in general, the assumption of a static, instantaneous primary X-ray source made in previous studies limits the applicability of such results. In the present work we relax these assumptions and consider the reverberation effects from corotating, off-axis, non-instantaneous flares above an accretion disc. We calculate the time-dependent iron line for different values of the accretion disc inclination, black hole spin parameter and position of the flare relative to the accretion disc.

2. RESULTS AND DISCUSSION

2.1. The signatures of black hole spin

The figures show the iron line variability corresponding to one full revolution of the flare above the accretion disc. There is a clear difference between the results for the two considered extremal values of the spin parameter (compare upper and lower panels). The iron line extends to lower frequencies in the Kerr case which is due to the cold iron line being produced very close to the black hole. In the case of the Schwarzschild black hole the disc may be highly ionized below the radius of marginal stability ($r_{\rm ms} = 6m$) and the line may be either destroyed by the Auger process or not produced because of the total ionization of the disc material. Other interesting features are the two 'bumps' seen for the inclination $i = 50°$ and the Schwarzschild black hole. These features are due to blueshifted 'hot' iron lines ($E^{(1)} = 6.67$ keV and $E^{(2)} = 6.97$ keV) from the ionized regions and *could be detected* by future X-ray satellites.

2.2. The manifestation of relativity in the variable Fe line

The most prominent feature seen in all diagrams is the drifting maximum of the flux. This sinusoid-shaped feature is due to the fluorescence from the most strongly illuminated part of the disc located just below the flare. Initially the flare moves in a direction perpendicular to the line of sight and the maximum is redshifted due to the transverse Doppler effect and gravitational redshift. Later it gradually begins to recede from the observer. The brightness of this maximum is suppressed at this stage because the flux from the highly-illuminated part of the disc below the flare decreases as a result of light beaming away from the observer. The emerging *high*

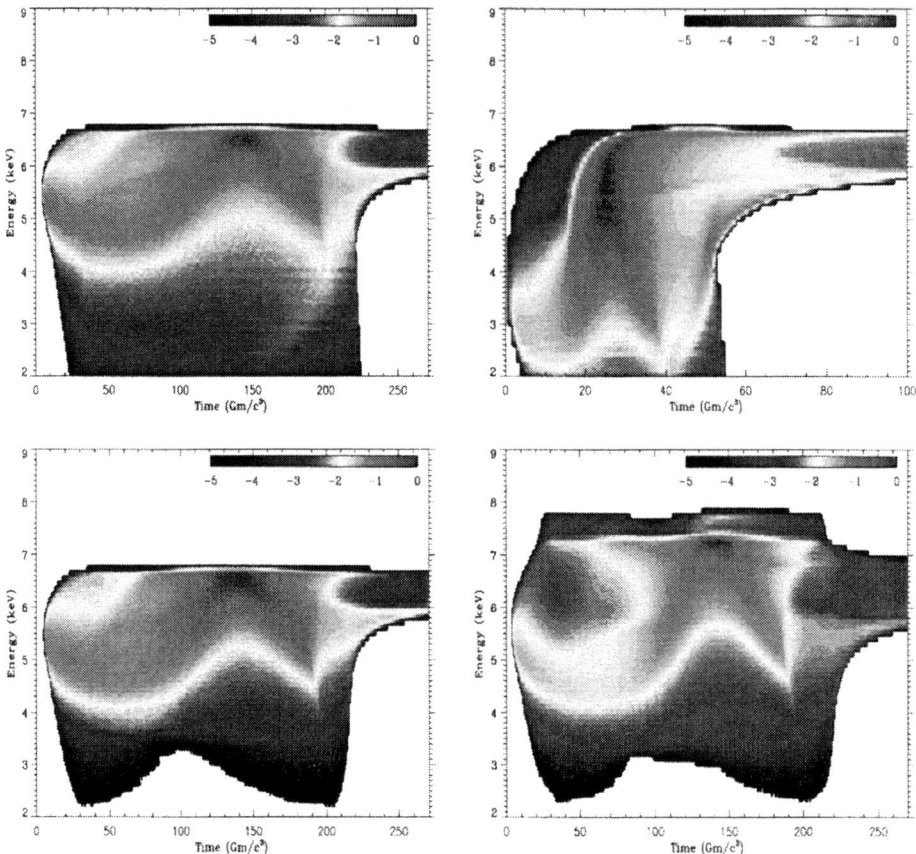

FIGURE 1. Normalized time sequence of Fe line spectrum ($\log(Flux/Flux_{\max})$) for: $a = 0.998, r = 10m, \theta = 70°$ (the inclination of the flare relative to the symmetry axis), $i = 30°$ (upper left panel); $a = 0.998, r = 3m, \theta = 70°, i = 30°$ (upper right panel); $a = 0.0, r = 10m, \theta = 70°, i = 30°$ (lower left panel); $a = 0.0, r = 10m, \theta = 70°, i = 50°$ (lower right panel);

energy maximum on the other hand (e.g. at 50-100 Gm/c^3 on lower right panel), is a consequence of the strong bending of light which is focused on the opposite side of the black hole relative to the actual position of the flare and leads to an enhanced illumination of the disc in this region. This radiation is then further amplified by Doppler boosting. This effect could be readily *observable* by future high throughput spectrometers. As the flare rotates and enters the approaching side of the disc the bulk of the iron line flux shifts towards high energies and the Doppler boosting of the radiation reflected by the strongly-illuminated regions of the disc leads to the rise (in energy) of the iron line maximum and its brightening. Note that during the whole revolution of the flare the gravitational and transverse Doppler redshifts may for some time overwhelm the blueshifts when the flare is on the approaching side of the disc. This effect leads to the sharpening of the bright Λ-shaped feature. An interesting feature can also be seen when the flare is receding ($\phi \approx 90°$). The main redshifted flux maximum *broadens* or even *splits* into two maxima (compare lower panels at 50-100 Gm/c^3). The middle maximum is a result of the strong bending of light which is focused from the back side of the disc relative to the observer. As in previous cases, such effect *could be observed* by future X-ray spectrometers.

The magnitude of the energy shift of the main maximum is related to the inclination of the accretion disc (compare lower panels) and also to the distance of the flare from the centre. The latter effect is demonstrated on the upper panels, where the bright Λ-shaped feature extends to lower energies for smaller distances r of the flare from the centre. The same figures also illustrate how the timescale of the sinusoidal variation changes as a function of r with shorter time scales corresponding to faster orbital motion of the closer flares. Note how the structures on the reverberation map change for a very close flare (see upper right panel). The middle and strong lower maxima on the right side of the Λ-shaped feature are due to strong gravitational focusing. The lack of the uppermost and middle maxima on the left hand side of this feature is related to the very large overall gravitational and transverse Doppler redshift. It is also caused by the faster motion of the flare (and thus the main flux maximum) combined with the relatively slow propagation of signals slowed by the Shapiro delay in the vicinity of the black hole.

ACKNOWLEDGEMENTS

MR acknowledges support from an External Research Studentship of Trinity College, Cambridge; an ORS Award; and the Stefan Batory Foundation.

REFERENCES

Fabian A.C., Rees M.J., Stella L., White N.E., 1989, MNRAS, 238, 729
Laor A., 1991, ApJ, 376, 90
Reynolds C.S., Young A.J., Begelman M.C., Fabian A.C, 1999, ApJ, 514, 164
Stella L., 1990, Nature, 344, 747

THE INFLUENCE OF RESONANT ABSORPTION ON THE FE EMISSION LINE PROFILES FROM ACCRETING BLACK HOLES

Mateusz Ruszkowski & Andrew C. Fabian

Institute of Astronomy, University of Cambridge, Madingley Road, Cambridge CB3 OHA, UK

ABSTRACT Diffuse plasma existing above an accretion disc can affect the X-ray spectrum by iron $K\alpha$ resonance absorption. This in turn can influence the estimation of the accretion disc and black hole parameters. We embark on a fully relativistic computation of this effect and calculate the iron line profile in the framework of a specific model in which rotating, highly ionized and resonantly-absorbing plasma occurs close to the black hole. This can explain the features seen in the iron $K\alpha$ line profile recently obtained by Nandra et al. (1999) for the Seyfert 1 galaxy NGC 3516. We show that the redshift of this feature can be mainly gravitational in origin and accounted for without the need to invoke fast accretion of the matter onto the black hole. New X-ray satellites such as XMM, ASTRO-E and Chandra provide excellent opportunities to test the model against high quality observational data.

KEYWORDS: accretion, accretion discs - black hole physics - galaxies: active - line: profiles - galaxies: individual (NGC 3516)

1. INTRODUCTION

We assume that the accretion disc is optically thick and geometrically thin and the primary source of X-ray radiation is an optically-thin corona located just above the accretion disc. The primary flux is incident on the disc and produces fluorescent photons. The line and continuum photons can then interact with the cooler($\sim 10^7$K) Thomson-thin plasma in which the disc, corona and the black hole are embedded. Under these conditions a significant fraction of iron ions in the diffuse plasma are in the hydrogen (Fe XXVI) and helium-like (Fe XXV) states. These ions can resonantly absorb the radiation from the accretion disc and corona. The allowed transitions which are of interest to us are $1s(^2S) - 2p(^2P)$ and $1s^2(^1S) - 1s2p(^1P)$ with the corresponding large oscillator strengths f_{lu} equal 0.416 for H-like and 0.794 for He-like ions. The rest energies of these transitions are $E_H = 6.9$ keV and $E_{He} = 6.7$ keV. This means that absorption lines can potentially change the shape of the observed broad iron line which is generated in the accretion disc with the rest energy $E_{disc} = 6.4$ keV. In order to calculate the optical depth, we first estimate the characteristic distance λ_{Sob} over which the energy of a photon, as seen by the observer who is at rest with the absorbing plasma, changes by the Doppler thermal width ΔE_D. We use the equation:

$$\left|\frac{dl}{dE}\right| = (|E_{abs}(p_\mu u^\mu)_{;\alpha} p^\alpha|)^{-1} \equiv \lambda = \frac{\lambda_{Sob}}{\Delta E_D}, \qquad (1)$$

where E_{abs} is the rest frequency of the absorption line, l is the proper length as measured in the local rest frame of the plasma flow (Novikov & Thorne, 1973). In the above equation p_μ is the 4-momentum of a photon and u^μ is the 4-velocity of the absorbing medium. We choose the velocity field $u^\mu = C(\partial_t + \Omega \partial_\phi)$, where $C = (-g_{tt} - 2\Omega g_{t\phi} - \Omega^2 g_{\phi\phi})^{-1/2}$ and $\Omega = u^\phi/u^t$ is the angular velocity of the absorbing medium as seen at infinity. From equation (1) we get $\lambda_{Sob} = 5.7 \times 10^{-4}(T/10^8 K)^{1/2}\lambda$. It has been checked numerically that for photon trajectories connecting the accretion disc and observer $\lambda_{Sob} \ll R_c$, where R_c is the radius of the absorbing cloud of constant density. The above constraint on λ_{Sob} ensures that we may use the Sobolev approximation to describe the resonant absorption in the plasma, i.e. we may assume that the absorption occurs locally and, for most of the trajectory of a photon with a given emission energy, the medium is transparent. Thus, in the further analysis, we model the absorption line as a Dirac delta function. The optical depth per absorption event is then given by the formula:

$$\tau = 5.5 \times 10^{-2} \left(\frac{E_{abs}}{6.7 \text{keV}}\right)^{-1} \left(\frac{A_{Fe}}{2A_{Fe\odot}}\right) \left(\frac{f_{lu}}{0.5}\right) \left(\frac{f_l N_H}{10^{23} \text{cm}^{-2}}\right) \left(\frac{\lambda}{R_c}\right),$$

where A_{Fe} is the abundance of iron, f_l is the fraction of ions in the H-like or He-like state and N_H is the parameter related to the column density. In our model the absorbing medium is optically thin for resonant and Thomson scattering.

We use the ray back-tracing method and search for the absorption regions corresponding to all energies and calculate the optical depth as a function of energy for each photon trajectory. We separately include an additional contribution to the iron line from the deexciting ions in the ionized cloud.

2. RESULTS AND DISCUSSION

The examples of the computed iron line profiles for the parameters similar to that obtained by Nandra et al. (1999) for NGC 3516 are shown in Fig. 1. The thick solid denotes the total iron line profile. The shape of the line on the left panel may provide an acceptable fit to the temporary line profile obtained by Nandra et al. (1999) (compare with the inset in their Fig. 1). It peaks around 6.2 keV, has a long redshifted tail and possesses the redshifted absorption feature below 6.0 keV. The observed redshift of the absorption feature can be mostly accounted for by the strong gravitational redshift alone. This has to be contrasted with the interpretation originally given by Nandra et al. (1999) who suggest that the redshift of the resonant absorption feature may be due to the scattering of radiation by the matter infalling onto the black hole. Such matter would have to fall in along the black hole rotation

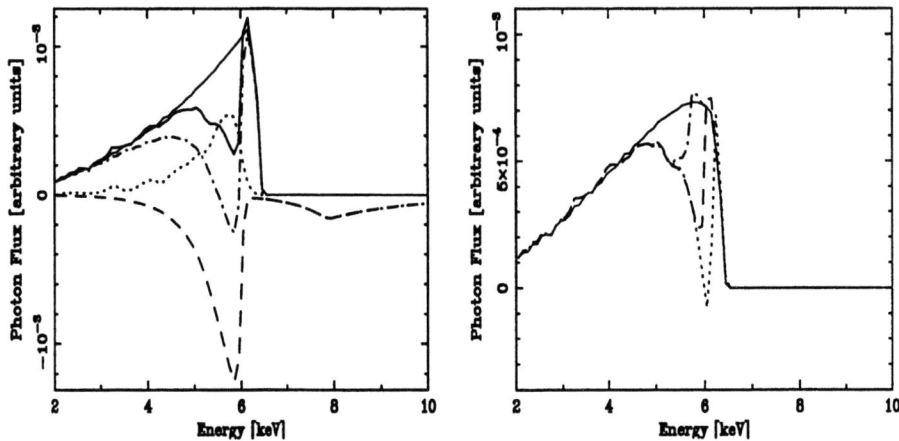

FIGURE 1. The Fe line profiles (see text for explanations)

axis in order to efficiently absorb radiation from the central parts of the disc. Thus their main interpretation implicitly implies accretion with low angular momentum which is less likely. Our results suggest that the current observational data do not necessarily give *direct* evidence for the accretion of matter onto the black hole in NCG 3516.

2.1. The components of the iron line profile

The other profiles seen on Fig. 1 (left panel) correspond to different contributions to the total line profile, namely: disc fluorescent emission (thin solid line), resonant absorption of the continuum (dashed line), deexcitation correction (dotted line) and the line absorbed by resonant and photoelectric absorption (dash-dotted line). Note that there is no intrinsic resonant absorption of the fluorescent disc emission line and only the continuum radiation is resonantly absorbed. This is the result of the low inclination of the disc. Emission line photons with rest energies at 6.4 keV leave the disc and continually move towards increasing gravitational potential and therefore their locally observed energies gradually decrease and can never exceed the rest energy of the resonant absorption line at 6.7 keV. The characteristic skewed shape of the profile of resonant absorption of the continuum is the result of the systematic effect of the gravitational redshift and is also due to the variations of the Sobolev length across the disc as seen by the observer. The coherence length is larger for photons on the approaching side of the disc and smaller on the opposite side because plasma and photons on the receding side locally travel in opposite directions. The left panel also shows the contribution to the line from the deexciting iron ions. As expected, the profile of this line has the shape similar to that of the disc fluorescent line.

2.2. The absorption edge

The depth of the accompanying photoelectric absorption edge depends on our 'column density' $N_{\rm H} \equiv n_{\rm H} R_{\rm c}$, where $n_{\rm H}$ is the hydrogen number density. The amount of ionized plasma which reprocesses the continuum and the emission line is constrained by a nondetection of the absorption edge of iron in the profiles obtained by Nandra et al. (1999). Our definition of $N_{\rm H}$ does not include the fact that the real column densities for geodesics intersecting the cloud further away from the centre are smaller which will reduce the iron edge. It also has to be stressed that any absorption edge created close to the black hole would be additionally smeared out by a factor of a few by the strong relativistic effects operating in this region. Therefore we calculate the iron edge using a fully relativistic treatment to check what the allowed values of $N_{\rm H}$ are. The strength of the calculated absorption edge is well within the acceptable range given the quality of the current data.

2.3. Geometry of the absorbing medium

The right panel shows the effect of the size of the cloud on the line profile. Large cloud sizes lead to stronger absorption because the coherence length over which resonant absorption can occur increases with distance from the black hole. Of course the smaller the cloud the greater the redshift of the continuum absorption feature, however the magnitude of absorption decreases as the size of the cloud gets smaller. It is plausible that the absorbing plasma exists in the form of small magnetically confined clumps. If the plasma is in photoionization equilibrium, then the ionization parameter necessary to produce a high ionization state, has to be of order $\xi \sim 10^4$. Assuming the X-ray luminosity, the size of the absorbing region and the column density to be respectively of the order $L_X \sim 10^{43}$ erg s^{-1}, $R_{\rm c} \sim 10^{14}$ cm and $N_{\rm H} \sim 10^{24}$ cm^{-2}, one can obtain a rough estimate of the filling factor $f \propto \xi N_{\rm H} R_{\rm c}/L_X \sim 0.1$ implying a clumpy structure to the plasma. Therefore, an ensemble of very small clouds or filaments distributed at a range of distances from the central black hole could in principle produce a number of absorption features superimposed on the emission line, an effect similar to the 'Lyα forest' observed in the spectra of distant quasars. In the present case such features are likely to be time variable.

ACKNOWLEDGEMENTS

MR acknowledges support from an ERS of Trinity College, Cambridge; an ORS Award; and the Stefan Batory Foundation. ACF thanks Royal Society for support.

REFERENCES

Nandra K., George I.M., Mushotzky R.F., Turner T.J., Yaqoob T., 1999, ApJ, 523, L17
Novikov I.D., Thorne K.S., 1993, in Black Holes [Les Houches Summer School 1972], DeWitt C. ed. (New York: Gordon and Breach Science Publishers)
Ruszkowski M., Fabian A.C., 2000, MNRAS, 315, 223

MORPHOLOGICAL ANALYSIS OF A STATISTICALLY COMPLETE X-RAY SELECTED SAMPLE OF SEYFERT GALAXIES.

M. Salvato [1], P. Böhm [1], J-U. Fischer [1], G. Hasinger [1], I. Lehmann [1], P. Rafanelli [2]

1) Astrophysikalisches Institut Potsdam Germany, 2) Dipartimento di Astronomia, Universitá di Padova, Italy

ABSTRACT   Galaxy interactions have been suggested to play an important role in triggering AGN activity. In this context we have started to analyze morphological properties of nearby AGNs ($z<0.1$) and their environment. We have chosen an unbiased statistical complete flux-limited AGN sample of 101 Seyfert galaxies from the ROSAT Bright Survey (RBS). Here we present the research project and show first results.

KEYWORDS: surveys, galaxies: interactions, galaxies: fundamental parameters, galaxies: Seyfert

1. INTRODUCTION

The hypothesis that galaxy interactions triggers AGN activity, originates from numerical simulations by Toomre & Toomre (1972) and theoretical suggestions by Gunn (1979). This idea was largely developed in the following years: several observations (e.g. Kennicutt et al. 1987; Keel & Van Soest 1992; Telesco et al. 1993) suggested that interaction can enhance star formation processes. This statement is supported also by numerical simulations of galaxy encounters (e.g. Barnes & Hernquist 1991, Hernquist & Mihos 1995). Petrosian (1982) and Dahari (1984) found an excess of active galaxies with close companions compared with a sample of normal galaxies and similar excesses have been reported by Keel et al. (1985), MacKenty (1989) and Rafanelli et al. (1995).

Studies of Fuentes-Williams & Stocke (1988) and Laurikainen & Salo (1995), instead, came to opposite results. They found in fact that Seyfert environments are not richer than the environments of a control sample of normal galaxies. Recently De Robertis et al. (1998) confirmed this result. Finally, Xanthopoulos (1996) looking for a connection between galaxy interaction and morphology, has found that in a sample of 27 Seyfert galaxies 52% are barred galaxies and half of these have nearby companions. However, in order to properly evaluate these results we must keep in mind the criteria used by each author to select their Seyfert and control samples, as well as the approach that was adopted in their analysis. The selection of the samples followed morphological or spectroscopical criteria in the optical band with

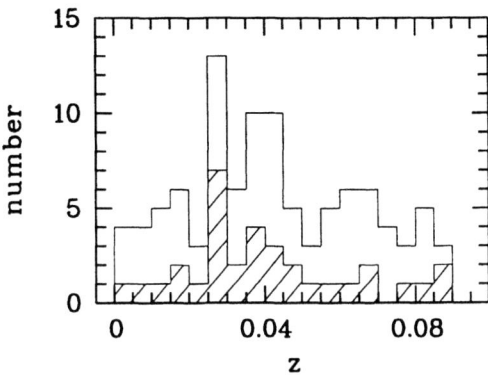

FIGURE 1. Sample redshift distribution. The solid line represents the total sample. The hatched line gives the fraction of Seyfert galaxies in groups (merging, pair, group) with concordant redshift resulting from NED.

possible biases depending on the selection. Even the definition of "companion" of the Seyfert is different from author to author.

We are choosing a different approach. First, we are using a complete X-ray selected sample of relatively local Seyfert galaxies derived form the ROSAT Bright Survey (RBS; Schwope et al. 2000). For Seyfert 1 galaxies this sample should have no selection biases in the optical band and the fraction of Seyfert2 as expected is very small. We then study the morphology distribution, the frequency of disturbed isophotes and the pairing fraction of this sample. A preliminary study of a subsample of the RBS has already revealed that 40% of Seyfert galaxies are in interacting systems (Fischer et al. 1997).

Since it is impossible to construct an unbiased X-ray selected comparison sample of non-Seyfert galaxies we plan to compare the morphological properties of our sample a.) with an optically selected sample of local Seyfert galaxies (Rafanelli et al. 1995) and b.) with a sample of 21 Seyfert galaxies at hight redshift ($0.9 < z < 1.9$) selected from the ROSAT Ultradeep Surveys (Hasinger et al. 1999), which will be observed with the WFPC2 at HST in AO8.

In the sections 2 and 3 we present the sample selection and the fundamental steps in data analysis. In the last section we present first results and discuss future works.

2. SAMPLE SELECTION

We extracted our sample from the RBS which comprises all the bright, high-galactic-latitude X-ray sources detected in the ROSAT All-Sky Survey (1RXS catalogue, Voges et al. 1996). Out of the RBS we selected a statistically complete, flux limited (ROSAT PSPC 0.5–2.0 keV countrate $> 0.3\ cts/sec$), local ($z < 0.1$) sample of 101

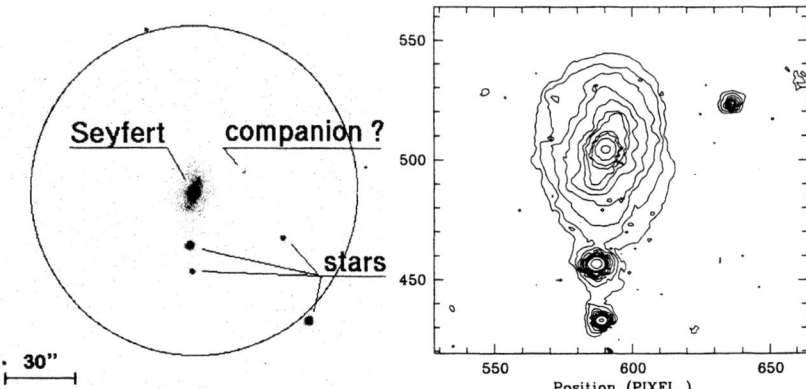

FIGURE 2. R-band image of the Seyfert galaxy Z0920+2308 (left panel) and isophotal contours (right panel) of the adaptively smoothed image. On the left panel the circle indicates the search radius for possible companions. A faint candidate companion is marked. On the right panel, the presence of the bar is suggested from the twist of the isophotes.

emission line AGNs. The luminosity of these objects is typicaly greater than 10^{42} erg/s, so we can safely assume that the X-ray emission is dominated by the nucleus. It turns out that all are Seyfert galaxies. Moreover from NED we found that 33 of them are in pairs or groups with concordant redshift (see Fig. 1).

3. OBSERVATIONS AND DATA ANALYSIS

We have started an imaging survey program to study the morphology of our sample using the CCD camera on the 1.23m Calar Alto telescope. The camera was equiped with a TEK # 7 (1024×1024) chip. The scale of the image is 0.5 arcsec per pixel. To track young stars and star-forming regions, concentrated especially in tidal tails and bridges, we use images in the B filter, while information about the old star distribution and H_α emission comes from images in the R filter.

After the application of standard reduction to the data we do the following steps:

1. We determine the apparent angular dimension of the major axis of the Seyfert to inspect for close companions. Following the criteria of Rafanelli et al. (1995), we search objects within 3 times the diameter of the Seyfert and differing not more than 3 mag from it (see Fig. 2). We will obtain optical spectra of the candidate companions to confirm their physical association with the Seyfert.

2. In parallel we started a detailed analysis of the Seyfert morphology. To reveal structures with low surface brightness we apply an smoothig adaptive filter to the image. In this way it is possible to take into account the fluctuation of the background in the image and to adapt to it the response (Lorenz et al. 1993).

From the resulting image, we get the isophotal contours. One example of the results is shown in Fig. 2 (right panel). The position angle of the major axes of the isophotes is changing as a function of radius of the Seyfert. This twist of the isophotes can be interpreted as evidence of presence of a bar.

3. The next step in the analysis will be to fit the isophotes with ellipses and to analyze the derived parameters. Moreover we will start to study the luminosity profiles and the color profiles of the galaxies.

4. SUMMARY

Using a new sample selection in the X-ray band with the purpose to avoid biases, and using a set of indicators of the morphology of the galaxies, we are trying to define the role of galaxy interactions and morphology assymmetries for AGN activity. The first promising results come from the analysis of the environment of the Seyferts. After analysis of 20 fields we found in 14 cases candidate companions using the criteria defined in section 3 and this is in accord with the idea of a connection between galaxy interaction and AGN activity. We will start the spectroscopic confirmation of their physical association with the Seyfert galaxies in April 2000. Then it will be possible to define the real fraction of interacting systems to compare with the results found in different samples as explained in section 1.

REFERENCES

Barnes, J.E., Hernquist, L., 1991, ApJ, 370, L65
Dahari, O., 1984, AJ, 89, 7
De Robertis, M.M., Yee, H.K.C., Hayhoe, K., 1998, ApJ, 496, 93
Fischer, J.-U. et al., 1998, Astron. Nachr., 319, 347
Fuentes-Williams, T., Stocke, J.T., 1988, AJ, 96, 1235
Gunn, J., 1979, Edited by Hazard and Mitton, CUP, 213
Hasinger, G., Lehmann, I., Giacconi, R., et al., 1999, astro-ph/9901103
Hernquist, L., Mihos, J.C., 1995, ApJ, 448, 41
Kell, W.C. et al., 1985, AJ, 90, 708
Kell, W.C., Van Soest, E.T.M., 1992, AAS, 94, 553
Kennicut, R.C. et al., 1987, AJ, 93, 469
Laurikainen, E., Salo, H., 1995, AAA, 293, 683
Lorenz, H., et al., 1993, AAA, 277, 321
MacKenty, J.W., 1989, ApJ, 343, 125
Petrosian, A.R, 1982, Afz, 18, 548
Rafanelli, P., Violato, M., Baruffolo, A., 1995, AJ, 109, 1546
Schwope, A., Hasinger, G., lehmann, I., et al., 2000, Astron. Nachr., 321, 1
Telesco, C.M., Dressel, L.L., Wolstencroft, R.D., 1993, ApJ, 414, 120
Toomre, A., Toomre, J., 1972, ApJ, 178, 623
Voges, W., 1999, A&A, 349, 389
Xanthopoulos, E., 1996, MNRAS, 280, 6

IRRADIATION OF THE SECONDARY STAR IN X-RAY NOVA SCORPII 1994 (=GRO J1655–40)

T. Shahbaz [1], P. Groot [2], S.N. Phillips [1], J. Casares [3], P.A. Charles [1], J. van Paradijs [2]

1) University of Oxford, Department of Physics, Oxford, UK
2) Astronomical Institute "Anton Pannekoek", University of Amsterdam and Center for High Energy Astrophysics, The Netherlands
3) Instituto de Astrofísica de Canarias, Tenerife, Spain

ABSTRACT

We have obtained intermediate resolution optical spectra of the black-hole candidate Nova Sco 1994 in June 1996, when the source was in an X-ray/optical active state ($R \sim 15.05$). We measure the radial velocity curve of the secondary star and obtain a semi-amplitude of 279±10 km s^{-1}; a value which is 30 per cent larger than the value obtained when the source is in quiescence. Our large value for K_2 is consistent with 60 per cent of the secondary star's surface being heated; compared to 35 per cent, which is what one would expect if only the inner face of the secondary star were irradiated. Effects such as irradiation-induced flows on the secondary star may be important in explaining the observed large value for K_2.

KEYWORDS: binaries: spectroscopic – black hole physics – X-rays: stars – stars: individual (GRO J1655–40).

1. INTRODUCTION

Nova Sco 1994 was discovered on July 27 1994 with BATSE on board the Compton Gamma Ray Observatory. It has been studied extensively during the past few years in X-rays and at optical and radio wavelengths. Strong evidence that the compact object in Nova Sco 1994 is a black hole was presented by Bailyn et al. (1995) who initially established a spectroscopic period of 2.601 ± 0.027 days, classified the secondary as an F2–F6IV type star and suggested $f(M)$=3.16±0.15 M$_\odot$. Shahbaz et al. (1999) using only quiescent data revsied $f(M)$ to 2.73±0.09 and the mass of the black hole to be 5.5–7.9 M$_\odot$.

The effect of heating of the secondary is to shift the 'effective centre' of the secondary, weighted by the strength of the absorption lines, from the centre of mass of the star. One expects that this results in a significant distortion of the radial velocity curve and renders a sinusoidal fit to be clearly inadequate, leading to a spuriously high radial velocity semi-amplitude. In order to quantify this effect we have determined the radial velocity variations of the secondary star in Nova Sco

1994, when it was in outburst and compare our results with others obtained using data taken when the source was in different X-ray states.

2. OBSERVATIONS AND DATA REDUCTION

Intermediate resolution optical spectra of Nova Sco 1994 were obtained on 1996 June 20–24 with the 1.54-m Danish Telescope at ESO in Chile using DFOSC. The spectral resolution was 7.6 Å and 5.5 Å for the first and other nights respectively. A total of 47 spectra were taken each having exposure times of 1800s.

Using the same setup as for the spectroscopy, we also obtained limited Bessell r-band images of Nova Sco 1994 every night. We applied aperture photometry to Nova Sco 1994 and several nearby comparison stars within the field of view and estimate $R \sim 15.05$ for Nova Sco 1994.

3. THE RADIAL VELOCITY CURVE

The radial velocities of the F-type secondary star in Nova Sco 1994 were measured from the spectra by the method of cross-correlation with a template star. Only regions of the spectrum devoid of emission lines (6400-6520Å) were used in the cross-correlation.

Using the orbital ephemeris given by van der Hooft et al., (1998) we phase-folded and binned the heliocentric radial velocities (see Figure 1). A sine wave fit to the data points (not including the point at phase 0.2) yields a χ_ν^2 of 1.5, a semi-amplitude $K_2 = 279 \pm 10$ km s^{-1}, systemic velocity $\gamma = -155 \pm 7$ and a phase shift of $-0.043 \pm 0.005\phi$.

4. IRRADIATION OF THE SECONDARY STAR

It has been known for some time, especially in studies of dwarf novae and polars, that substantial heating of the secondary star shifts the effective centre of the secondary, weighted by the strength of the absorption lines, from the centre of mass of the star. This results in a significant distortion of the radial velocity curve leading to a spuriously high semi-amplitude and a radial velocity curve that is eccentric. Davey & Smith (1992) describe a procedure for detecting the effects of irradiation on the radial velocity curve of the secondary star, whereby one tests the significance of an eccentricity in the orbital solution However, it should be noted that, although our data does not allow this eccentricity test, we can use the spuriously high radial velocity semi-amplitude to show that X-ray heating is present.

In order to investigate the effects of X-ray heating on the secondary star's radial velocity curve we used the model described by Phillips, Shahbaz & Podsiadlowski (1999). The maximum possible change that irradiation can have on K_{obs}, based purely on geometry, is 15 per cent. However, from our data presented in this paper, we observe $\Delta K_2/K_2 = 0.30 \pm 0.05$, which when compared with maximum possible value based on geometry, is significant at the 3-σ level.

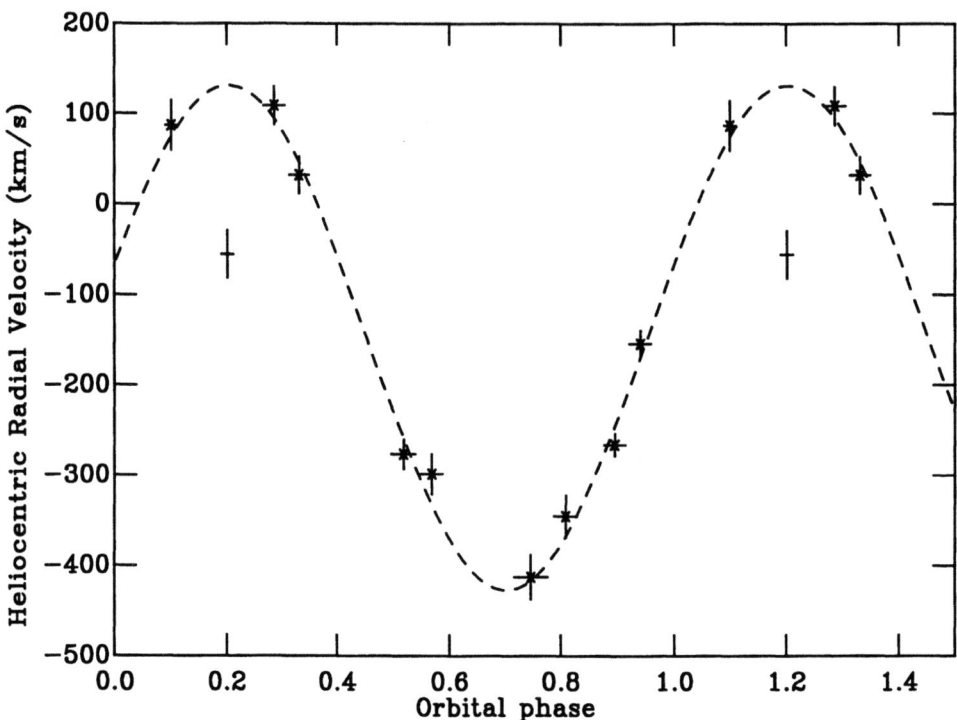

FIGURE 1. The radial velocity curve of the secondary star in Nova Sco 1994. The dashed curve is a sinusoidal fit to the data points marked with a star.

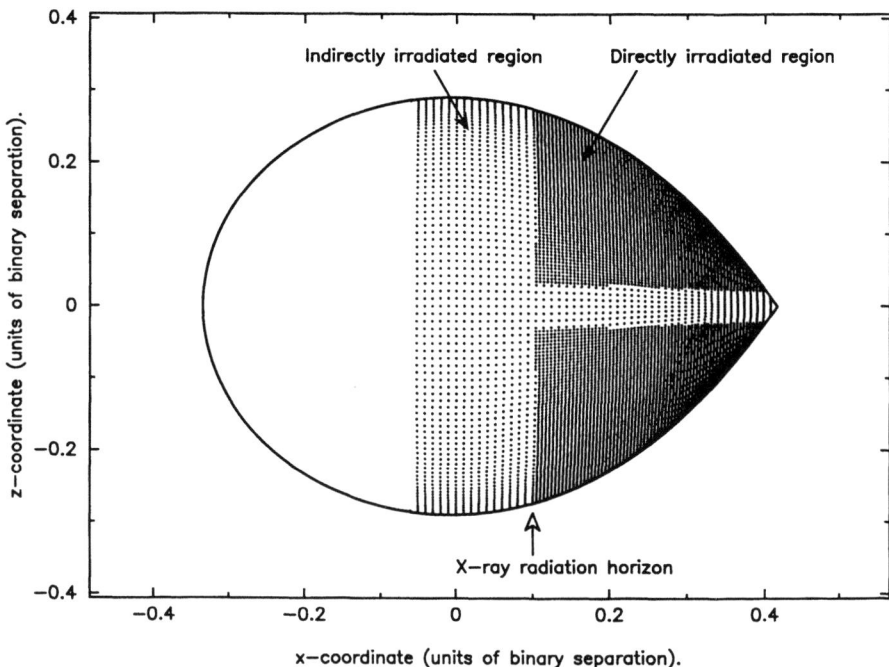

FIGURE 2. The irradiated secondary star's Roche lobe in the $(x-z)$ plane. The dense shaded region shows the area (35 per cent) that is irradiated directly by X-rays produced at the compact object; these regions do not contribute to the observed absorption line flux. The less dense region show the area (60 per cent) which must also be heated indirectly in order to produce the large observed radial velocity semi-amplitude.

In Figure 2 we show how much of the secondary star's surface needs to be heated in order to produce the observed radial velocity amplitude. We find that based purely on geometry 35 per cent of the secondary star's surface is directly heated by X-rays produced at the compact object. However, in order to produce the observed large radial velocity semi-amplitude, 60 per cent of the secondary star needs to be heated. This result may seem surprising at first, since one expects only the regions of the secondary star facing the compact object to be irradiated and yet our result implies that some of the regions not directly seen by the compact object are also affected by irradiation. However, one should note that effects such as asymmetric irradiation (e.g. from the bright spot), X-ray scattering and irradiation-induced flows on the surface of the secondary star (Phillips & Podsiadlowski 1999) can increase the fraction of the secondary star that responds to the X-ray source. Note that the regions on the secondary star that are shadowed by the accretion disc will be indirectly heated by such mechanisms. Therefore K_{obs} can be larger than that expected from heating the inner face of the secondary star alone.

In order to study the extent of irradiation of the secondary star one requires good quality spectra-photometric studies throughout an X-ray outburst. This will allow the surface intensity distribution across the secondary star to be mapped (see Rutten & Dhillon 1994 and Davey & Smith 1996), from which effects such as irradiation-induced circular flows or star-spots can be investigated.

REFERENCES

Bailyn, Orosz, J.A., McClintock, J.E., Remillard, R.A., 1995, Nat, 378, 157

Davey, S.C., Smith, R.C., 1992, MNRAS, 257, 476

Davey, S.C., Smith, R.C., 1996, MNRAS, 280, 481

Phillips, S.N., Shahbaz T., Podsiadlowski, Ph., 1999, MNRAS, 304, 839

Phillips, S.N., Podsiadlowski, Ph., 1999, MNRAS, in prep.

Rutten, R., Dhillon, V.S., 1994, A&A, 288, 773

Shahbaz, T., van der Hooft, F., Casares, J., Charles, P.A., van Paradijs, J., 1999, MNRAS, 306, 89

van der Hooft, F. et al., 1998, A&A, 329, 538

FAINT-SOURCE CONTRIBUTIONS TO THE EXTRAGALACTIC X-RAY BACKGROUND IN AN XMM DEEP FIELD [*]

R. Shirey, F. Cordova, J. Kennea, D. Pandel, T. Sasseen, & J. West

Department of Physics, University of California, Santa Barbara

ABSTRACT

Two 200-ks XMM Deep Field observations are planned to probe the faint-source populations that make up the hard X-ray background (2–10 keV). We use a simulated 200-ks XMM EPIC-PN Deep Field image, as well as simulated EPIC-PN spectra of potential types of faint-source populations, in order to explore the sensitivity of XMM for resolving the contributors to the hard X-ray background.

KEYWORDS: surveys — diffuse radiation — X-rays:galaxies — galaxies:active — galaxies:starburst

1. INTRODUCTION

Significant progress on resolving the 0.5–2.0 keV extragalactic background was achieved using deep *ROSAT* observations, which resolved 70–80% of the background at those energies into discrete sources (Hasinger et al. 1998). The majority of these sources were found to be active galactic nuclei (AGN) of type 1: classical quasars and Seyfert 1 galaxies (Schmidt et al. 1998). About 25% of the hard (2–10 keV) background has been detected as discrete sources with ASCA (Cagnoni et al. 1998; Ueda et al. 1999), but the majority of sources that give rise the hard background are as yet unresolved. Obscured (type 2) AGN, with harder spectra than type 1 AGN, are expected to be the primary contributors to the hard X-ray background (Setti & Woltjer 1989), with possible additional contributions from such sources as starburst galaxies (Moran, Lehnert, & Helfand 1999) and narrow emission line galaxies (McHardy et al. 1998).

Several programs are planned with *XMM* and *Chandra* to use deep exposures to study the faint contributors to X-ray background. In particular, we are participating in the *XMM* Optical Monitor consortium's guaranteed-time program of two 200-ks Deep Field exposures and ground-based follow-up. In this paper we use simulations of a Deep Field to explore XMM's sensitivity to faint-source contributors to the X-ray background.

---

[*]THE 200-KS XMM DEEP FIELDS ARE PART OF THE XMM OPTICAL MONITOR GUARANTEED-TIME PROGRAM, LED BY K. MASON WITH COLLABORATORS AT MSSL, UCSB, LANL, AND UNIV. OF SOUTHAMPTON.

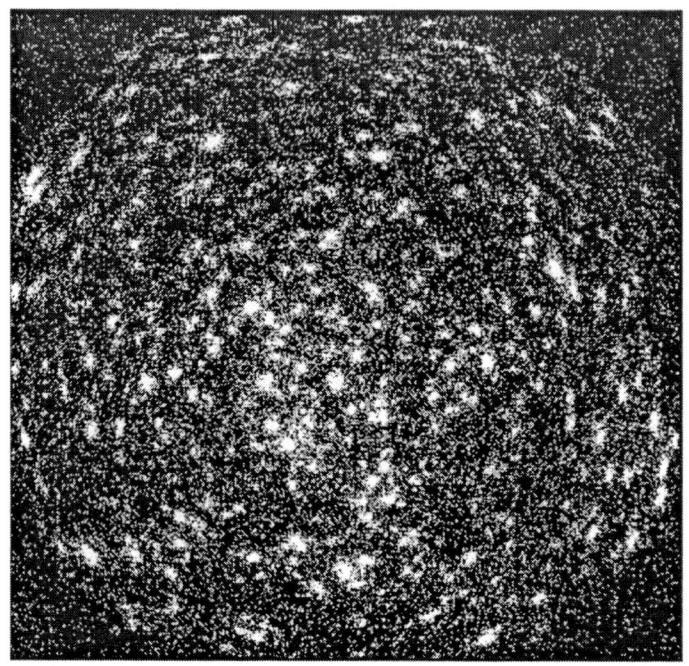

FIGURE 1. Simulated EPIC-PN image (2–10 keV, 30'x30') from a 200-ks exposure.

2. SIMULATIONS OF *XMM* DEEP FIELDS IMAGES AND SPECTRA

We simulated a 200-ks *XMM* Deep-Field observation using the *XMM* science simulator (SciSim). We generated the input sample of type 1 and type 2 AGN based on the X-ray background synthesis model of Comastri et al. (1995). Contributions due to diffuse Galactic X-ray emission and straylight have been included. The hard-band (2–10 keV) EPIC-PN image from our simulation is shown in Figure 1. The two EPIC-MOS cameras combined will provide comparable counts to the PN data we consider here. A simple source-detection algorithm detects $\sim$400 sources above a 3σ detection limit of $\sim 10^{-15}$ erg s^{-1} cm^{-2} ($\sim$23 counts in 200 ks, 2–10 keV) in this PN image. At this level, we expect to resolve $\sim$70% of the 2–10 keV extragalactic background flux.

A primary goal of the *XMM* Deep Fields is to identify the source populations that contribute to the X-ray background at faint flux levels. Observations at longer wavelengths (see below) will be used extensively for such identifications; however, for many faint sources the high throughput of *XMM* will provide X-ray spectra of sufficient quality to yield some information about the source type. To illustrate this capability, we simulated 200-ks EPIC-PN spectra of a luminous starburst galaxy and a Compton-thick Seyfert 2 galaxy of comparable luminosity (see Table 1). The input spectrum for each simulation was that of a nearby galaxy of the appropriate type.

Source Type	$L_{0.5-2\ keV}$	$L_{2-10\ keV}$	$f_{0.5-2\ keV}$	$f_{2-10\ keV}$	EPIC-PN counts
	(10^{41} erg s^{-1})		(10^{-16} erg s^{-1} cm^{-2})		
Starburst	2.2	2.3	7.4	59	170
Seyfert 2	2.0	5.7	3.5	27	200

TABLE 1. Intrinsic (unabsorbed) luminosity (L), observed flux (f), and 0.1–10 keV photon counts in 200 ks for the starburst galaxy and Seyfert 2 galaxy, both at a redshift of z=0.2, shown in Fig. 2. (The totally obscured primary radiation of the Seyfert 2 reflection component is not included in the unabsorbed luminosity.)

We used the spectral fitting package XSPEC to produce a spectrum with the same intrinsic spectral shape and luminosity as that of the nearby galaxy, but located at a redshift of $z = 0.2$, so that the 0.5–2 keV flux of each is fainter than the flux limit of the ROSAT Deep Survey (10^{-15} erg s^{-1} cm^{-2}; Hasinger et al. 1998). For the starburst galaxy, the input spectrum was that of NGC 3256, which Morand, Lehnert, & Helfand (1999) modeled as a hard power-law component of photon index 1.7 and a 0.8 keV thermal plasma, both moderately obscured ($N_H = 7.9 \times 10^{21}$ cm^{-2}), plus an additional soft (0.29 keV) thermal component. For the Seyfert 2 galaxy, the input spectrum was that of NGC 4939, which Maiolino et al. (1998) modeled as cold reflection of primary radiation (a power law of photon index 1.7, totally obscured), plus iron Kα lines and an unobscured 0.3 keV thermal component. The spectra have been background-subtracted, after first including diffuse Galactic emission, straylight, particle background, and unresolved sources in the simulated spectra.

Both spectra are dominated at low energy by their thermal components. The starburst galaxy is detected up to about 4 keV due to the central starburst and power-law component. The reflection component of the Seyfert galaxy is detected almost up to 10 keV, and the iron Kα emission is evident. At the flux levels of these sources, the starburst and Seyfert 2 models are clearly distinguishable in 200-ks EPIC-PN exposures.

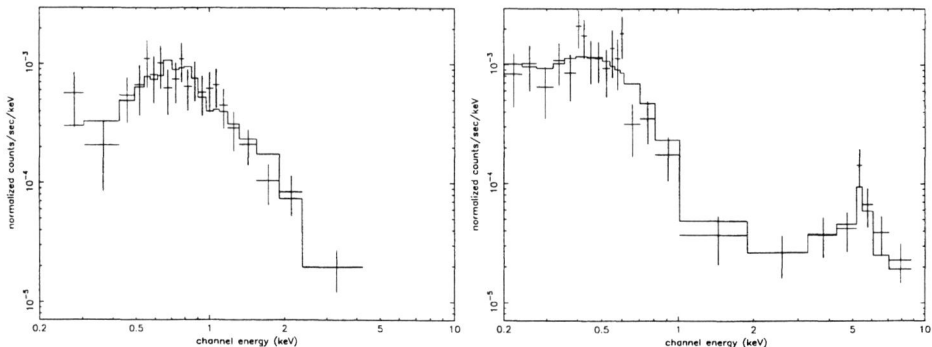

FIGURE 2. Simulated 200-ks EPIC-PN spectrum of a luminous starburst galaxy (left) and a Seyfert 2 galaxy of similar luminosity (right), each at a redshift of 0.2.

3. OPTICAL MONITOR AND GROUND-BASED SUPPORT OBSERVATIONS

We are undertaking a substantial multi-wavelength follow-up program designed to identify the X-ray sources detected in the *XMM* Deep Fields. The *XMM* Optical Monitor will concentrate on deep UV images that should reach a magnitude of ~23. Optical and IR photometric surveys that will enable R and B measurements of the X-ray sources (reaching planned limiting magnitudes of R~26 and B~27) are already in progress. The resulting colors measured in these surveys will be used to identify likely counterparts of the sources detected in the *XMM* exposures, and spectroscopic observations will be used to secure source types. A program of deep radio mapping of both fields (reaching a sensitivity of 5 μJy at 20 cm) is also underway to help differentiate between the source classes.

4. CONCLUSIONS

Current models of the 2–10 keV X-ray background (e.g., Wilman & Fabian 1999, Schmidt et al. 1999) predict almost complete saturation of the background at flux limits of 10^{-16} erg s^{-1} cm^{-2} and source counts of ~4000 deg^{-2}. We expect to detect about half of these sources at a 2–10 keV flux limit of 10^{-15} erg s^{-1} cm^{-2}, accounting for ~70% of the hard X-ray background flux. For sources a factor of a few brighter than our flux limit, yet undetected in previous surveys, we will obtain X-ray spectra of sufficient quality to begin to distinguish source types. The XMM Optical Monitor and ground-based observations will be used extensively in the identification process.

ACKNOWLEDGEMENTS

We would like to acknowledge the efforts of all the members of the XMM Optical Monitor team who are participating in the Deep Fields. This work was supported by NASA through contract NAS5-97119 and grant NAG5-6911.

REFERENCES

Cagnoni, I., Della Ceca, R., & Maccacaro, T. 1998, ApJ, 493, 54

Comastri, A., Setti, G., Zamorani, G., & Hasinger, G. 1995, A&A, 296, 1

Hasinger, G., Burg, R., Giacconi, R., Schmidt, M., Trumper, J., & Zamorani, G. 1998, A&A, 329, 482

Maiolino, R., et al. 1998, A&A, 338, 781

McHardy, I., et al. 1998, MNRAS, 295, 641

Moran, E., Lehnert, M., & Helfand, D. 1999, ApJ, 526, 649

Schmidt, M., et al., 1998 A&A, 329, 495

Setti, G., & Woltjer, L. 1989, A&A, L21

Schmidt, M., Giacconi, R., Hasinger, G., Trumper, J., & Zamorani, G. 1999, in "Highlights in X-ray Astronomy", in press, astro-ph/9908295

Ueda, Y., et al. 1999, ApJ, 518, 656

Wilman, R. J. & Fabian, A. C. 1999, MNRAS, 309, 862

BEPPOSAX OBSERVATIONS OF THE GALACTIC CENTER REGION: SOFT X–RAYS FROM THE RADIO HALO OF SGRA EAST

L. Sidoli [1,2], S. Mereghetti [1], A. Treves [3], L. Chiappetti [1], G.L. Israel [4] & M. Orlandini [5]

1) Istituto di Fisica Cosmica 'G. Occhialini", Milano, Italy
2) Dipartimento di Fisica, Università di Milano, Milano, Italy
3) Università dell'Insubria, Como, Italy
4) Osservatorio Astronomico di Roma, Monteporzio Catone, Roma, Italy
5) ITeSRE, Bologna, Italy

ABSTRACT   The $Beppo$-SAX satellite performed a survey of the Galactic Center Region in the 1-10 keV energy band with its Narrow Field Instruments. Several bright X–ray sources containing neutron stars and black holes have been observed and studied, including the possible counterpart of SgrA*. Here we report the results on the diffuse emission coming from the SgrA Complex.

The emission from within $8'$ from SgrA* has a double–temperature thermal spectrum ($kT_1 \sim$ 0.6 keV and $kT_2 \sim$ 8 keV) and an energy–dependent morphology: the hard emission (5–10 keV) is elongated along the galactic plane, while the soft one (2–5 keV) shows a triangular shape, very similar to the radio halo of SgrA East. This spatial correlation and the physical parameters of the lower temperature component support the interpretation of the radio halo of the Sgr A East shell as a SNR.

KEYWORDS: Galactic Center; X–rays; individual: SgrA East; supernova remnants.

1. INTRODUCTION

The BeppoSAX satellite performed a survey of the Galactic Center Region in the 1–10 keV energy band with its Narrow Field Instruments during 1997–1998.

A source positionally coincident with the Galactic Center (hereafter GC) was observed, together with strong diffuse emission and several point–like sources with luminosity $L_X \sim 10^{36}$ erg s^{-1}. The results on these sources, most of which are likely low mass X–ray binaries containing neutron stars and black holes, both with transient and persistent emission, are reported in detail by Sidoli et al. (1999). The spectral results for these sources are summarized in Fig. 1, where the photon index of the fits with a power law are plotted versus the hydrogen column density. An upper limit of $L_X \sim 10^{35}$ erg s^{-1} has also been placed to the 2–10 keV luminosity from the X–ray counterpart of SgrA* (see Sidoli et al. 1999 for details), confirming the underluminosity of this presumed supermassive black hole at high energies.

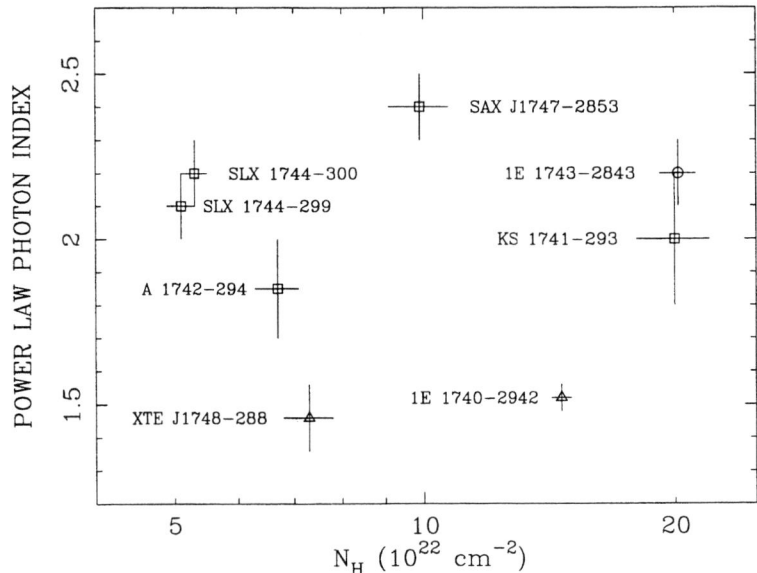

FIGURE 1. Spectral slope of the brightest sources observed in the GC region. Triangles mark the black hole candidates, the squares indicate LMXBs with neutron stars, while the circle marks a source the nature of which is still unknown.

Intense diffuse X-ray emission is also present in the GC region, the nature of which is still poorly known. Here we present the BeppoSAX results on the diffuse emission from the SgrA complex (Sidoli & Mereghetti 1999).

2. THE DIFFUSE EMISSION FROM THE SGRA COMPLEX

The Sgr A Region has been imaged with the MECS instruments (1.3–10 keV) with a spatial resolution of $\sim 1'$ (about 2.5 pc at the GC distance) in August 1997 (99.5 ksec net exposure time).

In order to study the spectral properties of the diffuse emission coming from the SgrA Complex, we extracted the MECS counts from four concentric annular regions ($0'-2', 2'-4', 4'-6', 6'-8'$) around SgrA*.

Several emission lines are present in all the spectra, with the K-lines from iron (E$\sim$ 6.7 keV) and sulfur (E$\sim$ 2.4 keV) particularly bright. The fit with a single temperature hot plasma model showed a nearly constant temperature ($\sim 7-8$ keV) at radii larger than $2'$, while in the innermost circle a softer spectrum (kT$\sim$ 4 keV) was found. This is probably due to the contribution from one (or more) point source(s) located close to SgrA* (Predehl & Trümper 1994, Maeda et al. 1996, Sidoli et al. 1999), that cannot be spatially resolved in our data. Since the temperature profile does not show spectral variations in the region from $2'$ to $8'$, we studied the overall spectrum from this entire corona. A single temperature plasma (MEKAL model) is not adequate to describe the spectrum, leaving positive residuals at low energy and especially around 6.4 keV. This can be due to the presence of fluorescent emission from neutral or weakly ionized iron in the nearby molecular

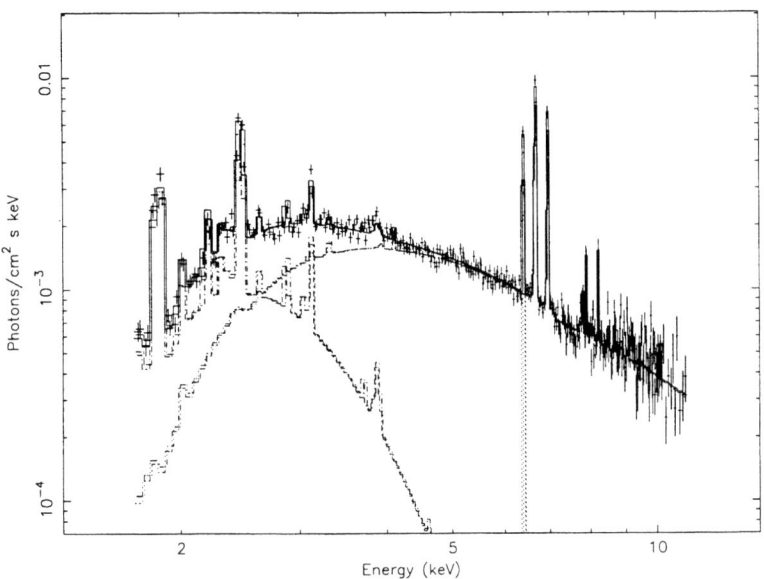

FIGURE 2. Best fit to the MECS spectrum ($2'-8'$ corona) from the Sgr A Complex.

clouds (Koyama et al. 1996). Thus we added a lower temperature plasma component plus a gaussian line at 6.4 keV. The resulting best fit is a double–temperature plasma, with $kT_1 \sim 0.6$ keV and $kT_2 \sim 8$ keV ($N_H = 8 \times 10^{22}$ cm^{-2}) and a gaussian line at 6.4 keV with an equivalent width of ~ 120 eV (see Fig. 2). The total flux corrected for the absorption is $F_X \sim 1.7 \times 10^{-10}$ erg cm^{-2} s^{-1} (2–10 keV), which translates into a luminosity of $L_X \sim 1.4 \times 10^{36}$ erg s^{-1}. About one third of the flux is contributed by the soft component.

2.1. Morphology of the Diffuse Emission

The spatial distribution of the diffuse emission was studied extracting two images in different energy ranges, below and above 5 keV (see Fig. 3). Both emissions are peaked at the GC position, but they have significantly different spatial distributions: the soft emission (2–5 keV) displays a triangular shape, while the hard one (5–10 keV) is elliptical and elongated in the direction of the galactic plane. While the hard emission can be simply part of the diffuse emission permeating the inner 60° of the galactic disk (e.g. Kaneda et al. 1997), the soft X–rays are spatially correlated with a structure observed in the radio band, known as the Sgr A East triangular halo (Pedlar et al. 1989). This is an extended non–thermal structure (probably a SNR) which surrounds in projection SgrA*. Since also our spectral data are well described by a two–temperature thermal model, it is tempting to give an interpretation in terms of two plasma components with different temperatures and spatial distributions. From the spectral analysis of the lower temperature plasma ($kT_1 \sim 0.6$ keV), we derive a luminosity $\sim 4.5 \times 10^{35}$ ergs s^{-1} (2–10 keV), an electron density $n_e \sim 3$ cm^{-3}, a total mass $M_g \sim 250\, M_\odot$ and an average thermal pressure

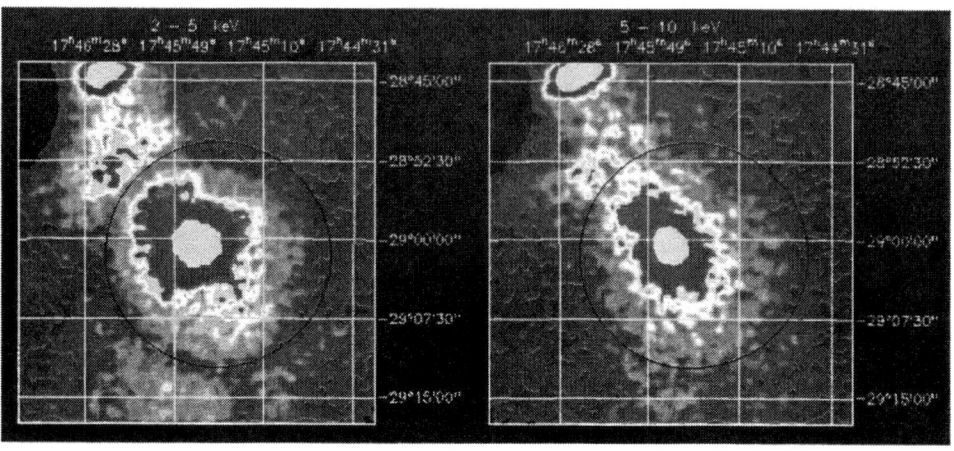

FIGURE 3. Spatial distribution of the diffuse emission from the Sgr A Complex below (left panel) and above (right panell) 5 keV. A smoothing with gaussian with FWHM=1' has been applied. The black circle marks the position of the circular strongback structure of the MECS instrument (about 20' diameter).

$P \sim 3 \times 10^{-9}$ ergs cm^{-3}, consistent with the pressure $P_{Sedov} \sim 4 \times 10^{-9}$ erg cm^{-3} derived for a SNR in a Sedov phase.

3. CONCLUSIONS

The BeppoSAX observation of the SgrA complex revealed the presence of at least two distinct components: a soft component with kT$\sim$0.5–1 keV spatially correlated with the SgrA East 7' triangular halo, and a hard one with kT$\sim$7–9 keV elongated along the Galactic Plane and possibly associated with the harder component of the Galactic Ridge emission.

The soft component, which accounts for about one third of the 2-10 keV diffuse luminosity from the SgrA complex, can be well explained as thermal emission from the SNR responsible for the radio halo of the SgrA East shell.

REFERENCES

Kaneda H. et al. 1997, ApJ 491, 638
Koyama K. et al. 1996, PASJ 48, 249
Maeda Y. et al. 1996, PASJ 48, 417
Pedlar A. et al. 1989, ApJ 342, 769
Predehl P. & Trümper J. 1994, A&A 290, L29
Sidoli L. & Mereghetti S. 1999, A&A, 347, L49
Sidoli L., Mereghetti S., Israel G.L., Chiappetti L., Treves A. & Orlandini M. 1999, ApJ 525, 215

ON PAIR CONTENT OF QUASAR JETS

M. Sikora [1], G. Madejski [2]

1) N. Copernicus Astronomical Ctr, Warsaw, Poland 2) NASA/GSFC, Greenbelt, USA

ABSTRACT

We use X-ray observations of blazars to infer that jets in quasars contain many more e^+/e^- pairs than protons, but dynamically, they are still dominated by protons. In particular, we show that pure pair jets can be excluded, as they overproduce soft X-ray radiation; likewise, pure proton-electron jets can be excluded, as they produce too little nonthermal X-ray radiation. We demonstrate that jets which initially consist of protons / electron plasma can be pair loaded due to interactions with 100 - 300 keV photons produced in the hot accretion disc coronae.

KEYWORDS: quasars; jets; pair plasma

1. INTRODUCTION

One of the central unresolved questions regarding the nature of blazar jets is that of their composition: are they dominated by plasma consisting mainly of protons and electrons, or electron-positron pairs, or a mixture of both?

Arguments in favor of proton-electron jets have been provided by Celotti & Fabian (1993). Using synchrotron-self-Compton constraints from radio-core observations and information about energetics of jets from radio-lobe studies, they showed that in the case of pure e^+ / e^- jets, the required number of pairs is too high to be delivered from the central engine. The limit is provided by the annihilation process (Guilbert, Fabian, & Rees 1983).

On the other hand, the recently discovered circular polarization in the radio cores of the GeV-emitting blazar 3C 279 and several other objects, and its interpretation in terms of the "Faraday-conversion" process suggests that the jet plasma is dominated by e^+ / e^- pairs (Wardle et al. 1998; Wardle & Homan 1999).

In this paper we derive constraints which are imposed on the pair content of quasar jets by the X-ray observations of blazars. Our results suggest that the pair content is high, but that dynamically the jets are still dominated by protons.

2. ELECTRON-POSITRON JETS

External UV photons, Comptonized by relativistic flux of cold electrons/positrons, are boosted in frequency by a square of a bulk Lorentz factor Γ and collimated along a jet axis. If the energy flux of quasar jets were dominated by cold pairs, this process should produce in the soft X–ray spectra of blazars a "bump" with luminosity

$$L_{BC} \geq 3 \times 10^{47} \, (\Gamma/10)^3 (\xi L_d)_{45} \, \text{erg s}^{-1}, \qquad (1)$$

where ξ is the fraction of the central radiation (with luminosity L_d) reprocessed / rescattered in the surrounding medium at a distance $r < 10^{17} L_{j,46}$ cm, and L_j is the kinetic luminosity of a jet (Sikora et al. 1999). Overall electromagnetic spectra of blazars show no such bumps. This imposes strong constraints on the number of cold pairs in a jet and suggests that the kinetic energy flux in a jet is dominated by protons.

It is of course possible that because of inefficient cooling of electrons/positrons below given energy, multiple reacceleration balances adiabatic losses in the conically diverging jet, and the pairs, once accelerated, remain relativistic forever. If relativistic electrons have Lorentz factors γ_0 that are narrowly distributed around $1 < \gamma_0 < 30$, then taking into account that $|dE_e/dt| \propto \gamma_0^2$ and that $n_e \propto L_j/\gamma_0$, the bulk-Compton luminosity, which is $\propto |dE_e/dt| n_e$, would be γ_0 times larger than the value given by Eq. (1), and would peak at $\sim h\nu \simeq \Gamma^2 \gamma_0^2 \nu_{UV} \simeq \gamma_0^2 (\Gamma/10)^2$ keV. No such bumps have been observed at keV energies. Alternatively, one can speculate that the blazar X–ray spectra are superposed from "thermal" peaks produced over several decades of distance (Sikora et al. 1997), but in this case, the X–ray luminosities are also predicted to be larger than those observed in blazars.

3. PROTON-ELECTRON JETS

Let us consider now another extreme, assuming that there are no pairs in jets at all. For a given energy flux in a jet, L_j, which now is proportional to $n_p m_p$, the number of electrons is m_e/m_p times smaller than the number of electrons plus positrons in the jet made from cold pairs. Thus, rescaling the formula (1) by a factor m_e/m_p, one can find that the proton-electron jets do not overproduce soft X–ray luminosities, which in blazars are $\sim 10^{46}$ ergs s^{-1} (Sambruna 1997).

But from the same reason – low number of electrons – the proton-electron jets are relatively inefficient producers of nonthermal radiation. The best way to study this is via analysis of the low energy tails of the nonthermal radiation components, where the requirement for a number of electrons is largest. In the case of synchrotron radiation, such tails are not observed because they are self-absorbed, and the only spectral band where the presence of such lower energy relativistic electrons can be verified are the soft and mid-energy X–rays, 0.1 – 20 keV.

Let us first investigate whether X–ray spectra of blazars can be produced by Comptonization of external diffuse radiation field, via the so–called External Radiation Compton process (ERC). Such a field in the jet comoving frame is dominated by broad emission lines (BEL) and near-IR radiation of hot dust. Assuming that at a distance r_{fl} all available electrons are accelerated and that energy flux in the jet is dominated by cold protons, i.e., that

$$n'_e = \int_{\gamma_{min}} n'_\gamma d\gamma = n'_p \simeq \frac{L_j}{m_p c^3 \Gamma^2 \pi a^2}, \qquad (2)$$

one can find that for $\Gamma \simeq 10$ Comptonization of broad emission lines gives

$$(L_\nu\nu)_{C(BEL)} \simeq 0.5 \times 10^{44} \frac{L_{BEL,45}}{r_{fl,17.5}} \left(\frac{h\nu}{1\text{keV}}\right)^{1-\alpha_x} \gamma_{min}^{2\alpha_x} L_{j,46}, \quad (3)$$

while Comptonization of near infrared radiation gives

$$(L_\nu\nu)_{C(IR)} \simeq 2 \times 10^{43} \left(\frac{\xi_{IR}}{0.1}\right) r_{fl,17.5} \left(\frac{h\nu}{1\text{keV}}\right)^{1-\alpha_x} \gamma_{min}^{2\alpha_x} L_{j,46}, \quad (4)$$

where L_{BEL} is the luminosity of broad emission lines, α_x is the X-ray spectal index, and ξ_{IR} is the fraction of central luminosity reprocessed by dust with temperature $T = 1000$ K.

Noting that in the case of Comptonization of BEL photons, radiation at 1 keV is produced by electrons with $\gamma \sim 1$, while in the case of Comptonization of near infrared radiation – by electrons with $\gamma \sim 50/\Gamma$, one can see that the predicted 1 keV ERC luminosities are about two orders of magnitude too small in comparison with typical observed soft X-ray luminosities.

The other candidate process – the synchrotron-self-Compton (SSC) – involves more relativistic electrons in the production of X–rays than ERC. Therefore, γ_{min} is not restricted to very low values, and assuming $\gamma_{min} \sim 100$ one can reproduce soft X–ray luminosities:

$$(L_\nu\nu)_{SSC} \sim 10^{46} \frac{L_{syn,47}}{r_{fl,17.5}} \left(\frac{h\nu}{1\text{keV}}\right)^{1-\alpha_x} \left(\frac{\gamma_{min}}{100}\right)^{2\alpha_x} L_{j,46}. \quad (5)$$

However the SSC spectra in the X–ray bands are much softer than the observed ones (Błażejowski et al. 1999).

Summarizing this section, we conclude that pure proton-electron jet models are inadequate, as they fail to produce the nonthermal X-ray luminosities observed in blazars. The number of pairs per proton required to produce the observed X-ray luminosities via the ERC process is

$$\frac{n_{pairs}}{n_p} \sim 50 \frac{L_{SX,46}}{L_{j,46}}. \quad (6)$$

4. PAIR PRODUCTION

We propose a scenario where jets are launched as proton-electron outflows ("proto-jets") in the innermost parts of the accretion flow and are loaded by pairs due to interactions with hard X-rays produced via thermal Comptonization in hot accretion disc coronae. Pairs are produced in two steps. The first step is Compton boosting of coronal 100 - 300 keV photons up to few MeV by cold electrons in the outflow propagating through the central region with $\Gamma \sim 10$ (Begelman & Sikora 1987). The opacity for such interactions is quite high,

$$\tau_{e\gamma} \simeq n_x r_{corona} \sigma_T \sim 60 \frac{L_{x,46}}{(h\nu/100\text{keV})r_{corona,15}}, \quad (7)$$

which means that each electron in the wind produces tens of 1 - 3 MeV photons. The second step is the absorption of MeV photons by the coronal 100 - 300 keV photons in the pair creation process. The pairs created in this manner are dragged by the jet, but before leaving the compact X-ray field they produce a second generation of MeV photons, and they in turn produce next generation of pairs. Such pair production can continue until the time when the "proto-jet" becomes opaque for coronal radiation, i.e., when $n_e r_{corona} \sigma_T \sim 1$. Within this limit, the e^+ / e^- flux can reach the value

$$\dot{N}_e \simeq n_e c \Omega_i r_{corona}^2 \simeq \frac{c \Omega_i r_{corona}}{\sigma_T}, \quad (8)$$

where Ω_i is the initial solid angle of the outflow. Comparing the e^+ / e^- flux with the proton flux

$$\dot{N}_p \sim L_j / \Gamma m_p c^2, \quad (9)$$

we find that proton-electron winds can be loaded by pairs in the central compact X-ray sources up to the value

$$\frac{n_{pairs}}{n_p} \simeq \frac{\dot{N}_e/2}{\dot{N}_p} \simeq \frac{m_p c^3}{2 \sigma_T} \frac{r_{corona} \Omega_i \Gamma}{L_j} \simeq 30 \frac{r_{corona,15} \Omega_i (\Gamma/10)}{L_{j,46}}. \quad (10)$$

This corresponds roughly to a pair content given by Eq. (6), provided Ω_i is not very small.

ACKNOWLEDGEMENTS

This project was supported by the ITP/NSF grant PHY94-07194 and the Polish KBN grant 2P03D00415.

REFERENCES

Begelman, M.C., & Sikora, M. 1987, ApJ, 322, 650
Błażejowski, M., Sikora, M, Moderski, R., & Bulik, T. 1999, this volume
Celotti, & Fabian, A.C. 1993, MNRAS, 264, 228
Ghisellini, G., Celotti, A., George, I.M., & Fabian, A.C. 1992, MNRAS, 258, 776
Guilbert, P.W., Fabian, A.C., & Rees, M.J. 1983, MNRAS, 205, 593
Sambruna, R.M. 1997, ApJ, 487, 536
Sikora, M., Madejski, G., Moderski, R., & Poutanen, J. 1997, ApJ, 484, 108
Wardle, J.F.C., & Homan, D.C. 1999, AAS Meeting 194
Wardle, J.F.C., Homan, D.C., Ojha, R., & Roberts, D. 1998, Nature, 395, 457

CCD PHOTOMETRY OF OUTBURSTS IN GK PER

V. Šimon [1], Z. Velič [2]

1) Astronomical Institute, 251 65 Ondřejov, Czech Republic, e-mail: simon@asu.cas.cz
2) L. Štúra 16/22-16, 01861 Beluša, Slovakia, e-mail: osbdpb@px.psg.sk

ABSTRACT Photometry of the dwarf nova type 1996 outburst of GK Per shows that the color indices are getting bluer during the late rise to the maximum brightness and a strong linear reddening is visible during the decline. The brightness fluctuations (cycles of 0.055 – 0.070 days and 0.028 – 0.045 days) were detected during the 1999 outburst. The amplitude of the fluctuations (intensity units) was increasing during the rise to the maximum light.

KEYWORDS: binaries: close; circumstellar matter; novae, cataclysmic variables; Stars: individual: GK Per

1. INTRODUCTION

GK Per (Nova Per 1901) is an intermediate polar (P_{orb} = 1.99 days (Crampton et al. 1986); P_{spin} = 351 sec (Watson et al. 1985)). Discrete brightenings by 2–3 mag recur on the time scale of about 3 years (e.g. Sabbadin and Bianchini 1983, Hudec 1981) and are accompanied by the X-ray outbursts (King et al. 1979). These events were modeled by Kim et al. (1992) using thermal instability disk model. Fluctuations of the luminosity (360–5000 sec) in the different spectral regions were observed during the outbursts (Duschl et al. (1985); Mazeh et al. (1985); Morales-Rueda et al. (1996)).

2. THE OBSERVATIONS AND ANALYSIS OF THE DATA

Two outbursts in 1996 and 1999 were observed with the 180/700 Newton telescope (CCD detector Texas Instruments TC 211; VRI filters corresponding to Kron-Cousins system) at Beluša Observatory by one of us (ZV). The star SAO 38899 (8.9 mag(V), A2) was used as the basic comparison star. One to three check stars were used to check the air and detector stability. The variable, the comparison and the check stars were always placed in the same image.

Detection of the color variations over the outburst was the aim of observing the outburst in 1996. The images (one in each filter) were secured in closely spaced intervals and are therefore unlikely to be largely affected by the flickering. The CCD observations in I-filter (differential with respect to SAO 38899) of the 1999 outburst were focused on detection of rapid variations.

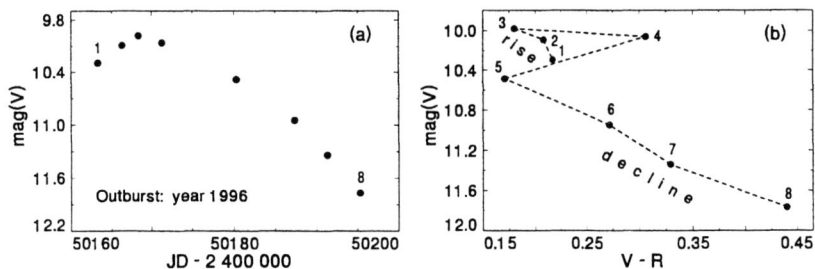

FIGURE 1. Color variations over the 1996 outburst. Numbers of the respective points in (b) can be identified from the light curve (a).

2.1. The outburst in 1996

Late rise and the decline were covered during 8 nights of VRI observation, numbered 1 to 8 in Fig. 1a. Variations of the color index $V-R$ versus mag(V) are displayed in Fig. 1b. The points are numbered according to Fig. 1a and allow to trace the evolution. $V-R$ is getting bluer during the late rise to maximum (nights 1–3). A strong linear reddening of $V-R$ with the fall of brightness occurs during the decline (nights 5–8). $R-I$ displays a larger scatter but the course is similar to $V-R$.

2.2. The outburst in 1999

The position of the CCD observations is illustrated by the one-day means of the AFOEV data in Fig. 2a. An arbitrary shift was applied to the differential CCD observations to match the visual curve. The short-term variations are clearly visible each night (Fig. 2b).

The autocorrelation method, described by Percy et al. (1981), was used for analysis of the fluctuations. This method allows to search for characteristic time scales or quasi-periods which may extend just for several cycles. A cycle-length of 0.055–0.070 days is always detected in the nights with sufficiently long coverage. Also another, less prominent cycle of 0.028–0.045 days was detected in most nights.

The two longest runs (JD = 2 451 250 and 2 451 255) allow to resolve the non-sinusoidal shape of the 0.06 day wave – the decreasing part is steeper than the rise (Fig 2b).

The energy output of the processes governing the fluctuations may be assessed by an analysis of the amplitude development of the fluctuations with the progress of the outburst. The respective light curves were transformed into intensity, normalized to unity at $\Delta \mathrm{mag}(I)=1.6$. The mean intensity and its standard deviation σ_{INT} of each run were calculated (Fig. 3). σ_{INT} was steadily increasing in the course of the rise to the maximum light. The projected dependence of σ_{INT} on intensity (Fig. 3) shows roughly linear increase of the intensity amplitude by a factor of three.

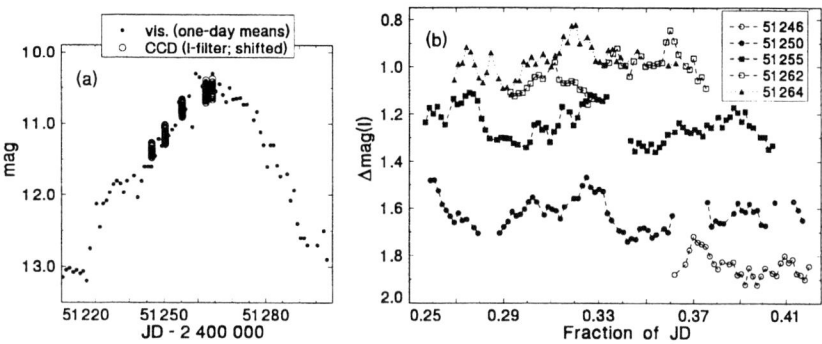

FIGURE 2. Fluctuations of brightness during the rise to the 1999 outburst. See text.

3. DISCUSSION

The color variations during the late rise and the decline of the 1996 outburst were obtained. The indices are getting bluer during the late rise to maximum. The index $V - R$ measures the slope of Paschen continuum and is free of Paschen and Balmer jumps. The strong linear reddening of $V - R$ during the decline therefore suggests a decrease of the temperature. Probably the evolved secondary becomes more prominent as the cooling front is bringing more and more disk back into the cool state.

Fluctuations of the brightness, consisting of the superposed cycles, persisted at least from the mid-rise till the maximum of the 1999 outburst. The range of the signals detected (both groups, but mainly the longer cycle of ≈0.06 days) mostly fits into the range found by Morales-Rueda et al. (1996) in the 1996 outburst. It therefore may be a repeating feature of outbursts in GK Per.

The ≈0.06 day fluctuations may be of the same origin as those observed spectroscopically in the previous outburst (1996) by Morales-Rueda et al. (1999). They offered an interpretation in terms of the model by Watson et al. (1985) where the observed periods are the beats between the spin frequency of WD and the Keplerian frequency of the blobs of gas at the inner rim of the disk. The phase of the minimum brightness would be caused by passing an accretion curtain in front of the blob.

I-filter measures Rayleigh-Jeans tail of the black body radiation where the luminosity increases roughly linearly with the temperature T_{eff}. Because luminosity of the inner disk in outburst is roughly proportional to the local mass flow $\dot{m}_{\text{inner}}$ one can infer that the amplitude (in the intensity units) of the fluctuations (represented by σ_{INT}) depends on the mass flow through the inner part of the disk. Increase of the luminosity of the accretion curtain (which may be expected due to increasing mass inflow into this region) is therefore smaller than that of the inner part of the disk where the blobs are suggested to form.

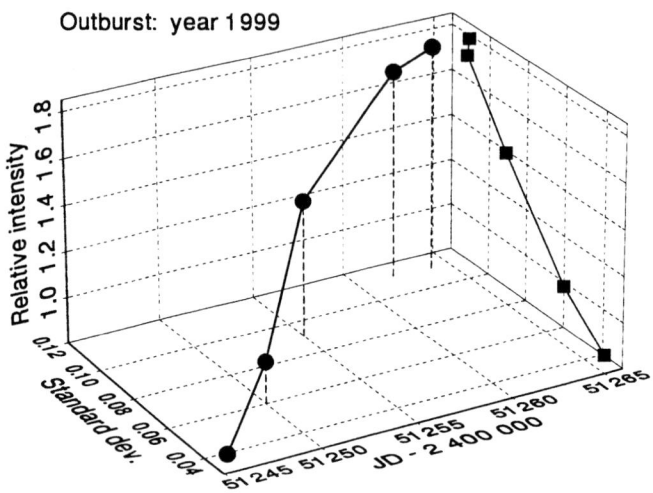

FIGURE 3. Evolution of the relative intensity (I-band) and σ_{INT} during the rise to the maximum brightness of the 1999 outburst. The projection shows the relation of intensity and σ_{INT}.

ACKNOWLEDGEMENTS

This research has made use of the AFOEV database, operated at CDS, France, and NASA's Astrophysics Data System Abstract Service. The investigation of cataclysmic variables with X-ray emission is partly supported by the project KONTAKT ME 137 by the Ministry of Education and Youth of the Czech Republic.

REFERENCES

Crampton, D., Cowley, A.P., Fisher, W.A., 1986, ApJ, 300, 788
Duschl, W. J.; Meyer-Hofmeister, E.; Meyer, F., 1985, ESA Recent Results on Cataclysmic Variables, p.221
Hudec, R., 1981, Bull. Astron. Inst. Czechosl., Vol. 32, 93
Kim, Soon-Wook; Wheeler, J. C.; Mineshige, S., 1992, ApJ, 384, 269
King, A.R., Ricketts, M.J., Warwick, R.S., 1979, MNRAS, 187, 77
Mazeh, T.; Tal, Y.; Shaviv, G.; Bruch, A.; Budell, R, 1985, A&A, 149, 470
Morales-Rueda, L.; Still, M. D.; Roche, P., 1996, MNRAS, 283, L58
Morales-Rueda, L.; Still, M. D.; Roche, P., 1999, MNRAS, 306, 753
Percy, J.R., Jakate, S.M., Matthews, J.M., 1981, AJ, 86, 53
Sabbadin, F., Bianchini, A., 1983, A&AS, 54, 393
Watson, M.G., King, A.R., Osborne, J., 1985, MNRAS, 212, 917

COLOR VARIATIONS DURING ACTIVITY STATES OF THE BINARY V SGE

V. Šimon [1], S. Shugarov [2]

1) Astronomical Institute, 251 65 Ondřejov, Czech Republic, e-mail: simon@asu.cas.cz
2) Sternberg Astronomical Institute, Moscow State University, Universitetsky Prospect 13, 119899 Moscow, Russia, e-mail: shugarov@sai.msu.ru

ABSTRACT Color variations in V Sge during various states of its activity in the course of 1995 – 1997 show that $B - V$ gets bluer while $U - B$ does not change significantly. Comparison of our data with the outburst observed by Herbig et al. (1965) shows that although the character of the activity in V Sge changed during the last decades (Šimon and Mattei 1999) the colors and their variations remained similar.

KEYWORDS: binaries: close; binaries: general; circumstellar matter; novae, cataclysmic variables; Stars: individual: V Sge

1. INTRODUCTION

V Sge is a peculiar eclipsing binary ($P_{\mathrm{orb}} = 0.514$ d) (Herbig et al. 1965 – HPSP). There have been accumulated several lines of evidence (the high-excitation emission lines (Patterson et al. 1998); X-ray variations (Greiner and van Teeseling 1998); the long-term activity (Šimon and Mattei 1999 – Paper I)) which strongly support the model of the mass accreting white dwarf (WD) primary from a massive companion, originally suggested by Williams et al. (1986). V Sge is a candidate for the super-soft X-ray source (SSXS) (Steiner and Diaz 1998). V Sge displays strong photometric activity (HPSP, Šimon 1996, Robertson et al. 1997, Paper I). The character of the activity changed from relatively isolated outbursts, seen in 1930's, to alternating high (HS) and low (LS) states, typical for the recent about 20 years (Paper I).

2. THE OBSERVATIONS AND ANALYSIS OF THE DATA

The photoelectric observations in Johnson UBV system were secured in three seasons (July – December 1995 and July 1996: 600/7500 mm Cassegrain tel. at Crimea Observatory; August 1997: 700/10500 mm Cassegrain tel. at Moscow Observatory). Diameter of the diaphragm (usually 14 arc sec.) at least largely minimized the contribution of the optical companion (14.4 mag(V), 9.7 arc sec.).

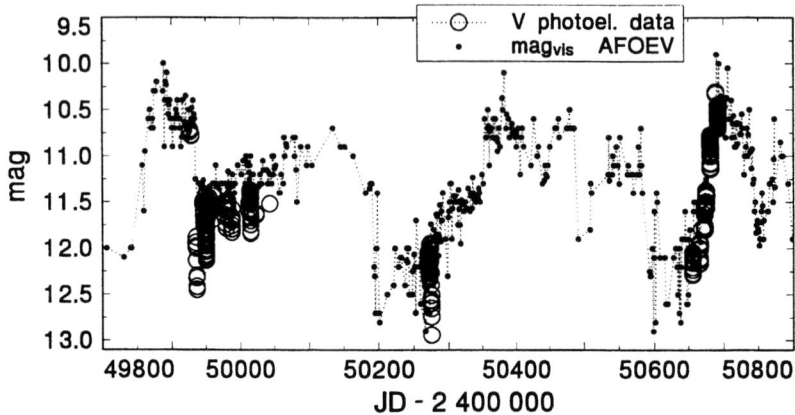

FIGURE 1. The light curve illustrating the character of activity in V Sge over the years 1995–1997.

2.1. B-V and U-B changes versus brightness

The superposed one-day means of the visual AFOEV observations (made in the same way as in Šimon 1996) illustrate that the UBV data fall into an active segment (following the notation of Paper I), that is a season of alternating high and low states (Fig. 1).

Figs. 2ab show $B - V$ and $U - B$ indices, plotted versus mag(V). The orbital phases are resolved (0.9–1.1: primary eclipse – empty circles) (0.1–0.9: filled circles).

Clustering of most points at some levels complicates the search for the mean course in Figs. 2ab. Moving averages did not yield a good fit. Arithmetic means (bins of 0.2 mag(V)) gave better results. Another method we applied was the code HEC13 (author Dr. P. Harmanec), based on the method of Vondrák (1969 and 1977). The input parameters were $\epsilon = 10^3$, Δmag = 0.01 for both $U - B$ and $B - V$. Both the fit to the whole set and just to the data within phases 0.1–0.9 were made to search for the phase color dependence. The medium state and LS have a very similar $B - V \approx +0.05$. $B - V$ becomes significantly smaller only in HS (about 11 mag(V) and brighter). Examination of the data just for the rise from LS in August 1997 (JD≈2 450 680) confirmed that the change of the slope of $B - V$ is not any product of mixing data from several seasons. Fig. 2b shows no prominent trend in variations of $U - B$.

Our data for 1995 – 1997 were compared with UBV out-eclipse observations, published by HPSP (large squares in Figs.2ab). The latter capture an outburst which occurred in the beginning of the sixties. The course of HPSP's data in $B - V$ $mag(V)$ diagram is similar to our new data – in both cases $B - V$ decreases as the systems brightens.

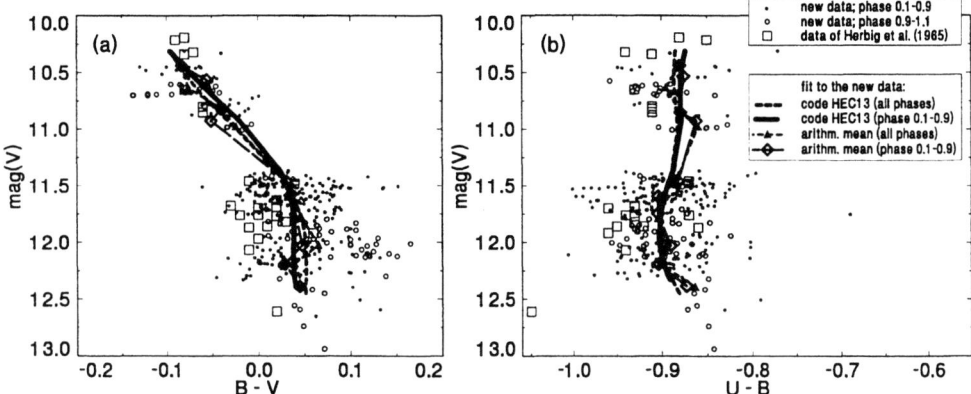

FIGURE 2. Color variations $B - V$ and $U - B$ versus brightness (V). The new data represent the years 1995–1997. Resolution of the orbital phases is made – the primary eclipse occurs at phase 0.0.

3. DISCUSSION

The observations of V Sge during the recent period of its activity (1995 – 1997), when compared with the old data of HPSP, show that although the character of activity has changed significantly during the recent decades (Paper I) the colors and their variations with the brightness level remained similar. This fact therefore strengthens the evidence (expressed already in Paper I on the base of the long-term trends in the activity) that the process which gives rise to the large brightness variations in V Sge is the same for the respective kinds of its activity.

Figs. 2ab show that the dominant part of the color variations must be ascribed to the long-term activity and not to the changes on the orbital scale. Decrease of $B - V$ at a roughly constant $U - B$ may imply that Balmer jump undergoes just minor changes over LS/HS transition and that the continuum gets hotter towards HS.

If HPSP's interpretation that the entire color variation was due to the increase in strength of the emission lines is true then the change of the slope in Fig. 2a suggests that the emission lines strengthen only when the system is brighter than about 11.5 mag(V).

In the framework of the super-soft X-ray binary model for V Sge (see sect. 1) it is reasonable to suppose that the mass-gaining primary is WD embedded in a modified disk with a high asymmetric rim (Meyer-Hofmeister et al. 1997). The variations of dimensions of the disk (its radius, height of the rim or both) can account for the brightening of the system without discernible color changes. The irradiation from WD which is thought to be substantial in SSXS (van den Heuvel et al. 1992, Schandl at al. 1997) flattens the radial temperature distribution. Variations of the outer disk radius therefore may not give rise to large color changes.

Change of the slope of $B - V$ above 11.5 mag(V) (Fig. 2a) suggests that some additional process begins to play a role above this level of brightness. It may be a radiatively driven wind from the luminous accretion disk (Hachiya et al. 1998), suggested also by Greiner and van Teeseling (1998) for explanation of the changes of the X-ray spectrum from soft to hard. Such a wind is also likely to be the source of the uneclipsed light which contaminates the light coming from both components of the binary and significantly lowers the amplitude of the orbital modulation in HS, as observed e.g. by HPSP and Patterson et al. (1998).

ACKNOWLEDGEMENTS

This research has made use of NASA's Astrophysics Data System Abstract Service and partly of the AFOEV database (operated at CDS, France). We are indebted to Dr. Harmanec for providing us with the program HEC 13. Investigation of CVs with X-ray emission is partly supported by the project KONTAKT ME 137 by the Ministry of Education and Youth of the Czech Republic. This study was also supported in part by the Russian Foundation for Basic Research and the Council of the Program for the "Support of Leading Scientific Schools" through grants No 99-02-17589 and 96-15-96489.

REFERENCES

Greiner, J., van Teeseling, A., 1998, A&A, 339, L21
Hachiya, M., Tajima, Y., Fukue, J., 1998, PASJ, 50, 367
Herbig, G.H., Preston, G.W., Smak, J., Paczynski, B., 1965, ApJ, 141, 617 (HPSP)
Meyer-Hofmeister, E., Schandl, S., Meyer, F., 1997, A&A, 321, 245
Patterson, J., Kemp, J., Shambrook, A., Thorstensen, J.R., Skillman, D.R., Gunn, J., Jensen, L.
    Vanmunster, T., Shugarov, S., Mattei, J.A., Shahbaz, T., Novák, R., 1998, PASP, 110, 380
Schandl, S., Meyer-Hofmeister, E., Meyer, F., 1997, A&A, 318, 73
Steiner, J.E., Diaz, M.P., 1998, PASP, 110, 276
Šimon, V., 1996, A&AS, 118, 421
Šimon, V., Mattei, J.A., 1999, A&AS (accepted, in press)
van den Heuvel, E.P.J., Bhattacharya, D., Nomoto, K., Rappaport, S.A., 1992, A&A, 262, 97
Vondrák, J., 1969, Bull. Astron. Inst. Czechosl., Vol. 20, 349
Vondrák, J., 1977, Bull. Astron. Inst. Czechosl., Vol. 28, 84
Williams, G.A., King, A.R., Uomoto, A.K., Hiltner, W.A., 1986, MNRAS, 219, 809

ACTIVITY OF THE DWARF NOVA CH UMA

V. Šimon

Astronomical Institute, 251 65 Ondřejov, Czech Republic, e-mail: simon@asu.cas.cz

ABSTRACT The outbursts in CH UMa can be divided into two classes (bright and faint), the division magnitude being 12 mag_{vis} and separation almost 2 mag_{vis}. The maximum brightness of the outbursts tends to decrease over 25 years. The bright outbursts fall into two subtypes according to their decay branches. The recurrence time of the bright outbursts 300 – 370 days displays variations with a strong secular trend.

KEYWORDS: Stars: activity; circumstellar matter; novae, cataclysmic variables; Stars: individual: CH UMa

1. INTRODUCTION

CH UMa is a U Gem type cataclysmic variable (CV) (Kukarkin et al. 1974). X-ray observations come from Becker et al. (1982) and Verbunt et al. (1997). In quiescence CH UMa is characterized by a hard X-ray spectrum and low X-ray to the optical flux ratio (0.04). CH UMa appears to exhibit several types of outbursts. Szkody and Mattei (1984) included it to a list of candidates for the SU UMa subgroup. However, P_{orb} = 8.28 hours (Thorstensen 1986) clearly rules out this classification. The contradiction between the outburst properties and the orbital period thus makes CH UMa an interesting object for a study of its activity.

2. THE SOURCES OF THE DATA

This analysis is based on the observations from the AFOEV database, operated at CDS, Strasbourgh, France. The positive observations (1756) were binned into one-day means (1366). The negative observations (4302) constrained the number of possible missing outbursts. The coverage by the upper limits is dense for magnitudes brighter than about 12.6 mag_{vis}. The peak brightness and the moment of the maximum light were determined by fitting a polynomial (typical error 1 – 2 days).

3. ANALYSIS OF THE DATA

Outburst parameters: The outburst history over the years 1973 – 1998 comprises about 21 outbursts, reaching up to 11.5 – 10.5 mag_{vis} from the quiescent level ($\approx$ 15 mag_{vis}). Several fainter outbursts (about 13 mag_{vis}) are apparent, too. Only outbursts defined by multiple observations are considered.

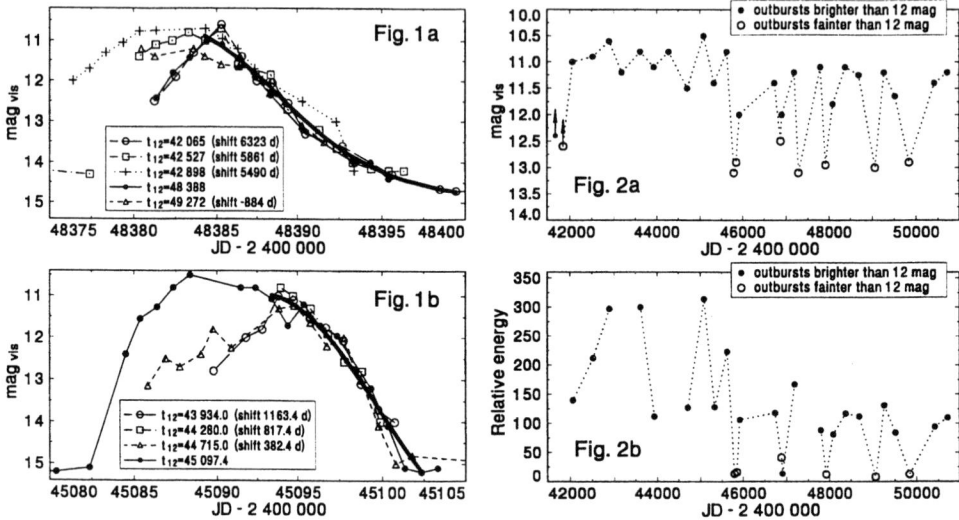

FIGURE 1 (left). Types of the decay branches of the bright outbursts ((a) slow; (b) fast). The respective light curves were aligned at 12 mag$_{vis}$. The thick lines represent the smoothed decay curves.

FIGURE 2 (right). The maximum brightness (a) and the relative energy (b) of the outbursts. The neighbouring outbursts are connected by lines to guide the eye. See text for details.

The findings can be summarized in the following way:

The statistical distribution of the outburst maximum brightness is bimodal (peaks at $\approx$ 11.1 mag$_{vis}$ and $\approx$ 13 mag$_{vis}$), hereafter "bright" and "faint" outbursts (division magnitude 12 mag$_{vis}$).

In regard to the decay branches, the bright outbursts fall into two types: *fast* and *slow*. They differ in both the rate and the course of the decay (Fig. 1ab). The decay branches were smoothed by the code HEC13 (author Dr. P. Harmanec), based on the method of Vondrák (1969 and 1977) (thick solid lines in Fig. 1ab).

The maximum brightness and the relative energy (see below) of the bright outbursts tend to decrease in the course of the covered interval (Fig. 2a).

The energy output of the respective outbursts was assessed using the relative energy (RE) of outburst (Fig. 2b). The whole light curve was transformed into intensity, normalized to unity at 14.5 mag$_{vis}$. RE of each outburst was then calculated by integration of the light curve – i.e. it represents the area outlined by the outburst light curve having intensity greater than unity. Since we are interested just in comparing the relative outputs of outbursts, RE may be expressed in dimensionless units. The mean RE of all measured bright and faint outbursts is $RE = 146 \pm 77$ and 17 ± 11, respectively. The bright and faint outbursts tend to alternate.

The outburst cycle-length and its variations: The method of the $O - C$

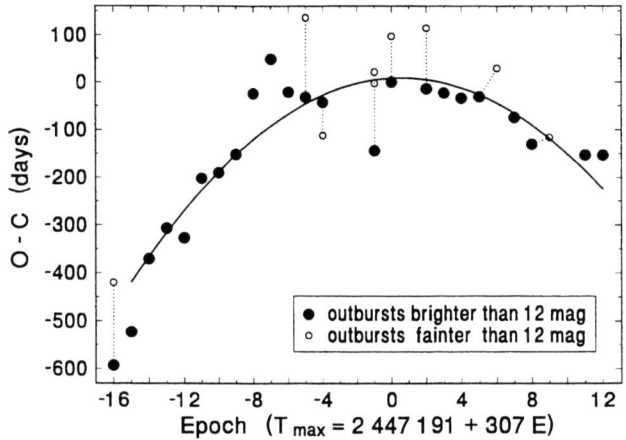

FIGURE 3. The $O-C$ diagram for the outbursts. The $O-C$ values of the outbursts brighter than 12 mag$_{vis}$ were fitted by parabola. See text for details.

residuals from a reference period was used for the search for the cycle-lengths T_C of outbursts and their changes (Fig. 3). The $O-C$ diagram can be constructed even if there are gaps in the data and allows to examine the $O-C$ curve for the respective types of outbursts. The $O-C$ values of the faint outbursts in Fig. 3 are negative (positive) if they occurred before (after) the following (preceding) bright outburst. The ephemeris $T_{max} = 2\,447\,191 + 307\,E$ was used. In most cases the error bars would be smaller than the symbols used. The respective bright outbursts keep the period from cycle to cycle much better than the faint ones but display a strong decrease of T_C which may be approximated by a parabola with superposed "fine" structure.

4. DISCUSSION

The analysis revealed that CH UMa exhibits two *largely* divergent alternating types of outburst. Although they may be analogous to the long and short outbursts in SS Cyg (e.g. Cannizzo and Mattei 1992), in the case of CH UMa all faint outbursts are not only more narrow than the bright ones but the difference in their maximum brightness is very large, approximately 2 mag$_{vis}$, and the ratio of their relative energies is almost 9:1.

The light curves of most outbursts (slow rise in some cases, often symmetric shape with a sharp top) bear a close resemblance to the type B (inside-out outburst). This is also supported by the occurrence of the alternating bright and faint outbursts, predicted for the B type (Smak 1984).

The two types of decay of the bright outbursts can be interpreted in terms of the different value of the viscosity parameter α for each type (the lower α, the slower decay), following the computations by Cannizzo (1994).

The long T_C in CH UMa implies a very low mass transfer rate $\dot{m}$ and/or α (Ichikawa and Osaki 1994), in accordance with the type B of outbursts.

The $O-C$ curve is plausibly defined only for the bright outbursts in which the entire disk is brought into the hot state. This is also seen for the long outbursts in SS Cyg and U Gem (Vogt 1980). Both RE of the outbursts and brightness of maxima in CH UMa tend to decrease (Fig. 2ab) while the secular term in the $O-C$ curve suggests the dominant shortening of T_C – this anticorrelation contradicts the variations of $\dot{m}$ (the model by Hameury et al. 1998). A variable removal of the angular momentum from the disk by the magnetic field from the spots on the secondary (Meyer- Hofmeister et al. 1996) is more promising. The general decrease of RE and maximum brightness (Fig. 2ab) with the trend of decreasing T_C are in accordance with this model.

ACKNOWLEDGEMENTS

This research has made use of the AFOEV database, operated at CDS, France, and NASA's Astrophysics Data System Abstract Service. I thank Dr. Hudec for the comments and Dr. Harmanec for providing me with the program HEC 13. My thanks also to the amateur observers. The investigation of cataclysmic variables with X-ray emission is partly supported by the Project KONTAKT ME 137 by the Ministry of Education and Youth of the Czech Republic.

REFERENCES

Becker, R.H., Chanan, G.A., Wilson, A.S., Pravdo, S.H., 1982, MNRAS, 201, 265
Cannizzo, J.K., 1994, ApJ, 435, 389
Cannizzo, J.K., Mattei, J.A., 1992, ApJ, 401, 642
Hameury, J.-M., Menou, K., Dubus, G., Lasota, J.-P., Huré, J.-M., 1998, MNRAS, 298, 1048
Ichikawa, S., Osaki, Y., 1994, Theory of Accretion Disks II, eds. W.J. Duschl, Kluwer Academic Publishers, Dordrecht, p.169
Kukarkin, B.V., et al., 1974, Second Suppl. to the Third Edition of GCVS, Nauka Publishing House, Moscow
Meyer-Hofmeister, E., Vogt, N., Meyer, F., 1996, A&A, 310, 519
Smak, J., 1984, Acta Astron., 34, 161
Szkody, P., Mattei, J.A., 1984, PASP, 96, 988
Thorstensen, J.R., 1986, AJ, 91, 940
Vogt, N., 1980, A&A, 88, 66
Vondrák, J., 1969, Bull. Astron. Inst. Czechosl., Vol. 20, 349
Vondrák, J., 1977, Bull. Astron. Inst. Czechosl., Vol. 28, 84

ROSAT REVEALS THE LARGE SCALE DISTRIBUTION OF MATTER

A. Sołtan[1], M. Freyberg[2], G. Hasinger[3], T. Miyaji[2], M. Treyer[4], J. Trümper[2]

[1] *Nicolaus Copernicus Astronomical Center, Warsaw,* [2] *MPI für extraterrestrische Physik, Garching,* [3] *Astrophysikalisches Institut Potsdam, Potsdam,* [4] *Laboratoire d'Astronomie Spatiale, Marseille*

ABSTRACT Fluctuations of the soft X-ray background (XRB) at angular scales of up to $\sim 6°$ are measured using the *ROSAT* All-Sky Survey (RASS). At $1°$ the amplitude of the autocorrelation function (ACF) amounts to 0.0035 ± 0.0007. In the range of $0°\!.3 - 4°$ the ACF slope $\gamma = -1.1 \pm 0.2$. These XRB variations are generated by the nonuniform spatial distribution of sources producing the XRB. The magnitude of the XRB fluctuations is used to estimate the level of clustering of X-ray sources. We show that the observed amplitude of the XRB fluctuations results from strong clustering of AGNs at redshifts where most of the XRB originates.

KEYWORDS: cosmology: large-scale structure of Universe – X-rays: general

1. THE AUTOCORRELATION FUNCTION OF THE SOFT XRB

The ACF is calculated using the RASS $12'$ resolution maps from which the non-cosmic diffuse contributions have been subtracted (Snowden et al. 1997). We selected a region in the north galactic hemisphere ($70° < l < 250°$, $b > 40°$), apparently free from the contamination by the emission of galactic plasma (Sołtan et al. 1996). The ACF of the extragalactic component is shown in Fig. 1. Error bars represent 1σ uncertainties resulting from stochastic nature of the count rate distribution. The solid line shows the best fit assuming the power law between $0°\!.3$ and $4°$:

$$w(\theta) = (\theta/0°\!.0058)^{-1.1}.$$

The ACF slope of -1.1 ± 0.2 is steeper than -0.8 found for nearby galaxies, but close to the ACF slope of radio source distribution in the FIRST survey. This coincidence is expected, since both surveys (X-ray and radio) are dominated by distant AGNs and probe similar samples of objects.

2. SOURCE CLUSTERING VS. THE XRB FLUCTUATIONS

Most, if not all, of the XRB is produced by discrete sources. Distinct classes of extragalactic objects (various types of active galactic nuclei, clusters of galaxies, "normal" galaxies) contribute to the XRB. The spatial distribution of these sources defines

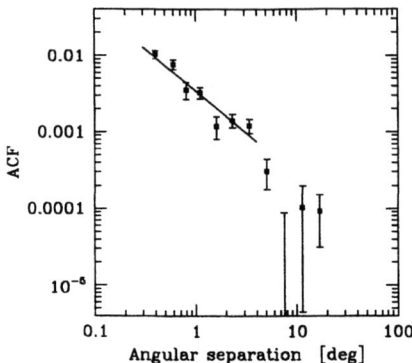

FIGURE 1. Points with error bars: the autocorrelation function (ACF) of the soft X-ray background (XRB) based on the *ROSAT* All-Sky Survey; Solid line: power-law best fit between $0°\!.3$ and $4°$.

surface brightness distribution of the XRB. The ACF measures the fluctuations of the XRB flux resulting from nonuniform spatial distribution of sources integrated along the line of sight. X-ray sources identified in the *ROSAT* Deep Survey have redshifts spread roughly uniformly between ~ 0.2 and ~ 2 with a tail extending up to ~ 3 (Schmidt et al. 1998). Such wide distribution of sources contributing to the XRB results from their strong evolution. Clustering of sources is also subject to the cosmic evolution. Thus, to reproduce observed angular fluctuations of the XRB, one should model both the luminosity and clustering evolution of X-ray sources over a wide redshift range.

2.1. Evolution of X-ray luminosity density

The largest contribution to the total XRB comes from AGNs, predominantly QSOs and Seyfert galaxies (Hasinger 1998). At low redshifts, however, a relatively large fraction of X-ray luminosity comes from normal galaxies and clusters of galaxies (Sołtan et al 1997). Quasars are subject to strong evolution. Although data on the rate of evolution of galaxies and clusters are scarce, it is certain that these objects exhibit at most weak evolutionary effects. For the purpose of the present analysis we divide the X-ray sources into two groups: (*a*) evolving with quasar-like rate of evolution, and (*b*) nonevolving - representing galaxies and clusters. The relative local contribution of both components is not well established. Various models of the evolution of the AGN luminosity function have been proposed in the literature. In the present calculations we have used the pure *luminosity evolution* (LE) model by Boyle et al. (1994) and the *luminosity dependent density evolution* (LDDE) by Miyaji et al (1999). Although these models assume basicly different types and rates of evolution, both predict very similar redshift distributions of the XRB flux. As an example, in Fig. 2 the redshift distribution of the XRB flux is shown for the LDDE model.

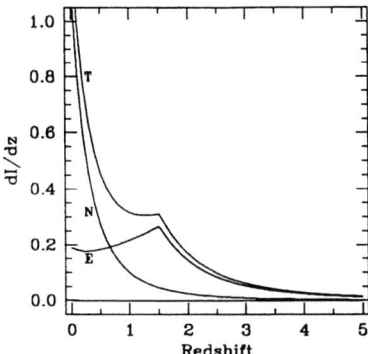

FIGURE 2. Redshift distribution of the soft X-ray background (i.e. relative contribution of consecutive redshift shells to the XRB) in the *luminosity dependent density evolution* (LDDE) model. Total (T) flux is a sum of contributions of nonevolving (N) and evolving (E) sources.

2.2. Evolution of spatial clustering

To describe the source clustering we use a spatial (3D) autocorrelation function (SCF). We tentatively assume that the spatial distribution of *nonevolving* sources resembles the distribution of the general population of galaxies. The SCF of galaxies at separations below $\sim 20\,\mathrm{Mpc}$ has a well defined power law shape $\xi = (r/r_o)^\gamma$ with characteristic clustering length $r_o \approx 5\,\mathrm{Mpc}$ ($H_o = 100\,\mathrm{km\,s^{-1}\,Mpc^{-1}}$) and a slope $\gamma \approx -1.8$ (e.g. Groth & Peebles 1977, Ratcliffe et al. 1998). At larger separations the SCF determinations are uncertain; it is generally accepted that the SCF steepens above $\sim 20\,\mathrm{Mpc}$, but there are significant differences between individual investigations. Evolution of clustering is usually described by means of the scaling factor $(1+z)^\varepsilon$:

$$\xi[r,z] = \xi[r/(1+z), 0](1+z)^{-(3+\varepsilon)},$$

where r denotes the *comoving* separation. Value of ε is not well constrained by observations but most investigations give $\varepsilon = 1 \pm 1$. Since contribution of the nonevolving objects to the XRB decreases rapidly with redshift (Fig. 2), the uncertainties of the clustering evolution rate of normal galaxies only weakly affect the present calculations.

Studies of the AGN clustering and clustering evolution provide results which are difficult to reconcile. It has been generally accepted that AGN are more strongly clustered than normal galaxies and that clustering amplitude is subject to evolution described by $\varepsilon < 0$. Some studies show that the AGN clustering is stable in *comoving* coordinates (e.g. Croom & Shanks 1996). In this case $\varepsilon \approx -1.2$, but La Franca et al. (1997) find $\varepsilon \approx -2.5$. However, Carrera et al. (1997) argued that AGN are subject to weaker clustering than previously reported and that both the local clustering amplitude and the clustering evolution of AGN is similar to that of 'normal' galaxies. To explore effects of AGN clustering on the XRB fluctuations we allow for stronger AGN clustering by using the ACF with high normalization of $r_o \approx 9\,\mathrm{Mpc}$.

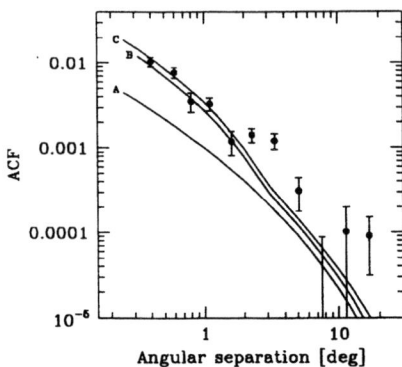

FIGURE 3. Points with error bars: the autocorrelation function of the XRB (same as Fig. 1). Curve A: example of model with weakly clustered X-ray sources; curves B and C: two examples of models with high clustering and strong clustering evolution, (B - LDDE, C - LE; see text for details).

3. MODELS THE XRB FLUCTUATIONS

Wide variety of evolutionary models for luminosity density and source clustering have been used to reproduce the observed ACF. It was found that models with low normalization of SCF ($r_o \approx 5$ Mpc) and clustering amplitude increasing with time ($\varepsilon \approx 1$) predict substantially weaker XRB fluctuations than are actually observed. As an example, curve A in Fig. 3 shows a model which assumes pure LE, 'standard' normalization of the SCF and $\varepsilon = 0.8$. Only models with high SCF normalization ($r_o \approx 9$ Mpc) and strong clustering at high redshifts produce adequate XRB fluctuations at seperations of up to $\sim 2°$. Discrepancies at larger separations will be discussed in a separate paper. Curves B (LDDE, $\varepsilon = -2.5$) and C (LE, $\varepsilon = -2.5$) in Fig. 3 show two such models.

ACKNOWLEDGEMENTS

This work has been partially supported by the Polish KBN grant 2 P03D 002 14.

REFERENCES

Boyle B.J., Shanks T., Georgantopoulos I., et al., 1994, MNRAS 271, 639
Carrera F.J., Barcons X., Fabian A.C. et al., 1997, MNRAS 285, 820
Croom S.M., Shanks T., 1996, MNRAS 281, 893
Groth E.J. & Peebles P.J.E., 1977, ApJ 217, 385
Hasinger G., 1998, Astron. Nachr. 319, 37
La Franca F., Andreani P. & Cristiani S., 1998, ApJ 497, 529
Miyaji T., Hasinger G., and Schmidt M., 1990, Highlights in X-ray Astronomy in honour of Joachim Trümper's 65th birthday, p. 222
Ratclife A., Shanks T., Parker Q.A., Fong R., 1998, MNRAS 296, 191
Schmidt M., Hasinger G., Gunn J., et al., 1998, A&A 329, 495
Snowden S.L., Egger R., Freyberg M.J., et al., 1997, ApJ 485, 125
Sołtan A.M., Hasinger G., Egger R., Snowden S. & Trümper J., 1996, AA 305, 17
Sołtan A.M., Hasinger G., Egger R., Snowden S. & Trümper J., 1997, AA 320, 705

THE IBIS DATA HANDLING SYSTEM

M. Stuhlinger, E. Goehler, C. Dreischer, R. Volkmer, R. Staubert, E. Kendziorra,
N. v.Krusenstiern, P. Risse, R. Weiss

IAAT-Astronomie, Univ. Tuebingen, Germany

ABSTRACT The Imager on Board of INTEGRAL (IBIS) has to cope with very high data rates (>15.000 events/sec) from two detector layers (ISGRI and PICsIT). Because the available telemetry rate is only about $1/30$ of the expected maximum data rate an on-board data reduction and pre-processing is neccessary. This task is performed by a tailored fast Hardware Event Preprocessor (HEPI) and a slow programmable Data Processing Electronics (DPE). Several processing modes can be combined to adapt the instrument IBIS to the aims of the observers. The functions and methods of the digital on-board data processing (hardware and software) are described.

1. INTRODUCTION

IBIS is one of four instruments of the INTEGRAL satellite to be launched in September 2001. It is designed to provide images in the energy range from 15 keV to 10 MeV with an angular resolution of about 12 arcmin.

This is achieved by a coded-mask aperture and a detector array. The structure of the mask allows a deconvolution even for many objects.

The detector consists of two detector layers:

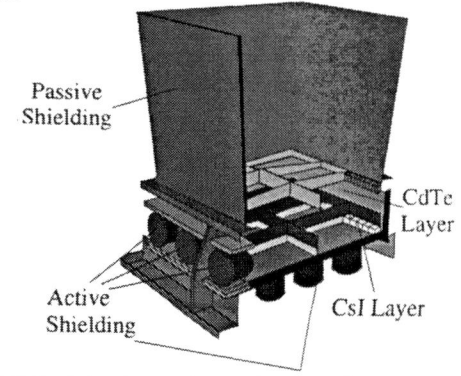

- The upper one consists of 128×128 Cadmium–Telluride semiconductor detectors (ISGRI) with an energy range from 15 keV to 700 keV.

- The lower one is made of 64×64 Caesium–Iodide–scintillation–pixels (PICsIT) with an energy range from 150 keV to 10 MeV.

Lateral radiation is shielded by an active BGO-detector veto system and a passive W-Pb foil.

2. CONSTRAINTS ON THE DATA HANDLING SYSTEM

The data handling system of IBIS has to cope with high input data rate and low telemetry bandwidth combined with comparatively slow on-board data processing unit:

- The ISGRI detector will generate about 98,000 bit/sec (ESA solar min scenario).
- The PICsIT detector may generate up to 966,400 bit/sec (ESA solar min scenario).
- The maximum avaliable telemetry rate for IBIS can only transmit 59,840 bit/sec.
- IBIS data handling will be performed in a Data Processing Electronics which runs at 13 Mhz at approximately 1,25 MIPS (DAIS).

3. THE HARDWARE EVENT PREPROCESSOR FOR IBIS (HEPI)

To handle these constraints the data handling system is split into two parts:

1. A fast tailored digital electronics – the Hardware Event Preprocessor for IBIS (HEPI).
2. The Data Processing Electronics (DPE) handles the more complex processing in a programmable computer.

4. FUNCTIONS OF HEPI

Energy Correction: HEPI performs a linear energy correction for all PICsIT events. A configurable look–up table inside HEPI memory provides linear correction curves for each of the 64 x 64 PICsIT detector elements.

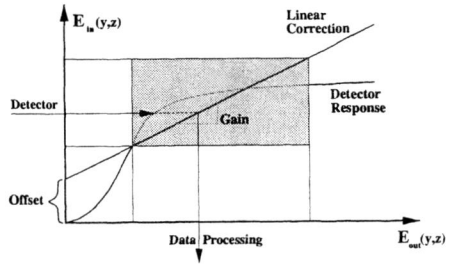

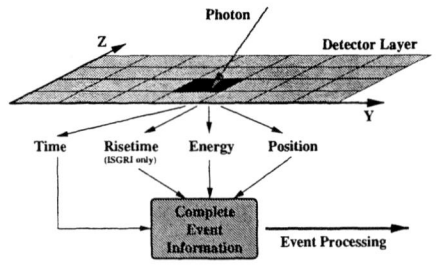

Photon by Photon: The Photon by Photon function passes all events from ISGRI and PICsIT to the DPE without any loss of information. The time tagging allows observations up to 72 hours with a time resolution of 61 μsec. (This is the only HEPI function for ISGRI data.)

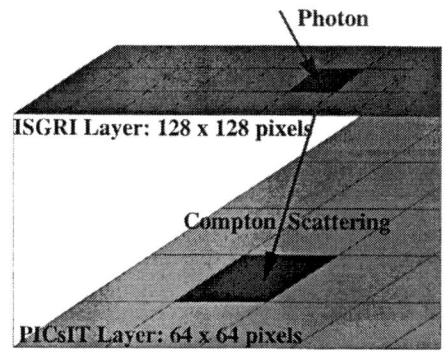

Compton Events: When both detector layers encounter an event simultaneously it is assumed that both events have their origin in the same photon which was Compton scattered in one of the detector layers before hitting the other one. The complete Compton event information from both detectors are available. This function is a special feature of the photon by photon function.

Imaging Histograms: The integration of histograms allows a constant telemetry rate for PICsIT data independently of the event rate (up to 15000 events/sec). For each pixel the events are counted in 256 configurable energy channels and stored in a HEPI memory area. The precise time information of the events is lost.

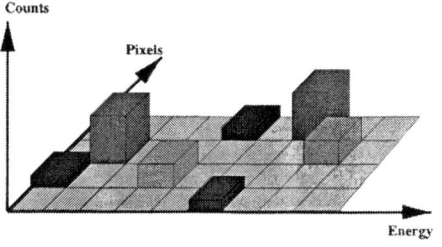

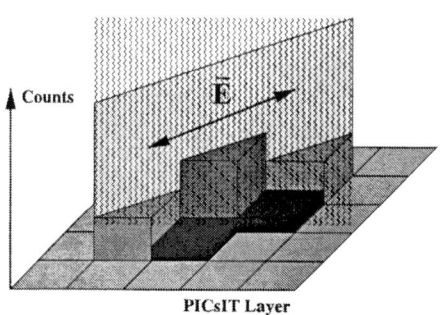

Polarimetry Histograms: This function analyses photons which are Compton scattered inside the PICsIT layer and deposit their energies inside two neighbouring pixels. It is assumed that the photons deposit the higher part of the energy inside the first pixel. For every pixel the energy is counted in 64 channels while the polarisation angle is kept in 8 different directions represented by the neighbouring pixel positions. The time information of the events is lost.

Spectral Timing Histograms: To analyze spectral time variability of sources HEPI provides the tool to generate histograms with a configurable integration time from 1/2 sec down to 1 msec. These spectral timing histograms consist of 8 configurable energy channels. The position information of the events is lost.

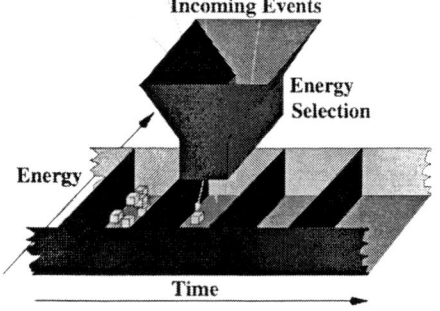

5. INSTRUMENT APPLICATION SOFTWARE (IASW)

The on-board software for IBIS (the IBIS IASW) running on the data processing electronics (DPE) is split in different software tasks which are implemented by a preemptive multitasking system (which is the same for all data processing electronics on INTEGRAL).

The tasks are priority controlled allowing maximal performance on data processing:

Apart from tasks processing telecommands received from ground and housekeeping data collection most performance is spent on fetching and compressing histogram blocks as well as the evaluation of single photon events supplied from HEPI:

- Single photon events encounter a loss–free compression by a time–difference method.

- Compton events recognized by HEPI which are out of the field of view will be rejected. The direction can be computed by the position and energy of both photon event counterparts.

- Histograms (Imaging and Timing) are compressed by exploiting the fact that most cells of the histogram are filled less than 1/8 of 256.

- The IASW distributes the available telemetry rate between different telemetry types according to their priority and the incoming event rate.

6. PERFORMANCE TESTS

Performance tests have shown that HEPI is able to handle events up to a rate of 4,096,000 bit/sec which is 5 times more than required.

Also the software running on the data processor can cope with 320,000 bit/sec which is >5 times more than the max. available telemetry rate.

The compression factor of 3:2 (single photons) and nearly 2:1 (histograms) will increase the amount of scientific data linked down.

ACKNOWLEDGEMENTS

The IBIS consortium consists of the following institutes:
CNR/IAS
CEA/SAP
CNR/TESRE
UNIVERSITY OF BERGEN
UNIVERSITY OF VALENCIA
COPERNICUS ASTR.CTR
NASA–MSFC
IAAT

THE POPULATION OF FAINT X-RAY SOURCES IN THE GALAXY AND THEIR CONTRIBUTION TO THE GALACTIC RIDGE X-RAY EMISSION

M. Sugizaki[1,2], K. Matsuzaki[2], H. Kaneda[2], S. Yamauchi[3], K. Mitsuda[2], and ASCA Galactic Plane Survey team

1) National Space Develop Agency of Japan, 2-1-1, Sengen, Tsukuba, 305-8505, Japan
2) ISAS, 3-7-1, Yoshinodai, Sagamihara, 229-8510, Japan
3) Facility of Humanities and Social Sciences, Iwate University, 3-18-34, Ueda, Morioka, 020-8550, Japan

ABSTRACT

The Galactic ridge X-ray emission (GRXE) is an enhanced X-ray emission along the Galactic plane, whose origin still remains unknown. The GRXE was studied for the first time in the energy band of 0.5–10 keV with a spatial resolution of $3'$ by the ASCA Galactic plane survey which covers the spatial area of $|l| \lesssim 45°$ and $|b| \lesssim 0.4°$ almost uniformly. We determined the large scale distribution of the GRXE after eliminating discrete X-ray sources with a flux above $10^{-12.5}$ ergs cm^{-2} s^{-1} and revealed that the volume emissivity of the GRXE is highly concentrated within the 4 kpc arm. We resolved 163 discrete X-ray sources by imaging analysis and obtained the LogN-LogS relations of those sources. In the hard (2–10 keV) band, the slope of the LogN-LogS relation is significantly smaller than 1. Considering a scale height of the Galactic plane covered by the ASCA Galactic plane survey, we conclude that the slope of the LogN-LogS relation represents that the spatial distribution of the sources has a scale height as small as ~ 10 pc, that the sources are distributed in arm-like structures, and/or that the relation reflects the luminosity function rather than the spatial distribution. We analyzed small-scale spatial intensity fluctuation of the GRXE after subtracting the large scale variations and the contributions of resolved discrete sources. The residual small-scale fluctuation is found to remain significantly over the photon-counting Poisson fluctuation. However, in the 2–10 keV band, the amplitude can be explained by the fluctuation of the cosmic X-ray background coming through the Galactic interstellar medium. From this, we can obtain a strong constraint to a flux and a number density of discrete sources if we are to explain the GRXE with a sum of discrete sources; more than 10^7 sources with the luminosity smaller than 10^{31} ergs s^{-1} must exist in the Galaxy. This source number density is larger by three orders of magnitude than that of CVs in the solar neighborhood, which would be a plausible candidate for the discrete source origin. Thus, We conclude that the diffuse emission origin should be much more probable even if it has large problems.

KEYWORDS: diffuse radiation — Galaxy: structure – X-rays: sources

1. INTRODUCTION

The Galactic ridge X-ray emission denotes an unresolved enhanced X-ray emission along the Galactic plane. The existence of the GRXE has been known from early stage of X-ray astronomy, however the origin still remains unknown (e.g. Warwick et al. 1985). The X-ray spectrum indicates an origin from hot plasma with a temperature of 10^8K (e.g. Kaneda et al. 1997). If such a hot plasma is freely floating in the Galactic plane, it cannot be constrained by the gravitational potential in the Galaxy. An alternative possible hypothesis for the origin of the GRXE is a superposition of unresolved faint sources in the Galaxy. However, the population of faint X-ray sources in the Galaxy has not been measured in the X-ray band above 2 keV, because a high sensitive imaging observation in the higher energy X-ray band has been very difficult. ASCA with a moderate spatial resolution of $3'$ in the wide energy band of 0.5–10 keV, gives a first opportunity to answer this problem. Therefore, we performed a large area survey on the Galactic plane by ASCA and analyzed those image data.

2. ANALYSIS

The ASCA Galactic plane survey, which covers the area of $|l| \lesssim 45°$ and $|b| \lesssim 0.4°$ by 170 successive pointing observations each with an exposure time of ~ 10 ks, was carried out from March, 1995 to April, 1999. We utilized data of the GIS for the image analysis, which has advantages to the SIS in the points of a larger field of view (FOV) and a better calibration accuracy of the response function. We analyzed the large data of the ASCA Galactic plane survey in the following steps:
    (1)   determine the large scale intensity variations averaged over each pointing,
    (2)   extract discrete X-ray sources,
    (3)   evaluate small scale fluctuations of the unresolved residual emission
Since the image response of the ASCA detectors is too complicated to allow deconvolution into an original sky image, we applied an image fitting method. Also, we eliminated data of pointings contaminated by stray light coming from bright X-ray sources outside the FOV.

    First, we determine the large-scale intensity variations. It is performed by fitting the image response for a uniform surface brightness to raw data in each GIS FOV. The derived large-scale intensity variations are shown in figure 1 for the soft (0.7–2 keV) and the hard (2–10 keV) band, respectively. The profile of the 2–10 keV band which shows an excess within the area of $l \lesssim 35°$ corresponding to the 4-kpc arm, is consistent with those of the past observations (e.g. Warwick et al. 1985).

    Next, we made a source survey. The procedure of the source survey is owed to that developed in the ASCA LSS (Ueda et al. 1999). We searched for peaks in the image and constructed the model for the observed image, which consists of the large scale intensity variations and the extracted peaks. We accepted peaks with a significance above 4σ in the image fitting as discrete X-ray sources. The source flux and the area surveyed for an arbitrary flux are estimated from the best fit model. We performed the source survey in the soft (0.7–2 keV) and the hard (2–10 keV)

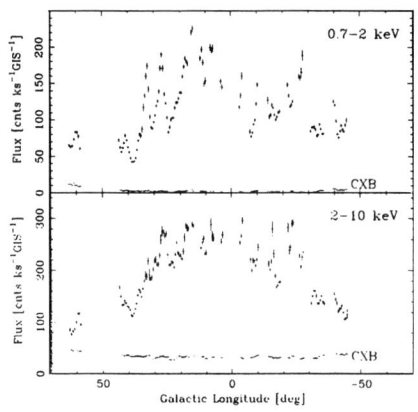

FIGURE 1. The large scale intensity variations of the GRXE averaged over each FOV of the GIS in the 0.7–2 keV band (upper) and the 2–10 keV band (lower). Contributions of the resolved discrete X-ray sources are eliminated. Inclusive contributions of the CXB coming through the Galactic ISM are shown together, which are calculated from the average spectrum of the CXB and the Galactic hydrogen column density obtained from the HI and CO line intensity.

band individually and detected in total 163 sources with a flux down to $10^{-12.5}$ ergs cm^{-2} s^{-1}. The LogN-LogS relation of these sources can be calculated from the number of detected sources and the area surveyed for each flux. In the soft band, we obtained the LogN-LogS relation a little different from the past results derived from the Einstein and the ROSAT Galactic plane surveys, which is reasonably explained from the difference of the area surveyed. In the hard band, the LogN-LogS relation of the faint Galactic sources was derived for the first time in this survey, and is shown in figure 2. It is significantly flatter than a power-law with an index of -1, expected for a two-dimensional source distribution. Thus, it implies that the spatial distribution of the X-ray sources shows a one-dimensional structure associated with the Galactic arm and/or the LogN-LogS relation reflect the luminosity function rather than the spatial distribution (Sugizaki 1999).

Finally, we analyzed the small scale spatial fluctuations of the unresolved emission in the 2–10 keV band, in which the Galactic ISM is almost transparent and the emission consists of the GRXE and the cosmic X-ray background (CXB). We fit a model of uniform surface brightness to the raw data after masking the resolved sources and subtracting the large scale intensity variations. The fit could not be accepted. We carefully estimated contributions of various unmodeled effects, like residuals of the large-scale intensity variations, difference of the effective exposure time, contamination from the extended PSF, stray light from the bright sources outside of the FOV, and uncertainty of the response function. As the result, we confirmed that significant residuals still remain. We evaluated the fluctuation of the CXB expected from the LogN-LogS relation of the extragalactic sources (Ueda et al. 1999), and found that it can explain the observed residual fluctuations (Sugizaki 1999). Therefore, small scale fluctuations in the GRXE are not necessarily required. It strictly constrain the unresolved faint sources as the origin of the GRXE. We construct a model of the source distribution on the 4-kpc arm of the Galaxy to explain the GRXE and derived a required condition: their averaged luminosity should be smaller than 1.6×10^{31} ergs s^{-1} and their number density is 0.7×10^{-3} pc^{-3} with a 90% confidence limit (Sugizaki 1999). That condition is illustrated in the figure 2.

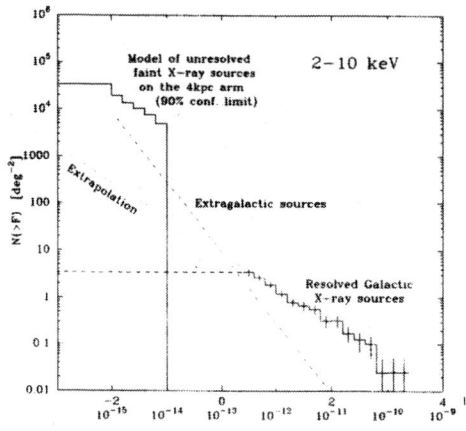

FIGURE 2. The LogN-LogS relation of the Galactic X-ray sources resolved by the ASCA survey in the 2–10 keV band and the allowed model for the unresolved faint sources within the 90% confidence limit from the fluctuation analysis. The LogN-LogS relation of the extragalactic sources observed through the Galactic absorption is shown together, which is estimated form the results of the ASCA LSS (Ueda et al. 1999).

3. DISCUSSION — ORIGIN OF THE RIDGE X-RAY EMISSION

From a similarity of the spectrum with the GRXE, young supernova remnant (SNR), cataclysmic variables (CVs) and massive stars are considered as the candidates. We can estimate the population of the young SNRs from the birth-rate and the evolution of the supernovae in the Galaxy. The number of SNRs with a temperature above $\sim 10^8$ K is at most ~ 100 (Koyama et al. 1986), which cannot explain the spatial uniformity of the GRXE. A population of CVs was estimated only for the solar neighborhood systems (Patterson 1985), which is $\sim 10^{-6}$ [pc^{-3}], then much smaller by three order of magnitude than that required for the 100% GRXE. We cannot reject a hypothesis that the CVs' density inside the 4 kpc arm is larger than that near the sun, however it seems very difficult from the stellar evolution in the Galaxy. Also, recent ASCA observations suggest that some massive stars have a hot plasma with a temperature of $\sim 10^8$ K (Matsuzaki 1999). If all massive stars in the Galaxy have such a hot plasma, the number of sources to explain the full GRXE is satisfied. However, the abundance of those massive stars is very small. Therefore, the hypothesis of diffuse hot plasma as the origin of the GRXE is most plausible even if it has a large problem: how is such a hot plasma produced and confined in the Galactic plane. Magnetic fields and low-energy cosmic rays, whose comprehensive properties in the Galaxy have not been acknowledged, might play an important role (Kaneda et al. 1997; Varinia & Marshall 1998).

REFERENCES

Kaneda, K., et al. ApJ, 491, 638-652, (1997)
Koyama, K., Ikeuchi, S., & Tomisaka,K. PASJ, 38, 503 (1986)
Matsuzaki, K. Ph.D Thesis Univ. of Tokyo (1999)
Patterson, J. & Raymond, J. C. ApJ, 292, 535-549 (1985)
Sugizaki, M. Ph.D Thesis Univ. of Tokyo (1999)
Ueda, Y., et al. ApJ, 518, 656 (1999)
Warwick, R. S. et al. Nature, 317, 218-221 (1985)

AN H-R DIAGRAM FOR AGN?

J. W. Sulentic [1], P. Marziani [2], M. Calvani [2]

1) University of Alabama, USA; 2) Astronomical Observatory, Padova, Italy

ABSTRACT We have been searching for a correlation space that provides an optimal discrimination between the diverse classes of AGN. A space involving FWHM Hβ, the ratio of EW FeII/EW Hβ and soft X-ray spectral index shows the strongest discrimination and intercorrelation. It represents a synthesis of several correlations that have emerged in the past decade. We interpret the correlations as driven primarily by orientation and L/M ratio of the central source with an uncertain role for Fe abundance.

KEYWORDS: Galaxies: active

1. INTRODUCTION

Correlations often drive the development of physical theories for astronomical phenomena. Boroson & Green 1992 (BG92) suggested that FeII strength was an important parameter for discriminating between different types of quasars (e.g. radio-loud=RL and radio-quiet=RQ). They identified a "1st eigenvector" from a Principal Component Analysis (PCA) of their BQS correlation matrix. FeII strength was found to anticorrelate with broad line FWHM Hβ and the strength of narrow line [OIII]λ5007. More recently the soft X-ray photon index Γ_{soft} has emerged as a correlate with both FeII and Hβ measures (Wang et al 1996). We combine BG92 measures with 18 RL sources from our own survey (Marziani et al 1996, MS96) to explore our expanded interpretation of an Eigenvector 1 (E1) correlation space involving: 1) FWHM Hβ_{BC}, 2) R_{FeII} = EW FeIIλ4570 blend/EW Hβ_{BC} and 3) Γ_{soft}.

2. E1 PARAMETER SPACE

Figure 1 shows the 2D projections of the E1 space. AGN plotted include: 1) the BG92 sample of 71 RQ (filled circles) and 16 RL sources and 2) 18 RL sources from MS96. FeII measures in MS96 have been converted to the BG92 system. Fig. 1 shows a total of 34 RL AGN (open symbols), 13 of which are flat spectrum (open circles). Soft X-ray photon indices are available for a large fraction of the above samples (e.g. Wang et al. 1996; Brinkmann et al. 1997; Siebert et al. 1998; Yuan et al. 1998).

Narrow-Line Seyfert 1 Galaxies (NLSy1) represent a narrow FWHM, strong EW FeII, weak EW Hβ and soft X-ray excess extremum in the parameter space. RL sources represent the opposite extremum in all parameters. BAL sources (4 in BG92) occupy a slightly larger domain than the NLSy1 but are displaced in: a) X-ray luminosity where they are X-ray quiet and b) optical luminosity where they are about 4× more luminous on average. E1 appears to be uncorrelated with optical luminosity as suggested by its appearance in the second (orthogonal) eigenvector of the PCA analysis in BG92. Steep-spectrum RL sources (open squares) represent the most extreme opposite population showing the weakest FeII strength and no evidence for a soft X-ray excess.

2.1. A RQ AGN Main Sequence?

E1 shows a continuity in all measured parameters for (mostly RQ) sources with FWHM H$\beta \leq$4000km s^{-1}. This challenges the idea that NLSy1 represent a population of AGN with unique properties. NLSy1 sources appear to form a smooth extremum in both EW FeII/Hβ ratio and Γ_{soft} relative to broader line profiles. The continuous "main sequence" in the R$_{\text{FeII}}$ vs. FWHM(Hβ) plane correlates with several other parameters including (see MS96): 1) Hβ_{BC} and CIV$_{BC}\lambda$1549 asymmetry increasing from blue- to red-ward as one proceeds from NLSy1 to broader profiles, 2) EW Hβ_{BC} and CIV$_{BC}\lambda$1549 increasing in the same sense and CIV$_{BC}\lambda$1549 centroid blueshift decreasing in sources with broader profiles, weaker FeII and harder X-ray spectra. We identify several physical parameters that are also involved in the correlation space including: a) ionization spectrum (lower in NLSy1), b) electron density (higher in NLSy1), c) decoupling of HIL and LIL (very strong in NLSy1) and d) as yet uncertain model-dependent parameters related to the soft X-ray excess (see Sulentic et al. 2000).

2.2. Two populations of RQ AGN?

There is a remarkable continuity and correlation in line parameters from NLSy1 to RQ AGN with FWHM$\leq$ 4000 km s^{-1}. Broader RQ sources show a larger scatter and overlap almost completely with the parameter space occupied by RL AGN. This suggests a possibly significant phenomenological distinction among RQ sources involving "Population A" with FWHM H$\beta \lesssim$ 4000km s^{-1} and a broader "Population B" with FWHM H$\beta \gtrsim$ 4000km s^{-1}. Pop. A includes 65% of BG92 RQ sources and very few RL. Pop. B involves RQ quasars that show optical properties most similar to RL AGN. A plot of EW FeII vs. FWHM Hβ for RQ sources reveals two disjoint populations of sources with no correlation within the individual populations. The correlations appear in the E1 parameter domain. Table 1 below presents representative values of important parameters for the two proposed populations.

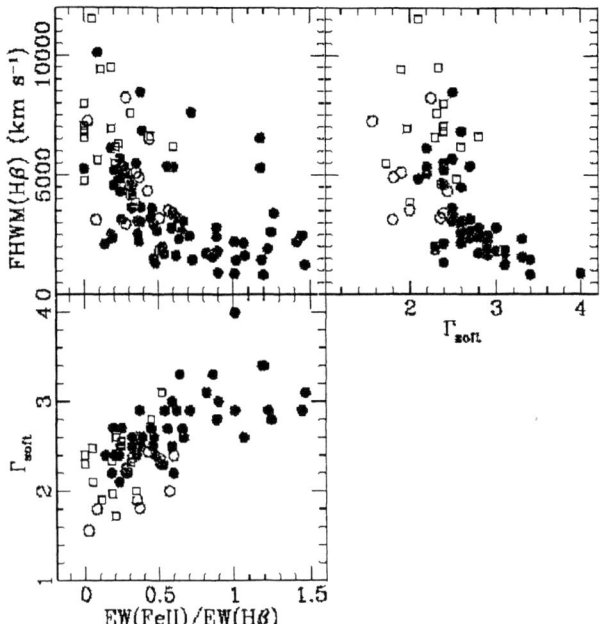

FIGURE 1. 2D projections of the 3D space correlating FWHM(Hβ), R_{FeII} and soft X-ray photon index Γ_{soft}. Radio quiet, radio loud steep and radio loud flat spectrum sources are shown as filled circles, open squares and open circles respectively.

Mean	Population A	Population B
EW FeII (Å)	65	30
R_{FeII}	0.7	0.3
FWHM(Hβ) (km s^{-1})	2000	5500
Γ_{soft}	2.9	2.4

3. DISCUSSION

Almost every currently fashionable model of the Broad Line Region is challenged by the statistical properties of AGN emission lines. E1 correlates AGN spectroscopic data in a way that removes much of the apparent "randomness" of line properties and redefines input parameters for photoionization and kinematical models. Interpretation of the correlations in E1 most likely involves at least three independent parameters: the ratio of AGN luminosity to black hole mass L/M, source orientation and Fe abundance.

The ratio L/M may be systematically higher in Pop. A AGN (i.e. higher accretion rate) relative to Pop. B and RL AGN. It is obvious that orientation cannot account for the systematically larger FWHM Hβ_{BC} observed for RL sources (MS96). A role for L/M is supported by: a) the observation of a soft X-ray excess, which has been related to a higher accretion rate (e.g. Pounds et al 1995) and b) the C$IV_{BC}\lambda$1549 profiles in NLSy1 which show the largest blue-shifts and lowest EW that are strongly suggestive of an outflow. At the same time reverberation mapping of Pop. B Sy1 nuclei suggests that the C$IV\lambda$1549 emitting gas is not outflowing. Redward asymmetries predominant in RL sources may signal the increasing role of gravitational redshift (MS96). The L/M ratio may only be high enough in Pop. A sources to trigger a radiation pressure driven outflow.

Orientation: NLSy1 show the narrowest FWHM Hβ_{BC} and sometimes almost completely blueward displaced C$IV_{BC}\lambda$1549 profiles. The short-term, high amplitude variability of the soft X-ray excess favors a preferential orientation for these sources. If FeII and Hβ_{BC} are emitted in a flattened configuration (an accretion disk), FWHM(Hβ) and FeII emission should be orientation dependent. The properties of core-dominated and superluminal sources, which are seen oriented close to the jet axis and which have FWHM(Hβ) $\leq$ 4000 km s^{-1}, at the low end of the radio-loud distribution, as well as of the recently-discovered "radio loud NLSy1" suggest that RL and RQ AGN may show parallel orientation-driven FWHM vs. FeII relations.

Iron abundance may not be able to account for the wide range of R$_{FeII}$ observed, but may account for at least part of the difference between RQ and RL sources.

The plane FWHM(Hβ)-R$_{FeII}$ may be the equivalent of an H-R diagram for AGN although Γ_{soft} is necessary for a full discrimination among AGN sub-types. Once the origin of the soft X-ray excess in Pop. A sources is understood the Γ_{soft} correlation may lead to further constraint on physical or aspect-related parameters.

ACKNOWLEDGEMENTS

JS acknowledges support and hospitality from Osservatorio Astronomico di Padova. MC and PM acknowledge support from the Italian Ministry for University and Research (Cofin 98–02–32).

REFERENCES

Boroson, T. A., Green, R. F. 1992, ApJS 80, 109
Brinkmann, W., Yuan, W., Siebert, J. 1997, A&A, 319, 413
Marziani, P., Sulentic, J.W., Dultzin-Hacyan, D., Calvani, M., Moles, M. 1996, ApJS 104, 37 (M96)
Pounds K. A., Done C., Osborne J. P. 1995, MNRAS 277, L5
Siebert, J., Brinkmann, W., Yuan, W. 1998, Astron. Nach. 319, 35
Sulentic, J. W., Marziani, P., Dultzin-Hacyan, D., 2000, Ann. Rev. Astr. Astroph., in press
Wang, T., Brinkmann, W., Bergeron, J. 1996, A&A, 309, 81
Yuan, W., Brinkmann, W., Siebert, J. & Voges, W. 1998, A&A 330, 108

ON THE MODELLING OF TIME-DEPENDENT PHENOMENA AROUND BLACK HOLES

E. Szuszkiewicz [1,2] and J. C. Miller [2,3]

1) Torun Centre for Astronomy, Nicolaus Copernicus University, Poland
2) International School for Advanced Studies, SISSA, Trieste, Italy
3) Nuclear and Astrophysics Laboratory, University of Oxford, England

ABSTRACT   We are carrying out a programme of calculations which is directed towards two specific goals. The first is to calculate light curves and spectra for individual accreting black holes; the second, is to determine the characteristic properties of populations of these objects. We highlight our most interesting results and report on the present status of the project.

KEYWORDS: accretion; accretion discs - instabilities

1. ACCRETING BLACK HOLES

One of the most characteristic properties of galactic black hole candidates (some of the X-ray binaries) and extragalactic black hole candidates (active galactic nuclei) is their variability. The possibility that intrinsic instabilities in the accretion discs thought to be present in these objects may be responsible for the observed variability has been extensively studied (see Kato, Fukue and Mineshige 1998 for a review).

Recent developments in observational techniques provide high quality data whose modelling requires detailed theoretical predictions. We are aiming to meet this need by calculating theoretical light curves and spectra corresponding to particular instabilities whose origin is well understood. Also, we plan to use our study of individual objects as a basis for considering whole populations of accreting black holes. The specific accretion scenarios characteristic of each instability will be taken into account for investigating the evolution of active galactic nuclei and black hole demography. Until now, such studies have been based on simple assumptions about the mass-luminosity relation (Salucci, Szuszkiewicz, Monaco and Danese, 1999).

2. THERMAL INSTABILITY DRIVEN BY RADIATION PRESSURE

We have started our programme of work by investigating the thermal instability driven by radiation pressure and have performed non-linear, time-dependent numerical calculations of non-stationary accretion onto black holes (taking $M = 10 M_\odot$) using the "slim disc" approach (Abramowicz, Czerny, Lasota, and Szuszkiewicz 1988) with vertically-integrated equations. Our strategy has been to start from the original version of the slim-disc model, with all of its associated approximations and assumptions, and then to add improvements to this systematically, one at a time, so that the effects of each can be clearly understood. In the first phase of the work, we used the original αp viscosity prescription. For $\alpha = 10^{-3}$ (Szuszkiewicz and Miller 1997) a violent instability was observed just outside the sonic point, the nature of which is under investigation. Removing this, we have found that models predicted to be stable by local analysis do indeed remain stable and stationary. The same applies for models which, according to local analysis, would have a potentially unstable region smaller than the minimum wavelength for unstable perturbations. In terms of luminosities, this means that all models with luminosity less than or equal to $0.08 L_E$ (L_E is the Eddington luminosity) are globally stable. For models with luminosities between $0.09 L_E$ and $1 L_E$ we see unstable behaviour near the sonic point. For $\alpha = 10^{-1}$ (Szuszkiewicz and Miller 1998) we have found the first complete limit-cycle behaviour (Figure 1) with successive evacuation and refilling of the inner parts of the disc, the evacuation being accompanied by a swelling up of the region of lowered surface density.

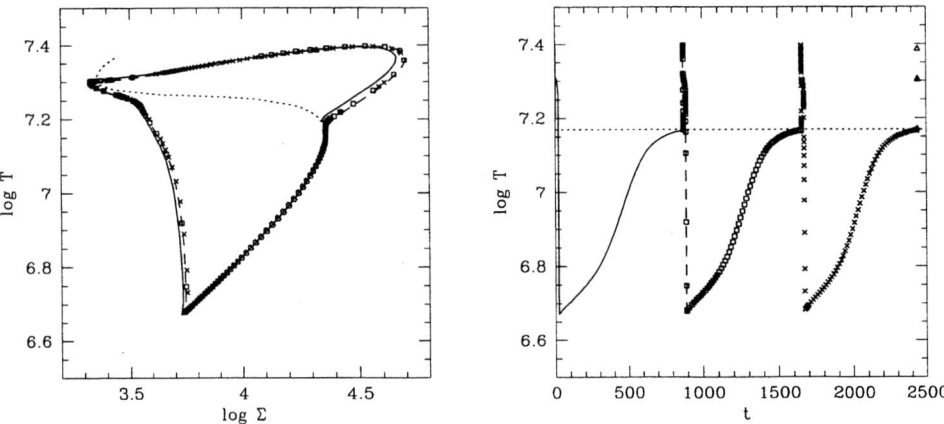

FIGURE 1. The phase portrait in the log T - log Σ plane (left panel) and the temperature time sequence (right panel) at 10 Schwarzschild radii. The temperature T is given in degrees K, Σ (surface density) is measured in g/cm^2 and time t in seconds.

Our next step in the work was to repeat the calculations for $\alpha = 10^{-1}$ with a more physical viscosity prescription (with the $r\varphi$ component of the viscous stress tensor being proportional to the radial gradient of the angular velocity). Also, we introduced a vertical acceleration equation rather than taking the disc to be in hydrostatic equilibrium in the vertical direction as was done previously.

We have obtained a limit cycle solution with this diffusion-type formulation of viscosity for the first time for transonic accretion discs. The solution is very similar

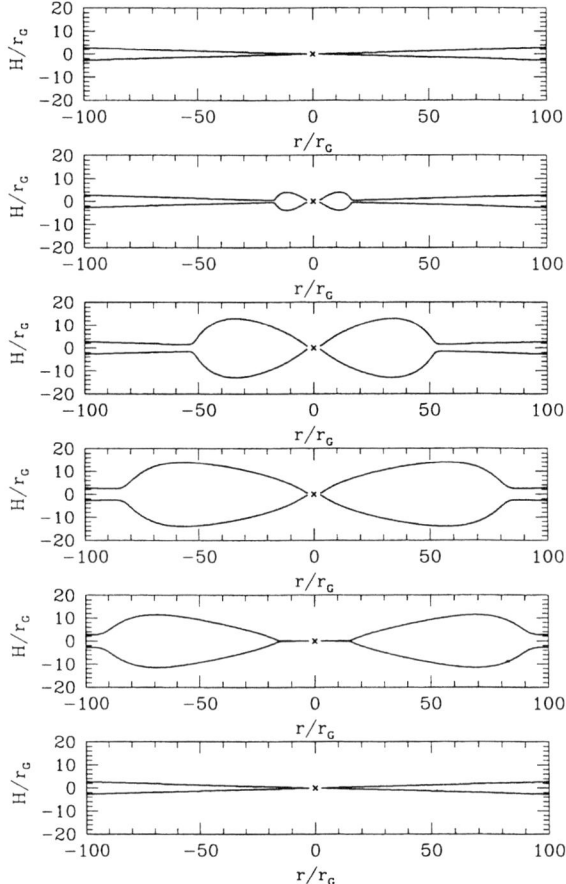

FIGURE 2. The shape of a transonic accretion disc during a thermal limit cycle shown at 0, 1, 6, 13, 16 and 760 seconds respectively.

to that obtained with the αp prescription for the same case but this is not surprising since the radial gradient of the angular velocity in the disc during the evolution is never very different from the Keplerian value. The structure of a transonic accretion disc during the thermal limit cycle, illustrating the characteristic swelling up of the region of low surface density, is shown in Figure 2.

3. OTHER INSTABILITIES

Thermal instability driven by radiation pressure might be relevant for intrinsically bright accreting sources. The most promising candidate where this instability seems to be at work is one of the two known superluminal objects in our galaxy, GRS 1915+105 (Belloni, Mendez, King, van der Klis and van Paradijs 1997). Although elegant, this scenario has not yet been confirmed by proper time-depending modelling. Recent studies seem to indicate several problems with such an interpretation (e.g. Nayakshin, Rappaport and Melia 1999). Further work is needed in order to get a definitive answer. This instability was also suggested as a mechanism for variability in NGC 5548 (Czerny, Schwarzenberg-Czerny, Loska 1999).

Another type of thermal instability, which is thought to be operating in a very well-known class of objects - cataclysmic variables - is driven not by radiation pressure but by partial ionization of hydrogen (Smak, 1982; Meyer and Meyer-Hofmeister, 1982). This type of instability might be present also in active galactic nuclei (Lin and Shields, 1986). Recently, Burderi, King and Szuszkiewicz (1998) have studied the effect of irradiation for this instability and concluded that it should indeed operate for most if not all accreting supermassive black holes with relevant masses and luminosities. This may have very interesting observational consequences (Siemiginowska and Elvis, 1999).

Accretion discs are also subject to pulsational instabilities, which might be responsible for quasi-periodic oscillations observed in accreting objects. This kind of instability has been seen in our calculations for models with relatively high values of α.

Future work will reveal the observational appearance and relevance of these instabilities.

ACKNOWLEDGEMENTS

We gratefully acknowledge financial support from the grant 2P03D01817 of Polish State Committee for Scientific Research (KBN). We thank also Ramesh Narayan for a very stimulating discussion.

REFERENCES

Abramowicz, M. A., Czerny, B., Lasota, J.-P., Szuszkiewicz, E., 1988, ApJ, 332, 646
Belloni, T., Mendez M., King, A. R., van der Klis, M., van Paradijs, J., 1997, ApJ, 479, L145
Burderi, L., King, A. R., Szuszkiewicz, E., 1998, ApJ, 509, 85
Czerny, B., Schwarzenberg-Czerny, A., Loska, Z., 1999, MNRAS, 303, 148
Kato, S., Fukue, J., Mineshige, S., 1998, " Black-hole accretion disks", Kyoto University Press, Japan
Meyer, F., Meyer-Hofmeister, E., 1982, A&A, 106, 34
Nayakshin, S., Rappaport S., Melia, F., 1999, preprint astro-ph/9905371
Salucci, P., Szuszkiewicz, E., Monaco, P., Danese, L., 1999, MNRAS, 307, 637
Lin, D. N. C., Shields, G. A., 1986, ApJ, 305, 28
Siemiginowska, A., Elvis, M., 1999, Nature, 397, 476
Smak, J., 1982, Acta Astron., 32, 199
Szuszkiewicz, E., Miller, J. C., 1997, MNRAS, 287, 165
Szuszkiewicz, E., Miller, J. C., 1998, MNRAS, 298, 888

FLARING BLAZARS WITH BEPPOSAX

G. Tagliaferri[1], G. Ghisellini[1], P. Giommi[2], L. Chiappetti[3], L. Maraschi[4],
A. Celotti[5], M. Chiaberge[5], A. Comastri[6], L. Costamante[1], G. Fossati[7], E.
Massaro[8], R. Nesci[8], E. Pian[9], C.M. Raiteri[10], M. Ravasio[1], F. Tavecchio[4],
G. Tosti[11], A. Treves[12], M. Villata[10], A. Wolter[4]

[1] *Osservatorio Astronomico di Brera, Via Bianchi 46, I-23807 Merate, Italy*
[2] *BeppoSAX Science Data Center, ASI, Via Corcolle 19, I-00131 Roma, Italy*
[3] *Istituto di Fisica Cosmica, CNR, Via Bassini 15, I-20133 Milano, Italy*
[4] *Osservatorio Astronomico di Brera, Via Brera 28, I-20121 Milano, Italy*
[5] *SISSA/ISAS, Via Beirut 2-4, I-34014 Trieste, Italy*
[6] *Osservatorio Astronomico di Bologna, Via Zamboni 33, I-40126 Bologna, Italy*
[7] *CASS, University of California, 9500 Gilman Drive, La Jolla, CA 92093-0424, USA*
[8] *Istituto Astronomico, Università "La Sapienza", Via Lancisi 29, I-00161 Roma, Italy*
[9] *TeSRE/CNR, Via Gobetti 101, I-40129 Bologna, Italy*
[10] *Osservatorio Astronomico Torino, St.da Osservatorio 20, 10025 Pino Torinese, Italy*
[11] *Osservatorio Astronomico, Università di Perugia, Via Pascoli, 06100 Perugia, Italy*
[12] *Dipartimento di Scienze, Università dell'Insubria, Via Lucini 3, I-22100 Como, Italy*

ABSTRACT   We observed three blazars in an active state with *Beppo*SAX. For ON 231 and BL Lac, we detected simultaneously both the synchrotron and the Compton component. Fast time variability was present in both sources, **only** for the synchrotron component. PKS 2005-489 was observed in a very high state and the 0.1-200 keV X-ray spectrum is well accounted for by the sole synchrotron emission. No fast time variability was detected. The SED of the three sources are well described by pure SSC models.

KEYWORDS: BL Lacertae objects: individual: ON 231, PKS 2005-489, BL Lac; X-rays: general

1. INTRODUCTION

The overall Spectral Energy Distribution (SED) of blazars shows two broad emission peaks, believed to be produced by the synchrotron and inverse Compton processes respectively. The position of the synchrotron peak is used to define different classes of blazars, i.e. HBL and LBL (high- and low-energy peak BL Lacs). The good *Beppo*SAX sensitivity and spectral resolution over a wide X-ray energy range (0.1–200 keV) are ideal to constrain existing models for the X-ray emission of blazars.

We successfully used the *Beppo*SAX satellite to perform observations of blazars that were known to be in a high state from other observations both in the X-ray and other bands (mainly optical and TeV). Here we present the preliminary analysis of the *Beppo*SAX observations of the three Blazars observed in an active state in 1998-99: ON 231, PKS 2005-489 and BL Lac.

2. THE DATA

2.1. ON 231 (W Com, B2 1219+28, $z = 0.102$)

It had an exceptional optical outburst in April–May 1998, reaching the most luminous state since the beginning of the century, and showed a continuing flaring activity. ON 231 has a SED that peaks in the optical, representing an intermediate object between HBL and LBL. It was observed in a high X-ray state by *Beppo*SAX in May and June, 1998, and was detected up to 100 keV. In both cases a single power law model plus absorption does not fit the data, while a broken power law model provides a good fit, with the second spectral index much flatter than the first one resulting in a concave shape (see Table 1)..

We therefore conclude that in both observations **both the synchrotron and the Compton component were detcted**. In May rapid X–ray variability of about a factor of three in few hours was clearly detected, but only at energies smaller than 3-4 keV, corresponding to the synchrotron component (Tagliaferri et al. 1999).

2.2. PKS 2005-489 ($z = 0.071$)

It is a bright BL Lac object with a SED that peaks in the UV-soft X-ray, typical of an HBL source. It has already been observed by *Beppo*SAX in September, 1996 (Padovani et al. 1998). Multiwavelength campaigns were organized in September and October, 1998 including *Rossi*XTE monitoring (Perlman et al. 1999). At the end of October, 1998, a flare alert was issued by the *Rossi*XTE team (Remillard 1998). This triggered also our *Beppo*SAX ToO observation and PKS2005-489 was observed by *Beppo*SAX on November 1-2, 1998.

We detected the source in a high state, up to 100-200 keV (see Fig. 1, left panel). Few days later the source was observed in an even higher state with the *Rossi*XTE satellite (Perlman et al. 1999). A single power law can not fit the 0.1-200 keV spectrum, while a broken power law provides a good fit. The spectrum is found to be convex, the spectral index increasing with energy by a small but significant amount with the break below 2 keV (see Table 1). This is consistent with the *Rossi*XTE results. No flux variability was detected during our observation.

As shown by the SED (see Fig. 1, left panel), for this HBL we see only the synchrotron component, that fits the X-ray data in the full 0.1-200 keV band. Recent *Beppo*SAX observations of MKN 501 and 1ES 2344+514 have shown that the peak of the synchrotron emission can move to very high energies during strong flares (Pian et al. 1998, Giommi et al. 1999). This does not seem to be the case for PKS 2005-489 where it definitely remained below few keV.

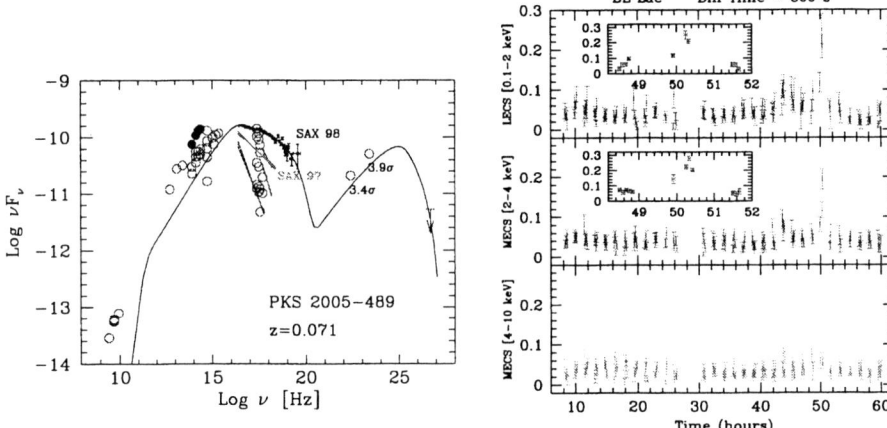

FIGURE 1. Left panel:The SED of PKS 2005-489, together with the SSC model used to fit the X-ray data. Right panel: MECS 2-10 keV (bottom panels) and LECS 0.1-2 keV (top panel) light curves of BL Lac during the June 1999 *Beppo*SAX observation. Note the clear flare detected at the end of the observations and better shown in the insert.

2.3. BL Lac ($z = 0.069$)

Its SED peaks in the optical-infrared band, i.e. it is typical of an LBL source. Also this source was already observed by *Beppo*SAX, in November 1997, in a high state with a 2-10 keV flux of $\sim 2 \times 10^{-11}$ erg s^{-1} cm^{-2} (Padovani et al. in preparation). This is comparable to the level observed by *Rossi*XTE during the 1997, July outburst. The X-ray spectrum of this object is quite hard with an energy spectral index of $\alpha = 0.4 - 0.9$, as measured by ASCA, *Rossi*XTE and *Beppo*SAX (Sambruna et al. 1999; Madejski et al. 1999; Padovani et al. in preparation).

On May, 1999, the source was again in an optically bright state, thus we triggered our *Beppo*SAX ToO and BL Lac was observed the 5-7 of June, 1999. The source was about a factor of two weaker than the first *Beppo*SAX observation. The 0.1-100 keV X-ray spectra is not fitted by a single power law. A broken power law gives a better fit with a concave shape (see Table 1). Thus, also for BL Lac we detected both the synchrotron and the Compton component.

During this observation we also find very short time variability, with the 0.1-10 keV flux increasing by a factor of two in about 20 minutes. As in the case of ON 231, this variability was detected only in the synchrotron part of the spectrum. In fact, the variability is seen by both the LECS and the MECS detectors only below the break (~ 4 keV) (see Fig. 1, rigth panel).

3. CONCLUSIONS

ON 231: the source was observed in a high state and we detected both the synchrotron and the Compton component in the 0.1-200 keV X-ray range. We detect

TABLE 1. Fit results for a broken power law model

source	date	Γ_1[a]	Γ_2	break keV	χ_r^2 (d.o.f.)	$F_{[2-10\text{ keV}]}$ erg cm^{-2} s^{-1}
ON 231	11-12/05/98	$2.60^{-0.08}_{+0.07}$	$1.13^{-0.20}_{+0.30}$	$4.0^{-0.70}_{+0.55}$	0.92 (63)	4.4×10^{-12}
	11-12/06/98	$2.68^{-0.12}_{+0.11}$	$1.49^{-0.26}_{+0.25}$	$2.6^{-0.60}_{+0.50}$	0.88 (47)	3.2×10^{-12}
PKS 2005	01-02/11/98	$2.02^{-0.04}_{+0.03}$	$2.21^{-0.02}_{+0.02}$	$1.87^{-0.35}_{+0.40}$	0.90 (212)	1.8×10^{-10}
BL Lac	5 - 7/06/99	$2.18^{-0.09}_{+0.08}$	$1.74^{-0.43}_{+0.22}$	$4.0^{-1.25}_{+2.35}$	1.17 (113)	6.5×10^{-11}

[a] photon spectral index; errors at 90% confidence level for 3 parameters of interest.

fast time variability, but only for the synchrotron component.

PKS 2005-489: the source was in a very high state. The 0.1-200 keV X-ray spectrum steepens with increasing energy and is accounted for by the synchrotron emission only. The synchrotron peak did not reach the hard X-rays as in the case of Mkn 501. No fast time variability was detected.

BL Lac: the source was not very bright during our observation. However, as for ON 231, we detected both the synchrotron and the Compton component and very fast time variability, again only for the synchrotron component.

We are able to explain the SEDs of the three sources with a pure homogeneous SSC model. The intrinsic luminosities that we derived are between $0.3 - 3 \times 10^{42}$ erg s^{-1}, the magnetic field B is between 0.7-1.5 Gauss, the size R of an assumed spherical region is between $7 - 10 \times 10^{15}$ cm, while the Doppler factor δ is 14-16. In Fig. 1, left panel, we show as an example the SED of PKS 2005-489 together with the SSC model used to fit the X-ray data.

ACKNOWLEDGEMENTS

We thank the *Beppo*SAX science data center for their support in the data analysis. This research is financially supported by the Italian Space Agency.

REFERENCES

Giommi P., Padovani P., Perlman E., 1999, MNRAS in press (astro-ph/9907377)
Madejski G.M., Sikora M., Jaffe T., et al., 1999, ApJ, 521, 145
Padovani P. et al., 1998, Nuclear Physics B (Proc. Suppl.) 69, 431
Perlman E.S., Madejski G.M., Stocke J.T., Rector T.A., 1999, ApJ, 523, L11
Pian E., Vacanti G., Tagliaferri G. et al., 1998, ApJ, 492, L17
Remillard R., 1998, IAU Circ. 7041
Sambruna R.M., Ghisellini G., Hooper E., et al., 1999, ApJ, 515, 140
Tagliaferri G., Ghisellini G., Giommi P., et al., 1999, A&A submitted

GAMMA-LOUD QUASARS: A VIEW WITH BEPPOSAX

Tavecchio F.[1], Maraschi L.[1] Ghisellini G.[1], Celotti A.[2], Chiappetti L.[3] Comastri A.[4], Fossati G.[5], Grandi P.[6], Haardt F.[7], Pian E.[8], Tagliaferri G.[1], Treves A.[7], Raiteri C.M.[9], Sambruna R.[10], Villata M.[9]

[1] *Osservatorio Astronomico di Brera, Milano, Italy,* [2] *SISSA/ISAS, Trieste, Italy,* [3] *IFC/CNR, Milano, Italy,* [4] *Osservatorio Astronomico di Bologna, Bologna, Italy,* [5] *CASS/UCSD, La Jolla, USA,* [6] *IAS/CNR, Roma, Italy,* [7] *Universita' dell'Insubria, Como, Italy,* [8] *ITESRE/CNR, Bologna, Italy,* [9] *Osservatorio Astronomico di Torino, Pino Torinese, Italy,* [10] *PennState University, USA*

ABSTRACT   We present $BeppoSAX$ observations of the γ-ray emitting quasars 0836+710, 1510-089 and 2230+114. All the objects have been detected in the PDS up to 100 keV and have extremely flat power-law spectra above 2 keV (α_x=0.3–0.5). 0836+710 shows absorption higher than the galactic value and marginal evidence for the presence of the redshifted 6.4 keV Iron line. 1510-089 shows a spectral break around 1 keV, with the low energy spectrum steeper (α_l=1.6) than the high energy power-law (α_h=0.3). The data are discussed in the light of current Inverse Compton models for the high energy emission.

KEYWORDS: quasars: individual (0836+710, 1510-089, 2230+114); radiation mechanisms: non-thermal; X-rays: galaxies

1. INTRODUCTION

Since the EGRET detection of about 60 blazars as strong γ-ray emitters the study of these extreme objects has received a renewed interest. The overall Spectral Energy Distribution (SED) of Blazars shows two broad components, the first one peaking at IR-up to soft X-rays, the second one in the γ-rays, from MeV up to TeV energies. The first peak is due to synchrotron radiation produced by relativistic electrons, while the high energy component is believed to be Inverse Compton scattered radiation. The seed photons for the IC scattering could be the synchrotron photons themselves (SSC model) or photons produced in the region external to the jet (EC model) which, especially for quasars with strong emission lines, is probably rich of optical-UV radiation. The subclass of quasar-like sources contains the most luminous sources, with apparent γ-ray luminosity up to 10^{48} erg s^{-1}.

In the following we present the $BeppoSAX$ observations of three gamma-loud quasars (0836+710, 2230+114 and 1510-089), detected up to 100 keV with the high energy instrument PDS, and we discuss the external Compton scenario. A full paper

Γ	E_b^* keV	Γ_h^*	N_H 10^{20} cm^{-2}	$F_{[2-10\,keV]}$ 10^{-12} erg cm^{-2} s^{-1}	χ^2/d.o.f.
			0836+710		
1.32 ± 0.04	-	-	$8.3_{5.7}^{13.0}$	26	63.47/63
$0.83_{0.3}^{1.17}$	1.2 ± 0.3	1.31 ± 0.03	2.98(fix)	26	63.14/62
			2230+114		
1.51 ± 0.04	-	-	$7.3_{4.6}^{11.1}$	6.05	51.11/51
			1510-089		
1.35 ± 0.07	-	-	0-2.44	5.3	60.24/64
$2.65_{2.05}^{3.28}$	1.3 ± 0.3	1.39 ± 0.08	7.8 (fix)	5.3	43.06/63

*: only for the broken power-law model

TABLE 1. Fits to *Beppo*SAX Data (LECS+MECS+PDS).

is in preparation (Tavecchio et al. 2000).

2. THE OBSERVED OBJECTS

• 0836+710: this is a distant quasar ($z = 2.172$), characterized by a very flat X-ray spectrum, observed with ROSAT and ASCA. The ASCA observation showed a column density greater than the galactic value (Cappi et al. 1997).

• 2230+114: this source ($z = 1.037$), observed with GINGA, ROSAT and ASCA (Lawson & Turner 1997, Brinkmann et al. 1994, Kubo et al. 1998) shows a flat spectrum extending smoothly in the gamma-ray band, as indicated by OSSE observations (Mc Naron-Brown et al. 1995).

• 1510-089: this interesting Highly Polarized Quasar ($z = 0.361$) shows a pronounced UV bump (Pian & Treves 1993). The EXOSAT observation (Singh et al. 1990) suggested the presence of a fluorescence iron line, not present in a more recent ASCA observation (Singh et al. 1997).

3. *BEPPO*SAX OBSERVATIONS AND RESULTS

We modelled the spectral data with either single or broken power-law models (with galactic and free absorption). Results of the spectral fits to the LECS+MECS+PDS data are shown in Table 1. In the following we discuss the results for each object.

0836+710: the data are consistent both with a power-law model with intrinsic absorption ($N_H \simeq 7 \times 10^{21}$ cm^{-2} in the QSO rest frame) and with a broken power-law model with fixed galactic absorption. With both models the residuals show an excess at about 2 keV, that could be interpreted as the redshifted fluorescence iron line (see Fig 1). Adding a gaussian line with energy as a free parameter the fit converges to an energy of $E = 1.99 \pm 0.1$ with an intrinsic equivalent width of $EW \simeq 110$ eV,

R_{16} (cm)	B (G)	δ	γ_{min}	n	$L_{inj,45}$ (erg s^{-1})	$\tau L_{ext,45}$ (erg s^{-1})	$R_{ext,18}$ (cm)
				0836+710			
4	5.3	18	50	3.05	14.8	3.2	1.5
				2230+114			
4	3.7	15.5	130	3	0.14	0.4	1
				1510-089			
2	3.1	17	65	2.9	0.01	0.45	1.

TABLE 2. Parameters used for the EC model

but the significance of the improvement of the χ^2, evaluated with the F-test, is not very high ($P \simeq 90\%$).

2230+114: A simple power-law model ($\alpha = 0.5$) with absorption consistent with the galactic value reproduces quite well the data in the whole range 0.1-100 keV.

1510-089: A simple absorbed power-law (although statistically acceptable, see Table 1) gives evident residuals at low energies. The F-test confirms that a broken power-law is a better model (with a probability $P > 99.99\%$). The PDS/MECS relative normalization (1.3-2.5 90% conf. level) is above the accepted range (0.77-0.93); this problem is possibly due to contamination by a source in the large FOV of PDS (~ 1 deg.). In the fit we fixed the normalization to 0.85.

4. DISCUSSION

• We constructed the SEDs of the observed sources using contemporaneous X-ray and optical observations and historical data taken from the literature (an example is reported in Fig.1) We have reproduced the observed spectrum using the homogeneous EC model discussed in detail in Ghisellini et al. (1998). In a spherical region with size R, a power-law electron distribution (with slope n and limits γ_{min} and γ_{max}) is continuously injected with luminosity L_{inj}. Electrons cool through the synchrotron and IC processes and are free to escape from the source at some velocity v_{esc}, forming a flat ($\alpha < 0.5$) power-law below γ_{min}. We assume that the external radiation field is described by a black body spectrum with luminosity L_{ext}, diluted in a spherical region with size R_{ext}. The parameter values for the models are reported in Tab. 2.

• It is interesting to note that the derived spectral indices in the medium to hard X-ray band are flatter than 0.5 in two out of three sources. A population of electrons cooling through synchrotron and IC forms a distribution which produce a spectrum with $\alpha = 0.5$. Therefore some additional mechanism is required in order to produce the observed flatter spectrum, e.g. escape or injection of an intrinsically flat distribution (see e.g. Ghisellini 1996).

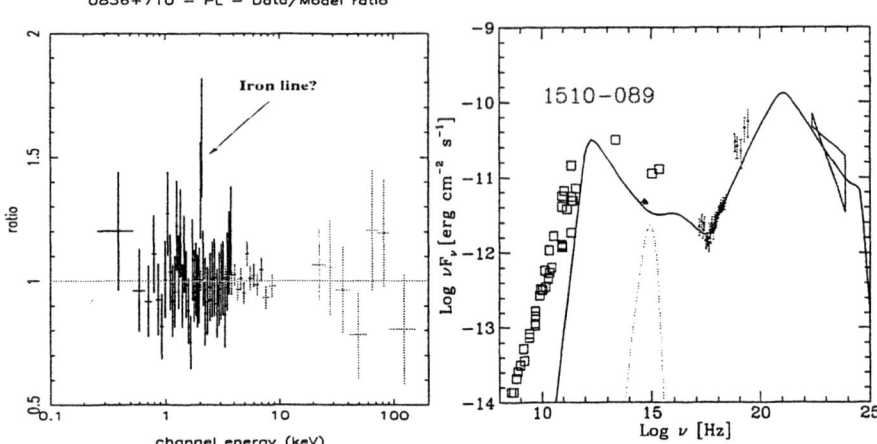

FIGURE 1. Left panel: Data/Model ratio for 0836+710 with the power-law model (N_H free). Right panel: Overall SED of 1510-089 with the spectrum calculated using the homogeneous EC model. Radio to UV (open squares) and gamma-ray data are taken from the literature.

- The excess of 1510-089 in the soft band is well understood as due to the SSC emission (see Fig.1), although another possible source is the tail of the strong UV bump.
- We confirm the presence of absorption higher than the galactic one in 0836+710. The origin of this absorption is likely intrinsic to the source. The fluorescence iron line suggested by our data could be produced through reprocessing by the same material responsible for the absorption. On the other hand a broken power-law continuum with galactic absorption can reproduce the data equally well: in this case the break could be due to the incomplete comptonization of the soft external photons (see Ghisellini 1996).

REFERENCES

Brinkmann, W. & Siebert, J. 1994, A&A, 285, 812
Cappi, M., et al. 1997, ApJ, 478, 49
Ghisellini, G. 1996, IAU Symposia, 175, 413
Ghisellini, G., et al.1998, MNRAS, 301, 451
Kubo, H., et al. 1998, ApJ, 504, 693
Lawson, A. J. & Turner, M. 1997, MNRAS, 288, 920
McNaron-Brown, K., et al. 1995, ApJ, 451, 575
Pian, E. & Treves, A. 1993, ApJ, 416, 130
Singh, K. P., et al. 1990, ApJ, 365, 455
Singh, K. P., et al. 1997, ApJ, 491, 515

CONSTRAINTS TO THE SSC MODEL FOR MKN 501

Fabrizio Tavecchio and Laura Maraschi

Osservatorio Astronomico di Brera, Milano, Italy

ABSTRACT We fit the SEDs of the TeV blazar Mkn 501 adopting the homogeneous Synchrotron-Self Compton model to simultaneous X-ray and TeV spectra recently become available. We present detailed model spectra calculated with the above constraints and taking into account the absorption of TeV photons by the IR background. We found that the curved TeV spectra can be naturally reproduced even without IRB absorption. Taking IRB absorption into account changes the required parameter values only slightly.

KEYWORDS: BL Lacertae objects: individual (Mkn 501); gamma rays: theory; radiation mechanisms: non-thermal

1. INTRODUCTION

The study of Blazars has been recently enriched by the detection of very high energy gamma rays (energy in the TeV range) from a handful of nearby sources. This discovery opens the possibility to effectively test and constrain the radiative mechanisms invoked to explain the emission from Blazars and to investigate the physical conditions in relativistic jets.

As discussed in Tavecchio et al. 1998 (hereafter T98), the knowledge of the simultaneous X-ray and TeV spectra of Blazars allows to univocally find the set of physical parameters necessary within a homogeneous Synchrotron–Self Compton model with an electron distribution described by a broken power-law. Here we apply the SSC model to recent high quality, simultaneous X-ray and TeV data of the well studied source Mkn 501 and discuss the results.

2. THE DATA

TeV spectra measured by the CAT team during 1997 were recently reported in Djannati-Atai et al (1999). In particular the TeV spectra associated with *Beppo*SAX observations of April 16 and 7 (reported in Pian et al 1998) are also discussed. For the large flare of April 16, given the high flux level, it has been possible to obtain a good quality TeV spectrum from a single observation which partially overlaps with the *Beppo*SAX observation. On Apr 7, Mkn 501 was less bright and the TeV spectrum is obtained using different observations with similar TeV fluxes and hardness ratios (for more details see Djannati-Atai et al 1999). We combine these data with

R_{16} (cm)	B (G)	δ	γ_{break}	K	n_1	n_2
\multicolumn{7}{c}{Mkn 501 High state (Apr 16)}						
1.8	0.02	10	10^7	10^4	2	8
\multicolumn{7}{c}{Mkn 501 High state with IRB}						
3.5	0.005	10	2×10^7	4×10^3	2	8
\multicolumn{7}{c}{Mkn 501 Low state (Apr 7)}						
2.5	0.01	10	6.3×10^6	2.3×10^3	2	8

TABLE 1. Values of the physical parameters used for the SSC model.

the X-ray spectra reported in Pian et al. (1998) for the same days constructing quasi simultaneous SEDs for two epoches. The TeV spectra of Mkn 501 are clearly curved. It was suggested that this curvature could be the "footprint" of the absorption of TeV photons by the Infra Red Background (IRB, e.g Konopelko et al. 1999, but see the detailed discussion in Vassiliev 1999)

3. SPECTRAL FITS WITH THE SSC MODEL

T98 obtained analytical relations for the IC peak frequency in Klein-Nishina (KN) regime adopting the step function approximation for the KN cross-section. A comparison between these approximate estimates and the detailed numerical calculations with the full KN cross section shows that the analytical formulae for the IC peak frequency given in T98 can overestimate its value by a factor of 3-10, depending on n_2, the index of the steeper part of the electron distribution. Therefore, although the analytical estimates are useful as a guideline, it is very important for the study of the TeV spectrum of Blazars to use precise numerical calculations, like those discussed in the following.

We reproduced the SED of Mkn 501 (reported in Fig.1) using the SSC model described in T98 and Chiappetti et al 1999. In a spherical region with radius R, magnetic field B and Doppler factor δ, a population of electrons with energy distribution of the form $N(\gamma) = K\gamma^{-n_1}(1 + \gamma/\gamma_{break})^{n_1-n_2}$ emits synchrotron and IC radiation. The IC spectrum is calculated with the full Klein-Nishina cross-section, using the formula derived by Jones (1968). The seed photons for the IC scattering are those produced by the same electrons through the synchrotron mechanism. For the case of the high state of Mkn 501 we considered possible intervening absorption by the IRB adopting the prescriptions for the *low model* discussed in Stecker & De Jager (1998). In the calculation reported here we assume a typical radius of the emitting region of $R \sim 10^{16}$ cm and a Doppler factor $\delta = 10$. With this choice the minimum variability timescale is of $\simeq 10$ h (see Tab.1 for the parameters used).

It is useful to calculate the IC emission from electrons with different energy

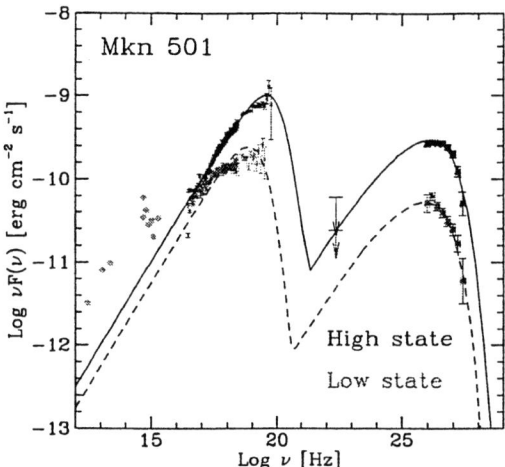

FIGURE 1. SED of Mkn 501 (High state and Low state) with the spectrum calculated with the SSC model. Upper limits at 100 MeV band are from Weekes et al. 1996 and Catanese et al 1997.

ranges. These "slices" are reported in Fig. 2 (see figure caption for the detailed energy ranges). It is interesting to note that, because of the KN cut at the high energy emission, the peak of the IC component is produced by electrons with Lorentz factor below γ_{break} while the peak of the Synchrotron component is produced by electrons at γ_{break}. This effect has important consequences for the study of the X-ray/TeV correlated variability (see also Maraschi et al. 1999).

4. DISCUSSION AND CONCLUSIONS

The parameters adopted for the model are reported in Table 1. Our model indicates a rather low value for the magnetic field and a high γ_{break} (see Tab.1). The transition from the low state to the high state in Mkn 501 is consistent with an increase of a factor of 2 in both γ_{break} and B and with an almost constant value of δ and R. These results are in agreement with a similar analysis of the April 16 flare by Bednarek & Protheroe (1999): for a minimum timescale of 2.5 h they found that the TeV spectrum can be well reproduced by $B \sim 0.03$ and $\delta \sim 15$.

Although some authors (see Konopelko et al. 1999) have recently proposed that the curvature of the TeV spectrum of Mkn 501 provides evidence for absorption by IRB, the curvature could be intrinsic and related to the curved electron distribution necessary to fit the X-ray data. Our models show that a curved spectrum is a plausible explanation for the curvature. The introduction of the IRB does not dramatically change the inferred physical parameters. As shown in Fig.2 IRB/no IRB TeV spectra are very similar up to 15 TeV, while for higher energy the power-law spectrum of the no IRB case is changed in an exponential profile. Therefore spectral data above 10 TeV are needed in order to understand if IRB absorption affects Blazar spectra.

Finally we note that in the theory of particle acceleration by shocks (for a recent critical discussion see Henri et al. 1999) the maximum Lorentz factor of accelerated electrons, obtained equating acceleration time and cooling time, is given

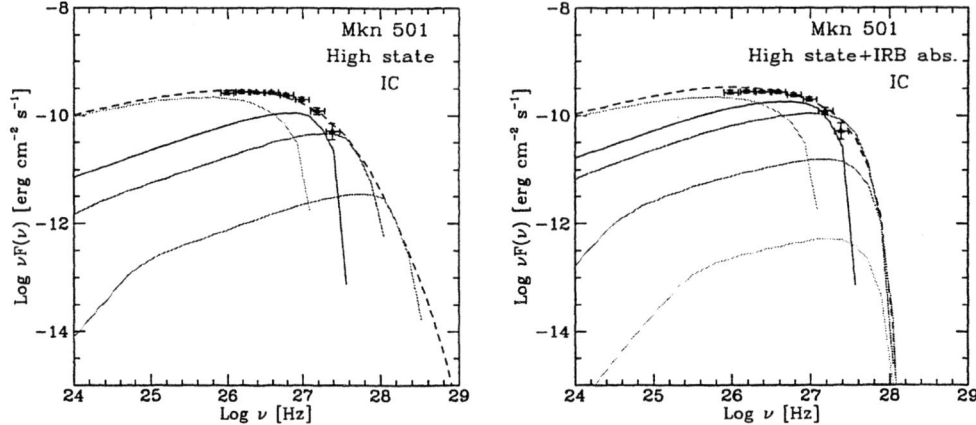

FIGURE 2. TeV spectrum of Mkn 501-High state without (left) and with (right) IRB absorption. We also show the contribution from electrons with different Lorentz factors. From the left the curves show the emission from electrons with Lorentz factor in the range: $1-10^6$, $10^6-3\times10^6$, $3\times10^6-10^7$, $10^7-3\times10^7$, $3\times10^7-10^8$

by $\gamma_{max} \simeq 10^7(B/0.02G)^{-1/2}$, where we used the value of B we found in the high state of Mkn 501. Therefore our fit suggests that during the high activity states of Mkn 501 the electrons can reach the maximum energy fixed by the balance between cooling and acceleration processes. In states of lower activity the acceleration time could be longer and the condition $t_{esc} < t_{acc}$ (where t_{esc} and t_{acc} are the escape time and the acceleration time respectively) could prevent the electrons for reaching the maximum energy.

ACKNOWLEDGEMENTS

We thank A. Djannati-Atai for sending us the TeV data of Mkn 501.

REFERENCES

Bednarek, W., & Protheroe, R.J., 1999, MNRAS, in press, astro-ph/9902050
Catanese, M., et al. 1997, ApJ, 487, L143
Chiappetti, L., et al. 1999, ApJ, 521,552
Djannati-Atai, A., et al. 1999, submitted to A&A
Henri, G., et al., Astropart. Phys, in press (astro-ph/9901051)
Jones, F., 1968, Phys. Rev., 167, 1159
Konopelko, A.K., et al. 1999, ApJ, 518, L13
Maraschi, L., et al. 1999b, ApJ Letters, in press
Pian, E., et al. 1998. ApJ, 491, L17
Stecker, F.W., & de Jager, O.C., 1998, A&A, 334, L85
Tavecchio, F., Maraschi, L. and Ghisellini, G., 1998, ApJ, 509, 608
Vassiliev, V.V., 1999, Astropart. Phys., in press
Weekes, T.C., et al. 1996, A&AS, 120, 603

VARIABILITY OF THE IRON LINE IN THE LOW-LUMINOSITY AGN NGC 4579

Y. Terashima[1], T. Yaqoob[1], P.J. Serlemitsos[1], H. Kunieda[2], K. Misaki[2,3], A.F. Ptak[4], and L.C. Ho[5]

1) NASA Goddard Space Flight Center, Greenbelt, MD 20771, USA
2) Institute of Space and Astronautical Science, Yoshinodai 3-1-1, Sagamihara, Kanagawa 229-8510, Japan
3) Nagoya University, Chikusa-ku, Nagoya 464-8602, Japan
4) Carnegie Mellon University, 5000 Forbes Ave., Pittsburgh, PA 15213, USA
5) The Observatories of the Carnegie Institution of Washington, 813 Santa Barbara St. Pasadena, CA 91101-1292, USA

ABSTRACT We present results of $ASCA$ observations of the low-luminosity AGN (LLAGN) NGC 4579 in 1995 July and 1998 December which show evidence for spectral variability. The X-ray luminosity in the 1998 observations is 2×10^{41} ergs s^{-1} in the 2–10 keV band and about 35% larger than in 1995. An Fe K emission line is clearly detected in both observations. The center energy is $6.73^{+0.13}_{-0.12}$ keV in 1995 and 6.39 ± 0.09 keV in 1998. If we model the Fe line with two Gaussians centered at 6.4 keV and 6.7 keV, the line intensity of the 6.7 keV line decreased while that of the 6.4 keV line increased within an interval of 3.5 years. This variability rules out thermal plasmas in the host galaxy as the origin of the ionized Fe line in this LLAGN. The detection of a variable 6.4 keV line suggests that an optically thick standard accretion disk is present in this object, whose Eddington ratio is only $L_{\rm Bol}/L_{\rm Edd} \sim 2 \times 10^{-3}$.

KEYWORDS: galaxies: nuclei, galaxies: individual (NGC 4579)

1. INTRODUCTION

Recent observations have shown that there exist many low luminosity AGNs (LLAGNs) with a typical X-ray luminosity of $10^{40} - 10^{41}$ ergs s^{-1} in the local universe (e.g., Ho 1999a; Terashima 1999). The origin of Fe lines in LLAGNs, however, is not well understood yet. One mysterious property is that several LLAGNs have Fe line centered at 6.7 keV, which is consistent with He-like Fe (e.g., Ishisaki et al. 1996; Terashima et al. 1998).

NGC 4579 is an LLAGN classified as a LINER 1.9/Seyfert 1.9 with an X-ray luminosity of 1.5×10^{41} ergs s^{-1}, and a Fe line was detected at 6.7 keV with $ASCA$ in 1995 July (Terashima et al. 1998). We observed NGC 4579 again with $ASCA$ in 1998 Dec. to search for spectral variability and to study the origin of the ionized Fe line. Here we report spectral variability between the 1995 and 1998 observations.

2. OBSERVATIONS AND RESULTS

ASCA observations of NGC 4579 were performed on 1995 June 25, 1998 Dec. 17, and 1998 Dec. 27. The effective exposure times are 32, 19, 18 ksec for SIS and 31, 20, 19 ksec for GIS, respectively. Detailed results on 1995 and 1998 observations are presented in Terashima et al. (1998) and Terashima et al. (2000), respectively. Since no significant spectral difference between the two observations in 1998 is seen, we present spectral results on a combined spectrum of these data.

2.1. 1995 Observation

The *ASCA* spectrum in 1995 is well represented by a model consisting of three component: (1) Raymond-Smith thermal plasma model with $kT = 0.90^{+0.11}_{-0.05}$ keV and abundance of 0.5 solar, (2) power-law with photon index $\Gamma = 1.72 \pm 0.05$, and (3) Gaussian with center energy $E_{\rm line} = 6.73^{+0.13}_{-0.12}$ keV (source rest frame), width $\sigma = 0.17^{+0.11}_{-0.12}$ keV, and equivalent width EW $= 490^{+180}_{-190}$ eV (errors are 90% confidence level for one interesting parameter). The X-ray luminosity is 1.5×10^{41} ergs s^{-1} in the 2–10 keV band at an adopted distance of 16.8 Mpc.

2.2. 1998 Observation

We fitted the 1998 spectrum with the same model as the 1995 data. We used an energy band of 1.0–10 keV band for SIS data to avoid calibration uncertainties in the low energy band. Because of the chosen energy band and a hard component which is brighter than that in the 1995 observation, we cannot constrain the spectral parameters of the soft thermal component. We therefore fixed kT and abundance at the best-fit values found in 1995. The best-fit parameters for the hard component are: $\Gamma = 1.85^{+0.07}_{-0.09}$, $E_{\rm line} = 6.39 \pm 0.09$ keV, $\sigma = 0(< 0.16)$ keV, and EW$=250^{+105}_{-95}$ eV. The X-ray luminosity is 2.0×10^{41} ergs s^{-1} in the 2–10 keV band, which is about 35% larger than that in 1995.

2.3. Spectral variability

The most intriguing spectral variability is the change of the Fe line center energy: the line energy $6.73^{+0.13}_{-0.12}$ keV decreased to 6.39 ± 0.09 keV within an interval of 3.5 yr. Confidence contours for the line center energy versus line normalization are shown in Fig. 1. The difference of the line center energy between the two occasions is significant at greater than 90% confidence level for two interesting parameters. This behavior may be due to a change of the line center energy caused by a change of the physical state of the line emitter, or it could be due to variability of multiple lines at 6.4 keV and 6.7 keV. In order to test the second possibility, we fitted the Fe line with two Gaussians, where the line center energy and width are fixed at the value in 1995 and in 1998 (Fig 2). We found that the intensity of the 6.4 keV (6.7 keV) line increased (decreased) significantly at more than 90% confidence level. The EW of the 6.7 keV line also decreased, while that of 6.4 keV marginally increased.

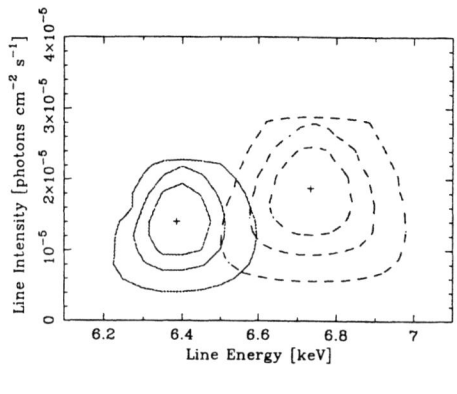
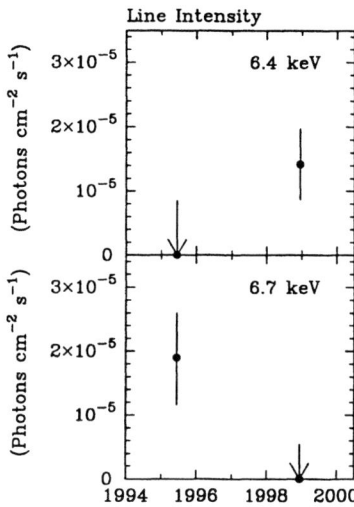

FIGURE 1. (left) Confidence contours ($\Delta\chi^2=2.3$, 4.6, and 9.1) for the line energy and normalization for the 1995 (dashed lines) and 1998 (solid lines) observations.

FIGURE 2. (right) Variability of the 6.4 and 6.7 keV line intensity for a double Gaussian fit.

3. DISCUSSION

We found that the Fe line center energy in NGC 4579 decreased in 3.5 years. This variability indicates that the Fe line emitter is located within ~ 1 pc from the nucleus, and thus rules out hot gas in the host galaxy (starburst activity, ridge emission) as the origin of the 6.7 keV Fe line. The observed Fe lines most probably originate from the AGN itself.

The relatively strong 6.4 keV line in 1998 indicates that cold matter subtends a large solid angle viewed from the nucleus. A part of the line could come from an obscuring torus. The torus contribution is at most 8.5×10^{-6} photons s^{-1} cm^{-2}, which is the 90% confidence upper limit on the unvariable 6.4 keV line, if the Fe line flux from the torus remains constant. This line flux corresponds to an EW of 155 eV for the continuum level in 1998. The rest of the 6.4 keV line probably comes from an optically thick accretion disk, which evidently exists even though the Eddington ratio of NGC 4579 is extremely low ($L_{\rm Bol}/L_{\rm Eddigton} = 2.0 \times 10^{-3}$; Ho 1999b). This is consistent with the estimate of the transition radius from an advection dominated accretion flow to a standard disk ($\sim 100 R_{\rm S}$) by Quartaert et al. (1999).

The origin of the 6.7 keV line is still puzling. If the Fe lines in 1995 and 1998 originated from the same region, the decrease of ionization state responding to the small flux increase of only 35% is difficult to be explained by photoionization. If most of the 6.4 keV line comes from the torus and the 6.7 keV line is from different photoionized matter, the disappearance of the 6.7 keV line might be understood by increasing the ionization state such that the gas is almost fully ionized. However, the small flux change, again, is not enough to change the ionization state so drastically. Therefore, it is conceivable that the physical conditions (density and geometry) of the ionized matter has changed between the two observations.

ACKNOWLEDGEMENTS

We thank all the *ASCA* team members who made these observations possible. YT and KM acknowledge JSPS for support.

REFERENCES

Ho, L. C. 1999a, Advances in Space Research, 23 (5–6), 813
Ho, L. C. 1999b, ApJ, 516, 672
Ishisaki, Y., et al. 1996, PASJ, 48, 237
Quataert, E., Di Matteo, T., Narayan, R., Ho, L. C. 1999, ApJ, 525, L89
Terashima, Y. 1999, Astrophysical Letters and Communications, 39, 149
Terashima, Y., Kunieda, H., Misaki, K., Mushotzky, R. F., Ptak, A. F., Reichert, G. A. 1998, ApJ, 503, 212
Terashima, Y., Ho, L. C., Ptak, A. F., Yaqoob, T., Kunieda, H., Misaki, K., Serlemitsos, P. J. 2000, ApJ, 535, L79

X-RAY PROPERTIES OF FR I RADIO GALAXIES

E. Trussoni[1], L. Feretti[2], A. Capetti[1], A. Celotti[3], M. Chiaberge[3]

1) Osservatorio Astronomico di Torino, Pino Torinese, Italy
2) Istituto di Radioastronomia del CNR, Bologna, Italy
3) SISSA, Trieste, Italy

ABSTRACT

We discuss the results of ROSAT observations of the FR I radio galaxies 3C 31, 3C 264, 3C 270 and 3C 465. The X-ray emission from the central regions of the first and the third source is fit by a composite spectrum with a thermal component (from the hot corona surrounding the galaxy) and a power law component (from the active nucleus). A non thermal emission is consistent with the whole flux from 3C 264, while only an upper limit to it is found in 3C 465. The implications for the unified schemes for AGN are shortly discussed.

KEYWORDS: Radio galaxies; Active galactic nuclei; X-rays

1. INTRODUCTION

The features of the X-ray emission from low brightness radio galaxies (Fanaroff-Riley I, FR I) are quite complex. The early-type galaxies associated with these objects are surrounded by hot halos ($T \sim 0.5 - 1$ keV), and most of them are members of clusters or groups with diffuse X-ray emission from the hot intergalactic gas ($T \sim 2 - 8$ keV). Furthermore X-ray cores coincident with the central active nuclei are present in several radio galaxies (Canosa et al. 1999). If FR I objects are identified as the parent population of BL Lac (unified schemes for AGN; Urry & Padovani 1995), we expect that the X-ray emission from the cores of these objects is dominated by the Doppler boosting, with non thermal spectrum and with some possible amount of absorption by an edge-on projected disk. We present here the X-ray properties of four FR I radio galaxies (3C 31, 3C 264, 3C 270 and 3C 465) deduced from ROSAT observations, and discuss the implications for the unified model for AGN.

2. OBSERVATIONS AND RESULTS

ROSAT data from HRI and PSPC pointed observations of the selected targets (only PSPC for 3C 264) are available in the public archive. The HRI observations provide information on the morphological structure of the inner regions around the AGN. Spectral fits to PSPC data require a composite model to disentangle the nuclear

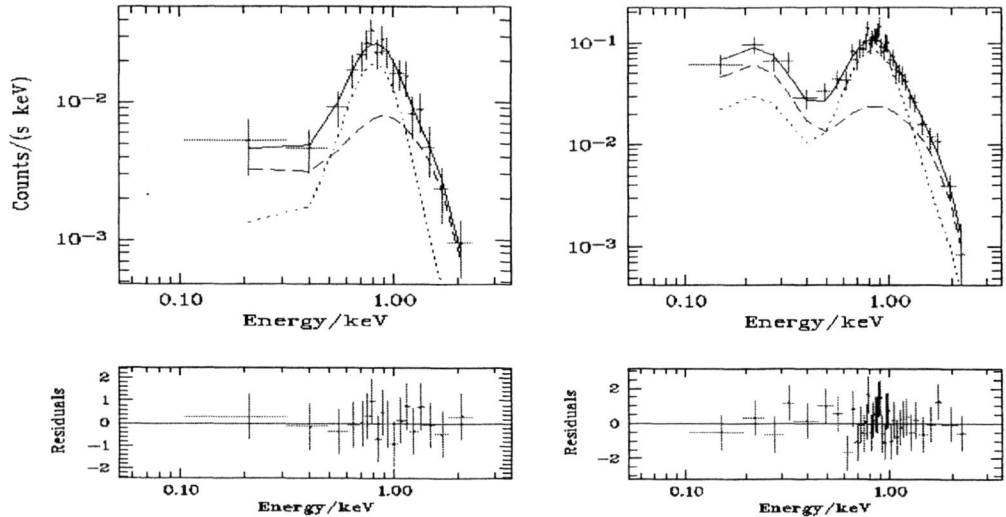

FIGURE 1. The composite spectra of 3C 31 (*left*) and 3C 270 (*right*). The dotted and dashed lines represent the thermal and the power law components, respectively

emission from the diffuse thermal component. Due to the low photon flux, in the thermal spectrum we have fixed the hydrogen column density to $N_{H,gal}$ and the relative metallicity to 0.5. We have tested that for 3C 31 and 3C 270 a composite spectral model improves the fit at a level of confidence of 90% and 99%, respectively (see Fig. 1). The main properties of each source are summarized below and the results are reported in Tab. 1.

3C 31. The spectral properties of this source (NGC 383, the main member of the group Arp 331) are consistent with a quite flat power law component (photon index $\Gamma \approx 1.8$) and a halo with low temperature ($T \approx 0.6$ keV). However the parameters are affected by large statistical uncertainties. The HRI map shows a compact component surrounded by a region of weak X-ray emission (radius ≈ 12 kpc).

3C 264. The spectrum of this radio galaxy (lying at $\approx 8'$ from the center of the cluster A 1367) is well fit by a steep power law model ($\Gamma \approx 2.46$). Only an upper limit is obtained for the thermal flux from a galactic corona.

3C 270. In this source, associated with the galaxy NGC 4261, the gas temperature is ≈ 0.8 keV and the non thermal spectrum is flat ($\Gamma \approx 1.7$). The X-ray image is consistent with an extended emission (radius ≈ 8.5 kpc).

3C 465. This target (NGC 7720) is the brightest member of the cluster A 2634, embedded in an extended X-ray region with $T \approx 4$ keV. The HRI image of the center is consistent with a pointlike source, coincident with the galaxy (radius < 5 kpc). The spectrum of the flux from this region is fitted by a thermal model ($T \approx 1.2$ keV) with an upper limit for the non thermal emission.

TABLE 1. Spectral fits from PSPC data (quantities without errors are fixed)

Source	Envir.	$N_{H,gal}^a$	T^b (keV)	Γ^b	$L_{X,th}^{b,c}$	$L_{X,pl}^{b,c}$
3C 31	Arp 331	5.23	$0.62^{+0.22}_{-0.24}$	$1.81^{+0.55}_{-1.12}$	$1.43^{+0.80}_{-0.33}$	$2.46^{+1.10}_{-0.98}$
3C 264	A 1367	2.55	1	$2.46^{+0.06}_{-0.06}$	≤ 6.6	106^{+2}_{-3}
3C 270	Group	1.63	$0.81^{+0.09}_{-0.13}$	$1.71^{+0.21}_{-0.18}$	$1.10^{+0.19}_{-0.13}$	$1.21^{+0.24}_{-0.22}$
3C 465	A 2634	5.22	$1.16^{+0.14}_{-0.18}$	1.7	$6.91^{+0.81}_{-1.10}$	≤ 4.8

a $\times 10^{20}$ cm^{-2}
b Errors are at 1σ, upper and lower limits are at 2σ
c $\times 10^{41}$ erg s^{-1} in the energy band 0.1 -2.4 keV; $H_\circ = 50$ km s^{-1} Mpc^{-1}

3. DISCUSSION

The thermal luminosity $L_{X,th}$ of 3C 31 and 3C 270, and the upper limit for 3C 264 are consistent with the optical magnitude/X-ray luminosity correlation holding for early type galaxies (Fig. 2; Trussoni et al. 1997). We remark anyway that the first two sources lie at the lower boundary of the region over which the data are scattered. In 3C 465 conversely $L_{X,th}$ is much lower than expected from the correlation. This result could be due to the difficulty to separate the coronal flux from the surrounding intracluster emission. On the other hand galactic haloes in clusters can undergo strong dynamical processes, and thus the small size and luminosity of the compact central region in 3C 465 could be related to the stripping of the corona from the outer medium. This galaxy is at the center of a cooling flow and its radio morphology (wide angle tailed) may indicate a strong interaction with the environment.

Referring to the non thermal emission from the active nuclei ($L_{X,pl}$), a correlation exists between the core spectral luminosities in the radio and X-ray bands (Canosa et al. 1999). Taking into account the high dispersion of the data, the luminosities found for 3C 31, 3C 270 and 3C 264 are basically consistent with this correlation, while for 3C 465, that has the brightest radio luminosity, only an X-ray upper limit has been found (Fig. 2). Concerning the high brightness of 3C 264, we point out that this radio galaxy has an optical jet oriented at a small angle to the line of sight ($\approx 15°$).

From the HST images Chiaberge et al. (1999) have shown that in FR I a correlation exists between the optical and radio brightness of the core. This correlation supports the connection between the relativistic beaming effects and the nuclear emission properties, while there is no evidence of obscuration from thick tori (contrary to the case of FR II radio galaxies). The X-ray observations are also consistent with the lack of absorption effects. The X-ray emission could originate along the jet far from the center, however the X-ray/radio correlation does not support this possibility, as VLBI data show that radio cores are unresolved on scales of ~ 0.1 pc.

The spectral energy distribution (SED) of BL Lac is quite complex, with two peaks of emission in the IR-optical- soft X-ray and hard X-gamma ray bands (Fossati

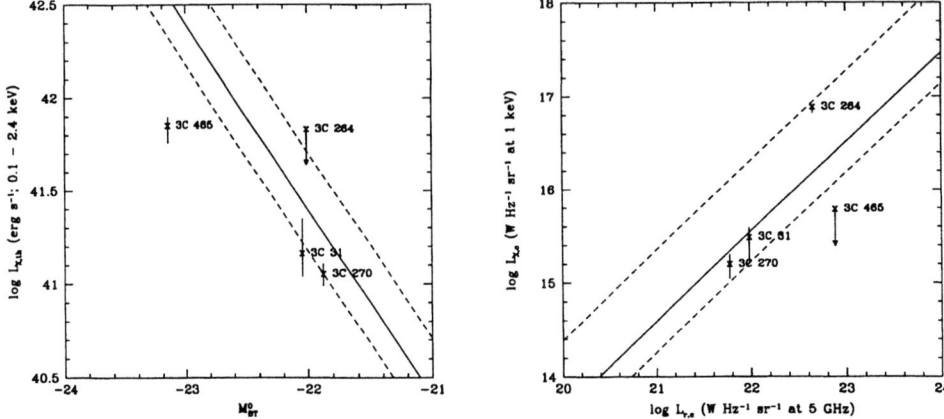

FIGURE 2. *Left:* X-ray thermal luminosity $L_{X,th}$ vs the optical magnitude M_{BT}^0. *Right:* X-ray spectral luminosity of the core $L_{X,c}$ vs the radio spectral luminosity of the core $L_{r,c}$. The solid lines are the correlations found by Trussoni et al. (1997) and Canosa et al. (1999) respectively, and the two pair of dotted lines limit the regions containing their data

et al. 1999). Therefore, in the framework of the unified schemes for AGN, these X-ray data provide further useful information to build the integral spectrum of FR I. The SED of these radio galaxies are qualitatively similar to those of BL Lac taking into account the beaming effects, consistently with the unified scheme for these two classes of objects. However it appears also that these sources are much brighter than expected if the relativistic beaming is considered. This could be related to a velocity structure across the jet (Capetti et al. 2000), even though extra contribution to the X-ray flux cannot in principle be excluded (e.g. from advection dominated disks).

ACKNOWLEDGEMENTS

This work was partially supported by the Italian Ministry for University and Research (MURST) under grant Cofin98-02-32.

REFERENCES

Canosa C.M., Worrall D.M., Hardcastle M.J., Birkinshaw M., 1999, 310, 30
Capetti A., Trussoni E., Celotti A., et al., 2000, MNRAS, in press
Chiaberge M., Capetti A., Celotti A., 1999, A&A, 349, 77
Fossati G., Maraschi L., Celotti A., et al., 1998, MNRAS, 299, 433
Trussoni E., Massaglia S., Ferrari R., et al., 1997, A&A, 327, 27
Urry C.M., Padovani P., 1995, PASP, 107, 803

TARTARUS – AN ASCA AGN DATABASE

T.J.Turner[1,2], K.Nandra[2,3], D.Turcan[2,4], I.M.George[2,3]

1) University of Maryland Baltimore County, 1000 Hilltop Circle, Baltimore, MD 21250; 2) Laboratory for High Energy Astrophysics, Code 660, NASA/Goddard Space Flight Center, Greenbelt, MD 20771; 3) Universities Space Research Association; 4) University of Maryland College Park

ABSTRACT

Tartarus is a database of products from $ASCA$ observations of AGN. We discuss access to and products available within the database, and outline the data-reduction process.

KEYWORDS: astronomical data bases; X-ray:galaxies; galaxies:active

1. INTRODUCTION

Version 1 of the Tartarus database was released 1999 July 27, and may be accessed via http://tartarus.gsfc.nasa.gov. The database contains reduced images, spectra and light curves for $ASCA$ observations of AGN and is currently complete up to the end of the AO-5 phase of the mission. We have provided significant improvements over standard mission products by including background spectra and light curves, optimizing the data-reduction for faint sources, performing a number of checks and including some supplementary information. In the initial release 444 reduced sequences were made available out of 473 in the archive. A few failed runs have been excluded until the problems can be isolated. We expect Version 2 of Tartarus to include these previously failed sequences and to have other minor existing problems fixed. We will continue to reduce AGN sequences as they enter the public domain, until all "70000000 series" sequences are available. For up-to-date information please see the Tartarus web pages.

2. SUMMARY OF AVAILABLE PRODUCTS AND INFORMATION

The Tartarus home page allows access to the database via objected-sorted and sequence-sorted lists, and a search facility is available. For each sequence there is a web page summarizing important information about the observation (e.g. exposure time, observation mode, date), as well as basic information about the source such as count rates, fluxes and a basic spectral fit. Data products (gzipped tar files) can be downloaded from web pages or via anonymous ftp to tartarus.gsfc.nasa.gov (cd to pub/tartarus/). Thumb gifs indicate whether problems occurred during processing.

Three thumbs up is obviously a good sign, if one or more thumbs is down please see the processing log for details of the problem. Comments and questions can be sent to tartarus@athena.gsfc.nasa.gov.

Data products include:

- **IMAGES** Exposure-corrected, smoothed images are available in sky co-ordinates from the summed instrument pairs. The web page shows the summed sky image from GIS2 and GIS3. The expected position of the target source is shown, based on the RA/DEC given by the observer. Other AGN lying within the field are noted.

- **DETECTOR IMAGES** Images are shown on the web page in detector coordinates with extraction regions overlaid.

- **LIGHT CURVES** Source and background light curves are available for each instrument, in 32 s bins. Selected light curves are available in 128 or 5760s bins.

- **SPECTRA** Integrated source and background spectra are available for all four detectors. The web page shows a fit to a powerlaw model along with the data and the residuals to the fit. Users can retrieve the spectral files alone in a gzipped tar file, the XSPEC command file `sequence_webfit.xcm` allows users to start their analysis at the fit shown on the web page.

3. DATA REDUCTION SUMMARY

The data reduction process is outlined here. The version of FTOOLS and Tartarus scripts used will change with time, so please see the web pages and log files for version numbers.

3.1. Examining the Observation Parameters

Observation catalogs are constructed for all four instruments (SIS0,1 and GIS2,3) using XSELECT. Filter files are merged into one and sorted by time (using FMERGE, FSORT). The datamode used is FAINT when all files are in FAINT mode, otherwise BRIGHT mode data are used. The BR_EARTH angle is determined as the angle from the earth where the SIS0 event rate in the nominal chip becomes lower than 200 c/s (then 5 degrees is added to be conservative, but never going lower than the constraints noted in the next section).

3.2. Extracting Events

The extraction command files are set up to include only data taken outside of the SAA, having angular offset from nominal pointing < 0.01 degrees, $RBM_{CONT} <$ 500 (radiation belt monitor) and $COR > 6$ (cut-off rigidity, GeV/c). For the SIS we require $ELV > 10°$ (angle from earths limb), BR_EARTH $> 20°$ (angle from bright Earth), time after day/night terminator $> 50, 100, 200$ for 1,2,4 CCD modes, respectively, SIS pixel threshold $< 100, 75, 50$ cts for 1,2,4 CCD modes. For the GIS we require $ELV > 5°$. A check is performed to determine whether good time exists for the observation. Events are extracted for all four instruments (SIS0,1 and GIS2,3) using XSELECT and the extraction command files.

3.3. Defining Regions

We create (XSELECT) and smooth (FGAUSS) sky and detector images for all detectors. The centroid is found by determining the pixel with the highest count in the smoothed image within radii $\sim 5'$ and $\sim 7'$ of the expected target position in the SIS and GIS, respectively (to minimize the risk of extracting a serendipitous source). Source extraction regions are centered on these centroid positions. Source regions are circles of radii 45 (SIS) or 27 (GIS) pixels (using XSELECT detector image defaults which have the SIS data spatially-binned by a factor 4), corresponding to $4.86'$ and $6.75'$, respectively. The background region for the SIS is everything outside of a circle of radius 47 pixels but inside the nominal chip (S0C1 or S1C3). For the GIS the background region is an annulus with inner radius 35 and outer radius of 55 pixels centered on the source. A check is made that SIS0, SIS1 centroids and GIS2, GIS3 centroids agree within 1 arcmin. A combined and smoothed sky image is created for the SIS0,1 pair and the GIS2,3 pair and then a check is made that the resulting SIS and GIS centroids agree within 1 arcmin.

3.4. Extracting Additional Products

SISPI is run on the SIS0 and SIS1 event files. Source and background spectra are extracted from the defined regions using XSELECT. Source and background mini-event files are extracted for SIS0 and SIS1 in the full (0.5–10 keV), hard (2–10 keV) and soft (0.5–2 keV) energy ranges from the defined regions using XSELECT. Source and background event files are extracted for GIS2 and GIS3 in the full and hard energy range.

3.4.1. Spatial Analysis

Four-channel spectra are temporarily created for use in making temporary efficiency maps (to speed up the run-time compared to using full resolution spectra for the energy-weighting). Exposure maps and sky images are created from the efficiency maps using the attitude information for the observation with ASCAEXPO. The maximum attitude deviation allowed is $15''$, and the image rebinning factor is 8. Combined, smoothed, exposure-corrected sky images are created for the SIS0,1 and GIS2,3 pairs using XIMAGE. The GIS image on the web page has overlaid a green circular marker showing the expected position of the target based on the RA and DEC *given by the observer* (taken from the attitude file header).

3.4.2. Spectral Analysis

If SIS datamode is BRIGHT then SIS spectra will have 512 pha channels. If the GISBITFIX problem is present then GIS3 spectra will have 128 channels. All other spectra have 1024 channels. Bad channels are then set such that for the SIS only data between 0.6 – 10 keV are denoted "good" and used in the spectral fitting. For GIS observations prior to June 1996 data between 0.7 – 10 keV are good, observations after that date use only data between 1.0 – 10.0 keV. The data are grouped to have at least 20 counts per spectral bin. Ancilliary response files (ARFs) are made using the spectral file, and assuming a point-source at the center of the region file. Redistribution matrix files (RMFs) are made using SISRMG.

A power law is fit to the data of all four instruments using XSPEC. Individual instrument normalizations were allowed to vary relative to SIS0 to allow for small uncertainties in the instrument cross-calibrations. The fit has three free parameters; photon index, normalization of the powerlaw and absorbing column and is performed on the 0.6-5 keV data (excluding data below 1 keV for the GIS) plus 7.5-10 keV data, i.e. excluding the iron Kα regime. The Galactic line-of-sight column is noted for comparison. The 5.0-7.5 keV data are read in after the fit has minimized, these data are included in the plots and contribute towards the fit statistic shown. The $\chi^2 + 4.61$ error ranges are calculated for N_H and Γ. The flux under the model is calculated in the 0.5–2 keV and 2–10 keV bands (from SIS0). For accuracy, the fluxes were obtained from a double powerlaw fit, not shown on the web pages (which will soon be available in BROWSE).

3.4.3. Timing Analysis

Combined SIS0,1 and GIS2,3 light curves are created using LCURVE. The instrument-pair light curves are the sum of the two individual instruments. Time series are created from the source region and background region for each instrument. Three energy ranges are used (full, hard, soft) and three binning intervals (32, 128, 5760 seconds). Combined SIS0,1 and GIS2,3 count rates are calculated from combined SIS and GIS light curves for the full, soft and soft energy ranges with background subtracted (The scaling parameter (BACKSCAL) is determined from source and background spectral files using FKEYPRINT).

ACKNOWLEDGEMENTS

This work was supported by NASA LTSA grant NAG 5-7385 (TJT) and ADP grant NAG 5-7067 (KN). This project has made use of data obtained through the HEASARC on-line service provided by NASA/GSFC. DT was supported by ASCA grants of TJT, KN, IMG.

X-RAY SPECTRAL COMPLEXITY IN NARROW-LINE SEYFERT 1 GALAXIES

S. Vaughan[1] J. Reeves[1] R. Warwick[1] R. Edelson[1,2] K. Pounds[1]

1) X-Ray Astronomy Group; Department of Physics and Astronomy; Leicester University; Leicester LE1 7RH; U.K.
2) Department of Physics and Astronomy; University of California, Los Angeles; Los Angeles, CA 90095-1562; U.S.A.

ABSTRACT

We present a systematic analysis of the X-ray spectral properties of a sample of 22 'narrow-line' Seyfert 1 galaxies for which data are available from the ASCA public archive. We also consider the spectrum of the bright NLS1 Ark 564 in detail using simultaneous ASCA and RXTE data.

KEYWORDS: galaxies: Seyfert; X-rays: galaxies; galaxies: active; galaxies: individual (Ark 564).

1. INTRODUCTION

Interest in Narrow-Line Seyfert 1 (NLS1) galaxies, defined as having FWHM(Hβ) < 2000 km/s (Osterbrock & Pogge 1985) and weak forbidden lines ([O III]/Hβ <3), has grown markedly since *ROSAT* observations showed NLS1s to exhibit unusually strong soft X-ray (<2 keV) emission (Boller, Brandt & Fink 1996). Subsequent observations with *ROSAT*, *ASCA* and *BeppoSAX* have confirmed other distinctive properties of NLS1s – compared with normal, broad-line Seyfert 1s (BLS1s). These include a generally steeper power law slope over the 2–10 keV band (Brandt, Mathur & Elvis 1997) and rapid and large amplitude variability (Boller et al. 1996; Turner et al. 1999). An intriguing absorption or emission feature near 1 keV has also been reported in recent *ASCA* and *BeppoSAX* observations of several NLS1s (e.g. Leighly et al. 1997; Turner, George & Nandra 1998). All these X-ray and optical properties are consistent with NLS1s (in terms of the widely accepted accreting black hole model) running at a higher accretion rate (than BLS1), a possibility first noted by analogy with Galactic black hole candidate spectra by Pounds, Done & Osborne (1995).

The X-ray spectrum should also bear the signature of 'reflection' from the disc surface, which is expected to be highly ionised in high accretion rate objects (e.g., Matt, Fabian & Ross 1993). The form of the reflection features, particularly the iron K line and absorption edge and the form of the soft X-ray continuum, should therefore differ in NLS1s from BLS1s if indeed NLS1s are accreting at a higher rate. Tentative support for this view has been presented by Comastri et al. (1998) and

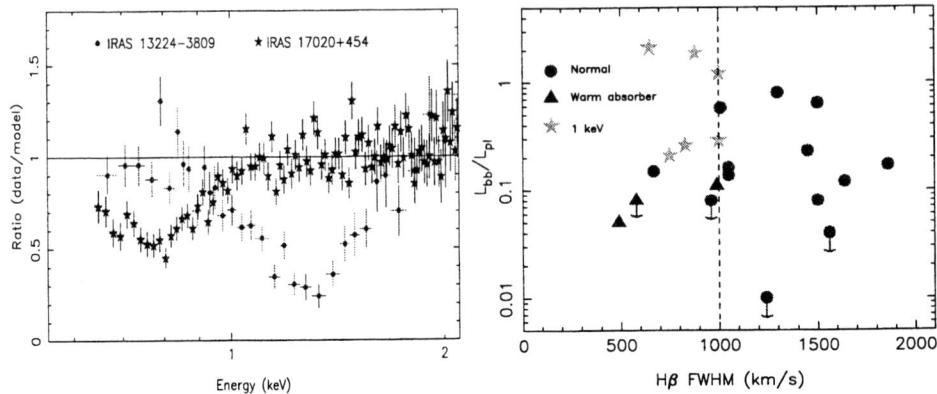

FIGURE 1. (a, left) The ratio of the *ASCA* SIS data to the best-fit continuum model for two NLS1s. IRAS 13224−3809 exhibits an anomalous absorption feature near 1.2 keV while IRAS 17020+454 probably a warm (O VII) edge near 0.74 keV. FIGURE 1(b, right) The relative luminosity of the soft excess (compared to the hard power law) against FWHM(Hβ). Objects with no evidence for absorption are marked by circles and those with normal warm absorbers by triangles. The sources exhibiting anomalous 1.1–1.4 keV features are marked by stars.

Turner et al. (1998) who find evidence for an emission line near 7 keV, consistent with K_α emission in hydrogen-like iron in the NLS1 Ton S180. If indeed a high accretion rate is the driver that distinguishes NLS1s, it underlines the importance of studying NLS1s as a means by which one of the most fundamental properties of AGN can be better understood.

2. THE ASCA SAMPLE

We have systematically analysed the X-ray spectra of a sample of 22 objects, identified in the literature as Seyfert 1 galaxies or quasars with FWHM(Hβ) ≤ 2000 km/s, for which *ASCA* data are publicly available (Vaughan et al. 1999a).

Their measured hard power-law continua have photon indices spanning the range $\Gamma = 1.6 - 2.5$ with a mean of 2.1, which is only slightly steeper than the norm for BLS1s. All but four of the NLS1s exhibit a soft excess, which can be modelled as blackbody emission ($kT \approx 90 - 250$ eV) superposed on the underlying power-law. This soft component is often so strong that, even in the relatively hard bandpass of *ASCA*, it contains a significant fraction, if not the bulk, of the X-ray luminosity, apparently ruling out models in which the soft excess is produced entirely through reprocessing of the hard continuum.

Emission lines at 6–7 keV are detected at high/modest significance levels in 9 objects and marginal significance in 2 others. However, the line properties (energy, equivalent width, intrinsic width) are in general only poorly constrained by the *ASCA* spectra. A point of interest, for follow-up when more sensitive X-ray spectra

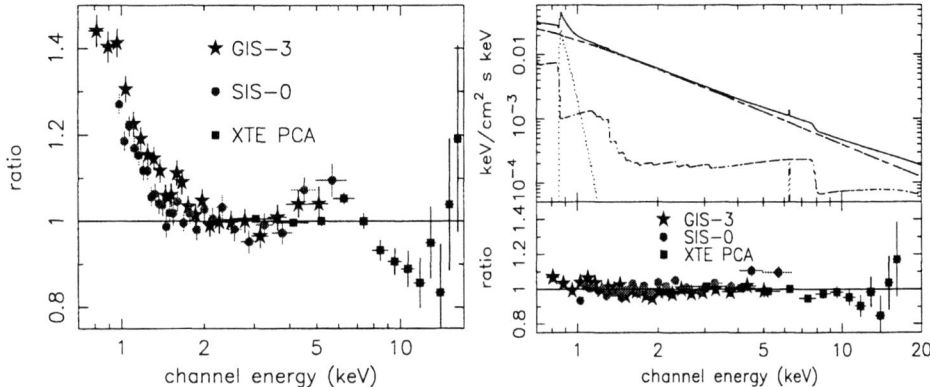

FIGURE 2. (a, left) The residuals from a simple power law fit to the $ASCA/RXTE$ data of Ark 564 indicating a 'soft excess' and Fe K shell features in the 6–10 keV range. FIGURE 2(b, right) The best fit ionised reflection model as fitted to the $ASCA/RXTE$ data.

are available, is the incidence of emission lines from highly ionised gas (corresponding to Fe XXII and above) in NLS1s, suggested by the current data.

Six NLS1s show evidence for the above mentioned absorption/emission feature near ~1.2 keV, whilst 3 others show more usual 'warm absorption' features below 0.9 keV (see Fig. 1a). Remarkably, these nine sources have extremely narrow permitted optical line widths, with FWHM(Hβ) below the sample mean of 1000 km/s (see Fig. 1b). The six NLS1s with the ~1.2 keV feature also tend to have the strongest soft excesses. Clearly they represent an anomalous group of NLS1s.

3. SIMULTANEOUS ASCA AND RXTE OBSERVATIONS OF ARK 564

Ark 564 is the brightest known NLS1 in the 2–10 keV band and hence can be considered an ideal object in which to study the spectral features described above. $ROSAT$ data revealed a complex spectrum in the soft X-ray band, well fitted with either a power law and strong soft excess, or a (steeper) power law and an absorption edge at 1.2 keV (Brandt et al. 1994). Ark 564 was observed simultaneously with $ASCA$ and $RXTE$, giving the first spectrum of an ultrasoft NLS1 measured out to 20 keV (Vaughan et al. 1999b).

Fig. 2(a) shows the $ASCA$ and $RXTE$ data, over the range 1–20 keV, compared to a steep power-law ($\Gamma = 2.6$) and Galactic column of 6.4×10^{20} cm^{-2}. Due to calibration problems with the $ASCA$ SIS we do not attempt to fit the spectrum below 0.8 keV. The residuals to the above fit reveal an apparent soft excess below ~2 keV and Fe K-shell features in the range 6–10 keV. Fitting these features suggests the presence of iron K$_\alpha$ emission (although the details are poorly constrained) and a relatively deep ($\tau \sim 0.2$) edge at ~8.6 keV, distinguished at high confidence from a neutral Fe K-edge at 7.1 keV. All these features can be naturally explained by reflection from ionised matter, and we find that the *RXTE/ASCA* data over the whole 0.8–20 keV band are well fitted by the ionised reflection model of Magdziarz & Zdziarski (1995). In these fits, the 'soft excess' in Fig. 2(a) is accounted for by the enhanced continuum reflection plus recombination emission from the ionised material, which together supply $\sim$ 16 per cent of the luminosity in the 0.8–2 keV band. Fig. 2(b) reproduces the ionised disk model fit including recombination emission from O VIII. Such reprocessed soft X-ray emission also provides an alternative explanation for the unusual spectral features near $\sim$1 keV seen in other NLS1s. Alternative spectral fits to the *ASCA* data of Ark 564 are given by Turner, George & Netzer (1999) and Ballantyne, Iawasawa & Fabian (2000), and the *BeppoSAX* spectrum is discussed in Pounds & Vaughan (2000).

We note, finally, that if the unusually steep power-law component is a result of Compton-cooling of a disc corona by an intense soft photon flux, then the implication is that the bulk of these soft photons lie in the unobserved extreme ultraviolet.

REFERENCES

Ballantyne, D. R., Iawasaw, K., Fabian, A. C., 2000, MNRAS, submitted
Boller, Th., Brandt, W. N. & Fink, H. 1996, A&A, 305, 53
Brandt, W. N., Fabian, A. C., Nandra, K., Reynolds, C. S., Brinkmann, W. 1994, MNRAS, 271, 958
Brandt, W. N., Mathur, S. & Elvis, M. 1997, MNRAS, 285, L25
Comastri, A. et al. 1998, A&A, 333, 31
Leighly, K., Mushotsky, R. F., Nandra, K., Forster, K. 1997, ApJ, 489, L25
Magdziarz, P., Zdziarski, A. A. 1995, MNRAS, 273, 837
Matt, G., Fabian, A. C., Ross, R. R. 1993, MNRAS, 262, 179
Osterbrock, D., Pogge, R. 1985, ApJ, 297, 166
Pounds, K. A., Done, C., Osborne, J. 1995, MNRAS, 277, L5
Pounds, K. A., Vaughan, S., 2000, in "Observational and Theoretical Progress in the Study of Narrow-Line Seyfert 1 Galaxies," eds. Boller Th., Brandt W.N., Leighly K.M., Ward M.J., New Astronomy Reviews, in press
Turner, T. J., George, I. M., Nandra, K. 1998, ApJ, 508, 648
Turner, T. J., George, I. M., Netzer, H. 1999, ApJ, 526, 52
Turner, T. J., George, I. M., Nandra, K., Turcan, D. 1999, ApJ, 524, 667
Vaughan, S., Reeves, J., Warwick, R., Edelson, R. 1999a, MNRAS, 309. 113
Vaughan, S., Pounds, K., Reeves, J., Warwick, R., Edelson, R. 1999b, MNRAS, 308, L34

ASCA VIEW ON HIGH-REDSHIFT RADIO-QUIET QUASARS

C. Vignali[1,2], A. Comastri[2], M. Cappi[3,4], G.G.C. Palumbo[1], M. Matsuoka[5]

1) Dipartimento di Astronomia, Università di Bologna, via Ranzani 1, I–40127, Bologna, Italy
2) Osservatorio Astronomico, via Ranzani 1, I–40127, Bologna, Italy
3) ITeSRE, C.N.R., via Gobetti 101, I-40129 Bologna, Italy
4) Harvard-Smithsonian Center for Astrophysics, 60 Garden Street, Cambridge MA 02138, USA
5) National Space Development Agency of Japan (NASDA), World Trade Center Bldg., 2-4-1, Hamamatsu-cho, Minato-ku, Tokyo 105-8060, Japan

ABSTRACT   We briefly discuss the latest $ASCA$ results on the X–ray spectral properties of high-redshift radio-quiet quasars.

KEYWORDS: active galaxies; quasars; nonthermal mechanism; X-rays

1. INTRODUCTION

The study of quasar X–ray properties as a function of redshift can address some important issues such as (a) the history of accretion processes over the cosmic time, (b) the evolution of quasar activity and (c) the condition of the Universe at the epoch of quasars formation. So far, while high-redshift radio-loud quasars (RLQs) have been widely studied in hard X–rays in the last few years (Cappi et al. 1997, Yamazaki et al. 1998), the X–ray properties of high-z radio-quiet quasars (RQQs) are still poorly known (even though they constitute about 90 % of the quasar population), due to their weaker X–ray flux with respect to radio-loud objects (Zamorani et al. 1981). In order to fill this gap we have performed a pilot study with $ASCA$ of a sample of high-redshift ($z > 1.85$) RQQs.

2. THE SAMPLE

The objects presented here have been chosen among the brightest ones found through a cross correlation of the Véron-Véron quasars catalogue (Véron-Cetty & Véron 1996) with the ROSAT All-Sky Survey catalogue of X–ray sources (Voges et al. 1999). We have obtained relatively good ASCA spectra for 9 sources (see Table 1 for the relevant data). The sample is clearly not complete and, due to its soft X-ray selection, may be biased toward less absorbed objects. Nonetheless, it may be considered adequate in order to obtain, for the first time and prior to XMM

Table. 1 - The Radio-Quiet Quasars sample

Object	z	$N_{H_{gal}}^a$	m_V	Exp.SIS(GIS) (ks)	CR$_{SIS(GIS)}$ (10^{-2} c/s)	α_{ox}^b
0151−4046	1.85	2.07	18.1	32 (36)	4.0 (2.3)	1.15
0040+0034	2.00	2.45	18.0	30 (36)	3.8 (3.2)	1.17
1352−2242	2.00	5.88	18.2	31.4	1.7 (1.5)	1.29
1247+267	2.04	0.9	15.6	36 (34)	1.3 (1.1)	1.69
1400+10	2.07	1.81		86.5	2.4 (1.8)	
1101−264	2.15	5.68	16.0	17.4 (19.7)	1 (0.6)	1.71
0059−304A	2.17	2.00	19.3	31 (37.4)	0.5 (0.6)	1.23
0300−4342	2.30	1.83	19.2	37.7	1.3 (1.2)	1.16
0130−4124	2.46	2.20	20.8	35.3 (38)	1.2 (1.1)	0.95

^{a} In units of 10^{20} cm^{-2}, Dickey & Lockman 1990; ^{b} $\alpha_{ox} = -\frac{Log(F_V/F_X)}{Log(\nu_V/\nu_X)}$

launch, a reliable measurement of the X–ray spectral properties of high-z RQQs. A standard analysis has been applied to the data, taking into account the most recent calibration uncertainties, and extensive checks on background subtraction have been performed. A detailed description of the data analysis and the first results can be found in Vignali et al. (1999) on a subsample of 4 objects, plus WEE 83, which is now excluded from discussion since its redshift has been re-measured and the revised value puts it fairly near (z=0.311, Wu et al. 1999).

3. RESULTS

The results of the spectral analysis are presented in Table 2 and summarized in the following.

- The spectra are well fitted by a single power law model over the ∼ 2–30 keV energy range (rest frame). The average spectral slope is $<\Gamma> = 1.75$, with dispersion $\sigma = 0.12$, which agrees very well with the value ($\Gamma = 1.72\pm0.03$) obtained from the co-added GIS+SIS spectra of all the sources in the overlapping rest-frame energy range (∼ 2.4–28.4 keV).

- A comparison with various samples of lower-z RQQs from *ASCA* (Reeves et al. 1997, George et al. 2000) and *BeppoSAX* (Costantini 1998) is showed in Fig. 1, suggesting either a possible flattening of the power law slope with redshift or toward high energies.

- There is no evidence of the signatures of cold matter either in transmission (the upper limits on intrinsic absorption ranging from 3 up to 8 × 10^{21} cm^{-2} rest frame) or in reflection, with upper limits on Fe Kα line EW of ∼ 70–200 eV (rest frame). The lack of intrinsic absorption in high-z RQQs is at variance with the findings by Elvis et al. (1994) and Cappi et al. (1997) for high-z RLQs. If this result will be confirmed by future observations, then a different evolution for RQQs and RLQs, possibly related to their environment, could be envisaged.

- A further check on the presence of a reflection component (peaking at 20–30 keV rest frame) has been performed on the co-added spectrum. The addition of

Table 2 - ASCA (SIS+GIS) 0.8–10 keV Spectral Results

Object	N_H (10^{21} cm^{-2})	Γ	R^a	χ^2/dof	$F_{2-10keV}$ (10^{-13} cgs)	$L_{2-10keV}$ (10^{46} cgs)
0040+0034	$\equiv N_{H_{gal}}$	$1.66^{+0.07}_{-0.06}$		240/237	15	2.7
	< 8.76	$1.69^{+0.10}_{-0.08}$		239/236		
	$\equiv N_{H_{gal}}$	$1.71^{+0.18}_{-0.14}$	< 2.34	239/236		
0300−4342	$\equiv N_{H_{gal}}$	$1.64^{+0.12}_{-0.15}$		168/148	5.0	1.2
	< 7.15	1.63 ± 0.14		168/147		
	$\equiv N_{H_{gal}}$	$1.96^{+0.19}_{-0.33}$	$3.62^{+6.38}_{-2.90}$	164/147		
1101−264	$\equiv N_{H_{gal}}$	$1.79^{+0.26}_{-0.25}$		29.3/23	2.9	0.7
	< 63.1	$2.05^{+0.62}_{-0.43}$		27.9/22		
	$\equiv N_{H_{gal}}$	$1.70^{+1.69}_{-0.23}$	unc.	29.2/22		
1352−2242	$\equiv N_{H_{gal}}$	1.66 ± 0.12		145/147	6.7	1.1
	< 7.53	$1.65^{+0.14}_{-0.10}$		145/146		
	$\equiv N_{H_{gal}}$	$1.84^{+0.80}_{-0.28}$	< 23	144/146		
1400+10	$\equiv N_{H_{gal}}$	$1.64^{+0.06}_{-0.05}$		292/256	9.4	1.8
	< 7.64	$1.64^{+0.09}_{-0.05}$		292/255		
	$\equiv N_{H_{gal}}$	$1.60^{+0.14}_{-0.08}$	< 1.13	292/255		
0130−4124	$\equiv N_{H_{gal}}$	1.78 ± 0.14		132/137	5.0	1.6
	< 54.3	$1.83^{+0.32}_{-0.18}$		131/136		
	$\equiv N_{H_{gal}}$	$1.69^{+0.38}_{-0.16}$	< 2.35	132/136		
0059-304 A	$\equiv N_{H_{gal}}$	$1.73^{+0.61}_{-0.55}$		88/60	2.3	0.6
	< 349	$1.74^{+1.79}_{-0.56}$		88/59		
	$\equiv N_{H_{gal}}$	$1.92^{+0.88}_{-0.77}$	unc.	87.8/59		
0151−4046	$\equiv N_{H_{gal}}$	$1.83^{+0.07}_{-0.06}$		163/197	11	2.0
	< 3.97	1.84 ± 0.07		163/196		
	$\equiv N_{H_{gal}}$	$1.88^{+0.19}_{-0.15}$	< 2.68	163/196		
1247+267	$\equiv N_{H_{gal}}$	$2.00^{+0.15}_{-0.13}$		130/124	4.8	1.3
	< 2.99	2.01 ± 0.14		130/123		
	$\equiv N_{H_{gal}}$	$2.23^{+0.40}_{-0.30}$	$2.87^{+7.42}_{-2.69}$	128/123		

Note: errors are quoted at 90 % confidence level for one interesting parameter.
^{a} R represents the normalization of reflected vs. direct continuum (R=1 means $\Omega=2\pi$ coverage).

this component results in a steeper spectrum ($\Gamma = 1.82^{+0.11}_{-0.09}$) and a value for R (the normalization of the reflected vs. the direct continuum) of $0.89^{+0.81}_{-0.56}$. However, this component statistically is not required by the data. This result, combined with the lack of any iron line, do indicate that in high-luminosity RQQs reprocessing, if present, is different with respect to nearby Seyfert galaxies.

• The lack of a reflection component and absorption features in the spectra of high-z RQQs is not surprising though. It can be explained as the result of a strong ionized reflection, which gives rise to a reprocessed spectrum similar to the incident one; alternatively, it may be caused by a low covering fraction of the reprocessing matter as seen from the X–ray source. The iron line may be weak due to resonant trapping and Auger effects or totally absent if the iron is fully stripped of electrons.

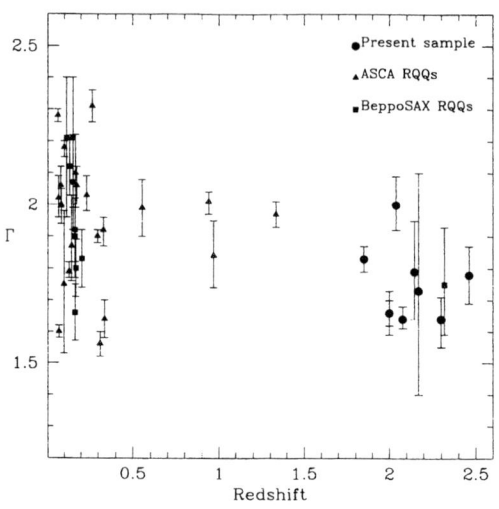

FIGURE 1. RQQs photon spectral indices as a function of redshift.

ACKNOWLEDGEMENTS

We thank the *ASCA* team, who operate the satellite and maintain the software and database. Financial support from Italian Space Agency under the contract ASI-ARS-98-119 and the Italian Ministry for University and Research (MURST) under grant Cofin98-02-32 are acknowledged by C. V. and A. C.

REFERENCES

Cappi M., Matsuoka M., Comastri A., Brinkmann W., Elvis M., Palumbo G. G. C., Vignali C. 1997, ApJ, 478, 492
Costantini E. 1998, Degree Thesis, University of Bologna
Elvis M., Fiore F., Wilkes B. J., McDowell J. C., Bechtold J. 1994, ApJ, 422, 60
George I. M., Turner T. J., Yaqoob T., Netzer H., Laor A., Mushotzky R. F., Nandra K., Takahashi T. 2000, ApJ, 531, 52
Nandra K., Pounds K. A. 1994, MNRAS, 268, 405
Reeves J. N., Turner M. J. L., Ohashi T., Kii T. 1997, MNRAS, 292, 468
Véron-Cetty M. P., Véron P. 1996, ESO Sci. Rep., 17, 1
Vignali C., Comastri C., Cappi M., Palumbo G. G. C., Matsuoka M., Kubo H. 1999, ApJ, 516, 582
Voges W., et al. 1999, A&A, 349, 389
Wu X.-B., Bade N., Beckmann V. 1999, A&A, 347, 63
Yamasaki N. Y., Miyazaki H., Ohashi T., Wilkes B. J. 1998 PASJ, 50, 19
Zamorani G., et al. 1981, ApJ, 245, 357

THE WARM SCATTERING MEDIUM IN NGC 4151

R. S. Warwick[1] & R. G. Griffiths[1]

1) Department of Physics and Astronomy, University of Leicester, Leicester, LE1 7RH, United Kingdom

ABSTRACT   Seyfert galaxies have been shown to exhibit a rich variety in terms of their X-ray spectra. One process which contributes in the soft X-ray band is the scattering of the continuum flux by highly ionized gas in the vicinity of the nucleus. This is particularly important in systems where the direct view of the continuum source is blocked below a few keV by line-of-sight absorption, in which case the scattered component is often apparent as a soft excess flux. A spectral feature observed near 0.9 keV in the X-ray spectra of several such Seyfert galaxies, including NGC 4151, can be interpreted as OVIII recombination originating in a warm, partially photoionized, scattering medium. We find that, in the case of NGC 4151, the electron scattering medium may be highly stratified with the warm ($T_{warm} \sim 3 \times 10^5$ K) plasma representing only a minor component in comparison to hotter ($T_{hot} \gtrsim 10^7$ K), more highly photoionized gas. Nevertheless it appears that the recombination continua of the hydrogen- and helium-like ions of elements such as oxygen, silicon and sulphur may, in future X-ray observations, provide an important new diagnostic of the temperature, the ionization state and conceivably the dynamics of the electron scattering media present in Seyfert nuclei.

KEYWORDS: galaxies: individual (NGC 4151) – active – galaxies:Seyfert – X-rays:galaxies.

1. INTRODUCTION

As a result of missions such as *Ginga*, *ROSAT* and *ASCA* and, more recently, *RXTE* and *BeppoSAX*, Seyfert galaxies have been shown to exhibit a rich phenomenology in terms of their X-ray spectra. A topic of particular recent interest has been the study of gas in the vicinity of the active nucleus which is strongly photoionized by the ultraviolet and X-ray flux emanating from the central source. Recent observations provide clear evidence for ionized absorption systems, more commonly known as warm absorbers, in many Seyfert 1 galaxies (e.g. Reynolds 1997; George et al. 1998). The presence of highly ionized gas has also been inferred in Seyfert 2 nuclei as the medium responsible for the scattering of the nuclear flux into our line of sight, even when our direct view of the nucleus is heavily obscured (e.g. Antonucci 1993).

Unless the scattering medium is completely photoionized by the incident nuclear flux, it will imprint absorption features on the emergent flux. There is also the possibility that the radiation processes intrinsic to the scatterer will be revealed through the superposition of sharp emission features. Here we discuss the presence of such features in the X-ray spectrum of NGC 4151.

2. THE X-RAY SPECTRUM OF NGC 4151

We have analysed the *ASCA* SIS spectra of NGC 4151 obtained from a deep (~ 100 ks) observation performed in May, 1995. We adopt a largely empirical model for the X-ray spectrum above 2 keV involving multiple partial covering of the continuum source plus an iron K_α line. Here we identify the soft excess as scattered nuclear continuum radiation and initially model it simply as a power-law component with the same spectral slope as the hard continuum. It is immediately obvious, from the count-rate residuals that pure electron scattering alone cannot adequately describe the soft X-ray spectrum of NGC 4151. The largest discrepancy between the data and the model is the presence of a line-like feature at ~ 0.9 keV, very similar to the situation in Mrk 3 (Griffiths et al. 1998). Evidence for this line-like feature has also been found by other authors in previous ASCA observations of NGC 4151 (e.g. Weaver et al. 1994). If the line-like feature is modelled by the addition of a narrow Gaussian component, then an energy and equivalent width of 0.89 ± 0.01 keV and 54 ± 8 eV respectively is obtained, in excellent agreement with the earlier measurements. The inclusion of the line component also leads to a huge improvement in the χ^2 of the fit ($\Delta\chi^2 \sim 128$).

Thus far it has been assumed that the scattering region is fully ionized and therefore superimposes no intrinsic features on the scattered nuclear signal. The next step is to consider a scenario in which the scattering region is only partially ionized. The photoionization code XSTAR (Kallman & Krolik 1997) was used to calculate the emission and absorption spectrum of a gas cloud subject to an intense hard ($\Gamma = 1.75$) photoionizing continuum flux. Further details of our use of XSTAR are given in Griffiths (1999). Our attempts to fit the soft excess flux in terms of a homogeneous partially ionized scattering region in both ionization and thermal equilibrium were not fully successful. The problem is that in order to obtain an appropriate equivalent width for the feature at ~ 0.9 keV, a rather high value is required for the ionization parameter (e.g. $\xi > 3$). However, the implied temperature of the plasma is then too high to be consistent with the observed intrinsic width of the 0.9 keV feature. Although this problem may be circumvented if the requirement for radiative (thermal) equilibrium is relaxed, a solution is also possible if the medium is stratified.

The best-fitting model shown in Fig. 1 has $\log \xi \approx 2.5$ and a corresponding plasma temperature of $T \approx 3 \times 10^5$ K. The spectral peak apparent near 0.9 keV in the *ASCA* SIS spectrum of NGC 4151 is attributed to recombination directly into the K-shell of fully stripped oxygen ions in the scattering medium. The overprediction of the equivalent width of the OVIII recombination feature is avoided by requiring a significant fraction of the scattered flux to be produced in plasma with at least a factor ten higher ionization parameter, at which point the temperature will have risen to close to the Compton temperature. A consistent picture is thus one in which very hot, fully ionized plasma, perhaps in the form of an outflow or wind

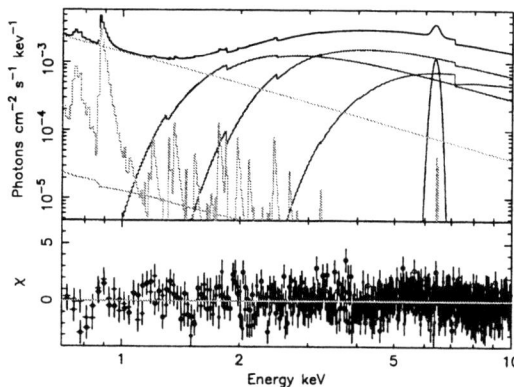

FIGURE 1. *Top panel:* The best-fitting photon spectrum of NGC 4151 with the soft excess modelled in terms of the scattered and intrinsic emission from a strongly photoionized and stratified scattering medium. *Bottom panel:* The corresponding count rate residuals to the fit of the *ASCA* SIS (S0 and S1) spectra.

from the nucleus of NGC 4151 (e.g. Krolik & Kriss 1995), is responsible for most of the scattering. However, regions of cooler, denser and less strongly photoionized gas give rise to the discernable OVIII recombination feature.

3. DISCUSSION

The above model for the scattering medium in NGC 4151 requires the co-existence of two distinct regions, in essence a warm region with $T_{warm} \approx 3 \times 10^5$ K which electron scatters 0.025% of the nuclear flux into our line of sight and a much hotter region, with $T_{hot} > 10^7$ K, which produces the bulk of the soft excess via a scattering fraction of 2.3%. There are two plausible settings in which these two scattering regions could co-exist as illustrated in Fig. 2. The first resembles a two phase medium of the type discussed by Krolik, McKee & Tarter (1981) in which clouds of warm gas are embedded in a pervasive hot medium. A rough pressure equilibrium may exist between these two phases with $n_{warm} \sim 10^9$ cm^{-3}, i.e. comparable to the density typical of broad-line region clouds. However, the cooler component lies on the part of the thermal equilibrium curve that is unstable to temperature variations and the expectation would be that such clouds would soon be heated by the surrounding gas and evaporate. An alternative possibility is that the two regions represent extremes within a highly stratified electron scattering region as depicted in Fig. 2. In this case the modelling requires $r \approx \Delta r_{hot} \sim 10^{16}$ cm^{-2} and $\Delta r_{warm} << \Delta r_{hot}$.

Finally we note that the OVIII recombination feature discussed here, will be a very prominent feature in the high quality X-ray spectra from missions such as Chandra and XMM. From such data it should be possible, for example, to use the intrinsic width of the recombination feature to determine the precise temperature of

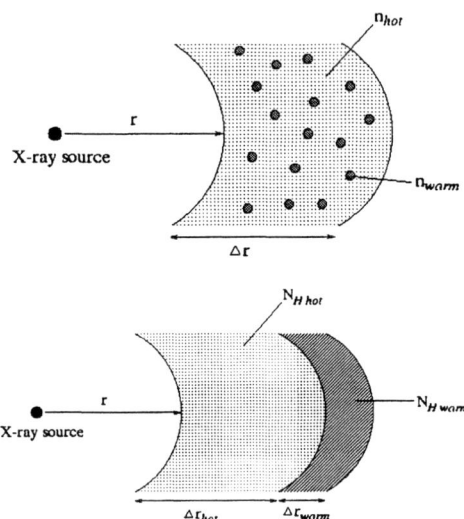

FIGURE 2. The distribution of strongly photoionized gas in the nucleus of NGC 4151. *Top panel:* The warm clouds are embedded in the hot medium. *Bottom panel:* The warm material is spatial separated from the hot component indicative of stratification in the electron scattering medium.

the scattering medium and, from the recombination edge energy, determine whether the material is in the process of outflow from (or conceivably in-fall towards) the central nuclear source.

ACKNOWLEDGEMENTS

RGG acknowledges support from PPARC in the form of a research studentship. The X-ray spectral data used in this work were obtained from the HEASARC facility at Goddard Space Flight Centre.

REFERENCES

Antonucci R.R.J., 1993, ARA&A, 31, 473
George I.M. et al., 1998, ApJS, 114, 73
Griffiths R.G. et al., 1998, MNRAS, 298, 1159
Griffiths R.G., 1999, Ph.D. Thesis, University of Leicester
Kallman T.R., Krolik J.H., 1997, XSTAR: A Spectral Analysis Tool. Version 1.40 of the User's Guide
Krolik J.H., Kriss G.A., 1995, ApJ, 447, 512
Krolik J.H., McKee C.F., Tarter C.B., 1981, ApJ, 249, 422
Reynolds C.S., 1997, MNRAS, 286, 513
Weaver K.A. et al., 1994, ApJ, 436, L27

THE X-RAY SPECTRA OF SYMBIOTIC STARS

Peter J. Wheatley

Dept. of Physics and Astronomy, University of Leicester, University Road, Leicester UK

ABSTRACT   Symbiotic stars are thought to show distinct X-ray emission from the accreting object and from the colliding winds of the two stars. I show that the colliding wind component is unnecessary. Instead, the spectra can be interpreted as emission only from the compact object that is strongly absorbed by the partially-ionised wind of the red giant. There remains no evidence of any X-ray emission from colliding winds, and thus no need for a substantial wind from the compact object.

KEYWORDS: binaries: symbiotic — stars: winds, outflows — white dwarfs — X-rays: stars

1. INTRODUCTION

Symbiotic stars are binary stars in which usually a white dwarf accretes from the wind of a red giant (e.g. Luthardt 1992). Their X-ray spectra are apparently dominated by distinct soft and hard X-ray components (I do not discuss the third "supersoft" case – usually interpreted as steady nuclear burning of accreted material). As an example I take the ASCA GIS2 spectrum of the bright symbiotic CH Cyg, plotted in Figure 1. The emission is seen to peak at 1 keV and at 5 keV. In Figure 1 I have also overlaid a simple 10 keV bremsstrahlung spectrum that has been folded through the response of the telescope. The dip in the observed spectrum (at 2 keV) corresponds to a maximum in the effective area of ASCA, and thus must be a true minimum in the X-ray emission of the CH Cyg.

The ASCA observation of CH Cyg was analysed by Ezuka, Ishida & Makino (1998). To achieve an acceptable fit they required three emission components (kT = 0.2, 0.7, 7.3 keV), each with a different absorption column, plus an additional partial-covering absorber. Usually the hard emission is attributed to the accreting compact object and the soft emission to colliding winds of the two stars. In this paper I demonstrate that the spectrum can be understood with a far more simple model, and that there is no need for a separate soft component.

2. IONISED ABSORPTION

The key to this new interpretation is to allow the absorbing medium to be partially ionised. This is reasonable because the wind of the red giant — the obvious candidate absorber — is strongly illuminated by ionising radiation from the accreting

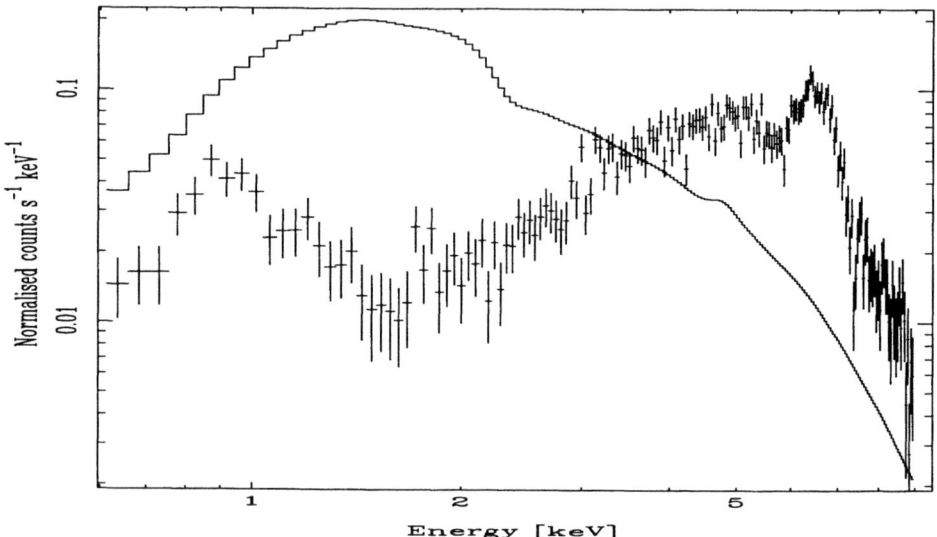

FIGURE 1. The ASCA GIS2 spectrum of the bright symbiotic star CH Cyg. The solid line is a simple 10 keV bremsstrahlung spectrum convolved with the response of ASCA, showing that the observed minimum at ~2 keV must be a true minimum in the X-ray emission of CH Cyg.

white dwarf. Fitting the ASCA spectrum with a single-temperature emission model (*mekal*) absorbed by a photoionised medium (*absori*) I readily achieve a fairly good fit (reduced χ^2=2.4 with 172 d.o.f.; top panel of Figure 2). The residuals are dominated by narrow features at 0.9 keV and 6.4 keV. Adding narrow lines to the model I find an acceptable fit (reduced χ^2=1.4 with 168 d.o.f.; bottom panel of Figure 2). Best fitting parameters are kT=11 keV, N_H = 4 × 10^{23} cm^{-2}, ξ=840 (ionisation parameter).

3. LINE EMISSION

The 6.4 keV emission line is most likely due to K_α fluorescence of weakly-ionised iron. Since the absorbing medium is strongly ionised, this fluorescence must arise elsewhere, probably through reflection from the surface of the compact object. The 0.9 keV line is more difficult to identify because its energy is less well constrained and there are a large number of emission lines in this portion of the X-ray spectrum. However, its proximity to the strong OVIII absorption edge (see Figure 3) suggests it is most likely the recombination continuum emission of OVIII. The emission spectrum of the absorbing medium is neglected in the ionised absorption model (*absori*).

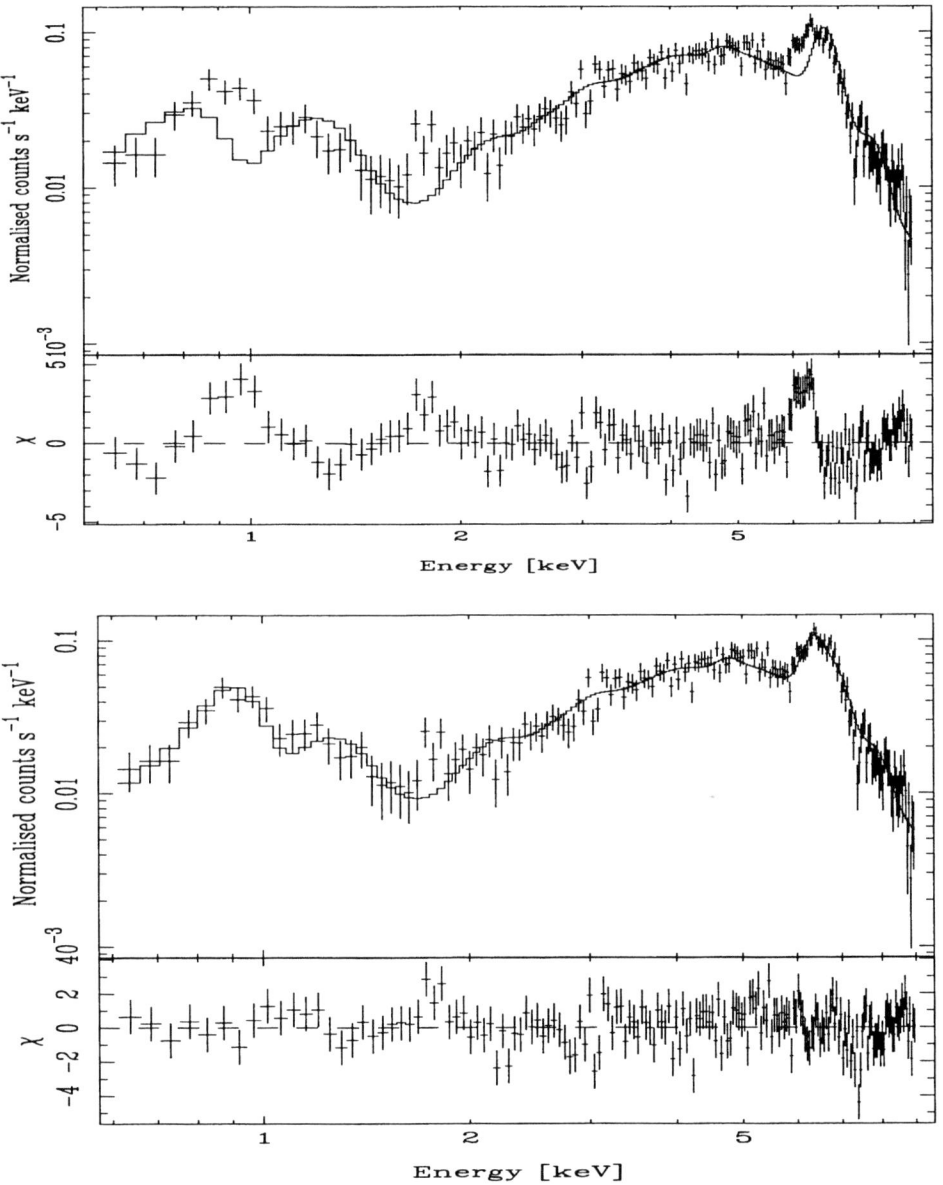

FIGURE 2. The ASCA GIS2 spectrum of CH Cyg fitted with an ionised absorption model and a single emission component (top panel). Narrow emission lines have been added to the model for the lower panel (at 0.9 & 6.4 keV), resulting in an acceptable fit to the spectrum (reduced χ^2=1.4 with 168 d.o.f.).

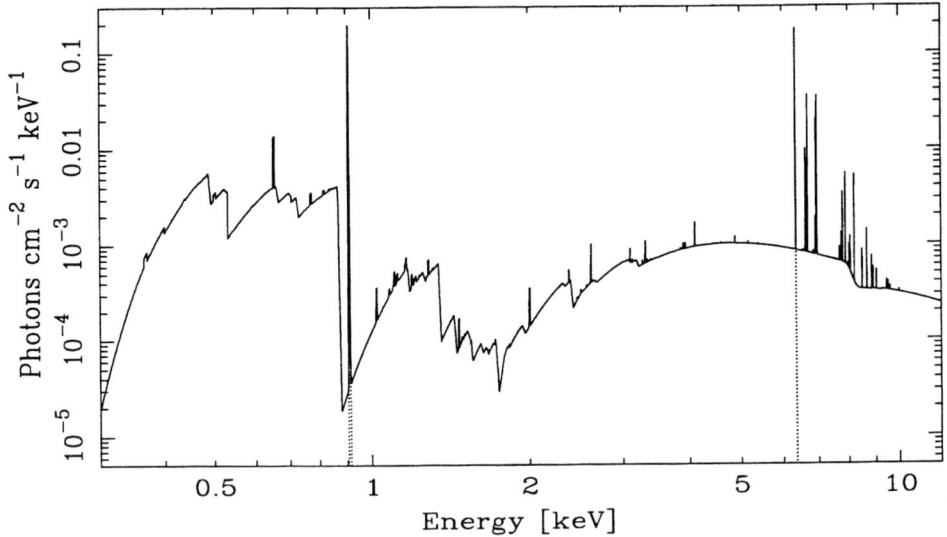

FIGURE 3. The model spectrum fitted to the ASCA spectrum in the lower panel of Figure 2.

4. NO NEED FOR COLLIDING WINDS

The consequence of my fit to the ASCA spectrum of CH Cyg is that a separate low-temperature emission component is no longer needed. Thus there is no need to invoke X-ray emission from colliding winds in this system, and indeed, no longer any need for a substantial wind from the white dwarf.

The model spectrum (Figure 3) shows how the partially ionised absorber cuts deeply at intermediate energies but allows soft photons to leak through. Most evidence taken to support X-ray emission from colliding winds has come from soft X-ray observations, e.g. ROSAT (Mürset, Wolff & Jordan 1997). Clearly the ROSAT spectrum (0.1-2.5 keV) of an absorbed system will reveal only the soft X-ray leak, and this could be mistaken for a low temperature emission spectrum. I believe that all the ROSAT spectra of symbiotic stars previously interpreted as emission from colliding winds may be reinterpreted as absorbed hard X-ray spectra.

REFERENCES

Luthardt, R. 1992, RvMA 5, 38
Ezuka, H., Ishida, M. & Makino, F. 1998, ApJ 499, 388
Mürset, U., Wolff, B. & Jordan, S. 1997, A&A 319, 201

HARD SYNCHROTRON BL LACS: THE CASE OF 1ES 1101-232

Anna Wolter[1], Gabriele Ghisellini[1], Gianpiero Tagliaferri[1], Fabrizio Tavecchio[1], Alessandro Caccianiga[2]

1) Osservatorio Astronomico di Brera, Milano, Italy 2) Observatório Astronómico de Lisboa, Lisboa, Portugal

ABSTRACT The bright X-ray selected BL Lac object 1ES1101-232 shows a flat X-ray spectrum, making it detectable with high statistics over the wide BeppoSAX energy range. We have observed it in two different epochs with BeppoSAX, and found a variation of the flux of about 30% that can be explained by a change in the spectral index above the synchrotron peak. We present here the data and infer limits on the strength of the magnetic field based on models of emission for High-frequency peaked BL Lacs.

KEYWORDS: (Galaxies:) BL Lacertae objects: general – X-rays: galaxies – BL Lacertae objects: individual: 1ES 1101-232

1. INTRODUCTION

Overall spectral energy distributions (SED) of BL Lacs and blazars in general show two broad peaks: the synchrotron one at low energies and the inverse Compton scattering peak at high energies. The position of the synchrotron peak defines different classes of BL Lacs: the HBL (High-peaked BL Lacs) and the LBL (Low-peaked BL Lacs). Ghisellini et al. (1998) and Fossati et al. (1998) propose a sequence for blazars in which the energy of the peak is anti-correlated with the bolometric luminosity, and fainter objects, as HBL, should have a peak in the UV–X-ray band.

1ES 1101-232 ($z=0.186$) is an extreme case of HBL in which the synchrotron component peaks in the X-ray band (~ 1 keV), as shown by our previous observation (Wolter et al. 1998). Even if not as extreme as that of the flaring states of Mkn 501 (Pian et al. 1998) and 1ES 2344+514 (Catanese et al. 1998), the SED of 1ES 1101-232 makes it a good candidate for TeV emission.

2. X-RAY DATA

BeppoSAX has observed 1ES1101-232 on two occasions, on 4 Jan 97 and 19 Jun 98. A single power law fit with Galactic absorption at low energy is rejected for both observations, while a broken power law yields an acceptable χ^2. In Wolter et al. (1999) all the details of the fits are reported. The broken power law model is preferred, from a statistical point of view besides for physical reasons, even over a single power law with intrinsic absorption. The PDS observations, being so short, are not of sufficient statistical significance to put a real constraint on the spectrum.

The position of the break energy (E_0) and the slope of the low energy part of spectrum (α_1) are the same in the two observations within the errors. On the contrary, the portion of the spectrum at higher energies (i.e. above E_0) has changed between the two observations. We therefore fit the two datasets together, by using an appropriate model; the best fit of a broken power law model, in which only the high energy index α_2 is untied between the two observations, is acceptable (see Table 1).

The fluxes are consistent with those obtained by the separate fits. Only the intensity above 2 keV changed (of $\sim$ 32%) between the two observations. Even if the flux variation is small, this result might bear an impact on spectral variability models in BL Lacs.

3. SPECTRAL ENERGY DISTRIBUTION

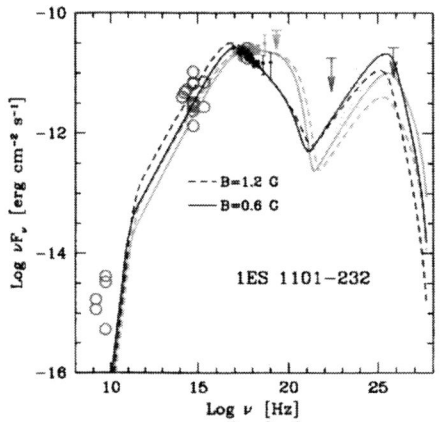

Figure 1: SED, points from literature and BeppoSAX observations. See text for an explanation of the model. Light gray line and dots refer to the Jan 1997 observation, while dark line and dots to the June 1998 BeppoSAX observation.

By using the same data as reported in Wolter et al. (1998) and adding the second *BeppoSAX* observation we construct the SED of Figure 1 and 2. Furthermore, 1ES1101-232 has been observed on the nights of 19-27 May 1998 with the Durham University Mark 6 atmospheric Čerenkov telescope (Chadwick et al. 1999). The source was not detected and an upper limit of f_{TeV} (> 300 GeV) $= 3.7 \times 10^{-11}$ photons cm^{-2} s^{-1} has been derived from the observation. This value also has been plotted in Figure 1.

We can reproduce the observed SED by using the homogeneous Synchrotron–Self Compton model described in detail in Ghisellini et al. (1998). A power-law distribution of electrons with slope n and minimum Lorentz factor γ_{min} is continuously injected in a spherical region with radius R. The source is in relativistic motion

TABLE 1. Broken power Law fit results for LECS+MECS data COMBINED

Date	α_1 En. index	α_2 En. index	E_0 keV	F[a]	χ^2 (dof)
Jan '97	0.64(0.51-0.76)	0.97(0.93-1.03)	1.28(1.16-1.41)	38.7	455.2(397)
Jun '98	same	1.31(1.27-1.35)	same	25.5	

Broken p.l. with $N_H = N_H^{Gal}$; α_1 and E_0 tied between the two datasets.
[a] Unabsorbed flux [2-10 keV] in 10^{-12} erg cm^{-2} s^{-1}.

toward the observer and relativistic effects are expressed by the Doppler factor δ. Electrons are free to cool and form the low energy flat spectrum with spectral index $\alpha = 0.5$.

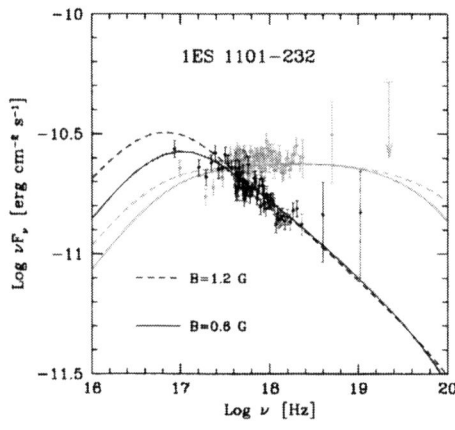

Figure 2: SED, enlargement of the X-ray band with the two observations and model. Light gray line and dots refer to the Jan 1997 observation, while dark line and dots to the June 1998 BeppoSAX observation. The agreement between the model and the X-ray points in the two observations is evident.

The model over-imposed on the SED is derived assuming a radius of $R = 1 \times 10^{16}$ cm, $\delta=15$, $L_{inj}^{intr.} = 9.3 \times 10^{41}$ erg/s; $\gamma_{max} = 4 \times 10^6$, with no external photons. The slope of the injected electrons is $s=2.7$ (1998) or $s=1.95$ (1997). $B = 0.6$ Gauss (and $\gamma_{min}^{inj} = 5. \times 10^4$) for the continuous line; $B = 1.2$ Gauss (and $\gamma_{min}^{inj} = 3. \times 10^4$) for the dashed line.

4. MAGNETIC FIELD

A small change in the magnetic field, while still consistent with the X-ray (BeppoSAX) observations (see Figure 1), produces a very different TeV emission. The TeV band data can therefore put stringent constraints on the magnetic field.

The TeV upper limit indicates that the Compton peak cannot be higher than the synchrotron peak ($L_C/L_S \leq 1$); using the analytical relations discussed in Tavecchio et al. (1998) we can calculate the minimum B allowed by the observed TeV upper limit for different values of ν_c and δ. The values that produce a SED in agreement with both the X-ray spectra and the TeV upper limit are very similar to those found for Mkn 501 (e.g. $\delta = 15$ and $B=0.2$ G; Kataoka et al. 1999) implying that the physical conditions of the two sources are also quite similar.

5. CONCLUSIONS

The X-ray spectrum of 1ES 1101-232 is fitted by a broken power law (a single or an absorbed power law are not statistically acceptable) with a break at 1.3 - 1.9 keV. From the first to the second observation, the spectrum varied at high energies, becoming softer (steeper). The flux has therefore decreased by about 32%, in the 2-10 keV band.

The TeV observation has not yielded a detection. However, since the TeV emission is largely sensitive to parameters like the magnetic field that produces the Synchrotron emission, interesting limits can be put on this quantity. Of course, more sensitive TeV instruments will produce more stringent constraints on the higher energy part of the spectrum and therefore on the emission mechanisms.

Multifrequency, simultaneous observations (e.g. optical, X-ray, TeV) will thus allow us to explain the variability of the sources, both from the energetic and the spectral distribution point of view.

ACKNOWLEDGEMENTS

This work has received partial financial support from the Italian Space Agency and from the European Commission, TMR Programme, Research Network Contract ERBFMRXCT96-0034 "CERES"

REFERENCES

Catanese M. et al., 1998, ApJ, 501, 616.
Chadwick P.M. et al., 1999, ApJ, 513, 161.
Fossati G. et al., 1998, MNRAS, 299, 433.
Ghisellini G. et al., 1998, MNRAS, 301, 451.
Kataoka, J., et al. 1999, ApJ, 514, 138.
Pian E. et al., 1998, ApJL, 492, L17.
Tavecchio F. et al. 1998, ApJ, 509, 608.
Wolter A. et al. 1998, A&A, 335, 899.
Wolter A. et al. 1999, A&A, submitted.

ASCA/ROSAT OBSERVATIONS OF PKS 2316-423: SPECTRAL PROPERTIES OF A LOW LUMINOSITY INTERMEDIATE-TYPE BL LAC OBJECT

S.-J. Xue[1] and Y.-H. Zhang[2]

1) Beijing Astronomical Observatory and Beijing Astrophysics Center of National Astronomical Observatories, Chinese Academy of Sciences. E-mail: xue@bac.pku.edu.cn

2) International School for Advanced Studies, SISSA/ISAS, via Beirut 2-4, I-34014 Trieste, Italy; E-mail: yhzhang@sissa.it

ABSTRACT We present the analysis of archival data from ROSAT and ASCA of a serendipitous source PKS 2316-423. According to its featureless non-thermal radio/optical continuum, the object has been assumed as a BL Lac candidate in the literature. PKS 2316-423 was evident variable over the multiple X-ray observations. Specially, a variable high-energy tail of the synchrotron radiation is revealed. X-ray spectral analysis provided further evidence of its synchrotron-nature broadband spectrum with steep and down-curved shape in the range of 0.1–10 keV, this is general signature of a HBL. The spectral energy distribution (SED) through radio-to-X-ray yields the synchrotron radiation peak at frequency $\nu_p = 7.3 \times 10^{15}$ Hz, with integrated luminosity of $L_{syn} = 2.1 \times 10^{44}$ ergs s^{-1}, this suggest that PKS 2316-423 is a low luminosity BL Lac object with high synchrotron peak frequency. Further SED analysis suggest that PKS 2316-423 is a very low luminosity "intermediate" or high energy peaked BL Lac object. Given the unusual low luminosity, the further studies of PKS 2316-423 might give clues on the evolution properties of BL Lacs.

KEYWORDS: BL Lac objects: individual (PKS 2316-423) – X-rays: galaxies

1. INTRODUCTION

Earlier studies of BL Lac objects have shown that the systematic differences between radio and X-ray selected BL Lac objects (RBLs vs XBLs) can be just attributed to orientation differences. Moreover, BL Lac objects have been reclassified by a more accurate way "low energy" and "high energy" peaked BL Lac objects (LBLs vs HBLs) based on the peak frequency of synchrotron radiation (e.g. Giommi and Padovani 1994). In general, RBLs and XBLs tend to be LBLs and HBLs, respectively. They generally represent two distinct extremes of BL Lacs. However, recent studies from deeper and larger X-ray survey have shown that BL Lac objects tend to exhibit more homogeneous distributions of the properties (Perlman et al. 1998; Caccianiga et al. 1999; Laurent-Muehleisen et al. 1999) rather than previously disparate ones. This has resulted in important roles of intermediate BL Lac objects

(IBLs) in revealing BL Lac mysteries.

In this paper, we present the X-ray spectral analysis (ROSAT and ASCA archival data) and spectral energy distribution (SED) of PKS 2316-423, aiming at showing its intermediate-BL Lac properties. It is a southern radio source at $z = 0.0549$, and was formerly classified as a BL Lac candidate on the base of its featureless non-thermal radio/optical continuum (Crawford & Fabian 1994; Padovani & Giommi 1996). We noticed this object as it has been the brightest contaminating source to the nearby narrow-line X-ray galaxy–NGC 7582 (Xue et al. 1998) in most of its historical X-ray records.

The ROSAT(PSPC) and ASCA satellites observed this object as a serendipitous source in April 1993 and November 1994 respectively. These observations, together with the two ROSAT/HRI observations made in 1992 and 1993, could not only extend our knowledge of the source SED properties to the X-ray domain ($\lesssim 10$ keV), but also provide a good opportunity for X-ray spectroscopic studies in the range of 0.1–10 keV. Which turn out to be very important for the unambiguous classification of the source.

2. SPECTRAL ANALYSIS

No evident variations in the source count rate were detected over both observations spanning about half a day and a little more than one day. Thus time-averaged spectra from both satellites were used for Spectral analyzing.

A simple power-law model, with photon index of $\Gamma \sim 2.0$ and absorption column density at the Galactic value, gives acceptable fit to the ROSAT/PSPC data (Figure 1). The inferred intrinsic luminosity is 5.7×10^{43} ergs s^{-1}, which is similar to that of other non-quasar AGN. The source was observed twice, and showed consistent fluxes, with ROSAT HRI in June 1992 and May 1993 respectively. However, the brightness decreased by 30% from the later HRI observation to the PSPC observation which was taken one week apart. These factors suggest the source is variable and thus there might be non-thermal origin for the X-ray flux.

A simple power-law model fails to well describe the ASCA data, mainly due to an abnormal excess absorption above the Galactic value is required. Consider that this excess absorption might be an artifact due to a false spectral model, we next fitted the data with a broken power-law with free break energy. Thus the fit to the data is notably improved at a $\sim 90\%$ level, yields the absorption in consistent with the Galactic value and two powerlaw components with a break-point at ~ 2.1 keV (see Table 1). The lower-energy part is flatter with a slope in good agreement with that of ROSAT spectrum; the higher-energy part is steeper with $\Gamma = 2.6^{+0.3}_{-0.3}$.

Comparison with the ROSAT/PSPC observation, the ASCA data indicate that the source brightness decreased by 33% in the 0.1–2.4 keV band in a 1.5 years interval. Meanwhile the broad-band X-ray spectrum remained the shape at the

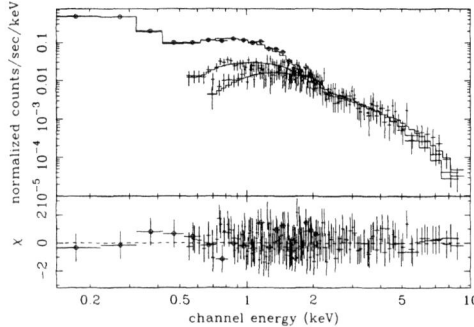

Table 1. Spectral fitting (with 90% errors).

Data	N_H [10^{20} cm^{-2}]	Γ_1	χ^2_ν/d.o.f.
ROSAT	$1.4^{+0.5}_{-0.4}$	$2.0^{+0.2}_{-0.2}$	1.1/17
ASCA	$2.2^{+2.2}_{-2.0}$	$2.0^{+0.4}_{-0.2}$	1.0/131
Γ_2	E_{break} keV	$F^{obs}_{0.1-2keV}$ (10^{-12}ergs cm^{-2} s^{-1})	
–	–	$2.63^{+0.15}_{-0.18}$	
$2.6^{+0.3}_{-0.3}$	2.1	$1.35^{+0.33}_{-0.20}$	

Figure 1. Folded ROSAT/PSPC and ASCA SIS/GIS spectra of PKS 2316-423.

lower-energy part, and hardened the slope in higher-energy range, which was likely in a manner of the prediction of the synchrotron radiation losses.

2.1. Spectral Energy Distribution

The composite SED (Figure 2), from both space and ground-based observations, provides further insights into the object. It is clear that the SED from radio to X-ray is possibly from only one radiation component (Synchrotron emission) and peaks at a higher frequency falling in the EUV/soft-X-ray band ($\nu_p = 7.3 \times 10^{15}$ Hz). The optical and ultraviolet radiation appear to be a continuation of the radio synchrotron spectrum; the X-ray data are likely from a common emission origin as the lower energy parts and represents a high energy tail of the synchrotron spectrum. Other relevant slope parameters from the SED are listed in Table 2.

For a comparison, we plotted in Figure 2 the EGRET sensitivity threshold as an upper limit to the GeV flux (marked by an arrow), since the source was never detected at γ-ray. It is shown that the source is well dominated by a synchrotron process.

3. DISCUSSIONS

Putting 2316-423 on the α_{ro} vs α_{ox} color-color diagram, we find it is in the intermediate range of BL Lacs. As we know, α_{XOX} can more precisely measure spectral changes from optical to soft X-ray bands, however, the values of α_{XOX} for PKS 2316-423 depend on the assumption of the X-ray spectral indices, being 0.18/0.26 and -0.42/-0.34 for $\alpha_x = 1.0$ and 1.6, respectively. These values should locate in the intermediate range of the α_{XOX} distribution of recent BL Lacs samples (Laurent-Muehleisen et al. 1999).

The importance of the frequency at which the synchrotron radiation peaks is that it provides a powerful diagnostics for the physical condition of the emitting region. Recent studies showed that among BL Lacs the synchrotron peak frequencies are inversely correlated with their luminosities (Fossati et al. 1998). We could put

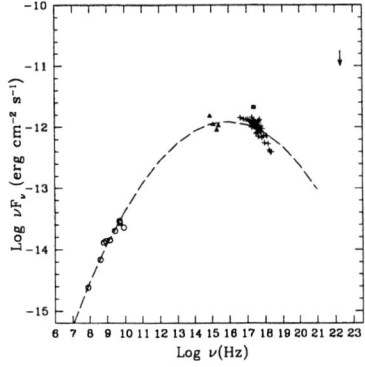

Table 2. Broad-band slopes of SED

α_{ro}	α_{ox}	α_x (0.1-2 keV)	α_x (2-10 keV)
0.56	1.18/1.26	1.0	1.6

Figure 2. The multifrequency SED of PKS 2316-423 and its parabolic fit. The X-ray points are data from this paper, plotted with solid squares for ROSAT/HRI and cross marks for ASCA and ROSAT/PSPC. The UV/optical points are data from Crawford and Fabian (1994) plotted with solid triangle. Circle symbols represent radio data from NASA/IPAC Extra-galactic Database (NED).

PKS 2316-423 on the Figure 7c of Fossati et al. (1998). Due to its lowest peak luminosity, PKS 2316-423 should locate the right-bottom end, this means that the peak frequency of PKS 2316-423 would be around $\sim 10^{18}$ Hz, however, our fit to the SED just gives $\nu_p \sim 10^{16}$ Hz. Therefore, we suggest that PKS 2316-423 might be a low luminosity "intermediate" object between HBLs and LBLs.

In a summary, the X-ray spectral and SED analysis of PKS 2316-423 point out its IBL or HBL attributes with very low luminosity compared with the most recent BL Lac samples. Because of its peculiar low luminosity, however, the more detailed studies of PKS 2316-423 will shed light on the evolution of BL Lac objects.

ACKNOWLEDGEMENTS

This research has made use of the NASA/IPAC Extra-galactic Database (NED) which is operated by the Jet Propulsion Laboratory, California Institute of Technology, under contract with the National Aeronautics and Space Administration. S.J.X. acknowledges the financial support from Chinese Post Doctoral Program.

REFERENCES

Caccianiga, A., Maccacaro, T., Wolter, A. et al., 1999, ApJ, 513, 51
Crawford C.S., Fabian A.C., 1994, MNRAS, 266, 669
Fossati, G., Maraschi, L., Celloti, A. et al., 1998, MNRAS, 299, 433
Giommi P., Padovani P., 1994, MNRAS, 268, L51
Laurent-Muehleisen et al., 1999, astro-ph/9905133
Perlman, E.S., Padovani, P., Giommi, P., 1998, AJ, 115, 1253
Padovani P., Giommi P., 1995, ApJ, 444, 567
Padovani P., Giommi P., 1996, MNRAS, 279, 526
Xue S.J., Otani C., Mihara T. et al. 1998, PASJ, 50, 519

ENERGY DEPENDENT X-RAY VARIABILITY OF THE TEV BLAZARS PKS 2155–304 AND MKN 421

Y.H. Zhang [1], A. Celotti [1], A. Treves [2], G. Fossati [3], L. Maraschi [4],
E. Pian [5], S. Paltani [6], F. Tavecchio [4], L. Chiappetti [7], G. Ghisellini [8],
G. Tagliaferri [8], M. Chiaberge [1]

1) International School for Advanced Studies, via Beirut 2-4, I-34014 Trieste, Italy
2) Dipartimento di Scienze, Università dell'Insubria, via Lucini 3, I-22100 Como, Italy
3) CASS, UCSD, 9500 Gilman Drive, La Jolla, CA 92093-0424, USA
4) Osservatorio Astronomico di Brera, via Brera 28, I-20121 Milano, Italy
5) Istituto TESRE/CNR, via Gobetti 101, I-40129 Bologna, Italy
6) ISDC, 16, ch. d'Écogia, 1200 Versoix, Switzerland
7) IFCTR/CNR, via Bassini 15, I-20133 Milano, Italy
8) Osservatorio Astronomico di Brera, via Bianchi 46, I-22055 Merate, Italy

ABSTRACT We present the X-ray variability properties of the TeV blazars PKS 2155–304 and MKN 421 as observed with $Beppo$SAX and ASCA. The minimum timescales of $\sim$ 1000s are suggested. For PKS 2155–304 we found that the soft X-ray lags (relative to higher energies) are inversely correlated with the source intensity, i.e. the lag is longer when the source is fainter. In April 1998 a flare of MKN 421 was detected, during which higher energy X-rays lagged lower energy ones by $\sim$ 2500s. The origin of these lags is briefly discussed.

KEYWORDS: BL Lacertae objects: individual (PKS 2155–304 and MKN 421); X-rays: variability.

1. INTRODUCTION

PKS 2155–304 (z=0.117) and MKN 421 (z=0.031) are among the brightest BL Lac objects at the X-ray wavelengths and two of the few extragalactic TeV sources (Chadwick et al. 1999, Punch et al. 1992). $Beppo$SAX 1997 (SAX97) is the brightest, $Beppo$SAX 1996 (SAX96) the faintest and ASCA 1994 (ASCA94) intermediate state among the three observations of PKS 2155–304 (Zhang et al. 1999a). MKN 421 1998 (SAX98) observation is the highest state recorded (Fossati et al. 1999a).

2. STRUCTURE FUNCTION (SF) ANALYSIS

The first order SF of a time series $F(t)$ is a function of a timescale "τ", and is defined as: $SF(\tau) = <(F(t+\tau) - F(t))^2>$ (Hughes, Aller & Aller 1992; Paltani

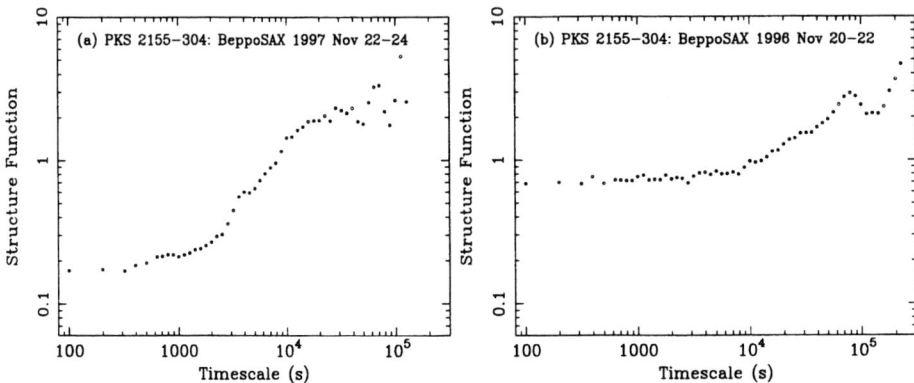

FIGURE 1. SFs (1.5−10 keV) of PKS 2155−304. The SF is normalized to the variance of the light curve. (a) SAX97; (b) SAX96.

1999). The SFs in the 1.5−10 keV band of the four light curves binned over 100s are presented in Figures 1 and 2, respectively.

Considering that observational length strongly affects SF at the largest timescales, the four SFs are compatible with a power law plus a constant representing the measurement uncertainties. However, they are flat up to different timescales possibly due to intrinsic or observational features from which we can not distinguish: if the source is brighter, measurement uncertainties will be comparatively lower, and the power law part of the SF will emerge from the noise at shorter timescale. Furthermore, the absence of steepening of the SFs prevent us from determining any physically meaningful minimum timescale of a process. However, as discussed by Paltani (1999), the absence of steepening of the SF suggests that the minimum timescale will be smaller than the timescale from which the power law emerge. Therefore, the upper limits of the minimum variability timescales would be $\sim$ 3000s, 4000s and 9000s for SAX97, ASCA94 and SAX96 observations of PKS 2155−304, and 3000s for MKN 421 SAX98, respectively. The minimum timescale down to $\sim$ 1000s is possible for the high state of PKS 2155−304 (SAX97) and MKN 421. In PKS 2155−304 whether these timescales correlate with the source brightness (a higher state corresponds to a smaller timescale) is worth studying. We mention this point because it has important consequences on the blazar radiation models.

3. CROSS CORRELATION ANALYSIS

We performed a detailed cross correlation analysis using two techniques suited to unevenly sampled time series: the Discrete Correlation Function (DCF, Edelson & Krolik 1988) and Modified Mean Deviation (MMD, Hufnagel & Bregman 1992). In addition, model-independent Monte Carlo simulations taking into account "flux randomization" (FR) and "random subset selection" (RSS) of the data sets (see

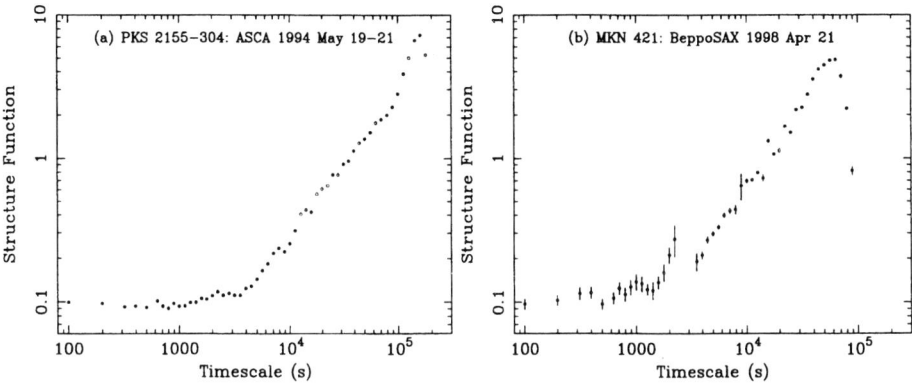

FIGURE 2. Same as Figure 1. (a) PKS 2155-304 (ASCA94) ; (b) MKN 421 (SAX98).

Peterson et al. 1998 for details) are used to statistically determine the significance of any time lag derived from DCF and MMD (see Zhang et al. 1999a for details). The cross correlated energy band is 0.1−1.5 keV vs 3.5−10 keV.

The analysis of PKS 2155-304 has been presented in Zhang et al. (1999a), the resulting soft time lags (which correspond to different brightness levels) are ∼ 0.4, 0.8, 4.0 hours for the SAX97, ASCA94 and SAX96 observations, respectively. These findings indicate that the inter-band soft lags are variable, and inversely correlate with both the source intensities and the ratios of the maximum to minimum fluxes (see Figure 10 of Zhang et al. 1999a). The suggestive trend is such that the lag is longer when the source is fainter and less variable.

On the contrary, we find a negative time lag for the flare of MKN 421 on April 21 1998, i.e. the higher energy X-ray photons lagged the lower energy ones. We notice that this behaviour is *opposite* to the previous findings for this source and other HBLs. The best Gaussian fits result in negative lags of -0.82 ± 0.06 (DCF) and -0.60 ± 0.08 (MMD) hours, confirmed with high significance by the FR/RSS Monte Carlo simulations (Figure 3).

4. DISCUSSION

One of the main results of the analysis for PKS 2155-304 is the inverse correlation between lags and source intensities. This finding sheds light on the brightness dependence of time-dependent emission models and origin of lags (e.g., cooling, light travel times, particle injection/acceleration mechanism, ratio of underlying and flare components).

The new result for MKN 421 is the significant detection of a hard lag, *opposite* to that previously detected in HBLs. Interestingly, we notice that the 1998 April 21 flare represents the brightest state of MKN 421, and thus suggests the possibility that the hard lag follows the trend found for PKS 2155-304. Work is in progress

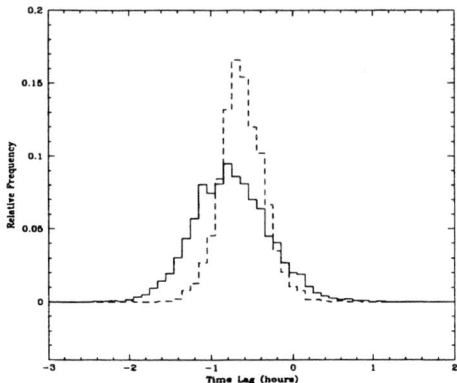

FIGURE 3. Cross Correlation Peak Distribution (CCPD) of FR/RSS simulations of MKN 421. The solid line represents the DCF, and the dashed line the MMD.

to examine this possibility (Zhang et al. 1999b), which would allow us to establish whether the acceleration/injection properties (e.g. Kirk, Rieger, & Mastichiadis, 1998) are related to the source brightness. The presence of a hard lag indicates that the flare evolution is driven by the acceleration/injection mechanism.

It is of primary importance to study together the evolution of time lags, X-ray spectra and energy shifts of the emission peaks. In at least a few HBLs when the source is more luminous, the spectra become harder, the synchrotron emission peak shifts to higher frequency and at least in the case of PKS 2155–304 the soft lag becomes shorter. The determination of the relative contribution of a flare and a steady emission component is also crucial to understand the underlying process at work (Fossati et al. 1999a; Fossati et al. 1999b).

REFERENCES

Chadwick, P.M., et al. 1999, ApJ, 513, 161
Edelson, R. A., & Krolik, J. H. 1988, ApJ, 333, 646
Fossati, G., et al. 1999a, ApJ, in press (astro-ph/0005066)
Fossati, G., et al. 1999b, these proceedings
Hufnagel, B. R., & Bregman, J. N. 1992, ApJ, 386, 473
Hughes, P.A., Aller, H.D., & Aller, M.F. 1992, ApJ, 396, 469
Kirk, J. G., Rieger, F. M., & Mastichiadis, A. 1998, A&A, 333, 452
Paltani, S. 1999, in Proc. of the BL Lac Phenomenon meeting, Turku, Finland, 22-26 June 1998, ed. L.O. Takalo, ASP Conf. Series, 159, 293
Peterson B. M., Wanders I., Horne K., Collier S., Alexander T., & Maoz D. 1998, PASP, 110, 660
Punch, M. et al. 1992, Nature, 358, 477
Zhang, Y.H. et al. 1999a, ApJ, 527, 719
Zhang, Y.H. et al. 1999b, in preparation

STRUCTURE OF THE CIRCUMNUCLEAR REGION OF SEYFERT 2 GALAXIES – CLUES FROM RXTE OBSERVATIONS OF NGC 4945

P. Życki[1], G. Madejski[2,3], C. Done[4]

[1] Nicolaus Copernicus Astronomical Center, Bartycka 18, 00-716 Warsaw, Poland
[2] LHEA, Code 662, NASA/GSFC, Greenbelt, MD 20771, USA
[3] with the Dept. of Astronomy, University of Maryland, College Park, MD 20742, USA
[4] University of Durham, Physics Dept., Durham DH1 3LE, England, UK

ABSTRACT NGC 4945 is a nearby (~ 3.7 Mpc) Seyfert 2 galaxy. Our line of sight to the center is heavily absorbed ($N_{\rm H} \approx 3 \times 10^{24}$ cm^{-2}, i.e. $\tau_{\rm T} \sim 2$), so that the primary hard X-rays are observed only above ≈ 10 keV (Done, Madejski & Smith 1996; Figure 1). The mass of the central black hole, $M \approx 1.4 \times 10^6 \, {\rm M}_\odot$ and the source inclination ($i \approx 90°$, i.e. edge-on) are known thanks to observations of the water maser emission (Greenhill et al. 1997).

According to unification schemes of Seyfert galaxies, the absorber could be envisioned as the molecular/dusty torus, geometrically thick (half-opening angle $\sim 45°$), extending on $\sim$parsec scale (Antonucci 1993). In such a geometry any intrinsic source variability would be smeared out due to distribution of path lengths of photons scattered within the absorber before reaching an observer. Our recent RXTE observations revealed, however, strong variability of the hard X-rays ($E > 10$ keV) on timescale of days (Figure 2).

This discovery rules out the 'classical' torus as the optically thick absorber in NGC 4945. The geometry of the absorber has to be such that the contribution from photons scattered towards our line of sight is negligible, or the spatial scale of the absorber is very small. This first possibility leads to a geometrically thin, disk-like absorber, i.e. a torus with a large opening (funnel) angle. Any geometrically thick, torus-like absorber would have to be located at distances $\ll 1$ light day $\sim 10^4 \, R_{\rm g}$, where its density would have to be $\gg 10^{13}$ cm^{-3} in order for the absorber to remain neutral, as observed.

Adopting the disk-like absorber geometry we estimate the true source luminosity to be $\geq 1.5 \times 10^{43}$ erg s^{-1}, i.e. $L \geq 0.08 \, L_{\rm Edd}$.

KEYWORDS: radiative transfer — galaxies: Seyfert — galaxies: individual (NGC 4945) — X-rays: galaxies

1. SPECTRAL MODELING

We use the model of Krolik, Madau & Życki (1994) for Monte Carlo radiative transfer through the torus. We assume a square cross-section of the torus, with

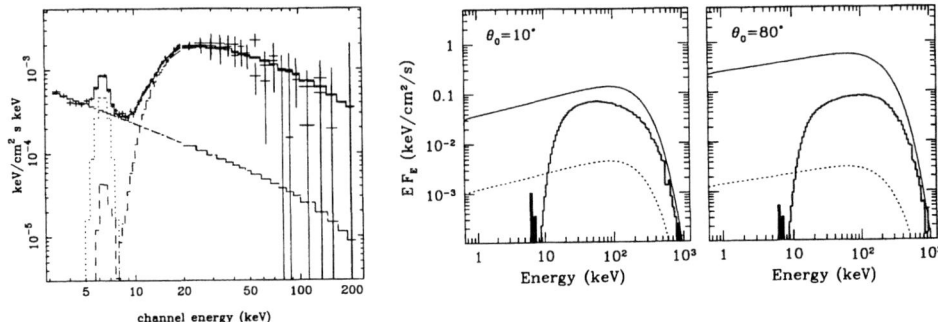

FIGURE 1. Best fit of the torus model for $\theta_0 = 80°$ (left panel), and primary spectra compared to the transmitted one (right panels)

Thomson thickness τ_T and the half-opening angle of the funnel, θ_0. Only the photons escaping in the equatorial plane are included in the model spectrum.

Results of fitting the model, for three values of $\theta_0 = 10°$, $60°$ and $80°$, are presented in Table 1. The quality of the three spectral fits is not very different, considering uncertainties in the underlying continuum. The best fit spectral index is $\Gamma = 1.7-1.8$, typical for Seyfert 1 galaxies. The torus optical thickness $\tau_T = 2.1-2.4$, corresponding to hydrogen column of $\approx 3 \times 10^{24}$ cm^{-2}. An additional emission line at 6.4 keV with EW$\sim$ 1.5 keV is still required. The data and unfolded best fit model for $\theta_0 = 80°$ are plotted in Figure 1.

The three cases of geometry predict rather different absorption-corrected X-ray luminosity. Figure 1 also shows the model spectra, as obtained from the Monte Carlo simulations, compared with the spectrum of radiation incident on the torus. For $\theta_0 = 10°$, the transmitted flux is ≈ 0.30 of the primary one in 0.1–1000 keV band, while this fraction is only ≈ 0.07 for $\theta_0 = 80°$. Thus, the total X-ray luminosity ($E > 0.1$ keV) is between 3.5 and 14 times larger than the observed X-ray luminosity.

The observed X-ray flux (above 1 keV but neglecting the scattered fraction) inferred from modeling is $\approx (2.6 - 3.2) \times 10^{-10}$ erg cm^{-2} s^{-1}, depending on θ_0. Correcting it for the absorption and adopting the distance of 3.7 Mpc gives L_X in the range $(1.5 - 7.3) \times 10^{42}$ erg s^{-1} (for $\theta_0 = 10°$ and $80°$, respectively). For $M =$

TABLE 1. Results of model fitting

θ_0	Γ	τ_T	EW (keV)	χ^2/d.o.f.
10°	1.7	2.1	1.2 ± 0.2	68.5/76
60°	1.8	2.3	1.5 ± 0.2	76.6/76
80°	1.8	2.4	1.6 ± 0.2	75.4/76

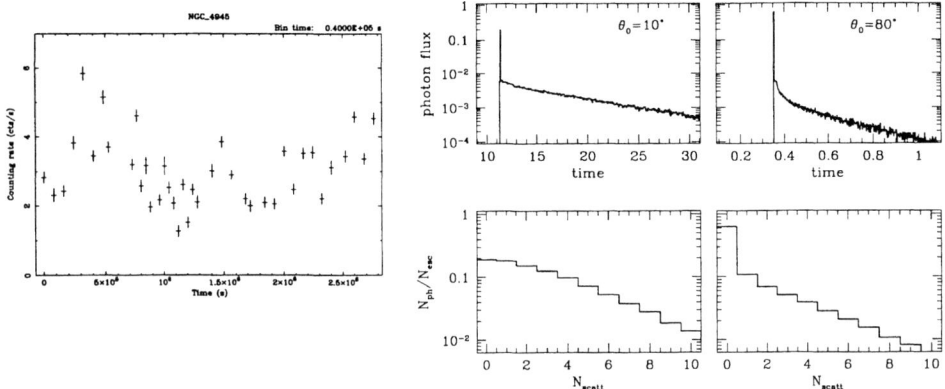

FIGURE 2. *RXTE* PCA light curve of NGC 4945 in 8–30 keV band, with all layers co-added (right panel), and model lightcurves corresponding to a $\delta(t)$ initial impulse (let, upper panels), and distributions of the number of scatterings the photons undergo before escape (left, lower panels). Geometrically thick torus smears out intrinsic variability more significantly than a thin one.

$1.4 \times 10^6 \, M_\odot$ (Greenhill et al. 1997), this corresponds to $L_X/L_{\rm Edd} = 0.008 - 0.041$. Since thermal emission from an accretion disk is at least $\sim L_X$, the bolometric luminosity should be $\geq 2L_X$.

2. TIME VARIABILITY

The PCA count rate clearly shows flux variations by a factor of up to 5 on time-scale of 4–5 days (Figure 2). Similar variability was recently also discovered by BeppoSAX (Guainazzi et al. 2000). *A priori* this could be an intrinsic source variability, or changes in the absorber. We analyzed the spectra from observations with high and low count rates, in order to determine the cause of variability. The hypothesis of constant absorption and changing intrinsic luminosity gives $\chi^2 = 143/110$ dof, while the hypothesis of changing absorption and constant luminosity gives $\chi^2 = 401/110$ dof. This clearly shows that the variability was intrinsic rather than caused by changes in absorption.

2.1. Simulations of the time variability of transmitted radiation

Any intrinsic source variability would be strongly affected by multiple scattering the photons undergo in the absorber. The smearing of the intrinsic variability is rather different in the three different torus geometries (θ_0). The larger the solid angle subtended by the torus from the central source, $\Omega_{\rm tor}$, (i.e. the *smaller* θ_0), the broader is the distribution of pathlengths of the transmitted photons, and the stronger the smearing of any initial variability. This is because for large $\Omega_{\rm tor}$ photons initially directed away from the equatorial plane can be scattered towards it. This effect is demonstrated in Figure 2 where we plot the 10–30 keV lightcurves resulting from a $\delta(t)$-impulse to the torus (i.e. Green's functions for this problem). Corresponding distributions of a number of photon scatterings before escape are also plotted.

For $\theta_0 = 10°$ only $\approx 17\%$ of the observed photons escape without scattering. All the remaining photons are scattered at least once, which lengthens their paths through the torus, leading to significant smearing of any variability. For $\theta_0 = 80°$ the unscattered fraction is $\approx 60\%$.

3. DISCUSSION AND CONCLUSIONS

The observed hard X-ray variability of NGC 4945 is intrinsic rather than due to changing absorption, and is incompatible with classical geometry of obscuring torus posited in the unification schemes of Seyfert galaxies.

The only way for the absorber to be geometrically thick is to postulate that it is located at a distance of $d = 10^2$–$10^3\,R_g \approx 10^{13}$–$10^{14}$ cm $\sim$several light hours, i.e. much less than the time scale of variability. For such an absorber to remain neutral (not ionized), its density would have to be quite high, $n = L_X/d^2\xi$, where ξ is the ionization parameter. Assuming $\xi < 10$, gives a rather high limit, $n > 10^{13}$ cm^{-3}, characteristic for an accretion disk rather than circumnuclear material. However, no theoretical suggestions as to the existence of such (sub)structures of accretion disks exist.

A more plausible geometry is that of a small scale height torus/disk (the half-opening angle θ_0 close to 90°). The location of such an absorber is then unconstrained from our analysis. One might speculate that it is a warped outer accretion disk, due e.g. to radiation pressure (Maloney, Begelman & Pringle 1996). The existence of an extended, optically thin ($N_H \leq 10^{23}$ cm^{-2}) absorber cannot then be ruled out, and is in fact supported by the observed strength of the 6.4 keV Fe Kα line.

Adopting the latter geometry we can estimate the true luminosity of the central source to be $\geq 1.5 \times 10^{43}$ erg s$^{-1} \approx 0.08 L_{\rm Edd}$. This is much larger than the value of $L/L_{\rm Edd} \sim 10^{-4}$ derived for NGC 4258 – another AGN with known central mass.

ACKNOWLEDGEMENTS

This work was supported in part by grant 2P03D01816 of the Polish State Committee for Scientific Research.

REFERENCES

Antonucci R. 1993 ARA&A, 31, 473
Done C., Madejski G. M., Smith D. A. 1996, ApJ, 463, L63
Greenhill L. J., Moran J. M., Herrnstein J. R. 1997, ApJ, 481, L23
Guainazzi et al. 2000, A&A, 356, 463
Krolik J. H., Madau P., Życki P. T. 1994, ApJ, 420, L20
Maloney P., Begelman M. C., Pringle J. 1996, ApJ, 472, 582

THE CONSTELLATION X-RAY MISSION

N.E. White [1], H. Tananbaum [2]

1) NASA's Goddard Space Flight Center, Code 662, Greenbelt, MD 20771, USA.
2) Harvard-Smithsonian Center for Astrophysics, 60 Garden St, Cambridge, MA 02138, USA.

ABSTRACT The Constellation-X mission is a large collecting area X-ray facility, emphasizing observations at high spectral resolution ($E/\Delta E \sim$ 300–3000) while covering a broad energy band (0.25–40 keV). This mission will achieve a factor of 100 increased sensitivity over current capabilities and is optimized to observe the effects of extreme gravity close to black holes, and test models for the formation of large scale structure in the Universe. It is apart of NASA's strategic plan for launch towards the end of the first decade of the 21st century.

KEYWORDS: instrumentation: detectors, techniques: spectroscopic, X-rays: general

1. INTRODUCTION

The Constellation-X mission science is driven by the study of *Gravity* under two extremes. The first is the study of the most extreme gravitational fields known, in the vicinity of black holes. The second is the evolution of the largest Structures in the Universe, clusters of galaxies, where gravity acts on the largest observable scales and is dominated by the effects of the dark matter. In essentially all cases where gravitational forces are important to the dynamics and energetics of the system, the presence of X-ray emitting gas is a natural consequence. As such, X-ray observations play a crucial role in addressing the origin, structure, and evolution of the Universe, and of its principal material constituents: galaxies, stars and dark matter.

The 0.25-10 keV X-ray band contains the K-shell lines for all of the abundant metals (carbon through zinc), and the L-shell lines of many. The detailed X-ray line spectra are rich in plasma diagnostics which also provide unambiguous constraints on physical conditions in the sources. A spectral resolving power ($R = E/\Delta E$) of at least 300 is required to separate the He-like density sensitive triplet. In the region near the iron K complex a resolving power exceeding 2000 is necessary to distinguish the lithium-like satellite lines from the overlapping helium-like transitions. Measurement of accurate radial velocities better than 100 km/s from X-ray emission lines is central for many astrophysical investigations. The current generation of X-ray observatories (*Chandra*, XMM, and Astro-E) will probe the X-ray sky to unprecedented depth with imaging using CCD detectors. These new observatories carry high resolution ($R > 300$) spectrometers that provide sufficient resolution

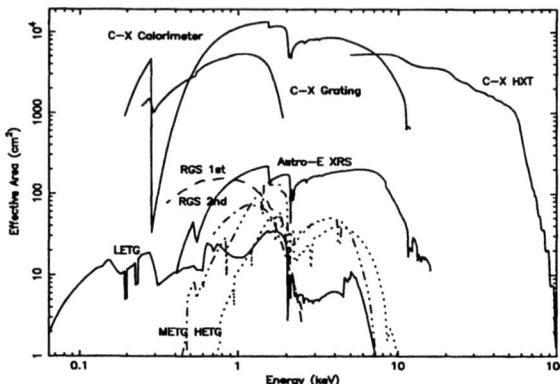

FIGURE 1. The effective area of the three Constellation-X coaligned telescope-instrument combinations. For comparison the lower curves represent the effective area of the high resolution spectrometers on *Chandra*, XMM and Astro-E.

to make detailed plasma diagnostics, but only on the very brightest sources. The Constellation-X observatory combines both the large collecting area and sensitive spectrometers required to obtain high spectral resolution, broad bandpass (0.25-40 keV) spectroscopy for all classes of X-ray sources, over a wide range of luminosity and redshift. It is the natural mission to follow after the current generation of X-ray observatories and will be able to obtain detailed spectroscopy of the faint X-ray source populations that we already know exist from the deepest ROSAT and ASCA surveys.

2. THE MISSION

Spectra of high statistical quality in an observing time of $\leq 10^5$ s for large populations of sources at the faintest flux levels reached in the ROSAT deep surveys require an effective area 20–100 more than provided by the high resolution spectrometers on *Chandra*, XMM, and Astro-E. This corresponds to an effective area of 15,000 cm^2 at 1 keV and 6,000 cm^2 at 6 keV (including the spectrometer efficiency). The spectrometer efficiency is of order 50% at 1 keV, which corresponds to a raw telescope area of 30,000 cm^2. The Constellation-X mission is designed to meet these requirements. The effective area curves of Constellation-X are shown in Figure 1, and compared with the equivalent curves from the high resolution spectrometers on *Chandra*, XMM and Astro-E. The spectacular increase in capability over the previous missions is apparent.

The Constellation-X design takes a radical new approach utilizing several smaller spacecraft and more modest launch vehicles (e.g. Delta-class) can cost less than one very large spacecraft and launcher (e.g. Titan-class). This approach also reduces risk by distributing the activity over several launches and spacecraft. The current baseline mission is four satellites that are carried in pairs on either two Atlas V or Delta IV launchers. The mission will be placed into a L2 orbit to facilitate high observing efficiency, provide an environment optimal for cyrogenic cooling, and simplify the spacecraft design. The interval between the two launches will be of order 1 year. The mission lifetime with all four satellites on orbit will be >4 years. There will be minimal consumables and once on orbit the mission lifetime can potentially last much longer.

To cover the entire 0.25–40 keV band Constellation-X will utilize a matched set of high throughput focusing telescope systems. A spectroscopy X-ray telescope (SXT) covers the 0.25–10 keV band, and is optimized to maintain a spectral resolving power of at least 300 across the bandpass. To maximize the collecting area per unit mass at minimum cost we have selected an angular resolution requirement of at least 15 arc sec half power diameter (HPD), with a goal of 5 arc sec. The 15 arc sec HPD meets the confusion limit for the limiting flux of the sources to be studied. The SXT uses two complementary spectrometer systems to achieve the desired energy resolution: an array of high efficiency quantum microcalorimeters with energy resolution of 2eV, and a set of reflection gratings. The gratings deflect part of the telescope beam away from the calorimeter array in a design similar to XMM. The calorimeter field of view is 2.5 arc min square, with 5 arc second pixels. The hard X-ray telescope (HXT) uses multi-layers to provide the first focusing optics system to operate in the 10–40 keV band. The improvement in the signal to noise results in a factor of 100 or more increased sensitivity over non-focussing methods used in this band. The HXT has an angular resolution of $\leq 1'$ HPD. The fine position resolution, along with the required efficiency and energy resolution (R > 10) can be met by a cadmium zinc telluride (CZT) detector system.

The Constellation-X technology development program is now underway and is making tremendous progress. The current schedule allows for a new start in 2005 and launch in 2008 & 2009. The goal is to have some overlap between Constellation-X and *Chandra* to allow for a science synergy similar to that between HST and the Keck Observatory.

3. MISSION SCIENCE

The Constellation-X high resolution spectrometers and hard X-ray sensitivity up to 40 keV will be able to reach AGN populations at high redshift and determine the total energy output, as well as the geometry, ionization state and density of the surrounding region. If most of accretion in the Universe is highly obscured, then the amount of emitted power per unit galaxy based on optical or UV quasar luminosity functions may have been under estimated (Fabian and Iwasawa 1998). The hard X-ray band is relatively immune to this obscuration and allows a unique view of this early activity. Observations of AGN in the X-ray band at $z \geq 1$ and beyond will hold important clues as to the overall evolution of the Universe. The supermassive black holes in active galactic nuclei (AGN) represent the most extreme gravitational environment known, where the distortion of time and space predicted by General Relativity is most pronounced. A broad iron K line discovered by the ASCA Observatory (Tanaka et al 1995) appears to be broadened by relativistic effects, indicating that it originates within a few gravitational radii of the black hole. This line provides an unexpected new diagnostic to determine not only the mass, but also the spin of the black hole. Constellation-X is optimized to study this line feature in order to probe the region where strong gravity dominates, and to measure the spin and mass of the black hole (Reynolds et al 1999) for a large sample of objects over a wide range of redshift.

Models of galaxy formation aim to account for the properties of all structure, from the first dwarf galaxies to the clusters of galaxies, that are virialising now. Clusters of Galaxies are the largest bound systems known, with their Baryonic mass dominated by X-ray emitting plasma, which is ten times the mass of the sum of the constituent galaxies (Mathiesen and Evrard 1998). When and how these largest of structures came into existance are crucial tests of Cosmological models. X-ray spectroscopy with Constellation-X will determine the internal velocity structure, gas fraction and abundances of clusters and groups of galaxies out to the highest redshift ($z \geq 1$). These observations can test hierarchical clustering models by observing the first clusters as they start to virialise and map the velocity and abundance distributions of nearby Clusters and groups of galaxies. The observed Baryons in the local universe fall far short of those predicted by standard big bang nucleosynthesis (Fukugita, Hogan and Peebles 1998). The baryons seen so clearly in the Lyman alpha forest, are not seen in the local universe. Numerical simulations (Cen and Ostriker 1998) predict that most of these missing Baryons are in a hot intergalactic medium (IGM). Constellation-X will detect the IGM by observing X-ray absorption lines imprinted by highly ionized metals on the spectrum of background quasars (Perna and Loeb 1998).

ACKNOWLEDGEMENTS

We thank Azita Valinia, Kim Weaver, Jean Grady, Oren Sheinman, Steve Kahn, Richard Mushotzky, Pat Tyler and the entire Constellation-X Facility Science Team for their help with the preparation of this paper.

REFERENCES

Cen, R. and Ostriker, J.P. 1998, ApJ, 519, 109.
Fabian, A.C. and Iwasawa, K. 1999, MNRAS, 303, 34.
Fukugita, M., Hogan, C.J. and Peebles, P.J.E. 1998, ApJ, 500, 79.
Hasinger, G., et al 1998, A&A, 329, 495.
Madau, P., Ghisellini, G. and Fabian, A.C., 1994, MNRAS, 270, L17.
Mathiesen, B. and Evrard, A.E. 1998, MNRAS, 295, 769.
Perna, R. and Loeb, A. 1998, ApJ, 503, 135.
Reynolds, C.S. et al 1999, ApJ 514, 614.
Tanaka, Y. et al 1995, Nature, 375, 659.

List of Participants

Arnaud Abrassart	arnaud.abrassart@obspm.fr
Lorenzo Amati	amati@tesre.bo.cnr.it
Xavier Barcons	barcons@ifca.unican.es
Didier Barret	didier.barret@cesr.fr
Loredana Bassani	bassani@tesre.bo.cnr.it
Werner Becker	web@mpe.mpg.de
Volker Beckmann	vbeckmann@hs.uni-hamburg.de
Mitchell C Begelman	mitch@jila.colorado.edu
Sara Benlloch	benlloch@astro.uni-tuebingen.de
Michal Blazejowski	blazejow@camk.edu.pl
Peter J Bleackley	pbl@star.le.ac.uk
Laurence Boirin	boirin@cesr.fr
Thomas Boller	bol@mpe.mpg.de
Hale Bradt	bradt@mit.edu
Niel Brandt	niel@astro.psu.edu
Albert C Brinkman	a.brinkman@sron.nl
Gianfranco Brunetti	gbrunetti@astbol.bo.cnr.it
Geoffrey Burbidge	gburbidge@ucsd.edu
Luciano Burderi	burderi@coma.mporzio.astro.it
Vadim Burwitz	burwitz@mpe.mpg.de
Alessandro Caccianiga	caccia@oal.ul.pt
Andrej Cadez	andrej.cadez@uni-lj.si
Ilaria Cagnoni	icagnoni@cfa.harvard.edu
Massimo Calvani	calvani@pd.astro.it
Sergio Campana	campana@merate.mi.astro.it
Milvia Capalbi	capalbi@gavi.sdc.asi.it
Massimo Cappi	mcappi@tesre.bo.cnr.it
Ruth Carballo	carballo@ifca.unican.es
Francisco J Carrera	fjc@mssl.ucl.ac.uk
Annalisa Celotti	celotti@sissa.it
Paula M Chadwick	p.m.chadwick@dur.ac.uk
Deepto Chakrabarty	deepto@space.mit.edu
Phil Charles	pac@astro.ox.ac.uk
Sylvain Chaty	sylvain.chaty@cesr.fr
James Chiang	chiangj@rocinante.colorado.edu
Andrea Comastri	comastri@astbo3.bo.astro.it
Remon Cornelisse	cornelis@sron.nl
Enrico Costa	costa@ias.rm.cnr.it
Elisa Costantini	elisa@head-cfa.harvard.edu
Daniele Dal Fiume	daniele@tesre.bo.cnr.it
Nichi D'amico	damico@bo.astro.it
Giovanni De Cesare	decesare@saturn.ias.rm.cnr.it
Roberto Della Ceca	rdc@brera.mi.astro.it
Stefano Del Sordo	delsordo@ifcai.pa.cnr.it
Domitilla De Martino	demartin@cerere.na.astro.it
Guido Di Cocco	dicocco@tesre.bo.cnr.it
Christine Done	chris.done@durham.ac.uk
Rick Edelson	rae@astro.ucla.edu
Dieter Engels	dengels@hs.uni-hamburg.de
Andy Fabian	acf@ast.cam.ac.uk
Silvia R Facondi	sfacondi@bo.astro.it
Yuxin Feng	fengyx@astrosv1.ihep.ac.cn
Fabrizio Fiore	fiore@quasar.mporzio.astro.it
Sharon Frigerio	frigerio@brera.mi.astro.it
Filippo Frontera	filippo@tesre.bo.cnr.it
Antonella Fruscione	afruscione@cfa.harvard.edu
Duncan K Galloway	duncan.galloway@utas.edu.au
Michael R Garcia	garcia@cfa.harvard.edu
Jonathan M Gelbord	jonathan@jhu.edu
Ioannis Georgantopoulos	ig@astro.noa.gr
Ian M George	ian.george@gsfc.nasa.gov
Gabriele Ghisellini	gabriele@merate.mi.astro.it
Marat Gilfanov	gilfanov@mpa-garching.mpg.de

Roberto Gilli	gilli@arcetri.astro.it
Paolo Giommi	giommi@sax.sdc.asi.it
Eckart Goehler	goehler@astro.uni-tuebingen.de
Paolo Goldoni	paolo@discovery.saclay.cea.fr
Eric V Gotthelf	evg@astro.columbia.edu
Paola Grandi	grandi@alphasax2.ias.rm.cnr.it
Duane E Gruber	dgruber@ucsd.edu
Matteo Guainazzi	mguainaz@astro.estec.esa.nl
Herbert Gursky	herbert.gursky@nrl.navy.mil
Frank Haberl	fwh@mpe.mpg.de
Volkmar Hable	drhable@alaska-scientific.com
Zoltan Haiman	zoltan@fnal.gov
Alice K Harding	harding@twinkie.gsfc.nasa.gov
Carole A Haswell	c.a.haswell@open.ac.uk
Kiyoshi Hayashida	hayasida@ess.sci.osaka-u.ac.jp
John Heise	j.heise@sron.nl
Rene Hudec	rhudec@asu.cas.cz
Philip J Humphrey	pjh@star.sr.bham.ac.uk
Kevin C Hurley	khurley@sunspot.ssl.berkeley.edu
John B Hutchings	john.hutchings@hia.nrc.ca
Stefan Immler	simmler@mpe.mpg.de
Hajime Inoue	inoue@astro.isas.ac.jp
Jean In 't Zand	jeanz@sron.nl
Gianluca Israel	gianluca.israel@oar.mporzio.astro.it
Kazushi Iwasawa	ki@ast.cam.ac.uk
Naoko Iyomoto	iyomoto@amalthea.phys.s.u-tokyo.ac.jp
Keith Jahoda	keith.jahoda@gsfc.nasa.gov
Philip Kaaret	pkaaret@cfa.harvard.edu
Peter Kahabka	ptk@astro.uva.nl
Peter M W Kalberla	pkalberla@astro.uni-bonn.de
Timothy Kallman	tim@xstar.gsfc.nasa.gov
Victoria M Kaspi	vicky@space.mit.edu
Nobuyuki Kawai	nkawai@postman.riken.go.jp
Eckhard Kendziorra	kendziorra@astro.uni-tuebingen.de
Juergen Kerp	jkerp@astro.uni-bonn.de
Richard, I Klein	klein@radhydro.berkeley.edu
Stefanie Komossa	skomossa@xray.mpe.mpg.de
Albert Kong	albertk@astro.ox.ac.uk
Taro Kotani	kotani@milkyway.gsfc.nasa.gov
Henric Krawczynski	henric.krawczynski@mpi-hd.mpg.de
Ingo Kreykenbohm	kreyken@astro.uni-tuebingen.de
Aya Kubota	aya@amalthea.phys.s.u-tokyo.ac.jp
Markus Kuster	kuster@astro.uni-tuebingen.de
Erik Kuulkers	e.kuulkers@sron.nl
Fabio La Franca	lafranca@fis.uniroma3.it
Georg Lamer	gl@astro.soton.ac.uk
Nicola La Palombara	nicola@ifctr.mi.cnr.it
Stefan Larsson	stefan@astro.su.se
Denis A Leahy	leahy@iras.ucalgary.ca
Ingo Lehmann	ilehmann@aip.de
Daniele Lentini	daniele@brera.mi.astro.it
Alexander Lutovinov	aal@hea.iki.rssi.ru
Tommaso Maccacaro	tommaso@brera.mi.astro.it
Daniel R Macdonald	danmac@mamacass.ucsd.edu
Giuseppe Malaguti	malaguti@tesre.bo.cnr.it
Angela Malizia	malizia@gavi.sdc.asi.it
Fabrizio Fiore	fiore@quasar.mporzio.astro.it
Julien Malzac	julien.malzac@cesr.fr
Laura Maraschi	maraschi@brera.mi.astro.it
Craig B Markwardt	craigm@lheamail.gsfc.nasa.gov
Andrea Martocchia	martok@sissa.it
Philippe B Marty	marty@ias.fr
Paolo Marziani	marziani@pd.astro.it

Nicola Masetti	masetti@tesre.bo.cnr.it
Keith O Mason	kom@mssl.ucl.ac.uk
Hironori Matsumoto	matumoto@space.mit.edu
Giorgio Matt	matt@haendel.fis.uniroma3.it
Katherine E Mcgowan	kem@astro.ox.ac.uk
Andrea Merloni	am@ast.cam.ac.uk
M Coleman Miller	miller@astro.umd.edu
Felix Mirabel	mirabel@discovery.saclay.cea.fr
Kazutami Misaki	misaki@u.phys.nagoya-u.ac.jp
Jonathan P Mittaz	jpdm@mssl.ucl.ac.uk
Sigenori Miyamoto	miyamoto@ouhs.ac.jp
Emi Miyata	miyata@ess.sci.osaka-u.ac.jp
Sandor M Molnar	sandor@pcasrv2.gsfc.nasa.gov
Raquel Morales	rm@ast.cam.ac.uk
James R Murray	jmu@star.le.ac.uk
Fumiaki Nagase	nagase@astro.isas.ac.jp
Kirpal Nandra	nandra@milkyway.gsfc.nasa.gov
Ludmila Nazarova	lsn@star.le.ac.uk
Hitoshi Negoro	negoro@postman.riken.go.jp
Ignacio Negueruela	ignacio@tocai.sdc.asi.it
Hagai Netzer	netzer@wise.tau.ac.il
Fabrizio Nicastro	fnicastro@cfa.harvard.edu
Werner Becker	web@mpe.mpg.de
Michael S Noble	mnoble@cfa.harvard.edu
Andrew J Norton	a.j.norton@open.ac.uk
Michael A Nowak	mnowak@rocinante.colorado.edu
Atsuo Okazaki	okazaki@elsa.hokkai-s-u.ac.jp
Mauro Orlandini	orlandini@tesre.bo.cnr.it
Astrid Orr	orr@astro.uni-tuebingen.de
Julian P Osborne	julo@star.le.ac.uk
Frits Paerels	fritsp@sron.nl
Mathew J Page	mjp@mssl.ucl.ac.uk
Eliana Palazzi	eliana@tesre.bo.cnr.it
Stéphane Paltani	stephane.paltani@obs.unige.ch
Giorgio G C Palumbo	ggcpalumbo@astbo3.bo.astro.it
Dirk Pandel	dpandel@eridanus.physics.ucsb.edu
Hara Papathanassiou	hara@sissa.it
Anastasia Pappa	apa@star.le.ac.uk
Arvind Parmar	aparmar@astro.estec.esa.nl
Biswajit Paul	bpaul@astro.isas.ac.jp
Silvia Pellegrini	pellegrini@astbo3.bo.astro.it
Eric S Perlman	perlman@stsci.edu
Matteo Perri	perri@gavi.sdc.asi.it
Massimo Persic	persic@ts.astro.it
Pierre-olivier Petrucci	petrucci@obs.ujf-grenoble.fr
Elena Pian	pian@tesre.bo.cnr.it
Michael Pivovaroff	mjp@space.mit.edu
Graziella Pizzichini	graziella@botes1.tesre.bo.cnr.it
Robert Popham	popham@mpa-garching.mpg.de
F Scott Porter	porter@milkyway.gsfc.nasa.gov
Juri Poutanen	juri@astro.su.se
Andrea Preite Martinez	andrea@ias.rm.cnr.it
Dimitrios Psaltis	dpsaltis@cfa.harvard.edu
Andrew F Ptak	ptak@astro.phys.cmu.edu
Elizabeth Puchnarewicz	emp@mssl.ucl.ac.uk
Hannah Quaintrell	h.quaintrell@open.ac.uk
A R Rao	arrao@tifr.res.in
Paul Ray	paul.ray@nrl.navy.mil
Andrew M Read	aread@mpe.mpg.de
Travis A Rector	rector@noao.edu
James Reeves	jnr@star.le.ac.uk
Ronald A Remillard	rr@space.mit.edu
Mikhail G Revnivtsev	revnivtsev@hea.iki.rssi.ru

Christopher S Reynolds	chris@rocinante.colorado.edu
Donatella Ricci	riccid@sax.sdc.asi.it
George R Ricker	grr@space.mit.edu
Guido Risaliti	risaliti@arcetri.astro.it
Patrick Risse	risse@astro.uni-tuebingen.de
Natale R Robba	robba@gifco.fisica.unipa.it
Timothy P Roberts	tro@star.le.ac.uk
Edward L Robinson	elr@astro.as.utexas.edu
Richard E Rothschild	rrothschild@ucsd.edu
Mateusz Ruszkowski	ruszkows@ast.cam.ac.uk
Samar R Safi-harb	samar@milkyway.gsfc.nasa.gov
Mara Salvato	msalvato@aip.de
Rita M Sambruna	rms@astro.psu.edu
Andrea Santangelo	andrea@ifcai.pa.cnr.it
Norbert Schartel	nscharte@xmm.vilspa.esa.es
Paola Severgnini	paolas@arcetri.astro.it
Tariq Shahbaz	tsh@astro.ox.ac.uk
I Chun Shih	icshih@astro.ox.ac.uk
Robert E Shirey	shirey@orion.physics.ucsb.edu
Lara Sidoli	sidoli@ifctr.mi.cnr.it
Aneta Siemiginowska	aneta@head-cfa.harvard.edu
Marek Sikora	sikora@camk.edu.pl
Vojtech Simon	simon@asu.cas.cz
Andrzej M Soltan	soltan@camk.edu.pl
Aldo Spizzichino	spizzichino@tesre.bo.cnr.it
Gordon C Stewart	gcs@star.le.ac.uk
Tod E Strohmayer	stroh@clarence.gsfc.nasa.gov
Martin Stuhlinger	stuhli@astro.uni-tuebingen.de
Mutsumi Sugizaki	sugizaki@oasis.tksc.nasda.go.jp
Ewa Szuszkiewicz	esz@astri.uni.torun.pl
Harvey Tananbaum	ht@cfa.harvard.edu
Makoto Tashiro	tashiro@phys.s.u-tokyo.ac.jp
Fabrizio Tavecchio	fabrizio@brera.mi.astro.it
Yuichi Terashima	terasima@olegacy.gsfc.nasa.gov
Edoardo Trussoni	trussoni@to.astro.it
Hiroshi Tsunemi	tsunemi@ess.sci.osaka-u.ac.jp
Tracey Jane Turner	turner@lucretia.gsfc.nasa.gov
Anastasios Tzioumis	atzioumi@atnf.csiro.au
Yoshihiro Ueda	ueda@astro.isas.ac.jp
Meg Urry	cmu@stsci.edu
Phil Uttley	pu@astro.soton.ac.uk
Maria Assunta Valenza	valenza@gifco.fisica.unipa.it
Rob L J Van Der Meer	robvdm@sron.nl
Simon A Vaughan	sav@star.le.ac.uk
Mario Vietri	vietri@corelli.fis.uniroma3.it
Cristian Vignali	vignali@kennet.bo.astro.it
Martin Ward	mjw@star.le.ac.uk
Robert Warwick	rsw@star.le.ac.uk
Michael G Watson	mgw@star.le.ac.uk
Kimberly A Weaver	kweaver@milkyway.gsfc.nasa.gov
Jennifer Rittenhouse West	jennifer@eridanus.physics.ucsb.edu
Peter J Wheatley	pjw@star.le.ac.uk
Nicholas White	nwhite@lheapop.gsfc.nasa.gov
Anna Wolter	anna@brera.mi.astro.it
Kent S Wood	wood@ssd0.nrl.navy.mil
Suijian Xue	xue@bac.pku.edu.cn
J S Yadav	jsyadav@tifr.res.in
Kumi Yoshita	kyoshita@ess.sci.osaka-u.ac.jp
Y H Zhang	yhzhang@sissa.it
Piotr Zycki	ptz@camk.edu.pl

AUTHOR INDEX

A

Abrassart, A., 489
Aharonian, F. A., 694, 866
Almaini, O., 834
Amati, L., 466, 493, 614, 754
Antonelli, L. A., 111
Arnaud, M., 842
Arons, J., 457
Awaki, H., 478

B

Bałucińska-Church, M., 658
Bandyopadhyay, R. M., 336
Barbanera, L., 582
Barbiellini, G., 582
Barcons, X., 3, 842
Barret, D., 518, 814, 842
Bartholdi, P., 826
Bassani, L., 497, 854
Bavdaz, M., 842
Beall, J., 336
Becker, W., 13
Beckmann, V., 502
Bell, J. F., 453, 598
Belloni, T., 606
Benjamin, R., 466
Benlloch, S., 506
Bernard, J.-P., 750
Beuermann, K., 606
Blaes, O., 578
Blanchard, A., 842
Błazejowski, M., 510
Bleackley, P. J., 514
Bloom, E. D., 336
Böhm, P., 918
Böhringer, H., 842
Boirin, L., 518
Boldt, E. A., 734
Boller, T., 25
Bonnet-Bidaud, J. M., 606
Boroson, B., 878
Boyle., B. J., 834
Bradt, H., 35
Braito, V., 602

Brandt, W. N., 53
Budini, G., 582
Bulik, T., 510
Burwitz, V., 522, 686

C

Caccianiga, A., 526, 726, 730, 1011
Čadež, A., 530, 534, 538
Cagnoni, I., 602, 874
Callanan, P. J., 433, 542
Calvani, M., 534, 538, 963
Camilo, F. M., 453, 598
Campana, S., 63, 546, 674
Capalbi, M., 858
Capetti, A., 987
Cappi, M., 550, 842, 854, 999
Caraveo, P., 582
Caroli, E., 497
Carrera, F. J., 3, 554, 782, 822
Casares, J., 922
Catanese, M., 866
Ceballos, M. T., 3
Celotti, A., 562, 586, 866, 971, 975, 987, 1019
Chadwick, P. M., 558, 566, 570
Chakrabarty, D., 336, 850
Charles, P. A., 690, 766, 922
Chaty, S., 574
Chiaberge, M., 562, 586, 971, 987, 1019
Chiang, J., 578
Chiappetti, L., 562, 586, 594, 866, 931, 971, 975, 1019
Christian, D., 542
Church, M. J., 658
Ciliegi, P., 111
Cinti, M. N., 493
Cocco, V., 582
Comastri, A., 73, 111, 590, 842, 874, 971, 975, 999
Cominsky, L., 336
Coppi, P. S., 694
Cordova, F., 927
Costa, E., 493, 582
Costamante, L., 586, 971
Costantini, E., 590

Courvoisier, T., 842
Covino, S., 674
Cowley, A. P., 662
Crampton, D., 662
Crawford, F., 453, 598
Cremonesi, D., 546
Cusumano, G., 614

D

Dadina, M., 550, 610
Dal Fiume, D., 283, 493, 594, 614, 674, 754, 838
D'Amico, N., 453, 598
Degrange, B., 866
Della Ceca, R., 526, 602, 726, 730
Della Valle, M., 466
Del Sordo, S., 614, 838
de Martino, D., 606
De Rosa, A., 470, 610
Di Cocco, G., 497, 550, 582
Di Giacomo, C., 534
Di Matteo, T., 83
Djannati-Atai, A., 866
Done, C., 1023
Dotani, T., 794
Drake, J. J., 542
Dreischer, C., 955
Dumont, A. M., 489

E

Edelson, R., 437, 894, 995
Edwards, P. G., 646
Ehle, M., 670
Elvis, M., 802
Endo, T., 794
Eracleous, M., 355

F

Fabian, A. C., 93, 770, 790, 842, 914
Fender, R. P., 101
Feretti, L., 987
Feroci, M., 493, 582
Filippenko, A. V., 433
Fiore, F., 111, 610, 738, 802, 858, 874

Fischer, J.-U., 918
Fossati, G., 562, 586, 866, 971, 975, 1019
Freyberg, M., 951
Fritz, G., 336
Frontera, F., 466, 493, 497, 614
Frontera, R., 754
Fruscione, A., 542, 590, 874
Frutti, M., 582

G

Gaensler, B. M., 445
Gallagher, S. C., 53
Galli, M. R., 594
Galloway, D. K., 618
Gänsicke, B. T., 606
Garcia, M. R., 433
Gelbord, J., 622
Georgantopoulos, I., 830, 834
George, I. M., 991
Ghisellini, G., 120, 441, 586, 866, 971, 975, 1011, 1019
Giacconi, R., 189
Giebels, B., 336
Gilli, R., 626, 898
Gioia, I. M., 526, 726, 730
Giommi, P., 111, 441, 586, 858, 866, 971
Godfrey, G., 336
Goehler, E., 955
Goldoni, P., 630
Goldwurm, A., 630
Gomboc, A., 530
Gorenstein, P., 654
Gotthelf, E. V., 445
Grandi, P., 130, 470, 610, 975
Greiner, J., 466
Griffiths, R., 842
Griffiths, R. E., 834
Griffiths, R. G., 1003
Grindlay, J., 678
Groot, P., 922
Gruber, D. E., 634, 698, 702, 706, 734
Grupe, D., 686
Guainazzi, M., 638
Gunn, J. E., 189

H

Haardt, F., 470, 594, 610, 975
Haberl, F., 449, 606
Haiman, Z., 140
Hamaguchi, K., 478
Harding, A. K., 150
Hasinger, G., 189, 842, 918, 951
Haswell, C. A., 642
Heindl, W. A., 506, 698, 702, 706
Heise, J., 493
Henri, G., 862
Hertz, P., 336
Hirabayashi, H., 646
Ho, L. C., 983
Homer, L., 690
Hudec, R., 650, 654
Humphrey, P. J., 658
Hurley, K., 160
Hutchings, J. B., 662

I

Immler, S., 666, 670
Inneman, A., 654
in't Zand, J., 493
Israel, G. L., 674, 931
Iwasawa, K., 169

J

Janek, M., 686
Jernigan, J. G., 457
Johnson, W. N., 336
Jourdain, E., 742

K

Kaaret, P., 179, 678
Kaastra, J. S., 610, 774
Kahabka, P., 682
Kaneda, H., 959
Karas, V., 746
Kaspi, V. M., 453, 598
Kawai, N., 462
Kendziorra, E., 955
Kennea, J., 927

Kim, Y., 336
King, A. R., 642
Kitamoto, S., 786
Klein, R. I., 457
Kobayashi, H., 646
Komossa, S., 686
Kong, A., 690
Kotani, T., 462, 794
Kowalski, M., 336
Koyama, K., 478
Krawczynski, H., 694, 866
Kretschmar, P., 698, 702
Kreykenbohm, I., 506, 698, 702
Kunieda, H., 778, 983
Kuster, M., 706
Kuulkers, E., 690

L

Labanti, C., 582
La Franca, F., 111, 762
Lamb, F. K., 678
Lamer, G., 710
Laor, A., 53
Lapidus, I., 614
Lapshov, I., 582
Laurent, P., 630
Law-Green, D., 894
Lazzati, D., 674
Leahy, D. A., 336, 714, 718, 722
Lebrun, F., 630
Lehmann, I., 189, 918
Lentini, D., 726
Levine, A., 35
Lingenfelter, R. E., 906
Longo, F., 582
Lovellette, M. N., 336
Lyne, A. G., 453, 598
Lyons, K., 558, 566, 570

M

Maccacaro, T., 526, 602, 726, 730
Maccarone, M. C., 774
Maccarone, T., 694
MacDonald, D. R., 734
Madejski, G., 578, 935, 1023
Magdziarz, P., 578

Maiolino, R., 111, 199, 898
Malaguti, G., 497, 550, 842, 854
Malizia, A., 497, 738
Malzac, J., 742
Manchester, R. N., 453, 598
Maraschi, L., 470, 586, 610, 866, 971, 975, 979, 1019
Marsden, D., 906
Marshall, H., 578
Martino, B., 582
Martocchia, A., 746
Marty, P., 750
Marziani, P., 534, 963
Masetti, N., 614, 754
Mason, K. O., 842, 874
Massaro, E., 971
Mastropietro, M., 582
Mateos, S., 3
Mathur, S., 590
Matsumoto, H., 758
Matsuoka, M., 462, 550, 999
Matsuzaki, K., 959
Matt, G., 111, 209, 470, 550, 606, 610, 746, 762, 802
McCaughrean, M., 189
McComb, T. J. L., 558, 566, 570
McGowan, K. E., 766
McHardy, I. M., 710
McKay, N. P. F., 453
Mereghetti, S., 219, 546, 582, 931
Merloni, A., 770
Michelson, P., 336
Mihara, T., 794
Miller, J. C., 967
Miller, M. Coleman, 229
Mineo, T., 466, 610, 774
Mirabel, I. F., 239, 574
Misaki, K., 778, 983
Mitsuda, K., 959
Mittaz, J. P. D., 554, 782, 822
Miyaji, T., 951
Miyamoto, S., 786
Moderski, R., 510
Molendi, S., 111
Morales, R., 790
Morelli, E., 582
Morgan, E. H., 678
Morris, D., 598
Morselli, A., 582
Motch, C., 244

Mouchet, M., 606
Mukai, K., 606
Muller, J. M., 493
Murakami, H., 478
Murata, Y., 646
Murray, N., 578

N

Nagase, F., 254, 462, 794, 850
Namiki, M., 462
Nandra, K., 264, 991
Negoro, H., 798
Negueruela, I., 810
Nesci, R., 971
Netzer, H., 274
Nicastro, F., 470, 590, 610, 802
Nicastro, L., 493
Nishiuchi, M., 478
Norton, A. J., 806
Nowak, M., 578

O

O'Brien, P. T., 514, 894
O'Donoghue, D., 766
Ögelman, H., 466
Ohashi, T., 842
Okazaki, A. T., 810
Olive, J.-F., 518, 814
Oosterbroek, T., 638, 674, 838
Orford, K. J., 558, 566, 570
Orio, M., 466
Orlandini, M., 283, 493, 614, 674, 754, 838, 931
Orr, A., 506, 818
Osborne, J. L., 558, 566, 570
Ozaki, M., 850
Ozawa, H., 794

P

Padovani, P., 441
Paerels, F., 842
Page, M. J., 554, 782, 822
Palazzi, E., 493, 614, 754, 866
Paltani, S., 826, 1019

Palumbo, G. G. C., 854, 999
Pandel, D., 927
Panzera, M. R., 674
Pappa, A., 830, 834
Pareschi, G., 614
Parkinson, P. Saz, 336
Parmar, A. N., 466, 614, 638, 658, 674, 838, 842
Paul, B., 846, 850
Peacock, T., 842
Pellegrini, S., 854
Pelletier, G., 862
Perola, G. C., 111, 470, 550, 610
Perotti, F., 582
Perri, M., 858
Persic, M., 854
Petrucci, P. O., 470, 610, 862
Phillips, S. N., 922
Pian, E., 754, 866, 971, 975, 1019
Picozza, P., 582
Pietrini, P., 818
Pietsch, W., 666, 670, 886
Pina, L., 654
Piro, L., 295, 470, 550, 610, 842
Pittori, C., 582
Pompilio, F., 762
Popham, R., 870
Possenti, A., 598
Pottschmidt, K., 506
Pounds, K., 894, 995
Poutanen, J., 310
Preite-Martinez, A., 774
Prest, M., 582
Ptak, A. F., 326, 983
Puchnarewicz, E. M., 874

Q

Quaintrell, H., 878

R

Rafanelli, P., 918
Raiteri, C. M., 866, 971, 975
Rao, A. R., 846
Rapisarda, M., 582
Ravasio, M., 971
Ray, P. S., 336

Rayner, S. M., 558, 566, 570
Read, A. M., 886
Rector, T. A., 890
Reeves, J., 894, 995
Reilly, K., 336
Reinsch, K., 522
Remillard, R., 35
Rephaeli, Y., 634
Reynolds, C. S., 346, 506, 578, 790
Ricci, D., 674
Risaliti, G., 626, 898
Risse, P., 955
Roberts, M., 336
Roberts, T. P., 474
Robinson, E. L., 902
Roche, P., 878
Rodríguez, L. F., 574
Ross, R. R., 770
Rothschild, R. E., 506, 698, 702, 706, 906
Rubini, A., 582
Ruszkowski, M., 910, 914

S

Sakano, M., 462, 478
Salvati, M., 626, 898
Salvato, M., 918
Salvini, C., 590
Sambruna, R. M., 355, 866, 975
Santangelo, A., 838
Sasseen, T., 927
Savaglio, S., 802
Scargle, J., 336
Schmidt, M., 189
Schmidtke, P., 662
Schmitt, J., 842
Schneider, D. P., 189
Scott, D. M., 722
Segreto, A., 838
Semionova, L., 718
Serlemitsos, P. J., 983
Shabad, G., 336
Shahbaz, T., 922
Shanks, T., 834
Shaw, S. E., 558, 566, 570
Shirey, R., 927
Shugarov, S., 943
Sidoli, L., 931

Siemiginowska, A., 874
Sikora, M., 510, 935
Šimon, V., 939, 943, 947
Simpson, C., 894
Smale, A. P., 690, 766
Smith, D. A., 35, 866
Soffitta, P., 493, 582
Sołtan, A., 951
Stairs, I. H., 598
Starrfield, S., 466
Staubert, R., 506, 698, 702, 706, 955
Stella, L., 365, 546, 614, 674
Stephen, J. B., 497
Stewart, G. C., 830, 834
Still, M. D., 878
Stirpe, G. M., 590
Stocke, J. T., 890
Strohmayer, T. E., 377
Stuhlinger, M., 955
Sugizaki, M., 959
Sulentic, J. W., 963
Sunyaev, R., 870
Swank, J. H., 678
Szuszkiewicz, E., 967

T

Tagliaferri, G., 441, 586, 674, 866, 971, 975, 1011, 1019
Takeshima, T., 462, 850
Tananbaum, H., 387, 1027
Tavani, M., 493, 582
Tavecchio, F., 586, 866, 971, 975, 979, 1011, 1019
Terashima, Y., 778, 983
Torii, K., 445
Torricelli-Ciamponi, G., 818
Tosti, G., 866, 971
Treves, A., 586, 594, 610, 866, 931, 971, 975, 1019
Treyer, M., 951
Trifoglio, M., 497
Trümper, J., 189, 951
Trussoni, E., 466, 987
Tsunemi, H., 478
Tsuru, T. G., 478, 758
Turcan, D., 991
Turner, T. J., 991
Turver, K. E., 558, 566, 570

U

Ueda, Y., 396, 462
Uno, S., 794
Urry, C. M., 866
Uttley, P., 710

V

v.Krusenstiern, N., 955
Vallazza, E., 582
van der Klis, M., 406, 842
van Paradijs, J., 922
Vasisht, G., 445
Vaughan, S., 894, 995
Velič, Z., 939
Vercellone, S., 582
Vietri, M., 416
Vignali, C., 111, 999
Villata, M., 866, 971, 975
Vink, J., 774
Vogler, A., 670
Volkmer, R., 955
Vrtilek, S. D., 878

W

Ward, M., 842, 894
Warwick, R. S., 474, 995, 1003
Weaver, K. A., 482, 622
Weiss, R., 955
Welsh, W. F., 902
West, J., 927
Wheatley, P. J., 1007
White, N. E., 1027
Wilkes, B., 590
Wills, B. J., 53
Wilms, J., 506, 698, 702, 706
Wolff, M. T., 336
Wolter, A., 502, 526, 586, 726, 730, 971, 1011
Wood, K. S., 336
Wu, K., 618

X

Xue, S.-J., 1015

Y

Yamauchi, S., 959
Yaqoob, T., 983
Yentis, D., 336
Yokogawa, J., 850
Young, P., 902

Z

Zamorani, G., 189
Zezas, A. L., 830
Zhang, W., 678
Zhang, Y.-H., 562, 1015, 1019
Życki, P., 1023